RADAR HANDBOOK

McGraw-Hill Reference Books of Interest

Handbooks

Avalone and Baumeister • STANDARD HANDBOOK FOR MECHANICAL ENGINEERS
Beeman • INDUSTRIAL POWER SYSTEMS HANDBOOK
Coombs • BASIC ELECTRONIC INSTRUMENT HANDBOOK
Coombs • PRINTED CIRCUITS HANDBOOK
Croft and Summers • AMERICAN ELECTRICIANS' HANDBOOK
DiGiacomo • VLSI HANDBOOK
Fink and Beaty • STANDARD HANDBOOK FOR ELECTRICAL ENGINEERS
Fink and Christiansen • ELECTRONICS ENGINEERS' HANDBOOK
Harper • HANDBOOK OF ELECTRONIC SYSTEMS DESIGN
Harper • HANDBOOK OF THICK FILM HYBRID MICROELECTRONICS
Hicks • STANDARD HANDBOOK OF ENGINEERING CALCULATIONS
Inglis • ELECTRONIC COMMUNICATIONS HANDBOOK
Johnson and Jasik • ANTENNA ENGINEERING HANDBOOK
Juran • QUALITY CONTROL HANDBOOK
Kaufman and Seidman • HANDBOOK FOR ELECTRONICS ENGINEERING TECHNICIANS
Kaufman and Seidman • HANDBOOK OF ELECTRONICS CALCULATIONS
Kurtz • HANDBOOK OF ENGINEERING ECONOMICS
Perry • ENGINEERING MANUAL
Stout • HANDBOOK OF MICROPROCESSOR DESIGN AND APPLICATIONS
Stout and Kaufman • HANDBOOK OF MICROCIRCUIT DESIGN AND APPLICATION
Stout and Kaufman • HANDBOOK OF OPERATIONAL AMPLIFIER DESIGN
Tuma • ENGINEERING MATHEMATICS HANDBOOK
Williams • DESIGNER'S HANDBOOK OF INTEGRATED CIRCUITS
Williams and Taylor • ELECTRONIC FILTER DESIGN HANDBOOK

Dictionaries

DICTIONARY OF COMPUTERS
DICTIONARY OF ELECTRICAL AND ELECTRONIC ENGINEERING
DICTIONARY OF ENGINEERING
DICTIONARY OF SCIENTIFIC AND TECHNICAL TERMS
Markus • ELECTRONICS DICTIONARY

Other Books

Boithias • RADIOWAVE PROPAGATION
Ewell • RADAR TRANSMITTERS
Gentili • MICROWAVE AMPLIFIERS AND OSCILLATORS
Johnson and Jasik • ANTENNA APPLICATIONS REFERENCE GUIDE
Milligan • MODERN ANTENNA DESIGN
Nathanson • RADAR DESIGN PRINCIPLES

---

*For more information about other McGraw-Hill materials,
call 1-800-2-MCGRAW in the United States. In other
countries, call your nearest McGraw-Hill office.*

RADAR HANDBOOK

Editor in Chief
MERRILL I. SKOLNIK

Second Edition

McGraw-Hill, Inc.
New York St. Louis San Francisco Auckland Bogotá
Caracas Lisbon London Madrid Mexico Milan
Montreal New Delhi Paris San Juan São Paulo
Singapore Sydney Tokyo Toronto

Library of Congress Cataloging-in-Publication Data

Radar handbook / editor in chief, Merrill I. Skolnik. — 2nd ed.
    p.   cm.
    Includes index.
    ISBN 0-07-057913-X
    1. Radar—Handbooks, manuals, etc.   I. Skolnik, Merrill
I. (Merrill Ivan), date.
TK6575.R262   1990
621.3848—dc20   89-35217
                                                     CIP

    7890   DOC/DOC   96

ISBN 0-07-057913-X

The editors for this book were Daniel A. Gonneau and Beatrice E.
Eckes, the designer was Naomi Auerbach, and the production supervisor
was Dianne Walber. It was set in Times Roman by the McGraw-Hill
Publishing Company Professional & Reference Division composition
unit.
Printed and bound by R. R. Donnelly & Sons Company.

CONTENTS

Chapter 4. Transmitters   *T. A. Weil*   4.1

Chapter 5 Solid-State Transmitters   *Michael T. Borkowski*   5.1

Chapter 20 Height Finding and 3D Radar *David J. Murrow* 20.1

Chapter 21 Synthetic Aperture Radar *L. J. Cutrona* 21.1

Index follows Chapter 25

CONTRIBUTORS

Lamont V. Blake, *Electronics Consultant* (CHAPTER 2)

Michael T. Borkowski, *Raytheon Company* (CHAPTER 5)

Leopold J. Cantafio, *Space and Technology Group TRW* (CHAPTER 22)

Theodore C. Cheston, *Naval Research Laboratory* (CHAPTER 7)

L. J. Cutrona, *Sarcutron, Inc.* (CHAPTER 21)

Daniel Davis, *Electronic Systems Group, Westinghouse Electric Corporation* (CHAPTER 6)

Gary E. Evans, *Electronic Systems Group, Westinghouse Electric Corporation* (CHAPTER 6)

A. Farina, *Radar Department, Selenia S.p.A., Italy* (CHAPTER 9)

Edward C. Farnett, *RCA Electronics Systems Department, GE Aerospace* (CHAPTER 10)

Joe Frank, *Technology Service Corporation* (CHAPTER 7)

V. Gregers-Hansen, *Equipment Division, Raytheon Company* (CHAPTER 15)

J. M. Headrick, *Naval Research Laboratory* (CHAPTER 24)

Dean D. Howard, *Locus, Inc., a subsidiary of Kaman Corp.* (CHAPTER 18)

Alex Ivanov, *Missile Systems Division, Raytheon Company* (CHAPTER 19)

Eugene F. Knott, *The Boeing Company* (CHAPTER 11)

William H. Long, *Westinghouse Electric Corporation* (CHAPTER 17)

David H. Mooney, *Westinghouse Electric Corporation* (CHAPTER 17)

Richard K. Moore, *The University of Kansas* (CHAPTER 12)

David J. Murrow, *General Electric Company* (CHAPTER 20)

William K. Saunders, *formerly of Harry Diamond Laboratories* (CHAPTER 14)

Helmut E. Schrank, *Electronics Systems Group, Westinghouse Electric Corporation* (CHAPTER 6)

Robert J. Serafin, *National Center for Atmospheric Research* (CHAPTER 23)

William W. Shrader, *Equipment Division, Raytheon Company* (CHAPTER 15)

William A. Skillman, *Westinghouse Electric Corporation* (CHAPTER 17)

Merrill I. Skolnik, *Naval Research Laboratory* (CHAPTER 1)

Fred M. Staudaher, *Naval Research Laboratory* (CHAPTER 16)

George H. Stevens, *RCA Electronics Systems Department, GE Aerospace* (CHAPTER 10)

John W. Taylor, Jr., *Westinghouse Electric Corporation* (CHAPTER 3)

G. V. Trunk, *Naval Research Laboratory* (CHAPTER 8)

T. A. Weil, *Equipment Division, Raytheon Company* (CHAPTER 4)

Lewis B. Wetzel, *Naval Research Laboratory* (CHAPTER 13)

Nicholas J. Willis, *Technology Service Corporation* (CHAPTER 25)

PREFACE

This edition has been thoroughly revised to reflect the advances made in radar over the past two decades. There are many new topics not found in the original, and over half of the 25 chapters were written by authors who did not participate in the first edition. The continued growth in radar capability and applications is reflected in much of the new material included in this second edition. The following are some of the many new radar advances that have occurred since the original edition (listed in no particular order):

- The use of digital techniques that allow sophisticated signal processing in MTI and pulse doppler radars, as well as digital data processing to perform automatic detection and tracking.

- The use of the doppler filter bank and the clutter map in MTI radar.

- The reduced dependency on operators for extracting information from a radar and the incorporation of CFAR in automatic detection and tracking systems.

- The emergence of the analog SAW dispersive delay line as the preferred technique for wideband (high-resolution) pulse compression; the use of digital processing for the pulse compression filter when the bandwidth permits; and the introduction of Stretch pulse compression, which allows high resolution, over a limited range interval with considerably reduced processing bandwidth.

- The increased use of 3D radar for military applications.

- The introduction of the ultralow-sidelobe antenna for airborne pulse doppler radar and, later, for ECCM.

- The replacement of the parabolic reflector antenna with the planar-aperture array antenna for 3D radar, ultralow-sidelobe antennas, and airborne radar.

- The high-power solid-state transmitter that consists of many transistor modules distributed on the rows of a 3D radar (such as the AN/TPS-59), or employed at the elements of a phased array (as in PAVE PAWS), or configured as a transmitter for a conventional radar (as in the AN/SPS-40 or the Canadian ATC radar known as RAMP).

- The serial production of phased arrays for the Patriot, Aegis, PAVE PAWS, and B1-B radar systems.

- The interest in the radar cross section of targets brought about by the attempts to reduce the cross section of military vehicles; and advances in computer methods for predicting the cross section of complex targets.

- The increased capability of military airborne radar (airborne intercept, AWACS, and AEW) due to advances in components and technology that permitted the application of AMTI and pulse doppler to the detection of aircraft in the midst of large clutter.

- The use of radar in space for rendezvous and landing, remote sensing of the earth's environment, planetary exploration, and the detection of targets on the oceans of the world.

- The use of semiactive radar for the guidance of military missile systems.

- The extraction of the doppler frequency shift in meteorological radars that permits the recognition of hazardous weather phenomena not possible with previous weather radars.

- The use of radar operating in the HF portion of the spectrum for long-range over-the-horizon detection of aircraft, ships, and missiles, as well as to provide the direction of the surface winds and the sea state over wide areas of the ocean.

- The development of electronic counter-countermeasures (ECCM) in military radars to thwart attempts to negate radar capability by hostile electronic radiations.

- The increased range resolution and doppler resolution in synthetic aperture radars (SAR) for the imaging of a scene, the use of inverse SAR (ISAR) for the imaging of targets, and the replacement of optical processing with digital processing for SAR imaging.

- The adaptive antenna for application in sidelobe cancelers (as an ECCM) and AMTI radar.

- The use of computers to reliably and quickly predict the capability and coverage of radar systems in the real environment.

The purpose of the above listing is to indicate that radar is dynamic. Not all the new advances made since the first edition of this handbook are listed, nor does the list include all the new material discussed in this edition. There continue to be significant advances in the application of new technology and in the appearance of new applications. Radar grows and is viable since it satisfies important societal, economic, and military needs. It has no serious competitor for most of its many applications.

The size of this edition of the handbook is smaller than the original edition. This is more an indication of the problems involved in technical book publishing rather than problems with the health of radar or the availability of material to include. It was not an easy task to constrain the chapter authors to a limited page budget, and I am appreciative of their efforts to keep the size of their chapters within the allocated number of pages. It would have been easy to double the size of this edition even without increasing the number of chapters. The limitation on size was one of the reasons a number of chapters found in the first edition do not appear here. Some of the omitted chapters were concerned with subjects for which there is not as much interest as there had been or whose technology has not advanced as much as other areas of radar. It is with some regret that 16 of the chapters in the original could not be included in this second edition.

As in the first edition, no attempt was made to utilize a standard notation throughout the book. Each particular subspecialty of radar seems to have developed its own nomenclature, and it is not appropriate in a book such as this to force authors to use notation that is foreign to their field even though it might be commonplace in some other aspect of radar.

Each chapter author was instructed to assume that the average reader has a general knowledge of radar but is not necessarily an expert in the particular subject covered by the chapter.

After a general introduction to radar in Chap. 1, there is a review of the methodology for predicting the range of a radar that has evolved over the years. This is followed by several chapters on the major subsystems of a radar: the receiver, transmitter, solid-state transmitters, reflector antennas, phased array antennas, data processing, ECCM, and pulse compression. Next are discussions of the target cross section and the nature of the radar echoes from the ground and the sea. The various types of radar systems are then discussed: CW and FM-CW, MTI, AMTI, pulse doppler, tracking, missile guidance, height finding and 3D, and synthetic aperture radar. This is followed by three specialized examples of radar that have their own unique character: radar in space, meteorological (weather) radar, and HF over-the horizon radar. The book closes with a treatment of the bistatic radar, which was the first type of radar explored during the onset of radar development in the 1930s.

I would hope that readers who refer in their own writings to material from this book do so by chapter author and title and not by citing just the "Radar Handbook." This will give proper credit to the individual authors who created the work.

It is with much pleasure that I acknowledge the contributions of the individual chapter authors. I enjoyed working with these talented radar engineers and having the opportunity to learn so much from my association with them. A handbook such as this exists only because of the dedicated efforts of the many experts who took the time and energy to prepare the individual chapters. I appreciate their hard work in committing to writing their knowledge and experience, and I am grateful that they have shared this with us.

Merrill Skolnik

CHAPTER 1
AN INTRODUCTION TO RADAR

Merrill I. Skolnik

1.1 DESCRIPTION OF RADAR

The basic concept of radar is relatively simple even though in many instances its practical implementation is not. A radar operates by radiating electromagnetic energy and detecting the echo returned from reflecting objects (targets). The nature of the echo signal provides information about the target. The range, or distance, to the target is found from the time it takes for the radiated energy to travel to the target and back. The angular location of the target is found with a directive antenna (one with a narrow beamwidth) to sense the angle of arrival of the echo signal. If the target is moving, a radar can derive its track, or trajectory, and predict the future location. The shift in frequency of the received echo signal due to the doppler effect caused by a moving target allows a radar to separate desired moving targets (such as aircraft) from undesired stationary targets (such as land and sea clutter) even though the stationary echo signal may be many orders of magnitude greater than the moving target. With sufficiently high resolution, a radar can discern something about the nature of a target's size and shape. Radar resolution may be obtained in range or angle, or both. Range resolution requires large bandwidth. Angle resolution requires (electrically) large antennas. Resolution in the cross-range dimension is usually not as good as the resolution that can be obtained in range. However, when there is relative motion between the individual parts of a target and the radar, it is possible to use the inherent resolution in doppler frequency to resolve in the cross-range dimension. The cross-range resolution of a synthetic aperture radar (SAR) for imaging a scene such as terrain can be explained as being due to resolution in doppler, although a SAR is usually thought of as generating a large "synthetic" antenna by storing received signals in a memory. The two views—doppler resolution and synthetic antenna—are equivalent. Resolution in the doppler domain is a natural way to envision the cross-range resolution achieved by the inverse synthetic aperture radar (ISAR) used for the imaging of a target.

Radar is an active device in that it carries its own transmitter and does not depend on ambient radiation, as do most optical and infrared sensors. Radar can detect relatively small targets at near or far distances and can measure their range with precision in all weather, which is its chief advantage when compared with other sensors.

The principle of radar has been applied from frequencies of a few megahertz

(HF, or high-frequency region of the electromagnetic spectrum) to well beyond the optical region (laser radar). This is a frequency extent of about 1 billion to 1. The particular techniques for implementing a radar differ greatly over this range of frequencies, but the basic principles remain the same.

Radar was originally developed to satisfy the needs of the military for surveillance and weapon control. Military applications have funded much of the development of its technology. However, radar has seen significant civil applications for the safe travel of aircraft, ships, and spacecraft; the remote sensing of the environment, especially the weather; and law enforcement and many other applications.

Radar Block Diagram.   The basic parts of a radar system are illustrated in the simple block diagram of Fig. 1.1. (Other examples of radar block diagrams can be found throughout the handbook.) The radar signal, usually a repetitive train of short pulses, is generated by the transmitter and radiated into space by the antenna. The duplexer permits a single antenna to be time-shared for both transmission and reception. Reflecting objects (targets) intercept and reradiate a portion of the radar signal, a small amount of which is returned in the direction of the radar. The returned echo signal is collected by the radar antenna and amplified by the receiver. If the output of the radar receiver is sufficiently large, detection of a target is said to occur. A radar generally determines the location of a target in range and angle, but the echo signal also can provide information about the nature of the target. The output of the receiver may be presented on a display to an operator who makes the decision as to whether or not a target is present, or the receiver output can be processed by electronic means to automatically recognize the presence of a target and to establish a track of the target from detections made over a period of time. With automatic detection and track (ADT) the operator usually is presented with the processed target track rather than the raw radar detections. In some applications, the processed radar output might be used to directly control a system (such as a guided missile) without any operator intervention.

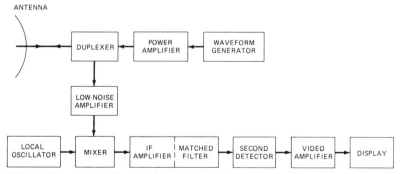

FIG. 1.1 Simple block diagram of a radar employing a power amplifier transmitter and a superheterodyne receiver.

The operation of the radar is described in more detail, starting with the transmitter.

Transmitter. The transmitter (Chap. 4) in Fig. 1.1 is shown as a power amplifier, such as a klystron, traveling-wave tube, crossed-field amplifier, or solid-state device (Chap. 5). A power oscillator such as a magnetron also can be used as the transmitter; but the magnetron usually is of limited average power compared with power amplifiers, especially the klystron, which can produce much larger average power than can a magnetron and is more stable. (It is the *average* power, rather than the *peak* power, which is the measure of the capability of a radar.) Since the basic waveform is generated at low power before being delivered to the power amplifier, it is far easier to achieve the special waveforms needed for pulse compression and for coherent systems such as moving-target indication (MTI) radar and pulse doppler radar. Although the magnetron oscillator can be used for pulse compression and for MTI, better performance can be obtained with a power amplifier configuration. The magnetron oscillator might be found in systems where simplicity and mobility are important and where high average power, good MTI performance, or pulse compression is not required.

The transmitter of a typical ground-based air surveillance radar might have an average power of several kilowatts. Short-range radars might have powers measured in milliwatts. Radars for the detection of space objects (Chap. 22) and HF over-the-horizon radars (Chap. 24) might have average powers of the order of a megawatt.

The radar equation (Sec. 1.2 and Chap. 2) shows that the range of a radar is proportional to the fourth root of the transmitter power. Thus, to double the range requires that the power be increased by 16. This means that there often is a practical, economical limit to the amount of power that should be employed to increase the range of a radar.

Transmitters not only must be able to generate high power with stable waveforms, but they must often operate over a wide bandwidth, with high efficiency and with long, trouble-free life.

Duplexer. The duplexer acts as a rapid switch to protect the receiver from damage when the high-power transmitter is on. On reception, with the transmitter off, the duplexer directs the weak received signal to the receiver rather than to the transmitter. Duplexers generally are some form of gas-discharge device and may be used with solid-state or gas-discharge receiver protectors. A solid-state circulator is sometimes used to provide further isolation between the transmitter and the receiver.

Antenna. The transmitter power is radiated into space by a directive antenna which concentrates the energy into a narrow beam. Mechanically steered parabolic reflector antennas (Chap. 6) and planar phased arrays (Chap. 7) both find wide application in radar. Electronically steered phased array antennas (Chap. 7) are also used. The narrow, directive beam that is characteristic of most radar antennas not only concentrates the energy on target but also permits a measurement of the direction to the target. A typical antenna beamwidth for the detection or tracking of aircraft might be about 1 or 2°. A dedicated tracking radar (Chap. 18) generally has a symmetrical antenna which radiates a pencil-beam pattern. The usual ground-based air surveillance radar that provides the range and azimuth of a target generally uses a mechanically rotated reflector antenna with a fan-shaped beam, narrow in azimuth and broad in elevation. Airborne radars and surface-based 3D air surveillance radars (those that rotate mechanically in azimuth to measure the azimuth angle but use some form of electronic steering or beamforming to obtain the elevation angle, as discussed in Chap. 20) often employ planar array apertures. Mechanical scanning of the radar antenna is usually quite acceptable for the vast majority of radar applications. When it is necessary to scan the beam more quickly than can be achieved with mechanical scanning and when high cost can be tolerated, the electronically steered phased array antenna can be employed. (Beam steering with electronically steered phased arrays can be accomplished in microseconds or less if necessary.)

The size of a radar antenna depends in part on the frequency, whether the radar is located on the ground or on a moving vehicle, and the environment in which it must operate. The lower the frequency, the easier it is to produce a physically large antenna since the mechanical (and electrical) tolerances are proportional to the wavelength. In the ultrahigh-frequency (UHF) band, a *large* antenna (either reflector or phased array) might have a dimension of 100 ft or more. At the upper microwave frequencies (such as X band), radar antennas greater than 10 or 20 ft in dimension can be considered large. (Larger antennas than the above examples have been built, but they are not the norm.) Although there have been microwave antennas with beamwidths as small as 0.05°, radar antennas rarely have beamwidths less than about 0.2°. This corresponds to an aperture of approximately 300 wavelengths (about 31 ft at X band and about 700 ft at UHF).

Receiver. The signal collected by the antenna is sent to the receiver, which is almost always of the superheterodyne type (Chap. 3). The receiver serves to (1) separate the desired signal from the ever-present noise and other interfering signals and (2) amplify the signal sufficiently to actuate a display, such as a cathode-ray tube, or to allow automatic processing by some form of digital device. At microwave frequencies, the noise at the receiver output is usually that generated by the receiver itself rather than external noise which enters via the antenna. The input stage of the receiver must not introduce excessive noise which would interfere with the signal to be detected. A transistor amplifier as the first stage offers acceptably low noise for many radar applications. A first-stage receiver noise figure (defined in Sec. 1.2) might be, typically, 1 or 2 dB. A low-noise receiver front end (the first stage) is desirable for many civil applications, but in military radars the lowest noise figure attainable might not always be appropriate. In a high-noise environment, whether due to unintentional interference or to hostile jamming, a radar with a low-noise receiver is more susceptible than one with higher noise figure. Also, a low-noise amplifier as the front end generally will result in the receiver having less dynamic range—something not desirable when faced with hostile electronic countermeasures (ECM) or when the doppler effect is used to detect small targets in the presence of large clutter. When the disadvantages of a low-noise-figure receiver are to be avoided, the RF amplifier stage is omitted and the mixer stage is employed as the receiver front end. The higher noise figure of the mixer can then be compensated by an equivalent increase in the transmitter power.

The mixer of the superheterodyne receiver translates the receiver RF signal to an intermediate frequency. The gain of the intermediate-frequency (IF) amplifier results in an increase of the receiver signal level. The IF amplifier also includes the function of the matched filter: one which maximizes the output signal-to-noise ratio. Maximizing the signal-to-noise ratio at the output of the IF maximizes the detectability of the signal. Almost all radars have a receiver which closely approximates the matched filter.

The second detector in the receiver is an envelope detector which eliminates the IF carrier and passes the modulation envelope. When doppler processing is employed, as it is in CW (continuous-wave), MTI, and pulse doppler radars, the envelope detector is replaced by a phase detector which extracts the doppler frequency by comparison with a reference signal at the transmitted frequency.

There must also be included filters for rejecting the stationary clutter and passing the doppler-frequency-shifted signals from moving targets.

The video amplifier raises the signal power to a level where it is convenient to display the information it contains. As long as the video bandwidth is not less than half of the IF bandwidth, there is no adverse effect on signal detectability.

A threshold is established at the output of the video amplifier to allow the detection decision to be made. If the receiver output crosses the threshold, a target is said to be present. The decision may be made by an operator, or it might be done with an automatic detector without operator intervention.

Signal Processing. There has not always been general agreement as to what constitutes the signal-processing portion of the radar, but it is usually considered to be the processing whose purpose is to reject undesired signals (such as clutter) and pass desired signals due to targets. It is performed prior to the threshold detector where the detection decision is made. Signal processing includes the matched filter and the doppler filters in MTI and pulse doppler radar. Pulse compression, which is performed before the detection decision is made, is sometimes considered to be signal processing, although it does not fit the definition precisely.

Data Processing. This is the processing done after the detection decision has been made. Automatic tracking (Chap. 8) is the chief example of data processing. Target recognition is another example. It is best to use automatic tracking with a good radar that eliminates most of the unwanted signals so that the automatic tracker only has to deal with desired target detections and not undesired clutter. When a radar cannot eliminate all nuisance echoes, a means to maintain a constant false-alarm rate (CFAR) at the input to the tracker is necessary.

The CFAR portion of the receiver is usually found just before the detection decision is made. It is required to maintain the false-alarm rate constant as the clutter and/or noise background varies. Its purpose is to prevent the automatic tracker from being overloaded with extraneous echoes. It senses the magnitude of the radar echoes from noise or clutter in the near vicinity of the target and uses this information to establish a threshold so that the noise or clutter echoes are rejected at the threshold and not confused as targets by the automatic tracker.

Unfortunately, CFAR reduces the probability of detection. It also results in a loss in signal-to-noise ratio, and it degrades the range resolution. CFAR or its equivalent is necessary when automatic tracking computers cannot handle large numbers of echo signals, but it should be avoided if possible. When an operator is used to make the threshold decision, CFAR is not a necessity as in limited-capacity automatic systems because the operator can usually recognize echoes due to clutter or to increased noise (such as jamming) and not confuse them with desired targets.

Displays. The display for a surveillance radar is usually a cathode-ray tube with a PPI (plan position indicator) format. A PPI is an intensity-modulated, maplike presentation that provides the target's location in polar coordinates (range and angle). Older radars presented the video output of the receiver (called *raw video*) directly to the display, but more modern radars generally display *processed* video, that is, after processing by the automatic detector or the automatic detector and tracker (ADT). These are sometimes called *cleaned-up displays* since the noise and background clutter are removed.

Radar Control. A modern radar can operate at different frequencies within its band, with different waveforms and different signal processing, and with different polarizations so as to maximize its performance under different environmental conditions. These radar parameters might need to be changed according to the local weather, the clutter environment (which is seldom uniform in azimuth and range), interference to or from other electronic equipment, and (if a military radar) the nature of the hostile ECM environment. The different parameters, optimized for each particular situation, can be programmed into the radar ahead of time in anticipation of the environment, or they can be chosen by an operator in real time according to the observed environmental conditions. On the other hand, a *radar control* can be made to automatically recognize when environmental conditions have changed and automatically select, without the aid of an operator, the proper radar operating parameters to maximize performance.

Waveform. The most common radar waveform is a repetitive train of short pulses. Other waveforms are used in radar when particular objectives need to be achieved that cannot be accomplished with a pulse train. CW (a continuous sine wave) is employed on some specialized radars for the measurement of radial velocity from the doppler frequency shift. FM/CW (frequency-modulated CW) is used when range is to be measured with a CW waveform (Chap. 14). Pulse compression waveforms (Chap. 10) are used when the resolution of a short pulse but the energy of a long pulse is desired. MTI radars (Chaps. 15 and 16) with low pulse repetition frequencies (PRFs) and pulse doppler radars (Chap. 17) with high PRFs often use waveforms with multiple pulse repetition intervals in order to avoid range and/or doppler ambiguities.

1.2 RADAR EQUATION

Perhaps the single most useful description of the factors influencing radar performance is the radar equation which gives the range of a radar in terms of the radar characteristics. One form of this equation gives the received signal power P_r as

$$P_r = \frac{P_t G_t}{4\pi R^2} \times \frac{\sigma}{4\pi R^2} \times A_e \qquad (1.1)$$

The right-hand side has been written as the product of three factors to represent the physical processes taking place. The first factor is the power density at a distance R meters from a radar that radiates a power of P_t watts from an antenna of gain G_t. The numerator of the second factor is the target cross section σ in square meters. The denominator accounts for the divergence on the return path of the electromagnetic radiation with range and is the same as the denominator of the first factor, which accounts for the divergence on the outward path. The product of the first two terms represents the power per square meter returned to the radar. The antenna of effective aperture area A_e intercepts a portion of this power in an amount given by the product of the three factors. If the maximum radar range $R_{\max}$ is defined as that which results in the received power P_r being equal to the receiver minimum detectable signal $S_{\min}$, the radar equation may be written

$$R^4{}_{\max} = \frac{P_t G_t A_e \sigma}{(4\pi)^2 S_{\min}} \qquad (1.2)$$

When the same antenna is used for both transmitting and receiving, the transmitting gain G_t and the effective receiving aperture A_e are related by $G_t = 4\pi A_e/\lambda^2$, where λ is the wavelength of the radar electromagnetic energy. Substituting into Eq. (1.2) gives two other forms of the radar equation:

$$R^4_{\text{max}} = \frac{P_t G_t^2 \lambda^2 \sigma}{(4\pi)^3 S_{\text{min}}} \qquad (1.3a)$$

$$R^4_{\text{max}} = \frac{P_t A_e^2 \sigma}{4\pi\lambda^2 S_{\text{min}}} \qquad (1.3b)$$

The examples of the radar equation given above are useful for rough computations of range performance but they are overly simplified and do not give realistic results. The predicted ranges are generally overly optimistic. There are at least two major reasons why the simple form of the radar equation does not predict with any accuracy the range of actual radars. First, it does not include the various losses that can occur in a radar. Second, the target cross section and the minimum detectable signal are statistical in nature. Thus the specification of the range must be made in statistical terms. The elaboration of the simple range equation to yield meaningful range predictions is the subject of Chap. 2. Although the range enters as the fourth power in Eq. (1.3), it can appear as the cube, as the square, or as the first power in specific situations, some of which are described later in this section and in other chapters.

In addition to its use for range prediction, the radar equation forms a good basis for preliminary system design by providing a guide to the possible tradeoffs among the various parameters that enter into radar performance.

The minimum detectable signal S_{min}, which appears in the radar equation, is a statistical quantity and must be described in terms of the probability of detection and the probability of a false alarm. This is discussed in more detail in Chap. 2; for present purposes it suffices to state that for a signal to be reliably detected it must be larger than noise (generally by 10 to 20 dB) at the point in the receiver where the detection decision is made. The minimum detectable signal can be expressed as the signal-to-noise ratio (S/N) required for reliable detection times the receiver noise. The receiver noise is expressed relative to the thermal noise that would be produced by an ideal receiver. The thermal noise is equal to kTB, where k is Boltzmann's constant, T is the temperature, and B is the receiver bandwidth. The receiver noise is the thermal noise multiplied by the factor F_n, the receiver noise figure. The receiver noise figure is measured relative to a reference temperature $T_0 = 290$ K (approximately room temperature), and the factor kT_0 becomes 4×10^{-21} W/Hz. The minimum detectable signal in the radar equation can be written

$$S_{\text{min}} = kT_0 B F_n \frac{S}{N} \qquad (1.4)$$

Sometimes the factor $T_0 F_n$ is replaced with T_s, the system noise temperature.

The above discussion of the radar equation was in terms of the signal power. Although power is a well-understood characteristic of the usual radar waveform consisting of a rectangular pulse, with more complicated waveforms the total signal energy is often a more convenient measure of waveform detectability. It also is more appropriate for theoretical reasons. The ratio of signal energy to noise energy, denoted E/N_0, is a more fundamental parameter than the signal-to-noise (power) ratio in theoretical analyses based on statistical detection theory. No matter what the shape of the received waveform, if the receiver is designed as a matched filter the peak signal-to-noise (power) ratio at the output of the matched filter is $2E/N_0$.

For a rectangular pulse of width τ the signal power is E/τ and the noise power is N_0B, where E = signal energy, N_0 = noise energy, or noise power per unit bandwidth (provided the noise is uniform with frequency), and B = receiver bandwidth. With these substitutions, S_{min} becomes $kT_0F_n(E/N_0)/\tau$. Substituting into Eq. (1.2) gives

$$R^4_{max} = \frac{E_tG_t A_e\sigma}{(4\pi)^2kT_0F_n(E/N_0)} \tag{1.5}$$

where $E_t = P_t\tau$ is the energy contained in the transmitted waveform. Although Eq. (1.5) assumes a rectangular pulse, it can be applied to any waveform provided that E_t is interpreted as the energy contained in the transmitted waveform and that the receiver of noise figure F_n is designed as a matched filter. Some of the published results of radar detection theory give the probability of detection and probability of false alarm in terms of S/N rather than E/N_0. When these results assume optimum (matched-filter) processing, the required values of E/N_0 for use in the radar equation can be obtained from the published results for S/N or the visibility factor as described in Chap. 2.

The radar equation can be manipulated into various forms, depending on the particular application. Several examples are given below.

Tracking. In this situation the radar is assumed to track continuously or "searchlight" a target for an interval of time t_0. Equation (1.5) applies, so that the tracking-, or searchlighting-, radar equation is

$$R^4_{max} = \frac{P_{av}t_0G_t A_e\sigma}{4\pi kT_0F_n(E/N_0)} \tag{1.6}$$

where $P_{av}t_0 = E_t$. Thus, in a tracking radar that must "see" to a long range, the average power must be high, the time on target must be long, and the antenna must be of large electrical size (G_t) and large physical size (A_e). The frequency does not enter explicitly. Since it is easier mechanically to move a small antenna than a large one, tracking radars are usually found at the higher frequencies, where small apertures can have high gain and thus an adequate $G_t A_e$ product.

The radar equation is based on detectability. A tracking radar must also be designed for good angular accuracy. Good angle accuracy is achieved with narrow beamwidths (large G_t) and with high E/N_0 (large A_e). Thus a large $G_t A_e$ product is consistent with good tracking accuracy as well as good detectability.

Volume Search. Assume that the radar must search an angular volume of Ω steradians in the time t_s. If the antenna beam subtends an angle of Ω_b steradians, the antenna gain G_t is approximately $4\pi/\Omega_b$. If the antenna beam dwells a time t_0 in each direction subtended by the beam, the total scan time is $t_s = t_0\Omega/\Omega_b$. Substituting these expressions into Eq. (1.5) and noting that $E_t = P_{av}t_0$,

$$R^4_{\text{max}} = \frac{P_{\text{av}} A_e \sigma}{4\pi k T_0 F_n (E/N_0)} \frac{t_s}{\Omega} \tag{1.7}$$

Thus for a volume search radar the two important parameters for maximizing range are the average transmitter power and the antenna aperture. Any decrease in time to scan the volume or any increase in the volume searched must be accompanied by a corresponding increase in the product $P_{\text{av}} A_e$. Note that the frequency does not enter explicitly.

Jamming. When the detection of the radar signal is limited by an external noise source, such as a deliberate noise jammer rather than by receiver noise, the parameters of importance in determining range performance are slightly different from those presented above (Chap. 9). The receiver noise power per unit bandwidth is now that determined by the jammer rather than the receiver noise figure. When a radar is performing volume search and jamming power enters from a particular direction via the sidelobes, the maximum range can be written

$$R^4_{\text{max}} = \frac{P_{\text{av}} t_s}{g_s} \frac{\sigma}{\Omega} \frac{\sigma}{E/N_0} \frac{R_J^2 B_J}{P_J G_J} \tag{1.8}$$

where g_s = sidelobe level relative to main beam (number less than unity)
  R_J = jammer range
  B_J = jammer bandwidth
  P_J = jammer power
  G_J = jammer antenna gain

and E/N_0 is the ratio of signal energy to noise power per unit bandwidth necessary for reliable detection. The parameter of importance is the average power. The antenna sidelobes are also important. This equation was derived by substituting for kT_0F_n in Eq. (1.7) the jamming noise power per unit bandwidth that would enter the radar receiving-antenna sidelobes. It applies only when the normal receiver noise is negligible compared with the jamming noise.

When the radar is searchlighting a target with a jammer, a mode of operation sometimes called *burnthrough*, the range becomes

$$R^2_{\text{max}} = \frac{P_{\text{av}} t_0 G_t}{4\pi} \frac{\sigma}{E/N_0} \frac{B_J}{P_J G_J} \tag{1.9}$$

The important radar parameters are the average power, the time of observation, and the transmitting-antenna gain. The maximum range is squared rather than raised to the fourth power as in other forms of the radar equation. Note that in neither jamming example does the antenna aperture area enter explicitly. A large aperture collects more signal, but it also collects more jamming noise. The receiver noise figure does not enter because it is assumed that the jamming noise is considerably larger than the receiver noise. Thus in a noisy environment one might not benefit from the effort to design a receiver with the ultimate in sensitivity. The above two examples of jamming radar equations are simplifications. Other variations are possible.

Clutter. When the radar must detect a small target located on the surface of the sea or land, the interfering unwanted clutter echoes can severely limit the detectability of the target. When clutter power dominates receiver noise power, the range equation simply reduces to an expression for the signal-to-clutter ratio. This ratio is equal to the ratio of the target cross section to the clutter cross section. If clutter is distributed more or less uniformly, the clutter echo will depend on the area illuminated by the radar resolution cell. Surface (ground or sea) clutter is described by the ratio of the clutter echo to the area illuminated by the radar. This normalized clutter coefficient is denoted σ^0.

Consider a pulse radar viewing the target and the clutter at low grazing angles. If single-pulse detection is assumed, the signal-to-clutter ratio is

$$\frac{S}{C} = \frac{\sigma}{\sigma^0 R \theta_b (c\tau/2) \sec \phi} \tag{1.10}$$

or

$$R_{\max} = \frac{\sigma}{(S/C)_{\min} \sigma^0 \theta_b (c\tau/2) \sec \phi}$$

where R = range to clutter patch
θ_b = azimuth beamwidth
c = velocity of propagation
τ = pulse width
ϕ = grazing angle

The clutter patch is assumed to be determined in azimuth by the width of the antenna beam and in the range coordinate by the pulse width. The ratio S/C takes a role similar to the ratio E/N_0 for thermal noise. It must be of sufficient magnitude to achieve reliable detection. The clutter statistics generally differ from the statistics of thermal noise but, as a first guess when no other information is available, the required values of S/C might be taken to be those of E/N_0. It is significant that the range dependence enters linearly rather than as the fourth power. Thus for detection of a target in clutter the radar beam should be narrow and the pulse width should be short. With assumptions other than those above, the important radar parameters for detection of targets in clutter might be different. If n hits are received per scan and if the clutter is correlated from pulse to pulse, no improvement in S/C is obtained as it would be if thermal noise, rather than clutter, were the limitation.

1.3 *INFORMATION AVAILABLE FROM THE RADAR ECHO*

Although the name *radar* is derived from *ra*dio *d*etection *a*nd *r*anging, a radar is capable of providing more information about the target than is implied by its name. *Detection* of a target signifies the discovery of its presence. It is possible to consider *detection* independently of the process of information extraction, but it is not often that one is interested in knowing only that a target is present without knowing something about its location in space and its nature. The extraction of useful target information is therefore an important part of radar operation.

The ability to consider *detection* independent of *information* extraction does not mean that there is no relation between the two. The extraction of information generally requires a matched filter, or its equivalent, for optimum processing. The more information that is known about the target a priori, the more efficient will be the detection. For example, if the target location were known, the antenna could be pointed in the proper direction and energy or time need not be wasted searching empty space. Or, if the relative velocity were known, the receiver could be pretuned to the correct received frequency, negating the need to search the frequency band over which the doppler shift might occur.

The usual radar provides the location of a target in range and angle. The rate of change of target location can also be measured from the change in range and angle with time, from which the track can be established. In many radar applications a *detection* is not said to occur until its track has been established.

A radar with sufficient resolution in one or more coordinates can determine a target's size and shape. Polarization allows a measure of the symmetry of a target. In principle, a radar can also measure the surface roughness of a target and determine something about its dielectric properties.

Range.   The ability to determine range by measuring the time for the radar signal to propagate to the target and back is probably the distinguishing and most important characteristic of conventional radar. No other sensor can measure range to the accuracy possible with radar, at such long ranges, and under adverse weather conditions. Surface-based radars can be made to determine the range of an aircraft to an accuracy of a few tens of meters at distances limited only by the line of sight, generally 200 to 250 nmi. Radar has demonstrated its ability to measure interplanetary distances to an accuracy limited only by the accuracy to which the velocity of propagation is known. At more modest distances, the measurement of range can be made with a precision of a few centimeters.

The usual radar waveform for determining range is the short pulse. The shorter the pulse, the more precise can be the range measurement. A short pulse has a wide spectral width (bandwidth). The effect of a short pulse can be obtained with a long pulse whose spectral width has been increased by phase or frequency modulation. When passed through a *matched filter*, the output is a compressed pulse whose duration is approximately the reciprocal of the spectral width of the modulated long pulse. This is called *pulse compression* and allows the resolution of a short (wide-bandwidth) pulse with the energy of a long pulse. A CW waveform with frequency or phase modulation can also provide an accurate range measurement. It is also possible to measure the range of a single target by comparing the phase difference between two or more CW frequencies. Range measurement with CW waveforms has been widely employed, as in aircraft radar altimeters and surveying instruments.

Radial Velocity.   From successive measurements of range the rate of change of range, or radial velocity, can be obtained. The doppler frequency shift of the echo signal from a moving target also provides a measure of radial velocity. However, the doppler frequency measurement in many pulse radars is highly ambiguous, thus reducing its utility as a direct measurement of radial velocity.

When it can be used, it is often preferred to successive range measurements since it can achieve a more accurate measurement in a shorter time.

Any measurement of velocity, whether by the rate of change of range or by the doppler frequency shift, requires time. The longer the time of observation, the more accurate can be the measurement of velocity. (A longer observation time also can increase the signal-to-noise ratio, another factor that results in increased accuracy.) Although the doppler frequency shift is used in some applications to measure radial velocity (as, for example, in such diverse applications as the police speed meter and satellite surveillance radars), it is more widely employed as the basis for sorting moving targets from unwanted stationary clutter echoes, as in MTI, AMTI (airborne MTI), pulse doppler, and CW radars.

Angular Direction. The direction of a target is determined by sensing the angle at which the returning wavefront arrives at the radar. This is usually accomplished with a directive antenna, i.e., one with a narrow radiation pattern. The direction in which the antenna points when the received signal is a maximum indicates the direction of the target. This, as well as other methods for measuring angle, assumes that the atmosphere does not perturb the straight-line propagation of the electromagnetic waves.

The direction of the incident waveform can also be determined by measuring the phase difference between two separated receiving antennas, as with an interferometer. Phase-comparison monopulse also is based on the phase measurement of signals in two separated antennas. The amplitude-comparison monopulse determines the angle of arrival by comparing the amplitudes of the signals received in two squinted beams generated by a single antenna.

The accuracy of the angle of arrival depends on the extent of the antenna aperture. The wider the antenna, the narrower the beamwidth and the better the accuracy.

The angle of arrival, or target direction, is not strictly a radar measurement (as are the range and radial velocity) if a radar measurement is defined as one obtained by comparing the reflected echo signal with the transmitted signal. The determination of angle basically involves only the one-way path. Nevertheless, the angle measurement is an integral part of most surveillance and tracking radars.

Size. If the radar has sufficient resolution, it can provide a measurement of the target's extent, or size. Since many targets of interest have dimensions of several tens of meters, resolution must be several meters or less. Resolutions of this order can be readily obtained in the range coordinate. With conventional antennas and the usual radar ranges, the angular resolution is considerably poorer than what can be achieved in range. However, target resolution in the cross-range (angle) dimension can be obtained comparable with that obtained in range by the use of resolution in the doppler frequency domain. This requires that there be relative motion between the various parts of the target and the radar. It is the basis for the excellent cross-range resolution obtained in a SAR in which the relative motion between target and radar occurs because of the travel of the aircraft or spacecraft on which the radar is mounted. In an ISAR (inverse synthetic aperature radar) the relative motion is provided by the movement of the target.

Shape. The size of a target is seldom of interest in itself, but its shape and its size are important for recognizing one type of target from another. A high-resolution radar that obtains the profile of a target in both range and cross range (as do SAR and ISAR) provides the size and shape of the target. The shape of an object can also be obtained by tomography, in which a two-dimensional image of a three-dimensional object is reconstructed from the measurement of phase and amplitude, at different angles of observation. (The radar might rotate around the fixed object, or the radar can be fixed and the object rotated about its own axis.) Range resolution is not necessary with the coherent tomographic radar method.

As mentioned earlier, comparison of the scattered fields for different polarizations provides a measure of target asymmetry. It should be possible to distinguish targets with different aspect ratios (shapes), as for example, rods from spheres and spheres from aircraft. The complete exploitation of polarization requires the measurement of phase, as well as amplitude of the echo signal at two orthogonal polarizations and a cross-polarization component. Such measurements (which define the polarization matrix) should allow in principle the recognition of one class of target from another, but in practice it is not easy to do.

One characteristic of target *shape* is its surface roughness. This measurement can be of particular interest for echoes from the ground and the sea. Rough targets scatter the incident electromagnetic energy diffusely; smooth targets scatter specularly. By observing the nature of the backscatter as a function of the incident angle it should be possible to determine whether a surface is smooth or rough. Surface roughness is a relative measure and depends on the wavelength of the illuminating signal. A surface that appears rough at one wavelength might appear smooth when illuminated with longer-wavelength radiation. Thus another method for determining surface roughness is by varying the frequency of the illuminating radiation and observing the transition from specular to diffuse scatter. A direct method for determining roughness is to observe the scatter from the object with a resolution that can resolve the roughness scale.

Other Target Measurements. Just as the radial velocity can be determined from the *temporal* doppler frequency shift, it is possible to measure the tangential (cross-range) component of velocity. This can be obtained from the analogous *spatial* doppler frequency shift that expands or compresses the apparent antenna radiation pattern (just as the radial component of velocity can expand or compress the time waveform of a radar signal reflected from a moving target to produce a temporal doppler frequency shift). A measurement of tangential velocity requires a wide-baseline antenna, such as an interferometer. The measurement of tangential velocity has not seen application because the required baseline is often too wide for practical purposes.

It is also possible to note the change of a complex target's radial projection from the change of the received-signal amplitude with time. (A change in the radial projection of a target usually results in a change of the radar cross section.)

Vibrations of the target, rotation of the propellers of an aircraft, or the rotation of a jet engine can induce distinctive modulation on the radar echo which can be detected by a spectral analysis of the radar echo signal.

1.4 RADAR FREQUENCIES

There are no fundamental bounds on radar frequency. Any device that detects and locates a target by radiating electromagnetic energy and utilizes the echo scattered from a target can be classed as a radar, no matter what its frequency.

Radars have been operated at frequencies from a few megahertz to the ultraviolet region of the spectrum. The basic principles are the same at any frequency, but the practical implementation is widely different. In practice, most radars operate at microwave frequencies, but there are notable exceptions.

Radar engineers use letter designations, as shown in Table 1.1, to denote the general frequency band at which a radar operates. These letter bands are universally used in radar. They have been officially accepted as a standard by the Institute of Electrical and Electronics Engineers (IEEE) and have been recognized by the U.S. Department of Defense. Attempts have been made in the past to subdivide the spectrum into other letter bands (as for waveguides and for ECM operations), but the letter bands in Table 1.1 are the only ones that should be used for radar.

The original code letters (P, L, S, X, and K) were introduced during World War II for purposes of secrecy. After the need for secrecy no longer existed, these designations remained. Others were later added as new regions of the spectrum were utilized for radar application. (The nomenclature *P band* is no longer in use. It has been replaced with *UHF*.)

Letter bands are a convenient way to designate the general frequency range of a radar. They serve an important purpose for military applications since they can describe the frequency band of operation without using the exact frequencies at which the radar operates. The exact frequencies over which a radar operates should be used in addition to or instead of the letter bands whenever proper to do so.

TABLE 1.1 Standard Radar-Frequency Letter-Band Nomenclature*

Band designation	Nominal frequency range	Specific frequency ranges for radar based on ITU assignments for Region 2
HF	3 MHz–30 MHz	
VHF	30 MHz–300 MHz	138 MHz–144 MHz
		216 MHz–225 MHz
UHF	300 MHz–1000 MHz	420 MHz–450 MHz
		890 MHz–942 MHz
L	1000 MHz–2000 MHz	1215 MHz–1400 MHz
S	2000 MHz–4000 MHz	2300 MHz–2500 MHz
		2700 MHz–3700 MHz
C	4000 MHz–8000 MHz	5250 MHz–5925 MHz
X	8000 MHz–12,000 MHz	8500 MHz–10,680 MHz
K_u	12.0 GHz–18 GHz	13.4 GHz–14.0 GHz
		15.7 GHz–17.7 GHz
K	18 GHz–27 GHz	24.05 GHz–24.25 GHz
K_a	27 GHz–40 GHz	33.4 GHz–36.0 GHz
V	40 GHz–75 GHz	59 GHz–64 GHz
W	75 GHz–110 GHz	76 GHz–81 GHz
		92 GHz–100 GHz
mm	110 GHz–300 GHz	126 GHz–142 GHz
		144 GHz–149 GHz
		231 GHz–235 GHz
		238 GHz–248 GHz

*From IEEE Standard 521-1984.

The International Telecommunications Union (ITU) assigns specific frequency bands for radiolocation (radar) use. These are listed in the third column of Table 1.1. They apply to ITU Region 2, which encompasses North and South America. Slight differences exist in the other two ITU regions. Although L band, for example, is shown in the second column of the table as extending from 1000 to 2000 MHz, in practice an L-band radar would be expected to be found somewhere between 1215 and 1400 MHz, the frequency band actually assigned by the ITU.

Each frequency band has its own particular characteristics that make it better for certain applications than for others. In the following, the characteristics of the various portions of the electromagnetic spectrum at which radars have been or could be operated are described. The divisions between the frequency regions are not as sharp in practice as the precise nature of the nomenclature.

HF (3 to 30 MHz). Although the first operational radars installed by the British just prior to World War II were in this frequency band, it has many disadvantages for radar applications. Large antennas are required to achieve narrow beamwidths, the natural ambient noise level is high, the available bandwidths are narrow, and this portion of the electromagnetic spectrum is widely used and restrictively narrow. In addition, the long wavelength means that many targets of interest might be in the Rayleigh region, where the dimensions of the target are small compared with the wavelength; hence, the radar cross section of targets small in size compared with the (HF) wavelength might be lower than the cross section at microwave frequencies.

The British used this frequency band, even though it had disadvantages, because it was the highest frequency at which reliable, readily available high-power components were then available. Ranges of 200 mi were obtained against aircraft. These were the radars that provided detection of hostile aircraft during the battle of Britain and were credited with allowing the limited British fighter resources to be effectively used against the attacking bomber aircraft. They did the job that was required.

Electromagnetic waves at HF have the important property of being refracted by the ionosphere so as to return to the earth at ranges from about 500 to 2000 nmi, depending on the actual condition of the ionosphere. This allows the over-the-horizon detection of aircraft and other targets. The long over-the-horizon ranges that are possible make the HF region of the spectrum quite attractive for the radar observation of areas (such as the ocean) not practical with conventional microwave radar.

VHF (30 to 300 MHz). Most of the early radars developed in the 1930s were in this frequency band. Radar technology at these frequencies represented a daring venture that pushed to the edge of technology known in the thirties. These early radars served quite well the needs of the time and firmly established the utility of radar.

Like the HF region, the VHF (very high frequency) region is crowded, bandwidths are narrow, external noise can be high, and beamwidths are broad. However, the necessary technology is easier and cheaper to achieve than at microwave frequencies. High power and large antennas are readily practical. The stable transmitters and oscillators required for good MTI are easier to achieve than at higher frequencies, and there is relative freedom from the blind speeds that limit the effectiveness of MTI as the frequency is increased. Reflections from rain are not a problem. With horizontal polarization over a good reflecting surface (such as the sea), the constructive interference between the direct wave and the wave reflected from the surface can result in a substantial increase in the maximum range against aircraft (almost twice the free-space range). However, a consequence of this increase in range due to constructive interference is that the accompanying destructive interference results in nulls in the coverage at other elevation angles and less energy at low angles. It is a good frequency for lower-cost radars and for long-range radars such as those for the detection of satellites. It is also the frequency region where it is theoretically difficult to reduce the radar cross section of most types of airborne targets.

In spite of its many attractive features, there have not been many applications of radar in this frequency range because its limitations do not always counterbalance its advantages.

UHF (300 to 1000 MHz). Much of what has been said regarding VHF applies to UHF. However, natural external noise is much less of a problem, and beamwidths are narrower than at VHF. Weather effects usually are not a bother. With a suitably large antenna, it is a good frequency for reliable long-range surveillance radar, especially for extraterrestrial targets such as spacecraft and ballistic missiles. It is well suited for AEW (airborne early warning), e.g., airborne radar that uses AMTI for the detection of aircraft. Solid-state transmitters can generate high power at UHF as well as offer the advantages of maintainability and wide bandwidth.

L Band (1.0 to 2.0 GHz). This is the preferred frequency band for land-based long-range air surveillance radars, such as the 200-nmi radars used for en route air traffic control [designated ARSR by the U.S. Federal Aviation Administration (FAA)]. It is possible to achieve good MTI performance at these frequencies and to obtain high power with narrow-beamwidth antennas. External noise is low. Military 3D radars can be found at L band, but they also are at S band. L band is also suitable for large radars that must detect extraterrestrial targets at long range.

S Band (2.0 to 4.0 GHz). Air surveillance radars can be of long range at S band, but long range usually is more difficult to achieve than at lower frequencies. The blind speeds that occur with MTI radar are more numerous as the frequency increases, thus making MTI less capable. The echo from rain can significantly reduce the range of S-band radars. However, it is the preferred frequency band for long-range weather radars that must make accurate estimates of rainfall rate. It is also a good frequency for medium-range air surveillance applications such as the airport surveillance radar (ASR) found at air terminals. The narrower beamwidths at this frequency can provide good angular accuracy and resolution and make it easier to reduce the effects of hostile main-beam jamming that might be encountered by military radars. Military 3D radars and height finding radars are also found at this frequency because of the narrower elevation beamwidths that can be obtained at the higher frequencies. Long-range airborne air surveillance pulse doppler radars, such as AWACS (Airborne Warning and Control System) also operate in this band.

Generally, frequencies lower than S band are well suited for air surveillance (detection and low-data-rate tracking of many aircraft within a large volume). Frequencies above S band are better for information gathering, such as high-data-rate precision tracking and the recognition of individual targets. If a single frequency must be used for both air surveillance and precision tracking, as in military air defense systems based on phased array multifunction radar, a suitable compromise might be S band.

C Band (4.0 to 8.0 GHz). This band lies between the S and X bands and can be described as a compromise between the two. It is difficult, however, to achieve long-range air surveillance radars at this or higher frequencies. It is the frequency where one can find long-range precision instrumentation radars used for the accurate tracking of missiles. This frequency band has also been used for multifunction phased array air defense radars and for medium-range weather radars.

X Band (8 to 12.5 GHz). This is a popular frequency band for military weapon control (tracking) radar and for civil applications. Shipboard navigation and piloting, weather avoidance, doppler navigation, and the police speed meter are all found at X band. Radars at this frequency are generally of convenient size and are thus of interest for applications where mobility and light weight are important and long range is not. It is advantageous for information gathering as in high-resolution radar because of the wide bandwidth that makes it possible to generate short pulses (or wideband pulse compression) and the narrow beamwidths that can be obtained with relatively small-size antennas. An X-band radar may be small enough to hold in one's hand or as large as the MIT Lincoln Laboratory Haystack Hill radar with its 120-ft-diameter antenna and average radiated power of about 500 kW. Rain, however, can be debilitating to X-band radar.

K_u, K, and K_a Bands (12.5 to 40 GHz). The original K-band radars developed during World War II were centered at a wavelength of 1.25 cm (24 GHz). This proved to be a poor choice since it is too close to the resonance wavelength of water vapor (22.2 GHz), where absorption can reduce the range of a radar. Later this band was subdivided into two bands on either side of the water-vapor absorption frequency. The lower frequency band was designated K_u, and the upper band was designated K_a. These frequencies are of interest because of the wide bandwidths and the narrow beamwidths that can be achieved with small apertures. However, it is difficult to generate and radiate high power. Limitations due to rain clutter and attenuation are increasingly difficult at the higher frequencies. Thus not many radar applications are found at these frequencies. However, the airport surface detection radar for the location and control of ground traffic at airports is at K_u band because of the need for high resolution. The disadvantages that characterize this band are not important in this particular application because of the short range.

Millimeter Wavelengths (above 40 GHz). Although the wavelength of K_a band is about 8.5 millimeters (a frequency of 35 GHz), the technology of K_a-band radar is more like that of microwaves than that of millimeter waves and is seldom considered to be representative of the millimeter-wave region. Millimeter-wave radar, therefore, is taken to be the frequency region from 40 to 300 GHz. The exceptionally high attenuation caused by the atmospheric oxygen absorption line at 60 GHz precludes serious applications in the vicinity of this frequency within the atmosphere. Therefore, the 94-GHz-frequency region (3-mm wavelength) is generally what is thought of as a "typical" frequency representative of millimeter radar.

The millimeter-wave region above 40 GHz has been further subdivided into letter bands in the IEEE Standard, as shown in Table 1.1. Although there has been much interest in the millimeter portion of the electromagnetic spectrum, there have been no operational radars above K_a band. High-power sensitive receivers and low-loss transmission lines are difficult to obtain at millimeter wavelengths, but such problems are not basic. The major reason for the limited utility of this frequency region is the high attenuation that occurs even in the "clear" atmosphere. The so-called propagation window at 94 GHz is actually of greater attenuation than the attenuation at the water-vapor absorption line at 22.2 GHz. The millimeter-wave region is more likely to be of interest for operation in space, where there is no atmospheric attenuation. It might also be considered for short-range applications within the atmosphere where the total attenuation is not large and can be tolerated.

Laser Frequencies. Coherent power of reasonable magnitude and efficiency, along with narrow directive beams, can be obtained from lasers in the infrared, optical, and ultraviolet region of the spectrum. The good angular resolution and range resolution possible with lasers make them attractive for target information-gathering applications, such as precision ranging and imaging. They have had application in military range finders and in distance measurement for surveying. They have been considered for use from space for measuring profiles of atmospheric temperature, water vapor, and ozone, as well as measuring cloud height and tropospheric winds. Lasers are not suitable for wide-area surveillance because of their relatively small physical aperture area. A serious limitation of the laser is its inability to operate effectively in rain, clouds, or fog.

1.5 RADAR NOMENCLATURE

Military electronic equipment, including radar, is designated by the Joint Electronics Type Designation System (JETDS), formerly known as the Joint Army-Navy Nomenclature System (AN System), as described in Military Standard MIL-STD-196D. The letter portion of the designation consists of the letters AN, a slant bar, and three additional letters appropriately selected to indicate where the equipment is installed, the type of equipment, and the purpose of the equipment. Table 1.2 lists the equipment indicator letters. Following the three letters are a dash and a numeral. The numeral is assigned in sequence for that particular combination of letters. Thus the designation AN/SPS-49 is for a shipboard surveillance radar. The number 49 identifies the particular equipment and indicates it is the forty-ninth in the SPS category to which a JETDS designation has been assigned. A suffix letter (A, B, C, etc.) follows the original designation for each modification where interchangeability has been maintained. A change in the power input voltage, phase, or frequency is identified by the addition of the letters X, Y, or Z to the basic nomenclature. When the designation is followed by a dash, the letter T, and a number, the equipment is designed for training. The letter V in parentheses added to the designation indicates variable systems (those whose functions may be varied through the addition or deletion of sets, groups, units, or combinations thereof). Experimental and developmental systems sometimes are assigned special indicators enclosed in parentheses, immediately following the regular designation, to identify the development organization; for example, (XB) indicates the Naval Research Laboratory, and (XW) indicates the

Rome Air Development Center. Empty parentheses, commonly called "bow-legs," are used for a developmental or series "generic" assignment.

TABLE 1.2 JETDS Equipment Indicators*

Installation (first letter)	Type of equipment (second letter)	Purpose (third letter)
A Piloted aircraft	A Invisible light, heat radiation	A Auxiliary assembly
B Underwater mobile, submarine	C Carrier	B Bombing
D Pilotless carrier	D Radiac (radioactive detection, indication, and computation devices)	C Communications (receiving and transmitting)
F Fixed ground		D Direction finder reconnaissance and/or surveillance
G General ground use		
K Amphibious	E Laser	
M Ground, mobile	G Telegraph or teletype	E Ejection and/or release
P Portable	I Interphone and public address	G Fire control or searchlight directing
S Water		
T Ground, transportable	J Electromechanical or inertial wire-covered	H Recording and/or reproducing (graphic meteorological)
U General utility	K Telemetering	
V Ground, vehicular	L Countermeasures	K Computing
W Water surface and underwater combination	M Meteorological	M Maintenance and/or test assemblies (including tools)
	N Sound in air	
Z Piloted and pilotless airborne vehicle combination	P Radar	N Navigational aids (including altimeters, beacons, compasses, racons, depth sounding, approach and landing)
	Q Sonar and underwater sound	
	R Radio	
	S Special types, magnetic, etc., or combinations of types	
	T Telephone (wire)	Q Special or combination of purposes
	V Visual and visible light	R Receiving, passive detecting
	W Armament (peculiar to armament, not otherwise covered)	S Detecting and/or range and bearing, search
	X Facsimile or television	T Transmitting
	Y Data processing	W Automatic flight or remote control
		X Identification and recognition
		Y Surveillance (search, detect, and multiple-target tracking) and control (both fire control and air control)

*From Military Standard Joint Electronics Type Designation System, MIL-STD-196D, Jan. 19, 1985.

In the first column of Table 1.2, the installation letter M is used for equipment installed and operated from a vehicle whose sole function is to house and transport the equipment. The letter T is used for ground equipment that is normally moved from place to place and is not capable of operation while being transported. The letter V is used for equipment installed in a vehicle designed for functions other than carrying electronic equipment (such as a tank). The letter G is used for equipment capable of being used in two or more different types of ground installations. Equipment specifically designed to operate while being carried by a person is designated by the installation letter P. The letter U implies use in a combination of two or more general installation classes, such as ground, aircraft, and ship. The letter Z is for equipment in a combination of airborne installations, such as aircraft, drones, and guided missiles.

The equipment indicator letter (second column of Table 1.2) that designates radar is the letter P; but it is also used for beacons which function with a radar, electronic recognition and identification systems, and pulse-type navigation equipment.

Canadian, Australian, New Zealand, and United Kingdom electronic equipment can also be covered by the JETDS designations. For example, a block of numbers from 500 to 599 and 2500 to 2599 is reserved for Canadian use.

The radars used in the air traffic control system of the FAA utilize the following nomenclature:

ASR   airport surveillance radar

ARSR   air route surveillance radar

ASDE   airport surface detection equipment

TDWR   terminal doppler weather radar

The numeral following the letter designation indicates the particular radar model of that type.

Weather radars in use by the U.S. National Weather Service employ the designation WSR, which is not associated with the JETDS nomenclature. The number following the designation indicates the year in which the radar was put into service. When a letter follows the number, it indicates the letter-band designation. Thus, the WSR-74C is a C-band weather radar introduced in 1974.

GENERAL BOOKS ON RADAR

1. Skolnik, M. I.: "Introduction to Radar Systems," 2d ed., McGraw-Hill Book Co., New York, 1980.

2. Skolnik, M. I.: "Radar Applications," IEEE Press, New York, 1988.

3. Barton, D. K.: "Modern Radar System Analysis," Artech House, Norwood, Mass., 1988.

4. Barton, D. K.: "Radar System Analysis," originally published by Prentice-Hall in 1964 and republished by Artech House, Norwood, Mass., in 1977.

5. Nathanson, F.: "Radar Signal Processing and the Environment," McGraw-Hill Book Co., New York, 1969.

6. Brookner, E.: "Aspects of Modern Radar," Artech House, Norwood, Mass., 1988.

7. Brookner, E.: "Radar Technology," Artech House, Norwood, Mass., 1977.

8. Levanon, N.: "Radar Principles," John Wiley & Sons, New York, 1988.

9. Eaves, J. L., and E. K. Reedy: "Principles of Modern Radar," Van Nostrand Reinhold Company, 1987.

10. Berkowitz, R.: "Modern Radar," John Wiley & Sons, New York, 1965.
11. Reintjes, J. F., and G. T. Coate: "Principles of Radar," McGraw-Hill Book Co., New York, 1952.
12. Ridenour, L. N.: "Radar System Engineering," McGraw-Hill Book Co., New York, 1947.

CHAPTER 2
PREDICTION OF RADAR RANGE

Lamont V. Blake
Electronics Consultant

2.1 INTRODUCTION

The basic physics governing the prediction of radar maximum detection range, for a specified target under free-space conditions with detection limited by thermal noise, has been well understood since the earliest days of radar. The term *free space* implies (in the present context) that a spherical region of space, centered at the radar and extending to considerably beyond the target, is empty except for the radar and the target. (*Considerably* as used here can be precisely defined for specific radars, but a general definition would be lengthy and not very useful.) It also implies that the only radar-frequency electromagnetic waves detectable within this region, other than those emanating from the radar itself, are from natural thermal and quasi-thermal noise sources, as described in Sec. 2.5.

Although this condition is never fully realized, it is approximated for some radar situations. Under many non-free-space conditions and with radically nonthermal forms of background noise, the prediction problem is considerably more complicated. Complications not considered in early analyses also result from modification of the signal and noise relationship by the receiving-system circuitry (*signal processing*).

In this chapter the free-space equation will be presented, basic signal processing will be discussed, and some of the most important non-free-space environments will be considered. The effect of some common nonthermal types of noise will be considered. Although it will not be feasible to consider all the possible types of radar situations, the methods to be described will indicate the general nature of the necessary procedures for environments and conditions not specifically treated here. Some of the specialized types of radar, for which special analyses are required, are described in later chapters of this handbook.

Definitions. The radar range equation contains many parameters of the radar system and its environment, and some of them have definitions that are interdependent. As will be discussed in Sec. 2.3, some definitions contain an element of arbitrariness, and it is common for different authors to employ different definitions of some of the range-equation factors. Of course, when generally accepted definitions do exist, they should be observed. But even more important, although some arbitrariness may be permissible for individual definitions, once a particular definition has been adopted for one of the range-equation factors, it will be found that definition of one or more of the other factors is no longer arbitrary.

As an example, for pulsed radar the definitions of pulse power and pulse length are highly arbitrary individually, but once a definition for either one of them has been adopted, the definition of the other is determined by the constraint that their product *must* equal the *pulse energy*. In this chapter, a set of definitions that are believed to conform to such rules of consistency, as well as to definitions adopted by standards organizations, will be presented.

Conventions. Because of the wide variability of propagation-path and other range-equation factors, certain *conventions* are necessary for predicting range under standard conditions when specific values of those factors are not known. A convention is a generally accepted *standard assumption*, which may never be encountered exactly in practice but which falls within the range of conditions that will be encountered, preferably somewhere near the middle of the range. An example is the conventional geophysical assumption, for calculating certain earth environment effects that depend on the earth's curvature, that the earth is a perfect sphere of radius 6370 km. The importance of conventions is that they provide a common basis for comparison of competing radar systems. To the extent that they are fairly representative of typical conditions, they also allow prediction of a realistic detection range. Commonly accepted conventions will be used in this chapter, and where needed conventions do not exist, appropriate ones will be suggested.

Range Prediction Philosophy. It is apparent from the foregoing discussion that a range prediction based on conventional assumptions will not necessarily be confirmed exactly by experimental results. This conclusion is further warranted by the statistical nature of the "noise" which is usually the limiting factor in the signal detection process. In other words, even if all the environmental factors were precisely known, a range prediction would not be likely to be verified exactly by the result of a single experiment. A statistical prediction refers to the average result of many trials. Therefore, radar range prediction is not an exact science. (In fact, the lesson of quantum mechanics seems to be that there is no such thing as an exact science in the strict sense.)

Nevertheless, calculations to predict radar range are useful. However inexact they may be on an absolute basis, they permit meaningful comparisons of the expected relative performance of competing system designs, and they indicate the relative range performance change to be expected if the radar parameters or environmental conditions are changed. They are therefore a powerful tool for the system designer. The predicted range is a figure of merit for a proposed radar system. It is not necessarily a complete one, since other factors such as target-position-measurement accuracy, data rate, reliability, serviceability, size, weight, and cost may also be important. Despite the inexactness of range predictions in the absolute sense, the error can be made small enough that the calculated range is a good indication of the performance to be expected under average environmental conditions. Section 2.10 is a more detailed discussion of prediction accuracy.

Attempts to evaluate range prediction factors accurately, to better than perhaps 1 dB, are sometimes disparaged on the grounds that some factors are unlikely to be known with accuracy in operational situations and that hence it is useless to seek better accuracy for any factor. Although there is some basis for this viewpoint, the overall accuracy will be unnecessarily degraded if the accuracy of all the factors in the equation is deliberately reduced. Therefore it is recommended that range predictions be based on as careful an evaluation of all the factors as is possible. A goal of 0.1 dB accuracy is perhaps reasonable, although admittedly it may be impossible to evaluate all the factors in the equation with that degree of precision.

Historical Notes. Possibly the first comprehensive treatise on radar maximum-range prediction was that of Omberg and Norton,[1] published first as a U.S. Army Signal Corps report in 1943. It presents a fairly detailed range equation and contains information on evaluating some of the more problematical factors, such as multipath interference and minimum detectable signal, within the limitations of the then-available knowledge. The signal detection process was assumed to be based on visual observation of a cathode-ray-tube display. The antenna was assumed to "searchlight" the target. Statistical aspects of signal detection were not considered.

D. O. North,[2] in a classical report published with a military security classification in 1943, outlined the basic theory of a statistical treatment of signal detection. (This report was republished in *Proceedings of the IEEE*, but not until 1963.) He introduced the concepts that are now called *probability of detection* and *false-alarm probability*, and he clearly delineated the role of integration in the detection of pulse signals. This report also introduced the concept of the *matched filter*, a contribution for which it had achieved some recognition prior to 1963. But except for the matched-filter concept, its contributions to signal detection theory were virtually unrecognized by radar engineers generally until the report was republished 20 years after its first appearance.

In a famous report[3] first published in 1948 and republished in *IRE Transactions on Information Theory* in 1960, J. I. Marcum extensively developed the statistical theory of detection with the aid of machine computation, referencing North's report. He computed probabilities of detection as a function of a range parameter related to signal-to-noise ratio, for various numbers of pulses integrated and for various values of a false-alarm parameter which he designated *false-alarm number*. He employed this type of computation to study the effects of various amounts and kinds of integration, different detector (demodulator) types, losses incurred by "collapsing" one spatial coordinate on the radar display, and various other effects. His results are presented as curves for probability of detection as a function of the ratio of the actual range to that at which the signal-to-noise ratio is unity, on the assumption that the received-signal power is inversely proportional to the fourth power of the range. Since this proportionality holds only for a target in free space, application of Marcum's results is sometimes complicated by this mode of presentation.

Marcum considered only *steady* signals (target cross section not varying during the period of observation), and most of his results assume the use of a square-law detector. Robertson[4] has published exceptionally detailed and useful steady-signal results applicable to the linear-rectifier detector, which is the type of detector almost universally used. (The square-law-detector results are also useful because they differ very little from the linear-detector results.) Swerling extended Marcum's work to include the case of fluctuating signals.[5] His report was republished in *IRE Transactions on Information Theory* in 1960. Fehlner[6] recomputed Marcum's and Swerling's results and presented them in the more useful form of curves with the signal-to-noise power ratio as the abscissa. The fluctuating-signal problem has subsequently been further treated by Kaplan,[7] Schwartz,[8] Heidbreder and Mitchell,[9] Bates,[10] and others.

Hall[11] published in 1956 a comprehensive paper on radar range prediction in which the concepts of probability of detection, false-alarm probability, the relative effects of predetection and postdetection integration, and the effects of scanning the antenna beam were considered. The range equation was formulated in terms of an ideal (matched-filter) utilization of the available received-signal power, with loss factors to account for departures from the ideal.

Blake[12] published an updating of the subject in 1961, applying recent advances in system-noise-temperature calculation, atmospheric absorption, plotting of coverage diagrams based on a realistic atmospheric refractive-index model, and multipath-interference calculation. This work was followed by Naval Research Laboratory (NRL) reports[13] and a book[14] in which further details were presented.

Contributions to the subject of range prediction have also been made by many others, far too numerous to mention by name. Only the major contributions can be recognized in this brief history. Special mention should be made, however, of the many contributions in two volumes (13 and 24) of the MIT Radiation Laboratory Series, edited by Kerr[15] and by Lawson and Uhlenbeck.[16] Much use is made in this chapter of results originally published in those volumes.

2.2 RANGE EQUATIONS

Radar Transmission Equation. The following equation, in the form given in Kerr,[15] is called the *transmission equation* for monostatic radar (one in which the transmitter and receiver are colocated):

$$\frac{P_r}{P_t} = \frac{G_t G_r \sigma \lambda^2 F_t^2 F_r^2}{(4\pi)^3 R^4} \tag{2.1}$$

where P_r = received-signal power (at antenna terminals)
  P_t = transmitted-signal power (at antenna terminals)
  G_t = transmitting-antenna power gain
  G_r = receiving-antenna power gain
  σ = radar target cross section
  λ = wavelength
  F_t = pattern propagation factor for transmitting-antenna-to-target path
  F_r = pattern propagation factor for target-to-receiving-antenna path
  R = radar-to-target distance (range)

This equation is not identical to Kerr's; he assumes that the same antenna is used for transmission and reception, so that $G_t G_r$ becomes G^2 and $F_t^2 F_r^2$ becomes F^4. The only factors in the equation that require explanation are the pattern propagation factors F_t and F_r. The factor F_t is defined as the ratio, at the target position, of the field strength E to that which would exist at the same distance from the radar in free space and in the antenna beam maximum-gain direction, E_0. The factor F_r is analogously defined. These factors account for the possibility that the target is not in the beam maxima (G_t and G_r are the gains in the maxima) and for any propagation gain or loss that would not occur in free space. The most common of these effects are absorption, diffraction and shadowing, certain types of refraction effects, and multipath interference.

For a target in free space and in the maxima of both the transmit and receive antenna patterns, $F_t = F_r = 1$. Detailed definitions of these and other range-equation factors are given in Secs. 2.3 to 2.7.

Maximum-Range Equation. Equation (2.1) is not a range equation as it stands, although it can be rewritten in the form

$$R = \left[\frac{P_t G_t G_r \sigma \lambda^2 F_t^2 F_r^2}{(4\pi)^3 P_r} \right]^{1/4} \tag{2.2}$$

This equation states that R is the range at which the received-echo power will be P_r if the transmitted power is P_t, target size σ, and so forth. It becomes a maximum-range equation by the simple expedient of attaching subscripts to P_r and R so that they become $P_{r,\text{min}}$ and R_{max}. That is, when the value of P_r in Eq. (2.2) is the minimum detectable value, the corresponding range is the maximum range of the radar.

However, this is a very rudimentary maximum-range equation, of limited usefulness. A first step toward a more useful equation is replacement of P_r by a more readily evaluated expression. This is done by first defining the signal-to-noise power ratio:

$$S/N = P_r/P_n \tag{2.3}$$

where P_n is the power level of the noise in the receiving system, which determines the minimum value of P_r that can be detected. This noise power, in turn, can be expressed in terms of a receiving-system noise temperature T_s:

$$P_n = kT_s B_n \tag{2.4}$$

where k is Boltzmann's constant (1.38×10^{-23} Ws/K) and B_n is the noise bandwidth of the receiver predetection filter, hertz. (These quantities are defined more completely in Secs. 2.3 and 2.5.[17]) Therefore,

$$P_r = (S/N) kT_s B_n \tag{2.5}$$

This expression can now be substituted for P_r in Eq. (2.2).

A further convenient modification is to redefine P_t as the transmitter power at the terminals of the transmitter, rather than [as in Eq. (2.1)] the usually somewhat smaller power that is actually delivered to the antenna terminals because of loss in the transmission line. This redefinition is desirable because when radar system designers or manufacturers specify a transmitter power, the actual transmitter output power is usually meant.

With this changed definition, P_t must be replaced by P_t/L_t, where L_t is a *loss factor* defined as the ratio of the transmitter output power to the power actually delivered to the antenna. (Therefore, $L_t \geqq 1$.)

It will later prove convenient to introduce additional loss factors similarly related to other factors in the range equation. These loss factors are multiplicative;

that is, if there are, for example, three loss factors L_1, L_2, and L_3, they can be represented by a single *system loss factor* $L = L_1L_2L_3$. The resulting maximum-range equation is

$$R_{max} = \left[\frac{P_tG_tG_r\sigma\lambda^2F_t^2F_r^2}{(4\pi)^3(S/N)_{min}kT_sB_nL} \right]^{1/4} \tag{2.6}$$

The quantities $(S/N)_{min}$ and T_s as here defined are to be evaluated at the antenna terminals, and that fact detracts from the utility of this form of the equation. As thus defined, $(S/N)_{min}$ is not independent of B_n, and the dependence is difficult to take into account in this formulation. If that dependence were ignored, this equation would imply that R_{max} is an inverse function of B_n; i.e., if all the other range-equation factors were held constant, R_{max} could be made as large as desired simply by making B_n sufficiently small. This is well known to be untrue. To remedy this difficulty, several factors must be considered. It is convenient to do this in terms of a particular transmitted waveform.

Pulse Radar Equation. Equation (2.6) does not specify the nature of the transmitted signal; it can be CW (continuous-wave), amplitude- or frequency-modulated, or pulsed. It is advantageous to modify this equation for the specific case of pulse radars and in so doing to remove the "bandwidth" difficulty encountered in using Eq. (2.6). Pulse radars are of course the most common type. As will be shown, although the equation thus modified will ostensibly be restricted to pulse radars, it can in fact be applied to other types of radar by appropriate reinterpretation of certain parameters.

D. O. North[2] demonstrated that the detectable signal-to-noise ratio $(S/N)_{min}$ will have its smallest possible value when the receiver bandwidth B_n has a particular (optimum) value and that this optimum value of B_n is inversely proportional to the pulse length τ. This implies that an equation can be written with pulse length in the numerator rather than with bandwidth in the denominator. North also showed that signal detectability is improved by *integrating* successive signal and noise samples in the receiver and that the detectability is a function of the total integrated signal energy. (The integration process is discussed in Sec. 2.4.) Finally, he showed that *when the receiver filter is matched to the pulse waveform*, the ratio of the received-pulse energy to the noise power spectral density at the output of the receiver filter is maximized *and is equal to the signal-to-noise power ratio at the antenna terminals*. The term *matched* in this context means, partially, that the filter bandwidth is optimum. The full meaning is that the filter transfer function is equal to the complex conjugate of the pulse spectrum.

Detectability Factor. An equation based on these facts can be derived by utilizing a parameter called *detectability factor*, defined by the Institute of Electrical and Electronics Engineers (IEEE)[18] as follows: "In pulsed radar, the ratio of single-pulse signal energy to noise power per unit bandwidth that provides stated probabilities of detection and false alarm, measured in the intermediate-frequency amplifier and using an intermediate-frequency filter matched to the single pulse, followed by optimum video integration." Deferring for the moment discussion of the meaning of some aspects of this definition, it can be expressed mathematically as follows:*

$$D_0 = E_r/N_0 = P_r\tau/kT_s \tag{2.7}$$

where D_0 is the detectability factor, E_r is the received-pulse energy, and N_0 is the noise power per unit bandwidth, both measured at the output of the receiver filter (i.e., at the demodulator input terminals).

The next step in this reformulation of the range equation is to define a bandwidth correction factor C_B, to allow for the possibility that the receiver filter bandwidth B_n may not be optimum. This factor is defined by the following relationship:

$$(S/N)_{\min}B_n = (S/N)_{\min(0)}B_{n,\,\mathrm{opt}}C_B = D_0B_{n,\,\mathrm{opt}}C_B \tag{2.8}$$

where $B_{n,\mathrm{opt}}$ is the optimum value of B_n. The factor C_B has been named the *bandwidth correction factor* because it was originally defined in terms of bandwidth optimization, but in actuality it is a *filter mismatch factor*, in the North matched-filter sense. As Eq. (2.8) implies, $C_B \geqq 1$. Evaluation of C_B is discussed in Sec. 2.3.

The quantity $(S/N)_{\min(0)}$ in Eq. (2.8) is the optimum-bandwidth (matched-filter) value of $(S/N)_{\min}$, which North showed to be equal to D_0. It is this fact that allows the range equation to be written, as desired, in terms of the signal-to-noise ratio at the detector input terminals (filter output) rather than the ratio at the antenna terminals.

North deduced that $B_{n,\mathrm{opt}} = 1/\tau$ exactly. As will be discussed later, some radar detection experiments with human observers have subsequently suggested that the constant of proportionality may not be exactly unity. However, North's analysis is theoretically correct for pulses of rectangular shape and for the definition to be given in Sec. 2.3 for the noise bandwidth B_n. For pulses of other shapes the pulse-length–bandwidth relationship is subject to the particular definition used for the pulse length. That definition is not an issue, of course, when the pulse shape is rectangular.

Based on that result, the range equation can be written with pulse length in the numerator by means of the following equivalence, in terms of the parameters of Eq. (2.8):

$$(S/N)_{\min}B_n = D_0C_B/\tau \tag{2.9}$$

The expression of the left-hand side of Eq. (2.9), where it occurs in the denominator of Eq. (2.6), can now be replaced by the expression of the right-hand side. The result is the desired pulse radar form of the range equation:

$$R_{\max} = \left[\frac{P_t\tau G_tG_r\sigma\lambda^2F_t^2F_r^2}{(4\pi)^3kT_sD_0C_BL}\right]^{1/4} \tag{2.10}$$

---

*In some of the literature it is stated that the matched-filter output signal-to-noise ratio is $2E_r/N_0$. That statement is based on defining peak signal power as the instantaneous value occurring not only at the peak of the output-pulse waveform but also at the peak of an RF cycle, where the instantaneous power is theoretically twice the average power. North's definition, based on the signal power averaged over an RF cycle, is consistent with the definition of noise power as the average over both the RF cycles and the random noise fluctuations.

A primary advantage of this formulation of the equation is that standard curves for the parameter D_0, as a function of the number of pulses integrated, are available, with the probabilities of detection and false alarm as parameters (Sec. 2.4). Calculation of these curves is necessarily done in terms of D_0, the signal-to-noise ratio at the demodulator input terminals.

The emphasis of this equation on the significance of the pulse *energy* (the product $P_t\tau$ in the numerator) is valuable to the system designer. It also provides a simple answer to the question of which pulse length to use in the range equation when the radar employs *pulse compression*, in which a coded pulse waveform of relatively long duration is transmitted and then "compressed" to a short pulse upon reception. The correct answer is deduced from the fact that the product $P_t\tau$ must equal the transmitted pulse energy. Therefore if the pulse power P_t is the power of the long (uncompressed) transmitted pulse, then τ must be the duration of that pulse.

A further advantage of this form of the equation, or more specifically of the definition of the detectability factor, is the indicated dependence of the radar detection range on the *integration* of successive pulses, if any, that takes place in the receiving system. Integration is discussed in Sec. 2.4.

Finally, as was mentioned earlier, this formulation of the range equation, although derived specifically in terms of pulse radar parameters, can be applied to CW radars and to radars that utilize forms of signal modulation other than pulses. Its application to these other radar types is accomplished by redefining the parameters τ and D_0. Details of this procedure are presented in Ref. 14, Chaps. 2 and 9.

Probabilistic Notation. It has been mentioned (Sec. 2.1) that the radar signal detection process is basically probabilistic or statistical in nature. This results from the nature of the noise voltage that is always present in the receiver circuits. This voltage is randomly varying or fluctuating, and when it is intermixed with a radar echo signal, it becomes impossible to tell with certainty whether a momentary increase of the receiver output is due to a signal or to a chance noise fluctuation. However, it is possible to define probabilities for these two possibilities and to discuss the detection process in terms of them in a quantitative manner. The probability that the signal, when present, will be detected is called the *probability of detection, P_d*, and the probability that a noise fluctuation will be mistaken for a signal is called the *false-alarm probability, P_{fa}*.

The notations R_{max}, $P_{r,min}$, and $(S/N)_{min}$ can then be replaced by more precise notation, using subscripts to denote the applicable values of P_d and P_{fa}. However, the *fa* subscript is ordinarily suppressed, though implied. Thus R_{50} can denote the range for 0.5 (i.e., 50 percent) probability of detection and some separately specified false-alarm probability.

If the target cross section σ fluctuates, this fluctuation will alter the signal-plus-noise statistics. As mentioned in Sec. 2.1, this problem has been analyzed by Swerling[5] and others.[6-10] Curves have been calculated that allow determining the appropriate value of D_0 for the fluctuating-signal case, for various probabilities of detection and false alarm (Sec. 2.4).

Automatic Detection. Detection* is said to be *automatic* if the decision concerning the presence or absence of a received signal is made by a purely

---

*A note on various meanings of the words *detect, detector,* and *detection* is desirable here. In radio usage, a detector has come to mean either a frequency converter (e.g., a superheterodyne first detector) or a demodulator (often the "second detector" of a superheterodyne receiver, which is usually a linear rectifier). Then, *detection* means the waveform modification produced by such a device. An *automatic*

physical device, without direct human intervention. Such a device, described by North,[2] establishes a threshold voltage level (for example, by means of a biased diode). If the processed (e.g., integrated) receiver output exceeds the threshold (as evidenced by diode current flow), some mechanism is actuated to indicate this fact in an unequivocal fashion. The mechanism could be the lighting of a light, the ringing of a bell, or more generally the setting of a bit to 1 in a binary data channel wherein a 0 corresponds to no signal. Various additional consequences may then automatically ensue. The analysis of radar detection can thus be regarded as a problem in statistical decision theory.

Bistatic Radar Equation. The foregoing equations assume that the transmitting and receiving antennas are at the same location (monostatic radar). A bistatic radar (Chap. 25) is one for which the two antennas are widely separated, so that the distance and/or the direction from the transmitting antenna to the target are not necessarily the same as the distance and/or direction from the receiving antenna to the target. Moreover, since the signal reflected from the target to the receiving antenna is not directly backscattered, as it is for monostatic radar, the target cross section is not usually the same (for a given target viewed in a given aspect by the transmitting antenna). A *bistatic radar cross section* σ_b is defined to apply for this situation. The symbol σ in the preceding equations implies a monostatic cross section. Range equations for bistatic radar are obtained from the foregoing monostatic equations by replacing the range R and the target cross section σ by the corresponding bistatic quantities. The bistatic equivalent of R is $\sqrt{R_t R_r}$, where R_t is the distance from the transmitting antenna to the target and R_r is the distance from the target to the receiving antenna.

Equations in Practical Units. The equations that have been given are valid when a consistent system of units is used, such as the rationalized meter-kilogram-second (mks) system. In many applications, however, it is convenient or necessary to employ "mixed" units. Moreover, it is usually more convenient to express the wavelength λ in terms of the equivalent frequency in megahertz. It is also desirable to combine all the numerical factors and the various unit-conversion factors into a single numerical constant. For a particular system of mixed units, the following equation is obtained from Eq. (2.10):

$$R_{max} = 129.2 \left[\frac{P_{t(kW)}\tau_{\mu s}G_t G_r \sigma F_t^2 F_r^2}{f_{MHz}^2 T_s D_0 C_B L} \right]^{1/4} \tag{2.11}$$

The subscript notation R_{max} is now meant to imply the range corresponding to specified detection and false-alarm probabilities. For this equation, the range is given in international nautical miles. (One international nautical mile is exactly 1852 m.) The target cross section σ is in square meters, transmitter power P_t in kilowatts, pulse length τ in microseconds, frequency f in megahertz, and system noise temperature T_s in kelvins. (All other quantities are dimensionless.)

If the range is desired in units other than nautical miles (all other units remaining the same), in place of the factor 129.2 the following numerical constants should be used in Eq. (2.11):

---

detector, however, is a decision-making device—for example, a device that replaces the human observer of a cathode-ray-tube display—and in that context *detection* is the making of a positive decision. In this chapter the meaning will ordinarily be evident from the context. Where confusion might otherwise result, the term *detection-decision device* may be used to denote an automatic detector.

Range units	Constant, Eq. (2.11)
Statute miles	148.7
Kilometers	239.3
Thousands of yards	261.7
Thousands of feet	785.0

A decibel-logarithmic form of the range equation is sometimes useful. An equation of that type, corresponding to Eq. (2.11), is readily obtained as the algebraic sum of the logarithms of the terms of that equation (with appropriate multipliers for the decibel format and for exponents), since it involves only multiplication, division, and exponentiation.

2.3 DEFINITION AND EVALUATION OF RANGE FACTORS

There is an element of arbitrariness in the definition of most of the factors of the radar range equation, and for some of them more than one definition is in common use. Since the definitions in these cases are arbitrary, one definition is in principle as good as another. However, once a definition has been chosen for one factor, there is no longer freedom of choice for one or more of the others. The factors are interdependent, and mutual compatibility is essential. A set of definitions that are believed to be mutually compatible will be given here. Also, information needed for evaluating these factors will be given insofar as is practicable. Certain range-equation factors that present special problems will be considered at greater length in subsequent sections of the chapter.

Transmitter Power and Pulse Length. The radar transmission equation, from which all the subsequent range equations are derived, is an equation for the dimensionless ratio P_t/P_r. Consequently, the most basic requirement on the definition of P_t is that it agree with the definition of P_r. For a CW radar, the power (averaged over an RF cycle) is constant, and there is no definition problem. For a pulse radar, both P_t and P_r are usually defined as the *pulse power*, which is the *average power during the pulse*. More precisely,

$$P_t = \frac{1}{\tau} \int_{T/2}^{-T/2} W(t)\, dt \qquad (2.12)$$

where $W(t)$ is the instantaneous power (a function of time, t). The definition of $W(t)$, however, excludes "spikes," "tails," and any other transients that are not useful for radar detection. The time interval T is the pulse period ($=1/\text{PRF}$, where PRF is the pulse repetition frequency in pulses per second). Because of the exclusion of nonuseful portions of the waveform (as it exists at the transmitter output terminals), P_t as thus defined may be called the *effective* pulse power. It is often referred to as the *peak power*. However, peak power more properly signifies the power level at the peak of the pulse waveform (averaged over an RF cycle), and *pulse power* is more appropriate.

In the transmission equation, Eq. (2.1), P_t and P_r are the transmitted and received powers at the antenna terminals. As was mentioned in Sec. 2.2, P_t is now defined at the *transmitter output* terminals, and any loss between these terminals and the antenna input terminals must be expressed as a loss factor L_t.

The pulse power P_t and the pulse length τ must be defined so that their product is the pulse energy. Any definition of τ will produce this result if the same definition is used in Eq. (2.12). The customary definition, and the one recommended here, is the time duration between the half-power points of the envelope of the RF pulse (0.707-V points). For some purposes, such as analyzing the range resolution or accuracy, this arbitrary definition of the pulse length is not permissible. But the half-power definition is customary and acceptable for use in the range equation.

The range equation can be written with the product $P_t\tau$ replaced by the pulse energy E_t. The more detailed notation is used here because, for ordinary pulse radars, P_t and τ are usually given explicitly and E_t is not. However, the use of E_t in the equation does have the advantage of avoiding the problems of defining P_t and τ, and it is especially useful when complicated waveforms are transmitted.

If *coherent integration for a fixed integration time* is assumed, the equation can also be written with the transmitted *average* power in the numerator. For simple pulse radars, the average power is the product of pulse power, pulse length, and pulse repetition frequency. In this average-power formulation, the average power $\bar{P}_t$ is multiplied by the integration time t_i (assumed to be long compared with the interpulse period) to obtain the transmitted energy. Then the value of D_0 used is that which would apply if detection were based on observation of a single pulse. (See Sec. 2.4, Fig. 2.3.) The average-power formulation is especially useful for CW or pulse doppler radars.

Antenna Gain, Efficiency, and Loss Factor. The gains G_t and G_r are defined as the power gains of the antennas *in the maximum-gain direction*. If a target of interest is at an elevation angle not in the beam maxima, that fact is accounted for by the pattern propagation factors F_t and F_r, discussed in Sec. 2.6. The maximum power gain of an antenna is equal to its directivity (maximum directive gain) multiplied by its radiation efficiency.[19] The directivity D is defined in terms of the electric-field-strength pattern $E(\theta,\phi)$ by the expression

$$D = \frac{4\pi E^2_{\max}}{\displaystyle\int_0^{2\pi}\int_0^{\pi} E^2(\theta,\phi) \sin\theta\, d\theta\, d\phi} \tag{2.13}$$

where θ and ϕ are the angles of a spherical-coordinate system whose origin is at the antenna and $E_{\max}$ is the value of $E(\theta,\phi)$ in the maximum-gain direction.

The radiation efficiency of the transmitting antenna is the ratio of the power input at the antenna terminals to the power actually radiated (including minor-lobe radiation). In terms of the receiving antenna, the equivalent quantity is the ratio of the total signal power extracted from the incident field by the antenna, with a matched-load impedance, to the signal power actually delivered to a matched load. The reciprocal of the radiation efficiency is the antenna loss factor L_a, which plays a part in the calculation of antenna noise temperature (Sec. 2.5).

Measured antenna gains are usually power gains, whereas gains calculated from pattern measurements or theory are directive gains. If the antenna gain figures supplied for use in the range equations of this chapter are of the latter type, they must be converted to power gains by dividing them by the appropriate loss factor. For many simple antennas the ohmic losses are negligible, and in those cases the power gain and the directive gain are virtually equal. However, this is by no means a safe assumption in the absence of specific knowledge. Array antennas in particular are likely to have significant ohmic losses in waveguides or coaxial lines used to distribute the power among the radiating elements.

If separate transmitting and receiving antennas are used and if their maximum gains occur at different elevation angles (this is a possible though not a common situation), appropriate correction can be made by means of the pattern factors $f_t(\theta)$ and $f_r(\theta)$, contained in the pattern propagation factors F_t and F_r (Sec. 2.6).

Antenna Beamwidth. This property of the antenna does not appear explicitly in the range equations, but it affects the range calculation through its effect on the number of pulses integrated when the antenna scans. The conventional definition is the angular width of the beam between the half-power points of the pattern. *Pattern* is used here in the usual antenna sense, for one-way transmission. It is not the two-way pattern of the radar echo signal from a stationary target as the antenna scans past it.

If a radar target, as viewed from the radar antenna, has an angular dimension that is appreciable compared with the beamwidth, the target cross section becomes a function of the beamwidth (see Sec. 2.8). For computing an effective value of σ in this case, in principle a special definition of beamwidth is needed (Ref. 15, p. 483). For practical work, however, the error that results from using the half-power beamwidth in this application is usually acceptable.

Target Cross Section. The definition of *radar target cross section* that applies for use in the foregoing radar range equations is given in Chap. 11, and the reader is referred to that chapter for a detailed discussion of the subject. Here mention will be made of a few aspects of the definitions that are of particular significance to the range prediction problem.

Targets can be classified as either *point targets* or *distributed targets*. A point target is one for which (1) the maximum transverse separation of significant scattering elements is small compared with the length of the arc intercepted by the antenna beam at the target range and (2) the maximum radial separation of scattering elements is small compared with the range extent of the pulse. At distance R from the antenna, the transverse dimension of the antenna beam is R times the angular beamwidth in radians. The range extent of the pulse is $c\tau/2$, where c is the speed of wave propagation in free space, 3×10^5 km/s, and τ is the pulse duration in seconds. Most of the targets for which range prediction is ordinarily of interest are point targets, e.g., aircraft at appreciable distances from the radar.

However, range predictions for distributed targets are sometimes wanted. The moon, for example, is a distributed target if the radar beamwidth is comparable to or less than 0.5° or if the pulse length is less than about 11.6 ms. A rainstorm is another example of a distributed target. Often, distributed targets are of interest because echoes from them (called *radar clutter*) tend to mask the echoes from the point targets whose detection is desired (see Sec. 2.8). Echoes from rain may be regarded as clutter when they interfere with detection of aircraft or other point targets, but they are themselves the signals of prime interest for weather radar.

The radar range equation is derived initially for a point target, and when that equation or the subsequent equations derived from it are used to predict the detection range for distributed targets, complications arise. In many cases, however, the point-target equation can be used for distributed targets by employing a suitable "effective" value of σ (Sec. 2.8).

The cross section of any nonspherical target is a function of the aspect angle from which it is viewed by the radar. It may also be a function of the polarization of the radar electromagnetic field. Therefore, in order to be wholly meaningful, a radar range prediction for a specific target, such as an aircraft, must stipulate the target aspect angle assumed and the polarization employed. Ordinarily, the nose aspect of an aircraft (approaching target) is of principal interest. The commonly used polarizations are horizontal, vertical, and circular. Tabulations of radar cross-section measurements of aircraft sometimes give nose, tail, and broadside values.

If the values are obtained from *dynamic* (moving-target) measurements, they are usually time averages of fluctuating values; otherwise they are *static* values for a particular aspect. Since the *instantaneous* cross section of a target is a function of the aspect angle, targets that are in motion involving random changes of aspect have cross sections that fluctuate randomly with time, as was mentioned in Sec. 2.2. This fluctuation must be taken into account in the calculation of probability of detection, as will be discussed in Sec. 2.4. When σ fluctuates, the value to be used in the range equation as formulated here is the time average.

Because of the wide variation of cross-section values of real targets, the range performance of a radar system is often stated for a particular target–cross-section assumption. A favorite value for many applications is 1 m². This represents the approximate cross section of a small aircraft, nose aspect, although the range for different "small" aircraft may be from less than 0.1 m² to more than 10 m². Radars are often performance-tested by using a metallic sphere, sometimes carried aloft by a free balloon, as the target because the cross section of a sphere can be accurately calculated and it does not vary with the aspect angle or the polarization.

A special definition problem arises when the target is large enough to be nonuniformly illuminated by the radar. A ship, for example, may be tall enough so that the pattern propagation factor F has different values from the waterline to the top of the mast. This matter is discussed in Ref. 15, p. 472 ff.

Wavelength (Frequency). There is ordinarily no problem in definition or evaluation (measurement) of the frequency to be used in the radar range equation. However, some radars may use a very large transmission bandwidth, or they may change frequency on a pulse-to-pulse basis, so that a question can exist as to the frequency value to be used for predicting range. Also, the presence of f (or λ) in the range equations makes it clear that the range can be frequency-dependent, but the exact nature of the frequency dependence is not always obvious because other factors in the range equation are sometimes implicitly frequency-dependent. Therefore an analysis of how the range depends on frequency can be rather complicated, and the answer depends partly on what factors are regarded as frequency-dependent and which ones are held constant as the frequency is changed. For example, most antennas have gain that is strongly frequency-dependent, but some antenna types are virtually frequency-independent over a fairly wide frequency band.

Bandwidth and Matching Factors. The frequency-response width (bandwidth) of the receiver selective circuits appears explicitly in Eqs. (2.4) to (2.6), but it is an implicit factor in the other range equations as well, through the factor C_B. From Eq. (2.4) it is clear that B_n directly affects the noise level in the receiver output. In general it also affects the signal, but not necessarily in the same manner as the noise is affected, because the signal spectrum is not usually uniform. There is a value of B_n that optimizes the output signal-to-noise ratio, as indicated by Eq. (2.8), and this optimum bandwidth is inversely proportional to the pulse length τ. (This statement applies to pulse compression radars as well as to others if τ is, *in this context*, the *compressed* pulse length, since it is the compressed pulse that is amplified in the receiver. However, as has been emphasized in Sec. 2.2, in the numerator of the range equation the *uncompressed* pulse length must be used along with the actual radiated pulse power, P_t.)

Since the range equation (2.6) and those subsequently derived from it incorporate the assumption of Eq. (2.4) (namely, that the noise output power of the receiver is equal to $kT_sB_nG_0$), the definition of B_n—the *noise bandwidth*—must conform to that assumption. The resulting correct definition, due to North,[20] is

$$B_n = \frac{1}{G_0}\int_0^\infty G(f)\,df \tag{2.14}$$

where G_0 is the gain at the nominal radar frequency and $G(f)$ is the frequency–power gain characteristic of the receiver predetection circuits, from antenna to detector.

The definition specifies $G(f)$ to be the response characteristic of the predetection circuits only. That is because for maximum postdetection signal-to-noise ratio the video bandwidth should be equal to at least half of the predetection bandwidth; and if it is of this width or wider, its exact width has little or no effect on signal detectability (Ref. 16, p. 211 ff.).

It is common practice to describe receiver bandwidth as the value between half-power points of the frequency-response curve. Fortunately, this value is usually very close to the true noise bandwidth, although the exact relationship of the two bandwidths depends on the particular shape of the frequency-response curve (Ref. 16, p. 177).

The bandwidth correction factor C_B in Eqs. (2.10) and (2.11) accounts for the fact that if B_n is not the optimum value, a value of signal-to-noise ratio larger than the optimum-bandwidth value D_0 is required. Therefore $C_B \geq 1$. From data obtained in signal detection experiments during World War II at the Naval Research Laboratory, Haeff[21] devised the following empirical expression:

$$C_B = \frac{B_n\tau}{4\alpha}\left[1 + \frac{\alpha}{B_n\tau}\right]^2 \tag{2.15}$$

where B_n is the noise bandwidth, τ is the pulse length, and α is the product of τ and $B_{n,\text{opt}}$ (optimum bandwidth). Figure 2.1 is a plot of Haeff's equation.

Actually, Haeff deduced from his experiments, as did North from theoretical analysis, that $B_{n,\text{opt}} = 1/\tau$; that is, $\alpha = 1$, for rectangular-shaped pulses. How-

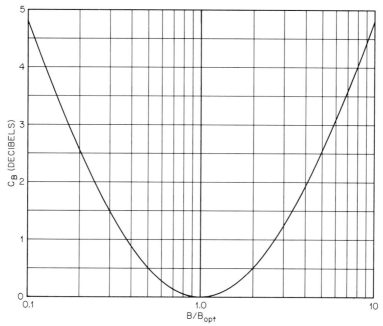

FIG. 2.1 Bandwidth correction factor C_B as a function of bandwidth B_n relative to optimum bandwidth B_{opt}; plotted from Haeff's empirical formula, Eq. (2.15).

ever, on the basis of experiments at the Massachusetts Institute of Technology (MIT) Radiation Laboratory conducted somewhat later (also with rectangular pulses), it was concluded (Ref. 16, p. 202) that $\alpha = 1.2$ for detection of signals by visual observation of cathode-ray-tube displays. Figure 2.2 is a plot of the Radiation Laboratory experimental results. The value 1.2 has subsequently been widely used for determining $B_{n,opt}$ in radar design and for computing C_B in radar range prediction.[11,12] However, North* has suggested that the $\alpha = 1.2$ figure may be based on a misinterpretation of the Radiation Laboratory data. Also, it has been noted† that the number of actual data points in Fig. 2.2 may be too few on which to base a good estimate of the optimum. Consequently, it is possible that the value of α for human observation of visual displays is much closer to 1 than was deduced by Lawson and Uhlenbeck. Fortunately, the minima of the curves (Fig. 2.2) are very broad and flat, and therefore the exact value of α does not make much difference for the usual range of values of $B_n\tau$.

The interpretation of C_B as a factor that accounts only for nonoptimum *width* of the predetection filter is permissible for simple pulse shapes and approximate results, but in principle it must also account for the complete amplitude-phase characteristic of the filter: its departure from a matched-filter characteristic. The

---

*In a private communication to the author in 1963.
†By M. I. Skolnik, editor of this handbook, in his review of this chapter.

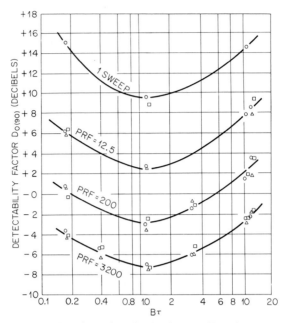

FIG. 2.2 Experimental results showing the effect of bandwidth (parameter $B\tau$) on 90 percent probability detectability factor $D_{0(90)}$, with pulse repetition frequency (PRF) as a parameter. The experiments were performed during World War II at the MIT Radiation Laboratory. (*From Ref. 15, Fig. 8.7.*)

matched-filter condition as stated by North[2] is that the receiver transfer characteristic must be the complex conjugate of the spectrum of the echo at the receiving-antenna terminals.

2.4 MINIMUM DETECTABLE SIGNAL-TO-NOISE RATIO

In Sec. 2.3, factors in the range equations were defined, and some information on how to evaluate them in typical cases was given. However, several very important factors were not covered because they are of sufficient importance to warrant more extensive treatment in separate sections. In this section and in Secs. 2.5 to 2.7, these additional factors will be discussed.

The quantities $P_{r,\min}$, $(S/N)_{\min}$, and D_0 are all related, as indicated in the development of Eqs. (2.3) to (2.9). Determination of the appropriate numbers to use for these quantities in their respective equations is a basic problem of radar range prediction. As will be seen, one of the problems is to define the meaning of *detectable*.

Integration of Signals. Detection of radar echo signals is usually (with some exceptions) accomplished by first integrating (e.g., adding) a sequence of received pulses and basing the detection decision on the resultant integrated signal voltage. Integrators that perform this operation will of course necessarily add noise as well as signals, but it is demonstrable that the ratio of the added signal voltages to the added noise voltages will be greater than the preintegration signal-to-noise ratio. Stated otherwise, the detectable signal-to-noise ratio evaluated ahead of an integrator will be smaller than when detection is performed by using single pulses.

There are many different methods of accomplishing integration. One method is the use of a feedback-loop delay line with a delay time equal to the interpulse period, so that signals (and noise) separated by exactly one pulse period will be directly added. Integration also occurs for visual detection by human radar operators if the phosphor of the cathode-ray-tube radar display (such as a PPI) has sufficient luminous persistence. In recent years, integration methods based on digital circuitry have become practical and are now perhaps the method of choice in many if not most cases.

The benefit of integration is a function of the number of pulses integrated. If integration is performed in the predetection stages of a receiver, ideally the addition of M equal-amplitude phase-coherent signal pulses will result in an output (integrated) pulse of voltage M times the single-pulse voltage. Adding M noise pulses, however, will result in an integrated noise pulse whose rms voltage is only $\sqrt{M}$ times as great as that of a single noise pulse if (as is true of ordinary receiver noise and many other types of noise) the added noise pulses are not phase-coherent. Therefore the signal-to-noise *voltage*-ratio improvement is, ideally, $M/\sqrt{M} = \sqrt{M}$. Consequently the signal-to-noise *power*-ratio improvement, and the reduction of the single-pulse minimum-detectable signal-to-noise power ratio, is equal to M.

Integration can also be performed after detection. In fact postdetection integration is used more commonly than is predetection integration, for reasons that will be explained, but the analysis of the resulting improvement is then much more complicated. After detection, the signals and the noise cannot be regarded as totally separate entities; the nonlinear process of detection produces an inseparable combination of signal and noise, so that one must then consider the comparison of signal-plus-noise to noise. As will be shown, the improvement that results from this type of integration is usually not as great as with ideal predetection integration of the same number of pulses. Nevertheless, postdetection integration produces worthwhile improvement. Moreover, "ideal" predetection integration is virtually unachievable because the echo fluctuation from most moving targets severely reduces the degree of phase coherence of successive received pulses. With rapidly fluctuating signals, in fact, postdetection integration will provide greater detectability improvement than does predetection integration, as discussed later in this section under the heading "Predetection Integration."

Number of Pulses Integrated. The number of pulses integrated is usually determined by the scanning speed of the antenna beam in conjunction with the antenna beamwidth in the plane of the scanning. The following equation can be used for calculating the number of pulses received between half-power-beam-width points for an azimuth-scanning radar:

$$M = \frac{\phi \, \overline{PRF}}{6 \, \overline{RPM} \cos \theta_e} \qquad (2.16)$$

where ϕ is the azimuth beamwidth, $\overline{PRF}$ is the radar pulse repetition frequency in hertz, $\overline{RPM}$ is the azimuthal scan rate in revolutions per minute, and θ_e is the target elevation angle. This formula strictly applies only if $\phi/\cos \theta_e$, the "effective" azimuth beamwidth, is less than 360°. (At values of θ_e for which $\phi/\cos \theta_e$ is greater than 360°, the number of pulses computed from this formula will obviously be meaningless. Practically, it is suggested that it be applied only for elevation angles such that $\phi/\cos \theta_e$ is less than about 90°.) This formula is based on the properties of spherical geometry. The formula also assumes that the beam maximum is tilted upward at the angle θ_e, but it can be applied with negligible error if θ_e is only approximately equal to the beam tilt angle.

The formula for the number of pulses within the half-power beamwidth for an azimuth- and elevation-scanning radar (which can be applied with minor modification to a radar scanning simultaneously in any two orthogonal angular directions) is

$$M = \frac{\phi\theta \ \overline{PRF}}{6\omega_v t_v \overline{RPM} \cos \theta_e} \qquad (2.17)$$

where ϕ and θ are the azimuth (horizontal-plane) and elevation (vertical-plane) beamwidths in *degrees*, θ_e is the target elevation angle, ω_v is the vertical scanning speed in degrees per second, and t_v is the vertical-scan period in seconds (including dead time if any). This formula should also be restricted to elevation angles for which $\phi/\cos \theta_e$ is less than about 90°. Here M is a function of the target elevation angle not only explicitly but also implicitly in that ω_v may be a function of θ_e.

Some modern radars, especially those capable of scanning by electronic means— i.e., without mechanical motion of the antenna—employ *step scanning*. In this method, the antenna beam is pointed in a fixed direction while a programmed number of pulses is radiated in that direction. Then the beam is shifted to a new direction, and the process is repeated. The number of pulses integrated in this scanning method is thus determined by the programming and not by the beamwidth. Also, the integrated pulses are then all of the same amplitude (except for the effect of target fluctuation), and so there is no *pattern loss* of the type described in Sec. 2.7. There is, however, a *statistical* loss if the target direction and the antenna beam maximum do not always coincide when the pulses are radiated.

Evaluation of Probabilities. As was mentioned in Sec. 2.2, if a threshold device is employed to make a decision as to the presence or absence of a signal in a background of noise, its performance can be described in terms of two probabilities: (1) the *probability of detection, P_d*, and (2) the *false-alarm probability, P_{fa}*. The threshold device is characterized by a value of receiver output voltage V_t (the *threshold*, equivalent to Marcum's[3] bias level), which, if exceeded, results in the decision report that a signal is present. If the threshold voltage is not exceeded at a particular instant, the detector reports "no signal."

There is always a definite probability that the threshold voltage will be exceeded when in fact no signal is present. The statistics of thermal random-noise voltage are such that there is a usually small but nonzero probability that it can attain a value at least equal to the saturation level of the receiver. (In the mathematical theory of thermal noise, there is a nonzero probability that it can attain any finite value, however large.) The probability that V_t is exceeded when no signal is present is the false-alarm probability. It is calculated from the equation

$$P_{fa} = \int_{V_t}^{\infty} p_n(v)dv \qquad (2.18)$$

where $p_n(v)$ is the probability density function of the noise. The probability of detection is given by the same expression, with the probability density function that of the signal-noise combination (usually called *signal-plus-noise*, but the "addition" is not necessarily linear):

$$P_d = \int_{V_t}^{\infty} p_{sn}(v)dv \qquad (2.19)$$

The signal-plus-noise probability density function $p_{sn}(v)$ depends on the signal-to-noise *ratio* as well as on the signal and noise statistics. Also, both p_n and p_{sn} are functions of the rectification law of the receiver detector and of any postdetection processing or circuit nonlinearities. Primarily, however, p_{sn} and therefore the probability of detection are functions of the signal-to-noise ratio. From Eq. (2.19), the variation of P_d with S/N can be determined. As would logically be assumed, it is a monotonic-increasing function of S/N for a given value of V_t. Similarly, the variation of P_{fa} as a function of V_t can be found from Eq. (2.18); it is a monotonic-decreasing function.

The method of applying these concepts to the prediction of radar range consists of four steps: (1) decide on a value of false-alarm probability that is acceptable (the typical procedure for making this decision will be described); (2) for this value of P_{fa}, find the required value of threshold voltage V_t, through Eq. (2.18); (3) decide on a desired value of P_d (in different circumstances, values ranging from below 0.5 up to as high as perhaps 0.99 may be selected); and (4) for this value of P_d and for the value of V_t found in step 2 find the required signal-to-noise ratio through Eq. (2.19). This requires evaluating the function $p_{sn}(v)$, taking into account the number of pulses integrated. Iteration is required, in this procedure, to find the value of D_0 corresponding to a specified probability of detection and number of pulses integrated. The value of D_0 thus found is the value to be used in the range equation [e.g., Eqs. (2.10) and (2.11)].

The process of finding the required value of D_0 for use in the range equation is greatly facilitated by curves that relate the number of pulses integrated to D_0 with P_{fa} and P_d as parameters. Many such curves have been published, and some representative ones are given as Figs. 2.3 through 2.7. The principal difficulty in computing them is determination of the probability density functions $p_n(v)$ and $p_{sn}(v)$ and in performing the requisite integrations. North[2] gives the exact functions that apply for single-pulse detection with a linear rectifier as detector and the approximations that apply when many pulses are integrated. The density functions appropriate to other situations, e.g., square-law detection and fluctuation of signals, are given by various authors.[3-10]

The decision as to the acceptable level of false-alarm probability is usually made in terms of a concept called *false-alarm time*, which will here be defined as the average time between false alarms. Other definitions are possible; Marcum[3] defines it as the time for which the probability of at least one false alarm is 0.5. However, the average time between false alarms seems a more practically useful concept. With it, for example, one can compute the average number of false alarms that will occur per hour, day, year, etc. With this definition, the false-alarm time is given by

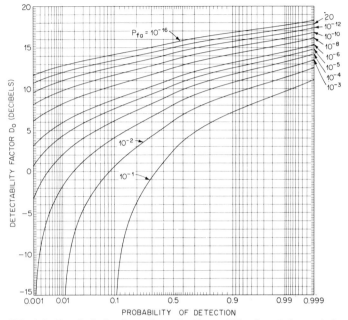

FIG. 2.3 Required signal-to-noise ratio (detectability factor) for a single-pulse, linear-detector, nonfluctuating target as a function of probability of detection with false-alarm probability (P_{fa}) as a parameter. *(From Ref. 13.)*

$$t_{fa} = \frac{M\,\tau}{P_{fa}} \tag{2.20}$$

where M is the number of pulses integrated and τ is the pulse duration.

This formula assumes that the integrator output is sampled at time intervals equal to τ. If range gates are employed and M pulses are integrated, if the ON time of the gate t_g is equal to or greater than the pulse length τ, and if there is some fraction of the time δ when no gates are open (*dead time*, e.g., just before, during, and after the occurrence of the transmitter pulse), then the formula is

$$t_{fa} = \frac{M\,t_g}{P_{fa}(1 - \delta)} \tag{2.21}$$

These false-alarm-time formulas assume that the receiver predetection noise bandwidth B_n is equal to or greater than the reciprocal of the pulse length and that the postdetection (video) bandwidth is equal to or greater than 0.5 B_n (as it usually is). These assumptions, usually met, amount to assuming that values of the noise voltage separated by the pulse duration are statistically independent; this independence occurs for times separated by $1/B_n$, sometimes called the Nyquist interval. Since ordinarily $B_n = 1/\tau$ and $t_g = \tau$, $1/B_n$ is sometimes used in place of τ or t_g in the false-alarm-time equations.

Marcum's false-alarm number n' is related to the false-alarm probability by the equation

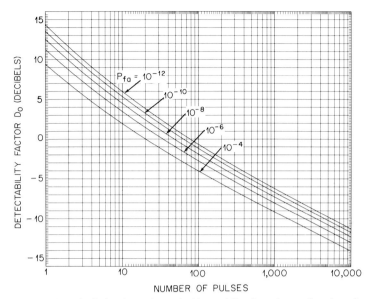

FIG. 2.4 Required signal-to-noise ratio (detectability factor) as a function of number of pulses noncoherently integrated, linear detector, nonfluctuating target, and 0.5 probability of detection. (*From Ref. 13.*)

$$1 - (1 - P_{fa})^{n'} = 0.5 \qquad (2.22a)$$

For the usual large values of n' that are of interest, a highly accurate approximate solution of this equation for P_{fa} is

$$P_{fa} = \frac{\log_e 0.5}{n'} = \frac{0.6931}{n'} \qquad (2.22b)$$

Fluctuating Target Cross Section. In general, the effect of fluctuation is to require higher signal-to-noise ratios for high probability of detection and lower values for low probability of detection than those required with nonfluctuating signals. Swerling has considered four cases, which differ in the assumed rate of fluctuation and the assumed statistical distribution of the cross section. The two assumed rates are (1) a relatively slow fluctuation, such that the values of σ for successive scans of the radar beam past the target are statistically independent but remain virtually constant from one pulse to the next, and (2) a relatively fast fluctuation, such that the values of σ are independent from pulse to pulse within one beamwidth of the scan (i.e., during the integration time).

The first of the two assumed distributions for the received-signal voltage is of the Rayleigh form,* which means that the target cross section σ has a probability density function given by

---

*The *Rayleigh* density function for a voltage v is

$$p(v) = \frac{2v}{r^2} e^{-v^2/r^2}$$

where r is the rms value of v.

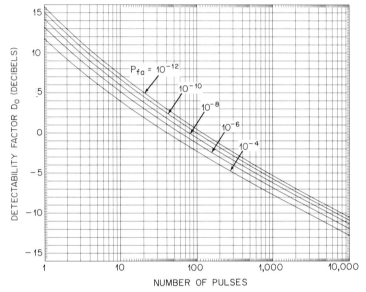

NUMBER OF PULSES

FIG. 2.5 Required signal-to-noise ratio (detectability factor) as a function of number of pulses noncoherently integrated, linear detector, nonfluctuating target, and 0.9 probability of detection. (*From Ref. 13.*)

$$p(\sigma) = \frac{1}{\overline{\sigma}} e^{-\sigma/\overline{\sigma}} \qquad (2.23)$$

where $\overline{\sigma}$ is the average cross section. (This is a *negative-exponential* density function, but a target having this distribution is called a *Rayleigh target* because this distribution of σ produces a received signal *voltage* which is Rayleigh-distributed.) The second assumed cross-section density function is

$$p(\sigma) = \frac{4\sigma}{\overline{\sigma}^2} e^{-2\sigma/\overline{\sigma}} \qquad (2.24)$$

The first distribution, Eq. (2.23), is observed when the target consists of many independent scattering elements of which no single one or few predominate. Many aircraft have approximately this characteristic at microwave frequencies, and large complicated targets are usually of this nature. (This result is predicted, for such targets, by the central limit theorem of probability theory.) The second distribution, Eq. (2.24), corresponds to that of a target having one main scattering element that predominates together with many smaller independent scattering elements. In summary, the cases considered by Swerling are as follows:

Case 1   Eq. (2.23), slow fluctuation
Case 2   Eq. (2.23), fast fluctuation
Case 3   Eq. (2.24), slow fluctuation
Case 4   Eq. (2.24), fast fluctuation

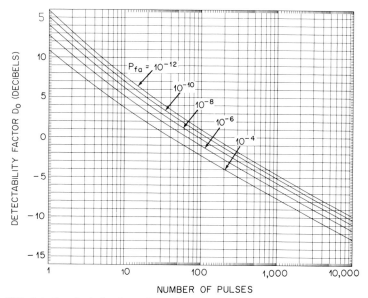

NUMBER OF PULSES

FIG. 2.6 Required signal-to-noise ratio (detectability factor) as a function of number of pulses noncoherently integrated, square-law detector, Swerling Case 1 fluctuating target, and 0.5 probability of detection. (*From Ref. 13.*)

The distribution of Eq. (2.24) is sometimes assumed for a small, rigid streamlined aircraft at the lower radar frequencies (e.g., below 1 GHz). Subsequent to Swerling's work, it has been found that many targets of the non-Rayleigh type are better represented by the so-called log-normal distribution, and analyses have been made for this case.[9]

Swerling's Case 1 is the one most often assumed when range prediction is to be made for a nonspecific fluctuating target. Results for this case are presented in Figs. 2.6 and 2.7. Curves for the other fluctuation cases and for additional values of detection probability are given in Refs. 13 and 14.

Detector Laws. A linear detector is a rectifier which has the rectification characteristic

$$\left. \begin{array}{ll} I_o = \alpha V_i & V_i \geq 0 \\ I_o = 0 & V_i < 0 \end{array} \right\} \tag{2.25}$$

where I_o is the instantaneous output current, V_i is the instantaneous input voltage, and α is a positive constant. Typical diodes approximate this law if V_i is larger than some very small value (e.g., a few millivolts). Such a diode is ordinarily used as the second detector of a superheterodyne radar receiver. Also, appreciable RF and IF gain usually precedes the second detector, so that the voltage applied to it is usually large enough (typically, an appreciable fraction of a volt) to ensure this "linear" type of operation.

A square-law detector is one that has the nonlinear characteristic

$$I_o = \alpha V_i^2 \tag{2.26}$$

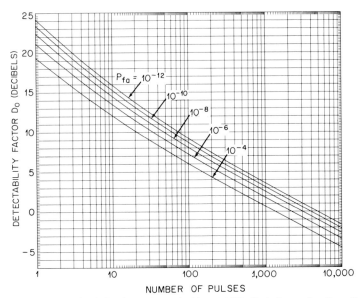

FIG. 2.7 Required signal-to-noise ratio (detectability factor) as a function of number of pulses noncoherently integrated, square-law detector, Swerling Case 1 fluctuating target, and 0.9 probability of detection. (*From Ref. 13.*)

Marcum[3] showed that a square-law detector is very slightly superior to a linear detector when many pulses are integrated, by about 0.2 dB. For a few pulses integrated, 10 or less, a linear detector is slightly superior—again, by about 0.2 dB or less. The mathematical analysis of probability of detection is somewhat more tractable when a square-law detector is assumed; this is probably its principal advantage.

The complexity of this matter is further compounded by the fact that, because of the statistics of the signal-noise superposition, in a *linear rectifier* there is a square-law relationship between the signal input voltage and the signal-plus-noise output voltage for small signal-to-noise ratios. This relationship becomes linear for large signal-to-noise ratios, as shown by Bennett,[22] North,[2] and Rice.[23] Because of this effect, it is sometimes erroneously thought that such a detector becomes square-law for small signal-to-noise *ratios*. But it is the input signal-plus-noise *voltage* V_i, and not the signal-to-noise ratio, that determines whether a diode rectifier is a linear or a square-law detector.

Curves for Visual Detection. The curves of Figs. 2.3 to 2.7 apply when the detection decision is based on an automatic threshold device as described. It is reasonable to suppose, however, that a human observer of a cathode-ray-tube display makes decisions in an analogous manner. That is, the equivalent of a threshold voltage (which would be a luminosity level for the PPI-scope type of display and a "pip-height" level for the A-scope display) exists somewhere in the observer's eye-brain system. This threshold, resulting in a particular false-alarm probability, is probably related to the observer's experience and personality: his or her innate cautiousness or daring. The probability of detection probably depends not only on the signal-to-noise ratio in relation to the threshold but on the observer's visual-mental acuity, alertness or fatigue, and experience. Consequently, curves calculated for an automatic threshold decision device cannot be assumed to apply accurately to the performance of a human observer of a cathode-ray-tube display. But such an assumption does not give grossly erroneous results, and it is justifiable when experimental human-observer data are not available or are of questionable accuracy.

Curves based on actual experiments with human observers, analogous to those of Figs. 2.4 through 2.7, are given in Ref. 14, Chap. 2, along with further discussion of visual detection.

Other Detection Methods. The discussion and results that have been presented have assumed perfect postdetection (video) integration of pulses prior to decision by an automatic threshold device. The noise statistics have been implicitly assumed to be those of ordinary receiver noise of quasi-uniform spectral density and (before detection) of gaussian probability density. A great many other detection procedures and signal-noise statistics are possible. Some of them are discussed in Ref. 14, Chap. 2.*

Predetection Integration. The results depicted in Figs. 2.3 through 2.7 apply for perfect *postdetection* (video) integration of a specified number of pulses. It was shown by North[2] that under ideal conditions *predetection* integration results in the smallest possible detectability factor and that for ideal predetection integration of M pulses the following relation holds:

$$D_0(M) = D_0(1)/M \qquad (2.27)$$

That is, the minimum detectable signal-to-noise power ratio at the demodulator input terminals is improved, relative to single-pulse detection, by a factor exactly equal to the number of pulses integrated, M. For perfect postdetection integration, the improvement factor is generally less than M, asymptotically approaching $M^{1/2}$ as M becomes indefinitely large.

An exception occurs in the $M < 10$ region for fast-fluctuating targets and high probabilities of detection. For those circumstances the postdetection integration improvement factor can actually exceed M, and under the same circumstances predetection integration yields little or no improvement. The result of adding successive fast-fluctuating signals before detection, with virtually uncorrelated phases, is practically the same as that of adding noise voltages. Therefore there is virtually no integration improvement.

Predetection integration is also called *coherent integration*, because of its dependence on phase coherence of the integrated pulses, and postdetection integration is called *noncoherent integration*.

When integration is not perfect, as is always the case practically, if the value of D_0 used in the range equation is based on perfect integration, an imperfect-integration loss factor or factors must be included in the system loss factor L, as discussed in Sec. 2.7.

Although the full benefits of predetection integration are realizable only for nonfluctuating targets, some benefit can be achieved by predetection integration of a moderate number of *slowly* fluctuating targets. For such targets, the phase fluctuation from pulse to pulse is small. This type of integration is being employed to an increasing extent in modern systems when the utmost sensitivity is important and when fast fluctuation is not expected.

---

*Chapter written by Lowell W. Brooks, Technology Service Corporation, Salida, Colo.

Since radial target motion produces a frequency shift of the received-echo signal (doppler effect) which is proportional to the target's radial velocity, this shift must be taken into account if predetection integration is used. This is done in *pulse doppler radar* (Chap. 17).

Some radar systems that integrate many pulses utilize a combination of coherent and noncoherent integration when the phase stability of the received pulses is sufficient for some coherent integration but not great enough to allow coherent integration of the entire pulse train during the antenna on-target dwell time. If the total number of received pulses is N and M of them (with $M < N$) are coherently integrated and if the coherent integrator is followed by a noncoherent integrator, then (assuming an appropriate implementation and ideal integrations) the detectability factor will be

$$D_{0(M,N)} = D_0(N/M)/M \qquad (2.28)$$

where $D_{0(M,N)}$ means the detectability factor for the assumed combination of coherent and noncoherent integration and $D_0(N/M)$ is the detectability factor for noncoherent integration of N/M pulses with no coherent integration (e.g., a value read from curves such as those of Figs. 2.4 through 2.7). As an example, if a train of $N = 24$ pulses is received and each set of $M = 8$ pulses is predetection- (coherently) integrated and if the predetection integrator is followed by a postdetection (noncoherent) integrator, the integration process produces at best a combined detectability-factor improvement corresponding to that of coherent integration of 8 pulses and noncoherent integration of 3 pulses.

2.5 SYSTEM NOISE TEMPERATURE

The concept of a *noise temperature* is derived from Nyquist's theorem,[24] which states that if a resistive circuit element is at temperature T (kelvins) there will be generated in it an open-circuit thermal-noise voltage given by

$$V_n = \sqrt{4kTRB} \qquad \text{volts} \qquad (2.29)$$

where k is Boltzmann's constant (1.38054×10^{-23} Ws/K), R is the resistance in ohms, and B is the bandwidth, in hertz, within which the voltage is measured (that is, the passband of an infinite-impedance voltmeter). The absence of the frequency in this expression implies that the noise is white—that the spectrum is uniform and extends to infinitely high frequency. But this also implies infinite energy, an obvious impossibility, indicating that Eq. (2.29) is an approximation. A more exact expression, which has frequency dependence, must be used if the ratio f/T exceeds about 10^8, where f is the frequency in hertz and T is the kelvin temperature of the resistor. Thus Eq. (2.29) is sufficiently accurate at a frequency of 30 GHz if the temperature is at least 300 K. The more accurate equations are given in Ref. 14 and in radio astronomy texts.

Available Power, Gain, and Loss.   As thus defined, V_n is the open-circuit voltage at the resistor terminals. If an external impedance-matched load of resistance $R_L = R$ is connected, the noise power delivered to it will be

$$P_n = kTB_n \qquad (2.30)$$

which does not depend on the value of R. This is of course also an approximation, but it is quite accurate at ordinary radar frequencies and temperatures. This
matched-load power is called the *available* power.[17]

The concepts of *available power*, *available gain*, and its reciprocal, *available
loss*, are assumed in all noise-temperature and noise-factor equations. These and
other noise-temperature concepts are explained fully in Refs. 14, 17, and 25.
Briefly, available power at an output port is that which would be delivered to a
load that matches (in the complex-conjugate sense) the impedance of the source.
Available gain of a two-port transducer or cascade of transducers is the ratio of
the available power at the output port to that available from the source connected
to the input port, with the stipulation that the available output power be measured with the actual input source (not necessarily impedance-matched) connected.

Noise Temperature.   The usual noise that exists in a radar receiving system
is partly of thermal origin and partly from other noise-generating processes.
Most of these other processes produce noise which, within typical receiver
bandwidths, has the same spectral and probabilistic nature as does thermal
noise. Therefore it can all be lumped together and regarded as thermal noise.
This is done, and the available-power level P_n is described by assigning to the
noise a semifictitious "noise temperature" T_n, which is

$$T_n = P_n/(kB_n) \tag{2.31}$$

This is of course simply an inversion of Eq. (2.30), except that T in Eq. (2.30)
refers to an actual (thermodynamic) temperature. The temperature defined by
Eq. (2.31) is semifictitious because of the nonthermal origin of some of the noise.
When this temperature represents the available-noise-power output of the entire
receiving system, it is commonly called the system noise temperature or operating noise temperature,[17] and it is then used to calculate the system noise power
and signal-to-noise ratio, as in Eqs. (2.4) to (2.6).

The Referral Concept.   A receiving system can be represented as a cascade
of two-port transducers, preceded by a source (the antenna) and terminated by
a load. [However, in the discussion of system noise temperature, only those
parts of the receiver that precede the detector (demodulator) are of
significance, for the noise level at that point determines the signal-to-noise ratio
for signal-detection-calculation purposes.]

Noise may arise at any and all points in this cascade, so that the noise level
changes from point to point. The important quantity is the *output* noise power
P_{no}. For purposes of signal-to-noise calculation, however, it is convenient to *refer* this output noise to the system input terminals. This is done by defining the
system noise temperature T_s so that it satisfies the relation

$$kT_sB_n = P_{no}/G_0 \tag{2.32}$$

where G_0 is the overall-system available gain and B_n is the noise bandwidth of the
system [Eq. (2.14)]. The output power P_{no} is thus "referred" to the system input
(the antenna terminals), and T_s is actually the system *input* noise temperature.
The product kT_sB_n is thus the system output noise power referred to the antenna
terminals.

Each two-port transducer of the receiving-system cascade can be regarded as having its own effective input noise temperature T_e, representing its intrinsic available output noise power referred to its own input terminals. Here *intrinsic* means the power that the transducer would generate with a *noise-free* input termination of the same impedance as the actual input termination. Transducer output power is referred to the input terminals by dividing the output power by the available gain of the transducer.

For an N-transducer cascade, the system input noise temperature (with the antenna terminals considered to be the system input terminals) is then given by

$$T_s = T_a + \sum_{i=1}^{N} \frac{T_{e(i)}}{G_i} \qquad (2.33)$$

Here T_a is the antenna noise temperature, representing the available noise power at the antenna terminals, and G_i is the available gain of the system between its input terminals and the input terminals of the ith cascaded component. (By this definition $G_1 = 1$ always.)

To illustrate these principles concretely, this formula will here be applied to a two-transducer cascade representing a typical receiving system (Fig. 2.8). The first transducer is the transmission line that connects the antenna to the receiver input terminals, and the second transducer is the predetection portion of the receiver itself. (As mentioned above, for purposes of signal-noise analysis subsequent portions of the receiver are not considered.) If desired, a many-transducer receiving system could be further broken down, with a preamplifier and possibly other units considered as separate elements of the cascade.

For this system, if the receiving-transmission-line noise temperature is represented by T_r and its loss factor is L_r $(=1/G_2)$ and if the receiver effective input noise temperature is T_e, Eq. (2.33) becomes

$$T_s = T_a + T_r + L_r T_e \qquad (2.34)$$

It now remains to discuss evaluation of T_a, T_r, L_r, and T_e.

Antenna Noise Temperature.   Antenna noise is the result of (1) noise in the form of electromagnetic waves received by the antenna from external radiating sources and (2) thermal noise generated in the ohmic components (resistive conductors and imperfect insulators) of the antenna structure. The product kT_aB_n is the noise power available at the antenna terminals within the receiver bandwidth.

This noise temperature is dependent in a somewhat complicated way on the noise temperatures of various radiating sources within the receiving-antenna pattern, including its sidelobes and backlobes. The concept of noise temperature of a radiating source is based on Planck's law or on the Rayleigh-Jeans approximation to it, analogous to the relationship of resistor noise temperature to Nyquist's theorem.

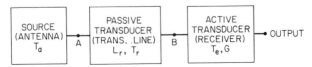

FIG. 2.8   Block diagram of a cascade receiving system.

The antenna noise temperature is not dependent on the antenna gain and beamwidth when a uniform-temperature source fills the beam. If the noise sources within the beam are of different temperatures, the resulting antenna temperature will be a solid-angle-weighted average of the source temperatures. The noise temperatures of most of the radiating sources that an antenna "sees" are frequency-dependent; therefore antenna temperature is a function of frequency. That is, antenna noise is not truly "white," but within any typical receiver passband it is virtually white.

In the microwave region, it is also a function of the antenna beam elevation angle, because in this region most of the "sky noise" is the result of atmospheric radiation. This radiation is related to atmospheric absorption, which is greater at low angles where the antenna beam sees a thicker slice of the lossy atmosphere than it does at higher angles.

Curves of antenna temperature for a lossless antenna are shown in Fig. 2.9, calculated for typical conditions.[14,25]

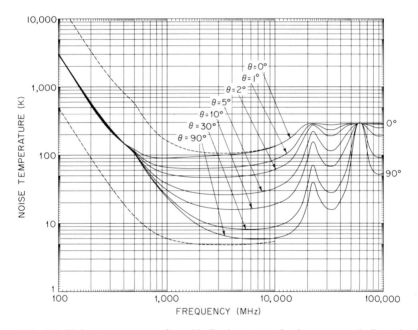

FIG. 2.9 Noise temperature of an idealized antenna (lossless, no earth-directed sidelobes) located at the earth's surface, as a function of frequency, for a number of beam elevation angles. Solid curves are for geometric-mean galactic temperature, sun noise 10 times quiet level, sun in unity-gain sidelobe, cool temperate-zone troposphere, 2.7 K cosmic blackbody radiation, and zero ground noise. The upper dashed curve is for maximum galactic noise (center of galaxy, narrow-beam antenna), sun noise 100 times quiet level, and zero elevation angle; other factors are the same as for the solid curves. The lower dashed curve is for minimum galactic noise, zero sun noise, 90° elevation angle. (The bump in the curves at about 500 MHz is due to the sun-noise characteristic. The curves for low elevation angles lie below those for high angles at frequencies below 400 MHz because of reduction of galactic noise by atmospheric absorption. The maxima at 22.2 and 60 GHz are due to water-vapor and oxygen absorption resonances; see Fig. 2.19.) (*From Ref. 13.*)

The curves of Fig. 2.9 apply to a lossless antenna that has no part of its pattern directed toward a warm earth. The lossless condition means that the curves represent only the noise received from external radiating sources. Therefore any thermal noise generated in the antenna must be added to the noise represented by these curves. In most practical cases, a ground noise-temperature component must also be added because part of the total antenna pattern is directed toward the ground. (This will be true because of sidelobes and backlobes even if the main beam is pointed upward.) But then also the sky-noise component given by Fig. 2.9 must be reduced somewhat because part of the total antenna pattern is not then directed at the sky. The reduction factor is $(1 - T_{ag}/T_{tg})$, where T_{ag} is the ground noise-temperature contribution to the total antenna temperature and T_{tg} is the effective noise temperature of the ground.

If α is the fraction of the solid-angle antenna power pattern subtended by the earth, then $T_{ag} = \alpha T_{tg}$. If the earth is perfectly absorptive (a thermodynamic blackbody), its effective noise temperature may be assumed to be approximately 290 K. A suggested conventional value for T_{ag} is 36 K, which would result if a 290 K earth were viewed over a π-steradian solid angle by sidelobes and backlobes averaging 0.5 gain (-3 dB). These sidelobes are typical of a "good" radar antenna but not one of the ultralow-noise variety.

Moreover, some practical antennas have appreciable ohmic loss, expressed by the loss factor L_a (Sec. 2.3). An additional thermal-noise contribution of amount $T_{ta}(1 - 1/L_a)$ then results, where T_{ta} is the thermal temperature of the lossy material of the antenna. However, the noise from external sources is then also reduced by the factor $1/L_a$. The total correction to the temperature values given by Fig. 2.9, to account for both ground-noise contribution and antenna loss, is then given by the following formula:

$$T_a = \frac{T_a'(1 - T_{ag}/T_{tg}) + T_{ag}}{L_a} + T_{ta}(1 - 1/L_a) \tag{2.35a}$$

where T_a' is the temperature given by Fig. 2.9. For $T_{ag} = 36$ K and $T_{tg} = T_{ta} = 290$ K, this becomes

$$T_a = \frac{0.876\, T_a' - 254}{L_a} + 290 \tag{2.35b}$$

and if $L_a = 1$ (lossless antenna), it further simplifies to

$$T_a = 0.876\, T_a' + 36 \tag{2.35c}$$

Transmission-Line Noise Temperature. Dicke[26] has shown that if a passive transducer of noise bandwidth B_n connected in a cascade system is at a thermal temperature T_t and if its available loss factor is L, the thermal-noise power available at its *output* terminals is

$$P_{no} = kT_t B_n(1 - 1/L) \tag{2.36}$$

A transmission line is a passive transducer. From Eq. (2.36) together with Eq. (2.31) and the definition of *input* temperature, it is deduced that the input noise temperature of a receiving transmission line of thermal temperature T_{tr} and loss factor L_r is

$$T_r = T_{tr}(L_r - 1) \tag{2.37}$$

(In this referral operation, multiplication by loss factor is equivalent to division by gain.) The receiving-transmission-line loss factor L_r is defined in terms of a CW signal received at the nominal radar frequency by the antenna. It is the ratio of the signal power available at the antenna terminals to that available at the receiver input terminals (points A and B, Fig. 2.8). A suggested conventional value for T_{tr} is 290 K.

Receiver Noise Temperature. The effective input noise temperature of the receiver T_e may sometimes be given directly by the manufacturer or the designer. In other cases, the *noise figure* F_n may be given. The relationship between the noise figure and the effective input noise temperature of the receiver or, in fact, of any transducer is given by[17]

$$T_e = T_0(F_n - 1) \tag{2.38}$$

where T_0 is, by convention, 290 K. In this formula F_n is a power ratio, not the decibel value that is usually given.

This formula is applicable to a *single-response* receiver (one for which a single RF input frequency corresponds to only one output or IF frequency and vice versa). Methods of computing noise temperatures when a double- or multiple-response receiver is used (e.g., for a superheterodyne receiver without preselection) are described in Refs. 17 and 25. Single-response receivers are ordinarily used in radar systems.

It is worth mentioning a point that has been well emphasized in the specialized literature of radio noise but is nevertheless easily overlooked. A receiver noise-temperature or noise-figure rating applies when a particular terminating impedance is connected at the receiver input. If this impedance changes, the noise temperature changes. Therefore, in principle, when a noise-temperature rating is quoted for a receiver, the source impedance should be specified, especially since the optimum (lowest) noise temperature does not necessarily occur when impedances are matched. However, when a receiver noise temperature is quoted without this impedance specification, it is presumable that the optimum source impedance is implied.

2.6 PATTERN PROPAGATION FACTOR

The pattern propagation factors F_t and F_r in the range equation account for the facts that (1) the target may not be in the beam maximum of the vertical-plane antenna pattern and (2) non-free-space wave propagation may occur. This single factor, rather than two separate factors, is designed to account for both of those effects. This is necessary because they become inextricably intertwined in the calculation of multipath interference, which is the most important non-free-space effect.

As will be seen, this effect can result in very considerable increase or decrease of the radar detection range compared with the free-space range. In this chapter, the basic ideas of pattern-propagation-factor calculation and some typical multipath-interference results will be presented. Additional details are given in Ref. 14, Chap. 6, and in Ref. 15.

General Discussion. As has been mentioned, the pattern propagation factors for transmitting and receiving are equal when the same antenna is used for transmitting and receiving, and it is then not necessary to denote them separately. In what follows, an unsubscripted quantity F will be used to denote the pattern propagation factor, with the understanding that it may sometimes be necessary to calculate separately a transmitting factor F_t and a receiving factor F_r. When that situation occurs, F^4 in the range equation is replaced by $F_t^2 F_r^2$. Similarly the vertical-plane antenna pattern factor $f(\theta)$, which is a component of the pattern propagation factor, can have a transmitting value $f_t(\theta)$ and a receiving value $f_r(\theta)$, but those are also equal when a common antenna is used. (The argument θ is the vertical-plane angle relative to zero elevation angle.) In what follows, a single direct-ray pattern factor $f(\theta_d)$ and a single reflected-ray pattern factor $f(\theta_r)$ will also be assumed, for simplicity, without serious loss of generality.

In principle, the pattern propagation factor can be used to account for all non-free-space propagation gains and losses, including atmospheric absorption and certain refraction losses (*lens effect*). But generally it is permissible and simpler to account for those atmospheric losses by introducing appropriate loss factors (Secs. 2.3 and 2.7).* That will be done here. Then the pattern propagation factor will account only for the vertical-plane antenna pattern, the multipath effect, and diffraction and shadowing for targets that are beyond the radar horizon.

Multipath interference occurs when a radar antenna overlooks a specular reflecting surface such as the sea. A *specular* reflector is one that obeys the physical laws of reflection for a smooth (mirrorlike) surface. When an antenna radiates toward a specular reflecting surface, the result is a reflected wave whose wavefront has a direction and phase that are predictable relative to the direction and phase of the incident wavefront, given the geometry and electrical properties of the reflecting surface. Figure 2.10 illustrates the geometry of the multipath effect. In this illustration, the reflecting surface is plane. This assumption is often permissible, even though the earth's curvature must sometimes be taken into account.

As shown, when there is specular reflection, it is possible for radar waves to travel from the radar antenna to the target and return by two different paths: a direct path (as in free space) and a reflected path. Although it is possible in principle to have more than one reflected ray (as discussed by Omberg and Norton[1]), the two-path case is the usual one, and this will be meant when the term *multipath* is used here.

As Fig. 2.10 indicates, the distances traveled by the two rays are not equal, and that fact results in a *phase difference* of the waves traveling via the direct- and reflected-ray paths. It is this phase difference that is primarily responsible for the multipath effect. In accordance with elementary principles of wave propagation, the phase difference corresponding to a path-length difference δ is equal to $2\pi\delta/\lambda$ rad, where λ is the radar wavelength.

Additional phase difference is contributed by the reflection coefficient of the reflecting surface and in some cases also by a phase difference of the antenna pattern factors in the directions of the direct and reflected waves. Because of the phase difference, the direct and reflected waves will interfere either constructively (additively) or destructively (subtractively) in vector-phasor fashion at the

---

*This is permissible if the losses for the direct ray and for the reflected ray are virtually equal. This is usually true for surface-based radars, because although the direct- and reflected-ray path lengths are unequal, the path difference is small as a percentage of the total path lengths.

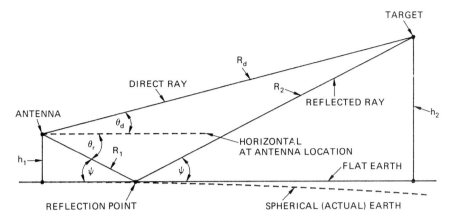

FIG. 2.10 Geometry of reflection from a plane earth. Parameters are antenna height (h_1), target height (h_2), direct-ray path length (R_d), reflected-ray path length ($R_1 + R_2$), direct-ray (target) elevation angle (θ_d), reflected-ray depression angle (θ_r), and grazing angle (ψ). *Flat earth* is tangent to actual spherical earth at subantenna point.

target. The two echo signals (direct and reflected) similarly interfere at the receiving antenna.

The electric fields of the interfering waves are usually essentially parallel or antiparallel; therefore the phasor aspect of the interference predominates, and any slight vector nonparallelism is usually ignored.

If the free-space range of the radar ($F = 1$) is denoted by R_0 and if the radar is monostatic with one antenna used for both transmitting and receiving so that $F_t = F_r = F$, then Eq. (2.10) can be written in the form

$$R_{max} = R_0 F \qquad (2.39)$$

(This assumes that there are no atmospheric losses.) Thus, except for the effect of atmospheric loss, the non-free-space range is directly proportional to F (or to $\sqrt{F_t F_r}$ in the more general case).

If the direct and reflected waves are of exactly equal amplitude and in phase, the resultant received voltage will be 4 times as great as it would be with free-space propagation. A 4-times increase of received signal voltage corresponds to a 16-times increase of signal power. Because of the fourth-root exponent on the right side of the range equation, this results in detection of a target at twice the free-space detection range. But if the direct and reflected waves are exactly out of phase, the resultant voltage and the "maximum range" will both be zero. In terms of Eq. (2.39), this means that the possible range of variation of F is from zero to 2 for the two-path multipath case. It is thus apparent that the multipath effect can produce drastic variation of the radar detection range relative to the free-space value.

Because of the equivalence (from the interference point of view) of phase differences separated by integral multiples of 2π rad, as a moving target approaches a radar at constant altitude (increasing elevation angle), the pattern propagation factor will undergo a cyclic variation of maxima and minima. Figure 2.11 is a plot illustrating this multipath effect. The quantity plotted is the ratio R_{max}/R_0 as a function of target altitude and/or elevation angle. For the assumed conditions, the

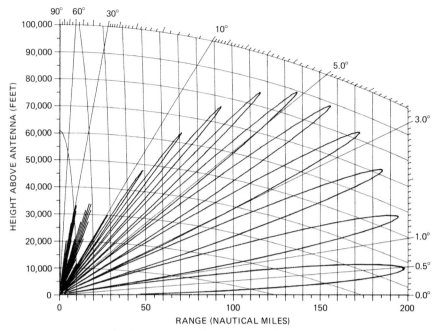

FIG. 2.11 A typical reflection-interference lobe pattern. It is a computer-generated detection-contour plot for frequency 1000 MHz, antenna height 30 ft above the sea surface, vertical beamwidth 10°, smooth sea, zero beam tilt, and horizontal polarization. The free-space range is 100 nmi.

detection range at the maxima of the interference lobes and in the antenna pattern maximum is $R_{max} = 2R_0$ (virtually total addition of the direct and reflected rays) and $R_{max} = 0$ in the minima (virtually total cancellation). Roughness of the sea surface, the curvature of the earth's surface, and atmospheric losses will usually somewhat modify this result, so that $R_{max} < 2R_0$ in the maxima and $R_{max} > 0$ in the minima.

The three following classes of multipath-interference problems require successively more elaborate procedures for their solution: (1) one terminal of the propagation path is low, and the reflection point is close enough to this terminal so that the earth's surface may be considered a plane reflector (*flat-earth case*); (2) the distance from the lower terminal of the path to the reflection point is appreciable, so that the earth's curvature is significant (*spherical-earth case*), but yet the grazing angle is great enough so that the path-length difference of the direct and reflected rays is at least an appreciable fraction of a half wavelength; and (3) the target is close to the radio horizon. A target beyond the horizon, where there is no multipath effect, is in the diffraction region.

In Cases 1 and 2 the target is said to be in the interference region. In Case 3 the separate existence of direct and reflected rays becomes questionable, and ray optics is no longer applicable. The target is then said to be in the intermediate region. The theoretical solution of this problem is difficult, but an approximate result can be obtained by straightforward calculation.[14,15]

At the usual radar frequencies (VHF and higher), a radar cannot see targets that are appreciably below the horizon, in the diffraction region. In this region, ordinarily $F \cong 0$ as a result of shadowing by the spherical earth. Exceptions occur for radar using frequencies below VHF at which ionospheric reflection takes place and at which a vertically polarized surface wave can result in significant signal strength at moderate distances beyond the horizon. (See Chap. 24.)

The sea surface is often a fairly good reflector. The usual roughness does not fully destroy the specularity of reflection, though it reduces it. Under special conditions, land surfaces can also act as specular reflectors.[1] The reflection characteristics of the sea are of special interest because it is a reflecting surface often encountered in practical radar applications.

Mathematical Definition. The pattern propagation factor is defined mathematically, for a point in space at a range R and elevation angle θ, as the magnitude of the ratio of the electric-field strength $E(R,\theta)$ at that point (e.g., volts per meter) to the field strength that would exist at the *same range in free space* and *in the antenna beam maximum*. Symbolically,

$$F(R,\theta) = E(R,\theta)/E_0(R) \qquad (2.40)$$

where $E_0(R)$ is the beam-maximum free-space field strength at range R. Thus F is defined as a voltage ratio. In the numerator on the right-hand side of the range equation, e.g., Eq. (2.10), quantities to which the received echo *power* is directly proportional occur to the first power (e.g., P_t, G_t, and G_r). The received *voltage* is proportional to F^2, and therefore F occurs in the range-equation numerator as F^4, or the transmitting and receiving factors occur as F_t^2 and F_r^2.

To solve any reflection-interference problem, it is necessary to know the value of the specular-reflection coefficient Γ of the surface and the characteristics of the vertical-plane antenna pattern, expressed as the pattern factor. The reflection coefficient is a complex number of magnitude ρ and phase angle ϕ. Similarly the pattern factor, a function of the vertical-plane angle θ, has a magnitude $|f(\theta)|$ and a phase angle β. In terms of these quantities, the general formula for F when multipath interference occurs is

$$F = |f_d + \rho f_r\, e^{-j\alpha}| \qquad (2.41)$$

where f_d is the magnitude of $f(\theta_d)$, f_r is the magnitude of $f(\theta_r)$, and α is the total phase difference of the direct and reflected waves at their point of superposition, i.e., at the target for the transmitted waves and at the receiving antenna for the returned echo. This total phase difference is the resultant of the phase difference $(\beta_r - \beta_d)$ of the pattern factors, the phase shift ϕ that occurs in the reflection process, and the phase difference due to the path-length difference. The absolute-value brackets indicate that F is a real number, although the reflection coefficient Γ and the pattern factor $f(\theta)$ are in general complex.

Pattern Factor. For a radar that has a broad vertical-plane antenna pattern with its maximum directed at the horizon, the effect of the vertical-plane pattern on the pattern propagation factor can be considered negligible for low-elevation-angle targets. But in general this pattern can affect the magnitudes and phases of both the direct and the reflected waves. The pattern factor $f(\theta)$ (a complex number) expresses the relative electric-field strength and phase of the wave radiated in a direction θ, relative to the magnitude and phase of the field in the direction of the beam maximum:

$$f(\theta_i) = f_i e^{-j\beta_i} \qquad (2.42)$$

where the subscripts can denote either the direct ray ($i = d$) or the reflected ray ($i = r$). This definition is stated in terms of a transmitting antenna, but an analogous definition applies for reception.

The magnitude of the pattern factor is a number in the range zero to 1; it is the ratio of the electric-field strength radiated in the direction θ to that radiated in the beam-maximum direction. This pattern information is often specified in terms of the antenna power gain factor $G(\theta)$. The pattern factor magnitude is related to $G(\theta)$ by the equation

$$f_i = \sqrt{G(\theta_i)/G_{max}} \qquad (2.43)$$

where G_{max} is the power gain in the beam-maximum direction.

Antenna pattern phase information is sometimes unavailable. For simple antenna types the phase can be considered constant within the main beam and within individual minor lobes, but with phase reversal (π-rad phase change) between adjacent lobes. Significant phase variation within a pattern lobe is most likely to occur with *shaped-beam* antennas, e.g., when the surface of a collimating reflector is deliberately shaped to depart from perfect collimation for beam-shaping purposes or when the phases of array elements are varied for a similar purpose.

For the purpose of computing the pattern propagation factor, the only important aspect of the pattern phase is the phase *difference* of the direct- and reflected-ray pattern factors. The difference will be zero if the pattern is symmetrical and if the beam maximum is midway between the directions of the direct and reflected rays. This will be the case if the antenna height is not too great, if the target is sufficiently distant, and if the beam maximum is at zero elevation angle.

Reflection Coefficient. The specular-reflection coefficient Γ is a complex number equal to the ratio of the reflected-electric-field phasor to the incident-field phasor, both evaluated at an infinitesimal distance from the point of reflection. Its magnitude ρ is therefore in the range between zero and 1, and its phase angle ϕ can vary from zero to π rad (180°). Equations exist for calculating the specular-reflection coefficient of a perfectly flat smooth surface of known permittivity, permeability, and conductivity.[14,15]

The magnitude and phase angle of the reflection coefficient of a smooth flat surface are determined entirely by the electromagnetic properties (permittivity, permeability, and conductivity) of the material of the surface, by the grazing angle ψ of the ray (Fig. 2.10), and by the wave polarization. They are calculable if those properties are known. In general, the electromagnetic properties of reflecting materials are functions of the frequency of the incident electromagnetic wave; therefore the reflection coefficient is a function of frequency. Results of reflection-coefficient calculations for seawater of average salinity and temperature are shown in Figs. 2.12 through 2.14 for horizontal and vertical polarizations. Details of the calculation procedure are given in Ref. 14.

Curves for the phase of the horizontal-polarization reflection coefficient are not given here because the phase is almost constant as a function of the grazing angle ψ. It is exactly π rad (180°) at $\psi = 0$, and it increases linearly to a maximum value that is less than 184° at $\psi = 90°$, for frequencies between 100 and 10,000 MHz.

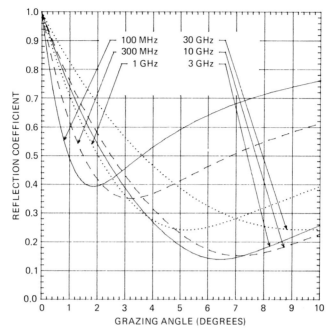

FIG. 2.12   Magnitude of the reflection coefficient for a smooth sea and vertical polarization at a number of frequencies.

Note that ϕ is also 180° at $\psi = 0$ for vertical polarization and that $\rho_0 = 1$ for both polarizations at this grazing angle. The result is that for targets near the horizon ($\psi = 0$) the reflected ray is almost exactly out of phase with the direct ray, and near cancellation occurs ($F = 0$). Actually, the earth curvature, together with the invalidity of ray optics theory at very small grazing angles, results in a more complicated behavior than this suggests, as will be discussed under the heading "The Intermediate Region." But it is nevertheless true that this cancellation effect makes radar detection for near-horizon targets very difficult.

The equations for elliptical (including of course circular) polarizations have been given by Shotland and Rollin,[27] and they are also summarized in Ref. 14 (pp. 264–265).

Total Phase Difference.   The angle α of Eq. (2.41) has been defined as the total phase difference of the direct and reflected rays at the radar target. It is numerically equal to the (algebraic) sum of the phase shifts due to the path difference δ, the phase angle of the reflection coefficient, and the phase difference (if any) of the antenna pattern factors in the directions of the direct and reflected rays. The pattern factor phase angles will be denoted β_d and β_r. Symbolically, therefore, the total phase difference is

$$\alpha = \frac{2\pi\delta}{\lambda} + \phi + (\beta_r - \beta_d) \qquad (2.44)$$

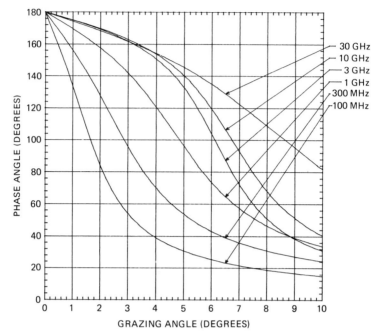

FIG. 2.13 Phase angle of the reflection coefficient for a smooth sea and vertical polarization at a number of frequencies.

Each of the terms on the right-hand side of Eq. (2.44) represents a phase lag of the reflected wave relative to the phase of the direct wave. Therefore β_r and β_d are defined as phase lags relative to the pattern phase at the beam maximum.

Reflection from a Rough Spherical Surface. The reflection coefficient that has been discussed thus far applies to specular reflection from a smooth flat surface. The magnitude of the specular-reflection coefficient for a partially rough surface is less than the value calculated for a perfectly smooth surface having the same electrical properties. It will be further reduced if the reflecting surface is curved rather than flat. Because the earth's surface is spherical, this effect is significant under some circumstances.

The reflection-coefficient reduction factor due to roughness will be denoted by r. The reduction due to the earth's curvature is called the divergence factor D. (The symbol D has also been used to denote the detectability factor. Both usages have become well established. Ordinarily no confusion of meaning will result.) Both r and D are numbers in the range zero to 1. If ρ_0 denotes the magnitude of the smooth-flat-earth reflection coefficient, the magnitude of the resulting complete reflection coefficient for a rough curved surface is

$$\rho = \rho_0 \, r \, D \qquad (2.45)$$

The phase angle ϕ of the specular-reflection coefficient is not affected by the roughness and divergence factors.

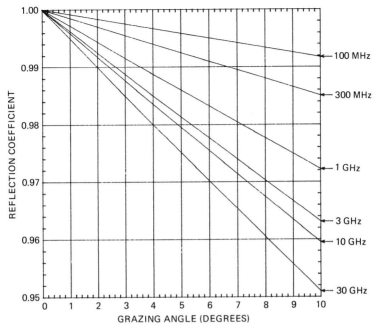

FIG. 2.14   Magnitude of the reflection coefficient for a smooth sea and horizontal polarization for a number of frequencies.

Specular-Reflection Roughness Factor.   Ament* gives the following formula for the specular-reflection-coefficient reduction factor due to surface roughness:

$$r = \exp\left[-2\left(\frac{2\,\pi\,H\,\sin\,\psi}{\lambda}\right)^2\right] \tag{2.46}$$

This formula is in fair agreement with rough-sea-reflection experiments reported by Beard et al.[29] The experimental results give somewhat larger values of r than predicted by Eq. (2.46) for values of H (sin ψ)/λ greater than about 0.1. A plot of this equation in terms of the parameter (fH sin ψ) is given in Fig. 2.15 for H in feet and f in megahertz. The dashed curve approximately represents the experimental data in the large-(fH sin ψ) region.

In this equation H is the standard deviation of waveheight. It is approximately equal to 0.25 times the so-called significant waveheight $H_{1/3}$; the latter quantity is defined as the average crest-to-trough height of the highest one-third of the waves. For use in the equation, the units in which H and $H_{1/3}$ are expressed must of course be the same as the units used for λ. In Fig. 2.15, however, the parameter H of the abscissa scale is in feet.

A modification of Eq. (2.46) has been proposed by Miller et al.[30] to conform more closely to the experimental results. Their formula, which is also based on theoretical considerations, is

---

*See Ref. 28. Ament derives this result but states that it was originally derived by Pekeris and, independently, by MacFarlane during World War II.

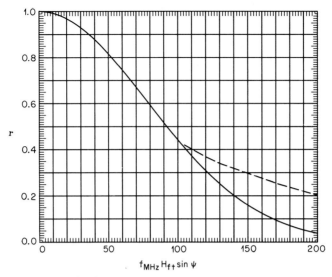

FIG. 2.15 Ratio of rough-sea to smooth-sea reflection coefficient r as a function of parameter (fH sin ψ); (f is the frequency, MHz, H is the standard deviation of waveheight distribution, feet, and ψ is the grazing angle). The solid curve is computed from Eq. (2.46). The dashed curve represents the experimental results of Beard et al., Ref. 29. (*From Ref. 13.*)

$$r = e^{-z} I_0(z) \qquad (2.47)$$

where $I_0(z)$ is the modified Bessel function of order zero [$=J_0(iz)$], and

$$z = 2\left[\frac{2\pi H \sin \psi}{\lambda}\right]^2 \qquad (2.48)$$

They state that this result is accurate, subject to their assumptions, for the range of the parameter (H sin ψ/λ) from zero to about 0.3.

When a surface is rough, there is in addition to the specular reflection a "diffuse" forward-scattered reflection component, which fluctuates as the sea moves. Its behavior in this respect is similar to that of the backscattered signal that causes the well-known phenomenon of *sea clutter* (Chap. 13). This diffusely forward-scattered fluctuating signal combines with the direct wave and with the specularly reflected wave as a third component of the total field at the target, and it will cause F to fluctuate.[29,31,32] Therefore, even though the target cross section is nonfluctuating, the received signal will then fluctuate. If the target cross section is fluctuating, the fluctuation of F will cause additional fluctuation of the received-echo signal.

The diffuse forward-scattered signal is not usually included in calculations of F. Because this signal is scattered over a large solid angle (whereas the specular reflection is concentrated into the solid angle of the antenna beam), its effect on the radar detection range is probably not significant in most situations.

Spherical-Earth Reflection Geometry. The principal problems of pattern-propagation-factor calculation are (1) finding the reflection coefficient Γ and (2) finding the path difference δ of the direct and reflected rays. Relatively simple formulas for the path difference can be derived from analysis of Fig. 2.10 for reflection from a plane surface.[14,15] The flat-earth assumption can be made without serious error if the antenna height is low (e.g., typical shipboard antenna heights) and the target elevation angle is sufficiently positive.

When this assumption is not valid, the path difference must be calculated by using the geometry of spherical-earth reflection as diagrammed in Fig. 2.16. The difficult part of this problem is to find the reflection point. The solution is described by Fishback[15]; it requires solving the following cubic equation:

$$2G_1^3 - 3GG_1^2 + [G^2 - 2a_e(h_1 + h_2)]G_1 + 2a_e h_1 G = 0 \qquad (2.49)$$

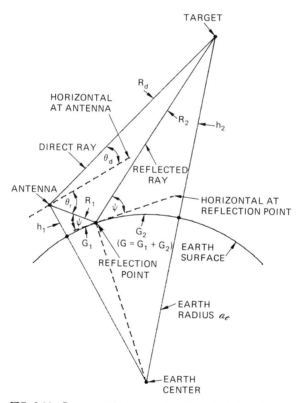

FIG. 2.16 Geometry of reflection from the spherical earth (compare with Fig. 2.10). Distances R_d, R_1, R_2, h_1, h_2, G_1, and G_2 are measured along solid lines between heavy dots. Antenna and target heights are exaggerated for diagrammatic clarity. (Analysis in the text assumes heights much smaller than the earth radius.)

where G, G_1, h_1, h_2, and a_e are as shown in Fig. 2.16. The arc length G_1 is the (unknown) distance from the subantenna point to the reflection point (a *ground range* measured on the curved-earth surface). The distance $a_e = ka$ is the *effective earth radius*, with $k = 4/3$, and a is the true earth radius, 6370 km. (Therefore $a_e = 8493.3$ km.) Use of this effective radius, as explained later, corrects for the curvature of rays due to atmospheric refraction in the normal atmosphere while allowing the analysis to assume that the ray paths are straight lines. This refraction correction is fairly accurate for altitudes up to 10 km, or 30,000 ft.

The solution described in Ref. 14 (slightly modified) is

$$G_1 = G/2 - p \sin (\xi/3) \tag{2.50}$$

with the parameters p and ξ given by

$$p = (2/\sqrt{3})\sqrt{a_e(h_1 + h_2) + (G/2)^2} \tag{2.51}$$

$$\xi = \sin^{-1} [2a_e G(h_2 - h_1)/p^3] \tag{2.52}$$

From Fig. 2.16, the remaining quantities needed for calculation of the path difference can be found by using simple geometric and trigonometric analysis. Although manual solution of these equations would require considerable labor, a rather short program can be written for solving them with a digital computer. The final formula for the path difference is

$$\delta = 4R_1R_2 \sin^2 \psi/(R_1 + R_2 + R_d) \tag{2.53}$$

Details are given in Refs. 14 and 15.

Divergence Factor. An additional quantity that must be calculated for spherical-earth multipath reflection is the *divergence factor*, which is an ingredient of the complete reflection coefficient, Eq. (2.45). This factor takes into account the weakening of the reflected-wave power density caused by the more rapid three-dimensional spreading of the wavefront when it is reflected from a spherical surface, as compared with the spreading rate before reflection. The following equation is an accurate approximation, using quantities already available in the foregoing calculation of δ:

$$D \cong \left[1 + \frac{2G_1G_2}{a_c G \sin \psi} \right]^{-1/2} \tag{2.54}$$

Spherical-Earth Pattern Propagation Factor. The pattern propagation factor can now be calculated from Eq. (2.41), which is the most general formula for that calculation. However, this complex-exponential expression can be written in a more convenient real-variable form as follows:

$$F = f_d |\sqrt{1 + x^2 + 2x \cos \alpha}| \tag{2.55}$$

where $$x = \rho\, f_r/f_d \qquad\qquad (2.56)$$

and the quantities f_r and f_d are the magnitudes of the pattern factors for the direct and reflected rays, respectively. The quantities α and ρ are defined by Eqs. (2.44) and (2.45). Equation (2.55), like Eq. (2.41), is completely general, applicable to either flat-earth or spherical-earth reflection.

The Intermediate Region. Two of the three previously defined multipath situations have now been considered: the flat-earth case and the spherical-earth case when the target is well above the horizon. In both of those cases the target is in what is called the *interference region*. Figure 2.17 illustrates the geometrical relationship of the third situation, the intermediate region, to the interference region and to the diffraction region.

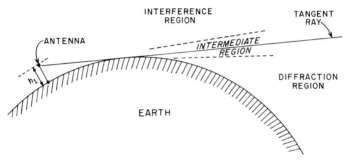

FIG. 2.17 Approximate locations of interference, intermediate, and diffraction regions. (*From Ref. 56.*)

In the interference region the pattern propagation factor can be calculated by the methods of interference theory that have been described above, which are based on the assumptions of ray optics. In the diffraction region, the assumptions of ray optics are not met, and it is necessary to employ a solution based on electromagnetic wave theory, i.e., Maxwell's equations. Fortunately this theory permits a relatively simple mathematical procedure in this region. But when a target is in the intermediate region, ray optics theory is not applicable, and the full-fledged electromagnetic wave theory becomes very complicated. Therefore it is desirable to find a method of solution that is more practical. The method commonly employed is to interpolate between points calculated at as great a range as is permissible in the interference region (i.e., within the region of validity of ray optics) and points in the diffraction region for which the wave-theory solution is relatively simple. This method was described by Fishback[15] and is discussed in some detail in Ref. 14.

Elevation-Angle Interference-Region Calculation. For purposes of plotting the vertical-plane coverage of a ground-based or shipboard radar system where one of the plot coordinates is elevation angle, it is convenient to be able to calculate the pattern propagation factor as a function of elevation angle, independent of range. It is demonstrable that if the target range and height are both very much larger than the antenna height, F can indeed be calculated as a function of the target elevation angle θ alone, independent of the target range and height. The method of making such calculations is described in detail in Ref. 14.

Refraction and Coverage Diagrams. The ultimate goal of pattern-propagation-factor calculations is usually a plot of the radar maximum range as a function of the elevation angle or a plot of signal-to-noise ratio as a function of the range, usually but not necessarily for a constant target altitude. If F has been calculated as a function of elevation angle, a vertical-plane coverage diagram can be plotted for an assumed free-space range R_0 by means of Eq. (2.39), with both R_{max} and F considered to be functions of elevation angle. (If it is desired to include the effect of atmospheric losses, an iterative solution is required, as discussed in Sec. 2.7.)

If the plot is to show the correct relationship between the target range, elevation angle, and height above the earth's surface (altitude), it is necessary to take into account the slight downward bending (refraction) of radio-frequency rays in the earth's atmosphere. For heights that are not too great, this refractive effect is approximately accounted for by the *effective-earth's-radius* method of Schelleng, Burrows, and Ferrell.[33] The refractive index n is assumed to decrease linearly with height h; that is, $dn/dh = C$, where C is a negative constant. The curvature of the rays is then such that, if plotted in a geometry where the earth has a radius greater than its true radius by a suitable factor k, they will appear to be straight lines. A standard value for k is 4/3. The range-height–angle relationship for these assumptions, for radar ranges R very much smaller than the earth's radius a, is

$$h_2 = h_1 + R \sin \theta + \frac{R^2 \cos^2 \theta}{2ka} \tag{2.57}$$

where θ is the target elevation angle, h_2 is the target height, and h_1 is the antenna height, with the assumption that $h_1 \ll h_2$ and h_2 is not greater than about 30,000 feet (10 km). (All units of length are assumed to be in the same system of units in this equation.)

A range-height–angle chart can be constructed on the basis of Eq. (2.57), in which the rays are straight lines. If the range and height coordinates are to the same scale, the constant-range lines are circles whose center is the radar antenna location. If the scales are not the same, the constant-range lines will be ellipses. The constant-height contours of this chart curve downward, with a radius of curvature, at height h, equal to $k(a + h)$. Plots of radar range as a function of elevation angle can be made on such charts; such plots are called *coverage diagrams*. When reflection interference occurs, these diagrams exhibit the lobe structure that has been described.

The curved-earth and intermediate-region calculation procedures that have been described are also based on the assumption that the refractive index is a linearly decreasing function of height. Since these procedures are meant to be used mainly for low- to moderate-altitude target problems, this assumption is not seriously in error. However, the error is significant at heights of about 30,000 ft or more for ray paths that have a low initial elevation angle. A more realistic refractive-index model that has been proposed and extensively studied is the exponential model:[34,35]

$$n(h) = 1 + (N_s \times 10^{-6})e^{-\gamma h} \tag{2.58}$$

where N_s is the surface refractivity and γ is a decay constant. Unfortunately, this model does not result in simple formulas for the range-height–angle relationship;

a numerical integration is required to trace a ray path.[36] However, machine-computed results can be used to plot a range-height–angle chart for this model.[37] A chart of this type is shown in Fig. 2.18, for N_s = 313 and γ = 0.04385, with h in Eq. (2.58) expressed in thousands of feet. The refractivity model for these values is called the CRPL reference atmosphere for N_s = 313.[35] This model and the method of constructing range-height–angle charts based on it are discussed in Ref. 14, Chap. 5.

The ray lines on the chart are plotted as straight lines even though in real space they have downward curvature due to refraction. (The straight ray lines of the chart are achieved by suitably distorting the lines of constant altitude.) Then if maximum-range contours based on Eq. (2.39) are plotted on this chart, with F calculated as a function of elevation angle, the plot will give the correct heights for the plotted points, for the refractivity profile used in constructing the chart. A sample plot of this type is shown in Fig. 2.11.

At frequencies below 1000 MHz, refraction by the ionosphere occurs for targets at great heights (e.g., space objects). (At higher frequencies ionospheric refraction is negligible.) This refraction is frequency-dependent and undergoes diurnal and longer-period time variations. It may also be, under some conditions, a complicated function of the direction of the ray in relation to the earth's magnetic field. Therefore no general chart can be made to represent the radar range-height–angle relationship in and above the ionosphere. Typical results are given by Millman.[38]

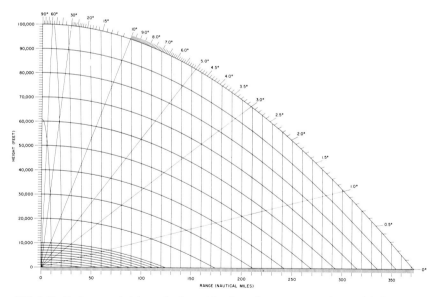

FIG. 2.18   Radar range-height–angle chart with refracted rays represented as straight lines. It is calculated for CRPL exponential reference atmosphere with N_s = 313, linear range and height scales, and nonlinear angle scale. (*From Ref. 37.*)

2.7 LOSS FACTORS

Loss is defined as the reciprocal of gain. The loss factor for a two-port (four-terminal) transducer is the ratio of the power input to the power output. If the power is defined as the available power (Sec. 2.5), the resultant loss factor is called the available loss (reciprocal of available gain). The available-loss concept must be used in calculations of noise temperature.

The general system loss factor of the radar range equations, L, represents the product of any and all individual loss factors that may occur in connection with the various radar range factors. A complete discussion of all the possible components of the total system loss factor would be impracticable here, and therefore this section will be limited to a discussion of principles, a listing of some of the more important loss factors, and equations and data for a few of the loss factors that are virtually always present. Additional information can be found in Ref. 14, Chap. 8, along with references to other sources.

The transmitting loss factor L_t has been mentioned in Sec. 2.3; it is the ratio of the transmitter power output to the power actually delivered to the antenna terminals. It represents the losses in the transmission line, duplexer, and any other components inserted between the transmitter and the antenna.

The receiving-transmission-line loss factor L_r does *not* appear as a component of the general system loss factor, nor does the antenna loss factor L_a, because these losses are fully accounted for by the definitions of T_s, G_t, and G_r.

In addition to L_t, two losses that are often present are the antenna pattern loss factor L_p (this applies to scanning radars only) and the atmospheric propagation loss factor L_α, which accounts for tropospheric-absorption loss and lens-effect loss. These two factors will be discussed in some detail, and several other important loss factors that sometimes occur will be discussed briefly. If these occasional loss factors are lumped together as L_x, the system loss factor in terms of the losses that have been mentioned is

$$L = L_t L_p L_\alpha L_x \qquad (2.59)$$

Loss factors are often expressed as decibel values. These must be converted to power ratios for use in Eqs. (2.10) and (2.11).

Antenna Pattern Loss. Search radars may detect targets by *searchlighting* (pointing the antenna in a fixed direction that is of special interest) or by scanning, i.e., moving the beam through an angular sector in a regular and repetitive fashion. In the searchlighting case, if the target happens to be in a direction other than that of the beam maximum, correction for this fact is made by means of the pattern factor $f(\theta,\phi)$ (Sec. 2.6).

In the scanning case, the target is in different parts of the beam for successive pulses. Therefore, as the beam traverses the target, the received train of pulses is modulated by the two-way pattern of the antenna. This amplitude variation must be taken into account in calculating the effect of integrating the received pulses. This is customarily done by using a value of D_0 in the range equation based on perfect integration of the *full-amplitude* pulses occurring between the half-power points of the beamwidth and then applying a loss factor to account for the effect of the beam pattern.

In the analysis by Blake,[39] the computed pattern loss for a typical beam shape is found to be $L_p = 1.6$ dB. This result is based on the assumption that there exists an optimum angular gate width* such that pulses are accepted for integration only within this gated angle centered on the beam maximum. It is also based on the assumption that the accepted pulses are postdetection-integrated with equal weighting. It may be assumed to apply at least approximately for human observers of visual displays, by postulating that the eye-brain combination somehow looks at an approximately optimum arc width (e.g., on a PPI display). This is a reasonable supposition.

Pencil-beam radars sometimes scan simultaneously in two angular directions, for example, in elevation and azimuth. The elevation scan is typically a sawtooth motion (in the sense that a graph of angle versus time has a sawtooth appearance), not necessarily linear, whereas the azimuth scan is usually a uniform rotation, although it can also be a sawtooth scan. The rates are usually adjusted so that the angular motion in one direction is at most a beamwidth during one complete scan in the other direction; this ensures no "holes" in the coverage. For such a bidirectional scan, the problem of analyzing the loss is more complicated and probably depends on the particular scanning pattern employed. In the absence of such analysis, an educated guess is $L_p = 3.2$ dB.

Radars may employ step scanning, in which the antenna beam remains stationary while pulses are transmitted in one direction, then moves to a new position (usually one or more beamwidths away), and remains there while another set of pulses is transmitted. The pulses transmitted in each position are all of equal amplitude, so that there is no pattern loss in the sense of the preceding discussion. However, there may be a pattern loss in the pattern factor sense, since a target may sometimes be exactly in the direction of one of the beam positions and sometimes not. Thus, some kind of statistical average loss can be computed.

Marcum[3] considered the unidirectional-scan pattern loss problem in his machine computation study of signal detectability. The details of his analysis are not given in his report, but he states that the loss is "in the neighborhood of 1.5 dB." Hall and Barton[40] extended the analysis to more general cases, calculating the loss in relation to the ideal utilization of the total signal energy within the entire beam pattern. They also pointed out that when the number of pulses per beamwidth is of the order of 1 or less (as has been the case with some very rapid scanning radars of high-volume coverage), the pattern loss factor for high probabilities of detection can become much larger than the customary values.[14]

Tropospheric-Absorption Loss.   Attenuation of radar signals by absorption in the troposphere is significant for propagation at low elevation angles and long ranges at frequencies above a few hundred megahertz. The absorption is due to quantum-mechanical resonances of the water-vapor and oxygen molecules. It can be serious even at short ranges in the vicinity of the water-vapor resonance frequency, 22.2 GHz, especially under conditions of high atmospheric water-vapor content. It can be even more serious near the oxygen resonance at 60 GHz. But also appreciable absorption at frequencies quite distant from these resonances can occur as a result of broadening the resonances by molecular collisions and because the oxygen absorption has a nonresonant component. Absorption at off-resonance frequencies is greater in the lower atmosphere than at higher altitudes, both because of the direct effect of the greater molecular density and because of the resultant greater frequency of collisions. The basic theory was published by Van Vleck[41] in 1945.

Direct radar measurement of atmospheric loss is very difficult in the fre-

---

*This optimum varies from 0.84 times the half-power beamwidth, for small signal-to-noise ratio, to 1.2 times the beamwidth when the signal-to-noise ratio is large.

quency region most commonly used for radar (i.e., at or below 10 GHz); in fact, the author is not aware of any such measurements. However, calculations based on accepted theory have been made for this frequency region (as well as for higher frequencies) by Bean and Abbott,[42] Hogg,[43] and Blake,[44] and their results are in good mutual agreement.

Figure 2.19 gives results applicable to detection by earth-surface-based radars of targets that are beyond the troposphere, which for practical purposes means targets at altitudes above about 30 km, or 100,000 ft; i.e., the ray paths are assumed to traverse the entire troposphere. (These results are useful for range calculations for "space" targets: satellites, ballistic missiles, and the like.) The curves are for the combined absorption by the oxygen and water vapor of the troposphere, for a standard atmosphere,[45] assuming no rain, snow, fog, or ice (hail). For targets that are within the troposphere (e.g., aircraft), curves such as Figs. 2.20 to 2.24, in which absorption loss is given as a function of the target range as well as frequency and elevation angle, are required.*

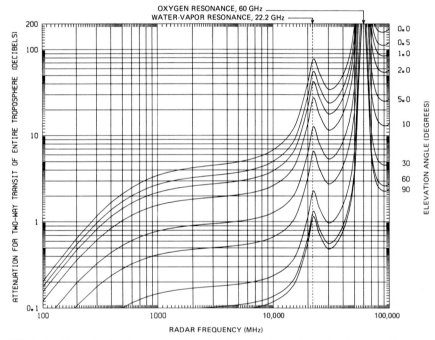

FIG. 2.19   Absorption loss for two-way transit of the entire troposphere at various elevation angles, calculated by using the Van Vleck theory for oxygen and water-vapor absorption. Ray paths are computed for the CRPL exponential reference atmosphere, $N_s = 313$. The pressure-temperature profile is based on the International Civil Aviation Organization (ICAO) standard atmosphere. Surface water-vapor content is 7.5 g/m^3. (*From Ref. 44, Fig. 101.*)

---

*The curves given here are similar but not identical to those published in the first edition of this handbook. They have been recalculated with improved equations for the residual (collision-broadened) effect of infrared water-vapor absorption, more exact calculation at frequencies in the vicinity of the oxygen resonances, and an improved model of the atmospheric pressure, temperature, and water-vapor content of the atmosphere as functions of altitude. The calculated absorption losses are slightly larger than the previous ones at some frequencies. Details of the calculation, together with plotted results for a more extensive set of frequencies, are given in Chap. 5 of Ref. 14 and in Ref. 44.

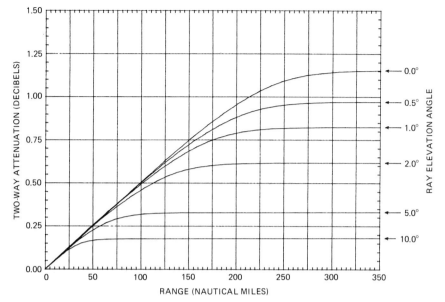

FIG. 2.20 Radar absorption versus range for targets within the troposphere; frequency 300 MHz.

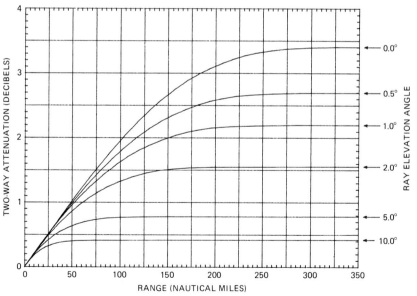

FIG. 2.21 Radar absorption versus range for targets within the troposphere; frequency 1000 MHz.

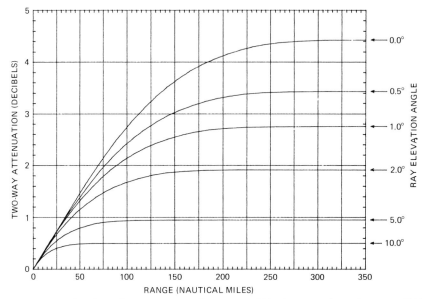

FIG. 2.22 Radar absorption versus range for targets within the troposphere; frequency 3000 MHz.

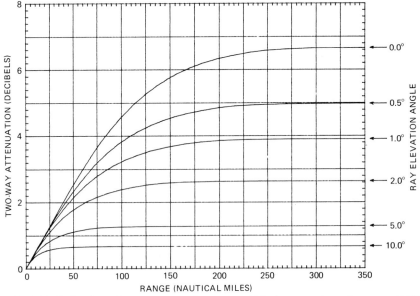

FIG. 2.23 Radar absorption versus range for targets within the troposphere; frequency 10,000 MHz.

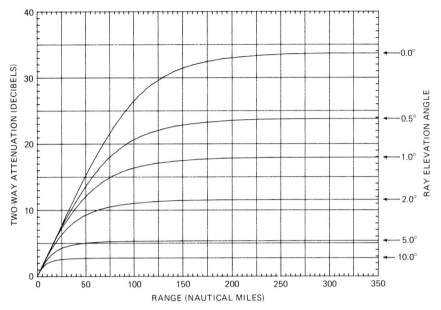

FIG. 2.24 Radar absorption versus range for targets within the troposphere; frequency 30,000 MHz.

These absorption-loss curves were calculated by taking atmospheric refraction into account. That is, the absorption was calculated by integrating the differential absorption along the actual refracted-ray path in the atmosphere, for a ray that has the indicated initial elevation angle. The refraction model used was the previously mentioned CRPL exponential-reference-atmosphere model[14] for N_s = 313.

The total atmospheric absorption in decibels is the sum of the decibel absorptions by water vapor and by oxygen. The oxygen component of the atmosphere is quite stable both temporally and spatially, but that is not true of the water-vapor component. It can vary in density from as little as about 2 g/m^3 in arctic regions in winter to as great as about 20 g/m^3 in tropical regions in summer. The curves given here are for a standard sea-level water-vapor content of 7.5 g/m^3, which is reasonably representative of average temperate-latitude conditions. It is a suitable "conventional" value according to the principles discussed in the introduction of this chapter.

In Ref. 14, curves of the type of Figs. 2.20 through 2.24 are given for selected frequencies in the range 100 to 32,500 MHz. Above 1200 MHz, separate curves are there plotted for oxygen and water-vapor absorption. This allows correcting the total absorption for water-vapor content other than the standard value 7.5 g/m^3, since the decibel water-vapor absorption is (with negligible error for practical purposes) directly proportional to the water-vapor content. (At or below 1200 MHz, water-vapor absorption is negligible.)

Reasonably good approximations at frequencies and angles intermediate to those for which curves are given can be made by judicious interpolation. Figure 2.19 can be used as a guide to whether or not a simple linear frequency interpolation will be satisfactory.

As mentioned above, these curves are for earth-surface-based radars and for targets at specified nonnegative elevation angles. *Earth-surface-based* can be interpreted, with negligible error, to mean antenna heights of up to a few hundred meters, or about 1000 ft. The curves can also be used for airborne or space-based radars looking downward at targets on or near the earth's surface, since the total absorption is the same when the positions of the radar and the target are interchanged. To use the curves in this way, however, it is necessary to know the elevation angle of the ray path at its point of intersection with the earth's surface. If the radar altitude and the target *slant range* (range measured along the ray path) are the known quantities, then the chart of Fig. 2.18 can be used to find the earth-surface elevation angle.

If a ray path does not conform to any of the above descriptions, various approximation methods may be employed. For example, if the radar-to-target path is between two points that are both at an appreciable altitude but the altitude variation over that path is not great, the absorption coefficient at the average ray altitude can be found or estimated using Fig. 2.25. This value multiplied by twice the distance between the radar and the target will give a reasonable approximation of the total radar absorption loss. This method can also be used for the loss between an earth-surface-based radar and a surface-based target. In both of these cases, the initial elevation angle of the ray may be negative so that part of the ray path is at a lower altitude than either the radar or the target.

Both Ref. 14 and Ref. 44 describe in detail the method of calculating absorption loss, and additionally Ref. 44 contains listings of the Fortran computer programs from which the absorption curves of this chapter were plotted.

Lens-Effect Loss. Since publication of the first edition of this handbook, a paper by Weil[46] has described an additional type of atmospheric loss. It is caused by the fact that rays emanating from a source near the earth's surface at different elevation angles are refracted (bent downward) by different amounts. The phenomenon of varying amounts of refraction as a function of elevation angle has long been known, but until Weil's paper was published it was not recognized by radar engineers that this could result in a signal-strength loss relative to the signal that would be received if there were no refraction.

Figure 2.26, from a more recent paper by Shrader and Weil,[47] shows the results of lens-effect loss calculations. In this paper, it is pointed out that the lens-effect loss (in decibels) is half as great for targets that are uniformly distributed in elevation angle as it is for point targets. This applies to targets which fill the radar vertical-plane beamwidth, such as rainstorms. Figure 2.26 gives the loss for both point targets and distributed targets.

Since atmospheric refraction is virtually frequency-independent in the frequency region used by conventional radars, this loss (unlike the absorption loss) can be considered frequency-independent, within the normal range of radar frequencies. As Fig. 2.26 shows, the two-way (radar) loss is about 1.5 dB at a range of 300 nmi for zero elevation angle. Thus the lens-effect loss is smaller than the absorption loss at microwave frequencies, but it is by no means negligible. It is notable that this loss is still increasing at that range, and in fact, unlike the absorption loss, it continues to increase even though the rays are virtually above the earth's atmosphere at that range. But it does approach a limit asymptotically. Weil has calculated this limit to be about 3 dB for a zero-elevation-angle ray, about 1.5 dB for 1° elevation angle, and 0.25 dB for 5° elevation.

The lens-effect loss and the absorption loss, in decibels, are directly additive, so that the total normal tropospheric radar loss is the sum of the water-vapor absorption loss, the oxygen absorption loss, and the lens-effect loss.

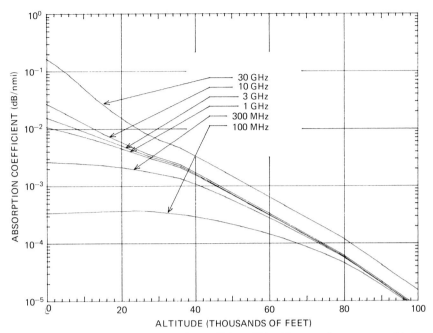

FIG. 2.25 Variation of the total absorption coefficient, dB/nmi (oxygen plus water vapor) in the standard atmosphere as a function of altitude for several frequencies. (Note that these values are for one-way propagation.)

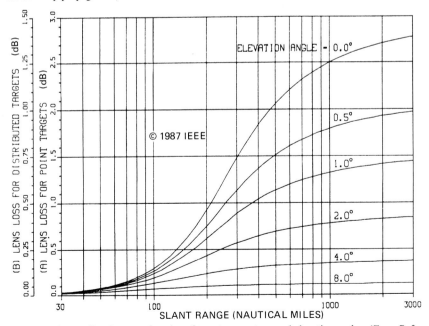

FIG. 2.26 Lens-effect loss as a function of target range at several elevation angles. (*From Ref. 47, courtesy of IEEE.*)

Other Tropospheric Losses.   The atmospheric losses thus far discussed are those that occur in the normal troposphere. The term *normal* here means that the atmosphere consists only of gases in molecular form, including the gaseous form of water (water vapor). As has been mentioned, the absorption losses are the result of collision-broadened quantum-molecular absorption resonances. Additional losses can occur when there is atmospheric water in liquid or frozen form: rain, fog, hail, or snow. The mechanisms of these losses are different from the molecular-resonance type of absorption. One mechanism is diffraction (scattering) of an incident electromagnetic wave by the liquid or frozen water droplets so that some of the wave energy is scattered in many directions, thus resulting in a reduction of the original-wavefront power density. The other is lossy-dielectric absorption, which converts electromagnetic energy into heat. These losses are discussed in Ref. 14. Additional information is given by Kerr,[15] Nathanson,[48] Crane,[49] and others.

The attenuation due to rain increases with increasing frequency, and it is a serious problem at frequencies above a few gigahertz. Typical values at 10 GHz are 0.01 dB/km for a light rain (1 mm/h) and 0.37 dB/km for a heavy rain (16 mm/h). Experimental verification of rain attenuation theory is difficult because of the variability of rainfall over a radar-to-target path.

Attenuation can also be caused by the presence of atmospheric water in the form of fog, mist, clouds, snow, or hail. Attenuation by dry snow and ice is much less than that of liquid water. However, as reported by Bell,[50] the attenuation of wet snow can be even larger than that of rain of equivalent water content.

Procedure for Range-Dependent Losses.   The radar range equation is a solution for the detection range when all the quantities on the right-hand side of the equation are range-independent. But atmospheric losses are in general range-dependent except when the target range is beyond the region of loss. A special procedure may therefore be necessary for solving the range equation when atmospheric losses are significant and the target is within the atmosphere.

The two possible procedures are iteration and graphical solution. Iteration is especially feasible for computer solution of the range equation because it is easily programmed. The graphical solution is also feasible with a computer if a computer-controlled plotter is available.

Collapsing Loss.   This term was initially applied to a loss that is related to the spatial dimensions of the radar data. The target position information received by a radar system is potentially three-dimensional, e.g., the range dimension and two orthogonal angle dimensions such as azimuth and elevation.

The total space surveyed by a particular radar can be partitioned into *resolution cells*, whose dimensions are determined by the radar range and angle resolution, that is, by the pulse length and beamwidths. Sometimes the received radar data is displayed or processed in such a way that some of the resolution cells overlap either partially or fully. In such cases the visual display (or the equivalent nonvisual processing) is said to be "collapsed" in one or more of the three dimensions. An example is the case of a radar that scans in both azimuth and elevation angles but presents the information on a PPI scope which can display only range and azimuth. The elevation dimension is then said to be collapsed.

A signal detectability loss then results because of the excess noise that is present due to the collapsing. For example, in the case of the above-cited azimuth- and elevation-scanning radar, the noise tending to obscure an echo signal received from a target at a particular elevation angle is increased by the ratio of the scanned elevation sector to the vertical-plane beamwidth, compared with the noise that would be present if there were no elevation scanning or if the display were gated in elevation angle so as to display only the elevation sector at the elevation angle of the target.

Marcum[3] defined a *collapsing ratio* ρ_c in terms of which this loss can be expressed:

$$\rho_c = \frac{p + q}{q} \tag{2.60}$$

where q is the number of resolution cells, in the collapsed dimension, that contain signals as well as noise, and p is the number that contain noise only. Thus in the above example of collapsing, q is proportional to the radar vertical beamwidth and $(p + q)$ is proportional to the angular size of the vertical-plane scanned sector.

As shown by Fehlner,[6] the collapsing-loss factor is related to the collapsing ratio by the following expression:

$$L_c = \frac{D_0(\rho_c N)}{D_0(N)} \tag{2.61}$$

where $D_0(x)$ is the detectability factor for integration of x pulses, from curves such as those of Figs. 2.4 through 2.7, and N is the number integrated in the case at hand. (Note, however, that D_0 here is a power ratio, not a decibel value.) For $\rho_c > 1$ the loss factor thus calculated will be smaller than the value of ρ_c.

The collapsing-loss concept is sometimes applied to situations which do not directly correspond to the foregoing description, for example, to the relationship between the size of the luminous spot of a cathode-ray tube used as a radar display, the speed of the spot in the range sweep, and the pulse duration, with the collapsing ratio given by

$$\rho_c = \frac{d + s\tau}{s\tau} \tag{2.62}$$

where d is the spot size in millimeters, s is the sweep speed in millimeters per second, and τ is the pulse duration in seconds. The collapsing-loss concept has also been applied to the mixing of video outputs of two or more receivers when only one receiver output contains the signal. The subject of collapsing loss is discussed at greater length in Ref. 14, pages 49–55 and 376–378.

Signal-Processing Losses. The term *signal processing* can broadly refer to any process that alters the waveform of the signal and the relationship of the signal to the receiver noise and other interfering signals. Such processing is always intended to be beneficial in some way, but it may also involve some loss relative to an ideal implementation of the processing. If the range equation is formulated to reflect the performance that would be ideally obtained from a particular form of processing, then any departure from that ideal must be accounted for by inclusion of an appropriate loss factor as a component of the total system loss factor.

A loss often incurred in modern radars is the result of *constant-false-alarm-rate* (CFAR) signal-noise processing. Its purpose is to adjust the detection threshold to maintain (as nearly as possible) the desired false-alarm probability when the noise level varies, as may occur under conditions of jamming or clutter (Sec. 2.8). This loss depends on the complexity of the circuitry used, the number of pulses integrated, the number of independent noise samples utilized, and the noise or clutter statistical parameters. It can range from a fraction of a decibel to as much as 20 or more dB, although of course a suitably designed processor would avoid such a large loss. Mitchell and Walker[51] give results (cited in Ref. 14, Chap. 7) for various conditions. For example, the CFAR loss with Rayleigh noise for 10 pulses integrated and five independent noise samples (cells) averaged per pulse and, for detection with $P_d = 0.9$ and $P_{fa} = 10^{-6}$, is 1.2 dB.

Other processes which are sometimes characterized by loss factors are MTI (moving-target indication), pulse compression processing, and processes occurring in special types of radar systems such as pulse doppler radar, CW radar, bistatic radar, and synthetic aperture radar, to name a few. Losses are also associated with the digital devices that are now used for many of the signal-processing tasks that were formerly implemented by analog circuitry. These losses are often the result of digital approximation to the exact analog procedure. Some of them are discussed in more detail by Nathanson[48] and in Ref. 14.

Miscellaneous Losses.   It is impossible to list all the possible losses that may occur in special situations, but a few that are known to occur at times will be mentioned.

A *polarization loss* can occur if for some reason the received-signal polarization does not agree with the polarization for which the receiving antenna is designed. This can happen when the target is beyond the ionosphere and the polarization is rotated by the ionospheric Faraday effect (which occurs at frequencies below 1000 MHz). If the receiving and transmitting antennas are both linearly polarized, the polarization of the received signal may be rotated by some angle ψ with respect to the antenna polarization. The applicable power loss factor in this case is $\sec^2 \psi$. (This loss can be avoided by using circular polarization, by adjusting the receiving-antenna polarization, or by receiving on two orthogonal polarizations.)

A *pulse-length loss* can occur when an array antenna with a certain type of feed system is used (e.g., the "serpentine" waveguide feed used in frequency-scanned arrays) such that a finite time elapses between the arrival of the leading edge of the RF pulse waveform at the first element of the array and its arrival at the last element. Since the beam is not fully formed until the latter event occurs, a portion of the pulse length is "lost" (actually, some radiation occurs during the "fill time" of the array, but at reduced antenna gain).

A *squint loss* occurs when a phase- or frequency-scanning array scans off the broadside direction owing to the loss of effective aperture size. If the gains G_t and G_r are defined to be those applicable to the broadside direction, then a squint loss factor should be included in the system loss for targets off the broadside direction. The loss factor is, to a first approximation, $\sec^2 \gamma$, where γ is the angle between the target direction and the broadside direction of the array. Element pattern effects may increase this loss at large squint angles.

2.8 JAMMING AND CLUTTER

Jamming.   The basic problem in radar detection, which was considered in Secs. 2.4 and 2.5, is the discernment of signals in a background of noise of the *thermal* type, which originates both in the receiving system and in the external environment as the result of natural phenomena. In the practical application of radar, the noise that competes with signals may originate in other ways. For military radars, deliberate radiation of jamming signals may introduce additional noise into the receiving system. When the jamming signals are either CW or have various forms of periodic modulation, much can be done to discriminate against them. But if the jammer uses "noise" modulation, so that the jamming signal essentially just adds to the already-present thermal noise, its effect is exactly the same as would be an increase in the system noise temperature due to the factors discussed in Sec. 2.5. Therefore the effect of this type of jamming is completely predictable.

Range equations for jamming situations are given in Chap. 9.

Clutter. In the discussion of signal detectability in Sec. 2.4, it was assumed that only one echo signal was present within a volume of space with dimensions of several resolution cells in all directions. If a few other targets are present within the total coverage volume of the radar, no two of them within the same or adjoining resolution cells, the detection of individual targets is not affected. But if there are so many targets that they run together on a cathode-ray-tube display or, more generally, if they fill all or most of the resolution cells of the radar within a volume of space containing the target of interest, detection of the desired target will be adversely affected. A profusion of echoes sufficient to produce this condition is called *clutter*.

A full coverage of the subject of radar detection in clutter is beyond the scope of this chapter. Additional information will be found in Chaps. 12 and 13 and in Refs. 14, 52, and 53. Here just a few of the basic aspects of the subject will be discussed, and the general nature of solutions to the problem will be described.

Clutter echoes generally fall into one of two categories: surface clutter and volume clutter. Surface clutter is caused by echoes from irregular surfaces, such as a rough sea or terrain, while volume clutter results from scatterers that are distributed within a volume of space, such as rain. Another source of volume clutter is the military radar countermeasure called chaff, in which an airborne cloud of lightweight strips of aluminum foil or similar reflecting objects produce clutter echoes. Each of these types of clutter has a different law of variation of amplitude with range, resulting in different range equations.

Targets in Clutter. The clutter echoes from a rough sea, from rain, from chaff, and from some types of terrain have many characteristics like those of thermal noise; they are randomly fluctuating in both amplitude and phase, and in many cases the probability density function (as observed after linear detection) is of the Rayleigh type, like that of thermal noise. But the *spectrum* is often much narrower than that of white or quasi-white (broadband) noise, and some types of clutter have decidedly non-Rayleigh probability density functions. This means that detectability curves such as those of Figs. 2.3 through 2.7 may not be directly applicable for targets in clutter.

The narrower spectral width of some types of clutter, compared with that of thermal or quasi-thermal noise, means that the clutter may be correlated, either partially or nearly totally, for times of the order of the typical period of signal integration. (The time, in seconds, required for virtually complete decorrelation of noise or clutter is of the order of the reciprocal of the spectral width, in hertz, of the clutter.)

The benefit of integration depends entirely on the statistical independence (lack of correlation) of noise samples separated by the same time interval as the signal samples being integrated. Consequently, the integration process may be considerably less beneficial for targets in clutter than it is for targets in ordinary noise.

When a train of echo pulses is integrated, if the noise or clutter is totally correlated for the entire integration time, there is no integration gain. The applicable detectability factor is that of a single pulse, as given by the curves of Fig. 2.3 for noise or clutter of the Rayleigh type. For *partial* pulse-to-pulse correlation of clutter, curves such as those of Figs. 2.4 through 2.7 can be applied by computing an *effective number of pulses integrated* (assuming that the clutter is Rayleigh-distributed). In other words, the integration gain obtainable with some actual number of integrated pulses in partially correlated clutter will be equivalent to that which would be obtained by integrating some lesser number of pulses in totally decorrelated noise. Formulas for computing an effective number of pulses integrated in this type of situation are given by Nathanson (Ref. 48, Secs. 3.5 and 3.6).

When the probability density function of the clutter is non-Rayleigh, then a set of detectability-factor curves for that function can in principle be calculated if the non-Rayleigh density function is a known and well-behaved mathematical function. In more difficult cases, results can be obtained by numerical integration using a digital computer.

When the clutter level is so much higher than the normal receiver noise level that the normal noise is insignificant, signal detection depends only on the signal-to-clutter ratio, S/C, in the same general way as it depends on the signal-to-noise ratio in nonclutter situations. A principal difference of the detection-in-clutter problem from the signal-in-noise problem is that receiver noise is the same at all target ranges while the clutter is highly range-dependent. Since both the target echo and the clutter echoes are equally affected by such radar parameters as the transmitter power and antenna gain, the primary factors affecting the signal-to-clutter ratio are the radar cross sections of the target and the clutter, σ_t and σ_c. However, the pattern propagation factors for the target and clutter may be different because, in general, their direction angles relative to the radar antenna beam maximum direction may be different. Therefore both the pattern factor or factors and the reflection-interference propagation factor or factors of the target and clutter may be different. The basic equation for S/C is consequently

$$\frac{S}{C} = \frac{\sigma_t F^4}{\sigma_c F_c^{\;4}} \qquad (2.63)$$

where F is the pattern propagation factor for the target and F_c is the corresponding quantity for the clutter. Here it has been assumed, without loss of generality, that $F_t = F_r = F$; when this assumption is not correct, F^4 in the clutter equations, as in the basic range equation, becomes $F_t^2 F_r^2$.

When the clutter is from a rough *surface*, σ_c is the product of the cross section per unit area, σ^O, and the area of the surface, A, illuminated by the radar at a particular instant. For a radar of horizontal beamwidth ϕ rad* and pulse length τ s, viewing the surface at a grazing angle ψ, this area is, for small values of ψ,

---

*As mentioned in Sec. 2.3, a special definition of beamwidth is required here; see Ref. 14, pp. 483–484. A similar modification of the pulse-length definition is also required.

$$A = R\phi\left(\frac{c\tau}{2}\right) \sec \psi \tag{2.64}$$

where R is the radar range to this clutter area and c is the radar wave-propagation velocity (3×10^8 m/s). The manner in which σ^O is defined and measured amounts to including the clutter pattern propagation factor as a part of σ^O. Therefore the assumption $F_c = 1$ can be made in any equation that contains σ^O.

Consequently, Eq. (2.63) can be rewritten as a range equation by expressing σ_c as the product $\sigma^O A$, rearranging terms, and attaching appropriate subscripts to S/C and R as follows:

$$R_{\max} = \frac{\sigma_t F^4}{\sigma^O \phi(c\tau/2) \sec \psi \, (S/C)_{\min}} \tag{2.65}$$

However, this allows a solution for the range only if the quantities σ^O, F, and ψ are all range-independent. Generally σ^O is not range-independent; it is a nonlinear function of R, and ψ is in turn a monotonic-decreasing function of the range. The variation of σ^O with range is frequency-dependent as well as dependent on such factors as the type of surface and the wave polarization. Therefore, in general, no simple range equation that takes these dependences into account can be written.

However, if data on the variation of σ^O with range—or equivalently, the grazing angle—is available, a graphical solution for the range can be obtained by plotting the right-hand side of Eq. (2.65) as a function of range. (An equation for the grazing angle as a function of range can be deduced from Fig. 2.16; or see Ref. 14, Chap. 6.) A range at which this quantity equals the minimum detectable signal-to-clutter ratio $(S/C)_{\min}$ is a boundary between the regions of detection and no detection.

If the clutter is from scatterers distributed within a volume (rain, chaff) rather than over a surface, the principle of range calculation is the same, except that $\sigma_c = \eta V$, where η is a cross section per unit volume and V is the instantaneously illuminated volume, given by

$$V = \frac{\Omega R^2 c\tau}{2} \tag{2.66}$$

Here Ω is the solid angle of the radar beam in steradians. In the volume clutter case, however, F_c of Eq. (2.63) is not included in the evaluation of η; therefore it must be included explicitly in the range equation. In some circumstances, F_c will vary within the region illuminated, and then an effective value must be determined. If η, F, and F_c are independent of range, a direct solution for detection range can be obtained by combining Eqs. (2.63) and (2.66), together with the relation $\sigma_c = \eta V$:

$$R_{\max} = \left[\frac{\sigma_t F^4}{\eta \Omega (c\tau/2) F_c^{\,4} (S/C)_{\min}}\right]^{1/2} \tag{2.67}$$

The detection of signals from *moving* targets in stationary or slow-moving clutter can be enhanced by suitable signal processing. The methods of accomplishing this improvement are called doppler filtering and MTI, discussed in Chaps. 17 to 19. The signal-to-clutter ratio calculated by the methods just described should then be modified to reflect the improvement produced by this signal processing.

The maximum detection range depends directly on the ratio S/C only as long as the clutter-to-noise ratio, C/N, is considerably greater than unity. If that is not true, target detectability then depends on the ratio $S/(C + N)$, where $(C + N)$ is the sum of the clutter and noise powers. Then Eqs. (2.65) and (2.67) are no longer applicable, and graphical or iterative solutions are necessary. This procedure is discussed in Ref. 14, Chap. 7.

If the radar employs pulse compression, the pulse length in Eqs. (2.64) through (2.67) refers to the compressed pulse rather than to the transmitted pulse. *Recall that the opposite rule holds for Eqs. (2.10) and (2.11).* As this implies, pulse compression is advantageous for detecting targets in clutter.

2.9 CUMULATIVE PROBABILITY OF DETECTION

In principle, if the probability of detecting the target on a single scan at range R_i is P_i, on the assumption that the target fluctuation is independently random from scan to scan, the cumulative probability $P_c(R)$ that the approaching target will be detected at least once by the time it reaches range R is

$$P_c(R) = 1 - \sum_{i=1}^{n} (1 - P_i) \tag{2.68}$$

where the scans occurring prior to the target reaching range R are numbered 1, 2, 3, ..., n. A more detailed formula is given by Mallett and Brennan.[54]

The assumption that the fluctuation is independently random from scan to scan may not be justified in all cases, and so caution must be used in applying this formula. Moreover, evaluation of the P_i's may be very difficult. If they are known as a tabulation of values from experimental data, calculation of $P_c(R)$ from this formula will require excessive labor unless a digital computer is employed.

There are many questions concerning the validity of assumptions necessary for computing cumulative probability. However, under some circumstances the necessary assumptions may be realized, and calculations of cumulative probability may then be of value. Also, criteria other than "probability of at least one detection" may be invoked. For example, it may be required that detection shall occur on at least two successive scans. Another criterion sometimes used is "at least two detections out of three successive scans." Equations to implement these requirements will obviously be more complicated than the at-least-one-detection criterion. Of course, all such criteria are somewhat arbitrary.

2.10 ACCURACY OF RADAR RANGE PREDICTION

Calculations of radar maximum range were notoriously unreliable in the early days of radar, for a number of reasons. Numerous losses that occur were not recognized. Propagation effects were often ignored (free-space propagation was often assumed when in fact there was multipath propagation). The sky noise temperature was often incorrectly assumed to be 290 K (the input termination temperature adopted as standard for receiver-noise-factor measurement). The probabilistic aspect of radar detection and the role of integration were not always properly taken into account.

In this chapter, methods of handling these factors correctly have been given, and if the factors in the equations are correctly determined, the predicted range will be correct in the probabilistic sense. But since the maximum range is incontrovertibly a stochastic variable, a precise agreement between prediction and the results of limited experiment cannot be expected.

Moreover, the factors in the range equation are never known with absolute accuracy (although some of them may be known fairly accurately). A quantity sometimes subject to inaccuracy is the radar cross section of the target, σ. Another factor that is not always known accurately is the pattern propagation factor F. Because of the strong dependence of the range on F compared with most of the other factors, this error may be especially damaging. It may arise especially through incorrect estimation of the magnitude of the earth- or sea-surface specular-reflection coefficient Γ. In some cases superrefractive effects may also cause unexpectedly large or small values of F. At microwave frequencies, excessive atmospheric moisture or precipitation may cause absorption losses much higher than those predicted for the normal atmosphere. Also, numerous unrecognized losses may occur within the radar system.

A measured antenna gain figure should be used in radar range calculation whenever possible; however, accurate gain measurement is sometimes fairly difficult to accomplish.

The system noise temperature contains two components that may be subject to appreciable error: the sky noise varies greatly over the celestial sphere, and the receiver noise temperature is not always accurately known.

In view of all these possible sources of error, it is of some interest to consider the relative effects of errors in the individual range-equation factors upon the total calculated-range error. The effect of definite increments of the independent variables is well known. For example, the range is proportional to the fourth root of the transmitter power and of several other quantities in the equation. Hence a change in one of these quantities by the factor x changes the range by the factor $x^{1/4}$. However, the range is directly proportional to the pattern propagation factor F and proportional to the square root of the frequency f. (But, as discussed in Sec. 2.3, this square root dependence on f is only the explicit part of the dependence. It does not hold unless all other factors in the equation are constant as the frequency changes.)

In considering the effect of sky noise variations upon the accuracy of the range calculation, it is necessary to realize that the fourth-root dependence is upon the *system* noise temperature, of which the sky noise is an additive component. Therefore, the sensitivity of the range to the value of the sky noise temperature depends partly upon the relative magnitudes of the sky noise temperature and the other components of system noise temperature. Similar statements are true of the other temperature components, such as the receiver noise temperature.

2.11 A SYSTEMATIC PROCEDURE FOR RANGE PREDICTION

The accompanying worksheet form has proved to be helpful in making the computations necessary for radar range prediction, including iteration with respect to atmospheric loss. It is self-explanatory. It ensures against omitting any of the factors in the range equation, and it simplifies the computation. It is based on Eq. (2.11). When many calculations of range for different values of the variables are required, programming of the range equation for a digital computer may be preferable, but when just a few calculations are required, this form is useful.

2.12 COMPUTER SOLUTION OF THE RANGE EQUATION

Manual calculation of radar detection range is entirely feasible for a specific set of radar parameters and for simple environments (no multipath interference, jamming, or clutter). The worksheet of Sec. 2.11 is helpful, as is a digital hand calculator with scientific function keys. But even with these aids manual calculation becomes very laborious if repeated calculations are required, as is true when a complete-coverage diagram is required, or if a graphical or iterative solution with many iterations is necessary. Then the use of a suitably programmed digital computer is very desirable or virtually necessary in many cases. For coverage diagrams, in fact, not only is the computer itself necessary, but a computer-controlled plotter (or some equivalent device such as a printer with graphics capability) is a virtual necessity.

It is a relatively simple matter to program the basic radar range equation, such as Eq. (2.11) for computer solution if calculations of such factors as multipath interference, antenna noise temperature, and atmospheric losses are not included; in fact, this task is easily done for a programmable pocket calculator. But the real advantage of a computer is that many of the truly complicated aspects of radar range calculation can be included in the calculation even when using a so-called personal (PC, or desktop) computer. A set of three such programs (written originally in the 1960s for a mainframe computer at the Naval Research Laboratory, using the Fortran language), is described in Refs. 55 and 56.

The first program of this set, named RGCALC, computes a quasi-free-space range, based on Eq. (2.11) of this chapter with $F_t = F_r = 1$. The term *quasi-free-space* denotes a free-space range modified by some of the simpler environmental factors such as atmospheric attenuation and external radiating noise sources (*antenna noise*) but not multipath propagation. A subroutine of this program calculates the detectability factor D_0 as a function of the required probability parameters and number of pulses integrated, so that it is unnecessary to consult curves of the type of Figs. 2.3 through 2.7. Other subroutines calculate the sky noise (for assumed "standard conditions"), the system noise temperature, and the atmospheric losses, so that Figs. 2.20 through 2.24 (or their equivalents) are also not needed.

The second program, called LOBEPLOT, computes and plots the vertical-plane coverage of a radar taking multipath interference into account, using the elevation-angle-only-dependence calculation of the pattern propagation factor. One optional input of this program is the free-space range R_0 calculated by RGCALC; alternatively, an arbitrary free-space range can be input. The quantity plotted is then the range calculated from Eq. (2.39), i.e., $R_{max}(\theta) = R_0 F(\theta)$, where θ is the elevation angle. The coverage diagram is plotted on a coordinate system which has been described in Sec. 2.4, there called a range-height–angle chart. It

PULSE-RADAR-RANGE CALCULATION WORKSHEET
Based on Eq. (2.11)

1. Compute system input noise temperature T_s, following outline in Sec. A below.
2. Enter range factors known in other than decibel form, in Sec. B below, for reference.
3. Enter logarithmic and decibel values in Sec. C below, positive values in plus column, negative in minus column. Example: If $D_{O(dB)}$ as given by Figs. 2.3 to 2.7 is negative, then $-D_{O(dB)}$ is positive and goes in plus column. For C_B, see Fig. 2.1. For definitions of range factors, see Eqs. (2.1) and (2.11).

Radar antenna height: h= ft Target elevation angle: $\theta=$ °. (See Figs. 2.16 and 2.18).

A. Computation of T_s: $T_s = T_a + T_r + L_r T_e$	B. Range factors	C. Decibel values	Plus (+)	Minus (−)
	P_t(kW)	10 log P_t (kW)	.	.
(a) Compute T_a. For $T_{tg} = T_{ta} = 290$ and $T_g = 36$, use Eq. (2.35b) Read $T_{a'}$ from Fig. 2.9. L_a(dB): _____ $L_{a'}$: _____ $T_a = (0.876\, T_{a}' - 254)/L_a + 290$ $\boxed{T_a =}$ K	τ_{μ}s	10 log τ_{μ}s	.	.
	G_t	G_t(dB)	.	.
	G_r	G_r(dB)	.	.
	σ(m2)	10 log σ	.	.
	f_{MHz}	− 20 log f_{MHz}	.	.
	T_s, K	− 10 log T_s	.	.
(b) Compute T_r, Eq. (2.37). L_r(dB): $\boxed{T_r =}$ K	D_O	$-D_O$(dB)	.	.
	C_B	$- C_B$ (dB)	//////	.
	L_t	$- L_t$(dB)	//////	.
(c) Compute T_e, Eq. (2.38).	L_p	$- L_p$ (dB)	//////	.
	L_x	$- L_x$(dB)	//////	.
F_n(dB): _____ T_e: _____ K L_r: _____ $\boxed{L_r T_e =}$ K	Range equation constant (40 log 1.292)		4.45	//////
	4. Obtain column totals ————▶		.	.
	5. Enter smaller total below larger ————▶		.	.
Add $\boxed{T_s =}$ K	6. Subtract to obtain net decibels (dB)—▶		+ .	− .

7. Calculate free-space range R_0 from the equation
   $R_0 = 100 \times$ antilog (dB/40)
   where dB is the decibel value obtained in Step 6. (Observe sign of dB.)
8. Multiply R_0 by pattern propagation factor F to obtain non-free-space range $R' = R_0 F$. Use equations (2.41) through (2.56) as appropriate.
9. Correct R' for atmospheric attenuation, using Figs. 2.19 through 2.26 as appropriate. Iteration or graphical method may be used. Range correction factor for x-dB attentuation is antilog (x/40).

takes into account atmospheric refraction based on the CRPL exponential-refractive-index model. Figure 2.11 is a sample plot made by this program.

The third of these programs, named SIGPLOT, produces a plot of received signal power expressed in decibels relative to the minimum-detectable-signal power as a function of range. The received power is calculated from the following equation:

$$S_{dB}(R) = 40\log_{10}(R_0 F/R) - L_{dB} \qquad (2.69)$$

where $S_{dB}(R)$ is the signal power in decibels at range R, taking into account multipath interference, antenna pattern factor, and atmospheric loss, relative to the signal power at the free-space range R_0, assuming no atmospheric loss, no multipath interference, and target in beam maximum; and L_{dB} is the actual atmospheric loss in decibels at range R. The target, in this calculation, is assumed to be approaching the radar at a constant altitude, although constant altitude is not a fundamental restriction. The calculation of F for this program is based on the spherical-earth equations, not restricted (as is LOBEPLOT) to distant targets, to positive elevation angles, and to the interference region. The plot continues to a range beyond the normal radar horizon into the intermediate and diffraction regions, where the target is actually at a negative elevation angle. Figure 2.27 is a sample plot of this type.

The advantage of the LOBEPLOT format is that it portrays the complete elevation-angle coverage (except for elevation angles near and below zero). The SIGPLOT format, on the other hand, provides important information as to the radar performance against low-altitude and low-angle targets without some of the approximation necessarily used in LOBEPLOT. Also, the SIGPLOT algorithm automatically does a graphical solution of the radar detection range when the range equation includes range-dependent variables such as noise from a self-screening jammer, sea or land clutter, and atmospheric attenuation. This feature eliminates the need for iteration, which is necessary with the RGCALC and LOBEPLOT programs, under those conditions. However, the refraction correction in SIGPLOT uses the effective-earth's-radius method (because the exponential-refractive-index model is usable only when the pattern propagation

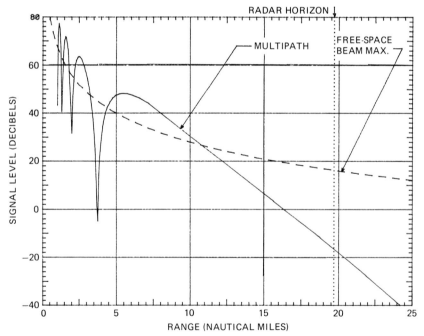

FIG. 2.27 Computer-generated plot of signal relative to minimum detectable signal, as a function of range, for frequency 3000 MHz, antenna height 100 ft, target height 200 ft, and free-space detection range 100 nmi.

factor is elevation-angle-only-dependent). But this is not a serious defect because the effective-earth's-radius correction is quite good for targets up to about 30,000-ft altitude, and SIGPLOT is primarily used for (though not restricted to) targets below that altitude.

Enhancements of these programs have been made in the ensuing years at many research and development organizations. Some of those enhanced programs are also available commercially on diskettes for desktop computers of the IBM PC type.* Some of them have the capability for handling jamming and clutter, MTI, and pulse doppler and for (optionally) plotting probability of detection versus range instead of signal power level, to mention a few. Also, the original programs have been modified to use more efficient and, in some cases, more accurate equations.

REFERENCES

1. Omberg, A. C., and K. A. Norton: The Maximum Range of a Radar Set, *Proc. IRE*, vol. 35, pp. 4–24, January 1947. (Originally published as *Rept. ORG-P-9-1*, Operational Research Group, Office of Chief Signal Officer, U.S. Army, February 1943.)

2. North, D. O.: An Analysis of the Factors Which Determine Signal/Noise Discrimination in Pulsed Carrier Systems, *Proc. IEEE*, vol. 51, pp. 1015–1028, July 1963. (Originally published as *RCA Lab. Rept. PTR-6C*, June 1943.)

3. Marcum, J. I.: A Statistical Theory of Detection by Pulsed Radar, and Mathematical Appendix, *IRE Trans.*, vol. IT-6, pp. 59–267, April 1960. (Originally published as *RAND Corp. Res. Mem. RM-754*, December 1947, and *RM-753*, July 1948.)

4. Robertson, G. H.: Operating Characteristics for a Linear Detector of CW Signals in Narrow-Band Gaussian Noise, *Bell Syst. Tech. J.*, vol. 46, pp. 755–774, April 1967.

5. Swerling, P.: Probability of Detection for Fluctuating Targets, *IRE Trans.*, vol. IT-6, pp. 269–308, April 1960. (Originally published as *RAND Corp. Res. Mem. RM-1217*, March 1954.) Also, Swerling, P.: Detection of Radar Echoes in Noise Revisited, *IEEE Trans.*, vol. IT-12, pp. 348–361, July 1966.

6. Fehlner, L. F.: Marcum and Swerling's Data on Target Detection by a Pulsed Radar, *Johns Hopkins Univ. Appl. Phys. Lab. Rept. TG-451*, July 1962, and *Suppl., TG-451A*, September 1964.

7. Kaplan, E. L.: Signal Detection Studies with Applications, *Bell Syst. Tech. J.*, vol. 34, pp. 403–437, March 1955. (Also published as *Bell Telephone Lab. Monog.* 2394.)

8. Schwartz, M.: Effects of Signal Fluctuation on the Detection of Pulsed Signals in Noise, *IRE Trans.*, vol. IT-2, pp. 66–71, June 1956.

9. Heidbreder, G. R., and R. L. Mitchell: Detection Probabilities for Log-Normally Distributed Signals, *IEEE Trans.*, vol. AES-3, pp. 5–13, January 1967.

10. Bates, R. H. T.: Statistics of Fluctuating Target Detection, *IEEE Trans.*, vol. AES-2, pp. 137–138, January 1966; also, Discussion, vol. AES-2, pp. 621–622, September 1966.

11. Hall, W. M.: Prediction of Pulse Radar Performance, *Proc. IRE*, vol. 44, pp. 224–231, February 1956.

---

*One such product, known as the Radar Workstation, is marketed by Technology Service Corporation, Silver Spring, Md. It consists of three modules, called RADARCALC, CONTOUR PLOT, and SIGNAL PLOT, which correspond in their general nature to the three above-mentioned NRL programs. RADARCALC utilizes the well-known Lotus 1-2-3 spreadsheet program, while CONTOUR PLOT and SIGNAL PLOT are graphics-oriented programs written in the C language. Additional products of this type are listed in Refs. 57 through 59.

12. Blake, L. V.: Recent Advancements in Basic Pulse-Radar Range Calculation Technique, *IRE Trans.*, vol. MIL-5, pp. 154–164, April 1961.

13. Blake, L. V.: A Guide to Basic Pulse-Radar Maximum-Range Calculation, *Naval Res. Lab. Rept.* 5868, December 1962. (Reprinted with revisions, December 1963.) Second edition, pt. 1, *Naval Res. Lab. Rept.* 6930, 1969; and pt. 2, *Rept.* 7010, 1969.

14. Blake, L. V.: "Radar Range-Performance Analysis," D. C. Heath and Company (Lexington Books), Lexington, Mass., 1980; 2d ed., Artech House, Norwood, Mass., 1986.

15. Kerr, D. E. (ed.): "Propagation of Short Radio Waves," MIT Radiation Laboratory Series, vol. 13, McGraw-Hill Book Company, New York, 1951. Republished as EW024 of IEE Electromagnetic Wave Series, 1987; available from IEEE Service Center, 445 Hoes Lane, Piscataway, N.J. 08855.

16. Lawson, J. L., and G. E. Uhlenbeck (eds.): "Threshold Signals," MIT Radiation Laboratory Series, vol. 24, McGraw-Hill Book Company, New York, 1950.

17. IEEE Standard 161-1971, Institute of Electrical and Electronics Engineers, New York (revision of Standards 57IRE 7.S2, 59IRE 20.S1, and 62IRE 7.S2); and tutorial papers published in *Proc. IRE*, vol. 45, pp. 938–1010, July 1957; vol. 48, pp. 61–68, January 1960; and vol. 51, pp. 434–442, March 1963.

18. IEEE Standard 686-1982, "IEEE Standard Radar Definitions," Institute of Electrical and Electronics Engineers, New York, 1982.

19. IEEE Standard 149, "IEEE Test Procedure for Antennas," Institute of Electrical and Electronics Engineers, New York, January 1965; also published in *IEEE Trans.*, vol. AP-13, pp. 437–466, May 1965.

20. North, D. O.: The Absolute Sensitivity of Radio Receivers, *RCA Rev.*, vol. 6, pp. 332–343, January 1942.

21. Haeff, A. V.: Minimum Detectable Radar Signal and Its Dependence on Parameters of Radar Systems, *Proc. IRE*, vol. 34, pp. 857–861, November 1946. (Originally published as a Naval Research Laboratory report in 1943.)

22. Bennett, W. R.: The Response of a Linear Rectifier to Signal and Noise, *J. Acoust. Soc. Am.*, vol. 15, pp. 164–172, January 1944; also, *Bell System Tech. J.*, vol. 23, p. 97, January 1944.

23. Rice, S. O.: Mathematical Analysis of Random Noise, *Bell System Tech. J.*, vol. 23, pp. 282–332, July 1944, and vol. 24, pp. 45–156, January 1945. (See especially vol. 23, p. 119, Eq. 4.2-3.)

24. Nyquist, H.: Thermal Agitation of Electric Charge in Conductors, *Phys. Rev.*, vol. 32, pp. 110–113, July 1928.

25. Blake, L. V.: Antenna and Receiving System Noise-Temperature Calculation, *Proc. IEEE* (Correspondence), vol. 49, pp. 1568–1569, October 1961. (This correspondence summarizes *Naval Res. Lab. Rept.* 5868, same title, September 1961.)

26. Dicke, R. H., et al.: Atmospheric Absorption Measurements with a Microwave Radiometer, *Phys. Rev.*, vol. 70, p. 340, 1946.

27. Shotland, E., and Rollin, R. A., "Complex Reflection Coefficient over a Smooth Sea in the Micro and Millimeter Wave Bands for Linear and Circular Polarizations," *Rept. FS-76-060*, Johns Hopkins University Applied Physics Laboratory, Laurel, Md., May 1976.

28. Ament, W. S.: Toward a Theory of Reflection by a Rough Surface, *Proc. IRE*, vol. 41, pp. 142–146, January 1953.

29. Beard, C. I., I. Katz, and L. M. Spetner: Phenomenological Vector Model of Microwave Reflection from the Ocean, *IRE Trans.*, vol. AP-4, pp. 162–167, April 1956; and C. I. Beard: Coherent and Incoherent Scattering of Microwaves from the Ocean, *IRE Trans.*, vol. AP-9, pp. 470–483, September 1961.

30. Miller, A. R., R. M. Brown, and E. Vegh, "New Derivation for the Rough-Surface

Reflection Coefficient and for the Distribution of Sea-Wave Elevations," *IEE Proc.*, vol. 131, pp. 114–116, April 1984; also, *NRL Repts.* 7705 (1974), 8744 (1983) and 8898 (1985).

31. Blake, L. V.: Reflection of Radio Waves from a Rough Sea, *Proc. IRE*, vol. 38, pp. 301–304, March 1950.

32. Beckmann, P., and A. Spizzichino: "The Scattering of Electromagnetic Waves from Rough Surfaces," The Macmillan Company, New York, 1963.

33. Schelleng, J. C., C. R. Burrows, and E. B. Ferrell: Ultra-Short-Wave Propagation, *Proc. IRE*, vol. 21, pp. 427–463, March 1933.

34. Bauer, J. R., W. C. Mason, and F. A. Wilson: Radio Refraction in a Cool Exponential Atmosphere, *MIT Lincoln Lab. Rept.* 186, August 1958. (ASTIA No. 202331.)

35. Bean, B. R., and G. D. Thayer: Models of the Atmospheric Refractive Index, *Proc. IRE*, vol. 47, pp. 740–755, May 1959; also, Bean, B. R., and E. J. Dutton: Radio Meteorology, *Natl. Bur. Std. (U.S.) Monog.* 92, 1966.

36. Blake, L. V.: Ray Height Computation for a Continuous Nonlinear Atmospheric Refractive-Index Profile, *Radio Sci.*, vol. 3 (new ser.), pp. 85–92, January 1968.

37. Blake, L. V.: Radio Ray (Radar) Range-Height–Angle Charts, *Naval Res. Lab. Rept.* 6650, Jan. 22, 1968; also *Microwave J.*, vol. 4, October 1968.

38. Millman, G. H.: Atmospheric Effects on VHF and UHF Propagation, *Proc. IRE*, vol. 46, pp. 1492–1501, August 1958.

39. Blake, L. V.: The Effective Number of Pulses per Beamwidth for a Scanning Radar, *Proc. IRE*, vol. 41, pp. 770–774, June 1953; Addendum, vol. 41, p. 1785, December 1953.

40. Hall, W. M., and D. K. Barton: Antenna Pattern Loss Factor for Scanning Radars, *Proc. IEEE* (Correspondence), vol. 53, pp. 1257–1258, September 1965.

41. Van Vleck, J. H.: The Absorption of Microwaves by Oxygen, and the Absorption of Microwaves by Uncondensed Water Vapor, *Phys. Rev.*, vol. 71, pp. 413–433, Apr. 1, 1947.

42. Bean, B. R., and R. Abbott: Oxygen and Water-Vapor Absorption of Radio Waves in the Atmosphere, *Rev. Geofis. Pura Appl. (Milano)*, vol. 37, pp. 127–144, 1957.

43. Hogg, D. C.: Effective Antenna Temperatures Due to Oxygen and Water Vapor in the Atmosphere, *J. Appl. Phys.*, vol. 30, pp. 1417–1419, September 1959.

44. Blake, L. V.: Radar/Radio Tropospheric Absorption and Noise Temperature, *Naval Res. Lab. Rept.* 7461, October 1972; also, preceding work summarized in Tropospheric Absorption Loss and Noise Temperature in the Frequency Range 100–10,000 Mc, *IRE Trans.*, vol. AP-10, pp. 101–102, January 1962.

45. Minzner, R. A., W. S. Ripley, and T. P. Condron: U.S. Extension to the ICAO Standard Atmosphere, U.S. Department of Commerce, Weather Bureau, and USAF ARDC Cambridge Research Center, Geophysics Research Directorate, 1958.

46. Weil, T. A.: Atmospheric Lens Effect; Another Loss for the Radar Range Equation, *IEEE Trans.*, vol. AES-9, pp. 51–54, January 1973.

47. Shrader, W. W., and T. A. Weil: Lens-Effect Loss for Distributed Targets, *IEEE Trans.*, vol. AES-23, pp. 594–595, July 1987.

48. Nathanson, F. E., "Radar Design Principles," McGraw-Hill Book Company, New York, 1969.

49. Crane, R. K.: Comparative Evaluation of Several Rain Attenuation Prediction Models, *Radio Sci.*, vol. 20, pp. 843–863, July–August 1985.

50. Bell, J.: Propagation Measurements at 3.6 and 11 Gc/s over a Line-of-Sight Radio Path, *IEE Proc. (British)*, vol. 114, pp. 545–549, May 1967.

51. Mitchell, R. L., and J. F. Walker: Recursive Methods for Computing Detection Probabilities, *IEEE Trans.*, vol. AES-7, pp. 671–676, July 1971.

52. Barton, D. K., "Radars," vol. 5: "Radar Clutter" (selected reprints from the literature), Artech House, Norwood, Mass., 1975 (second printing 1977).

53. Long, M. W., "Radar Reflectivity of Land and Sea," 2d ed., Artech House, Norwood, Mass., 1983.

54. Mallett, J. D., and L. E. Brennan: Cumulative Probability of Detection for Targets Approaching a Uniformly Scanning Search Radar, *Proc. IRE*, vol. 51, pp. 596–601, April 1963.

55. Blake, L. V.: A Fortran Computer Program to Calculate the Range of a Pulse Radar, *NRL Rept.* 7448, August 1972. (AD No. 749686.)

56. Blake, L. V.: Machine Plotting of Radio/Radar Vertical-Plane Coverage Diagrams, *NRL Rept.* 7098, June 1970. (AD No. 709897.)

57. Fielding, J. E., and G. D. Reynolds: "RGCALC: Radar Range Detection Software and User's Manual," Artech House, Norwood, Mass., 1987.

58. Fielding, J. E., and G. D. Reynolds: "VCCALC: Vertical Coverage Plotting Software and User's Manual," Artech House, Norwood, Mass., 1988.

59. Skillman, W. A.: "SIGCLUT: Radar Calculation Software and User's Manual," Artech House, Norwood, Mass., 1988.

CHAPTER 3
RECEIVERS

John W. Taylor, Jr.
Westinghouse Electric Corporation

3.1 THE CONFIGURATION OF A RADAR RECEIVER

The function of a radar receiver is to amplify the echoes of the radar transmission and to filter them in a manner that will provide the maximum discrimination between desired echoes and undesired interference. The interference comprises not only the noise generated in the radar receiver but also energy received from galactic sources, neighboring radars and communication equipment, and possibly jammers. The portion of the radar's own radiated energy that is scattered by undesired targets (such as rain, snow, birds, insects, atmospheric perturbations, and chaff) may also be classed as interference. Where airborne radars are used for altimeters or mapping, other aircraft are undesired targets, and the ground is the desired target. More commonly, radars are intended for detection of aircraft, ships, surface vehicles, or personnel, and the reflection from sea or ground is classified as clutter interference.

The boundaries of the radar receiver must be defined arbitrarily. This chapter will consider those elements shown in Fig. 3.1 as the receiver. The input signal is derived from the duplexer, which permits a single antenna to be shared between transmitter and receiver. Some radar antennas include low-noise amplifiers prior to forming the receive beams; although these are generally considered to be antenna rather than receiver elements, they will be discussed in this chapter.

The receiver filters the signal to separate desired echoes from interference in many ways, but some elements of such processing are covered by other chapters because of the depth of treatment required:

Chapter 14 describes continuous-wave (CW) and FM-CW radars; the discussion here will be confined to receivers for pulse radars, the dominant form. Low-pulse-repetition-frequency (PRF) pulse radars transmit a burst of energy and listen for echoes between transmissions. Their outstanding advantage is that neither leakage from the transmitter nor the very strong echoes from close-range clutter occur at the same instant as reception of weak echoes from long-range targets. The delay of the pulsed echo also provides an instantaneous measurement of range.

Chapters 15 to 17 relate to discrimination of desired targets from interference on the basis of velocity or the change in phase from one pulse to the next; the

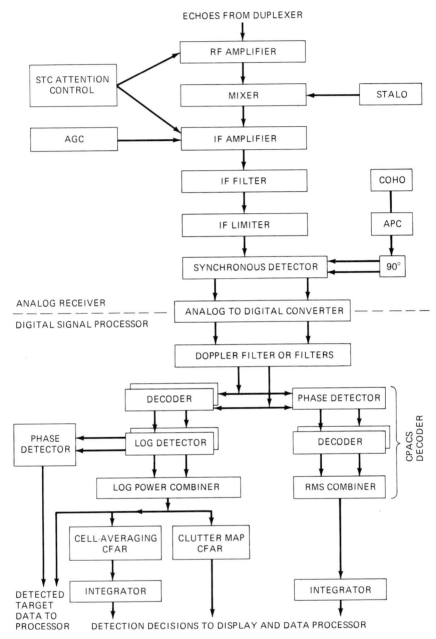

FIG. 3.1 General configuration of a radar receiver.

receiver discussed here serves only to provide the individual pulse signals in proper form for such doppler filtering. Section 3.12 will discuss data distortions in the synchronous detector or analog-to-digital (A/D) converter which affect these doppler filters.

Chapter 10 deals with pulse compression, and only brief mention of its ability to aid in the discrimination process will be included here. Physically, decoding such echoes may occur as part of the intermediate-frequency (IF) filter, typically with surface acoustic-wave (SAW) devices or with digital correlators, either preceding or following doppler filtering. In Fig. 3.1, pulse compression is provided by the cascade combination of IF filter and digital decoders (correlators) after doppler filtering.

Similarly, Chaps. 18 and 20 describe tracking radars and height finding radars, but the peculiarities of the receivers required to perform these functions will be mentioned briefly.

The purpose of Fig. 3.1 is to illustrate the usual sequence of processing functions that may occur in any radar receiver and the variety of possible outputs, although no radar receiver will include all these functions or provide all these outputs.

Virtually all radar receivers operate on the superheterodyne principle shown in Fig. 3.1. The echo, after modest amplification, is shifted to an intermediate frequency by mixing with a local-oscillator (LO) frequency. More than one conversion step may be necessary to reach the final IF, generally between 0.1 and 100 MHz, without encountering serious image- or spurious-frequency problems in the mixing process. Not only is amplification at IF less costly and more stable than at microwave frequency, but the wider percentage bandwidth occupied by the desired echo simplifies the filtering operation. In addition, the superheterodyne receiver can vary the LO frequency to follow any desired tuning variation of the transmitter without disturbing the filtering at IF. These advantages have been sufficiently powerful that competitive forms of receivers have virtually disappeared; only the superheterodyne receiver will be discussed in any detail.

Other receiver types include the superregenerative, crystal video, and tuned radio frequency (TRF). The superregenerative receiver is sometimes employed in radar-beacon applications because a single tube may function as both transmitter and receiver and because simplicity and compactness are more important than superior sensitivity. The crystal video receiver also is simple but of poor sensitivity. The TRF receiver uses only RF and video amplification; although its noise temperature may be low, its sensitivity is poor because optimum-bandwidth filtering of the usual radar echo spectrum is generally impractical to achieve. Only for radars that radiate a relatively wide-percentage-bandwidth signal is filtering practical.

3.2 NOISE AND DYNAMIC-RANGE CONSIDERATIONS

Receivers generate internal noise which masks weak echoes being received from the radar transmissions. This noise is one of the fundamental limitations on the radar range, and for this reason the subject has been treated in Sec. 2.5. The analysis of radar sensitivity is facilitated if the noise contribution of each element of the system is expressed as a *noise temperature* rather than as a *noise factor* or *noise figure*; these terms are defined and interrelated in Sec. 2.5.

The noise temperature of the radar receiver has been reduced to the point that it no longer represents a dominant influence in choosing between available alternatives. It is a paradox that a noise parameter is usually the first characteristic specified for a radar receiver, yet few radars employ the lowest-noise receiver available because such a choice represents too great a sacrifice in some other characteristic.

Cost is rarely a consideration in rejecting a lower-noise alternative. A reduction of requirements for antenna gain or transmitter power invariably produces cost savings far in excess of any added cost of a lower-noise receiver. More vital performance characteristics generally dictate the choice of receiver front end:

1. Dynamic range and susceptibility to overload
2. Instantaneous bandwidth and tuning range
3. Phase and amplitude stability
4. Cooling requirements

A direct compromise must be made between the noise temperature and the dynamic range of a receiver. The introduction of an RF amplifier in front of the mixer necessarily involves raising the system noise level at the mixer to make the noise contribution of the mixer itself insignificant. Even if the RF amplifier itself has more than adequate dynamic range, the mixer dynamic range has been compromised, as indicated below:

Ratio of front-end noise to mixer noise	6 dB	10 dB	13.3 dB
Sacrifice in mixer dynamic range	7 dB	10.4 dB	13.5 dB
Degradation of system noise temperature by mixer noise	1 dB	0.4 dB	0.2 dB

Definitions. Section 2.5 defined the noise parameters of a receiver in simple terms. Dynamic range, which represents the range of signal strength over which the receiver will perform as expected, is more difficult to define. It requires the specification of three parameters:

1. *Minimum signal of interest:* This is usually defined as the input signal that produces unity signal-to-noise ratio (*SNR*) at the receiver output. Occasionally, a minimum-detectable-signal definition is employed.

2. *Allowable deviation from expected characteristic:* The maximum signal is one that will cause some deviation from expected performance. Linear receivers usually specify a 1 dB decrease in incremental gain (the slope of the output-versus-input curve). Limiting or logarithmic receivers must define an allowable error in their outputs. Gain-controlled receivers must distinguish between instantaneous dynamic range and that achieved partly as a result of programmed gain variation.

3. *Type of signal:* Three types of signals are of general interest in determining dynamic-range requirements: distributed targets, point targets, and wideband-noise jamming. If the radar employs a phase-coded signal, the elements of the receiver preceding the decoder will not restrict the dynamic range of a point target as severely as they will distributed clutter; the bandwidth-time product of the coded pulse indicates the added dynamic range that the decoder will extract from point targets. Conversely, if the radar incorporates an excessively wide-bandwidth RF amplifier, its dynamic range to wideband-noise interference may be severely restricted.

When low-noise amplifiers (LNAs) are included in the antenna, prior to forming the receive beams, the sidelobe levels achieved are dependent upon the degree to which gain and phase characteristics are similar in all LNAs. Dynamic range has an exaggerated importance in such configurations because matching nonlinear characteristics is impractical. The effect of strong interference [mountain clutter, other radar pulses, or electronic countermeasures (ECM)], entering through the sidelobes, will be exaggerated if it exceeds the dynamic range of the LNAs because sidelobes will be degraded. The LNAs are wideband devices, vulnerable to interference over the entire radar operating band and often outside this band; although off-frequency interference is filtered in subsequent stages of the receiver, strong interference signals can cause clutter echoes in the LNA to be distorted, degrading the effectiveness of doppler filtering and creating false alarms. This phenomenon is difficult to isolate as the cause of false alarms in such radars owing to the nonrepetitive character of many sources of interference.

Evaluation. A thorough evaluation of all elements of the receiver is necessary to prevent unanticipated degradation of noise temperature or dynamic range. Inadequate dynamic range makes the radar receiver vulnerable to interference, which can cause saturation or overload, masking or hiding the desired echoes. A tabular format for such a computation (a typical example of which is shown as Table 3.1) will permit those components that contribute significant noise or restrict the dynamic range to be quickly identified. "Typical" values are included in the table for purposes of illustration.

One caution is required in using Table 3.1. The dynamic range of each component is computed by comparing the maximum signal and system noise levels at the output of each component. The assumption inherent in this method is that all filtering (bandwidth reduction and decoding) by this component is accomplished prior to any saturation. It is important to treat those stages of the receiver that provide significant filtering as separate elements; if multiple stages are lumped into a single filter, this assumption may be grossly in error.

3.3 BANDWIDTH CONSIDERATIONS

Definitions. The instantaneous bandwidth of a component is the frequency band over which the component can simultaneously amplify two or more signals to within a specified gain (and sometimes phase) tolerance. The tuning range is the frequency band over which the component may operate without degrading the specified performance if suitable electrical or mechanical controls are adjusted.

Important Characteristics. The environment in which a radar must operate includes many sources of electromagnetic radiation, which can mask the relatively weak echoes from its own transmission. The susceptibility to such interference is determined by the ability of the receiver to suppress the interfering frequencies if the sources have narrow bandwidth or to recover quickly if they are more like impulses in character. One must be concerned with the response of the receiver in both frequency and time domains.

Generally, the critical response is determined in the IF portion of the receiver; this will be discussed in Sec. 3.7. However, one cannot ignore the RF portion of the receiver merely by making it have wide bandwidth. Section 3.2 discussed

TABLE 3.1 Noise and Dynamic-Range Characteristics

		Antenna	Trans-line	RF amplifier	Mixer	Filter	Log detector
Noise temperature of component	K			520	1300	300	24K
Gain of component *	dB		-1.0	25	-6	15	
Total gain to input	dB			-1.0	24	18	33
Noise-temperature contribution referred to antenna	System 838 K	80	75	660	6	5	12
	29.3 dB K						
Overall RX bandwidth	63.0 dBHz						
	92.3						
Boltzmann's constant	-198.6						
Narrowband noise level †	-106.3 dBm	-106	-107	-82	-88	73	(-73)
Maximum signal capability †	dBm			-5	-16	+5	(+7)
Dynamic range to distributed targets †	dB			77	72	78	(80)
Bandwidth x time of point target †	dB	11	11	11	11	0	0
Dynamic range to point target †	dB			88	83	78	(80)
Bandwidth of receiver †	MHz			200	100	2	2
Ratio to overall receiver bandwidth †				100	50	1	1
Wideband-noise vulnerability †	dB			20	17	0	0
Dynamic range to wideband noise †	dB			57	55	78	(80)

*CW output–CW input on the center frequency, not coded pulse.

†At the output terminal of the designated component except where indicated by parentheses (at the input terminal of a nonlinear device).

how excessively wide bandwidth can penalize dynamic range if the interference is wideband noise. Even more likely is an out-of-band source of strong interference (TV station or microwave communication link) which, if allowed to reach this point, can either overload the mixer or be converted to IF by one of the spurious responses of the mixer.

Ideal mixers in a superheterodyne receiver act as multipliers, producing an output proportional to the product of the two input signals. Except for the effect of nonlinearities and unbalance, these mixers produce only two output frequencies, equal to the sum and the difference of the two input frequencies. Product mixers, although common at intermediate frequencies, are not generally available for RF conversion down to IF, and diode mixers are most commonly employed. The frequency-conversion properties of the diode are produced by its nonlinear characteristics. If its characteristic is defined by a power series, only the square-law term produces the desired conversion. The other terms produce spurious products, which represent an unwanted ability to convert off-frequency signals to the IF of the receiver. The efficiency of conversion of these unwanted frequencies, except for the image frequency, is sufficiently poor that the system noise temperature is not significantly degraded, but the mixer is vulnerable to strong out-of-band interference. The best radar receiver is one with the narrowest RF instantaneous bandwidth commensurate with the radiated spectrum and hardware limitations, and with good frequency and impulse responses.

A wide tuning range provides a flexibility to escape interference, but if the interference is intentional (jamming), change in frequency on a pulse-to-pulse basis may be required. Such frequency agility can be achieved by using switchable microwave filters or electronically tuned yttrium iron garnet (YIG) filters to restrict the instantaneous bandwidth. Each involves some insertion loss, another sacrifice in noise temperature to achieve more vital objectives.

3.4 RECEIVER FRONT END

Configuration. The radar *front end* consists of a bandpass filter or bandpass amplifier followed by a downconverter. The radar frequency is downconverted to an intermediate frequency, where filters with suitable bandpass characteristics are physically realizable. The mixer itself and the preceding circuits are generally relatively broadband. Tuning of the receiver, between the limits set by the preselector or mixer bandwidth, is accomplished by changing the LO frequency.

Effect of Characteristics on Performance. Noncoherent pulse radar performance is affected by front-end characteristics in three ways. Noise introduced by the front end restricts the maximum range. Front-end saturation on strong signals may limit the minimum range of the system or the ability to handle strong interference. Finally, the front-end spurious characteristic affects the susceptibility of off-frequency interference.

Coherent radar performance is even more affected by spurious mixer characteristics. Range and velocity accuracy is degraded in the pulse doppler radar; stationary-target cancellation is impaired in MTI (moving-target indication) radar; and range sidelobes are raised in high-resolution pulse compression systems.

Spurious Distortion of Radiated Spectrum. It is a surprise to many radar engineers that components of the radar receiver can cause degradation of the radiated transmitter spectrum, generating harmonics of the carrier frequency or spurious doppler spectra, both of which are often required to be 50 dB or more below the carrier. Harmonics can create interference in other electronic equipment, and their maximum levels are specified by the National Telecommunications and Information Administration (NTIA) and MIL-STD-469. Spurious doppler spectra levels are dictated by requirements to suppress clutter interference through doppler filtering.

Harmonics are generated by any component which is nonlinear at the power level created by the transmitter and which passes those harmonics to the antenna. Gaseous or diode receiver-protectors are designed to be nonlinear during the transmitted pulse and reflect the incident energy back toward the antenna. Isolators or circulators are often employed to absorb most of the reflected fundamental, but they are generally much less effective at the harmonics. Moreover, these ferrite devices are nonlinear in themselves and can generate harmonics.

Harmonic filters are included in most radars but often are improperly located to perform adequately. It is useless to locate the harmonic filter between the transmitter and the duplexer if the latter generates unacceptable harmonic levels itself; the filter must be located between the antenna and the duplexer.

Spurious doppler spectra are created by any process which does not reoccur precisely on each transmitted pulse. Gaseous receiver-protectors ionize under transmitter power levels, but there is some small statistical variation in the initiation of ionization on the leading edge of the pulse and in its subsequent development. In radars demanding high clutter suppression (in excess of 50 dB), it has sometimes been found necessary to prevent this variable reflected power from being radiated by use of both a circulator and an isolator in the receive path.

Spurious Responses of Mixers

Mathematical Mixer Model. The power-series representation of the mixer is perhaps the most useful in predicting the various spurious effects that are often noted. The current i flowing in a nonlinear resistance may be represented by a power series in the voltage V across the resistor terminals:

$$i = a_0 + a_1 V + a_2 V^2 + a_3 V^3 + \cdots + a_n V^n \tag{3.1}$$

The voltage applied to the mixer is the sum of the LO voltage $V_1 e^{j\omega_1 t}$ and the signal voltage $V_2 e^{j\omega_2 t}$:

$$V = V_1 e^{j\omega_1 t} + v_2 e^{j\omega_2 t} \tag{3.2}$$

When V from Eq. (3.2) is substituted into Eq. (3.1) and the indicated operations performed, the spectral characteristics are predicted.

Mixer Spurious-Effects Chart. The results of these calculations have been tabulated in several forms to show the system designer at a glance which combinations of input frequencies and bandwidths are free of strong low-order spurious components. The most useful form of the mixer chart[1] is shown in Fig. 3.2. The heavy line shows the variation of normalized output frequency $(H - L)/H$ with normalized input frequency L/H. This response is caused by the first-order mixer product $H - L$, which originates mainly from the square-law term in the power-series representation. All other lines on the chart define spurious effects arising from the cubic and higher-order terms in the power series. To simplify use of the chart, the higher input frequency is designated by H and the lower input frequency by L.

Seven particularly useful regions have been outlined on the chart. Use of the chart is illustrated by means of the region marked A, which represents the widest available spurious-free bandwidth centered at $L/H = 0.63$. The available RF passband is from 0.61 to 0.65, and the corresponding IF passband is from 0.35 to 0.39. However, spurious IF frequencies of 0.34 ($4H - 6L$) and 0.4 ($3H - 4L$) are generated at the extremes of the RF passband. Any extension of the instantaneous RF bandwidth will produce overlapping IF frequencies, a condition that is not corrected by IF filtering. The $4H - 6L$ and $3H - 4L$ spurious frequencies,

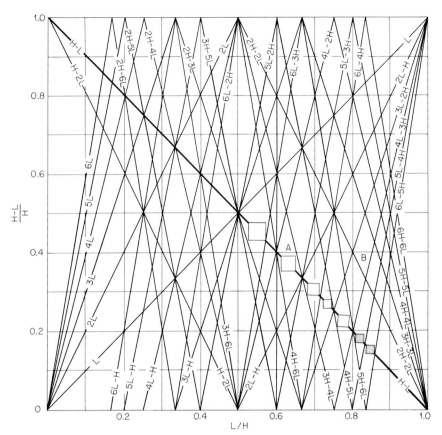

FIG. 3.2 Downconverter spurious-effects chart. H = high input frequency; L = low input frequency.

like all spurious IF frequencies, arise from cubic or higher-order terms in the power-series model of the mixer.

The available spurious-free bandwidth in any of the designated regions is roughly 10 percent of the center frequency or $(H - L)/10H$. Thus receivers requiring a wide bandwidth should use a high IF frequency centered in one of these regions. For IF frequencies below $(H - L)/H = 0.14$ the spurious frequencies originate from extremely high-order terms in the power-series model and are consequently so low in amplitude that they can usually be ignored. For this reason, single-conversion receivers generally provide better suppression of spurious responses than double-conversion receivers. The rationale for a choice of double conversion should always be validated.

The spurious-effects chart also demonstrates spurious input responses. One of the stronger of these occurs at point B, where the $2H - 2L$ product causes a mixer output in the IF passband with an input frequency at 0.815. All the products of the form $N(H - L)$ produce potentially troublesome spurious responses. These frequencies must be filtered at RF to prevent their reaching the mixer.

A spurious input response not predicted by the chart occurs when two or more off-frequency input signals produce by intermodulation a third frequency that lies within the RF passband. This effect is caused by quartic and higher-order even terms in the series. Its effect will be noted, for example, when

$$\frac{2H - L_1 - L_2}{H} = \frac{H - L}{H} \tag{3.3}$$

Intermodulation is reduced in some mixer designs by forward-biasing the mixer diodes to reduce the higher-order curvature.

The Balanced Mixer. The mixer model and the spurious-effects chart predict the spectral characteristics of a single-ended mixer. In the balanced-mixer configuration these characteristics are modified by symmetry. The two most common forms of balanced-mixer configuration are shown by Fig. 3.3a and b.

The configuration of Fig. 3.3a suppresses all spurious IF frequencies and spurious RF responses derived from even harmonics of the *signal* frequency. For the case where the subtraction is not obtained by a time delay, the LO frequency and *all* its harmonics are suppressed at the signal input port. Also of importance, noise sidebands of the LO which are converted to IF frequency are suppressed at the mixer IF port.

The configuration of Fig. 3.3b suppresses all spurious IF frequencies and spurious responses derived from even harmonics of the LO frequency. For the case where the RF phase shift is not obtained by a time delay, the LO frequency and

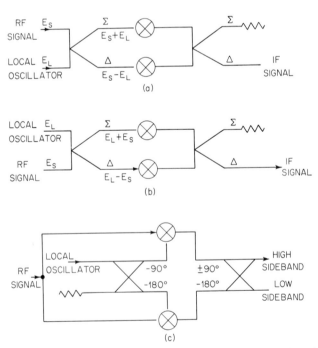

FIG. 3.3 (*a*) Balanced mixer with an inverted signal. (*b*) Balanced mixer with an inverted LO. (*c*) Image-reject mixer.

its *odd* harmonics are suppressed at the signal input port. Noise sidebands of the LO converted to IF *are not* rejected by this configuration, however.

Image-Reject Mixer. The single-ended mixer has two input responses which are derived from the square-law term in the power series. The responses occur at points above and below the LO frequency where the frequency separation equals the IF. The unused response, known as the image, is suppressed by the image-reject or single-sideband mixer shown in Fig. 3.3c. The RF hybrid produces a 90° phase differential between the LO inputs to the two mixers (which may be balanced mixers). The effect of this phase differential on the IF outputs of the mixers is a +90° shift in one sideband and a −90° shift in the other. The IF hybrid, adding or subtracting another 90° differential, causes the high-sideband signals to add at one output port and to subtract at the other. Where wide bandwidths are involved, the IF hybrid is of the all-pass type.

Characteristics of Amplifiers and Mixers

Noise Temperature. The most frequently cited figure of merit for a mixer or amplifier is its noise figure. However, the concept of noise temperature has proved more useful. Chapter 2 defines the usage of these parameters in determining the detectability of signals in a noise background.

Dynamic Range. A second useful figure of merit of the front-end device is the dynamic range from rms noise to the signal level that causes 1 dB compression in dynamic gain. Since the rms noise is dependent on the IF bandwidth, the effective dynamic range decreases with increasing IF bandwidth.

The balanced-diode mixer exhibits the largest dynamic range for a given IF bandwidth. However, the dynamic range of the mixer preceded by a low-noise amplifier will be reduced in proportion to the gain of the amplifier. Thus noise performance and dynamic range cannot be simultaneously optimized. A solution to this problem may come in the form of an active converter.[2,3]

3.5 LOCAL OSCILLATORS

Functions of the Local Oscillator. The superheterodyne receiver utilizes one or more local oscillators and mixers to convert the echo to an intermediate frequency that is convenient for filtering and processing operations. The receiver can be tuned by changing the first LO frequency without disturbing the IF section of the receiver. Subsequent shifts in intermediate frequency are often accomplished within the receiver by additional LOs, generally of fixed frequency.

Pulse-amplifier transmitters also use these same LOs to generate the radar carrier with the required offset from the first local oscillator. Pulsed oscillator transmitters, with their independent "carrier" frequency, use automatic frequency control (AFC) to maintain the correct frequency separation between the carrier and first LO frequencies.

In many early radars, the only function of the local oscillators was conversion of the echo frequency to the correct intermediate frequency. The majority of modern radar systems, however, coherently process a series of echoes from a target. The local oscillators act essentially as a timing standard by which the echo delay is measured to extract range information, accurate to within a small fraction of a wavelength. The processing demands a high degree of phase stability throughout the radar. Although these processing techniques are described elsewhere (Chaps. 15 to 17 and 21), they determine the basic stability requirements of the receiver.

The first local oscillator, generally referred to as a stable local oscillator (stalo), has a greater effect on processing performance than the transmitter. The final local oscillator, generally referred to as a coherent local oscillator (coho), is often utilized for introducing phase corrections which compensate for radar platform motion or transmitter phase variations.

Stalo Instability. The stability requirements of the stalo are generally defined in terms of a tolerable phase-modulation spectrum. Sources of unwanted modulation are mechanical or acoustic vibration from fans and motors, power supply ripple, and spurious frequencies and noise generated in the stalo. In general, the tolerable phase deviation decreases with increasing modulation frequency because the doppler filter is less efficient in suppressing the effects. In a radar having two-pulse MTI, there is a linear relationship between the tolerable phase deviation and the period of the modulation. Their ratio is the allowable FM (frequency modulation) or *short-term frequency stability* sometimes encountered in the literature. This parameter does not adequately define the phase-stability requirements for pulse doppler or MTI radars where more than two pulses are coherently processed.

The phase-modulation spectrum of the stalo may be measured and converted into the MTI improvement factor limitation, which is dependent on range to the clutter and the characteristics of the two cascaded filters in the radar receiver. This conversion process involves three steps, described below.

It should be noted that some spectrum analyzers do not distinguish between frequencies below the desired stalo frequency and those above; their response is the sum of the power in the two sidebands at each designated modulation frequency. This is of no consequence in MTI radars which have equal response to positive and negative doppler frequencies. In radars using doppler filters unsymmetrical about zero doppler, it is necessary to assume that the stalo spectrum measured is symmetrical, generally a valid assumption. The examples shown subsequently employ measured data from this type of double-sideband (DSB) spectrum analyzer. If a single-sideband (SSB) spectrum analyzer is available, positive and negative modulation-frequency components can be measured separately and analyzed without any assumption of symmetry. It is essential that the measured data be defined as SSB or DSB, since there is a 3 dB difference in the two forms of data.

Range Dependence. Most modern radars use the stalo to generate the transmitted pulse as well as to shift the frequency of the received echoes. The transmitters are power amplifiers (traveling-wave tubes, klystrons, twystrons, crossed-field amplifiers, solid-state amplifiers, etc.) rather than oscillators (magnetron, etc.). It is this double use of the stalo that introduces a dependence on range of the clutter and exaggerates the effect of certain unintentional phase-modulation components by 6 dB, the critical frequencies being those which change phase by odd multiples of 180° during the time period between transmission and reception of the clutter echo from a specified range. At these critical frequencies, a maximum positive phase deviation on transmission changes to a maximum negative deviation at the time of reception, doubling the undesired phase modulation of the echo at IF.

Figure 3.4 shows this range-dependent filter characteristic, which may be expressed mathematically as

$$dB = 10 \log 4 \sin^2 (2\pi f_m R/c) = 10 \log 4 \sin^2 (\pi f_m t) \tag{3.4}$$

where f_m = modulation frequency, Hz
R = range, m
c = propagation velocity, 3×10^8 m/s
t = time delay = $2R/c$

A short time delay can tolerate much higher disturbance at low modulation frequencies, as illustrated by the two cases in Fig. 3.4. Consequently, stalo stability needs to be computed for several time delays.

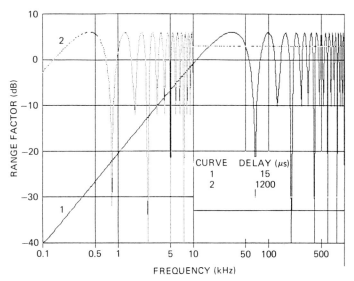

FIG. 3.4 Effect of range delay on clutter cancellation.

For example, stalo phase modulation caused by power supply ripple at 120 Hz creates nearly equivalent phase modulation of a clutter echo from nearly 100-nmi range (delay of 1200 µs resulting in 0 dB range factor). Phase modulation of a clutter echo with a range delay of 15 µs is about 38 dB less than the stalo phase modulation because the stalo phase has changed only slightly in this short time interval; the phase added by the stalo to the transmitted pulse is nearly the same as the phase subtracted from the received echo in the mixer.

Adding the decibel values of the measured stalo spectrum and the range-dependent effect at each modulation frequency provides the spectrum of undesired doppler modulation at the output of the mixer.

Receiver Filtering. Subsequent stages of the radar receiver have responses which are functions of the doppler modulation frequency; so the output residue spectrum can be obtained by adding the decibel responses of these filters to the preceding spectrum at the mixer.

The receiver contains two cascaded filters: an optimum-bandwidth filter at IF and a doppler filter, generally implemented digitally in modern radars. The example illustrated in Fig. 3.5 includes a gaussian filter at IF with a 3 dB bandwidth of 1.6 MHz and a four-pulse MTI with variable interpulse periods and time-

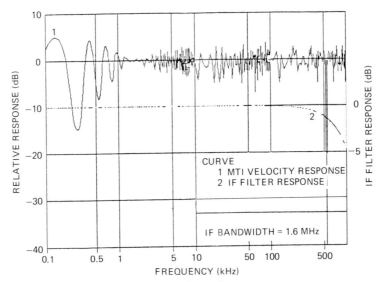

FIG. 3.5 Frequency responses of radar receiver.

varying weights. This MTI exaggerates certain stalo modulation frequencies by as much as 5 dB more than average. Note that for this analysis to be correct the MTI velocity response must be scaled to have 0 dB gain to noise, not to an optimum doppler frequency.

Integration of Residue Power. The MTI improvement factor limitation of the stalo may be expressed as the ratio of the stalo power to the total power of the echo modulation spectrum it creates at the output of the cascaded filters (Figs. 3.4 and 3.5).

Figure 3.6 shows an example of the measured modulation spectrum of a stalo (curve 1) and the effect of 15-μs delay on the clutter residue (curve 2). A computer program can be utilized to alter the measured spectral data, using the filters of Fig. 3.4 and 3.5, and to integrate the total power in the doppler residue spectrum, with the exception of modulation frequencies below 100 Hz, which cannot be measured. The result (51.8 dB) is the MTI improvement factor limitation due to stalo instability.

Figure 3.7 shows the same measured stalo modulation spectrum, subjected to a range delay of 1200 μs. The clutter residue spectrum contains more power at the lower modulation frequencies than Fig. 3.6, but the MTI residue is increased by only 1 dB. Long-range clutter is suppressed to nearly the same degree as short-range clutter.

If the radar utilizes more than one doppler filter, the effect of stalo instability should be calculated for each individually. If an individual filter's doppler response is unsymmetrical, residues from positive and negative doppler bands must be computed separately and added in power.

It should be noted that many textbooks analyze only a simple two-pulse MTI, and the resulting equations for the limitation on MTI improvement factor cannot be employed for more sophisticated doppler filters. This important fact is generally overlooked by the casual reader.

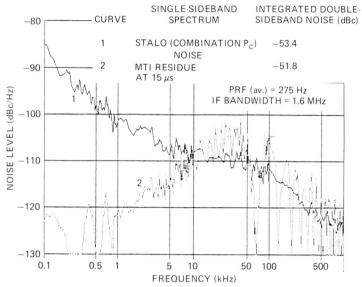

FIG. 3.6   Effect of 15-μs range delay on stalo MTI limitation.

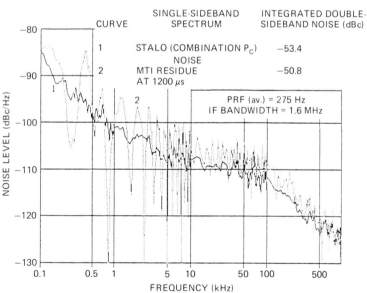

FIG. 3.7   Effect of 1200-μs range delay on stalo MTI limitation.

It should be noted that most textbook analyses assume that the stalo instability is due either to a single modulation frequency or to a combination of white gaussian-noise modulations. Rarely are these assumptions valid for real stalos; so a different method of analysis must be employed.

The computer analysis described measures the colored modulation spectrum of the stalo, modifies it in conformance with the range-dependent effect and receiver filters of the radar, and integrates the output residue power.

It is a valid procedure for determining stalo stability that requires no assumptions, but the resulting value must be considered merely a figure of merit to compare *different* stalos for a *given* application. The same stalo would have a different stability value in another application with different receiver filters.

Coho and Timing Instability. In modern radars using pulsed amplifier transmitters, the coho is rarely a significant contributor to receiver instability. However, in older radars using pulsed oscillator transmitters, the coho must compensate for the random phase of each individual transmitter pulse, and imperfect compensation results in clutter residue at the output of the doppler filters. Compensation is possible only for echoes from the most recent transmission; echoes from prior transmissions (multiple-time-around clutter) cannot be suppressed by doppler filtering in pulsed oscillator radars, and for this reason alone such radars are no longer popular. Those readers interested in the methods employed to compensate the coho in these older radars are referred to Sec. 5.5 of the 1970 edition of this handbook.

When the radar is on a moving platform or when the clutter is moving rain or sea, the frequency of the coho is sometimes varied to try to compensate for this motion, shifting the clutter spectrum back to zero doppler. The servo which accomplishes this task, if properly designed, will introduce insignificant instability under ideal environmental conditions (solely clutter echoes and receiver noise typical of laboratory tests), but the effect of strong moving targets and pulsed interference from other radars can sometimes be serious, shifting the coho frequency from the proper compensation value.

Timing signals for the transmitter and A/D converter are usually generated from the coho, and timing jitter can cause clutter attenuation to be degraded. However, the effect of timing jitter is too complex to predict accurately; so it is rarely measured separately.

Total Radar Instability. The primary sources of radar instability are usually the stalo and the transmitter. If the doppler spectra of these two components are available, either through measurements or through predictions based on similar devices, the convolution of the two-way stalo spectrum (modified by the range-dependent effect) and the transmitter doppler spectrum provides an estimate of the spectrum of echoes from stable clutter, which is then modified by the two receiver filters and integrated to obtain the residue power caused by these two contributors. This power can be larger than the sum of the residues created by each contributor alone. These procedures are employed to diagnose the source of radar instability in an existing radar or to predict the performance of a radar in the design stage.

Measurement of total radar instability can be conducted with the radar antenna searchlighting a stable point clutter reflector which produces an echo close to (but below) the dynamic-range limit of the receiver and doppler filter. Suitable clutter sources are difficult to find at many radar sites, and interruption of rotation of the antenna to conduct such a test may be unacceptable at others; in this case, a microwave delay line can be employed to feed a delayed sample of the transmitter pulse into the receiver. All sources of instability are included in this single measurement except for any contributors outside the delay-line loop. It is important to recognize that timing jitter does not produce equal impact on all parts of the echo pulse and generally has minimal effect on the center of the pulse; so it is essential to collect data samples at a multiplicity of points across the echo, including leading and trailing edges. The total radar instability is the ratio of the sum of the multiplicity of residue powers at the output of the doppler filter to the sum of the powers at its input, divided by the ratio of receiver noise at these locations. Stability is the inverse of this ratio; both are generally expressed in decibels.

In radars with phase-coded transmission and pulse compression receivers, residue may be significant in the range sidelobe region as well as in the compressed pulse, caused by phase modulation during the long transmitted pulse rather than solely from pulse to pulse. Measurement of stability of such radars must employ a very large number of data points to obtain an answer valid for clutter distributed in range.

Radar instability produces predominantly phase modulation of the echoes, and the scanning antenna produces predominantly amplitude modulation; so the combined effect is the sum of the residue powers produced by each individually.

3.6 GAIN-CONTROLLED AMPLIFIERS

Sensitivity Time Control (STC). The search radar detects echoes of widely differing amplitudes, typically so great that the dynamic range of any fixed-gain receiver will be exceeded. Differences in echo strength are caused by differences in radar cross sections, in meteorological conditions, and in range. The effect of range on radar echo strength overshadows the other causes, however.

The radar echo power received from a reflective object varies inversely with the fourth power of the range or propagation time of the radar energy. The effect of range on signal strength impairs the measurement of target size. Yet determination of target size is needed to discriminate against radar echoes from insects, atmospheric anomalies, or birds (which in some cases have radar cross sections only slightly less than that of a jet fighter). Also, many radar receivers exhibit objectionable characteristics when signals exceed the available dynamic range. These effects are prevented by a technique known as sensitivity time control, which causes the radar receiver sensitivity to vary with time in such a way that the amplified radar echo strength is independent of range.

Search radars often employ a cosecant-squared antenna pattern whose gain diminishes with increasing elevation angle. The pattern restricts the power at high elevation angles, since an aircraft at a high elevation angle is necessarily at close range and little power is required for detection. At the high elevation angles, however, the echo power becomes independent of range and varies instead with the inverse fourth power of the altitude. An STC characteristic that is correct for the low-angle radar echoes restricts high-angle coverage. This incompatibility of the STC requirements at the elevation extremes severely limits the usefulness of STC.

The restriction on STC imposed by the cosecant-squared antenna pattern can be reduced by a more realistic radar design philosophy. It is recognized that the antenna must radiate more energy at the high angles than is provided by the cosecant-squared pattern. There are two reasons for this. First, high-angle coverage is limited by clutter from the stronger low-altitude section of the beam rather than from system noise. Second, ECM reduces both the maximum range and the altitude coverage of the radar. Of these, the loss of altitude coverage is the more serious. Both factors have caused the cosecant-squared pattern to be abandoned in favor of one that directs more energy upward.

The advent of stacked-beam radars, which achieve their coverage pattern by use of multiple beams, has liberated STC from the restrictions of the antenna pattern. In these systems, there is one receiver channel for each beam, and STC may be applied to the receiver channels independently. Consequently, the upper-beam receivers may be allowed to reach maximum sensitivity at short ranges, whereas the lower-beam receiver reaches maximum sensitivity only at long range.

Most modern radars generate STC waveforms digitally. The digital commands may be used directly by digital attenuators or converted to voltage or current for control of diode attenuators or variable-gain amplifiers. Digital control permits calibration of each attenuation to determine the difference between the actual attenuation and the command, by injecting a test pulse during *dead time*. This is essential in monopulse receivers which compare the echo amplitudes received in two or more beams simultaneously to accurately determine the target's position in azimuth or elevation. Accurate measurements depend on compensation for any difference in gains of the monopulse receivers.

Readers interested in the methods of generating the analog STC waveforms used in older radars may find descriptions of various methods in Sec. 5.6 of the 1970 edition of this handbook.

Clutter Map Automatic Gain Control.   In some radars, mountain clutter can create echoes which would exceed the dynamic range of the subsequent stages of the receiver (A/D converter, etc.) if the STC attenuation at that range allows detection of small aircraft. The spatial area occupied by such clutter is typically a very small fraction of the radar coverage; so AGC is sometimes considered as an alternative to either boosting the STC curve (a performance penalty affecting detectability of small aircraft in areas of weaker clutter or no clutter) or increasing the number of bits of the A/D converter and subsequent processing (an economic penalty).

Clutter map AGC is controlled by a digital map which measures the mean amplitude of the strongest clutter in each map cell of many scans and adds attenuation where necessary to keep the mean amplitude well below saturation. One disadvantage of clutter map AGC is that it degrades detectability of small aircraft over clutter which, in the absence of AGC, would be well below saturation. The scan-to-scan fluctuation of clutter requires a 6 to 10 dB safety margin between the maximum mean level controlled and saturation. Another problem is the vulnerability of the map to pulsed interference from other radars.

Clutter map AGC can serious degrade other critical signal-processing functions, and the following fundamental incompatibilities prevent its successful application to many types of radars:

• Suppression of clutter by doppler filtering is degraded by change of attenuation from one interpulse period to the next.

• Control of false alarms in distributed clutter (rain, sea) can be degraded by change of attenuation from one range sample to the next (see Sec. 3.13).

• Time sidelobes of compressed pulses in radars which transmit coded waveforms are degraded by attenuation variation in range prior to compression. Gradual STC variations can be tolerated, but not large step changes.

Automatic Noise-level Control. AGC is widely employed to maintain a desired level of receiver noise at the A/D converter. As will be described in Sec. 3.11, too little noise relative to the quantization increment of the A/D converter causes a loss in sensitivity; too much noise means a sacrifice of dynamic range. Samples of noise are taken at long range, often beyond the instrumented range of the radar (in dead time), to control the gain by means of a slow-reaction servo. If the radar has RF STC prior to any amplification, it can achieve meaningful dead time by switching in full attenuation; this minimizes external interference with minimal (and predictable) effect on system noise temperature. Most radars employ amplifiers prior to STC; so they cannot attenuate external interference without affecting the noise level which they desire to sense, and the servo must be designed to tolerate pulses from other radars and echoes from rainstorms or mountains at extreme range. This interference occasionally can be of high amplitude but generally has a low duty cycle during a 360° scan; so the preferred servo is one which increments a counter when any sample is below the desired median noise level and decrements the counter when the sample is above that level, independently of how great the deviation is. The most significant bits of the counter control the gain, and the number of bits of lesser significance in the counter dictates the sluggishness of the servo.

3.7 FILTERING

Filtering of the Entire Radar System. The filter provides the principal means by which the receiver discriminates between desired echoes and interference of many types. It may approximate either of two forms: a matched filter, which is a passive network whose frequency response is the complex conjugate of the transmitted spectrum, or a correlation mixer, an active device which compares the received signals with a delayed replica of the transmitted signal. Receiver filters are assumed to have no memory from one transmission to the next; their response is to a single transmission.

Actually, most radars direct a multiplicity of pulse transmissions at a target before the antenna beam is moved to a different direction, and the multiplicity of echoes received is combined in some fashion. The echoes may be processed by an integrator, which is analogous to a matched filter in that ideally its impulse response should match the echo modulation produced by the scanning antenna. The echoes may be applied directly to a PPI, with the viewer visually integrating the dots in an arc which he or she associates with the antenna beamwidth. Various doppler processes (including MTI) may be applied to separate desired from undesired targets. From the radar system standpoint, these are all filtering functions, but they are treated in other chapters of this handbook. The receiver filtering to be discussed here is that associated with separating a single pulse from interference, although the subsequent problem of filtering the train of echoes from a single target dictates the stability of the receiver filter.

At some point in the radar receiver, a detector produces an output voltage which is some function of the envelope of the IF signal. If it provides a linear function, it is termed an envelope detector; logarithmic detectors will be described in Sec. 3.8. The response of a linear detector to weak signals which do not greatly exceed noise level has been extensively analyzed.[4] Various pairs of frequency components of input noise, which may be far removed from the spec-

trum of the desired echo, can intermodulate to produce a beat-frequency component at the detector output that is within the desired band. Similarly, the noise
intermodulation smears some of the signal energy outside the desired band. As a
consequence, filtering after envelope detection is less efficient than filtering prior
to detection. All postdetection circuitry should have several times the bandwidth
of the echo, and predetection filtering should be optimized, as will be described.

Definitions. The reader is cautioned that there are no universally accepted
definitions of the terms *pulse duration* and *spectral bandwidth* of the
transmitted signal, *impulse response* and *bandwidth* of filters, or the equivalent
antenna parameters, *beamwidth* and *spectral bandwidth* caused by scanning.
These terms should always be used with clarifying adjectives to define their
meaning.

Energy Definitions. For detection of radar echoes against a noise background, the only fundamental parameters are the energy content of the transmitted signal, of the receiver noise, and of the echoes received as the antenna scans
past the target. These energy parameters define the width of a rectangular function that has the same peak response and same energy content as the real function. Their only purpose is to relate the peak value of the function to the more
vital energy content.

Of these parameters, only the energy of the pulse is easy to measure (average
power/PRF), and this may be employed in the radar range equation directly,
without distinguishing peak power and "energy" pulse width. The noise or energy bandwidth of a receiver is often employed in theoretical analyses but rarely
stated in the tabulation of radar parameters; bandwidth need not even be included
in the radar range equation if the receiver approximates a matched filter.

3 dB Definitions. In the interest of making possible direct measurement of
parameters from oscilloscope waveforms or pen recordings of these functions, it
has been customary to utilize widths measured either at half-power (3 dB) or half-
voltage (6 dB) points. For functions that resemble a gaussian pattern, the 3 dB
width is a close approximation to the energy width; the receiver bandpass generally fulfills this condition sufficiently to make the 3 dB bandwidth meaningful.
Transmitter pulse shapes and spectra generally deviate significantly from
gaussian.

6 dB Definitions. Although antenna beamwidths (and number of echoes received) are often specified between the 3 dB points, this actually represents a
6 dB definition of the echo response as the radar antenna scans past the target; in
those radars whose transmitting and receiving beams are not identical, the 6 dB
points of the two-way pattern are usually specified. Most definitions of pulse
shapes include voltage parameters, with rise and fall times being represented by
10 and 90 percent points and pulse duration by 50 percent (6 dB) points. Likewise, filter bandpass characteristics are often defined by their widths at the 6 and
60 dB points. The 6 dB definitions will be the dominant definitions employed in
this chapter.

Entirely aside from custom, there are several valid arguments favoring the use
of 6 dB parameters. As indicated in Table 3.2, the optimum bandwidth–time
product for detection of a pulse in white gaussian noise, with each defined at the
6 dB points, does not deviate significantly from unity for most practical functions. The 3 dB or energy definitions yield widely variable optimum bandwidths,
dependent upon the shape of the pulse and the bandpass of the filter; there can be
no quick estimation of optimum bandwidth if these parameters are utilized.

Distributed clutter, rain or chaff, is often a more serious interference with target detection than noise. In passing through an optimum-bandwidth filter and optimum-bandwidth integrator, the echo is stretched in both range and angle; the clutter spectrum, being the product of the transmitted spectrum and the bandpass of the receiver, is narrower than either. As a result, the 6 dB two-way beamwidth and the 6 dB pulse width closely approximate the extent of the radar cell from which the "optimum" receiver accumulates clutter energy.

To summarize the general utility of 6 dB definitions: (1) The range equation for detection in noise need not include peak power, pulse width, or receiver bandwidth; only the efficiency of the integrator requires a definition of the number of pulses being received, and a 6 dB echo definition is universally employed. (2) The optimum bandwidth is close to the inverse of the echo duration if both are 6 dB definitions; this applies to both the receiver filter and the integrator. (3) The energy of the interference from clutter, rain or chaff, that is accepted by an approximately matched receiver is well defined by the 6 dB pulse duration and 6 dB two-way beamwidth.

Approximations to Matched Filters. The most efficient filter for discriminating between white gaussian noise and the desired echoes is a matched filter, a passive network whose frequency response is the complex conjugate of the transmitted spectrum. It can process echoes from all ranges. The correlation mixer, an active device which compares the received signals with a delayed replica of the transmitted signal, is mathematically equivalent to a matched filter, but it is responsive only to echoes from one specific range; consequently its use in radar systems is more limited.

Table 3.2 illustrates the degree of sacrifice in detectability that results in approximating a matched filter, either to simplify the hardware or to achieve better filtering of other forms of interference. The optimum bandwidths of these filters

TABLE 3.2   Approximations to Matched Filters

Transmitted pulse shape	Receiver filter	Optimum bandwidth–time product			Mismatch loss, dB
		6 dB	3 dB	Energy	
Gaussian	Gaussian bandpass	0.88	0.44	0.50	0
Gaussian	Rectangular bandpass	1.05	0.74	0.79	0.51
Rectangular	Gaussian bandpass	1.05	0.74	0.7	0.51
Rectangular	5 synchronously tuned stages	0.97	0.67	0.76	0.51
Rectangular	2 synchronously tuned stages	0.95	0.61	0.75	0.56
Rectangular	Single-pole filter	0.70	0.40	0.63	0.88
Rectangular	Rectangular bandpass	1.37	1.37	1.37	0.85
Phase-coded					
Biphase	Gaussian	1.05	0.74	0.79	0.51
Quadriphase	Gaussian	1.01	0.53	0.5	0.09

also are tabulated in terms of product of the filter bandwidth and pulse duration. Typically, the bandwidth may deviate 30 to 50 percent from the optimum value before the detectability is degraded by an additional 0.5 dB. This rather broad "optimum" region is centered near unity bandwidth-time product for virtually all practical filters if one uses the 6 dB definitions.

Sometimes the bandwidth of a radar receiver is in excess of the optimum to allow for some offset between the echo spectrum and the filter bandpass, caused by target velocity and receiver tuning tolerances. Although this makes the radar more susceptible to off-frequency narrowband interference (Fig. 3.8), it reduces the time required to recover from impulse interference (Fig. 3.9). These figures also illustrate that, to provide good suppression of both forms of interference, the shape of the filter bandpass characteristic is even more important than its bandwidth. Rectangular bandpass or impulse responses should be avoided; the closer one approximates a gaussian filter, the better the skirts in both frequency and time domains.

In the case of phase-coded transmissions, the duration of the subpulses is the time parameter. This is equivalent to the spacing of the subpulses with one exception: the quadriphase code[5,6] employs half-cosine subpulses with 6 dB width equal to four-thirds of subpulse spacing. One of the virtues of the quadriphase code is the unusually low mismatch loss, owing to the fact that the impulse response of the gaussian filter is an excellent approximation to the subpulse shape. A digital correlator in a later stage of the receiver completes the matched filter.

Filtering Problems Associated with Mixer Spurious Responses. The approximation of a matched filter is generally most easily accomplished at some

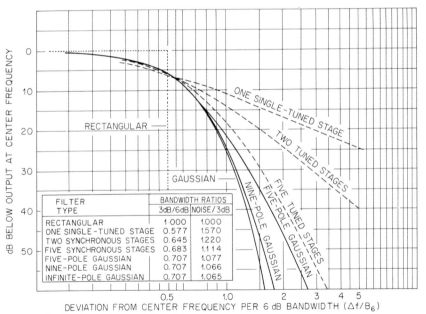

FIG. 3.8 Bandpass characteristics of filters.

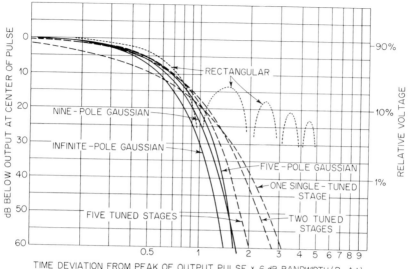

FIG. 3.9   Impulse characteristics of filters.

frequency other than that radiated by the radar. The optimum filtering frequency is a function of the bandwidth of the echo and the characteristics of the filter components. Consequently, it is necessary for the radar receiver to translate the frequency of the echo to that of the filter, in one or more steps, using local oscillators and mixers.

Section 3.4 described how spurious responses are generated in the mixing process. Unwanted interference signals can be translated to the desired intermediate frequency even though they are well separated from the echo frequency at the input to the mixer. The ability of the radar to suppress such unwanted interference is dependent upon the filtering preceding the mixer as well as on the quality of the mixer itself.

The image frequency is the most serious of the spurious-response bands, but an image-rejection mixer can readily suppress these signals by 20 dB. A filter can further attenuate image-frequency signals before they reach the mixer unless the ratio of input to output frequencies of the mixer exceeds the loaded Q of the available filters. This image-suppression problem is the reason why some receivers do not translate from the echo frequency directly to the final intermediate frequency in a single step.

The other spurious products of a mixer generally become more serious if the ratio of input to output frequencies of the downconverter is less than 10. The spurious-effects chart (Fig. 3.2) shows that there are certain choices of frequency ratio that provide spurious-free frequency bands, approximately 10 percent of the intermediate frequency in width. By the use of a high first IF, one can eliminate the image problem and provide a wide tuning band free of spurious effects. Filtering prior to the mixer remains important, however, because the neighboring spurious responses are of relatively low order and may produce strong outputs from the mixer.

In addition to external sources of interference, the radar designer must be concerned with internal signal sources. MTI and pulse doppler radars are particularly susceptible to any such internal oscillators that are not coherent, i.e., that do not have the same phase for each pulse transmission. The effect of the spurious signal then is different for each echo, and the ability to reject clutter is degraded.

A truly coherent radar generates all frequencies, including its interpulse periods, from a single stable oscillator. Not only all the desired frequencies but also all the internally generated spurious signals are coherent, and they do not affect the clutter rejection.

More commonly, MTI and pulse doppler radars are pseudo-coherent, as illustrated in Fig. 3.10. The coho is the master reference for the phase detector and may be the clock from which the interpulse periods are determined. The coho also is employed in generating the transmitter frequency, offset from a noncoherent local oscillator. Neither the local oscillator nor the transmitter is co-

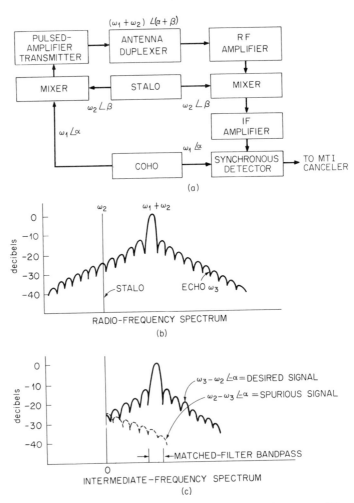

FIG. 3.10 Pseudo-coherent radar. (*a*) Block diagram. (*b*) RF spectrum of the echo. (*c*) IF spectrum of the echo.

herent; their phases are different for each pulse transmitted. Only in the IF portion of the receiver is the desired echo from a stationary target coherent (α = constant, where α is the phase of the echo when the transmitter is turned on); virtually all the spurious outputs of the mixer are noncoherent and produce a fluctuating signal at IF.

The only spurious component that may be coherent in the pseudo-coherent radar is the image frequency. As illustrated in Fig. 3.10, a pulsed spectrum can have fairly wide skirts and can overlap into the image band. The folded spectrum at intermediate frequency has two components within the bandpass of the matched filter, and if the percentage bandwidth at IF is large, the undesired image component can be significant. Only when it is coherent is a degradation of clutter cancellation impossible.

Consider a pseudo-coherent radar in which the interpulse periods are generated from an independent source. Now the phase of the coho (α) changes from pulse to pulse, and although the phase of the desired echo from a stationary target changes equally, the phase of the spurious image echo changes oppositely. The cancellation of the stationary-target echo is limited to the level of the spurious echo.

It is clear that the overall filtering capability of the radar, its ability to enhance the desired echoes and suppress undesired interference, may be degraded by the spurious responses of the various mixing stages. Particularly susceptible to degradation are MTI and pulse doppler radars, which may not provide the expected improvement in the rejection of clutter if the coho is not at the same phase condition each time that the transmitter pulses. All radars are susceptible to off-frequency interference, which, unless it is filtered before reaching the mixer, may create a detectable output in the desired IF band.

The ability of MTI or pulse doppler processing to suppress clutter may be degraded if the receiver filter is not perfectly stable. The receiver's transfer characteristic (gain, time delay, and bandpass or impulse response) must be constant so that its effect on each echo pulse is identical.

The mixer spurious responses just discussed and the stalo and coho problems of Sec. 3.5 represent only the most likely sources of instability to be encountered. Other elements of the receiver require attention to avoid instability problems. Vibration or power supply ripple can cause gain and phase modulation, particularly in RF amplifiers. Such modulation will degrade clutter attenuation unless the ripple frequency is a harmonic of the PRF.

3.8 LOGARITHMIC DEVICES

Characteristics

Accuracy. Logarithmic devices and IF amplifiers are devices whose output is proportional to the logarithm of the envelope of the IF input. They often approximate the logarithmic characteristic by multiplicity of linear segments. Normally linear segments of equal *length ratio* and varying slope are joined to give a best fit to a logarithmic curve. Each segment will be correct at two points and will have a maximum error at the ends and center. The magnitude of the error[7] increases with the length ratio of the segment. Figure 3.11 shows how this error changes with segment length ratio (also called gain per segment). In practice, the joints between the "linear" segments are not abrupt, and the best fit to the logarithmic curve may have less than the theoretical error.

Logarithmic detectors and amplifiers are frequently designed with adjustments in each stage. This allows for adjustment of the slope and/or length of the segment for a better fit. A precision exponentially decaying IF waveform from a test set is applied to the unit under test. The unit is adjusted for a linearly decaying output, which indicates the correct adjustment.

Dynamic Range. The dynamic range of a logarithmic detector or amplifier is dependent on the number of linear segments N and on the length ratio G of the segments:

$$\text{Dynamic range} = 20N \log G \qquad (3.5)$$

A well-designed logarithmic detector may have a dynamic range of 80 dB derived from nine stages with an error as low as ± 0.2 dB.

Bandwidth. The bandwidth of a logarithmic detector or amplifier generally varies with signal level. For this reason the logarithmic device is usually designed with excess bandwidth and is preceded by filters which establish the receiver bandwidth. However, the large signal bandwidth of the logarithmic device itself may be measured by using the method indicated in Fig. 3.12. The input voltage is increased from V_i to $V_i + 3$ or 6 dB, changing the operating point from A to B. The frequency is now changed in both directions to find those two frequencies that place the operating point at C. The same result also may be obtained by changing the frequency until the output is reduced by $3S$ or $6S$, where S is the volts-per-decibel characteristic. This places the operating point at C', which lies on the same curve as C. Because of the dependence of bandwidth on signal level, if possible the logarithmic amplifier should be aligned with signals below the threshold and its pulse response used as a criterion of performance.

The pulse response is measured with a pulsed IF signal having much faster rise and fall times than those of the logarithmic device being tested. The rise time is the time required for the output to rise from the $-20S$ to the $-S$ point, and the fall time is the time required for the output to fall from the $-S$ point to the $-20S$ point. Because of the logarithmic characteristic, the fall time tends to be a straight line and to exceed the rise time.

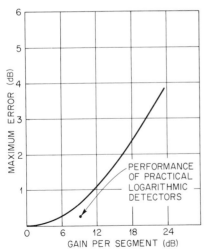

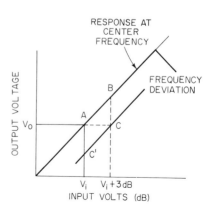

FIG. 3.11 Error in the approximation of a logarithmic curve by linear approximation.

3.12 Method of measuring logarithmic-amplifier bandwidth.

Analog Logarithmic Devices

Logarithmic Detector. A well-known form of the logarithmic detector uses successive detection,[6] wherein the detected outputs of N similar limiter stages are summed as shown by Fig. 3.13. If each stage has a small signal gain G and a limited output level E, the intersections of the approximating segments fall on a curve described by

$$E_0(M) = n\left[E\frac{\log E_i(M)G^{N+1}}{\log G} + E\left(\frac{1}{G^{M-1}} + \cdots + \frac{1}{G^2} + \frac{1}{G}\right)\right] \quad (3.6)$$

where n is the detector efficiency and $E_i(M)$ represents the particular input levels that correspond to the intersections of the line segments,

$$E_i(M) = \frac{E}{G^M} \quad (3.7)$$

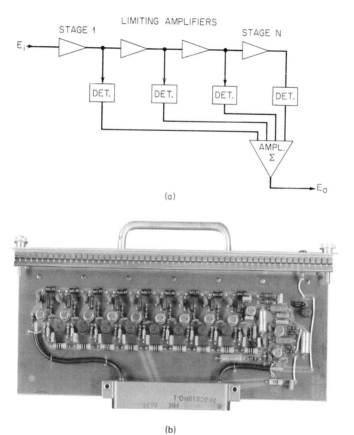

(a)

(b)

FIG. 3.13 (*a*) Logarithmic detector. (*b*) Nine-stage logarithmic detector. Gain adjustment for each stage is shown below the transistor.

The independent variable M is the order of the stage at incipient saturation and will take on only integer values from 1 to N.

A nonlogarithmic term in the form of a power series in $1/G$ has only a minor effect on logarithmic accuracy. The successive difference in this term as M is varied from 2 to N is E/G^{M-1}. The overall tendency of this term, therefore, is to produce an offset in the output with only a minor loss in logarithmic accuracy at the highest signal levels.

A typical logarithmic detector may have an accuracy of ± 4 dB, a dynamic range of 80 dB, and a bandwidth of 5 to 10 MHz at 30 MHz, derived from nine stages (Fig. 3.13b). However, Rubin[9] describes a four-stage successive detection design having a bandwidth of 640 MHz centered at 800 MHz. Pulse response time of 2.5 ns is claimed.

Logarithmic Amplifier. A logarithmic IF amplifier may be implemented with N identical cascaded dual-gain stages. In this case a precise logarithmic response results if each amplifier has a threshold E_T below which the gain is a fixed value G and above which the incremental gain is unity. The intersections of the approximating segments fall on a curve described by

$$E_0(M) = E_T\, G\frac{\log E_i(M)G^N}{\log G} \tag{3.8}$$

where $E_i(M)$ represents the input levels corresponding to the intersection of the line segments:

$$E_i(M) = E_T\, G^{1-M} \tag{3.9}$$

The independent variable M is the order of the stage that is at incipient saturation; M will have integer values between 1 and N. In the case of the dual-gain logarithmic amplifier, all intersections of the line segments fall on a logarithmic curve.

A typical logarithmic amplifier may have a dynamic range of 80 dB, derived from nine stages, with an overall bandwidth of 5 MHz or more. Typical accuracy is $\pm\frac{1}{4}$ dB over 70 dB dynamic range and ± 1 dB over the full dynamic range.

Details of a typical stage are shown in Fig. 3.14, and the voltage characteristics in Fig. 3.15. The stage consists of an attenuator bridged by a series-diode limiter and followed by an amplifier. In the absence of an input voltage, the current divides equally between the limiter diodes. The thresholds are reached when diode $D2$ carries either all or none of the current. If the voltage drops across the diodes are neglected, this occurs when

$$E_i = E_T = \pm\frac{V}{2}\left(\frac{R}{R + R_s}\right) \tag{3.10}$$

Signals above the threshold in magnitude are attenuated by a factor $R/(R + R_f)$, the reciprocal of the amplifier gain. The incremental gain in this region is therefore unity. The blocking capacitors are used because of the dc offset voltage that is characteristic of this form of limiter.

Digital Logarithm. The trend toward digital processing requires mention of a piecewise linear digital approximation which may be accomplished after analog-to-digital conversion and digital doppler filtering to suppress clutter

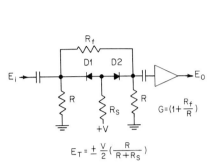

FIG. 3.14 Dual-gain IF stage.

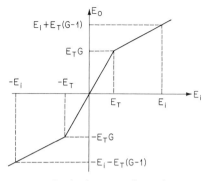

FIG. 3.15 Dual-gain-stage voltage characteristic.

interference prior to any nonlinear operation. A digital word in the power-of-2 binary format may be written

$$E = 2^{N-M}\left(b_{N+1-M} + \frac{b_{N-M}}{2} + \frac{b_{N-1-M}}{4} + \cdots + \frac{b_0}{2^{N-1}} \right) \qquad (3.11)$$

where M is the place beyond which all coefficients to the left are zero. Note that M has essentially the same significance as in previous sections. The logarithm of E to the base 2 is

$$\log_2 E = N - M + \log_2 \left(1 + \frac{b_{N-M}}{2} + \frac{b_{N-1-M}}{4} + \cdots + \frac{b_0}{2^{N-1}} \right) \qquad (3.12)$$

$$\log_2 E \approx N - M + \left(\frac{b_{N-M}}{2} + \frac{b_{N-1-M}}{4} + \cdots + \frac{b_0}{2^{N-1}} \right) \qquad (3.13)$$

The whole number $(N - M)$ becomes the characteristic, and the series, a fractional number, is a line approximation of the mantissa. The approximation is accurate to ± 0.25 dB if the mantissa contains at least 4 bits. The accuracy may be improved to any desired degree by use of a programmable read-only memory (PROM) to convert the linear approximation to the true mantissa.

Digital Log Power Combiner. In log format fewer bits of data need to be stored or manipulated, and many arithmetic computations are simplified. For example, the rms combination of the voltages (I and Q) is complicated in linear format, and approximations are generally employed which introduce some error. In log format the process is simple and more accurate:

$$\log_2 I^2 = 2 \log_2 |I| \qquad \log_2 Q^2 = 2 \log |Q| \qquad (3.14)$$

$$\log_2 (I^2 + Q^2) = \log_2 I^2 + \log_2 (1 + Q^2/I^2) \qquad (3.15)$$

The latter term of Eq. (3.15) is the output of a PROM, using $\log_2 I^2 - \log_2 Q^2$ as the address.

The size of the PROM can be substantially reduced by comparing the two variables to determine the larger (*L*) and smaller (*S*):

$$\log_2 (I^2 + Q^2) = \log_2 L + \log_2 (1 + S^2/L^2) \tag{3.16}$$

The latter term of Eq. (3.16) requires fewer bits than that of Eq. (3.15) because its maximum value is 3 dB. The PROM address also is unipolar and may be limited at a ratio where the PROM output drops to zero.

The log power combiner makes power (square-law) integration feasible. A number of variables may be weighted by addition of logarithmic scaling factors and successively accumulated, using the log power combiner to combine each with the prior partial power summation. Square-law integration of multiple echoes from a target provides better sensitivity than prior methods, but this was impossible with analog signal processing and costly when using conventional digital processing.

3.9 IF LIMITERS

Applications. When signals are received that saturate some stage of the radar receiver which is not expressly designed to cope with such a situation, the distortions of operating conditions can persist for some time after the signal disappears. Video stages are most vulnerable and take longer to recover than IF stages; so it is customary to include a limiter in the last IF stage, designed to quickly regain normal operating conditions immediately following the disappearance of a limiting signal. The limiter may be set either to prevent saturation of any subsequent stage or to allow saturation of the A/D converter, a device which is usually designed to cope with modest overload conditions.

The IF phase detector described in Sec. 3.10 requires a limiter to create an output dependent on phase and independent of amplitude. It is employed in phase-lock servos and phase-monopulse receivers.

An IF limiter is sometimes employed prior to doppler filtering to control the false-alarm rate when the clutter echo is stronger than the filter can suppress below noise level. This was widely used in early two-pulse MTI, but it has drastic impact on the performance of the more complex doppler filters of modern radars. It is only compatible with phase-discrimination constant false-alarm rate (CFAR; Sec. 3.13), but it serves a useful purpose in radars utilizing this CFAR process after doppler filtering.

Characteristics. The limiter is a circuit or combination of like circuits whose output is constant over a wide range of input signal amplitudes. The output waveform from a bandpass limiter is sinusoidal, whereas the output waveform from a broadband limiter approaches a square wave.

There are three basic characteristics of limiters whose relative importance depends upon the application. They are performance in the presence of noise, amplitude uniformity, and phase uniformity. When the input signal varies over a sufficiently wide range, all these characteristics become significant. Amplitude uniformity and phase uniformity are dependent largely on the design of the limiter and are a direct measure of its quality.

Noise. Limiter performance in the presence of noise is characterized by a failure to limit signals buried in noise and by an output signal-to-noise ratio that differs from the input signal-to-noise ratio. Gardner[10] gives an approximate but useful relation for bandpass limiters which demonstrates the effect of limiting on the signal-to-noise power ratio when both are present simultaneously.

$$\left(\frac{S}{N}\right)_0 = \left(\frac{S}{N}\right)_i \frac{1 + 2(S/N)_i}{4/\pi + (S/N)_i} \tag{3.17}$$

$$\left(\frac{S}{N}\right)_0 = 2\left(\frac{S}{N}\right)_i \quad \text{when} \quad \left(\frac{S}{N}\right)_i \gg 1 \tag{3.18}$$

This is the result of limiter suppression of the in-phase noise component, leaving only the quadrature component to compete with the signal.

$$\left(\frac{S}{N}\right)_0 = \frac{\pi}{4}\left(\frac{S}{N}\right)_i \quad \text{when} \quad \left(\frac{S}{N}\right)_i \ll 1 \tag{3.19}$$

This slight suppression of weak signals by the noise approximates the degradation of detectability that can be attributed to the limiter in the phase-discrimination CFAR techniques described in Sec. 3.13. Strictly speaking, these equations do not apply to the detection of pulsed radar echoes; they do not relate to the probability distributions of signal plus noise compared with noise alone. However, as the bandwidth-time product at the limiter increases, Eq. (3.19) defines the effect on the minimum detectable signal more and more closely.

Gardner also demonstrates the relation between output power and $(S/N)_i$:

$$S_0 \approx \frac{(S/N)_i}{4/\pi + (S/N)_i} \tag{3.20}$$

When $(S/N)_i \gg 1$, the output signal power is seen to be constant; however, when $(S/N)_i \ll 1$, the output signal power is seen to be a linear function of input signal power. This is of considerable importance in the design of phase-lock loops.

Amplitude Uniformity. No single-stage limiter will exhibit a constant output over a wide range of input signal amplitude. One cause is apparent if one considers a single-stage limiter having a perfectly symmetrical clipping at $\pm E$. The rms output at the threshold of limiting is $E/2$, rising to the value $(4/\pi)(E/2)$ when the limiter is fully saturated and the output waveform is rectangular.

In a practical case, the amplitude performance is also affected by capacitive coupling between input and output of each limiting stage, charge storage in transistors and diodes, and RC time constants which permit changes in bias with signal level. For these reasons, two or more limiter stages are cascaded when good amplitude uniformity is required.

Phase Uniformity (Phamp). The change of phase with signal amplitude (phamp) of a limiter is more readily measured than analyzed. Calaway[11] bases some very useful conclusions on a series of experiments with five common limiter circuits. He demonstrates that a transistor provides maximum phase uniformity when used in the current-mode switching configuration and that the diode provides better overall performance in the series mode when charge storage effects are not evident.

In conventional limiter types known for their phase uniformity, there are two common denominators worth noting. In each case the peak-to-peak output can be expressed as the product of a switched fixed current and a resistance; these limiters have been described as *available-power* switching types. Second, the phamp of any particular limiter circuit is generally directly proportional to the frequency at which it is operated; its variation with signal amplitude is better characterized by nanoseconds per decibel than by degrees per decibel.

3.10 PHASE DETECTORS AND SYNCHRONOUS DETECTORS

Definitions and Characteristics. The distinction between a phase detector, a synchronous or phase-sensitive detector, and a balanced mixer is sometimes unclear. This results from the similarity of analog circuits that perform these functions. It is generally agreed, however, that a particular circuit is used as a phase detector when only phase information is present in the output, as a synchronous detector where both phase and amplitude information is present in the output, and as a mixer when phase, amplitude, and frequency information is present in the output. Doppler frequency shifts are excepted in this convention.

Phase-detector output characteristics generally fall into one of the three classes shown in Fig. 3.16. Peculiarly, the characteristic of a given detector is not invariant. Gardner[10] shows three cases where the characteristic of a particular detector depends upon the shape of the applied signals. Certain types of diode detectors exhibit a sinusoidal characteristic with sinusoidal inputs and a triangular characteristic with square-wave inputs. In some cases a shift from the triangular to the sinusoidal characteristic accompanies a reduction in signal level or the introduction of noise.

In certain high-performance systems where maximum information is to be retained, a pair of synchronous detectors may be operated in quadrature. Their operation is described by the following diagram:

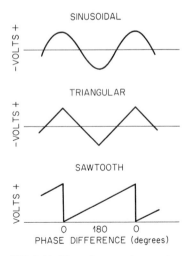

FIG. 3.16 Phase-detector characteristics.

$$E_S \sin (\omega t + \theta) \longrightarrow \otimes \longrightarrow \frac{E_S E_R}{2} \cos \theta \qquad (3.21)$$
$$\uparrow$$
$$E_R \sin \omega t$$

$$E_S \sin (\omega t + \theta) \longrightarrow \otimes \longrightarrow \frac{E_S E_R}{2} \sin \theta \qquad (3.22)$$
$$\uparrow$$
$$E_R \cos \omega t$$

The in-phase (I) and quadrature (Q) operation is described by the first and second lines, respectively. If the detector produces a triangular rather than sinusoidal function of phase angle, the two outputs may still be described as in-phase and quadratic components, but they will be distinguished by quotation marks ("I" and "Q"). In this case, the detector operates as a phase detector rather than as a synchronous detector.

Applications

MTI Radar. A phase-sensitive detector is a key element in nearly all MTI radar systems. It is used to detect the echo vector phase shift produced by target motion between pulses. The reference signal to the detector is supplied by a second local oscillator, known as a coho because of its coherence with the transmitted signal. The phase detector allows the echo signal to be stored in the form of a video signal or arithmetic number representing the echo vector.

To reduce cost and maintenance, early MTI cancelers often possessed only one phase component. In such a canceler the triangular "I" component is used for large echoes, as this minimizes blind phase effects that occur in the I characteristic, at the point where $dE/d\theta$ is near zero. For echoes below the limit level, the "I" characteristic converts to an I characteristic, and blind phase effects are accepted. The "I" characteristic is unsuitable for cancelers using three or more pulses. This results from the dependency of the weighting on a smoothly varying doppler vector. In any case, the best performance results when both I and Q are used.

Phase and Phase-Rate Trackers. The phase detector is used to measure accurately the velocity of targets that have been isolated by other means. High accuracy results from the long time base used in the measurement. The $(N + 1)$st measurement of radial velocity is given by

$$V_{N+1} = \frac{c}{2\omega} \left(\frac{\phi_{N+1} - \phi_N}{\text{PRT}} \right) \qquad (3.23)$$

where c = velocity of light
ω = radar angular frequency
ϕ_N = Nth phase measurement
PRT = interpulse time

Extraction of this information in a practical case requires a filter, such as the tracking loop.

Figure 3.17a shows a phase tracker in which the predicted phase ϕ_P is continuously compared with the measured phase ϕ_M. The velocity is extracted from the derivative of the predicted phase ϕ_P. Because the tracker is measuring phase,

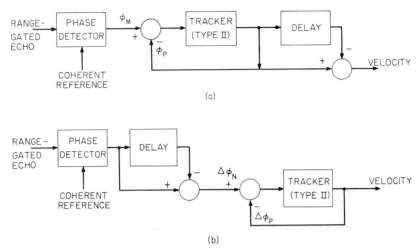

FIG. 3.17 (*a*) Phase tracker. (*b*) Phase-rate tracker.

which is equivalent to range, the system is acceleration-limited and therefore does not have wide application.

Figure 3.17*b* shows a phase-rate tracker in which the predicted phase difference $\Delta\phi_P$ is continuously compared with the measured phase difference $\Delta\phi_M$. This is accomplished by taking the derivative of the phase ahead of the tracker. In this case the velocity is directly proportional to the predicted phase difference $\Delta\phi_P$. Because the tracker is measuring phase rate, which is equivalent to velocity, the system is not acceleration-limited. Lag errors are introduced only by the next higher derivative, jerk.

Monopulse Angle Measurement. In the monopulse radar, a pair of antenna horns are used to form two like beams, which point in slightly different directions. The two antenna ports are connected to a hybrid with sum and difference outputs. The antenna pattern of the difference port will exhibit a null directly between the beams, whereas the pattern of the sum port will exhibit a peak. These hybrid outputs are referred to as the difference signal (Δ) and the sum signal (Σ).

Figure 3.18*a* illustrates the most common amplitude-comparison monopulse configuration. Automatic gain control derived from the amplified signal in the Σ channel causes matched variable-gain amplifiers in both channels to have a gain proportional to Σ^{-1}. The amplified difference-channel signal Δ/Σ is synchronously detected to preserve phase polarity. This monopulse receiver is restricted to radars and targets that create only small variation in echo amplitude from one pulse to the next.

Figure 3.18*b* illustrates a monopulse receiver without this limitation. In this receiver, the amplitude information is converted into phase information by a quadrature hybrid. The hybrid outputs $A \pm jB$ and $A \mp jB$ are translated to IF frequency and ultimately phase-detected against each other. The phase detector for this application should have a sawtooth characteristic, in which case its output is

$$2 \sin^{-1}\frac{\pm B}{K}$$

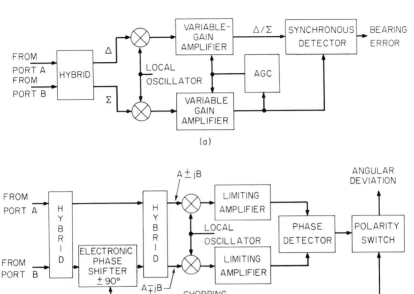

FIG. 3.18 (*a*) Amplitude-comparison monopulse receiver. (*b*) Phase-comparison monopulse receiver.

where K is a constant equal to $\sqrt{A^2 + B^2}$. The alternating polarity of B is introduced by a phase chopper to allow for correction of receiver phase errors. It is removed by a reversing switch following the detector.

Recording. Radar echoes are sometimes recorded for subsequent analysis, usually by a digital computer. Digital recording prevents degradation of data in recording and also provides direct access to the computer. If the dynamic range is reasonably low, I and Q signals may be recorded. Where the dynamic range is large, phase and log amplitude may be used to minimize the number of bits. As an example, consider a system requiring 3 percent recording increments over a 72 dB dynamic range. I and Q recording requires two channels of 17 bits each, whereas phase- and log-amplitude recording for the same performance requires only 8 bits in both channels. Details of the comparison are given in Table 3.3.

TABLE 3.3 Comparison of I and Q with Phase- and Log-Amplitude Recording

Parameter	I and Q recording	Recordings of phase and log amplitude	
Least significant bit (LSB)	$\frac{1}{32}$	0.03 rad	0.28 dB
Most significant bit (MSB)	4000	2 rad	72 dB
MSB/LSB	128,000	256	256
Number of bits	17 each	8	8

Examples of Phase Detectors

Multiplier Detector. The gated-beam tube and, to some extent, the beam-deflection tube have been used as analog multipliers to obtain the product of the signal and reference waveforms. They are self-limiting and produce a gradual transition from the *I* to "*I*" characteristics about the saturation level. When they are used as synchronous detectors, the dynamic range is restricted by the high noise level of these devices. This type of multiplier detector also may be implemented by using a field-effect transistor (FET) multiplier as suggested by Highleyman and Jacob.[12]

Balanced-Diode Detector. The balanced-diode detector of Fig. 3.19*a* is widely used because of its unusually favorable characteristics. When two sinusoids of frequency ω and phase difference θ are applied to this detector, the output is given by

$$E_{out} = K(\cos\theta - \cos 2\omega t + \text{higher-order terms}) \qquad (3.24)$$

Under these conditions, the characteristic is sinusoidal, and the ripple is free from a fundamental component. When bandwidth permits, the detector will operate with square-wave inputs to give a triangular characteristic.

This circuit can be purchased in modular form, containing a pair of balanced wideband transformers and matched hot-carrier diodes. The detector can be obtained with 35 dB isolation between ports over a frequency range of 3 to 100 MHz. Units having a maximum frequency limit of 1 GHz are available.

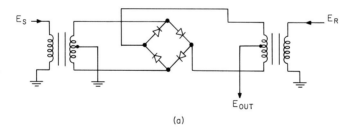

(a)

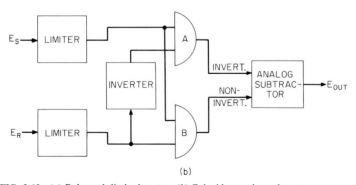

(b)

FIG. 3.19   (*a*) Balanced-diode detector. (*b*) Coincidence phase detector.

In theory, the dynamic range is determined by the maximum signal-to-noise ratio at the output. In practice, however, it tends to be limited by unbalanced residuals and their fluctuation. Very precise detectors of this type may have a usable dynamic range of 50 dB.

Coincidence Phase Detector. The coincidence detector of Fig. 3.19b provides a triangular output characteristic. When E_S and E_R are in phase, AND gate B registers coincidence half the time, and AND gate A registers no coincidence. This condition leads to maximum negative output. When E_S and E_R are out of phase, the reverse condition exists, and maximum positive output results. Normally the triangular characteristic exhibits some rounding of the peaks. However, the detector has been built with very sharp peaks, using a tunnel-diode threshold in each channel.

The coincidence phase detector has a fundamental ripple-frequency component. A higher ripple frequency results from exclusive OR-gate logic, but this introduces voltage offsets that may be troublesome.

Analog-to-Digital Phase Detector. The phase detector of Fig. 3.20 measures the time interval between positive (or negative) zero crossings of the signal and reference waveforms. A pair of zero-crossing detectors generate sharp spikes at their respective points of crossing. The reference-channel spike sets a RESET-SET (*R-S*) flip-flop, and the signal-channel spike resets it. A gated oscillator generates a clock waveform while the flip-flop is in the SET state, and a counter measures the length of this interval. Filtering is accomplished by a buffer register.

The resolution of this phase detector is determined by the ratio of the clock frequency to the intermediate frequency ω_2 of the signal. In a tracking radar with a range gate, the signal bandwidth may be narrowed drastically after gating with-

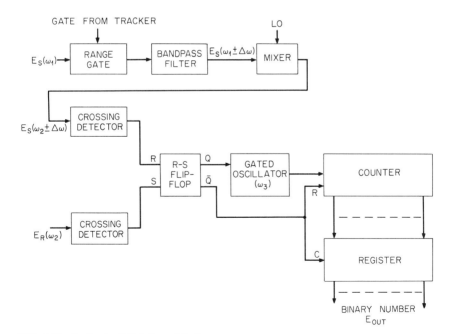

FIG. 3.20   Analog-to-digital phase detector.

out loss in signal-to-noise ratio. The resultant signal may then be translated to a very low IF frequency without spectral foldover problems.

In this system, the filter receives a signal of short duration compared with its response time. It then rings at its own natural frequency, whose deviation ($\Delta\omega$) from the nominal frequency (ω_1) must be very small if the sawtooth phase detector is to produce correspondence between the 0 and 360° points to within the least significant bit. Given the filter tolerance and the counter speed, the best accuracy is achieved by making the frequency at the crossing detector equal to their geometric mean.

This type of phase detector can be made fairly precise. However, it was noted by Gardner[10] in connection with a similar device that the signal quality must be high to ensure reliable operation of the flip-flop. The narrowband filter, which in effect samples and holds the phase, overcomes this problem.

Digital Phase Detector. In most modern radars, digital doppler filters are utilized to attenuate clutter interference prior to extracting the phase (and amplitude) of the desired target echo. The data is in digital form, with real (I) and imaginary (Q) components; so the phase detectors which have been discussed previously (requiring IF inputs) are inappropriate.

The phase is extracted most simply by converting the data to logarithmic format, by the methods described in Sec. 3.8. The phase-detector inputs then consist of polarities of I and Q, $\log_2 I^2$, and $\log_2 Q^2$. The two polarities define the quadrant, and the phase within the quadrant is a PROM function of $\log_2 I^2 - \log_2 Q^2$.

Phase-discrimination CFAR (Sec. 3.13) tolerates crude phase information, and insignificant benefit results from extracting phase to better than 3 bits. A 3-bit digital phase detector may utilize either linear or log data to extract the octant merely by comparing the magnitudes of I and Q. The PROM address consists of nine states:

$$I < 0 \quad I = 0 \quad I > 0$$

$$Q < 0 \quad Q = 0 \quad Q > 0$$

$$|I| < |Q| \quad |I| = |Q| \quad |I| > |Q|$$

It is important in this application[13] to maintain an equal probability of each phase output in noise; so the data corresponding to the boundary conditions [$I = 0$, $Q = 0$, or $|I| = |Q|$] must be assigned to the neighboring octant in a consistent rotational sense (either clockwise or counterclockwise). A pseudorandom phase output is necessary at the origin [$I = 0$, $Q = 0$].

The coded-pulse anticlutter system (CPACS) decoder[13] requires data in the form of $k \cos \phi$ and $k \sin \phi$ rather than phase (ϕ). The conversion can be made by a second PROM. A 3-bit output is provided when $k = 7$; output states are 0, ± 5, and ± 7. A 2-bit output is provided when $k = 3$; output states are 0, ± 2, and ± 3, which introduces tolerable error.

3.11 ANALOG-TO-DIGITAL CONVERTER

Applications. Analog-to-digital converters find numerous applications in modern radar systems. The trend toward digital processing of radar data has resulted in a demand for fast converters that are able to convert data in real time.

Digital MTI is an example of a technique requiring such high-speed converters. Here, the synchronous-detector output is sampled at a rate not less than the receiver bandwidth, and the digital result is stored in a large digital memory. Data is read from the memory to allow comparison with corresponding returns from subsequent radar "looks." The flexibility of this method has permitted MTI velocity response characteristics previously unobtainable with analog memory devices.

Many tracking radars use a converter to encode the echo in the tracking gate. In this case, a general-purpose computer provides all computations required to track a target and to provide range and velocity outputs. Precise data-smoothing and stabilizing characteristics are provided by the computer.

High-speed converters have been used to encode the height information from a stacked-beam radar. This permits an arithmetic interpolation of target position. Errors following conversion are, of course, eliminated.

Another application of converters is in the field of digital recording. This is used where vast quantities of data are to be analyzed or where an isolated event is to be analyzed. In this case, the encoded data is stored on magnetic tape. The results are then analyzed in nonreal time with arithmetic accuracy.

Formats. The most frequently used digital format is the power-of-2 binary forms:

$$E = K(b_N 2^{N-1} + b_{N-1} 2^{N-2} + b_{N-2} 2^{N-3} + \cdots + b_1 2^0) \qquad (3.25)$$

where E = analog voltage
N = number of binary digits
b_i = state of ith binary digit

The encoded word usually is applied to a general-purpose computer in serial form but is applied to special-purpose high-speed computers in parallel form.

The Grey code is used in certain types of asynchronous converters where encoded data is read out of the converter continuously. This code allows all adjacent transitions to be accomplished by the change of a single digit only. Use of the Grey code greatly reduces the magnitude of transient errors in such cases.

Converters in radar systems normally have a complemented power-of-2 format for negative inputs. This simplifies both the converter and subsequent computations. In the complemented format, the converter counts up from the most negative value to zero and then continues to count up from zero to the most positive value. A sign bit indicates which half of the range applies. The process for the 2's complement may be described by

$$E = k(-b_N 2^{N-1} + b_{N-1} 2^{N-2} + b_{N-2} 2^{N-3} + \cdots + b_1 2^0) \qquad (3.26)$$

and for the 1's complement by

$$E = k[-b_N(2^{N-1} - 1) + b_{N-1} 2^{N-2} + b_{N-2} 2^{N-3} + \cdots + b_1 2^0] \qquad (3.27)$$

The 1's complement is seen to have two binary values for 0, but confusion may be eliminated by suppressing the one where all b_i's are unity.

Synchronization. Converters operate in either a synchronous or an asynchronous mode. In the synchronous mode, the converter samples the analog data, decodes, and stores the result on command. An asynchronous converter, however, constantly tracks the changing analog input, and the digital output is in effect sampled by the terminal equipment. The asynchronous converter should employ the Grey-code format to avoid the possibility of a large error should the output be sampled at the instant when digits are changing.

Performance Characteristics

Signal Bandwidth. The digital data used by the terminal equipment is always sampled. The bandwidth of this "digital signal" is limited to half the sampling frequency.

Resolution. The resolution of a converter is determined by the number of bits. For an N-bit converter, the resolution is $E_{max}/(2^N - 1)$ if the converter is truly monotonic, that is, if its response to an analog ramp is a uniform progression of binary numbers. This characteristic is usually realized with a slowly changing analog input but must be verified under pulsed conditions.

Dynamic Range. If the A/D converter is sampling the two components of the echo vector (I and Q), each component contains half of the noise power and up to 100 percent of the signal power. The dynamic range is the maximum IF signal-to-noise ratio which can be handled by the A/D converter without saturating at any phase condition.

$$\text{Dynamic range (dB)} = 6N - 9 - 20 \log (\sigma/\text{LSB}) \qquad (3.28)$$

where N = number of bits including sign

σ = rms noise in I or Q

LSB = least-significant-bit voltage

Quantization Noise. The conversion of sampled noise voltages into integer numbers introduces an added random error which can be considered as an additional source of noise, requiring an increase in echo strength to achieve the desired detection probability if false-alarm probability is maintained constant.

$$\text{Quantization loss (dB)} = 10 \log \left[1 + \left(\frac{\text{LSB}}{\sigma} \right)^2 \right] \qquad \sigma \geq \text{LSB} \qquad (3.29)$$

Sampling. When the signal bandwidth is so great that the analog voltage changes significantly from sample to sample, the instantaneous signal may become distorted by the sampling process.[15] A slewing error results when the exponential charging is incomplete. An entirely separate lag error results from changes in signal amplitude during the sampling interval. Current flows in the storage capacitor, causing an IR drop, which is still present when the switch is opened.

An additional error is introduced by the finite opening time of the switch. The signal tends to be averaged over this interval, and the sampled voltage does not correspond exactly to the voltage at the instant when the switch starts to open. The time required to open the switch is sometimes called the *aperture time*.

Design data specifying the slewing and lag errors is presented in Fig. 3.21. Practical circuits having RC time constants of 3 ns and sampling intervals of 50 ns have been used in high-speed A/D conversion. The resultant slewing error is seen to be less than 0.001 percent, and, at a signal frequency of 0.5 MHz, the lag error is 0.46 percent. It should be emphasized that large sampling errors are not always fatal in a radar system. For example, in an MTI radar the error will repeat from one interpulse period to the next in stationary clutter, and it is therefore removed by subtraction in the canceler.

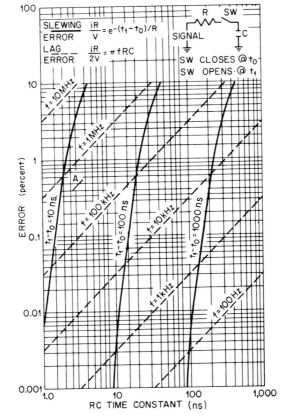

FIG. 3.21   Sampling-circuit errors.

Multiplexing. I and Q data may be processed by the same A/D converter if its conversion rate limit is more than double the required sampling rate. It is essential that I and Q signals be sampled at the same instant of time; so such A/D converters generally employ a pair of sample-and-hold circuits, one to sample I and the other to sample Q. Their stored voltages are converted to digital numbers sequentially prior to the next sample time. Alternatively, a delay line can be employed for either I or Q to allow a single sample-and-hold to be employed, but pulse shape distortion in the delay line and timing error restrict this approach to applications which can tolerate such distortion (see Sec. 3.12).

3.12 I/Q DISTORTION EFFECTS AND COMPENSATION METHODS

Gain or Phase Unbalance. If the gains of the I and Q channels are not exactly equal or if their coho phase references are not exactly $\pi/2$ rad apart, an input signal at a single doppler frequency will create an output at the image doppler frequency. If the ratio of voltage gains is $(1 \pm \Delta)$ or if the phase references differ by $(\pi/2 \pm \Delta)$, the ratio of the spurious image at $-f_d$ to the desired output of f_d is $\Delta/2$ in voltage, $\Delta^2/4$ in power, or 20 (log Δ) $- 6$ in decibels.

In simple MTI, which creates a symmetrical rejection notch to suppress clutter near zero doppler, the image frequency is not a problem; it is suppressed to the same degree as the true doppler. Early MTI radars exploited this tolerance of unbalance to simplify the hardware, processing only I or Q, not both.

Modern radars often provide doppler filters to suppress moving clutter caused by movement of the radar platform or rain or sea clutter, and the image of the clutter in the rejection band can appear in the passband; so a high degree of balance must be preserved. Figure 3.22 shows the measured response of such a doppler filter with a large unbalance in I/Q gain before and after compensation. Note that at those frequencies where the doppler sidelobes of the ideal filter and the image of the passband are of comparable magnitude, the response fluctuates widely, depending on the phase of the input signal.

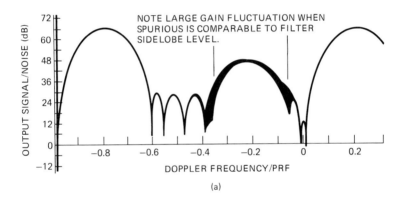

(a)

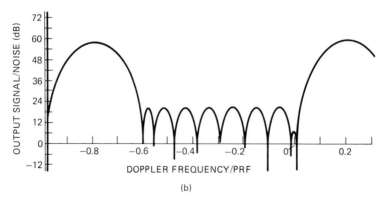

(b)

FIG. 3.22   Measured doppler filter response with (*a*) 2 dB I/Q gain unbalance generating image spurious ($-f_D$) down 18 dB and (*b*) all spurious below noise.

The fluctuating response is the result of the vector addition of the two components: the image frequency and the sidelobe response of the filter to the input doppler. It represents a gain variation as a function of phase of the input signal. If echoes from a rainstorm have a velocity where this fluctuation is occurring, the amplitude statistics of the residue at the filter output will differ markedly from the input statistics, which are similar to noise. Amplitude-discrimination CFAR processes described in Sec. 3.13 are vulnerable to such a change in statistics; false-alarm rate in the rain area will increase significantly. To hold gain fluctuation to ± 1 dB, the image must be suppressed 18 dB below the doppler filter sidelobes; so extremely low doppler sidelobes may not be a desirable characteristic. Phase-discrimination CFAR ignores all amplitude information and is more tolerant of spurious responses.

Error in gain and phase unbalance can be measured by injecting a doppler test tone into the receiver beyond the maximum range of interest, at a frequency corresponding to a null in the sidelobe response of an ideal filter (Fig. 3.22b) and near the peak response to the image (Fig. 3.22a).[16] The output of the filter represents the error, and a phase reference derived from a conjugate filter allows the error to be separated into gain and phase unbalance components. Two servos are used to drive this error down to the tolerable level of Fig. 3.22b.

Phase error may be corrected by a vernier phase shifter in one or both of the coho lines feeding the synchronous detector. Gain error may be corrected by a vernier change in gain in the IF, video, or digital stage of either or both I and Q channels. Video gain control often exaggerates the nonlinearity of those stages; digital control is preferable. It should be implemented by choosing a set of filter-weighting coefficients from a PROM appropriate for the measured gain unbalance.[16] It should not be implemented by scaling either I or Q data with a separate multiplier because this drastically increases the number of bits in the scaled input to the doppler filter. Truncation of the bits of lesser significance in this scaled data effectively causes the scaling factor to vary with signal level, resulting in a variety of unexpected and undesirable characteristics.

A measurement of the doppler spectrum at the center of an echo pulse, such as shown in Fig. 3.23a, indicates the degree of gain and phase unbalance compensation provided by the servos. However, as the following discussion will explain, the suppression of image doppler energy from moving clutter interference may be substantially less than indicated by this measurement at pulse center.

Time Delay and Pulse Shape Unbalance. If the responses of the I and Qchannels to the spectrum of the echo pulse are not identical, the two pulses will have slightly different pulse shapes or time delays. The gain and phase servos compensate for errors at the center of the receiver passband if they use a doppler test-tone pulse of substantially longer duration than the radar echo, but extreme care must be exercised in design of the video stages to ensure that this compensation is appropriate across the entire receiver passband. Optimum bandpass filtering should be at IF, where it affects I and Q channels identically, not at baseband. Video bandwidth should be many times wider than IF and controlled by precision components.

The test for this problem is a measurement of the desired output and the spurious image from a doppler-offset test pulse identical to the transmitted waveform. The measurement is performed at a multiplicity of sample points across the pulse until the image is no longer discernible on the leading and trailing edges. The image suppression of moving clutter echoes is

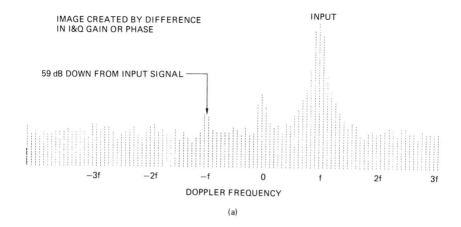

IMAGE CREATED BY DIFFERENCE
IN I&Q GAIN OR PHASE

INPUT

59 dB DOWN FROM INPUT SIGNAL

−3f   −2f   −f   0   f   2f   3f

DOPPLER FREQUENCY

(a)

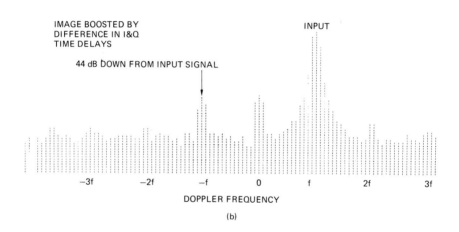

IMAGE BOOSTED BY
DIFFERENCE IN I&Q
TIME DELAYS

INPUT

44 dB DOWN FROM INPUT SIGNAL

−3f   −2f   −f   0   f   2f   3f

DOPPLER FREQUENCY

(b)

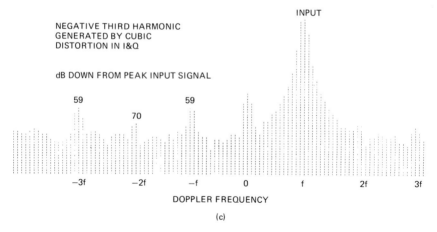

INPUT

NEGATIVE THIRD HARMONIC
GENERATED BY CUBIC
DISTORTION IN I&Q

dB DOWN FROM PEAK INPUT SIGNAL

59   59

70

−3f   −2f   −f   0   f   2f   3f

DOPPLER FREQUENCY

(c)

FIG. 3.23 Spurious doppler frequencies generated in receiver. (*a*) Image created by difference in the *I* and *Q* gain or phase with sample at pulse center 49 dB above noise. (*b*) Image boosted by difference in *I* and *Q* time delays with sample at pulse edge 48 dB above noise. (*c*) Negative third harmonic generated by cubic distortion in *I* and *Q* with sample at pulse center 58 dB above noise.

$$10 \log \sum_{n=1}^{N} P_{Dn} - 10 \log \sum_{n=1}^{N} P_{In} \qquad (3.30)$$

where P_{Dn} = doppler power at sample point n
P_{In} = image power at sample point n
N = number of sample points

The problem is apparent in Fig. 3.23b with the sample point on the edge of the echo. The echo amplitude has been boosted so that the amplitude at the sample point is close to that in Fig. 3.23a, but the image response is 15 dB worse.

Nonlinearity in I and Q Channels. It is more difficult to achieve high dynamic range in the video stages (I and Q) than at IF or RF; so nonlinearity generally shows up first in these stages. Component tolerances often lead to somewhat different nonlinearities in I and Q, which can generate the variety of spurious doppler components shown in Fig. 3.23c, which involves a pulse within a few decibels of A/D saturation.
    The ideal doppler input signal is

$$V = Ae^{j2\pi ft} = I + jQ \qquad (3.31)$$

$$V = \frac{A}{2}(e^{j2\pi ft} + e^{-j2\pi ft}) - j\frac{A}{2}e^{j2\pi ft} - e^{j2\pi ft}) \qquad (3.32)$$

Each video channel response can be expressed as a power series. For simplicity, only symmetrical distortion distortion will be considered. The A/D output, including a residual gain unbalance of Δ, is

$$V' = I' + jQ' \qquad (3.33)$$

$$I' = I - aI^3 - cI^5 \qquad (3.34)$$

$$Q' = (1 + \Delta)Q - bQ^3 - dQ^5 \qquad (3.35)$$

Substitution of Eqs. (3.34) and (3.35) into Eq. (3.32) yields the amplitudes of the doppler spectral components listed in Table 3.4. Note that if the nonlinearities in I and Q were identical ($a = b$; $c = d$), spurious components at $-5f$ and $+3f$ would not be present and image would be proportional to input signal amplitude. Spurious at zero doppler is not due to dc offset; it is the result of even-order nonlinearities which were omitted from the above equations. The negative third harmonic is the dominant spurious produced by nonlinearity.

TABLE 3.4   Spurious Doppler Components Generated by I/Q Nonlinearity

Doppler frequency	Amplitude of spectral component
$-5f$	$A^5(c-d)/32$
$-3f$	$A^3(a+b)/8 + 5A^5(c+d)/32$
$-f$	$A(\Delta/2) + 3A^2(a-b)/8 + 5A^5(c-d)/16$
(Input) f	$A(1+\Delta/2) - 3A^2(a+b)/8 - 5A^5(c+d)/16$
$+3f$	$A^3(a-b)/8 + 5A^5(c-d)/32$
$+5f$	$A^5(c+d)/32$

DC Offset. Small signals and receiver noise can be distorted by an offset in the mean value of the A/D converter output unless the doppler filter suppresses this component. Actually, it is preferable to operate the A/D with an intentional offset because all bits switch between 0 and −1 and their accumulated errors make this voltage increment less predictable than others. An intentional offset can be subtracted at the A/D output, and it is the output of this subtractor which should be maintained close to zero.

False-alarm control in receivers without doppler filters is sometimes degraded by errors of a small fraction of the LSB; so correction is preferably applied at the analog input to the A/D rather than at the point where the error is sensed. An up-down counter reacting to polarity of noise (and ignoring zero) is an ideal error sensor; its most significant bits are converted to an analog voltage and added to the A/D input. A number of unused bits of lesser significance dictate the reaction time of the servo and prevent an interference pulse, no matter how strong, from having significant effect on the compensation.

3.13 CFAR DETECTION PROCESSES

Application. Many radars operate in an environment where the noise generated within its own receiver is not the dominant source of interference. Undesired echoes from rain and clutter and undesired signals from other radiating sources often exceed the receiver noise level. These sources of interference may completely obliterate the radar display, or they may overload either a computer that is attempting to track valid targets of interest or narrow-bandwidth data links to remote users.

The digital decision process usually involves threshold criteria both at the input and at the output of the digital processor. At each point, one can define the probability of detection of a desired target and the probability of false alarm from noise or one of the above sources of interference. An operator viewing a PPI display makes a somewhat similar decision, and so the concept of false alarms is applicable to most radar systems.

Either an operator or a digital processor would like to keep the false-alarm rate reasonably constant, having the radar automatically adjust its sensitivity as the intensity of interference varies. Receivers that have this property are called CFAR (constant false-alarm rate) receivers.

In older radars, the CFAR process was implemented in analog hardware, described in Sec. 5.8 of the 1970 edition of the handbook. In modern radars, digital implementation is virtually universal because it provides both greater accuracy and sophistication impractical by prior methods.

The CFAR processes described below are applied in the radar receiver's digital signal processor at every range sample point, independently of whether a target is present. This high processing rate makes special-purpose computers more economical than general-purpose programmable types. Implementation of two of the most widely employed CFAR processes will be described, and their effect on radar sensitivity, range resolution, and azimuth accuracy will be defined. Additional CFAR processes are discussed in Sec. 8.2.

If the target of interest is small in physical size compared with the range resolution cell of the radar and if the interference signals are distributed reasonably uniformly over a large area relative to the range resolution, one may exploit this fact to discriminate between them.

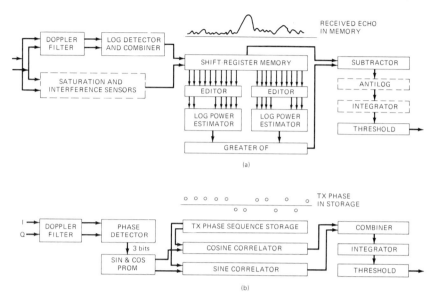

FIG. 3.24 Modern digital signal processing for discriminating between point targets and distributed interference. (*a*) Amplitude-discrimination CFAR (cell-averaging). (*b*) Phase-discrimination CFAR (CPACS).

Amplitude-Discrimination CFAR (Cell-Averaging CFAR). The process shown in Fig. 3.24*a*, often called *range-averaging CFAR* or *cell-averaging CFAR*, is the digital descendant of the *log CFAR* described in the 1970 edition of the handbook (Sec. 5.8). It estimates the mean intensity of signals in neighboring range cells (the CFAR zone) and computes the signal-to-interference ratio of the central range sample. Because this estimate is based on a finite number of samples, it will fluctuate about the true value; the increased signal-to-noise ratio required to achieve a given detection probability, relative to that required if the estimated value did not fluctuate, is known as the *CFAR loss*.

Multiple interpulse periods of data may be integrated as the antenna beam scans past the target; so the estimate of interference can actually be based on a two-dimensional array of data. The CFAR loss decreases as the quantity of *independent* data samples increases, but adjacent data samples are rarely independent; the IF filter partially correlates adjacent noise samples in range, and a doppler filter (MTI) correlates noise samples from one interpulse period to the next. Equation (3.36), derived from Hansen's data,[17] shows the effect of these two filters on CFAR loss.

Amplitude-discrimination CFAR loss (dB) =

$$\frac{5.5\chi}{(B_6 t_s)\left(\dfrac{M + h_M}{1 + h_M}\right)\left[\left(1 - \dfrac{D_6}{\mathrm{PRF}}\right)\left(\dfrac{N + h_N}{1 + h_N}\right)\right]^i} \tag{3.36}$$

where $\chi = - \log P_{FA}$; $P_{FA} = 10^{-\chi}$
M = number of range samples integrated
N = number of interpulse periods integrated
h = inefficiency factor of data integrated ($h = 0$ for power averaging, 0.09 for voltage averaging, and 0.65 for log averaging.)
i = efficiency factor of postdetection integration ($i = 0$ for ratio following integration of N pulses, and $i = 0.7$ for ratio preceding integration of N pulses.)
B_6 = 6 dB bandwidth of IF filter, MHz
t_s = sample spacing, μs, or $1/B_6$, whichever is smaller
D_6 = 6 dB width of rejection band of doppler filter, Hz
PRF = average pulse repetition frequency of data in doppler filter, Hz

Power averaging provides the minimum fluctuation in the estimate of interference. Averaging of voltage or logarithmic data produces greater fluctuation in the estimate, as indicated by the h factors. The log data format is generally advantageous for four reasons:

1. Fewer bits cover the large dynamic range of radar echoes with adequate accuracy.

2. Signal-to-interference-ratio calculation requires only subtraction of log data rather than the more difficult division of voltage or power data.

3. Power summation of data in the CFAR zone is readily accomplished, using the log power combiner described in Sec. 3.8.

4. Integration of multiple interpulse periods may be accomplished in either power or voltage with equal facility.

Although integration prior to computing the ratio of signal to interference provides a lower CFAR loss ($i = 1$), the sequence shown in Fig. 3.24a is generally preferable because it can cope with variation in the level of interference from one interpulse period to the next. Intermittent interference (jamming or long pulses from other radars) produces no increase in false alarms and less sacrifice in sensitivity if integration follows the ratio calculation. If a 1-bit moving window is utilized as an integrator, the bracketed term in the denominator of Eq. (3.36) is omitted; the false-alarm rate at the *input* to the moving window defines χ. When a doppler filter is involved, it must be remembered that the number of independent noise inputs is reduced; an m-out-of-N detection criterion must be considered as an equivalent moving window with a criterion of

$$m\left(1 - \frac{D_6}{PRF}\right) \text{ out of } N\left(1 - \frac{D_6}{PRF}\right) \tag{3.37}$$

The size of the CFAR zone is limited by the size of the area where interference can be considered to have reasonably constant intensity. To maintain control of alarms in the center of severe rainstorms with typical diameters (3 dB) of 2 to 3 nmi, the CFAR zone should extend less than 1 nmi from the cell being examined. To prevent excessive alarms at the leading and trailing edges of the storm, it is advantageous to estimate interference separately in the "early" and "late" CFAR zones and use the "greater of" the two estimates rather than the average. The CFAR loss of the greater-of CFAR is only 0.1 to 0.3 dB more than one averaging the two estimates.[18]

The pulse spectrum of rain echoes is narrower than the noise spectrum; it is the product of the IF bandwidth and the spectrum of the transmitted pulse. With reasonably optimum IF bandwidth, the reduction factor is $\sqrt{2}$. This causes more fluctuation in the estimate of rain echo level than in noise level, and the false-alarm rate in the rainstorm is higher due to this factor alone. If wind shear is inadequate to create an interference spectrum (rain plus noise) wider than noise alone at the output of a doppler filter, use of an integrator can cause another boost in the rain alarm rate due to this factor. Consequently, integrators should be used only in radars whose doppler filters create noise bandwidths narrower than the typical doppler spectrum of rain (15 to 30 kn).

If no postdetection integration is employed, the control of false alarms from sea clutter is very good with vertical or circular polarization but very poor with horizontal polarization, owing to a drastic difference in amplitude statistics of the echoes, which also depend upon compressed pulse width, beamwidth, etc. If the doppler filter is able to keep sea clutter residue equal to or lower than noise level, an integrator may be employed without creating false-alarm problems. The sea clutter residue spectrum is in part of the doppler band different from the output noise, and the composite spectrum is wider than noise alone. The false-alarm rate drops under conditions where sea clutter residue is comparable with noise because the composite interference has less correlation from one interpulse period to the next than noise alone.

The amplitude distribution of echoes from rain and the sea (with vertical or circular polarization and compressed pulse width exceeding 1 μs) does not differ significantly from noise, and the mean value varies slowly with range. Ground clutter varies much more rapidly with range; so the average clutter in neighboring cells is a poor estimate of the clutter in the center cell. Where the doppler filter is unable to suppress ground clutter well below noise, this CFAR process must be supplemented by other techniques to control alarms. These are discussed in Sec. 8.2 and in the "Clutter Map CFAR" subsection in this chapter.

Phase-Discrimination CFAR (CPACS). Another class of CFAR receivers completely obliterates all amplitude information in the echoes by employing limiters, with the normal noise level well above limit level.[19] These receivers discriminate between desired echoes and interference solely by the variation with time of the phase pattern at the limiter output, on the basis of how well it correlates with the phase code that was transmitted. This technique is often called the coded-pulse anticlutter system (CPACS) and comprises the elements so labeled in Fig. 3.1. Figure 3.24b explains the elements in greater detail.

In contrast to amplitude-discrimination techniques, phase-discrimination techniques have no problems with speed of response to changing intensity of interference. They sense only how well the echo matches a predetermined pattern of phase, and they can tolerate rapid and wide variation in signal strength: pulse interference from another radar, for example.

Fundamentally, limiting destroys some information and results in some degradation in performance. Detectability in noise is degraded by the limiting process,[20] but this loss drops to about 1 dB if the bandwidth-time product at the limiter exceeds 20. Similarly, limiting degrades the clutter attenuation capability of MTI radars that compare more than two pulses but have fewer than 20 hits per beamwidth. Consequently, doppler filtering must precede limiting to avoid this performance sacrifice in clutter. The digital phase detector described in Sec. 3.10 provides the required data for this CFAR process from a digital doppler filter; 3 bits of phase are adequate, while 2 bits add over 0.6 dB to CFAR loss. A PROM converts the phase (ϕ) into $k \sin \phi$ and $k \cos \phi$ with $k = 7$ or 3.

Transmitter pulse length should not exceed the size of rainstorms (≈ 25 μs), and for these modest pulse lengths biphase or quadriphase codes are preferable to FM. The quadriphase code[5,6] is particularly advantageous, in that the impulse response of a gaussian IF filter is a close reproduction of the half-cosine subpulse of the code; the combination of the IF filter and digital correlator provides a nearly perfect matched filter. Range sampling loss and spectral splatter of the quadriphase code also are significantly superior to binary codes.

Both types of codes share the advantage of a simple CPACS decoder (a digital correlator requiring no multiplication). The CPACS decoder[13] merely adds M_C adjacent data samples in range after rotating the phase of each sample $0°$, $\pm 90°$, or $180°$, corresponding to the transmitted phase code at that tap location but with opposite polarity. When the echo from a point target is centered in the digital correlator, these phase rotations cause data at all taps to have the same phase, creating the maximum output. Noise or distributed clutter produces random phase conditions, creating a mean power output lower by a factor of M_C, the number of subpulses in the code.

CFAR loss is described by Eq. (3.38), which is similar in form to that of amplitude-discrimination CFAR. The effect of an integrator on false-alarm control in rain or sea clutter is also similar.

Phase-discrimination CFAR loss (dB) =

$$1 + \frac{5.5\chi}{(M_C - 1)\left[\left(1 - \dfrac{D_6}{PRF}\right)\left(\dfrac{N + h_N}{1 + h_N}\right)\right]^i} \qquad (3.38)$$

where all parameters are defined in the same manner as in amplitude-discrimination CFAR, except that M_C = the number of subpulses in biphase or quadriphase code.

In contrast to amplitude-discrimination techniques, CPACS with a transmitter pulse length of 3.25 μs has been able to control alarms from clutter residue from the Swiss Alps without supplementary CFAR processes. The IF limiter is adjusted to begin limiting at a level where the clutter-to-noise ratio approaches the MTI improvement factor. The limiter splatters the scan modulation spectrum of the clutter so that the resulting residue has a broader doppler spectrum than the noise output of the MTI. CPACS ignores the extremely different amplitude statistics of the residue from limiting clutter, and the integrator reacts favorably to the broader residue spectrum.

Effect on Range Resolution and Azimuth Accuracy. The ability to detect two targets separated only in range can be seriously degraded by the CFAR processes described. Range resolution is generally dictated by the size of the CFAR zone, which is many times the width of the echo at the receiver output.

Probability of resolution (*Pr*) is the ratio of the probability of detecting both targets at specified separation to the probability of detecting both targets when widely separated. Resolution is the separation of equal Swerling Case 1 targets providing *Pr* = 50 percent, if not otherwise specified.

Amplitude-discrimination CFAR is particularly vulnerable when the mean echo strengths of the two targets differ. The stronger target's echo located in the CFAR zone of the weaker, if included in the estimate of interference, will cause the weaker target to disappear. The effect is exaggerated by power averaging and absence of postdetection integration; under these conditions, Swerling 1 echoes of even equal mean strength provide *zero* probability of detecting both targets on the same scan when the number of range samples integrated is less than or equal to 24.

The best solution to this problem is editing the strongest sample in the CFAR zone[21] and the neighboring samples before estimating the interference. Editing a single sample[21] is inadequate because the echo will generally affect a pair of samples; a strong echo will affect three or four samples.

Figure 3.25 shows the excellent probability of resolution provided by such editing. Without this editing, probability of resolution is *zero* with power averaging. Voltage averaging without editing provides poor probability of resolution; even the two echoes have equal mean strengths (dotted curve). Note that editing three samples provides greater immunity from degradation when one target has a considerably larger mean radar cross section than the other. Range samples are spaced 75 percent of the transmitted pulse width or 60 percent of the 6 dB width of the echo after IF filtering. Amplitude-discrimination CFAR with editing provides probability of resolution which depends primarily on the ratio of mean echo powers of the fluctuating targets; signal-to-noise ratio and range separation (within the CFAR zone) have little effect.

Phase-discrimination CFAR (CPACS) provides probability of resolution which depends primarily on range separation; signal-to-noise ratio and the ratio of mean echo powers have little effect. The larger echo provides the dominant phase information in the region where echoes overlap. The weaker echo provides only a portion of its code; in the extreme, its information content in the overlap region is completely obliterated. Even when targets have equal mean echo powers, their independent fluctuation characteristics generally signify that one echo is substantially stronger than the other; so probability of resolution is only moderately better than when target sizes are drastically different.

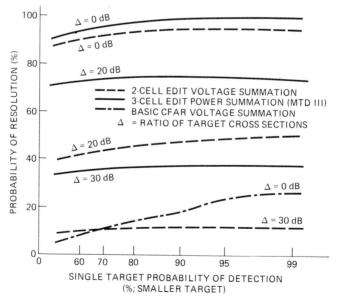

FIG. 3.25 Three-cell editing improves the range resolution of two Swerling Case 1 targets within each other's CFAR zones (pair of 12-cell zones).

The effect of truncation of a portion of the weaker echo by a much stronger one is illustrated in Fig. 3.26 for three different overlap conditions b, c, and d. Quadriphase code 28B is employed in this example.[5,6] Loss of echo data in the overlap region causes a proportional reduction in the peak amplitude and some degradation in range sidelobes. When the peak falls below the detection threshold in Fig. 3.26d, the weaker echo is no longer detectable. Noise and the sidelobes of the stronger echo are not included in the presentation of Fig. 3.26; although their effects are small, one must select a code whose truncation sidelobes are well below the detection threshold to allow for their contribution. This is a factor in selecting good code which is often overlooked.

Range resolution of phase-discrimination CFAR can be estimated from Eq. (3.39).

$$\frac{\Delta t}{T} \approx \frac{4}{\sqrt{M_C}\left[\left(1 - \frac{D_6}{PRF}\right)\left(\frac{N + h_N}{1 + h_N}\right)\right]^i} \tag{3.39}$$

where Δt = time separation of two echoes to achieve 50 percent probability of resolution

T = transmitted-pulse duration

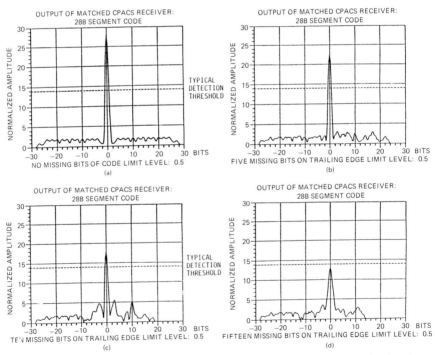

FIG. 3.26 Effect of CPACS truncation of a weaker echo by a stronger target overlapping the trailing edge by a varying number of code segments: (a) 0; (b) 5; (c) 10; (d) 15.

Typically the answer is close to half of the transmitted-pulse duration. Azimuth accuracy can be degraded if the azimuth of each target is calculated as the average of the first and last azimuths where the CFAR process detects the target. When two targets are separated in both range and azimuth, the effect of CFAR is to delay the initial detection of one target and advance the final detection of the other. The targets appear to be more widely separated in azimuth than they truly are. This problem can be avoided by using the CFAR process only for a detection decision, which initiates the transfer of a multiplicity of echo amplitudes to the data processor for determination of azimuth, range, and decision as to whether two targets are present with slightly different azimuths and/or ranges.

Clutter Map CFAR. A clutter map may be employed to provide better detectability of aircraft on near-tangential flight paths over clutter. The near-zero doppler echoes are suppressed by the doppler filter used to attenuate ground clutter, but terrain clutter does not cover large areas solidly. Screening of hills creates shadow areas where clutter is absent, and ability to detect aircraft in these areas, called *interclutter visibility*, is the primary reason for use of a clutter map CFAR.

Map cells must be spaced no farther apart than the azimuth beamwidth and compressed pulse width to provide maximum interclutter visibility. The mean echo voltage or power in each map cell over many scans is determined by exponential weighting, or integration; and the detection threshold is boosted above this value by a factor necessary to provide the desired probability of false alarm in noise. Clutter fluctuation from scan to scan is assumed to have statistics similar to noise *if no postdetection integration is employed*, as might be caused by windblown foliage.

Nitzberg[22] has calculated the CFAR loss of an exponentially weighted integrator with power input and feedback factor of $(1 - w)$. His results may be expressed as

$$\text{Clutter map CFAR loss (dB)} = \frac{5.5\chi}{1 + 2/w} \qquad (3.40)$$

where χ is the exponent of the probability of false alarm. Map data is actually stored in logarithmic format to minimize the number of bits of data, but either voltage or power integration may be achieved by converting the equation to log format.

Clutter map CFAR is based on the assumption that clutter statistics are stationary. Moving rainstorms, jamming, pulses from other radars, and similar dynamically changing clutter conditions can cause an excessive alarm rate. Provisions to turn off this channel automatically in beams where excessive alarms occur are a necessity.

The clutter map may also be employed to sense locations where clutter echoes are too strong to be suppressed below noise by the doppler filter. To be effective, it must apply appropriate boost in the detection threshold at these locations and at neighboring azimuths as well;[23] the scanning radar beam creates residue on the leading and trailing edges of the beam, where echoes are weak but changing rapidly, as well as near the beam center.

3.14 DIPLEX OPERATION

Benefits. Diplex operation consists of two receivers which simultaneously process echoes from transmissions on two frequencies. Transmissions are usually nonoverlapping to avoid a 6 dB increase in peak power, but simultaneous reception of their echoes requires duplicate receivers. Although this doubles the cost of receivers and signal processors, the required average power of the transmitter or transmitters is reduced substantially; in most cases diplex operation reduces total cost.

Some transmitters, particularly solid-state, run closer to the peak power limit of the device than to the average power limit; their cost can be reduced if longer pulse durations can be tolerated. Diplex operation with unequal pulse lengths allows pulse duration to be more than doubled.

Minimum range dictates the shorter of the two pulse durations. Echoes cannot be received while transmission is occurring on either frequency; so the shorter pulse is the second transmission.

The effect of the highest-speed target on the compressed pulse restricts the longer-pulse duration. Digital phase codes and nonlinear FM provide low mismatch loss for targets of low doppler, but mismatch loss and range sidelobes of the compressed pulse degrade at maximum doppler; the longer the pulse, the more the degradation. Linear FM has a higher mismatch loss, but doppler has little effect on it or on range sidelobes. However, both linear and nonlinear FM produce a range offset as a function of doppler; they measure where the target was a fraction of a second ago or where it will be a fraction of a second in the future. These range displacements must match in the two receivers to within a small fraction of the compressed pulse width; otherwise, the sensitivity benefits of diplex operation are not totally achieved. Also, range accuracy may be degraded.

The sensitivity benefit of diplex operation for detecting Swerling 1 targets is shown in Fig. 3.27, increasing with P_D. For example, diplex operation achieves 90 percent P_D with 2.6 dB less total signal power than simplex. The ability at least to double the duration of transmission provides a further benefit of a 3 dB or more reduction of peak power requirements. Assumptions made in deriving Fig. 3.27 are:

1. Echoes on the two frequencies are added in voltage or power prior to the detection decision rather than being subjected to individual detection decisions.

2. Separation of the two frequencies is sufficient to make their Swerling 1 fluctuations independent. This depends on the physical length of the target in the range dimension ℓ_r. The minimum frequency separation is 150 MHz/ℓ_r (m); 25 MHz will maintain the diplex benefit for aircraft longer than 6 m (20 ft).

3. Equal energy is transmitted in both pulses. A 2:1 unbalance sacrifices only 0.2 dB of the benefit at 90 percent P_D.

Recommended Implementation. The echoes can be amplified in a wideband low-noise RF amplifier but should be separated into two individual channels prior to the mixer, using RF filters. Where rapid tuning or frequency agility is required, a switched filter bank is the preferred method. YIG filters are sometimes employed in radars which do not aim for high clutter attenuation. If both echoes were processed by the same mixer, with separation occurring at IF, the spurious signals generated in this nonlinear device could seriously degrade clutter attenuation. The number of spurious frequencies is much larger than those shown in Fig. 3.2, and they can be intolerably strong.

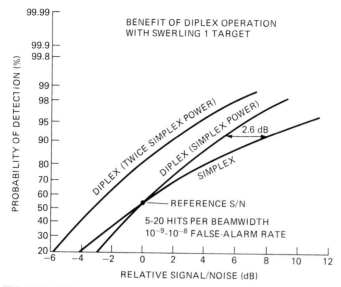

FIG. 3.27 Diplex operation improves the sensitivity of the receiver.

Although separation prior to mixing solves the largest spurious problem, it does not eliminate similar nonlinearity effects in the shared RF amplifier. To avoid problems here, the separation of the two frequencies must be an integer multiple of the clock frequency used for timing the transmissions. This ensures that overlapping clutter echoes from stationary objects have the same relative phase after each transmission; so the effect of distortions in the RF amplifier is repeated.

Two operating frequencies can be provided by a single stalo, using both upper and lower sidebands.[24] By making the IF frequencies of the two receivers differ by a few megahertz, each stalo frequency can provide four operating frequency choices. This reduces the number of stalo choices which must be provided to meet specified tuning requirements. If each IF channel is designed to operate on a different sideband, the transmitted frequencies are separated by the *sum* of the two IFs. Use of a single sideband would separate the transmitted frequencies by the *difference* in the two IFs; this may not provide the full benefit of diplex operation with small aircraft, since condition 2 above might not be satisfied.

REFERENCES

1. Brown, T. T.: Mixer Harmonic Chart, *Electron. Buyer's Guide*, pp. R46, R47, June 1954.

2. Barber, M. R.: A Numerical Analysis of the Tunnel Diode Frequency Converter, *IEEE Trans.*, vol. MTT-13, pp. 663–670, September 1965.

3. Gambling, W. R., and S. B. Mallick: Tunnel-Diode Mixers, *Proc. IEEE*, pp. 1311–1318, July 1965.

4. Rice, S. O.: Response of a Linear Rectifier to Signal and Noise, *J. Acoust. Soc. Am.*, vol. 16, p. 164, 1944.

5. Taylor, J. W., Jr., and H. J. Blinchikoff: The Quadriphase Code—A Radar Pulse Compression Signal with Unique Characteristics, *IEE Conf. Publ.* 281, pp. 315–319, London, October 1987.

6. Taylor, J. W., Jr., and H. D. Blinchikoff: Quadriphase Code—A Radar Pulse Compression Signal with Unique Characteristics, *IEEE Trans.*, vol. AES-24, pp. 156–170, March 1988.

7. Solms, S. J.: Logarithmic Amplifier Design, *IRE Trans.*, vol. I-8, pp. 91–96, 1959.

8. Croney, J.: A Simple Logarithmic Receiver, *Proc. IRE*, vol. 39, pp. 807–813, July 1951.

9. Rubin, S. N.: A Wide-Band UHF Logarithmic Amplifier, *IEEE J. Solid State Circuits*, vol. 1, pp. 74–81, December 1966.

10. Gardner, F. M.: "Phaselock Techniques," John Wiley & Sons, New York, 1967.

11. Calaway, W.: The Design of Wideband Limiting Circuits, *Electron. Design News*, vol. 10, pp. 42–53, December 1965.

12. Highleyman, W. H., and E. S. Jacob: An Analog Multiplier Using Two Field Effect Transistors, *IRE Trans.*, vol. CS-10, pp. 311–317, September 1962.

13. Taylor, J. W., Jr.: Constant False Alarm Rate Radar System and Method of Operating the Same, U.S. Patent 4,231,005, October 1980.

14. Daley, F. D.: Analog-to-Digital Conversion Techniques, *Electro-Technol. (New York)*, vol. 79, pp. 34–39, May 1987.

15. Barr, P.: Voltage to Digital Converters and Digital Voltmeters, *Electromech. Design*, vol. 9, pp. 301–310, January 1965.

16. Taylor, J. W., Jr.: System and Method of Compensating a Doppler Processor for Input Unbalance and an Unbalance Sensor for Use Therein, U.S. Patent 4,661,229, October 1986.

17. Hansen, V. G.: Constant False Alarm Rate Processing in Search Radars, "Radar—Present and Future," *IEE Conf. Publ.* 105, London, October 1973.

18. Hansen, V. G., and J. H. Sawyers: Detectability Loss Due to "Greatest of" Selection in a Cell-Averaging CFAR, *IEEE Trans.*, vol. AES-16, pp. 115–118, January 1980.

19. Bogotch, S. E., and C. E. Cook: The Effect of Limiting on the Detectability of Partially Time-Coincident Pulse Compression Signals, *IEEE Trans.*, vol. MIL-9, pp. 17–24, January 1965.

20. Bello, P., and W. Higgins: Effect of Hard Limiting on the Probabilities of Incorrect Dismissal and False Alarm at the Output of an Envelope Detector, *IRE Trans.*, vol. IT-7, pp. 60–66, April 1961.

21. Richard, J. T., and G. M. Dillard: Adaptive Detection Algorithms for Multiple-Target Situations, *IEEE Trans.*, vol. AES-13, pp. 338–343, July 1977.

22. Nitzberg, R.: Clutter Map CFAR Analysis, *IEEE Trans.*, vol. AES-22, pp. 419–421, July 1986.

23. Taylor, J. W., Jr.: Point Clutter Threshold Determination for Radar Systems, U.S. Patent 4,713,664, December 1987.

24. Taylor, J. W., Jr.: Double Sideband Pulse Radar, U.S. Patent 4,121,212, October 1978.

CHAPTER 4
TRANSMITTERS

T. A. Weil
Equipment Division
Raytheon Company

4.1 INTRODUCTION

The Transmitter as Part of a Pulsed Radar System. Figure 4.1 shows a block diagram of a typical pulsed radar system. Of these dozen blocks, the public news media generally show only the antenna and displays. The rest of the blocks are "unsung heroes," but they are equally important to the system and can be equally interesting from a design standpoint.

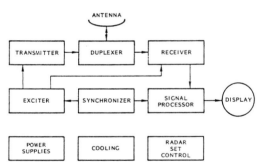

FIG. 4.1 Block diagram of a typical radar system.

The transmitter is usually a large fraction of radar system cost, size, weight, and design effort, and it typically requires a major share of system prime power and maintenance. It generally ends up being a big box that sits in the corner of the radar equipment room, hums to itself, and has a big sign on it that says, "Danger, High Voltage"; so most people prefer to keep away from it. Its insides tend to look peculiar, more like a brewery than a TV set or a computer. This chapter will try to explain why transmitters have to be what they are and hopefully will make them appear a little less peculiar to the reader.

Why So Much Power?   Transmitters are big, heavy, and costly and draw so much prime power because they are required to generate so much RF power output; and that requirement, in turn, comes from the radar system design tradeoffs.

The useful range of a search radar varies as the fourth root of the product of average RF power, antenna aperture area (which determines antenna gain), and the time allowed to scan the required solid angle of coverage (which limits how long the signal in each direction can be collected and integrated to improve signal-to-noise ratio):

$$R^4 \propto P \times A \times T \tag{4.1}$$

The range varies as the fourth root of power because both the outgoing transmitted power density and the returning echo energy density from the target become diluted as the square of the distance traveled. Trying to increase range by increasing transmitter power is costly: a 16-fold increase in power is needed to double the range. Conversely, negotiating a reduced range requirement can produce remarkable savings in system cost.

Power-aperture product as a measure of radar performance is fundamental. It is so fundamental that it was explicitly mentioned in the SALT-I Treaty as the basis for limiting the capabilities of antiballistic missile (ABM) radars.

Receiver sensitivity is not shown as a factor in Eq. (4.1) because thermal noise sets a definite limit on receiver sensitivity, and this simplified range equation merely assumes that the receiver is always made as sensitive as possible.

Since average transmitter power is only one of the factors in the range equation and is so costly, why does power usually end up being so high? Wouldn't it be better to use less power and to make up for it with more aperture or more scan time? The flaw in this argument is that increasing the antenna aperture increases its cost quickly because its weight, structural complexity, dimensional tolerance problems, and pedestal requirements grow rapidly with antenna size. The only other factor, scan time, is usually set by some definite system operational requirement: to look at all aircraft within 100 mi every 4 s, for example, to permit prompt recognition of changes in aircraft direction of travel; so scan time is usually not flexible (which probably explains why everyone talks about the "power-aperture product" of a radar rather than its "power-aperture–scan-time product").

It obviously would not make sense for a radar to have a huge, costly antenna and a tiny, inexpensive transmitter, or vice versa, because doubling the tiny part would allow cutting the huge part in half, which would clearly reduce total system cost. Thus, minimizing total system cost requires a reasonable *balance* between the costs of these two subsystems. The result, for any nontrivial radar task, is that significant transmitter power is always demanded by the system designers.

The same result occurs when the system design is based on a required range coverage in the presence of standoff jammers (rather than just thermal noise).

For detection of a target carrying a self-screening jammer, the range equation becomes

$$R^2 \propto (P_r \times A_r) / (P_j \times A_j) \tag{4.2}$$

where P_r and A_r are the power and aperture of the radar and P_j and A_j are those of the jammer. The result is very similar: power and aperture are still the driving factors, and a balanced system design again results in significant transmitter power.

The inescapable conclusion is that "It's watts up front that count." The desire to attain maximum radar performance capability thus means, more often than not, that *both* the antenna size and the transmitter power are pushed to the maximum affordable.

Pushing the transmitter design to the maximum affordable power is not without its problems, however. Historically, this pressure has often led to problems in development time, unexpected costs, and other risks, especially when a new RF tube had to be developed for the application. The AN/FPN-10 L-band beacon radar system development, for example, was never completed because the tube vendor was unable to make the magnetron stable enough over the wide range of duty cycle. The ballistic missile early-warning system (BMEWS) radar development was in similar danger until a second (backup) tube development contract was placed that used integral vacuum cavities rather than external cavities for the high-average-power klystron development.[1] Even a "successful" RF tube development may end up with a design that is marginal in arcing rate and/or in cooling design, leading to reliability problems, excessive maintenance and logistics costs, and an unhappy customer.

As a result of the risks of pushing RF tube developments to (or unwittingly beyond) the state of the art, and especially if the desired power is known to be beyond the capabilities of a single tube, it becomes attractive to use more than a single RF tube and to combine their RF outputs; this turns out to be a very practical approach, as will be discussed later (Sec. 4.5). This ability to combine, readily and reliably, is also what makes solid-state transmitters practical, since individual solid-state RF devices have much lower power-handling capability than single RF tubes. Combining a few RF tubes to obtain a needed high-power level adds complexity to a transmitter, of course; but, on the other hand, combining a *large* number of RF devices, as must typically be done for solid-state transmitters, leads to certain advantages, such as graceful degradation and improved reliability, as described in Chap. 5.

Why Pulsed? Radar transmitters would be much less complex and costly if they could simply operate CW (continuous-wave) like broadcast stations. Having to generate very high pulsed RF power leads to much higher operating voltages (both dc and RF), energy storage problems, and the necessity for high-power switching devices. Some RF devices, like Class C amplifiers (tube or solid-state), are *self-pulsing* and draw current only when RF drive is applied, but most microwave tubes require some type of pulse modulator (Sec. 4.8) so they won't waste power and so they won't generate interfering noise during the receiving period between pulses.

Basically, pulsing is used because it's hard to hear while you're talking (not everyone at meetings seems to understand that point). In a radar system, if the transmitter is always on, it is very hard to keep the transmitter from interfering with the receiver that is trying to hear faint echoes from distant targets. CW radars have been made to work by using separate transmit and receive antennas to isolate the receiver from the transmitter. When the two antennas cannot be widely enough separated to reduce transmitter leakage into the receiver below the receiver noise level (such as when both antennas have to be on the same vehicle), the residual transmitter leakage can be reduced by *feedthrough nulling*, which works by using negative feedback at the receiver input to cancel whatever transmitter carrier signal may appear there. The feedback loop must be selective enough, however, to cancel only the carrier, since signals offset from the carrier include the desired target doppler signals. As a result, a fundamental limitation in CW radar system sensitivity is that leakage into the receiver of transmitter noise sidebands (resulting from imperfect transmitter stability) sets a limit below which small moving-target signals cannot be seen; the maximum range capability of CW radars is often limited by this factor.

A pure CW radar system can detect moving targets by their doppler offset, but no range information is obtained. The normal solution to that problem is to use an FM–CW system, in which the transmitted frequency is swept (usually linearly versus time) so that both range and doppler information can be extracted by proper interpretation of the received signals;[2] the frequency of the echo determines how long ago the signal was transmitted and thus the range to the target. Nevertheless, one fundamental limitation in such CW radars is that weak echoes from distant targets must compete with strong echoes from short-range clutter. This requires superb clutter cancellation, which in turn is limited by transmitter instabilities (which produce noise sidebands). In other words, strong short-range clutter effectively adds more transmitter leakage into the receiver.

In pulsed radar systems, short-range and long-range echoes arrive at different times, and receiver sensitivity can be adjusted accordingly with STC (sensitivity time control). Note that high-PRF pulse doppler systems, which also receive signals from multiple ranges simultaneously, suffer the same type of limitation as CW radars, so long-range pulse radars seldom use a continuous pulse doppler waveform. However, much of the same benefit of wide unambiguous doppler coverage can be obtained by a compromise waveform called *burst*, in which a finite group of high-PRF pulses is transmitted; the duration of the burst is made short enough to avoid making long-range target echoes compete with short-range clutter echoes.

A further disadvantage of CW radar is that it requires two antennas, which effectively "wastes" 3 dB of range-equation performance that could be gained if that total aperture area were combined into a single antenna and used for both transmit and receive. Pulsed radar does exactly that; it shares a single antenna for both the transmitter and the receiver by using a *duplexer*,[2] as shown in Fig. 4.2.

A gas-tube duplexer (Fig. 4.2a) uses the presence of high power during transmit to fire the gas-filled T/R (transmit/receive) "tubes," which are actually just sections of transmission line filled with a low-breakdown-voltage gas, to direct the transmitter power to the antenna. The T/R tubes recover (deionize) quickly after the transmitted pulse, which then allows received signals to flow to the receiver. A limiter is also used, as shown, to pro-

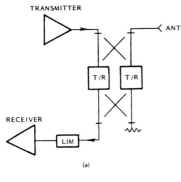

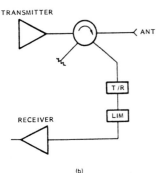

FIG. 4.2   Duplexers. (*a*) Gas-tube duplexer.
(*b*) Ferrite circulator duplexer.

tect the receiver from power leakage through the T/R tubes during transmit. The limiter also protects the receiver from signals from nearby radars that may not be strong enough to fire the T/R tubes but could be large enough to hurt the receiver. A ferrite duplexer (Fig. 4.2*b*) uses a ferrite circulator,[3] instead of T/R tubes, to send the transmitter power to the antenna and the received signals to the receiver. However, in this case reflected power from the antenna voltage–standing-wave ratio (VSWR) during transmit also is directed to the receiver, so a T/R tube and limiter are still required to protect the receiver during transmit.

In either case, the duplexer accomplishes the objective of letting the transmitter and the receiver share a single antenna in pulsed radar systems.

4.2 MAGNETRON TRANSMITTERS

Historically, the invention of the microwave magnetron during World War II made pulsed radar practical, and early radar systems were undoubtedly tailored around what magnetrons could do. The 5J26, for example, has been used in search radars for over 40 years. It operates at L band and is mechanically tunable from 1250 to 1350 MHz. It is typically operated at 500-kW peak power with 1-μs pulse duration and 1000 pulses per second (pps), or 2 μs and 500 pps, either of which is 0.001 duty cycle and provides 500 W of average RF power. Its 40 percent efficiency is typical for magnetrons. The 1- to 2-μs pulse duration provides 150- to 300-m range resolution and is "convenient" for magnetrons, which simply oscillate at the resonant frequency of their mechanical cavities and are subject to frequency instabilities that would be unacceptable compared with the narrower signal bandwidth of longer pulse widths.

Magnetron transmitters are well described in the literature.[2] They readily produce high peak power; and they are quite small, simple, and low in cost. Pulsed magnetrons vary from a 1-in[3], 1-kW peak-power beacon magnetron up to several megawatts peak and several kW average power, and CW magnetrons have been made up to 25 kW for industrial heating. All commercial marine radars have used magnetrons.

Magnetron transmitters have been widely used for moving-target indication (MTI) operation, typically allowing 30 to 40 dB of clutter cancellation. It is remarkable that magnetrons are stable enough for MTI operation at all, considering that it requires the self-excited magnetron to repeat its frequency, pulse to pulse, within about 0.00002 percent. The starting RF phase, however, is arbitrary on each pulse as the magnetron starts to oscillate, so a locked coho (coherent oscillator)[4] or an equivalent (which measures phase on transmit and corrects in a signal processor on receive) must always be used. The high-voltage power supply (HVPS) and pulse modulator must provide very stable (repeatable) pulsing to the magnetron, as well, in order not to spoil MTI performance. Modulation of magnetron frequency by microphonics, from ambient vibration, has also been a limiting factor in some cases.

Automatic frequency control (AFC) is typically used to keep the receiver tuned to the transmitter as the magnetron slowly drifts with ambient temperature and self-heating. The AFC can be applied to the magnetron instead to keep it operating on an assigned frequency, within the accuracy limits of its tuning mechanism.

Limitations. In spite of their wide capabilities, magnetrons may not be suitable for various reasons:

1. If precise frequency control is needed, better than can be achieved through the magnetron tuner after allowing for backlash, warmup drift, pushing, pulling, etc.

2. If precise frequency jumping is required, or frequency jumping within a pulse or within a pulse group.

3. If the best possible stability is required. Magnetrons are not stable enough to be suitable for very long pulses (e.g., 100 μs), and starting jitter limits their use at very short pulses (e.g., 0.1 μs), especially at high power and lower frequency bands.

4. If coherence is required from pulse to pulse for second-time-around clutter cancellation, etc. Injection locking has been tried but requires too much power to be attractive. For the same reason, combining the power outputs of magnetrons has not been attractive.

5. If coded or shaped pulses are required. A range of only a few decibels of pulse shaping is feasible with a magnetron, and even then frequency pushing may prevent obtaining the desired benefits.

6. If lowest possible spurious power levels are required. Magnetrons cannot provide a very pure spectrum but instead produce considerable electromagnetic interference (EMI) across a bandwidth much wider than their signal bandwidth (coaxial magnetrons are somewhat better in this respect).

Magnetron Features. Where a magnetron is suitable, it can be obtained with features that have broadened considerably since its early days.

Tuners. High-power magnetrons can be mechanically tuned over a 5 to 10 percent frequency range routinely, and in some cases as much as 25 percent.

Rotary Tuning. The rotary-tuned ("spin-tuned") magnetron was developed around 1960.[5,6] A slotted disk is suspended above the anode cavities as shown in Fig. 4.3 and, when rotated, alternately provides inductive and capacitive loading of the cavities to raise and lower the frequency. Very fast tuning rates are feasible because each revolution of the tuner disk tunes the tube across the band and back a number of times equal to the number of cavities around the anode. The disk is mounted on bearings inside the vacuum (developed first for rotating-anode x-ray tubes) and is magnetically coupled to a shaft outside the vacuum. At 1800 r/min, a tube with 10 cavities tunes across the band 300 times per second. By ensuring that the modulator pulse rate is not synchronous with the tuning rate, the transmitted frequency will vary from pulse to pulse in a regular pattern as the PRF beats with the tuning rate. Irregular (pseudorandom) jumping of the frequency can be obtained by varying the modulator PRF or by varying the motor speed rapidly. First-order tracking information for the receiver local oscillator (LO) is obtained from an

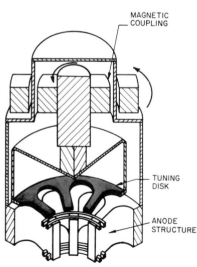

FIG. 4.3 Magnetron rotary tuner. (*Courtesy of Raytheon Company.*)

internal transducer, usually capacitive, on the same shaft as the tuning disk.
The use of rotary tuners involves some penalties besides higher cost and
weight. Less average power output is feasible than for tubes with conventional
tuners, since cooling the rotary tuner is more difficult. Precise band-edge tuning
is not assured; since the entire tuning range is always covered on each cycle and
since system operation outside the assigned band is usually not permissible, tolerances on tuning range must be absorbed within the band. When used for MTI
(with the tuner stopped), stability is less good than with other tuners.

   Stabilized Magnetrons.   The most common form of stabilized magnetron is
the coaxial magnetron, in which a high-Q annular cavity is intimately coupled to
the anode vanes inside the inner cylinder, as shown in Fig. 4.4. At the higher
frequencies (above X band) an inside-out version, called an inverted coaxial
magnetron, as shown in Fig. 4.5, is more suitable because the cavity becomes
very small and the normal construction would leave inadequate room for the
cathode and the anode structure.

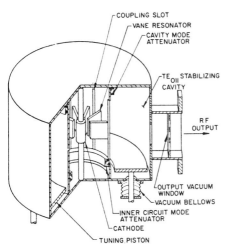

FIG. 4.4 Coaxial magnetron. (*From Ref. 8.*)

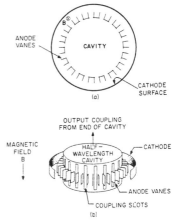

FIG. 4.5 Inverted coaxial magnetron.
(*a*) Simplified cross section. (*b*) Simplified perspective.

   These techniques[5,7,8] permit an increase in stability by a factor of 3 to 10 in
pushing figure and pulling figure (defined later in this section). This is of most
importance at high frequency (X and K bands), where the effects of pushing and
pulling are more significant compared with the bandwidth occupied by typical
system pulse widths. Stabilization is also most practical at these frequencies because a high-Q cavity is then of acceptable size. MTI performance may be improved over conventional magnetrons because the pulse-to-pulse and intrapulse
frequency stability is better. However, the expected benefit may not be realizable
unless jitter and noise during the starting time of each pulse are low enough, and
these characteristics vary considerably among the different types of stabilized
magnetrons.

   Common Problems.   The classic problems of magnetron operation still exist,
but they now tend to be better understood, specified, and controlled. Briefly,
the most common problems are as follows:

1. *Sparking:* Especially when a magnetron is first started, it is normal for anode-to-cathode arcing to occur on a small percentage of the pulses. Sometimes this also applies to moding and/or misfiring. The modulator must tolerate this for brief periods without tripping off and must deliver normal output immediately following sparking.

2. *Moding:* If other possible operating-mode conditions exist too close to the normal-mode current level, stable operation is difficult to achieve. Starting in the proper mode requires the proper rate of rise of magnetron cathode voltage, within limits that depend on the tube starting time and the closeness of other modes. Too fast a rate of rise may even result in failure to start at all. Since starting time is roughly $4Q_L/f_o$, where Q_L is the loaded Q, it is more difficult and inefficient to operate high-power, low-frequency magnetrons at short pulse lengths. Techniques to minimize this problem include *despiking*, usually a simple series-RC network to slow down the modulator voltage rate of rise, or a *pulse bender*, which uses a diode and a parallel-RC network to load down only the last portion of the rate of rise.[9] On the other hand, too slow a voltage rate of rise (or too slow a fall time at the end of the pulse) can also excite a lower-current mode if that tube has one.[10]

3. *Noise rings:* Excessive inverse voltage following the pulse, or even a small forward "postpulse" of voltage applied to the magnetron, may make it produce sufficient noise to interfere with short-range target echoes. The term *noise ring* is used because this noise occurs at a constant delay after the transmitted pulse and produces a circle on a plan position indicator (PPI). This can also occur if the pulse voltage on the magnetron does not fall fast enough after the pulse.[11]

4. *Spurious RF output:* In addition to their desired output power, magnetrons generate significant amounts of spurious noise. The kinds and amounts are similar to those listed under crossed-field amplifiers (CFAs) in Table 4.2 below and discussed in Secs. 4.3 and 4.4, but the resonant nature of the magnetron tends to suppress noise that is far from the operating frequency, except for harmonics.

5. *RF leakage out of the cathode stem:* Typically, an S-band tube may radiate significant VHF and UHF energy as well as fundamental and harmonics out of its cathode stem. This effect varies greatly among different magnetrons, and when it occurs, it also varies greatly with lead arrangements, filament voltage, magnetic field, etc. Although it is preferable to eliminate cathode-stem leakage within the tube, it has sometimes been successfully trapped, absorbed, or tolerated outside the tube.

6. *Drift:* Magnetron frequency varies with ambient temperature (of the cooling air or water) according to the temperature coefficient of its cavities, and it may also vary significantly during warmup. Even during continuous operation, a change in tuner setting may result in drifting again after the change if cavity or tuner heating varies with tuner setting. Temperature-compensated designs are available in some cases.

7. *Pushing:* The amount by which a magnetron's frequency varies with changes in anode current is called its *pushing figure*,[10] and the resulting pulse-to-pulse and intrapulse frequency changes must be kept within system requirements by proper modulator design.

8. *Pulling:* The amount by which a magnetron's frequency varies as the phase of a mismatched load is varied is called its *pulling figure*.[10] Thanks to the ready availability of ferrite isolators, pulling is seldom a problem in modern radar

transmitters. For the same reason, long-line effect[12] is a problem of the past, since isolators readily reduce the mismatch seen by the magnetron to a value low enough to guarantee freedom from frequency skipping.

9. *Life:* Although some magnetrons have short wear-out life, many others have short life because of mishandling by inexperienced personnel. Dramatic increases in average life have been obtained by improved handling procedures and proper operator training (Sec. 4.4).[13,14]

10. *Tuner life:* Because of cost and size tradeoffs, tube life may be limited by the finite fatigue life of the bellows required to allow actuating tuners inside the vacuum. Tuners that operate outside the vacuum must still have adequate gear and bearing design if they are not to limit tube life; in particular, backlash may be a limit.

4.3 AMPLIFIER CHAIN TRANSMITTERS

It was the limitations of magnetrons that eventually pushed radars to use amplifier chain transmitters, which are more capable but also more complex. The key difference is that the transmitter signal is generated at low level, as precisely as desired, and is amplified all the way from there to the required peak power level. As shown in Figs. 15.44 and 15.45, the change in the system block diagram is small, just the direction of two arrows and the change from oscillator to amplifier for the high-power RF source; but there is a *huge* difference in the hardware required to implement the amplifier chain system, including many stages of RF amplification, each with its own power supplies, modulator, and controls; and *all* these stages must be stable to achieve good system MTI performance (Chap. 15).

Oscillator versus Amplifier. Amplifier chain systems can readily achieve full coherence from pulse to pulse and can provide all the features that pulsed oscillator systems (usually magnetrons) cannot provide: coded pulses, true frequency agility, and combining and arraying. The price is higher system complexity and cost. Thus, "oscillator versus amplifier" is one of the basic choices that must be made early in radar system design. Some of the factors entering into this choice are given below.

Accuracy and Stability of Carrier Frequency. In an oscillator-type transmitter, the RF power tube determines its own operating frequency, as opposed to having it determined by a separate low-power stable oscillator. The frequency may thus be affected by tube warmup drift, temperature drift, pushing, pulling, tuner backlash, and calibration error. In an amplifier chain type of transmitter, the frequency accuracy is essentially equal to that of its low-level stable crystal (or other) oscillator. Furthermore, the frequency of the amplifier chain can be changed instantaneously by electronic switching among several oscillators, at a rate faster than that of any mechanical tuner.

Coherence. An amplifier chain system can generate its LO (local oscillator) and coho [coherent intermediate-frequency (IF) oscillator] signals with precision, whereas an oscillator-type transmitter requires manual tuning or an automatic frequency control (AFC) servosystem to tune the LO to the correct frequency. Since an oscillator-type transmitter starts each pulse at an arbitrary phase angle with respect to the coho and LO, coho locking must be provided; in an amplifier system, coho locking is inherent in the signal generation process. Furthermore, since phase coherence can be maintained over a train of pulses in an amplifier-type transmitter, second-time-around clutter can also be canceled, whereas in an oscillator-type system, second-time-around clutter will be noise-modulated by the random starting phase of the oscillator tube. Amplifier chains also allow full coherence, in which the PRF, IF, and RF frequencies are all locked together; this is sometimes necessary to keep PRF harmonics out of IF doppler bands.

Instabilities. As discussed in Secs. 4.6 and 15.11, different kinds of instabilities are associated with a pulsed oscillator system and a pulsed amplifier chain. For the oscillator, pulse-to-pulse *frequency* stability depends on HVPS ripple, and intrapulse *frequency* changes depend on modulator droop and ringing. Tolerable limits are shown in Table 15.4, but these limits may be loosened if coho locking is based on an effective average of the transmitter frequency during the pulse length. For an amplifier chain, pulse-to-pulse *phase* stability depends on HVPS ripple, and intrapulse *phase* variations depend on droop and ringing; tolerable limits are also shown in Table 15.4.

An interesting compromise is also feasible: if a locked coho is used with an amplifier chain, then pulse-to-pulse phase variations in the chain are not significant (except on second-time-around clutter). This technique is particularly convenient when a CFA power booster is added to an existing pulsed oscillator MTI radar system; simply by changing the point at which the RF sample is taken for locking the existing coho, the added CFA is not required to have tight pulse-to-pulse phase stability.

A digital equivalent of the locked coho has also been used. The phase of the transmitter is simply measured on each pulse, and the proper correction is made on the received signals in the signal processor. Like the locked coho, this technique is not effective on second-time-around targets.

Amplifier Chains: Special Considerations. The decision to use an amplifier chain, most often for coherence and agility, introduces many complications, some of which are noted here.

Timing. Because modulator rise times differ, triggers to each amplifier stage must usually be separately adjusted to provide proper synchronization without excessive wasted beam energy. In CFA chains, allowance must also be made for the pulse-width shrinkage that occurs because of the necessary overlap of RF drive, as noted in Sec. 4.4.

Isolation. Each intermediate stage of a chain must see proper load match even if the following stage has high VSWR input, as in a typical broadband klystron, or even if it has significant reverse-directed power coming back from it, as is the case with CFAs. This reverse-directed power results from mismatch at the CFA output that sends power back through the low-loss structure of the CFA. For example, a load with 1.5:1 VSWR reflects power 14 dB down. At certain frequencies, this reflected power will combine with reflections inside the tube and may typically return to the input of the CFA at a power level only 8 dB down from full output power. This amount of reverse-directed power is 2 dB *greater* than the RF input power arriving at that point even if the CFA has only 10 dB of gain. Although this does not seem to interfere with normal CFA operation, it does require an isolator at the CFA input with 16 dB isolation, in this case, just to bring the VSWR seen by the previous stage down to 1.5:1.

Matching. RF tubes used in amplifier chains are often more "fussy" about the match they see than oscillator tubes. Because good isolators are now generally available, improved amplifier ratings are sometimes available if the tube is guaranteed to see a good match, such as 1.1:1. Furthermore, CFAs and traveling-wave tubes (TWTs) generally require that the match they see be controlled over a much wider range than the specified operating frequency band to ensure that the amplifier tube will remain stable.

Signal-to-Noise Ratio. The noise power output of an amplifier tube may be significant. When several tubes are connected as a chain, the output signal-to-noise ratio cannot be better than that of the worst stage. For this reason the input stage, especially, must be checked to see if it has an adequately low noise figure; otherwise, it may prevent the entire chain from achieving a satisfactory signal-to-noise ratio. For example, a low-level TWT with 0.5-mW RF signal input and 35 dB noise figure will limit signal-to-noise ratio of the amplifier chain to 74 dB in a 1-MHz bandwidth. Conventional CFAs have higher noise levels than linear-beam tubes, and their signal-to-noise ratio is typically only 55 dB in a 1-MHz bandwidth; the low-noise CFA, however, can be 70 dB or better (Sec. 4.4).

Leveling. In a multistage chain of linear-beam tubes, the performance of each tube depends in part on the performance of the tubes preceding it. In particular, power flatness (constant power output across the frequency band) requires careful specification of flatness for each stage in the presence of a suitable allowance of nonflatness of the stage preceding it. For example, the *saturated gain* of a tube may be constant across the band, and yet the power output may vary considerably across the band with constant RF drive. Saturated gain is measured by varying the drive at each frequency until the point of maximum power output is found; at that point, saturated gain is the ratio of RF power output divided by RF power input. Unless the saturated power output is constant over the band, the saturated gain bears little relationship to power flatness across the band with constant RF drive. Nor does flat small-signal gain indicate power flatness at large-signal conditions. Therefore, it is usually best to specify that the tube be tested in a way that will ensure proper performance in the system, including adequate tolerances on the RF drive.

Naturally, the transmitter gain and leveling plan must cover all losses and tolerances of components between the stages as well as the tube tolerances. It is also feasible to consider passive frequency-shaping networks to compensate for known deviations from flatness in the RF tube characteristics.

In CFA chains, leveling is far simpler because excess drive power is harmless (it just feeds through and adds to the output),[15] and it is only necessary to ensure that there is always adequate drive power.

Stability Budgets. In a multistage chain, each stage must have better stability than the overall requirement on the transmitter, since the contributions of all stages may add. They may add randomly or directly, or in certain cases they may be arranged to cancel, depending on the nature and source of the instabilities. Normally it is necessary to subdivide the transmitter stability requirement into several smaller numbers that are then allocated to each stage according to *degree of difficulty.* Such *stability budgets* are usually required for pulse-to-pulse variations, for intrapulse variations, and sometimes for phase linearity. Jitter is usually dependent primarily on a single stage and is therefore usually not budgeted among stages.

RF Leakage. A typical amplifier chain may have 90 dB of gain at the transmitter frequency in one shielded room or one location. In order to keep the chain from oscillating, leakage from the output of the chain back to its input must clearly be at least 90 dB down. However, a more stringent requirement is that RF leakage into the input stage of the chain must be kept below the desired level of MTI *purity* with respect to the signal level at that point, since the leakage path might conceivably be modulated by fan blades, cabinet vibration, etc. The leakage feedback will also affect pulse compression sidelobe levels. Since a typical level of purity desired for MTI or pulse compression might be 50 dB, this leads to an isolation requirement of 140 dB from chain output to input. Since typical waveguide joints and coaxial-cable connectors may have leakage levels of the order of -60 dB, 140 dB of isolation can be difficult. Other contributors to amplifier chain RF leakage problems often include collector seals on linear-beam tubes and cathode stems on CFAs. Successful amplifier chain design therefore requires conscious and careful control of RF leakage.

Reliability. The complexity of transmitter amplifier chains often makes it difficult to achieve the desired reliability. Solutions usually involve the use of redundant stages or a whole redundant chain, and many combinations of switching are feasible. Careful analysis and restraint are usually necessary; otherwise, the complexity and cost of fault monitoring and automatic switching very quickly grow out of bounds. Appropriate design for acceptable reliability involves trading off various serial and redundant transmitter chain and switching alternatives, but such system-reliability calculations are beyond the scope of this handbook.

RF Amplifiers. Successful amplifier chain transmitter design depends upon the availability of suitable RF amplifier devices or the feasibility of developing them. Since solid-state transmitters are covered in Chap. 5, we will limit this chapter to discussions of RF tubes for radar systems, as described in the next section.

4.4 RF AMPLIFIER TUBES

Until the mid-1970s, radar transmitters used only vacuum tubes of one kind or another for microwave power generation. The earliest systems all used magnetrons, as has been noted, and amplifier chain system development had to await the development of suitable high-power pulsed-amplifier tubes. Although many varieties were developed, the successful kinds were klystrons, TWTs, and CFAs. Triodes and tetrodes[2] have also been used in radars at frequencies below 600 MHz.

Klystrons and TWTs are called *linear-beam tubes* because the direction of the dc electric field that accelerates the electron beam coincides with the axis of the magnetic field that focuses and confines the beam. This is in contrast to *crossed-field tubes*, such as magnetrons and CFAs, in which the electric and magnetic fields are at right angles to each other.

Since there are a number of excellent references that describe the theory and operation of RF amplifier tubes,[2–4,16–18] this discussion will be limited primarily to system considerations in selecting and using microwave amplifier tubes in radar transmitters.

Crossed-Field Amplifiers (CFAs). High efficiency, small size, and relatively low-voltage operation make CFAs especially attractive for lightweight systems for transportable or airborne use, from UHF to K band. Having relatively low gain, CFAs are generally used only in the one or two highest-power stages of an amplifier chain, where they may offer an advantage in efficiency, operating voltage, size, and/or weight compared with linear-beam tubes. The output-stage CFA is usually preceded by a medium-power TWT that provides most of the chain gain. CFAs have also been used to boost the power output of previously existing radar systems.

The dominant types of CFAs are all reentrant, distributed emission

CFAs.[4,16-18] The high-gain CFA[19] was not developed until 1987, but it is very attractive, both because it requires less drive power and because the presence of RF drive at the cathode results in lower noise levels.

Backward-wave CFAs were developed first and were applied first (the amplitron).[15] In backward-wave devices, the voltage required for a given peak current is essentially proportional to frequency, but this can be accommodated by the inherent contant-power characteristic of a line-type modulator or by a hard-tube modulator operated in the constant-current region. The constant-current switch tube also helps to regulate CFA current against HVPS capacitor-bank-voltage droop. Forward-wave CFAs, which were developed later, operate at nearly constant voltage across their frequency band and can therefore be considered for *dc operation*, which requires only a control electrode (see below) instead of a full-power pulse modulator.

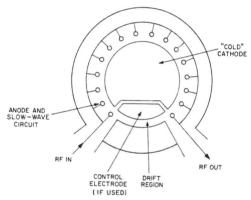

FIG. 4.6 Drift region and control electrode in a reentrant CFA.

Some CFAs have cold cathodes and are started by applying RF drive. The RF drive power must be applied in time to permit the tube to start drawing current before the cathode voltage pulse overshoots the proper operating voltage. However, even when there is a drift region in the CFA, as shown in Fig. 4.6, the tube will not stop when RF drive is removed; the reentrant electrons still carry enough energy that secondary emission from the cathode is maintained, and the tube will oscillate near a band edge or generate broadband noise until the cathode voltage pulse ends. In addition, once operation has been started by RF drive, back bombardment heats the cathode, and on following pulses the cathode current may start from thermionic emission even before RF drive is applied. Since this would also produce noise output, it is customary to make the RF drive pulse straddle the modulator voltage pulse to prevent this. Allowance must be made for the resulting pulse-width shrinkage in an amplifier chain with one or more CFAs. The output pulse will also have "pedestals" owing to feedthrough of the wider RF drive pulse length, as shown in Fig. 4.7.

A control electrode[20,21] usually consists of a segment of the cathode structure in the drift region, as shown in Fig. 4.6. The control electrode is pulsed positive

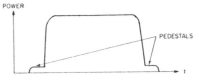

POWER
PEDESTALS

FIG. 4.7 Pedestals on a CFA RF output pulse.

(with respect to the cathode) at the end of the RF pulse to collect the electrons passing through the drift region and thereby make the tube turn off even though high voltage is still applied.[22] Turnoff control electrodes in CFAs thus make possible dc operation, which eliminates the high-power modulator. In dc operation, the high voltage is continuously present between anode and cathode, and the current is turned on by applying RF drive and turned off by pulsing the control electrode. To prevent the tube from starting without RF drive, the cathode must be kept cool enough to prevent thermionic emission. The control electrode requires only a short, medium-power pulse, typically one-third of the anode voltage and one-third of the anode peak current. The greatly reduced modulator requirements for dc operation make it practical to use more complex pulse coding. However, some energy is dissipated on the control electrode each time it is pulsed, and since it is an insulated electrode, it is difficult to cool. Control-electrode heating can therefore be a limitation on the maximum PRF that may be used.

In practice, dc operation has seldom been used[23] because it requires a much larger capacitor bank to limit droop (as opposed to using a switch tube in the constant-current mode) and because an arc in a dc-operated CFA requires crowbarring (Sec. 4.9), which interrupts operation for a few seconds instead of only for a single pulse; it has not been possible to make CFAs arc-free. Problems have also occurred as a result of adjacent radars injecting enough RF energy into the system antenna and back into the transmitter to turn on a dc-operated CFA at the wrong times.

The low insertion loss of CFAs from RF input to RF output without modulator voltage applied permits convenient programming of CFA amplifier chain power output in steps.[24] For example, in an amplifier chain transmitter with two CFAs preceded by a TWT, three power-output levels can be selected simply by choosing which modulators to pulse. Power programming, also called *feedthrough operation*, is especially useful in 3D radar applications, since it permits conserving average power by reducing the peak power output at high scan angles.

The low insertion loss of a CFA also allows power reflected at the output to be passed back through the tube to its input; in many cases the reflected output power coming back out of the input may even exceed the incoming drive power. A properly rated isolator[3,25] is thus a necessity between stages of a CFA chain.

Certain additional problems long identified with magnetron operation are also common to operation of CFAs. For details, see paragraphs on sparking, moding, noise rings, spurious RF output, and RF leakage, discussed under "Common Problems" in magnetrons (Sec. 4.2). One difference is that because RF drive is present during the voltage rise time, many (but not all) cathode-pulsed CFAs allow a much faster voltage rate of rise than magnetrons. For the same reason (i.e., RF drive is already present) there is little starting-time delay in the desired CFA operating mode; but the π-mode oscillation has a finite starting time and will produce little energy if the voltage passes quickly enough through the range in which it can occur. In a dc-operated CFA, the π-mode oscillation should not occur at all because the cathode voltage is at full value all the time.

Klystrons. The multicavity klystron has always been known for its high-gain and high-power capability. However, its bandwidth during the 1950s tended to be 1 percent or less, with wider ranges being covered by mechanical tuning of the cavities; *gang tuning* (tuning all cavities at once in response to rotation of a single tuning knob or motor drive) is often used. Although a tradeoff between klystron gain and bandwidth was always known to be feasible, the stagger tuning of a klystron is far more complex than that of an IF strip. The overall frequency response of a klystron contains intermediate gain products as well as the total product of the individual cavity responses; certain tuning combinations produce excessive harmonic output; and broadband small-signal gain does not ensure broadband saturated gain. Klystron bandwidth capability increases with power level[16] because the stronger beam provides heavier loading of the cavities.

Modern digital computers made it possible to determine improved cavity-tuning arrangements, and klystron bandwidth improved greatly; 8 percent bandwidth 3 dB down) has been obtained with fixed cavity settings, and even 11 percent in rare cases (Varian VA-812C). The achievement of this bandwidth in klystrons also depended partly on improvements in beam perveance, but, more important, it required progress in output-cavity design, because the *power bandwidth* can be no better than the ability of the output cavity alone to extract the energy from the beam, regardless of the gain or drive power available preceding it. Single-cavity output circuits are therefore replaced in broadband klystrons by double-tuned and triple-tuned cavities, sometimes called an *extended-interaction circuit,*[26,27] which uses more than one interaction gap to extract energy from the beam, as shown in Fig. 4.8. This technique of grouping cavities was later extended to the prior cavities as well, and by discovering that the cavities in each group need not be coupled to each other, the *clustered-cavity klystron*[28] has achieved as much as 20 percent bandwidth. Although more complex and expensive than a normal klystron, the clustered-cavity klystron is still less complex and costly than a comparable TWT or Twystron.

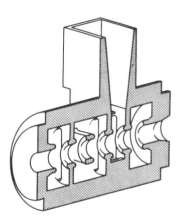

FIG. 4.8 Extended-interaction output circuit. (*From Ref. 26.*)

Traveling-Wave Tubes (TWTs).   The low-power helix TWT is still the king of bandwidth. Because it has virtually constant phase velocity at all frequencies, the helix permits TWTs to have bandwidths in excess of an octave. However, the helix TWT has not been used in high-power radars because high power requires a high-voltage beam, and the resulting electron velocity is too fast to synchronize with the low velocity of the RF wave on a helix slow-wave circuit. The limit of helix tubes is about 10 kV and a peak RF power output of a few kilowatts. For higher power levels, other kinds of slow-wave circuits with a higher RF velocity must be used, and the bandpass characteristics of those circuits can lead to band-edge oscillation problems. Furthermore, both forward waves and backward waves may propagate on the RF structures, leading to the possibility of backward-wave oscillations. Depending on the circuit used, other kinds of oscillations can also occur. For

FIG. 4.9   Cloverleaf slow-wave circuit.   (*Courtesy of Varian Associates.*)

these reasons, high-power TWT development lagged that of klystrons and is still more expensive. By 1963, however, Varian had produced multimegawatt pulsed TWTs using the cloverleaf circuit, as shown in Fig. 4.9,[29] a structure that can be made heavy and rugged enough to handle power comparable with that of klystrons.

Slow-wave structures for high-power TWTs include helix-derived structures (contrawound helix, or ring-bar circuit) and coupled-cavity circuits, of which the cloverleaf is one example, and the ladder network.[30] Below 100 kW, ring-bar circuits usually have broader bandwidth and higher efficiency than coupled-cavity circuits. Above 200 kW or even below that if average power is a limitation, coupled-cavity circuits dominate.[31]

If a TWT using a coupled-cavity circuit is cathode-pulsed, there is an instant during the rise and fall of voltage when the beam velocity becomes synchronous with the cutoff frequency (π mode) of the RF circuit, and the tube usually oscillates. These oscillations at the leading and trailing edges of the RF output pulse have a characteristic appearance that has given them the name *rabbit ears*, as shown in Fig. 4.10. Only in rare cases has it been possible to suppress these oscillations completely. However, since this particular oscillation depends on electron velocity, which in turn depends on beam voltage, the problem is avoided by the use of mod-anode or grid pulsing (described later in this section). In this case, it is only necessary to be sure not to let the modulator begin pulsing the beam current during turn-on of the HVPS until the voltage is safely above the oscillation range, which is typically somewhere between 60 and 80 percent of full operating voltage.

FIG. 4.10   Rabbit-ear oscillations on the envelope of the RF output from a cathode-pulsed TWT amplifier.

Discontinuities, called *severs*, are necessary in the slow-wave structure of high-power TWTs to prevent oscillation due to reflections at input and output of the structure. Although oscillation could also be prevented by distributing loss along the structure, this would result in lower efficiency, which is unattractive in high-power tubes. Typically, a sever is used for each 15 to 30 dB of tube gain. At each sever, the modulated beam carries the signal forward, while the power traveling in the slow-wave circuit at that point is dissipated in the sever loads; thus reverse-directed power is stopped at each sever. The sever loads may be placed external to the tube to reduce dissipation within the RF structure itself.

TWTs tend to be less efficient than klystrons because of the necessity for loading the structure for stability and because relatively high RF power is present in an appreciable fraction of the entire structure. One important technique for improving the efficiency of high-power TWTs is called *velocity tapering*. This technique consists of tapering the length of the last few circuit sections of the slow-wave circuit to take into account the slowing down of the beam as the energy is extracted from it. Velocity tapering permits extracting more of the energy from the beam and significantly improves the power-bandwidth performance of the tube.[31] Nevertheless, high-power TWTs generally show an appreciable falloff of power output toward the band edges, so that the rated bandwidth depends very much on how much power falloff can be tolerated by the system.

To improve the efficiency of TWTs (or klystrons), the use of depressed collectors[16,32] has been developed to a remarkably successful degree. The use of multiple collector sections at intermediate voltages allows catching each spent electron at a voltage near optimum. Up to 10 collector sections have been used in some communications tubes, but 3 sections, as shown in Fig. 4.11, are more typical for modern high-power TWTs for radar systems. The several different voltages needed for the depressed collectors add complexity to the HVPS, but fortunately these voltages need not be as well regulated as the main beam voltage.

Twystrons. In 1963 Varian assembled a hybrid tube consisting of klystron cavities in all but the output section, while a cloverleaf traveling-wave circuit was used for the output section. The purpose was to produce a more efficient version of the VA-125 broadband S-band TWT, based on the more effective beam-bunching action of the cavities. The result was not only slightly higher efficiency but also a significant improvement in bandwidth as a result of the flexibility in tuning of the cavities combined with the broad power-bandwidth capability of the TWT output section. To compensate for the inherent droop in gain of the TWT output section at the edges of the band, the klystron cavities were purposely tuned to boost the gain at those frequencies. Because it is part klystron and part TWT, Varian named the hybrid tube a Twystron.[33] A 3 dB bandwidth of 14 percent has been demonstrated in the VA-145, or a 1 dB bandwidth of 12 percent; 48 percent efficiency has been shown at midband with 41 dB gain. Although more complex and expensive than most klystrons, the Twystron appears capable of equally high power with broader bandwidth than all but perhaps the clustered-cavity klystron.

RF Tube Selection. Table 4.1 summarizes the main differences among the leading RF tube types. The factors that most often dominate in tube selection are cost, bandwidth, spurious noise level, control electrodes, gain, size, voltage, and availability (not shown in the table). Linear-beam tubes are quieter and higher-gain and can be grid-pulsed, but CFAs are smaller, lighter, lower-voltage, and less costly. Sometimes one of these factors is overwhelmingly important, and the transmitter designer is forced to accommodate all the less

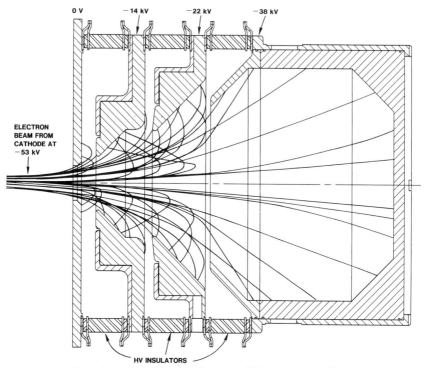

0 V   −14 kV   −22 kV   −38 kV

ELECTRON
BEAM FROM
CATHODE AT
−53 kV

HV INSULATORS

FIG. 4.11 Multiple depressed collectors. (*Courtesy of Hughes Aircraft Electron Dynamics Division.*)

significant disadvantages. Often, several choices are feasible, and tradeoffs are then considered among system cost, schedule, and performance. Further comments on Table 4.1 are given below.

Voltage. Voltage affects the size and cost of the HVPS and modulator as well as x-ray severity.

Gain. Gain strongly affects the number of stages and therefore the complexity (parts count, control, fault monitoring, and maintenance) required in an amplifier chain.

Bandwidth. The bandwidth listed here is the instantaneous bandwidth of the tube, i.e., without tuning adjustments. The tube bandwidth must be compatible with system requirements; in turn, system bandwidth tends to be based on available or presumed tube capabilities.

X-rays. X-rays affect transmitter weight because of the shielding required to protect personnel (and sensitive semiconductors).

Efficiency. This strongly affects transmitter weight, cost, and cooling requirements as well as prime power. The numbers shown do *not* include heater power, solenoid power, or cooling-system power, which may be significant.

Ion Pump. Residual outgassing in a microwave tube can spoil its vacuum and cause RF or dc breakdown. A Vac-Ion pump (trademark of Varian Associates) can be used to maintain a good vacuum, even during storage, and to indicate the quality of the vacuum. Most crossed-field tubes don't require ion pumps because they will pump themselves down when operated.

TABLE 4.1 High-Power Pulsed Amplifiers Compared for Same Frequency and Peak and Average Power Output

	Linear-beam tubes		Crossed-field tubes*	
	Klystron	TWT	Conventional	High-gain
Voltage	High voltage (1 MW requires approx. 90 kV)		Low voltage (1 MW requires approx. 40 kV)	
Gain	30–70 dB		8–15 dB | 15–30 dB	
Bandwidth	1–8%† | 10–15%		10–15%	
X-rays	Severe, but lead is reliable		Not usually a problem	
Efficiency				
Basic	15–30%		35–45%	
With depressed collectors	40–60%		N.A.	
Ion pump	Required on large tubes		Self-pumping	
Weight	Higher		Lower	
Size	Larger		Smaller	
Cost	Medium | High		Medium	
Spurious noise‡	−90 dB		−55 dB | −70 dB	
Spurious modes (typical)	None	π mode during rise and fall if cathode-pulsed; none if mod-anode- or grid-pulsed	π mode during rise and fall if cathode-pulsed; full-power noise output if turned on without RF drive	
Usable dynamic range	40–80 dB		A few decibels	
Control electrode	None, or mod-anode, or grid		None, or turnoff electrode	
Magnetic field	PPM up to 1 MW at S band; solenoid otherwise; barrel magnet rare (SLAC); none for ESFK		Permanent magnet	
Dynamic/static impedance	0.8		0.05–0.2	
Phase-modulation sensitivity	5–40° per 1% $\Delta E/E$		0.5–3.0° per 1% $\Delta I/I$	

*Distributed emission, reentrant, circular.
†Clustered-cavity klystron can achieve 10 to 15 percent bandwidth at higher cost.
‡In 1-MHz bandwidth.

Weight. The weights of the tubes alone are compared in Table 4.1; however, the solenoid required for high-power linear-beam tubes and their higher voltages usually make linear-beam-tube transmitters considerably heavier than CFA transmitters.

Size. The same comments apply as noted under "Weight."

Cost. The costs listed here refer only to tube costs. Also see comments under "Control Electrodes," below. The comments listed apply roughly to both development costs and unit costs.

Spurious Noise. This has become a more significant factor as radar bands become more crowded, receiver sensitivities increase, and electromagnetic compatibility (EMC) receives more attention. Spurious noise should be considered in four parts:

1. *Harmonics:* Both linear-beam tubes and CFAs produce harmonic power output of typically −25 dB (with respect to fundamental power output) at the second harmonic, −30 dB at the third harmonic, and considerably less at higher harmonics. Although these figures vary greatly from one tube design to another, there is no strong difference between linear-beam tubes and CFAs in general. If harmonic power output is a problem, excellent high-power microwave filters are practical.

2. *Adjacent-band spurious noise:* Problems of this type may occur from adjacent modes in CFAs or in TWTs, typically appearing a few percent above (in forward-wave tubes) or below (in backward-wave tubes) the desired operating band. The problem may be severe in cathode-pulsed tubes but is avoided in tubes pulsed by a control electrode (dc-operated CFA, or mod-anode or gridded TWT). If present, adjacent-band spurious noise is usually easy to filter out unless it is too close to the desired operating band.

3. *In-band spurious noise:* This is the factor noted in Table 4.1, since it is most serious to system operation and usually cannot be filtered out. In-band spurious noise may interfere with other systems or may prevent achieving desired MTI cancellation or pulse compression sidelobe levels in the tube's own system. In-band spurious noise also sets a limit on the spectrum improvement that may be obtained by pulse shaping (Sec. 4.7). In-band spurious noise may also be degraded by the source of RF drive (Sec. 4.3).

4. *Interpulse noise:* Unlike the above three factors, interpulse noise is the noise produced by an RF tube when it is supposed to be completely off; that is, between pulses. Noise generated at this time is of concern because, in almost all radar system configurations, this noise will feed directly into the radar receiver and may either produce false targets or mask real targets. In cathode-pulsed tubes, the high voltage is removed from the RF tube between pulses, and no interference with receiver operation is normally encountered unless there is excessive modulator fall time or backswing, which can produce *noise rings* (Sec. 4.2). With dc-operated CFAs or with mod-anode or gridded linear-beam tubes, the high voltage remains on the tube between pulses, and serious noise may be generated if even a small amount of beam current is allowed to flow through the tube. Since all dc-operated CFAs use cold cathodes, no current can flow until RF drive is applied. With linear-beam tubes, beam current must be well enough cut off to keep noise output (and amplified input signals) small enough. Despite the nearly 200 dB between typical RF peak power output and typical receiver noise levels, most RF tubes readily meet the interpulse-noise requirements. Problems occurred mostly with older-style intercepting-grid linear-beam tubes because the hot grid may emit and produce residual beam current even if cathode current is cut off.

Spurious Modes. The spurious modes listed in Table 4.1 are the ones most commonly present. In some cases tubes have been made in which these modes are fully suppressed, whereas in poorly designed tubes other modes may also appear, such as band-edge oscillations, harmonic oscillations, etc.

Usable Dynamic Range. Dynamic range and linearity may be of importance for pulse shaping, as discussed in Sec. 4.7.

Control Electrodes. These determine the type of modulator required, which in turn affects transmitter size, weight, complexity, and cost. Control electrodes avoid the need for a full-power cathode-pulse modulator. A *mod-anode* (modulating anode) can be used in any linear-beam tube; it acts like a control grid with a mu of 1 and is inexpensive and reliable. A high-mu grid is also feasible for all but the very highest power linear-beam tubes, and it greatly simplifies the modulator requirements; but it raises the cost and may lower reliability and life of the tube.

Magnetic Field. Except for a few electrostatically focused klystrons,[34,35] all magnetrons, CFAs, klystrons, and TWTs require a magnetic field to control the path of the electron beam. Virtually all CFAs use permanent magnets. Periodic permanent magnet (PPM) focusing is used on all but very high power linear-beam tubes, which still require solenoids. Use of a solenoid affects transmitter size, weight, efficiency, cost, servicing, and tube protection.

Dynamic Impedance. This indicates how rapidly the tube current changes for a given change in applied voltage (also see Table 4.2). The significance of this factor depends on the type of modulator used, or it may affect the HVPS capacitor bank size required for a given permissible power droop during the pulse length (Sec. 4.8).

Phase-Modulation Sensitivity. This indicates how hard the transmitter designer must work to ensure that system phase-stability requirements will be met (Sec. 4.6). Large as the difference may seem between linear-beam tubes and CFAs in this characteristic, modulation sensitivity is seldom a dominant factor in RF tube selection. However, it does enter into the size, weight, and cost tradeoffs by its effect on HVPS filter size or on modulator complexity.

Historically, it has always been feasible to obtain the extremely low ripple levels desired for MTI systems, limited only by inherent noise levels (including jitter and starting-time noise) in the tubes. For pulse compression systems, the necessary freedom from ringing during the pulse length has been fairly easy to achieve with hard-tube modulators and difficult to achieve with line-type modulators, with either linear-beam tubes or CFAs.[36]

Life. Linear-beam tubes and CFAs have both shown the feasibility of long life (over 40,000 h in some cases), but both have also shown very short life when the wrong tube-design compromises were made, when development problems remained, or when tubes were misapplied or carelessly handled in the field. Attainment of very long life, in the region of 10,000 h or more, requires judicious selection of power ratings, conservative cathode-current density, and conscientious counseling of the marriage between tube and transmitter.

RF Tube Power Capabilities. The peak power capabilities of RF tubes have progressed sufficiently far that the limitation has become breakdown in practical waveguide systems, even with 20 lb/in^2 of SF$_6$ in the waveguide. Therefore, since the early 1960s there has been a tendency for radar systems to employ increased duty cycle, by the use of techniques such as pulse compression, to achieve higher average power without a further increase in peak power. Although in many cases a single RF tube can produce so much average power that even pure-copper waveguide requires water cooling, the limit in system average power may still be the RF tube. Furthermore, asking for the ultimate power capability in one RF tube has a high risk of leading to an unsuccessful development program or to an unreliable tube even if the development is "successful." From a reliability standpoint as well, multiple smaller tubes are often preferable to a single very large tube. Therefore, it has become quite common for high-power radar systems to use more than a single RF tube, as described in the next section.

4.5 COMBINING AND ARRAYING

It is often necessary to use more than one RF tube or solid-state device to produce the required radar transmitter RF power output. Since the mid-1950s, two
or more microwave tubes have often been used to achieve more total power output than can be obtained from a single tube. Since about 1960, there has been
interest in using more than one RF device, especially if it can then be solid-state,
to provide increased system reliability from the greatly lowered probability of
multiple failures. Although failure of any one device reduces total power output
and may also degrade performance in other ways (e.g., increased array sidelobes
if the outputs are combined in space, as discussed below), the system can be designed to perform adequately with a reasonable number of devices out of operation (referred to as *graceful degradation*). Calculations and tradeoffs in this area
are well covered in references on reliability.

Hybrid Combining (or Magic T). For combining the power output of two
devices, a hybrid or magic T works very nicely in waveguide, coax, stripline,
or slabline. Multiple devices are usually combined with corporate structures by
using hybrid or magic-T combiners in each stage, but a multibranch Wilkinson
combiner[3] or a replicated combiner[37] can also be considered for minimum
combining loss. For a 2:1 combiner, the choice between magic T and hybrid
and between types of construction depends on power level and physical
convenience; they all function the same way. If the power outputs of two
devices are equal and if their outputs arrive in proper phase at the combiner,
all the power (less only a small insertion loss) will appear at the output arm of
the combiner. The less perfectly these two criteria are met, the more power
appears at the fourth (dummy-load) arm of the combiner, as noted below:

$$\text{Output-arm power: } \left(\frac{P_1 + P_2}{2}\right) + \sqrt{P_1 P_2} \, \cos \theta \qquad (4.3)$$

$$\text{Fourth-arm power: } \frac{P_1 + P_2}{2} - \sqrt{P_1 P_2} \, \cos \theta \qquad (4.4)$$

where θ is the relative phase of the two signals arriving at the combiner. If the
relative phase is purposely changed by 90°, the combiner will deliver half the
power to the output arm and half to the fourth arm; or if the phase is changed by
180°, the roles of the output arm and the fourth arm may be interchanged, such as
for testing into a dummy load. Conversely, the ability of a pair of tubes to be
efficiently combined into the desired output arm over a range of frequencies without adjustment depends on the *phase tracking* of the two RF devices being combined. That is, although the phase delay through the two devices may vary with
frequency, the two devices must vary together within a tolerance dependent on
the allowable power loss to the fourth arm.

For convenience in planning allowable tolerances, fourth-arm power (wasted
power) is plotted in Fig. 4.12. It can be seen that phase differences are much
more significant than amplitude differences. For example, note that the relative
phase of the two signals arriving at the combiner can differ by as much as 32°
before the combined output will be reduced by 10 percent (0.5 dB). Amplitude
differences are even less critical; a 2:1 difference (3.0 dB) in the power level of
the two signals arriving at the combiner reduces their combined total by only 2.9
percent (0.13 dB).

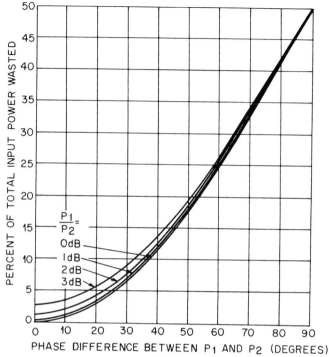

FIG. 4.12 Fourth-arm (wasted) power versus phase and amplitude difference between two tubes (or other devices) being combined.

Since a hybrid or magic T may also be used as a power splitter, a common way of operating two identical devices in parallel is shown in Fig. 4.13a. Figure 4.13b shows another alternative, in which the outputs are recombined only in space but the devices are still effectively operating in parallel. Figure 4.13c shows two whole chains operating in parallel; but the greater the number of items that are included in each of the two paths, the more chance exists for phase differences to occur between the two paths as a function of frequency, temperature, or component tolerances. Therefore, combining chains is more difficult than combining single stages and is usually avoided. However, the combined chains allow use of a low-power 180° diode or ferrite switch to provide very fast switching of the combined power output between the normal output arm and the fourth arm of the combiner for pulse coding, for pulse shaping, or for switching between two antennas or array faces.

Phased Arrays. From the RF tube standpoint, phased arrays may be of three types:

Case 1. *Single spigot:* The whole array may be fed from a single RF power source by using a corporate divider structure to divide the single source to drive all phased array elements or by using an equivalent optical-feed system (Sec. 7.8) for the same purpose. To avoid having to use two separate phase shifters in cascade at each element to respond to simple row-and-column steering commands, these two commands are usually summed for each ele-

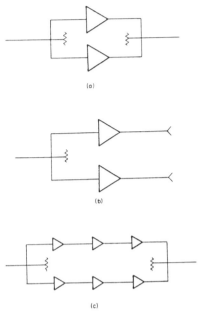

FIG. 4.13 Parallel operation of RF amplifier devices or chains. (*a*) Power splitter and power combiner. (*b*) Power splitter and recombination in space. (*c*) Parallel operation of two chains.

ment into a single phase command. This can be done for each element in a central computer, which then requires that a different command be delivered to every element in the array; or the summing can be done by a controller chip at each element, in which case the number of commands to be sent to the array is reduced to the number of rows and columns. The ultimate is just to send all elements in the array the same three pieces of data: azimuth angle, elevation angle, and frequency. Each element must then contain a controller chip smart enough to compute its own proper phase setting based on its location in the array. The central computer or smart controller chip at each element can also absorb the tasks of collimation and error correction to eliminate the need for mechanical trimming of nominal phase at each element.

Case 1 is nearly conventional from the tube-type transmitter standpoint, except for a likely emphasis on very high average power capability. Tube-type systems of this type usually use multiple tubes in parallel for increased reliability. Physical size is usually no problem.

Case 2 *Subarrays:* One RF source may be used per group of array elements, called a *subarray.* Power dividers (small corporate structures) distribute the power to the subarray elements, but the dividers are identical for all subarrays. The high-power phase shifters in each subarray require separate steering commands, but the commands to all subarrays can be identical. Low-power phase shifters (or true-time-delay elements if the pulse waveform bandwidth is wide enough compared with array dimensions to require it) precede each RF device and require separate steering commands, but there are many fewer separate steering commands required than in Case 1.

Case 2, like Case 1, usually requires high average power per RF source or group, but if there are enough subarrays to allow adequate system performance with one or more subarrays failed, then the reliability required of the RF source for each subarray can be reduced.

Case 2 imposes similarity requirements on the RF sources comparable with those listed below for Case 3. Size may or may not be a problem, depending on the physical size of the subarray, which in turn depends on the operating frequency band and the number of elements per subarray. For tube systems, the optimum number of elements per subarray depends on the power capability of a single RF tube, on the economics of tube quantity versus power per tube, and possibly on true-time-delay considerations.

Case 3 *One tube per element:* An arrangement of one RF source per array element allows all phase shifting for beam steering to be done at a low power level ahead of the final-stage RF amplifiers. Several arrangements for using row-and-column steering commands are feasible, allowing a simpler beam-steering computer.

Case 3 ordinarily does not require high average power per source, nor does the achievement of very high system reliability require extremely high reliability of each source. The tolerable number of RF sources that may be in a failed condition at one time depends instead primarily on the tolerable degradation in array-antenna-pattern sidelobe levels.[38,39] However, the mean life of all the RF devices obviously has a very strong impact on operating costs, both because of the cost of replacement devices and because of the maintenance cost to perform the replacements. The relatively high maintenance required in Case 3 makes it imperative that systems of this type be designed to permit replacing failed devices without shutting down the system. Physical size can be very important in Case 3 because the construction becomes very unwieldy if the RF sources cannot be placed immediately behind the array elements.

Case 3 requires that the phase delay through all the RF sources, as well as their power outputs, be similar and remain similar in order to prevent excessive degradation of antenna pattern sidelobes as a result of these errors. Besides similarity of the RF devices themselves, it is also necessary to ensure similarity of power supply voltages to all tubes, of modulator voltages (such as for gridded TWTs), of magnetic fields if solenoids are used, of RF drive levels, and of associated RF components such as ferrite isolators. It is usually necessary to develop an error budget for these similarity factors, similar to the stability budget required for pulse-to-pulse and intrapulse phase-stability requirements discussed under "Amplifier Chains: Special Considerations" (Sec. 4.3).

4.6 TRANSMITTER STABILITY REQUIREMENTS

Pulsed MTI Systems. Table 15.2 lists the tolerable amounts of instabilities of various kinds in an MTI system. Several conclusions may immediately be drawn from that table from the transmitter standpoint:[40]

1. Time-jitter limits are harder to meet in short-pulse systems.

2. The pulse-to-pulse frequency stability required of a pulsed oscillator transmitter becomes harder to meet at high RF frequency and in long-pulse systems; so does the intrapulse frequency-stability requirement.

3. The pulse-to-pulse phase stability required of a pulsed amplifier transmitter is independent of RF frequency and pulse length; so is the intrapulse phase-stability requirement.

Other comments on oscillator versus amplifier systems are given in Sec. 4.3.
Modulator and HVPS Effects. In order to calculate tolerable modulator ripple and tolerable intrapulse voltage or current variation, the frequency (or phase) pushing versus anode current or voltage applied to the RF tube must be known, together with the factor by which anode current or voltage varies with a change in HVPS voltage. Frequency (or phase) pushing versus anode current or voltage is known as the frequency- (or phase-) modulation sensitivity, and typical values are tabulated in Table 4.2. Since the rate at which anode current or voltage varies

with HVPS voltage depends on the relative dynamic impedances of the RF tube and its modulator, several alternatives are listed in Table 4.2. Note that although the dynamic impedance of crossed-field devices is very low, typically one-tenth of their static impedance, the nominally matched internal impedance of a line-type modulator reduces the amount of current variation that results from a 1 percent change in HVPS voltage to about 2 percent instead of the 10 percent that might be expected.

For a grid-pulsed or mod-anode-pulsed linear-beam tube, the beam voltage is determined directly by the dc HVPS voltage, and the phase-modulation sensitivity of the tube applies directly to any HVPS ripple or droop. In addition, the tube will have a phase-modulation sensitivity to its control-electrode voltage (grid or mod-anode voltage), but this sensitivity is usually much lower (in degrees per percent change) than the sensitivity to beam-voltage changes. Phase-modulation sensitivities to filament voltage and to magnetic field may also be significant in some cases. Attainment of satisfactory transmitter stability for MTI purposes requires adequate control of *all* sources of instability.

With proper care, adequate transmitter stability to permit 40 to 60 dB MTI improvement factor has been achieved with a wide variety of RF tubes and operating conditions, including staggered PRF (discussed below) and intermittent bursts of pulses. However, the development of stealth aircraft and missiles continues to force radar designers to handle smaller cross-section targets, even in clutter, which requires even higher levels of MTI performance. In some cases, at least with uniform-PRF operation, transmitter stability consistent with a 60 to

TABLE 4.2 Stability Factors

		Impedance	Current or voltage change for 1 percent change in HVPS voltage	
	Frequency- or phase-modulation sensitivity	Dynamic / Static	Line-type modulator	Low-impedance* hard-tube modulator, or dc operation
Magnetron	$\dfrac{\Delta f}{f} = \left(\begin{matrix}0.001 \\ to \\ 0.003\end{matrix}\right)\dfrac{\Delta I}{I}$	0.05–0.1	$\Delta I = 2\%$	$\Delta I = 10$–20%
Stabilotron or stabilized magnetron	$\dfrac{\Delta f}{f} = \left(\begin{matrix}0.0002 \\ to \\ 0.0005\end{matrix}\right)\dfrac{\Delta I}{I}$	0.05–0.1	$\Delta I = 2\%$	$\Delta I = 10$–20%
Backward-wave CFA	$\Delta\phi = 0.4$ to $1°$ for 1% $\Delta I/I$	0.05–0.1	$\Delta I = 2\%$	$\Delta I = 10$–20%
Forward-wave CFA	$\Delta\phi = 1.0$ to $3.0°$ for 1% $\Delta I/I$	0.1–0.2	$\Delta I = 2\%$	$\Delta I = 5$–10%
Klystron	$\dfrac{\Delta\phi}{\phi} = \dfrac{1}{2}\dfrac{\Delta E}{E}$ $\quad\phi\approx5\lambda$ $\Delta\phi \approx 10°$ for 1% $\Delta E/E$	0.67	$\Delta E = 0.8\%$	$\Delta E = 1\%$
TWT	$\dfrac{\Delta\phi}{\phi} \approx \dfrac{1}{3}\dfrac{\Delta E}{E}$ $\quad\phi\approx15\lambda$ $\Delta\phi \approx 20°$ for 1% $\Delta E/E$	0.67	$\Delta E = 0.8\%$	$\Delta E = 1\%$
Triode or tetrode	$\Delta\phi = 0$ to $0.5°$ for 1% $\Delta I/I$	1.0	$\Delta I = 1\%$	$\Delta I = 1\%$

*A high-impedance modulator is not listed because its output would (ideally) be independent of HVPS voltage.

80 dB improvement factor has been demonstrated; other portions of the system may become the limiting factor in such cases.

RF Tube Effects. Linear-beam tubes are inherently capable of very high MTI improvement factors, limited only by power supply and modulator effects, as noted above, or by possible microphonic effects (rare). The inherent noise level in CFAs, however, is much higher than in linear-beam tubes and can set a limit to the achievable MTI improvement factor. For example, in a system with a compressed (or actual) pulse width of 0.1 μs, the typical noise level of −55 dB/MHz (Secs. 4.3 and 4.4) of conventional CFAs would limit the MTI improvement factor to less than 45 dB. Fortunately, the low-noise CFA is at least 15 dB better in this respect.

Staggered PRF. An especially difficult requirement, from the transmitter standpoint, is the use of nonuniform interpulse periods, usually called *staggered-PRF* MTI (Sec. 15.9). Although the stability requirements on the transmitter are usually the same as discussed above, it is much more difficult for the transmitter to meet them because the output from the power supply and/or modulator tends to vary when the interpulse period varies.

The most significant source of transmitter instability in staggered-PRF operation is the variation in HVPS voltage from pulse to pulse that results from the nonuniform power drain on the HVPS. In the region of small output-voltage changes considered here, the power supply current I into the filter capacitor C can be considered essentially constant at its long-term average value (rectification ripple effects can be considered separately). The power-supply-voltage variation ΔV induced by the staggered PRF can therefore be determined from $\Delta V = I\Delta t/C$, where Δt is the variation in interpulse periods. Use of a large enough filter capacitor can make this voltage variation as small as desired, but in typical cases this "brute-force" approach is usually impractical. Some kind of fast regulator (Sec. 4.9) capable of reducing pulse-to-pulse variations is therefore usually required in modulators for staggered-PRF systems. Even if the HVPS voltage is held constant, variations can occur in other portions of the transmitter and system as a result of nonuniform interpulse periods.

Essentially, the change from uniform to nonuniform pulse periods makes it unlikely rather than likely that uniform modulator output will be obtained, as a result of a variety of factors that can rightfully be ignored in uniform-PRF systems. The detailed solutions to these problems vary, depending on the pulse-to-pulse stability level being sought. In extreme cases it may be necessary to inject an appropriate compensation waveform into the regulator circuit. In an amplifier chain transmitter, it may be easier instead to correct for transmitter pulse-to-pulse phase variations in the receiver (with a locked coho, or "measure on transmit and correct on receive," for example), unless second-time-around clutter must also be well canceled.

Pulse Compression Systems. In a linear-FM (chirp) pulse compression system, the RF portions of the system can distort the chirp signal in ways that will cause the compressed-pulse output of the receiver to be broadened or to be accompanied by undesired *paired echoes*, also called *time sidelobes*.[41–43] These distortions may be the result of reflections or of amplitude and phase nonlinearities (dispersion) in the RF path, or they may be the result of modulator effects on the RF tubes. An echo will be tolerable if it is small enough, depending on the system time-sidelobe level being sought; a greater echo may be tolerable if its effect is only to broaden the pulse; and it can be allowed to be quite large if it falls completely within the compressed-pulse width. An echo resulting from reflections will not appear fully outside the compressed pulse unless it is delayed by a whole compressed-pulse width; and nonlinearities will not produce comparably located echoes unless the distortion is a rapid enough function of frequency to produce at least two cycles of variation within the chirp-pulse duration.

Although similar echoes may be produced in the receiver in both cases, it is convenient to separate the discussion of fixed distortions in the RF path from the time-varying effects produced by the modulator.

Reflections and Nonlinearities. An RF tube may introduce serious distortion either internally or in conjunction with the waveguide system with which it connects. Internal effects may be of concern if the tube contains sufficiently high-Q circuits to have appreciable amplitude or phase nonlinearity (versus frequency) within the chirp bandwidth, but this is unlikely if the bandwidth of the RF tube is much larger than the chirp bandwidth. Internal effects may also be of concern if the time delay through the tube is comparable to a compressed-pulse width; in this case, RF reflections in the tube or RF leakage around the tube can produce echoes outside the compressed-pulse width. By vector addition,[43] these reflections or leakage paths will also cause a tube to exhibit amplitude and phase nonlinearities as a function of frequency. Therefore, adequate amplitude and phase linearity of a tube assure freedom from such echoes.

As noted above, a nonlinearity will be tolerable if it is small enough or if it is a slow enough function of frequency. For two cycles of phase variation to occur within a chirp bandwidth, the round-trip path length of the reflection (or leakage signal) must change by two wavelengths over that bandwidth. Therefore, a broadband RF tube probably cannot exhibit two cycles of amplitude or phase variation within a chirp bandwidth unless the length of the tube in RF wavelengths is at least equal to the reciprocal of the fractional chirp bandwidth. For example, a 50-MHz chirp at 5 GHz is only a 1 percent (1/100) chirp bandwidth, and a broadband RF tube would have to be at least 100 wavelengths long to contribute a nonlinearity versus frequency having two cycles of variation within a 1 percent bandwidth. Correspondingly, a 50-MHz chirp results in a compressed-pulse width (with typical weighting) of 40 ns, and 100 wavelengths at 5 GHz has a round-trip time delay of 40 ns. Therefore, since RF tubes are seldom over 40 wavelengths long, significant time sidelobes should not be expected from broadband RF tubes for chirp bandwidths less than about 2.5 percent. Measurements tend to confirm this conclusion; for example, phase nonlinearities in S-band TWTs tend to have a periodicity of about 80 MHz, and in C-band TWTs they tend to have a periodicity of about 150 MHz.

For chirp bandwidths over a few percent, RF tubes do show significant nonlinearities,[44] as do other microwave portions of the transmitter. To the extent that these nonlinearities are constant, a transversal equalizer (or equivalent) in the pulse compression receiver can be adjusted to compensate for them. However, large changes in system operating frequency or large doppler shifts caused by very fast targets can make this compensation ineffective.[41]

Even when an RF tube by itself may not be long enough to cause distinct time sidelobes, its mismatches must be considered in conjunction with the system waveguide runs with which it connects. These runs often are long enough to produce noticeable sidelobes if the reflections are large enough. This is of greater than normal concern with CFAs, because a reflected signal will pass back through the CFA, be re-reflected by the mismatch at its input, and be amplified when it passes forward through the CFA. System waveguide reflections arriving at the CFA output thus see the mismatch at the input of the CFA magnified by the CFA gain. The obvious solution is a good isolator close to the CFA output so that echoes from this effect will fall within the compressed pulse.

Especially in a wideband chirp system, a direct measurement of the effects of an RF tube on a short pulse may be more useful than linearity tests in the frequency domain. If a very short pulse (equal to or less than the compressed-pulse width) is passed through the RF tube, any significant reflections or nonlinearities in the tube will result in echoes immediately following the short test pulse. This test may be difficult to perform satisfactorily, however, because of the limited broadband signal-to-noise ratio in a practical amplifier chain, as discussed in Secs. 4.3 and 4.4. A linear-beam tube may also be beam-pulsed without RF drive, and a very short test pulse fed into its *output* with a circulator; this is essentially a time-domain reflectometer test[45] and can be used to determine the magnitude and location of individual mismatches within the tube, but only as far back as the nearest sever or drift space. Since the input of the tube may be terminated for this test, the signal-to-noise-ratio problem is reduced. If the tube is a TWT, the results must be interpreted with allowance for the fact that reflected signals traveling forward through the output section of the tube will be amplified. Other input and output connections of the tube can also be tested. However, useful and interesting as these tests may be, it is important to note that they do not take into account any modulator effects (or any time-related effects within the tube[46]).

Modulator Effects.   Regardless of the chirp bandwidth β, several cycles of amplitude or phase modulation during a pulse length τ may easily occur because of modulator or power supply ringing. It is therefore necessary that ringing and ripple on the modulator pulse shape meet the requirement that, when divided into its Fourier components during the pulse length, those components between two and βτ cycles during the pulse must have amplitudes small enough to allow system sidelobe levels to be met,[41] and components of lower frequency must be small enough not to broaden the compressed pulse excessively. For typical cases seeking 30- to 40-dB sidelobe levels, the one-cycle-per-pulse-length component may be allowed to be 4 times as great as the two-cycle-component limit. Droop or curvature having less than one cycle in the pulse length can be allowed to be at least 30° phase modulation or 1-dB amplitude change.

If the ringing or curvature lasts for less than the full duration of the pulse, it may be allowed to have correspondingly greater amplitude. Thus, fortunately, ringing or curvature that typically occurs at the leading edge of a pulse is allowed greater limits than effects that persist for the entire pulse length.

For the same reason, the pedestals associated with CFA chains (Secs. 4.3 and 4.4) are usually of such short duration that they have very little effect. For minimum effect, the frequency used during the pedestals should be a smooth continuation of the chirp (or other coding) used during the main portion of the pulse. In any case, any disturbance that is as short as a compressed-pulse duration will have its energy well dispersed in time by the receiver's pulse compression network and will thereby be reduced in amplitude by the compression ratio. For these reasons, the pedestals in CFA chains can usually be tolerated in pulse compression systems. Similarly, a finite rise and fall time on the RF pulse, comparable to a compressed-pulse duration, is not only tolerable but actually desirable.[47]

Hard-tube modulators can normally provide adequately "clean" pulse shapes to be compatible with pulse compression systems seeking sidelobe levels of 30 to 40 dB, whereas line-type modulators have great difficulty in doing so because of pulse transformer ringing and the lumped-section nature of pulse-forming networks.

Stability Improvement by Feedback or Feedforward. Feedback loops have long been used to improve the stability of CW radar transmitters and CW illuminators; in those applications, it is usually called *noise degeneration*. Basically, a sample of the transmitter RF output is compared with a sample of the RF drive signal, using a phase detector; the resulting phase-error signal is used to drive a fast phase shifter in the transmitter to reduce the error. The phase shifter is usually at a low power level for convenience, but it must follow the point where the RF drive sample is taken. The drive sample must also be suitably delayed, by an amount equal to the phase delay through the transmitter, so the phase detector will not show a phase "error" as a result of carrier frequency changes.

The success of noise degeneration in CW transmitters has led to its use in pulsed transmitters as well to provide increased phase stability for MTI[48] and improved phase linearity for wideband pulse compression. However, the feedback bandwidths of interest in a pulsed transmitter are usually much wider than for a CW transmitter, which may limit the effectiveness that can be achieved. Pulsed transmitters also generally have higher peak power output, which generally means there are more stages around which the feedback must be applied.

Each stage in a chain, as well as the microwave interconnections between them, has some time delay, and the total time delay around the feedback loop limits the achievable loop bandwidth. Time delay adds extra (lagging) phase shift to the other factors in the feedback loop's open-loop gain characteristic. The amount of phase shift added is proportional to frequency and equals 360° at the frequency where the time delay equals one whole cycle. Reasonable feedback-loop design can only tolerate an extra 45° or so from this factor at 0-dB crossover and still retain satisfactory gain and phase margins for loop stability, especially over the expected range of variations in loop gain. Furthermore, it must be recognized that useful error correction occurs only at frequencies well below crossover, where the loop gain becomes appreciably greater than 0 dB.

As an example, a typical tube-type L-band amplifier chain, including the RF tube slow-wave circuits and the feedback return path, might easily have a total path length of 50 wavelengths, resulting in a time delay of 40 ns (time delay in a typical solid-state transmitter is several times greater, as a result of the phase delay introduced by the input and output impedance transformation networks used in each of the many stages required). As a result, useful feedback-loop bandwidth would be limited to about 3 MHz, and useful error correction, such as 10:1 (20 dB), could be achieved only up to about 300 kHz. This may be sufficient if the required noise degeneration bandwidth is no more than about 1 MHz. On the other hand, if the loop must correct disturbances over a greater bandwidth, the effectiveness of the feedback will be quite limited. Noise degeneration should work better at the higher frequency bands, where there is less time delay per wavelength.

As a result of the limitations in using feedback, feedforward (which has been successfully used in microwave relay links[49,50]) has sometimes also been considered. Because feedforward is open-loop, it will try to correct not only the small errors it is intended to correct but also any gross phase changes versus time or versus frequency across the band unless these are automatically programmed out, which adds to the complexity. The even more severe "catch" to feedforward is that it is linear and will automatically try to correct amplitude errors as well, which forces the correction amplifier to be excessively large and powerful even if it is intended only to make very small phase-error corrections.

4.7 TRANSMITTER SPECTRUM CONTROL

As the RF spectrum becomes more and more crowded, interest in EMC continues to grow. Although EMC covers all aspects of compatibility, such as optimal geographical assignment of frequencies, the aspects of concern here are those which affect RF tube selection and achievement of minimum spectrum occupancy by the transmitter.

Reduction of Spurious Outputs. RF tube spurious outputs may be grouped into three kinds: harmonics, adjacent-band, and in-band.

All RF tubes produce some harmonic output (Sec. 4.4). In general, little can be done in tube design to reduce harmonic outputs, but it is feasible to filter out harmonics very well (30 to 60 dB reduction) with modern high-power filters.

Adjacent-band spurious output also occurs in most cathode-pulsed TWTs and CFAs (Sec. 4.4). Adjacent-band spurious output is affected by tube and modulator selection, but it can also be filtered by a high-power microwave filter if necessary.

All RF tubes produce some in-band background noise level. In a 1-MHz bandwidth, this noise is typically 50 to 60 dB down in conventional CFAs, 70 to 80 dB down in the low-noise high-gain CFA, and 90 dB down or better in linear-beam tubes. In-band spurious cannot normally be improved with filters, since it occurs within the same frequency range as the desired signal spectrum. Attempts to use noise degeneration to reduce inherent RF tube noise levels are subject to the limits described in Sec. 4.6. In-band spurious signals can also result from power supply and modulator instabilities, as described previously.

Reduction of Spectrum Amplitude Exceeding (sin x)/x. The spectrum of a perfectly rectangular pulse has the familiar (sin x)/x form, where x is $\pi(f_0 - f)\tau$, f_0 is the operating frequency, and τ is the pulse width. If $1/\tau$ is called the nominal bandwidth of the signal, the envelope of the spectrum peaks falls off 6 dB per bandwidth octave, and this reduction will continue until the envelope reaches the inherent noise output level of the transmitter. This rate of spectrum falloff is too slow to meet most system requirements. Nevertheless, without special care the actual spectrum envelope may be even worse than this, depending on tube characteristics, as a result of phase modulation during the finite rise and fall of practical modulator and RF drive pulse shapes. In these cases, either the leading and trailing edges must be appropriately tailored, or else (in linear-beam tubes) the RF drive may be withheld during the rise and fall time, as shown in Fig. 4.14. Although this may slightly reduce the apparent efficiency, it should be noted that energy outside approximately $1/\tau$ generated during the rise and fall with RF drive present is not utilized by the receiver anyway.[51]

Improvement over (sin x)/x by Means of Shaped Pulses. Since the energy in the spectrum beyond plus or minus $1/\tau$ from f_0 is not used by the receiver, it is desirable for EMC purposes to avoid transmitting energy beyond those limits; this objective may be approached by using a pulse shape different from the convenient and conventional rectangular pulse, as noted in the literature.[52–56] However, idealized pulse shapes have rarely been used in radar systems because of the complexity and loss of efficiency that result.

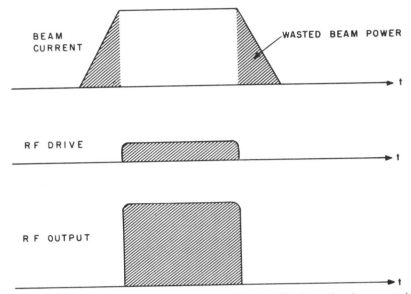

FIG. 4.14 Use of gated RF drive to avoid phase distortion. The figure also shows wasted beam power during modulator rise and fall times.

Pulses with Shaped Edges. A simpler and yet effective approach to spectrum improvement is to shape only the rise and fall of a rectangular pulse.[57] This attenuates the spectrum at frequencies far from f_0, while the flat-topped center portion of the pulse retains high transmitter efficiency for most of the pulse duration. Since a rectangular pulse has the best transmitter efficiency but the widest spectrum, whereas a gaussian pulse has the narrowest spectrum but very poor transmitter efficiency, the fraction of the pulse length to be used for the shaped rise and fall is a crucial decision.

Although the improvement attainable in practice is limited by phase modulation in the transmitter during the rise and fall,[54,56,58] significant improvements can be obtained. In a linear-beam-tube transmitter with properly shaped RF drive, for example, the spectrum width at 60 dB down can usually be narrowed by about an order of magnitude at a cost of about 1 dB in transmitter efficiency.

In practice, most amplifier chain radar systems, whether tube or solid-state, now use at least some shaping of the edges of the transmitted RF pulse to reduce RF spectrum width. This is usually done simply by slowing the rise and fall times of the exciter signal to the transmitter; this has generally been adequate to satisfy MIL-STD-469 and related system requirements.

4.8 PULSE MODULATORS

Since pulse-modulator design is well covered in existing literature,[59–61] this section will mainly summarize and compare available modulator techniques. The type of modulator required is usually determined by the available type of RF tube. A grid pulser, for example, is the smallest, easiest, and least expensive type of modulator, but it can only be used if the RF tube has a grid. Although grids have become more common, a grid may still not be feasible in a very high power RF tube. On the other hand, several types of modulators may be suitable for a given application; Table 4.3 compares some of the performance advantages and disadvantages of various modulator techniques. The final choice is then based on tradeoffs among cost, size, weight, efficiency, and life. The conclusions vary greatly as a function of system requirements and the type of RF tube to be pulsed, as proved by the wide variety of modulators in use.

Line-Type Modulators. The classic line-type modulator is shown in Fig. 4.15. In this type of modulator, the switching device $V1$ (thyratron, ignitron, silicon controlled rectifier, reverse-switching rectifier,[62] or spark gap) merely initiates and carries the discharge of the pulse-forming network (PFN); the actual shape and duration of the pulse are determined entirely by the PFN and other passive circuit elements. The pulse ends when the passive elements have discharged sufficiently that current in the switch stops and allows the switch to recover its voltage hold-off capability.

The self-terminating nature of the pulse discharge is what permits the use of simple switching devices (fully on or fully off only), but this characteristic is also the greatest weakness of line-type modulators. The switching device merely times the pulse discharge and has no control on the pulse shape. Although the PRF can be varied if a series diode is used in the resonant-charging circuit,[59] pulse length can only be changed by switching the connections to multiple PFNs or PFN sections, which requires high-voltage switches. For similar reasons, the trailing edge of the pulse is usually not sharp, since it depends on the energy stored in multiple reactive elements all going to zero at the same time. Furthermore, a well-matched condition is difficult to achieve into nonlinear loads such as RF tubes and all their stray circuit impedances. Achieving the desired pulse shape into very nonlinear loads such as magnetrons often requires despiking or damping circuits.[9,59]

Heater power (if required) for the RF tube load is usually supplied either by a low-capacity high-voltage-insulated heater transformer or by a bifilar secondary winding on the pulse transformer, as shown in Fig. 4.15.

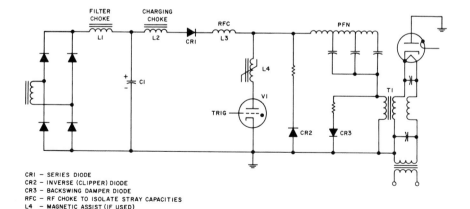

CRI — SERIES DIODE
CR2 — INVERSE (CLIPPER) DIODE
CR3 — BACKSWING DAMPER DIODE
RFC — RF CHOKE TO ISOLATE STRAY CAPACITIES
L4 — MAGNETIC ASSIST (IF USED)
TI — PULSE TRANSFORMER (SHOWN WITH BIFILAR SECONDARY)

FIG. 4.15 Line-type modulator.

TABLE 4.3 Comparison of Modulators

Modulator		Fig.	Flexibility		Pulse-length capability		Pulse flatness	Crowbar required		Modulator voltage level
			Duty cycle	Mixed pulse lengths	Long	Short		Load arc	Switch arc	
Line type	Thyratron / SCR	15	Limited by charging circuit	No	Large PFN	Good	Ripples	No		Medium / Low
Magnetic modulator		...	Limited by reset and charging time	No	Large C's and PFN	Good	Ripples	No		Low
Hybrid SCR magnetic modulator		...	Limited by reset and charging time	No	Large C's and PFN	Good	Ripples	No		Low
Active switch	Series switch	16a	No limit	Yes	Excellent; large capacitor bank	Good	Good	Maybe	Yes	High
	Capacitor-coupled	16b	Limited	Yes	Large coupling capacitor	Good	Good	Maybe	Yes	High
	Transformer-coupled	16c	Limited	Yes	Difficult; XF gets big; large capacitor bank	Good	Fair	Maybe	Yes	Medium-high
	Mod-anode	17	No limit	Yes	Excellent; large capacitor bank	OK, but efficiency low	Excellent	Yes	Yes	High
	Grid	...	No limit	Yes	Excellent; large capacitor bank	Excellent	Excellent	Yes	...	Low

4.34

Operation of a line-type modulator into a mismatched load results in residual energy in the PFN at the end of the desired pulse length. If the load is lower than match (i.e., if the load impedance is lower than the PFN impedance as seen through the pulse transformer), the energy remains as a voltage of reverse polarity on the PFN. Within limits, this allows additional time for the switch device to recover, but an inverse clipper diode (CR2 in Fig. 4.15) is required to discharge this energy so that the charging voltage on the next pulse will not be affected.

A well-designed clipper circuit[59,63,64] will prevent the charging voltage from rising more than a few percent on the next charging cycle even when the load arcs and presents a short circuit to the modulator. Observation of the peak charging voltage[65] while a grounding stick is used to simulate arcing of the RF tube will quickly show how effective a particular clipper circuit may be, and all line-type modulators should be subjected to this test. Crossed-field tubes must be allowed to arc occasionally without tripping off the modulator, especially during "burn-in," and modern modulators are usually designed not to trip off unless the arcing continues excessively.

When a line-type modulator is operated into a load impedance higher than match, a train of pulses of exponentially decaying amplitude is theoretically expected, leading to concern that the thyratron will not deionize before the next charging cycle begins. This indeed occurs on a pure resistive load and results in "hangfire" of the thyratron and trip-off of the modulator. However, the presence of the pulse transformer in typical line-type-modulator circuits ensures that the modulator can operate properly into a load higher than match. The buildup of magnetizing current in the pulse transformer continues to discharge the PFN until its voltage reverses (perhaps after several pulse lengths), just as if the load were lower than match.

Since a line-type modulator ordinarily runs at or near match, moderate variations in load impedance can be analyzed on the basis that constant power will be delivered by the modulator. This is true as long as the PFN charging voltage is constant, which depends on having an effective clipper circuit. Similarly, a 1 percent increase in peak charging voltage will produce a 2 percent increase in power delivered to the load, regardless of its dynamic impedance (Table 4.2).

Various line-type-modulator arrangements using more than one PFN are feasible in order to produce an output pulse at a different impedance level than that switched by the thyratron.[59,66] This may offer advantages in certain cases, but in general these techniques increase the PFN cost and make it harder to achieve good pulse shape. Since in these cases the voltage on the PFN has both polarities during normal operation, it is awkward or impossible to achieve effective clipper-circuit action. The result is that most multiple-network modulators tend to have multiple postpulses (extraneous pulse outputs following the desired pulse output) due to multiple internal reflections of the residual energy in the several PFNs. For these reasons multiple-PFN modulators are relatively uncommon in radar usage.

Throughout the history of line-type modulators, high-power applications have grown faster than switching devices, so two or more of the largest devices have often been used in series or parallel to handle higher peak or average power.[61,67]

Active-Switch Modulators. There is such a variety of active-switch pulse modulators that it is useful to categorize them as cathode pulsers, mod-anode pulsers, and grid pulsers. Cathode pulsers must control the full beam power of the RF tube, either directly or through a coupling circuit. Mod-anode pulsers must usually provide a voltage swing equal to the full beam voltage of the tube, but the current required is only that needed to charge and discharge the circuit capacitances at the beginning and end of the pulse, since a mod-anode usually draws very little current during the pulse. Pulsers for gridded RF tubes perform the same kind of task as mod-anode pulsers, but since the term *grid* is used here to describe a high-mu control electrode, the voltage swing required from a grid pulser is far smaller and permits the use of lower-voltage components and techniques.

Before semiconductors became available, these types of modulators were all called *hard-tube modulators* because they used vacuum tubes exclusively. Active-switch modulators require switching devices that can be both turned on and turned off at will, since, unlike the line-type modulator, the switching device controls both the beginning and the end of the pulse. In active-switch modulators, the pulse is terminated when only a fraction of the stored energy available in the HVPS or modulator has been delivered to the load.

Transistors and gate-turnoff silicon-controlled rectifiers (SCRs) are the only semiconductors inherently capable of being turned off at will, but their power-handling capabilities are much lower than those of conventional SCRs. Therefore, because interest in using semiconductors for high power has been so great, special commutating circuits have been developed to make SCRs turn off at a desired time by means of other SCRs. Although the same techniques could be applied to hydrogen thyratrons and other switching devices normally limited to line-type modulators, it has not been done, probably because a multiplicity of hot cathodes is less palatable than a multiplicity of semiconductor devices.

In general, active-switch modulators provide great flexibility in pulse length and PRF, including mixed pulse lengths and bursts of pulses, since the pulse length is generated by low-level circuits. Maximum-pulse-length capability, within some allowable droop limit, is determined by the size of the energy storage capacitor used (and pulse transformer, if used). Since the energy stored in a capacitor is $CE^2/2$, a 5 percent voltage-droop limit (for example) means that the capacitor must store 10 times the energy delivered to the load in a single pulse. In high-power transmitters with long pulse lengths, the capacitor becomes very large, requiring series and/or parallel combinations of many capacitors, since the maximum practical energy in a single capacitor case is a few thousand joules. The collection of capacitors is then usually known as a capacitor bank (where the joules are stored), and such banks are reasonably common in the range of 10,000 to 1 million J. For example, a transmitter delivering 10 MW of peak power to an RF tube for 100 μs (1000 J per pulse) requires at least a 10,000-J capacitor bank to limit droop to 4 percent, which will produce about a 13 percent droop on the RF output power of a linear-beam tube (unless droop compensation is used, as discussed below). The problem is about 4 times as severe for CFAs because of their low dynamic impedance (Table 4.2).

Special circuits can be used to reduce the effective droop for a given capacitor bank size or to reduce the capacitor bank size for a given allowable droop. Droop can be eliminated (although some ripple is likely to be added) by adding inductors to make the capacitor bank appear as a low-impedance PFN,[68] but this works well only for a fixed pulse length. Droop compensation is less critical and can be accomplished by inserting a parallel *RL* network in series between the capacitor bank and the pulsed load.[69] The drop across the *RL* network is highest at the start of the pulse and gradually decreases during the pulse, which tends to cancel the droop; but some energy is lost in the *RL* network. As an example, a 5 percent droop can be reduced to 2 percent with an efficiency loss of 2 percent.

In general, active-switch modulators are capable of excellent pulse shape if careful attention is paid to stray circuit inductances and capacitances, since there is no lumped-section PFN to limit rise time and to introduce ripple during the pulse length.

Like line-type modulators, active-switch modulators must be designed to tolerate occasional load arcing without damage. Since the RF tube is connected directly to the energy storage capacitor bank in the case of dc-operated CFAs or in the case of a linear-beam tube using mod-anode or grid pulsing, a crowbar (Sec. 4.9) is required to protect the tube from being damaged by the discharge of all that energy when a load arc occurs. With a cathode pulser, the switching device should ordinarily be able to interrupt the load arc current, and firing the crowbar should not be necessary unless the switching device itself arcs.

Cathode Pulsers. The basic types of active-switch cathode pulsers are shown in Fig. 4.16. The triode shown represents any suitable active switch, either a hard tube or a string of solid-state devices, and the linear-beam tube shown as the load represents any cathode-pulsed RF tube, whether crossed-field or linear-beam and whether oscillator or amplifier. Table 4.3 provides a comparison of the features of cathode pulsers.

There are two basic types of cathode pulsers. Most often, the switching device is driven hard enough to bring its voltage as low as possible during the pulse; the device is said to be *bottomed*. This approach minimizes switching-device dissipation and maximizes efficiency, but variations in power supply voltage due to rectification ripple, line-voltage variations, or energy-storage-capacitor droop are passed directly to the load. The alternative is to operate the switching device as a *constant-current* device by limiting its drive signal. The effects of capacitor-bank-voltage droop and of power-supply-voltage variations on the load are then reduced by $(Rp + R_L)/R_L$, where Rp is the dynamic resistance of the device and R_L is the dynamic resistance of the load (Table 4.2).

Tetrode switch tubes are better suited to constant-current service than triodes because of their higher plate resistance. However, in constant-current operation, any fluctuations in grid drive affect load current directly, whereas if the switch tube is bottomed, variations in grid drive have relatively little effect. In constant-current operation, the grid drive may also be programmed to provide even better droop reduction than is provided by constant grid drive; for example, a rising *ramp* on the grid drive can be adjusted to compensate fully for the droop on the energy storage capacitor during the pulse.[68]

As the power ratings of metal-oxide-semiconductor field-effect transistors (MOSFETs) have grown, series strings of these devices have become attractive for use in active-switch modulators at increasingly high power levels, both as bottomed switches and as constant-current switches.[70]

Mod-Anode Pulsers. A basic modulating-anode pulser, sometimes called a floating-deck modulator, is shown in Fig. 4.17.[71–73] The klystron shown represents any linear-beam tube having a mod-anode, and the triodes shown represent any suitable active-switch device. During the pulse the ON tube holds the mod-anode near ground potential to turn on the klystron, and between pulses $R3$ holds the mod-anode negatively biased with respect to the klystron cathode to keep the klystron beam current cut off. The ON tube carries significant current only during the leading edge of the pulse when it is charging up the mod-anode stray capacitance C_s (including all associated stray capacitances, such as that of the ON deck), and the OFF tube similarly carries significant current only to discharge C_s at the end of the pulse. The OFF tube may be thought of as an end-of-pulse *tailbiter*, which is vital in this case since the load on the modulator is primarily capacitive.

Extremely good pulse flatness during the pulse can be obtained with mod-anode pulsers because the klystron is directly across the capacitor bank and because variations in grid drive to the ON tube during the pulse can readily be

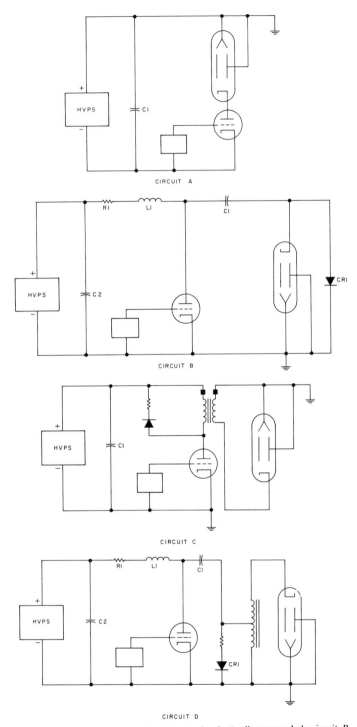

FIG. 4.16 Active-switch cathode pulsers: circuit *A*, direct-coupled; circuit *B*, capacitor-coupled; circuit *C*, transformer-coupled; circuit *D*, capacitor- and transformer-coupled.

4.38

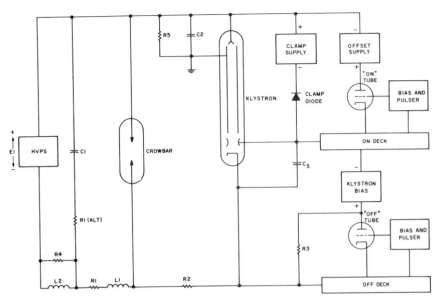

FIG. 4.17 Modulating-anode pulser: direct-coupled.

clamped to produce a flat-topped pulse. Except for capacitor bank size, there is no limit on maximum pulse length, but with very short pulses efficiency drops because of the finite time and energy it takes to turn the mod anode on and off. The ON and OFF tubes can be considerably smaller than a switch tube for cathode-pulsing the same klystron, since they carry less current and carry it only briefly. However, power and triggers must be coupled to two decks floating at high voltage, one of which is at the dc power supply voltage $E1$ and one of which pulses up and down with the mod anode. Since the dissipation in each switch tube is essentially $C_s (E1)^2/2$ times the PRF, it is important to minimize C_s, especially if the PRF is high.

Grid Pulsers. When the RF tube has a high-mu grid, the pulse required for it becomes quite small, comparable with the pulse required for the grid of a switch tube in other kinds of hard-tube modulators, and will not be described here. All the types of modulators previously mentioned may be considered, except that the voltage excursion required for grid pulsing is much less than full beam voltage. Because stray-capacity charging losses are reduced by the square of the mu, grid modulators can more readily handle high-PRF and/or burst-mode operation of radars and are often called on to do so.

4.9 HIGH-VOLTAGE CROWBARS, REGULATORS, AND POWER SUPPLIES

Providing the power needed by RF tubes and their pulse modulators involves a number of considerations that are peculiar to radar transmitters, as described in the following subsections.

Crowbars. Virtually without exception, microwave tubes and their high-voltage switch tubes all arc occasionally, and this effectively places a short circuit across the power supply and/or modulator delivering power to the tube. Since roughly 50 J of energy can usually cause damage to an RF tube (or switch tube) and since the HVPS capacitor bank must often store far more than 50 J, a "crowbar" circuit must be provided to divert the stored energy when an arc occurs. Line-type modulators do not usually require crowbars because their short-circuit load current is inherently limited to about twice normal and because their pulse length is limited to normal duration no matter what the switch device does. Furthermore, the stored energy in the PFN of a line-type modulator is only that needed for a single pulse, whereas in active-switch modulators the stored energy must be many times that much (Sec. 4.8). Crowbars are usually required for all active-switch modulators; they are also required for dc-operated CFAs, which are connected directly across a capacitor bank just like a mod-anode-pulsed linear-beam tube. The term *crowbar* originated[74] because the action is, in concept, equivalent to placing a heavy conductor (like a crowbar) directly across the power supply. These circuits and devices are also called *energy diverters* to describe their function more clearly (although less colorfully). Typical crowbar devices[74–78] are listed and compared in Table 4.4.

In one special case, a crowbar may not be required. Some gridded linear-beam tubes can now be obtained with a *control anode*, which is similar to a modulating anode but is simply left tied to the body (ground) through a resistor. The most likely place for an arc to occur is from cathode to control anode, and in this case it simply results in the tube turning off. The anode is specially made so that an arc there is not likely to cause any other arcing when it suddenly goes to cathode voltage. The arc current is limited to a small enough value by the resistor that the arc will soon extinguish, and normal operation can then continue. Some additional arrangement must still be made to interrupt excessively long pulses if there is a modulator failure, but this can simply be a triggered spark gap to bring the control anode to cathode *as if* there had been an arc; this triggered gap can be far smaller and simpler than a full crowbar.

TABLE 4.4 Crowbar Devices

	Hydrogen thyratron	Mercury pool ignitron	EG & G-triggered gas gap	Energy Systems ball gap	Multigap crowbar	Plasma-triggered vacuum gap
Voltage rating (maximum)	40–125 kV	50 kV	70 kV	No limit	No limit	75 kV
Current and joule ratings	5–10 kA	50 kA	100 kA, 4 kJ	No limit	No limit	70 kA
Firing range	10:1	50:1	3:1	"Infinite"*	"Infinite"	"Infinite"
Self-firing	No	No	No	No	Yes	No
Triggered end	Negative	Negative	Either	Either	Either	Negative
Trigger *V* as fraction of voltage rating	1/10	1/50 (several)	1/3	1/2	1/2	5 kV (all sizes)
Size	Small	Small	Small	Large	Large	Small
Cost	Low	Low	Medium	Medium	High	Medium

*With some inductance in series with load, which limits effectiveness.[77]

Regulators.   RF tubes must usually be operated at a very specific value of peak pulse current, within a tolerance of a few percent or better, in order to ensure proper performance and long life. This is also desirable from the system standpoint, because less power output reduces system performance, while excess output is wasteful of prime power, overheats components, and raises the cooling burden. In crossed-field tubes, other modes may lie near the normal operating point and must be avoided. In linear-beam tubes, gain and bandwidth can change drastically with small changes in beam voltage. In most transmitters, therefore, some kind of regulator is necessary to keep HVPS and modulator output constant in spite of power-line voltage variations (including transients) and PRF and pulse-length changes. A regulator may also be useful or necessary to reduce HVPS ripple to the extremely small values required for satisfactory MTI performance, since passive LC components for this purpose may be impractically large. Table 4.5 shows the variety of regulators[79–81] that have been used for these purposes and compares their features.

High-Voltage Power Supplies.   Besides the usual considerations for high-voltage power supply design, the fact that the HVPS is to be used with a pulsed transmitter imposes some special requirements, as noted below.

The power output of a pulsed transmitter is delivered, obviously, in pulses; yet it is imperative that the power drawn from the power line be essentially smooth and continuous. To meet this requirement, the HVPS filter must often be made much larger than is required just for ripple considerations. In addition, although a multisection filter may appear attractive for high ripple attenuation, a large single-section filter will do a better job of smoothing the prime power drain at low PRFs. A minimum number of filter sections may also be more attractive when lower-frequency ripple components from line unbalance and generator subharmonics are considered.

In modern computer-controlled radar systems, especially multifunction phased array radars, the PRF and duty cycle are seldom constant. This fluctuating drain on the HVPS results in similar fluctuations in prime power drain, since this type of fluctuation lasts far too long (many milliseconds) to be filtered out by any reasonable filter. Power-line drain variations of this type are called *thump* and may be a serious problem if the transmitter power drain is a significant fraction of the power system capacity, as is often true in shipboard or field systems. The effects of thump are an increase in peak power demand from the power line, an increase in rms voltamperes drawn by the transmitter, and a decrease in its power factor. Fluctuating duty requirements must thus be carefully negotiated before a transmitter design is begun.

HVPSs for transmitters must be designed to withstand many short circuits during their lifetime. In line-type modulators, the thyratron (or other switch) may "hang fire"; and in active-switch modulators, load arcing and crowbar firing are common. Short-circuit current from the power supply continues to flow until the prime power circuit is interrupted, and all components in the HVPS must be rated accordingly. Similar current surges occur briefly if the power supply is "snapped on," i.e., is started instantly rather than being "run up" by a Variac, an SCR regulator, or other "soft-start" device. Similarly, the power line must be rated to tolerate these conditions, and the prime power contactor must be rated to interrupt the short-circuit current for thousands of operations.

The advent of high-power solid-state switching devices and high-frequency power-conditioning techniques (inverters, choppers, resonant converters, etc.[82]) has greatly reduced the size and weight and improved the performance of high-

TABLE 4.5 Regulators for HVPSs and Modulators

Type of regulator	Input or output control range	Efficiency	Speed	Accuracy (typical), per cent	Ripple reduction	Notes
Primary control						
Motor-driven Variac*	Full	Very high	Very slow	0.3	No	Moving parts, heavy.
Ferroresonant regulator	Fixed output	Good	Fast	1.0	No	Heavy, fixed frequency, single phase only. Output may be square-wave or sine-wave.
SCR or ignitron phase control	Full	Very high†	Medium	0.1	No	Small, raises ripple, generates some radio-frequency interference.
High-frequency power conditioning (10–100 kHz)	Full	Good	Fast	0.1	Yes	Initial development costly; then can be standard modules. Lightweight, compact. SCRs or transistors. Some radio-frequency interference.
DC high-voltage regulators						
Series regulator	Full	Poor	Very fast	0.01	Yes	Can use transistors or tubes.
Constant-current hard-tube modulator	Full	Poor	Very fast	0.1	Yes	See Sec. 4.8.
For line-type modulators only						
De-Q-ing	30%	Poor	Fast	1.0	Yes	Tubes or SCRs.
Series triode and return diode	Full	Good	Fast	1.0	Yes	Tubes. Clipper circuit can be omitted.
Postcharge regulator	30%	Poor	Fast	0.1	Yes	Tubes or transistors.

*Or equivalent Powerstat, Inductrol, etc.
†But lowers power factor.

voltage power supplies for radar (and other) systems. Vacuum-tube series regulators, required previously for very high stability systems, have been replaced by solid-state power conversion systems, operating typically at 10 to 100 kHz, whose high-frequency operation allows sufficient closed-loop regulator bandwidth to satisfy MTI system requirements even with staggered PRF operation (Sec. 15.9). Similarly, static-interrupter techniques (e.g., force-commutated SCRs) can now be used to interrupt the prime power circuit so rapidly, even in the case of inverter hang-up or crowbar operation, that no surge current above normal is drawn from the power line.

REFERENCES

1. "Radar Handbook," 1st ed., 1970, p. 7-33.
2. Skolnik, M. I.: "Introduction to Radar Systems," 2d ed., McGraw-Hill Book Company, New York, 1980.
3. Chatterjee, R.: "Elements of Microwave Engineering," John Wiley & Sons, New York, 1986.
4. Ewell, G. W.: "Radar Transmitters," McGraw-Hill Book Company, New York, 1981.
5. Edwards, R.: High Power Magnetrons: A State-of-the-Art Report, *Electron. Design News*, September 1965.
6. Edwards, R.: New Magnetron Shifts Frequency Fast, *Electronics*, vol. 37, pp. 76–81, Apr. 6, 1964.
7. Okress, E.: "Crossed-Field Microwave Devices," vols. 1 and 2, Academic Press, New York, 1961.
8. How to Speak Coaxial Magnetrons, *SFD Laboratories, Inc., Bull.* 1770, 2.5*M*, September 1967.
9. Main, J. H.: A Self-Biased Solid-State Diode Despiker, *Proc. Eighth Symp. Hydrogen Thyratrons and Modulators*, pp. 416–422, May 1964, AD454-991.
10. Collins, G. B. (ed.): "Microwave Magnetrons," MIT Radiation Laboratory Series, vol. 6, McGraw-Hill Book Company, New York, 1948.
11. "Radar Handbook," 1st ed., 1970, p. 7-12.
12. Pritchard, W. L.: Long-Line Effect and Pulsed Magnetrons, *IRE Trans.*, vol. MTT-4, pp. 97–110, April 1956.
13. Microwave and Power Tube Reliability in the Field, *Raytheon Microwave and Power Tube Div., Bull.* AEB-14, August 1961.
14. Neal, R. B.: "The Stanford Two-Mile Accelerator," W. A. Benjamin, New York, 1968.
15. Brown, W. C.: Crossed Field Microwave Tubes, *Electronics*, vol. 33, pp. 75–79, Apr. 29, 1960.
16. Gilmour, Jr., A. S.: "Microwave Tubes," Artech House, Norwood, Mass., 1986.
17. Special section in March 1973 issue of *Proc. IEEE*, pp. 279–381: Clampitt, L. L.: High Power Microwave Tubes; Mendel, J. T.: Helix and Coupled-Cavity Traveling-Wave Tubes; Staprans, A., E. W. McCune, and J. A. Reutz: High-Power Linear-Beam Tubes; Skowron, J. F.: The Continuous-Cathode Crossed-Field Amplifier; Yingst, T. E., D. R. Carter, J. A. Eshleman, and J. M. Pawlikowski: High-Power Gridded Tubes.
18. Smith, W. A.: Types of Crossed-Field Microwave Tubes and Their Applications, chap. 23 in Brookner, E.: "Radar Technology," Artech House, Norwood, Mass., 1977.
19. McMaster, G. H.: Current Status of Crossed-Field Devices, *IEEE Int. Electron. Devices Meet.*, December 1988.
20. Introduction to Pulsed Crossed-Field Amplifiers, *SFD Laboratories, Inc., Publ.*, 2.5*M*, June 1967.
21. Handy, R. A.: A dc-Operated CFA Chain for 1 MW, 5 kW at S-Band, *Third Symp. High-Power Microwave Tubes*, Fort Monmouth, N.J., May 1967.
22. Cutoff Electrode Operation in CFA's, *Raytheon Bull.* ENG-9, June 1967.
23. Weil, T. A.: Comparison of CFA's for Pulsed-Radar Transmitters, *Microwave J.*, pp. 51–72, June 1973.
24. Clampitt, L. L.: S-Band Amplifier Chain, *Raytheon Company*, Waltham, Mass., NATO Conf. Microwave Techniques, Paris, Mar. 5, 1962.
25. "Radar Handbook," 1st ed., 1970, sec. 8.5.

26. Luebke, W., and G. Caryotakis: Development of a One Megawatt CW Klystron, *Microwave J.*, pp. 43–47, August 1966.

27. Chodorow, M., and T. Wessel-Berg: A High Efficiency Klystron with Distributed Interaction, *IRE Trans.*, vol. ED-8, pp. 44–55, January 1961.

28. Symons, R. S., B. Arfin, R. E. Boesenberg, P. E. Ferguson, M. Kirshner, and J. R. M. Vaughan: An Experimental Clustered Cavity Klystron, *IEDM Dig.*, pp. 153–156, Dec. 6, 1987.

29. Ruetz, J. A., and W. H. Yocom: High-Power Traveling-Wave Tubes for Radar Systems, *IRE Trans.*, vol. MIL-5, pp. 39–45, April 1961.

30. James, B. G.: Coupled-Cavity TWT Designed for Future mm-Wave Systems, *Microwave Syst. News*, December 1986.

31. Yocom, W. H.: High Power Traveling-Wave Tubes: Their Characteristics and Some Applications, *Microwave J.*, vol. 8, pp. 73–78, July 1965.

32. Kosmahl, H. G.: Modern Multistage Depressed Collectors—A Review, *Proc. IEEE*, vol. 70, pp. 1325–1334, November 1982.

33. LaRue, A. D.: The Twystron Hybrid TWT, *Electron. Design News*, vol. 9, pp. S-17–S-23, September 1964.

34. Prommer, A. J.: New Method for Focusing Klystrons, *Electron. Ind.*, vol. 23, pp. 152–153, 156–158, June 1964.

35. Hechtel, J. R., and A. Mizuhara: A New Type of High Power Microwave Tube: The Electrostatically Focused Klystron Amplifier, *Microwave J.*, vol. 8, pp. 78–83, September 1965.

36. "Radar Handbook," 1st ed., 1970, p. 7–67.

37. O'Shea, R. L.: Radio Frequency Power Divider/Combiner Networks, U.S. Patent 4,583,061, Apr. 15, 1986.

38. Allen, J. L.: Array Radars: A Survey of Their Potential and Limitations, *Microwave J.*, vol. 5, pp. 67–69, May 1962.

39. Hevesh, A. H., and D. J. Harrahy: Effects of Failure on Phased-Array Radar Systems, *IEEE Trans.*, vol. R-15, pp. 22–32, May 1966.

40. Weil, T. A.: Applying the Amplitron and Stabilotron to MTI Radar Systems, *IRE Nat. Conv. Rec.*, vol. 6, pt. 5, pp. 129–130, 1958.

41. Klauder, J. R., A. C. Price, S. Darlington, and W. J. Albesheim: The Theory and Design of Chirp Radars, *Bell Syst. Tech. J.*, vol. 39, pp. 745–808, July 1960.

42. DiFranco, J. W., and W. L. Rubin: Analysis of Signal Processing Distortion in Radar Systems, *IRE Trans.*, vol. MIL-6, pp. 219–227, April 1962.

43. Reed, J.: Long Line Effect in Pulse Compression Radar, *Microwave J.*, vol. 4, pp. 99–100, September 1961.

44. Anderson, L. B., W. A. Janvrin, C. Rowe, and M. E. Schwartz: Determination of Microwave-Tube Transfer Functions, *Proc. IEE (London)*, vol. 114, pp. 873–877, July 1967.

45. Time Domain Reflectometry, *Hewlett-Packard Co.*, *Appl. Note* 62, 1964.

46. Clampitt, L. L., M. Huse, and W. Smith: Measurement of Phase Characteristics of High Power Microwave Tubes, *IEEE Conv. Rec.*, vol. 11, pt. 3, pp. 147–153, 1963.

47. Cook, C. E., and J. Paolillo: A Pulse Compression Predistortion Function for Efficient Sidelobe Reduction in a High-Power Radar, *Proc. IEEE*, vol. 52, pp. 337–389, April 1964.

48. Stover, J. V.: Recent Impacts upon Ground and Shipboard Microwave Transmitter Design, *Power Modulator Symp. Rec.*, IEEE CH2056-0/84, 1984.

49. Seidel, H.: A Microwave Feed-Forward Experiment, *Bell Syst. Tech. J.*, pp. 2879–2916, November 1971.

50. Jurgen, R. K.: Feedforward Correction: A Late-Blooming Design, *IEEE Spectrum*, pp. 41–43, April 1972.

51. Miller, S. N.: The Source of Spectrum Asymmetry in High Power RF Klystrons, *IEEE MIL-E-CON-9*, pp. 18–24, Sept. 22, 1965.

52. Ashley, J. R., and A. D. Sutherland: Microwave Spectrum Conservation by Means of Shaped Pulse Transmitters, *Sperry Electronic Tube Div., Rept. NJ*-2761-0 168, July 1964.

53. Cumming, R. C., M. Perry, and D. H. Preist: Calculated Spectra of Distorted Gaussian Pulses, *Microwave J.*, vol. 8, pp. 70–75, April 1965.

54. Cumming, R. C.: The Influence of Envelope-Dependent Phase Deviation on the Spectra of RF Pulses, *Microwave J.*, vol. 8, pp. 100–105, August 1965.

55. Goldbarb, E. M., and R. C. Cumming: Interference Reduction Techniques for Pulsed Transmitters, *Energy Systems, Final Rept. ECOM*-01444-F, July 1967, ASTIA AD820216.

56. "Radar Handbook," 1st ed., 1970, Chap. 29.

57. Weil, T. A.: Efficient Spectrum Control for Pulsed Radar Transmitters, chap. 27 in Brookner, E.: "Radar Technology," Artech House, Norwood, Mass., 1977.

58. Brookner, E., and Bonneau, R. J.: Spectra of Rounded Trapezoidal Pulses Having AM/ PM Modulation and Its Application to Out-of-Band Radiation, *Microwave J.*, vol. 16, pp. 49–51, December 1983.

59. Glasoe, G. N., and J. V. Lebacqz (eds.): "Pulse Generators," MIT Radiation Laboratory Series, vol. 5, McGraw-Hill Book Company, New York, 1948.

60. Lewis, I. A. D., and F. H. Wells: "Millimicrosecond Pulse Techniques," McGraw-Hill Book Company, New York, 1954.

61. Gilmour, A. S.: Conference Records of the Power Modulator Symposia (1950–1986), reprinted by State University of New York at Buffalo, Amherst, N.Y. 14260.

62. Hill, R. A., and W. R. Olson: Lightweight, High-Power Modulator Uses RSR Switch Device, *Proc. Tenth Modulator Symp.*, pp. 155–163, May 1968.

63. Zinn, M. H.: Analysis of Clipper Diode Conditions, *U.S. Army Signal Corps. Eng. Lab. Mem. M*1649, May 2, 1955.

64. Watrous, W. W., and J. J. McArtney: A Gas Clipper Tube for High Power Radar Modulators, *Proc. Fifth Symp. Hydrogen Thyratrons and Modulators*, May 1958.

65. Weil, T. A.: Pulse Measurement Technique, *Proc. Eighth Symp. Hydrogen Thyratrons and Modulators*, pp. 120–142, May 1964, AD454991.

66. Woodrow, G. V.: The Multiple Line Modulator, *Third Hydrogen Thyratron Symp. Tech. Minutes*, pp. 171–175, May 1953.

67. "Radar Handbook," 1st ed., 1970, p. 7-75.

68. Martinovitch, V. N.: Hard Tube Modulator Techniques That Permit Utilization of Minimum Size Capacitor Banks, *Proc. Seventh Symp. Hydrogen Thyratrons and Modulators*, pp. 415–420, May 1962, AD296-002.

69. Weil, T. A.: Design Charts for Droop-Compensation Networks, *11th Modulator Symp. Rec.*, September 1973, IEEE CHO 773-2 ED.

70. Conrad, G. M.: High-Voltage Constant-Current Solid-State Modulator, *17th Power Modulator Symp.*, June 1986, IEEE 86CH2262-4.

71. Pappas, C.: Some Advances in the Technology of Modulating Anode Pulsers for High-Power Klystrons, *Proc. Seventh Symp. Hydrogen Thyratrons and Modulators*, pp. 361–387, May 1962, AD296002.

72. Grotz, G.: Design Consideration for 180 kV Floating Deck Modulator, *Proc. Ninth Modulator Symp.*, pp. 368–384, May 1966, AD651694.

73. Main, J. H.: A Simplified Floating Deck Modulator, *Proc. Ninth Modulator Symp.*, pp. 385–389, May 1966, AD651-694.

74. Parker, W. N., and M. V. Hoover: High-Speed Electronic Fault Protection for Power Tubes and Their Circuitry, *IRE Conv. Rec.*, vol. 3, pt. 9, pp. 10–15, 1955.

75. Schneider, S., M. H. Zinn, and A. J. Buffa: A Versatile Electronic Crowbar System, *Proc. Seventh Symp. Hydrogen Thyratrons and Modulators*, pp. 482–506, May 1962, AD296002.

76. Morris, A. J., and J. P. Swanson: The High Speed Protection of Microwave Tubes and Systems, *Proc. Seventh Symp. Hydrogen Thyratrons and Modulators*, pp. 436–453, May 1962, AD296002.

77. Taylor, G. W., and S. Schneider: Inductance Effects in Energy-Diverter Circuits, *Proc. Eighth Symp. Hydrogen Thyratrons and Modulators*, pp. 72–81, May 1964, AD454-991.

78. "Radar Handbook," 1st ed., 1970, Sec. 7.16.

79. Smalley, K. M.: A Post-Charge Regulator, *11th Modulator Symposium*, IEEE 73 CHO 773-2 ED, pp. 184–188, 1973.

80. Weil, T. A.: Looking Back, Looking Forward, *17th Power Modulator Symp.*, IEEE 86CH2262-4, pp. 1–8, 1986.

81. "Radar Handbook," 1st ed., 1970, sec. 7.17.

82. Tarter, R. E.: "Principles of Solid-State Power Conversion," Howard W. Sams & Co., Indianapolis, 1985.

CHAPTER 5
SOLID-STATE TRANSMITTERS

Michael T. Borkowski
Raytheon Company

5.1 INTRODUCTION

Solid-state devices have largely superseded vacuum tubes in logic and other low-power circuits and even in some very high power applications such as power supplies and power converters below 1 MHz. The only exception seems to be cathode-ray tubes (CRTs), which are less costly than large plasma displays. In radar transmitters, the transition from high-power klystrons, traveling-wave tubes (TWTs), crossed-field amplifiers (CFAs), and magnetrons to solid-state has been more gradual because the power output of individual solid-state devices is quite limited. However, compared with tubes, solid-state devices offer many advantages:

1. No hot cathodes are required; therefore, there is no warmup delay, no wasted heater power, and virtually no limit on operating life.

2. Device operation occurs at much lower voltages; therefore, power supply voltages are on the order of volts rather than kilovolts. This avoids the need for large spacings, oil filling, or encapsulation, thus saving size and weight and leading to higher reliability of the power supplies as well as of the microwave power amplifiers themselves.

3. Transmitters designed with solid-state devices exhibit improved mean time between failures (MTBF) in comparison with tube-type transmitters. Module MTBFs greater than 100,000 h have been measured.

4. No pulse modulator is required. Solid-state microwave devices for radar generally operate Class-C, which is *self-pulsing* as the RF drive is turned on and off.

5. Graceful degradation of system performance occurs when modules fail. This results because a large number of solid-state devices must be combined to provide the power for a radar transmitter, and they are easily combined in ways that degrade gracefully when individual units fail. Overall power output, in decibels, degrades only as $20 \log r$, where r is the ratio of operating to total amplifiers.

6. Extremely wide bandwidth can be realized. While high-power microwave radar tubes can achieve 10 to 20 percent bandwidth, solid-state transmitter modules can achieve up to 50 percent bandwidth or more with good efficiency.

7. Flexibility can be realized for phased array applications. For phased array

systems, an active transceiver module can be associated with every antenna element. RF distribution losses that normally occur in a tube-powered system between a point-source tube amplifier and the face of the array are thus eliminated. In addition, phase shifting for beam steering can be implemented at low power levels on the input feed side of an active array module; this avoids the high-power losses of the phase shifters at the radiating elements and raises overall efficiency. Also, peak RF power levels at any point are relatively low since the outputs are combined only in space. Furthermore, amplitude tapering can be accomplished by turning off or attenuating individual active array amplifiers.

The general replacement of high-power microwave tubes by solid-state devices has proceeded more slowly than was once forecast. With hindsight, the reason for this is that it is usually too costly to use solid-state devices to replace a pulsed RF tube directly while operating at the same peak power and duty cycle. This is true because microwave semiconductor devices have much shorter thermal time constants than RF tubes (milliseconds rather than seconds). The result is that a microwave transistor that is capable of perhaps 50-W average power cannot handle much more than 100 to 200 W of peak power without overheating during the pulse. The short pulse lengths and low duty cycles typical of older tube-type radars would thus make very inefficient use of the average power capabilities of microwave transistors. For example, to replace the old, well-proven 5J26 L-band magnetron that develops 500 W of average RF power at 0.1 percent (typical) duty cycle would require 2500 to 5000 of the 50-W transistors just mentioned. In other words, microwave transistors are much more cost-effective when the required radar system average power can be provided by a lower peak power at a higher duty cycle. As a result, there have been relatively few direct replacements of older low-duty-cycle transmitters by solid-state transmitters; the AN/SPS-40 is an interesting exception to this rule and will be discussed later. For new radar systems, the system designers have been motivated by these considerations to choose as high a duty cycle as possible, both to reduce the peak power required and to permit using solid-state devices at a reasonable cost. With a 10 percent duty cycle, for example, the 500-W average power mentioned earlier in the paragraph could be provided by only 25 to 50 of the 50-W transistors.

The decision to use a high transmitter duty cycle, however, has significant impact on the rest of the radar system. Operation at a high duty cycle generally requires the use of pulse compression to provide the desired unambiguous range coverage together with reasonably small range resolution. Other consequences follow in turn: the wide transmitted pulse used with pulse compression blinds the radar at short ranges, so a "fill-in" pulse must also be transmitted and processed. To prevent points of strong clutter from masking small moving targets, the signal processor must achieve low pulse compression time sidelobes and high clutter cancellation ratio. As a result, it is much easier to design a solid-state transmitter as part of a new system than it is to retrofit one into an old system that usually does not have all these features.

High-power microwave transistors have been developed more quickly at HF through L band than at higher-frequency bands, so the widest use of solid-state transmitters has been at these lower bands, as shown in Table 5.1. Note, also, that solid-state transmitters at UHF and below have generally been much higher in peak and average power than those at L band.

The use of solid-state does not eliminate all the problems of transmitter design, of course. The RF combining networks must be designed with great care and skill to minimize combining losses in order to keep transmitter efficiency

TABLE 5.1   Fielded Solid-State Transmitters

System	Contractor	Fre-quency, MHz	Peak power, kW	Duty cycle	Average power, kW	No. of mod-ules	Peak power per mod-ule, W	Year fielded
ROTHR	Raytheon	5–30	210	CW	210	84	3000	1986
NAVSPASUR*	Raytheon	218	850	CW	850	2666	320	1986
SPS-40*	Westinghouse	400–450	250	1.6%	4	112	2500	1983
PAVE PAWS†	Raytheon	420–450	600	25.0%	150	1792	340	1978
BMEWS*	Raytheon	420–450	850	30.0%	255	2500	340	1986
TPS-59	GE	1200–1400	54	18.0%	9.7	1080	50	1975
TPS-59‡	GE	1200–1400	54	18.0%	9.7	540	100	1982
SEEK IGLOO	GE	1200–1400	29	18.0%	5.2	292	100	1980
MARTELLO*	Marconi	1250–1350	132	3.75%	5	40	3300	1985
RAMP	Raytheon	1250–1350	28	6.8%	1.9	14	2000	1986
SOWRBALL	Westinghouse	1250–1350	30	4.0%	1.2	72	700	1987

*Solid-state replacements of tube-type transmitters.
†Parameters per array face.
‡Upgraded with 100-W peak power modules.

high. Suitable isolation from excessive voltage–standing-wave ratio (VSWR) must be provided to protect the microwave transistors, and their harmonic power output must be properly filtered to meet MIL-STD-469 and other specifications on RF spectrum quality. Because most microwave transistors operate Class-C, no pulse modulators are required, but Class-C operation makes it more difficult to provide controlled shaping of the RF rise and fall time for spectrum control. Also, just as in tube-type transmitters, energy management is still crucial. Each dc power supply must have a capacitor bank large enough to supply the energy drawn by its solid-state modules during an entire pulse, and each power supply must recharge its capacitor bank smoothly between pulses without drawing an excessive current surge from the power line. While the required power supply is generally not a "catalog" power supply, there are plenty of solid-state devices and circuits available to satisfy these requirements.

As a result of unavoidable losses in combining the outputs of many solid-state devices, it is especially tempting to avoid combining before radiating, since combining in space is essentially lossless. For this reason, many solid-state transmitters consist of modules that feed either rows, columns, or single elements of an array antenna. Especially in the last-named case, it is necessary to build the modules (and probably their power supplies) into the array structure. Furthermore, locating the modules at the antenna avoids the losses of long waveguide runs. Nevertheless, there are cases where building that much equipment weight into the antenna is undesirable, which may force designers to stay with conventional combining schemes. One such case is shipboard radars; the antenna is always mounted as high on the ship as possible, where weight must be minimized to maintain roll stability, and where access for maintenance is extremely difficult.[1]

Because of the large number of individual modules in a typical solid-state transmitter, failure of an individual or a few modules has little effect on overall transmitter performance. However, the module outputs add as voltage vectors, so that loss of 2 of 10 modules (or 20 percent of 1000 modules) results in a reduction to 80 percent of voltage output, which is 64 percent of power output; but even this is only a 2-dB reduction (the difference between 64 and 80 percent of the power ends up in the combiner loads or in sidelobes if the combining is in space). As a result of this "graceful degradation," overall reliability of solid-state transmitters is very high even if maintenance is delayed until convenient scheduled periods; however, this advantage should not be abused. Consider a case where 20 percent of 1000 modules are allowed to fail before output power falls below requirements, and assume that maintenance occurs at scheduled 3-month intervals. In this case, module MTBF need only be 22,000 h to provide 90 percent confidence that the transmitter will not "fail" in less than 3 months; but the cost of replacement modules and labor would be very unattractive, since nearly 40 percent of the transmitter would have to be replaced every year. Higher MTBFs are thus essential to ensure that the transmitter is not only available but also affordable. Fortunately, solid-state module reliability has proved to be even better than the MIL-HDBK-217 predictions; AN/FPS-115 (PAVE PAWS), for example, actual transceiver module MTBF, including the receiver transmit/receiver (T/R) switches and phase shifters as well as the power amplifiers, has grown to 141,000 h, which is 2.3 times the predicted value. In fact, MTBF for the output power transistors measures better than 1.1 million h.[2]

5.2 SOLID-STATE MICROWAVE POWER GENERATION

Although the RF power-generating capability of single solid-state devices is small with respect to the overall peak and average power requirements of a radar transmitter, solid-state devices can be used quite effectively. Large peak and average powers can be attained by combining the outputs of hundreds or thousands of identical solid-state amplifiers. The power output level from a particular device is a function of the exact operating frequency and the operating conditions, namely, the pulse width and duty cycle, and within normal operating constraints bipolar transistors can provide power outputs in the 50-W through 500-W range. These devices have been used for successful designs in the UHF through L-band frequency ranges, as noted in Table 5.1. Bipolar devices satisfy system requirements of reliability, electrical performance, packaging, cooling, availability, and maintainability. In fact, these devices offer an attractive alternative to tube operation at the lower frequencies.

At higher frequencies, and especially for microwave phased array applications where a physically small module with transmit and receive functions is necessary, the gallium arsenide field-effect transistor (GaAs FET) and its associated batch-processed monolithic microwave integrated circuitry (MMIC) can be used. GaAs FETs are well established as low-noise devices up to 60 GHz;[3] and, with individual cell-combining techniques,[4] GaAs FETs can be used as power amplifiers in the 1- to 20-GHz range. In general, the attribute that makes GaAs an attractive technology is that the GaAs FET can be fully integrated with the passive circuitry that is necessary to provide the biasing, loading, filtering, and switching functions that are necessary for multistage transceiver module designs. As a result of fundamental device limitations, however, this approach is not envisioned

as a cost-effective alternative for module designs that require power outputs exceeding 25 to 30 W.
For the upper end of the solid-state microwave spectrum, i.e., the millimeter-wave range, the single-port microwave diode can be used as a low-power oscillator. Unfortunately, the power output and efficiency of these devices are in general very low; in fact, the efficiency is significantly lower than that of their tube counterparts. However, CW and pulsed power output are attainable up to 300 GHz. Gunn and IMPATT diodes are the devices that offer the most promise for millimeter-wave solid-state operation.

Brief descriptions of these device types and their associated technologies are given in the following subsections.

Microwave Bipolar Power Transistors. The silicon bipolar power transistor is a common device choice for a solid-state system. At the lower frequencies, especially below 3 GHz, this component provides adequate performance at the lowest cost among competing solid-state technologies. Amplifier design is realizable for frequencies up through S Band, where the tradeoff between device performance and overall system cost begins to reach a point of diminishing returns. The silicon bipolar transistor technology is very mature, and, with the continuing advances in device processing, packaging, and circuit design techniques, manufacturers should be able to continue demonstrating increased levels of power output, bandwidth, and reliability for these transistors. In addition, as a relative figure of merit, the cost per watt of device output power has been decreasing as a result of improvements in processing yields and as a result of increased automated or semiautomated assembly techniques.

Microwave power transistors can be considered complex hybrid circuits and are generally single-chip or multichip devices. For devices with very high power output, several transistor dice are always combined in parallel within a small hermetic ceramic package. In addition, some form of internal impedance prematching circuitry is often included in order to preserve the high intrinsic bandwidth of the semiconductor chip and to make the task of external impedance matching easier. The internal matching also increases the terminal impedances of the packaged device to a level where the component losses of the circuitry external to the transistor become less critical.

The processing and planar layout of these chips is somewhat standardized among manufacturers. Figure 5.1 shows a partial cross section of a typical microwave bipolar power transistor chip. The structure is an NPN silicon device with a vertical diffusion profile; i.e., the collector contact forms the bottom layer of the chip. The P-type base region has been diffused or implanted into the collector, the N-type emitter has been diffused or implanted into the base, and both base and emitter regions are accessible from the top surface of the chip. The collector region consists of an N-doped, low-resistivity epitaxial layer that is grown on a very low resistivity silicon substrate. The characteristics of the epitaxial layer, i.e., thickness and resistivity, can determine the upper limit of performance of the device in terms of ruggedness, efficiency, and saturated power output.

The fundamental limitation on high-frequency transistor performance is the overall collector-to-emitter delay time. If a signal is introduced to either the base or the emitter, four separate regions of attenuation or time delay are encountered: the emitter-base junction capacity charging time, the base transit time, the col-

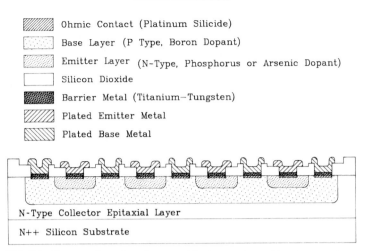

FIG. 5.1 Cross section of a microwave bipolar transistor chip. (*Reprinted with permission from E. D. Ostroff et al., "Solid-State Transmitters," Artech House, Norwood, Mass., 1985.*)

lector depletion-layer transit time, and the collector capacitance-resistance charging time. High-frequency transistor design is concerned with optimizing the physical parameters that contribute to the time-delay components.[5]

For high-power chips, the design challenge is to maintain a uniform high current density over a large emitter area with a minimum temperature rise. High-frequency devices require shallow, narrow, high-resistance base regions under the emitter region. This causes most of the current carried in the device to be crowded along the periphery of the emitter. Thus, in order to maximize the current-handling capability of the device and hence the power output capability of the device, the emitter periphery is maximized. Since the capacitance of the collector-base junction appears as a deleterious parasitic electrical component, the emitter-periphery to base-area ratio, or Ep/Ba, is maximized where possible. Generally, higher-frequency devices exhibit higher Ep/Ba ratios; and to obtain a high Ep/Ba ratio very fine line geometries are required, where the term *geometry* refers to the surface construction details of the transistor.

One limit on the RF power output capability of the transistor is the breakdown voltage of the collector-base junction. Within that limit, the maximum practical level of power output that can be obtained from a single transistor over a given bandwidth is governed by two further limitations: the thermal-dissipation limit of the device and the terminal input or output impedance limit of the device. These latter two limitations are somewhat related by virtue of the physical construction of typical devices.

Active transistor area on the surface of the chip is divided into cells, where the cell size is most often custom-designed for a particular application or range of applications. Pulse width and duty cycle or, as a result, the peak and average dissipated power are the parameters that determine the cell size and arrangement of cells on a chip. As devices become larger and the dissipative heat flux from the top surface of the die to the bottom layer of the transistor increases, the junction temperature increases to the point where the transistor becomes thermally limited.

The ultimate operating junction temperature of the transistor is largely dependent on the transient heating that will be encountered and the layout and area of the individual cells. For devices that are designed to operate for long pulses or CW, an increase in the average power capability of the transistor can be achieved by dividing the active area of a transistor into small, thermally isolated cell areas.

There is a thermal time constant associated with the numerous thermally resistive layers between the transistor junction and the heat sink or cold plate to which the device is attached. This occurs because each layer (silicon, ceramic, transistor flange) not only has a thermal resistance but also exhibits a thermal capacity. Since the overall thermal time constant for a typical L-band power transistor may be on the order of hundreds of microseconds, the tradeoff between peak and average power versus device size can be significant for typical radar pulse widths in the 20- to 1000-μs range. Devices that operate for short-pulse and low-duty-cycle applications, such as DME (distance-measuring equipment), Tacan, and IFF (identification, friend or foe) systems, differ in design from the devices that have been designed for the longer pulse widths and moderate-duty-cycle waveforms that are more typical for surveillance radars. Very high duty cycles or CW operation dictates careful thermal design. An illustration of the thermal-time-constant effect, as it relates to a train of RF pulses, is shown in Fig. 5.2. Table 5.2 illustrates some reported device applications and their general performance characteristics. A photograph of the 115-W UHF transistor used for the PAVE PAWS transmitter is shown in Fig. 5.3, and the schematic, shown in Figure 5.4, may be considered typical for a packaged multichip 100-W L-band transistor.

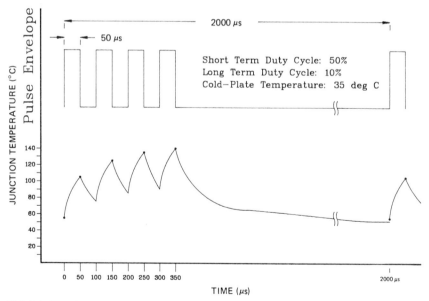

FIG. 5.2   Transient thermal response of a Class-C-biased silicon power transistor for a pulsed RF input.

TABLE 5.2   System Applications for Microwave Power Transistors*

System	Frequency, MHz	Pulse/duty	Transistor performance Peak power, W	Gain, dB	Efficiency, percent
OTH	5–30	CW	130	14.0	60
NAVSPASUR	217	CW	100	9.2	72
AN/SPS-40	400–450	60 µs at 2%	450	8.0	60
PAVE PAWS	420–450	16 ms at 20%	115	8.5	65
BMEWS	420–450	16 ms at 20%	115	8.5	65
AN/TPS-59	1215–1400	2 ms at 20%	55	6.6	52
RAMP	1250–1350	100 µs at 10%	105	7.5	55
MARTELLO S723	1235–1365	150 µs at 4%	275	6.3	40
MATCALS	2700–2900	100 µs at 10%	63	6.5	40
AN/SPS-48	2900–3100	40 µs at 4%	55	5.9	32
AN/TPQ-37	3100–3500	100 µs at 25%	30	5.0	30
HADR	3100–3500	800 µs at 23%	50	5.3	35

*Reprinted with permission from E. D. Ostroff et al., "Solid-State Transmitters," Artech House, Norwood, Mass., 1985.

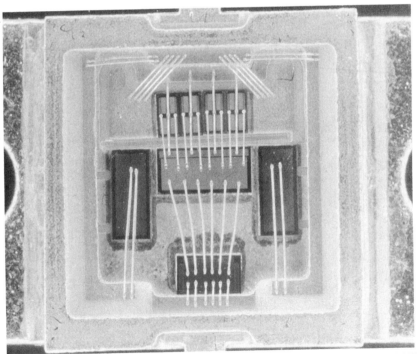

FIG. 5.3   A UHF 115-W power transistor for long pulse and high duty cycle, used in the PAVE PAWS transmitter. (*Photograph courtesy of M/A-COM, PHI.*)

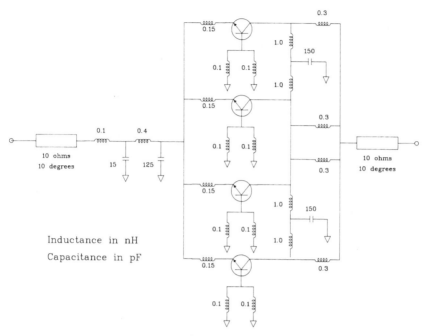

FIG. 5.4 Circuit schematic of an internally input- and output-matched 100-W L-band power transistor.

Microwave Field-Effect Transistors (FETs). Silicon power FETs have demonstrated power output characteristics comparable to the silicon bipolar transistor in the 0.1- to 1.0-GHz frequency range[6]; however, power FETs that are made from gallium arsenide (GaAs) are limited in power output capability, primarily because of the poor thermal conductivity of GaAs and lower typical breakdown voltages; but they are capable of much higher frequency operation than silicon devices.[7] In addition, the GaAs FET can be utilized in a monolithic format, where the remaining passive circuitry of the amplifier is totally integrated on the same substrate with the active device.

GaAs Power FETs. In an FET, the flow of charge carriers between the source and drain electrodes is modulated by one or more gate electrodes, and when the FET is operated as a power device, it may be considered as a simple current switch. Power devices in GaAs are commonly built as metal-semiconductor field-effect transistors (MESFETs), so called because the gate metal is deposited directly onto the semiconductor channel region, forming a Schottky barrier. Power FETs that are fabricated on GaAs exhibit substantially higher frequency performance than similar devices fabricated on silicon because GaAs has a higher electron bulk mobility and a greater maximum electron drift velocity than silicon. In addition, the electron mobility in epitaxial GaAs approaches the bulk value; hence GaAs FETs exhibit lower parasitic series resistances and higher transconductances than silicon FETs with similar geometries. Electrons in GaAs travel at approximately twice the speed that is possible in silicon. In addition, the electron mobility in GaAs is a factor of 3 higher than in silicon. Thus, for comparable geometries, the intrinsic gain for a GaAs device will be substantially higher than for a silicon device.

The cell design and geometry configuration of power GaAs FETs follow design rules similar to those used by designers of silicon bipolar devices. Figure 5.5 shows the cross section of a power GaAs FET that uses air bridge construction and *via holes* to ground the source terminal of the FET. The gate length, seemingly a misnomer because it is shorter than the gate width, is the major parameter that determines device gain and hence operating frequency. Gate width is sometimes also referred to as the periphery. In general, it is desirable to use the largest gate length that permits sufficient gain at the operating frequency. This maximizes processing yield and hence minimizes component cost. Gate lengths for devices in the 1- to 30-GHz range may vary from 2.0 μm to as little as 0.25 μm. While frequency limits can be increased to some degree by decreasing the gate length, increases in power output require greater transistor gate width to support the increased current flow; however, if gate fingers are made too wide, the microwave signal will accumulate a phase shift and will be attenuated while propagating down the gate metal; consequently, the overall gain will be degraded. Total effective increases in gate width can be achieved by paralleling several adjacent gates in order to increase the total channel area per given area. This is similar to increasing emitter periphery per unit base area in the design of bipolar transistors. In addition, the structure must be designed to maintain as high a source-drain breakdown voltage as is possible in order to maximize power output capability. The output power capability of GaAs FETs increases almost linearly with increases in the total gate width, while the power gain decreases slowly with increasing total gate width. The maximum practical total gate width that can be accommodated on a single chip is limited by the following factors:

1. *Yield of the device:* A typical dc processing yield may be 0.995 per 100 μm of gate periphery. A 24-mm chip would therefore have an expected yield of only 30 percent.

2. *Difficulty of impedance matching:* Increases in power output are the result of adjacent channels being connected in parallel. For higher power levels, overall device impedances become lower and lower. For example, the real part of the input impedance for a 24-mm chip would be less than 1 ohm.

3. *Decrease of total power gain:* Uniform current distribution among gate fingers becomes increasingly difficult to manage as the number of paralleled gates increases. Combining inefficiencies result, and the overall device gain decreases.

4. *Physical device size:* The size of a 24-mm periphery chip would be approximately 3000 square mils (75 mils by 40 mils). Larger chip areas decrease the probable assembly yield because assembly difficulty increases with larger chips, greater numbers of wires, and larger packages.

5. *Dissipated power:* The thermal conductivity of GaAs is poor. The dissipated power from larger devices will result in extremely high channel temperatures on the die surface, and reliability will be impacted.

Silicon Power FETs. Silicon bipolar power transistors have been under development far longer than their silicon FET equivalents; consequently, many of the earliest solid-state transmitter designs utilized the bipolar devices. However, the silicon power FET is a viable alternative device type for the amplifier designer. Unlike bipolar transistors, which are minority carrier devices, FETs are

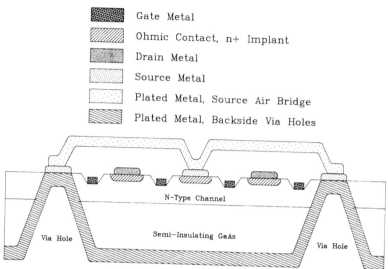

Gate Metal

Ohmic Contact, n+ Implant

Drain Metal

Source Metal

Plated Metal, Source Air Bridge

Plated Metal, Backside Via Holes

FIG. 5.5   Cross section of a GaAs power MESFET chip.

majority carrier devices and exhibit inherent thermal stability. In contrast, bipolar devices are minority carrier devices, and since minority carriers are thermally generated, bipolar power transistors tend to generate localized hot spots and can become thermally unstable. Resistive ballasting techniques used in both the collector and the emitter leads of a bipolar transistor reduce the intrinsic device gain and efficiency but offset the thermal-instability problem. In the power FET, however, large active cell areas can be combined without using these ballasting techniques and without experiencing the problems of thermal runaway.

In general, there are advantages that silicon FETs exhibit, namely,[8]

1. *Thermal stability:* This results from a negative temperature coefficient of power gain.

2. *Gain control ability:* Pulse shaping can be accomplished by using low power gate bias modulation.

3. *Ease of impedance matching:* This is particularly true for the static induction transistor (SIT), which can operate from dc supply voltages as high as 100 V and hence can provide higher impedance levels than other device types for a given power output level.

Millimeter-Wave Solid-State Power Sources.   Solid-state power in the millimeter-wave frequency range is generated from low-power oscillators or negative-resistance amplifiers. The most promising results have been obtained from IMPATT diodes or Gunn diodes. However, TUNNETT (tunnel injection transit time) devices and BARITT (barrier injection transit time) devices are also used. When an extremely short gate length is used, a MESFET construction can be employed at these frequencies, but fundamental limitations on charge carrier velocities and processing tolerances on physically short gate lengths limit the practical operation of MESFET oscillators to below 50 GHz.

IMPATT diodes have been made from silicon, gallium arsenide (GaAs), or indium phosphide (InP) and operate as millimeter-wave oscillators. Silicon devices offer the most promise because silicon provides the most efficient junction heat removal. Overall performance of the diode depends upon the doping density and the thicknesses of the epitaxial, junction, and interface layers. The level of power output from the devices depends upon whether the device is operated pulsed or CW but can range from 1 W CW at 80 GHz to approximately 20 W pulsed at 80 GHz.[9] In addition to IMPATT diodes, transferred electron devices (more commonly named Gunn diodes) made from GaAs or InP can be used up to about 100 GHz. At the lower part of the millimeter-wave band, CW power levels up to 2 W with 15 percent efficiency and pulsed power levels up to 5 W with 20 percent efficiency have been reported.

5.3 SOLID-STATE MICROWAVE DESIGN

The solution to the radar range equation for most applications invariably requires high peak and average radiated power from the antenna in order to ultimately maintain some minimum signal-to-noise ratio on receive. The impact on the solid-state transmitter designer of the requirement for high radiated power is fundamental: high power must be achieved by combining the outputs of lower-power amplifiers in order to develop the required radiated levels. The amplifier-combining approach generally takes one of two different configurations: space-combined or corporate-combined structures, as shown in Fig. 5.6; however, there are also hybrid approaches in which corporate-combined modules feed

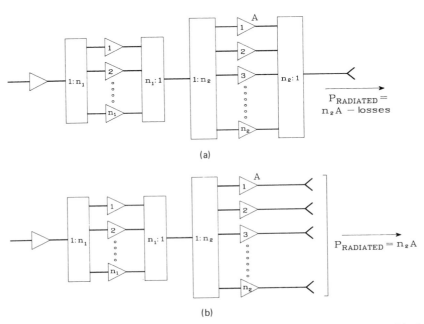

(a)

(b)

FIG. 5.6 Block diagram of (a) a corporate-combined power amplifier and (b) a space-combined power amplifier.

rows of a space-combined array. The phased array configuration is an example of the space-combined approach wherein each radiating antenna element is fed by an amplifier module and the wavefront is formed in space. An example of the corporate-combined design is a parabolic reflector antenna illuminated from a single feedhorn. The power radiated from the horn has been combined from the outputs of many amplifier modules. Solid-state transmitter designs have been built around each of these generic forms, and the components that are required in the implementation of each share similar characteristics and devices.

High-Power Amplifier Design.   In the corporate-combined system, high power levels are generated at a single point by combining the outputs of many low-power amplifiers. The amplifier module is usually partitioned such that the required electrical performance is achieved while the constraints imposed by the mechanical, cooling, maintenance, repair, and reliability disciplines are simultaneously achieved. In general, a module, as shown in Fig. 5.7, consists of a number of identical amplifier stages that are parallel-combined and isolated from one another through the use of microwave combining and isolating techniques. Drive power for this parallel group is obtained from driver or predriver stages, using microwave power dividers. A circulator at the module output port is commonly used to protect the amplifier from the damaging effects of high-load VSWR, most notably from the antenna. Also, ancillary circuitry such as energy storage capacitance, built-in-test (BIT) sensors, or adaptive control components may be included.[10]

Single-Stage Characteristics.   Transistors that operate in the HF through S-band frequency ranges are commonly biased either Class-B or Class-C. Class-C

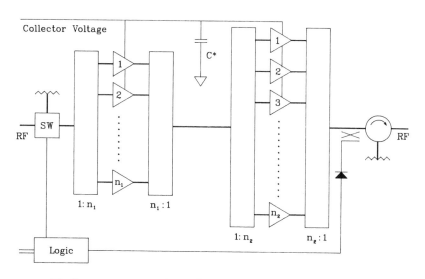

C* : Energy storage capacitance for pulsed amplifiers

$$C = \frac{I \times dt}{dv}$$

where   I = peak current
        dt = pulse length
        dv = pulse voltage droop

FIG. 5.7   Block diagram of a high-power solid-state amplifier module for a corporate-combined transmitter.

operation is the preferred mode since the RF output power of the amplifier is maximized for a given prime power input.[11,12] In general, the base-emitter junction is reverse-biased, and collector current is drawn for less than half of an RF cycle. Collector current is drawn only when the input voltage exceeds the reverse bias across the input and the output voltage is developed across a resonant-tuned load. The net result is high amplifier efficiency. The practical implications of a Class-C-biased amplifier stage are as follows:

   1. No quiescent dc current is drawn while the device is not being driven, such as in the radar receive mode. Hence there is no power dissipation in the amplifier while the transmitter is operating in this mode.

   2. Only one power supply voltage is necessary for the collector terminal of the transistor. The Class-C operation is a *self-bias*, wherein the transistor draws collector current only when the RF voltage swing on the input exceeds the built-in potential of the emitter-base junction. Additional reverse biasing may be introduced as a result of the voltage drop induced by current flow across parasitic resistance of the base or emitter bias return, and in common-base operation this will result in degraded power gain.

   Unlike Class-A linear amplifiers, there are peculiar operating characteristics of the Class-C-biased amplifier that must be recognized in the overall amplifier design. Among these are the RF characteristics of the device as a function of varying RF input drive levels, varying collector voltage supply levels, or varying load impedance. As the RF input drive level of a Class-C-biased device is increased from zero, the dc potential of the reverse-biased base-emitter junction is surpassed and the device begins to draw collector current from the fixed dc supply voltage. The amplifier shows somewhat "linear" transfer characteristics as the drive is increased until the device begins to saturate. Eventually a point of saturated power output capability of the device is attained, and further increases in RF input drive level will actually produce a degraded power output level. At this point the device may also be thermally limited at a device junction. One of the characteristics of this mode of operation is that devices will continue to draw collector current as the amplifier is driven harder; consequently, there exists an optimum operating point with regard to RF drive level. This generally occurs as the transistor is driven approximately 0.5 to 0.75 dB into saturation.

   At the chosen operating point and under the conditions of fixed RF input drive and fixed dc supply voltage, amplifiers of this type also exhibit sensitivities of insertion phase and power output to changes in the input drive level, collector voltage, temperature, and load impedance.[13] Some of the common sensitivities of a Class-C-biased amplifier are given in Table 5.3. Although a nominal 50-ohm load impedance may be assumed, the typical loading effect from the microwave power combiners and the circulator will produce variations in the load impedance presented to the transistor stage that may vary by ±50 percent from the nominal level. Depending on the phase of this mismatch, which can vary among adjacent devices, the port-to-port characteristics of an amplifier can vary dramatically. An important facet of Class-C design is that the response of the single-stage amplifier to these external perturbations must be addressed. Proper selection of the nominal load impedance directly affects the power output, gain, insertion phase, efficiency, and peak junction temperature of the single stage. Changes in the port-to-port insertion phase may result in combining inefficiencies among adjacent amplifiers since the RF power that is lost to the fourth-port termination of a microwave combiner, when adjacent amplifiers are combined, is given by

TABLE 5.3 Performance Sensitivities for a Class-C-Biased Amplifier

Parameter	Value
Amplitude sensitivity to RF drive	0.2–0.9 dB/dB*
Amplitude sensitivity to collector voltage	0.2–04 dB/V†
Phase sensitivity to RF drive	10–13°/dB
Phase sensitivity to collector voltage	0.5–1.5°/V
Phase runout	5–20°‡

*Function of the saturation level.
†Function of the collector voltage level.
‡Function of pulse length and transistor geometry.

$$\text{Power lost (dB down)} = 10 \log [(1 + \cos \theta)/2] \tag{5.1}$$

where θ is the phase difference between amplifiers. Finally, the long-term reliability of the amplifier may be affected by the choice of nominal load impedance since this affects the operating junction temperature of the transistors.

Module Design. In a very simple sense, the design of an amplifier module consists of matching the power transistors to the proper impedance level and then combining the power levels at these impedances. A typical packaged power transistor has low input and output impedances that must be transformed up to higher level, usually 50 ohms. Thus, the typical amplifier design task must address both low-loss and inexpensive reactive impedance-transforming networks that can provide the proper source and load impedances to the transistor. The common medium for providing this function is a microstrip transmission line. Microstrip is a quasi-TEM mode transmission-line medium that requires photolithographically defined lines on a low-loss, high-quality dielectric substrate. Reactive components that are necessary as impedance-matching elements can be approximated in the microstrip format. An inexpensive reactive matching network can be formed by using an interconnected pattern of microstrip elements. Shunt- and series-connected inductive reactances as well as shunt capacitive reactances are the most easily fabricated and most frequently used matching elements up through 10 GHz.

The outputs of identical single-stage power amplifiers are commonly summed by using splitting and combining techniques that also provide isolation between paralleled amplifiers. It is important to note that isolation is necessary between adjacent combined amplifier stages. Should one device fail, the power combiner must provide a constant load impedance to the remaining device. Half the power of the remaining active device, however, will be dissipated in the isolation resistor of the combiner.

A splitting-combining network must also provide serial isolation among amplifier stages as well as parallel isolation. As a Class-C-biased transistor is pulsed, it passes through its cutoff, linear, and saturation regions. Consequently, the input and output impedances are dynamically varying, and the input impedance changes most dramatically. The input impedance match may change from a near-infinite VSWR in the OFF state to a well-matched condition in the ON state. When amplifier stages are serially cascaded without any means of isolating successive stages, the changing input impedance will appear as a varying load impedance to the previous stage. This may very well send the previous stage into oscillation. Figure 5.8 illustrates splitting-combining configurations that provide the necessary isolation by utilizing reflected signal phase cancellation techniques.

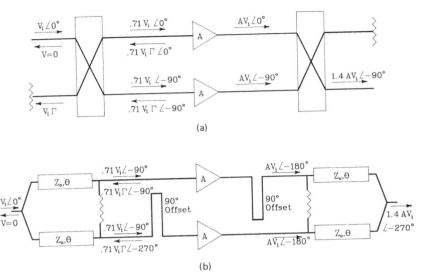

(a)

(b)

FIG. 5.8 Power amplifier combining configurations that provide minimum input port reflected power. (*a*) Quadrature-coupled amplifier pair. (*b*) Split-T amplifier pair with a 90° offset. (*Reprinted with permission from E. D. Ostroff et al., "Solid-State Transmitters," Artech House, Norwood, Mass., 1985.*)

Phased Array Transceiver Module Design. In contrast to the design of a corporate-combined output, where significant losses can accrue in the combining circuitry, the solid-state phased array system uses individual transceiver modules to feed antenna elements on an array face. Consequently, the phase shifting can be done at a low power level, where dissipated power levels in the phase shifters can be much smaller. The transceiver module, regardless of complexity, has four fundamental functions: (1) to provide gain and power output in the transmit mode, (2) to provide gain and low-noise figure in the receive mode, (3) to switch between transmit and receive states, and (4) to provide phase shift for beam steering in the transmit and receive states. A block diagram of a typical transceiver module is shown in Fig. 5.9.

Microwave Monolithic Integrated Circuits. The use of integrated circuits in transceiver designs has enabled bold new module configurations, and hence phased array systems, to be envisioned. Since some of the more complex functions in the generic transceiver block diagram can be fabricated by using MMIC technology, the components that can be realized through the use of this technology can be employed to create system architectures that are difficult if not impossible to design with other, less integrated technologies. The MMIC design approach utilizes active and passive devices that have been manufactured by using a single process. Active and passive circuit elements are formed on a semi-insulating semiconductor substrate, commonly GaAs, through various deposition schemes. The monolithic approach to circuit design inherently offers several advantages:

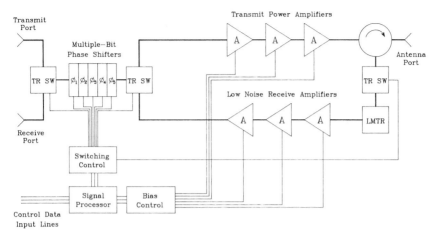

FIG. 5.9   Block diagram of a generic transceiver module for phased array radar.

1. *Low-cost circuitry:* Component assembly is eliminated since complex circuit configurations using both active and passive components are batch-processed on the same substrate.

2. *Increased reliability:* Batch-processed components lead to a reduced number of parts, from the reliability standpoint, and hence to increased reliability.

3. *Increased reproducibility:* Circuitry that is batch-processed or circuits that originate from the same wafer exhibit consistent electrical characteristics from component to component.

4. *Small size and weight:* Integration of active and passive components onto a single chip results in high-density circuitry with multiple functions on a single chip. Overall, the transceiver module can be made much smaller than with discrete components.

However, there are inherent drawbacks to the MMIC approach to component design. The nonrecurring engineering design cost is very high. A typical multistage low-noise amplifier chip design, for example, may easily consume a year's worth of effort before a final design is realized. The cost of wafers and the processing costs in general are very high, such that low-yielding circuit designs may still result in high-cost components. Little circuit trimming can be accommodated in the MMIC approach; consequently, designs must be made tolerant to processing variations or lower performance standards must be accepted for a given design.

The typical processing construction sequence for a GaAs MMIC chip is fairly similar among the GaAs foundries.[14] The active channel region of an FET is delineated by any of several patterning techniques on a semi-insulating GaAs substrate. A combination of deposited dielectric films and metal layers is used to form the passive components and also to interconnect all the elements of the circuit. Standard libraries of circuit elements may include FETs (used as linear amplifiers, low-noise amplifiers, saturating power amplifiers, or switches), resistors, capacitors, inductors, diodes, transmission lines, interconnects, and plated ground vias.

Monolithic circuit elements can be viewed as consuming real estate on a GaAs substrate, and the processing complexity of each step determines the relative yield for individual elements. Although typical processing yields for FET devices may exceed 95 percent per millimeter of gate periphery and greater than 99 percent per picofarad of capacitance, the net yield for a complex circuit may be quite low. For example, a four-stage power amplifier chip that is capable of 3 W of power output at 2 GHz may require more than 9 mm of total gate periphery and may have a total of 75 pF of blocking, bypassing, and matching capacitance. The overall dc yield for this device may be as low as 30 percent when the processing yields for each step are accounted for; high-reliability screening that addresses visually detected imperfections may reduce that yield again by half.

Transceiver Module Performance Characteristics. The partitioning of transceiver module circuit functions onto GaAs chips usually represents a tradeoff among several design issues, and the resultant circuit configurations represent a compromise among the goals of optimum RF performance, high levels of integration, and fabrication yields that are consistent with processing capabilities of GaAs MMICs. Among the single-chip circuit designs that have been reported from UHF through millimeter-wave frequencies are power amplifiers, low-noise amplifiers, wideband amplifiers, phase shifters, attenuators, T/R switches, and other special function designs.

Component Characteristics. Performance characteristics for monolithic circuits vary significantly and are the result of processing variations, layout considerations, yield optimization, or circuit complexity. However, the design tradeoffs have resulted in commonly partitioned circuits. Some of the design criteria or characteristics peculiar to amplifiers and other circuits as follows:

LOW-NOISE AMPLIFIERS. (1) Multiple-stage linear designs require proper device sizing of successive stages in order to maintain low intermodulation distortion products. (2) Circuit losses on the input, before the first stage, degrade the noise figure of the design; therefore some designs utilize off-chip matching. (3) A low noise-figure requires a bias condition that is close to the pinch-off voltage of the FET. Both gain and noise figure are highly dependent on the pinch-off voltage when the FET is biased close to pinch-off. Since the pinch-off voltage can vary significantly for devices from the same wafer, the bias condition must be chosen carefully. Gain and noise figures are usually traded off against repeatable performance.

POWER AMPLIFIERS. (1) Total gate periphery is usually at a premium. For high-power design, the load impedance presented to the final device must be carefully chosen such that power output and efficiency are maximized. (2) Losses in the output circuit of the final stage can significantly reduce power output and efficiency. Off-chip matching may be necessary to maximize power output for a given design. (3) GaAs is a poor thermal conductor. Power FET design that addresses thermal management is required. Adequate heat sinking of the chip is mandatory. (4) For efficient multiple-stage designs, it is necessary that the final stage of the amplifier reach saturation before the preceding stages. This must be addressed in the circuit design.

T/R SWITCHING. (1) For switching applications, the FET design should be chosen such that the ratio of OFF-ON resistance of the FET is kept as high as possible. The channel length largely determines the ON resistance and hence the insertion loss of the device. The tradeoff between short gate length (thus lower processing yield) and insertion loss must be examined. (2) The value of the parasitic drain-source capacitance will affect the OFF-state isolation of the device. This capacitance depends largely on the source-drain spacing of the FET geometry. Critical applications are usually only the front-end switching configurations in a transceiver module, i.e., before the receive low-noise amplifier or after the transmit amplifier.

PHASE SHIFTERS. (1) Phase-shifter designs generally utilize either a switched-line or a loaded-line circuit configuration, using either distributed transmission-line components or lumped-element equivalent circuits, to achieve multiple-bit phase shifting. Switched-line configurations rely on FET switches to switch lengths of transmission line in and out of the circuit and are typically used for higher frequencies where less chip area is needed. Loaded-line configurations use the switched FET parasitics as circuit elements to introduce the necessary phase changes.

The photograph of a representative MMIC chip, shown in Fig. 5.10, is a 12-W power amplifier chip pair that operates at S band. The final stages of this particular design use FETs with a total of 30-mm gate periphery.

Module Characteristics. Performance data exists for modules that use GaAs MMIC technology in the 1- to 10-GHz frequency range. Animated interest in lightweight, adaptive array applications will continue to push this technology. The reported performance characteristics of transceiver modules are generally a combination of multiple MMIC chip configurations and/or multiple-chip configurations with additional hybrid components to augment the performance of the GaAs components. In addition, the complexity of the transmit/receive functions varies among the module configurations. Data for various modules is enumerated in Table 5.4.[15,16,17] An integrated transceiver module that operates in the X-band frequency range is shown in Fig. 5.11.

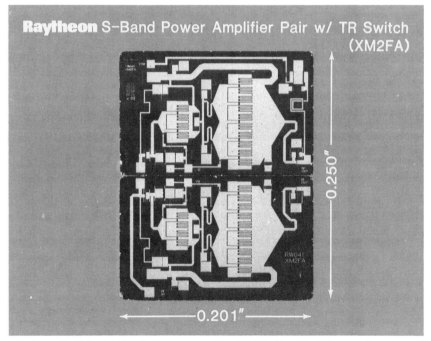

FIG. 5.10   An S-band 12-W GaAs power amplifier MMIC chip pair. (*Photograph courtesy of Raytheon Company.*)

TABLE 5.4   Integrated Transceiver Module Performance Characteristics and Sensitivities

Performance characteristics

	Transmit mode			Receive mode					
Frequency	RF power, W	Gain, dB	Efficiency, percent	Gain, dB	Noise figure, dB	rms phase error Gain, dB	rms phase error Phase, deg	Size, in³	weight, oz
L band*	11	35	30	30	3.0	0.8	5.0	4.0	4.0
S band	10	31	16	25	4.1	0.5	4.0	2.4	2.4
S band	2	23	22	27	3.8	N.A.	4.6	2.9	3.6
S/X band†	2	30	25	...	...	...	...	0.25	N.A.
X band	2.5	30	15	22	4.0	0.6	6.0	0.7	0.7

Performance sensitivities

Parameter	Transmit	Receive
Gain sensitivity to drain voltage	1 dB/V	1 dB/V
Phase sensitivity to drain voltage	4°/V	2°/V

*Includes hybrid stage on transmit output and receive input.
†Transmit amplifier only.

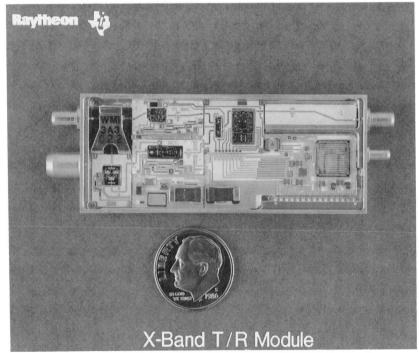

FIG. 5.11   An integrated X-band transceiver module. (*Photograph courtesy of Raytheon Company/Texas Instruments; development work sponsored by RADC, Griffiss Air Force Base.*)

5.4 TRANSMITTER SYSTEM DESIGN

Solid-state amplifiers are usually used in space-combined configurations, corporate-combined configurations, or a hybrid combination of the two. The advantages of solid-state in transmitter applications have already been noted, and some of the performance nuances are noted in the following paragraphs.

Performance Sensitivities

Peak and Average Power Differences. A significant difference between the optimum operating characteristics of solid-state amplifiers and tube amplifiers is that transistors are average-power devices while tubes typically are peak-power devices. The operating characteristics that result from this difference have tremendous impact on the overall system design. The peak power output of microwave transistors is limited by electrical characteristics, while the average power output capability is determined by thermal layout of the transistor. Transistors can be designed to operate for short ($<$ 10 μs), medium (10 to 150 μs), or long ($>$ 150 μs) pulse widths and for duty cycle ranges to CW. Unlike tubes, where the ratio of peak to average power capability can be very high, the peak power output capability of a transistor operated in a short-pulse low-duty mode may only be 2 to 3 times greater than the peak power capability when that device is operated CW. Thus, solid-state operation favors long pulse widths and high duty cycles. For retrofit systems, where high peak power from a short pulse width is required, the acquisition cost of a solid-state transmitter may be prohibitive. The cost of the transmitter may ultimately be reduced if the waveform can be altered to favor higher duty cycle, since the cost in dollars per average watt of the transistor becomes lower as the duty cycle is increased. As long as this cost differential is large enough to offset the added cost in signal processing to accommodate long pulses, a less costly transmitter may be realized.

Amplitude and Phase Sensitivities. Transistor amplifiers that utilize Class-C-biased devices exhibit sensitivity to RF drive level that may degrade the output pulse characteristics. The single-stage amplifier will typically exhibit a very narrow "linear" transfer characteristic; the linear region may exist over only a narrow 1- to 3-dB window. This becomes strikingly critical when several Class-C-biased stages are cascaded in series, as is common in most amplifier configurations. The final tier of output transistors in a serial amplifier chain must be driven into saturation by the preceding stages, and the drive level must be held relatively constant as a function of time and temperature. Since these devices show a narrow operating range, small decreases in the input RF drive level to a multistage amplifier may bring the final tier of devices out of saturation. The net result is an unacceptably degraded output pulse fidelity. One solution has involved a feedback path from a drive-level monitoring point to a variable power supply voltage that maintains the drive level within a specified range for various operating conditions.[10]

In addition to the problems associated with pulse envelope distortions, the phase and amplitude sensitivity of transistor amplifiers to power supply ripple may impact the MTI improvement factor that can be attained. The sensitivities of amplifier stages have already been described. In a multistage amplifier the phase errors due to power supply sensitivity of serially cascaded stages will add, and the limit on MTI improvement factor is

$$I = 20 \log d\theta \qquad (5.2)$$

where $d\theta$ is the magnitude of the insertion phase ripple. The corresponding limit on improvement factor caused by amplifier amplitude instability is

$$I = 20 \log \, (dA/A) \qquad (5.3)$$

where dA and A are the amplitude ripple and the magnitude of the amplitude voltage, respectively.

Time jitter of the RF output pulse envelope can also result from power supply ripple, as a result of Class-C operation of the module, and will also limit the MTI improvement factor. The limitation from this effect is

$$I = 20 \log \, (dt/T) \qquad (5.4)$$

where dt is the time jitter and T is the pulse width. If pulse compression is used, T is still the transmitted-pulse width, not the compressed-pulse width.

In addition, careful design must take into account interactions that can occur as a result of the many cascaded stages of solid-state amplification. These include the following:

1. Phase errors in cascaded stages simply add. However, it may also be possible to arrange them to cancel by proper phasing of power supply ripples for different stages. Similarly, in a stage with N modules in parallel, each with its own high-frequency power-conditioned power supply, the overall phase ripple can usually be assumed to be reduced by a factor equal to the square root of N if the power supply clocks are purposely not synchronized.

2. Because of saturation effects amplitude errors in cascaded stages do not simply add. However, amplitude errors in driving stages will cause drive-induced phase variations in the following stages, as noted above, all of which must be counted.

3. Time jitter in cascaded stages simply adds unless they are arranged to cancel or to be root-sum-squared, as discussed in Par. 1. In addition, amplitude fluctuations in the RF drive will also cause drive-induced jitter, which may even exceed power-supply-ripple-induced jitter, so this factor must be carefully measured.

Spectral Emissions. As a result of Class-C-biased amplifier operation, when a rectangular RF drive pulse is applied to a module, the amplifier will typically show rise and fall times that are on the order of nanoseconds. The output signal spectrum of this pulse shape may not meet spectral emissions requirements, and it may be necessary to slow the rise and fall times. This becomes very difficult when stages are serially cascaded. Because of the highly nonlinear effect described in the preceding subsection, each Class-C stage tends to speed up the rise and fall times of the driving pulse. Consequently, an input pulse shape with slow rise and fall times may be necessary to achieve the desired output-pulse spectral composition.

Control of the rise and fall times is complicated by the necessary use of external bias injection networks. The pulse fall times are generally very fast, on the order of nanoseconds. However, the rise time for a high-power transistor amplifier may be slower, on the order of 100 nanoseconds, and is the result of the reverse biasing that may be encountered by instantaneous current flow in the emitter bias return as the transistor is turning on. This is a design problem for common-base operation, but a common-base configuration is often necessary since it provides more power gain than common-emitter operation at frequencies above approximately 1 GHz.

Power Combining. To achieve very high levels of output power from a single port, combining the outputs of a large number of modules is required, and therefore a complex combining design is necessary. A power combiner coherently adds together the RF output voltages of individual modules and delivers to a single port the sum total of the modules' output power, minus the losses of the combiner. There are several power combiner-splitter configurations available to the module circuit designer, and all display somewhat varied characteristics.[18] In general, the requirements for a power combiner are:

1. The combiner should have low insertion loss, such that transmitter power output and efficiency are not compromised.

2. The combiner should have RF isolation among ports, such that failed modules do not affect the load impedances or combining efficiency for the remaining functioning modules.

3. The combiner should provide a controlled RF impedance to the amplifier modules, such that the amplifier characteristics are not degraded.

4. The combiner reliability should far exceed the reliability of other transmitter components.

5. The dissipated power capability of the power combiner terminations should be sufficient to accommodate any combination of power amplifier failures.

6. The mechanical packaging of the power combiner should allow modules to be repaired easily. The packaging should also provide short, equal phase and low-loss interconnections between the amplifier modules and the combiner.

High-power combiners may be either reactive or hybrid (or equivalent magic-T) designs. In the hybrid design, any imbalance or difference between the phase and amplitude of the voltages that are being combined is directed to a resistive termination. The net result is that a constant load impedance is presented to the amplifier module under all conditions even when an adjacent module in a combining tier has failed. In a reactive combiner design, any imbalance in power or phase between two input signals results in reflected power and increased VSWR to the module driving it. Power amplifier modules that are not protected from high-load VSWR can be damaged by reflected power from the combiner. In addition, frequency-dependent phase and amplitude ripple may result from this configuration.

Typical RF transmission media that are used in the construction of high-power combiners include coaxial transmission lines, microstrip or stripline transmission lines, or waveguide. The choice of transmission medium is generally a function of many parameters, including peak and average power-handling capability, operating frequency and bandwidth, mechanical packaging constraints, and, of course, the overall loss that can be tolerated. More often than not a combiner design utilizes a hierarchy of cascaded designs to sum the outputs of many modules;[10] however, unique configurations that sum many ports to a single port have been built.[19,20,21]

Solid-State Transmitter Design Examples

AN/TPS-59. The AN/TPS-59 (Sec. 20.1) is a solid-state, L-band, long-range, 3D air defense surveillance radar developed for the Marine Corps by the Electronic Systems Division of the General Electric Company.[22] The radar is tactically mobile and consists of a trailer-mounted rotating antenna and two radar shelters. The shelters house the digital signal processor, waveform generator, preprocessor, computer, peripherals, and display consoles. One additional shelter is provided for maintenance aids. Radar surveillance coverage encompasses a volume out to 300 nmi and up to 100,000 ft with a 90 percent probability of detection within 200 nmi for a 1-m^2 fluctuating target. The search volume is scanned mechanically in azimuth through 360° and electronically in elevation through 20°. The rotating 15-ft by 30-ft antenna structure houses 54 row transceivers, each of which drives an RF distribution board that feeds one row of 24 dipole antenna elements. The peak power output capability of the system is 54 kW at an average duty cycle of 18 percent.

The row electronics circuitry feeds each of 54 row feed networks. The row electronics consists of transmit preamplifiers, transmit amplifiers, phase shifters, circulator, and logic control. There are ten 100-W transmit amplifier modules in the final output stage of the row electronics circuitry. Each power amplifier consists of two 55-W silicon bipolar power transistors driven by a smaller 25-W device. The 55-W transistors provide a minimum of nearly 7.0-dB gain from 1215 through 1400 MHz from a 28-V dc power supply. All three devices are beryllia-based transistors that are soldered into a metal-ceramic hermetic enclosure, a photograph of which is shown in Fig. 5.12.

Variants of this system were sold by General Electric to the North Atlantic Treaty Organization (NATO) as the GE-592 and to the U.S. Air Force as the AN/

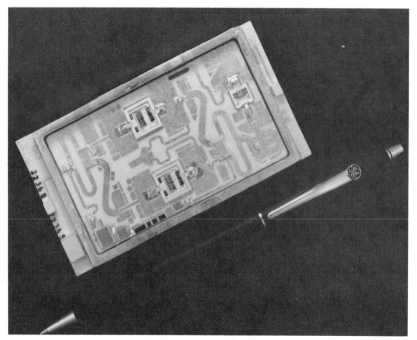

FIG. 5.12 AN/TPS-59 transmitter amplifier module. (*Photograph courtesy of General Electric Company.*)

FPS-117. The AN/FPS-117 radars are operational for the U.S. Air Force in Alaska (SEEK IGLOO Program), northern Canada (North Warning Program), and Berlin. The radar is also in use in Korea, for NATO in Iceland, and for Saudi Arabia as part of the Peace Shield Program. MTBF for these radars has been demonstrated to exceed the 1076-h specification, and more than 750,000 h MTBF has been demonstrated for the 100-W solid-state RF power amplifier modules.

PAVE PAWS. The PAVE PAWS (AN/FPS-115) system is a UHF solid-state active aperture phased array radar that was built for the Electronic Systems Division of the U.S. Air Force by the Equipment Division of the Raytheon Company.[23] The radar is a long-range system with a primary mission to detect and track sea-launched ballistic missiles. The radar uses 1792 active transceiver modules per face to feed dipole antenna elements. Extra elements and a narrow beam are used on receive, and upgrade capability has been included for the future installation of up to 5354 transceiver modules per array face. The peak power output from each face of the baseline system is 600 kW, and the average power output is 150 kW.

Among the 1792 modules per face, groups of 32 transceiver modules are operated as a subarray. In transmit, a high-power array predriver is used to drive 56 subarray driver amplifiers. Each of these power amplifiers provides enough RF drive for all 32 modules in one subarray. In receive, the signal from each of the 56 subarrays is fed into a receive beamforming network.

The transceiver module contains predriver, driver, and final transmit output stages, transmit/receive switching, low-noise amplifiers, limiter, phase shifters, and logic control. The transceiver module block diagram is shown in Fig. 5.13, and a photograph is shown in Fig. 5.14. The transmitter portion of the T/R module contains seven silicon bipolar power transistors, operated Class-C from a +31-V dc power supply. The amplifier is a 1-2-4 configuration, and each of the four final stages delivers 110 W peak for 16-ms pulse widths at duty cycles up to 25 percent. Table 5.5 enumerates some of the salient measured performance

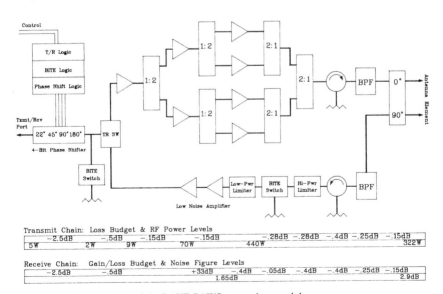

FIG. 5.13 Block diagram of the PAVE PAWS transceiver module.

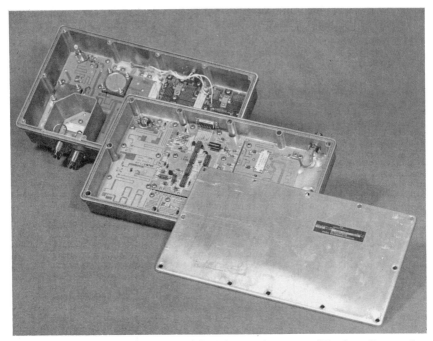

FIG. 5.14   PAVE PAWS transceiver module. (*Photograph courtesy of Raytheon Company.*)

TABLE 5.5   PAVE PAWS Solid-State Transceiver Module Performance*

Parameter	Performance	Specification
Peak power output	330 W	322 W
Output power tracking	0.24 dB rms	0.58 dB rms
Transmit phase tracking	6.7° rms	14° rms
Transmit phase-shifter error	2.52° rms	4.6° rms
Receive phase-shifter error	2.30° rms	4.6° rms
Efficiency	37.9 percent	36 percent
Receive gain	34 dB	27 dB
Receive gain tracking	0.57 dB rms	0.81 dB rms
Receive phase tracking	5.56° rms	10° rms
Noise figure	2.71 dB rms	2.9 dB rms

*Reprinted with permission from D. J. Hoft, Solid-State Transmit/Receive Module for the PAVE PAWS Phase Array Radar, *Microwave J.*, Horizon House, Norwood, Mass., October 1978.

NOTE: Frequency: 420–450 MHz; pulse width: 0.25–16 ms; duty cycle: 0–25 percent.

characteristics of the module. More than 180,000 transistors have been built into more than 25,000 modules.

AN/SPS-40.   The AN/SPS-40 was an existing UHF, tube-type, long-range, 2D shipboard search radar system, for which a new solid-state transmitter, built for the Naval Sea Systems Command by the Westinghouse Electric Corporation, replaced the tube transmitter.[10] The existing waveform from the original transmitter was not changed, and the solid-state unit was installed as a direct retrofit. This was not quite as difficult as usual, because the tube-type system already used long pulses and pulse compression, with a duty cycle of nearly 2 percent, which is a lot higher than older 0.1 percent duty cycle systems. Although it may have been desirable to go to a higher duty cycle and lower peak power to make the solid-state retrofit easier, the Navy preferred not to have to modify the rest of the system.

The 250-kW peak power transmitter uses a total of 128 high-power amplifier modules, which, along with power combining, predrivers, drivers, and control circuitry, are housed in three separate cabinets. There are 112 final power output modules arranged in two groups of 56. Each module produces 2500 W peak and 50 W average for a 60-μs pulse width at a 2 percent duty cycle. Drive power for the two banks of final output modules, 17.5 kW, is provided from the combined outputs of 12 more identical modules in the driver group. Predrivers and a redundant preamplifier are used as preceding drive stages.

The power amplifier module consists of 10 identical silicon bipolar power transistors arranged in a 2-driving-8 amplifier configuration to develop more than 2500-W peak power output over the 400- to 450-MHz frequency bandwidth. A photograph of the transmitter module is shown in Fig. 5.15. Each transistor is a 400-W peak-power device that is operated in a balanced push-pull circuit design. By using a push-pull configuration, the circuit designers have alleviated some of the low-impedance-matching problems normally associated with very high power transistors. The RF input drive to the module is 120 W peak and is used to drive two devices. A combined power level of greater than 600 W is split eight ways and drives the eight identical output stages. Losses in the output circulator, final power combining, and the fault detection circuitry reduce the combined power level to 2500 W. Output modules are liquid-cooled for normal operation, but an emergency backup forced-air cooling is provided in the event of a primary-cooling-system failure. The dissipated heat can be tolerated because the system operates at a low duty cycle.

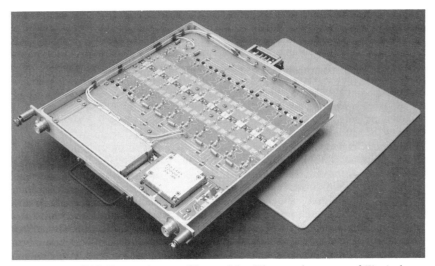

FIG. 5.15   AN/SPS-40 transmitter amplifier module. (*Photograph courtesy of Westinghouse Electric Corporation.*)

The power combining for each output cabinet consists of 56:1 combiners. The reactive power combiner consists of seven groups of 8:1 combiners fabricated in air stripline using 0.5-in ground-plane spacing. The seven outputs are combined by using a single 7:1 air stripline combiner with 1.0-in ground-plane spacing. The 130-kW outputs of the two 56:1 combiners are combined in a single 2:1 isolated hybrid that is manufactured by using 3⅛-in coaxial transmission line. The advertised losses of the 2:1 and 56:1 combiners are 0.1 dB and 0.25 dB, respectively.

Other features of the system include a self-monitoring and self-adjusting driver group power output level. The output level is monitored, and changes to the programmable 24- to 40-V dc power supplies maintain a constant driver group output level as a function of time and temperature.

NAVSPASUR. The solid-state transmitter for the Naval Space Surveillance System (NAVSPASUR) was designed and built by the Equipment Division of the Raytheon Company.[24] The NAVSPASUR is a CW radar, operating at 217 MHz, and is used to provide detection and track data on satellites and other objects as they pass over the continental United States. The solid-state transmitter was procured as a direct replacement for the prior tube-type version. This system produces a very high average power output (850 kW), and with the antenna gain of the system it produces an effective radiated power of over 98 dBW.

There is one main transmitter site at Lake Kickapoo, Texas, with smaller auxiliary sites in Gila River, Arizona, and Jordan Lake, Alabama. The transmitter sites are phased array dipole antennas driven through a coaxial corporate-feed system. The main transmitter site at Lake Kickapoo consists of 2556 antenna elements, each driven by a 300-W solid-state module located directly below the antenna. The most apparent advantages that the solid-state system has over the former tube version are:

1. Much lower dissipation is experienced in the corporate feed because the modules are colocated with the antenna. As a result, the overall site efficiency has been nearly doubled, thus contributing to lower life-cycle costs.

2. A fault-tolerant architecture has led to a system availability of near unity. With a module MTBF of 100,000 h, the maintenance for failed modules can be neglected for nearly 2 years before the transmitter power output degrades by 1 dB. Figure 5.16 illustrates the comparison between the original tube system and the solid-state retrofit.

The solid-state module is a 300-W CW amplifier that uses a 1-driving-4 configuration of silicon bipolar transistors operated common-emitter by using Class-C bias and a 28-V dc power supply. The operating characteristics of the module are delineated in Table 5.6, and Figs. 5.17 and 5.18 are a photograph and a block diagram, respectively. An input power level of 6.2 W is required to drive the module into saturation at the 300-W level. The module dissipates 200 W to the baseplate, which is convection- and radiation-cooled in the outdoor environment. Other module features include automatic fault detection and shutoff, harmonic filtering, and factory-adjustable delay line for module insertion phase matching.

RAMP. The RAMP (Radar Modernization Project) radar system is an L-band system built by the Raytheon Company to replace the earlier primary and secondary surveillance radars used for air traffic control by Canada's Ministry of Transport.[25,26] The primary surveillance radar consists of a rotating reflector, horn-fed by a solid-state transmitter, and redundant receive channels with receiver-exciters and signal processors. The primary surveillance radar operates

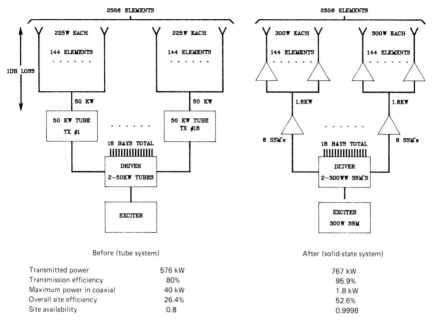

	Before (tube system)	After (solid-state system)
Transmitted power	576 kW	767 kW
Transmission efficiency	80%	95.9%
Maximum power in coaxial	40 kW	1.8 kW
Overall site efficiency	26.4%	52.6%
Site availability	0.8	0.9998

FIG. 5.16 NAVSPASUR transmitter design tradeoffs.

TABLE 5.6 NAVSPASUR Solid-State Power Amplifier Module Performance

Parameter	Performance	Specification
Frequency	216.98 MHz	216.98 MHz
RF power output (CW)	320.0 W	300 W + 0.5 dB
Gain	17.1 dB	16.8 dB
Spurious RF output (near-in)	−75 dBc	−70 dBc maximum
Spurious RF output (far-out)	−85 dBc	− 80 dBc maximum
RF dc efficiency	61.5 percent	58 percent minimum
Input return loss	14 dB	14 dB
Power output similarity (1σ)	0.29 dB	0.5 dB
Phase similarity (1σ)	3.0°	3°
DC voltages	28 V/16.5 A and 8.9 V/0.18 A	28 V/19 A and 8.9 V/0.2 A
Size	21 × 16 × 4.3 in	21 × 26 × 4.3 in
Weight	47 lb	47 lb
Operating ambient temperatures	0–116°F	0–116°F
Pressurization	5 lb/in^2	5 lb/in^2

between 1250 and 1350 MHz with a 25-kW peak power output and provides radar coverage to 80 nmi and to an altitude of 23,000 ft with an 80 percent probability of detection for a 2-m^2 target; with azimuth and range resolution to 2.25° and 0.25 nmi, respectively. The receiver-exciter efficiently utilizes the transmitter solid-state devices with a high-duty-cycle waveform. A double-pulse pair is used in the frequency-agile system, and target returns are processed by a four-pulse moving-target detector. The pulse pair consists of a 1-μs CW pulse that provides cover-

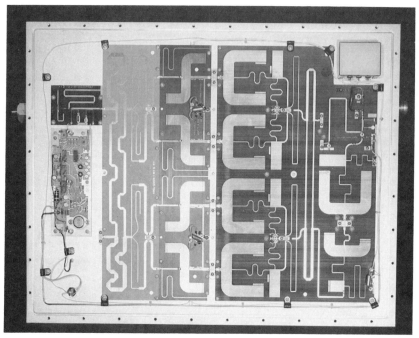

FIG. 5.17   NAVSPASUR transmitter amplifier module. (*Photograph courtesy of Raytheon Company.*)

age to 8 nmi and a 100-μs nonlinear chirp pulse that provides coverage to 80 nmi. The 100-μs pulse is compressed to 1 μs such that high duty cycle is achieved without compromising range resolution.

The transmitter consists of 14 modules, each capable of 2000-W power output, that are combined to produce the greater than 25-kW peak level. Two modules and a 33-V dc power supply make up a single transmitting group. The module consists of a 2-8-32 amplifier configuration of silicon bipolar power transistors. The 32 final output devices and the eight driver devices are 100-W transistors capable of operating up to a 10 percent duty cycle over the 100-MHz bandwidth at collector efficiencies greater than 52 percent. Each module is air-cooled, and the measured efficiency is greater than 25 percent when the module is operating at an 8.2 percent average duty cycle. Module power gain is greater than 16 dB. Figure 5.19 shows a photograph of the 50-lb module and the lineup of power transistors down the center spine of the module. A circulator is used on the output port to protect the 100-W devices from antenna-generated reflections, and control circuitry has been included to switch off modules in the event of cooling-system failure. A 14:1 high-power replicated combiner,[21] built by using a combination of reactive and resistive power-combining techniques in air dielectric stripline, is employed to sum the module outputs to the 25-kW level.

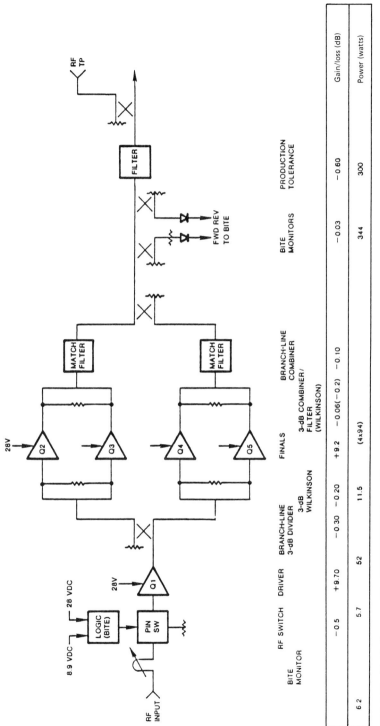

	BITE MONITOR	RF SWITCH	DRIVER	BRANCH-LINE 3-dB DIVIDER		FINALS	3-dB COMBINER/ FILTER (WILKINSON)	BRANCH-LINE COMBINER	BITE MONITORS	PRODUCTION TOLERANCE	Gain/loss (dB)
Gain/loss (dB)		− 0 5	+ 9.70	− 0.30	− 0 20	+ 9.2	− 0.06(− 0.2)	− 0.10	− 0.03	− 0 60	
Power (watts)	6 2	5.7	52		11.5	(4×94)			344	300	

FIG. 5.18   Block diagram of the NAVSPASUR transmitter amplifier module.

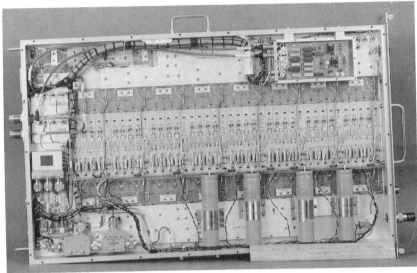

FIG. 5.19   RAMP transmitter amplifier module. (*Photograph courtesy of Raytheon Company.*)

REFERENCES

1. Skolnik, M. I.: The Application of Solid-State RF Transmitters to Navy Radar, *NRL Memo. Rept.* 3074, Naval Research Laboratory, Washington, June 1975.

2. Considine, B.: personal communication, Raytheon Company, January 1988.

3. Watkins, E. T., et al.: A 60 GHz GaAs FET Amplifier, *IEEE MTT-S Int. Microwave Symp. Dig.*, pp. 145–147, 1983.

4. Soares, R., J. Graffeuil, and J. Obregon, "Applications of GaAs MESFET's," Artech House, Norwood, Mass., 1983, pp. 157–207.

5. Cooke, H.: Microwave Transistors: Theory and Design, *Proc. IEEE*, vol. 59, pp. 1163–1181, August 1971.

6. Ostroff, E., M. Borkowski, H. Thomas, and J. Curtis: "Solid-State Radar Transmitters," Artech House, Norwood, Mass., 1985.

7. DiLorenzo, J. V., and D. D. Khandelwal: "GaAs FET Principles and Technology," Artech House, Norwood, Mass., 1982, pp. 201–276.

8. Browne, J.: RF Devices Gain Higher Power Levels, *Microwaves & RF*, p. 148, November 1987.

9. Bhartia, P., and I. J. Bahl: "Millimeter Wave Engineering and Applications," John Wiley & Sons, New York, 1984, pp. 75–152.

10. Lee, K., C. Corson, and G. Mols: A 250 kW Solid-State AN/SPS-40 Radar Transmitter, *Microwave J.*, pp. 93–105, July 1983.

11. Krauss, H. L., C. W. Bostian, and F. H. Raab: "Solid State Radio Engineering," John Wiley & Sons, New York, 1980, pp. 394–431.

12. Pitzalis, O., and R. Gilson: Broadband Microwave Class-C Transistor Amplifiers, *IEEE Trans.*, vol. MTT-21, pp. 660–668, November 1973.

13. Lavallee, L. R.: Two-Phased Transistors Shortchange Class-C Amps, *Microwaves*, pp. 48–54, February 1978.

14. Ferry, D. K.: "Gallium Arsenide Technology," Howard W. Sams & Co., Indianapolis, 1985, pp. 189–300.

15. Laighton, D. G.: private communication, Raytheon Company, January 1988.

16. Pierro, J., and R. Clouse: An Ultraminiature 5–10 GHz, 2-Watt Transmit Module for Active Aperture Application, *IEEE MIT Symp. Dig.*, pp. 941–944, June 1987.

17. Green, C. R., et al.: A 2-Watt GaAs TX/RX Module with Integral Control Circuitry for S-Band Phased Array Radars, *IEEE MTT Symp. Dig.*, pp. 933–936, June 1987.

18. Howe, H.: "Stripline Circuit Design," Artech House, Norwood, Mass., 1974, pp. 77–180.

19. Sanders, B. J.: A 110-Way Parallel Plate RF Divider/Combiner Network & Solid-State Module, *Military Microwaves Conf.*, 1980.

20. Beltran, F.: "Multi-Port Radio Frequency Networks for an Antenna Array," U.S. Patent 4,612,548, Raytheon Company, Sept. 16, 1986.

21. O'Shea, R. L.: "Radio Frequency Power/Divider Combiner Networks," U.S. Patent 4,583,061, Raytheon Company.

22. Perkins, W. B.: personal communication, January 1988.

23. Hoft, D. J.: Solid-State Transmit/Receive Module for the PAVE PAWS Phased Array Radar, *Microwave J.*, pp. 33–35, October 1978.

24. Francoeur, A.: private communication, Raytheon Company, January 1988.

25. Dyck, J. D., and H. R. Ward: RAMP's New Primary Surveillance Radar, *Microwave J.*, p. 105, December 1984.

26. Ward, H. R.: The RAMP PSR, A Solid State Surveillance Radar, *IEE Int. Radar Conf.*, London, October 1987.

CHAPTER 6
REFLECTOR ANTENNAS

Helmut E. Schrank
Gary E. Evans
Daniel Davis
Electronic Systems Group
Westinghouse Electric Corporation

6.1 INTRODUCTION

Role of the Antenna. The basic role of the radar antenna is to provide a transducer between the free-space propagation and the guided-wave propagation of electromagnetic waves. The specific function of the antenna during transmission is to concentrate the radiated energy into a shaped directive beam which illuminates the targets in a desired direction. During reception the antenna collects the energy contained in the reflected target echo signals and delivers it to the receiver. Thus the radar antenna is used to fulfill reciprocal but related roles during its transmit and receive modes. In both of these modes or roles, its *primary purpose is to accurately determine the angular direction of the target.* For this purpose, a highly directive (narrow) beamwidth is needed, not only to achieve angular accuracy but also to resolve targets close to one another. This important feature of a radar antenna is expressed quantitatively in terms not only of the beamwidth but also of *transmit gain* and *effective receiving aperture.* These latter two parameters are proportional to one another and are directly related to the detection range and angular accuracy. Many radars are designed to operate at microwave frequencies, where narrow beamwidths can be achieved with antennas of moderate physical size.

The above functional description of radar antennas implies that a single antenna is used for both transmitting and receiving. Although this holds true for most radar systems, there are exceptions: some monostatic radars use separate antennas for the two functions; and, of course, bistatic radars must, by definition, have separate transmit and receive antennas. In this chapter, emphasis will be on the more commonly used single antenna and, in particular, on the widely used reflector antennas. Phased array antennas are covered in Chap. 7.

Beam Scanning and Target Tracking. Because radar antennas typically have directive beams, coverage of wide angular regions requires that the narrow beam be scanned rapidly and repeatedly over that region to assure detection of targets wherever they may appear. This describes the *search* or *surveillance function* of a radar. Some radar systems are designed to follow a target once it has been detected, and this *tracking function* requires a specially designed antenna different from a surveillance radar antenna. In some radar systems, particularly airborne radars, the antenna is designed to perform both search and track functions.

Height Finding. Most surveillance radars are two-dimensional (2D), measuring only range and azimuth coordinates of targets. In early radar systems, separate height finding antennas with mechanical rocking motion in elevation were used to determine the third coordinate, namely, elevation angle, from which the height of an airborne target could be computed. More recent designs of three-dimensional (3D) radars use a single antenna to measure all three coordinates: for example, an antenna forming a number of stacked beams in elevation in the receive mode and a single broad-coverage elevation beam in the transmit mode. The beams are all equally narrow in azimuth, but the vertically stacked receive beams allow measuring echo amplitudes in two adjacent overlapping beams to determine the elevation angle of the target.

Classification of Antennas. Radar antennas can be classified into two broad categories, *optical antennas* and *array antennas*. The optical category, as the name implies, comprises antennas based on optical principles and includes two subgroups, namely, *reflector antennas* and *lens antennas*. Reflector antennas are still widely used for radar, whereas lens antennas, although still used in some communication and electronic warfare (EW) applications, are no longer used in modern radar systems. For that reason and to reduce the length of this chapter, lens antennas will not be discussed in detail in this edition of the handbook. However, references on lens antennas from the first edition will be kept in the list at the end of the chapter.

6.2 BASIC PRINCIPLES AND PARAMETERS

This section briefly reviews basic antenna principles with emphasis on definitions of terms useful to a radar system designer. In order to select the best type of antenna for a radar system, the system designer should have a clear understanding of the basic performance features of the wide variety of antenna types from which he or she must choose.[1] This includes the choice between reflector antennas, covered in this chapter, and phased arrays, covered in Chap. 7. Another alternative is a reflector fed by a phased array.

Although the emphasis in this chapter is on reflectors, many of the basic principles discussed in this section apply to all antennas. The three basic parameters that must be considered for any antenna are:

- Gain (and effective aperture)
- Radiation pattern (including beamwidth and sidelobes)
- Impedance (voltage–standing-wave ratio, or VSWR)

Other basic considerations are *reciprocity* and *polarization*, which will be briefly discussed in this section.

Reciprocity. Most radar systems employ a single antenna for both transmitting and receiving, and most such antennas are reciprocal devices, which means that their performance parameters (gain, pattern, impedance) are identical for the two functions. This reciprocity principle[2] allows the antenna to be considered either as a transmitting or as a receiving device, whichever is more convenient for the particular discussion. It also allows the antenna to be tested in either role (Sec. 6.10).

Examples of *nonreciprocal* radar antennas are phased arrays using nonreciprocal ferrite components, active arrays with amplifiers in the transmit/receive (T/R) modules, and height finding antennas for 3D (range, azimuth, and elevation) radars. The last-named, typified by the AN/TPS-43[3] radar, uses several overlapping beams stacked in elevation for receiving with a broad elevation beam for transmitting. All beams are equally narrow in the azimuth direction. These nonreciprocal antennas must be tested separately for their transmitting and receiving properties.

Gain, Directivity, and Effective Aperture. The ability of an antenna to concentrate energy in a narrow angular region (a directive beam) is described in terms of antenna gain. Two different but related definitions of antenna gain are *directive gain* and *power gain*. The former is usually called *directivity*, while the latter is often called *gain*. It is important that the distinction between the two be clearly understood.

Directivity (directive gain) is defined as the maximum radiation intensity (watts per steradian) relative to the average intensity, that is,

$$G_D = \frac{\text{maximum radiation intensity}}{\text{average radiation intensity}} = \frac{\text{maximum power per steradian}}{\text{total power radiated}/4\pi} \qquad (6.1)$$

This can also be expressed in terms of the maximum radiated-power density (watts per square meter) at a far-field distance R relative to the average density at that same distance:

$$G_D = \frac{\text{maximum power density}}{\text{total power radiated}/4\pi\ R^2} = \frac{p_{max}}{P_t/4\pi\ R^2} \qquad (6.2)$$

Thus the definition of directivity is simply how much stronger the actual maximum power density is than it would be if the radiated power were distributed isotropically. Note that this definition does not involve any dissipative losses in the antenna but only the concentration of *radiated power*.

Gain (power gain) does involve antenna losses and is defined in terms of *power accepted* by the antenna at its input port P_0 rather than radiated power P Thus gain is given by

$$G = \frac{\text{maximum power density}}{\text{total power accepted}/4\pi\ R^2} = \frac{p_{max}}{P_0/4\pi\ R^2} \qquad (6.3)$$

For realistic (nonideal) antennas, the power radiated P_t is equal to the power accepted P_0 times the *radiation efficiency* factor η of the antenna:

$$P_t = \eta P_0 \qquad (6.4)$$

As an example, if a typical antenna has 1.0 dB dissipative losses, $\eta = 0.79$, and it will radiate 79 percent of its input power. The rest, $(1 - \eta)$ or 21 percent, is converted into heat. For reflector antennas, most losses occur in the transmission line leading to the feed and can be made less than 1 dB.

By comparing Eqs. (6.2) and (6.3) with (6.4), the relation between gain and directivity is simply

$$G = \eta G_D \qquad (6.5)$$

Thus antenna gain is always less than directivity except for ideal lossless antennas, in which case $\eta = 1.0$ and $G = G_D$.

Approximate Directivity—Beamwidth Relations.   An approximate but useful relationship between directivity and antenna beamwidths (see Sec. 2.3) is

$$G_D \simeq \frac{40,000}{B_{az}B_{el}} \tag{6.6}$$

where B_{az} and B_{el} are the principal-plane azimuth and elevation half-power beamwidths (in degrees), respectively. This relationship is equivalent to a 1° by 1° pencil beam having a directivity of 46 dB. From this basic combination, the approximate directivities of other antennas can be quickly derived: for example, a 1° by 2° beam corresponds to a directivity of 43 dB because doubling one beamwidth corresponds to a 3 dB reduction in directivity. Similarly, a 2° by 2° beam has 40 dB, a 1° by 10° beam has 36 dB directivity, and so forth. Each beamwidth change is translated into decibels, and the directivity is adjusted accordingly. This relation does not apply to shaped (e.g., cosecant-squared) beams.

Effective Aperture.   The aperture of an antenna is its physical area projected on a plane perpendicular to the main-beam direction. The concept of effective aperture is useful when considering the antenna in its receiving mode. For an ideal (lossless), uniformly illuminated aperture of area A operating at a wavelength λ, the directive gain is given by

$$G_D = 4\pi A/\lambda^2 \tag{6.7}$$

This represents the *maximum available gain* from an aperture A and implies a perfectly flat phase distribution as well as uniform amplitude.

Typical antennas are not uniformly illuminated but have a tapered illumination (maximum in the center of the aperture and less toward the edges) in order to reduce the sidelobes of the pattern. In this case, the directive gain is less than that given by Eq. (6.7):

$$G_D = 4\pi A_e/\lambda^2 \tag{6.8}$$

where A_e is the effective aperture or *capture area* of the antenna, less than the physical aperture A by a factor ρ_a usually called the *aperture efficiency*.

$$A_e = \rho_a A \tag{6.9}$$

This aperture efficiency would better be called *aperture effectiveness* because it does not involve RF power turned into heat; i.e., it is not a dissipative effect but only a measure of how effectively a given aperture is utilized. An antenna with an aperture efficiency of, say, 50 percent ($\rho_a = 0.5$) has a gain 3 dB below the uniformly illuminated aperture gain but does not dissipate half the power involved. The effective aperture represents a smaller, uniformly illuminated aperture having the same gain as that of the actual, nonuniformly illuminated aperture. It is the area which, when multiplied by the incident power density P_i, gives the power received by the antenna:

$$P_r = P_i A_e \tag{6.10}$$

Radiation Patterns. The distribution of electromagnetic energy in three-dimensional angular space, when plotted on a relative (normalized) basis, is called the *antenna radiation pattern*. This distribution can be plotted in various ways, e.g., polar or rectangular, voltage intensity or power density, or power per unit solid angle (radiation intensity). Figure 6.1 shows a typical radiation pattern for a circular-aperture antenna plotted isometrically in terms of the logarithmic power density (vertical dimension in decibels) versus the azimuth and elevation angles in rectangular coordinates. The *main lobe* (or *main beam*) of the pattern is a *pencil beam* (circular cross section) surrounded by *minor lobes*, commonly referred to as *sidelobes*. The angular scales have their origins at the peak of the main lobe, which is generally the *electrical reference axis* of the antenna.

This axis may or may not coincide with the *mechanical axis* of the antenna, i.e., the axis of symmetry, sometimes called the *boresight axis*. If the two do not coincide, which usually happens unintentionally, the angular difference is referred to as a *boresight error* and must be accounted for in the measurement of target directions.

Figure 6.1*a* shows the three-dimensional nature of all antenna patterns, which require extensive data to plot in this form. This same data can also be plotted in the form of constant-power-level contours, as shown in Fig. 6.1*c*. These contours are the intersections of a series of horizontal planes through the 3D pattern at various levels and can be quite useful in revealing the distribution of power in angular space.

More frequently, however, 2D plots are sufficient and more convenient to measure and plot. For example, if the intersection of the 3D pattern of Fig. 6.1*a* with a vertical plane through the peak of the beam and the zero azimuth angle is taken, a 2D slice or "cut" of the pattern results, as shown in Fig. 6.1*b*. This is called the *principal-plane elevation pattern*. A similar cut by a vertical plane or

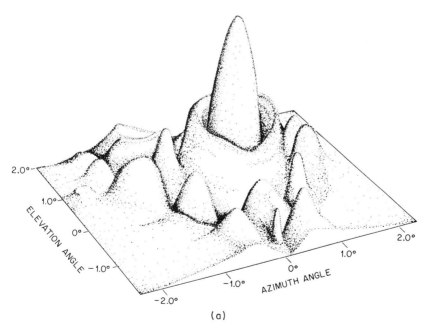

(*a*)

FIG. 6.1 Typical pencil-beam pattern. (*a*) Three-dimensional cartesian plot of complete pattern.

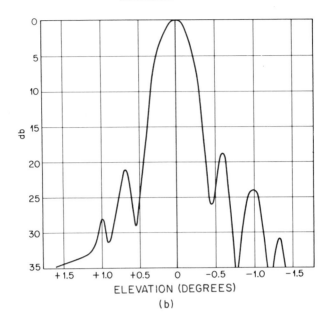

ELEVATION (DEGREES)

(b)

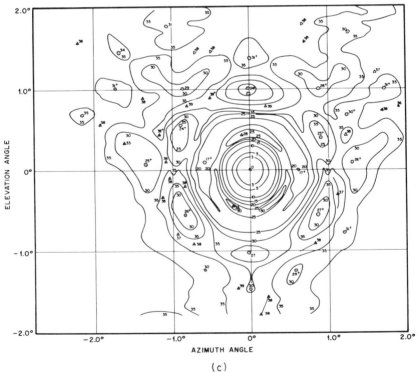

(c)

FIG. 6.1 (*Continued*) (*b*) Principal-plane elevation pattern. (*c*) Contours of constant intensity (isophotes). (*Courtesy of D. D. Howard, Naval Research Laboratory.*)

thogonal to the first one, i.e., containing the peak and the 0° elevation angle, re-sults in the so-called *azimuth pattern*, also a principal-plane cut because it con-tains the peak of the beam as well as one of the angular coordinate axes.

These principal planes are sometimes called *cardinal planes*. All other vertical planes through the beam peak are called *intercardinal planes*. Sometimes pattern cuts in the ±45° intercardinal planes are measured and plotted, but most often it is sufficient to plot only the azimuth and elevation patterns to describe the pat-tern performance of an antenna. In other words, it is sufficient (and much less costly) to *sample* the 3D pattern with two planar cuts containing the beam axis.

The terms *azimuth* and *elevation* imply earth-based reference coordinates, which may not always be applicable, particularly to an airborne or space-based (satellite) system. A more generic pair of principal planes for antennas in general are the so-called *E* and *H* plane of a linearly polarized antenna. Here the *E*-plane pattern is a principal plane containing the direction of the electric (*E*-field) vector of the radiation from the antenna, and the *H* plane is orthogonal to it, therefore containing the magnetic (*H*-field) vector direction. These two principal planes can be independent of earth-oriented directions such as azimuth and elevation and are widely used.

It should be noted that sampling 3D antenna patterns is not limited to planar cuts as described above. Sometimes it is meaningful and convenient (from a measuring-technique viewpoint) to take *conical cuts*, i.e., the intersections of the 3D pattern with cones of various angular widths centered on the electrical (or me-chanical) axis of the antenna.

The typical 2D pattern shown in Fig. 6.1*b* is plotted in rectangular coordi-nates, with the vertical axis in decibels. This is by far the most widely used form of plotting patterns because it provides a wide dynamic range of pattern levels with good visibility of the pattern details. However, other forms of plotting-pattern data are also used, as illustrated in Fig. 6.2. This shows four forms of plotting the same sin *x*/*x* pattern, including (*a*) a polar plot of relative voltage (in-tensity), (*b*) a rectangular plot of voltage, (*c*) a rectangular plot of relative power (density), and (*d*) a rectangular plot of logarithmic power (in decibels). The linear voltage and power scales in *a*, *b*, and *c* leave much to be desired in showing lower-level pattern details, while *d* provides good ''visibility'' of the entire pat-tern. Of course, polar patterns can also be plotted by using decibels in the radial dimension, but lower-level details are compressed near the center of the pattern chart and visibility is very poor. Figure 6.2 shows the reason why *rectangular-decibel pattern plots* are most often preferred.

Beamwidth. One of the main features of an antenna pattern is the *beamwidth* of the main lobe, i.e., its angular extent. However, since the main beam is a con-tinuous function, its width varies from the peak to the nulls (or minima). The most frequently expressed width is the *half-power beamwidth* (HPBW), which occurs at the 0.707-relative-voltage level in Fig. 6.2*a* and *b*, at the 0.5-relative-power level in *c*, and at the 3 dB level in *d*. Sometimes other beamwidths are specified or measured, such as the one-tenth power (10 dB) beamwidth, or the width between nulls, but unless otherwise stated the simple term *beamwidth* im-plies the half-power (3 dB) width. This half-power width is also usually a measure of the *resolution* of an antenna, so that two identical targets at the same range are said to be resolved in angle if separated by the half-power beamwidth.

The beamwidth of an antenna depends on the size of the antenna aperture as well as on the amplitude and phase distribution across the aperture. For a given distribution, the beamwidth (in a particular planar cut) is inversely proportional to the size of the aperture (in that plane) expressed in wavelengths. That is, the half-power beamwidth is given by

$$\text{HPBW} = K/(D/\lambda) = K\lambda/D \qquad (6.11)$$

where D is the aperture dimension, λ is the free-space wavelength, and K is a proportionality constant known as the *beamwidth factor*. Each amplitude distribution (assuming a linear phase distribution) has a corresponding beamwidth factor, expressed either in radians or in degrees.

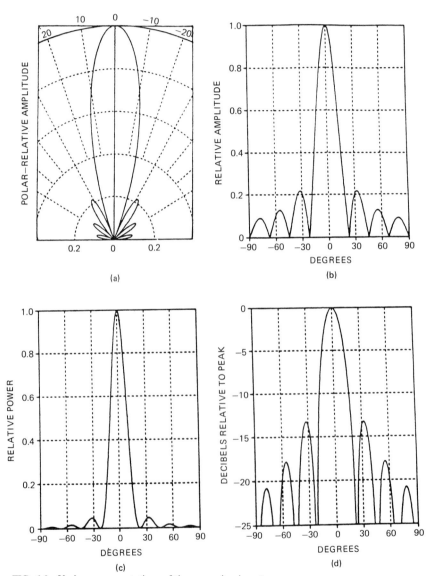

FIG. 6.2 Various representations of the same sin x/x pattern.

Sidelobes. The lobe structure of the antenna radiation pattern outside the major-lobe (main-beam) region usually consists of a large number of *minor lobes*, of which those adjacent to the main beam are *sidelobes*. However, it is common usage to refer to all minor lobes as sidelobes, in which case the adjacent lobes are called *first sidelobes*. The minor lobe approximately 180° from the main lobe is called the *backlobe*.

Sidelobes can be a source of problems for a radar system. In the transmit mode they represent wasted radiated power illuminating directions other than the desired main-beam direction, and in the receive mode they permit energy from undesired directions to enter the system. For example, a radar for detecting low-flying aircraft targets can receive strong ground echoes (*clutter*) through the sidelobes which mask the weaker echoes coming from low radar cross-section targets through the main beam. Also, unintentional interfering signals from friendly sources (electromagnetic interference, or EMI) and/or intentional interference from unfriendly sources (jammers) can enter through the minor lobes. It is therefore often (but not always) desirable to design radar antennas with sidelobes as low as possible (consistent with other considerations) to minimize such problems. (NOTE: There are systems in which the lowest possible sidelobes are not desirable; for example, to minimize main-beam clutter or jamming, it may be better to tolerate higher sidelobes in order to achieve a narrower main-beam null width).

To achieve low sidelobes, antennas must be designed with special tapered amplitude distributions across their apertures. For a required antenna gain this means that a larger antenna aperture is needed. Conversely, for a given size of antenna, lower sidelobes mean less gain and a correspondingly broader beamwidth. The optimum compromise (tradeoff) between sidelobes, gain, and beamwidth is an important consideration for choosing or designing radar antennas. Figure 7.23 of Chap. 7 shows these tradeoff relations for the optimum Taylor amplitude distribution[4,5] being widely used for sidelobe suppression in radar antennas. One set of curves is for rectangular (linear) apertures; the other, for circular Taylor distributions.

The *sidelobe levels* of an antenna pattern can be specified or described in several ways. The most common expression is the *relative sidelobe level*, defined as the peak level of the highest sidelobe relative to the peak level of the main beam. For example, a ''-30 dB sidelobe level'' means that the peak of the highest sidelobe has an intensity (radiated power density) one one-thousandth (10^{-3} or -30 dB) that of the peak of the main beam. Sidelobe levels can also be quantified in terms of their *absolute level* relative to isotropic. In the above example, if the antenna gain were 35 dB, the absolute level of the -30 dB relative sidelobe is $+5$ dBi, i.e., 5 dB above isotropic. For some radar systems, the peak levels of individual sidelobes are not as important as the *average level* of all the sidelobes. This is particularly true for airborne ''down-look'' radars like the Airborne Warning and Control System, or AWACS (E-3A), which require very low (ultralow) average sidelobe levels in order to suppress ground clutter. The average level is a *power average* (sometimes referred to as the *rms level*) formed by integrating the power in all minor lobes outside the main lobe and expressing it in decibels relative to isotropic (dBi). For example, if 90 percent of the power radiated is in the main beam, 10 percent (or 0.1) is in all the sidelobes: this corresponds to an average sidelobe level of -10 dBi. If the main beam contains 99 percent of the radiated power, the average sidelobe level is 0.01, or -20 dBi, etc. Ultralow average sidelobe levels, defined as better than −20 dBi, have been achieved with careful design and manufacturing processes.

One other way to describe sidelobe levels (not often used but sometimes meaningful) is by the *median level*; this is the level such that half of the angular space has sidelobe levels above it and the other half has them below that level.

Polarization. The direction of polarization of an antenna is defined as the direction of the electric-field (*E*-field) vector. Many existing radar antennas are *linearly polarized*, usually either *vertically* or *horizontally*; although these designations imply an earth reference, they are quite common even for airborne or satellite antennas.

Some radars use *circular polarization* in order to detect aircraftlike targets in rain. In that case, the direction of the *E* field varies with time at any fixed observation point, tracing a circular locus once per RF cycle in a fixed plane normal to the direction of propagation. Two senses of circular polarization (CP) are possible, right-hand (RHCP) and left-hand (LHCP). For RHCP, the electric vector appears to rotate in a clockwise direction when viewed as a wave receding from the observation point. LHCP corresponds to counterclockwise rotation. These definitions of RHCP and LHCP can be illustrated with hands, by pointing the thumb in the direction of propagation and curling the fingers in the apparent direction of *E*-vector rotation. By reciprocity, an antenna designed to radiate a particular polarization will also receive the same polarization.

The most general polarization is *elliptical polarization* (EP), which can be thought of as imperfect CP such that the *E* field traces an ellipse instead of a circle. A clear discussion of polarizations can be found in Kraus.[6,7]

Another increasingly important consideration for radar antennas is not only what polarization they radiate (and receive) but how pure their polarization is. For example, a well-designed vertical-polarization antenna may also radiate small amounts of the orthogonal horizontal polarization in some directions (usually off the main-beam axis). Similarly, an antenna designed for RHCP many also radiate some LHCP, which is mathematically orthogonal to RHCP. The desired polarization is referred to as the *main polarization* (COPOL), while the undesired orthogonal polarization is called cross polarization (CROSSPOL). Polarization purity is important in the sidelobe regions as well as in the main-beam region. Some antennas with low COPOL sidelobes may, if not properly designed, have higher CROSSPOL sidelobes, which could cause clutter or jamming problems. A well-designed antenna will have CROSSPOL components at least 20 dB below the COPOL in the main-beam region, and 5 to 10 dB below in the sidelobe regions. Reflecting surfaces near an antenna, such as aircraft wings or a ship superstructure, can affect the polarization purity of the antenna, and their effects should be checked.

6.3 TYPES OF ANTENNAS

Reflector antennas are built in a wide variety of shapes with a corresponding variety of feed systems to illuminate the surface, each suited to its particular application. Figure 6.3 illustrates the most common of these, which are described in some detail in the following subsections. The paraboloid in Fig. 6.3*a* collimates radiation from a feed at the focus into a pencil beam, providing high gain and minimum beamwidth. The parabolic cylinder in Fig. 6.3*b* performs this

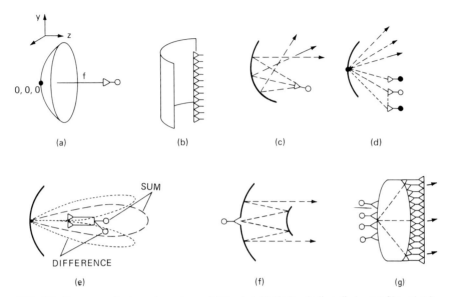

FIG. 6.3 Common reflector antenna types. (*a*) Paraboloid. (*b*) Parabolic cylinder. (*c*) Shaped. (*d*) Stacked beam. (*e*) Monopulse. (*f*) Cassegrain. (*g*) Lens.

collimation in one plane but allows the use of a linear array in the other plane, thereby allowing flexibility in the shaping or steering of the beam in that plane. An alternative method of shaping the beam in one plane is shown in Fig. 6.3*c*, in which the surface itself is no longer a paraboloid. This is a simpler construction, but since only the phase of the wave across the aperture is changed, there is less control over the beam shape than in the parabolic cylinder, whose linear array may be adjusted in amplitude as well.

Very often the radar designer needs multiple beams to provide coverage or to determine angle. Figure 6.3*d* shows that multiple feed locations produce a set of secondary beams at distinct angles. The two limitations on adding feeds are that they become defocused as they necessarily move away from the focal point and that they increasingly block the aperture. An especially common multiple-beam design is the monopulse antenna of Fig. 6.3*e*, used for angle determination on a single pulse, as the name implies. In this instance the second beam is normally a difference beam with its null at the peak of the first beam.

Multiple-reflector systems, typified by the Cassegrain antenna of Fig. 6.3*f*, offer one more degree of flexibility by shaping the primary beam and allowing the feed system to be conveniently located behind the main reflector. The symmetrical arrangement shown has significant blockage, but offset arrangements can readily be envisioned to accomplish more sophisticated goals.

Lenses (Fig. 6.3*g*) are not as popular as they once were, largely because phased arrays are providing many functions that lenses once fulfilled. Primarily they avoid blockage, which can become prohibitive in reflectors with extensive feed systems. A very wide assortment of lens types has been studied.[8–13]

In modern antenna design, combinations and variations of these basic types are widespread, with the goal of minimizing loss and sidelobes while providing the specified beam shapes and locations.

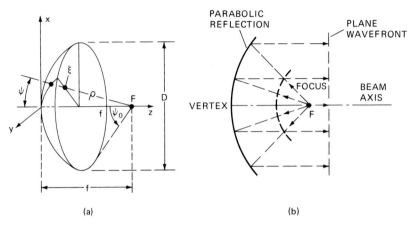

(a)   (b)

FIG. 6.4 Geometrical representation of a paraboloidal reflector. (*a*) Geometry. (*b*) Operation.

Paraboloidal Reflector Antennas. The theory and design of paraboloidal reflector antennas are extensively discussed in the literature.[2–4,14,15] The basic geometry is that of Fig. 6.4*a*, which assumes a parabolic conducting reflector surface of focal length f with a feed at the focus F. It can be shown from geometrical optics considerations that a spherical wave emerging from F and incident on the reflector is transformed after reflection into a plane wave traveling in the positive z direction (Fig. 6.4*b*).

The two coordinate systems that are useful in analysis are shown in Fig. 6.4*a*. In rectangular coordinates (x, y, z), the equation of the paraboloidal surface with a vertex at the origin $(0, 0, 0)$ is

$$z = (x^2 + y^2)/4f \tag{6.12}$$

In spherical coordinates (ρ, ψ, ξ) with the feed at the origin, the equation of the surface becomes

$$\rho = f \sec^2 \frac{\psi}{2} \tag{6.13}$$

This coordinate system is useful for designing the pattern of the feed. For example, the angle subtended by the edge of the reflector at the feed can be found from

$$\tan \frac{\psi_0}{2} = D/4f \tag{6.14}$$

The aperture angle $2\psi_0$ is plotted as a function of f/D in Fig. 6.5. Reflectors with the longer focal lengths, which are flattest and introduce the least distortion of polarization and of off-axis beams, require the narrowest primary beams and therefore the largest feeds. For example, the size of a horn to feed a reflector of $f/D = 1.0$ is approximately 4 times that of a feed for a reflector of $f/D = 0.25$. Most reflectors are chosen to have a focal length between 0.25 and 0.5 times the diameter.

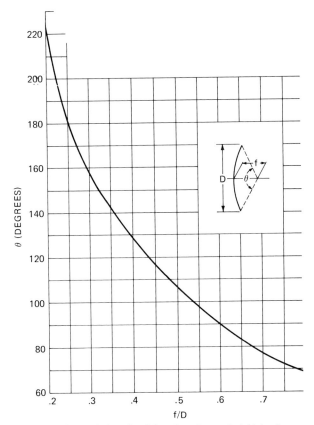

FIG. 6.5 Subtended angle of the edge of a paraboloidal reflector.

When a feed is designed to illuminate a reflector with a particular taper, the distance ρ to the surface must be accounted for, since the power density in the spherical wave falls off as $1/\rho^2$. Thus the level at the edge of the reflector will be lower than at the center by the product of the feed pattern and this "space taper." The latter is given in decibels as

$$\text{Space taper (dB)} = 20 \log \frac{(4f/D)^2}{1 + (4f/D)^2} \qquad (6.15)$$

Equation (6.15) is graphed in Fig. 6.6, showing a significant contribution at the smaller focal lengths. In low-sidelobe applications this amplitude variation may be used in conjunction with the feed pattern to achieve a specific shaping to the skirts of the distribution across the aperture.

Although this reflector is commonly illustrated as round with a central feed point, a variety of reflector outlines are in use, as shown in Fig. 6.7. Often, the

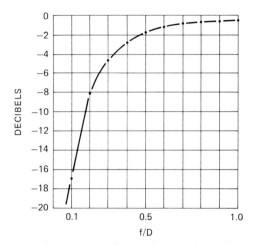

FIG. 6.6 Edge taper due to the spreading of the spherical wave from the feed ("space loss").

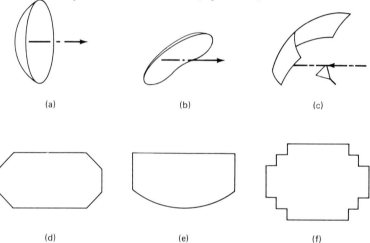

(a)   (b)   (c)

(d)   (e)   (f)

FIG. 6.7 Paraboloidal reflector outlines. (*a*) Round. (*b*) Oblong. (*c*) Offset feed. (*d*) Mitered corner. (*e*) Square corner. (*f*) Stepped corner.

azimuth and elevation beamwidth requirements are quite different, requiring the "orange-peel," or oblong, type of reflector of Fig. 6.7*b*.

As sidelobe levels are reduced and feed blockage becomes intolerable, offset feeds (Fig. 6.7*c*) become necessary. The feed is still at the focal point of the portion of the reflector used even though the focal axis no longer intersects the reflector. Feeds for an offset paraboloid must be aimed beyond the center of the reflector area to account for the larger space taper on the side of the dish away from the feed. The result is an unsymmetrical illumination.

The corners of most paraboloidal reflectors are rounded or mitered as in Fig. 6.7*d* to minimize the area and especially to minimize the torque required to turn the antenna. The deleted areas have low illumination and therefore least contribution to the gain. However, circular and elliptical outlines produce sidelobes at all angles from the principal planes. If low sidelobes are specified away from the principal planes, it may be necessary to maintain square corners, as shown in Fig. 6.7e.

Parabolic reflectors still serve as a basis for many radar antennas in use today, since they provide the maximum available gain and minimum beamwidths with the simplest and smallest feeds.

Parabolic-Cylinder Antenna.[2,16,17] It is quite common that either the elevation or the azimuth beam must be steerable or shaped while the other is not. A parabolic cylindrical reflector fed by a line source can accomplish this flexibility at a modest cost. The line source feed may assume many different forms ranging from a parallel-plate lens to a slotted waveguide to a phased array using standard designs.[2-4]

The parabolic cylinder has application even where both patterns are fixed in shape. The AN/TPS-63 (Fig. 6.8) is one such example in which elevation beam

FIG. 6.8 AN/TPS-63 parabolic-cylinder antenna. (*Courtesy Westinghouse Electric Corporation.*)

shaping must incorporate a steep skirt at the horizon to allow operation at low elevation angles without degradation from ground reflection. A vertical array can produce much sharper skirts than a shaped dish of equal height can, since a shaped dish uses part of its height for high-angle coverage. The array can superimpose high and low beams on a common aperture, thereby benefiting from the full height for each.

The basic parabolic cylinder is shown in Fig. 6.9, in which the reflector surface has the contour

$$z = y^2/4f \tag{6.16}$$

The feed is on the focal line F–F', and a point on the reflector surface is located with respect to the feed center at x and $\rho = f \sec^2 \psi/2$. Many of the guidelines for paraboloids except space taper can be carried over to parabolic cylinders. Since the feed energy diverges on a cylinder instead of a sphere, the power density falls off as ρ rather than ρ^2. Therefore, the space taper of Eq. (6.15) is halved in decibels.

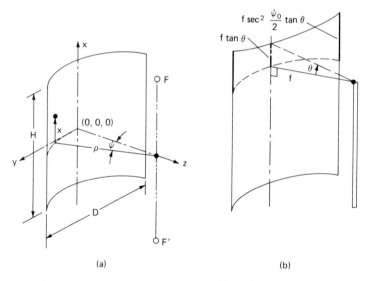

FIG. 6.9 Parabolic cylinder. (*a*) Geometry. (*b*) Extension.

The height or length of the parabolic cylinder must account for the finite beamwidth, shaping, and steering of the linear feed array. As Fig. 6.9 indicates, at angle θ from broadside the primary beam intercepts the reflector at $f \tan \theta$ past the end of the vertex. Since the peak of the primary beam from a steered line source lies on a cone, the corresponding intercepts on the right and left corners of the top of reflector are farther out at $f \sec^2 \psi_0/2 \tan \theta$. For this reason, the corners of a parabolic cylinder are seldom rounded in practice.

Parabolic cylinders suffer from large blockage if they are symmetrical, and they are therefore often built offset. Properly designed, however, a cylinder fed by an offset multiple-element line source can have excellent performance[18] (Fig. 6.10). A variation on this design has the axis of the reflector horizontal, fed with

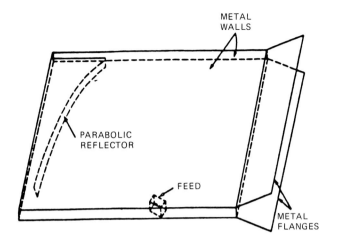

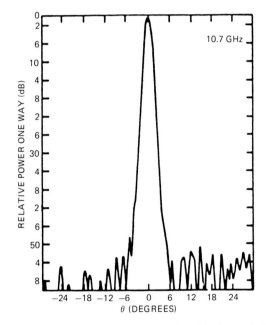

FIG. 6.10 Pillbox structure used to test a low-sidelobe parabolic cylinder and its measured pattern. (*Courtesy Ronald Fante, Rome Air Development Center.*)

a linear array for low-sidelobe azimuth patterns and shaped in height for elevation coverage. It is an economical alternative to a full two-dimensional array.

Shaped Reflectors. Fan beams with a specified shape are required for a variety of reasons. The most common requirement is that the elevation beam provide coverage to a constant altitude. If secondary effects are ignored and if the transmit and receive beams are identical, this can be obtained with a power radiation pattern proportional to $\csc^2\theta$, where θ is the elevation angle.[2,19] In practice, this well-known cosecant-squared pattern has been supplanted by similar but more specific shapes that fit the earth's curvature and account for sensitivity time control (STC).

The simplest way to shape the beam is to shape the reflector, as Fig. 6.11 illustrates. Each portion of the reflector is aimed in a different direction and, to the extent that geometric optics applies, the amplitude at that angle is the integrated sum of the power density from the feed across that portion. Silver[2] describes the procedure to determine the contour for a cosecant-squared beam graphically. However, with modern computers arbitrary beam shapes can be approximated accurately by direct integration of the reflected primary pattern. In so doing, the designer can account for the approximations to whatever accuracy he or she chooses. In particular, the azimuth taper of the primary beam can be included,

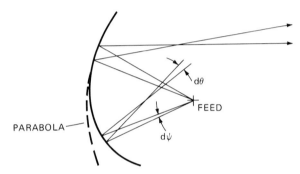

FIG. 6.11   Reflector shaping.

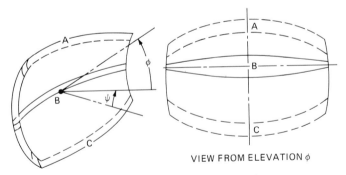

VIEW FROM ELEVATION ϕ

FIG. 6.12   Three-dimensional shaped-reflector design.

and the section of the reflector aimed at elevation θ can be focused in azimuth and have a proper outline when viewed from elevation θ (Fig. 6.12). Without these precautions off-axis sidelobes can be generated by banana-shaped sections.

Most shaped reflectors take advantage of the shaping to place the feed outside the secondary beam. Figure 6.13 shows how blockage can be virtually eliminated even though the feed appears to be opposite the reflector.

The ASR-9 (Fig. 6.14) typifies shaped reflector antennas designed by these procedures. The elevation shaping, azimuth beam skirts, and sidelobes are closely controlled by the use of the computer-aided design process.

A limitation of shaped reflectors is that a large fraction of the aperture is not used in forming the main beam. If the feed pattern is symmetrical and half of the power is directed to wide angles, it follows that the main beam

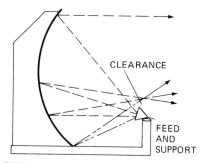

FIG. 6.13 Elimination of blockage.

will use half of the aperture and have double the beamwidth. This corresponds to shaping an array pattern with phase only and may represent a severe problem if sharp pattern skirts are required. It can be avoided with extended feeds.

Multiple Beams and Extended Feeds.[19–21] A feed at the focal point of a parabola forms a beam parallel to the focal axis. Additional feeds displaced from the focal point form additional beams at angles from the axis. This is a powerful capability of the reflector antenna to provide extended coverage with a modest increase in hardware. Each additional beam can have nearly full gain, and adjacent beams can be compared with each other to interpolate angle.

A parabola reflects a spherical wave into a plane wave only when the source is at the focus. With the source off the focus, a phase distortion results that in-

FIG. 6.14 ASR-9 shaped reflector with offset feed and an air traffic control radar beacon system (ATCRBS) array mounted on top. (*Courtesy Westinghouse Electric Corporation.*)

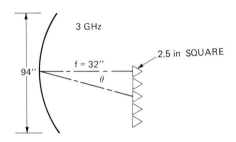

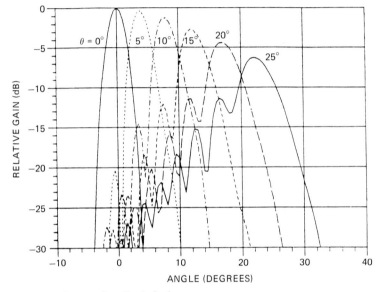

FIG. 6.15   Patterns for off-axis feeds.

creases with the angular displacement in beamwidths and decreases with an increase in the focal length. Figure 6.15 shows the effect of this distortion on the pattern of a typical dish as a feed is moved off axis. A flat dish with a long focal length minimizes the distortions. Progressively illuminating a smaller fraction of the reflector as the feed is displaced accomplishes the same purpose.

Two secondary effects are influential in the design of extended feeds. If an off-axis feed is moved parallel to the focal axis, the region of minimum distortion moves laterally in the reflector. At the same time, if the reflector is a paraboloid of revolution, the focus in the orthogonal plane (usually the azimuth) is altered. For the reflector region directly in front of the displaced feed, it has been found that both planes are improved by moving progressively back from the focal plane. This is clearly illustrated in the side view of the AN/TPS-43 antenna of Fig. 6.16. If that feed is examined carefully, one can also see that off-axis feeds become progressively larger so as to form progressively wider elevation beams that maintain a nearly constant number of beamwidths off axis. This is often made possible by radar coverage requirements having reduced range at wide elevation angles.

FIG. 6.16 AN/TPS-43 multiple-beam antenna. (*Courtesy Westinghouse Electric Corporation.*)

For some purposes the extended feed is not placed about the focal plane at all. If we consider the reflector as a collector of parallel rays from a range of angles and examine the converging ray paths (Fig. 6.17) it is evident that a region can be found that intercepts most of the energy. A feed array in that region driven with suitable phase and amplitude can therefore efficiently form beams at any of the angles. This ability has been used in various systems as a means of forming agile beams over a limited sector. It has also been used as a means of shaping beams and of forming very low sidelobe illumination functions. One such antenna is illustrated in Fig. 6.18.

Monopulse Feeds.[22–25] Monopulse is the most common form of multiple-beam antenna, normally used in tracking systems in which a movable antenna keeps the target near the null and measures the mechanical angle, as opposed to a surveillance system having overlapping beams with angles measured from RF difference data.

Two basic monopulse systems, phase comparison and amplitude comparison, are illustrated in Fig. 6.19. The amplitude system is far more prevalent in radar antennas, using the sum of the two feed outputs to form a high-gain, low-sidelobe beam, and the difference to form a precise, deep null at boresight. The sum beam is used on transmit and on receive to detect the target. The difference port provides angle determination. Usually both azimuth and elevation differences are provided.

If a reflector is illuminated with a group of four feed elements, a conflict arises

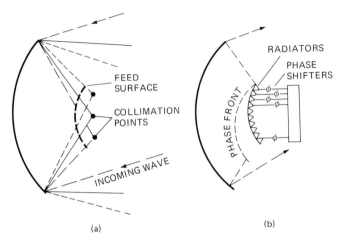

FIG. 6.17 Extended feeds off the focal plane. (*a*) Geometry. (*b*) Feed detail.

FIG. 6.18 Low-sidelobe reflector using an extended feed. (*Courtesy Westinghouse Electric Corporation.*)

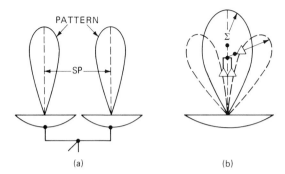

FIG. 6.19 Monopulse antennas. (*a*) Phase. (*b*) Amplitude.

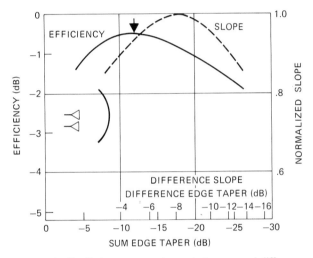

FIG. 6.20 Conflicting taper requirements for sum and difference horn designs (*H* plane illustrated).

between the goals of high sum-beam efficiency and high difference-beam slopes. The former requires a small overall horn size, while the latter requires large individual horns (Fig. 6.20). Numerous methods have been devised to overcome this problem, as well as the associated high-difference sidelobes. In each case the comparator is arranged to use a different set of elements for the sum and difference beams. In some cases this is accomplished with oversized feeds that permit two modes with the sum excitation. Hannan[24] has tabulated results for several configurations, as summarized in Table 6.1.

Multiple-Reflector Antennas.[26–31] Some of the shortcomings of paraboloidal reflectors can be overcome by adding a secondary reflector. The contour of the added reflector determines how the power will be distributed across the primary reflector and thereby gives control over amplitude in addition to phase in the aperture. This can be used to produce very low spillover or to produce a specific low-sidelobe distribution. The secondary reflector may also be used to

TABLE 6.1 Monopulse Feedhorn Performance

Type of horn	H plane Efficiency	H plane Slope	E plane Slope	Sidelobes, dB Sum	Sidelobes, dB Difference	Feed shape
Simple four-horn	0.58	1.2	1.2	19	10	
Two-horn dual-mode	0.75	1.6	1.2	19	10	
Two-horn triple-mode	0.75	1.6	1.2	19	10	
Twelve-horn	0.56	1.7	1.6	19	19	
Four-horn triple-mode	0.75	1.6	1.6	19	19	

relocate the feed close to the source or receiver. By suitable choice of shape, the apparent focal length can be enlarged so that the feed size is convenient, as is sometimes necessary for monopulse operation.

The Cassegrain antenna (Fig. 6.21), derived from telescope designs, is the most common antenna using multiple reflectors. The feed illuminates the hyperboloidal subreflector, which in turn illuminates the paraboloidal main reflector. The feed is placed at one focus of the hyperboloid and the paraboloid focus at the other. A similar antenna is the gregorian, which uses an ellipsoidal subreflector in place of the hyperboloid.

The parameters of the Cassegrain antenna are related by the following expressions:

$$\tan \psi_r/2 = 0.25 D_m/f_m \tag{6.17}$$

$$1/\tan \psi_v + 1/\tan \psi_r = 2f_s/D_s \tag{6.18}$$

$$1 - 1/e = 2 L_r/f_c \tag{6.19}$$

where the eccentricity e of the hyperboloid is given by

$$e = \sin [(\psi_v + \psi_r)/2]/\sin [(\psi_v - \psi_r)/2] \tag{6.20}$$

The equivalent-paraboloid[4,26] concept is a convenient method of analyzing the radiation characteristics in which the same feed is assumed to illuminate a virtual reflector of equal diameter set behind the subreflector. The equation

$$f_e = D_m/(4 \tan \psi_r/2) \tag{6.21}$$

defines the equivalent focal length, and the magnification m is given by

$$m = f_e/f_m = (e + 1)/(e - 1) \tag{6.22}$$

Thus the feed may be designed to produce suitable illumination within subtended angles $\pm\psi_r$ for the longer focal length. Typical aperture efficiency can be better than 50 to 60 percent.

Aperture blocking can be large for symmetrical Cassegrain antennas. It may

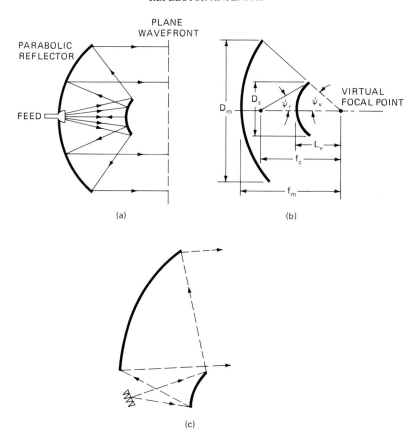

FIG. 6.21 The Cassegrain reflector antenna. (*a*) Schematic diagram. (*b*) Geometry. (*c*) Offset dual reflector.

be minimized by choosing the diameter of the subreflector equal to that of the feed.[26] This occurs when

$$D'_s \simeq \sqrt{2f_m\lambda/k} \qquad (6.23)$$

where k is the ratio of the feed-aperture diameter to its effective blocking diameter. Ordinarily k is slightly less than 1. If the system allows, blocking can be reduced significantly by using a polarization-twist reflector and a subreflector made of parallel wires.[3,26] The subreflector is transparent to the secondary beam with its twisted polarization.

In the general dual-reflector case, blockage can be eliminated by offsetting both the feed and the subreflector (Fig. 3.21*c*). With blockage and supporting struts and spillover virtually eliminated, this is a candidate for very low sidelobes.[32] It can be used in conjunction with an extended feed to provide multiple or steerable beams.[33]

Special-Purpose Reflectors. Several types of antennas are occasionally used for special purposes. One such antenna is the spherical reflector,[34] which can be scanned over very wide angles with a small but fixed phase error known as spherical aberration. The basis of this antenna is that, over small regions, a spherical surface viewed from a point halfway between the center of the circle and the surface is nearly parabolic. If the feed is moved circumferentially at constant radius $R/2$, where R = the radius of the circular reflector surface, the secondary beam can be steered over whatever angular extent the reflector size permits. In fact, 360° of azimuth steering may be accomplished if the feed polarization is tilted 45° and the reflector is formed of conducting strips parallel to the polarization. The reflected wave is polarized at right angles to the strips on the opposite side. This antenna is known as a Helisphere.[35]

If the scanning is in azimuth only, the height dimension of the reflector may be parabolic for perfect elevation focus. This is the parabolic torus,[3,4] which has been used in fixed radar installations.

6.4 FEEDS

Because most radar systems operate at microwave frequencies (L band and higher), feeds for reflector antennas are typically some form of flared waveguide horn. At lower frequencies (L band and lower) dipole feeds are sometimes used, particularly in the form of a linear array of dipoles to feed a parabolic-cylinder reflector. Other feed types used in some cases include waveguide slots, troughs, and open-ended waveguides, but the flared waveguide horns are most widely used.

Paraboloidal reflectors (in the receive mode) convert incoming plane waves into spherical phase fronts centered at the focus. For this reason, feeds must be point-source radiators; i.e., they must radiate spherical phase fronts (in the transmit mode) if the desired directive antenna pattern is to be achieved. Other characteristics that a feed must provide include the proper illumination of the reflector with a prescribed amplitude distribution and minimum spillover and correct polarization with minimum cross polarization; the feed must also be capable of handling the required peak and average power levels without breakdown under all operational environments. These are the basic factors involved in the choice or design of a feed for a reflector antenna. Other considerations include operating bandwidth and whether the antenna is a single-beam, multibeam, or monopulse antenna.

Rectangular (pyramidal) waveguide horns propagating the dominant TE_{01} mode are widely used because they meet the high power and other requirements, although in some cases circular waveguide feeds with conical flares propagating the TE_{11} mode have been used. These single-mode, simply flared horns suffice for pencil-beam antennas with just one linear polarization.

When more demanding antenna performance is required, such as polarization diversity, multiple beams, high beam efficiency, or ultralow sidelobes, the feeds become correspondingly more complex. For such antennas segmented, finned, multimode, and/or corrugated horns are used. Figure 6.22 illustrates a number of feed types, many of which are described in more detail in antenna references.[3,36,37]

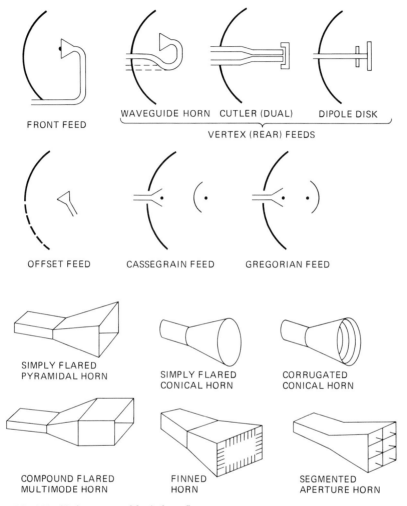

FRONT FEED

WAVEGUIDE HORN   CUTLER (DUAL)   DIPOLE DISK

VERTEX (REAR) FEEDS

OFFSET FEED   CASSEGRAIN FEED   GREGORIAN FEED

SIMPLY FLARED
PYRAMIDAL HORN

SIMPLY FLARED
CONICAL HORN

CORRUGATED
CONICAL HORN

COMPOUND FLARED
MULTIMODE HORN

FINNED
HORN

SEGMENTED
APERTURE HORN

FIG. 6.22   Various types of feeds for reflector antennas.

6.5   REFLECTOR ANTENNA ANALYSIS

In calculating the antenna radiation pattern, it is assumed that the reflector is a distance from the feed such that the incident field on it has a spherical wavefront. There are two methods[2,38] for obtaining the radiation field produced by a reflector antenna. The first method, known as the current-distribution method, calculates the field from the currents induced on the reflector because of the primary field of the feed. The second method, known as the aperture-field method, obtains the far field from the field distribution in the aperture plane. Both the current and aperture-field distributions are obtained from geometrical optics considerations. The two methods predict the same results in the limit $\lambda/D \to ?$. However, in contrast to the aperture-field method, the current-distribution method can explain the effect of antenna surface curvature on the sidelobe levels and on the polarization. While the aperture-field method is handy for approximations and estimates, another problem is that it assumes that the reflection from the surface forms a planar wavefront. This is true for a paraboloidal reflector fed at its focal point, but otherwise it is not true. For this reason, the analysis that follows is devoted to the more general current-distribution method, or induced-current approach.

Although most antennas are reciprocal devices (have the same patterns in receive and transmit), analysis typically follows the transmit situation in which the signal begins at the feed element and its progress is tracked to the far field. Also at the feed, the polarization is in its purest form, so the vector properties are best known at this point and are described in many textbooks. In the analyses that follow, the constants are usually stripped away from the textbook versions since the antenna designer's primary goal in analysis is normally to determine the antenna's gain and pattern for main and cross polarizations. Therefore, the designer will normally integrate the power radiated from a feed into a sphere to determine the normalization factors needed. The magnetic field $\bar{H}$ of the feed is chosen because it leads to the reflector surface current $\bar{J}$ via the normal to the surface $\hat{n}$, by $\bar{J} = \hat{n} \times \bar{H}$.

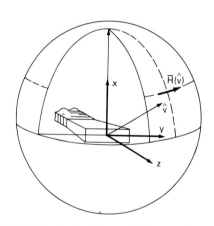

FIG. 6.23 Feed and normalization geometry.

The primary feed is assumed to radiate with the H field, $\bar{H}(\hat{v})$, perpendicular to the direction unit vector $\hat{v}$ (Fig. 6.23). It is dependent on feed type, say, horn or dipole. This H field is normalized such that the total power into a surrounding sphere (the magnitude of the tangent field squared) is equal to 1 W. This may be done by numerical integration using as much symmetry as possible to reduce computation time.

Radar reflectors are normally used to shape and distribute energy, which is more complicated than the case of the symmetrical paraboloidal reflector. Thus, where the focused paraboloid reflects into a common plane over the entire reflector, the shaped reflector focuses into many planes and the most general analysis is to treat the problem as an incremental summation of E fields. Another advantage of this analysis method is that the reflector outline can also be most general.

The surface of the reflector is divided into rectangular grid regions of area dA (Fig. 6.24), which intercept the feed-radiated field. The surface current then is the cross product of the H field with the normal to the surface $\hat{n}$ modified by differential area and a phase term,

$$\bar{J} = \hat{n} \times \bar{H}(\hat{v}) \, (\hat{v} \cdot \hat{n}) \, e^{-ikr} \, dA/4\pi r \qquad (6.24)$$

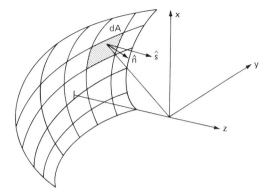

FIG. 6.24 Reflector geometry.

where r = the distance from the feed to the reflecting surface and $k = 2\pi/\lambda$ = the wavenumber.

Each reflector grid region represents the reflection of a small uniformly illuminated section. It has a gain factor and also a direction of reflection, which follows Snell's law. The direction of reflection $\hat{s}$ can be written

$$\hat{s} = \hat{v} - 2\,(\hat{n} \cdot \hat{v})\,\hat{n} \qquad (6.25)$$

and the differential surface reflection at each grid region is modified by a pattern factor represented by a uniformly illuminated reflection steered in the direction of unit vector $\hat{s}$ and determined in the pattern direction unit vector $\hat{p}$ by

$$\text{Pattern factor} = \frac{4\pi dA}{\lambda^2 |\hat{n} \cdot \hat{s}|} \; \frac{\sin \pi \Delta x(s_x - p_x)}{\pi \Delta x(s_x - p_x)} \; \frac{\sin \pi \Delta y(s_y - p_y)}{\pi \Delta y(s_y - p_y)} \qquad (6.26)$$

This factor modifies the surface current $\bar{J}$ as seen in the far field, projected in the direction of interest. At a distant spherical surface, vector $\hat{p}$ is normal to the surface. Two vectors are determined at the surface for the polarizations of interest, both perpendicular to p and perpendicular to each other, considered main and cross-polarized directions. The dot product of each of these unit vectors with $\bar{J}$ gives the field in the main and cross-polarized directions.

This pattern solution is found by a compromise between the number of grid regions and the time to compute all the parameters for each grid point. When the pattern is desired far off the pattern peak, one must consider the artificial grating lobes created by the computation method itself, in which case the grid density must be increased. With grid size Δx, the artificially induced grating lobe will appear at the angle found from $\sin \theta = \lambda/\Delta x$. Frequently, the user of such computational tools will trade off grid density in the orthogonal plane to enhance computation accuracy in the plane of interest.

Typically, the computer time-consuming operations in this type of pattern computation are trigonometric functions, sines and cosines, and square roots used in length and thus phase calculations. Extensive techniques are usually devised to minimize repetitive calculations through the use of arrays containing the unit vectors in the pattern directions of interest, polarization vectors for those directions, and symmetry.

The literature contains many articles showing how *geometric theory of diffraction* (GTD) techniques can be used to compute reflector patterns.[7,39] The problem one encounters with GTD is in generalizing the situation, such as an irregular antenna outline. The primary use one finds for GTD is in defining the antenna's operation in the back hemisphere, but for many antennas the irregularity of the edges requires an agonizingly complex description of the antenna. This is often found to be impractical to implement into a GTD analysis. Sometimes simple analyses are performed at special angles of interest.

6.6 SHAPED-BEAM ANTENNAS

Rotating search radars typically require antenna patterns which have a narrow azimuth beamwidth for angular resolution and a shaped elevation pattern designed to meet multiple requirements. When circular polarization is also one of the system requirements, a shaped reflector is almost always the practical choice, since circularly polarized arrays are quite expensive.

A typical range coverage requirement might look like that shown in Fig. 6.25. At low elevation angles, the maximum range is the critical requirement. Above the height–range limit intersection, altitude becomes the governing requirement, resulting in a cosecant-squared pattern shape. At still higher elevation angles,

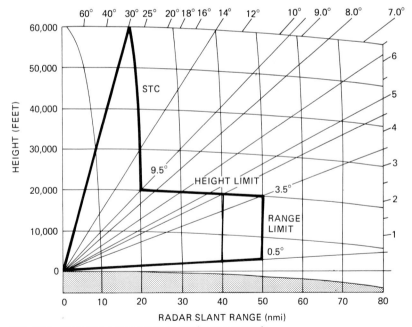

FIG. 6.25 Typical two-way coverage requirement example.

many systems use STC (sensitivity time control), which causes an increased range requirement for those angles. Small, close targets such as birds and insects at lower elevation angles can cause clutter. STC is an effective method for eliminating these unwanted radar echoes. By plotting range against angle as in Fig. 6.26, this two-way coverage requirement example is converted into an antenna pattern requirement by recognizing that range is directly proportional to pattern amplitude. With this simple example, the radar engineer can determine many system requirements relative to the antenna, and these are discussed below.

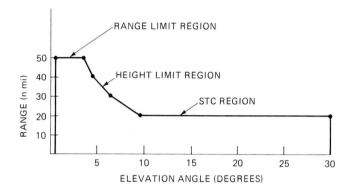

FIG. 6.26   Example of range versus elevation angle.

Gain Estimation. To estimate the antenna gain resulting from the beam-shaping requirements, the designer recognizes that the range versus angle plot represents the minimum range requirement; thus the pattern must cover the plot from above but may not dip below the curve. At the same time, one fundamental concern is to minimize power into the antenna while obtaining the coverage requirements, and this is done by coming as close to the requirement as possible.

A second requirement relates to rotational drive power which limits the size of the antenna. This will be discussed later, but for gain purposes a reasonable situation is achieved by assuming that the size will be determined so that a section of reflector is shaped such that the −3 dB points of the resulting pattern will hit the two corners of the range limit portion of the curve. In the cosecant-squared and STC coverage regions, some ripple must be anticipated, normally about ±1 dB; so the design curve will have to be raised 1 dB.

Gain of an antenna is the ratio of power density at the pattern peak relative to the power density when the same power is radiated isotropically. For the shaped-beam gain considered here, consider the azimuth and elevation beam shapes as independent power patterns, FAZ and FEL. In terms of azimuth angle ϕ and elevation angle θ, the gain can be written as

$$G = \frac{720 \text{ deg}^2}{\iint FAZ(\phi)\, FEL(\theta)\, d\phi\, d\theta} = \frac{720 \text{ deg}^2}{\int FAZ(\phi)\, d\phi \int FEL(\theta)\, d\theta} \qquad (6.27)$$

The integration can be handled separately. For simplicity, the azimuth power pattern can be approximated by a gaussian function,

$$FAZ(\phi) = \exp [- 2.7726 \ (\phi/BWAZ)^2] \qquad (6.28)$$

where BWAZ is the azimuth half-power beamwidth. This function has a peak amplitude of unity, and when integrated over infinite limits (using the normal distribution as a guide),

$$\int FAZ(\phi) \ d\phi = 1.0645 \ BWAZ \qquad (6.29)$$

Now to treat the elevation pattern, the amplitude pattern in Fig. 6.26 must be converted to a power pattern. A parabolic power function approximates the lowest, range-limited portion of the curve, fitted at the two corners. The peak is set to unity to normalize the curve. Renormalizing the coverage region also, with a 1 dB increase, results in the plot in Fig. 6.27, which is easily integrated.

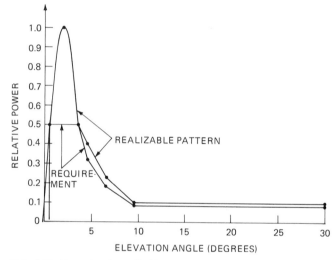

FIG. 6.27   Example of a realizable pattern.

Shaped-Beam Antenna Design. A shaped-beam design process that has worked successfully is that of joining sections of offset paraboloidal reflectors, each successive section pointing slightly higher than the preceding section. An example of this type of design is the ASR-9 antenna, shown in Fig. 6.14. Fig. 6.28 shows the optical projection of each section. It is seen that the projections miss the feed; so blockage is not a problem.

The antenna design is performed in two dimensions by setting up two curves on the same graph, both relating to the fraction of the total power. One curve relates power emanating from the feed striking reflector sections to the reflector section. The other relates the fraction of power in the pattern to the elevation angle. These are shown in Fig. 6.29. From the curves, the direction in which each reflector section is to steer is selected. The process uses a computer to iterate, based on an initial guess at the solution.

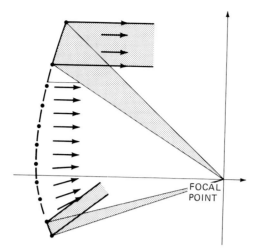

FIG. 6.28 Shaped-beam design process.

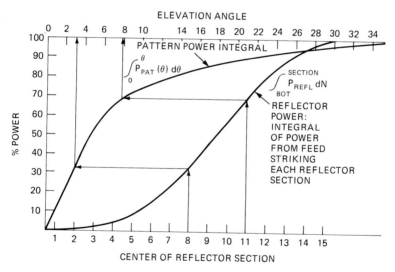

FIG. 6.29 Matching reflector-elevation power integrals.

Antenna Size. The antenna's width is practically determined by the beamwidth. Feedhorn patterns combine with space attenuation to provide a distribution taper which generally gives about 30 dB sidelobes. Thus the width can be estimated as

$$\text{Width} = 65 \ \lambda/\text{BWAZ} \qquad (6.30)$$

where width is the antenna width, λ is the wavelength, and BWAZ is the azimuth half-power beamwidth.

The antenna's height is another consideration. In the subsection "Gain Estimation," it was suggested that a section or several sections of the reflector be shaped to cover the lowest region. This minimizes the reflector height but also causes the slowest falloff at low elevation, which may or may not be of great concern to the designer. The falloff affects the power below the horizon, causing ground reflections which cause lobing, and the pattern overshoots above the range limit.

An estimate of the antenna height may be derived from the realizable power pattern FEL(θ), described in Fig. 6.27, by using the relation

$$\text{Height} = 65 \ (\lambda/\text{BWEL}) \ \frac{\displaystyle\int_{\text{entire pattern}} \text{FEL}(\theta) \ d\theta}{\displaystyle\int_{\text{range-limited region}} \text{FEL}(\theta) \ d\theta} \tag{6.31}$$

where height is the antenna height and BWEL is the 3-dB beamwidth in the elevation plane.

Accuracy. By using computers to compute the shaped-reflector patterns, reasonable accuracy, say, ± 1 dB to ± 2 dB, can be achieved in the specified coverage region. Some of the difficulties in practice are achieving the desired feedhorn pattern and minimizing the pattern effects caused by the feed-support structure.

Feedhorn patterns, regardless of the amount of study completed to date, are still an art form and, as such, have to be tailored to the application. The usual practice is to estimate a horn size for all computations in the design and, when acceptable patterns are achieved, to configure the horn to achieve the theorized patterns. A difficult area to deal with is the effect of edge taper caused by the feedhorn since it will rely on accurate control of the horn pattern at the -15 dB to -25 dB levels, where the slope is steepest. Shaped-beam patterns are very seriously influenced by reflector edge diffraction; so feed-pattern control is a major design problem.

The support structure also has a drastic effect on the pattern. Deflectors and diffusers may be needed to scatter energy away from the feed or reflector. Examples of these devices can be seen on the plate supporting the reflector and feed assembly and on the feed tower in the photograph of the ASR-9 antenna, Fig. 6.14.

6.7 DESIGN CONSIDERATIONS

Feed Blockage.[40,41] Most reflector systems suffer feed and feed-support blockage to some extent if the reflecting area includes the vertex of the parabola. The magnitude of the resulting sidelobes depends on the square of the blocked area, so that extended feeds, monopulse feeds, and subreflectors can have a large effect.

Unless the reflector is electrically small, feed blockage can generally be ignored anywhere outside the reflector outline projected in the direction of the beam. Thus the approximate illumination intensity on the feed is usually calculated by projecting the reflector illumination along the beam direction. The feed and its supports have two effects: they put a shadow or hole in the desired distribution, and they scatter the intercepted power. The effective size of the hole can be larger than the projected obstacle area and depends on the polarization. For conducting obstacles, the effective size is the actual dimension in the E plane, while in the H plane it is approximately $W + \sqrt{0.5\lambda d}$ for a strut W wide and d deep. Thus the blockage of a strut that is 0.125λ square in cross section and parallel to the E field appears to be 3 times larger than the same strut orthogonal to the E field, and its effect on the pattern is 9 times larger.

The effect of the hole on the forward pattern can be estimated by subtracting the pattern of the feed shape from the unperturbed antenna pattern, as Fig. 6.30 illustrates. If a different feed support is used, the orientation of the ridge of sidelobes is altered as in Fig. 6.31. It is to be noted that the tripod arrangement produces sidelobe ridges that are broader but 6 dB lower than those of the quadrapods.

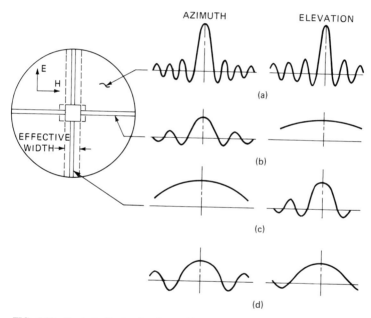

FIG. 6.30 Typical effects of reflector blockage. The unblocked pattern (*a*) has sidelobes that alternate in sign. Each strut (*b* horizontal and *c* vertical) contributes a pattern that is narrow in the direction along its axis. The feed (*d*) contributes a broad pattern corresponding to its small dimensions.

The diffraction sidelobes of typical antennas have sidelobes of alternating phase (the sidelobe adjacent to the main beam is π rad if the main beam phase is 0 rad). Consequently, blockage effects can usually be identified as a sequence of high, low, and high sidelobes. Figure 6.32 shows the levels to be expected for various feed sizes in front of a circular reflector.

Most of the scattering from the feed returns toward the reflector. The portion that misses the reflector is generally of less interest since it has low gain and the sidelobes that it affects tend to be lower. The portion returned to the reflector can be a problem in that it is collimated by the reflector into a secondary beam su-

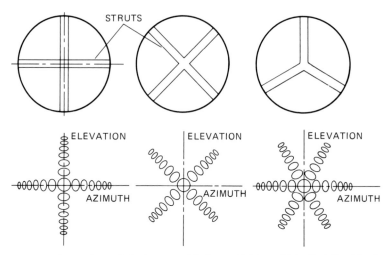

FIG. 6.31 Comparison of feed-support blockage sidelobes. Note that the tripod sidelobes are 6 dB lower than those of either quadrapod.

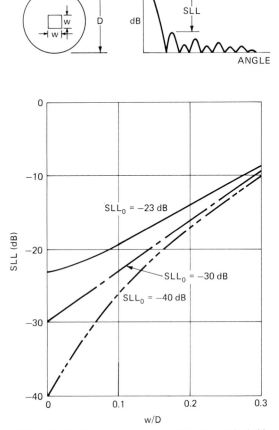

FIG. 6.32 Feed blockage estimates with unperturbed sidelobe levels (SLL$_o$) of -23 dB, -30 dB, and -40 dB.

6.36

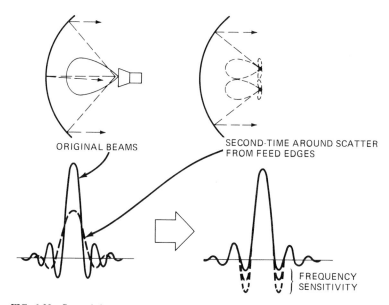

FIG. 6.33 Second-time-around scatter often reflects from the feed edges to produce a broadened contribution to the secondary pattern.

perimposed on the original beam. The reflected beam is generally wider than the original, causing a frequency-sensitive first sidelobe as shown in Fig. 6.33.

Feed and feed-support blockage can be reduced by several techniques besides offsetting the reflector. The support can be made of dielectric tubes, which rephase the intercepted signal. If the phase of the added delay is kept less than 60°, this has the additional advantage that it adds a level in quadrature with the sidelobes, primarily affecting the nulls rather than the peaks. If the delay is more than 60°, the effects of a dielectric strut can be more serious than that of a metal support. Furthermore, since they have to be thicker than metal supports, their blockage may be equivalent, although they may be tuned. If the feed is deep in the focal axis direction, the wave spreads out to produce an exaggerated blockage. This can be significantly reduced by loading the wall of the feed with a capacitive layer $\lambda/4$ off the surface. Figure 6.34 shows how this can be done, and Fig. 6.35 shows the etched panel used to accomplish it in the extended AN/TPS-63 feed array.

Feed Spillover and Diffraction. The edge illumination of a reflector determines both the diffraction sidelobes and the spillover lobes, as Fig. 6.36 shows. The spillover does not have the gain of the reflector, and it is diffracted at the edge; so it is down by approximately $G_s - G_p$ + edge taper + X, where $G_s - G_p$ is the difference between secondary and primary gains and X is the increase at the peak of the diffraction ripple and is up to 2 dB depending on the slope of the illumination. However, if the feed has large sidelobes, they can predominate since they are not diffracted.

Spillover reduction is accomplished by shaping the feed pattern to cut off sharply at the edge of the dish and by having low sidelobes in the feed pattern. Figure 6.37 illustrates the extent to which this can be accomplished in a low-sidelobe design.

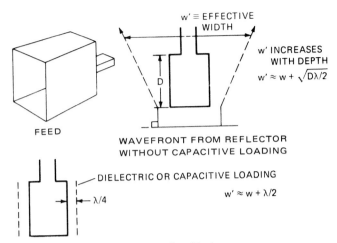

FIG. 6.34 Capacitive loading to reduce blockage.

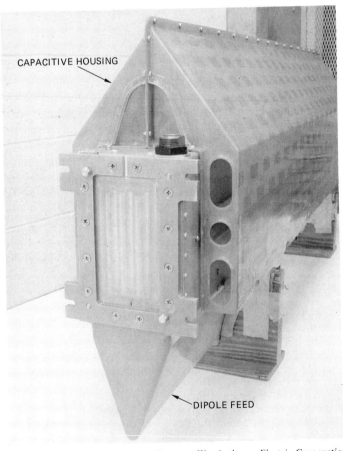

FIG. 6.35 AN/TPS-63 feed array. (*Courtesy Westinghouse Electric Corporation.*)

A good view on the total pattern is achieved by plotting the feedhorn pattern on the same diagram as the secondary pattern, being sure to adjust the gains (Fig. 6.38). The angle at which the feedhorn pattern is blocked by the reflector is easy to calculate. At this angle, the feedhorn power is expected to be reduced by an extra 6 dB due to diffraction, and the variation in this region can normally be estimated by considering the reflector as offering knife-edge diffraction to the feedhorn and using the Cornu spiral[42] to determine the ripple function. The interference pattern between primary and secondary patterns will be difficult to estimate. Usually the secondary pattern will be negligibly low in this region.

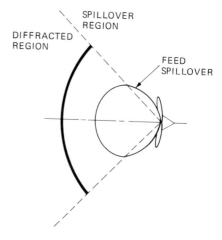

FIG. 6.36 Spillover lobes.

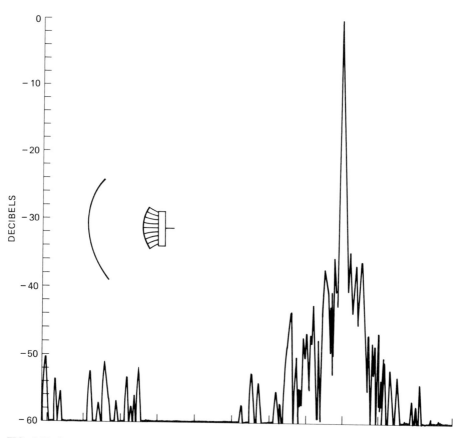

FIG. 6.37 Pattern measured using a low-spillover feed design.

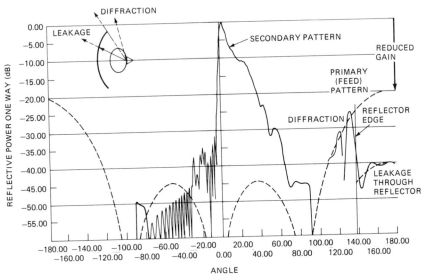

FIG. 6.38 Primary feed and secondary shaped-beam patterns superimposed on the same diagram to allow estimates of spillover. The dashed primary pattern is naturally directed opposite the solid secondary pattern and has substantially less gain. It is also blocked by the reflector over the back 80°.

Surface Leakage. The backlobe is determined by the combination of spillover and leakage through the reflector. Many reflectors are designed with either a tubular grid or a wire-mesh surface in order to minimize wind resistance and weight. To accomplish this, the opening size is chosen to be as large as possible, especially if icing may occur. Some of the common reflector surfaces are shown in Fig. 6.39. In every case, the gap S between conductors must be substantially less than $\lambda/2$ in the H plane for the polarization in use, so that the passage of electromagnetic energy will be below cutoff. If this is the case, the attenuation in decibels through the passage depth t is approximately $27\ t/S$, plus fringing losses, which are approximately $27(\lambda/2S - 1)$.

For specific cases, Fig. 6.40 shows the leakage levels to be expected. The difference between primary (feedhorn) and secondary (reflector) gain must be added to this to obtain the actual backlobe level.

6.8 MECHANICAL CONSIDERATIONS

All radar antennas, and particularly mechanically scanned reflector antennas, require careful consideration of mechanical design details. First, the reflecting surface must be designed and built so that it remains within close tolerances (typically ± 0.03 wavelength) of the ideal surface even under dynamic operating and environmental conditions. Also, the feed must be mounted with its phase center at the focal point of the reflector and accurately aligned with respect to the reflector axis. The feed-support structure and the stiffening structure behind the reflector must maintain the critical dimensions while the antenna is being scanned or rotated, with minimum degradation of the radiation performance caused by

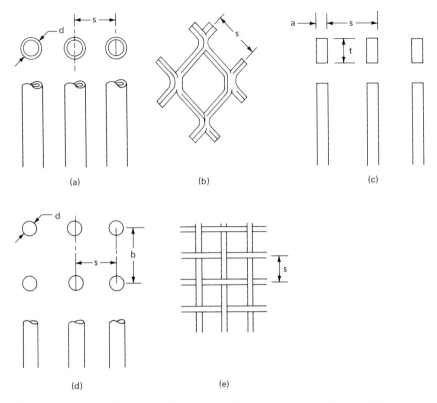

FIG. 6.39 Common reflector materials for reduced wind resistance. (*a*) Tubing. (*b*) Expanded metal. (*c*) Rectangular. (*d*) Double layer. (*e*) Screen.

blockage or dimensional distortions of the structure from wind loads, temperature variations, or other environmental factors. In most cases, the mechanical design of an antenna requires greater engineering effort and innovation than does the relatively simple electrical (RF) design. This is especially true when reliable high-performance servo-drive mechanisms are involved.

Antenna tolerance theory has been discussed by Ruze[43,44] from the statistical point of view. The loss of gain due to small phase error is approximately[44]

$$\frac{G}{G_0} \approx 1 - \overline{\delta^2} \tag{6.32}$$

where $\overline{\delta^2}$ = mean square phase error
  G_0 = gain without phase error
  G = gain with phase error

Equation (6.32) indicates that for 1 dB loss of gain the rms phase variation about the mean-phase plane must be less than 0.45 rad, or for shallow reflectors the surface error must be less than $\lambda/28$.

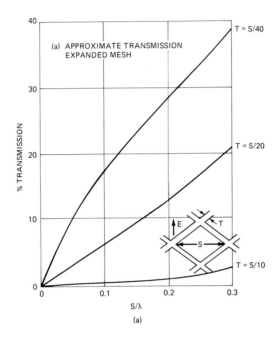

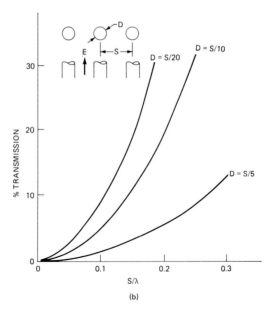

FIG. 6.40 Expected leakage levels of various reflectors. (*a*) Transmission through expanded mesh. (*b*) Transmission through parallel tubes.

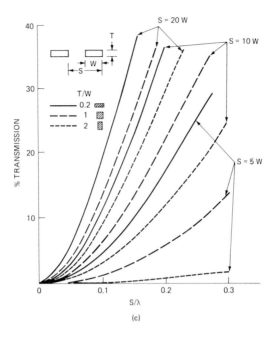

FIG 6.40   *(Continued)* (*c*) Transmission through a rectangular grid.

The permissible limits in the systematic errors in the design of the antenna assembly may be found in the literature.[2,20,38] The total error is arbitrarily restricted to $\lambda/8$ or $\pm\lambda/16$. On the basis of this criterion the maximum deviation from the parabolic profile should be $\pm\lambda/32$. The displacement of the feed along the axis and away from the focus produces even-order phase error on the aperture illumination. The maximum deviation $d_\parallel$ of the feed along the axis and away from the focus should be

$$d_\parallel \simeq \frac{\lambda}{8(1 - \cos \psi_0)} \tag{6.33}$$

The displacement of the feed normal to the axis and away from the focus produces odd-order phase error in the aperture. If the displacement $d_\perp$ is $<< f$, the phase error is essentially linear and the beam is displaced from the axis. The tolerance on $d_\perp$

$$d_\perp \lesssim \frac{\lambda f}{4D} \frac{(16f^2 - D^2)}{D^2} \tag{6.34}$$

and the beam will be shifted undistorted and away from the axis by

$$\theta_0 \approx \frac{\lambda}{4D} \frac{(16f^2 - D^2)}{D^2} \tag{6.35}$$

where f is the focal length, D is the reflector aperture dimension (diameter), and ψ_0 is the half angle subtended by the reflector at the focal point (Fig. 6.4a).

6.9 RADOMES*

Radomes are used when it is necessary to protect antennas from adverse environmental effects. Ideally, a radome should be perfectly transparent to the RF radiation from (or to) the antenna and yet be able to withstand such environmental effects as wind, rain, hail, snow, ice, sand, salt spray, lightning, and (in the case of high-speed airborne applications) thermal, erosion, and other aerodynamic effects. In practice, these environmental factors determine the mechanical design of the radome, and the desire for ideal RF transparency must be compromised because mechanical and electrical requirements are often in conflict.

Radomes cause four major electrical effects on antenna performance. *Beam deflection* is the shift of the electrical axis, which is critical for tracking radar. *Transmission loss* is the measure of energy lost by reflection and absorption. The *reflected power* causes antenna mismatch in small radomes and sidelobes in larger radomes. *Secondary effects* include depolarization and antenna noise additions.

Radome design is a specialized art, and many books[3,4,45] are devoted to its intricate details. This section makes no attempt to provide radome design information as such but instead is aimed at making the radar system designer aware of the basic concepts behind the types of radomes available for various applications.

There are two main categories of radomes for radar antennas: radomes for ground-based or shipboard systems and radomes for airborne or missile systems. Although these two differ significantly in size and form, they have some general characteristics in common.

Types of Radomes and General Considerations. Three general classes of radome are of interest: *feed covers*, which often have to endure pressure, high voltage, and heating; *covers attached to the reflector*, which alter the pattern in a fixed manner; and *external radomes*, within which the antenna moves. External radomes are the most common and will therefore be emphasized. Within each class, a variety of skin and skin-supporting designs is available to minimize the electrical effects under the constraints of the environment. The radome skin may be rigid, supported by a framework, or air-supported.

The most common rigid radome-wall structures are shown in Fig. 6.41 and are known as homogeneous single layer, A-sandwich, B-sandwich, C-sandwich, multiple-layer sandwich, and dielectrics with metal inclusions.

Single Layer. The homogeneous single-layer radome has been used in many radome applications. Materials for this type have included fiberglass-reinforced plastics, ceramics, elastomers, and monolithic foam. The optimum thickness for a single layer is a multiple of a half wavelength in the dielectric material at the

---

*Some of the material in this section has been taken directly from Chap. 14, "Radomes," of the first (1970) edition of this handbook, which was authored by Vincent J. DiCaudo.

appropriate incidence angle, but many single-layer radomes are simply thin-wall approximations to the zero-thickness case.

A-Sandwich. A commonly used radome-wall cross section is the A-sandwich, which consists of two relatively dense thin skins and a thicker low-density core. This configuration exhibits high strength-to-weight ratios. The skins are generally fiberglass-reinforced plastics, and the core is a foam or honeycomb. Inorganic skin and core sandwiches[2] also have been developed for high-temperature applications. As a rule, the skins of the sandwich are made symmetrical or of equal thickness to allow midband cancellation of reflections.

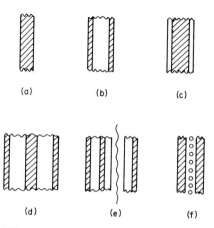

FIG. 6.41 Common radome-wall cross sections. (*a*) Single layer. (*b*) A-sandwich. (*c*) B-sandwich. (*d*) C-sandwich. (*e*) Multiple-layer sandwich. (*f*) Dielectrics with metal inclusions.

B-Sandwich. In contrast to the A-sandwich, the B-sandwich is a three-layer configuration whose skins have a dielectric constant lower than that of the core material. This wall cross section is heavier than that of the A-sandwich because of the relatively dense core. The B-sandwich is not commonly used because the core dielectric constant is quite high for a proper match.

C-Sandwich. The C-sandwich is a five-layer design consisting of outer skins, a center skin, and two intermediate cores. The symmetrical C-sandwich can be thought of as two back-to-back A-sandwiches. This configuration is used when the ordinary A-sandwich will not provide sufficient strength, or for certain electrical performance characteristics, or when one layer is to serve as a warm-air duct for deicing.

Multiple-Layer Sandwich. Multiple-layer sandwiches of 7, 9, 11, or more layers are sometimes considered when great strength, good electrical performance, and relatively light weight are required. Some of these designs[46,47] have used thin layers of fiberglass laminates and low-density cores to attain high transmission performance over large frequency bands.

Dielectric Layers with Metal Inclusions. Metal inclusions have been considered for use with dielectric layers to achieve frequency filtering, broad-frequency-band performance, or reduced-thickness radomes. Thin layers of metal inclusions exhibit the characteristics of lumped circuit elements shunted across a transmission line. For example, a grid of parallel metal wires exhibits the properties of a shunt-inductive susceptance.

Environmental Effects. Environmentally, wind loads generally produce the most severe stresses that radomes must withstand, whether due to surface winds or to aircraft motion. Associated with the wind are both heating and erosion. Erosion due to impinging rain or hail can be a problem, especially with airborne radomes. Typical fiberglass-reinforced organic-resin radomes used for subsonic-speed aircraft exhibit poor rain-erosion properties. Neoprene coatings, about 0.010 in thick, generally are applied to the outer surface of these radomes and provide the necessary protection. At higher speeds, the neoprene coatings do not survive the temperatures, owing to embrittlement and loss of adhesion. Ceramic radomes or ceramic coatings over plastic radomes generally are used at supersonic speeds to achieve rain-erosion resistance. At these higher speeds, the shape of the radome should be streamlined. Rocket-powered-sled tests[48] indicate that blunt-shaped radomes of either ceramic or plastic construction are not suitable for speeds greater than Mach 2.0.

Ice and snow can be degrading for both structural and electrical performance. In some airborne applications, ice is removed by a pulsating pneumatic boot. A more satisfactory solution is to heat the outer surface of the radome by passing hot air through plastic ducts embedded in the radome wall, thus preventing ice from forming. For ground radomes, the accumulation of ice and snow increases the loads that the radome must withstand and also degrades the electrical performance.

Airborne and Missile Radomes. Streamlined radomes are sometimes necessary to satisfy aerodynamic design requirements. If the antenna scans within the radome, the incidence angles may vary from normal incidence to as high as 70 to 80° from the normal. This large range of incidence angles leads to greater degradation of antenna performance with the streamlined radome than with the normal-incidence (spherical) radome. Variation of insertion phase difference and transmission across the antenna aperture due to the radome shape results in beam deflection, pattern distortion, and loss in gain.

The rigorous solution of antenna performance in the presence of a radome has been attained for a few simple shapes such as spheres and cylinders by a boundary-value approach. However, this technique is not applicable to the general radome shape. If the radome radius of curvature is large with respect to wavelength, the surface may be considered locally flat and plane-wave theory can be used. This approach becomes less accurate as the radome radius of curvature gets smaller with respect to wavelength, as in the nose area of streamlined radomes.

Numerous approximation techniques have been developed to estimate the beam deflection caused by a radome.[46,49–53] Several methods use ray-tracing techniques to determine the phase and amplitude distribution over a new aperture plane outside the radome.

High-Temperature Radomes. The advent of high-speed aircraft, missiles, and space vehicles has introduced the need to consider high temperatures, the radome shape necessary for aerodynamic requirements, and, in some instances, the rain-erosion characteristics of the radome surface.

High temperatures can produce changes in the structural and electrical characteristics of the radome material. The temperature over the surface and through the cross section of the radome may vary considerably. Supersonic-aircraft nose radomes must withstand temperatures in the 500 to 1000°F range. Laminates using organic resins such as silicone, polyimide (PI), and polybenzimidazole (PBI) can withstand temperatures of 500 to 700°F. An inorganic laminate composed of fiberglass reinforcement and aluminum phosphate binder shows promise for use in the 1000 to 1200°F range.[54]

Supersonic missiles are characterized by high heating rates that cause the radome to be subjected to thermal shock with surface temperatures up to 1600°F. Glass-ceramic radomes have found application here because of good thermal-shock resistance. Hypersonic vehicles present severe problems in the use of radomes because of extremely high temperatures and/or thermal shock. Ablation materials are used when high heating rates are encountered for short time periods. The ablation process results in a cooling effect because a large amount of heat energy is used in subliming the radome material. High-density ceramic materials are used for radomes whose location on the hypersonic vehicle eliminates the thermal-shock problem but requires a high-temperature material.

Ground and Shipboard Radomes. Ground and shipboard radomes are usually truncated spheres mounted on a tower or directly on the ground. They drastically reduce the structural and drive requirements on the antenna that is enclosed. For example, the exposed 16-ft-wide ASR-9 antenna requires more drive horsepower than the enclosed 42-ft-wide ARSR-3. A variety of construction techniques are used. Ground radomes are generally coated with a paint such as Hypalon, Radolon, or epoxy, as well as a Tedlar film. These are often loaded with titanium dioxide to reflect ultraviolet radiation.

Inflatable Radomes. The first large radomes were air-supported structures. The inflatable radome still is used where transportability, light weight, and minimum packaging volume are important factors. A Hypalon-coated Dacron material whose thickness results in a thin-wall structure is often used. Various sizes of inflatable radomes have been built, the largest being 210 ft in diameter.[55] The transmission performance of thin-wall air-supported radomes may be estimated from the flat-panel characteristics. One of the problems with inflatable radomes is the need for a continuous internal air pressure and the auxiliary equipment that must provide the source of power. Failure of the pressurization system can result in loss of both the radome and the antenna.

Foam Radomes. The foam-shell radome offers great promise of excellent electrical performance characteristics over large frequency bands. The low dielectric constant and loss tangent of foam materials permit the use of relatively thick-wall cross sections required to meet the structural loads. The individual panels comprising the radome may be joined by an adhesive joint or by foaming the joint in place with the basic panel materials.

Sandwich Radomes. A-sandwich-type construction has been used in large-radome applications to achieve excellent electrical performance characteristics at frequencies under 6 GHz.[56–58] At the lower frequencies, large cross sections may be used for structural purposes while maintaining good electrical performance. The core is generally a low-density paper or fiberglass honeycomb material with a low dielectric constant. The skins are fiberglass that permit the bolting of panels together. Sandwich radomes have been built up to 140 ft in diameter for the AN/FPS-24 radar and the AN/FPS-49 Ballistic Missile Early-Warning System (BMEWS) radar. Scale-model tests indicate that a transmission efficiency greater than 95 percent and a maximum boresight error of 0.3 mrad are achieved. Pattern distortion introduced by the radome is small. Theoretical and scale-model studies conducted for large dielectric-sandwich and metal-space-frame radomes indicate that the sandwich radome has the best electrical performance for application in the ultrahigh-frequency (UHF) region. The sandwich radome is more expensive than the metal space frame, and this factor must be weighed against the superior electrical performance that can be achieved.

Metal Space Frames. The electrical performance of the metal-space-frame radome is dominated by the scattering characteristics of the metal beams and hubs. The skin material enclosing the structure is generally an electrically thin wall that can operate over a wide frequency band. Depending on the particular site requirements, the skin thickness may range from 0.030 to 0.100 in. The transmission losses of the laminate skin material may be estimated from the flat-panel performance. The electrical performance of metal space frames may be estimated from a knowledge of the rib and hub geometry and orientation. The general technique[59,60] is to determine the far-field radiation scattered from the metal members that are illuminated by energy from the antenna aperture. This scattered field is combined with the field that would exist if the radome were not present, to give the total far field of the antenna in the presence of the radome.

Other Radome Effects

Effect of Radome Panels.   Whenever an obstacle is placed in front of an antenna, there is a possibility of a change in performance. This subsection describes how systems parameters are determined from radome junction measurements and window-area analysis. The following discussion will focus on the sandwich type of radome.

A sandwich type of radome affects radar performance in two ways:

1. The window areas have heat loss and reflection loss that reduce the transmitted power.

2. The junctions scatter energy uniformly in space, which adds or cancels from the normal pattern as taken without a radome.

The total loss is the loss due to the window area and the loss that is scattered due to the junctions. The junctions affect many other system parameters, and they are discussed in detail later on.

Junction Effects.   Junction effects can be estimated by using a combination of measurements and analyses developed and related through private communication with Essco Corporation, a leading radome manufacturer.

A typical panel junction is shown in Fig. 6.42. Since the junctions affect the pattern shape of sidelobes, boresight error, and height accuracy, their effect is critical. The junctions are arranged in random fashion. In this way the scattering from each member is random, and the scattered energy may be visualized as a spherical pattern of constant level that adds or cancels from the normal pattern. The determination of this scattered-energy level is the first step in the analysis.

The determination of the junction's effect on patterns is accomplished by the induced field ratio (IFR) method, that is, by first examining a single junction and its field disturbance, then relating it to the total scattered energy, and finally relating it to antenna patterns and system performance. To determine the junction's IFR, a panel measurement such as that shown in Figure 6.42b is made. In this measurement, two flat panels are built, bolted together, and mounted on a rail so that they may be slid between radiating and receiving horns. As they move, the insertion amplitude and phase are measured. Figure 6.43 shows an example of the measurements. By using these measurements, the calculation of IFR described by Rusch, Hansen, Klein, and Mittra in Forward Scattering from Square Cylinders in the Resonance Region with Application to Aperture Blockage (published in the March 1976 issue of the *IEEE Transactions on Antennas and Propagation*) is performed. In this article they define induced field ratio by the equation

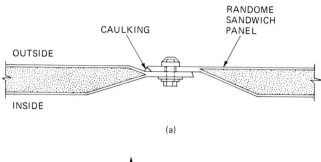

(a)

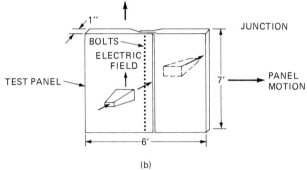

(b)

FIG. 6.42 (*a*) Panel construction, sandwich radome. (*b*) Panel measurements.

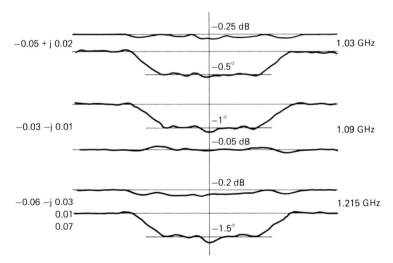

FIG. 6.43 Experimental panel measurement of a tuned joint, showing amplitude and phase at three frequencies for vertical polarization. Onset of the panel is seen by a large step in the phase, while the joint effect is seen as a small ripple in amplitude. (*Courtesy of Essco Corp.*)

$$\text{IFR} = (10^{-0.05\Delta dB}e^{-j\Delta\phi} - 1)e^{-j45°}\ \frac{\sqrt{\lambda\ D1\ D2}}{W\ \sqrt{D1 + D2}} \qquad (6.36)$$

where ΔdB = amplitude change due to junction
$\Delta\phi$ = phase change due to junction
λ = wavelength
$D1$ = distance from transmit horn to panel
$D2$ = distance from receive horn to panel
W = width of junction

From the IFR calculation, the junction loss is computed by using the absolute value of the real part of the IFR and the fraction of area blocked relative to the antenna area, using the equation

$$\text{Junction loss} = 20 \log \left[1 + B_{\text{ratio}} |\text{REAL(IFR)}| \right] \qquad (6.37)$$

where B_{ratio} = area blocked by the junction's antenna aperture area.

Window-Area Loss and Total Transmission Loss. The transmissivity through the radome depends upon the construction of the radome wall and also upon the antenna look angle, the field distribution over the antenna aperture, the location of the antenna inside the radome, and the radome-to-antenna-size ratio. By using ray-tracing methods, the transmissivity of the radome can be determined by summing the losses from all rays which have different weight functions and look angles. The calculation of power radiated, reflected, and lost in heating follows the analysis presented by H. Leaderman and L. A. Turner in "Theory of the Reflector and Transmission of Electromagnetic Waves by Dielectric Media," Chap. 12, MIT Radiation Laboratory Series, vol. 26. In this analysis, the direction of incidence is not taken into account because the antennas are close to the center and so the change due to different angles of incidence across the aperture is slight.

Scattered Energy. The energy loss due to the junctions is scattered spherically. Its level is computed relative to the pattern peak by the equation

$$\text{Scattered energy} = 10 \log (1 - 10^{\text{junction loss}/10}) - \text{gain} \qquad (6.38)$$

It is used to calculate the radome's effect on sidelobe level, beamwidth change, and height accuracy. By using the gains and junction losses at each frequency, the scattered-energy levels are computed.

Sidelobe-Level Error. The scattered-energy level determined above adds vectorially with the unobstructed sidelobe level. The change in sidelobe level with the radome is given by the equation

$$\text{Sidelobe-level change} = 20 \log (1 + 10^{(\text{scattered energy} - SL)/20}) \qquad (6.39)$$

where SL is the sidelobe level measured without the radome. By using the scattered-energy level above, the change in sidelobe level is computed.

Difference-Pattern Null Depth. The way in which the radome affects the difference pattern near its null is the same as the way it affects a sidelobe; that is, the scattered energy adds vectorially with the pattern as measured without the radome. The principal difference is that the gain of the difference pattern is normally about 2.5 dB below the gain of the sum pattern; so the scattered-energy level relative to the nominal pattern is that much higher. On this assumption the change in the difference-pattern null depth is computed.

Cross-Polarization Ratio. The way in which the radome affects cross polarization is also the same as the way it affects a sidelobe; that is, the scattered energy adds vectorially with the pattern as measured without the radome. The principal difference here is that the energy scattered into the crossed polarization must be less than half of that scattered into the normal polarization; thus the scattered-energy level used for the effect on cross-polarized energy is 3 dB below that assumed for the sidelobe-level-change calculation. On this assumption, the change in the cross-polarization level is computed.

Boresight Error Level. A spherically homogeneous sandwich radome will not introduce any boresight error if the enclosed antenna axis is coincident with the radome center. In large panelized radomes, some boresight error is introduced by the panel junction since at different rotation angles there is one more or less junction on each side of the radiating aperture. A. Kay and D. Patterson, in Design of Metal Space Frame Radome (*Rept. RADC-TDR*-64, Rome Air Development Center, Griffiss Air Force Base, N.Y., pp. 36–55, June 1964), expressed the rms boresight error (BSE) as

$$BSE_{rms} = 0.27 \; \lambda \; L \; W \; |IMAG(IFR)|/A^3 \tag{6.40}$$

where
$$
\begin{aligned}
\lambda &= \text{wavelength} \\
L &= \text{length of panel junction} \\
W &= \text{junction width} \\
IMAG(IFR) &= \text{imaginary part of IFR} \\
A &= \text{radius of antenna}
\end{aligned}
$$

This expression is for a round pencil-beam antenna. For shaped antennas, the expression is modified. The antenna radius can be related to the beamwidth of a typical 25 to 40 dB antenna pattern by

$$\lambda/(2A) = BW/65 \tag{6.41}$$

where BW = beamwidth in degrees. Also, πA^2 is the area of a round antenna and is related to the effective aperture area via the antenna gain:

$$\pi A^2 = \lambda^2 \; 10^{gain/10}/(4\pi) \tag{6.42}$$

By replacing terms when applying the above assumptions, the rms boresight error is rewritten as

$$BSE_{rms} = \frac{2.16\pi^2}{65} \; \frac{L \; W}{\lambda^2 \; 10^{gain/10}} \; BW \; |IMAG(IFR)| \tag{6.43}$$

Computing the maximum boresight error from the rms level only requires multiplying the above by 3.

Boresight Error Slope. The rate of change of boresight error is dependent on the change in the amount of dielectric material in front of the antenna. The shortest period of variation is one-half panel, so the deflection shifts from positive to negative over this period. The boresight error slope is computed as a percentile, from the angle that the panel subtends, nominally 9°, by using the equation

$$BSE \text{ slope} = 100 \; 2 \; BSE_{max}/9 \quad \% \tag{6.44}$$

Antenna Beamwidth Error. The effect of the radome on the antenna beamwidth may be computed by assuming that the scattered energy due to the junctions adds a small constant level all across the main lobe. Thus by relating the change in level to the slope at the -3 dB point, the change in beamwidth is determined. Near the peak of the beam, the pattern shape can be expressed in decibels as

$$dB = -3(2\theta/BW)^2 = -12\theta^2/BW^2 \qquad (6.45)$$

Differentiating with respect to θ gives

$$\Delta dB = -24(\theta/BW^2)\Delta\theta \qquad (6.46)$$

and solving for $\Delta\theta$ at $\theta = BW/2$,

$$\Delta\theta = -BW\ \Delta dB/12 \qquad (6.47)$$

in which the change in decibel level at the half-power point can be substituted. Since this affects both sides of the beam, it is doubled to get the beamwidth error:

$$\text{Beamwidth error} = 2\ BW\ 20\ \log\ (1 + 10^{(\text{scattered energy } + 3)/20})/12 \qquad (6.48)$$

6.10 ANTENNA TESTING

Pattern Test Ranges.[61–63] Antenna pattern testing requires special techniques and extensive test facilities. The goal of the test is almost always the determination of free-space patterns and gain, and the accuracy of the test is limited by the reflections and distortions of the test range. Four basic approaches are taken to minimize these errors, as Fig. 6.44 illustrates:

a. Elevated range: In this conventional approach, the ground reflections are reduced by using a narrow elevation beam on the source antenna.

b. Ground range: In this case, the ground is held flat to provide specular reflection, eliminating any ground scatter that could reach the test antenna.

c. Compact range: A compact range uses a large reflector to produce a region of uniform phase and amplitude.

d. Near-field range: Rather than illuminate the entire antenna simultaneously, a near-field range uses a movable source several wavelengths from the aperture, with a computer to collect and store the data and to reconstruct the far-field pattern.

For far-field ranges, the most common criterion for minimum range length is $2D^2/\lambda$, where D is the largest dimension of the aperture under test. At this distance, the spherical phase front from the source deviates from the ideal plane phase front by a quadratic error reaching $22\frac{1}{2}°$. The gradual nature of this error confines the pattern impact to be close to the main beam. Figure 6.45 indicates that for most range lengths only the first two nulls and sidelobes on either side of the main beam are significantly affected. However, extremely long ranges would be necessary to make the effect on the first sidelobe level negligible for low-sidelobe antennas. Lengthening the range to reduce this error must be considered carefully, since the transmit power and acreage increase by the square of the length. At the same time the subtended angle to obstacles is reduced in inverse proportion to the length, making it

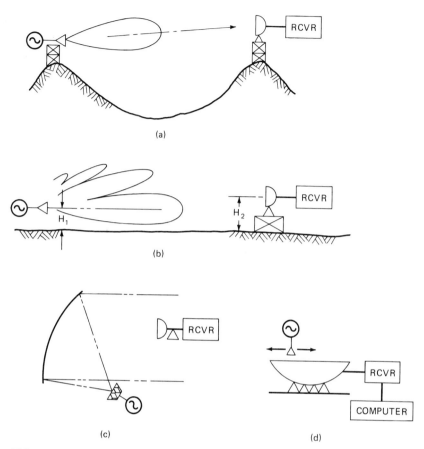

FIG. 6.44 Range types. (*a*) Elevated. (*b*) Ground. (*c*) Compact. (*d*) Near field. The antenna under test (AUT) is shown connected to a receiver; the antenna illuminating the AUT is connected to an RF power source and is called the source antenna.

more difficult to avoid sidelobes at wide angles. Figure 6.46 shows typical range lengths in the major radar bands.

The majority of ranges are elevated, requiring minimum modification of the site. Ground ranges become necessary when low sidelobes are required at the same time that ground illumination becomes unavoidable. Compact ranges require a reflector several times larger than the test antenna, so usually are restricted to antennas up to perhaps 10 ft across. In addition to saving real estate, they are valuable as a means of conducting tests indoors. Near-field ranges go one step further in reducing range size. The point-by-point measurement of the field usually adds a precision mechanism for probe motion but eliminates one or more axes of rotation on the test antenna mount. The computer processing of the data may be programmed to produce far-field patterns or, in the case of an array, an element-by-element tabulation of phase and amplitude.

Elevated Ranges. Figure 6.47 shows the equipment at a conventional elevated-range setup to test a receive antenna. The power source must generate

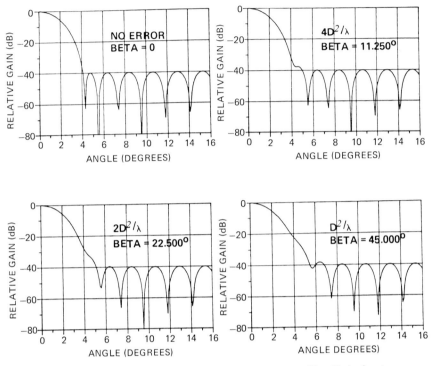

FIG. 6.45 Effect of range length on patterns of a 48-element array with a Chebyshev taper applied. BETA is the maximum phase deviation from a planar phase front.[69]

sufficient power to keep the receive sidelobes well above noise, typically several watts. The transmitting antenna should have a wide enough beam to provide nearly constant amplitude across the test antenna width D.

A generally accepted taper is 0.25 dB. Approximately, this corresponds to a 3 dB beamwidth no less than $(3/0.25)^{1/2} D/R$ rads, or 200 $D/R°$. A typical source antenna will have a null-to-null beamwidth of at least 3 times the 3 dB beamwidth. Therefore, if the taper is less than 0.25 dB, the main beam will illuminate a region at least 10 times the width of the antenna, and ordinarily a much larger area should be cleared of obstacles. Similarly, the towers should be tall enough to avoid illuminating the specular image of the test antenna in the ground. This image occurs at approximately $2H_R/R$ rad from the boresight of the source antenna. The receive tower height should therefore be at least 3 times the vertical size of the test antenna. In many cases, the source dimensions are not optimal, and the clearances are not sufficient. The test antenna mounting must then be arranged to avoid illuminating the scatterers with the main beam. For example, if the source has a wide elevation beam, the test antenna beam should be kept off the ground between the towers as the pedestal rotates.

Pattern range receivers are designed to reject interference, with a wide dynamic range and exceptional linearity over a 40 to 80 dB range of inputs. Most modern receivers provide both amplitude and phase information. Phase is measured by comparing the signal with a reference signal. Either hard lines between the towers or reference pickup radiators are used to obtain the reference.

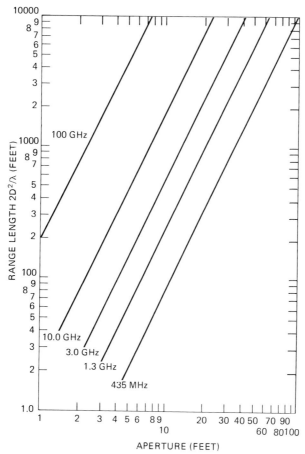

FIG. 6.46 Typical range lengths for far-field antenna ranges.

Pedestals and their angle encoders provide precise angular information to the recorder or computer. The simplest pedestals use an azimuth rotator or an elevation rotator mounted on (i.e., over) an azimuth rotator. However, the most common pedestals add an "upper" azimuth rotator on the elevation mount. Figure 6.48 shows the coordinate system in which a combination operates. Since planar arrays steer in conical coordinates, it is often necessary to produce patterns in such cuts, as is indicated.

Ground Ranges. The equipment for ground ranges is almost identical to that for elevated ranges, except that the path loss is 6 dB less. The source antenna may have a broad elevation beam, but it must be positioned at height $H_t \simeq \lambda R/4H_r$, where R = range length and H_r = test antenna height above ground, for each frequency (see Fig. 6.49). This maintains the first peak of the ground lobing pattern at the height of the receive antenna. For a typical receiver height of one-twentieth of the range length, the source height is 5 wavelengths.

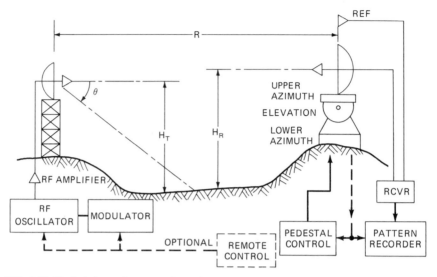

FIG. 6.47 Typical elevated range equipment.

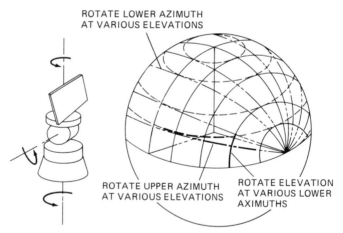

FIG. 6.48 Pedestal axes.

With a small source antenna, the level at the receive tower increases with height h as 20 log [sin $(\pi h/2H_r)$]. This limits the minimum receive tower height for a given amplitude flatness. As Fig. 6.50 indicates, the antenna must be mounted at least 3.3 times as high as it is tall for a 0.25 dB edge taper.

The surface of the range is illuminated at near grazing incidence; so fairly large gradual deformations are permissible. Depending on the tower height, a surface that is held level to within 1 to 2 wavelengths is satisfactory for most purposes.

Compact Ranges.[64]   Most of the constraints on a far-field range are involved with approximately uniform amplitude and phase over a region in space. A

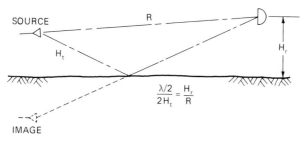

FIG. 6.49 Ground-range geometry.

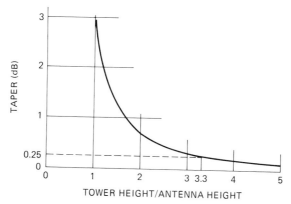

FIG. 6.50 Ground-range tower height requirements.

large reflector or lens can produce the same result without extensive real estate. In many respects, the reflector design is similar to that of an antenna, but certain characteristics are special.

1. The feed is normally offset out of the beam.

2. The extent of the flat-amplitude region is maximized by the feed design. Space attenuation must be accounted for as well.

3. Direct radiation from the feed must be minimized.

4. Diffraction from the reflector edge must be minimized both by feed-pattern shaping and by scalloping or rolling the edge.

5. The walls of the room must absorb the stray radiation to quite low levels.

To accomplish all of this requires extensive use of absorber and special feed designs. Dual-reflector design can accomplish much of the feed-pattern shaping. When all these steps are taken, a usable test region of about one-third of the reflector dimensions can be produced.

Near-Field Ranges.[65] A standard calibration technique for a far-field range involves probing the field over the region in which the test antenna will be placed. It is a logical step to reverse the probe and pick up the field radiated from the test antenna. Given the total phase and amplitude in the vicinity of the test antenna, the far-field pattern can be reconstructed from well-known electromagnetic equations. Alternatively, the data can be used to compute the fields on the antenna surface as a means of locating errors and failures.

Three technologies are critical to a successful near-field measurement: (1) the precise movement and location of the probe, (2) probe design, and (3) pattern analysis software.

The probe is generally moved in a spherical, cylindrical, or planar coordinate system (Fig. 6.51). The spherical system is convenient in that existing pedestals can be used, but implementation has been delayed by difficulties in the analysis of the spherical functions. Cylindrical near-field analysis is simpler, but in recent years planar scanning has predominated. Scanner design requires extreme precision and involves special foundations for the building, temperature control to several degrees Fahrenheit, and laser location of the probe. The probe must move beyond the antenna edge at least to the projected scanned aperture limits. At S band, a tolerance of 0.005 in. on probe locations over a 20-ft-square aperture is not an unusual requirement. If the antenna is too large for the available scan-

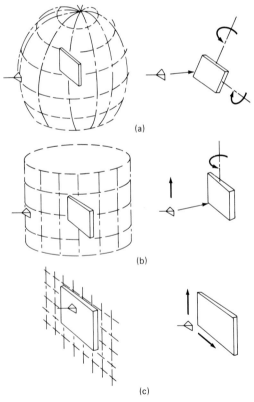

FIG. 6.51 Near-field scanning. (*a*) Spherical. (*b*) Cylindrical. (*c*) Planar.

ner, it is practical to shift the antenna in stages, merging the data into the equivalent of a single set taken on a larger scanner.

Probe design and probe correction are critical in a near-field range design. It can be shown that the probe pattern is a direct factor in the computed antenna pattern. If the probe is too close to the surface, it will reflect significant levels back into the antenna, causing frequency-sensitive variations in the pattern variations. If it is too distant, the track must extend far beyond the edge of the antenna. Approximately 3 to 5 wavelengths off the aperture is a common spacing.

Real probes have distinct patterns in both amplitude and phase and distinct polarization characteristics, all of which impact directly on the field measurements. In a planar scan, the bulk of the pattern information outside the main beam is coupled to the probe in the off-boresight direction. Consequently, almost all modern testing includes a calibration for probe correction.

Reflections from the probe and probe carriage are a major source of error, so extensive use of absorber is made around these objects. Likewise, walls and ceilings that are visible from both the antenna and the probe must be covered if the space attenuation is insufficient for the sidelobe level being measured.

Errors result from a variety of sources: (1) multiple reflections and room reflections, (2) probe position, (3) scan truncation, (4) probe calibration, and (5) instrumentation linearity. These have been analyzed in detail by various authors,[66,67] and techniques have been developed to reduce each error. For example, multiple reflections can be resolved by scanning at multiple spacings from the antenna.

A near-field range involves a major investment and skilled operators. Both amplitude and phase must be measured. Significant rework is necessary to change frequency bands, including new probes, probe calibrations, and possibly absorber. Planar measurements do not give backlobes. Nevertheless, most of the current developments in antenna testing are related to near-field methods. The capability to work indoors and to obtain detailed information on individual elements of an array are powerful incentives to use near-field testing. In some cases, such as space applications, the antenna is too fragile to be mounted on a pedestal and rotated in the conventional manner.

Miscellaneous Tests. Several antenna tests deserve special mention, including gain, circular polarization, field probing, and pattern analysis.

Gain. The most critical antenna parameter is usually gain, and its measurement is seldom trivial. The gain of the antenna results from a combination of its spatial directivity and its losses. In principle, the directivity is available from an integration of measured patterns over all space, but this is rarely practical. Instead, the level at the pattern peak is compared with that of a known or calculable gain standard. Gain standards[68] are usually short pyramidal horns with a deliberately nonplanar phase front so that the gain varies slowly with frequency. They typically have between 10 and 20 dB of gain. Because they have a wide beamwidth, they may illuminate much more of the range than the high-gain antenna under test, with significant errors. A reflection 30 dB down can result in ±0.3 dB of gain error. Consequently, it is wise to move the gain standard several wavelengths in each direction to average out ripples due to wide-angle reflections.

Circular Polarization. When circular polarization is involved, the purity of the polarization is commonly of interest. This polarization purity can be expressed either in terms of ellipticity (axial ratio) or as the cancellation ratio (the fraction of cross-polarized radiation). These are related in the graph of Fig. 6.52.

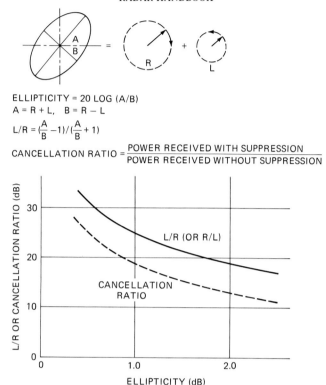

ELLIPTICITY = 20 LOG (A/B)
A = R + L,   B = R − L

$$L/R = (\frac{A}{B} - 1)/(\frac{A}{B} + 1)$$

$$\text{CANCELLATION RATIO} = \frac{\text{POWER RECEIVED WITH SUPPRESSION}}{\text{POWER RECEIVED WITHOUT SUPPRESSION}}$$

FIG. 6.52   Circular-polarization characteristics.

Polarization properties can be measured with circularly polarized source antennas, but it is usual to use rotatable linear-polarized sources. Often circular polarization is used for rejection of weather echoes, in which case average performance over an extended area is of significance. The generally accepted figure of merit is the integrated cancellation ratio (ICR). It is a measure of total return to the radar for raindrops uniformly distributed throughout space. It is measured with a spinning, linearly polarized source. Preferably, the entire source antenna is spun rather than the feed alone to avoid polarization errors of the reflector. Feed supports can be a source of such errors.

The round-trip nature of the ICR calculation makes performance near the peak dominant. Contributions outside the −10 dB point on the main beam are generally negligible.

Field Probing.   The standard test of pattern range adequacy has been a field probe in which the source illuminates the receive mount and a small radiator is scanned across the region of interest, usually with a horizontal and a vertical scan. The finite range phase taper can be detected as well as the amplitude curvature due to the source beamwidth and pointing error. Off-axis reflections show up as periodic ripples. The period of the ripples determines the angle off axis as $\alpha = \arcsin(\lambda/s)$. The peak-to-peak magnitude D of the ripple is determined by the amplitude of the reflection R as approximately

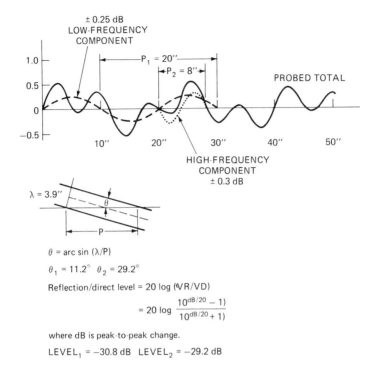

θ = arc sin (λ/P)

$\theta_1 = 11.2°$ $\theta_2 = 29.2°$

Reflection/direct level = 20 log (VR/VD)

$$= 20 \log \frac{10^{dB/20} - 1)}{10^{dB/20} + 1)}$$

where dB is peak-to-peak change.

LEVEL$_1$ = −30.8 dB LEVEL$_2$ = −29.2 dB

FIG. 6.53 Field probe data showing two reflections.

$$D \text{ (dB)} = 20 \log [(1 + R)/(1 - R)] \qquad (6.49)$$

or the reflected level in decibels is

$$R \text{ (dB)} = 20 \log [(10^{D/20} - 1)/(10^{D/20} + 1)] \qquad (6.50)$$

Figure 6.53 shows a typical recorded field probe result. From the data we can determine that sources exist at 11.2° and 29.2° off axis with magnitudes of −30.8 dB and −29.2 dB, respectively. In principle, this analysis can be carried to the point that subsequent patterns are corrected for the measured reflections.

Probe results must be used with care. Reflections outside the probe beamwidth are not detected. Thus a large reflection to the side of the test tower would show up in an azimuth cut but not in a field probe. On the other hand, the probe can also illuminate objects outside the path of a high-gain test antenna, thereby giving pessimistic data.

REFERENCES

1. Skolnik, M. I.: Antenna Options in Radar System Design, *Microwave J.*, vol. 26, pp. 75–82, December 1983.

2. Silver, S. (ed.): "Microwave Antenna Theory and Design," MIT Radiation Laboratory Series, vol. 12, McGraw-Hill Book Company, New York, 1949.

3. Johnson, R. C., and H. Jasik (eds.): "Antenna Engineering Handbook," 2d ed., McGraw-Hill Book Company, New York, 1984, pp. 32-11, 32-12.

4. Hansen, R. C.: "Microwave Scanning Antennas," Academic Press, New York, 1966; Peninsula Publishing, Los Altos, Calif., 1985.

5. Barton, D. K., and H. R. Ward: "Handbook of Radar Measurement," Prentice-Hall, Englewood Cliffs, N.J., 1969; Artech House, Norwood, Mass., 1984.

6. Kraus, J. D.: "Antennas," 2d ed., McGraw-Hill Book Company, New York, 1988, sec. 2-34.

7. Kraus, J. D.: "Electromagnetics," 3d ed., McGraw-Hill Book Company, New York, 1984, sec. 11-5.

8. Brown, J.: "Microwave Lenses," Methuen & Co., Ltd., London, 1953.

9. Kock, W. E.: Metal Lens Antennas, *Proc. IRE*, vol. 34, pp. 828–836, November 1946.

10. Luneburg, R. K.: "Mathematical Theory of Optics," University of California Press, Berkeley, 1964, chap. 3.

11. Braun, E. H.: Radiation Characteristics of the Spherical Luneburg Lens, *IRE Trans.*, vol. AP-4, pp. 132–138, April 1956.

12. Warren, F. G. R., and S. F. A. Pinnel: The Tin Hat Scanning Antenna, *RCA Victor Co., Ltd. (Montreal, Canada) Tech. Rept.* 6, 1951. Contract DRBS-2-1-44-4-3. Also, The Mathematics of the Tin Hat Scanning Antenna, *RCA Victor Co., Ltd. (Montreal, Canada) Tech. Rept.* 7, 1951. Contract DRBS-2-1-44-4-3.

13. Rotman, W., and R. F. Turner: "Wide-Angle Microwave Lens for Line Source Applications," *IEEE Trans.*, vol. AP-11, pp. 623–632, November 1963.

14. Sletten, C. J.: The Theory of Reflector Antennas, *Air Force Cambridge Res. Lab., AFCRL*-66-761, *Phys. Sci. Res. Paper* 290, 1966.

15. Kelleher, K. S.: High-Gain Reflector-Type Antennas, in H. Jasik (ed.), "Antenna Engineering Handbook," McGraw-Hill Book Company, New York, 1961, chap. 12.

16. Kiely, D. G.: Parabolic Cylinder Aerials, *Wireless Eng.*, vol. 28, pp. 73–78, March 1951.

17. Sengupta, D. L.: Parabolic Cylindrical Reflector Antennas, *J. Telecommun. Eng. (India)*, vol. 11, pp. 248–255, July 1965.

18. Fante, R. L. et al.: A Parabolic Cylinder Antenna with Very Low Sidelobes, *IEEE Trans.*, vol. AP-28, pp. 53–59, January 1980.

19. Kelleher, K. S., and H. P. Coleman: Off-Axis Characteristics of the Paraboloidal Reflector, *Naval Res. Lab. Rept.* 4088, 1952. Also see Ref. 4, pp. 15-19–15-20.

20. Lo, Y. T.: On the Beam Deviation Factor of a Parabolic Reflector, *IRE Trans.*, vol. AP-8, pp. 347–349, May 1960.

21. Sandler, S. S.: Paraboloidal Reflector Patterns for Off-Axis Feed, *IRE Trans.*, vol. AP-8, pp. 368–379, July 1960.

22. Cohen, W., and C. M. Steinmetz: "Amplitude and Phase Sensing Monopulse System Parameters," *Microwave J.*, pp. 27–33, October 1959.

23. Rhodes, D. R.: "Introduction to Monopulse," McGraw-Hill Book Company, New York, 1959.

24. Hannan, P. W., and P. A. Loth: A Monopulse Antenna Having Independent Optimi-

zation of the Sum and Difference Modes, *IRE Int. Conv. Rec., pt. 1*, pp. 57–60, March 1961.

25. Ricardi, L. J., and L. Niro: Design of a Twelve-Horn Monopulse Feed, *IRE Int. Conv. Rec.*, pt. 1, pp. 49–56, March 1961.

26. Hannan, P. W.: Microwave Antennas Derived from the Cassegrain Telescope, *IRE Trans.*, vol. AP-9, pp. 140–153, March 1961.

27. Potter, P. D.: Aperture Illumination and Gain of a Cassegrainian System, *IEEE Trans.*, vol. AP-11, pp. 373–375, May 1963.

28. Potter, P. D.: Application of Spherical Wave Theory to Cassegrainian-Fed Paraboloids, *IEEE Trans.*, vol. AP-15, pp. 727–736, November 1967.

29. Rusch, W. V. T.: Scattering from a Hyperboloidal Reflector in a Cassegrainian Feed System, *IEEE Trans.*, vol. AP-11, pp. 414–421, July 1963.

30. Wilkinson, E. J., and A. J. Applebaum: Cassegrain Systems, *IRE Trans.*, vol. AP-9, pp. 119–120, January 1961.

31. Skolnik, M. I.: "Introduction to Radar Systems," 2d ed., McGraw-Hill Book Company, New York, 1980.

32. Sletten, C. J. et al.: Offset Dual Reflector Antennas for Very Low Sidelobes, *Microwave J.*, pp. 221–240, May 1986.

33. Pearson, R. A., and M. S. Smith: Electronic Beam Scanning Using an Array-Fed Dual Offset Reflector Antenna, *IEEE AP-S Int. Symp. Dig.*, pp. 263–266, 1986.

34. Ashmead, J., and A. B. Pippard: The Use of Spherical Reflectors as Microwave Scanning Aerials, *J. IEE (London)*, vol. 93, pt. 3A, pp. 627–632, January 1946.

35. Schrank, H. E., and R. D. Grove: The Helisphere Passive Beacon, *Sixteenth U.S.A.F. Antenna Symp.*, University of Illinois, October 1966. (Also *Microwaves*, August 1968.)

36. Love, A. W.: "Electromagnetic Horn Antennas," IEEE Press, New York, 1976.

37. Barton, D. K.: "Modern Radar System Analysis," Artech House, Norwood, Mass., 1988.

38. Fradin, A. Z.: "Microwave Antennas," Pergamon Press, New York, 1961. (Translated from Russian by M. Nadler.)

39. Wood, P. J.: "Reflector Antenna Analysis and Design," IEE/P. Peregrinus, Ltd., London and New York, 1980.

40. Rusch, W. V. T.: Scattering from a Hyperboloidal Reflector in a Cassegrain Feed System, *IEEE Trans.*, vol. AP-11, pp. 414–421, July 1963.

41. Gray, C. L.: Estimating the Effect of Feed Support Member Blocking on Antenna Gain and Sidelobe Level, *Microwave J.*, pp. 88–91, March 1964.

42. Jenkins, F. A., and H. E. White: "Fundamentals of Optics," McGraw-Hill Book Company, New York, 1937, pp. 365–370.

43. Ruze, J.: The Effect of Aperture Errors on the Antenna Radiation Pattern, *Nuovo Cimento, Suppl.*, vol. 9, no. 3, pp. 364–380, 1952.

44. Ruze, J.: Antenna Tolerance Theory—A Review, *Proc. IEEE*, vol. 54, pp. 633–640, April 1966.

45. Walton, J. D., Jr. (ed): "Radome Engineering Handbook," Marcel Dekker, New York, 1970.

46. Tice, T. E. (ed.): Techniques for Airborne Radome Design, *AFAL-TR*-66-391, vol. 1, pp. 239, 273–275, December 1966.

47. Walton, J. D. (ed.): Techniques for Airborne Radome Design, *AFAL-TR*-66-391, vol. 2, pp. 121–128, December 1966.

48. Guarini, J. F.: Rain Erosion Sled Tests, *Proc. OSU-RTD Symp. Electromagnetic Windows*, vol. 3, sess. 5, pp. A1–A10, June 2–4, 1964.

49. Pressel, P. I.: Boresight Prediction Techniques, *Proc. OSU-WADC Radome Symp.*, vol. 1, pp. 33–40, August 1956.

50. Pressel, P. I., and H. F. Mathis: Improved Boresight Prediction Techniques, *Proc. OSU-WADC Radome Symp.*, vol. 1, pp. 126–133, June 1957.

51. Pressel, P. I., and H. F. Mathis: Effect of Conical Scan on Boresight Error, *Proc. OSU-WADC Radome Symp.*, vol. 1, pp. 134–137, June 1957.

52. Tricoles, G.: A Radome Error Prediction Method Based on Aperture Fields and Rays: Formulation and Application, *Proc. OSU-WADD Symp. Electromagnetic Windows*, vol. 1, pp. 544–564, June 1960.

53. Tricoles, G.: Application of a Ray Tracing Method for Predicting Radome Errors to a Small Radome, *Proc. OSU-RTD Symp. Electromagnetic Windows*, vol. 4, June 2–4, 1964.

54. Chase, V. A., and R. L. Copeland: Development of a 1200°F Radome, *Proc. OSU-RTD Symp. Electromagnetic Windows*, vol. 1, sess. 2, June 2–4, 1964.

55. Bird, W. W.: Large Air Supported Radomes for Satellite Communications Ground Stations, *Proc. OSU-RTD Symp. Electromagnetic Windows*, vol. 5, June 2–4, 1964.

56. Beal, C., J. R. Gruber, and D. J. Driscoll: Design and Performance of 60 Foot C-Band Rigid Radome CW-424/RPS-26, *Proc. OSU-RTD Symp. Electromagnetic Windows*, vol. 2, sess. 4, June 2–4, 1964.

57. Curtis, R. B.: Survey of Ground Radomes, *Rome Air Development Center, RADC-TDR*-64-127, pp. 2, 3, 6, 7, May 1964.

58. Technical Data on Five Types of Ground Radomes Manufactured by Goodyear Aircraft Company, *Goodyear Aircraft Co., GER*-10064, Nov. 23, 1960.

59. Kay A. L.: Electrical Design of Metal Space Frame Radomes, *IEEE Trans.*, vol. AP-13, pp. 188–202, March 1965.

60. Design of Metal Space Frame Radomes, *Rome Air Development Center, RADC-TDR*-64-334, November 1964.

61. Clutter, C. C., A. P. King, and W. E. Kock: Microwave Antenna Measurements, *Proc. IRE*, vol. 35, pp. 1462–1471, December 1947.

62. Hansen, R. C.: Test Procedure for Antennas, *IEEE Trans.*, vol. AP-13, pp. 437–466, May 1965.

63. Hollis, J. S.: Antenna Measurements—Part I, *Microwave J.*, vol. 3, pp. 39–46, February 1960.

64. Johnson, R. C., H. A. Ecker, and R. A. Moore: Compact Range Techniques and Measurements, *IEEE Trans.*, vol. AP-17, pp. 568–576, 1969.

65. Yaghjian, A. D.: An Overview of Near-Field Antenna Measurement, *IEEE Trans.*, vol. AP-34, pp. 30–45, January 1986.

66. Joy, E. B.: Maximum Near-Field Measurement Error Specification, *Dig. Int. Symp. Antennas Propag.*, Stanford, Calif., pp. 390–393, June 1973.

67. Wang, J. J. H.: An Examination of the Theory and Practices of Planar Near-Field Measurement, *IEEE Trans.*, vol. AP-36, pp. 746–753, June 1988.

68. Braun, E. H.: Gain of Electromagnetic Horns, *Proc. IRE*, vol. 14, pp. 109–115, January 1953.

69. Hacker, P.S., and H. E. Schrank: Range Distance Requirements for Measuring Low and Ultralow Sidelobe Antenna Patterns, *IEEE Trans.*, vol. AP-30, pp. 956–966, September 1982.

CHAPTER 7
PHASED ARRAY RADAR ANTENNAS

Theodore C. Cheston
Naval Research Laboratory

Joe Frank
Technology Service Corporation

7.1 INTRODUCTION

Phased Array Radars

Multifunction Radar. Early radar systems used antenna arrays formed by the combination of individual radiators. Such antennas date back to the turn of the twentieth century.[1,2,3] Antenna characteristics are determined by the geometric position of the radiators and the amplitude and phase of their excitation. As radars progressed to shorter wavelengths, arrays were displaced by simpler antennas such as parabolic reflectors. For modern radar applications the advent of electronically controlled phase shifters and switches has once more directed attention to array antennas. The aperture excitation may now be modulated by controlling the phase of the individual elements to give beams that are scanned electronically. This chapter will be devoted to arrays of this type.

The capability of rapidly and accurately switching beams permits multiple radar functions to be performed, interlaced in time or even simultaneously. An electronically steered array radar may track a great multiplicity of targets, illuminate a number of targets with RF energy and guide missiles toward them, perform complete hemispherical search with automatic target selection, and hand over to tracking. It may even act as a communication system, directing high-gain beams toward distant receivers and transmitters. Complete flexibility is possible; search and track rates may be adjusted to best meet particular situations, all within the limitations set by the total use of time. The antenna beamwidth may be changed to search some areas more rapidly with less gain. Frequency agility is possible with the frequency of transmission changing at will from pulse to pulse or, with coding, within a pulse. Very high powers may be generated from a multiplicity of amplifiers distributed across the aperture. Electronically controlled array antennas can give radars the flexibility needed to perform all the various functions in a way best suited for the specific task at hand. The functions may be programmed adaptively to the limit of one's capability to exercise effective automatic management and control.

Phased array theory was studied intensively in the 1960s, bringing understanding. Technology advanced and led to a series of operational systems in the 1980s; many publications became available.[4-15] In terms of performance improvement, ultralow sidelobes (less than −40 dB) were demonstrated first in the 1970s by Westinghouse Electric Corporation's AWACS (Airborne Warning and Control System) and brought about tight tolerances in construction and phase settings. The advent of more and better computer modeling and sophisticated test equipment such as network analyzers has led to improved methods of designing well-matched apertures. Better components such as radiating elements, phase shifters, and power dividers are now available. More economical solid-state devices and memory chips have led to precision aperture phase control with corrections for frequency and temperature variations. Solid-state microwave devices hold great promise for future systems where a solid-state module is associated with each radiating element; improvements in terms of aperture control, reliability, and efficiency continue. Phased arrays can be controlled adaptively, particularly for sidelobe cancellation. This is an area where theory and understanding have advanced much. Also great progress has been made with indoor near-field antenna ranges,[16] where computer-controlled precision two-dimensional radiation patterns are derived at multiple frequencies, and with scanning.

Phased arrays are very expensive. As technology advances, costs are reduced, particularly in the areas of phase shifters and drivers. At the same time, the quest for better performance with lower sidelobes and wider bandwidth keeps the costs high. The greatest potential for cost reductions is believed to lie in the application of solid-state systems with a transmit/receive module at each element.

Phased Array Antennas. The phased array antenna has an aperture that is assembled from a great many similar radiating elements, such as slots or dipoles, each element being individually controlled in phase and amplitude. Accurately predictable radiation patterns and beam-pointing directions can be achieved.

The general planar array characteristics are readily obtained from a few simple equations, given here but discussed later in greater detail. With the elements spaced by $\lambda/2$ (λ = wavelength) to avoid the generation of multiple beams (grating lobes), the number of radiating elements N for a pencil beam is related to the beamwidth by

$$N \approx \frac{10,000}{(\theta_B)^2}$$

$$\theta_B \approx \frac{100}{\sqrt{N}}$$

where θ_B is the 3 dB beamwidth in degrees. The corresponding antenna gain, when the beam points broadside to the aperture, is

$$G_0 \approx \pi N \eta \approx \pi N \eta_L \eta_a$$

where η accounts for antenna losses (η_L) and reduction in gain due to unequal weighting of the elements with a nonuniform amplitude distribution (η_a). When scanning to an angle θ_0, the gain of a planar array is reduced to that of the projected aperture:

$$G(\theta_0) \approx \pi N \eta \cos \theta_0$$

Similarly, the scanned beamwidth is increased from the broadside beamwidth (except in the vicinity of endfire, $\theta_0 = 90°$):

$$\theta_B \text{ (scanned)} \approx \frac{\theta_B(\text{broadside})}{\cos \theta_0}$$

The total number of beams M (with broadside beamwidth and square stacking) that fit into a sphere is approximately equal to the gain and with $\eta \approx 1$ is thus simply related to N by

$$M \approx \pi N$$

In a planar array where the beamwidth changes with the scan angle, the number of beams that can actually be generated and fitted into a sphere is

$$M' \approx \frac{\pi}{2}N$$

An array where the elements are fed in parallel (Sec. 7.8) and which is scanned by phase shift, modulo 2π, has limited bandwidth since for wideband operation constant path lengths rather than constant phases are required. The limit is given by

$$\text{Bandwidth } (\%) \approx \text{beamwidth (deg)}$$

This is equivalent to limitations given by

$$\text{Pulse length} = 2 \times \text{aperture size}$$

With these criteria, the scanned radiation pattern at 60° is steered by $\pm$ one-fourth of the local scanned beamwidth as the frequency is changed over the band. If all the frequencies in the band are used with equal weighting, then twice the bandwidth (half the pulse length) becomes acceptable. At a scan angle θ_0 the beam steers with frequency through an angle $\Delta\theta$ so that

$$\delta\theta \approx \frac{\delta f}{f} \tan \theta_0 \qquad \text{rad}$$

For wider bandwidths, time-delay networks have to be introduced to supplement the phase shifters.

Conformal Arrays.[17,18]   Phased arrays may conform to curved surfaces as required, for example, for flush-mounting on aircraft or missiles. If the surface has a large radius of curvature so that all the radiating elements point to substantially the same direction, then the characteristics are similar to those of a planar array even though the exact 3D position of the element has to be taken into account to calculate the required phase. A small radius of curvature is found with cylindrical (or spherical) arrays used for 360° coverage. Elements are switched to avoid sections of the antenna where they point away from the desired beam direction. Difficulties may be encountered in matching the radiating elements and in maintaining polarization purity. This geometry has not yet found use in radar systems.

The discussions in this chapter will concentrate on planar phased arrays.

3D Volumetric Search. 3D volumetric radar search is possible with electronic scanning in both azimuth and elevation; important regions (e.g., the horizon) may be emphasized at will and searched more frequently. The radar may operate with a higher than normal false-alarm rate since targets can easily be confirmed by repeated interrogation. Phase control allows beams to be widened, for example, to reduce search time for the more elevated regions, where reduced ranges need less antenna gain. A separate rotating surveillance radar system may be added for extra coverage (at a second frequency) and to allow more emphasis on tracking.

Monopulse Track. Phased array radars are well suited for monopulse tracking. The radiating elements of the array can be combined in three different ways to give the sum pattern and the azimuth and elevation difference patterns. Contradictory requirements in optimum amplitude distribution for sum and difference patterns exist,[19] but, as with other antenna systems, they may be independently satisfied. The sum and difference patterns are scanned simultaneously.

The difference-pattern null in a phased array system gives good beam-pointing accuracy. Absolute beam-pointing accuracies to within less than one-fiftieth of a (scanned) beamwidth have been measured with scans up to 60°.[20] The accuracy is limited by phase and amplitude errors. Since phase shift rather than time delay is used, as the frequency is changed, the direction of the null of the scanned beam is also changed, and the beam moves toward broadside with an increase in frequency. The amplitude at boresight of the difference-pattern output then increases linearly with a change in frequency. With a scan angle of 60° this change is from a null at the design frequency to a value of about −9 dB relative to the sum pattern, at the edge of the band, where the band is defined by bandwidth (percent) = beamwidth (deg). This is discussed more fully in Sec. 7.7.

Shaped Beams. The radiation pattern of an array may be shaped by modifying the aperture distribution. Good pattern approximations can be obtained by using phase only. In particular, the beam may be broadened by applying a spherical phase distribution to the aperture or by approximating it with a gable (triangular) phase distribution. Beams of this type are of particular interest since they are easily generated. They may be used for transmission in a system where the receiving antenna has a cluster of simultaneous beams, or, as previously discussed, they may be used in a search system to reduce the number of angular cells in regions of shorter range.

Monitoring. Electronically scanned arrays are composed of very many parts and include electronic circuitry to drive the phase shifters or switches that steer the beam. The overall reliability of such arrays can be great; graceful degradation has been claimed, since the failure of as much as 10 percent of the components leads to a loss in gain of only 1 dB. There is, however, a degradation of (low) sidelobes. Nevertheless, the functioning of the antenna is complex, and there is need for providing test or monitoring circuitry. The decision to point a beam in a certain direction is made somewhere in the radar control system and is normally defined by two direction cosines. A test or monitoring circuit should establish the correct functioning of all components, including all beam-pointing computations, electronic drivers and phase shifters or switches, and all their interconnections. Frequent indications that the antenna system is functioning or is capable of functioning should be available. In one possible method the phase shifters are programmed to focus on a nearby monitor probe and scan past it.[21] This will yield a close approximation of the complete radiation pattern, where gain and sidelobes can be measured and compared with previous results. The contribution of individual elements and their phase shifters (and drivers) can also be checked with this configuration. The phase at each element is sequentially rotated at some low frequency; the amplitude and phase of this modulation as received by the probe relate directly to both the relative amplitude excitation of the element and its relative phase setting.[22] Other methods have been proposed[23] where measurements are compared with previously recorded ones.

Deployment of Apertures. With planar arrays, scanning is limited by the loss in gain and the increase in beamwidth corresponding to the reduction of the aperture to its projected area. Practical extreme values of scanning are therefore in the region of 60 to 70°. A minimum of three planar array apertures is then necessary for hemispherical coverage. For shipborne use, a minimum of four apertures appears desirable since, with pitch and roll, more than hemispherical coverage is necessary. The antennas may be positioned as shown in Fig. 7.1, permitting a view that is unimpeded by the central superstructure. The apertures would normally be tilted back from the vertical to balance the scan angles.

Radiating Elements. The most commonly used radiators for phased arrays are dipoles, slots, open-ended waveguides (or small horns), and printed-circuit "patches" (originally called *Collings radiator* after their inventor[24]). The element has to be small enough to fit in the array geometry, thereby limiting the element to an area of a little more than $\lambda^2/4$. In addition, many radiators are required, and the radiating element should be inexpensive and reliable and have identical characteristics from unit to unit.

Since the impedance and pattern of a radiator in an array are determined predominantly by the array geometry (Sec. 7.4), the radiating element may be chosen to suit the feed system and the physical requirements of the antenna. For example, if the radiator is fed from a stripline phase shifter, a stripline dipole

FIG. 7.1 Guided missile cruiser showing two out of four phased array antennas. (*Courtesy of Ingalls Shipbuilding Division of Litton.*)

would be a logical choice. If a waveguide phase shifter is used, an open-ended waveguide or a slot might be convenient. At the lower frequencies, where coaxial components are prevalent, dipoles have been favored. A ground plane is usually placed about λ/4 behind an array of parallel dipoles so that the antenna forms a beam in only one hemisphere. At the higher frequencies open-ended waveguides and slots are frequently used. Considerable bandwidth (perhaps 50 percent) can be obtained, even with patch radiators, provided they are fed similarly to a dipole.[25]

For limited scanning (say, less than 10°), it is possible to use directive radiators having dimensions of height and width of several wavelengths. With such separation, the mutual coupling effects can be small, and the pattern and impedance of an element in the array approach those of the isolated element.

The element must be chosen to give the desired polarization, usually vertical or horizontal. The special case of circular polarization is discussed below.

If polarization diversity is required or if an array is required to transmit one polarization and receive the orthogonal or both polarizations, either crossed dipoles or circular or square radiators seem suitable. With appropriate feed systems, both are capable of providing vertical and horizontal polarization independently and may be combined to provide any desired polarization, including circular. Such polarization diversity adds considerable complexity, requiring two feed systems or switches at the radiating element level.

Circular Polarization.   From the point of view of the antenna designer, circular polarization is possible, though difficulties may be encountered in matching for large scan angles. On scanning, a component of the undesired orthogonal polarization will be generated,[26] and some provision should be made to absorb that energy.[27] With a conventional circularly polarized antenna, such as a parabolic dish with a circularly polarized feed, good circularity may be obtained over part of the main beam, with rapid deterioration over the rest of the pattern. With a planar array the relevant beamwidth is the beamwidth of the element in the array rather than the array beamwidth. The element beamwidth is broad, and good circularity may be expected over wide angles, including the main beam and sidelobes.

With circular polarization, the signal returned from a single-bounce target will require an antenna matched to the opposite sense of circular polarization from that transmitted. If the same antenna is used, then single-bounce targets are rejected. Such a system can therefore give a measure of suppression of rain echoes,[28] ideally amounting to

$$20 \log (e^2 + 1)/(e^2 - 1) \qquad \text{dB}$$

where e is the voltage-ellipticity ratio. An early model of a Raytheon reflectarray gave an ellipticity ratio of less than 1.5 dB with scans up to 30°, corresponding to a theoretical rain rejection of at least 15 dB. At the same time, an aircraft target would typically lose approximately 3 dB, leaving a relative net improvement of 12 dB of rain rejection.

Phased Arrays with Very Wide Bandwidth.   A radar system that has the capability of changing frequency over a very wide band can, with advantage, adapt its transmission to take into account frequency-dependent multipath characteristics, target response, environmental conditions, interference, and jamming. Further, wideband processing can give fine range resolution.

Phased arrays have the potential of operating over very wide bandwidths. Some ferrite phase shifters operate over two octaves,[29] and digital diode phase shifters that switch line lengths may function over even wider bands. The high end of the frequency band is limited by the physical size of the elements, which must be spaced close enough in the array to avoid the generation of grating lobes. For wide instantaneous bandwidth (rather than tunable bandwidth), time delays have to be added to prevent the beam from being scanned as the frequency is changed.

The impedance of the radiating element at the aperture (with closely spaced elements) is approximately independent of frequency, but the element must be matched over the wide band. This is difficult to achieve without exciting harmful surface waves when scanning. Nevertheless, matching with octave bandwidth for scanning to $\pm 60°$ appears possible.

Limited Scanning.[30] If scanning is limited to a small angular volume, considerable simplifications become possible. The total number of active phase-shifter controls can be reduced to about equal the total number of beams. Subarrays may be formed, each with only one phase control and of a size such that its beamwidth includes all the scan angles. Alternatively, a small phased array could be placed in the focal region of a large reflector to scan the narrow beamwidth of the reflector over a limited scan angle.

Scanning of Arrays

Phase Scanning. The beam of an antenna points in a direction that is normal to the phase front. In phased arrays, this phase front is adjusted to steer the beam by individual control of the phase of excitation of each radiating element. This is indicated in Fig. 7.2a. The phase shifters are electronically actuated to permit rapid scanning and are adjusted in phase to a value between 0 and 2π rad. With an interelement spacing s, the incremental phase shift ψ between adjacent elements for a scan angle θ_0 is $\psi = (2\pi/\lambda)s \sin \theta_0$. If the phase ψ is constant with frequency, the scan angle θ_0 is frequency-dependent.

Time-Delay Scanning. Phase scanning was seen to be frequency-sensitive. Time-delay scanning is independent of frequency. Delay lines are used instead of

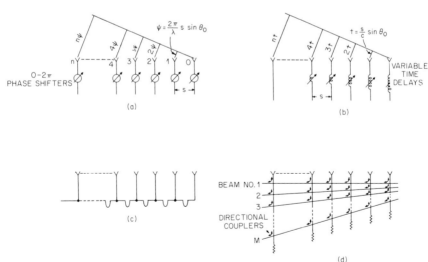

FIG. 7.2 Generation of scanned beams. (a) Phased array. (b) Time-delay array. (c) Frequency-scanned array. (d) Blass-type array.

phase shifters, as shown in Fig. 7.2*b*, providing an incremental delay from element to element of $t = (s/c) \sin \theta_0$, where c = velocity of propagation. Individual time-delay circuits (Sec. 7.7) are normally too cumbersome to be added to each radiating element. A reasonable compromise may be reached by adding one time-delay network to a group of elements (subarray) where each element has its own phase shifter.

Frequency Scanning.[31] Frequency rather than phase may be used as the active parameter to exploit the frequency-sensitive characteristics of phase scanning. Figure 7.2*c* shows the arrangement. At one particular frequency all radiators are in phase. As the frequency is changed, the phase across the aperture tilts linearly, and the beam is scanned. Frequency-scanning systems are relatively simple and inexpensive to implement. They have been developed and deployed in the past to provide elevation-angle scanning in combination with mechanical horizontal rotation for 3D radars. A chapter in the first edition of this handbook was devoted to this approach, which since then has received much less attention; frequency is usually considered too important a parameter to give up for scanning.

IF Scanning. For receiving, the output from each radiating element may be heterodyned (mixed) to an intermediate frequency (IF). All the various methods of scanning are then possible, including the beam-switching system described below, and can be carried out at IF, where amplification is readily available and lumped constant circuits may be used.

Digital Beamforming.[32-34] For receiving, the output from each radiating element may be amplified and digitized. The signal is then transferred to a computer for processing, which can include the formation of multiple simultaneous beams (formed with appropriate aperture illumination weighting) and adaptively derived nulls in the beam patterns to avoid spatial interference or jamming. Limitations are due to the availability and cost of analog-to-digital (A/D) converters and to their frequency and dynamic-range characteristics. Partial implementation is possible by digitizing at subarray levels only.

Beam Switching. With properly designed lenses or reflectors, a number of independent beams may be formed by feeds at the focal surface. Each beam has substantially the gain and beamwidth of the whole antenna. Allen[35] has shown that there are efficient equivalent transmission networks that use directional couplers and have the same collimating property. A typical form, after Blass,[36] is shown in Fig. 7.2*d*. The geometry can be adjusted to provide equal path lengths, thus providing frequency-independent time-delay scanning. Another possible configuration providing multiple broadband beams uses parallel plates containing a wide-angle microwave lens[37,38] (Gent, Rotman). Each port corresponds to a separate beam. The lens provides appropriate time delays to the aperture, giving frequency-invariant scanning. The beams may be selected through a switching matrix requiring $M-1$ single-pole–double-throw (SPDT) switches to select one out of M beams. The beams are stationary in space and overlap at about the 4 dB points. This is in contrast to the previously discussed methods of scanning, where the beam can be steered accurately to any position. The beams all lie in one plane. The system may be combined with mechanical rotation of the antenna, giving vertical switched scanning for 3D coverage. Much greater complexity is required for a system switching beams in both planes.

Multiple Simultaneous Beams. Instead of switching the beams, as described in the preceding paragraph, all the beams may be connected to separate receivers, giving multiple simultaneous receive beams. The transmitter radiation pattern would need to be wide to cover all the receive beams. Such multibeam systems have found application in combination with mechanical rotation for 3D coverage.

Multiple Independently Steered Beams. Independent multiple beams may be generated with a single beamformer by modifying both amplitude and phase at the aperture. This can be seen from Fig. 7.3, where, for example, two independent beams are generated. Both beams have the same amplitude (voltage) distribution $F(x)$ but differently inclined linear phased fronts. The total aperture excitation with both beams is

$$F(x,\psi) = F(x)e^{j2\psi_1(x/a)} + F(x)e^{j2\psi_2(x/a)} = 2F(x)\left[\cos(\psi_1 - \psi_2)\frac{x}{a}\right]e^{j(\psi_1 + \psi_2)(x/a)}$$

That is, the aperture amplitude distribution required for two separate beams varies cosinusoidally, and the phase distribution is linear and has the average inclination.

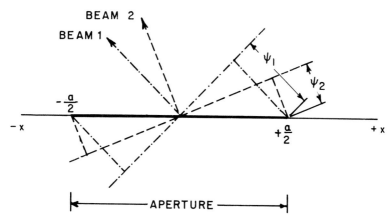

FIG. 7.3 Aperture distribution giving two beams.

In most phased array systems only the phase can be controlled. Ignoring the required amplitude variations still leads to good approximations for forming multiple beams, by superimposing the various required phase-shifter settings (modulo 2π). In the case of two beams, the aperture phase slope has the average inclination and varies periodically from 0 to π.

It should be noted that when a multilobed radiation pattern is received or transmitted in one channel, the gain is shared between the lobes. When the beams are contained in separate (beamforming) channels, however, each channel has the full gain of the aperture.

Vertical Scan Only. A greatly simplified phased array system becomes possible if there is no need for multifunction capabilities, including fire control, where a beam may have to be pointed in any given direction at any time. The array is scanned in the vertical plane only and mechanically rotated to give azimuth coverage. The number of phase control points is then reduced to the number of horizontal rows. In the case of a ship's surveillance radar, the antenna should be positioned as high as possible to avoid shadowing by the superstructure, but the pedestal need not be stabilized since stabilization can be achieved by electronic beam steering. Scanning can be in the form of phase scanning or beam switching, or multiple simultaneous beams may be used on receive with a wide antenna pattern on transmit. Many systems of this type have been developed for both naval and land-based use (Sec. 7.11).

7.2 ARRAY THEORY

Array with Two Elements. Figure 7.4 shows two isotropic radiators which are spaced by a distance s and excited with equal amplitude and phase. With unity input power, the vector sum of their contributions, added at a great distance as a function of θ, is the radiation pattern

$$E_a(\theta) = \frac{1}{\sqrt{2}} [e^{j(2\pi/\lambda)(s/2)\sin\theta} + e^{-j(2\pi/\lambda)(s/2)\sin\theta}]$$

where θ is measured from the broadside direction. Normalizing, to give unity amplitude when $\theta = 0$, and simplifying give

$$E_a(\theta) = \cos\left[\pi \frac{s}{\lambda} \sin\theta\right] \qquad (7.1)$$

The absolute value of $E_a(\theta)$ is plotted in Fig. 7.4 as a function of $\pi(s/\lambda)\sin\theta$. If the plot had been in terms of the angle θ, the lobes would have been found to increase in width as $|\theta|$ increased. The main lobe occurs when $\sin\theta = 0$. The other lobes have the same amplitude as the main lobe and are referred to as *grating lobes*. They occur at angles given by $\sin\theta = \pm [m/(s/\lambda)]$, where m is an integer. For the half space given by $-90° < \theta < +90°$, there are $2m'$ grating lobes, where m' is the largest integer smaller than s/λ. If $s < l$, grating-lobe maxima do not occur, and the value at $\pm90°$ is $\cos(\pi s/\lambda)$. This value is for isotropic radiators and is reduced if the radiators have directivity.

Linear Array.[39] With a linear array of N isotropic radiators, excited with equal amplitudes and phase and separated by distances s, as shown in Fig. 7.5, the condition for the occurrence of grating lobes is unchanged from the simpler case just considered. They occur for the same values of $\pi(s/\lambda)\sin\theta$, but the width of the lobes is reduced, and they are separated by minor lobes. Summing the vector contributions from all elements, with element 0 as phase reference, gives

$$E_a(\theta) = \frac{1}{\sqrt{N}} \sum_{n=0}^{n=N-1} e^{j(2\pi/\lambda)\,ns\sin\theta}$$

The factor $1/\sqrt{N}$ shows that each element is energized with $1/N$ of the (unity) input power. Normalizing the gain to unity at broadside, $\theta = 0$, gives the pattern

$$E_a(\theta) = \frac{\sin[N\pi(s/\lambda)\sin\theta]}{N\sin[\pi(s/\lambda)\sin\theta]} \qquad (7.2)$$

$E_a(\theta)$ gives the radiation pattern for isotropic radiators and is known as the *array factor*. It is shown in Fig. 7.6 for $N = 10$. The pattern is repetitive, and the lo-

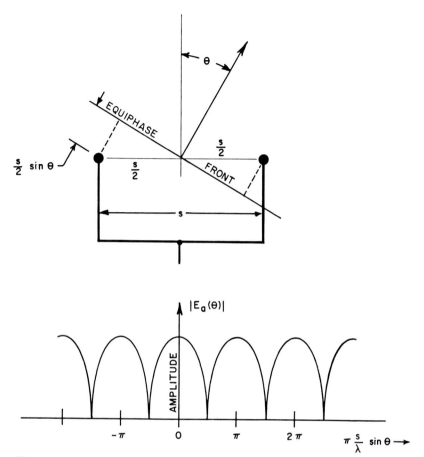

FIG. 7.4 Radiation pattern of two isotropic radiators.

cations of the adjacent grating lobes at angles θ_1 and θ_2 are separated by $\pi(s/\lambda)$ $(\sin \theta_1 - \sin \theta_2) = \pi$.

The radiating elements are not isotropic but have a radiation pattern $E_e(\theta)$, known as the *element factor* or *element pattern*; then the complete radiation pattern $E(\theta)$ is the product of the array factor and the element pattern:

$$E(\theta) = E_e(\theta)E_a(\theta) = E_e(\theta)\frac{\sin [N\pi(s/\lambda) \sin \theta]}{N \sin [\pi(s/\lambda) \sin \theta]} \qquad (7.3)$$

An approximation to the pattern of Eq. (7.2) is in the form

$$E(\theta) = \frac{\sin [\pi(a/\lambda) \sin \theta]}{\pi(a/\lambda) \sin \theta]} \qquad (7.4)$$

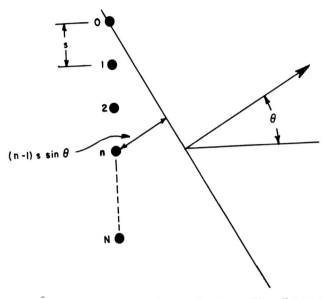

FIG. 7.5 Linear array with N radiators uniformly spaced by a distance s.

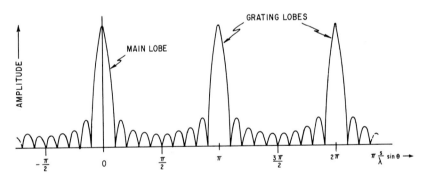

FIG. 7.6 Array factor with $N = 10$ elements.

where the effective aperture is $a = Ns$, which extends by $s/2$ beyond the centers of the end elements. In contrast to the array factor, this pattern has only one maximum and is nonrepetitive. It is the well-known Fourier transform of a continuous constant-amplitude distribution and is a reasonable approximation for small values of θ when the aperture is greater than several wavelengths. The *half-power beamwidth* is obtained from Eq. (7.4):

$$\theta_B = \frac{0.886}{a/\lambda} \text{ (rad)} = \frac{50.8}{a/\lambda} \text{ (deg)} \qquad (7.5)$$

The first sidelobe is 13.3 dB down from the peak of the main beam.

For larger values of θ the pattern of a continuous aperture is modified from Eq. (7.4) by the obliquity factor[40,41] $\frac{1}{2}(1 + \cos \theta)$, which arises from the definition of a Huygens source. This gives

$$E(\theta) = \frac{1}{2} (1 + \cos \theta) \frac{\sin [\pi(a/\lambda) \sin \theta]}{\pi(a/\lambda) \sin \theta} \qquad (7.6)$$

For closely spaced elements the obliquity factor is very similar to the amplitude pattern of a well-designed (matched) radiating element, $\sqrt{\cos \theta}$ for values up to some 60 or 70°. At greater angles the element pattern has values that are greater than those given by $\sqrt{\cos \theta}$ and that are a function of the total number of elements.[42]

Scanned Linear Arrays. The pattern of the array may be steered to an angle θ_0 by applying linearly progressive phase increments from element to element so that the phase between adjacent elements differs by $2\pi(s/\lambda) \sin \theta_0$. Equation (7.2) is then modified, giving the normalized array factor of a uniformly illuminated array as

$$E_a(\theta) = \frac{\sin N\pi(s/\lambda)(\sin \theta - \sin \theta_0)}{N \sin \pi(s/\lambda)(\sin \theta - \sin \theta_0)} \qquad (7.7)$$

and the pattern is

$$E(\theta) = E_e(\theta) \frac{\sin N\pi(s/\lambda)(\sin \theta - \sin \theta_0)}{N \sin [\pi(s/\lambda \sin \theta]} \qquad (7.8)$$

Equation (7.8) describes the fundamental response of a scanned array system. The array factor will have only one single major lobe, and grating-lobe maxima will not occur for $-90° < \theta < +90°$ as long as

$$\pi \frac{s}{\lambda} |\sin \theta - \sin \theta_0| \setminus \pi$$

or

$$\frac{s}{\lambda} < \frac{1}{1 + |\sin \theta_0|} \qquad (7.9)$$

which is always true if $s/\lambda < \frac{1}{2}$. When scanning is limited, the value of s/λ may be increased, for example, to $s/\lambda < 0.53$ for scanning to a maximum of 60° or to $s/\lambda < 0.59$ for scanning to a maximum of ±45°.

For larger values of s/λ, grating lobes occur at angles θ_1, given by

$$\sin \theta_1 = \sin \theta_0 \pm \frac{n}{s/\lambda} \qquad (7.10)$$

where n is an integer.

In the limit, the inequality (7.9) does allow a grating-lobe peak to occur at 90° when scanning to θ_0. Even though the grating lobe is reduced when multiplied by the element pattern, it may be prudent to space the elements such that the first null of the grating lobe, rather than the peak, occurs at 90°. With N elements this more restrictive condition is given by

$$\frac{s}{\lambda}\frac{N-1}{N} \times \frac{1}{1 + |\sin \theta_0|} \qquad (7.11)$$

Equation (7.8) may again be approximated by the Fourier transform of the illumination across the continuous aperture:

$$E(\theta) = \frac{1}{2}(1 + \cos \theta)\frac{\sin \pi(a/\lambda)(\sin \theta - \sin \theta_0)}{\pi(a/\lambda)(\sin \theta - \sin \theta_0)} \qquad (7.12)$$

The Fourier-transform solutions for continuous apertures[19,43] may be used to approximate patterns for practical amplitude and phase distributions as long as the element-to-element spacing is small enough to suppress grating lobes.[44] *Monopulse difference patterns* may be approximated in the same way from the Fourier transforms of the corresponding continuous odd aperture distributions. For example, with a constant amplitude distribution, the difference-pattern array factor calculated by the exact vector addition of all radiating elements is

$$E_a(\theta) = \frac{1 - \cos N\pi(s/\lambda)(\sin \theta - \sin \theta_0)}{N \sin \pi(s/\lambda)(\sin \theta - \sin \theta_0)}$$

The Fourier transform gives the same expression with the *sine* in the denominator replaced by its argument, giving (in the denominator) $\pi(a/\lambda)(\sin \theta - \sin \theta_0)$, where $a = Ns$.

For small scan angles θ_0 and small values of θ the expression $\sin \theta - \sin \theta_0$ may be approximated by $\theta - \theta_0$. For larger values of θ_0, the expression $\sin \theta - \sin \theta_0$ may be expanded to give the response in the general direction of the (narrow) scanned beam in terms of the *small angle* $(\theta - \theta_0)$:

$$\sin \theta - \sin \theta_0 \approx a(\theta - \theta_0) \cos \theta_0 \qquad (7.13)$$

This gives, with Eq. (7.12),

$$E(\theta) \not{=} \frac{1}{2}(1 + \cos \theta)\frac{\sin [(\pi a \cos \theta_0)/\lambda](\theta - \theta_0)}{[(\pi a \cos \theta_0)/\lambda](\theta - \theta_0)} \qquad (7.14)$$

Equation (7.14) measures the angle $\theta - \theta_0$ from the scanned direction. It shows that the effect of scanning is to reduce the aperture to the size of its projected area in the direction of scan. Correspondingly, the beamwidth is increased to

$$\theta_B \text{ (scanned)} \not{=} \frac{\theta_B \text{ (broadside)}}{\cos \theta_0} = \frac{0.886}{(a/\lambda) \cos \theta_0} \text{ (rad)} = \frac{50.8}{(a/\lambda) \cos \theta_0} \text{ (deg)} \qquad (7.15)$$

When the beam is scanned from broadside by an angle $\theta_0 < 60°$ and the aperture

$a/\lambda \hbar 5$, Eq. (7.15) gives a beamwidth that is too narrow, the error being less than 7 percent.

When the beam is scanned to very large scan angles, toward endfire, more exact calculations become necessary.[42,45] Equation (7.8) still applies and gives, for endfire with isotropic radiators,

$$\theta_B \text{ (endfire)} = 2\sqrt{\frac{0.886}{a/\lambda}} \text{ rad} \tag{7.16}$$

Element Factor and Gain of Planar Arrays. The gain of a uniformly illuminated and lossless aperture of area A, with a broadside beam, is $G_{(o)} = 4\pi A/\lambda^2$. With a nonuniform aperture distribution and with losses present, the gain is reduced by the efficiency term η to

$$G_{(0)} = 4\pi \frac{A}{\lambda^2} \eta \tag{7.17}$$

If the aperture is considered as a matched receiver, then the amount of energy arriving from a direction θ_0 is proportional to its projected area. The gain with scanning therefore is

$$G(\theta_0) = 4\pi \frac{A \cos \theta_0}{\lambda^2} \eta \tag{7.18}$$

The variation of gain with the cosine of the scan angle agrees with the equivalent variation in beamwidth given by Eq. (7.15). The gain may be expressed in terms of the actual beamwidth, giving, from Eqs. (7.15) and (7.18),

$$G(\theta_0) \not{\approx} \frac{32,000}{\theta_B \phi_B} \eta \tag{7.19}$$

where θ_B and ϕ_B are the beamwidths in degrees in the two principal planes with the beam scanned to θ_0.

If the aperture is made up of N equal radiating elements and is matched to accept the incident power, then the contribution to the overall gain is the same from all elements, whence

$$G(\theta) = NG_e(\theta)\eta \tag{7.20}$$

where G_e is the gain per element. It follows from Eq. (7.18) that the matched-element power pattern is

$$G_e(\theta) = 4\pi \frac{A}{N\lambda^2} \cos \theta \tag{7.21}$$

and the normalized radiation amplitude pattern of the (matched) element or (matched) *element pattern* is

$$E_e(\theta) = \sqrt{\cos \theta} \tag{7.22}$$

It has already been noted that the matched-element pattern is very similar to the obliquity factor $\frac{1}{2}(1 + \cos \theta)$ and differs markedly only near endfire, where the number of elements begins to matter.[42]

For a given element spacing s the total number of radiators N in the area A is $N = A/s^2$, and Eq. (7.21) gives

$$G_e(\theta) = 4\pi \left[\frac{s}{\lambda}\right]^2 \cos \theta$$

When the element spacing is $s = \lambda/2$, then the power pattern of an element that is perfectly matched at all scan angles is

$$G_e(\theta) = \pi \cos \theta \qquad (7.23)$$

and the peak antenna gain in the direction of scan, θ_0, is

$$G(\theta_0) = \pi N \eta \cos \theta_0 \qquad (7.24)$$

where the efficiency term η accounts for losses and for a nonuniform aperture distribution. For a broadside beam $\theta_0 = 0$ and

$$G_0 = \pi N \eta \qquad (7.25)$$

and the element gain is $G_e = \pi$.

The effects of the element pattern are most marked with wide beams. Figure 7.7 shows the array and element factors and the resulting pattern for a 10-element array, with element spacing $s = \lambda/2$, scanned to 60°. The pattern maximum is noted to occur at less than 60° because the gain of the element pattern increases toward broadside. The pattern value at 60° is $\cos 60° = 0.5$ in power or 0.707 in amplitude, relative to the maximum at broadside, as expected. The sidelobes in the general region of broadside are not reduced since in that region the element

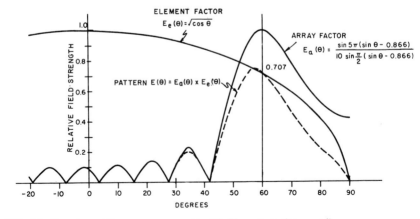

FIG. 7.7 Ten-element linear array scanned to 60°. Element spacing $s = \lambda/2$.

pattern is approximately unity. Relative to the beam maximum, therefore, the sidelobes near broadside are increased by approximately 3 dB.

7.3 PLANAR ARRAYS AND BEAM STEERING

Planar Arrays. A planar array is capable of steering the beam in two dimensions. In a spherical-coordinate system the two coordinates θ and ϕ define points on the surface of a unit hemisphere. As shown in Fig. 7.8, θ is the angle of scan measured from broadside and ϕ is the plane of scan measured from the x axis. Von Aulock[46] has presented a simplified method for visualizing the patterns and the effect of scanning. He considers the projection of the points on a hemisphere onto a plane (Fig. 7.9); the axes of the plane are the direction cosines $\cos \alpha_x$, $\cos \alpha_y$. For any direction on the hemisphere the direction cosines are

$$\cos \alpha_x = \sin \theta \cos \phi$$
$$\cos \alpha_y = \sin \theta \sin \phi$$

The direction of scan is indicated by the direction cosines $\cos \alpha_{xs}$, $\cos \alpha_{ys}$. Here the plane of scan is defined by the angle ϕ measured counterclockwise from the $\cos \alpha_x$ axis and is given by

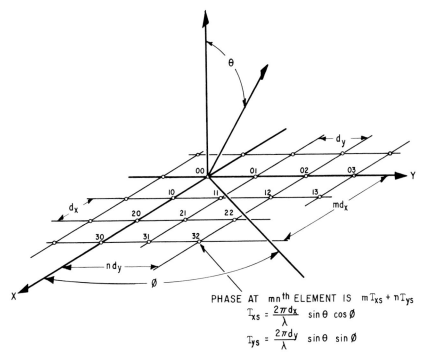

$$\text{PHASE AT } mn^{th} \text{ ELEMENT IS } m\,T_{xs} + n\,T_{ys}$$

$$T_{xs} = \frac{2\pi d_x}{\lambda} \sin \theta \cos \phi$$

$$T_{ys} = \frac{2\pi d_y}{\lambda} \sin \theta \sin \phi$$

FIG. 7.8 Planar-array-element geometry and phasing.

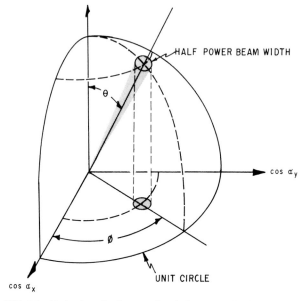

FIG. 7.9   Projection of points on a hemisphere onto the plane of the array.

$$\phi = \tan^{-1}\frac{\cos \alpha_{ys}}{\cos \alpha_{xs}}$$

The angle of scan θ is determined by the distance of the point ($\cos \alpha_{xs}$, $\cos \alpha_{ys}$) from the origin. This distance is equal to $\sin \theta$. For this reason a representation of this sort is called $\sin \theta$ space. A feature of $\sin \theta$ space is that the antenna pattern shape is invariant to the direction of scan. As the beam is scanned, every point on the plot is translated in the same direction and by the same distance as is the beam maximum.

The region inside the unit circle where

$$\cos^2 \alpha_x + \cos^2 \alpha_y \leq 1$$

is defined as *real space*, the hemisphere into which energy is radiated. The infinite region outside the unit circle is referred to as *imaginary space*. Although no power is radiated into imaginary space, the concept is useful for observing the motion of grating lobes as the array is scanned. In addition, the pattern in imaginary space represents stored energy and contributes to the element impedance in the array.

The most common element lattices have either a rectangular or a triangular grid. As shown in Fig. 7.8, the mnth element is located at (md_x, nd_y). The triangular grid may be thought of as a rectangular grid where every other element has been omitted. The element locations can be defined by requiring that $m + n$ be even.

Calculations for the element-steering phases are greatly simplified by the adoption of the direction cosine coordinate system. In this system the linear-phase tapers defined by the beam-steering direction ($\cos \alpha_{xs}$, $\cos \alpha_{ys}$) may be summed at each element so that the phasing at the mnth element is given by

$$\psi_{mn} = mT_{xs} + nT_{ys}$$

where T_{xs} = $(2\pi/\lambda)d_x \cos \alpha_{xs}$ = element-to-element phase shift in the x direction
T_{ys} = $(2\pi/\lambda)d_y \cos \alpha_{ys}$ = element-to-element phase shift in the y direction

The array factor of a two-dimensional array may be calculated by summing the vector contribution of each element in the array at each point in space. For an array scanned to a direction given by the direction cosines $\cos \alpha_{xs}$ and $\cos \alpha_{ys}$, the array factor of an $M \times N$ rectangular array of radiators may be written

$$E_a(\cos \alpha_{xs}, \cos \alpha_{ys}) = \sum_{m=0}^{M-1} \sum_{n=0}^{N-1} |A_{mn}| e^{j[m(T_x - T_{xs}) + n(T_y - T_{ys})]}$$

where T_x = $(2\pi/\lambda) d_x \cos \alpha_x$
T_y = $(2\pi/\lambda) d_y \cos \alpha_y$
A_{mn} = amplitude of mnth element

An array may be visualized as having an infinite number of grating lobes only one of which (namely, the main beam) is desired in real space. It is convenient to plot the position of the grating lobes when the beam is phased for broadside and observe the motion of these lobes as the beam is scanned. Figure 7.10 shows the grating-lobe locations for both rectangular and triangular spacings. For a rectangular array the grating lobes are located at

$$\cos \alpha_{xs} - \cos \alpha_x = \pm \frac{\lambda}{d_x} p$$

$$\cos \alpha_{ys} - \cos \alpha_y = \pm \frac{\lambda}{d_y} q$$

$$p, q = 0, 1, 2, \ldots$$

The lobe at $p = q = 0$ is the main beam. A triangular grid is more efficient for suppressing grating lobes than a rectangular grid,[47] so that for a given aperture size fewer elements are required. If the triangular lattice contains elements at (md_x, nd_y), where $m + n$ is even, the grating lobes are located at

$$\cos \alpha_{xs} - \cos \alpha_x = \pm \frac{\lambda}{2d_x} p$$

$$\cos \alpha_{ys} - \cos \alpha_y = \pm \frac{\lambda}{2d_y} q$$

where $p + q$ is even.

Since only one main lobe is normally desired in real space, an appropriate design will place all but one maximum in imaginary space for all angles of scan. With scanning, lobes that were originally in imaginary space may move into real space if the element spacing is greater than $\lambda/2$. As the array is scanned away from broadside, each grating lobe (in $\sin \theta$ space) will move a distance equal to the sine of the angle of scan and in a direction determined by the plane of scan. To ensure that no grating lobes enter real space, the element spacing must be

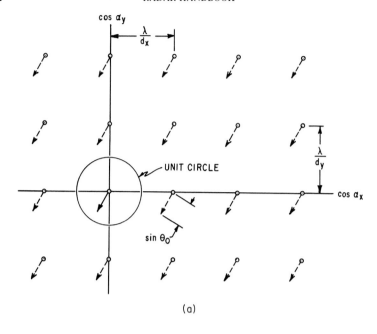

(a)

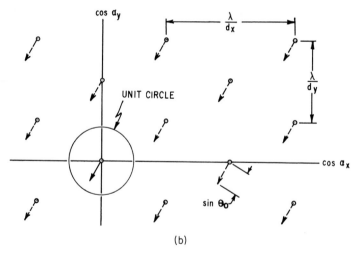

(b)

FIG. 7.10 Grating-lobe positions for (*a*) rectangular and (*b*) triangular grids, showing the motion of the lobes as the beam is scanned an angle θ_o.

chosen so that for the maximum scan angle θ_m the movement of a grating lobe by $\sin \theta_m$ does not bring the grating lobe into real space. If a scan angle of 60° from broadside is required for every plane of scan, no grating lobes may exist within a circle of radius $1 + \sin \theta_m = 1.866$. The square grid that meets this requirement has

$$\frac{\lambda}{d_x} = \frac{\lambda}{d_y} = 1.866 \quad \text{or} \quad d_x = d_y = 0.536\lambda$$

Here, the area per element is

$$d_x d_y = (0.536\lambda)^2 = 0.287\lambda^2$$

For an equilateral-triangular array, the requirement is satisfied by

$$\frac{\lambda}{d_y} = \frac{\lambda}{\sqrt{3} \, d_x} = 1.866 \quad \text{or} \quad d_y = 0.536\lambda \quad d_x = 0.309\lambda$$

Since elements are located only at every other value of mn, the area per element is

$$2d_x d_y = 2(0.536\lambda)(0.309\lambda) = 0.332\lambda^2$$

For the same amount of grating-lobe suppression, the square geometry requires approximately 16 percent more elements.

Element-Phasing Calculations. A computer is usually required to perform the steering computations for a phased array antenna. It can compensate for many of the known phase errors caused by the microwave components, the operating environment, and the physical placement of the elements. For example, if the insertion and differential phase variations (which may occur from phase shifter to phase shifter) are known, they may be taken into account in the computations. Known temperature variations across the array that would induce phase errors may be compensated for. Finally, many feeds (e.g., optical and series feeds) do not provide equal phase excitation at the input to each phase shifter. The relative phase excitation caused by these feeds is a known function of frequency. In these cases, the computer must provide a correction based on the location of the element in the array and on the frequency of operation.

For a large array with thousands of elements, many calculations are required to determine the phasing of the elements. These calculations must be performed in a short period of time. The use of the orthogonal phase commands mT_{xs}, nT_{ys} helps to minimize these calculations. Once the element-to-element phase increments T_{xs}, T_{ys} have been computed for a given beam-pointing direction, the integral multiples of T_{ys} may be used to steer the columns (Fig. 7.8). If an adder is located at each element, the row-and-column values mT_{xs} and nT_{ys} may be summed at the element. It is also possible to put two phase shifters in series so that the summation can be done at microwave frequencies. This may be implemented with the use of a series feed, as shown in Fig. 7.11. Here the row steering commands apply equally to all rows. An amplifier between a row phase shifter and a series feed is desirable so that the generated power does not have to take the loss of two phase shifters in series. In addition, the row phase shifters must be capable of accurate phasing. Since relatively few row phase shifters are required, it is reasonable to make them considerably more accurate than the phase shifters in the array. Any corrections for a phase taper

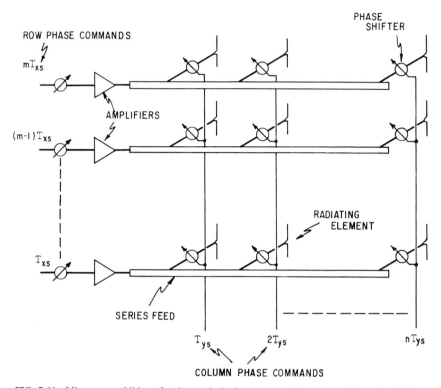

FIG. 7.11   Microwave addition of orthogonal phasing commands by means of a series feed.

across the series feed may be applied for an entire column if all the feeds have the same phase characteristics.

A large array requires many electronic phase-shifter drivers and very complex wiring to provide control signals and energy. The problem is complicated by the relatively close spacing of elements in the array. Further, many phase shifters are of the digital type and require a driver and control signal for each bit. The problems are eased somewhat in the system described above with two RF phase shifters in series, since many elements may use the same steering commands and the same drivers. In other systems it may be necessary to provide each element in the array with an independent phase command. The command may include a phase correction for the correction of errors due to component tolerances. An adder at each element provides rapid steering through the use of row-and-column steering commands. If high-speed phase shifting is not required, the computer may compute sequentially and store the phases for each of the elements. All the phase commands can then be delivered simultaneously.

7.4 APERTURE MATCHING AND MUTUAL COUPLING [48]

Significance of Aperture Matching.   An antenna is a device that acts as a transformer to provide a good match between a source of power and free space. If the antenna is not matched to free space, power will be reflected back toward the generator, resulting in a loss in radiated power. In addition, a mismatch produces standing waves on the feed line to the antenna. The voltage at the peaks of these standing waves is $(1 + |\Gamma|)$ times greater than the voltage of a matched line, where Γ is the voltage reflection coefficient. This corresponds to an increased power level that is $(1 + |\Gamma|)^2$ times as great as the actual incident power. Therefore, while the antenna is radiating less power, individual components must be designed to handle more peak power. With antennas that do not scan, the mismatch may often be tuned out by conventional techniques, preferably at a point as close to the source of the mismatch as possible.

In a scanning array the impedance of a radiating element varies as the array is scanned, and the matching problem is considerably more complicated. Unlike a conventional antenna, where mismatch affects only the level of the power radiated and not the shape of the pattern, spurious lobes in the scanning array may appear as a consequence of the mismatch. Further, there are conditions where an antenna that is well matched at broadside may have some angle of scan at which most of the power is reflected.

The variation in element impedance and element pattern is a manifestation of the mutual coupling between radiating elements that are in close proximity to one another. For a practical design, two empirical techniques are of great value:

1. Waveguide simulators provide a means for determining the element impedance in an infinite array with the use of only a few elements. The effectiveness of a matching structure based on these measurements may also be determined in the simulator.

2. A small array is the best technique for determining the active element pattern. The active element pattern, obtained by exciting one element and terminating its neighbors, is the best overall measure of array performance other than the full array itself. If a large reflection occurs at some angle of scan, it can be recognized by a null in the element pattern. The small array can also provide data on the coupling between elements. This data can be used to calculate the variation in impedance as the array is scanned.

Both these techniques will be discussed later in this section.

Effects of Mutual Coupling. When two antennas (or elements) are widely separated, the energy coupled between them is small and the influence of one antenna on the current excitation and pattern of the other antenna is negligible. As the antennas are brought closer together, the coupling between them increases. In general, the magnitude of the coupling is influenced by the distance between the elements, the pattern of the elements, and the structure in the vicinity of the elements. For example, the radiation pattern of a dipole has a null in the $\theta = \pm 90°$ direction and is omnidirectional in the $\theta = 0°$ plane. Therefore it can be expected that dipoles in line will be loosely coupled and parallel dipoles will be tightly coupled. When an element is placed in an array of many elements, the effects of coupling are sufficiently strong that the pattern and impedance of the element in the array are drastically altered.

The terms *active element pattern* and *element impedance* refer to an element in its operating environment (i.e., in an array with its neighboring elements excited). In the array, each excited element couples to every other element. The coupling from several elements to a typical central element, element 00, is shown in Fig. 7.12. The $C_{mn,pq}$ are mutual-coupling coefficients relating the voltage (amplitude and phase) induced in the mnth element to the voltage excitation at the

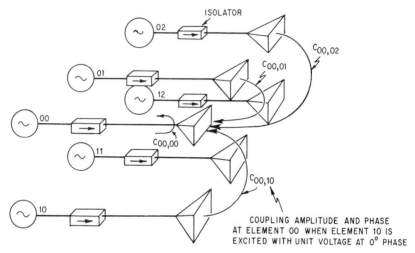

FIG. 7.12   Coupled signals to a central element from neighboring elements.

pqth element. The coupled signals add vectorially to produce a wave traveling toward the generator of element 00 that appears to be a reflection from the radiator of element 00. As the phases of the neighboring elements are varied to scan the beam, the vector sum of the coupled signals changes and causes an apparent change in the impedance of element 00. For some scan angles the coupled voltages tend to add in phase, causing a large reflection and possibly the loss of the main beam. Large reflections often occur at scan angles just prior to the emergence of a grating lobe into real space, but in some instances such reflections may occur at smaller scan angles.

The description of the impedance variation given above made no reference to the feed network or the phase shifters and assumed that the only coupling between elements is via the radiating aperture. The coupling coefficients would be measured, and by superposition the phased-voltage contributions from every element in the array (or at least those in the immediate vicinity) would be added vectorially to produce the voltage reflected back toward the generator. In a practical array the impedance variation depends upon the feed system and the phase shifter. If these are taken into account, the impedance variation may be different from what the above model might predict. In most analyses only the coupling at the aperture is considered. This description provides insight into the intrinsic impedance variation of the aperture when it is isolated from other effects, as in the case where each element has an independent feed (e.g., its own generator and isolator). In this case it is a simple matter to measure the voltage–standing-wave ratio (VSWR) in any line and determine exactly the extent of the impedance and mismatch variation. For many feed systems this is not possible, and a measurement of the reflected energy will provide erroneous information and a false sense of security. Unless all the reflections are collimated back at some central point (or independent feeds are used), some of the reflected energy will generally be re-reflected and contribute to large undesirable sidelobes.

For large arrays the impedance of an element located near the center of the array is often taken as typical of the impedance of every element in the array. As might be expected, this element is most strongly influenced by elements in its immediate vicinity. When the array is scanned, the influence of elements several wavelengths distant is also significant. For dipoles above a ground plane the magnitude of the coupling between elements decays rapidly with distance. For a reasonable indication of array performance, an element in the center of a 5 by 5 array may be taken as typical of an element in a large array. For dipoles with no ground plane (or the dual, slots in a ground plane) the coupling between elements does not decay so rapidly, and a 9 by 9 array appears reasonable. For an array of open-ended waveguides, a 7 by 7 array should suffice (see Fig. 7.22, below). If accurate prediction of the array performance is required, many more elements are needed than are indicated above.[49,50]

It is often convenient to assume that the array is infinite in extent and has a uniform amplitude distribution and a linear-phase taper from element to element. In this manner every element in the array sees exactly the same environment, and the calculations for any element apply equally to all. These assumptions provide a significant simplification in the calculation of the element impedance variations. In addition, impedance measurements made in simulators correspond to the element impedance in an infinite array. In spite of the assumptions, the infinite-array model has predicted with good accuracy the array impedance and the impedance variations. Even arrays of modest proportions (less than 100 elements) have been in reasonable agreement with the results predicted for an infinite array.[51]

Element Pattern. From energy considerations the directional gain of a perfectly matched array with constant amplitude distribution ($\eta = 1$) will vary as the projected aperture area, from Eq. (7.18)

$$G(\theta_0) = \frac{4\pi A}{\lambda^2} \cos \theta_0$$

If it is assumed that each of the N elements in the array shares the gain equally, the gain of a single element is [Eq. (7.21)]

$$G_e(\theta) = \frac{4\pi A}{N\lambda^2} \cos \theta_0$$

If the element is mismatched, having a reflection coefficient $\Gamma(\theta, \phi)$ that varies as a function of scan angle, the element gain pattern is reduced to

$$G_e(\theta) = \frac{4\pi A}{N\lambda^2} (\cos \theta)[1 - |\Gamma(\theta, \phi)|^2]$$

The element pattern is seen to contain information pertaining to the element impedance.[52–55] The difference between the total power radiated in the element pattern and the power delivered to the antenna terminals must equal the reflected power. In terms of the radiation patterns of the scanning array, this means that since the scanned antenna patterns trace out the element pattern, it follows that the average power lost from the scanned patterns is equal to the power lost from the element pattern because of reflections. It is not enough to match one element in the presence of all its terminated neighbors. The element will deliver power to its neighbors, and this loss in power corresponds to the average power lost when scanning. An ideal although not necessarily realizable element pattern would place all the radiated power into the scan region, giving a pattern like a cosine on a pedestal and thereby providing maximum antenna gain for the number of elements used.

Thinned Arrays. The number of radiating elements in an array may be reduced to a fraction of those needed completely to fill the aperture without suffering serious degradation in the shape of the main beam. However, average sidelobes are degraded in proportion to the number of elements removed. The element density may be thinned so as to effectively taper the amplitude distribution, and the spacing is such that no coherent addition can occur to form grating lobes. A thinned aperture, where elements have been removed randomly from a regular grid,[56] is shown in Fig. 7.13. The gain is that due to the actual number of elements $NG_e(\theta)$, but the beamwidth is that of the full aperture. For example, if the array has been thinned so that only 10 percent of

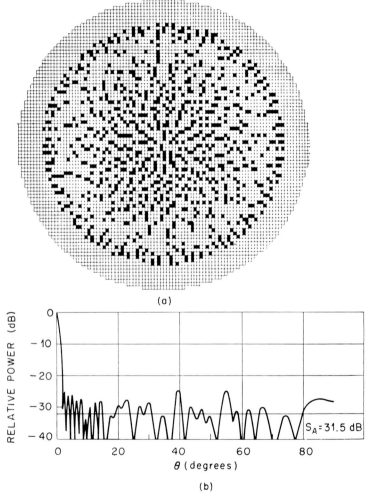

(a)

(b)

FIG. 7.13 (a) Thinned array with a 4000-element grid containing 900 elements. (b) Typical pattern for a thinned array. S_A is the average sidelobe level. (*From Willey*,[56] *courtesy of Bendix Radio.*)

the elements are used, the gain of the array will drop by 10 dB. However, since the main beam is virtually unchanged, about 90 percent of the power is delivered to the sidelobe region.

If the removed elements (in a regular thinned array) are replaced with elements with matched loads, the element pattern is identical to that of one in the regular array with all elements excited. The element pattern is independent of the array excitation, and the same fractional amount of power will be lost (because of mismatch) whether the array is thinned, tapered, or uniformly illuminated. It should be noted that the concept of an element pattern that applies equally to every element is valid only when isolating feeds are used and edge effects are ignored.

A thinned array may also be implemented with an irregular element spacing, although this is not common. In this case the element gain (and impedance) will vary from element to element, depending upon the environment of a given element. To obtain the gain of the array, it is necessary to sum all the different element gains $G_{en}(\theta)$. Thus

$$G(\theta) = \sum_n G_{en}(\theta)$$

Impedance Variation of Free Space. It is of interest to examine the case of a large continuous aperture which may be considered to be the limiting case of an array of many very small elements.[57] The free-space impedance E/H varies as $\cos \theta$ for scanning in the E plane and as $\sec \theta$ for scanning in the H plane. The impedance of a medium is thus dependent upon the direction of propagation, and the impedance variation of a scanning aperture is a natural consequence of this dependence. The continuous aperture appears to represent a lower limit to the impedance variation with scanning. This is indicated by Allen's results,[58] where the impedance variation with scanning was calculated for dipoles above a ground plane. In spite of increased mutual coupling, or perhaps because of it, the more closely the dipoles were spaced, the smaller the impedance variation with scanning (Fig. 7.14). Although the impedance variation decreased, the absolute impedance of the dipoles also decreased, making them more difficult to match at broadside. It is expected that to obtain an impedance variation smaller than that of free space some impedance compensation must be employed.

Element Impedance. The simplest and most straightforward method for computing the variation in reflection coefficient (and impedance) is by means of the scattering matrix of mutual-coupling coefficients. The mutual-coupling coefficients may be easily measured for elements of all types by exciting one element and terminating each of the other elements in a matched load. The ratio of the induced voltage at element mn to the excitation voltage at element pq gives the amplitude and phase of the coupling coefficient $C_{mn,pq}$. Once these coefficients are determined, it is a simple matter to compute the mismatch for any set of phasing conditions.

Consider the two-element array shown in Fig. 7.15 where each element is provided with an isolating feed. The incident wave in each element is represented by V_1, V_2, and the total reflected wave in each element is represented by V_1', V_2'. It should be apparent that the total reflected wave in any element

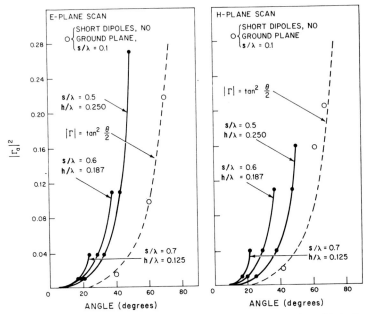

FIG. 7.14 Scanned mismatch variation for different element spacings (h/λ is the dipole spacing above a ground plane). (*After Allen.*[58])

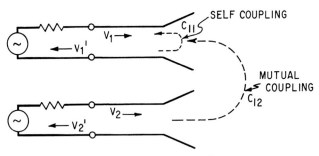

FIG. 7.15 Scattering-matrix model for a two-element array.

is the vector sum of the couplings from all elements, including its own reflection as a self-coupling:

$$V_1' = C_{11} V_1 + C_{12} V_2$$

$$V_2' = C_{21} V_1 + C_{22} V_2$$

The reflection coefficient in each element is obtained by dividing the reflected voltage by the incident voltage in the channel:

$$\Gamma_1 = \frac{V_1'}{V_1} = C_{11} \frac{V_1}{V_1} + C_{12} \frac{V_2}{V_1}$$

$$\Gamma_2 = \frac{V_2'}{V_2} = C_{21} \frac{V_1}{V_2} + C_{22} \frac{V_2}{V_2}$$

Note that all the quantities must contain both phase and amplitude and that as the phases of V_1, V_2 are varied to scan the beam, the reflection coefficients (Γ_1, Γ_2) will vary. Although only two elements have been used in this example, the technique is completely general. For a large array the reflection coefficient of the mnth element is given by

$$\Gamma_{mn} = \sum_{\text{all } pq} C_{mn,pq} \frac{V_{pq}}{V_{mn}}$$

The general case is treated in more detail by Oliner and Malech.[59] No restrictions need be placed on either the amplitude or the phase of the excitation at each element. There are also no restrictions on the spacing between the elements as long as the coupling coefficients are measured for the spacing and environment to be used. A considerable simplification is obtained by assuming that each of the elements sees the same environment and has the same voltage excitation. Then $|V_{pq}|/|V_{mn}|$ will always be unity, and the reflection coefficient at element mn is merely the sum of the mutual-coupling coefficients with the excitation phase at each element taken into account:

$$\Gamma_{mn} = \sum_{\text{all } pq} C_{mn,pq} e^{j(m-p)T_{xs}} e^{j(n-q)T_{ys}}$$

where $e^{j(m-p)T_{xs}}$ and $e^{j(n-q)T_{ys}}$ give the relative phase excitations in the x and y directions of the pqth element with respect to the mnth element. The impedance variation relative to a matched impedance at broadside may be obtained immediately from

$$\frac{Z_{mn}(\theta,\phi)}{Z_{mn}(0,0)} = \frac{1 + \Gamma_{mn}(\theta,\phi)}{1 - \Gamma_{mn}(\theta,\phi)}$$

Analytical Techniques. Stark[60] presents a thorough description of analytical techniques and insight to the problem of mutual coupling. He derives necessary and sufficient conditions for array blindness (i.e., nulls in the active element pattern). Evidence is provided to demonstrate that more closely spaced elements reduce the variation of impedance due to scanning in spite of the increase in mutual coupling.

Elliott and coworkers[61,62] have developed design procedures for slot arrays which are fed by either air-filled or dielectric-filled waveguide. These procedures take into account the differences in mutual coupling for central elements in an array as well as for the edge elements. This is particularly of interest when designing a small array.

Munk and colleagues[63,64] have developed a procedure for reducing the variations in impedance with scanning by matching elements with dielectric slabs. Their analytical results show a VSWR of less than 1.5 for scan angles of $\pm 80°$ in each of the principal planes. The penalty they incur for placing a thick (0.4λ) and potentially heavy dielectric slab in front of the aperture is mitigated by the low dielectric constant of the slab (ϵ_r = 1.3). This suggests that a lightweight loaded foam material could be used. Dielectric slabs on the surface of an array can often create surface waves and array blindness. In this case, however, the dielectric constant is quite low. The surface wave can be avoided by spacing the elements more closely. Here, as in some other cases, it is seen that closely spaced elements are beneficial for improving impedance matching for scanning arrays. Obviously, reducing the element spacing increases the number of elements and hence the cost without any increase in gain or reduction in beamwidth.

Another very useful tool for the computation of the impedance variation with scanning is the grating-lobe series,[65,66] which describes the impedance variation of an infinite array of regularly spaced elements.

Nonisolating Feeds. When nonisolating feeds are used, the mutual-coupling effects become dependent on whether the phase-shifting element is reciprocal or nonreciprocal. Figure 7.16 shows a space-fed array where it is assumed that the initial excitations at the input aperture are of equal amplitude and phase. If the phase shifters are nonreciprocal, the round-trip phase of a signal reflected from the radiating aperture is independent of phase-shifter setting. Therefore, when reflections occur at the radiating aperture, the reflected signal is phased so that the input aperture appears as a mirror as seen from the feed side, with the magnitude of the reflection being determined by the radiating aperture. Since the reflected beam does not scan, it should be possible to provide a good match at the input aperture. Matching the input aperture, in this case, is equivalent to providing the radiating aperture with independent feeds. Any secondary reflection from the input aperture will radiate in the original direction of scan.

If reciprocal phase shifters are used, the energy reflected from the radiating aperture will pick up equal additional phase shift on reflection from the radiating aperture, resulting in a beam at the input aperture that is phased to scan to twice the original scan angle (in sin θ space). Some energy will undergo secondary re-

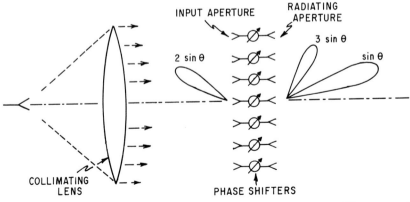

FIG. 7.16 Nonisolating feed showing spurious lobes when reciprocal phase shifters are used.

flection from the input aperture (which now has a mismatch corresponding to a scan of 2 sin θ) and will be phase-shifted once again to produce a beam at the radiating aperture in the direction 3 sin θ. Additional reflections acquire twice the original phase shift for each round trip, resulting in radiated beams at positions of 5 sin θ, 7 sin θ, etc. The magnitude of these beams is approximately equal to the product of the voltage reflection coefficients if the number of bounces is taken into account. For example, let $\Gamma_r(\sin\theta)$, $\Gamma_i(2\sin\theta)$ denote the reflection coefficients corresponding to scan angles of sin θ at the radiating aperture and 2 sin θ at the input aperture. If $\Gamma_r(\sin\theta) = 0.2$ and $\Gamma_i(2\sin\theta) = 0.5$, the magnitude of the radiated lobe directed at 3 sin θ would be $\Gamma_r(\sin\theta)\ \Gamma_i(2\sin\theta) = 0.1$, or 20 dB down from the main lobe. Similar results can be expected from series feeds[67] and from reactive parallel feeds.[48]

Mutual Coupling and Surface Waves. The mutual coupling between two small isolated dipoles[68] should decrease as $1/r$ in the H plane and $1/r^2$ in the E plane (E and H planes are interchanged for slots). Coupling measurements[69] have shown that in the array environment the rate of decay is slightly greater than predicted above, indicating that some of the energy is delivered to other elements in the array and may be dissipated and reradiated from these elements. The same measurements have shown that the phase difference of the energy coupled to elements is directly proportional to their distance from the excited elements, indicative of a surface wave traveling along the array, leaking energy to each of the elements. For best performance the velocity of the surface wave should be very close to that of free space. If the array contains waveguides or horns loaded with dielectric, the velocity will decrease slightly. Further, if the dielectric protrudes from the radiators or if a dielectric sheet is used in front of the array, the velocity of the surface wave may decrease dramatically. This surface wave is important since it can cause a large reflection (and an accompanying loss of the beam) for some angles of scan. This can best be seen by examining the condition of phasing for which the couplings from many elements will add in phase to cause a large reflection in a typical element.

Consider an array in which the velocity of the surface wave is that of free space. The difference in the phase of the voltages coupled from an adjacent pair of elements to element 00 (Fig. 7.17) is related to the scan angle by

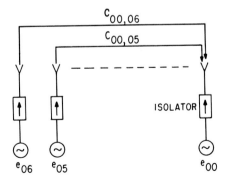

FIG. 7.17 Two adjacent elements coupling to another element in the same row.

$$\Delta\chi = \frac{2\pi s}{\lambda} + \frac{2\pi s}{\lambda}\sin\theta_0 = \frac{2\pi s}{\lambda}(1 + \sin\theta_0)$$

The couplings will be in phase when $\Delta\psi = 2\pi$ or when

$$\frac{s}{\lambda} = \frac{1}{1 + \sin\theta_0}$$

This is seen to be exactly the same condition as previously determined for the emergence of a grating lobe into real space. Therefore, it may be expected that when a grating lobe is about to emerge into real space, the coupled voltages tend to add in phase and cause a large mismatch. If the dielectric protrudes from the aperture or if a dielectric sheet covers the aperture (this is one technique for scan compensation discussed below), a large reflection may occur well before the grating lobe reaches real space.[70] If a surface-wave velocity of v_s is assumed, the couplings will add in phase when

$$\frac{s}{\lambda} = \frac{1}{c/v_s + \sin\theta_0}$$

For the purposes of mutual coupling, a slow wave across the aperture may be envisioned as being equivalent to spacing the elements farther apart in free space.

A phenomenon that produces similar effects can come about without dielectric in front of the aperture, e.g., by using a periodic structure of baffles. Under certain conditions an array of open-ended waveguides will perform as though a slow wave were propagating across the aperture. This effect has been analyzed and studied experimentally,[71,72] and it has been shown that as the array is scanned, higher-order modes are excited in the waveguides. Even though these modes are cut off, they contribute to the active impedance. At certain angles, almost all the energy is reflected, causing a null in the element pattern. To guard against these reflections it is best to design the radiators so that higher-order modes are well into the cutoff region. Figure 7.18 shows the results obtained by Diamond,[72] using a waveguide array. When only the dominant TE_{10} mode was taken into account, the null could not be explained. When the TE_{20} mode (which was only slightly cut off) was taken into account, the null showed up clearly. Since a null in the element pattern indicates in-phase addition of the mutually coupled signals, it appears that the array of open-ended waveguides causes the couplings between elements to have a phase variation corresponding to a velocity slower than that of free space.

Regardless of the cause of the null, it will show up in the element pattern. If only a few elements surround the central element, the null will normally be shallow and broad. If many elements are used, the null will be deep and sharp. The null will also show up if the mutual-coupling coefficients are measured and used to calculate the reflection coefficient.

Array Simulators. A good deal of effort has gone into matching a radiator in the presence of an array of radiators. The use of waveguide simulators as developed by Wheeler Laboratories has made it possible to determine the matching structure experimentally without the need of building an array. A waveguide, operating in a TE_{10} mode, may be considered to contain two inclined-plane waves propagating down the guide. The angle that each of the plane waves makes with the longitudinal direction (Fig. 7.19) is determined by

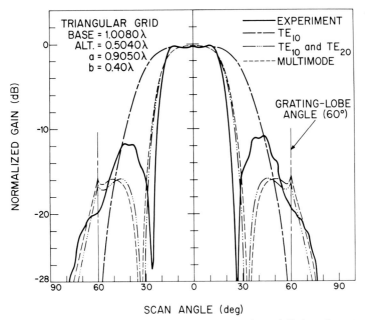

FIG. 7.18 Comparison of the theoretical and an experimental *H*-plane element pattern: triangular array of waveguides with a 2:1 ratio of width to height. (*From Diamond.*[72])

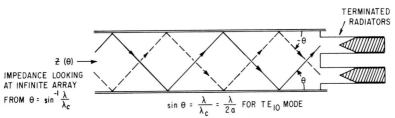

FIG. 7.19 Array simulator terminated with two dummy elements.

the *H* dimension of the waveguide and simulates the angle of scan of an infinite array:

$$\sin \theta = \frac{\lambda}{\lambda_c} \qquad (7.26)$$

where θ = scan angle
  λ = free-space wavelength
  λ_c = cutoff wavelength of guide

Additional scan angles may be simulated by exciting other modes. The waveguide dimensions are chosen so that a radiating element or elements placed in the waveguide sees mirror images in the walls of the waveguide that appear to be at the same spacing as the array to be simulated. Both rectangular and triangular arrays may be simulated, as shown in Fig. 7.20. The impedance measurements are made by looking into a waveguide simulator that is terminated with dummy elements. This is equivalent to looking at an infinite array from free space at a scan angle given by Eq. (7.26). A matching structure, designed from the simulator impedance data, may be placed into the simulator to measure its effectiveness. Several simulator designs, results, and a complete discussion of the topic have been presented by Hannan and Balfour.[73] The technique is limited in that only discrete scan angles can be simulated. Several scan angles in both planes of scan give a general idea of the array impedance but may miss a large reflection of the type described in this section under "Mutual Coupling and Surface Waves." Nevertheless, the array simulator is the best method available for empirically determining the array impedance without building an array.

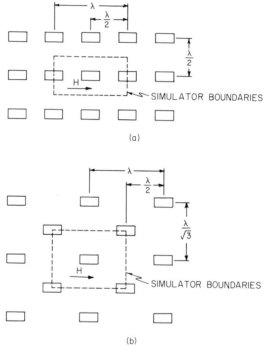

FIG. 7.20 Rectangular- and triangular-array geometries with simulator boundaries superimposed. (*a*) Square array with simulator superimposed. (*b*) Triangular array with simulator superimposed.

Under certain conditions, the matched element impedance may be determined analytically by envisioning a simulator.[74] For example, if the waveguide simulator is terminated by a single dummy element, the matched impedance of the element is exactly the impedance of the waveguide. Since waveguide impedances are well known, the matched impedance at the simulator scan angle is also known.

Compensation for Scanned Impedance Variation. The impedance of an element in an array has been discussed and has been shown to vary as the array is scanned. An array that is matched at broadside can be expected to have at least a 2:1 VSWR at a 60° angle of scan. To compensate for the impedance variation, it is necessary to have a compensation network that is also dependent on scanning.

One method uses a thin sheet of high-dielectric-constant material (e.g., alumina) spaced a fraction of a wavelength from the array, as shown in Fig. 7.21.This is in contrast to the method used by Munk[63,64] and described under "Analytical Techniques." The properties of a thin dielectric sheet (less than a quarter wave in the medium) are such that to an incident plane wave it appears as a susceptance that varies with both the plane of scan and the angle of scan. Magill and Wheeler[75] describe the technique in greater detail and present the results of a particular design using simulators. An alumina matching sheet is attractive because it simultaneously provides a natural radome. It should be cautioned that a dielectric sheet in front of the aperture may produce a slow surface wave and a possible null in the element pattern. However, a thin dielectric matching sheet has been used for a 400-element array,[51] and some compensation has been achieved without any noticeable slow-wave phenomenon.

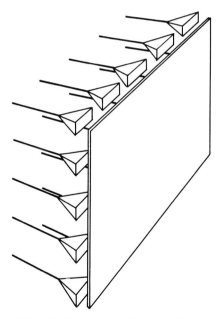

FIG. 7.21 Planar array with a thin dielectric sheet spaced a fraction of a wavelength from the radiators.

Small Arrays. The element pattern is the best single indicator of impedance matching in a scanning array. One way of determining the element pattern is to build a small array. A central element is excited, and all other elements are terminated. The pattern of this central element is the active element pattern.

Diamond[76] has examined the number of elements required in a small array to provide a reasonable approximation to an element in an infinite array. He concludes that 25 to 37 elements are required to provide a good indication. Figure 7.22 shows the change in the measured active element pattern as the number of elements is increased. For a 41-element array the null is very pronounced. Even for the 23-element array it is clear that the gain variation with scanning is dramatically greater than cos θ.

The small array can also be used to measure coupling coefficients as demonstrated by Fig. 7.12. These coupling coefficients can be used to calculate the impedance variation as the array is scanned. Grove, Martin, and Pepe[77] have noted that for the element to be matched in its operating environment the self-coupling C_{11} must exactly cancel the coupling from all other elements. They have used this technique to provide a good match on an ultralow-sidelobe, wideband phased array. After adjusting C_{11} to cancel the sum of the other mutuals, they further adjusted the amplitude of C_{11} to compensate for the variation in mismatch loss due to the resultant change in C_{11}. Results were obtained for an array of greater than two-octave bandwidth. In this design grating lobes were permitted to enter real space. At this point a high mismatch was observed. The mismatch improved as the grating lobe moved further into real space.

The combination of waveguide simulators and small arrays provides powerful empirical tools to supplement the analytical techniques. Experience has demonstrated that a large antenna should not be built until the element pattern has been verified with a small array.

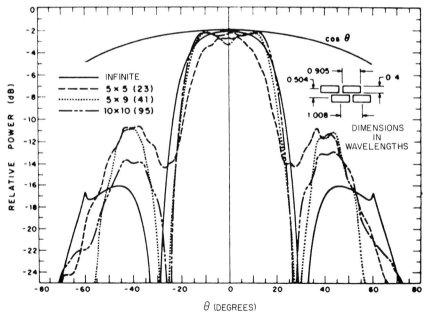

FIG. 7.22 Experimental *H*-plane patterns of the center elements of waveguide arrays. (*After Diamond.*[76])

7.5 LOW-SIDELOBE PHASED ARRAYS

Low sidelobes have long been of interest to antenna designers. This interest has been heightened by the jamming which threatens most military radars. The requirement for low sidelobes for clutter rejection in the AWACS radar resulted in technology which now supports sidelobe levels of more than 50 dB below the main-beam peak.[78,79] The price that must be paid to achieve these low sidelobes includes (1) a reduction in gain, (2) an increase in beamwidth, (3) increased tolerance control, (4) increased cost, and (5) the need to operate in an environment free from obstructions that can readily increase the sidelobes.[80] In spite of these drawbacks, the trend to low-sidelobe antennas has accelerated since low sidelobes provide an excellent counter to electronic countermeasures (ECM).

Antenna sidelobes can be controlled by the aperture amplitude distribution. For phased arrays the amplitude of each element may be controlled individually, and therefore good sidelobe control can be achieved. The process of designing a low-sidelobe antenna can be considered in two parts:

1. Choose the correct illumination function to achieve the desired design (error-free) sidelobes.

2. Control the phase and amplitude errors which are the contributors to the random sidelobes.

Of the two it is the control of errors which fundamentally limits sidelobe performance. The effects of illumination function and errors are discussed below.

Illumination Functions. The relation between aperture illumination and the far-field pattern has been studied extensively and is well documented in the literature.[81-84] For a continuous aperture, the far-field pattern is the Fourier transform of the distribution across the aperture. For an array, samples are taken of the continuous distribution at each of the discrete locations. Some typical illumination functions are given in Table 7.1. It is seen that uniform illumination (constant amplitude) results in the highest gain and the narrowest beamwidth but at the cost of high sidelobes. As the amplitude is tapered, the gain drops, the beam broadens, and the sidelobes may be reduced. It is important for the antenna designer to choose an efficient and realizable illumination function that provides low sidelobes at a minimum loss in gain. For low-sidelobe radars, the Taylor illumination[85,86] for the sum patterns and the Bayliss illumination[87] for the difference patterns have almost become an industry standard. The Taylor illumination is somewhat similar to a cosine squared on a pedestal and is readily implemented. The Bayliss illumination is a derivative form of the Taylor illumination and is also readily implemented. It should be noted that in many phased arrays the sidelobe performance for the difference pattern is comparable to that of the sum pattern. For both sum and difference patterns the sidelobes are referenced to the peak of the sum pattern.

Figure 7.23 gives the approximate loss in gain and the beamwidth factor for the Taylor illumination as the sidelobes change. For a more comprehensive treatment, see Barton and Ward.[88] The sidelobes predicted by Table 7.1 are for antennas which have perfect phase and amplitude across the aperture. None of these have yet been built, and none are anticipated. To allow for errors, aperture illuminations are often chosen to provide peak sidelobes below those required. For example, if the antenna specification calls for −40 dB sidelobes, a Taylor illumination that provides −45 dB design sidelobes might be chosen.

TABLE 7.1   Illumination Functions

Illumination function	Efficiency, η	Peak sidelobe, dB	Beamwidth factor, k
Linear illumination functions: beamwidth = $k\lambda/a$ (degrees); a = length of antenna			
Uniform	1	− 13.3	50.8
Cosine	0.81	− 23	68.2
Cosine squared (Hanning)	0.67	− 32	82.5
Cosine squared on 10 dB pedestal	0.88	− 26	62
Cosine squared on 20 dB pedestal	0.75	− 40	73.5
Hamming	0.73	− 43	74.2
Dolph-Chebyshev	0.72	− 50	76.2
Dolph-Chebyshev	0.66	− 60	82.5
Taylor $\bar{n} = 3$	0.9	− 26	60.1
Taylor $\bar{n} = 5$	0.8	− 36	67.5
Taylor $\bar{n} = 8$	0.73	− 46	74.5
Circular illumination functions: beamwidth = $k\lambda/D$ (degrees); D = diameter of antenna			
Uniform	1	− 17.6	58.2
Taylor $\bar{n} = 3$	0.91	− 26.2	64.2
Taylor $\bar{n} = 5$	0.77	− 36.6	70.7
Taylor $\bar{n} = 8$	0.65	− 45	76.4

It should be noted that for a rectangular array a different illumination may be chosen for each plane. This is appropriate if the sidelobe requirements in each of the planes are different. The resultant loss in gain is then the sum (in decibels) of the losses in each plane.

Effect of Errors.   When errors occur in phase or amplitude, energy is removed from the main beam and distributed to the sidelobes. If the errors are purely random, they create random sidelobes which are considered to be radiated with the gain and pattern of the element. When the errors are correlated, the sidelobe energy will be lumped at discrete locations in the far field. The correlated errors will therefore provide higher sidelobes, but only at a limited number of locations. Both correlated and random sidelobes are of concern to antenna designers. Correlated errors are discussed in Sec. 7.6.

Analyses of the far-field effect of errors are based on the fact that antennas are linear devices. That is, the far-field pattern is the sum of the voltages (amplitude and phase) of each radiating element in the antenna. For this reason the far-field voltage pattern can be considered to be the sum of the design pattern plus the pattern created solely by the errors.

$$E_T(\theta,\phi) = E_{\text{design}}(\theta,\phi) + E_{\text{error}}(\theta,\phi)$$

In general, three regions will be recognized in the total resultant pattern: a low noise floor generated by random errors which follows the element pattern, a few peak sidelobes due to correlated errors, and the main beam with its sidelobes due to the design distribution.

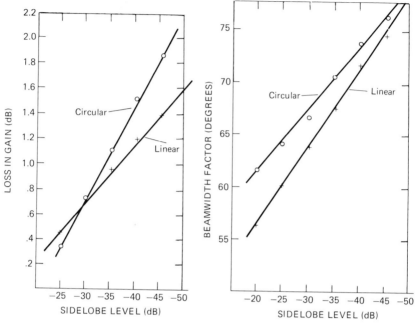

FIG. 7.23 Taylor illumination: loss and beamwidth factor.

Random Errors. Allen[89] and Ruze[90] have made detailed analyses of the effects of random errors on antennas. This discussion will follow the analysis performed by Allen. As previously mentioned, amplitude and phase errors take a fraction of the energy from the main beam and distribute this energy to the sidelobes. This fraction is, for small independent random errors,

$$\sigma_T^2 = \sigma_\phi^2 + \sigma_A^2$$

where σ_ϕ = rms phase error, rad
 σ_A = rms amplitude error, volts/volt (V/V)

This energy is radiated into the far field with the gain of the element pattern. To determine the mean-squared-sidelobe level (MSSL), it is necessary to compare this energy with the peak of the pattern of an array of N elements so that the mean-squared-sidelobe level is

$$\text{MSSL} = \frac{\sigma_T^2}{\eta_a N(1 - \sigma_T^2)} \qquad (7.27)$$

Note that in the denominator of this expression the gain due to the array factor N is reduced by the aperture efficiency η_a and by the errored power lost from the main beam $(1 - \sigma_T^2)$. As an example, consider an array of 5000 elements with an aperture efficiency of 70 percent, $\sigma_a = 0.1$ v/v, and $\sigma_\phi = 0.1$ rad. Then $\sigma_T^2 = (0.1)^2 + (0.1)^2 = 0.02$.

$$\text{MSSL} = \frac{\sigma_T^2}{\eta N(1 - \sigma_T^2)} = \frac{0.02}{(0.7)(5000)(0.98)} = 5.8 \times 10^{-6} = -52 \text{ dB}$$

The result is that this array has a mean floor of random sidelobes which on the average is 52 dB below the peak of the beam. It also illustrates that to achieve low sidelobes very tight tolerances are required. The amplitude of 0.1 v/v is equivalent to a total amplitude standard deviation of 0.83 dB rms. The total rms phase error is 5.7°. It should be noted that there are numerous sources of phase and amplitude errors which are induced by the phase shifters, the feed network, the radiating elements, and the mechanical structure. The task of building a low-sidelobe antenna is one of reducing each of the amplitude errors to a few tenths of a decibel and the phase errors to a few degrees. The fewer the number of elements used, the tighter the tolerances become.

The individual effects of phase and amplitude errors and of failed elements are summarized in Fig. 7.24.[91] The resultant rms sidelobes are referenced to the gain of a single element so that the curve can be used for any number of elements with independent errors. For example, a 5° rms phase error will produce an rms sidelobe level which is approximately 21 dB below the gain of an element. If 1000

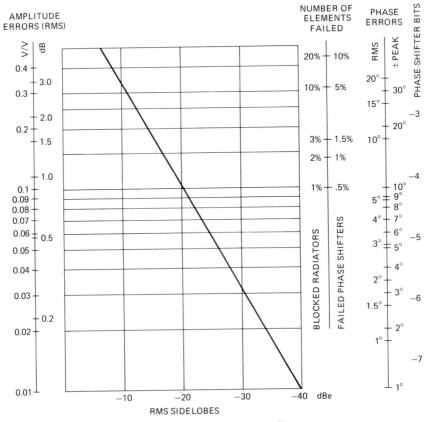

FIG. 7.24 Random errors and rms sidelobes. (*After Cheston.*[91])

elements (30 dB) are used, the rms sidelobe level is 51 dB below the gain of the array. This is the effect of only the random phase errors. The effects due to amplitude errors and failed elements must also be included.

The previous discussion applies to the rms sidelobe level. This analysis has been extended by Allen[89] to apply the probability of keeping a single sidelobe below a given level and then to the probability of keeping a number of sidelobes below a given level. By ignoring the element pattern, the MSSL including failed elements is given by

$$\text{MSSL} = \frac{(1 - P) + \sigma_A{}^2 + P\sigma_\phi{}^2}{\eta_a PN}$$

where $1 - P$ = probability of a failed element. Note that if $P = 1$ (no failed elements), this equation becomes

$$\text{MSSL} = \frac{\sigma_A{}^2 + \sigma_\phi{}^2}{\eta_a N} = \frac{\sigma_T{}^2}{\eta_a N}$$

This is the same as equation (7.27) except for $(1 - \sigma_T{}^2)$ in the denominator, which is not significant for low-sidelobe antennas. For the case in which the design sidelobes are well below the sidelobes caused by errors, Allen has developed the set of curves shown in Fig. 7.25. An example will illustrate the use of these curves. If you desire to hold the sidelobe at a given point in space to less than 40 dB below the peak of the beam with a probability of 0.99, draw a vertical line from -40 dB on the abscissa until it intersects the 0.99 curve. From this intersection draw a horizontal line and read the value of MSSL, in this case -47 dB. Then

$$\text{MSSL} = -47 \text{ dB}$$

or

$$\text{MSSL} = 2 \times 10^{-5} = \frac{(1 - P) + \sigma_A{}^2 + P\sigma_\phi{}^2}{\eta_a PN}$$

For a 10,000-element array

$$0.2 = \frac{(1 - P) + \sigma_A{}^2 + P\sigma_\phi{}^2}{\eta_a P}$$

For $\eta_a = 1$, this array can tolerate $P = 0.83$, or $\sigma_A = 3.2$ dB, or $\sigma_\phi = 25.6°$. Naturally, each type of error must be anticipated, and one must allow a budget for failed elements, amplitude errors, and phase errors.

For a number of independent sidelobes, the probability that n sidelobes can be kept below a given level R_T is equal to the product of the probabilities that each sidelobe can be held below this level.

$$P\,[n \text{ sidelobes} < R_T] = \prod_{i=1}^{n} P\,[R(\theta_i) < R_T]$$

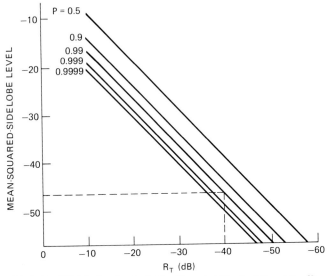

FIG. 7.25   Sidelobe level to be held with probability P. (*After Allen.*[89])

$$R(\theta_i) = \text{sidelobe level at } \theta_i$$

Assuming the same sidelobe requirement at each θ_i,

$$P[n \text{ sidelobes} < R_T] = \{1 - P[\text{one sidelobe} > R_T]\}^n$$

and for P [one sidelobe $> R_T$] $< \!\!\!< 1$,

$$P[n \text{ sidelobes} < R_T] = 1 - n P[\text{one sidelobe} > R_T]$$

A simple example will illustrate the process. If it is necessary to keep all 100 sidelobes in a sector below -40 dB with a probability of 0.9, determine the required probability on any one given sidelobe:

$$0.9 = P[100 \text{ sidelobes} < 40 \text{ dB}]$$

Then

$$0.9 = 1 - 100 P[\text{one given sidelobe} > 40 \text{ dB}]$$
$$0.001 = P[\text{one given sidelobe} > 40 \text{ dB}]$$
$$0.999 = P[\text{one given sidelobe} < 40 \text{ dB}]$$

That is, to keep all 100 sidelobes below -40 dB with a probability of 0.9, it is necessary to keep any given sidelobe below -40 dB with a probability of 0.999. The process of controlling each and every sidelobe is thus seen as formidable, since the total number of sidelobes is approximately equal to the number of elements in a phased array. For a 5000-element array and a probability of 0.999 that a single sidelobe will not exceed R_T at any single location, it will still be expected that 5 sidelobes will exceed R_T when all 5000 sidelobe locations are taken into account.

For very low sidelobe arrays, it is reasonable to allow a few sidelobes to ex-

ceed the MSSL value by as much as 10 to 12 dB to account for random variations. This can be seen from Fig. 7.25 as the difference between P = 0.5 and
P = 0.999 or P = 0.9999. If this allowance is not granted, the antenna will be
overdesigned. It is worthwhile to do some probability calculations before specifying the exact sidelobe requirements.

7.6 QUANTIZATION EFFECTS

Of concern here are errors peculiar to phased arrays, which are due to the quantization of amplitude and phase and to the lobes that occur when these errors are
repeated periodically. The effect on the gain and radiation pattern of random errors in the antenna excitation function is discussed in Sec. 7.5.

Phase Quantization

Phase Errors. Phase shifters suitable for steering phased arrays are described in Sec. 7.9. Most of these phase shifters are digitally controlled and can
be set with an accuracy that is a function of the number of bits. A small number
of bits is desired for simplicity of phase-shift computation and operation, for minimal insertion loss in the case of diode phase shifters, and for minimal cost. On
the other hand, a large number of bits is required for best performance in terms of
gain, sidelobes, and beam-pointing accuracy.

The phase of a phase shifter having P bits can be set to the desired value with
a residual error:

$$\text{Peak phase error} = \alpha = \pm \frac{\pi}{2^P} \qquad (7.28)$$

$$\text{RMS phase error} = \sigma_\phi = \frac{\pi}{\sqrt{3}\, 2^P} \qquad (7.29)$$

Loss in Gain. As discussed in Sec. 7.5, the loss in gain is $\sigma_\phi{}^2$, which gives
with Eq. (7.29)

$$\Delta G \approx \sigma_\phi{}^2 = \frac{1}{3} \frac{\pi^2}{2^{2P}} \qquad (7.30)$$

With many array elements this result is statistically independent of the amplitude
distribution. An enumeration of Eq. (7.30) gives

Number of phase-shifter bits, P	2	3	4
Loss in gain ΔG, dB	1.0	0.23	0.06

From the point of view of gain, therefore, 3 or 4 bits would appear ample.

RMS Sidelobes. Phase quantization decreases the gain of the main beam, as
shown above. The energy that has been lost is distributed to the sidelobes. The
resulting rms sidelobes are, therefore, σ_ϕ^2 relative to the gain of a single element,
as shown in Fig. 7.24.

Beam-Pointing Accuracy. The accurate determination of the direction of targets is made with the monopulse difference pattern. The accuracy of the null position of the difference pattern is therefore of interest. With quantized phase

shifters the position of this null can be moved with a granularity that is a function
of the bit size.

Following the analysis of Frank and Ruze,[92] Fig. 7.26 shows an aperture with
an even number of elements N, separated by distances s. All elements are excited
with equal amplitude and antisymmetric phase to give a difference pattern

$$D(\theta) = 2j\left\{\sin\left(\frac{\pi s}{\lambda}\sin\theta - \psi_1\right) + \sin\left(\frac{3\pi s}{\lambda}\sin\theta - \psi_2\right) + \cdots \right.$$
$$\left. + \sin\left[\frac{(N-1)\pi s}{\lambda}\sin\theta - \psi_{N/2}\right]\right\}$$

For a null of the difference pattern at θ_0

$$\sum_{n=1}^{n=N/2} \sin\left[\frac{(2n-1)\pi s}{\lambda}\sin\theta_0 - \psi_n\right] = 0$$

If the phases are nearly those required to produce a null at θ_0, the *sine* can be
replaced by its argument, and the equation simplifies to

$$\sin\theta_0 = \frac{4}{\pi(s/\lambda)N^2}\sum_{n=1}^{n=N/2}\psi_n \tag{7.31}$$

If a phase increment equal to the smallest bit size $2\pi/2^P$ is applied
antisymmetrically to a pair of radiators, the null will shift by $\delta\theta$ rad, so that

$$\delta\theta\cos\theta_0 = \frac{4}{\pi(s/\lambda)N^2}\frac{2\pi}{2^P} \tag{7.32}$$

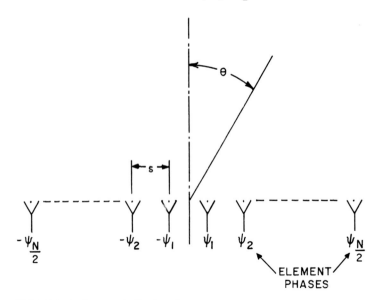

FIG. 7.26 Antisymmetrically phased array.

It should be noted that the movement of the null is independent of the position of the radiators and that the phase on any symmetric pair of radiators may be changed with equal effectiveness. Equation (7.32) can be expressed in terms of beamwidth from Eq. (7.15), with aperture $a = Ns$:

$$\frac{\delta\theta}{\theta_B(\text{scanned})} = \frac{9}{N2^P} \tag{7.33}$$

Small steering increments are possible. For example, with an array with 100 elements and 3-bit phase shifters, the beam may be steered in increments of about 1 percent of a beamwidth. If only one element rather than a symmetric pair has the phase change, then the beam is moved by half of the amount indicated by Eq. (7.33).

The sum $\sum_{n\,=\,1}^{n\,=\,N/2} \psi_n$ in Eq. (7.31) represents the total phase increments to all elements. This may be used in performing the beam-steering computations by distributing the total phase increments to all the elements in a linear manner. The monopulse null will then have an absolute accuracy of one-half of the value given by Eq. (7.33).

With tapered amplitude distributions the phase bits become proportionately weighted.

The exact pointing direction of the sum-pattern peak may be derived by similar means. Surprisingly, it does not move synchronously with the difference pattern. Its motion is dependent on both the weighting and the position of the elements.

Periodic Errors

Periodic Amplitude and Phase Modulation.   Both amplitude and phase quantization lead to discontinuities which may be periodic and give rise to *quantization lobes* that are similar to grating lobes.

Amplitude or phase errors that vary cosinusoidally may be analyzed simply, after Brown.[93] Figure 7.27a shows an original amplitude distribution $F(x)$ disturbed by a cosinusoidal ripple $q \cos (2\pi x/s)$, giving a new distribution $F'(x)$ such that

$$F'(x) = F(x) + qF(x) \cos \frac{2\pi x}{s}$$
$$= F(x) + \frac{q}{2}[F(x)e^{j(2\pi x/s)} + F(x)e^{-j(2\pi x/s)}]$$

In addition to the unperturbed aperture distribution there are two other distributions with positive and negative slopes, giving rise to two beams (amplitude quantization lobes) or relative amplitudes $q/2$ and offset by an angle θ_1, where $\sin \theta_1 = \pm 1/(s/\lambda)$.

Similarly, small phase ripples $\beta \cos (2\pi x/s)$, as shown in Fig. 7.27b, modify the aperture distribution to

$$F'(x) = F(x)e^{j\beta \cos (2\pi x/s)} \approx F(x)(1 + j\beta \cos 2\pi x/s)$$
$$\approx F(x) + j\frac{\beta}{2}[F(x)e^{j(2\pi x/s)} + F(x)e^{-j(2\pi x/s)}]$$

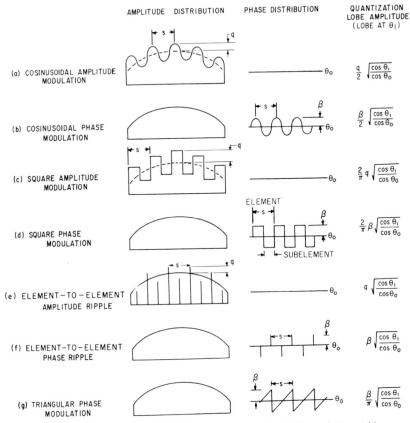

FIG. 7.27 Effects of periodic amplitude and phase modulation ($\sin \theta_1 = \sin \theta_0 \pm \lambda/s$).

Phase-quantization lobes occur as before at $\sin \theta_1 = \pm 1/(s/\lambda)$, this time with an amplitude $\beta/2$.

When the beam is scanned to θ_0, the quantization lobes occur at an angle θ_1, where

$$\sin \theta_1 = \sin \theta_0 \pm \frac{1}{s/\lambda}$$

The gain of the aperture varies as $\cos \theta$, and the relative amplitude of the quantization lobe is modified by the factor $\sqrt{(\cos \theta_1)/(\cos \theta_0)}$. Figure 7.27c–g shows the effects of various other periodic aperture modulations.

Phase-Shifter Quantization Lobes. Peak quantization sidelobe values will be derived by considering the actual aperture phase distribution. Figure 7.28, after Miller,[94] shows this distribution for some scan angle θ_0 and the resulting errors due to phase quantization. Although a continuous curve has been drawn, only points corresponding to integral values of M are meaningful.

The greatest phase-quantization lobes seem to occur with a phase slope such that the elements are spaced, as indicated in the figure, by a distance exactly one-

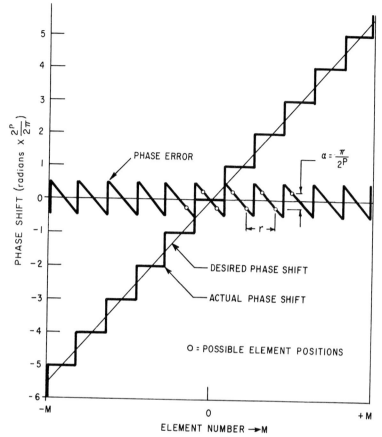

FIG. 7.28 Aperture phase error due to phase quantization. (*From Miller.*[94])

half of the period r or an exact odd multiple thereof. Since the distance between elements is about $\lambda/2$, the value of r is about 1 wavelength, and the array has to be scanned from broadside before the quantization lobe emerges. It will appear at an angle θ_1, given [Eq. (7.10)] by

$$\sin \theta_1 = \sin \theta_0 - \frac{1}{r/\lambda} = \sin \theta_0 - 1 \qquad (7.34)$$

The phase error under these conditions has element-to-element phase ripples with a peak-to-peak value $\alpha = \pi/2^P$. From Fig. 7.27f, this is seen to give

$$\text{Peak phase-quantization lobe} = \frac{\pi}{2} \frac{1}{2^P} \sqrt{\frac{\cos \theta_1}{\cos \theta_0}} \qquad (7.35)$$

These values are shown in Fig. 7.29 and are greater than those given by Miller. The triangular repetitive phase error shown in Fig. 7.28 appears at the aperture if

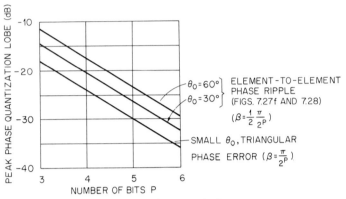

FIG. 7.29   Peak sidelobes due to phase quantization.

the beam points close to broadside and there are many radiators within the distance r. The same shape of phase error is obtained, although spanning several sawtooth cycles, if the element spacing corresponds to a distance slightly greater than r or multiples thereof. Figure 7.27g gives, for this triangular phase error of $\pm \pi/2^P$,

$$\text{Peak phase-quantization lobe} = \frac{1}{2^P} \qquad (7.36)$$

This value is shown in Fig. 7.29 for small scan angles, when $\sqrt{(\cos \theta_1)/(\cos \theta_0)} = 1$. In this case the position (θ_1) of the quantization lobe is given by $\sin \theta_1 = \sin \theta_0 - [1/(r/\lambda)]$, where, from Fig. 7.28, $(2\pi r/\lambda) \sin \theta_0 = 2\alpha = 2\pi/2^P$. This simplifies to

$$\sin \theta_1 = (1 - 2^P)\theta_0$$

An examination of Fig. 7.29 shows that peak phase-quantization lobes are significant, and attempts should be made to reduce them.

Reduction of Peak Phase-Quantization Lobes.   Miller[94] points out that the peak quantization lobes can be reduced by decorrelating the phase-quantization errors. This may be done by adding a constant phase shift in the path to each radiator, with a value that differs from radiator to radiator by amounts that are unrelated to the bit size. The variable phase shifter is then programmed to account for this additional insertion phase. With a spherical or quadratic law of insertion phase variation, as obtained with optical-feed systems (Sec. 7.8), the reductions in peak quantization lobes are equivalent to adding 1 bit to the phase shifters in a 100-element array, 2 bits in a 1000-element array, and 3 bits in a 5000-element array.

Amplitude Quantization.   When the aperture of a phased array is divided into equal subarrays, then the amplitude distribution across each subarray is constant. Aperture taper for the antenna is approximated by changing amplitude from subarray to subarray, and quantization lobes arise from these discontinuities. The value of these lobes may be estimated from the various results shown in Fig. 7.27

or actually computed by summing all contributions at the known quantization (grating) lobe angles. The distribution becomes smoother as the number of subarrays is increased or as they are interlaced.

7.7 BANDWIDTH OF PHASED ARRAYS

The phenomenon of focusing an array is the result of the energy from each element adding in phase at some desired point within the antenna. When energy is incident normal to the array, each element receives the same phase independent of frequency. When energy is incident from some angle other than normal, the phase difference from the planar phase front to each element is a function of frequency and most phased arrays with phase shifters become frequency-dependent. This same phenomenon can be viewed in the time domain. As shown in Fig. 7.30, when a pulse of energy is incident at an angle other than normal, the energy is received earlier at one edge of the array than at the other edge and a period of time must elapse before energy appears in all elements. The concept of aperture fill time $T = L/c \sin \theta_0$ is just another way of explaining the bandwidth of a phased array.

The bandwidth of phased arrays is described by Frank[95] as being composed of two effects: the aperture effect and the feed effect. In both effects, it is the path-length differences that contribute to the bandwidth sensitivity of a phased array. For a parallel-fed array (equal line length), the feed network does not contribute to a change in phase with frequency, and so only the aperture effect remains. This will be discussed first, and the effect of the feed will follow.

Aperture Effects.   When energy is incident on an array at an angle other than broadside (Fig. 7.30), the phase required on the edge element is $\psi = (2\pi L)/\lambda \sin \theta_0$. Note the λ in the denominator. This indicates that the required phase is frequency-dependent. If frequency is changed and the phase shifters are not changed, the beam will move. For an equal-line-length feed, the beam shape will be undistorted (in sine space) and the beam will move toward broadside as the frequency is increased. If the phase shifters are replaced by time-delay networks, then the phase through the time-delay networks will change with frequency and the beam will remain stationary.

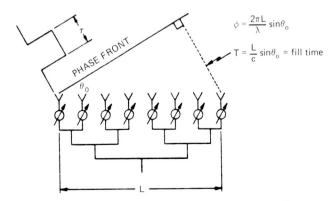

$$\psi = \frac{2\pi L}{\lambda} \sin\theta_0$$

$$T = \frac{L}{c} \sin\theta_0 = \text{fill time}$$

FIG. 7.30   Aperture-fill-time parallel-fed array. (*After Frank.*[95])

When phase shift (independent of frequency) is used to steer the beam, the beam is steered to a direction $\Delta\theta_0$ with a phase of

$$\psi = \frac{2\pi x}{\lambda_1} \sin \theta_0 = \frac{2\pi x}{c} f_1 \sin \theta_0$$

on the element located a distance x from the array center. At a frequency f_2 this same phase setting steers the beam to a new direction, $\theta_0 + \Delta\theta_0$, so that

$$\frac{2\pi x}{c} f_1 \sin \theta_0 = \frac{2\pi x}{c} f_2 \sin (\theta_0 + \Delta\theta_0) \qquad (7.37)$$

With a small change in frequency, the change in direction $\Delta\theta_0$ is small, and

$$\sin (\theta_0 + \Delta\theta_0) \not\approx \sin \theta_0 + \Delta\theta_0 \cos \theta_0$$

which gives, with Eq. (3.37),

$$\Delta\theta_0 = \frac{\Delta f}{f} \tan \theta_0 \qquad \text{(rad)} \qquad (7.38)$$

As the frequency is increased, the beam scans toward broadside by an angle that is independent of the aperture size or beamwidth. However, the permissible amount that the beam may scan with frequency is related to the beamwidth, since pattern and gain deteriorations are a function of the fractional beamwidth scanned. The angle that the beam actually scans, on the other hand, is related to the percentage bandwidth. A bandwidth factor may, therefore, be defined in terms of the broadside beamwidth:

$$\text{Bandwidth factor} = K = \frac{\text{bandwidth } (\%)}{\text{beamwidth (deg)}}$$

A reasonable criterion is to limit the bandwidth so that the beam never scans by more than $\pm$ one-fourth of a local beamwidth with frequency, i.e.,

$$\text{Criterion:} \qquad \left| \frac{\Delta\theta_0}{\theta_B \text{ (scanned)}} \right| \leq \frac{1}{4}$$

With a scan of 60° this gives $K = 1$, and in terms of broadside beamwidth the limit is

$$\text{Bandwidth } (\%) = \text{beamwidth (deg)} \qquad \text{(CW)}$$

For example, if the array has a beamwidth of 2°, this criterion permits a 2 percent change in frequency prior to resetting the phase shifters. This allows the beam to move from one-fourth beamwidth on one side of the desired direction to one-fourth beamwidth on the other side as the frequency changes by 2 percent. At smaller scan angles the effect is reduced as given by Eq. (7.38), and broader-band operation is possible.

The explanation given above applies to an antenna operating at a single (CW) frequency and describes how the beam moves as this frequency is changed.

However, most radars are pulsed and radiate over a band of frequencies. For a beam scanned from broadside, each spectral component is steered to a slightly different direction. To determine the composite effect of the components, it is necessary to add the far-field patterns of all spectral components. This analysis has been performed,[96,97] and it is apparent that the overall antenna gain of the pulse will be less than that of a single spectral component which has maximum gain in the desired direction. As with the CW situation the greatest loss occurs at the maximum scan angle, which is assumed to be 60°. For this situation the criterion chosen is to allow a spectrum which loses 0.8 dB of energy on target due to frequency-scanned spectral components. For a beam scanned to 60° this becomes

$$\text{Bandwidth (\%)} = 2 \text{ beamwidth (deg)} \quad \text{(pulse)}$$

Note that this is twice the bandwidth permitted for the CW situation. Another way of analyzing this phenomenon is in terms of aperture fill time. As shown in Fig. 7.30, the time it takes to fill the aperture with energy is

$$T = \frac{L}{c} \sin \theta_0$$

It has been shown that if the radar pulse is chosen to be equal to the aperture fill time $\tau = L/c \sin \theta_0$, this is equivalent to beamwidth (%) = 2 beamwidth (deg). Hence, a loss of 0.8 dB can be expected if the pulse width is equal to the aperture fill time. Longer pulses will have less loss. The exact amount of the loss will depend on the specific spectrum transmitted, but the variation will amount to less than 0.2 dB for most waveforms. Rothenberg and Schwartzman[98] provide details and also treat the problem as a matched filter.

The preceding discussion assumed an equal-path-length feed. However, it is unlikely that a feed will provide exactly equal path lengths. It will suffice to have the path lengths kept within 1 wavelength of one another. The phase errors introduced can then be corrected by programming the phase shifters. This will have the beneficial effect of breaking up the quantization errors and thereby reducing quantization lobes. As frequency is changed over a 10 to 15 percent bandwidth, the phase errors introduced will then be only a few degrees.

Feed Effect. When an equal-path-length feed is not in use, the feed network will produce a change in phase with frequency. In some cases (Rotman lens[99] or equal-length Blass matrix[100]) the feed can actually compensate for the aperture effect and produce a beam direction which is independent of frequency. However, the more conventional feeds tend to reduce the bandwidth of the array.

End-Fed Series Feed. An end-fed series array is shown in Fig. 7.38a below. The radiating elements are in series and progressively further and further removed from the feed point. When the frequency is changed, the phase at the radiating elements changes proportionally to the length of feed line so that the phase at the aperture tilts in a linear manner and the beam is scanned. This effect is useful for frequency-scanning techniques, but in the case of phased arrays it is undesirable and reduces the bandwidth. It has been shown previously that with phased arrays the pointing direction of a scanned beam also changes with frequency [Eq. (7.38)] since phase rather than time delay is adjusted. These two changes in beam-pointing direction may add or subtract, depending on the direction in which the beam has been scanned. The worst case will be considered here.

With a change in frequency, a nondispersive transmission line having free-space propagation characteristics and a length L equal to the size of the aperture that it feeds will produce a linear-phase variation across the aperture with a maximum value at the edges of

$$\Delta\psi_{max} = \frac{\Delta f}{f} \frac{2\pi L}{\lambda} \quad \text{rad}$$

where $\Delta f/f$ is the fractional change in frequency. This linear-phase progression across the aperture will translate the beam by

$$\Delta\theta_0 = \frac{\Delta f}{f} \frac{1}{\cos\theta_0} \quad \text{rad}$$

For one direction of scanning, this effect will add to the aperture effect; for the opposite direction it will tend to cancel the aperture effect. In waveguide, with the guide wavelength denoted λg, the effect is more pronounced and the resultant change in beam position is

$$\Delta\theta_0 = \frac{\lambda g}{\lambda} \frac{\Delta f}{f} \frac{1}{\cos\theta_0} \quad \text{rad}$$

When analyzing an end-fed series feed it is necessary to consider both the feed effect and the aperture effect. The total frequency scan of this feed will be

$$\Delta\theta_0 = \frac{\Delta f}{f} \tan\theta_0 \pm \frac{\lambda g}{\lambda} \frac{\Delta f}{f} \frac{1}{\cos\theta_0} \quad \text{rad} \quad \text{(CW)}$$

Center-Fed Series Feed.[101] A center-feed array (Fig. 7.31) can be considered as two end feeds. Each feed controls an aperture which is half the total and therefore has twice the beamwidth. As the frequency is changed, each half of the aperture scans in the opposite direction. This initially creates a broader beam with reduced gain. As frequency continues to change, the two beams will eventually split apart. At broadside, the center-fed antenna has poorer performance than a parallel feed since each half scans. However, at 60° scan the compensation on one-half of the array assists in keeping the gain comparable to that of a parallel feed. From the viewpoint of gain reduction, the criterion for a center-fed array is

$$\text{Bandwidth (\%)} = \frac{\lambda}{\lambda g} \text{ beamwidth (deg)} \quad \text{(CW)}$$

However, from the viewpoint of sidelobes this criterion may be unacceptable. For a low-sidelobe design, either CW or pulsed, the sidelobes for this feed should be calculated since a change in frequency no longer produces a translated beam but rather produces a broader beam composed of two translated beams.

Space Feed. The space (optical) feed can be considered to be somewhat between a parallel feed and a center-fed series feed. With a very long focal length, the space feed approximates a parallel feed. With a very short focal length it approximates a center-fed series feed. Since the bandwidth performance of each is comparable, the optical-feed criterion is

$$\text{Bandwidth (\%)} = \text{beamwidth (deg)} \quad \text{(CW)}$$

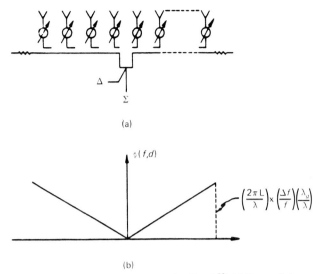

(a)

(b)

FIG. 7.31   Center-fed series array. (*After Frank.*[95]) (*a*) Center-fed series feed. (*b*) Gabled phase due to feed.

Table 7.2 summarizes the bandwidth criteria for the various feed networks. Both the CW criterion and the comparable pulse criterion are given.

Broad Instantaneous Bandwidth.   For a stationary beam in space, independent of frequency, it is necessary to use time delay rather than phase steering. It is not practical to provide a time-delay network at each element in a phased array since these networks are expensive and lossy and contain errors. An alternative is to use a broadband beam-switching technique such as the equal-line-length Blass matrix[100] or a Rotman lens.[99] For 2D scanning these techniques become quite complex.

Another technique for improving the bandwidth by a considerable factor is to use an array of subarrays. The radiating elements of a phased array may be grouped into subarrays where time-delay elements are added. This is shown in Fig. 7.32. The antenna may be regarded as an array of subarrays. The subarray

TABLE 7.2   Bandwidth Criteria for Several Feed Networks*

Feed	CW bandwidth, (%)	Pulse bandwidth, (%)
Equal line length	Beamwidth	$2 \times$ beamwidth
End-fed series	$\dfrac{1}{\left(1 + \dfrac{\lambda_g}{\lambda}\right)} \times$ beamwidth	$\dfrac{2}{\left(1 + \dfrac{\lambda_g}{\lambda}\right)} \times$ beamwidth
Center-fed series	$\dfrac{\lambda}{\lambda_g} \times$ beamwidth	$2\dfrac{\lambda}{\lambda_g} \times$ beamwidth
Space-fed (optical)	Beamwidth	$2 \times$ beamwidth

*After Frank.[95]
NOTE: All beamwidths are in degrees and refer to the broadside beamwidth. λ_g = guide wavelength; λ = free-space wavelength.

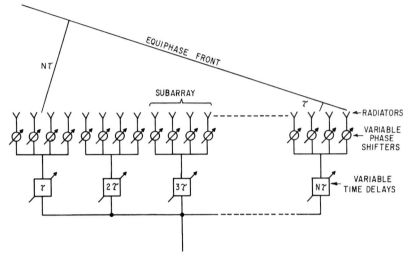

FIG. 7.32 Phased array using subarrays with time delay.

pattern forms the element factor; it is steered by phase shifters in the desired direction, but it scans with frequency as indicated by Eq. (7.38). The array factor is scanned by adjusting the frequency-independent time-delay elements. All subarrays are steered in the same manner. The total radiation pattern is the product of the array factor and the element factor. A change in frequency gives rise to grating lobes rather than shifts of the beam position. This can be seen from Fig. 7.33. The subarray pattern is shown at the design frequency f_0 and is seen to have a null at the position of the grating lobe. As the frequency is changed by δf, the

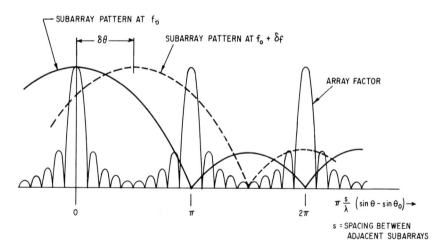

FIG. 7.33 Generation of grating lobes by a change of frequency.

pattern is scanned. It is shown dashed in a position where it has been scanned by a little more than half of its beamwidth. This is clearly too much, for the product of array and element factors gives two beams of equal amplitude.

The loss in gain and the magnitude of the grating lobe are functions of the fractional subarray beamwidth that has been scanned as a result of the change in frequency. The results may be expressed in terms of the bandwidth factor K (referred to the *subarray broadside beamwidth*). At the edges of the band

$$\text{Relative grating-lobe amplitude} \approx \frac{1}{(4/K \sin \theta_0) - 1}$$

$$\text{Loss in gain } \Delta G \approx 1 - \left(\frac{\sin [(\pi/4) K \sin \theta_0]}{(\pi/4) K \sin \theta_0} \right)^2$$

Figure 7.34 shows these values as functions of K for a scan of 60°.

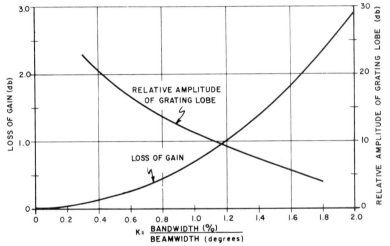

FIG. 7.34 Loss of gain and grating-lobe amplitude as functions of bandwidth (phased subarrays with time delay, scanned 60°). The value of the grating lobe will be modified by the element pattern.

$K = 1$ was the value previously used for scanning 60° and appears acceptable here, where the relevant beamwidth is the broadside beamwidth of the subarray. Thus, if the aperture is split into N subarrays in one plane, with time-delay networks at each subarray level, the bandwidth is increased by a factor of N. This same bandwidth criterion leads to a reduction in gain of about 0.7 dB and a grating lobe of about −11 dB at the edges of the band with 60° of scan. Interlacing of subarrays can reduce the grating lobes.

The monopulse null position is unaffected by the behavior of the subarrays as long as they all respond in the same way. The null position is determined only by the time-delay networks behind the subarray and corresponds to the array-factor null, which is unaffected when multiplied by the subarray pattern.

Time-Delay Networks. Figure 7.35a shows a time-delay network that is digitally controlled by switches. The total delay path length that has to be provided nondispersively amounts to $L \sin \theta_{max}$, where θ_{max} is the maximum scan angle for the aperture L. The smallest bit size is about $\lambda/2$ or λ, with the precise setting adjusted by an additional variable phase shifter. A 1° beam scanned 60°, for example, requires a time delay of 6 or 7 bits, the largest being 32 wavelengths, as well as an additional phase shifter. The tolerances are tight, amounting in this case to a few degrees out of about 20,000, and are difficult to meet. Problems may be due to leakage past the switch, to a difference in insertion loss between the alternate paths, to small mismatches at the various junctions, to variations in temperature, or to the dispersive characteristics of some of the components. Painstaking design is necessary. The switches may be diodes or circulators. Leakage past the switches may be reduced by adding another switch in series in each line. The difference in insertion loss between the two paths may be equalized by padding the shorter arm. The various problems have been comprehensively assessed and analyzed by Lincoln Laboratory.[48]

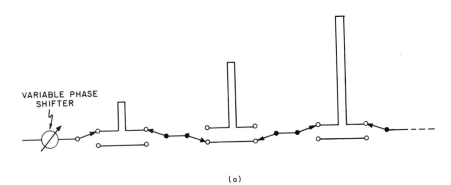

(a)

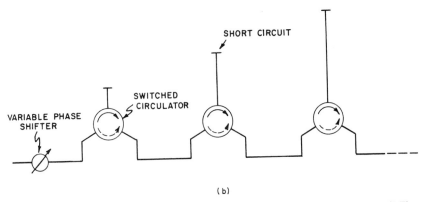

(b)

FIG. 7.35 Time-delay configurations. (a) Time delay by choosing upper or lower paths. (b) Time delay by using switched circulators.

On transmitting, the tolerances are less severe, since the requirements are usually for power on target rather than for accurate angle determination or low sidelobes.

Figure 7.35*b* shows another configuration that has the advantage of simplicity. Each of the switchable circulators connects either directly across (counterclockwise) or via the short-circuited length. Isolation in excess of 30 dB is required. It is clear that the insertion loss of time-delay circuits is very high for most practical systems. They would, therefore, precede a final power amplifier for transmitting and follow a preamplifier for receiving.

A further method of providing delay is possible by translating the problem from the microwave domain and delaying at IF.

A wide variety of time-delayed subarray techniques are available. Tang[102] describes interlaced and overlapped subarrays along with their implementation. One such configuration is shown in Fig. 7.36. Here a Butler or Blass matrix is used to feed a space-fed array. The array contains conventional phase shifters. The matrix, however, is fed with time-delay networks. If there are N inputs to the matrix, N beams will be formed to illuminate the space-fed aperture. Each of these N beams can be time-steered to provide a measure of time steering to the aperture. As with other subarray techniques, the bandwidth will increase in proportion to the number of time-delay networks added. If two-dimensional scanning is required, then a similar set of time-delay networks is required in the orthogonal

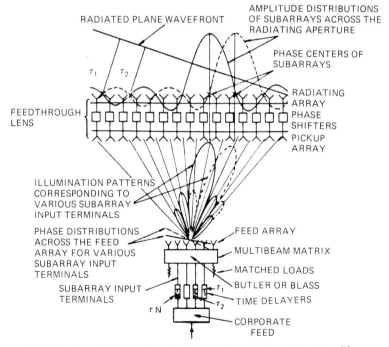

FIG. 7.36 Completely overlapped subarray antenna system. (*After Tang.*[102])

plane. For example, if a factor-of-10 improvement in bandwidth is required and scanning is required in each plane, then 100 time-delay networks will be needed.

7.8 FEED NETWORKS (BEAMFORMERS)

Optical-Feed Systems. Phased arrays may be in the form of lens arrays or reflectarrays, as shown in Fig. 7.37, where an optical-feed system provides the proper aperture illumination. The *lens* has input and output radiators coupled by phase shifters. Both surfaces of the lens require matching. The primary feed should be designed with care and can be complex to give the desired aperture amplitude distribution with low spillover losses. The transmitter feed can be separated from the receiver feed by an angle α as shown. The phase shifters are then reset between transmitting and receiving so that in both cases the beam points in the same direction. This method allows flexibility in optimizing the transmitter aperture distribution, perhaps for maximum power on the target, and separately optimizing the receiver sum and difference patterns for low sidelobes. Since a change in feed position corresponds to scanning with time delay, additional feeds may be added to provide several time-delay-compensated directions of scan for a corresponding increase in bandwidth.

The phasing of the antenna has to include a correction for the spherical phase front. This can be seen (Fig. 7.37*a*) to amount to

$$\frac{2\pi}{\lambda}\left(\sqrt{f^2 + r^2} - f\right) = \frac{\pi}{\lambda}\frac{r^2}{f}\left[1 - \frac{1}{4}\left(\frac{r}{f}\right)^2 + \ldots\right]$$

With a sufficiently large focal length, the spherical phase front may be approximated by that of two crossed cylinders, permitting the correction to be applied simply with row-and-column steering commands.

Correction of a spherical phase front with the phase shifters reduces peak phase-quantization lobes (Sec. 7.6). Space problems may be encountered in assembling an actual system, especially at higher frequencies, since all control circuits have to be brought out at the side of the aperture.

Multiple beams may be generated by adding further primary feeds. All the beams will be scanned simultaneously by equal amounts in sin θ.

The reflect array shown in Fig.7.37*b* has general characteristics similar to those of the lens. However, the same radiating element collects and radiates after reflection. Ample space for phase-shifter control circuits exists behind the reflector. To avoid aperture blocking, the primary feed may be offset as shown. As before, transmitting and receiving feeds may be separated. Multiple beams are again possible with additional feeds.

The phase shifter must be reciprocal so that there is a net controllable phase shift after passing through the device in both directions. This rules out frequently used nonreciprocal ferrite or garnet phase shifters.

Constrained Feeds. The optical-feed system divides power very simply in one step from the feed to the many elements on the aperture. In contrast, the constrained-feed system uses many steps. For a high-performance low-sidelobe system, each of these power-dividing steps has to be well matched over the band. If mismatches are present, they will be separated by many wavelengths and will add to give frequency-sensitive phase and amplitude perturbations at the aperture. In general, these perturbations will be different for the monopulse sum and

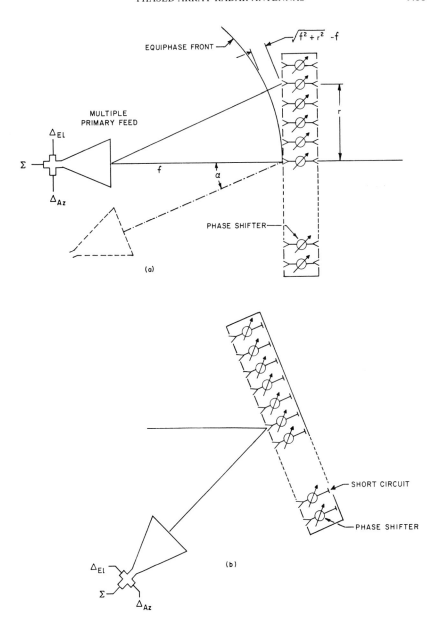

FIG. 7.37 Optical-feed systems. (*a*) Lens array. (*b*) Reflectarray.

difference patterns, so that only a common average can be compensated by calibration.

Series Feeds. Figure 7.38 shows several types of series feeds. In all cases except *d*, the electrical path length to each radiating element has to be computed as a function of frequency and taken into account when setting the phase shifters. Type *a* is an end-fed array. It is frequency-sensitive and leads to more severe bandwidth restrictions than most other feeds. Type *b* is center-fed and has almost the same bandwidth as a parallel-feed network. Sum- and difference-pattern outputs are available, but they have contradictory requirements for optimum amplitude distributions that cannot both be satisfied. As a result, either good sum or good difference patterns can be obtained. At the cost of some additional complexity, the difficulty can be overcome by the method shown in Fig. 7.38c. Two separate center-fed feed lines are used and combined in a network to give sum- and difference-pattern outputs.[103] Independent control of the two amplitude distributions is possible. For efficient operation the two feed lines require distributions that are *orthogonal*, that is, that give rise to patterns where the peak value of one coincides with a null from the other and aperture distributions are respectively even and odd.

A very wideband series feed with equal path lengths is shown in Fig. 7.38d. If the bandwidth is already restricted by phase scanning, then very little advantage is obtained at the cost of a considerable increase in size and weight. The network in Fig. 7.38e gives simplicity in programming since each phase shifter requires the same setting. The insertion loss increases for successive radiators, and the tolerances required for setting the phases are high. This type is not commonly used.

Parallel Feeds. Figure 7.39 shows a number of parallel-feed systems. They frequently combine a number of radiators into subarrays, and the subarrays are then combined in series or in parallel to form sum and difference patterns.

Figure 7.39a shows a *matched corporate feed*, which is assembled from

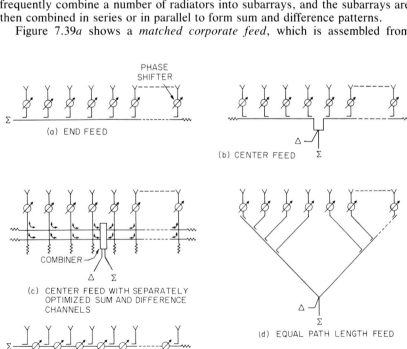

FIG. 7.38 Series-feed networks.

matched hybrids. The out-of-phase components of mismatch reflections from the aperture and of other unbalanced reflections are absorbed in the terminations. The in-phase and balanced components are returned to the input. To break up periodicity and reduce peak quantization lobes (Sec. 7.6), small additional fixed phase shifts may be introduced in the individual lines and compensated by corresponding readjustments of the phase shifters.

The reactive feed network of Fig. 7.39*b* is simpler than the matched configuration. It has the disadvantage of not terminating unbalanced reflections which are likely to be at least partly reradiated and thus contribute to the sidelobes. A stripline power divider is shown in Fig. 7.39*c*, and a constrained optical power divider using an electromagnetic lens is shown in Fig. 7.39*d*. The lens may be omitted and the correction applied at the phase shifters. With nonreciprocal phase shifters, a fraction of the power reflected from the aperture will then be reradiated (as sidelobes) rather than returned to the input. The amplitude distribution across the horn is given by the waveguide mode. It is constant with an *E*-plane horn as shown.

Subarrays. The phased-array aperture may be divided into subarrays, all similar to simplify manufacturing and assembly. Beamforming now requires the combination of the subarrays for suitable sum and difference patterns. Figure 7.40*a* shows a method of doing this by combining opposite subarrays into their sums and differences. All sum channels are then added with proper weighting to obtain the desired amplitude distribution. The difference channels are treated similarly with independent amplitude weighting. This method may be extended to include combination in the other plane.

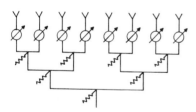

(a) MATCHED CORPORATE FEED

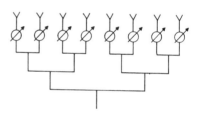

(b) REACTIVE CORPORATE FEED

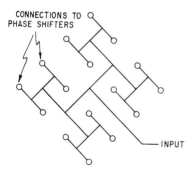

(c) STRIPLINE REACTIVE FEED

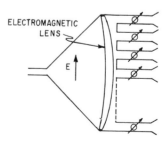

(d) MULTIPLE REACTIVE POWER DIVIDER

FIG. 7.39 Parallel-feed networks.

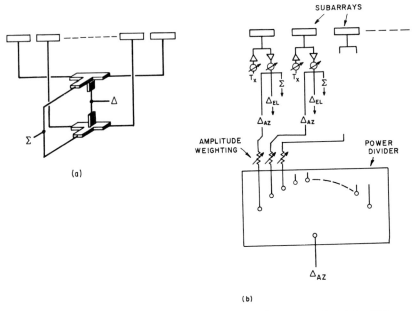

FIG. 7.40 Combination of subarrays to form sum and difference channels. (*a*) Combining opposite subarrays. (*b*) Combining subarrays after amplification.

Amplification on receiving or on both receiving and transmitting may be convenient at the subarray level. On receiving, the noise figure is established by the preamplifier so that further processing may include lossy circuits. The receiving channel may be split three ways, as indicated in Fig. 7.40*b*, into the sum and the elevation and azimuth difference channels. These are then weighted and summed with corresponding outputs from the other subarrays. On transmitting, all separate power amplifiers may be energized equally to give maximum power on the target. The addition of phase shifters shown at the subarray level simplifies the beam-steering computation, permitting all subarrays to receive identical steering commands. They may be replaced by time-delay circuits giving a wide instantaneous bandwidth (Sec. 7.7).

A simple method of providing TR (transmit-receive) switching can be obtained by combining two halves of a subarray with a four-port hybrid junction. The transmitter input into one port energizes both halves of the aperture, say, in phase. The receiver, connected to the remaining port, then requires that the phase shifters of half the aperture give an additional phase shift of π during the receiving period. This is easily programmed.

Multiple receive beams may be formed by combining the subarrays after amplification in as many different ways as separate beams are required. The limitation is that the beams have to lie within the beamwidth of the subarray in order to avoid excessive grating lobes. A cluster of such simultaneous receiver beams requires a wider transmitter beam. This may be obtained efficiently from the same antenna with a gabled or spherical phase distribution.

With identical subarrays, a desired aperture amplitude taper (for low sidelobes) is applied with a granularity that depends on their size and shape. The resulting amplitude steps will cause quantization lobes (Sec. 7.6).

The subarrays can be in the form of line sources. In particular, horizontal line sources can be stacked for vertical-only phase scanning of an antenna that is mechanically rotated in azimuth. Only one phase shifter per line is required.

7.9 PHASE SHIFTERS

The three basic techniques for electronic beam steering are (1) frequency scanning, (2) beam switching, and (3) phase scanning with phase shifters. Of the three techniques, the use of phase shifters is the most popular, and considerable effort has gone into the development of a variety of phase shifters. Phase shifters can be separated into two categories: reciprocal and nonreciprocal. The reciprocal phase shifter is not directionally sensitive. That is, the phase shift in one direction (e.g., transmit) is the same as the phase shift in the opposite direction (e.g., receive). Therefore, if reciprocal phasers are used, it is not necessary to switch phase states between transmit and receive. With a nonreciprocal phaser it is necessary to switch the phaser (i.e., change phase state) between transmit and receive. Typically it takes a few microseconds to switch nonreciprocal ferrite phasers. During this time the radar is unable to detect targets. For low-pulse-repetition-frequency (PRF) radars [e.g., 200 to 500 pulses per second (pps)] this may not cause a problem. For example, if the PRF is 200 pps (or Hz), the time between pulses is 500 μs. If the switching time for the phaser is 10 μs, only 2 percent of the time is wasted and less than 1.0 nmi of minimum range is lost. On the other hand, if the PRF = 50 kHz, the time between pulses is 20 μs, and it would not be possible to tolerate 10 μs of dead time for the switching of phasers.

All diode phasers are reciprocal along with certain types of ferrite phasers. It is worth noting that, owing to losses associated with their magnetic properties, ferrite phasers are almost never used at frequencies below 3 GHz. Diode phasers, in contrast, improve as the frequency gets lower.

At the present time there are three basic types of phasers which typically compete for use in a phased array. They are (1) the diode phaser, (2) the nonreciprocal ferrite phaser, and (3) the reciprocal (dual-mode) ferrite phaser. Each has its strengths, and the choice of which to use is highly dependent on the radar requirements. Each will be discussed in turn. A fourth, less frequently used phaser, the reciprocal ferrite rotary-field phaser, will also be discussed.

Diode Phasers.[104-107] Diode phasers are typically designed by using one of three techniques: (1) switched-line, (2) hybrid-coupled, and (3) loaded-line. These are shown diagrammatically in Fig. 7.41. The switched-line technique simply switches in lengths of line in binary increments (e.g., 180°, 90°, 45°) and requires a set of diodes for each bit. The diodes are used as switches to control which bits are activated to achieve a particular phase state.

The hybrid-coupled technique uses a microwave hybrid and effectively changes the distance at which the reflection takes place. This technique is usually used in binary increments, and an additional set of diodes is required for each phase state.

The diode phasers described above are limited in their ability to handle high peak power. Depending on their size and frequency, they are normally restricted to power levels of less than 1 kW. For higher power levels, the loaded-line technique is used. The diodes are used to switch in increments of capacitance and inductance which provide small changes in phase. Since the diodes are decoupled

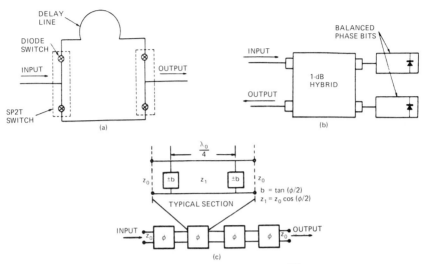

FIG. 7.41 Diode phase-shifter configurations. (*After Stark et al.*[104]) (*a*) Switched-line phase bit. (*b*) Hybrid-coupled phase bit. (*c*) Loaded-line phase bit.

from the main transmission line, they need handle only modest amounts of power in each diode. Very high power configurations are possible. The technique does require many diodes, and the phasers are typically large and bulky as compared with the switched-line and hybrid-coupled techniques.

Diode phasers have the advantage of being small and light in weight (except for high-power devices). They are suitable for stripline, microstrip, and monolithic configurations. Typical performance characteristics are given in Table 7.3. The main disadvantage of the diode phaser is that an additional set of diodes is normally required for each additional bit. As lower-sidelobe antennas are required, the number of bits increases. For very low sidelobe antennas 5, 6, or 7 bits may be required. As the number of bits increases, both cost and loss of the diode phasers are also increased. This need not be the case with ferrite devices.

Ferrite Phasers. Most ferrite phasers are nonreciprocal,[108–110] and their early versions used discrete lengths of ferrite (as shown in Fig. 7.42) to

TABLE 7.3 Diode Phase Shifters*

Frequency range	S band	C band	X band
Insertion range	1–1.2 dB	2.0 dB maximum	3.0 dB maximum
VSWR	1.4:1 maximum	1.3:1 maximum	1.5:1 maximum
RF power	50 W CW	50 W peak	50 W peak
Number of bits	6	6	6
Switching time	0.5 μs	1.0 μs	1.0 μs
Size†	2.5 by 1.5 by 0.5 in	2.5 by 1.5 by 0.5 in	2.5 by 1.5 by 0.5 in

*Courtesy of Anghel Laboratories, Inc.
†This size includes connectors. For an integrated phase shifter, much smaller sizes are achievable.

implement each of the bits (180°, 90°, 45°, etc.). In this configuration a current pulse is passed through each bit, and the ferrite toroid is saturated. When the current is removed, the ferrite toroid is said to be *latched* and retains its magnetization owing to its hysteresis properties. If the current is in a forward direction, the ferrite is latched with a particular phase (e.g., 180°). The ferrite maintains the phase until a current pulse in the opposite direction is applied. The ferrite phaser is then latched to the reference phase (0°).

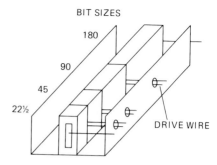

FIG. 7.42 Digital ferrite phase shifter using toroids. (*After Stark.*[104])

This change in phase with a change in current direction is due to the nonreciprocal nature of the device. As mentioned, early devices saturated each bit so that a ferrite toroid and electronic driver were required to control each bit individually. Other phasers use a single toroid and a single driver.[111,112] In this configuration the phaser is latched on a minor hysteresis loop by only partially magnetizing the ferrite. The distinct advantage of this technique is that any number of bits may be implemented while using only a single toroid. Table 7.4 shows the characteristics of typical nonreciprocal phasers. They have the advantage of low loss and relatively high power operation. Devices which handle up to 100 kW of peak power have been built. They are amenable to waveguide construction and are heavier and bulkier than comparable diode devices.

Reciprocal ferrite phasers, including dual-mode phasers,[113, 114] are a more recent addition to the family of phasers. Since the ferrite phase-shift mechanism is inherently nonreciprocal, it is necessary in this device to add another

TABLE 7.4 Nonreciprocal Ferrite Phase Shifters*

Frequency, GHz	Phase shift available	RF power (peak/average)	Insertion loss, dB maximum	Maximum VSWR	Phase-shift accuracy, rms	Maximum switching time, μs
S band (5%) bandwidth	360° 5–8 bits	40 kW/800 W	1.0	1.2:1	2–5°	20
C band (5%) bandwidth	360° 5–8 bits	10 kW/400 W	0.8	1.2:1	2–5°	8
X band (5%) bandwidth	360° 5–8 bits	500 W/100 W	0.6	1.2:1	2–5°	3
K_u band (5%) bandwidth	360° 5–8 bits	250 W/50 W	0.5	1.2:1	2–5°	2
6.5–18 inclusive	360° 4–6 bits	5 kW/100 W	1.8	1.5:1	6°	3
K_a band (5%) bandwidth	360° 4–6 bits	50 W/20 W	1.2	1.2:1	2–5°	<2

*Courtesy of Electromagnetic Sciences, Inc.
NOTE: The above parameters represent nominal performance and can be optimized for specific applications.

nonreciprocal device to the circuit as shown diagrammatically in Fig. 7.43. As physically implemented, a nonreciprocal quarter-wave plate is introduced to create circular polarization within the phaser. As with the nonreciprocal phaser, the phase is shifted by latching the ferrite toroid to a nonsaturated magnetic state with a pulse of current. After the microwave energy is phase-shifted, it goes through another nonreciprocal quarter-wave plate, which returns the polarization to linear. If energy is traveling in the opposite direction, the procedure is reversed, with the net result that the overall device is reciprocal. Typical characteristics are given in Table 7.5.

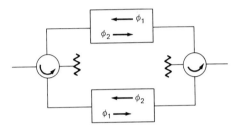

BASIC CONCEPT: DUAL-CHANNEL PHASER

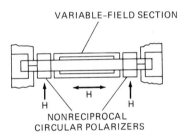

FIG. 7.43 Dual-mode phaser configuration. (*After Boyd.*[113])

    The dual-mode phasers tend to be low in loss and can be accurately controlled. As with the other ferrite phasers, they are typically latched on a minor hysteresis loop by using a current pulse and are therefore capable of implementing any number of bits with a single toroid. Dual-mode phasers are usually larger than diode devices and smaller than nonreciprocal devices. Similarly, they handle more power than diodes and less power than nonreciprocal ferrite devices.
    The largest drawback of the dual-mode phaser is the slow switching speed, which ranges from dozens to hundreds of microseconds. The faster speeds are associated with the smaller high-frequency devices. However, since these devices are reciprocal, it is not necessary to switch between transmit and receive, and high-speed switching may not be necessary.
    The rotary-field ferrite phase shifter[115] is also capable of reciprocal phase shift. It is less commonly used but is of considerable value for systems which require high power and high accuracy. Its requirement for considerable holding power and its slow switching speed have limited its application.
    In summary, diodes and nonreciprocal ferrite and reciprocal ferrite phasers

TABLE 7.5 Dual-Mode Reciprocal Latching Ferrite Phase Shifter*

Frequency range	C band, 10% bandwidth	X band, 10% bandwidth	K_u band, 12% bandwidth
Insertion loss	1.0 dB typical	1.0 dB typical	1.0 dB typical
VSWR	1.5:1 maximum	1.5:1 maximum	1.5:1 maximum
RF power	250 W peak, 10 W average	250 W peak, 10 W average	200 W peak, 10 W average
Phase error	4° rms	4° rms	4° rms
Switching time	250 μs	150 μs	50 μs
Weight (with driver)	4 oz	1.5 oz	0.8 oz
Size	0.815-in diameter, 4 in long	0.55-in diameter, 3 in long	0.34-in diameter, 2.5 in long

*Courtesy of Microwave Applications Group, CA.

are all viable competitors. Diode phasers, owing to improvements in the diodes, will continue to improve more rapidly than the ferrite devices. At L band and lower, diode phasers are an obvious choice. At S band and higher, ferrites should continue to hold an edge in higher-power systems and where additional bits are needed for the low phase errors required for low-sidelobe antennas. Ferrite phase shifters are more temperature-sensitive than diodes, and the phase will change with a change in temperature. This can be controlled by keeping the temperature constant (within a few degrees) across the array. A more common technique is to sense the temperature at several locations in the array and then correct the phase commands to the phase shifters.

7.10 SOLID-STATE MODULES[116–118]

A solid-state module may be connected to every radiating element or to every subarray of a phased array antenna, forming what is sometimes called an *active aperture*. Applications range from ultrahigh frequency (UHF) for surveillance to X band and above for airborne systems. Chapter 5 discusses the solid-state transmitter, but this section highlights some aspects specifically related to phased array radar.

A typical module is shown in simple schematic in Fig. 7.44. It consists of a transmitter amplifier chain, a preamplifier for receiving, a shared phase shifter with driver, and circulators and/or switches to separate the transmit and receive paths (the gain around the loop must be less than unity to avoid oscillations).

Power amplifiers for transmitting at the element level would typically have a gain of 30 dB or more to compensate for the loss of power dividing in the beamformer. Transistors are capable of generating high average power but only relatively low peak power. High-duty-cycle waveforms (10 to 20 percent) aretherefore required to transmit enough energy efficiently. This lack in high peak power is the main disadvantage of the solid-state modules in phased array radars. To a great extent, it can be compensated for by using more pulse com-

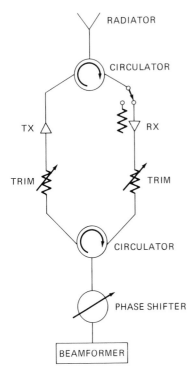

FIG. 7.44 Typical solid-state module.

pression in the receiver and extra-wide bandwidth to counteract jamming, but at the cost of increased signal processing. An important advantage of transistors is their potential for wide bandwidth.

The receiver requires a gain of typically 10 to 20 dB to adequately establish a low noise figure and allow for phase shifting and beamforming losses. The module also receives interfering signals from all directions within the element pattern (not just the antenna pattern), at all frequencies in the band. Low receiver gain may therefore be advantageous to conserve dynamic range. Tight tolerances between modules are needed to provide amplitude and phase tracking over the band for low-sidelobe performance. An electronically programmable gain adjustment may be helpful to correct module-to-module variations, allowing less demanding specifications. The trimmer would also provide a degree of freedom in aperture control for special situations. Since the noise figure has been established, the feed network may be split to give separate optimum aperture amplitude distributions for transmitting as well as for receiving on sum and difference channels. In an alternative configuration, the feed network could be designed for a constant amplitude aperture distribution to give highest transmit power on the target, and the receiver gain control be used to provide an amplitude taper for the sum channel. Perhaps a secondary feed system could be added for the difference channels. Poirier[117] has analyzed this case, the effect on noise, and degradation due to amplitude quantization steps.

The module phase shifter operates at low signal power levels. Its insertion loss may be high, since it is followed by amplification on transmit and preceded by amplification on receive. Diode phase shifters are therefore quite permissible even with many bits (e.g., 5, 6, or 7 bits for low sidelobes). Variations in insertion loss may be compensated for dynamically with the gain adjustment.

The circulator at the high-power side provides an impedance match to the power amplifier and may be adequate by itself for protecting the receiver. In Fig. 7.44 it is shown augmented by a switch which causes absorption of power reflected from antenna mismatches and also gives added protection to the receiver during transmission. If weight is an important consideration, as it would be in a space-based system, then the circulator could be replaced by diode switches requiring additional logic and driving circuitry.

Further information about the characteristics of solid-state transmitters, their advantages, and their disadvantages is given in Chap. 5.

7.11 PHASED ARRAY SYSTEMS

A number of phased array radar systems have been built, and a representative selection is briefly reviewed below.

AN/SPY-1.[119, 120] This S-band phased array radar is part of the Navy Aegis weapon system and was developed by RCA. It has four phased array apertures to give unobstructed hemispherical coverage (Fig. 7.1). In its early configuration it used a simple feed system with, on receive, 68 subarrays, each containing 64 waveguide-type radiators, for a total of 4352 elements. On transmit, subarrays were combined in pairs, and 32 such pairs gave a transmit aperture of 4096 radiators. The phase shifters fed directly into waveguide radiators, had 5 bits, and were of the nonreciprocal, flux-drive, latching garnet configuration. A later version was designed for low sidelobes. The subarray size had to be reduced to 2 elements to avoid quantization lobes, and, similarly, the phase shifter had to be refined by driving it with 7-bit accuracy. The resulting phased array has an aperture with a constrained-feed structure and 4350 waveguide-type radiators. Monopulse sum and difference receive patterns and the transmit pattern are separately optimized.

Figure 7.45 shows the antenna being tested in a near-field facility[16] where the field near the aperture is measured with a probe for far-field pattern calculation.

Thomson CSF Disk Antenna.[13, 14, 121] Thomson CSF has developed an S-band multifunction phased array radar, the TRS 3501, for naval use (Fig. 7.46). The planar antenna has a number of unusual features. It is mounted on a turntable and either rotated for 360° surveillance or pointed for sector coverage.

The antenna aperture contains circularly polarized helical radiators (for rain rejection) that are wound on a tulip-shaped dielectric form and arranged in roughly concentric rings. The elements also contain diode phase shifters for beam steering. The phase-shifter drivers are housed along the periphery of the aperture.

The feed network is of very simple construction. It consists of a parallel-plate medium forming a disk behind the radiating aperture. The medium is energized at its center with a monopulse feed. The radiating elements simply couple to the medium with probes, and the probe depth determines the amount of coupling. Since the same circularly polarized antenna is used for both transmitting and receiving, single-bounce targets, including rain, are rejected.

PATRIOT.[122] PATRIOT is a multifunction phased array radar system developed by Raytheon for the Army in the form of a lens array using an optical feed as shown in Fig. 7.37. Sum and difference patterns are separately optimized with a monopulse feed. The aperture is round and contains about 5000 elements. It uses 4-bit flux-driven nonreciprocal ferrite phase shifters and waveguide-type radiators at both apertures. The antenna is shown in Fig. 7.47. It is mounted on a vehicle and folds flat for transportation.

Electronically Agile Radar (EAR).[15, 22, 123] The Westinghouse Electric Corporation developed this X-band phased array radar for airborne application. The system later evolved into the AN/APQ-164 radar of the B-1B bomber, shown in Fig. 7.48 during assembly of its elliptical aperture with 1526 elements. The aperture of EAR is 39 in in diameter and is made up from 1818 radiating elements, each with its own reciprocal ferrite latching phase shifter. The system has many

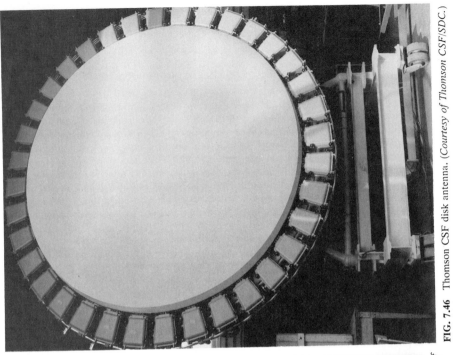

FIG. 7.46 Thomson CSF disk antenna. (*Courtesy of Thomson CSF/SDC.*)

FIG. 7.45 SPY-1 antenna undergoing tests in a near-field range. (*Courtesy of RCA Electronic Systems Department.*)

FIG. 7.47 PATRIOT multifunction phased array. (*Courtesy of Raytheon Company.*)

FIG. 7.48 AN/APQ-164 phased array during assembly. (*Courtesy of Westinghouse Electric Corporation.*)

flexible features. The aperture phase can be distorted to change the beam shape from pencil to cosecant-squared to vertical fan. Polarization can also be changed, from vertical to one-sense circular. This is accomplished with a switchable Faraday rotator in combination with a ferrite quarter-wave plate. A fault location and isolation system is included.[22] During tests on the antenna range, the phase of each individual element is rotated in sequence, and the resulting variations in signal are used to determine the proper relative phase setting (correcting for feed system errors), to check the amplitude of excitation, and to check the proper functioning of the polarization diversity.

AWACS (Airborne Warning and Control System).[13, 14, 124] The first operational radar antenna with very low sidelobes was developed by the Westinghouse Electric Corporation for the AWACS system. Its success, once established, was followed by others, and sidelobe levels well below what once was thought achievable are being routinely specified. The AWACS antenna is a slotted-waveguide array with over 4000 slots that is used for aerial surveillance (Fig. 7.49). It is rotated mechanically in a rotor dome and scanned electronically in the vertical plane with 28 reciprocal ferrite precision phase shifters.

MESAR (Multifunction Electronically Scanned Adaptive Radar).[125] This is an ambitious S-band solid-state air defense phased array radar development undertaken jointly by the Admiralty Research Establishment and Plessey Radar in Great Britain. The array is designed to give an aperture of 1.8 by 1.8 m and contains 918 waveguide-type radiating elements, constructed from metallized glass-fiber-reinforced plastic to reduce weight and cost (Fig. 7.50).

The solid-state modules are of the general form shown in Fig. 7.44. The phase shifter has 4 bits. The power amplifier is a discrete device and gives 2 W with 20 percent bandwidth and a power-added efficiency of 22 percent. On receive, the signal is preamplified and phase-shifted in the module and condensed in a beamformer to 16 subarrays, each with its own receiver. The outputs of these receivers are digitized with an 8-bit A/D converter, offering a powerful adaptive nulling capability.

AN/TPS-70.[14, 126] The AN/TPS-70 shown in Fig. 7.51, developed by the Westinghouse Electric Corporation, is a low-sidelobe array which does not use phase shifters. The array is an upgrading of the AN/TPS-43 and uses multiple simultaneous beams to provide elevation coverage while the antenna rotates in azimuth. The antenna employs 36 horizontal waveguide *sticks*. Each stick has 94 slots to provide a 1.6° beamwidth and low azimuth sidelobes, as previously developed for AWACS. Twenty-two of the sticks are excited on transmit to produce an elevation beamwidth of 20°. The transmit excitation is designed to provide higher gain at low elevation angles and lower gain at high elevation angles. On receive, the energy from all 36 sticks is combined to produce six simultaneous beams to cover elevations from 0 to 20°. The elevation beamwidth of the six beams varies from 2.3° for the lowest beam to 6° for the highest beam. Each of the six beams has a separate receiver, and monopulse information is available in elevation by comparing the energy in these beams. The advantage of a simultaneous-beam system is that it provides the time needed to perform signal-processing functions in a heavy-clutter environment.

The AN/TPS-70 is fully transportable, provides 240 nmi of range, and uses 3 MW of peak power and 5 kW of average power. The radar and its variants have been used widely throughout the world.

FIG. 7.49 AWACS antenna. (*Courtesy of Westinghouse Electric Corporation.*)

FIG. 7.50 MESAR, a solid-state phased array development. (*Courtesy of Plessey Radar Ltd. and the Admiralty Research Establishment.*)

FIG. 7.52 AN/TPQ-37 weapon-locating radar. (*Courtesy of Hughes Aircraft Company.*)

FIG. 7.51 AN/TPS-70 multibeam array. (*Courtesy of Westinghouse Electric Corporation.*)

7.74

AN/TPQ-37.[127]   The U.S. Army has developed the Firefinder, AN/TPQ-37, designed by Hughes Aircraft, to detect artillery shells and trace them back to their source of origin (Fig. 7.52). The radar employs a limited-scan phased array[30,128] which provides wide scan angles in azimuth together with limited scan angles in elevation. The limited-scan coverage reduces the number of phase shifters substantially. The system uses only 360 diode phasers, with each phaser controlling six radiating elements in a vertical line. Each phaser is capable of handling 4 kW of peak power and 165 W of average power.

A monopulse feed network is used to form sum, azimuth difference, and elevation difference beams. The beamformer is composed of air stripline and waveguide power dividers. The overall antenna size is 8 by 12 by 2 ft. Dozens of these systems have been built for the United States and international markets.

PAVE PAWS.[11, 13]   The Raytheon Company developed the PAVE PAWS radar system, Fig. 7.53, which has been deployed to provide early warning of ballistic missiles and to perform satellite tracking. It is a UHF solid-state phased array that is described in Sec. 5.4. An installation contains two radars with apertures inclined to each other by 120°, thereby providing a total field of view of 240°. A 10-m^2 target can be detected at 3000 nmi.

FIG. 7.53 PAVE PAWS ultrahigh-frequency solid-state phased array. (*Courtesy of Raytheon Company.*)

FIG. 7.54   COBRA DANE L-band phased array. (*Courtesy of Raytheon Company.*)

COBRA DANE.[13]   Raytheon's COBRA DANE radar is a very large L-band phased array, designed and positioned for gathering intelligence of foreign ballistic missile tests. Figure 7.54 shows the radar, which has some unusual features. It is a thinned array, 95 ft in diameter, with 34,768 elements, of which 15,360 are active; the rest are dummies which could be replaced by active elements at a later date. The active elements are divided into 96 subarrays, each with 160 radiators, fed, on transmit, by a traveling-wave tube (TWT). The total power at the antenna is 15.4 MW peak and 0.92 MW average. The radar has a wide instantaneous bandwidth, 200 MHz, which gives 2.5-ft range resolution for the measurement of target size and shape. Since the effective antenna aperture fill time is much greater than the pulse length, even with modest scan angles, time-delay beam steering becomes necessary. Time delays are applied to the subarrays, in addition to the phase steering applied to the individual radiating elements.

COBRA JUDY.[14]   This unique, large phased array was developed by Raytheon for the Air Force to collect data on foreign ballistic missile tests. It is mounted on a turntable on the USS *Observation Island* as shown in Fig. 7.55. The array is 22.5 ft in diameter and contains 12,288 elements fed by 16 TWTs for transmission.

FIG. 7.55 COBRA JUDY large rotatable phased array on the USS *Observation Island*. (*Courtesy of Raytheon Company*.)

REFERENCES

1. Blondel, A.: Belgian Patent 163,516, 1902; British Patent 11,427, 1903.

2. Brown, S. G.: British Patent 14,449, 1899.

3. Foster, R. M.: Directive Diagrams of Antenna Arrays, *Bell System Tech. J.*, vol. 5, pp. 292–307, April 1926.

4. Hansen, R. C.: "Microwave Scanning Antennas," vols. I, II, and III, Academic Press, New York, 1964.

5. Amitay, N., R. C. Pecina, and C. P. Wu: Radiation Properties of Large Planar Arrays, *Bell Teleph. Syst. Monog.* 5047, February 1965.

6. Oliner, A. A., and G. H. Knittel: "Phased Array Antennas," Artech House, Norwood, Mass., 1972.

7. Stark, L. I.: Microwave Theory of Phased Array Antennas—A Review, *Proc. IEEE*, vol. 62, pp. 1661–1701, December 1974.

8. Mailloux, R. J.: Phased Array Theory and Technology, *Proc. IEEE*, vol. 70, March 1982.

9. Rudge, A. W., K. Milne, A. D. Olver, and P. Knight: "The Handbook of Antenna Design," vol. 2, Peter Peregrinus, Ltd., London, 1983.

10. Johnson, R. C., and H. Jasik (eds.): "Antenna Engineering Handbook," 2d ed., McGraw-Hill Book Company, New York, 1984.

11. Brookner, E.: Phased Array Radars, *Sci. Am.*, vol. 252, pp. 94–102, February 1985.

12. Steyskal, H. P.: Phased Arrays 1985 Symposium, *RADC Rept.* TR-85-171, Rome Air Development Center, Bedford, Mass., August 1985.

13. Brookner, E.: A Review of Array Radars, *Microwave J.*, vol. 24, pp. 25–114, October 1981.

14. Brookner, E.: Radar of the 80's and Beyond, *IEEE Electro 84*, May 1984.

15. Brookner, E.: Array Radars: An Update, *Microwave J.*, vol. 30, pt. I, pp. 117–138, February 1987; pt. II, pp. 167–174, March 1987.

16. Harmening, W. A.: A Laser-Based, Near-Field Probe Position Measurement System, *Microwave J.*, pp. 91–102, October 1983.

17. Mailloux, R. J.: Chap. 21 in Johnson, R. C., and H. Jasik (eds.): "Antenna Engineering Handbook," 2d ed., McGraw-Hill Book Company, New York, 1984.

18. Borgiotti, G. V.: Chap. 11 in Rudge, A. W., K. Milne, A. D. Olver, and P. Knight: "The Handbook of Antenna Design," vol. 2, Peter Peregrinus, Ltd., London, 1983.

19. Schroeder, K. G.: Near Optimum Beam Patterns for Phase Monopulse Arrays, *Microwaves*, pp. 18–27, March 1963.

20. Frank, J.: Phased Array Antenna Development, *Johns Hopkins University, Appl. Phys. Lab. Rept. TG 882*, pp. 114–117, March 1967.

21. Scharfman, W. E., and G. August: Pattern Measurements of Phased Arrayed Antennas by Focussing into the Near Zone, in Oliner, A. A., and G. H. Knittel: "Phased Array Antennas," Artech House, Norwood, Mass., 1972, pp. 344–349.

22. Alexander, D. K., and R. P. Gray, Jr.: Computer-Aided Fault Determination for an Advanced Phased Array Antenna, *Proc. Antenna Application Symp.*, Allerton, Ill., September 1979.

23. Ronen, J., and R. H. Clarke: Monitoring Techniques for Phased-Array Antennas, *IEEE Trans.*, vol. AP-33, pp. 1313–1327, December 1985.

24. Collings, R. H.: U.S. Patent 3,680,136, 1972.

25. Byron, E. V.: A New Flush Mounted Antenna Element for Phased Array Application, in Oliner, A. A., and G. H. Knittel: "Phased Array Antennas," Artech House, Norwood, Mass., 1972, pp. 187–192.

26. Oliner, A. A., and R. G. Malech: Chaps. 2–4 in Hansen, R. C.: "Microwave Scanning Antennas," vol. II, Academic Press, New York, 1964.

27. Parad, L. I., and R. W. Kreutel: Mutual Effects between Circularly Polarized Elements, *Symp. USAF Antenna Res. Develop., Antenna Arrays Sec., Abstr.*, University of Illinois, Urbana, 1962.

28. Skolnik, M. I.: "Introduction to Radar Systems," 2d ed., McGraw-Hill Book Company, New York, 1980, pp. 504–506.

29. Querido, H., J. Frank, and T. C. Cheston: Wide Band Phase Shifters, *IEEE Trans.*, vol. AP-15, p. 300, March 1964.

30. Howell, J. M.: Limited Scan Antennas, *IEEE AP-5 Int. Symp.*, 1972.

31. Ajioka, J. S.: Frequency-Scan Antennas, chap. 19 in Johnson, R. C., and H. Jasik (eds.): "Antenna Engineering Handbook," 2d ed., McGraw-Hill Book Company, New York, 1984.

32. Gabriel, W. F.: Guest editor, special issue on adaptive antennas, *IEEE Trans.*, vol. AP-24, September 1976.

33. Forrest, J. R.: Guest editor, special issue on phased arrays, *Proc. IEE (London)*, vol. 127, pt. F, August 1980.

34. Steyskal, H.: Digital Beamforming Antennas—An Introduction, *Microwave J.*, vol. 30, pp. 107–124, January 1987.

35. Allen, J. L.: A Theoretical Limitation on the Formation of Lossless Multiple Beams in Linear Arrays, *IRE Trans.*, vol. AP-9, pp. 350–352, July 1961.

36. Blass, J.: The Multidirectional Antenna: A New Approach to Stacked Beams, *IRE Int. Conv. Rec.*, vol. 8, pt. 1, pp. 48–50, 1960.

37. Gent, H.: The Bootlace Aerial, *Royal Radar Estab. J. (U.K.)*, pp. 47–57, October 1957.

38. Rotman, W., and R. F. Turner: Wide Angle Microwave Lens for Line Source Application, *IEEE Trans.*, vol. AP-11, pp. 623–632, November 1963.

39. Schelkunoff, S. A.: A Mathematical Theory of Linear Arrays, *Bell Syst. Tech. J.*, vol. 22, pp. 80–107, January 1943.

40. Ramsay, J. F., J. P. Thompson, and W. D. White: Polarization Tracking of Antennas, *IRE Int. Conv.*, sess. 8, Antennas I, 1962.

41. Woodward, P. M.: A Method of Calculating the Field over a Planar Aperture Required to Produce a Given Polar Diagram, *J. IEE (London)*, vol. 93, pt. 3A, pp. 1554–1558, 1946.

42. Hansen, R. C.: Aperture Theory, in "Microwave Scanning Antennas," vol. I, Academic Press, New York, 1964, pp. 18–21.

43. Ramsay, J. F.: Lambda Functions Describe Antenna Diffraction Pattern, *Microwaves*, pp. 70–107, June 1967.

44. Ksienski, A.: Equivalence between Continuous and Discrete Radiating Arrays, *Can. J. Phys.*, vol. 39, pp. 35–349, 1961.

45. Bickmore, R. W.: Note on Effective Aperture of Electrically Scanned Arrays, *IRE Trans.*, vol. AP-6, pp. 194–196, April 1958.

46. Von Aulock, W. H.: Properties of Phased Arrays, *IRE Trans.*, vol. AP-9, pp. 1715–1727, October 1960.

47. Sharp, E. D.: Triangular Arrangement of Planar-Array Elements That Reduces Number Needed, *IRE Trans.*, vol. AP-9, pp. 126–129, March 1961.

48. Allen, J. L., et al.: Phased Array Radar Studies, *MIT Lincoln Lab. Tech. Rept.* 381, March 1965.

49. Allen, J. L., and B. L. Diamond: Mutual Coupling in Array Antennas, *MIT Lincoln Lab. Tech. Rept.* 424, October 1966.

50. Diamond, B. L.: Chap. 3 in Allen, J. L., and B. L. Diamond: Mutual Coupling in Array Antennas, *MIT Lincoln Lab. Tech. Rept.* 424, pt. III, October 1966.

51. Frank, J.: Phased Array Antenna Development, *Johns Hopkins University, Appl. Phys. Lab. Rept. TG* 882, March 1967.

52. Hannan, P. W.: Element-Gain Paradox for a Phased-Array Antenna, *IEEE Trans.*, vol. AP-12, pp. 423–433, July 1964.

53. Wasylkiwskyj, W., and W. K. Kahn: Element Patterns and Active Reflection Coefficient in Uniform Phased Arrays, *IEEE Trans.*, vol. AP-22, March 1974.

54. Wasylkiwskyj, W., and W. K. Kahn: Element Pattern Bounds in Uniform Phased Arrays, *IEEE Trans.*, vol. AP-25, September 1977.

55. Kahn, W. K.: Impedance-Match and Element-Pattern Constraints for Finite Arrays, *IEEE Trans.*, vol. AP-25, November 1977.

56. Willey, R. E.: Space Tapering of Linear and Planar Arrays, *IRE Trans.*, vol. AP-10, pp. 369–377, July 1962.

57. Wheeler, H. A.: Simple Relations Derived from a Phased-Array Antenna Made of an Infinite Current Sheet, *IEEE Trans.*, vol. AP-13, pp. 506–514, July 1965.

58. Allen, J. L.: On Array Element Impedance Variation with Spacing, *IEEE Trans.*, vol. AP-12, p. 371, May 1964.

59. Oliner, A. A., and R. G. Malech: "Microwave Scanning Antennas," vol. II, Academic Press, New York, 1964, chaps. 2–4.

60. Stark, L. I.: Microwave Theory of Phased Array Antennas—A Review, *Proc. IEEE*, vol. 62, pp. 1661–1701, December 1974.

61. Elliott, R. S., and L. A. Kurtz: The Design of Small Slot Arrays, *IEEE Trans.*, vol. AP-26, pp. 214–219, March 1978.

62. Elliott, R. S.: An Improved Procedure for Small Arrays of Shunt Slots, *IEEE Trans.*, vol. AP-31, pp. 48–53, January 1983.

63. Munk, B. A., T. W. Kornband, and R. D. Fulton: Scan Independent Phased Arrays, *Radio Sci.*, vol. 14, pp. 979–990, November 1979.

64. Shubert, K. A., and B. Munk: Matching Properties of Arbitrarily Large Dielectric Covered Arrays, *IEEE Trans.*, vol. AP-31, pp. 54–59, January 1983.

65. Wheeler, H. A.: The Grating-Lobe Series for the Impedance Variation in a Planar Phased-Array Antenna, *IEEE Trans.*, vol. AP-13, pp. 825–827, September 1965.

66. Stark, L.: Radiation Impedance of Dipole in Infinite Planar Phased Array, *Radio Sci.*, vol. 1, pp. 361–377, March 1966.

67. Kurtz, L. A., and R. S. Elliott: Systematic Errors Caused by the Scanning of Antenna Arrays: Phase Shifters in the Branch Lines, *IRE Trans.*, vol. AP-4, pp. 619–627, October 1956.

68. Hannan, P. W.: The Ultimate Decay of Mutual Coupling in a Planar Array Antenna, *IEEE Trans.*, vol. AP-14, pp. 246–248, March 1966.

69. Debski, T. T., and P. W. Hannan: Complex Mutual Coupling Measured in a Large Phased Array Antenna, *Microwave J.*, pp. 93–96, June 1965.

70. Lechtrek, L. W.: Cumulative Coupling in Antenna Arrays, *IEEE Int. Symp. Antennas Propag. Dig.*, pp. 144–147, 1965.

71. Farrell, G. F., Jr., and D. H. Kuhn: Mutual Coupling Effects of Triangular Grid Arrays by Modal Analysis, *IEEE Trans.*, vol. AP-14, pp. 652–654, September 1966.

72. Diamond, B. L.: Resonance Phenomena in Waveguide Arrays, *IEEE Int. Symp. Antennas Propag. Dig.*, 1967.

73. Hannan, P. W., and M. A. Balfour: Simulation of a Phased-Array Antenna in a Waveguide, *IEEE Trans.*, vol. AP-13, pp. 342–353, May 1965.

74. Balfour, M. A.: Active Impedance of a Phased-Array Antenna Element Simulated by a Single Element in Waveguide, *IEEE Trans.*, vol. AP-15, pp. 313–314, March 1967.

75. Magill, E. G., and H. A. Wheeler: Wide-Angle Impedance Matching of a Planar Array Antenna by a Dielectric Sheet, *IEEE Trans.*, vol. AP-14, pp. 49–53, July 1966.

76. Diamond, B. L.: Small Arrays—Their Analysis and Their Use for the Design of Array Elements, in Oliner, A. A., and G. H. Knittel: "Phased Array Antennas," Artech House, Norwood, Mass., 1972.

77. Grove, C. E., D. J. Martin, and C. Pepe: Active Impedance Effects in Low Sidelobe and Ultra Wideband Phased Arrays, *Proc. Phased Arrays Symp.*, pp. 187–206, 1985.

78. Evans, G. E., and H. E. Schrank: Low Sidelobe Radar Antennas, *Microwave J.*, pp. 109–117, July 1983.

79. Evans, G. E., and S. G. Winchell: A Wide Band, Ultralow Sidelobe Antenna, *Antenna Applications Symp.*, Allerton, Ill., September 1979.

80. Winchell, S. G., and D. Davis: Near Field Blockage of an Ultralow Sidelobe Antenna, *IEEE Trans.*, vol. AP-28, pp. 451–459, July 1980.

81. Barton, D. K., and H. R. Ward: "Handbook of Radar Measurement," Prentice-Hall, Englewood Cliffs, N.J., 1969, pp. 242–338.

82. Ramsey, J. F.: Lambda Functions Describe Antenna/Diffraction Patterns, *Microwaves*, p. 60, June 1967.

83. Yarnall, W. M.: Twenty-seven Design Aids for Antennas, Propagation Effects and Systems Planning, *Microwaves*, pp. 47–73, May 1965.

84. Harris, F. J.: On the Use of Windows for Harmonic Analysis with the Discrete Fourier Transform, *Proc. IEEE*, vol. 66, pp. 51–83, January 1978.

85. Taylor, T. T.: Design of Line Source Antennas for Narrow Beamwidth and Low Sidelobes, *IEEE Trans.*, vol. AP-3, pp. 16–28, 1955.

86. Taylor, T. T.: Design of Circular Apertures for Narrow Beamwidth and Low Sidelobes, *IEEE Trans.*, vol. AP-8, pp. 17–22, 1960.

87. Bayliss, E. T.: Design of Monopulse Antenna Difference Patterns with Low Sidelobes, *Bell Syst. Tech. J.*, pp. 623–650, May–June 1968.

88. Barton, D. K., and H. R. Ward: "Handbook of Radar Measurement," Prentice-Hall, Englewood Cliffs, N.J., 1969, pp. 256–266.

89. Allen, J. L.: The Theory of Array Antennas, *MIT Lincoln Lab. Rept.* 323, July 1963.

90. Ruze, J.: Physical Limitations on Antennas, *MIT Res. Lab. Electron. Tech. Rept.* 248, Oct. 30, 1952.

91. Cheston, T. C.: Effect of Random Errors on Sidelobes of Phased Arrays, *IEEE APS Newsletter—Antenna Designer's Notebook,* pp. 20–21, April 1985.

92. Frank, J., and J. Ruze: Steering Increments for Antisymmetrically Phased Arrays, *IEEE Trans.,* vol. AP-15, pp. 820–821, November 1967.

93. Brown, J.: unpublished communication, 1951.

94. Miller, C. J.: Minimizing the Effects of Phase Quantization Errors in an Electronically Scanned Array, *Proc. Symp. Electronically Scanned Array Techniques and Applications, RADC-TDR*-64-225, vol. 1, pp. 17–38, 1964.

95. Frank, J.: Bandwidth Criteria for Phased Array Antennas, in Oliner, A. A., and G. H. Knittel: "Phased Array Antennas," Artech House, Norwood, Mass., 1972, pp. 243–253.

96. Adams, W. B.: Phased Array Radar Performance with Wideband Signals, *AES Conv. Rec.,* pp. 257–271, November 1967.

97. Sharp, C. B., and R. B. Crane: Optimization of Linear Arrays for Broadband Signals, *IEEE Trans.,* vol. AP-14, pp. 422–427, July 1966.

98. Rothenberg, C., and L. Schwartzman: Phased Array Signal Bandwidth, *IEEE Int. Symp. Antennas Propag. Dig.,* pp. 116–123, December 1969.

99. Rotman, W., and R. F. Turner: Wide Angle Lens for Line Source Applications, *IEEE Trans.,* vol. AP-11, pp. 623–632, 1963.

100. Blass, J.: The Multi-Directional Antenna: A New Approach to Stacked Beams, *Proc. IRE Conv.,* vol. 8, pt. I, pp. 48–51, 1960.

101. Kinsey, R. F., and A. Horvath: Transient Response of Center-Series Fed Array Antennas, in Oliner, A. A., and G. H. Knittel: "Phased Array Antennas," Artech House, Norwood, Mass., 1972, pp. 261–271.

102. Tang, R.: Survey of Time-Delay Beam Steering Techniques, in Oliner, A. A., and G. H. Knittel: "Phased Array Antennas," Artech House, Norwood, Mass., 1972, pp. 254–260.

103. Lopez, A. R.: Monopulse Networks for Series Feeding an Array Antenna, *IEEE Int. Symp. Antennas Propag. Dig.,* 1967.

104. Stark, L., R. W. Burns, and W. P. Clark: Chap. 12 in Skolnik, M. I. (ed.): "Radar Handbook," 1st ed., McGraw-Hill Book Company, New York, 1970.

105. White, J. F.: "Semiconductor Control," Artech House, Norwood, Mass., 1977.

106. Ince, W. J.: Recent Advances in Diode and Phaser Technology for Phased Array Radars, pts. I and II, *Microwave J.,* vol. 15, no. 9, pp. 36–46, and no. 10, pp. 31–36, 1972.

107. White, J. F.: Diode Phase Shifters for Array Antennas, *IEEE Trans.,* vol. MTT-22, pp. 658–674, June 1974.

108. Fruchaft, M. A., and L. M. Silber: Use of Microwave Ferrite Toroids to Eliminate External Magnets and Reduce Switching Power, *Proc. IRE,* vol. 46, p. 1538, August 1958.

109. Frank, J., C. A. Shipley, and J. H. Kuck: Latching Ferrite Phase Shifter for Phased Arrays, *Microwave J.,* pp. 97–102, March 1967.

110. Ince, W. J., and D. H. Temme: Phasers and Time Delay Elements, "Advances in Microwaves," vol. 4, Academic Press, New York, 1969.

111. Whicker, L. R., and C. W. Young: The Evolution of Ferrite Control Components, *Microwave J.,* vol. 2, no. 11, pp. 33–37, 1978.

112. DiBartolo, J., W. J. Ince, and D. H. Temme: A Solid State "Flux Drive" ControlCircuit for Latching-Ferrite-Phaser Applications, *Microwave J.*, vol. 15, pp. 59–64, September 1972.

113. Boyd, C. R., Jr.: A Dual-Mode Latching Reciprocal Ferrite Phase Shifter, *IEEE Trans.*, vol. MTT-18, pp. 1119–1124, December 1970.

114. Hord, W. E., and C. R. Boyd, Jr.: A New Type of Fast-Switching Dual-Mode Ferrite Phase Shifter, *IEEE Trans.*, vol. MTT-35, pp. 1221–1226, December 1987.

115. Boyd, C. R., Jr.: Analog Rotary-Field Ferrite Phase Shifter, *Microwave J.*, pp. 41–43, December 1977.

116. Pucel, R. A. (ed.): "Monolithic Microwave Integrated Circuits," IEEE Press, Order No. PCO1867, New York, 1985.

117. Poirier, J. L.: An Analysis of Simplified Feed Architectures for MMIC T/R Module Arrays, *Rome Air Development Center Rept. RADC-TR*-86-236 (AD A185474), February 1987.

118. Perkins, W. H., and T. A. Midford: MMIC Technology: Better Performance at Affordable Cost, *Microwave J.*, vol. 31, pp. 135–143, April 1988.

119. Scudder, R. M., and W. H. Sheppard: AN/SPY-1 Phased Array Antenna, *Microwave J.*, vol. 17, pp. 51–55, May 1974.

120. Britton, R. L., T. W. Kimbrell, C. E. Caldwell, and G. C. Rose: AN/SPY-1 Planned Improvements, *Conf. Rec. Eascon '82*, pp. 379–386, September 1982.

121. Drabowitch, P. S., and F. Gautier: Antennes-réseaux phasées: Des principes aux réalisations, *Rev. Tech. THOMSON-CSF*, vol. 12, March 1980.

122. Carey, D. R., and W. Evans: The PATRIOT Radar in Tactical Air Defense, *Microwave J.*, vol. 31, pp. 325–332, May 1988.

123. Klass, P. J.: B-1 Pioneers Airborne Phased Array, *Aviat. Week Space Technol.*, pp. 84–89, Apr. 9, 1984.

124. Walsh, B.: An Eagle in the Sky, *Countermeasures—The Military Electron. Mag.*, pp. 30–63, July 1976.

125. Billam, E. R., and D. H. Harvey: MESAR—An Advanced Experimental Phased Array Radar, *Proc. RADAR-87 (London)*, pp. 19–26, October 1987.

126. Barton, D. K.: "Modern Radar System Analyses," Artech House, Norwood, Mass., 1988, pp. 196–198.

127. Ethington, D. E.: The AN/TPQ-36 and AN/TPQ-37 Fire Finder Radar Systems, *Conf. Rec. Eascon '77*, pp. 4-3A–4-F, 1977.

128. Mailloux, R. J., et al.: Grating Lobe Control in Limited Scan Arrays, *IEEE Trans.*, vol. AP-27, pp. 79–85, 1979.

CHAPTER 8
AUTOMATIC DETECTION, TRACKING, AND SENSOR INTEGRATION

G. V. Trunk
Naval Research Laboratory

8.1 INTRODUCTION

Since the invention of radar, radar operators have detected and tracked targets by using visual inputs from a variety of displays. Although operators can perform these tasks very accurately, they are easily saturated and quickly become fatigued. Various studies have shown that operators can manually track only a few targets. To correct this situation, automatic detection and tracking (ADT) systems were attached to many radars. As digital processing increases in speed and hardware decreases in cost and size, ADT systems will become associated with almost all but the simplest radars.

In this chapter, automatic detection, automatic tracking, and sensor integration systems for air surveillance radar will be discussed. Included in this discussion are various noncoherent integrators that provide target enhancement, thresholding techniques for reducing false alarms and target suppression, and algorithms for estimating target position and resolving targets. Then, an overview of the entire tracking system is given, followed by a discussion of its various components such as tracking filter, maneuver-following logic, track initiation, and correlation logic. Next, multiscan approaches to automatic tracking such as maximum likelihood are discussed. Finally, the chapter concludes with a discussion of sensor integration and radar netting, including both colocated and multisite systems.

8.2 AUTOMATIC DETECTION

The statistical framework necessary for the development of automatic detection was applied to radar in the 1940s by Marcum,[1] and later Swerling[2] extended the work to fluctuating targets. They investigated many of the statistical problems associated with the noncoherent detection of targets in Rayleigh noise. (NOTE: If the quadrature components are gaussian-distributed, the envelope is Rayleigh-distributed and the power is exponentially distributed.) Marcum's most important result was the generation of curves of probability of detection (P_D) versus signal-to-noise ratio (S/N) for a detector which sums N envelope-detected samples (either linear or square-law) under the assumption of equal signal amplitudes. However, in a search radar, as the beam sweeps over the target, the returned signal amplitude is modulated by the antenna pattern. Many authors investigated various detectors (weightings), comparing detection performance and angular estimation results with optimal values; and many of these results are presented later in this section.

In the original work on detectors, the environment was assumed known and homogeneous, so that fixed thresholds could be used. However, a realistic environment (e.g., containing land, sea, and rain) will cause an exorbitant number of false alarms for a fixed-threshold system that does not utilize excellent coherent processing. Three main approaches, adaptive thresholding, nonparametric detectors, and clutter maps, have been used to solve the false-alarm problem. Both adaptive thresholding and nonparametric detectors are based on the assumption that homogeneity exists in a small region about the range cell that is being tested. The adaptive thresholding method assumes that the noise density is known except for a few unknown parameters (e.g., the mean and the variance). The surrounding reference cells are then used to estimate the unknown parameters, and a threshold based on the estimated density is obtained. Nonparametric detectors obtain a constant false-alarm rate (CFAR) by ranking the test samples (ordering the samples from smallest to largest), usually with the reference cells. Under the hypothesis that all the samples (test and reference) are independent samples from an unknown density function, the test sample has a uniform density function, and, consequently, a threshold which yields CFAR can be set. Clutter maps store an average background level for each range-azimuth cell. A target is then declared in a range-azimuth cell if the new value exceeds the average background level by a specified amount.

Optimal Detector. The radar detection problem is a binary hypothesis-testing problem in which H_0 denotes the hypothesis that no target is present and H_1 is the hypothesis that the target is present. While several criteria (i.e., definitions of optimality) can be used to solve this problem, the most appropriate for radar is the Neyman-Pearson.[3] This criterion maximizes the probability of detection P_D for a given probability of false alarm P_{fa} by comparing the likelihood ratio L [defined by Eq. (8.1)] to an appropriate threshold T which determines the P_{fa}. A target is declared present if

$$L(x_1, \ldots, x_n) \frac{p(x_1, \ldots, x_n | H_1)}{p(x_1, \ldots, x_n | H_0)} \geq T \qquad (8.1)$$

where $p(x_1, \ldots, x_n | H_1)$ and $p(x_1, \ldots, x_n | H_0)$ are the joint probability density functions of the n samples x_i under the conditions of target presence and target absence, respectively. For a linear envelope detector the samples have a Rayleigh density under H_0 and a ricean density under H_1, and the likelihood ratio detector reduces to

$$\prod_{i=1}^{n} I_0\left(\frac{A_i x_i}{\sigma^2}\right) \geq T \qquad (8.2)$$

where I_0 is the Bessel function of zero order, σ^2 is the noise power, and A_i is the target amplitude of the ith pulse and is proportional to the antenna power pattern. For small signals ($A_i < \!\!\!\!\! \phi\ s$), the detector reduces to the square-law detector

$$\sum_{i=1}^{n} A_i^2 x_i^2 \geq T \tag{8.3}$$

and for large signals ($A_i >\!\!\!\!\$\ s$), it reduces to the linear detector

$$\sum_{i=1}^{n} A_i x_i > T \tag{8.4}$$

For constant signal amplitude (i.e., $A_i = A$) these detectors were first studied by Marcum[1] and were studied in succeeding years by numerous other people. Detection curves for both linear and square-law detectors are given in Chap. 2. The most important facts concerning these detectors are the following:

- The detection performances of the linear and square-law detectors are similar, differing only by less than 0.2 dB over wide ranges of P_D, P_{fa}, and n.

- Since the signal return of a scanning radar is modulated by the antenna pattern, to maximize the S/N when integrating a large number of pulses with no weighting (i.e., $A_i = 1$) only 0.84 of the pulses between the half-power points should be integrated, and the antenna beam-shape factor (ABSF) is 1.6 dB.[4] The ABSF is the number by which the midbeam S/N must be reduced so that the detection curves generated for equal signal amplitudes can be used for the scanning radar.

- The collapsing loss for the linear detector can be several decibels greater than the loss for a square-law detector[5] (see Fig. 8.1). The collapsing loss is the additional signal required to maintain the same P_D and P_{fa} when unwanted noise samples along with the desired signal-plus-noise samples are integrated. The number of signal samples integrated is N, the number of extraneous noise samples integrated is M, and the collapsing ratio $\rho = (N + M)/N$.

- Most automatic detectors are required not only to detect targets but to make angular estimates of the azimuth position of the target. Swerling[6] calculated the standard deviation of the optimal estimate by using the Cramer-Rao lower bound. The results are shown in Fig. 8.2, where a normalized standard deviation is plotted against the midbeam S/N. This result holds for a moderate or large number of pulses integrated, and the optimal estimate involves finding the location where the correlation of the returned signal and the derivative of the antenna pattern is zero. Although this estimate is rarely implemented, its performance is approached by simple estimates.

Practical Detectors. Many different detectors (often called *integrators*) are used to accumulate the radar returns as a radar sweeps by a target. A few of the most common detectors[7] are shown in Fig. 8.3. Though they are shown in the figure as being constructed with shift registers, they would normally be implemented with random-access memory. The input to these detectors can be linear, square-law, or log video. Since linear is probably the most commonly used, the advantages and disadvantages of the various detectors will be stated for this video.

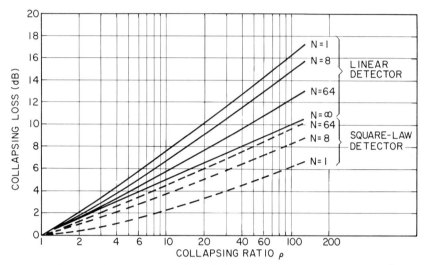

FIG. 8.1   Collapsing loss versus collapsing ratio for a probability of false alarm of 10^{-6} and a probability of detection of 0.5. (*Copyright 1972, IEEE; from Ref. 5.*)

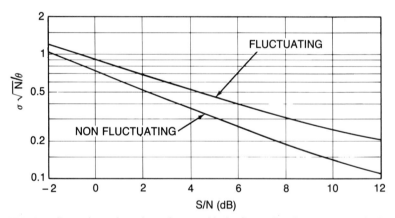

FIG. 8.2   Comparison of angular estimates with the Cramer-Rao lower bound. σ is the standard deviation of the estimation error, and N is the number of pulses within the 3-dB beamwidth, which is v. The S/N is the value at the center of the beam. (*Copyright 1956, IEEE; after Ref. 6.*)

Moving Window.   The moving window in Fig. 8.3a performs a running sum of n pulses in each range cell;

$$S_i = S_{i-1} + x_i - x_{i-n} \qquad (8.5)$$

where S_i is the sum at the ith pulse of the last n pulses and x_i is the ith pulse. The performance[8] of this detector for $n \approx 10$ is only 0.5 dB worse than the optimal detector given by Eq. (8.3). The detection performance can be obtained by using

INTEGRATORS

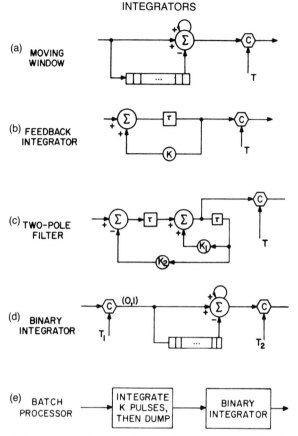

FIG. 8.3   Block diagrams of various detectors. The letter C indicates a comparison, τ is a delay, and loops indicate feedback. (*From Ref. 7.*)

an ABSF of 1.6 dB and the detection curves in Chap. 2. The angular estimate that is obtained by either taking the maximum value of the running sum or taking the midpoint between the first and last crossings of the detection threshold has a bias of $n/2$ pulses, which is easily corrected. The standard deviation of the estimation error of both estimators is about 20 percent higher than the optimal estimate specified by Cramer-Rao bound. A disadvantage of this detector is that it is susceptible to interference; that is, one large sample from interference can cause a detection. This problem can be minimized by using limiting. A minor disadvantage is that the last n pulses for each range cell must be saved, resulting in a large storage requirement when a large number of pulses are integrated. However, because of the availability of large memories of reduced size and cost, this is a minor problem.

The detection performance discussed previously is based on the assumption that the target is centered in the moving window. In the real situation the radar scans over the target, and decisions which are highly correlated are made at every pulse. Hansen[9] analyzed this situation for $N = 2, 4, 8,$ and 16 pulses and

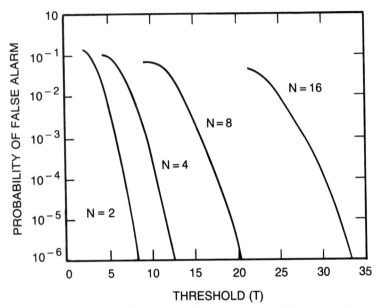

FIG. 8.4 Single-sweep false-alarm probability P_{fa} versus threshold for moving window. The noise is Rayleigh-distributed with $\sigma = 1$. (*Copyright 1970, IEEE; after Ref. 9.*)

calculated the detection thresholds shown in Fig. 8.4, the detection performance shown in Fig. 8.5, and the angular accuracy shown in Fig. 8.6. Comparing Hansen's scanning calculation with the single-point calculation, one concludes that 1 dB of improvement is obtained by making a decision at every pulse. The angular error of the beam-splitting procedure is about 20 percent greater than the optimal estimate. For large signal-to-noise ratios, the accuracy (rms error) of the beam-splitting and maximum-return procedures will be limited by the pulse

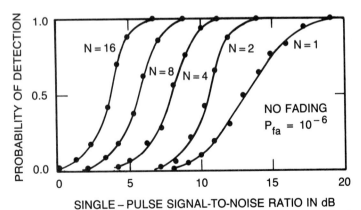

FIG. 8.5 Detection performance of the analog moving-window detector for the no-fading case. (*Copyright 1970, IEEE; after Ref. 9.*)

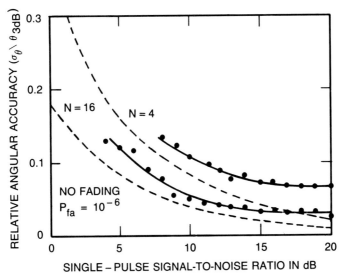

FIG. 8.6 Angular accuracy obtained with beam-splitting estimation procedure for the no-fading case. Broken-line curves are lower bounds derived by Swerling,[6] and points shown are simulation results. (*Copyright 1970, IEEE; after Ref. 9.*)

spacing[10] and will approach

$$\sigma(\hat{\theta}) = \Delta\theta/\sqrt{12} \tag{8.6}$$

where $\Delta\theta$ is the angular rotation between transmitted pulses. Consequently, if the number of pulses per beamwidth is small, the angular accuracy will be poor. For instance, if pulses are separated by 0.5 beamwidth, $\sigma(\hat{\theta})$ is bounded by 0.14 beamwidth. However, improved accuracy can be obtained by using the amplitudes of the radar returns. An accurate estimate of the target angle is given by

$$\hat{\theta} = \theta_1 + \frac{\Delta\theta}{2} + \frac{1}{2a\Delta\theta}\ln(A_2/A_1) \tag{8.7}$$

where

$$a = 1.386/(\text{beamwidth})^2 \tag{8.8}$$

and A_1 and A_2 are the two largest amplitudes of the returned samples and occur at angles θ_1 and $\theta_2 = \theta_1 + \Delta\theta$ respectively. Since the estimate should lie between θ_1 and θ_2 and Eq. (8.7) will not always yield such an estimate, $\hat{\theta}$ should be set equal to θ_1 if $\hat{\theta} < u_1$ and $\hat{\theta}$ should be set equal to θ_2 if $\hat{\theta} > u_2$. The accuracy of this estimator is given in Fig. 8.7 for the case of $n = 2$ pulses per beamwidth. This estimation procedure can also be used to estimate the elevation angle of a target in multibeam systems where θ_1 and θ_2 are the elevation-pointing angles of adjacent beams.

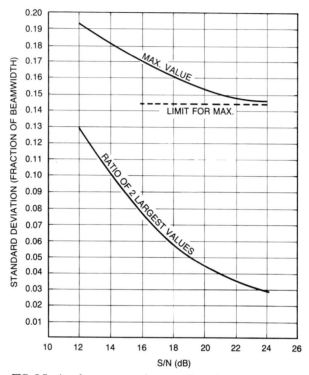

FIG. 8.7 Angular accuracy using two-pulse estimates.

Feedback Integrator. The amount of storage required can be reduced significantly by using a feedback integrator shown in Fig. 8.3*b*:

$$S_i = KS_{k-1} + x_i \qquad (8.9)$$

For a feedback value of K, the effective number of pulses integrated M is $M = 1/(1 - K)$, and for optimal (maximum P_D) performance $M = 0.63\ N$, where N is the number of pulses between the 3-dB antenna beamwidth.[11] The detection performance is given by the detection curves for M pulses with ABSF = 1.6 dB. Although the feedback integrator applies an exponential weighting into the past, its detection performance is only 1 dB less than that of the optimal integrator.[8] Unfortunately, difficulties are encountered when using the feedback integrator to estimate the azimuth position.[11] The threshold-crossing procedure yields estimates only 20 percent greater than the lower bound, but the bias is a function of S/N and must be estimated. On the other hand, the maximum value, though it has a constant bias, has estimates that are 100 percent greater than the lower bound. Furthermore, the exponential weighting function essentially destroys the radar antenna sidelobes. Because of these problems, the feedback integrator has limited utility.

Two-Pole Filter. The two-pole filter in Fig. 8.3*c* requires the storage of an intermediate calculation in addition to the integrated output and is described mathematically by

$$y_i = x_i - k_2 z_{i-1} \tag{8.10}$$

and

$$z_i = y_{i-1} + k_1 z_{i-1} \tag{8.11}$$

where x_i is the input, y_i is the intermediate calculation, z_i is the output, and k_1 and k_2 are the two feedback values. The values[12,13] which maximize P_D are given by

$$k_1 = 2 \exp\left(-\xi \omega_d \tau / \sqrt{1 - \xi^2}\right) \cos\left(\omega_d \tau\right) \tag{8.12}$$

and

$$k_2 = \exp\left(-2\xi \omega_d \tau / \sqrt{1 - \xi^2}\right) \tag{8.13}$$

where $\xi = 0.63$, $N\omega_d \tau = 2.2$, and N is the number of pulses between the 3-dB points of the antenna. With this rather simple device a weighting pattern similar to the antenna pattern can be obtained. The detection performance is within 0.15 dB of the optimal detector, and its angular estimates are about 20 percent greater than the Cramer-Rao lower bound. If the desired number of pulses integrated is changed (e.g., because of a change in the antenna rotation rate of the radar), it is only necessary to change the feedback values k_1 and k_2. The problems with this detector are that (1) it has rather high detector sidelobes, 15 to 20 dB, and (2) it is extremely sensitive to interference (i.e., the filter has a high gain resulting in a large output for a single sample that has a high value).

Binary Integrator.   The binary integrator is also known as the dual-threshold detector, M-out-of-N detector, or rank detector (see "Nonparametric Detectors" later in this section), and numerous individuals have studied it.[14–18] As shown in Fig. 8.3d, the input samples are quantized to 0 or 1, depending on whether or not they are less than a threshold T_1. The last N zeros and ones are summed and compared with a second threshold $T_2 = M$. For large N, the detection performance of this detector is approximately 2 dB less than the moving-window integrator because of the hard limiting of the data, and the angular estimation error is about 25 percent greater than the Cramer-Rao lower bound. Schwartz[16] showed that within 0.2 dB the optimal value of M for maximum P_D is given by

$$M = 1.5\sqrt{N} \tag{8.14}$$

when $10^{-10} < P_{fa} < 10^{-5}$ and $0.5 < P_D < 0.9$. The optimal value of P_n, the probability of exceeding T_1 when only noise is present, was calculated by Dillard[18] and is shown in Fig. 8.8. The corresponding threshold T_1 is

$$T_1 = \sigma(-2 \ln P_N)^{1/2} \tag{8.15}$$

A comparison of the optimal (best value of M) binary integrator with various other procedures is given in Figs. 8.9 and 8.10 for $P_D = 0.5$ and 0.9, respectively.

The binary integrator is used in many radars because (1) it is easily implemented, (2) it ignores interference spikes which cause trouble with integrators that directly use signal amplitude, and (3) it works extremely well when the noise has a non-Rayleigh density.[19] For $N = 3$, comparison of the optimal binary integrator (3 out of 3), another binary integration (2 out of 3), and the moving-window detector in log-normal interference (an example of a non-Rayleigh density) is shown in Fig. 8.11. The optimal binary integrator is much better than the moving-

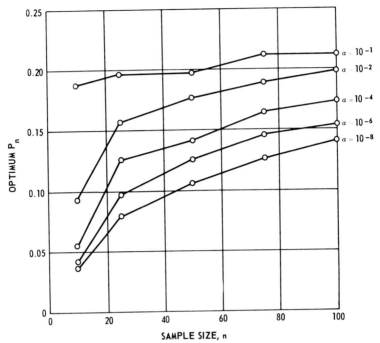

FIG. 8.8 Optimum values of P_N as a function of the sample size n and the probability of false alarm α; Ricean distribution with $S/N = 0$ dB per pulse. (*Copyright 1967, IEEE; from Ref. 18.*)

window integrator. The optimal values for log-normal interference were calculated by Schleher[19] and are $M = 3, 8,$ and 25 and $N = 3, 10,$ and 30, respectively.

A modified version of binary integration is sometimes used when there is a large number of pulses. It also has flexibility to integrate a different number of pulses. The modified binary moving window (MBMW) differs from the ordinary binary moving window (OBMW) by the introduction of a third threshold. When the second threshold is reached, one counts the number of consecutive pulses for which the second threshold is exceeded. When this number equals the third threshold, a target is declared. The performance of the MBMW and a comparison with the OBMW were given in Ref. 20. The major conclusion to be drawn is that the larger the value of N, the larger the difference in performance between the MBMW and OBMW detectors. For instance, with respect to the OBMW, the MBMW incurs losses of 0.15, 0.53, 1.80, and 2.45 dB for $N = 8, 16, 24,$ and 32 pulses, respectively.

Batch Processor.   The batch processor (Fig. 8.3*e*) is very useful when a large number of pulses are in the 3-dB beamwidth. If KN pulses are in the 3-dB beamwidth, K pulses are summed (batched) and either a 0 or a 1 is declared, depending on whether or not the batch is less than a threshold T_1. The last N zeros and ones are summed and compared with a second threshold M. An alternative version of this detector is to put the batches through a moving-window detector.

The batch processor, like the binary integrator, is easily implemented, ignores interference spikes, and works extremely well when the noise has a non-Rayleigh

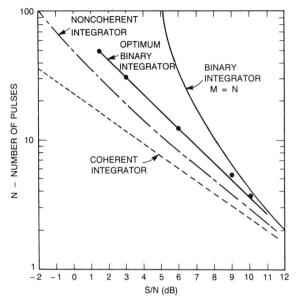

FIG. 8.9 Comparison of binary integrator (M out of N) with other integration methods ($P_{fa} = 10^{-10}$; $P_D = 0.5$). (*Copyright 1956, IEEE; after Ref. 16.*)

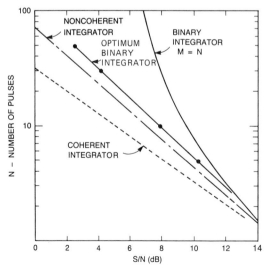

FIG. 8.10 Comparison of binary integrator (M out of N) with other integration methods ($P_{fa} = 10^{-10}$; $P_D = 0.90$). (*Copyright 1956, IEEE; after Ref. 16.*)

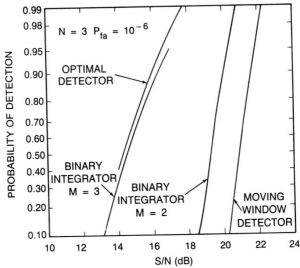

FIG. 8.11 Comparison of various detectors in log-normal ($\sigma = 6$ dB) interference ($N = 3$; $P_{fa} = 10^{-6}$). (*Copyright 1975, IEEE; after Ref. 19.*)

density. Furthermore, the batch processor requires less storage, detects better, and estimates angles more accurately than the binary integrator. For instance, if there were 80 pulses on target, one could batch 16 pulses, quantize this result to a 0 or a 1, and declare a target with a 3-out-of-5 (or 2-out-of-5) binary integrator. With an 8-bit analog-to-digital converter, the storage requirement per range cell is 17 bits (12 bits for the batch and 5 for the binary integrator) for the batch processor as opposed to 80 bits for the binary integrator and 640 bits for the moving window. The detection performance of the batch processor for a large number of pulses integrated is approximately 0.5 dB worse than the moving window. The batch processor has been successfully implemented by the Applied Physics Laboratory[21] of Johns Hopkins University. To obtain an accurate azimuth estimate $\hat{\theta}$, approximately 20 percent greater than the lower bound,

$$\hat{\theta} = \frac{\Sigma B_i \theta_i}{\Sigma B_i} \tag{8.16}$$

is used, where B_i is the batch amplitude and θ_i is the azimuth angle corresponding to the center of the batch.

False-Alarm Control.   In the presence of clutter, if fixed thresholds are used with the previously discussed integrators, an enormous number of detections will occur and will saturate and disrupt the tracking computer associated with the radar system. Four important facts should be noted:

- A tracking system should be associated with the automatic detection system (the only exception is when one displays multiple scans of detections).
- The P_{fa} of the detector should be as high as possible without saturating the tracking computer.

- Random false alarms and unwanted targets (e.g., stationary targets) are not a problem if they are removed by the tracking computer.

- Scan-to-scan processing can be used to remove stationary point clutter or moving-target indication (MTI) clutter residues.

One can limit the number of false alarms with a fixed-threshold system by setting a very high threshold. Unfortunately, this would reduce target sensitivity in regions of low noise (clutter) return. Three main approaches—adaptive threshold, nonparametric detectors, and clutter maps—have been used to reduce the false-alarm problem. Adaptive thresholding and nonparametric detectors assume that the samples in the range cells surrounding the test cell (called *reference cells*) are independent and identically distributed. Furthermore, it is usually assumed that the time samples are independent. Both kinds of detectors test whether the test cell has a return sufficiently larger than the reference cells. Clutter maps allow variation in space, but the clutter must be stationary over several (typically 5 to 10) scans. Clutter maps store an average background level for each range-azimuth cell. A target is then declared in a range-azimuth cell if the new value exceeds the average background level by a specified amount.

Adaptive Thresholding.   The basic assumption of the adaptive thresholding technique is that the probability density of the noise is known except for a few unknown parameters. The surrounding reference cells are then used to estimate the unknown parameters, and a threshold based on the estimated parameters is obtained. The simplest adaptive detector, shown in Fig. 8.12, is the cell-averaging CFAR (constant false-alarm rate) investigated by Finn and Johnson.[22] If the noise has a Rayleigh density, $p(x) = x \exp(-x^2/2\sigma^2)/\sigma^2$, only the parameter σ (σ^2 is the noise power) needs to be estimated, and the threshold is of the form $T = K\Sigma x_i = Kn\sqrt{\pi/2}\hat{\sigma}$, where $\hat{\sigma}$ is the estimate of σ. However, since T is set by an estimate $\hat{\sigma}$, it has some error and must be slightly larger than the threshold that one would use if σ were known exactly a priori. The raised threshold causes a loss in target sensitivity and is referred to as a CFAR loss. This loss has been

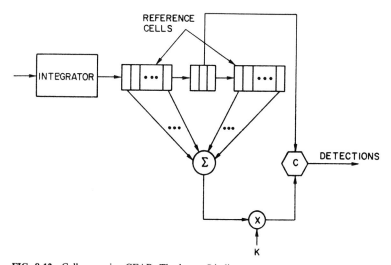

FIG. 8.12 Cell-averaging CFAR. The letter C indicates a comparison. (*From Ref. 7.*)

TABLE 8.1 CFAR Loss for $P_{fa} = 10^{-6}$ and $P_D = 0.9^*$

Number of pulses integrated	Loss for various numbers of reference cells, dB					
	1	2	3	5	10	∞
1	. . .	15.3	7.7	3.5	0	
3	. . .	7.8	5.1	3.1	1.4	0
10	6.3	3.3	2.2	1.3	0.7	0
30	3.6	2.0	1.4	1.0	0.5	0
100	2.4	1.4	1.0	0.6	0.3	0

*After Ref. 23.

calculated[23] and is summarized in Table 8.1. As can be seen, for a small number of reference cells the loss is large because of the poor estimate of σ. Consequently, one would prefer to use a large number of reference cells. However, if one does this, the homogeneity assumption (i.e., all the reference cells are statistically similar) might be violated. A good rule of thumb is to use enough reference cells so that the CFAR loss is below 1 dB and at the same time not let the reference cells extend beyond 1 nmi on either side of the test cell. For a particular radar this might not be feasible.

If there is uncertainty about whether or not the noise is Rayleigh-distributed, it is better to threshold individual pulses and use a binary integrator as shown in Fig. 8.13. This detector is tolerant of variations in the noise density because by setting K to yield a 1 with probability 0.1, a $P_{fa} \approx 10^{-6}$ can be obtained by using a 7-out-of-9 detector. While noise may be non-Rayleigh, it will probably be very Rayleigh-like out to the tenth percentile. Furthermore, one can use feedback based on several

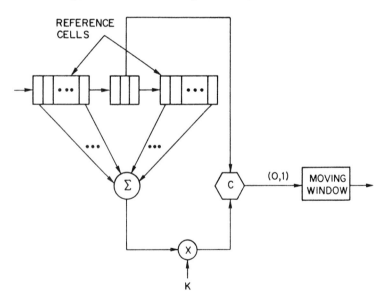

FIG. 8.13 Implementation of a binary integrator. The letter C indicates a comparison. (*From Ref. 7.*)

scans of data to control K in order to maintain a desired P_{fa} either on a scan or a sector basis. This demonstrates a general rule: to maintain a low P_{fa} in various environments, adaptive thresholding should be placed in front of the integrator.

If the noise power varies from pulse to pulse (as it would in jamming when frequency agility is employed), one must CFAR each pulse and then integrate. While the binary integrator performs this type of CFAR action, analysis[24,25] has shown that the ratio detector in Fig. 8.14 is a better detector. The ratio detector sums signal-to-noise ratios and is specified by

$$\sum_{i=1}^{n} \frac{x_i^2 (j)}{\dfrac{1}{2m} \sum_{k=1}^{m} [x_i^2(j+1+k) + x_i^2(j-1-k)]} \tag{8.17}$$

where $x_i(j)$ is the ith envelope-detected pulse in the jth range cell and $2m$ is the number of reference cells. The denominator is the maximum-likelihood estimate of σ_i^2, the noise power per pulse. It will detect targets even though only a few returned pulses have a high signal-to-noise ratio. Unfortunately, this will also cause the ratio detector to declare false alarms in the presence of narrow-pulse interference. To reduce the number of false alarms when narrow-pulse interference is present, the individual power ratios can be soft-limited[25] to a small enough value so that interference will cause only a few false alarms. A comparison of the ratio detector with other commonly used detectors is shown in Figs. 8.15 and 8.16 for nonfluctuating and fluctuating targets. A typical performance in sidelobe jamming when the jamming level varies by 20 dB per pulse is shown in Fig. 8.17. By employing a second test to identify the presence of narrow-pulse interference, a detection performance approximately halfway between the limiting and nonlimiting ratio detectors can be obtained.

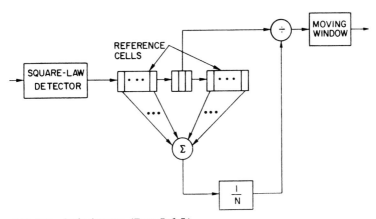

FIG. 8.14   Ratio detector. (*From Ref. 7.*)

If the noise samples are dependent in time or have a non-Rayleigh density such as the chi-square density or log-normal density, it is necessary to estimate two parameters and the adaptive detector is more complicated. Usually several pulses are integrated so that one can assume the integrated output has a gaussian probability density. Then the two parameters that must be estimated are the

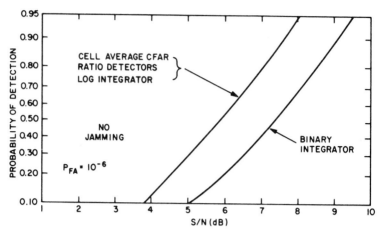

FIG. 8.15 Curves of probability of detection versus signal-to-noise ratio per pulse for the cell-averaging CFAR, ratio detectors, log integrator, and binary integrator: nonfluctuating target, $N = 6$, and probability of false alarm $= 10^{-6}$. (*From Ref. 25.*)

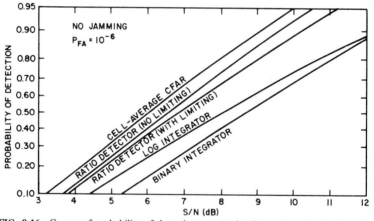

FIG. 8.16 Curves of probability of detection versus signal-to-noise ratio for the cell-averaging CFAR, ratio detectors, log integrator, and binary integrator: Rayleigh, pulse-to-pulse fluctuating target, $N = 6$, and probability of false alarm $= 10^{-6}$. (*From Ref. 25.*)

mean and the variance, and a threshold of the form $T = \hat{\mu} + K\hat{\sigma}$ is used. Though the mean is easily obtained in hardware, the usual estimate of the standard deviation

$$\hat{\sigma} = \left[\frac{1}{N}\Sigma(x_i - \bar{x})^2\right]^{1/2} \tag{8.18}$$

where

$$x = \frac{1}{N}\Sigma x_i \tag{8.19}$$

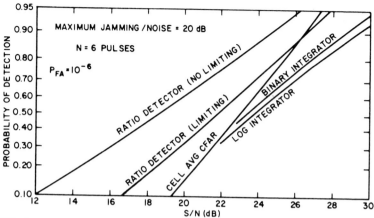

FIG. 8.17 Curves of probability of detection versus signal-to-noise ratio for the cell-averaging CFAR, ratio detectors, log integrator, and binary integrator: Rayleigh, pulse-to-pulse fluctuations, probability of false alarm = 10^{-6}, and maximum jamming-to-noise ratio = 20 dB. (*From Ref. 25.*)

is more difficult to implement. Consequently, the mean deviate defined by

$$\sigma = A\Sigma|x_i - \bar{x}| \qquad (8.20)$$

is sometimes used because of its ease of implementation. Nothing can be done to the binary integrator to yield a low P_{fa} if the noise samples are correlated. Thus, it should not be used in this situation. However, if the correlation time is less than a batching interval, the batch processor will yield a low P_{fa} without modifications.

Target Suppression.   Target suppression is the loss in detectability caused by other targets or clutter residues in the reference cells. Basically, there are two approaches to solving this problem: (1) remove large returns from the calculation of the threshold,[26-28] or (2) diminish the effects of large returns by either limiting or using log video. The technique that should be used is a function of the particular radar system and its environment.

Rickard and Dillard[27] proposed a class of detectors D_K, where the K largest samples are censored (removed) from the reference cells. A comparison of D_0 (no censoring) with D_1 and D_2 for a Swerling 2 target and a single square-law detected pulse is shown in Fig. 8.18, where N is the number of reference cells, β is the ratio of the power of the interfering target to the target in the test cell, and the bracketed pair (m, n) indicates the Swerling models of the target and the interfering target, respectively. As shown in Fig. 8.18, when one has an interfering target, the P_D does not approach 1 as S/N increases. Another approach[26] which censors samples in the reference cell if they exceed a threshold is briefly discussed in the subsection "Nonparametric Detectors."

Finn[28] investigated the problem of the reference cells spanning two continuous different "noise" fields (e.g., thermal noise, sea clutter, land clutter, etc.). On the basis of the samples, he estimated the statistical parameters of the two noise fields and the separation point between them. Then, only those reference cells which are in the noise field containing the test cell are used to calculate the adaptive threshold.

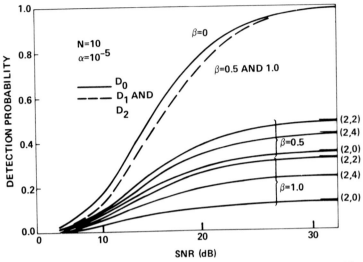

FIG. 8.18　Detection probability versus *SNR* for Swerling Case 2 primary target. (*Copyright 1977, IEEE; from Ref. 27.*)

An alternative approach for interfering targets is to use log video. By taking the log, large samples in the reference cells will have little effect on the threshold. The loss associated with using log video is 0.5 dB for 10 pulses integrated and 1.0 dB for 100 pulses integrated.[29] An implementation of the log CFAR[30] is shown in Fig. 8.19. In many systems the antilog shown in Fig. 8.19 is not taken. To maintain the same CFAR loss as for linear video, the number of reference cell $M_{\log}$ for the log CFAR should equal

$$M_{\log} = 1.65 \, M_{\lin} - 0.65 \tag{8.21}$$

where $M_{\lin}$ is the number of reference cells for linear video. The effect of target suppression with log video is discussed later in this section (Table 8.2).

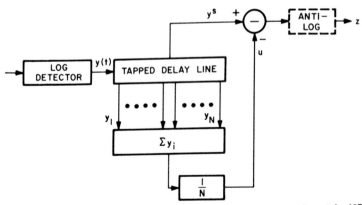

FIG. 8.19　Block diagram of cell-averaging log-CFAR receiver. (*Copyright 1972, IEEE; from Ref. 30.*)

Nonparametric Detectors. Usually nonparametric detectors obtain CFAR by ranking the test sample with the reference cells.[31,32] Ranking means that one orders the samples from the smallest to the largest and replaces the smallest with rank 0, the next smallest with rank 1,..., and the largest with rank $n-1$. Under the hypothesis that all the samples are independent samples from an unknown density function, the test sample has equal probability of taking on any of the n values. For instance, referring to the ranker in Fig. 8.20, the test cell is compared with 15 of its neighbors. Since in the set of 16 samples, the test sample has equal probability of being the smallest sample (or equivalently any other rank), the probability that the test sample takes on values 0, 1,..., 15 is 1:16. A simple rank detector is constructed by comparing the rank with a threshold K and generating a 1 if the rank is larger, a 0 otherwise. The 0s and 1s are summed in a moving window. This detector incurs a CFAR loss of about 2 dB but achieves a fixed P_{fa} for any unknown noise density as long as the time samples are independent. This detector was incorporated into the ARTS-3A postprocessor used in conjunction with the Federal Aviation Administration airport surveillance radar (ASR). The major shortcoming of this detector is that it is fairly susceptible to target suppression (e.g., if a large target is in the reference cells, the test cell cannot receive the highest ranks).

If the time samples are correlated, the rank detector will not yield CFAR. A modified rank detector, called the modified generalized sign test (MGST),[26] maintains a low P_{fa} and is shown in Fig. 8.21. This detector can be divided into three parts: a ranker, an integrator (in this case a two-pole filter), and a threshold (decision process). A target is declared when the integrated output exceeds two thresholds. The first threshold is fixed (equals $\mu + T_1/K$ in Fig. 8.21) and yields $P_{fa}=10^{-6}$ when the reference cells are independent and identically distributed. The second threshold is

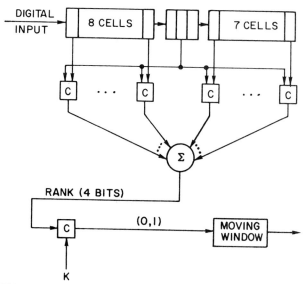

FIG. 8.20   Rank detector: output of a comparator C is either a zero or a one. (*From Ref. 7.*)

adaptive and maintains a low P_{fa} when the reference samples are correlated. The device estimates the standard deviation of the correlated samples with the mean deviate estimator, where extraneous targets in the reference cells have been excluded from the estimate by use of a preliminary threshold T_2.

The rank and MGST detectors are basically two-sample detectors. They decide that a target is present if the ranks of the test cell are significantly greater than the ranks of the reference cells. Target suppression occurs at all interfaces (e.g., land, sea) where the homogeneity assumption is violated. However, some tests exist, such as the Spearman Rho and Kendall Tau tests,[33] that depend only on the test cell. These tests use the fact that as the antenna beam sweeps by a point target, the signal return increases and then decreases. Thus, for the test cell the ranks should follow a pattern, first increasing and then decreasing. Although these detectors do not require reference cells and hence have the useful property of not requiring homogeneity, they are not generally used because of the large CFAR loss that occurs for moderate sample sizes. For instance, the CFAR losses are approximately 10 dB for 16 pulses on target and 6 dB for 32 pulses on target.[33]

The basic disadvantages of all nonparametric detectors are that (1) they have relatively large CFAR losses, (2) they have problems with correlated samples, and (3) one loses amplitude information, which can be a very important discriminant between target and clutter.[34] For example, a large return ($\sigma \hbar 1000$ m^2) in a clutter area is probably just clutter breakthrough. See "Contact Entry Logic" in Sec. 8.3.

Clutter Mapping. A clutter map uses adaptive thresholding where the threshold is calculated from the return in the test cell on previous scans rather than from the surrounding reference cells on the same scan. This technique has

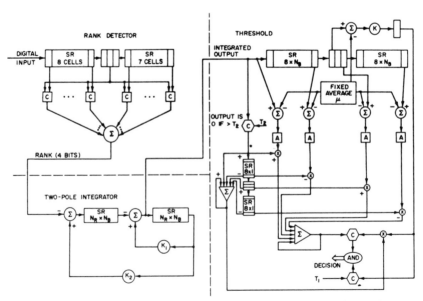

FIG. 8.21   Modified generalized sign test processor. (*Copyright 1974, IEEE; from Ref. 26.*)

the advantage that for essentially stationary environments (e.g., land-based radar against ground clutter), the radar has interclutter visibility—it can see between large clutter returns. Lincoln Laboratory[35] in its moving-target detector (MTD) used a clutter map for the zero-doppler filter very effectively. The decision threshold T for the ith cell is

$$T = A\ S_{i-1} \tag{8.22}$$

where
$$S_i = K\ S_{i-1} + X_i \tag{8.23}$$

S_i is the average background level, X_i is the return in the ith cell, K is the feedback value which determines the map time constant, and A is the constant which determines the false-alarm rate. In the MTD used for ASR application K is 7:8, which effectively averages the last eight scans. The main utility of clutter maps is with fixed-frequency land-based radars. While clutter maps can be used with frequency-agile radars and on moving platforms, they are not nearly as effective in these environments.

Target Resolution.   In automatic detection systems, a single large target will probably be detected many times, e.g., in adjacent range cells, azimuth beams, and elevation beams. Therefore, automatic detection systems have algorithms for merging the individual detections into a single centroided detection. Most algorithms have been designed so that they will rarely split a single target into two targets. This procedure results in poor range resolution capability. A merging algorithm[36] often used is the adjacent-detection merging algorithm, which decides whether a new detection is adjacent to any of the previously determined sets of adjacent detections. If the new detection is adjacent to any detection in the set of adjacent detections, it is added to the set. Two detections are adjacent if two of their three parameters (range, azimuth, and elevation) are the same and the other parameter differs by the resolution element: range cell ΔR, azimuth beamwidth θ, or elevation beamwidth γ.

A simulation[36] was run to compare the resolving capability of three common detection procedures: linear detector with $T = \hat{\mu} + A\hat{\sigma}$, linear detector with $T = B\hat{\mu}$, and log detector with $T = C + \hat{\mu}$. The constants $A, B,$ and C are used to obtain the same P_{fa} for all detectors. The estimates $\hat{\mu}$ and $\hat{\sigma}$ of μ and σ were obtained from either (1) all the reference cells or (2) the leading or lagging half of the reference cells, choosing the half with the lower mean value. The simulation involved two targets separated by 1.5, 2.0, 2.5, or 3.0 range cells and a third target 7.0 range cells from the first target. When the two closely spaced targets were well separated, either 2.5 or 3.0 range cells apart, the probability of detecting both targets (P_{D2}) was < 0.05 for the linear detector with $T = \hat{\mu} + A\hat{\sigma}$; $0.15 < P_{D2} < 0.75$ for the linear detector with $T = B\hat{\mu}$; and $P_{D2} > 0.9$ for the log detector. A second simulation, involving only two targets, investigated the effect of target suppression on log video, and the results are summarized in Table 8.2. One notes an improved performance for small S/N (10 to 13 dB) when one calculates the threshold using only the half of the reference cells with the lower mean value. The resolution capability of the log detector which uses only the half of the reference cells with the lower mean is shown in Fig. 8.22. The probability of resolving two equal-amplitude targets does not rise above 0.9 until they are separated in range by 2.5 pulse widths.

By assuming that the target is small with respect to the pulse width and that the pulse shape is known, the resolution capability can be improved by fitting

TABLE 8.2 Probability of Detecting with Log Video Two Targets Separated by 1.5, 2.0, 2.5, or 3.0 Range Cells*

Thresholding technique	Target separation	S/N of target no. 2				
		10	13	20	30	40
All reference cells	1.5	0.0	0.04	0.0	0.00	0.00
	2.0	0.0	0.22	0.54	0.14	0.10
	2.5	0.04	0.24	0.94	0.62	0.32
	3.0	0.0	0.24	0.88	0.92	0.76
Reference cells with minimum mean value	1.5	0.0	0.0	0.00	0.0	0.02
	2.0	0.10	0.32	0.44	0.12	0.04
	2.5	0.18	0.58	0.98	0.46	0.28
	3.0	0.22	0.66	0.98	0.82	0.74

*S/N of target 1 is 20 dB. S/N of target 2 is 10, 13, 20, 30, or 40 dB. After Ref. 36.

the known pulse shape to the received data and comparing the residue square error with a threshold.[37] If only one target is present, the residue should be only noise and hence should be small. If two or more targets are present, the residue will contain signal from the remaining targets and should be large. The results of resolving two targets with $S/N = 20$ dB are shown in Fig. 8.23. These targets can be resolved at a resolution probability of 0.9 at separations

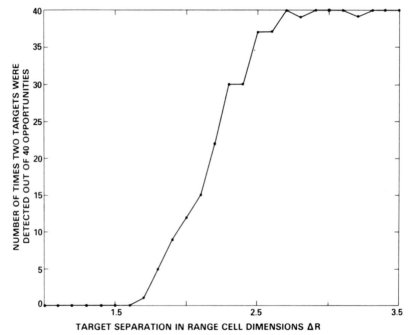

FIG. 8.22 Resolution capability of a log detector which uses the half of the reference cells with the lower mean. (*Copyright 1978, IEEE; from Ref. 36.*)

varying between one-fourth and three-fourths of a pulse width, depending on the relative phase difference between the two targets. Furthermore, this result can be improved further by processing multiple pulses.

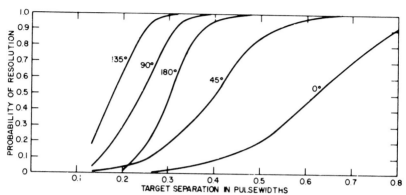

FIG. 8.23 Probability of resolution as a function of range separation: sampling rate $\Delta R = 1.5$ samples per pulse width; target strengths—nonfluctuating, $A_1 = A_2 = 20$ dB; phase differences = 0°, 45°, 90°, 135°, and 180°. (*Copyright 1984, IEEE; from Ref. 37.*)

Detection Summary. When only 2 to 4 samples (pulses) are available, a binary integrator should be used to avoid false alarms due to interference. When a moderate number of pulses (5 to 16) are available, a binary integrator, a rank detector, or a moving-window integrator should be used. If the number of pulses is large (greater than 20), a batch processor or a two-pole filter should be used. If the samples are independent, a one-parameter (mean) threshold can be used. If the samples are dependent, one can either use a two-parameter (mean and variance) threshold or adapt a one-parameter threshold on a sector basis. These rules should serve only as a general guideline. It is *highly recommended* that before a detector is chosen the radar video from the environment of interest be collected and analyzed and that various detection processes be simulated on a computer and tested against the recorded data.

8.3 AUTOMATIC TRACKING

Track-while-scan (TWS) systems are tracking systems for surveillance radars whose nominal scan time (revisit time) is from 4 to 12 s for aircraft targets. If the probability of detection (P_D) per scan is high, if accurate target location measurements are made, if the target density is low, and if there are only a few false alarms, the design of the correlation logic (i.e., associating detections with tracks) and tracking filter (i.e., filter for smoothing and predicting track positions) is straightforward. However, in a realistic radar environment these assumptions are seldom valid, and the design of the automatic tracking system is complicated. In actual situations one encounters target fades (changes in signal strength due to multipath propagation, blind speeds, and atmospheric conditions), false alarms (due to noise, clutter, interference, and jamming), and poor radar parameter estimates (due to noise, unstabilized antennas, unresolved targets, target splits, multipath propagation, and propagation effects). The tracking system must deal with all these problems.

A generic TWS system will be considered first. This will be followed by a discussion of maximum-likelihood multiple-scan approaches to automatic tracking.

Track-While-Scan System.   A general outline of a TWS system is presented. Next, since a major portion of any tracking system must deal with manipulating a large amount of data efficiently, a typical file system and pointer system are described. Then the contact entry logic, coordinate systems, tracking filter, maneuver-following logic, track initiation, and correlation logic are discussed.

System Organization.   Almost all TWS systems operate on an azimuthal sector basis which provides basic system timing. A typical series of operations is shown in Fig. 8.24. For instance, if the radar has reported all the detections in sector 11 and is now in sector 12, the tracking program would start by correlating (trying to associate) the clutter points (stationary tracks) in sector 10 with detections in sectors 9, 10, and 11. Those detections that are associated with clutter points are deleted (are not used for further correlations) from the detection file and are used to update the clutter points. Next, firm tracks in sector 8 are correlated with detections in sectors 7, 8, and 9. By this time all clutter points have been removed from sectors 9 and below. Those detections that are associated with firm tracks are deleted from the detection file and are used to update the appropriate track.

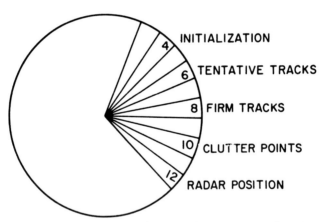

FIG. 8.24   Various operations of a track-while-scan system performed on a sector basis. (*From Ref. 7.*)

Usually, some provision is made for giving preference to firm tracks (instead of tentative tracks) in the correlation process. By performing the tentative-track correlation process two sectors behind firm-track correlations (Fig. 8.24), it is impossible for tentative tracks to steal detections belonging to firm tracks. In some other tracking systems the correlation for firm and tentative tracks is performed in the same sector; however, the generalized distance D, which quantifies how close tracks are to detections [see Eq. (8.47)], is incremented by ΔD if the track is tentative.

Finally, detections that are not associated with either clutter points or tracks are used for initiation of new tracks. The most common initiation procedure is to initiate a tentative track. Later the tentative track is dropped or else made a firm track or a clutter point, depending on its velocity. An alternative approach[38] is to establish both a clutter point and a tentative track. If the detection came from a stationary target, the clutter point will be updated and the tentative track will eventually be dropped. On the other hand, if the detection came from a moving target, the tentative track will be made firm and the clutter point will be dropped. This method requires less computer computation time when most of the detections are clutter residues.

File System. When a track is established in the computer, it is assigned a track number. All parameters associated with a given track are referred to by this track number. Typical track parameters are the smoothed and predicted position and velocity, time of last update, track quality, covariance matrices if a Kalman-type filter is being used, and track history (i.e., the last n detections). Each track number is also assigned to a sector so that the correlation process can be performed efficiently.[39] In addition to the track file, a clutter file is maintained. A clutter number is assigned to each stationary or very slowly moving echo. All parameters associated with a clutter point are referred to by this clutter number. Again, each clutter number is assigned to a sector in azimuth for efficient correlation. The basic files and pointer systems are described below in a Fortran format. Several higher-level languages, such as Pascal, have pointer systems which permit efficient implementation.

TRACK AND CLUTTER NUMBER FILES. Only the operation of the track number file is described, since the operation of the clutter number file is identical. The development in Ref. 39 is used. The track number parameters are listed in Table 8.3. The track number file is initialized by setting LISTT(I) = $I + 1$ for $I = 1$ through M. LISTT(M) is set equal to zero (denoting the last available track number in the file), NEXTT = 1 (the next available track number), LASTT = M (the last track number not being used), and FULLT = $M - 1$ (indication that $M - 1$ track numbers are available). A flowchart of the operation is shown in Fig. 8.25. When a new track number is requested, DROPT is set equal to one, and the system checks to see if FULLT is zero. If FULLT is not equal to zero, the routine is called. Since DROPT = 1, the new track is assigned the next available track number; i.e., NT = NEXTT. The next available track number in the list is found, and NEXTT is set equal to LISTT(NT). FULLT is decremented, indicating that one less track number is available. Finally, LISTT(NT) is set equal to 512, a number larger than M, as an aid in debugging the program.

TABLE 8.3 Track Number Parameters.*

Parameters	Description
NT	Track number
DROPT	1 (obtain) or 0 (drop) a track number NT
FULLT	Number of available track numbers
NEXTT	Next track number available
LASTT	Last track number not being used
LISTT(M)	File whose M locations correspond to track numbers
M	Maximum number of tracks

*From Ref. 39.

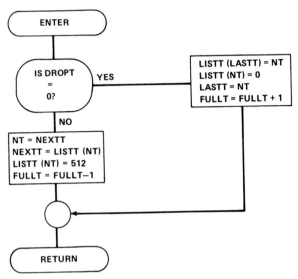

FIG. 8.25   Flowchart for track number file. (*From Ref. 39.*)

When DROPT = 0, a track number NT is dropped by setting the last available track number LISTT(LASTT) equal to the track number NT. LISTT(NT) is set equal to zero to denote the last track number, and LASTT is then set equal to the track number being dropped, LASTT = NT. The parameter FULLT is incremented, indicating that one more track number is available. Thus, the track (and clutter) number files maintain a linkage from one number to the next, thereby eliminating searching techniques.

TRACK NUMBER ASSIGNMENT TO AZIMUTH SECTOR FILES.   The range-azimuth plane is usually separated into 64 or 128 equal azimuth sectors. After a track has been updated or initiated, the predicted position of the target is checked to see which sector it occupies and the track is assigned to this sector. If the track is dropped or moves to a new sector, it is dropped out of the sector in which it was previously located. The parameters associated with the track sector files are listed in Table 8.4. Only the assignment of track numbers to azimuth sectors is described, since the clutter number assignment is identical. The TBX(I) file contains the first track number in sector *I*. If TBX(I) = 0, no tracks are in sector *I*. The IDT(M) file has storage locations corresponding to each of the possible *M* track numbers. The first track number in sector *I* is obtained from FIRST = TBX(I). The second track number in the sector is obtained by NEXT1 = IDT(FIRST). The next track number in the sector is obtained by NEXT2 = IDT(NEXT1). The process is continued until a zero is encountered, indicating that no more track numbers are in the sector.

When a new track is added or a track moves from one sector to another, a track number must be added to the new sector. To accomplish this, the first track number in the sector is stored, the track number NT being added is made the first track number in the sector, and the pointer associated with track NT is made equal to the original first track in the sector. The corresponding computer code requires the three Fortran statements: NL = TBX(I), TBX(I) = NT, and

TABLE 8.4 Track Sector Files.*

Parameter	Description
TBX(I)	First track number in sector I (a subscript of array IDT)
IDT(M)	Pointer file (each location corresponds to a track number, and the location contains the next track number in sector I or a zero)

*From Ref. 39.

IDT(NT) = NL. This procedure is essentially a push-down stack, pushing the older track numbers further down in the file.

When a track is dropped or moves out of the sector, the track number must be removed from the sector. The flow diagram for this is shown in Fig. 8.26. It is first determined whether the first track number in the sector TBX(I) is the one being dropped. If it is, the first track number in the sector is set equal to the second track number in the sector, and the location in the IDT file corresponding to

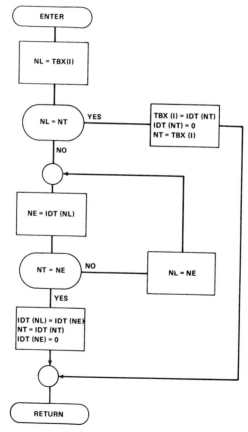

FIG. 8.26 Dropping a track number NR from sector I. (*From Ref. 39.*)

the track number NT being dropped is set to zero. NT is set equal to the track number in the file following the one just dropped, so that we now have the next available track number. If the track number being dropped is not the first one in the file, then the push-down stack IDT(NL) is searched sequentially until the track number is found. The variable containing IDT(NL) containing NT as the next track is replaced by the next track number following NT, and the variable in the IDT file corresponding to NT is set equal to zero. Again, NT is set equal to the track number in the file after the one being dropped.

Contact Entry Logic. When the radar system has either no or ineffective coherent processing, not all the detections declared by the automatic detector are used in the tracking process. Rather, many of the detections (contacts) are filtered out in software using a process called *contact entry logic.*[34] The basic idea is to use the target amplitude in connection with the type of environment from which the detection comes to eliminate some detections. The first step is for the operator to identify the environment in various regions. Types of environments which can be identified include land clutter, rain clutter, sea clutter, and interference. Histograms of detections in various regions are developed, and an example for rain clutter is shown in Fig. 8.27. For low-frequency radars (ultrahigh frequency and L band), usually the amplitudes of rain clutter detections are lower than those of targets. Consequently a threshold can be set to inhibit low-amplitude detections. This threshold can be controlled by the local detection density: raised in high detection densities and lowered in low detection densities. In no circumstances would a detection be inhibited if it fell within a track gate (i.e., a gate centered on the predicted position of a firm track). In land clutter areas, high-amplitude detections would be inhibited.

Coordinate Systems. The target location measured by the radar is in spherical coordinates: range, azimuth, and elevation. Thus, it may seem natural to perform tracking in spherical coordinates. However, this causes difficulties since the motion of constant-velocity targets (straight lines) will cause acceleration terms in all coordinates. A simple solution to this problem is to track in a cartesian coordinate system. Although it may appear that tracking in cartesian coordinates will seriously degrade the accurate range track, it has been shown[40] that the inherent accuracy can be maintained. Another approach[41] noted that since maneuvering targets cause a large range error but a rather insignificant azimuth error, a target-oriented cartesian coordinate system could be used. Specifically, the x axis is taken along the azimuth direction of the target, and the y axis is taken in the cross-range direction. Finally, no matter what coordinate system is used for tracking, the correlation of detections with tracks must be performed in spherical coordinates.

Tracking Filters. The simplest tracking filter is the α-β filter described by

$$x_s(k) = x_p(k) + \alpha[x_m(k) - x_p(k)] \tag{8.24}$$

$$V_s(k) = V_s(k - 1) + \beta[x_m(k) - x_p(k)]/T \tag{8.25}$$

$$x_p(k + 1) = x_s(k) + V_s(k)T \tag{8.26}$$

where $x_s(k)$ is the smoothed position, $V_s(k)$ is the smoothed velocity, $x_p(k)$ is the predicted position, $x_m(k)$ is the measured position, T is the scanning period (time between detections), and α and β are the system gains.

The minimal mean-square-error (MSE) filter for performing the tracking when the equation of motion is known is the Kalman filter, first discussed by Kalman[42] and later by Kalman and Bucy.[43] The Kalman filter is a popular filter for radar

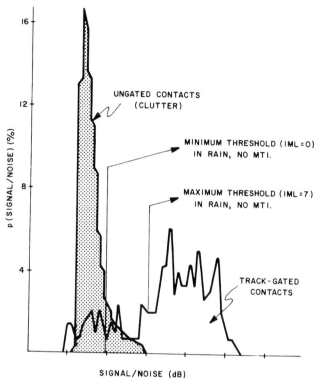

FIG. 8.27 Effectiveness of the signal-to-noise-ratio test in rain clutter. (*From Ref. 34.*)

and is a recursive filter which minimizes the MSE. The state equation in *xy* coordinates for a constant-velocity target is

$$X(t + 1) = \phi(t)\, X(t) + \Gamma(t)\, A(t) \tag{8.27}$$

where

$$X(t) = \begin{vmatrix} x(t) \\ \dot{x}(t) \\ y(t) \\ \dot{y}(t) \end{vmatrix} \quad \phi(t) = \begin{vmatrix} 1 & T & 0 & 0 \\ 0 & 1 & 0 & 0 \\ 0 & 0 & 1 & T \\ 0 & 0 & 0 & 1 \end{vmatrix} \tag{8.28}$$

$$\Gamma(t) = \begin{vmatrix} T^2/2 & 0 \\ T & 0 \\ 0 & T^2/2 \\ 0 & T \end{vmatrix} \quad \text{and } A(t) = \begin{vmatrix} a_x^{(t)} \\[6pt] a_y^{(t)} \end{vmatrix} \tag{8.29}$$

with $X(t)$ being the state vector at time t, consisting of position and velocity components $\dot{x}(t)$, $x(t)$, $\dot{y}(t)$, and $y(t)$; $t + 1$ being the next observation time; T being the time between observations; and $a_x(t)$ and $a_y(t)$ being random accelerations with covariance matrix $Q(t)$. The observation equation is

$$Y(t) = M(t)X(t) + V(t) \tag{8.30}$$

where
$$Y(t) = \begin{vmatrix} x_m(t) \\ y_m(t) \end{vmatrix} \qquad M(t) = \begin{vmatrix} 1 & 0 & 0 & 1 \\ 0 & 0 & 0 & 0 \end{vmatrix} \text{ and } V(t) = \begin{vmatrix} v_x(t) \\ v_y(t) \end{vmatrix} \tag{8.31}$$

with $Y(t)$ being the measurement at time t, consisting of positions $x_m(t)$ and $y_m(t)$, and $V(t)$ being a zero-mean noise whose covariance matrix is $R(t)$.

The problem is solved recursively by first assuming that the problem is solved at time $t - 1$. Specifically, it is assumed that the best estimate $\hat{X}(t - 1|t - 1)$ at time $t - 1$ and its error covariance matrix $P(t - 1|t - 1)$ are known, where the caret in the expression of the form $\hat{X}(t|s)$ signifies an estimate and the overall expression signifies that $X(t)$ is being estimated with observations up to $Y(s)$. The six steps involved in the recursive algorithm are:

1. Calculate the one-step prediction

$$\hat{X}(t|t - 1) = \phi(t - 1)\, \hat{X}(t - 1|t - 1) \tag{8.32}$$

2. Calculate the covariance matrix for the one-step prediction

$$P(t|t - 1) = \phi(t - 1)P(t - 1|t - 1)\phi^T(t - 1) + \Gamma(t - 1)Q(t - 1)\Gamma^T(t - 1) \tag{8.33}$$

3. Calculate the predicted observation

$$\hat{Y}(t|t - 1) = M(t)\hat{X}(t|t - 1) \tag{8.34}$$

4. Calculate the filter gain matrix

$$\Delta(t) = P(t|t - 1)M^T(t)\, [M(t)P(t|t - 1)M^T(t) + R(t)]^{-1} \tag{8.35}$$

5. Calculate the new smoothed estimate

$$\hat{X}(t|t) = \hat{X}(t|t - 1) + \Delta(t)[Y(t) - \hat{Y}(t|t - 1)] \tag{8.36}$$

6. Calculate the new covariance matrix

$$P(t|t) = [I - \Delta(t)M(t)]P(t|t - 1) \tag{8.37}$$

In summary, starting with an estimate $\hat{X}(t - 1|t - 1)$ and its covariance matrix $P(t - 1|t - 1)$, after a new observation $Y(t)$ has been received and the six quantities in the recursive algorithm have been calculated, a new estimate $\hat{X}(t|t)$ and its covariance matrix $P(t|t)$ are obtained.

It was shown[44] that, for a zero random acceleration $Q(t)$ 0 and a constant measurement covariance matrix $R(t) = R$, the $\alpha - \beta$ filter can be made equivalent to the Kalman filter by setting

$$\alpha = \frac{2(2k - 1)}{k(k + 1)} \tag{8.38}$$

and
$$\beta = \frac{6}{k(k + 1)} \tag{8.39}$$

on the kth scan. Thus as time passes, α and β approach zero, applying heavy smoothing to the new samples. Usually it is worthwhile to bound α and β from zero by assuming a random acceleration $Q(t) \neq 0$ corresponding to approximately a 1-g maneuver. The previously stated Kalman-filter method is optimal (with respect to MSE) for straight-line tracks but must be modified to enable the filter to follow target maneuvers.

Maneuver-Following Logic. Benedict and Bordner[45] noted that in TWS systems there is a conflicting requirement between good tracking noise reduction (implying small α and β) and good maneuver-following capability (implying large α and β). Although some compromise is always required, the smoothing equations should be constructed to give the best compromise for a desired tracking-noise reduction. Benedict and Bordner defined a measure of transient-following capability and showed that α and β should be related by

$$\beta = \frac{\alpha^2}{2 - \alpha} \qquad (8.40)$$

Thus, an (α, β) pair satisfying Eq. (8.40) can be chosen so that the tracking filter will follow a specified g turn. Cantrell[46] calculated the probability that a target detection will fall within a correlation region centered at the predicted target position when the target is doing a specified g turn. He suggested using the (α, β) pair satisfying Eq. (8.40) which yields the smallest correlation region. However, if high g turns must be followed, the noise performance is poor.

An alternative approach uses a turn detector which consists of two correlation regions, as shown in Fig. 8.28. The inner nonmaneuvering gate is usually set so that the probability of the target detection being in the gate is greater than 0.99 when the target is doing a 1-g maneuver. Then, if the detection is in the nonmaneuvering correlation region, the filter operates as usual, the filter gains being reduced according to Eq. (8.35) or Eqs. (8.38) and (8.39). When the target detection falls outside the inner gate but within the maneuver gate, a maneuver is declared and the filter bandwidth is increased (α and β are increased); Quigley and Holmes[41] increase the bandwidth by lowering the value of k in Eqs. (8.38) and (8.39). To avoid the problem of a target's fading causing a missed detection and a false alarm appearing in the large maneuver gate, the track should be bifurcated when a maneuver is declared. That is, two tracks are generated: (1) the continuation of the old track but not updated with the new detection and (2) a new maneuvering track with the new detection and increased filter bandwidth. The next scan is used to resolve the ambiguity and remove one of the tracks. While the turn detector is the most common approach for maneuver following, other solu-

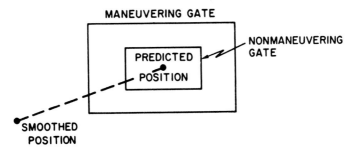

FIG. 8.28 Maneuver and nonmaneuver gates centered at the target's predicted position. (*From Ref. 7.*)

tions are to adjust the bandwidth as a function of the measurement error[38] or to use the Kalman filter with a realistic target-maneuvering model.[47]

Specifically, Cantrell et al.[38] suggested that the α-β filter described by Eqs. (8.24), (8.25), and (8.26) be made adaptive by adjusting α and β by

$$\alpha = 1 - e^{-2\xi\omega_0 T} \tag{8.41}$$

and

$$\beta = 1 + e^{-2\xi\omega_0 T} - 2e^{-\xi\omega_0 T} \cos(\omega_0 T\sqrt{1 - \xi^2}) \tag{8.42}$$

in which

$$\omega_0 = 0.5|p_1(k)/p_2(k)| \tag{8.43}$$

where

$$p_1(k) = e^{-\omega_a T}p_1(k - 1) + (1 - e^{-\omega_a T}) \epsilon(k)\epsilon(k - 1) \tag{8.44}$$

$$p_2(k) = e^{-\omega_b T}p_2(k - 1) + (1 - e^{-\omega_b T}) \epsilon(k)\epsilon(k) \tag{8.45}$$

ξ is the damping coefficient (nominally 0.7), T is the time since the last update, ω_a and ω_b are weighting constants, and $\epsilon(k)$ is the error between the measured and predicted positions on the kth update. The basic principle of the filter is that $p_1(k)$ is an estimate of the covariance of successive errors and $p_2(k)$ is an estimate of the error variance. When the target trajectory is a straight line, $p_1(k)$ approaches 0, since the expected value of $\epsilon(k)$ is 0. Thus ω_0 approaches 0, and the filter performs heavy smoothing. When the target turns, $p_1(k)$ grows, since the error $\epsilon(k)$ will have a bias. Thus ω_0 grows, and the filter can follow the target maneuver.

Singer[47] suggested using the Kalman filter with a realistic target-maneuvering model. He assumed that the target was moving at a constant velocity but was perturbed by a random acceleration. The target acceleration is correlated in time; and it was assumed that the covariance of the correlation was

$$r(\tau) = E[a(t)a(t + \tau)] = \sigma_m^2 e^{-\gamma\tau} \tag{8.46}$$

where $a(t)$ is the target acceleration at time t, σ_m^2 is the variance of the target acceleration, and γ is the reciprocal of the maneuver time constant. The density function for target acceleration consists of delta functions at $\pm A_{max}$, a delta function at 0 with probability P_0, and a uniform density between $-A_{max}$ and A_{max}. For this target motion Singer then calculated the state transition matrix $\phi(t)$ and the covariance matrix $Q(t)$, thereby specifying the Kalman-filter solution. He generated curves which give the steady-state performance of the filter for various data rates, single-look measurement accuracies, encounter geometry, and class of maneuvering targets.

Track Initiation. Detections that do not correlate with clutter points or tracks are used to initiate new tracks. If the detection does not contain doppler information, the new detection is usually used as the predicted position (in some military systems, one assumes a radially inbound velocity); and a large correlation region must be used for the next observation. The correlation region must be large enough to capture the next detection of the target, assuming that it could have the maximum velocity of interest. Since the probability of obtaining false alarms in the large correlation region is sometimes large, one should generally disregard the initial detection if no correlation is obtained on the second scan. Also one should not declare a track firm until at least a third detection (falling within a smaller correlation region) is obtained. A common track initiation criterion is four out of five, although one may require only three detections out of five opportunities in regions with a low false-alarm rate and a low target density. The possible exceptions for using only two detections are when doppler information is available (so that a small correlation region can be used immediately and the range rate can be used as an additional correlation parameter) or for *pop-up* targets (i.e., targets that suddenly appear at a close range) in a military situation.

An alternative track initiation logic[41] uses a sequential hypothesis-testing scheme. When a correlation is made on the ith scan, Δ_i is added to a score function; when a correlation opportunity is missed, Δ_i' is subtracted from the score function. The increments are a function of the state of the tracking system, the closeness of the association, the number of false alarms in the region, the a priori probability of targets, and the probability of detection. When the score function exceeds a particular value, the track is made firm. Although this method will inhibit false tracks in dense detection environments, it will not necessarily establish the correct tracks.

To initiate tracks in a dense detection environment, the technique known as *retrospective processing*[48] uses the detections over the last several scans to initiate straight-line tracks by employing a collection of filters matched to different velocities. An example of the processing is illustrated in Fig. 8.29, where one is looking for surface targets in the presence of sea spikes: large, targetlike sea clutter returns. It is clear from this figure that detections 1, 4, 6, 10, 12, and 14 form a track. Even though the false-alarm rate per scan is relatively high (approximately 10^{-3}), the processing can be implemented within a microprocessor.

Track Drop.   Firm tracks that are not updated in several scans corresponding to 40 to 60 (for aircraft) are usually dropped. In some systems, before a track is dropped, the track symbol is blinked, indicating that a track is about to be dropped. This gives an operator the chance to update the track with a manual detection to keep it from being dropped.

Correlation Logic.   To limit the number of detections that need to be tested in updating a track, correlation gates are used. A detection can never update a track unless it lies within the correlation gate that is centered at the track's predicted position. The correlation gate should be defined in $r - \theta$ coordinates regardless of what coordinate system is used for tracking. Furthermore, the gate size should be a function of the measurement accuracy and prediction error [specified by Eq. (8.33)] so that the probability of the correct detection lying within the gate is high (at least 0.99). In some tracking systems,[49] the location of the correlation gate is fed back to the automatic detector, and the detection threshold is lowered in the gate to increase the probability of detection P_D. The gate also disables the contact entry logic described in this section.

When several detections are within the correlation region, the usual and simplest solution is to associate the closest detection with the track. Specifically, the measure of closeness is the statistical distance

$$D^2 = \frac{(r_p - r_m)^2}{\sigma_r^2} + \frac{(\theta_p - \theta_m)^2}{\sigma_\theta^2} \tag{8.47}$$

where (r_p, θ_p) is the predicted position, (r_m, θ_m) is the measured position, σ_r^2 is the variance of $r_p - r_m$, and σ_θ^2 is the variance of $\theta_p - \theta_m$. These variances are

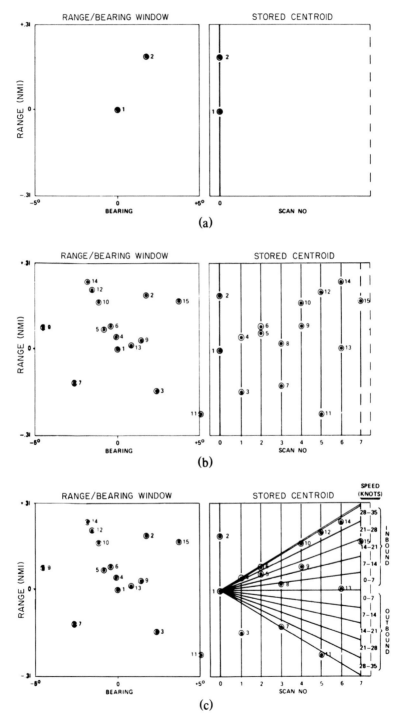

FIG. 8.29 The retrospective process. (*a*) A single scan of data. (*b*) Eight scans of data. (*c*) Eight scans of data with trajectory filters applied. (*From Ref. 48.*)

a by-product of the Kalman filter. Since the prediction variance is proportional to the measurement variance, σ_θ^2 and θ_r^2 are sometimes replaced by the measurement variances. Statistical distance rather than euclidean distance must be used because the range accuracy is usually much better than the azimuth accuracy.

Problems associated with multiple detections and tracks are illustrated in Fig. 8.30; two detections are within gate 1, three detections are within gate 2, and one detection is within gate 3. Table 8.5 lists all detections within the tracking gate, and the detections are entered in the order of their statistical distance from the track. Tentatively, the closest detection is associated with each track, and then the tentative associations are examined to remove detections that are used more than once. Detection 8, which is associated with tracks 1 and 2, is paired with the closest track (track 1 in this case); then all other tracks are reexamined to eliminate all associations with detection 8. Detection 7 is associated with tracks 2 and 3; the conflict is resolved by pairing detection 7 with track 2. When the other association with detection 7 is eliminated, track 3 has no associations with it and consequently will not be updated on this scan. Thus, track 1 is updated by detection 8, track 2 is updated by detection 7, and track 3 is not updated.

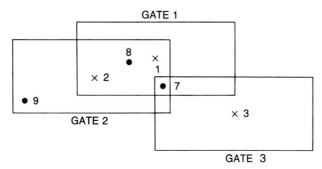

● DETECTION

× PREDICTED POSITION (CENTER OF GATE)

FIG. 8.30 Examples of the problems caused by multiple detections and tracks in close vicinity. (*From Ref. 7.*)

TABLE 8.5 Association Table for Example Shown in Fig. 8.30*

Track number	Closest association		Second association		Third association	
	Detection number	D^2	Detection number	D^2	Detection number	D^2
1	8	1.2	7	4.2		
2	8	3.1	7	5.4	9	7.2
3	7	6.3				

*From Ref. 7.

An alternative strategy is to always pair a detection with a track if there is only one correlation with a track. As before, ambiguities are removed by using the smallest statistical distance. Thus, track 3 in the example is updated by detection

7, track 1 is updated by detection 8, and track 2 is updated by detection 9. This procedure yields better results when P_D is close to one and the probability of false alarm is very low.

The optimal strategy which maximizes the probability of correct correlation is a joint maximum-likelihood approach. This involves examining all possible combinations and selecting the combination that is most probable in a statistical sense. This procedure requires that one knows the probability of detection and the probability of false alarms. A discussion of the joint maximum-likelihood approach can be found later in this section.

Singer and Sea[50] recognized and characterized the interaction between the correlation and track update functions. Specifically, three distinct situations can occur: the track is not updated, the track is updated with the correct return, and the track is updated with an incorrect return. The authors generalized the tracking filter's error covariance equations [i.e., generalized Eq. (8.33)] to account for the a priori probability of incorrect returns being correlated with the track. This permits the analytical evaluation of tracking accuracy in a multitarget environment which produces false correlations. Furthermore, by using the generalized tracking-error covariance equation, they optimized the filter gain matrix [generated a new equation to replace (Eq. 8.35)], which yielded a new minimum-error tracking filter for multitarget environments. Also, they generated a suboptimal fixed-memory version of this filter to reduce computation and memory requirements.

A subsequent paper by Singer et al.[51] uses a posteriori correlation statistics based on all reports in the vicinity of the track. The mathematical structure is similar to the Kalman filter. The estimation error is denoted by $\tilde{X}(t|t') = \hat{X}(t) - X(t|t')$ and has mean and covariance matrices denoted by $b(t|t')$ and $P(t|t')$. It is assumed that n_k detections fall within the correlation gate on scan k. Included in the number n_k are extraneous reports whose number obeys a Poisson distribution and whose positions are uniformly distributed within the gate. The smooth estimate is given by

$$\hat{X}(t|t) = \hat{X}(t|t-1) + A(t) \tag{8.48}$$

where the correction vector $A(t)$ is chosen to minimize the noncentral second moment of the filter estimation error. The problem is solved by using track histories. A track history α at scan k is defined by selecting, for each scan, a detection or a miss (a miss corresponds to the hypothesis that none of the reports belong to the track). The number of such track histories is

$$L(k) = \prod_{i=1}^{k} (1 + n_i) \tag{8.49}$$

Associated with each history α is the probability $p_\alpha(t)$ that the history α is the correct one, given observations through time t (scan k). The terms $b_\alpha'(t|t-1)$ and $P_\alpha'(t|t-1)$ are the bias and covariance of the estimation error $\tilde{X}(t|t-1)$, 7 given observations through time $t-1$ and given that track history α' at time $t-1$ is the (only) correct one. Then it is shown that the optimal correction vector is given by

$$A(t) = \sum_{\alpha=1}^{L(k)} P_\alpha(t) b_\alpha(t|t-1) \tag{8.50}$$

The a posteriori tracking filter in essence considers all possible track histories and calculates a probability of each history being correct, and the target track is a weighted (by the calculated probabilities) sum of all track histories. In contrast, the ordinary Kalman filter considers only one detection on each scan (usually the closest one to the predicted position) and assumes with probability 1 that this detection is the correct one to update the track.

The trouble with the optimal a posteriori filter is that it requires a growing memory [i.e., the number of track histories grows exponentially with K; see Eq. (8.49)]. Hence several suboptimal filters were investigated. The first suboptimal filter considers only the last N scans; track histories which are identical for the last N scans are merged. The second suboptimal filter considers only the L nearest neighbors in the correlation gate; essentially the gate size is changed to limit the number of reports to L. The last method uses both techniques: it considers only the last N scans and restricts the number of reports on any scan to L. In Fig. 8.31 the filter variance normalized by the theoretical (perfect-correlation) Kalman-filter variance is plotted for several filters. As a class the a posteriori filters provide better performance than the other filters. For high density of false reports ($4\beta\sigma_R^2 = 0.1$), the variance of the a posteriori filter is 30 times larger than predicted by the standard Kalman-filter approach. Thus, by ignoring the real possibility that the filter can be updated by incorrect detections, the variance of the standard Kalman filter is 30 times smaller than it should be. The consequences of this are that (1) a false detection can capture the track and (2) the gate size which is directly propor-

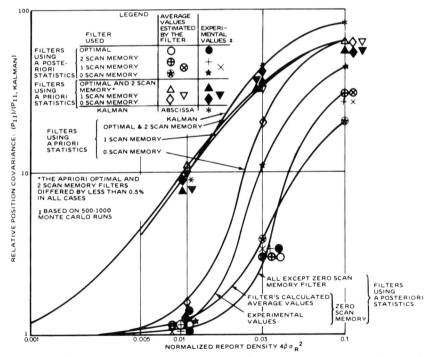

FIG. 8.31 Simulated and calculated variances in filtered position errors for optimal and suboptimal a posteriori and a priori tracking filters. β is the extraneous report density, and σ_R^2 is the measurement error variance. (*Copyright 1974, IEEE; from Ref. 51.*)

tional to the calculated variance is much too small, resulting in missed detections or incorrect maneuver declarations. Thus, the standard approach should never be used in dense-target (or false-target) environments.

If one considers only the last scan ($N = 0$ case), one obtains the probabilistic data association filter (PDAF).[52,53] In essence, each detection is used to update a Kalman filter, and the final estimate is the weighted sum of all the estimates of each detection, where the weight is the probability that each detection is the proper update. This procedure does not require a growing memory, requires slightly higher computational requirements than the ordinary Kalman filter, and solves the correlation problem by using all detections within the track gate to update the filter. As originally formulated, track initiation and termination were not considered. However, Colegrove et al.[54,55] have proposed solutions to these problems. Furthermore, they have implemented the PDAF on an over-the-horizon (OTH) radar. It should be noted that this is a different type of tracking problem because of the nature of the data (OTH radar has very accurate doppler but inaccurate range).

Maximum-Likelihood Approaches. The previously discussed procedures updated a single track with zero, one, or multiple detections; however, each track was operated on individually; i.e., each track was considered by itself. Better performance can be obtained by using all the detections, on all the previous scans, by employing a maximum-likelihood approach. This better performance is obtained at the cost of an enormous computational requirement.

Sittler[56] formulated the maximum-likelihood approach to tracking. In his approach he assumed that (1) objects appear in the surveillance region according to a Poisson model and are uniformly distributed in space, (2) the persistence of objects is independent with a duration that follows an exponential density, (3) the track positions have a known density, (4) the detection model is a Poisson process, (5) the measurement errors have a known density (later assumed to be gaussian), and (6) the false alarms follow a Poisson process. The approach of Stein and Blackman[57] is similar to that of Sittler.[56] It differs in the following ways: (1) a sample data system is assumed, and the detection process is specified by the detection probability; (2) the development eliminates the concept of undetected targets and of distinguishing between true targets generating a single detection and false alarms; (3) it uses realistic target maneuver models;[45] and (4) it uses the Kalman filter.[42,43] Later, Morefield[58] developed a similar approach fairly suitable for implementation. It is based upon likelihood functions and converts the association of detections to tracks into an integer programming problem. Finally, Trunk and Wilson[59] considered the effects of resolution in the track formation process.

Since the general approach of all these maximum-likelihood procedures is similar, the more comprehensive approach[59] will be discussed in detail. The maximum-likelihood method calculates the likelihood that a set of tracks correctly represents a given set of detections. The statement that a set of tracks represents a given set of detections means that each detection is either declared a false alarm or assigned to one or more tracks and that only one detection is assigned to a track on any scan. In the calculation of the likelihood, the probability of detection P_D, the probability of false alarm, the measurement error characteristics, and the probability of target resolution are all taken into account. For simplicity, the likelihood is written only in terms of the range measurement. Reference 59 indicates how the angle measurements are included.

The likelihood of an N_T track combination is given by

$$L(N_T) = \left[\prod_{j=1}^{N_T} \binom{N_s}{N_j} P_D^{N_j} (1 - P_D)^{N_s - N_j} \right]$$

$$\times [(1/R_I)^{N_{fa}} N_s! \prod_{i=1}^{N_D} P_i^{M_i}/M_i!]$$

$$\times \left\{ \prod_{j=1}^{N_T} \left[\prod_{i=1}^{N_s} f[(X_j(i), \hat{X}_j(i)]/(R_I)^2 (2\pi\sigma^2)(N_j - 2)/2] \right] \right.$$

$$\times \left[\prod_{k=1}^{N_u} P_R(X_k) \right] P_E(X_k) \tag{8.51}$$

where the first term gives the probability of associating detections with the N_T tracks. For the jth track, the probability of obtaining N_j detections in N_s scans is binomially distributed; and the probability of obtaining the detections associated with a set of N_T tracks is just the product of the binomial probabilities associated with each track. The second term represents the probability of obtaining a specified number of false alarms in a range interval R_I on each of the N_s scans. This probability is given by a multinomial probability where N_D is the total number of detections, M_i is the number of scans which contain exactly i false alarms, and the Poisson probability P_i of obtaining i false alarms on any scan is

$$P_i = (\lambda R_I)^i \exp(-\lambda R_I)/i! \tag{8.52}$$

where λ is the false-alarm rate per unit range. Finally, the probability of a false alarm occurring at a particular range is given by the uniform density $1/R_I$, and the total number of false alarms N_{fa} is

$$N_{fa} = \sum_{i=1}^{N_D} i\, M_i \tag{8.53}$$

Thus, the probability of the location of all the false alarms is $1/R_I$ raised to the N_{fa} power. The third term is the likelihood of obtaining a given set of measurements on the jth track, where σ^2 is the measurement error variance, $X_j(i)$ is the range of the associated detection on the ith scan for the jth track, $\hat{X}_j$ is the minimum mean square estimate of the track on the ith scan, and the function $f(\cdot, \cdot)$ is defined by

$$f(x, y) = \begin{cases} 1, & \text{if there is no detection associated with track on the } i\text{th scan or if the associated detection is also associated with any other track} \\ \\ \exp[-(x - y)^2/2\sigma^2] & \text{otherwise} \end{cases} \tag{8.54}$$

The last term represents the probability of N_u unresolved detections. When several targets are close to one another, merging algorithms may yield a single detection from the closely spaced targets.[36] The probability of obtaining a single (unresolved) detection X_k from N_k closely spaced targets is calculated by first ordering the predicted positions so that $\hat{X}_1 \leq \hat{X}_2 \leq \ldots \leq \hat{X}_{N_k}$ where for notational convenience the scan identifier has been dropped. Letting $D_m = \hat{X}_m - \hat{X}_{m-1}$, the probability of not resolving the N_k targets is given by

$$P_R(X_k) = \prod_{m=2}^{N_k} P(D_m) \tag{8.55}$$

where the probability of not resolving two targets separated by distance D is given in Ref. 36 by

$$P(D) = \begin{cases} 1 & D \leq 1.7R \\ (2.6R - D)/0.9R & 1.7R \leq d \leq 2.6R \\ 0 & D \geq 2.6R \end{cases} \tag{8.56}$$

where R is the 3-dB pulse width (range-cell dimension). The density of the position of X_k can be approximated by

$$P_E(X_k) = \{[\exp(-\epsilon_k^2/2\sigma^2)]/\max[1, (\hat{X}_{i,N,k} - \hat{X}_{i1})/(2\pi\sigma^2)^{1/2}]\} \tag{8.57}$$

where $$\epsilon_k = \max(0, X_k - \hat{X}_k - \hat{X}_{N_k}, \hat{X}_1 - X_k) \tag{8.58}$$

is the distance from X_k to the nearest predicted position if X_k lies outside the interval defined by the predicted positions; otherwise ϵ_k is zero.

Usually, the likelihood includes the a priori probability of having N_T tracks in the local region. Since in some situations this can be the result of a deterministic decision (e.g., a wing of aircraft in a military situation), the a priori probabilities will be assumed to be equal, and hence not included, rather than the commonly used Poisson arrival model.[56,57]

An example of this procedure applied to recorded data of a flight of five aircraft is shown in Fig. 8.32. The five tracks identified are (1, M, 1, 1, M, 1), (M, 1, 2, 2, 1, 2), (2, 2, 2, M, 1, 2), (3, 3, 3, 3, 2, 3), and (4, 4, 4, 4, 3, 4), where each track is represented by a sextuple with the number specifying a detection number and an M specifying a miss. Note that tracks 2 and 3 have common detections on scans 3, 5, and 6. By using Eq. (8.51), this sequence of tracks was found to be over 100 times more likely than the best four-track combination.

8.4 MULTISENSOR INTEGRATION

Multisensor integration will be divided into three topics: (1) colocated radar integration, (2) multisite radar integration, and (3) integration of data from unlike sensors [e.g., radar, identification, friend or foe (IFF), etc.). Radars are considered to be colocated if it is not necessary to take into account that the radars are located at different sites.

There are several methods of integrating data from multiple radars into a single system track file. The type of radar integration that should be used is a function of the radar's performance, the environment, and whether or not the radars

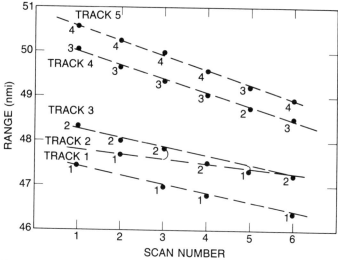

FIG. 8.32 Detections per scan shown versus range. Dots indicate detections, arcs unresolved detections, and dashed lines maximum-likelihood tracks.

are colocated. Several integration methods which have been used in various systems are:

1. *Track selection:*   Generate a track with each radar, and choose one of the tracks as the system track.

2. *Average track:*   Generate a track with each radar and weight, according to the Kalman filter's covariance matrices, the individual tracks to form a system track.

3. *Augmented track:* Generate a track with each radar, choose one of the tracks as the system track, and use selected detections from the other radars to update the system track.

4. *Detection-to-track:*   Use all radar detections to update the system track; tracks may or may not be initiated by using all detections from all radars.

Theoretically, the detection-to-track method of integration yields the best tracks because all the available information is used. However, the detections must be weighted properly, and care must be taken so that bad data does not corrupt good data.

There are many advantages of radar integration. Probably the most important is that it provides a common surveillance picture to all users so that decisions can be made more effectively. Radar integration will also improve track continuity and tracking of maneuvering targets because of the higher effective data rate. Improvement in track initiation times is a function of the target trajectory. For instance, long-range targets are usually detected by only one radar so that little or no improvement in initiation time is achieved. However, there could be an appreciable reduction in the initiation time for pop-up targets. Finally, the general tracking performance is improved in an electronic countermeasures (ECM) environment because of the integration of radars in different frequency bands located at different positions, providing both spatial and frequency diversity.

The main advantage of integrating different sensors with radar is to provide classification and/or identification information on radar tracks. In general, the other sensors do not provide position data of an accuracy comparable with the radar data. The sensors can also alert each other to conditions which can cause the mode of operation to be changed. For instance, a strong direction-finding (DF) bearing strobe on a noise (jammer) source or emitter which cannot be correlated with any radar track may cause the radar to use burnthrough, lower its detection thresholds, or change its initiation criterion in the sector containing the DF bearing strobe.

Colocated Radar Integration. The fundamental work[38,60,61] on colocated radar integration has been performed aboard military combatant ships, where a typical ship has two or three radars within several hundred feet of one another. Various radar integration techniques have been investigated; however, the one implemented on most ships is the detection-to-track integration philosophy. The typical functions of the detection-to-track integration philosophy are shown in Fig. 8.33. Associated with each radar is an automatic detector which performs a thresholding operation to control the false-alarm rate. After detections (contacts) have been declared, some further tests (contact entry logic) are applied to see whether the detections should be made available to the tracking system or discarded. (For a description of the contact entry logic see Sec. 8.3.) As detections are made available to the tracking system, they are associated in turn with the stationary-track filter (clutter map), the track file, and the saved-detection file (previously uncorrelated detections). Thus, a detection from one radar can be correlated with a saved detection from another radar, resulting in a track entry sooner than possible with any individual radar.

The tracking algorithms for a multiple-radar tracking system are quite similar to those for a single-radar tracking system. Consequently, only the differences will be discussed. First, a time is associated with each detection. Then, the detection with the oldest detection time is subjected to the correlation process. In a single-radar tracking system, time is not required since all the timing can take place in terms of the radar scan time. The second difference is that the tracking

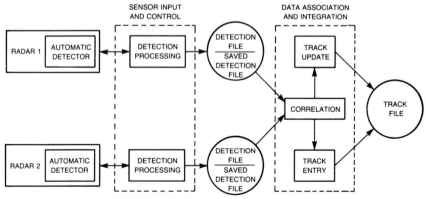

FIG. 8.33 Detection-to-track processing. (*Copyright 1977, IEEE; after Ref. 60.*)

filter updates the target tracks with detections whose accuracy varies with the radar and which arrive "randomly" in time. The optimal tracking filter (with respect to the minimum mean square error) is the Kalman filter.[42,43] Consequently, either a Kalman filter or an approximation of it should be used.

Another problem unique to the multiradar tracking system is the azimuthal and range alignment between radars. This problem is solved by adaptively aligning all the radars to one radar which is arbitrarily selected to be the accurate radar. This is accomplished by examining radar tracks which are being updated by multiple radars and which meet other specified criteria (e.g., frequency of update). Then the average error for the ith radar (over all targets) between the measured and predicted positions is weighted and subtracted from future measurements made by the ith radar (over all targets). This feedback arrangement will drive the bias errors toward zero. The residue bias error is approximately one-tenth of the accuracy of the measurement and prediction errors.[60,61]

Multisite Radar Integration. The methods used for exchanging tracking information between noncolocated sites are a function of whether the sites are fixed or mobile and what communication links are available. The most common multisite tracking system philosophy is that one site controls a radar track (has reporting responsibility) and transmits the tracking parameters over the communication link and that receiving units display the remote track. The track receives only updates from one site and thus does not benefit from the mutual support available from other radars in different frequencies and spatial locations. This philosophy is usually employed in any multisite system when the communication link has limited capacity.

Multisite integration with large gridlock and sensor misalignment errors is a difficult problem. If the targets from the multiple sites can be correlated, one can use Kalman filtering techniques to estimate the gridlock and misalignment errors.[62] However, if the gridlock and misalignment errors are large, one cannot perform the necessary correlation. Bath[63] solved this problem by solving the correlation, gridlock, and misalignment problems simultaneously. One first defines the maximum errors in latitude, longitude, and azimuth misalignment. Next, one defines error bins of size Δx in latitude, Δy in longitude, $\Delta\theta$ in azimuth, and divides the error space into error bins of this size. The error between one track from site 1 and one track from site 2 is calculated, and a 1 is entered into the bin corresponding to the calculated error. If there are n tracks at site 1 and m tracks at site 2, $m \times n$ errors are calculated and $m \times n$ 1s are entered into the error bins. For tracks that are different, the 1s will be randomly distributed throughout the error space. However, for tracks that are the same, all the 1s will appear in the same error cell. An example of this two-dimensional cross-correlation function is shown in Fig. 8.34, where the azimuth error is zero. The spike, which results from all the 1s adding up, not only gives the gridlock biases but also identifies the proper correlation of tracks from multiple sites. Then, Kalman-filter techniques could be used with the correlated track pairs to obtain more accurate estimates of the biases.

The netted radar program (NRP)[64] is a netted radar system for ground-target surveillance developed by Lincoln Laboratory. The basic idea is to associate automatic detection and tracking systems (similar to the MTD system) with each radar and then transmit tracks to a central location (radar control center), where the tracks are integrated into a single track file. Tracks are transmitted rather than detections because the bandwidth required to transfer tracks is significantly lower than that required to transfer detections. The bandwidth used in the NRP

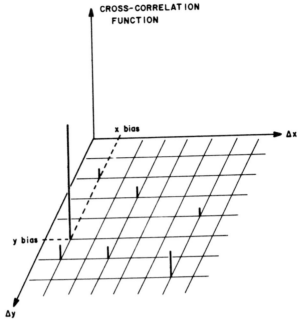

FIG. 8.34 Two-dimensional cross-correlation function (for two-track bases with x and y biases only). (*From Ref. 63.*)

system is approximately 2400 Bd. The NRP system was tested at Fort Sill, Oklahoma. The automatic integration at the radar control center provided continuity in track updating and identity despite missing detections caused by terrain masking and occasional misses caused by a variety of sources. It should be noted that gridlock was obtained by surveying the radar sites. This is the usual and preferred solution to gridlock when the sites are fixed and there are the time and the resources to perform the survey.

Unlike-Sensor Integration. A number of sensors can be integrated: radar, IFF, air traffic control radar beacon system (ATCRBS), infrared, optical, and acoustic. The sensors that are most easily integrated are the electromagnetic sensors, i.e., radar, IFF, and strobe extractors of noise sources or emitters.

IFF Integration. The problem of integrating radar and military IFF data is less difficult than that of integrating two radars. The question of whether detections or tracks should be integrated is a function of the application. In a military situation, by integrating detections one could interrogate the target only a few times, identify it, and then associate it with a radar track. From then on there would be little need for reinterrogating the target. However, in an air traffic control situation using ATCRBS, targets would be interrogated at every scan, and consequently either detections or tracks could be integrated.

Radar–DF Bearing Strobe Integration. Correlating radar tracks with DF bearing strobes on emitters has been considered by Coleman[65] and later by Trunk and Wilson.[66,67] Trunk and Wilson considered the problem of, given K DF angle tracks, each specified by a different number of DF measurements, associating each DF track with either no radar track or one of m radar tracks, again with each radar track being specified by a different number of radar measurements. Since each target can carry multiple emitters (i.e., multiple DF tracks can be associated with each radar track), each DF track association can be considered by itself, resulting in K disjoint association problems. Consequently, an equivalent problem is, given a DF track specified by n DF bearing measurements, to associate the DF track with no radar track or one of m radar tracks, the jth radar track being specified by m_j radar measurements. Using a combination of Bayes and Neyman-Pearson procedures and assuming that the DF measurement errors are usually independent and gaussian-distributed with zero mean and constant variance σ^2 but with occasional outliers (i.e., large errors, not described by the gaussian density), Trunk and Wilson argued that the decision should be based on the probability

$$P_j = \text{probability } (Z \geq d_j) \tag{8.59}$$

where Z has a chi-square density with n_j degrees of freedom and d_j is given by

$$d_j = \sum_{i=1}^{n_j} \min\{4, [\theta_e(t_i) - \theta_j(t_i)]^2/\sigma^2\} \qquad j = 1,\ldots,m \tag{8.60}$$

where n_j is the number of DF measurements overlapping the time interval for which the jth radar track exists, $\theta_e(t_i)$ is the DF measurement at time t_i, $\theta_j(t_i)$ is the predicted azimuth of radar track j for time t_i, and the factor 4 limits the square error to $4\sigma^2$ to account for DF outliers. By using the largest P_j's designated $P_{\max}$ and P_{next} and thresholds T_L, T_H, T_M, and R, the following decisions and decision rules were generated:

1. *Firm correlation:* DF signal goes with radar track having largest P_j (i.e., $P_{\max}$) when $P_{\max} \geq T_H$ and $P_{\max} \geq P_{\text{next}} + R$.

2. *Tentative correlation:* DF signal probably goes with radar track having largest P_j (i.e., $P_{\max}$) when $T_H > P_{\max} \geq T_M$ and $P_{\max} \geq P_{\text{next}} + R$.

3. *Tentative correlation with some track:* DF signal probably goes with some radar track (but cannot determine which) when $P_{\max} \geq T_M$ but $P_{\max} < P_{\text{next}} + R$.

4. *Tentatively uncorrelated:* DF signal probably does not go with any radar track when $T_M > P_{\max} > T_L$.

5. *Firmly uncorrelated:* DF signal does not go with any radar track when $T_L \geq P_{\max}$.

The lower threshold T_L determines the probability that the correct radar track (i.e., the one associated with the DF signal) will be incorrectly rejected from further consideration. If one desires to keep the rejection rate for the correct track at P_R, one can obtain this by setting $T_L = P_R$. The threshold T_H is set equal to P_{fa}, defined as the probability of falsely associating a radar track with a DF signal when the DF signal does not belong with the radar track. The threshold T_H is a function of the azimuth difference μ between the true (DF) position and the radar track under consideration. The threshold T_H was found for $\mu = 1.0\sigma$ and $\mu = 1.5\sigma$ by simulation techniques, and the results for $P_{fa} = 0.01$ are shown in Fig. 8.35. Between the high and low thresholds there is a tentative region. The middle threshold divides the "tentative" region into a tentatively correlated region and a tentatively uncorrelated region. The rationale in setting the threshold

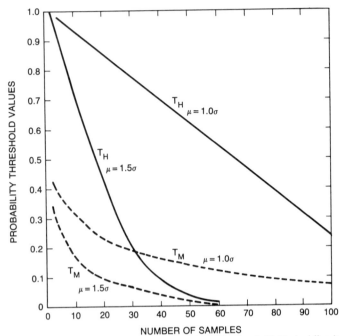

FIG. 8.35 High threshold (solid lines) and middle threshold (dashed lines) versus number of samples for two different separations. (*Copyright 1987, IEEE; after Ref. 66.*)

is to set the two associated error probabilities equal for a particular separation. The threshold T_M was found by using simulation techniques and is also shown in Fig. 8.35.

The probability margin R ensures the selection of the proper DF-radar association (avoiding false-target classifications) when there are two or more radar tracks close to one another. The correct selection is reached by postponing a decision until the two discriminant probabilities differ by R. The value for R is found by specifying a probability of an association error P_e according to $P_e = P_R(P_{max} \geq P_{next} + R)$, where P_{max} corresponds to an incorrect association and P_{next} corresponds to the correct association. The probability margin R is a function of P_e and the separation μ of the radar tracks. The probability margin R was found for $\mu = 0.25\sigma$, 0.50σ, and 1.00σ by using simulation techniques, and the results for $P_e = 0.01$ are shown in Fig. 8.36. Since the curves cross one another, we can ensure that $P_e \leq 0.01$ for any μ by setting R equal to the maximum value of any curve for each value of n.

The algorithm was evaluated by using simulations and recorded data. When the radar tracks are separated by several standard deviations of the measurement error, correct decisions are made rapidly. However, if the radar tracks are close to one another, errors are avoided by postponing the decision until sufficient data is accumulated. An interesting example with recorded data is shown in Figs. 8.37 and 8.38. Figure 8.37 shows the radar (azimuth) measurements of the control aircraft, the radar measurements of four aircraft of opportunity in the vicinity of the

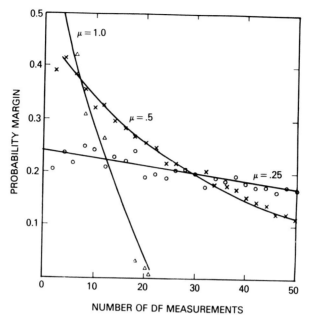

FIG. 8.36 Probability margin verses number of DF measurements for three different target separations. The o's, x's, and Δ's are the simulation results for $\mu = 0.25$, $\mu = 0.5$, and $\mu = 1.0$, respectively. (*Copyright 1987, IEEE; from Ref. 66.*)

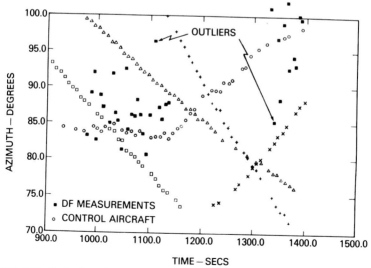

FIG. 8.37 Radar detections o and DF measurements collected on the control aircraft. The o's, Δ's, $+$'s, and x's are radar detections on four aircraft of opportunity in the vicinity of the control aircraft. (*Copyright 1987, IEEE; from Ref. 66.*)

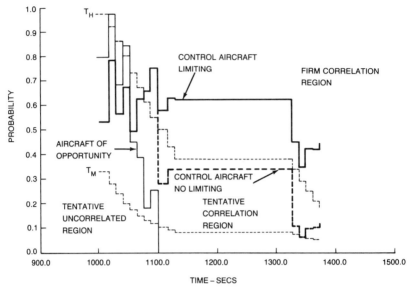

FIG. 8.38 Probability discriminants for experimental data. The bold lines are discriminants for control aircraft; the solid line, for limiting; the dashed line, for no limiting; the thin line, the discriminant for the aircraft of opportunity; and the thin dashed lines, the thresholds T_M and T_H. (*Copyright 1987, IEEE; after Ref. 66.*)

control aircraft, and the DF measurements from the radar on the control aircraft. The probability discriminants, one without limiting and one with limiting, are shown in Fig. 8.38. Initially, an aircraft of opportunity has the highest discriminant probability; however, a firm decision is not made since $P_{\max}$ does not exceed P_{next} by the probability margin. After the fourteenth DF measurement, the emitter is firmly correlated with the control aircraft. However, at the eighteenth DF measurement a very bad measurement (outlier) is made, and the firm correlation is downgraded to a tentative correlation if limiting is not used. If limiting is employed, however, the correct decision remains firm.

In a complex environment where there are many radar tracks and DF signal sources, it is quite possible that many DF signals will be assigned the category that the DF signal probably goes with some radar track. To remove many of these ambiguities, multisite DF operation can be considered. The extension of the previous procedures to multisite operation is straightforward. Specifically, if $\theta_{e1}(t_i)$ and $\theta_{e2}(t_k)$ are the DF angle measurements with respect to sites 1 and 2 and if $\theta_{j1}(t_i)$ and $\theta_{j2}(t_k)$ are the estimated angular positions of radar track j with respect to sites 1 and 2, the multisite squared error is simply

$$d_j = \sum_{i=1}^{n_{1j}} \min\{4\,[\theta_{e1}(t_i) - \theta_{j1}(t_i)]^2/\sigma_1^2\} + \sum_{k=1}^{n_{2j}} \min\{4\,[\theta_{e2}(t_k) - \theta_{j2}(t_k)]^2/\sigma_2^2\} \qquad (8.61)$$

Then, the previously described procedure can be used with d_j being defined by Eq. (8.61) instead of Eq. (8.60).

REFERENCES

1. Marcum, J. I.: A Statistical Theory of Target Detection by Pulsed Radar, *IRE Trans.*, vol. IT-6, pp. 59–267, April 1960.

2. Swerling, P.: Probability of Detection for Fluctuating Targets, *IRE Trans.*, vol. IT-6, pp. 269–300, April 1960.

3. Neyman, J., and E. S. Pearson: On the Problems of the Most Efficient Tests of Statistical Hypotheses, *Philos. Trans. R. Soc. London*, vol. 231, ser. A, p. 289, 1933.

4. Blake, L. V.: The Effective Number of Pulses per Beamwidth for a Scanning Radar, *Proc. IRE*, vol. 41, pp. 770–774, June 1953.

5. Trunk, G. V.: Comparison of the Collapsing Losses in Linear and Square-Law Detectors, *Proc. IEEE*, vol. 60, pp. 743–744, June 1972.

6. Swerling, P.: Maximum Angular Accuracy of a Pulsed Search Radar, *Proc. IRE*, vol. 44, pp. 1146–1155, September 1956.

7. Trunk, G. V.: Survey of Radar ADT, *Naval Res. Lab. Rept.* 8698, June 30, 1983.

8. Cooper, D. C., and J. W. R. Griffiths: Video Integration in Radar and Sonar Systems, *J. Brit. IRE*, vol. 21, pp. 420–433, May 1961.

9. Hansen, V. G.: Performance of the Analog Moving Window Detection, *IEEE Trans.*, vol. AES-6, pp. 173–179, March 1970.

10. Trunk, G. V.: Comparison of Two Scanning Radar Detectors: The Moving Window and the Feedback Integrator, *IEEE Trans.*, vol. AES-7, pp. 395–398, March 1971.

11. Trunk, G. V.: Detection Results for Scanning Radars Employing Feedback Integration, *IEEE Trans.*, vol. AES-6, pp. 522–527, July 1970.

12. Trunk, G. V., and B. H. Cantrell: Angular Accuracy of a Scanning Radar Employing a 2-Pole Integrator, *IEEE Trans.*, vol. AES-9, pp. 649–653, September 1973.

13. Cantrell, B. H., and G. V. Trunk: Corrections to "Angular Accuracy of a Scanning Radar Employing a Two-Pole Filter," *IEEE Trans.*, vol. AES-10, pp. 878–880, November 1974.

14. Swerling, P.: The "Double Threshold" Method of Detection, *Project Rand Res. Mem.* RM-1008, Dec. 17, 1952.

15. Harrington, J. V.: An Analysis of the Detection of Repeated Signals in Noise by Binary Integration, *IRE Trans.*, vol. IT-1, pp. 1–9, March 1955.

16. Schwartz, M.: A Coincidence Procedure for Signal Detection, *IRE Trans.*, vol. IT-2, pp. 135–139, December 1956.

17. Cooper, D. H.: Binary Quantization of Signal Amplitudes: Effect for Radar Angular Accuracy, *IEEE Trans.*, vol. ANE-11, pp. 65–72, March 1964.

18. Dillard, G. M.: A Moving-Window Detector for Binary Integration, *IEEE Trans.*, vol. IT-13, pp. 2–6, January 1967.

19. Schleher, D. C.: Radar Detection in Log-Normal Clutter, *IEEE Int. Radar Conf.*, pp. 262–267, Washington, 1975.

20. Mao, Y.-H.: The Detection Performance of a Modified Moving Window Detector, *IEEE Trans.*, vol. AES-17, pp. 392–400, May 1981.

21. Radar Processing Subsystem Evaluation, vol. 1, *Johns Hopkins University, Appl. Phys. Lab. Rept.* FP8-T-013, November 1975.

22. Finn, H. M., and R. S. Johnson: Adaptive Detection Mode with Threshold Control as a Function of Spacially Sampled Clutter-Level Estimates, *RCA Rev.*, vol. 29, pp. 414–464, September 1968.

23. Mitchell, R. L., and J. F. Walker: Recursive Methods for Computing Detection Probabilities, *IEEE Trans.*, vol. AES-7, pp. 671–676, July 1971.

24. Trunk, G. V., and J. D. Wilson: Automatic Detector for Suppression of Sidelobe Interference, *IEEE Conf. Decision & Control*, pp. 508–514, December 7–9, 1977.

25. Trunk, G. V., and P. K. Hughes II: Automatic Detectors for Frequency-Agile Radar, *IEE Int. Radar Conf.*, pp. 464–468, London, 1982.

26. Trunk, G. V., B. H. Cantrell, and F. D. Queen: Modified Generalized Sign Test Processor for 2-D Radar, *IEEE Trans.*, vol. AES-10, pp. 574–582, September 1974.

27. Rickard, J. T., and G. M. Dillard: Adaptive Detection Algorithms for Multiple-Target Situations, *IEEE Trans.*, vol. AES-13, pp. 338–343, July 1977.

28. Finn, H. M.: A CFAR Design for a Window Spanning Two Clutter Fields, *IEEE Trans.*, vol. AES-22, pp. 155–168, March 1986.

29. Green, B. A.: Radar Detection Probability with Logarithmic Detectors, *IRE Trans.*, vol. IT-4, March 1958.

30. Hansen, V. G., and J. R. Ward: Detection Performance of the Cell Average Log/CFAR Receiver, *IEEE Trans.*, vol. AES-8, pp. 648–652, September 1972.

31. Dillard, G. M., and C. E. Antoniak: A Practical Distribution-Free Detection Procedure for Multiple-Range-Bin Radars, *IEEE Trans.*, vol. AES-6, pp. 629–635, September 1970.

32. Hansen, V. G., and B. A. Olsen: Nonparametric Radar Extraction Using a Generalized Sign Test, *IEEE Trans.*, vol. AES-7, September 1981.

33. Hansen, V. G.: Detection Performance of Some Nonparametric Rank Tests and an Application to Radar, *IEEE Trans.*, vol. IT-16, May 1970.

34. Bath, W. G., L. A. Biddison, S. F. Haase, and E. C. Wetzlar: False Alarm Control in Automated Radar Surveillance Systems, *IEE Int. Radar Conf.*, pp. 71–75, London, 1982.

35. Muehe, C. E., L. Cartledge, W. H. Drury, E. M. Hofstetter, M. Labitt, P. B. McCorison, and V. J. Sferrino: New Techniques Applied to Air-Traffic Control Radars, *Proc. IEEE*, vol. 62, pp. 716–723, June 1974.

36. Trunk, G. V.: Range Resolution of Targets Using Automatic Detectors, *IEEE Trans.*, vol. AES-14, pp. 750–755, September 1978.

37. Trunk, G. V.: Range Resolution of Targets, *IEEE Trans.*, vol. AES-20, pp. 789–797, November 1984.

38. Cantrell, B. H., G. V. Trunk, F. D. Queen, J. D. Wilson, and J. J. Alter: Automatic Detection and Integrated Tracking, *IEEE Int. Radar Conf.*, pp. 391–395, Washington, 1975.

39. Cantrell, B. H., G. V. Trunk, and J. D. Wilson: Tracking System for Two Asynchronously Scanning Radars, *Naval Res. Lab. Rept.* 7841, Dec. 5, 1974.

40. Cantrell, B. H.: "Description of an α-β Filter in Cartesian Coordinates, *Naval Res. Lab. Rept.* 7548, March 1973.

41. Quigley, A. L., and J. E. Holmes: The Development of Algorithms for the Formation and Updating of Tracks, *Admiralty Surface Weapons Estab., ASWE-WP-XBC*-7512, Portsmouth P06 4AA, November 1975.

42. Kalman, R. E.: A New Approach to Linear Filtering and Prediction Problems, *J. Basic Eng. (ASME Trans., ser. D)*, vol. 82, pp. 35–45, March 1960.

43. Kalman, R. E., and R. S. Bucy: New Results in Linear Filtering and Prediction Theory, *J. Basic Eng. (ASME Trans., ser. D)*, vol. 83, pp. 95–107, March 1961.

44. Quigley, A. L.: Tracking and Associated Problems, *IEE Int. Conf. Radar—Present & Future*, pp. 352–357, London, 1973.

45. Benedict, T. R., and G. W. Bordner: Synthesis of an Optimal Set of Radar Track-While-Scan Smoothing Equations, *IRE Trans.*, vol. AC-7, pp. 27–32, July 1962.

46. Cantrell, B. H.: Behavior of α-β Tracker for Maneuvering Target under Noise, False Target, and Fade Conditions, *Naval Res. Lab. Rept.* 7434, August 1972.

47. Singer, R. A.: Estimating Optimal Tracking Filter Performance for Manned Maneuvering Targets, *IEEE Trans.*, vol. AES-6, pp. 473–484, July 1970.

48. Prengaman, R. J., R. E. Thurber, and W. B. Bath: A Retrospective Detection Algorithm for Extraction of Weak Targets in Clutter and Interference Environments, *IEE Int. Radar Conf.*, pp. 341–345, London, 1982.

49. Cook, S. R.: Development of IADT Tracking Algorithm, *Johns Hopkins University, Appl. Phys. Lab. Rept.* F3C-1-061, September 1974.

50. Singer, R. A., and R. G. Sear: New Results in Optimizing Surveillance System Tracking and Data Correlation Performance in Dense Multi-Target Environments, *IEEE Trans.*, vol. AC-18, pp. 571–582, December 1973.

51. Singer, R. A., R. G. Sear, and K. B. Housewright: Derivation and Evaluation of Improved Tracking Filters for Use in Dense Multi-Target Environments, *IEEE Trans.*, vol. IT-20, pp. 423–432, July 1974.

52. Bar-Shalom, Y., and A. Jaffer: Adaptive Nonlinear Filtering for Tracking with Measurements of Uncertain Origin, *Proc. IEEE Conf. Decision & Control*, pp. 243–247, New Orleans, December 1972.

53. Bar-Shalom, Y., and E. Tse: Tracking in a Cluttered Environment with Probabilistic Data Association, *Proc. Fourth Symp. Nonlinear Estimation*, University of California, San Diego, September 1973; *Automatica*, vol. 11, pp. 451–460, September 1975.

54. Colegrove, S. B., and J. K. Ayliffe: An Extension of Probabilistic Data Association to Include Track Initiation and Termination, *20th IREE Int. Conv. Dig.*, pp. 853–856, Melbourne, September 1985.

55. Colegrove, S. B., A. W. Davis, and J. K. Ayliffe: Track Initiation and Nearest Neighbors Incorporated into Probabilistic Data Association, *J. Elec. Electron. Eng. (Australia), IE Aust.* and *IREE Aust.*, vol. 6, pp. 191–198, September 1986.

56. Sittler, R. W.: An Optimal Association Problem in Surveillance Theory, *IEEE Trans.*, vol. MIL-8, pp. 125–139, April 1964.

57. Stein, J. J., and S. S. Blackman: Generalized Correlation of Multi-Target Track Data, *IEEE Trans.*, vol. AES-11, pp. 1207–1217, November 1975.

58. Morefield, C. L.: Application of 0-1 Integer Programming to Multi-Target Tracking Problems, *IEEE Trans.*, vol. AC-22, pp. 302–312, June 1977.

59. Trunk, G. V., and J. D. Wilson: Track Initiation of Occasionally Unresolved Radar Targets, *IEEE Trans.*, vol. AES-17, pp. 122–130, January 1981.

60. Casner, P. G., Jr., and R. J. Prengaman: Integration and Automation of Multiple Co-Located Radars, *IEEE EASCON '77*, pp. 10-1A–10-1E, Washington, 1977.

61. Casner, P. G., Jr., and R. J. Prengaman: Integration and Automation of Multiple Co-Located Radars, *IEE Int. Radar Conf.*, pp. 145–149, London, 1977.

62. Miller, J. T.: Gridlock Analysis Report, vol. I: Concept Development, *Johns Hopkins University, Appl. Phys. Lab. Rept.* FS-82-023, February 1982.

63. Bath, W. G.: Association of Multiside Radar Data in the Presence of Large Navigation and Sensor Alignment Errors, *IEE Int. Radar Conf.*, pp. 169–173, London, 1982.

64. Mirkin, M. I., C. E. Schwartz, and S. Spoerri: Automated Tracking with Netted Ground Surveillance Radars, *IEEE Int. Radar Conf.*, pp. 371–379, Washington, 1980.

65. Coleman, J. O.: Discriminants for Assigning Passive Bearing Observations to Radar Targets, *IEEE Int. Radar Conf.*, pp. 361–365, Washington, 1980.

66. Trunk, G. V., and J. D. Wilson: Association of DF Bearing Measurements with Radar Tracks, *IEEE Trans.*, vol. AES-23, pp. 438–447, July 1987.

67. Trunk, G. V., and J. D. Wilson: Correlation of DF Bearing Measurements with Radar Tracks, *IEE Int. Radar Conf.*, pp. 333–337, London, 1987.

CHAPTER 9
ELECTRONIC COUNTER-COUNTERMEASURES

A. Farina
Radar Department
Selenia S.p.A.

9.1   INTRODUCTION

Since World War II both radar and electronic warfare (EW) have achieved a very high state of performance.[1,2] Modern military forces depend heavily on electromagnetic systems for surveillance, weapon control, communication, and navigation. Electronic countermeasures (ECM) are likely to be taken by hostile forces to degrade the effectiveness of electromagnetic systems. As a direct consequence, electromagnetic systems are more and more frequently equipped with so-called electronic counter-countermeasures (ECCM) to ensure effective use of the electromagnetic spectrum despite an enemy's use of EW actions.

This chapter is devoted to the description of the ECCM techniques and design principles to be used in radar systems when they are subject to an ECM threat. Section 9.2 starts with a recall of the definitions pertaining to EW and ECCM. The topic of radar signals interception by EW devices is introduced in Sec. 9.3; the first strategy to be adopted by radar designers is to try to avoid interception by the opponent electronic devices. Section 9.4 is dedicated entirely to the analysis of the major ECM techniques and strategies. It is important to understand the ECM threat to a radar system in order to be able to efficiently react to it. To facilitate the description of the crowded family of ECCM techniques (Secs. 9.6 through 9.10), a classification is attempted in Sec. 9.5. Then, the techniques are introduced according to their use in the various sections of radar, namely, antenna, transmitter, receiver, and signal processing. A key role is also played by those ECCM techniques which cannot be classified as electronic, such as human factors, methods of radar operation, and radar deployment tactics (Sec. 9.10).

The ensuing Sec. 9.11 shows the application of the aforementioned techniques to the two most common radar families, namely, surveillance and tracking radars. The main design principles (e.g., selection of transmitter power, frequency, and antenna gain) as dictated by the ECM threat are also discussed in some detail.

The chapter ends with an approach to the problem of evaluating the efficacy of ECCM and ECM techniques (Sec. 9.12). There is a lack of theory to properly quantify the endless battle between ECCM and ECM techniques. Nevertheless, a commonly adopted approach to determine the ECM effect on a radar system is based on evaluation of the radar range under jamming conditions. The advantage of using specific ECCM techniques can also be taken into account by calculating the radar range recovery.

9.2 TERMINOLOGY

Electronic warfare is defined as a military action involving the use of electromagnetic energy to determine, exploit, reduce, or prevent radar use of the electromagnetic spectrum.[3–6] EW is organized into two major categories: electronic warfare support measures (ESM) and electronic countermeasures (ECM). Basically, the EW community takes as its job the degradation of radar capability. The radar community takes as its job the successful application of radar in spite of what the EW community does; the goal is pursued by means of ECCM techniques. The definitions of ESM, ECM, and ECCM are listed below.[3,6,7]

ESM is that division of electronic warfare involving actions taken to search for, intercept, locate, record, and analyze radiated electromagnetic energy for the purpose of exploiting such radiations in the support of military operations. Thus, electronic warfare support measures provide a source of electronic warfare information required to conduct electronic countermeasures, threat detection, warning, and avoidance. ECM is that division of electronic warfare involving actions taken to prevent or reduce a radar's effective use of the electromagnetic spectrum. ECCM comprises those radar actions taken to ensure effective use of the electromagnetic spectrum despite the enemy's use of electronic warfare.

The topic of EW is extremely rich in terms, some of which are also in general use in other electronic fields. A complete glossary of terms in use in the ECM and ECCM fields is found in the literature.[3,6,8]

9.3 ELECTRONIC WARFARE SUPPORT MEASURES

ESM is based on the use of intercept or warning receivers and relies heavily on a previously compiled directory of both tactical and strategic electronic intelligence (ELINT).[4,9,10] ESM is entirely passive, being confined to identification and location of radiated signals. Radar interception, which is of particular interest in this section, is based on the information gleaned from analysis of the signals transmitted by radar systems. The scenario in which ESM should operate is generally crowded with pulsed radar signals: figures of 500,000 to 1 million pulses per second (pps) are frequently quoted in the literature.[4] The train of interleaved pulses is processed in the ESM receiver to identify for each pulse the center frequency, amplitude, pulse width, time of arrival (TOA), and bearing. This information is then input to a pulse-sort processor which deinterleaves the pulses into the pulse repetition interval (PRI) appropriate to each emitter. Further comparison against a store of known radar types permits the generation of an emitter list classified with its tactical value. The ESM receiver is used to control the deployment and operation of ECM; the link between ESM and ECM is often automatic.

A single received radar pulse is characterized by a number of measurable parameters. The availability, resolution, and accuracy of these measurements must all be taken into account when designing the deinterleaving system because the approach used depends on the parameter data set available. Obviously, the better the resolution and accuracy of any parameter measurement, the more efficiently the pulse-sort processor can carry out its task. However, there are limitations on the measurement process from outside the ESM system (e.g., multipath), from inside the system (e.g., timing constraints, dead time during reception), and from cost-effectiveness considerations. Angle of arrival is probably the most important sorting parameter available to the deinterleaving process since the target bearing does not vary from pulse to pulse. A rotating directional antenna could be used for direction finding (DF); however, an interferometric system with more than one antenna is preferred because the probability of interception is higher than with the system having only one antenna.

The carrier frequency is the next most important pulse parameter for deinterleaving. A common method of frequency measurement is to use a scanning superheterodyne receiver that has the advantage of high sensitivity and good frequency resolution.[4] Unfortunately, this type of receiver has a poor probability of intercept for the same reasons as the rotating bearing measurement system. The situation is much worse if the emitter is also frequency-agile (random variation) or frequency-hopping (systematic variation). One method of overcoming this problem would be to use banks of contiguous receiver channels. This approach is today feasible owing to the availability of accurate surface acoustic-wave (SAW) filters and the integrated optic spectrum analyzer which utilizes the Bragg refraction of an optical guided beam by an SAW to perform spectral analysis.[4] Pulse width is an unreliable sorting parameter because of the high degree of corruption resulting from multipath transmission. Multipath effects can severely distort the pulse envelope, for example, by creating a long tail to the pulse and even displacing the position of the peak.

The TOA of the pulse can be taken as the instant that a threshold is crossed, but in the presence of noise and distortion this becomes a very variable measurement. Nevertheless, the TOA is used for deriving the PRI of the radar. The amplitude of the pulse is taken as the peak value. Dynamic-range considerations must take into account at least some three orders of magnitude for range variation and three orders of magnitude for scan pattern variations. In practice, 60-dB instantaneous dynamic range sounds like a minimum value; in many applications it should be larger. The amplitude measurement is used (along with TOA) for deriving the scan pattern of the emitter.[4]

Radar intercept receivers are implemented at varying levels of complexity. The simplest is the radar warning receiver (RWR), which in an airborne installation advises of the presence of threats such as a missile radar, supplying the relative bearing on a cockpit-based display. It is an unsophisticated low-sensitivity equipment which is preset to cover the bandwidth of expected threats, and it exploits the range advantage to indicate the threat before it comes into firing range. Receivers then increase in complexity through tactical ESM to the full ELINT (intelligence-gathering) capability. The specification of an ideal ELINT receiver for today's applications demands an instantaneous frequency coverage of 0.01 to 40 GHz, a sensitivity of better than −60 dBm, an instantaneous dynamic range greater than 60 dB, and a frequency resolution of 1 to 5 MHz. A diversity of signals, such as pulsed, CW, frequency-agile, PRI-agile, and intrapulse-modulated (chirp, multiphase-shift-keyed, etc.), must all be accommodated with a high probability of intercept (POI) and a low false-alarm rate (FAR).[4]

The range at which a radar emission is detected by an RWR depends primarily on the sensitivity of the receiver and the radiated power of the victim radar. The calculation of the warning range can be obtained by the basic *one-way beacon equation*, which provides the signal-to-noise ratio at the RWR:

$$\left(\frac{S}{N}\right)_{\text{at RWR}} = \left(\frac{P}{4\pi R^2}\right) G_t \left(\frac{G_r \lambda^2}{4\pi}\right) \left(\frac{1}{kT_S B}\right) \frac{1}{L} \tag{9.1}$$

where P is the radar radiated power, R is the range from the RWR to the radar, G_t is the transmitting-antenna gain of the radar, G_r is the receiving-antenna gain of the RWR, λ is the radar wavelength, the quantity $kT_S B$ is the total system noise power of the RWR, and L is the losses.

Equation (9.1) is the basis of performance calculation for an RWR. It is noted that the RWR detection performance is inversely proportional to R^2 rather than to R^4 of the radar target detection equation. For this reason, the RWR can detect a radiating radar at distances far beyond those of radar's own target detection capability. The radar-versus-interceptor problem is a battle in which the radar's advantage lies in the use of matched filtering, which cannot be duplicated by the interceptor (it does not know the exact radar waveform), while the interceptor's advantage lies in the fundamental R^2 advantage of one-way versus two-way radar propagation.[10]

9.4 ELECTRONIC COUNTERMEASURES

The objectives of an ECM system are to deny information (detection, position, track initiation, track update, and classification of one or more targets) that the radar seeks or to surround desired radar echoes with so many false targets that the true information cannot be extracted.

ECM tactics and techniques may be classified in a number of ways, i.e., by main purpose, whether active or passive, by deployment-employment, by platform, by victim radar, or by a combination of them.[8,11] An encyclopedia of ECM tactics and techniques can be found in the literature.[8,12] Here it is intended to limit description to the most common types of ECM.

ECM includes both jamming and deception. *Jamming* is the intentional and deliberate transmission or retransmission of amplitude, frequency, phase, or otherwise modulated intermittent, CW, or noiselike signals for the purpose of interfering with, disturbing, exploiting, deceiving, masking, or otherwise degrading the reception of other signals that are used by radar systems.[8] A jammer is any ECM device that transmits a signal of any duty cycle for the sole or partial purpose of jamming a radar system.[8]

Radio signals by special transmitters intended for interfering with or precluding the normal operation of a victim radar system are called *active jamming*. They produce at the input of a victim system a background which impedes the detection and recognition of useful signals and determination of their parameters. The most common forms of active noise jamming are spot and barrage noises. Spot noise is used when the center frequency and bandwidth of the victim system to be jammed are known and confined to a narrow band. However, many radars are frequency-agile over a wide band as an ECCM against spot jamming. If the rate of frequency agility is slow enough, the jammer can follow the frequency changes and maintain the effect of spot jamming. Barrage or broadband jamming is simultaneously radiated across the entire band of the radar spectrum of interest. This method is used against frequency-agile systems whose rates are too fast to follow or when the victim's frequency parameters are imprecisely known.

Jammer *size* is characterized by the *effective radiated power*; ERP = $G_j P_j$, where G_j is the transmit antenna gain of the jammer and P_j is the jammer power.

Passive ECM is synonymous with chaff, decoys, and other reflectors which require no prime power. The chaff is made of elemental passive reflectors or absorbers which can be floated or otherwise suspended in the atmosphere or exoatmosphere for the purpose of confusing, screening, or otherwise adversely affecting the victim electronic system. Examples are metal foils, metal-coated dielectrics (aluminum, silver, or zinc over fiberglass or nylon being the most common), string balls, rope, and semiconductors.[8] The basic properties of chaff are effective scatter area, the character and time of development of a chaff cloud, the spectra of the signals reflected by the cloud, and the width of the band that conceals the target.[4,12–14] Chaff consists of dipoles cut to approximately a half wavelength of the radar frequency. It is usually packaged in cartridges which contain a broad range of dipole lengths designed to be effective over a wide frequency band. From a radar viewpoint, the properties of chaff are very similar to those of weather clutter, except that its broadband frequency capability extends down to VHF. The mean doppler frequency of the chaff spectrum is determined by the mean wind velocity, while the spectrum spread is determined by wind turbulence and a shearing effect due to different wind velocities as a function of altitude.[12]

Decoys, which are another type of passive ECM, are a class of physically small radar targets whose radar cross sections are generally enhanced by using reflectors or a Luneburg lens to simulate fighter or bomber aircraft. The objective of decoys is to cause a dilution of the assets of the defensive system, thereby increasing the survivability of the penetrating aircraft.

The other major type of active jammer is deceptive ECM (DECM). *Deception* is the intentional and deliberate transmission or retransmission of amplitude, frequency, phase, or otherwise modulated intermittent or continuous-wave (CW) signals for the purpose of misleading in the interpretation or use of information by electronic systems.[8] The categories of deception are manipulative and imitative. *Manipulative* implies the alteration of friendly electromagnetic signals to accomplish deception, while *imitative* consists of introducing radiation into radar channels which imitates a hostile emission. DECM is also divided into *transponders* and *repeaters*.[12] Transponders generate noncoherent signals which emulate the temporal characteristics of the actual radar return. Repeaters generate coherent returns which attempt to emulate the amplitude, frequency, and temporal characteristics of the actual radar return. Repeaters usually require some form of memory for microwave signals to allow anticipatory returns to be generated; this is usually implemented by using a microwave acoustic memory or a digital RF memory (DRFM).[12]

The most common type of deception jammer is the range-gate stealer, whose function is to pull the radar tracking gate from the target position through the introduction of a false target into the radar's range-tracking circuits. A repeater jammer sends back an amplified version of the signal received from the radar. The deception jammer signal, being stronger than the radar's return signal, captures the range-tracking circuits. The deception signal is then progressively delayed in the jammer by using an RF memory, thereby "walking" the range gate off the actual target (range-gate pull-off, or RGPO, technique). When the range gate is sufficiently removed from the actual target, the deception jammer is turned off, forcing the tracking radar into a target reacquisition mode.[12]

Another DECM technique is called *inverse-gain jamming*; it is used to capture the angle-tracking circuits of a conical-scan tracking radar.[8] This technique repeats a replica of the received signal with an induced amplitude modulation which is the inverse of the victim radar's combined transmitting and receiving antenna scan patterns. Against a conically scanning tracking radar, an inverse-gain repeater jammer has the effect of causing positive feedback, which pushes the tracking-radar antenna away from the target rather than toward the target. Inverse-gain jamming and RGPO are combined in many cases to counter conical-scan tracking radars.[12]

A different form of DECM used against the main beam of surveillance radar attempts to cover the target's skin return with a wide pulse in order to confuse the radar's signal-processing circuitry into suppressing the actual target return.

In the deployment-employment of ECM, five classes can be singled out.[12] In the standoff jammer (SOJ) case, the jamming platform remains close to but outside the lethal range of enemy weapon systems and jams these systems to protect the attacking vehicles. Standoff ECM systems employ high-power noise jamming which must penetrate through the radar antenna receiving sidelobes at long ranges. *Escort jamming* is another ECM tactic in which the jamming platform accompanies the strike vehicles and jams radars to protect the strike vehicles.

Mutual-support, or *cooperative*, ECM involves the coordinated conduct of ECM by combat elements against acquisition and weapon control radars. One advantage of mutual-support jamming is the greater ERP available from a collection of platforms in contrast with a single platform. However, the real value of mutual-support jamming is in the coordinated tactics which can be employed. A favorite tactic employed against tracking radars, for example, is to switch between jammers located on separate aircraft within the radar's beamwidth. This blinking has the effect of introducing artificial glint into the radar tracking circuits, which, if introduced at the proper rate (typically 0.1 to 10 Hz), can cause the radar to break angle track. In addition, blinking has the desirable effect of confusing radiation homing missiles which might be directed against the jammer radiations.[12]

A *self-screening jammer* (SSJ) is used to protect the carrying vehicle. This situation stresses the capability of an ECM system relating to its power, signal-processing, and ESM capabilities.

Stand-forward jamming is an ECM tactic in which the jamming platform is located between the weapon systems and the strike vehicles and jams the radars to protect the strike vehicles. The stand-forward jammer is usually within the lethal range of defensive weapon systems for a considerable time. Therefore, only the use of relatively low-cost remotely piloted vehicles (RPVs) might be practical. RPVs can assist strike aircraft or missiles in penetrating radar-defended areas by jamming, ejecting chaff, dropping expendable jammers or decoys, acting as decoys themselves, and performing other related ECM tasks.

According to the platform, the jammer can be classified as space-borne, air-borne, missile-borne, based on the ground, or based on the sea surface.

A special class of missile-borne threat is the antiradiation missile (ARM), having the objective of homing on and destroying the victim radar. The sorting and acquisition of radar signals is preliminarily made by an ESM system; afterwards it cues the ARM, which continues homing on the victim radar by means of its own antenna, receiver, and signal processor. Acquisition depends on the direction of arrival, operating band, carrier frequency, pulse width, PRI, scan rate, and other parameters of the victim radar. An ARM homes on the continuous radiation from the radar sidelobes or on the flash of energy from the main beam. ARM benefits from the one-way-only radar signal attenuation. However, ARM receiver sensitivity is affected by mismatching losses, and accuracy in locating the victim radar is affected by the limited dimension of the ARM antenna.

9.5 OBJECTIVES AND TAXONOMY OF ECCM TECHNIQUES

The primary objective of ECCM techniques when applied to a radar system is to allow the accomplishment of the radar intended mission while countering the effects of the enemy's ECM. In greater detail, the benefits of using ECCM techniques may be summarized as follows: (1) prevention of radar saturation, (2) enhancement of the signal-to-jamming ratio, (3) discrimination of directional interference, (4) rejection of false targets, (5) maintenance of target tracks, (6) counteraction of ESM, and (7) radar system survivability.[3]

There are two broad classes of ECCM: (1) electronic techniques (Secs. 9.6–9.9) and (2) operational doctrines (Sec. 9.10). Specific electronic techniques take place in the main radar subsystems, namely, the antenna, transmitter, receiver, and signal processor. Suitable blending of these ECCM techniques can be implemented in the surveillance and tracking radars, as discussed in Sec. 9.11.

The ensuing description is limited to the major ECCM techniques; the reader should be aware that an alphabetically listed collection of 150 ECCM techniques and an encyclopedia of ECCM tactics and techniques can be found in the literature.[3,15] Many other references describe the ECCM problem, among which Refs. 13, 16, and 17 are worth noting.

9.6 ANTENNA-RELATED ECCM

Since the antenna represents the transducer between the radar and the environment, it is the first line of defense against jamming. The directivity of the antenna in the transmission and reception phases allows space discrimination to be used as an ECCM strategy. Techniques for space discrimination include antenna coverage and scan control, reduction of main-beam width, low sidelobes, sidelobe blanking, sidelobe cancelers, and adaptive array systems. Some of these techniques are useful during transmission, while others operate in the reception phase. Additionally, some are active against main-beam jammers, and others provide benefits against sidelobe jammers.

Blanking or turning off the receiver while the radar is scanning across the azimuth sector containing the jammer or reducing the scan sector covered are means to prevent the radar from looking at the jammer. Certain deception jammers depend on anticipation of the beam scan or on knowledge or measurement of the antenna scan rate. Random electronic scanning effectively prevents these deception jammers from synchronizing to the antenna scan rate, thus defeating this type of jammer. A high-gain antenna can be employed to spotlight a target and burn through the jammers. An antenna having multiple beams can also be used to allow deletion of the beam containing the jammer and still maintain detection capabilities with the remaining beams. Increased angular resolution of jammers in the main beam can be reached by resorting to spectral analysis algorithms, commonly referred to as *superresolution* techniques. Although they add complexity, cost, and possibly weight to the antenna, reduction of main-beam width and control of coverage and scan are valuable and worthwhile ECCM features of all radars.

If an air defense radar operates in a severe ECM environment, the detection range can be degraded because of jamming entering the sidelobes. On transmit, the energy radiated into spatial regions outside of the main beam is subject to being received by enemy RWRs or ARMs. For these reasons, low sidelobes are desirable on both receive and transmit.[18] Sometimes the increase in main-beam width that results from low sidelobes worsens the problem of main-beam jamming; this consequence should be carefully considered in specifying the antenna radiation pattern.

Usually, specification of the sidelobes as a single number (e.g., -30 dB) means that the peak of the highest sidelobe is 30 dB below the peak of the main beam. The average, or root-mean-square (rms), sidelobe level is often more important. For example, if 10 percent of the radiated power is in the sidelobes, the average sidelobe level is -10 dB, where dB refers to the number of decibels by which the average sidelobe level is below the gain of an isotropic (ideal) radiator. In theory, extremely low sidelobes can be achieved with aperture illumination functions that are appropriately tapered. This leads to the well-known tradeoffs among gain, beamwidth, and sidelobe level.[19] In order to keep the beamwidth small with low sidelobes, a larger and more costly antenna is needed. Other design principles involved in low antenna sidelobes are the use of radar-absorbent material (RAM) about the antenna structure, the use of a fence on ground installations, and the use of polarization screens and reflectors. This means that very low sidelobe antennas are costly in terms of size and complexity when compared with conventional antennas of similar gain and beamwidth characteristics. Second, as the design sidelobes are pushed lower and lower, a point is reached where minor error contributions to scattered energy (random errors) or misdirected radiation (systematic errors) become significant. In practice, peak sidelobe levels as low as -30 to -35 dB (average level, -5 to -20 dB) can be readily realized with phased array antennas which electronically scan. To obtain sidelobes at levels -45 dB down from the main beam (average level, below -20 dB), the total phase-error budget is required to be in the order of 5° rms or less. This is extremely difficult in arrays which electronically scan: the errors induced by phase shifters, active components, and feed elements must be included in this budget. Arrays have been realized in practice which have peak sidelobes in the vicinity of the -45 dB level; however, these are generally mechanically scanned, and the low error budgets are achieved by using all-passive feed components. Future antenna development will yield -45 dB sidelobe antennas which do scan electronically.[12]

Two additional techniques to prevent jamming from entering through the radar's sidelobes are the so-called sidelobe blanking (SLB) and sidelobe cancelers (SLC). An example of the practical effectiveness of the SLB and SLC devices is presented in the literature, where the plan position indicator (PPI) display is shown for a radar, subject to an ECM, equipped with and without the SLB and SLC systems.[17]

Other discrimination means are based on polarization. The polarization characteristics of a radar can be exploited as ECCM techniques in two ways. First, the cross-polarized pattern (i.e., the orthogonal polarization to the main plane of polarization) of a radar antenna should be kept as low as possible consistent with radar system cost. Ratios of copolarized main-beam peak gain to cross-polarized gain anywhere in the antenna pattern should be greater than 25 dB to provide protection against common cross-polarized jamming. This is thought of as an ECCM technique, but it is really no more than good antenna design. The cross-polarized jamming in this case attacks a design deficiency in the radar. The requirement for good cross-polarization design practice in a radar antenna system extends to any auxiliary ECCM antennas as well. If their cross-polarized gains are high, ECCM techniques such as SLC and SLB may not be effective against cross-polarized noise or repeater jammers.[15]

In the second use of polarization the radar antenna system purposely receives the cross-polarization component of the radar wave in addition to the copolarized component. The two orthogonally polarized components can be used to discriminate the useful target from chaff and jammer on the basis of their different polarizations.[20] However, limited benefits (few decibels of cancellation ratio) can be obtained at the expense of a more complex antenna system (consider, for example, a phased array with radiating elements able to separately receive and possibly transmit the two orthogonal components of a radar wave) and of a duplication of the receiver and signal processing.

Sidelobe-Blanking (SLB) System. The purpose of an SLB system is to prevent the detection of strong targets and interference pulses entering the radar receiver via the antenna sidelobes.[21–24] A method of achieving this is to employ an auxiliary antenna coupled to a parallel receiving channel so that two signals from a single source are available for comparison. By suitable choice of the antenna gains, one may distinguish signals entering the sidelobes from those entering the main beam, and the former may be suppressed. Figure 9.1*a* illustrates the radiation pattern of the main antenna together with a low-gain auxiliary antenna. An implementation of the SLB processor is shown in Fig. 9.1*b*, where the square-law-detected outputs of the two channels, ideally identical except for the antenna patterns, are compared. The comparison is made at each range bin for each pulse received and processed by the two parallel channels. Thus, the SLB decides whether or not to blank the main channel on a single-sweep basis and for each range bin. A target A in the main beam will result in a large signal in the main receiving channel and a small signal in the auxiliary receiving channel. A proper blanking logic allows this signal to pass. Targets and/or jammers J situated in the sidelobes give small main but large auxiliary signals so that these targets are suppressed by the blanking logic. It is assumed that the gain G_A of the auxiliary antenna is higher than the maximum gain G_{sl} of the sidelobes of the radar antenna.

The performance of the SLB may be analyzed by looking at the different outcomes obtained as a consequence of the pair (u, v) of the processed signals (see Fig. 9.1*b*). Three hypotheses have to be tested: (1) the null hypothesis H_0 corresponding to the presence of noise in the two channels, (2) the H_1 hypothesis pertaining to the target in the main beam, and (3) the H_2 hypothesis corresponding to target or interference signal in the sidelobe region. The null and H_1 hypotheses correspond to the usual decisions of "no detection" and "target detection," respectively. The blanking command is delivered when H_2 is detected.

SLB performance can be expressed in terms of the following probabilities: (1) The probability P_B of blanking a jammer in the radar sidelobes, which is the probability of associating the received signals (u, v) with H_2 when the same hypoth-

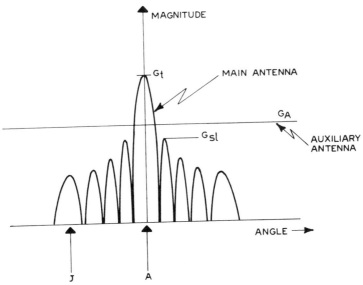

FIG. 9.1a Main and auxiliary antenna patterns for the SLB. (*From Ref. 21.*)

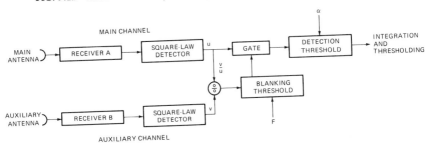

FIG. 9.1b Scheme of sidelobe-blanking system. (*From Ref. 21.*)

esis is true; P_B is a function of the jammer-to-noise ratio (*JNR*) value, the blanking threshold F, and the gain margin $\beta = G_A/G_{sl}$ of the auxiliary antenna with respect to the radar antenna sidelobes. (2) The probability P_{FA} of false alarm, which is the probability of associating the received signals (u, v) with the hypothesis H_1 when the true hypothesis is H_0; P_{FA} is a function of the detection threshold α normalized to the noise power level and of the blanking threshold F. (3) The probability P_D of detecting a target in the main beam, which is the probability of associating the received signal (u, v) with H_1 when the same hypothesis is true; P_D depends, among other things, on the signal-to-noise power ratio *SNR*, P_{FA}, and the blanking threshold F. (4) The probability P_{FT} of detecting a false target produced by a jammer entering through the radar sidelobes. P_{FT} is the probability of associating (u, v) with H_1 when H_2 is true; it is a function of *JNR*, the thresholds α and F, and the gain margin β. (5) The probability P_{TB} of blanking a target received in the main beam. This is the probability of associating (u, v) with H_2 when H_1 is the true hypothesis. P_{TB} is related to *SNR*, F, and the auxiliary gain $w = G_A/G_t$ normalized to the gain G_t of the main beam. To complete the list of parameters needed to describe the SLB performance, the last figure to consider is the detection loss L on the main-beam target. This can be found by comparing the SNR values required to achieve a specified P_D value for the radar system with and without the SLB. L is a function of many parameters such as P_D, P_{FA}, F, G_A, JNR, and β. A numerical evaluation of some of these performance parameters can be found in the literature.[21,24]

The SLB design requires the selection of suitable values for the following parameters: (1) the gain margin β and then the gain w of the auxiliary antenna, (2) the blanking threshold F, and the normalized detection threshold α. The a priori known parameters are the radar sidelobe level G_{sl} and the values of SNR and JNR. The design parameters can be selected by trying to maximize the detection probability P_D while keeping at prescribed values the probabilities P_B and P_{FA} and trying to minimize P_{FT}, P_{TB}, and L.

Sidelobe Canceler (SLC) System.   The objective of the SLC is to suppress high duty cycle and noiselike interferences (e.g., SOJ) received through the sidelobes of the radar. This is accomplished by equipping the radar with an array of auxiliary antennas used to adaptively estimate the direction of arrival and the power of the jammers and, subsequently, to modify the receiving pattern of the radar antenna to place nulls in the jammers' directions. The SLC was invented by P. Howells and S. Applebaum.[25,26]

The conceptual scheme of an SLC system is shown in Fig. 9.2. The auxiliary antennas provide replicas of the jamming signals in the radar antenna sidelobes. To this end the auxiliary patterns approximate the average sidelobe level of the radar receiving pattern. In addition, the auxiliaries are placed sufficiently close to the phase center of the radar antenna to ensure that the samples of the interference which they obtain are statistically correlated with the radar jamming signal. It is also noted that as many auxiliary antennas are needed as there are jamming signals to be suppressed. In fact, at least N auxiliary patterns properly controlled in amplitude and phase are needed to force to zero the main-antenna receiving pattern in N given directions. The auxiliaries may be individual antennas or groups of receiving elements of a phased array antenna.

The amplitude and phase of the signals delivered by the N auxiliaries are controlled by a set of suitable weights: denote the set with the N-dimensional vector $\mathbf{W} = (W_1, W_2, \ldots, W_N)$. Jamming is canceled by a linear combination of the signals from the auxiliaries and the main antenna. The problem is to find a suitable means of controlling the weights $\mathbf{W}$ of the linear combination so that the maximum possible cancellation is achieved. Owing to the stochastic nature of the jamming signals in the radar and in the auxiliary channels and the hypothesized linear combination of signals, it is advisable to resort to the techniques of linear prediction theory for stochastic processes.[27] Denote with V_M the radar signal at a certain range bin and with $\mathbf{V} = (V_1, V_2, \ldots, V_N)$ the N-dimensional vector containing the set of signals, at the same range bin, from the N auxiliary antennas. It is assumed that all the signals have bandpass frequency spectra; therefore, the signals can be represented by their complex envelopes, which modulate a common carrier frequency that does not appear explicitly. The jamming signals in the channels may be regarded as samples of a stochastic process having zero mean value and a certain time autocorrelation function. For linear prediction problems, the set of samples $\mathbf{V}$ is completely described by its N-dimensional covariance matrix $\mathbf{M} = E(\mathbf{V}^*\mathbf{V}^T)$, where $E(.)$ denotes the statistical expectation, the asterisk * indicates the complex conjugate, and $\mathbf{V}^T$ is the transpose vector of $\mathbf{V}$. The statistical relationship between V_M and $\mathbf{V}$ is mathematically represented by the N-

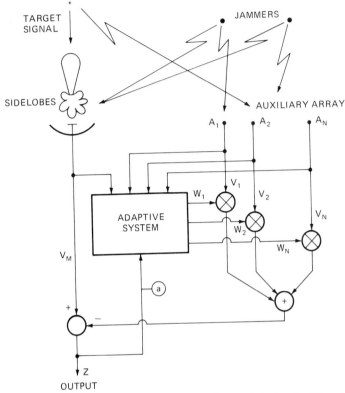

FIG. 9.2 Principle of SLC operation (connection *a* only in the closed-loop implementation techniques).

dimensional covariance vector $\mathbf{R} = E(V_M \mathbf{V}^*)$. The optimum weight vector $\hat{\mathbf{W}}$ is determined by minimizing the mean square prediction error which equals the output residual power:

$$P_Z = E\{ |Z|^2 \} = E\{|V_M - \hat{\mathbf{W}}^T \mathbf{V}|^2\} \tag{9.2}$$

where Z is the system output. It is found that the following fundamental equation applies:[27]

$$\hat{\mathbf{W}} = \mu \mathbf{M}^{-1} \mathbf{R} \tag{9.3}$$

where μ is an arbitrary constant value.

The benefit of using the SLC can be measured by introducing the jammer cancellation ratio (JCR), defined as the ratio of the output noise power without and with the SLC:

$$JCR = \frac{E\{|V_M|^2\}}{E\{|V_M - \hat{\mathbf{W}}^T \mathbf{V}|^2\}} = \frac{E\{|V_M|^2\}}{E\{|V_M|^2\} - \mathbf{R}^T \mathbf{M}^{-1} \mathbf{R}^*} \tag{9.4}$$

By applying Eqs. (9.3) and (9.4) to the simple case of one auxiliary antenna and one jammer, the following results are easily found:

$$\hat{W} = \frac{E\{V_M V_A^*\}}{E\{|V_A|^2\}} \stackrel{\Delta}{=} \rho \qquad JCR = \frac{1}{1 - |\rho|^2} \qquad (9.5)$$

It is noted that the optimum weight is related to the correlation coefficient ρ between the main signal V_M and the auxiliary signal V_A; high values of the correlation coefficient provide high values of JCR.

The problem of implementing the optimum-weight set [Eq. (9.3)] is essentially related to the real-time estimation of **M** and **R** and to the inversion of **M**. Several processing schemes have been conceived which may be classified in two main categories: (1) closed-loop techniques, in which the output residue (connection *a* of Fig. 9.2) is fed back into the adaptive system; and (2) direct-solution methods, often referred to as *open-loop*, which operate just on the incoming signals V_M and **V**. Broadly speaking, closed-loop methods are cheaper and simpler to implement than direct-solution methods.[27,28] By virtue of their self-correcting nature, they do not require components which have a wide dynamic range or a high degree of linearity, and so they are well suited to analog implementation. However closed-loop methods suffer from the fundamental limitation that their speed of response must be restricted in order to achieve a stable and not noisy steady state. Direct-solution methods, on the other hand, do not suffer from problems of slow convergence but, in general, require components of such high accuracy and wide dynamic range that they can only be realized by digital means.[27,29] Of course, closed-loop methods can also be implemented by using digital circuitry, in which case the constraints on numerical accuracy are greatly relaxed and the total number of arithmetic operations is much reduced by comparison with direct-solution methods.

Practical considerations often limit the SLC nulling capabilities to JCR of about 20 to 30 dB, but their theoretical performance is potentially much higher.[30,31] Examples of possible limitations are listed below:[27]

1. Mismatch between the main and auxiliary signals including the propagation paths, the patterns of the main and auxiliary antennas, the paths internal to the system up to the cancellation point, and the crosstalk between the channels.[32,33]

2. The limited number of auxiliary channels adopted in a practical system as compared with the number of jamming signals.[32]

3. The limited bandwidth of the majority of the schemes implementing Eq. (9.3) as compared with the wide band of a barrage jammer which can be regarded as a cluster, spread in angle, of narrowband jammers.[28,30,34]

4. The pulse width which limits the reaction time of the adaptive system, in order to avoid the cancellation of target signal.[33]

5. The target signal in the auxiliary array which may result in nonnegligible steering of the auxiliaries toward the main-beam direction.[33]

6. The presence of clutter which, if not properly removed, may capture the adaptive system, giving rise to nulls along directions different from those of the jammers.

7. The tradeoff which has to be sought between the accuracy of weights estimation and the reaction time of the adaptive system.

8. The quantization and processing accuracy in the digital implementation.

Adaptive Arrays.   An adaptive array (Fig. 9.3) is a collection of N antennas, feeding a weighting and summing network, with automatic signal-dependent weight adjustment to reduce the effect of unwanted signals and/or emphasize the desired signal or signals in the summing network output. Output signal z is envelope-detected and compared with a suitable threshold α to detect the presence of a useful target.[28,34–40] The adaptive array is a generalization of the SLC system concept described in the preceding subsection. The basic theory of jammer cancellation and target enhancement is considered first, and attention is then focused on the use of adaptive arrays to obtain superresolution capabilities which can be of help for ECCM. The implementation of the adaptive array concept is more and more related to digital beamforming technology.[41–43]

Jammer Cancellation and Target Signal Enhancement.   Adaptive array principles have found a thorough mathematical treatment since the early 1970s.[40] The basic result is given by the expression of the optimum set of weights:

$$\hat{\mathbf{W}} = \mu\mathbf{M}^{-1}\mathbf{S}^* \tag{9.6}$$

where $\mathbf{M} = E(\mathbf{V}^*\mathbf{V}^T)$ is the N-dimensional covariance matrix of the overall disturbance (noise and jammer) $\mathbf{V}$ received by the array and $\mathbf{S}$ is the N-dimensional vector containing the expected signal samples in the array from a target along a certain direction of arrival. The similarity of Eq. (9.6) to Eq. (9.3) governing the SLC is immediately recognized.

With respect to SLC, adaptive array techniques offer the capability of enhancing the target signal while canceling the disturbance. The adaptive system allocates in an optimum fashion its degrees of freedom (i.e., the set of received

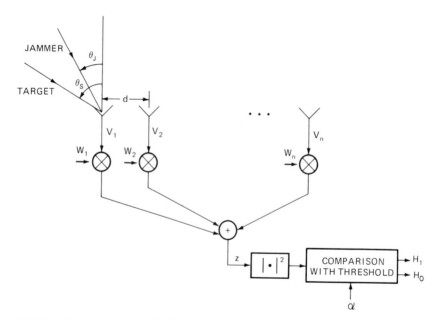

FIG. 9.3   The adaptive array scheme.

pulses from each antenna of the array) to the enhancement of the target signal and to the cancellation of clutter, chaff, and jammer.

Several generalizations of the basic theory have been considered, including: (1) the target model **S** is not known a priori, as it is assumed in deriving Eq. (9.6); (2) in addition to spatial filtering, doppler filtering is performed to cancel clutter and chaff; and (3) the radar platform is moving as in airborne and shipborne applications.

The detection probability P_D for the optimum filter of Eq. (9.6) is:[40]

$$P_D = Q\left(\sqrt{\mathbf{S^T M^{-1} S^*}}, \sqrt{2 \ln 1/P_{FA}}\right) \tag{9.7}$$

where $Q(.,.)$ is the Marcum Q function and P_{FA} is the prescribed probability of false alarm. It is also shown that the set of weights of Eq. (9.6) provides the maximum value of the improvement factor I_f, which is defined as follows:

$$I_f = \frac{\text{signal-to-interference power ratio at the output}}{\text{signal-to-interference power ratio at the input}} \tag{9.8}$$

The signal-to-interference power ratio $(SNR)_I$ at the input is measured at the input of an antenna of the array and refers to one echo pulse. The I_f value corresponding to the optimum set of weights of Eq. (9.6) is[40]

$$I_f = \frac{\mathbf{S^T M^{-1} S^*}}{(SNR)_I} \tag{9.9}$$

The I_f is better suited than the cancellation ratio, adopted in the SLC, to represent the performance of the adaptive array. In fact, in the latter case the useful signal is integrated while the interference is canceled.

The implementation of an adaptive array has been limited to experimental systems with a small number of antennas (say, 10), so that the matrix inversion can be handled by practical computing systems.[44,45] Arrays with a large number of receiving elements need some form of processing reduction. One method of partial adaptivity is to arrange the array elements in subgroups which form the inputs of the adaptive processor. Careful selection of the subgroup elements is necessary to avoid grating lobes.[46,47]

Other simplifications of the fully adaptive array are deterministic spatial filtering and phase-only nulling techniques. In the first case, a fixed reduction of the sidelobes is operated in those directions or solid angles from which the interferences are expected to come. As an example, a probable region with interferences is the horizon or part of it because jammers are mostly ground-based or at long range. The weights are computed offline, by assuming an a priori known covariance matrix **M**, and stored in a memory where a "menu" of weights is available to an operator or an automatic decision system.[48] The idea of phase-only nulling in phased array antennas is appealing because the phase shifters are already available as part of the beam-steering system. Hence, if the same phase shifters can be employed for the dual purposes of beam steering and adaptive nulling of unwanted interferences, costly retrofitting could be unnecessary. However, phase-only null synthesis presents analytic and computational difficulties not present when both the amplitude and the phase of the element weights can be freely perturbed.[49,50] Nevertheless, experimental systems have been tested with success.[51–53]

Superresolution. The resolution of a conventional antenna is limited by the well-known Rayleigh criterion, which states that two equal-amplitude noise sources can be resolved if they are separated in angle by 0.8 λ/L, in radians, where λ is the wavelength and L is the aperture length. When the incident wave is received with a high signal-to-thermal noise ratio, an adaptive array antenna may achieve a narrower *adaptive beamwidth*, giving a sharper bearing estimation of the incident wave. This is important for ECCM purposes: very accurate strobes of the jammers can be obtained. It is also possible to measure the source strength and to obtain a spatial spectrum pattern without sidelobes. The estimated angles of the jammers can be used to form beams in the jammer directions, which are used as auxiliary channels for adaptive interference suppression.[54] The interference directions can also be used for deterministic nulling, which is of special interest for main-beam nulling.[55] In addition to the interference source directions and source strengths, this technique can provide other information as to the number of sources and any cross correlations (coherence) between the sources. Such information can be used to track and catalog the interference sources in order to properly react to them.

The superresolution concept was mainly developed and analyzed by W. F. Gabriel.[56] Different methods for bearing estimation were described by Gabriel and, subsequently, by other authors.[39,57-59] One is the maximum-entropy method (MEM). It works well with a Howells-Applebaum adaptive beamformer, which has an omnidirectional receiving pattern except where signals are present. The presence of signals is indicated by nulls in the receiving pattern. Since nulls are always sharper than antenna lobes, signal bearings can be obtained more accurately from the adaptive beam pattern, and superresolution is the result. The desired spatial spectrum pattern is obtained as simply the inverse of the adapted pattern. As Gabriel points out, there is not a true antenna pattern because there is no linear combination of the signals from an array that could produce such a peaked spatial pattern. It is simply a function computed from the reciprocal of a true adapted antenna pattern. Superresolution and adaptive antennas are identical mathematically, use the same algorithms, and have identical hardware. Roughly speaking, the difference is that one produces a pattern with the nulls down (adaptive antenna for jammer cancellation) and the other with the nulls up (superresolution of jammers).

The achievable degree of superresolution depends heavily on the way in which the algorithms are implemented. The required accuracy of signal quantization and the matching of channels are comparable with those of adaptive nulling. The heavy computational task required by the algorithms can be handled by resorting to systolic array processors.[60] Experiments indicate that the resolution limit is determined rather more by implementation factors like channel mismatching errors than by the pure *SNR*. Two incoherent sources separated by a quarter beamwidth seem to be the lower limit for superresolution with the current technology for achieving equality between the channels, offset compensation, equality of I and Q channel amplification, etc. Resolution is worse for more than two sources.[58]

9.7 TRANSMITTER-RELATED ECCM

The different types of ECCM are related to the proper use and control of the power, frequency, and waveform of the radiated signal. One brute-force approach to defeating noise jamming is to increase the radar's transmitter power. This technique, when coupled with "spotlighting" the radar antenna on the target, results in an increase of the radar's detection range. Spotlighting or burnthrough modes might be effective, but a price must be paid. As the radar dwells in a particular direction, it is not looking elsewhere, where it is supposed to look. In addition, the burnthrough mode is not effective against chaff, decoys, repeaters, spoofers, and so on.

More effective is the use of complex, variable, and dissimilar transmitted signals which place a maximum burden on ESM and ECM. Different ways of operation refer to the change of the transmitted frequency in frequency-agility or frequency-diversity modes or to the use of wide instantaneous bandwidth.[61–64] *Frequency agility* usually refers to the radar's ability to change the transmitter frequency on a pulse-to-pulse or batch-to-batch basis. The batch-to-batch approach allows doppler processing, which is not compatible with frequency agility on a pulse-to-pulse basis. In a waveform with pulse-to-pulse frequency agility, the center frequency of each transmitted pulse is moved, in either a random or a programmed schedule, between a large number of center frequencies on a pulse-to-pulse basis. The frequency of the next pulse cannot generally be predicted from the frequency of the current pulse.[65] *Frequency diversity* refers to the use of several complementary radar transmissions at different frequencies, either from a single radar (e.g., a radar having stacked beams in elevation by employing different frequencies on each elevation beam) or from several radars. The objective of frequency agility and diversity is to force the jammer to spread its energy over the entire agile bandwidth of the radar; this corresponds to a reduction of the jammer density and resulting ECM effectiveness.[15] Signals with wide instantaneous bandwidth exhibit considerable variation of the frequency within each transmitted pulse. A spread of about 10 percent of the transmitter center frequency can be proper. Three of the more common coded-pulse waveforms are recalled: (1) the linear frequency-modulated signal, where the carrier frequency is varied linearly within the pulse; (2) the frequency-shift-coded signal, where the carrier frequency is changed in a stepwise fashion within the pulse; and (3) the phase-coded signal, in which the phase of the RF carrier is shifted at a rate equal to the bandwidth of the waveform.

Frequency agility, diversity, and instantaneous wideband techniques represent a form of ECCM in which the information-carrying signal is spread over as wide a frequency (or space, or time) region as possible to reduce detectability by ESM and/or ARM and make jamming more difficult. This ECCM technique pertains to the realm of waveform coding.[12,29] Waveform coding includes pulse-repetition-frequency (PRF) jitter, PRF stagger, and, perhaps, shaping of the transmitted radar pulse. All these techniques make deception jamming or spoofing of the radar difficult, since the enemy should not know or anticipate the fine structure of the transmitted waveform; as a consequence, they give assurance of maximum range performance against jamming. Intrapulse coding to achieve pulse compression may be particularly effective in improving target detection capability by radiation of enough average radar power without exceeding peak power limitations within the radar and by improving range resolution (larger bandwidth), which in turn reduces chaff returns and resolves targets to a higher degree.

Some advantage can be gained by including the capability to examine the jammer signals, find holes in their transmitted spectra, and select the radar frequency with the lowest level of jamming. This approach is particularly useful against pulsed ECM, spot noise, and nonuniform barrage noise; its effectiveness depends primarily on the extent of the radar agile bandwidth and the acquisition speed and frequency tracking of an "intelligent" jammer. A technique suited to this purpose is referred to as automatic frequency selection (AFS).[64,66]

Another method to reduce the effect of main-beam noise jamming is to increase the transmitter frequency (as an alternative means to the use of a larger antenna) in order to narrow the antenna's beamwidth. This restricts the sector which is blanked by main-beam jamming and also provides a strobe in the direction of the jammer. Strobes from two or three spatially separated radars allow the jammer to be located.

9.8 RECEIVER-RELATED ECCM

Jamming signals that survive the antenna ECCM expedients can, if large enough, saturate the radar processing chain. Saturation results in the virtual elimination of information about targets. Wide dynamic range (i.e., log and lin-log) receivers are normally used to avoid saturation.

Other special processing circuits can be used in the radar to avoid saturation, i.e., fast-time-constant (FTC) devices, automatic gain control (AGC), and constant-false-alarm rate (CFAR).[3,15,17] However, they cannot be said to be ECCM techniques. For example, FTC allows the detection of signals that are greater than clutter by preventing the clutter from saturating the display. FTC does not provide subclutter visibility. AGC keeps the radar receiver operating within its dynamic range, preventing system overload and providing proper normalization so as to furnish signals of standardized amplitude to radar range, velocity, and angle processing-tracking circuits. CFAR is a technique made necessary because of the limitations of the computer in automatic systems. It prevents the computer from being overloaded by lowering the capability of the radar to detect desired targets. In conclusion, these devices have a place in the radar but not as means for fighting the ECM battle.

A log (logarithmic) receiver is a device whose video output is proportional to the logarithm of the envelope of the RF input signal over a specified range. It is useful in preventing receiver saturation in the presence of variable intensities of jamming noise, rain, clutter, and chaff. Log receivers have the ECCM advantage of permitting the radar receiver to detect target returns that are larger than jamming noise, chaff, or clutter levels. By comparison with a linear receiver of low dynamic range, moderate jamming noise levels will normally cause the display to saturate so that the target signal will not be detected. However, the disadvantage lies in the fact that low-level jamming signals will be amplified more than higher-level target signals, thereby reducing the signal-to-jamming ratio and allowing a low-level noise jammer to be more effective. Another disadvantage is that a log characteristic causes spectral spreading of the received echoes. It would not be possible to maintain clutter rejection in an MTI (moving-target indicator) or pulse doppler radar if the spectrum of clutter echoes were to spread into the spectral region in which target returns were expected.[13,15]

In a lin-log (linear-logarithmic) receiver the output signal amplitude is closely proportional to the logarithm of the envelope of the RF input signal amplitude for high input signal amplitudes, while the output signal amplitude is directly proportional to the envelope of the RF input signal amplitude for low input signal amplitudes. The lin-log receiver can minimize the disadvantage of the log receiver when the signal is higher than the jamming and when signal and jamming are both at relatively low levels so that both are in the linear region of the lin-log receiver.[15]

Hard or soft limiters can also be used to counter jamming signals. They are nonlinear memoryless devices which chop jamming signals having wide amplitudes. The Dicke-Fix receiver counters high rates of swept-frequency CW jamming and swept spot noise jammers.[15,67] In a radar receiver, the Dicke-Fix uses a wideband intermediate-frequency (IF) amplifier and a limiter ahead of the narrow-bandwidth IF amplifier. The wideband amplifier allows a rapid recovery time from the effects of the swept jammer, and the limiter chops the jamming signal. The narrowband target signal, after transit through the wideband amplifier and the limiter without remarkable degradation, is integrated by the narrowband filter matched to the signal.

9.9 SIGNAL-PROCESSING-RELATED ECCM

Digital coherent signal processing greatly alleviates the effects of clutter and chaff.[68] This is motivated by the use of coherent doppler processing techniques such as fixed or adaptive MTI. Noncoherent devices are also required because of the limited degree of clutter, chaff, and jammer suppression practically achieved by coherent devices, so that the residual interference is still a significant source of false alarm. Among the noncoherent devices it is worth mentioning the CFAR detector (Chaps. 3 and 8 of this handbook) and the pulse-width and PRF discriminator, which are effective against pulsed jammers. The pulse-width discrimination circuit measures the width of each received pulse. If the received pulse is not of approximately the same width as the transmitted pulse, it is rejected and not passed to the signal processor or display. Similar concepts apply to discrimination based on PRF. A pulse-width discrimination technique can help in rejecting chaff; in fact, echo returns from chaff corridors are much wider than the transmitted pulse.

The first type of doppler processing to consider is the MTI, which allows the detection of moving targets embedded in stationary clutter (Chap. 15 of this handbook). The basic principle involves filtering the undesired clutter return by exploiting the differential doppler frequency shift which exists where there is relative radial motion between the target scatterers and the clutter scatterers. It is possible to follow the conventional MTI with either a coherent integrator [fast Fourier transform (FFT), or transversal filter bank] or an incoherent (after detection) integrator. The MTI-coherent integrator is a key element of the MTD (moving-target detector) processor.

The characteristics of chaff are similar to those of weather clutter, except that the chaff-scattering elements are cut to respond to a broad spectrum of radar frequencies. Weather clutter and chaff differ from ground clutter in that both the mean doppler shift and the spread cannot be predicted, being determined by wind velocity and wind shear, the latter arising from the variation of wind velocity with height. To reduce these returns, the MTI filter can be modified so that one or more of its zeros can be shifted from zero doppler to the mean doppler of the weather and chaff. Unfortunately, this mean doppler value is a priori unknown and varying along range, azimuth, and elevation. This situation motivates the use of the adaptive MTI. Briefly speaking, the mean doppler value of the clutter-chaff spectrum is derived by online estimation of the cross correlation between consecutive radar echoes pertaining to the same range cell. The MTI filtering band is then centered on the estimated mean doppler value of the clutter spectrum.[29,69] However, the adaptive capability of this MTI is confined to the estimation of the mean doppler frequency value of the clutter and/or chaff. In other words, if two interferences have the same mean doppler values but different spectra, they are processed in the same manner by the MTI. In addition, this type of MTI does not properly work against more than one clutter source. To overcome these limitations resort is made to optimum detection theory.[40] The target to be detected is represented as a vector S containing the sequence of N samples from a sinusoidal tone with random initial phase and a prescribed doppler frequency. The disturbance (e.g., chaff, clutter, and noise) is modeled as a stochastic sequence of N samples having a gaussian probability density function with zero mean value and a prescribed N-dimensional covariance matrix M. The problem refers to the maximization of the detection probability P_D for a given probability of false alarm P_{FA}. It is found that the optimum processor is the cascade of an optimum filter, an envelope detector, and a comparison with threshold. The optimum filter is of the finite impulse response (FIR) type, which linearly combines the batch of N received azimuthal samples pertaining to the same range cell. The weights of the combination are formally given by the same Eq. (9.6) found for the adaptive array of antennas, but the meaning of the symbols is different. This result is not surprising, owing to the duality between the processing in the doppler domain of a signal sampled in time and the processing in the angular domain of a radar wave sampled by an array of antennas. The generalized adaptive MTI rejects any interference which is highly correlated in time, while the adaptive array of antennas rejects any directive spatial interference. In many respects, the implementation of the generalized adaptive MTI is less critical than the implementation of an adaptive array of antennas. In fact, in the former case the processing is limited to just one digital channel, in contrast with many digital channels linked to the many antennas of the adaptive array. Nevertheless, the problem of real-time estimation and inversion of the interference covariance matrix still remains. This topic has been widely considered in the technical literature,[70] and a number of viable solutions have been offered and tested.[29,71–73] A successful approximation of the optimum signal processing is the so-called moving-target detection used in surveillance radars, provided that a sufficient number of echo pulses (e.g., 16 to 20) are available in the radar beamwidth.[74]

A disadvantage of the MTI and MTD processors is the relatively large amount of pulses (up to 10 or more) which must be transmitted at a stable frequency and PRF. A responsive jammer could measure the frequency of the first transmitted pulse and then center the jammer to spot-jam the following pulses. Also, the requirement for a stable PRF precludes the use of pulse-to-pulse jitter, which is one of the most effective techniques against deception and camouflage jammers that rely on anticipating the radar transmitter pulse. MTI and MTD are also generally vulnerable to asynchronous pulse interference because a specific sequence of pulse returns is required in normal operation to achieve the desired cancellation.[75] In addition, high-performance MTI and MTD operation requires a wide-dynamic-range linear receiver, which precludes the use of the hard or soft limiting associated with a number of ECCM fixes (e.g., Dicke-Fix) against certain forms of jamming.[12]

The digital coherent implementation of the Dicke-Fix receiver concept requires the use of a coherent hard limiter which preserves the phase of the signal while keeping the amplitude at a constant value. The coherent limiter is inserted upstream of the pulse compression filter in a radar which uses phase-coded signals. In reception, the jammer and target signals are chopped in amplitude. The preservation of the target signal phase coding allows the integration of target energy by means of the pulse compression filter matched to the phase code. The Dicke-Fix processing scheme suffers from two limitations. The first is related to the detection loss experienced when the target does not compete with the jammer. The other disadvantage refers to the masking effect of a weak target signal sufficiently close in range (compared with the spatial extension of the code) to a strong target.

A minor ECCM technique to be considered is pulse compression; it is intimately related to the waveform coding discussed in Sec. 9.7. Pulse compression is a pulse radar technique in which long pulses are transmitted to increase the energy on target while still retaining the target range resolution of a short pulse transmission. It is almost always used in radar for achieving high range resolution or reducing the peak power. Pulse compression offers some ECCM advantage that is discussed hereafter.[12,76] When the pulse compression search radar is compared, from an ESM standpoint, with a conventional search radar with the same wide pulse, the enemy receiver on a jamming platform will not know (in the general case) the pulse compression reference code and will be at a disadvantage. Compared with a radar that uses an uncompressed wide pulse, the pulse compression technique increases the radar's capability against extended signal returns like chaff and clutter. In addition, noise from a common jammer does not pulse-compress. Extended clutter tends to be noiselike and will not pulse-compress, which keeps down interference displayed to operators.[15] The disadvantages of pulse compression are related to the long duration of the coded pulse, which gives more time for the ECM equipment to process the pulse. In many cases pulse compression can provide the means for easy radar jamming for the enemy ECM operator. Pulse compression is also vulnerable to cover-pulse jamming, in which the ECM pulse is returned to the radar with a high jammer-to-signal power ratio such that the normal target return is covered by the jamming pulse. The width of the ECM pulse is normally wider than the radar skin return.[8] This type of deception can be counteracted by an ECCM technique such as the cover-pulse channel, where the tracking is on the ECM transmission rather than on the skin return from the target.[15]

9.10 OPERATIONAL-DEPLOYMENT TECHNIQUES

To this point in the chapter only electronic ECCM techniques have been considered. However, radar operational philosophy and deployment tactics may also have a significant effect on the radar's resistance to ECM, ESM, and ARM. This group of techniques can be subdivided into those involving the operator, the methods of operation, the radar deployment tactics, and the friend ESM in support to ECCM.[3]

The role of the operator in the ECCM chain pertains to the more general topic of *human-factors* ECCM.[15] This is a generic ECCM technique that covers the ability of an air defense officer, a radar operator, a commanding officer, and/or any other air defense associated personnel to recognize the various kinds of ECM, to analyze the effect of the ECM, to decide what the appropriate ECCM should be, and/or to take the necessary ECCM action within the framework of the person's command structure. However, the human operator is less effective against a simultaneous attack of many enemy vehicles supported by a strong ECM force. An operator confronted with a large mix of ECM types and a large number of ECCM techniques is likely to do the wrong thing and/or react too slowly. In this situation it might be proper to resort to automatically applied ECCM techniques.

The operational methods include emission control (EMCON), the appropriate assignment of operating frequencies to various radars, the use of combined ECCMs to meet combined ECMs, the use of dummy transmitters to draw ECM to other frequencies, and so on. EMCON is a technique for the management of all electromagnetic radiations of a friendly system, force, or complex to obtain maximum advantages in the areas of intelligence data reception, detection, identification, navigation, enemy missile guidance, etc., over the enemy in a given situation. EMCON permits essential operations while minimizing the disclosure of location, identification, force level, or operational intentions to enemy intelligence receptors. It includes the authorization to radiate, the control of radiation parameters such as amplitude, frequency, phase, direction, and time, the prohibition of radiation, and the scheduling of such actions for all units and equipment of a complex.[15] The on-off scheduling of the radar's operation, to include only those time intervals when surveillance is required, can reduce the probability of the radar location being found by DF (direction finding) equipment or radar homing and warning receivers. Radar blinking (using multiple radars with coordinated on-off times) can confuse an ARM seeker and guidance or a DF receiver. Decoy antennas may also be employed to confuse DF receivers and ARMs; these decoys can also operate in conjunction with the radars in a blinking mode.

Proper site selection for ground-based radars in fixed installations can provide a degree of natural signal masking to prevent, for example, detection by ground-based ESM equipment. A high degree of mobility for tactical systems allows "radiate and run" operations which are designed to prevent the radar from being engaged by DF location techniques and associated weapons.[12] The deployment of a radar network with overlapping coverage could provide some ECCM benefits. In the netted monostatic case, the radars have different frequencies for interference reduction purposes; consequently, the ECM has to consider jamming all radars in the overlapping zone, thus reducing its efficacy. This is a kind of frequency diversity discussed in Sec. 9.7.

Finally, it is worth noting that friendly ESM can support ECCM action by warning of possible hostile activity, providing angular locations of hostile jammers and information characteristics of jammers. This information is helpful in the selection of a suitable ECCM action.

9.11 APPLICATION OF ECCM TECHNIQUES

This section shows the application of the previously described ECCM techniques to surveillance and tracking radars. The use of ECCM techniques in such other types of radars as mortar location radars, missile guidance radars, and navigation and mapping radars is considered in the literature.[12]

Surveillance Radars. The function of a surveillance radar is to search a large volume of space and locate the position of targets within the search coverage. The radar range and the azimuth-elevation coverage depend on the specific radar applications. The target reports generated by a surveillance radar are processed to form target tracks. The key features of a surveillance radar are the detection range in clear, clutter, and jamming environments, the accuracy and rate of the extracted data, and the false-alarm rate. In the ensuing discussion, the design principles, driven by the requirements forced by the threat, are mainly addressed.[12]

Detection in a clear environment is a feature of early-warning radars, which look primarily for high-altitude targets at long ranges beyond the surface horizon, where the effects of clutter can be ignored. Under these conditions, a simplified analysis states that radar performance is relatively insensitive to transmitter frequency and waveform shape (see Sec. 9.12). The maximum detection range on a target with a certain radar cross section (RCS) σ in free space, for a surveillance radar which must uniformly search a specified volume in a given time period, depends on the product of the average transmitter power ($\bar{P}$) and the effective antenna aperture (A_r). It also depends on the inverse of the system noise temperature. The situation is more complex when the target to be detected is of the stealth type; in this case the radar operating frequency, the waveform, and many other factors (e.g., type of signal processing and technology) play a key role.[70]

Waveform design and operating frequency are relevant parameters in tactical surveillance radars, which must be able to detect low-flying penetrating targets, which attempt to use terrain shielding effects to escape radar detection. In this case, the selection of waveform and frequency is made to tackle the problems of masking, multipath, chaff, and clutter.

The major threats to a surveillance radar are (1) noise jamming, (2) chaff, (3) deception jamming, (4) decoys and expendables, and (5) ARM.

The most common types of jamming are main-beam noise jamming and sidelobe noise jamming. Against this threat, good radar ECCM performance is achieved by increasing the product ($\bar{P}A_r$) of average transmitter power by the effective antenna aperture. A military radar should always have 20 dB more power-aperture product than given by standard designs, yet this is seldom allowed. The request for low sidelobe level has to be traded off with the corresponding degradation of the main-beam width; the widening of the main-beam width may make the radar more vulnerable to main-beam jamming.

The noise jammer situation is basically an energy battle between the radar and the jammer. In the main-beam noise-jamming situation, the advantage is with the jammer because the radar experiences a two-way propagation loss of its energy as contrasted with the one-way propagation loss between the radar and the jammer. With sidelobe jamming the radar designer can reduce the jammer's advantage by low-sidelobe design coupled with the use of sidelobe cancellation techniques. With main-beam noise jamming, the radar can maximize the received-target energy by transmitting more average power, dwelling longer on the target, or increasing the antenna gain. If the radar's data rate is fixed and a uniform angular search rate is dictated by mechanical or search strategy, then the only option for the radar is to increase its average transmitter power. The next option is to manage the data rate, thereby allowing a longer dwell time on the target (burnthrough mode) along specific spatial sectors where needed. The ability to vary the data rate in an optimal manner is one of the principal advantages of phased array radars.[12]

Another principle of ECCM design against main-beam noise jamming is to minimize the amount of jamming energy accepted by the radar. This is accomplished by spreading the transmitted frequency range of the radar over as wide a band as possible, thus forcing the jammer into a barrage-jamming mode. This can be obtained by resorting to frequency agility and/or frequency diversity. Some radars incorporate an automatic frequency selection (AFS) device which allows the radar frequency to be tuned to that part of the spectrum which contains the minimum jamming energy.

In accordance with the search-radar equation (see Sec. 9.12), ECCM performance appears (explicitly) to be insensitive to frequency. Increasing the radar frequency does not affect the signal-to-interference energy ratio within a radar resolution cell when the antenna aperture and the radar data rate are held constant. The increased frequency increases both the antenna gain and the number of radar resolution cells which must be searched by equivalent amounts; the net effect is that the target return power is increased by the same amount by which the target dwell time is decreased, thereby holding the target-to-jamming energy ratio constant. Nevertheless, in practice the effect of main-beam noise jamming can be reduced with high radar frequency. Higher-frequency radars tend to have narrower antenna beamwidths and larger operating frequency bandwidths (5 to 10 percent of radar center frequency) than lower-frequency radars. Thus, main-beam jammers will blank smaller sectors of high-frequency radars than of low-frequency radars. In addition, main-beam jamming of a narrow-beam radar tends to provide a strobe in the direction of the jammer, which can be used to triangulate and reveal the jammer's location. Wider radar bandwidth, with appropriate coding, forces the jammer to spread its energy over a wider band, thereby diluting the effective jamming energy.[12]

ECCM design principles for main-beam noise jamming also apply to sidelobe noise jamming, with the addition that the sidelobe response in the direction of the jammer must be minimized. Ultralow sidelobes in the order of 45 dB below the antenna's main-beam peak response are feasible by using currently available advanced technology. Sometimes the control of sidelobe noise jamming by using ultralow-sidelobe antennas is not proper; this is true because the main-beam width might be increased 2 to 3 times. In addition, most operational radars do not use ultralow (less than −40 dB) or low (−30 to −40 dB) sidelobe antennas and have antenna sidelobes in the −20 to −30 dB region with average sidelobes of 0 to 5 dB below isotropic. SLC has the potential of reducing noise jamming through the antenna sidelobes, and it is used for this purpose in operational radars.[12]

ECCM techniques against chaff are those based on coherent doppler processing (e.g., adaptive MTI, and MTD).

Another class of ECCM techniques is aimed at contrast with deceptive ECM. Deception jammers have a number of specific characteristics which can be used by radars to identify their presence. The most prominent is that false-target returns must usually follow the return from the jammer-carrying target and must all lie in the same direction within a radar PRI. If the deception jammer uses a delay which is greater than a PRI period to generate an anticipatory false-target return, then pulse-to-pulse PRI jitter identifies the false-target returns. The generation of false targets in directions different from that of the jammer-carrying aircraft requires injecting pulse-jamming signals into the radar's sidelobes. Many radars employ the SLB (see Sec. 9.6) to defeat this type of ECM.

True-target returns tend to fluctuate from scan to scan with fixed-frequency radars and from pulse to pulse with frequency-agility radars. Transponder jammers generally send the same amplitude reply to all signals they receive above a threshold and hence do not simulate actual target fluctuation responses. In addition, they usually appear wider in azimuth than real targets owing to the modulation effect of the radar scanning antenna's response on the real target. Repeater jammers can be made to simulate the actual amplitude response of real targets and hence are more effective over transponder-type jammers from an ECM viewpoint. An operating mode to be included in a radar to distinguish useful targets from transponder and repeater jammers is based on a doppler spectrum analyzer implemented in one of two embodiments: (1) an FFT device able to process several tens of pulse echoes to achieve the required frequency resolution or (2) a device built around a superresolution scheme operating on few echo pulses but each having a high signal-to-noise ratio. Additional expensive techniques against deceptive jamming are based on the measurement and analysis of the angular and polarization spectra of the echo signals.

The same ECCM considerations apply with decoy targets which have the general attributes of real targets and are very difficult to identify as false targets. A method sometimes employed is to test the scintillation characteristics of the detected targets to determine whether they follow those of real targets. Expendables which tend to be designed under stringent economic constraints often return only a steady signal to the radar. With doppler spectrum analysis it is possible to look for returns from rotating components of the target that any form of powered target must possess. Examples are jet engine or propeller modulation returns associated with aircraft targets.

Antiradiation missiles pose a serious threat to a surveillance radar. The survivability of a surveillance radar to an ARM attack relies upon waveform coding (to dilute the energy in the frequency range), the management of radiated energy in time and along the angular sectors, and the adoption of low sidelobes in transmission. These actions make it more difficult for an ARM to home on radar. When an ARM attack is detected, it may be useful to turn on spatially remote decoy transmitters to draw the ARM away from the radar site. Blinking with a network of radars achieves better results. The ARM trajectory is usually selected to attack the radar through the zenith hole region above the radar, where its detection capability is minimal. Thus, a supplemental radar which provides a high probability of detection in the zenith hole region is required. There are certain advantages in choosing a low transmitting frequency (UHF or VHF) for the supplemental radar. The radar cross section of the ARM becomes greater as the wavelength of the radar approaches the missile dimensions, causing a resonance effect. A low-frequency radar is somewhat less vulnerable to an ARM attack owing to the difficulty of implementing a low-frequency antenna with the limited aperture available in the missile.[12] However, the use of low frequency might increase radar vulnerability to ECM. In addition, low-frequency radar has poor angular resolution.

Tracking Radars.   Tracking radars provide good resolution and precise measurement of the kinematic parameters (position, velocity, and acceleration) of targets. The estimation, update with measurements, and prediction of the kinematic parameters as the time runs are the processing steps used to build up the *tracks* of targets. Tracks allow guidance and control of friendly forces, threat assessment, and enemy target engagement by weapons. Tracking can be accomplished in three ways: (1) The dedicated radar tracker continuously points its antenna at a single target by sensing errors from the true target position and correcting these errors by a servo control system. (2) The track-while-scan (TWS) system generates tracks of more than one target by using a series of scan-to-scan target measurements taken as the antenna samples the target paths. (3) The multifunctional phased array radar tracks multiple targets by multiple independent beams, formed by the same array aperture, that are allotted to different targets. This subsection is limited to the design principles, driven by the threat requirements, of the dedicated radar tracker.[12,77]

Good ECCM performance is achieved by radiating as large an average transmitter power at the highest transmitter frequency practicable, coupled with as low a sidelobe level as achievable. Increasing the transmitter frequency increases the antenna gain G_t, which, in turn, increases the received target power as G_t^2. For mainbeam noise jamming, the received jamming power increases directly as G_t, resulting in a net increase in signal-to-jamming power by a factor proportional to the antenna gain G_t. Here there is noted a basic difference between surveillance and tracking radar: the detection range of a tracking radar improves as the frequency is increased for a fixed-size antenna. The reason for this improvement is that the antenna gain is directly increased with frequency, thereby focusing more power on the target. This increased power is integrated for a time which is inversely proportional to the bandwidth of the servo control loop. For a surveillance radar, this increased power is collected for a proportionally shorter time, since the radar must search more cells in the same time because of the narrower antenna beamwidth.

With sidelobe jamming, the received jamming power is proportional to the sidelobe antenna gain, resulting in a net increase in signal-to-jamming power ratio by the factor $G_t G_{sl}^{-1}$. As with surveillance radars, sidelobe jamming can be further attenuated by the use of sidelobe cancelers in conjunction with sidelobe blankers.[12]

The use of higher transmission frequencies and the long target dwell times for tracking radars generally make them less susceptible to noise jamming than surveillance radars. In addition, many tactical tracking radars make provision for angle-tracking noise jammers for jammer self-protection. Tracking a noise jammer in angle from two spatially dispersed radars provides enough information to locate a jammer with sufficient accuracy.

A more threatening ECM against tracking radars is DECM. These measures require considerably less energy than noise jamming (a feature particularly important on tactical aircraft, where available space is limited). Nevertheless, they are very effective in capturing and deceiving the range gate (with the range-gate pull-off, or RGPO technique), the speedgate (with the velocity-gate pull-off, or VGPO technique), and the angle-tracking circuits. A primary ECCM defense against RGPO is the use of a leading-edge range tracker. The assumption is that the deception jammer needs time to react and that the leading edge of the return pulse will not be covered by the jammer. Pulse repetition interval (PRI) jitter and frequency agility both help to ensure that the jammer will not be able to anticipate the radar pulse and lead the actual skin interval. Alternatively, the tracking radar might employ a multigate range-tracking system to simultaneously track both the skin and false-target returns. This approach utilizes the fact that both the jamming signals and the target return come from the same angular direction, so that the radar's angle-tracking circuits are always locked onto the real target.[12]

The methodology of introducing VGPO into the radar's tracking circuits is analogous to the method used with RGPO. The frequency shift is initially programmed so that the repeated signal is within the passband of the doppler filter containing the target return. This is needed to capture the doppler filter containing the target, through the radar's AGC action. The repeater jammer signal is then further shifted in frequency to the maximum expected doppler frequency of the radar. The repeated signal is then switched off, forcing the victim radar to reacquire the target.[12] Coherent tracking radars can check the radial velocity derived from doppler measurements with that derived from differentiated range data. Anomalous differences provide a warning of the probable presence of a deception jammer. When RGPO and VGPO operate simultaneously, the best defense is the contemporary tracking of true and false targets in both range and doppler dimensions. The use of multimode (high, low, and medium PRF) radars can also be an effective ECCM measure helping to counter range-gate and velocity-gate stealers by switching radar modes.

Angle-gate stealing is particularly effective against conical-scanning or sequential-lobing tracking radars. It is for this reason that such trackers should not be used in military applications. The fundamental problem with these radars is that angle tracking is accomplished by demodulating the amplitude modulation imposed on the target return pulses over a complete scanning or lobing cycle. To jam this type of radar effectively, the radar's angle-tracking–error-sensing circuits must be captured with a false amplitude-modulated signal, at the scanning or lobing rate, which is significantly out of phase with that from the target return. When the conical-scan or lobing modulation is imposed on both the transmitter and the receiver beams, it is relatively simple for a jammer to synthetize the appropriate jamming signal by inverting and repeating the transmitter modulation (inverse-gain repeater).[78] This can be partially overcome by a conical-scan-on-receiver-only (COSRO) system, where the tracking radar radiates a nonscanning transmitting beam but receives with a conical-scan beam. The jammer then has no knowledge of the phase of the conically scanned receiving beam and must adopt a trial-and-error method of scanning the jamming modulation until a noticeable reaction occurs in the tracking radar beam. (This jamming technique is called *jog detection*.[8]) A sequential lobing-on-receive-only (LORO) system conceals the lobing rate from a potential jammer.[12]

Monopulse tracking is inherently insensitive to angle deceptive jamming from a single point source. This is a result of the monopulse angle-error-sensing mechanism, which forms an error proportional to the angle between the target and the antenna's boresight on each return pulse. This is accomplished by comparing signals received simultaneously in two or more antenna beams, as distinguished from techniques such as lobe switching or conical scanning, in which angle information requires multiple pulses. Effective monopulse jamming techniques generally attempt to exploit the monopulse radar's susceptibility to target glint or multipath signals.[8]

One jamming approach, known as *cross-eye*, used against monopulse radars generates artificial glint into the monopulse tracking loop.[8] Cross-eye is basically a two-source interferometer whose antennas usually are mounted on the aircraft's wingtip as far apart as possible. The signals received in each wingtip antenna are repeated in the opposite wingtip antenna, except for a 180° phase shift which is inserted in one line to direct an interferometric null toward the victim radar. In effect, this creates an apparent change of target direction as viewed from the radar. A large repeater gain is required to generate a high jammer-to-signal ratio; otherwise, the skin echo will overwhelm the jamming signals in the interferometer pattern nulls. The maximum effectiveness of the techniques implies a considerable delay (on the order of 100 ns) in the repeated signal owing to the transmission line and amplifier between the receiver and transmitter antennas. Thus, leading-edge or multigate range tracking should be an effective ECCM technique against cross-eye jamming.[8,12]

Terrain bounce is another monopulse jamming technique which is used against semiactive missile seekers and airborne tracking radars. With this technique, the jammer aircraft illuminates the earth's surface in front of and below it, so that the semiactive missile homes on the illuminated ground spot and not on the jammer aircraft. The uncertainty of the terrain scattering parameters and the possible depolarizing effects of surface reflection are some of the problems associated with this technique.[12] Monopulse radars which use parabolic reflector antennas are susceptible to jamming through cross-polarization lobes generated by the reflector surface.[8,12] This occurs because the angle-error-sensing discriminator has an inverse slope for a cross-polarized signal, which causes the angle-tracking servo to have positive feedback instead of the negative feedback required for tracking. Planar array antennas usually have a high resistance to cross-polarization jamming.

9.12   ECCM AND ECM EFFICACY

There is a need for a quantitative measurement of the efficacy of one or more ECCM electronic techniques when a radar equipped with these devices is subject to an ECM threat. One performance measure generally used for an unjammed search radar is the detection range of a certain target against a system noise background; this situation is referred to as *detection in clear environment*. When the radar is jammed, it is of interest to calculate the degradation of the detection range with respect to self-protection, standoff, and escort jammers. These calculations apply to both search and tracking radars. For tracking radars it is also worthwhile to consider the degradation of measurement accuracy and resolution. The benefits of using ECCM techniques such as frequency agility, MTI, very low sidelobe antennas, and SLC can be easily assessed at a first approximation by properly modifying the parameters involved in the radar equation. If, for instance, an SLC is adopted against an SOJ, its net effect is to reduce jamming power by the amount of jammer cancellation ratio that the SLC can offer.

The prediction of radar range is difficult because of the many factors which are hard to represent with models of the required accuracy. The factors involve the target to be detected (target returns of a statistical nature, distributed rather than point targets), the natural environment in which the target is embedded (clutter returns, unintentional interference, uncontrollable environmental refraction and absorption, noise radiation from extraterrestrial sources, and reflections from the earth's surface), the random nature of the interference, and the radar itself (system noise temperature, signal distortions, etc.). Nevertheless, radar range prediction made under average conditions provides a preliminary and useful indication of performance under ECM threat and ECCM design effectiveness that produces baseline values prior to simulation and operational tests.

The calculation of radar range based on the radar equation is covered in Chap. 2. Many papers deal with the application of the radar equation to the jamming threat.[8,79,80] Here a short review of the basic radar equations under jamming conditions for search and tracking radars is given. Also, it is worth mentioning an approximate method to compute radar detection performance in accordance with the Marcum-Swerling theory. The method approximates the Marcum-Swerling detection curves to the order of 0.5-dB accuracy; the advantage gained is in simplified detection calculations.[12]

Of course, the radar equation is a simplification in assessing ECM-ECCM interactions; a measure of ECCM effectiveness should involve the whole weapon system in which the radar operates. The measure of effectiveness should be expressed in terms of the number of attackers killed or the probability of radar survival. A tentative step in this direction can be recognized in the introduction of the ECCM improvement factor (EIF).[81] Other tentatives also are discussed in the literature.[82,83] However, the whole topic is still in its infancy since it is difficult to achieve a realistic measure of ECCM effectiveness.

Simulation is another means to assess the ECCM benefits in the radar and weapon systems. An advantage of this approach resides in the capability to artificially generate different types of threats and to look at the radar and weapon system reactions. However, the simulation of such a complex system is a difficult, time-consuming task which sometimes involves the use of ad hoc programming languages suitable for simulation.

Simulation of a complex system on a digital computer is a technique used for the analysis, design, and testing of a system whose behavior cannot be easily evaluated by means of analysis or computation. The procedure essentially consists of reproducing the algorithms of a suitable model of the examined system by means of computer programs. Proper inputs to the model, corresponding to the most relevant operational conditions for the real system, can be generated by the same computer programs. The outputs obtained are compared with some reference values (expected or theoretical) to assess system performance. When random inputs are provided, a number of statistically independent trials are performed to achieve a significant sample of the output values from which reliable statistics can be estimated.

The accuracy and detail of the model may vary from a coarse functional description of the system to a very accurate one, according to the purpose of the simulation and the required accuracy of the results. However, it is desirable to limit the complexity of the simulation tools in order to have manageable programs, giving results which are easily interpreted. The accuracy in representing each system function depends upon its relevance with respect to system performance. When a very complex system is to be simulated, it is generally preferred to resort to several programs of limited complexity in lieu of a single bulky simulation. This approach corresponds to partitioning the whole system into subsystems separately modeled in detail. From each partial simulation, a limited number of relevant features are extracted and employed to build a simplified model of the overall system.

Unfortunately, even though the theory of simulation is well established, few papers are available in the open literature dealing with the simulation of ECM and ECCM in radar and weapon systems.[3,84]

The Radar Equation in Jamming and Chaff Conditions. The range equation for a search radar in clear environment is[79,85]

$$R_{\max}^4 = \left[\frac{2\sigma T_f}{4\pi\Omega(SNR)kT_S L_T L_r L_{ar}} \right] (\bar{P} A_r) \qquad (9.10)$$

where σ is the target radar cross section, T_f is the frame time required to search a given solid angle Ω, SNR is the peak signal-to-noise ratio required for a given level of detectability (generally 90 percent), k is Boltzmann's constant, T_S is the system noise temperature, L_T is the transmit loss factor, L_r is the receiver loss factor, L_{ar} is the round-trip atmospheric loss, $\bar{P}$ is the average transmit power, and A_r is the receiving aperture. When all the quantities in the first brackets are specified, the maximum radar range R_{max} is proportional to the one-fourth power of the power-aperture product. On a first-order approximation, R_{max} is independent of transmitter frequency; in a more accurate design the quantities T_S, L_T, L_r, L_{ar}, and σ have to be considered as a function of frequency.

Consider now a barrage noise jammer at range R_j from the radar. The total noise power density at the radar input is given by

$$P_N = kT_S + \frac{P_j G_j A_r g_{sl}}{4\pi R_j^2 B_j L_r L_{aj}} \tag{9.11}$$

where $P_j G_j/B_j$ is the radiated power spectral density of the jammer over the bandwidth B_j, L_{aj} is the atmospheric loss over the path length R_j (one way), and g_{sl} is the relative sidelobe gain factor of the receiving radar antenna at the pointing angle of the jammer to the antenna main-beam pointing direction.

By assuming that the average noise power density is dominated by the barrage noise density level, Eq. (9.10) becomes

$$R_{max}^4 = \left[\frac{2\sigma T_f}{\Omega(SNR)L_T L_{ar}}\right]\left(\frac{B_j R_j^2 L_{aj}}{P_j G_j}\right)\left(\frac{\bar{P}}{g_{sl}}\right) \tag{9.12}$$

Thus for a given set of requirements in the first brackets and for a given threat in the second brackets, the range performance is determined by the average transmit-power sidelobe-level ratio. This equation, which applies to the SOJ case, is also applicable to the SSJ situation by taking the detection range equal to R_j and $g_{sl} = 1$.

For tracking radar, the parameter estimation accuracy is of the form

$$\sigma_\theta^2 = C\theta^2/SNR \tag{9.13}$$

where C is a constant, θ is the resolution element for the parameter of interest (elevation angle, azimuth, range, range rate), and σ_θ^2 is the error variance. The radar equation (9.10) in clear environment can be rewritten as

$$R_{max}^4 = \left[\frac{2\sigma T_d}{(4\pi)^2 CkT_S L_T L_r L_{ar}}\right]\left(\frac{\sigma_\theta}{\theta}\right)^2 (\bar{P}G_t A_r) \tag{9.14a}$$

where T_d is the dwell time:

$$T_d = T_f \xi/\Omega \tag{9.14b}$$

and ξ is the solid-angle antenna beamwidth:

$$\xi = 4\pi/G_t \tag{9.14c}$$

with G_t being the radar antenna gain. For a given fractional accuracy requirement, which is established by weapon system considerations such as missile launch accuracy requirements, the maximum track range is proportional to the one-fourth root of the power-aperture-gain product. The radar range in a jamming dominated noise environment is

$$R_{max}^4 = \left(\frac{2\sigma T_d}{4\pi CL_T L_{ar}}\right)\left(\frac{R_j^2 B_j L_{aj}}{P_j G_j}\right)\left(\frac{\sigma_\theta}{\theta}\right)^2\left(\frac{\bar{P}G_t}{g_{sl}}\right) \tag{9.15}$$

The above results reveal the strong dependence of radar performance on the sidelobe level in a jamming environment. As for search radars, in tracking radars the aperture A_r (present in clear environment) is replaced by g_{sl}^{-1} in presence of SOJ.

Let us now briefly consider radar performance under chaff.[85-87] The range equation for a search radar equipped with an MTI device in presence of chaff is

$$R^2 = \frac{\sigma I_f}{C_d \Theta \Phi \, r(S/C)L}\gamma \tag{9.16}$$

where C_d is the backscatter coefficient (in square meters per cubic meter) of chaff, Θ and Φ are the azimuth and elevation beamwidths, r is the range cell size, S/C is the ratio of target echo power to the power received from the chaff cell for detection, L is the overall processing losses, I_f is the improvement factor provided by the MTI, and γ is a suitable constant value. I_f depends on the dwell time T_d. The exact relationship is determined by the type of MTI and the various assumptions made about the chaff doppler spectrum. By assuming a multiple-beam receiving antenna with N parallel receiving channels, the mean dwell time available in the search mode is obtained by Eq. (9.14b):

$$T_d = (NT_f\Theta\Phi)/\Omega \tag{9.17}$$

Substitution of practical values to Eq. (9.16) shows that to obtain adequate detection performance the MTI must produce a substantial improvement. This implies a sufficiently large T_d for a number of pulses to be coherently processed, and this in turn implies several parallel receiving channels if the beams have to be reasonably narrow. If the chaff is confined to a limited sector, an alternative solution is to have an adaptive radar in which extra dwell time can be allocated to the chaff at the expense of the rest of the coverage.

The above-mentioned equations can be applied in analysis and design problems. In the first case, having the parameters of the target, jammer, and the radar, the equations allow us to evaluate the corresponding maximum detection-range values. The design case generally requires selection of the key radar parameters (i.e., transmitter power, antenna aperture, sidelobe level, transmitter frequency) in such a way as to limit detection-range degradation under jamming conditions. An additional application concerns performance comparison of different system solutions. The effects of several ECCM strategies (e.g., SLC) can also be accounted for.

REFERENCES

1. Johnston, S. L.: World War II ECCM History, suppl. to *IEEE Int. Radar Conf. Rec.*, pp. 5.2–5.7, May 6–9, 1985.

2. Hoffmann-Heiden, A. E.: Anti-Jamming Techniques at the German AAA Radars in World War II, suppl. to *IEEE Int. Radar Conf. Rec.*, pp. 5.22–5.29, May 6–9, 1985.

3. Johnston, S. L. (ed.): "Radar Electronic Counter-Countermeasures," Artech House, Norwood, Mass., 1979.

4. Special issue on electronic warfare, *IEE Proc.*, vol. 129, pt. F, pp. 113–232, June 1982.

5. Davis, W. A.: Principles of Electronic Warfare: Radar and EW, *Microwave J.*, vol. 33, pp. 52–54, 56–59, February 1980.

6. Van Brunt, L. B.: "The Glossary of Electronic Warfare," EW Engineering, Inc., Dunn Loring, Va., 1984.

7. Department of Defense, Joint Chiefs of Staff, "Dictionary of Military and Associated Terms," JCS Pub-1, September 1974.

8. Van Brunt, L. B.: "Applied ECM," vol. 1, EW Engineering, Inc., Dunn Loring, Va., 1978.

9. Wiley, R. G.: "Electronic Intelligence: The Analysis of Radar Signals," Artech House, Norwood, Mass., 1985.

10. Wiley, R. G.: "Electronic Intelligence: The Interception of Radar Signals," Artech House, Norwood, Mass., 1986.

11. Johnston, S. L.: Philosophy of ECCM Utilization, *Electron. Warfare*, vol. 7, pp. 59–61, May–June, 1975.

12. Schleher, D. C.: "Introduction to Electronic Warfare," Artech House, Norwood, Mass., 1986.

13. Maksimov, M. V., et al.: "Radar Anti-Jamming Techniques," Artech House, Norwood, Mass., 1979. (Translated from Russian, Zaschita at Radiopomekh, Soviet Radio, 1976.)

14. Clifford Bell, D.: Radar Countermeasures and Counter-Countermeasures, *Mil. Technol.*, pp. 96–111, May 1986.

15. Van Brunt, L. B.: "Applied ECM," vol. 2, EW Engineering, Inc., Dunn Loring, Va., 1982.

16. Gros, P. J., D. C. Sammons, and A. C. Cruce: ECCM Advanced Radar Test Bed (E/ARTB) Systems Definition, *IEEE Nat. Aerosp. Electron. Conf. NAECON 1986*, pp. 251–257, May 19–23, 1986.

17. Johnson, M. A., and D. C. Stoner: ECCM from the Radar Designer's View Point, *Microwave J.*, vol. 21, pp. 59–63, March 1978.

18. Patton, W. T.: Low Sidelobe Antennas for Tactical Radars, *IEEE Int. Radar Conf. Rec.*, pp. 243–254, April 28–30, 1980.

19. Harrys, F. J.: On the Use of Windows for Harmonic Analysis with the Discrete Fourier Transform, *Proc. IEEE*, vol. 66, pp. 51–83, January 1978.

20. Giuli, D.: Polarization Diversity in Radars, *Proc. IEEE*, vol. 74, pp. 245–269, February 1986.

21. Maisel, L.: Performance of Sidelobe Blanking Systems, *IEEE Trans.*, vol. AES-4, pp. 174–180, March 1968.

22. Arancibia, P. O.: A Sidelobe Blanking System Design and Demonstration, *Microwave J.*, vol. 21, pp. 69–73, March 1978; reprinted in Ref. 3, 1979.

23. Harvey, D. H., and T. L. Wood: Designs for Sidelobe Blanking Systems, *IEEE Int. Radar Conf. Rec.*, pp. 41–416, April 1980.

24. O'Sullivan, M.: A Comparison of Sidelobe Blanking Systems, *IEE Int. Conf. Radar-87*, Conf. Publ. 281, pp. 345–349, London, Oct. 19–21, 1987.

25. Howells, P. W.: Intermediate Frequency Sidelobe Canceller, U.S. Patent 3,202,990, Aug. 24, 1965.

26. Applebaum, S. P., P. W. Howells, and C. Kovarik: Multiple Intermediate Frequency Side-Lobe Canceller, U.S. Patent 4,044,359, Aug. 23, 1977.

27. Monzingo, R. A., and T. W. Miller: "Introduction to Adaptive Arrays," John Wiley & Sons, New York, 1980.

28. Gabriel, W. F.: Adaptive Arrays: An Introduction, *Proc. IEEE*, vol. 64, pp. 239–272, 1976.

29. Lewis, B. L., F. F. Kretschmer, and W. W. Shelton: "Aspects of Radar Signal Processing," Artech House, Norwood, Mass., 1986.

30. Old, J. C.: Multiple Open-Loop Interference Canceller for a Rotating Search Radar, *IEE Proc.*, vol. 131, pt. F, pp. 203–207, April 1984.

31. Farina, D. J.: Adaptive Array Performance Analysis, *IEE Int. Conf. Radar-87*, Conf. Publ. 281, pp. 264–268, London, October 1987.

32. Farina, A., and R. Giusto: Performance of Side Lobe Cancellation Techniques, *Int. Conf. Digital Signal Processing*, pp. 922–929, Florence, Sept. 2–5, 1981.

33. Reis, G., and K. Krucker: Problems of Adaptive Sidelobe Suppression, AGARD CP-197 on "New Devices, Techniques and Systems in Radar," 1976.

34. White, W. D.: Wideband Interference Cancellation in Adaptive Sidelobe Cancellers, *IEEE Trans.*, vol. AES-19, pp. 915–925, November 1983.

35. Applebaum, S. P.: Adaptive Arrays, *Syracuse University Research Corporation Rept. SPL TR* 66-1, 1966. This report is reproduced in *IEEE Trans.*, vol. AP-24, pp. 585–598, September 1976.

36. Marr, J. D.: A Selected Bibliography on Adaptive Antenna Arrays, *IEEE Trans.*, vol. AES-22, pp. 781–798, November 1986.

37. Special issue on adaptive antennas, *IEEE Trans.*, vol. AP-24, pp. 573–764, September 1976.

38. Special issue on adaptive arrays, *IEE Proc. (London)*, vol. 130, pts. F and H, pp. 1–151, February 1983.

39. Special issue on adaptive processing antenna systems, *IEEE Trans.*, vol. AP-34, pp. 273–462, March 1986.

40. Brennan, L. E., and I. S. Reed: Theory of Adaptive Radar, *IEEE Trans.*, vol. AES-9, pp. 237–252, March 1973.

41. Wardrop, B.: The Role of Digital Processing in Radar Beamforming, *GEC J. Res.*, vol. 3, no. 1, pp. 34–45, 1985.

42. Valentino, P.: Digital Beamforming: New Technology for Tomorrow's Radars, *Def. Electron.*, pp. 102–107, October 1984.

43. Steyskal, H.: Digital Beamforming Antennas: An Introduction, *Microwave J.*, pp. 107–124, January 1987.

44. Baldwin, P. J., E. Denison, and S. F. O'Connor: An Experimental Adaptive Array for Radar Applications, *Proc. Ninth European Microwave Conf., Brighton (England)*, pp. 667–671, Sept. 17–20, 1979.

45. Wardrop, B.: Experimental Linear Phased Array with Partial Adaptivity, *IEE Proc.*, vol. 130, pt. F, pp. 118–124, February 1983.

46. Chapman, D. J.: Partial Adaptivity for the Large Array, *IEEE Trans.*, vol. AP-24, pp. 685–695, September 1976.

47. Morgan, D. R.: Partial Adaptive Array Techniques, *IEEE Trans.*, vol. AP-26, pp. 823–833, November 1978.

48. Giusto, R., and P. De Vincenti: Phase-Only Optimization for the Generation of Wide Deterministic Nulls in the Radiation Pattern of Phased-Arrays, *IEEE Trans.*, vol. AP-31, pp. 814–817, September 1983.

49. Baird, C. A., and G. G. Rassweiler: Adaptive Nulling Using Digitally Controlled Phase-Shifters, *IEEE Trans.*, vol. AP-24, pp. 638–649, September 1976.

50. Oestrich, E. T., and E. Mendelovicz: Phase Only Adaptive Nulling with Discrete Values, *Proc. Ninth European Microwave Conf., Brighton (England)*, pp. 164–168, Sept. 17–20, 1979.

51. Turner, R. M.: Null Placement and Antenna Pattern Synthesis by Control of the Element Steering Phases of a Phased-Array Radar, *IEE Int. Conf. Radar-77*, Conf. Publ. 155, pp. 222–225, London, Oct. 25–28, 1977.

52. Turner, R. M., et al.: A Steering-Phase Control Architecture and Its Use for Null Steering in a Phased-Array Radar, *Proc. Int. Conf. Radar*, pp. 441–449, Paris, Dec. 4–8, 1978.

53. Haupt, R. L., M. J. O'Brien, and R. A. Shore: Using the Phase Shifters in an Experimental Array for Adaptive Nulling, *Proc. Int. Symp. Noise and Clutter Rejection in Radars and Imaging Sensors*, pp. 579–584, Tokyo, 1984.

54. Brookner, E., and J. M. Howells: Adaptive-Adaptive Array Processing, *IEE Int. Conf. Radar-87*, Conf. Publ. 281, pp. 257–263, London, Oct. 19–21, 1987.

55. Dicken, L. W.: The Use of Null Steering in Suppressing Main Beam Interference, *IEE Int. Conf. Radar-77*, Conf. Publ. 155, pp. 226–231, London, Oct. 25–28, 1977.

56. Gabriel, W. F.: Spectral Analysis and Adaptive Array Superresolution Techniques, *Proc. IEEE*, vol. 68, pp. 654–666, June 1980.

57. Nickel, U.: Angle Estimation with Adaptive Arrays and Its Relation to Superresolution, *IEE Proc.*, vol. 134, pt. H, pp. 77–82, February 1987.

58. Nickel, U.: Angular Superresolution with Phased Array Radar: A Review of Algorithms and Operational Constraints, *IEE Proc.*, vol. 134, pt. F, pp. 53–59, February 1987.

59. Walach, E.: On Superresolution Effects in Maximum-Likelihood Adaptive Arrays, *IEEE Trans.*, vol. AP-32, pp. 259–263, March 1984.

60. Ward, C. R., et al.: Application of a Systolic Array to Adaptive Beamforming, *IEE Proc.*, vol. 131, pt. F, pp. 638–645, October 1984.

61. Barton, D. K.: "Radar," vol. 6, "Frequency Agility and Diversity," Artech House, Norwood, Mass., 1977.

62. Bergkvist, B.: Jamming Frequency Agile Radars, *Def. Electron.*, vol. 12, pp. 75.78–81.83, January 1980.

63. Nitzberg, R.: Losses for Frequency Diversity, *IEEE Trans.*, vol. AES-14, pp. 474–486, May 1978.

64. Strappaveccia, S.: Spatial Jammer Suppression by Means of an Automatic Frequency Selection Device, *IEE Int. Conf. Radar-87*, Conf. Publ. 281, pp. 582–587, London, Oct. 19–21, 1987.

65. Gager, C. H.: The Impact of Waveform Bandwidth upon Tactical Radar Design, *IEE Int. Radar Conf. 82*, pp. 278–282, London, Oct. 18–20, 1982.

66. Petrocchi, G., S. Rampazzo, and G. Rodriguez: Anticlutter and ECCM Design Criteria for a Low Coverage Radar, *Proc. Int. Conf. Radar*, pp. 194–200, Paris, Dec. 4–8, 1978.

67. Hansen, V. G., and A. J. Zottl: The Detection Performance of the Siebert and Dicke-Fix CFAR Detectors, *IEEE Trans.*, vol. AES-7, pp. 706–709, July 1971.

68. Johnston, S. L.: Radar Electronic Counter-Countermeasures against Chaff, *Proc. Int. Conf. Radar*, pp. 517–522, Paris, May 1984.

69. Galati, G., and P. Lombardi: Design and Evaluation of an Adaptive MTI Filter, *IEEE Trans.*, vol. AES-14, pp. 899–905, November 1978.

70. Farina A. (ed.): "Optimised Radar Processors," Peter Peregrinus, Ltd., London, 1987.

71. Sawyers, J. H.: Adaptive Pulse-Doppler Radar Signal Processing Using the Maximum Entropy Method, *IEEE Int. Radar Conf. Rec.*, pp. 454–461, Apr. 28–30, 1980.

72. Metford, P. A. S., and S. Haykin: Experimental Analysis of an Innovation Based Detection Algorithm for Surveillance Radar, *IEE Proc.*, vol. 132, pt. F, pp. 18–26, February 1985.

73. Nitzberg, R.: Implementation of an Adaptive Processor by Modern Spectral Analysis Techniques, *IEEE Int. Conf. Commun. Rec.*, pp. 10.3.1–10.3.4, June 14, 1981.

74. O'Donnel, R., C. Muehe, M. Labitt, and L. Cartledge: Advanced Signal Processing for Airport Surveillance Radars, *IEEE Electron. Aerosp. Syst. Conv. Rec. EASCON '74*, pp. 71A–71F, Oct. 7–9, 1974.

75. Fong, E., J. A. Walker, and W. G. Bath: Moving Target Indication in the Presence of Radio Frequency Interference, *IEEE Int. Radar Conf. Rec.*, pp. 292–296, May 6–9, 1985.

76. Van Brunt, L. B.: Pulse-Compression Radar: ECM and ECCM, *Def. Electron.*, vol. 16, pp. 170–185, October 1984.

77. Leonov, A. I., and K. J. Fomichev: "Monopulse Radar," Artech House, Norwood, Mass., 1987.

78. Johnston, S. L.: Tracking Radar Electronic Counter-Countermeasures against Inverse Gain Jammers, *Int. Conf. Radar-82*, Conf. Publ. 216, pp. 444–447, London, October 1982.

79. Di Franco, J. V., and G. Kaiteris: Radar Performance Review in Clear and Jamming Environments, *IEEE Trans.*, vol. AES-17, pp. 701–710, September 1981.

80. Barton, D. K.: Radar Equations for Jamming and Clutter, *IEEE Trans.*, vol. AES-3, pp. 340–355, November 1967.

81. Johnston, S. L.: The ECCM Improvement Factor (EIF): Illustration Examples, Applications, and Considerations in Its Utilization in Radar ECCM Performance Assessment, *Int. Conf. Radar*, pp. 149–154, Nanjing (China), Nov. 4–7, 1986.

82. Clarke, J., and A. R. Subramanian: A Game Theory Approach to Radar ECCM Evaluation, *IEEE Int. Radar Conf. Rec.*, pp. 197–203, May 6–9, 1985.

83. Li Nengjing: Formulas for Measuring Radar ECCM Capability, *IEE Proc.*, vol. 131, pt. F, pp. 417–423, July 1984.

84. Cook, D. H.: ECM/ECCM Systems Simulation Program, Electronic and Aerospace Systems Record, *IEEE Conv. Rec. EASCON '68*, pp. 181–186, Sept. 9–11, 1968.

85. Skillman, W. A.: "SIGCLUT: Surface and Volumetric Clutter-to-Noise, Jammer and Target Signal-to-Noise Radar Calculation Software and User's Manual," Artech House, Norwood, Mass., 1987.

86. Radford, M. E.: Radar ECCM: A European Approach, *IEEE Trans.*, vol. AES-14, pp. 194–198, January 1978.

87. Nathanson, F. E.: "Radar Design Principles," McGraw-Hill Book Company, New York, 1969.

CHAPTER 10
PULSE COMPRESSION RADAR

Edward C. Farnett
George H. Stevens
RCA Electronic Systems Department
GE Aerospace

10.1 INTRODUCTION

Pulse compression involves the transmission of a long coded pulse and the processing of the received echo to obtain a relatively narrow pulse. The increased detection capability of a long-pulse radar system is achieved while retaining the range resolution capability of a narrow-pulse system. Several advantages are obtained. Transmission of long pulses permits a more efficient use of the average power capability of the radar. Generation of high peak power signals is avoided. The average power of the radar may be increased without increasing the pulse repetition frequency (PRF) and, hence, decreasing the radar's unambiguous range. An increased system resolving capability in doppler is also obtained as a result of the use of the long pulse. In addition, the radar is less vulnerable to interfering signals that differ from the coded transmitted signal.

A long pulse may be generated from a narrow pulse. A narrow pulse contains a large number of frequency components with a precise phase relationship between them. If the relative phases are changed by a phase-distorting filter, the frequency components combine to produce a stretched, or expanded, pulse. This expanded pulse is the pulse that is transmitted. The received echo is processed in the receiver by a compression filter. The compression filter readjusts the relative phases of the frequency components so that a narrow or compressed pulse is again produced. The pulse compression ratio is the ratio of the width of the expanded pulse to that of the compressed pulse. The pulse compression ratio is also equal to the product of the time duration and the spectral bandwidth (time-bandwidth product) of the transmitted signal.

A pulse compression radar is a practical implementation of a matched-filter system. The coded signal may be represented either as a frequency response $H(\omega)$ or as an impulse time response $h(t)$ of a coding filter. In Fig. 10.1a, the coded signal is obtained by exciting the coding filter $H(\omega)$ with a unit impulse. The received signal is fed to the matched filter, whose frequency response is the complex conjugate $H^*(\omega)$ of the coding filter. The output of the matched-filter section is the compressed pulse, which is given by the inverse Fourier transform of the product of the signal spectrum $H(\omega)$ and the matched-filter response $H^*(\omega)$:

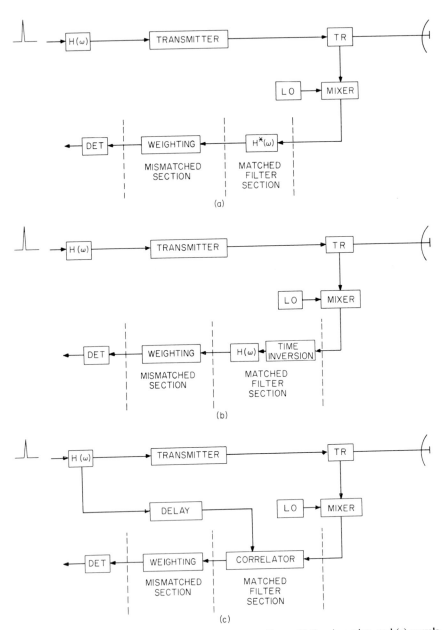

FIG. 10.1 Pulse compression radar using (*a*) conjugate filters, (*b*) time inversion, and (*c*) correlation.

$$y(t) = \frac{1}{2\pi}\int\limits_{-\infty}^{\infty} |H(\omega)|^2 e^{j\omega t} d\omega$$

The implementation of Fig. 10.1*a* uses filters which are conjugates of each other for the expansion and compression filters.

A filter is also matched to a signal if the signal is the complex conjugate of the time inverse of the filter's response to a unit impulse. This is achieved by applying the time inverse of the received signal to the compression filter, as shown in Fig. 10.1*b*. Identical filters may be used for both expansion and compression, or the same filter may be used for both expansion and compression with appropriate switching between the transmitting and receiving functions. The output of this matched filter is given by the convolution of the signal *h*(*t*) with the conjugate impulse response *h**(− *t*) of the matched filter:

$$y(t) = \int\limits_{-\infty}^{\infty} h(\tau)h^*(t - \tau)d\tau$$

The matched filter results in a correlation of the received signal with the transmitted signal. Hence, correlation processing as shown in Fig. 10.1*c* is equivalent to matched filtering. In practice, multiple delays and correlators are used to cover the total range interval of interest.

The output of the matched filter consists of the compressed pulse accompanied by responses at other ranges, called time or range sidelobes. Frequency weighting of the output signals is usually employed to reduce these sidelobes. This results in a mismatched condition and leads to a degradation of the signal-to-noise output of the matched filter. In the presence of a doppler frequency shift, a bank of matched filters is required, with each filter matched to a different frequency so as to cover the band of expected doppler frequencies.

10.2 FACTORS AFFECTING CHOICE OF PULSE COMPRESSION SYSTEM

The choice of a pulse compression system is dependent upon the type of waveform selected and the method of generation and processing. The primary factors influencing the selection of a particular waveform are usually the radar requirements of range coverage, doppler coverage, range and doppler sidelobe levels, waveform flexibility, interference rejection, and signal-to-noise ratio (*SNR*). The methods of implementation are divided into two general classes, active and passive, depending upon whether active or passive techniques are used for generation and processing.

Active generation involves generating the waveform by phase or frequency modulation of a carrier without the occurrence of an actual time expansion. An example is digital phase control of a carrier. Passive generation involves exciting a device or network with a short pulse to produce a time-expanded coded waveform. An example is an expansion network composed of a surface-acoustic-wave (SAW) delay structure. Active processing involves mixing delayed replicas of the transmitted signal with the received signal and is a correlation-processing approach. Passive processing involves the use of a compression network that is the conjugate of the expansion network and is a matched-filtering approach. Although a combination of active and passive techniques may be used in the same radar system, most systems employ the same type for generation and processing; e.g., a passive system uses both passive generation and passive processing.

The performance of common types of pulse compression systems is summarized in Table 10.1. The systems are compared on the assumption that information is extracted by processing a single waveform as opposed to multiple-pulse processing. The symbols B and T are used to denote, respectively, the bandwidth and the time duration of the transmitted waveform. Ripple loss refers to the SNR loss incurred in active systems because of the fluctuation or ripple in the SNR that occurs as a target moves from range cell to range cell. Clutter rejection performance of a single waveform is evaluated on the basis of doppler response rather than range resolution; pulse compression provides a means for realizing increased range resolution and, hence, greater clutter rejection. In applications where an insufficient doppler frequency shift occurs, range resolution is the chief means for seeing a target in clutter.

10.3 LINEAR FM

The linear-FM, or chirp, waveform is the easiest to generate. The compressed-pulse shape and SNR are fairly insensitive to doppler shifts. Because of its great popularity, more approaches for generating and processing linear FM have been developed than for any other coded waveform.[1] The major disadvantages are that (1) it has excessive range-doppler cross coupling which introduces errors unless either range or doppler is known or can be determined (i.e., a shift in doppler causes an apparent change in range and vice versa); and (2) weighting is usually required to reduce the time sidelobes of the compressed pulse to an acceptable level. Time and frequency weighting are nearly equivalent for linear FM and cause a 1 to 2 dB loss in SNR. Passive linear-FM generation and processing may be used as in Fig. 10.1a or b, where conjugate networks or a single network is employed. Active linear-FM generation and processing may be used as in Fig. 10.1c.

10.4 NONLINEAR FM

The nonlinear-FM waveform has attained little acceptance although it has several distinct advantages. The nonlinear-FM waveform requires no time or frequency weighting for range sidelobe suppression since the FM modulation of the waveform is designed to provide the desired amplitude spectrum. Matched-filter reception and low sidelobes become compatible in this design. Thus, the loss in signal-to-noise ratio associated with weighting by the usual mismatching techniques is eliminated. If a symmetrical FM modulation is used with time weighting to reduce the frequency sidelobes, the nonlinear-FM waveform will have a near-ideal ambiguity function. A symmetrical waveform typically has a frequency that increases (or decreases) with time during the first half of the pulse and decreases (or increases) during the last half of the pulse. A nonsymmetrical waveform is

TABLE 10.1 Summary of Performance of Various Pulse Compression Implementations

	Linear FM		Nonlinear FM		Phase-coded	
	Active	Passive	Active	Passive	Active	Passive
Range coverage	Limited range coverage per active correlation processor.	Provides full range coverage.	Limited range coverage per active correlation processor.	Provides full range coverage.	Limited range coverage per active correlation processor.	Provides full range coverage.
Doppler coverage	Covers any doppler up to $\pm B/10$, but a range error is introduced. SNR and time-sidelobe performance poor for larger doppler.		Multiple doppler channels required, spaced by $(1/T)$ Hz.			
Range sidelobe level	Requires weighting to reduce the range sidelobes below $(\sin x)/x$ falloff.		Good range sidelobes possible with no weighting. Sidelobes determined by waveform design.		Good range sidelobes. $N^{-1/2}$ for an N-element code.	
Waveform flexibility	Bandwidth and pulse width can be varied.	Limited to one bandwidth and pulse width per compression network.	Bandwidth and pulse width can be varied.	Limited to one bandwidth and pulse width per compression network.	Bandwidth, pulse width, and code can be varied.	
Interference rejection	Poor clutter rejection.		Fair clutter rejection.		Fair clutter rejection.	
SNR	Reduced by weighting and by ripple loss versus range.	Reduced by weighting.	Reduced by ripple loss versus range.	No SNR loss.	Reduced by ripple loss versus range.	No SNR loss.
Comments	1. Very popular with the advent of high-speed digital devices. 2. Extremely wide bandwidths achievable.	1. Widely used in past. 2. Well-developed technology.	1. Limited use. 2. Waveform generation by digital means most popular.	1. Limited use. 2. Extremely limited development.	1. Widely used. 2. Waveform very easy to generate.	1. Limited use. 2. Waveform moderately difficult to generate.

10.5

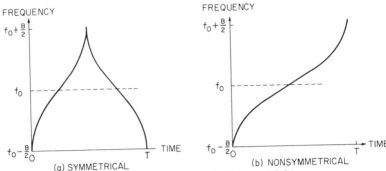

FIG. 10.2   Nonlinear-FM waveforms with 40 dB Taylor weighting.

obtained by using one-half of a symmetrical waveform (Fig. 10.2). However, the nonsymmetrical waveform retains some of the range-doppler cross coupling of the linear-FM waveform.

The disadvantages of the nonlinear-FM waveform are (1) greater system complexity, (2) limited development of nonlinear-FM generation devices, and (3) the necessity for a separate FM modulation design for each amplitude spectrum to achieve the required sidelobe level. Because of the sharpness of the ambiguity function, the nonlinear waveform is most useful in a tracking system where range and doppler are approximately known.

To achieve a 40 dB Taylor time-sidelobe pattern, the frequency-versus-time function of a nonsymmetrical transmitted pulse of bandwidth W is[2]

$$f(t) = W\left(\frac{t}{T} + \sum_{n=1}^{7} K_n \left|\sin\left|\frac{2\pi n t}{T}\right|\right.\right)$$

where $K_1 = -0.1145$
$K_2 = +0.0396$
$K_3 = -0.0202$
$K_4 = +0.0118$
$K_5 = -0.0082$
$K_6 = +0.0055$
$K_7 = -0.0040$

For a symmetrical frequency-versus-time function based on the above waveform, the first half ($t \leq T/2$) of the frequency-versus-time function will be the $f(t)$ given above, with T replaced with $T/2$. The last half ($t \geq T/2$) of the frequency-versus-time function will be the $f(t)$ above, with T replaced with $T/2$ and t replaced with $T/2 - t$.

10.5 PULSE COMPRESSION DEVICES

Major advances are continually being made in the devices used in pulse compression radars. Significant advances are evident in the digital and SAW techniques.

These two techniques allow the implementation of more exotic signal waveforms such as nonlinear FM. The digital approach has blossomed because of the manyfold increase in the computational speed and also because of the size reduction and the speed increase of the memory units. SAW technology has expanded because of the invention of the interdigital transducer,[3] which provides efficient transformation of an electrical signal into acoustic energy and vice versa. In spite of these advanced technologies, the most commonly used pulse compression waveforms are still the linear-FM and the phase-coded signals. Improved techniques have enhanced the processing of these "old standby" waveforms.

Digital Pulse Compression. Digital pulse compression techniques are routinely used for both the generation and the matched filtering of radar waveforms. The digital generator uses a predefined phase-versus-time profile to control the signal. This predefined profile may be stored in memory or be digitally generated by using appropriate constants. The matched filter may be implemented by using a digital correlator for any waveform or else a "stretch" approach for a linear-FM waveform.

Digital pulse compression has distinct features that determine its acceptability for a particular radar application. The major shortcoming of a digital approach is that its technology is restricted in bandwidths under 100 MHz. Frequency multiplication combined with stretch processing would increase this bandwidth limitation. Digital matched filtering usually requires multiple overlapped processing units for extended range coverage. The advantages of the digital approach are that long-duration waveforms present no problem, the results are extremely stable under a wide variety of operating conditions, and the same implementation could be used to handle multiple-waveform types.

Figure 10.3 shows the digital approach[4] for generating the radar waveform. This technique is normally used only for FM-type waveforms or polyphase-coded waveforms. Biphase coding can be achieved in a simpler manner, as shown in Sec. 10.6. The phase control element supplies digital samples of the in-phase component I and the quadrature component Q, which are converted to their analog equivalents. These phase samples may define the baseband components of the desired waveform, or they may define the waveform components on a low-frequency carrier. If the waveform is on a carrier, the balanced modulator is not required and the filtered components would be added directly. The sample-and-hold circuit is to remove the transients due to the nonzero transition time of the digital-to-analog (D/A) converter. The low-pass filter smooths (or interpolates) the analog signal components between waveform samples to provide the equivalent of a much higher waveform-sampling rate. The $I(t)$ component modulates a 0° carrier signal, and the $Q(t)$ component modulates a 90° phase-shifted carrier signal. The desired waveform is the sum of the 0°-modulated carrier and the 90°-modulated carrier. As mentioned earlier, when the digital phase samples include the carrier component, the I and Q components are centered on this carrier fre-

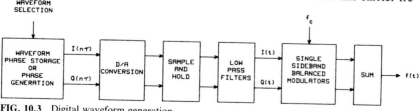

FIG. 10.3 Digital waveform generation.

quency and the low-pass filter can be replaced with a bandpass filter centered on the carrier.

Digital waveform generators are very stable devices with a well-defined distortion. As a result, the generated waveform may be frequency-multiplied to achieve a much wider waveform bandwidth. With multiplication, the distortion components are increased in magnitude by the multiplication factor, and tighter control of the distortion is required.

When a linear-FM waveform is desired, the phase samples follow a quadratic pattern and can be generated by two cascaded digital integrators. The input digital command to the first integrator defines this quadratic phase function. The digital command to the second integrator is the output of the first integrator plus the desired carrier frequency. This carrier may be defined by the initial value of the first integrator. The desired initial phase of the waveform is the initial value of the second integrator or else may be added to the second-integrator output.

Figure 10.4 illustrates two digital approaches to providing the matched filter for a pulse compression waveform. These approaches provide only limited range coverage, and overlapped processors are needed for all-range performance. Figure 10.4a shows a digital implementation of a correlation processor that will provide matched-filter performance for any radar waveform. Figure 10.4b shows a

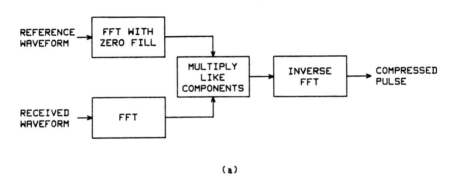

(a)

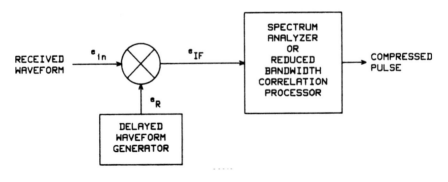

(b)

FIG. 10.4 Digital matched filter. (a) Correlation processor. (b) Stretch processor.

stretch processor for a linear-FM waveform. The delayed waveform has a bandwidth that is equal to or somewhat less than the transmitted waveform and a length that exceeds the duration of the transmitted waveform. This excess length equals the range window coverage.

The digital correlation processor[5] operates on the principle that the spectrum of the time convolution of two waveforms is equal to the product of the spectrum of these two signals. If M range samples are to be provided by one correlation processor, the number of samples in the fast Fourier transform (FFT) must equal M plus the number of samples in the reference waveform. These added M samples are filled with zeros in the reference waveform FFT. For extended range coverage, repeated correlation processor operations are required with range delays of M samples between adjacent operations. This correlation processor can be used with any waveform, and the reference waveform can be offset in doppler to achieve a matched filter at this doppler.

A stretch processor[6] can expand or contract the time scale of the compressed-pulse waveform within any defined time window. This general technique can be applied to any waveform, but it is much easier to use with a linear-FM waveform. For any waveform other than linear FM, an all-range pulse expansion approach is required in the received waveform path ahead of the mixer of Fig. 10.4b. Time contraction has not been applied to radar situations, as it requires an increased bandwidth for the compressed pulse. The stretch processing consideration will be restricted to time expansion of a linear-FM waveform.

Figure 10.4b shows the basic configuration of a time-expansion stretch processor for a linear-FM waveform. Let the received waveform be given by

$$e_{in} = A \, \text{rect}\left(t - \frac{\tau_{in}}{T_{in}}\right) \sin \left[2\pi(f_0 + f_d)(t - \tau_{in}) + \pi\alpha_{in} (t - \tau_{in})^2 + \phi\right]$$

where rect (X/T) is a unit amplitude pulse of duration T for $|X| \leq T/2$; τ_{in}, T_{in}, and α_{in} are the target time delay, the time pulse length, and the input frequency slope, respectively. The delayed waveform generator output will be

$$e_R = 2 \, \text{rect}\left(t - \frac{\tau_r}{T_R}\right) \sin \left[2\pi f_R (t - \tau_R) + \pi\alpha_R(t - \tau_R)^2 + \phi\right]$$

where the constants are the reference waveform equivalent of the received waveform constants. The intermediate-frequency (IF) input to the pulse compressor can easily be shown to be

$$e_{IF} = A \, \text{rect}\left(t - \frac{\tau_{in}}{T}\right) \text{rect}\left(t - \frac{\tau_R}{T_R}\right)$$

$$\cos \left[2\pi(f_o + f_d - f_R)(t - \tau_{in}) + \pi(\alpha_{in} - \alpha_R)\right.$$

$$\left. (t - \tau_{in})^2 + 2\pi\alpha_R(\tau_R - \tau_{in})(t - \tau_{in}) + \psi\right]$$

The resultant waveform is a reduced-frequency-slope linear-FM waveform with a target-range-dependent frequency offset riding on the doppler-shifted IF carrier frequency. Note that the frequency slope of the received waveform will be modified by the target's velocity.

For the special case where the two frequency slopes are equal, the IF waveform is a constant-frequency pulse with an offset of $f_d + \alpha_R (\tau_R - \tau_{in})$. A spectrum analysis of this IF signal will yield the relative target range $(\tau_R - \tau_{in})$ information. This frequency offset (exclusive of the target doppler) can be rewritten as $B (\Delta T/T)$, where B is the transmitted waveform bandwidth and ΔT is the time separation between the two waveforms. If the waveform bandwidth is 1 GHz and the analyzer can process only a 10-MHz bandwidth, the range coverage is restricted to under 1 percent of the transmitted waveform length. To increase the range coverage, a wider processing bandwidth is required. This stretch approach allows the full range resolution of a wide-bandwidth waveform to be realized with a restricted bandwidth processor. Note that the duration of the reference waveform should exceed the duration of the received waveform by the range processed interval, or else an S/N loss will occur.

A stretch processor with unequal-frequency-slope waveforms requires pulse compression of the residual linear FM. A linear FM with a frequency slope of $\alpha_{in} - \alpha_R$ occurs at the target's range. This linear FM will be offset in frequency by $\alpha_R \Delta_T$. With the range-doppler coupling of the linear-FM waveform, the apparent range of this target will be

$$\tau_{app} = - \alpha_R \Delta T/(\alpha_{in} - \alpha_R)$$

This results in a time-expansion factor of $\alpha_R/(\alpha_{in} - \alpha_R)$ for the compressed pulse. Again the range coverage capability of the system depends on the processing bandwidth that can be implemented.

Surface-Wave Pulse Compression. A SAW pulse compression unit consists of an input transducer and an output transducer mounted on a piezoelectric substrate. These transducers are usually implemented as interdigital devices which consist of a metal film deposited on the surface of the acoustic medium. This metal film is made of fingers (see Fig. 10.5) that dictate the frequency characteristic of the unit. The input transducer converts an electrical signal into a sound wave with over 95 percent of the energy traveling along the surface of the medium. The output transducer taps a portion of this surface sound wave and converts it back into an electric signal.

The SAW device[7-9] has unique features that dictate its usefulness for a given radar application. The major shortcomings of the SAW approach are that the waveform length is restricted to under 200 μs by the physical size of available crystals and that each waveform requires another design. The advantages of the SAW device are its compact size, the wide bandwidths that can be attained, the ability to tailor the transducers to a particular waveform, the all-range coverage of the device, and the low cost of reproducing a given design.

SAW pulse compression devices depend on the interdigital transducer finger locations or else the surface-etched grating to determine its bandpass characteristic. Figure 10.5 shows three types of filter determination approaches. Figure 10.5a has a wideband input transducer and a frequency-selective (dispersive) output transducer. When an impulse is applied to the input, the output signal is initially a low frequency that increases (based on the output transducer finger spacings) at later portions of the pulse. This results in an up-chirp waveform which would be a matched filter for a down-chirp transmitted waveform. In Figure 10.5b, both the input transducer and the output transducer are dispersive. This would result in the same impulse response as that of Fig. 10.5a. For a given crystal length and material, the waveform duration for approaches in Fig. 10.5a and b would be the same and is limited to the time that it takes an acoustic wave to

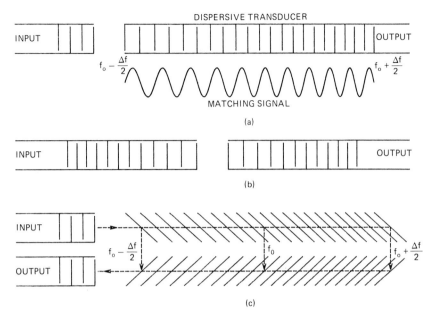

FIG. 10.5 SAW transducer types. (*a*) Dispersive output. (*b*) Both input and output dispersive. (*c*) Dispersive reflections.

traverse the crystal length. Figure 10.5*c* shows a reflection-array-compression (RAC) approach[10] which essentially doubles the achievable pulse length for the same crystal length. In an RAC, the input and output transducers have a broad bandwidth. A frequency-sensitive grating is etched on the crystal surface to reflect a portion of the surface-wave signal to the output transducer. This grating coupling does not have a significant impact on the surface-wave energy. Except for a 2:1 increase in the waveform duration, the impulse response of the RAC is the same as for approaches in Fig. 10.5*a* and *b*. Thus, these three approaches yield a similar impulse response.

Figure 10.6 shows a sketch of a SAW pulse compression device with dispersive input and output transducers. As the energy in a SAW device is concentrated in its surface wave, the SAW approach is much more efficient than bulk-wave devices, where the wave travels through the crystal. The propagation velocity of the surface wave is in the range of 1500 to 4000 m/s, depending on the crystal material, and allows a large delay in a compact device. Acoustic absorber material is required at the crystal edges to reduce the reflections and, hence, the spurious responses. Figure 10.7 shows the limit that can be expected from an SAW device and shows that bandwidths up to 1 GHz and delays up to 200 μs are achievable. The upper frequency limit depends on the accuracy that can be achieved in the fabrication of the interdigital transducer. The SAW device must provide a response that is centered on a carrier, as the lowest frequency of operation is about 20 MHz and is limited by the crystal. A matched-filter SAW pulse compression device can use variable finger lengths to achieve frequency weighting, and this internal weighting can correct for the Fresnel wiggles[11] in the FM spectrum. With this correction, 43 dB time-sidelobe levels can be achieved

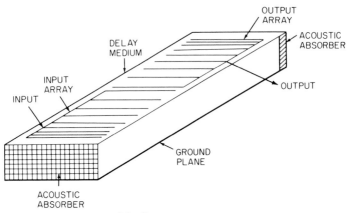

FIG. 10.6   Surface-wave delay line.

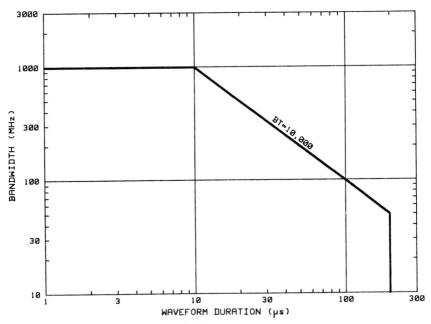

FIG. 10.7   Waveform limits for a SAW device.

for a linear-FM waveform with a BT as low as 15. The dynamic range is limited to under 80 dB by nonlinearities in the crystal material. The most common SAW materials are quartz and lithium niobate.

Other Passive Linear-FM Devices.   Table 10.2 summarizes the general characteristics of several other passive devices that are used for linear-FM

TABLE 10.2 Characteristics of Passive Linear-FM Devices

	B, MHz	T, μs	BT	f_0, MHz	Typical loss, dB	Typical spurious, dB
Aluminum strip delay line	1	500	200	5	15	−60
Steel strip delay line	20	350	500	45	70	−55
All-pass network	40	1000	300	25	25	−40
Perpendicular diffraction delay line	40	75	1000	100	30	−45
Surface-wave delay line	40	50	1000	100	70	−50
Wedge-type delay line	250	65	1000	500	50	−50
Folded-tape meander line	1000	1.5	1000	2000	25	−40
Waveguide operated near cutoff	1000	3	1000	5000	60	−25
YIG crystal	1000	10	2000	2000	70	−20

pulse compression. These passive devices fall into two broad classes: (1) bulk ultrasonic devices in which an electrical signal is converted into a sonic wave and propagates through the medium and (2) electrical devices that use the dispersive characteristic of an electrical network. The main objectives in designing and selecting a device are (1) a flat-amplitude characteristic over the bandwidth B, (2) a linear delay slope with a differential delay T across the bandwidth B, (3) minimum spurious responses and minimum distortion to achieve low sidelobes, and (4) a low insertion loss.

In a bulk ultrasonic device the input electrical signal is transformed into an acoustic wave, propagates through a medium at sonic speeds, and is then converted back to an electrical signal at the output. Since the wave propagates at sonic speeds, longer delays are achieved than with an electrical device of comparable size. A major disadvantage of ultrasonic devices is that the transducers required for coupling electrically to the acoustic medium are inefficient energy converters and hence cause high insertion losses. The most common types of bulk ultrasonic dispersive devices are (1) strip delay lines, (2) perpendicular diffraction delay lines, (3) wedge delay lines, and (4) yttrium iron garnet (YIG) crystals. The strip delay line and the YIG crystal depend on the dispersive nature of the medium for their operation. The other two types use a nondispersive medium and depend upon the diffraction characteristics of the input and output transducers for their operation; hence they are called grating-type delay lines.

A strip delay line[12–15] is made of a long, thin strip of material with transducers at opposite ends. Since the strips must be extremely thin (of the order of a few milli-inches), metal is selected because of its ruggedness. Aluminum and steel are the only metals that have found wide application. The dispersive strip delay line uses the phenomenon that if acoustic energy is propagated through a medium as a longitudinal wave, the medium exhibits a nearly linear delay-versus-frequency characteristic over an appreciable frequency range. The strip width is not critical as long as it is greater than 10 acoustic wavelengths. The thickness, however, is very critical and must be about one-half of an acoustic wavelength at a frequency equal to the center of the linear delay-versus-frequency characteristic. The length of the strip is a linear function of the differential delay required, but the bandwidth is independent of length. The differential delay corresponds to the time separation between the initial frequency and the final frequency of the waveform and is usually equal to the expanded pulse width T.

Because the thickness is very critical and cannot be controlled adequately, the stripline is placed in an oven whose temperature is adjusted to control the final operating frequency. One side of the strip is treated with an absorbing material to prevent reflections which could excite a wave that is not longitudinal and could thus introduce spurious signals.

Aluminum strip delay lines have the lowest losses, but their center frequency and bandwidth must be kept low. It is necessary to operate these lines below about 5 MHz if differential delays of over 50 μs are required. Aluminum lines have a midband delay of 7 to 10 μs/in.

Steel strip delay lines have high losses but operate at higher center frequencies, permitting wider bandwidths. Steel lines have typical losses of 70 to 80 dB and operating frequencies between 5 and 45 MHz. Steel lines have midband delays of 9 to 12 μs/in.

The perpendicular diffraction delay line[13,14,16] uses a nondispersive delay medium, such as quartz, with nonuniform input and output array transducers arranged on adjacent, perpendicular faces of the medium to produce the dispersion. The array element spacings decrease with increasing distance from the vertex of the right angle between the arrays. Thus only a positive slope of delay versus frequency can be produced. The bandwidth of the device is dictated by the array designs, and the delay is controlled by the size of the device. Errors in the array spacings produce phase errors which generate amplitude ripples and delay nonlinearities. Since many paths exist at a given frequency, these delay and amplitude errors tend to average out. Because of the averaging of the phase errors, the best delay linearity is achieved when the maximum number of grating lines is used. The center-frequency delay is limited to less than 75 μs for normal lines and 225 μs for polygonal lines because of limitations on the size of the quartz. In polygonal lines, the acoustic wave reflects off several reflecting faces in traveling from the input to the output array.

The wedge-type dispersive delay line[14] uses a wedge of quartz crystal and a frequency-selective receiver array to produce a linear delay-versus-frequency characteristic. The input transducer has a wide bandwidth, and the receiving-array elements are spaced in a quadratic manner. Reversal of the spacing of the output-array elements will change the output from an up-chirp waveform to a down-chirp waveform. The delay slope is dependent on the output-array configuration and the wedge angle. This device is fairly sensitive to grating phase errors since there is only one delay path per frequency.

YIG crystals[15,17] provide a dispersive microwave delay. YIG devices do not have a linear delay-versus-frequency characteristic, but their delay characteristic is very repeatable. The crystals require an external magnetic field, and the bandwidth and center frequency increase with the field strength. The delay of a YIG is determined by the crystal length. The maximum crystal length is limited to about 1.5 cm, corresponding to a delay of about 10 μs.

In the electrical-network class of linear-FM waveform generators, a signal is passed through an electrical delay network designed to have a linear delay-versus-frequency characteristic. The most common electrical networks that are used to generate linear-FM waveforms are (1) all-pass networks, (2) folded-tape meander lines, and (3) waveguide operated near its cutoff frequency. The all-pass network is a low-frequency device that uses lumped constant elements. The other two networks operate at very high frequencies and depend upon distributed parameters for delay.

An all-pass time-delay network[18,19] is ideally a four-terminal lattice network with constant gain at all frequencies and a phase shift that varies with the square of the frequency to yield a constant delay slope. The networks have equal input and output impedances so that several networks can be cascaded to increase the differential delay.

The folded-tape meander line[20] is the UHF or microwave analog of the low-frequency, all-pass network. A meander line consists of a thin conducting tape extending back and forth midway between two ground planes. The space between tape meanders and between the tape and the ground plane is filled with dielectric material. The center frequency of a meander loop is the frequency at which the tape length is $\lambda/4$. The time delay per meander loop is a function of the dimensions of the loop and the distance from the ground plane. To achieve a linear delay-versus-frequency curve, several loops with staggered delay characteristics are used in series. The number of meander loops required is greater than $B\Delta T$.

Other microwave dispersive networks include a waveguide operated near its cutoff frequency and stripline all-pass networks. If a section of rectangular waveguide is operated above its cutoff frequency, the time delay through the waveguide decreases with frequency. Over a limited frequency band, delay is a linear function of frequency. The usable frequency band and the delay linearity are significantly improved by employing a tapered-waveguide structure. Since stripline all-pass networks are microwave counterparts of the low-frequency all-pass networks, the synthesis of these networks is usually based on the low-frequency approach.

Voltage-Controlled Oscillator. A voltage-controlled oscillator (VCO) is a frequency generation device in which the frequency varies with an applied voltage. Ideally, the frequency is a linear function of the applied voltage, but most devices have a linearity error of over 1 percent. If a linear voltage ramp is applied to an ideal VCO, a linear-FM waveform is generated. A linear voltage ramp can be generated by applying a voltage step to an analog integrator. The integrator must be reset at the end of the generated pulse. If the VCO has a defined nonlinearity characteristic, the voltage into the integrator can be varied during the pulse so that the voltage ramp compensates for the VCO nonlinearity. Precompensation of this type is often employed. The characteristics of several common VCO devices are given in Table 10.3. The frequency-versus-voltage characteristic of the backward-wave oscillator is exponential; all the others have a linear characteristic. If coherent operation of the VCO is required, the output signal must be phased-locked to a coherent reference signal.

10.6 PHASE-CODED WAVEFORMS

Phase-coded waveforms differ from FM waveforms in that the pulse is subdivided into a number of subpulses. The subpulses are of equal duration, and each has a particular phase. The phase of each subpulse is selected in accordance with a given code sequence. The most widely used phase-coded waveform employs two phases and is called binary, or biphase, coding. The binary code consists of a sequence of either 0s and 1s or +1s and −1s. The phase of the transmitted signal alternates between 0° and 180° in accordance with the sequence of elements,

TABLE 10.3 Characteristics of VCO Devices

VCO device	Center-frequency range	Maximum frequency deviation as percent of center frequency, %	Maximum linearity as percent of deviation, %	Maximum center-frequency stability	Comments
LC oscillator	Up to 50 MHz	± 15	± 0.5	± 10 to ±100 ppm	
Crystal oscillator	100 kHz to 300 MHz	± 0.25	± 1	± 1 to ±10 ppm	
Three-terminal gallium arsenide oscillator	60 to 2500 MHz	± 2	± 2	±1%	
Voltage-tunable magnetron	100 to 10,000 MHz	± 50	± 1	±0.2%	Requires anode-voltage-control range of 750 to 3000 V.
Backward-wave oscillator	2 to 18 GHz	± 20	± 0.3*	±0.2%	Requires helix-voltage-control range of 400 to 1500 V.

*Deviation from an exponential frequency-versus-voltage curve.

0s and 1s or +1s and −1s, in the phase code, as shown in Fig. 10.8. Since the transmitted frequency is not usually a multiple of the reciprocal of the subpulse width, the coded signal is generally discontinuous at the phase-reversal points.

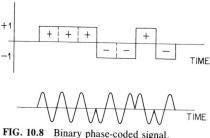

FIG. 10.8 Binary phase-coded signal.

Upon reception, the compressed pulse is obtained by either matched filtering or correlation processing. The width of the compressed pulse at the half-amplitude point is nominally equal to the subpulse width. The range resolution is hence proportional to the time duration of one element of the code. The compression ratio is equal to the number of subpulses in the waveform, i.e., the number of elements in the code.

Optimal Binary Sequences. Optimal binary sequences are binary sequences whose peak sidelobe of the aperiodic autocorrelation function (see Fig. 10.10b below) is the minimum possible for a given code length. Codes whose autocorrelation function, or zero-doppler responses, exhibit low sidelobes are desirable for pulse compression radars. Responses due to moving targets will differ from the zero-doppler response. However, with proper waveform design the doppler/bandwidth ratio can usually be minimized so that good doppler response is obtained over the target velocities of interest. The range-doppler response, or ambiguity diagram, over this velocity region then approximates the autocorrelation function.

Barker Codes. A special class of binary codes is the Barker[21] codes. The peak of the autocorrelation function is N, and the magnitude of the minimum peak sidelobe is 1, where N is the number of subpulses or length of the code. Only a small number of these codes exist. All the known Barker codes are listed in Table 10.4 and are the codes which have a minimum peak sidelobe of 1. These codes would be ideal for pulse compression radars if longer lengths were available. However, no Barker codes greater than 13 have been found to exist.[22–24] A pulse compression radar using these Barker codes would be limited to a maximum compression ratio of 13.

Allomorphic Forms. A binary code may be represented in any one of four allomorphic forms, all of which have the same correlation characteristics. These forms are the code itself, the inverted code (the code written in reverse order), the complemented code (1s changed to 0s and 0s to 1s), and the inverted complemented code. The number of codes listed in Table 10.4 is the number of codes, not including the allomorphic forms, which have the same minimum peak sidelobe. For example, the following 7-bit Barker codes all have the same

TABLE 10.4 Optimal Binary Codes

Length of code N	Magnitude of minimum peak sidelobe	No. of codes	Code (octal notation* for $N > 13$)
2	1	2	11,10
3	1	1	110
4	1	2	1101,1110
5	1	1	11101
6	2	8	110100
7	1	1	1110010
8	2	16	10110001
9	2	20	110101100
10	2	10	1110011010
11	1	1	11100010010
12	2	32	110100100011
13	1	1	1111100110101
14	2	18	36324
15	2	26	74665
16	2	20	141335
17	2	8	265014
18	2	4	467412
19	2	2	1610445
20	2	6	3731261
21	2	6	5204154
22	3	756	11273014
23	3	1021	32511437
24	3	1716	44650367
25	2	2	163402511
26	3	484	262704136
27	3	774	624213647
28	2	4	1111240347
29	3	561	3061240333
30	3	172	6162500266
31	3	502	16665201630
32	3	844	37233244307
33	3	278	55524037163
34	3	102	144771604524
35	3	222	223352204341
36	3	322	526311337707
37	3	110	1232767305704
38	3	34	2251232160063
39	3	60	4516642774561
40	3	114	14727057244044

*Each octal digit represents three binary digits:

0	000	4	100
1	001	5	101
2	010	6	110
3	011	7	111

autocorrelation peak value and the same minimum peak sidelobe magnitude: 1110010, 0100111, 0001101, 1011000. For symmetrical codes, the code and its inverse are identical.

Other Optimal Codes. Table 10.4 lists the total number of optimal binary codes for all N up through 40 and gives one of the codes for each N. As an example, the minimum peak sidelobe for a 19-bit code is 2. There are two codes having this minimum peak sidelobe, one of which is 1610445 = 1 110 001 000 100 100 101. Computer searches are generally used to find optimal codes.[25] However, the search time becomes excessively long as N increases, and recourse is often made to using other sequences which may not be optimal but possess desirable correlation characteristics.

Maximal-Length Sequences. The maximal-length sequences are of particular interest. They are the maximum-length sequences that can be obtained from linear-feedback shift-register generators. They have a structure similar to random sequences and therefore possess desirable autocorrelation functions. They are often called pseudorandom (PR) or pseudonoise (PN) sequences. A typical shift-register generator is shown in Fig. 10.9. The n stages of the shift register are initially set to all 1s or to combinations of 0s and 1s. The special case of all 0s is not allowed, since this results in an all-zero sequence. The outputs from specific individual stages of the shift register are summed by modulo-2 addition to form the input.

Modulo-2 addition depends only on the number of 1s being added. If the number of 1s is odd, the sum is 1; otherwise, the sum is 0. The shift register is pulsed at the clock-frequency, or shift-frequency, rate. The output of any stage is then a binary sequence. When the feedback connections are properly chosen, the output

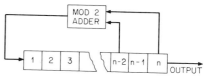

FIG. 10.9 Shift-register generator.

is a sequence of maximal length. This is the maximum length of a sequence of 1s and 0s that can be formed before the sequence is repeated.

The length of the maximal sequence is $N = 2^n - 1$, where n is the number of stages in the shift-register generator. The total number M of maximum-length sequences that may be obtained from an n-stage generator is

$$M = \frac{N}{n}\Pi\left(1 - \frac{1}{p_i}\right)$$

where p_i are the prime factors of N. The fact that a number of different sequences exist for a given value of n is important for applications where different sequences of the same length are required.

The feedback connections that provide the maximal-length sequences may be determined from a study of primitive and irreducible polynomials. An extensive list of these polynomials is given by Peterson and Weldon.[26]

Table 10.5 lists the length and number of maximal-length sequences obtainable from shift-register generators consisting of various numbers of stages. A feedback connection for generating one of the maximal-length sequences is also given for each. For a seven-stage generator, the modulo-2 sum of stages 6 and 7 is fed back to the input. For an eight-stage generator, the modulo-2 sum of stages 4, 5, 6, and 8 is fed back to the input. The length N of the maximal-length sequence is

TABLE 10.5 Maximal-Length Sequences

Number of stages, n	Length of maximal sequence, N	Number of maximal sequences, M	Feedback-stage connections
2	3	1	2,1
3	7	2	3,2
4	15	2	4,3
5	31	6	5,3
6	63	6	6,5
7	127	18	7,6
8	255	16	8,6,5,4
9	511	48	9,5
10	1,023	60	10,7
11	2,047	176	11,9
12	4,095	144	12,11,8,6
13	8,191	630	13,12,10,9
14	16,383	756	14,13,8,4
15	32,767	1,800	15,14
16	65,535	2,048	16,15,13,4
17	131,071	7,710	17,14
18	262,143	7,776	18,11
19	524,287	27,594	19,18,17,14
20	1,048,575	24,000	20,17

equal to the number of subpulses in the sequence and is also equal to the time-bandwidth product of the radar system. Large time-bandwidth products can be obtained from registers having a small number of stages. The bandwidth of the system is determined by the clock rate. Changing both the clock rate and the feedback connections permits the generation of waveforms of various pulse lengths, bandwidths, and time-bandwidth products. The number of zero crossings, i.e., transitions from 1 to 0 or from 0 to 1, in a maximal-length sequence is 2^{n-1}.

Periodic waveforms are obtained when the shift-register generator is left in continuous operation. They are sometimes used in CW radars. Aperiodic waveforms are obtained when the generator output is truncated after one complete sequence. They are often used in pulsed radars. The autocorrelation functions for these two cases differ with respect to the sidelobe structure. Figure 10.10 gives the autocorrelation functions for the periodic and aperiodic cases for a typical 15-element maximal-length code obtained from a four-stage shift-register generator. The sidelobe level for the periodic case is constant at a value of -1. The periodic autocorrelation function is repetitive with a period of $N\tau$ and

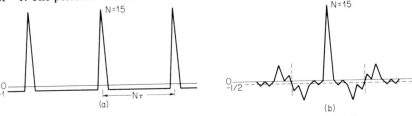

FIG. 10.10 Autocorrelation functions for (a) the periodic case and (b) the aperiodic case.

a peak value of N, where N is the number of subpulses in the sequence and τ is the time duration of each subpulse. Hence the peak-sidelobe–voltage ratio is N^{-1}.

For the aperiodic case, the average sidelobe level along the time axis is $-\frac{1}{2}$. The sidelobe structure of each half of the autocorrelation function has odd symmetry about this value. The periodic autocorrelation function may be viewed as being constructed by the superposition of successive aperiodic autocorrelation functions, each displaced in time by $N\tau$ units. The odd symmetry exhibited by the aperiodic function causes the sidelobe structure for the periodic function to have a constant value of -1. When the periodic waveform is truncated to one complete sequence, this constant sidelobe property is destroyed. For large N the peak-sidelobe–voltage ratio is approximately $N^{-1/2}$ for the aperiodic case.

Maximal-length sequences have characteristics which approach the three randomness characteristics ascribed to truly random sequences,[27] namely, that (1) the number of 1s is approximately equal to the number of 0s; (2) runs of consecutive 1s and 0s occur with about half of the runs having a length of 1, a quarter of length 2, an eighth of length 3, etc.; and (3) the autocorrelation function is thumbtack in nature, i.e., peaked at the center and approaching zero elsewhere. Maximal-length sequences are of odd length. In many radar systems it is desirable to use sequence lengths of some power of 2. A common procedure is to insert an extra 0 in a maximal-length sequence. This degrades the autocorrelation function sidelobes somewhat. An examination of sequences with an inserted 0 will yield the sequence with the best autocorrelation characteristics.

Quadratic Residue Sequences. Quadratic residue (p. 254 of Ref. 26), or Legendre, sequences offer a greater selection of code lengths than are available from maximal-length sequences. Quadratic residue sequences satisfy two of the randomness characteristics: the periodic autocorrelation function is as shown in Fig. 10.10a having a peak of N and a uniform sidelobe level of -1, and the number of 1s is approximately the same as the number of 0s.

A quadratic residue sequence of length N exists if $N = 4t - 1$, with N a prime and t any integer. The code elements a_i for $i = 0, 1, 2, \ldots, N - 1$ are 1 if i is a quadratic residue modulo N and -1 otherwise. Quadratic residues are the remainders where x^2 is reduced modulo N for $x = 1, 2, \ldots, (N - 1)/2$. As an example, the quadratic residues for $N = 11$ are 1, 3, 4, 5, 9. Hence the code elements a_i for $i = 1, 3, 4, 5, 9$ are 1, and the sequence is -1, 1, -1, 1, 1, 1, -1, -1, -1, 1, -1, or 10100011101. The periodic autocorrelation function of this sequence has a peak of 11 and a uniform sidelobe level of -1. Also, the numbers of 1s and 0s are approximately equal; the number of 1s is one more than the number of 0s.

Complementary Sequences. Complementary sequences consist of two sequences of the same length N whose aperiodic autocorrelation functions have sidelobes equal in magnitude but opposite in sign. The sum of the two autocorrelation functions has a peak of $2N$ and a sidelobe level of zero. Figure 10.11 shows the individual autocorrelation functions of the complementary sequences for length 26 and also the sum of the two autocorrelation functions. Golay[28,29] and Hollis[30] discuss general methods for forming complementary codes. In general, N must be an even number and the sum of two squares. In a practical application, the two sequences must be separated in time, frequency, or polarization, which results in decorrelation of radar returns so that complete sidelobe cancellation may not occur. Hence they have not been widely used in pulse compression radars.

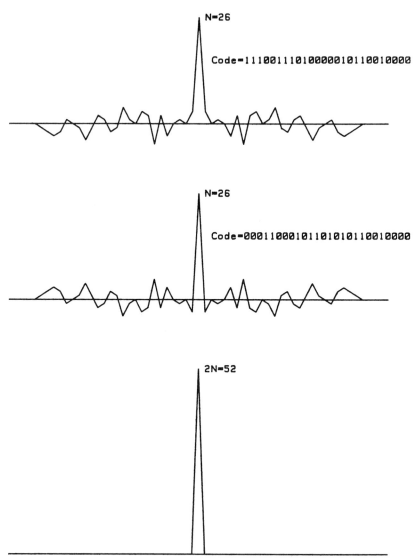

FIG. 10.11   Complementary-code aperiodic autocorrelation function.

Implementation of Biphase-Coded Systems. Digital implementation is generally used to perform the pulse compression operation in biphase-coded systems. A block diagram of a digital pulse compression system is given in Fig. 10.12. The code generator generates the binary sequence, which is sent to the RF modulator and transmitter and to the correlators. Received IF signals are passed through a bandpass filter matched to the subpulse width and are demodulated by *I* and *Q* phase detectors. The *I* and *Q* detectors compare the phase of the received IF signal with the phase of a local-oscillator (LO) signal at the same IF frequency. The LO signal is also used in the RF modulator to

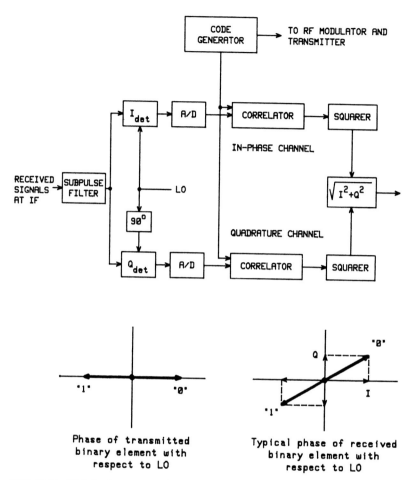

FIG. 10.12 Digital pulse compression for phase-coded signals.

generate the biphase-modulated transmitted signal. The phase of each transmitted binary element is 0° or 180° with respect to the LO signal. The phase of the received signal with respect to the LO signal, however, is shifted by an amount depending upon the target's range and velocity. Two processing channels are used, one which recovers the in-phase components of the received signal and the other which recovers the quadrature components. These signals are converted to digital form by analog-to-digital (A/D) converters, correlated with the stored binary sequence and combined, e.g., by the square root of the sum of the squares. A processing system of this type, which contains an in-phase and quadrature channel and two matched filters or correlators, is called a homodyne or zero IF system. There is an average loss in signal-to-noise ratio of 3 dB if only one channel is implemented instead of both I and Q channels. Each correlator may actually consist of several correlators, one for each quantization bit of the digitized signal.

Two methods of implementing the correlators are shown in Fig. 10.13. Figure 10.13a shows a fixed reference correlator; i.e., only one binary sequence is used. The received input sequence is continuously clocked into a shift register whose number of stages is equal to the number of elements in the sequence. The output of each stage is multiplied by weight a_i, which is either $+1$ or -1 in accordance with the reference sequence. The summation circuit provides the output correlation function or compressed pulse.

Figure 10.13b shows an implementation where the reference may be changed for each transmitted pulse. The transmitted reference sequence is fed into the reference shift register. The received input sequence is continuously clocked into the signal shift register. In each clock period the comparison counter forms the sum of the matches minus the sum of the mismatches between corresponding stages of the two shift registers, which is the output correlation function. In some systems, only the sum of the matches is counted and an offset of $-N/2$ is added to the sum.

Doppler Correction. In many applications the effect of doppler is negligible over the expanded pulse length, and no doppler correction or compensation is required. These applications transmit a short-duration phase-coded pulse, and

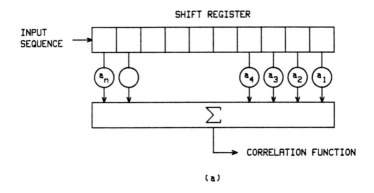

(a)

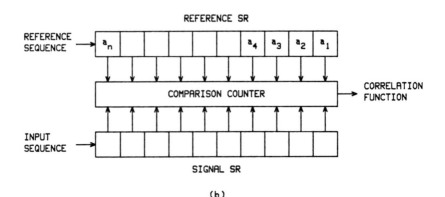

(b)

FIG. 10.13 Digital correlation with (a) fixed and (b) variable references.

the phase shift due to doppler over each expanded pulse width is negligible. Pulse compression is performed on each pulse. When the doppler shift over the expanded pulse width is not negligible, multiple doppler channels are required to minimize the loss in *SNR*. The received signals may be mixed with multiple LO signals (see Fig. 10.12), each offset in frequency by an amount corresponding to a doppler resolution element which is the reciprocal of the expanded pulse length. The processing following the subpulse filter in Fig. 10.12 is then duplicated for each doppler channel.

An alternative technique is to use a single LO signal and single-bit A/D converters in Fig. 10.12. Doppler compensation is performed on the outputs of the A/D converters prior to the correlators. This doppler compensation is in the form of inverting data bits, i.e., changing 1s to 0s and 0s to 1s, at time intervals corresponding to 180° phase shifts of the doppler frequency. As an example, the first doppler channel corresponds to a doppler frequency which results in a 360° phase shift over the pulse width. The bits are inverted after every half pulse width and remain inverted for a half pulse width. Bit inversion occurs at intervals of a quarter pulse width for the second doppler channel, an eighth pulse width for the third doppler channel, etc. Negative doppler frequency channels are handled in the same manner as for positive doppler frequency channels, but bits that were inverted in the corresponding positive channel are not inverted in the negative channel, and bits that were not inverted in the positive channel are inverted in the negative channel. No bit inversion occurs in the zero doppler channel. Each doppler channel consists of the single-bit *I* and *Q* correlators and the combiner, e.g., square root of the sum of the squares. After initial detection occurs, linear doppler processing may then be used to reduce the SNR loss. For example, the LO signal in Fig. 10.12 would then correspond to the doppler which resulted in the initial detection, and full A/D conversion is used. Some radar systems use long-duration pulses with single-bit doppler compensation to obtain initial detection and then switch to shorter-duration pulses which require no doppler compensation.

Polyphase Codes.   Waveforms consisting of more than two phases may also be used.[31,32] The phases of the subpulses alternate among multiple values rather than just the 0° and 180° of binary phase codes. The Frank polyphase codes[33] derive the sequence of phases for the subpulses by using a matrix technique. The phase sequence can be written as $\phi_n = 2\pi i(n - 1)/P^2$, where P is the number of phases, $n = 0, 1, 2, \ldots, P^2 - 1$, and $i = n$ modulo P. For a three-phase code, $P = 3$, and the sequence is 0, 0, 0, 0, $2\pi/3$, $4\pi/3$, 0, $4\pi/3$, $2\pi/3$.

The autocorrelation function for the periodic sequence has time sidelobes of zero. For the aperiodic sequence, the time sidelobes are greater than zero. As P increases, the peak-sidelobe–voltage ratio approaches $(\pi P)^{-1}$. This corresponds to approximately a 10 dB improvement over pseudorandom sequences of similar length. The ambiguity response over the range-doppler plane grossly resembles the ridgelike characteristics associated with linear-FM waveforms, as contrasted with the thumbtack characteristic of pseudorandom sequences. However, for small ratios of doppler frequency to radar bandwidth, good doppler response can be obtained for reasonable target velocities.

Lewis and Kretschmer[34] have rearranged the phase sequence to reduce the degradation that may occur by receiver band limiting prior to pulse compression. The rearranged phase sequence is

$$\phi_n = \frac{n\pi}{P}\left[1 - P + \frac{2(n - i)}{P}\right] \qquad \text{for } P \text{ odd}$$

$$\phi_n = \frac{\pi}{2P}(P - 1 - 2i)\left[P - 1 - \frac{2(n - i)}{P}\right] \quad \text{for } P \text{ even}$$

where P, n, and i are as defined above for the Frank code. For $P = 3$, the phase sequence is 0, $-2\pi/3$, $-4\pi/3$, 0, 0, 0, 0, $2\pi/3$, $4\pi/3$.

Generation and processing of polyphase waveforms use techniques similar to those for the FM waveforms of Sec. 10.5.

10.7 TIME-FREQUENCY-CODED WAVEFORMS

A time-frequency-coded waveform (Fig. 10.14) consists of a train of N pulses with each pulse transmitted at a different frequency. The ambiguity response for a periodic waveform of this type consists of a central spike plus multiple spikes or ridges displaced in time and frequency. The objective is to create a high-resolution, thumbtacklike central spike with a clear area around it; measurement is then performed on the high-resolution central spike. The range resolution or compressed pulse width is determined by the total bandwidth of all the pulses, and the doppler resolution is determined by the waveform duration T. For example, a typical waveform in this class has N contiguous pulses of width τ, whose spectra of width $1/\tau$ are placed side by side in frequency to eliminate gaps in the composite spectrum. Since the waveform bandwidth is now N/τ, the nominal compressed-pulse width is τ/N. Relationships are summarized in Table 10.6.

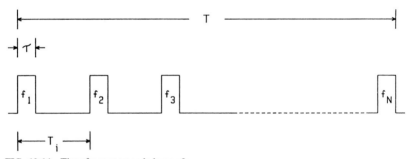

FIG. 10.14 Time-frequency-coded waveform.

TABLE 10.6 N Pulses Contiguous in Time and Frequency

Waveform duration, T	$N\tau$
Waveform bandwidth, B	N/τ
Time-bandwidth product, TB	N^2
Compressed pulse width, $1/B$	$\tau/N = T/N^2$

Shaping of the high-resolution central spike area as well as the gross structure of the ambiguity surface can be accomplished by variations of the basic waveform parameters such as amplitude weighting of the pulse train, staggering of the pulse repetition interval, and frequency or phase coding of the individual pulses.[35]

10.8 WEIGHTING AND EQUALIZATION

The process of shaping the compressed-pulse waveform by adjustment of the amplitude of the frequency spectrum is known as *frequency weighting*. The process of shaping the doppler response by control of the waveform envelope shape is called *time weighting*. The primary objective of weighting in either domain is to reduce sidelobes in the other domain. Sidelobes can severely limit resolution when the relative magnitudes of received signals are large.

Paired Echoes and Weighting. A description of the weighting process is facilitated by the application of paired-echo theory.[36–39] The first seven entries in Table 10.7 provide a step-by-step development of Fourier transforms useful in frequency and time weighting, starting with a basic transform pair. The last entry pertains to phase-distortion echoes. The spectrum $G(f)$ of the time function $g(t)$ is assumed to have negligible energy outside the frequency interval $-B/2$ to $+B/2$, where B is the bandwidth in hertz. The transform pairs of Table 10.7 are interpreted as follows:

Pair 1. Cosinusoidal amplitude variation over the passband creates symmetrical paired echoes in the time domain in addition to the main signal $g(t)$, whose shape is uniquely determined by $G(f)$. The echoes are replicas of the main signal, delayed and advanced from it by n/B s and scaled in amplitude by $a_n/2$.

Pair 2. The rectangular frequency function $W_0(f)$, that is, uniform weighting over the band, leads to a $(\sin x)/x$ time function $w_0(t)$ with high-level sidelobes, which can be objectionable in some cases. A normalized logarithmic plot of the magnitude of this time function is shown by curve A in Fig. 10.15. (All functions illustrated are symmetrical about $t = 0$.) The sidelobe adjacent to the main lobe has a magnitude of -13.2 dB with respect to the main-lobe peak. The sidelobe falloff rate is very slow.

Pair 3. Taper is applied by introducing one amplitude ripple ($n = 1$) in the frequency domain to form $W_1(f)$. By pairs 1 and 2, the time function is the superposition of the three time-displaced and weighted $(\sin x)/x$ functions.[39] Low time sidelobes are attainable in the resultant function $w_1(t)$ by the proper choice of the coefficient F_1. In particular, $F_1 = 0.426$ corresponds to Hamming weighting[40–42] and to the time function whose magnitude is represented by the solid curve B in Fig. 10.15.

Pair 4. The frequency-weighting function includes a Fourier series of $\bar{n} - 1$ cosine terms, where the selection of $\bar{n}$ is determined by the required compressed pulse width and the desired sidelobe falloff. By pairs 1 and 2, the time function includes the superposition of $2(\bar{n} - 1)$ echoes that occur in $\bar{n} - 1$ symmetrical pairs. If the coefficients F_m are selected to specify the Taylor weighting function[39,42,43] $W_{Tay}(f)$, the corresponding resultant time function $w_{Tay}(t)$ exhibits good resolution characteristics by the criterion of small main-lobe width for a specified sidelobe level. Taylor coefficients chosen for a -40 dB sidelobe level, with $\bar{n}$ selected as 6, lead to the main-sidelobe structure indicated by curve C of Fig. 10.15.

TABLE 10.7 Paired-Echo and Weighting Transforms

$g(t)=\int_{-\infty}^{\infty}G(f)\exp(j2\pi ft)df$	$G(f)=\int_{-\infty}^{\infty}g(t)\exp(-j2\pi ft)dt$
PAIRED ECHOES:	n AMPLITUDE RIPPLES:
1. $\dfrac{a_n}{2}g(t+\dfrac{n}{B})+g(t)+\dfrac{a_n}{2}g(t-\dfrac{n}{B})$	$G(f)\left[1+a_n\cos 2\pi n\dfrac{f}{B}\right]$ (REFS. 36–39)
HIGH SIDELOBES (–13.2db):	UNIFORM WEIGHTING:
2. $w_0(t)=B\dfrac{\sin \pi Bt}{\pi Bt}$	$W_0(f)=\begin{cases}1 & \|f\|<\dfrac{1}{2}B \\ 0 & \|f\|>\dfrac{1}{2}B\end{cases}$
LOW SIDELOBES: 3. $w_1(t)=$ $F_1 w_0(t+\dfrac{1}{B})+w_0(t)+F_1 w_0(t-\dfrac{1}{B})$	TAPER: $W_1(f)=$ $W_0(f)\left[1+2F_1\cos 2\pi\dfrac{f}{B}\right]$ (REFS. 39–42)

Pairs 5 to 7. The duality theorem 5 permits the interchange of time and frequency functions in each of the preceding pairs. Functions may be interchanged if the sign of the parameter t is reversed. Examples are pairs 6 and 7 obtainable from pairs 2 and 4 with the substitution of T s for B Hz. Taylor time weighting is applied in pair 7 to achieve good frequency resolution when the coefficients are selected for a specified sidelobe level.

Pair 8. Similarly to the amplitude variations of pair 1, sinusoidal phase variation over the passband creates symmetrical paired echoes in the time domain in addition to the main signal $g(t)$. The echoes are replicas of the main signal, de-

TABLE 10.7 Paired-Echo and Weighting Transforms (*Continued*)

	TAYLOR WEIGHTING:
4. $w_{Tay}(t) = \sum_{m=-\infty}^{\infty} F_m w_0(t - \frac{m}{B})$ where $F_0 = 1$, $F_m = 0$ for $\|m\| \geq \bar{n}$ and $F_m = F_{-m}$	$W_{Tay}(f) =$ $W_0(f)\left[1 + 2\sum_{m=1}^{\bar{n}-1} F_m \cos 2\pi m \frac{f}{B}\right]$ (REFS. 39,42,43)
DUALITY THEOREM:	
5. $\quad\quad\quad\quad G(-t)$	$g(f)$
6. $W_0(t) = \begin{cases} 1 & \|t\| < \frac{T}{2} \\ 0 & \|t\| > \frac{T}{2} \end{cases}$	$w_0(f) = T\dfrac{\sin \pi f T}{\pi f T}$
7. $W_{Tay}(t) =$ $W_0(t)\left[1 + 2\sum_{m=1}^{\bar{n}-1} F_m \cos 2\pi m \frac{t}{T}\right]$	$w_{Tay}(f) = \sum_{m=-\infty}^{\infty} F_m w_0(f - \frac{m}{T})$ (SEE PAIR No. 4)
PAIRED ECHOES:	n PHASE RIPPLES:
8. $\dfrac{b_n}{2} g(t + \frac{n}{B}) + g(t) - \dfrac{b_n}{2} g(t - \frac{n}{B})$ 	$G(f)e^{\,jb_n \sin 2\pi n \frac{f}{B}} \cong$ $\left[1 + jb_n \sin 2\pi n \frac{f}{B}\right]G(f)$ $\|b_n\| < 0.4$ radian (REFS. 36–39)

layed and advanced from it by n/B s, scaled in amplitude by $b_n/2$, and opposite in polarity.

Comparison of Weighting Functions. The performance achieved with various frequency-weighting functions is summarized in Table 10.8. With a change in parameter, the table also applies to time weighting (or weighting of the aperture distribution of an antenna). Pedestal height H is defined in all cases as the weighting-function amplitude at the band edge ($f = \pm B/2$) when the function has been normalized to unit amplitude at the band center ($f = 0$). The loss in the signal-to-noise ratio is based on the assumption that the transmitted amplitude spectrum is rectangular.

Item 1, uniform weighting, thus provides matched-filter operation with no

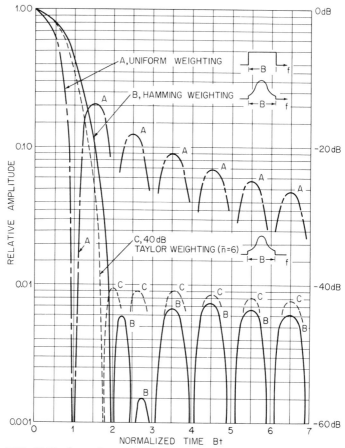

FIG. 10.15 Comparison of compressed-pulse shapes for three frequency-weighting functions.

SNR loss. Weighting in other cases is applied by a mismatch of the receiver amplitude characteristic. Item 2, Dolph-Chebyshev[44] weighting, is optimum in the sense of producing the minimum main-lobe width for a specified sidelobe level. However, the Dolph-Chebyshev function is physically unrealizable[39,41,42] for the continuous spectra under discussion. Item 3, Taylor weighting, provides a realizable approximation to Dolph-Chebyshev weighting. Time sidelobes have little decay in the region $B|t| \setminus \bar{n} - 1$ but decay at 6 dB per octave when $B|t| \hbar \bar{n}$. Item 4, cosine-squared-plus-pedestal weighting, becomes equivalent, after normalization and use of a trigonometric identity, to the weighting function $W_1(f)$ of pair 3 in Table 10.7. The normalized pedestal height H is related to the taper coefficient F_1 by $H = (1 - 2F_1)/(1 + 2F_1)$. The Hamming function produces the lowest sidelobe level attainable under category 4 of Table 10.8. Item 4b, 3:1 *taper ratio* (that is, $1/H = 3$), is analogous to a typical antenna distribution with power tapering to about 10 percent at the aperture edges.[45] Cosine-squared weighting without pedestal

TABLE 10.8 Performance for Various Frequency-Weighting Functions

Weighting function	Pedestal height H, %	SNR loss, dB	Main-lobe width, -3 dB	Peak sidelobe level, dB	Far sidelobe falloff
1 Uniform	100	0	$0.886/B$	-13.2	6 dB/octave
2 Dolph-Chebyshev			$1.2/B$	-40	No decay
3 Taylor ($\bar{n} = 8$)	11	1.14	$1.25/B$	-40	6 dB*/octave
4 Cosine-squared plus pedestal:$H + (1 - H)$ $\cos^2 (\pi f/B)$					
a. Hamming	8	1.34	$1.33/B$	-42.8	6 dB/octave
b. 3:1 "taper ratio"	33.3	0.55	$1.09/B$	-25.7	6 dB/octave
5 $\cos^2 (\pi f/B)$	0	1.76	$1.46/B$	-31.7	18 dB/octave
6 $\cos^3 (\pi f/B)$	0	2.38	$1.66/B$	-39.1	24 dB/octave
7 $\cos^4 (\pi f/B)$	0	2.88	$1.94/B$	-47	30 dB/octave
8 Triangular: $1 - 2|f|/B$	0	1.25	$1.27/B$	-26.4	12 dB/octave

*In the region $|t| ≥ 8/B$.

($H = 0$, $F_1 = 1/2$), listed as item 5, achieves a faster decay in far-off sidelobes and may simplify implementation. Entries 6 to 8 are of interest primarily because of the sidelobe falloff rate. The falloff rate can be shown to be related to the manner in which the frequency function and its derivatives behave at cutoff points, $f = \pm B/2$.[46,47]

Taylor versus Cosine-Squared-Plus-Pedestal Weighting. Figure 10.16a plots the taper coefficient F_1 and pedestal height H versus the peak sidelobe level for cosine-squared-plus-pedestal weighting. Table 10.9 lists Taylor coefficients F_m and main-lobe widths for various sidelobe levels and selections of $\bar{n}$.[48] The table illustrates that, for low design sidelobe levels, F_1 is much greater than $|F_m|$ when $m > 1$, indicating that Taylor weighting is closely approximated by the cosine-squared-plus-pedestal taper. A larger value of F_1 is required, however, in the latter case to yield the same sidelobe level. $F_1 = 0.426$ ($H = 0.08$), corresponding to Hamming weighting, produces the lowest level, -42.8 dB, attainable with this function. As indicated in Fig. 10.16a, larger values of $F_1(H < 0.08)$ increase the sidelobe level. For a given peak sidelobe level, Taylor weighting offers theoretical advantages in pulse width and SNR performance, as illustrated in Fig. 10.16b and c.

Taylor Weighting with Linear FM. The spectrum of a linear-FM pulse with a rectangular time envelope is not exactly rectangular in amplitude, nor is its phase exactly matched by the linear group delay of the compression filter.[2,39,42] The discrepancy is particularly severe for small time-bandwidth products. Therefore, the use of 40 dB Taylor weighting based on a simplified model which assumes a rectangular amplitude spectrum and a parabolic phase spectrum (that can be matched by the linear group delay) fails to achieve a -40 dB sidelobe level. Further degradation results when there is a doppler shift. Figure 10.17 plots the peak sidelobe level versus the target's doppler

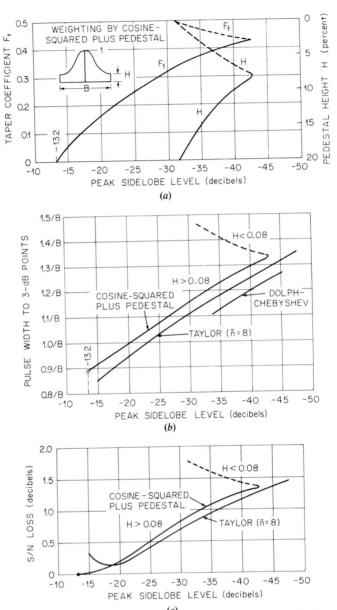

FIG. 10.16 (a) Taper coefficient and pedestal height versus peak side-lobe level. (b) Compressed-pulse width versus peak sidelobe level. (c) SNR loss versus peak sidelobe level.

TABLE 10.9 Taylor Coefficients F_m *

Design sidelobe ratio, dB	-30	-35	-40	-40	-45	-45	-50
$\bar{n}$	4	5	6	8	8	10	10
Main lobe width, -3 dB	1.13/B	1.19/B	1.25/B	1.25/B	1.31/B	1.31/B	1.36/B
F_1	0.292656	0.344350	0.389116	0.387560	0.428251	0.426796	0.462719
F_2	-0.157838(-1)	-0.151949(-1)	-0.945245(-2)	-0.954603(-2)	0.208399(-3)	-0.682067(-4)	0.126816(-1)
F_3	0.218104(-2)	0.427831(-2)	0.488172(-2)	0.470359(-2)	0.427022(-2)	0.420099(-2)	0.302744(-2)
F_4		-0.734551(-3)	-0.161019(-2)	-0.135350(-2)	-0.193234(-2)	-0.179997(-2)	-0.178566(-2)
F_5			0.347037(-3)	0.332979(-4)	0.740559(-3)	0.569438(-3)	0.884107(-3)
F_6				0.357716(-3)	-0.198534(-3)	0.380378(-5)	-0.382432(-3)
F_7				-0.290474(-3)	0.339759(-5)	-0.224597(-3)	0.121447(-3)
F_8						0.246265(-3)	-0.417574(-5)
F_9						-0.153486(-3)	-0.249574(-4)

*$F_0 = 1$; $F_{-m} = F_m$; floating decimal notation: $-0.945245(-2) = -0.00945245$.

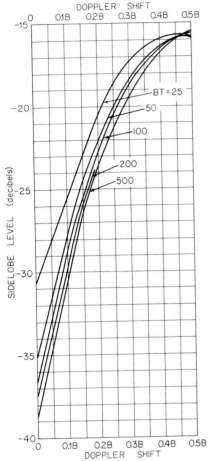

FIG. 10.17 Peak sidelobe level versus doppler shift for linear FM.

frequency. As the time-bandwidth product is increased, the model rectangular spectrum with parabolic phase is approached, and the sidelobe level in the absence of doppler shift approaches −40 dB. Unless SAW compression networks that compensate for the nonideal spectrum are employed, equalization techniques described later in this section are needed when sidelobe levels lower than about −30 dB are required. In Fig. 10.18 the loss in signal-to-noise ratio is plotted as a function of doppler shift. To obtain the total *SNR* loss with respect to that achieved with matched-filter reception, it is necessary to add 1.15 dB (see Fig. 10.16c for Taylor weighting) to the loss of Fig. 10.18.

Discrete Time Weighting[2]. A stepped-amplitude function for the reduction of doppler sidelobes is shown in Fig. 10.19. It is symmetrical about the origin, with N denoting the number of steps on each side. Table 10.10 lists stepped-amplitude functions optimized to yield minimum peak sidelobes for $N = 2, 3, 4$, and 5. $N = 1$, corresponding to the rectangular time envelope, is included for comparison. For $N = 2, 3$, and 4, the list corresponds very closely to stepped-antenna-aperture distributions[49] optimized by the criterion of maximizing the percentage energy included between the first nulls of the antenna radiation pattern.

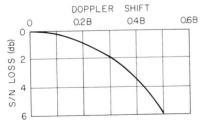

FIG. 10.18 Loss in signal-to-noise ratio versus doppler shift for linear FM.

Amplitude and Phase Distortion. The ideal compressed pulse has an amplitude spectrum that exactly matches the frequency-weighting function chosen to meet time-sidelobe requirements. Its phase spectrum is linear, corresponding to constant group delay over the band. Amplitude and phase distortion represent a departure of the actual spectrum from this ideal. All radar components are potential sources of distortion which can

TABLE 10.10 Optimum Stepped-Amplitude Time-Weighting Functions

N	Peak sidelobe, dB	Main-lobe width, −3 dB	a_1	a_2	a_3	a_4	a_5	b_1	b_2	b_3	b_4	b_5
1	− 13.2	0.886/T	1					1				
2	− 20.9	1.02/T	0.5	0.5				1	0.55			
3	− 23.7	1.08/T	0.35	0.35	0.30			1	0.625	0.350		
4	− 27.6	1.14/T	0.25	0.25	0.25	0.25		1	0.78	0.56	0.34	
5	− 29.6	1.16/T	0.300	0.225	0.235	0.170	0.070	1	0.72	0.54	0.36	0.18

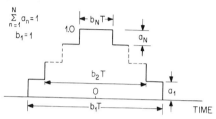

FIG. 10.19 Stepped-amplitude time weighting.

contribute to cumulative radar system distortion. Distortion degrades system performance usually by increasing the sidelobe level and, in extreme cases, by reducing the *SNR* and increasing the pulse width.

The paired-echo concept is useful in estimating distortion tolerances necessary to achieve a required time-sidelobe level.[50] Pair 1 of Table 10.7 shows

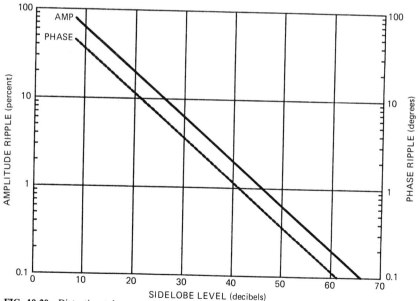

FIG. 10.20 Distortion tolerances versus time sidelobes.

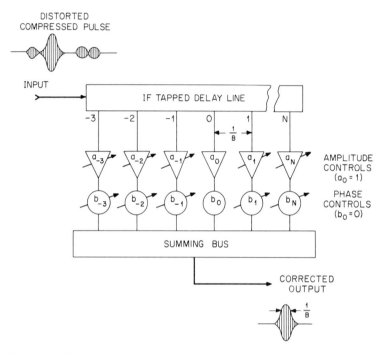

FIG. 10.21 Transversal filter.

that an amplitude ripple results in time sidelobes around the compressed pulse. Pair 8 of Table 10.7 shows that a phase ripple also results in time sidelobes around the compressed pulse. Figure 10.20 shows the amplitude and phase tolerances versus sidelobe level. To obtain time sidelobes of 40 dB below the compressed pulse, the amplitude and phase tolerances are 2 percent and 1.15°, respectively.

Equalization. The transversal filter[51,52] is widely used in the equalization of cumulative amplitude and phase distortion. One version of the transversal filter is shown in Fig. 10.21. It consists of a wideband, dispersion-free IF tapped delay line connected through each of its taps to a summing bus by amplitude and phase controls. The zeroth tap couples the distorted compressed pulse, unchanged except for delay, to the bus. The other taps make it possible to "buck out" distortion echoes of arbitrary phase and amplitude over a compensation interval equal to the total line delay. Reducing time sidelobes to an acceptable level is in effect synthesizing an equalizing filter, which makes the spectrum of the output pulse approach the ideal one described above. Because the transversal filter provides the means for reducing time sidelobes, it eliminates the need for a separate weighting filter since frequency weighting (see pairs 3 and 4 of Table 10.7) can be incorporated in the filter.

REFERENCES

1. Delay Devices for Pulse Compression Radar, *IEE (London) Conf. Publ.* 20, February 1966.

2. Murakami, T.: Optimum Waveform Study for Coherent Pulse Doppler, *RCA Final Rept.*, prepared for Office of Naval Research, Contract Nonr 4649(00)(x), Feb. 28, 1965. AD641391.

3. Morgan, D. P.: Surface Acoustic Wave Devices and Applications, *Ultrasonics*, vol. 11, pp. 121–131, 1973.

4. Eber, L. O., and H. H. Soule, Jr.: Digital Generation of Wideband LFM Waveforms, *IEEE Int. Radar Conf. Rec.*, pp. 170–175, 1975.

5. Hartt, J. K., and L. F. Sheats: Application of Pipeline FFT Technology in Radar Signal and Data Processing, *EASCON Rec.*, pp. 216–221, 1971; reprinted in David K. Barton, *Radars*, vol. 3, Books on Demand UMI, Ann Arbor, Michigan, 1975.

6. Caputi, W. J., Jr.: Stretch: A Time-Transformation Technique, *IEEE Trans.*, vol. AES-7, pp. 269–278, March 1971.

7. Gautier, H., and P. Tournois: Signal Processing Using Surface-Acoustic-Wave and Digital Components, *IEE Proc.*, vol. 127, pt. F, pp. 92–93, April 1980.

8. Slobodnik, A. J., Jr.: Surface Acoustic Waves and SAW Materials, *Proc. IEEE*, vol. 64, pp. 581–594, May 1976.

9. Bristol, T. W.: Acoustic Surface-Wave-Device Applications, *Microwave J.*, vol. 17, pp. 25–27, January 1974.

10. Williamson, R. C.: Properties and Applications of Reflective-Array Devices, *Proc. IEEE*, vol. 64, pp. 702–703, May 1976.

11. Judd, G. W.: Technique for Realizing Low Time Sidelobe Levels in Small Compression Ratio Chirp Waveforms, *Proc. IEEE Ultrasonics Symp.*, pp. 478–481, 1973.

12. Coquin, G. A., T. R. Meeker, and A. H. Meitzler: Attenuation of Longitudinal and Flexural Wave Motions in Strips, *IEEE Trans.*, vol. SU-12, pp. 65–70, June 1965.

13. May, J. E., Jr.: Ultrasonic Traveling-Wave Devices for Communications, *IEEE Spectrum*, vol. 2, pp. 73–85, October 1965.

14. Eveleth, J. H.: A Survey of Ultrasonic Delay Lines Operating Below 100 Mc/s, *Proc. IEEE*, vol. 53, pp. 1406–1428, October 1965.

15. Improved Delay Line Technique Study, *RADC Tech. Rept. RADC-TR*-65-45, May 1965. ASTIA AD617693.

16. Coquin, G. A., and R. Tsu: Theory and Performance of Perpendicular Diffraction Delay Lines, *Proc. IEEE*, vol. 53, pp. 581–591, June 1965.

17. Rodrigue, G. P.: Microwave Solid-State Delay Line, *Proc. IEEE*, vol. 53, pp. 1428–1437, October 1965.

18. O'Meara, T. R.: The Synthesis of "Band-Pass," All-Pass, Time Delay Networks with Graphical Approximation Techniques, *Hughes Aircraft Co. Res. Rept.* 114, June 1959.

19. Peebles, P. Z., Jr.: Design of a 100:1 Linear Delay Pulse Compression Filter and System, master thesis, Drexel Institute of Technology, Philadelphia, December 1962.

20. Hewett, H. S.: Highly Accurate Compression Filter Design Technique, *Stanford University, Electron. Lab. Tech. Rept.* 1965-3, November 1967. See also H. S. Hewett: A Computer Designed, 720 to 1 Microwave Compression Filter, *IEEE Trans.*, vol. MTT-15, pp. 687–694, December 1967.

21. Barker, R. H.: Group Synchronization of Binary Digital Systems, in Jackson, W. (ed.): "Communication Theory," Academic Press, New York, 1953, pp. 273–287.

22. Turyn, R., and J. Stover: On Binary Sequences, *Proc. Am. Math. Soc.*, vol. 12, pp. 394–399, June 1961.

23. Luenburger, D. G.: On Barker Codes of Even Length, *Proc. IEEE*, vol. 51, pp. 230–231, January 1963.

24. Turyn, R.: On Barker Codes of Even Length, *Proc. IEEE* (correspondence), vol. 51, p. 1256, September 1963.

25. Lindner, J.: Binary Sequences Up to Length 40 with Best Possible Autocorrelation Function, *Electron. Lett.*, vol. 11, p. 507, October 1975.

26. Peterson, W. W., and E. J. Weldon, Jr.: "Error Correcting Codes," app. C, M.I.T. Press, Cambridge, Mass., 1972.

27. Golomb, S. W.: "Shift Register Sequences," Holden-Day, Oakland, Calif., 1967, chap. 3.

28. Golay, M. J. E.: Complementary series, *IRE Trans.*, vol. IT-7, pp. 82–87, April 1961.

29. Golay, M. J. E.: Note on complementary series, *Proc. IRE*, vol. 50, p. 84, January 1962.

30. Hollis, E. E.: Another type of complementary series, *IEEE Trans.*, vol. AES-11, pp. 916–920, September 1975.

31. Golomb, S. W., and R. A. Scholtz: Generalized Barker Sequences, *IEEE Trans.*, vol. IT-11, pp. 533–537, October 1965.

32. Somaini, U., and M. H. Ackroyd: Uniform Complex Codes with Low Autocorrelation Sidelobes, *IEEE Trans.*, vol. IT-20, pp. 689–691, September 1974.

33. Frank, R. L.: Polyphase Codes with Good Nonperiodic Correlation Properties, *IEEE Trans.*, vol. IT-9, pp. 43–45, January 1963.

34. Lewis, B. L., and F. F. Kretschmer, Jr.: A New Class of Polyphase Pulse Compression Codes and Techniques, *IEEE Trans.*, vol. AES-17, pp. 364–372, May 1981. (See correction, *IEEE Trans.*, vol. AES-17, p. 726, May 1981.)

35. Rihaczek, A. W.: "Principles of High-Resolution Radar," McGraw-Hill Book Company, New York, 1969, chap. 8.

36. Wheeler, H. A.: The Interpretation of Amplitude and Phase Distortion in Terms of Paired Echoes, *Proc. IRE*, vol. 27, pp. 359–385, June 1939.

37. MacColl, L. A.: unpublished manuscript referred to by H. A. Wheeler (see Ref. 36, p. 359, footnote 1).

38. Burrow, C. R.: Discussion on Paired Echo Distortion Analysis, *Proc. IRE*, vol. 27, p. 384, June 1939.

39. Klauder, J. R., A. C. Price, S. Darlington, and W. J. Albersheim: The Theory and Design of Chirp Radars, *Bell Syst. Tech. J.*, vol. 39, pp. 745–808, July 1960.

40. Blackman, R. B., and J. W. Tukey: "The Measurement of Power Spectra," Dover Publications, New York, 1958.

41. Temes, C. L.: Sidelobe Suppression in a Range Channel Pulse-Compression Radar, *IRE Trans.*, vol. MIL-6, pp. 162–169, April 1962.

42. Cook, C. E., and M. Bernfield: "Radar Signals: An Introduction to Theory and Application," Academic Press, New York, 1967.

43. Taylor, T. T.: Design of Line-Source Antennas for Narrow Beamwidth and Low Sidelobes, *IRE Trans.*, vol. AP-3, pp. 16–28, January 1955.

44. Dolph, C. L.: A Current Distribution for Broadside Arrays Which Optimizes the Relationship between Beam Width and Sidelobe Level, *Proc. IRE*, vol. 34, pp. 335–348, June 1946.

45. Ramsay, J. F.: Fourier Transforms in Aerial Theory, *Marconi Rev.*, vol. 9, October–December 1946.

46. Cummings, R. D., M. Perry, and D. H. Preist: Calculated Spectra of Distorted Gaussian Pulses, *Microwave J.*, pp. 70–75, April 1965.

47. Mason, S. J., and H. J. Zimmerman: "Electronic Circuits, Signals and Systems," John Wiley & Sons, New York, 1960, p. 237.

48. Spellmire, R. J.: Tables of Taylor Aperture Distributions, *Hughes Aircraft Co., Syst. Dev. Lab. Tech. Mem.* 581, October 1958.

49. Nash, R. T.: Stepped Amplitude Distributions, *IEEE Trans.*, vol. AP-12, pp. 515–516, July 1964.

50. DiFranco, J. V., and W. L. Rubin: Signal Processing Distortion in Radar Systems, *IRE Trans.*, vol. MIL-6, pp. 219–225, April 1962.

51. Kallmann, H. E.: Transversal Filters, *Proc. IRE*, vol. 28, pp. 302–310, July 1940.

52. Pratt, W. R.: Transversal Equalizers for Suppressing Distortion Echoes in Radar Systems, *Proc. Symp. Pulse Compression Techniques*, pp. 119–128, *Rome Air Dev. Center, RADC-TDR*-62-580, April 1963.

CHAPTER 11
RADAR CROSS SECTION

Eugene F. Knott
The Boeing Company

11.1 INTRODUCTION

A radar detects or tracks a target, and sometimes can identify it, only because there is an echo signal. It is therefore critical in the design and operation of radars to be able to quantify or otherwise describe the echo, especially in terms of such target characteristics as size, shape, and orientation. For that purpose the target is ascribed an effective area called the *radar cross section*. It is the projected area of a metal sphere which would return the same echo signal as the target had the sphere been substituted for the target.

Unlike the echo of the sphere, however, which is independent of the viewing angle, the echoes of all but the simplest targets vary significantly with orientation. As such, one must mentally allow the size of this fictitious sphere to vary as the aspect angle of the target changes. As will be shown, the variation can be quite rapid, especially for targets many wavelengths in size.

The echo characteristics depend in strong measure on the size and nature of the target surfaces exposed to the radar beam. The variation is small for electrically small targets (targets less than a wavelength in size) because the incident wavelength is too long to resolve target details. On the other hand, the flat, singly curved and doubly curved surfaces of electrically large targets each give rise to different echo characteristics. Reentrant structures like jet engine intakes and exhausts generally have large echoes, and even the trailing edges of airfoils can be significant echo sources. The characteristics of some common targets and target features are discussed in Sec. 11.2.

The radar cross sections of simple bodies can be computed exactly by a solution of the wave equation in a coordinate system for which a constant coordinate coincides with the surface of the body. The exact solution requires that the electric and magnetic fields just inside and just outside the surface satisfy certain conditions that depend on the electromagnetic properties of the material of which the body is made.

While these solutions constitute interesting academic exercises and can, with some study, reveal the nature of the scattering mechanisms that come into play, there are no known tactical targets that fit the solutions. Thus, exact solutions of the wave equation are, at best, guidelines for gauging other (approximate) methods of computing scattered fields.

An alternative approach is the solution of the integral equations governing the distribution of induced fields on target surfaces. The most useful approach at solution is known as the *method of moments*, in which the integral equations are reduced to a system of linear homogeneous equations. The attraction of the method is that the surface profile of the body is unrestricted, allowing the computation of the scattering from truly tactical objects. Another is that ordinary methods of solution (matrix inversion and gaussian elimination, for example) may be employed to effect a solution. The method is limited by computer memory and execution time, however, to objects a few dozen wavelengths in size at best.

Alternatives to these exact solutions are several approximate methods that may be applied with reasonable accuracy to electrically large target features. They include the theories of geometrical and physical optics, the geometrical and physical theories of diffraction, and the method of equivalent currents. These approximations are discussed in Sec. 11.3. Other approximate methods not discussed here are explored in detail in some of the references listed at the end of this chapter.

The practical engineer cannot rely entirely on predictions and computations and must eventually measure the echo characteristics of some targets. This may be done by using full-scale test objects or scale models thereof. Small targets often may be measured indoors, but large targets usually must be measured on an outdoor test range. The characteristics of both kinds of test facilities are described in Sec. 11.4.

Control of the echo characteristics of some targets is of vital tactical importance. There are only two practical ways of doing so: shaping and radar absorbers. Shaping is the selection or design of surface profiles so that little or no energy is reflected back toward the radar. Because target contours are difficult to change once the target has become a production item, shaping is best implemented in the concept definition stage before production decisions have been made. Radar-absorbing materials actually soak up radar energy, also reducing the energy reflected back to the radar. However, the application of such materials can be expensive, whether gauged in terms of nonrecurring engineering costs, lifetime maintenance, or reduced mission capabilities. The two methods of echo control are discussed in Sec. 11.5.

Unless otherwise noted, the time convention used in this chapter is exp $(-i\omega t)$, with the time dependence suppressed in all equations. Readers who prefer the exp $(j\omega t)$ time convention may replace i by $-j$ wherever it appears.

11.2 THE CONCEPT OF ECHO POWER

Definition of RCS.   An object exposed to an electromagnetic wave disperses incident energy in all directions. This spatial distribution of energy is called *scattering*, and the object itself is often called a *scatterer*. The energy scattered back to the source of the wave (called *backscattering*) constitutes the *radar echo* of the object. The intensity of the echo is described explicitly by the radar cross section of the object, for which the abbreviation RCS has been generally recognized. Early papers on the subject called it the *echo area* or the *effective area*, terms still found occasionally in contemporary technical literature.

The formal definition of radar cross section is

$$\sigma = \lim_{R \to \infty} 4\pi R^2 \frac{|E_s|^2}{|E_0|^2} \qquad (11.1)$$

where E_0 is the electric-field strength of the incident wave impinging on the target and E_s is the electric-field strength of the scattered wave at the radar. The derivation of the expression assumes that a target extracts power from an incident wave and then radiates that power uniformly in all directions. Although the vast majority of targets do *not* scatter energy uniformly in all directions, the definition assumes that they do. This permits one to calculate the scattered power density on the surface of a large sphere of radius R centered on the scattering object. R is typically taken to be the range from the radar to the target.

The symbol σ has been widely accepted as the designation for the RCS of an object, although this was not so at first.[1,2] The RCS is the projected area of a metal sphere which is large compared with the wavelength and which, if substituted for the object, would scatter identically the same power back to the radar. The RCS of all but the simplest scatterers fluctuates greatly with the orientation of the object. As such, this imaginary sphere would have to expand and contract with changing target orientation to represent the amplitude fluctuations displayed by most objects.

The limiting process in Eq. (11.1) is not always an absolute requirement. In both measurement and analysis, the radar receiver and transmitter are usually taken to be in the far field of the target (discussed in Sec. 11.4), and at that distance the scattered field E_s decays inversely with the distance R. Thus, the R^2 term in the numerator of Eq. (11.1) is canceled by an identical but implicit R^2 term in the denominator. Consequently the dependence of the RCS on R, and the need to form the limit, usually disappears.

Radar cross section is therefore a comparison of the scattered power density at the receiver with the incident power density at the target. An equally valid definition of the RCS results when the electric-field strengths in Eq. (11.1) are replaced with the incident and scattered magnetic-field strengths. It is often necessary to measure or calculate the power scattered in some other direction than back to the transmitter, a *bistatic* situation. A bistatic RCS may be defined for this case as well as for backscattering, provided it is understood that the distance R is measured from the target to the receiver. *Forward scattering* is a special case of bistatic scattering in which the bistatic angle is 180°, whence the direction of interest is along the shadow zone behind the target.

The shadow itself can be regarded as the sum of two fields of nearly equal strength but 180° out of phase. One is the incident field, and the other is the scattered field. The formation of the shadow implies that the forward scattering is large, which is indeed the case. The fields behind the target are hardly ever precisely zero, however, because some energy usually reaches the shadow zone via diffraction from the sides of the target.

While there are few two-dimensional (infinite cylindrical) objects in the physical world, analyses of the scattering from two-dimensional structures are very useful. A two-dimensional object is, by definition, a cylinder formed by the pure translation of a plane curve to plus and minus infinity along an axis perpendicular to the plane of that curve. Many scattering problems become analytically tractable when there is no field variation along the cylindrical axis, such as when the infinite structure is illuminated by a plane wave propagating at right angles to the cylinder axis.

In this case, one defines a scattering *width* instead of a scattering area,

$$\sigma_{2D} = \lim_{\rho \to \infty} 2\pi\rho \frac{V_s^{\,2}}{V_0^{\,2}} \qquad (11.2)$$

where ρ is the distance from the cylindrical body to a remove receiver, measured perpendicularly to the cylindrical axis. We have appended the subscript 2D to distinguish the scattering width of Eq. (11.2), whose dimension is length, from the scattering cross section of Eq. (11.1), whose dimension is the square of length.

By virtue of the linear properties of electromagnetic fields, the solutions of two-dimensional problems may be resolved into two cases, one each for the electric field or the magnetic field parallel to the cylindrical axis. The ratio $|V_s|/|V_0|$ thus represents either the incident and scattered electric fields or the incident and scattered magnetic fields, depending on the case at hand. These two cases are often called E and H polarizations, respectively. They are also known as TM and TE polarizations.

Practical three-dimensional problems often involve truncated segments of two-dimensional structures, such as shown in Fig. 11.1. In the practical world, those segments may be viewed at angles other than incidence perpendicular to the cylindrical axis, as implied in the solution of two-dimensional problems. The three-dimensional RCS of a truncated two-dimensional structure may be found from the approximate relationship

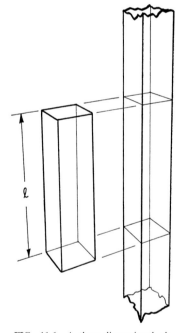

$$\sigma = \frac{2\ell^2 \sigma_{2D}}{\lambda} \left| \frac{\sin (k\ell \sin \tau)}{k\ell \sin \tau} \right|^2 \qquad (11.3)$$

where ℓ is the length of the truncated structure, σ_{2D} is its two-dimensional scattering width (obtained for the infinite structure), and τ is the tilt angle of the segment measured from broadside incidence. This approximation assumes that the amplitudes of the fields induced on the three-dimensional body are identically those induced on the corresponding two-dimensional structure and that the tilt angle influences only the phase of the surface fields induced on the body. The expression should not be used for large tilt angles, for which the amplitudes obtained from the two-dimensional solution no longer apply to the three-dimensional problem.

Examples of RCS Characteristics

FIG. 11.1 A three-dimensional object whose profile does not vary along its length, such as the truncated rectangular cylinder on the left, is a finite chunk of an infinite (two-dimensional) structure having the same profile, such as the one on the right. Equation (11.3) relates the RCS of the two structures.

Simple Objects. Because of its pure radial symmetry, the perfectly conducting sphere is the simplest of all three-dimensional scatterers. Despite the simplicity of its geometrical surface, however, and the invariance of its echo with orientation, the RCS of the sphere varies considerably with electrical size.

The exact solution for the scattering by a conducting sphere is known as the Mie series,[3] illustrated in Fig. 11.2.

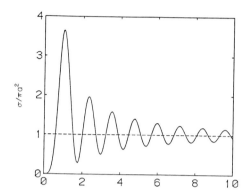

FIG. 11.2 RCS of a perfectly conducting sphere as a function of its electrical size ka.

The parameter $ka = 2\pi a/\lambda$ is the circumference of the sphere expressed in wavelengths, and the RCS is shown normalized with respect to the projected area of the sphere. The RCS rises quickly from a value of zero to a peak near $ka = 1$ and then executes a series of decaying undulations as the sphere becomes electrically larger. The undulations are due to two distinct contributions to the echo, one a *specular reflection* from the front of the sphere and the other a *creeping wave* that skirts the shadowed side. The two go in and out of phase because the difference in their electrical path lengths increases continuously with increasing ka. The undulations become weaker with increasing ka because the creeping wave loses more energy the longer the electrical path traveled around the shadowed side.

The log-log plot of Fig. 11.3 reveals the rapid rise in the RCS in the region $0 < ka < 1$, which is known as the *Rayleigh region*. Here the normalized RCS increases with the fourth power of ka, a feature shared by other electrically small or thin structures. The central region characterized by the interference between the specular and creeping-wave contributions is known as the *resonance region*. There is no clear upper boundary for this part of the curve, but a value near $ka = 10$ is generally accepted. The region $ka > 10$ is dominated by the specular return from the front of the sphere and is called the *optics region*. For spheres of these sizes the geometric optics approximation πa^2 is usually an adequate representation of the magnitude of the RCS.

The echoes of all scattering objects, and not just the perfectly conducting sphere, can be grouped according to the electrical-size characteristics of the object. The dimensions of a Rayleigh scatterer are much less than a wavelength, and the RCS is proportional to the square of the volume of the body. Resonant scatterers are generally of the order of one-half to 10 wavelengths in size, for which neither Rayleigh nor optics approximations may be very accurate. In the optics region several approximations are available for making estimates or predictions (see Sec. 11.3).

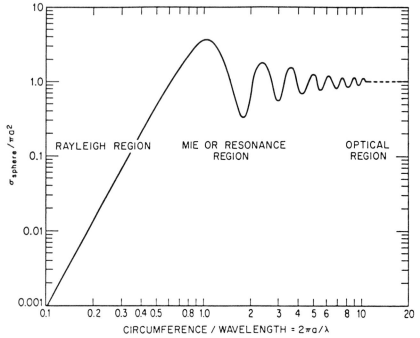

FIG. 11.3 Log-log version of the data displayed in Fig. 11.2.

The echo characteristics of permeable (dielectric) bodies can be more complicated than those of perfect conductors because energy may enter the body and suffer several internal bounces before emerging. An example is the dielectric sphere whose RCS is plotted in Fig. 11.4. Because the dielectric material is slightly lossy, as indicated by the nonzero imaginary component of the index of refraction, the RCS of the sphere decays gradually with increasing electrical size. The RCS of small dielectric bodies does not exhibit this complexity, on the other hand, because the sources of reflection are too close to each other to be resolvable by the incident wave. An example is the two-dimensional Rayleigh region RCS of a thin dielectric cylinder, plotted in Fig. 11.5. The thin dielectric cylinder has been used to model the target support lines sometimes employed in RCS measurements.[5] Note that the H-polarized echo is barely 6 dB less than that for E polarization for this particular dielectric constant.

The thin wire (a metal dipole) can have a complicated pattern, as shown in Fig. 11.6. The RCS of the wire varies with the wire length, the angle subtended by the wire and the line of sight, and on that component of the incident electric field in the plane containing the wire and the line of sight. The wire diameter has only a minor influence if it is much smaller than the wavelength. In addition to the prominent broadside lobe at the center of the pattern, there are traveling-wave lobes near the left and right sides. The traveling-wave lobes tend to disappear as the dipole becomes shorter and are closely related to those excited on traveling-wave antennas.

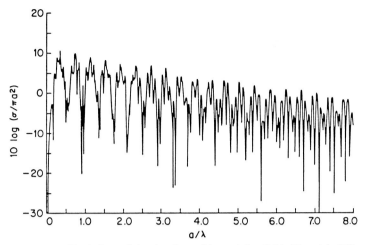

FIG. 11.4 RCS of a lossy dielectric sphere with $n = 2.5 + i0.01$. (*Copyright 1968, IEEE.*[4])

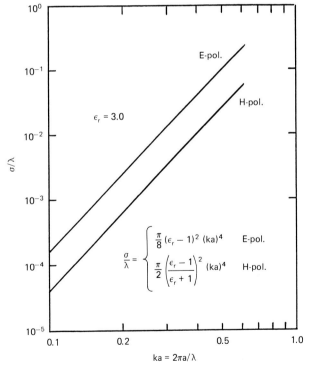

FIG. 11.5 RCS of a slender dielectric cylinder with $\epsilon_r = 3.0$.

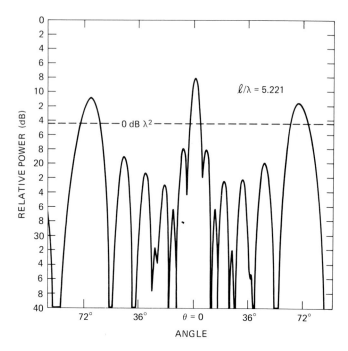

FIG. 11.6 Measured RCS pattern of a dipole 5.221λ long. (*Courtesy of University of Michigan Radiation Laboratory.*[6])

Figure 11.7 shows the broadside resonances of a wire dipole as a function of dipole length. The first resonance occurs when the dipole is just under a half wavelength long, and its magnitude is very nearly λ^2. Other resonances occur near odd multiples of a quarter wavelength, with plateaus of nearly constant re-

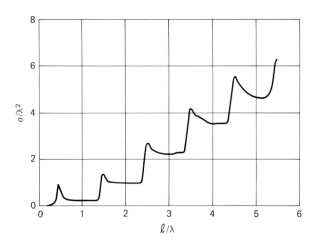

FIG. 11.7 Measured broadside returns of a thin dipole. (*Courtesy of University of Michigan Radiation Laboratory.*[6])

turn between the resonant peaks. These plateaus rise as the dipole becomes thicker, and the resonances eventually disappear.

Bodies considerably thicker than the thin wire also support surface traveling waves that radiate power in the backward direction. An example is the ogive, a spindle-shaped object formed by rotating an arc of a circle about its chord. Figure 11.8 is the RCS pattern of a 39-wavelength 15° half-angle ogive recorded for horizontal polarization (incident electric field in the plane of the ogive axis and the line of sight). The large lobe at the right side of the pattern is a specular echo in the broadside sector, and the sequence of peaks at the left side is the contribution of the surface traveling wave near end-on incidence. Note that the RCS is extremely small (not measurable in this case) at precisely end-on incidence. Theoretical predictions in the end-on region closely match the measured pattern for this particular body.

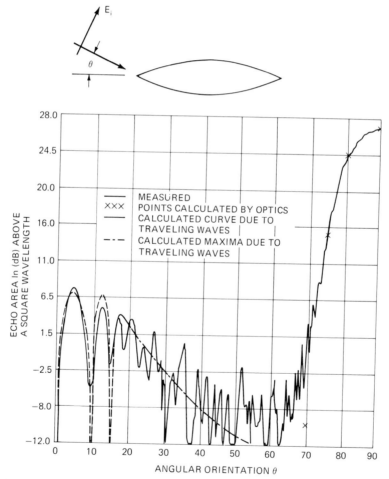

FIG. 11.8 Measured RCS pattern of a 39-wavelength 15° half-angle metal ogive. (*Copyright 1958, IEEE.*[7])

The dominant scattering mechanisms for the right circular conducting cone are the tip and the base. The return from the tip is very small in the nose-on region, and the RCS pattern is dominated by the echo from the base. Figures 11.9 and 11.10 are patterns of the RCS of a 15° (half-angle) cone with a base circumference of 12.575λ. Both patterns were measured as the cone was rotated about a vertical axis parallel to the base of the cone. The transmitted and received electric polarization was in the plane swept out by the cone axis (horizontal polarization) for Fig. 11.9 and was perpendicular to that plane (vertical polarization) for Fig. 11.10.

Nose-on incidence lies at the center of the patterns, and the sharp peaks near the sides are the specular returns from the slanted sides of the cone, also called specular *flashes*. The RCS formula for singly curved surfaces given in Table 11.1 may be used to predict the amplitudes of the specular flash within a fraction of a decibel. At precisely nose-on incidence the RCS must be independent of polarization because the cone is a body of axial symmetry. This may be verified by comparing the nose-on values in the two figures. At this angle the entire ring of the base of the cone is excited, but as the aspect angle swings away from nose-

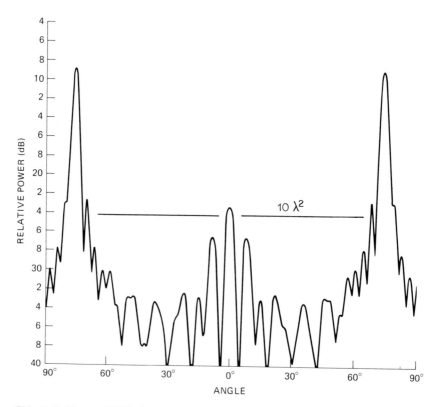

FIG. 11.9 Measured RCS of a 15° half-angle cone (horizontal polarization). The base circumference is 12.575λ. The heavy horizontal line indicates $10\lambda^2$. (*Courtesy of University of Michigan Radiation Laboratory.*[8])

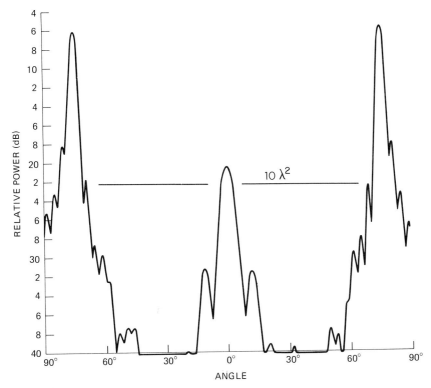

FIG. 11.10 Measured RCS of a 15° half-angle cone (vertical polarization). The base circumference is 12.575λ. The heavy horizontal line indicates 10λ². (*Courtesy of University of Michigan Radiation Laboratory.*[8])

TABLE 11.1 RCS Approximations for Simple Scattering Features

Scattering feature	Orientation (1)	Approximate RCS	
Corner reflector	Axis of symmetry along LOS	$4\pi A^2_{\text{eff}}/\lambda^2$	(2)
Flat plate	Surface perpendicular to LOS	$4\pi A^2/\lambda^2$	(3)
Singly curved surface	Surface perpendicular to LOS	$2\pi a\ell^2/\lambda$	(4)
Doubly curved surface	Surface perpendicular to LOS	$\pi a_1 a_2$	(5)
Straight edge	Edge perpendicular to LOS	ℓ^2/π	(6)
Curved edge	Edge element perpendicular to LOS	$a\lambda/2$	(7)
Cone tip	Axial incidence	$\lambda^2 \sin^4 (\alpha/2)$	(8)

NOTES:
1. LOS = line of sight.
2. A_{eff} = effective area contributing to multiple internal reflections.
3. A = actual area of the plate.
4. a = mean radius of curvature; ℓ = length of slanted surface.
5. a_1, a_2 = principal radii of surface curvature in orthogonal planes.
6. ℓ = edge length.
7. a = radius of edge contour.
8. α = half angle of the cone.

on, the scattering from the base degenerates to a pair of flash points. They lie at opposite ends of a diameter across the base in the plane containing the direction of incidence and the cone axis.

The echoes from the flash points at the sides of the base weaken as the aspect angle moves away from nose-on incidence, and the sidelobes seen at $+13°$ in Fig. 11.10 are actually due to an interaction between the two flash points *across the shadowed side* of the base. (The sidelobes disappear when a pad of absorber is cemented to the base.[8]) The flash point at the far side of the base disappears when the aspect angle moves outside the backward half cone, but the near flash point remains visible, and its echo decays with increasing aspect angle. Trailing-edge contributions like these are excited by that component of the incident electric field perpendicular to the edge; therefore they are stronger for horizontal incident polarization than for vertical polarization.

A flat plate also can support multiple diffraction from one side of the plate to the other, as shown in Fig. 11.11. The axis of rotation was in the plane of the plate parallel to one edge; normal incidence to the incident wave is $0°$, at the left side of each chart, with edge-on incidence at $90°$ near the right side. The specular return from the plate is the large peak at $0°$, which is predicted with quite good accuracy by the flat-plate formula given in Sec. 11.3. The edge-on return for vertical polarization is well predicted by the straight-edge formula given in Table 11.1.

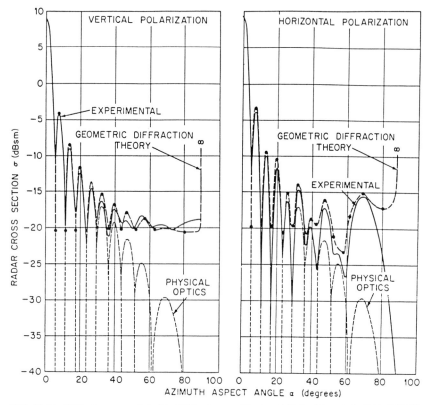

FIG. 11.11 RCS of a square flat plate 6.5 in along a side; $\lambda = 1.28$ in. (*Copyright 1966, IEEE.*[9])

These undulating patterns follow a sin x/x variation quite closely for aspect angles out to about 30°, but beyond that angle the two patterns differ by progressively wider margins. The sin x/x behavior is characteristic of a uniformly illuminated aperture, but unlike the one-way illumination function encountered in antenna work, the argument x for a flat plate includes a two-way (round-trip) illumination function. Thus, the beamwidth of the echo response of a flat plate is half the beamwidth of an antenna aperture of the same size. The prominent lobe in the horizontal pattern at 68° is a surface traveling-wave lobe closely related to the one appearing at nearly the same angle in the dipole pattern of Fig. 11.7.

In contrast to the pattern of a flat plate, the RCS pattern of a corner reflector is quite broad. This is true because the corner reflector is a reentrant structure, and no matter what its orientation (within limits, of course), internally reflected waves are directed back toward the source of the incident wave. A corner reflector is formed by two or three flat plates intersecting at right angles, and waves impinging on the first face are reflected onto the second; if there is a third face, it receives waves reflected by the first two faces. The mutual orthogonality of the faces ensures that the direction taken by waves upon final reflection is back toward the source.

The individual faces of the corner reflector may be of arbitrary shape, but the most common is an isosceles triangle for the trihedral corner; dihedral corners typically have rectangular faces. The RCS of a corner reflector seen along its axis of symmetry is identically that of a flat plate whose physical area matches the effective area of the corner reflector. The magnitude of the echo may be determined by finding the polygonal areas on each face of the corner receiving waves reflected by the other faces, and from which the final reflection is back toward the source. The effective area is determined by summing the projections of the areas of those polygons on the line of sight;[10] the RCS is then found by squaring that area, multiplying by 4π and dividing by λ^2.

Figure 11.12 is a collection of RCS patterns of a trihedral corner reflector with triangular faces. The reflector was fabricated of three triangular plywood panels, metallized to enhance their surface reflectivities. The aperture exposed to the radar was therefore an equilateral triangle, as shown in Fig. 11.13. The eight patterns in Fig. 11.12 were measured with the plane of the aperture tilted above or below the line of sight by the angle ϕ.

The broad central part of these patterns is due to a triple-bounce mechanism between the three participating faces, while the "ears" at the sides of the patterns are due to the single-bounce, flat-plate scattering from the individual faces. Along the axis of symmetry of the trihedral reflector in Fig. 11.13 ($\theta = 0°$, $\phi = 0°$), the RCS is $\pi\ell^4/3\lambda^2$, where ℓ is the length of one of the edges of the aperture. Not shown are the echo reductions obtained when the trihedral faces are angled other than at 90° from each other. The reductions resulting from changes in the angles of the corner faces depend on the size of the faces expressed in wavelengths.[11,12]

The RCS of most of the simple scattering features discussed above may be estimated by using the simple formulas listed in Table 11.1. The RCS of some complicated targets may be estimated by representing the target as a collection of features like those listed in Table 11.1, calculating the individual contributions, and then summing the contributions coherently or noncoherently. More detailed formulas are given in Sec. 11.3 that account for surface orientations not included in Table 11.1.

Complex Objects. Objects like antennas, insects, birds, airplanes, and ships can be much more complex than those discussed above, either because of the multiplicity of scatterers on them or because of the complexity of their surface profiles and dielectric constants. Insects are examples of the latter.

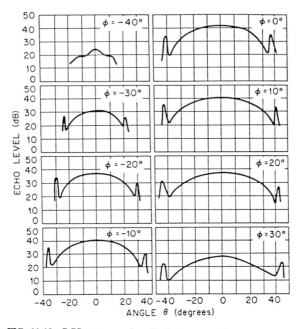

FIG. 11.12 RCS patterns of a trihedral corner reflector. Edge of aperture = 24 in; λ = 1.25 cm. (*Reprinted with permission from the* AT&T Technical Journal, *copyright 1947, AT&T.*[2])

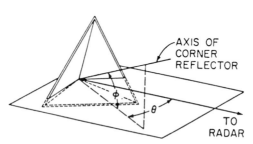

FIG. 11.13 Coordinate system for the RCS patterns in Fig. 11.12. (*Reprinted with permission from the* AT&T Technical Journal, *copyright 1947, AT&T.*[2])

Measured values for a dozen species are listed in Table 11.2. (The spider is an arachnid, not an insect, of course.) The animals were live for the measurements but had been drugged to immobilize them. Figure 11.14 shows the relationship between the RCS and the mass of an insect, with the variation of a water droplet shown for comparison. Similar comparisons have been made for both birds and insects.[15] The following values have been reported for the RCS of a man:[16]

TABLE 11.2   Measured Insect RCS at 9.4 GHz[13]

Insect	Length, mm	Width, mm	Broadside RCS, dBsm	End-on RCS, dBsm
Blue-winged locust	20	4	− 30	− 40
Armyworm moth	14	4	− 39	− 49
Alfalfa caterpillar butterfly	14	1.5	− 42	− 57
Honeybee worker	13	6	− 40	− 45
California harvester ant	13	6	− 54	− 57
Range crane fly	13	1	− 45	− 57
Green bottle fly	9	3	− 46	− 50
Twelve-spotted cucumber beetle	8	4	− 49	− 53
Convergent lady beetle	5	3	− 57	− 60
Spider (unidentified)	5	3.5	− 50	− 52

NOTE: Original values reported in square centimeters have been converted here to dBsm.

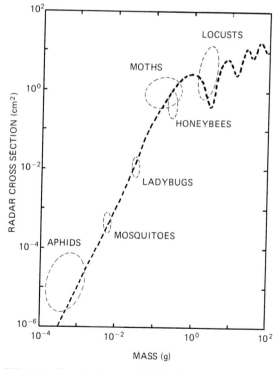

FIG. 11.14   Sample of measured RCS of insects as a function of insect mass at 9.4 GHz, based on Riley's summary. The solid trace is the calculated RCS of water droplets for comparison. (*Copyright, 1985 IEEE.*[14])

Frequency, GHz	RCS, m^2
0.41	0.033–2.33
1.12	0.098–0.997
2.89	0.140–1.05
4.80	0.368–1.88
9.375	0.495–1.22

Examples of the RCS of aircraft are shown in Figs. 11.15 through 11.17. The B-26 pattern in Fig. 11.15 was measured at a wavelength of 10 cm (frequency of about 3 GHz); the polar format is useful for display purposes but is not as convenient for detailed comparisons as a rectangular format is. The RCS levels shown in the scale model Boeing 737 patterns of Fig. 11.16 are those at the measurement frequency. To obtain the corresponding full-scale values, one must add 23.5 dB (10 log 225); the full-scale frequency is one-fifteenth of the measurement frequency in this case, or 667 MHz. The patterns shown in Fig. 11.17 are medians of RCS averages taken in cells 10° square. With modern data-collecting and -recording equipment, it is feasible to plot measured results at much finer intervals than are plotted in this figure. Note that the data is relative to 1 ft^2; to convert the displayed results to dBsm, one must subtract 10.3 dB (10 log 10.76 ft^2/m^2).

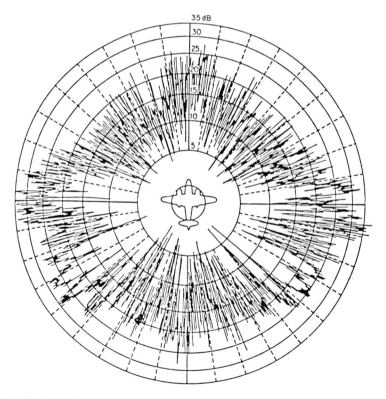

FIG. 11.15 Measured RCS pattern of a B-26 bomber at 10-cm wavelength. (*Copyright 1947, McGraw-Hill Book Company.*[17])

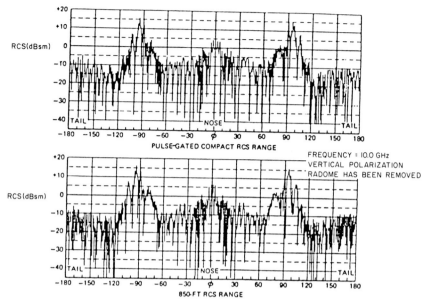

FIG. 11.16 Measured RCS of a one-fifteenth scale model Boeing 737 commercial jetliner at 10 GHz and vertical polarization. (*Copyright 1970, IEEE.*[18])

Figure 11.18 charts the RCS of a ship measured at 2.8 and 9.225 GHz at horizontal polarization. The data was collected by a shore-based radar instrumentation complex as the ship steamed in a large circle on Chesapeake Bay. The three traces in these charts are the 80, 50, and 20 percentile levels of the signals collected over aspect angle "windows" 2° wide. The patterns are not symmetrical, especially at the higher frequency. Note that the RCS can exceed 1 mi² (64.1 dBsm).

An empirical formula for the RCS of a naval ship is

$$\sigma = 52 f^{1/2} D^{3/2} \qquad (11.4)$$

where f is the radar frequency in megahertz and D is the full-load displacement of the vessel in kilotons.[20,21] The relationship is based on measurements of several ships at low grazing angles and represents the average of the median RCS in the port and starboard bow, and quarter aspects, but excluding the broadside peaks. The statistics include data collected at nominal wavelengths of 3.25, 10.7, and 23 cm for ship displacements ranging from 2 to 17 kilotons.

Figure 11.19 summarizes the general RCS levels of the wide variety of targets discussed in this section, with the RCS of a metallic sphere shown as a function of its volume for comparison. The ordinate is the RCS in square meters, and the abscissa is the volume of the target in cubic feet. Because the chart is intended only to display the wide range in RCS that may be encountered in practice, the locations of targets on the chart are approximate at best. Within given classes of target the RCS may be expected to vary by as much as 20 or 30 dB, depending on frequency, aspect angle, and specific target characteristics. The reader requiring

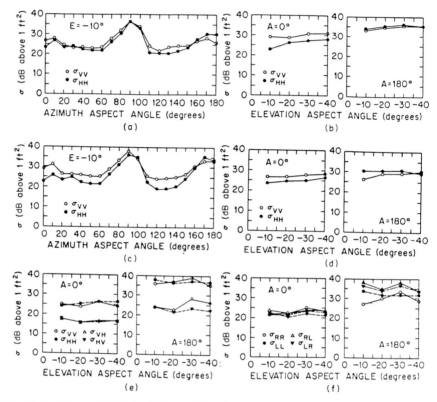

FIG. 11.17 Measured RCS of a C-54 aircraft in azimuth and elevation planes for linear and circular polarizations. Plotted values are the average RCS in a cell 10° in azimuth by 10° in elevation. Azimuth patterns *a* and *c* are for a fixed elevation angle of −10°. The remaining patterns are in the elevation plane for fixed nose-on or tail-on azimuths. The first and second subscripts give transmitted and received polarizations; *H* and *V* indicate horizontal and vertical polarizations, and *R* and *L* indicate right circular and left circular polarizations. (*Courtesy of I. D. Olin and F. D. Queen,*[19] *Naval Research Laboratory.*)

more explicit detail than this should consult referenced material at the end of this chapter.

11.3 RCS PREDICTION TECHNIQUES

Although the complexity and size of most scattering objects preclude the application of exact methods of radar cross-section prediction, exact solutions for simple bodies provide valuable checks for approximate methods. The exact methods are restricted to relatively simple or relatively small objects in the Rayleigh and resonant regions, while most of the approximate methods have been developed for the optics region. There are exceptions to these general limitations, of course; the exact solutions for many objects can be used for large bodies in the optics region if one uses arithmetic of sufficient precision, and many of the optics ap-

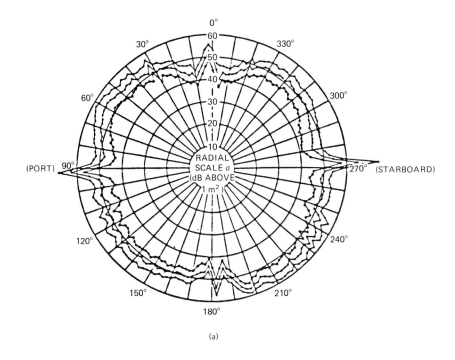

(a)

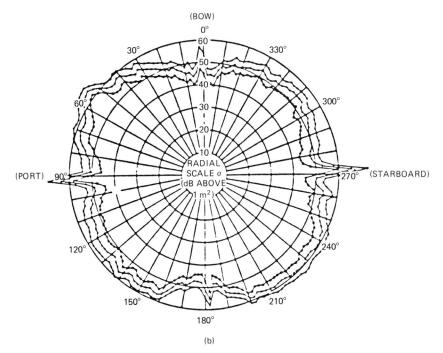

(b)

FIG. 11.18 Measured RCS of a large naval auxiliary ship for horizontal incident polarization. Upper pattern (a) is for 2.8 GHz and the lower (b) for 9.225 GHz. Shown are the 80, 50, and 20 percentile levels based on the statistics of the data over 2° aspect angle windows.

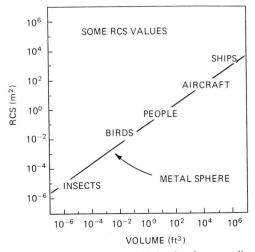

FIG. 11.19 Summary of RCS levels of targets discussed in this section. The locations of targets on the chart are general indications only.

proximations can be extended to bodies of modest electrical size in the resonance region. Low-frequency approximations developed for the Rayleigh region can extend nearly into the resonance region.

Exact Methods

Differential Equations. The exact methods are based on either the integral or differential form of Maxwell's equations. Maxwell's four differential equations constitute a succinct statement of the relationship between electric and magnetic fields produced by currents and charges and by each other.[22] The four equations may be manipulated for isotropic source-free regions to generate the wave equation

$$\nabla^2 \mathbf{F} + k^2 \mathbf{F} = 0 \qquad (11.5)$$

where **F** represents either the electric field or the magnetic field. Equation (11.5) is a second-order differential equation which may be solved as a boundary-value problem when the fields on the surface of the scattering obstacle are specified. The fields are typically represented as the sum of known and unknown components (incident and scattered fields), and the boundary conditions are the known relationships that must be satisfied between the fields (both electric and magnetic) just inside and just outside the surface of the obstacle exposed to the incident wave. Those boundary conditions are particularly simple for solid conducting or dielectric objects.

The boundary conditions involve all three components of the vector fields, and the surface of the body must coincide with a coordinate of the geometrical system in which the body is described. The solution of the wave equation is most useful for those systems in which the equation is separable into ordinary differential equations in each of the variables. The scattered fields are typically expressed in terms of infinite series, the coefficients of which are to be determined in the actual solution of the problem. The solution allows the fields to be calculated at any point in space, which in RCS problems is the limit as the distance from the obstacle becomes infinite. The product implied in Eqs. (11.1) and (11.2) is then formed from the solution of the wave equation, yielding the scattering cross section or the scattering width.

An example of a solution of the wave equation is the following infinite series for a perfectly conducting sphere:

$$\frac{\sigma}{\pi a^2} = \left| \sum_{n=1}^{\infty} \frac{(-1)^n (2n + 1)}{f_n(ka)[ka\, f_{n-1}(ka) - nf_n(ka)]} \right|^2 \tag{11.6}$$

The function $f_n(x)$ is a combination of spherical Bessel functions of order n and may be formed from the two immediately lower order functions by means of the recursion relationship

$$f_n(x) = \frac{2n - 1}{x} f_{n-1}(x) - f_{n-2}(x) \tag{11.7}$$

An efficient computational algorithm may be developed by using the two lowest orders as starting values,

$$f_0(x) = 1$$

$$f_1(x) = (1/x) - i$$

Equation (11.6) was used to compute the RCS characteristics plotted in Figs. 11.2 and 11.3. The infinite summation is truncated at the point where additional terms are negligible. The number of terms N required to compute the value of the bracketed term in Eq. (11.6) to six decimal places for $ka < 100$ is approximately

$$N = 8.53 + 1.21(ka) - 0.001(ka)^2 \tag{11.8}$$

The constants in Eq. (11.8) are slightly different for $ka > 100$ and are lower in value for fewer decimal places in the required accuracy.

The solution of the wave equation for the infinite, perfectly conducting circular cylinder can be resolved into two cases, one each for the incident electric or magnetic field parallel to the cylinder axis. The expressions are slightly simpler than Eq. (11.6) and involve cylindrical Bessel functions of the first and second kinds.[23] Figures 11.20 and 11.21 illustrate the backscattering behavior for the two principal polarizations as a function of the electrical circumference of the cylinder.

The response for E polarization (Fig. 11.20) is much larger than geometric optics value, πa, when the cylinder is less than a fraction of a wavelength in circumference, but it approaches the geometric optics value within a few percent for cylinders larger than about 2 wavelengths in circumference. The backscattering is markedly different for H polarization (Fig. 11.21), exhibiting the same kind of undulations noted earlier in the case of the metallic sphere. These undulations are caused by creeping waves that propagate around the rear of the cylinder just as they do around a sphere. However, the peaks and nulls of the sphere and cylinder interference patterns are not perfectly aligned with each another, suggesting that the relative phase angles between the creeping waves and specular contributions are slightly different for the two geometries.

The exact expression for the RCS of the dielectric cylinder is more complicated than for the conducting cylinder, but it accounts for the fact that energy penetrates the interior of the body. Unless the cylinder material is a perfect in-

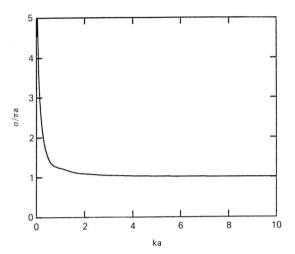

FIG. 11.20 Normalized scattering width of an infinite, perfectly conducting cylinder for E polarization (incident electric field parallel to the cylinder axis). The normalization is with respect to the geometric optics return from the cylinder.

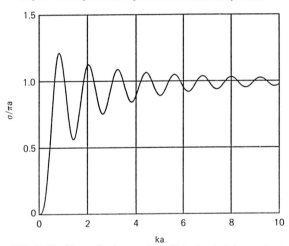

FIG. 11.21 Normalized scattering width of an infinite, perfectly conducting cylinder for H polarization (incident magnetic field parallel to the cylinder axis). The normalization is with respect to the geometric optics return from the cylinder.

sulator, its index of refraction is a complex function whose imaginary part gives rise to losses in the material. This in turn requires the computation of Bessel functions of complex argument, not an insignificant undertaking. Quite simple formulas for the scattering width may be obtained in the Rayleigh region, however, for which the cylinder diameter is much smaller than the incident wavelength. Figure 11.5 illustrates the scattering behavior of very thin dielectric cylinders.

Integral Equations.   Maxwell's equations may also be manipulated to generate a pair of integral equations (known as the Stratton-Chu equations[24]),

$$\mathbf{E}_s = \oint \{ ikZ_0(\mathbf{n} \times \mathbf{H})\psi + (\mathbf{n} \times \mathbf{E}) \times \nabla\psi + (\mathbf{n} \cdot \mathbf{E})\nabla\psi \} dS \qquad (11.9)$$

$$\mathbf{H}_s = \oint \{ -ikY_0(\mathbf{n} \times \mathbf{E})\psi + (\mathbf{n} \times \mathbf{H}) \times \nabla\psi + (\mathbf{n} \cdot \mathbf{H})\nabla\psi \} dS \qquad (11.10)$$

where **n** is the unit surface normal erected at the surface patch dS and the Green's function ψ is

$$\psi = e^{ikr}/4\pi r \qquad (11.11)$$

The distance r in Eq. (11.11) is measured from the surface patch dS to the point at which the scattered fields are desired. These expressions state that if the total electric and magnetic field distributions are known over a closed surface S, the scattered fields anywhere in space may be computed by summing (integrating) those surface field distributions.

The surface field distributions may be interpreted as induced electric and magnetic currents and charges, which become unknowns to be determined in a solution. The two equations are coupled because the unknowns appear in both. Unknown quantities also appear on both sides of the equations because the induced fields include the known incident field intensity and the unknown scattered field intensity. The method of solution is known as the *method of moments* (MOM),[25] reducing the integral equations to a collection of homogeneous linear equations which may be solved by matrix techniques.

The solution of the integral equations begins with the specification of the relation between the incident and scattered fields on the surface S, as governed by the material of which the object is made. If the body is perfectly conducting or if the electric and magnetic surface fields can be related by a constant (the surface impedance boundary condition), the equations become decoupled, and only one or the other need be solved. If the body is not homogeneous, the fields must be sampled at intervals within its interior volume, complicating the solution.

Once the boundary conditions have been specified, the surface S is split into a collection of small discrete patches, as suggested in Fig. 11.22. The patches must be small enough (typically less than 0.2λ) that the unknown currents and charges on each patch are constant or at least can be described by simple functions. A weighting function may be assigned to each patch, and the problem is essentially solved when the amplitude and phase of those functions have been determined.

The point of observation is forced down to a general surface patch, whereupon the fields on the left sides of Eqs. (11.9) and (11.10) are those due to the coupling of the fields on all other patches, plus the incident fields and a "self-field." The self-field (or current or charge) is moved to the right side of the equations, leaving only the known incident field on the left side. When the process is repeated for each patch on the surface, a system of $2n$ linear homogeneous equations in $2n$ unknowns is generated. If the boundary conditions permit the decoupling of the equations, the number of unknowns may be halved (n equations in n unknowns). The coefficients of the resulting matrix involve only the electrical distances (in wavelengths) between all patches taken by pairs and the orientation of the patch surface normals. The unknown fields may be found by inverting the resulting matrix and multiplying the inverted matrix by a column matrix representing the incident field at each patch. The surface fields are then summed in integrals like Eqs. (11.9) and (11.10) to obtain the scattered field, which then may be inserted in

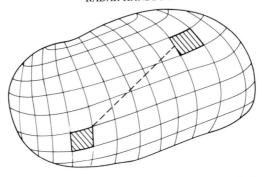

FIG. 11.22 The method of moments divides the body surface into a collection of discrete patches.

Eq. (11.1) to compute the RCS. Equation (11.2) and the two-dimensional counterparts of Eqs. (11.9) and (11.10) must be used for two-dimensional geometries, of course.

The method of moments has become a powerful tool in the prediction and analysis of electromagnetic scattering, with applications in antenna design as well as RCS prediction. The method has three limitations, however.

First, because computer memory and processing time both increase rapidly with the electrical size of the object, MOM is economically restricted to objects not much more than a few wavelengths, or perhaps a few dozen wavelengths, in size. As such, MOM is not a useful tool for predicting the RCS of, say, a jet fighter in the beam of a radar operating at 10 GHz. The second limitation is that MOM yields numbers, not formulas, and is therefore a numerical experimental tool. Trends may be established only by running a numerical experiment repeatedly for small parametric changes in the geometry or configuration of an object or in the angle of arrival or the frequency of the incident wave. Third, the solutions for some objects may contain spurious resonances that do not actually exist, thereby reducing the confidence one may have in applying the method to arbitrary structures.

Figure 11.23 traces the broadside RCS of a perfectly conducting cube computed by means of the method of moments. Spurious resonances were suppressed in the computations by forcing the normal surface component of the magnetic field to zero. The surface of the cube was divided into 384 patches (64 patches per face), which was about the limit of the central memory of the Cyber 750 computer used in the computations. It required more than 2 h for the Cyber 750 to generate the data plotted in the figure.[26]

Approximate Methods. Approximate methods for computing scattered fields are available in both the Rayleigh and the optics regions. Rayleigh region approximations may be derived by expanding the wave equation (11.5) in a power series of the wavenumber k.[27] The expansion is quasi-static for small wavenumbers (long wavelengths compared with typical body dimensions), and higher-order terms become progressively more difficult to obtain. The RCS pattern of a Rayleigh scatterer is very broad, especially if the object has similar transverse and longitudinal dimensions. The magnitude of the echo is proportional to the square of the volume of the object and varies as the fourth power of the frequency of the incident wave.[28] Because the method of moments is well suited to the solution of Rayleigh region problems, approximate methods for predicting the RCS of electrically small objects are not presented here.

Several approximate methods have been devised for the optics region, each with its particular advantages and limitations. The most mature of the methods are *geometric optics* and *physical optics*, with later methods attacking the problem of diffraction from edges and shadow boundaries. While the general accuracy of the optics region approximations improves as the scattering obstacle becomes electrically larger, some of them give reasonably accurate results (within 1 or 2 dB) for objects as small as a wavelength or so.

The theory of geometric optics is based on the conservation of energy within a slender fictitious tube called a *ray*. The direction of propagation is along the tube, and contours of equal phase are perpendicular to it. In a lossless medium, all the energy entering the tube at one end must come out the other, but energy losses within the medium may also be accounted for. An incident wave may be represented as a collection of a large number of rays, and when a ray strikes a surface, part of the energy is reflected and part is transmitted across the surface. The amplitude and phase of the reflected and transmitted rays depend on the properties of the media on either side of the surface. The reflection is perfect if the surface is perfectly conducting, and no energy is transmitted across the boundary. When energy can pass through the surface, transmitted rays are bent toward the surface normal in crossing a surface into an electrically denser medium (higher index of refraction) and away from the surface normal into a less dense medium. This bending of rays is known as *refraction*.

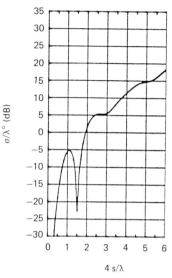

FIG. 11.23 Broadside RCS of a perfectly conducting cube (s = edge length). (*Copyright 1985, IEEE.*[26])

Depending on surface curvature and body material, reflected and transmitted rays may diverge from one another or they may converge toward each other. This dependence is the basis for the design of lenses and reflectors at radar wavelengths as well as optical wavelengths. The variation of the refractive index of the water molecule with wavelength is responsible for the rainbow, the result of two refractions near the front of a spherical water droplet and a single internal reflection from the rear. Secondary and tertiary rainbows are due to double and triple internal reflections.

The reduction in intensity as the rays diverge (spread away) from the point of reflection can be calculated from the curvatures of the reflecting surface and the incident wave at the *specular point*, that point on the surface where the angle of reflection equals the angle of incidence. The principal radii of curvature of the surface are measured in two orthogonal planes at the specular point, as shown in Fig. 11.24. When the incident wave is planar and the direction of interest is back toward the source, the geometric optics RCS is simply

$$\sigma = \pi a_1 a_2 \qquad (11.12)$$

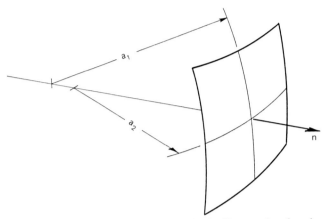

FIG. 11.24 The geometric optics RCS of a doubly curved surface depends on the principal radii of curvature at the specular point. The specular point is that point on the surface where the surface normal points toward the radar.

where a_1 and a_2 are the radii of curvature of the body surface at the specular point.

This formula becomes exact in the optical limit of vanishing wavelength and is probably accurate to 10 or 15 percent for radii of curvature as small as 2 or 3 wavelengths. It assumes that the specular point is not close to an edge. When applied to dielectric objects, the expression should be multiplied by the square of the voltage reflection coefficient associated with the material properties of the object. Internal reflections should also be accounted for, and the phase of internally reflected rays adjusted according to the electrical path lengths traversed within the body material. The net RCS then should be computed as the coherent sum of the surface reflection plus all significant internal reflections. Equation (11.12) fails when one or both surface radii of curvature at the specular point become infinite, yielding infinite RCS, which is obviously wrong. This occurs for flat and singly curved surfaces.

The theory of physical optics (PO) is a suitable alternative for bodies with flat and singly curved surface features. The theory is based on two approximations in the application of Eqs. (11.9) and (11.10), both of which are reasonably effective approximations in a host of practical cases. The first is the *far-field approximation*, which assumes that the distance from the scattering obstacle to the point of observation is large compared with any dimension of the obstacle itself. This allows one to replace the gradient of Green's function with

$$\nabla \psi = ik\psi_o \mathbf{s} \qquad (11.13)$$

$$\psi_o = e^{-ik\mathbf{r} \cdot \mathbf{s}} e^{ikR_o}/4\pi R_o \qquad (11.14)$$

where $\mathbf{r}$ is the position vector of integration patch dS and $\mathbf{s}$ is a unit vector pointing from an origin in or near the object to the far-field observation point, usually back toward the radar.[29] R_o is the distance from the origin of the object to the far-field observation point.

The second is the *tangent plane approximation*, in which the tangential field components $\mathbf{n} \times \mathbf{E}$ and $\mathbf{n} \times \mathbf{H}$ are approximated by their geometric optics values. That is, a tangent plane is passed through the surface coordinate at the patch dS, and the total surface fields are taken to be precisely those that would have existed had the surface at dS been infinite and perfectly flat. Thus the unknown fields in the integrals of Eqs. (11.9) and (11.10) may be expressed entirely in terms of the known incident field values. The problem then becomes one of evaluating one of the two integrals and substituting the result into Eq. (11.1) to obtain the RCS.

If the surface is a good conductor, the total tangential electric field is virtually zero and the total tangential magnetic field is twice the amplitude of the incident tangential magnetic field:

$$\mathbf{n} \times \mathbf{E} = 0 \tag{11.15}$$

$$\mathbf{n} \times \mathbf{H} = \begin{cases} 2\mathbf{n} \times \mathbf{H}_i & \text{illuminated surfaces} \\ 0 & \text{shaded surfaces} \end{cases} \tag{11.16}$$

Note that the tangential components of both the electric and the magnetic fields are set to zero over those parts of the surface shaded from the incident field by other body surfaces. Other approximations may be devised for nonconducting surfaces; if the incident wavelength is long enough, for example, the surface of a soap bubble or the leaf of a tree may be modeled as a thin membrane, on which neither the electric nor the magnetic fields are zero.

The integral is easy to evaluate for flat metallic plates because the phase is the only quantity within the integral that varies, and it varies linearly across the surface. The result for a rectangular plate viewed in a *principal plane* is

$$\sigma = 4\pi \left| \frac{A \cos \theta}{\lambda} \cdot \frac{\sin (k \ell \sin \theta)}{k \ell \sin \theta} \right|^2 \tag{11.17}$$

where A is the physical area of the plate, θ is the angle between its surface normal and the direction to the radar, and ℓ is the length of the plate in the principal plane containing the surface normal and the radar line of sight. A more general physical optics formula is available for the bistatic scattering of a polygonal plate with an arbitrary number of sides.[30,31]

A rectangular plate has a pair of orthogonal principal planes, and the edge length ℓ in Eq. (11.17) is that lying in the plane of measurement. If we designate w as the width of the plate in the opposite plane, the area of the plate is $A = \ell w$. To evaluate the maximum sidelobe levels of the plate RCS in the principal plane of measurement, we may replace the numerator of the sin $(x)/x$ term in Eq. (11.17) by unity. Normalizing with respect to the square of the width of the plane in the plane orthogonal to the measurement plane, we find the maximum sidelobe levels to be

$$\frac{\sigma}{w^2} = \frac{1}{\pi \tan^2 \theta} \tag{11.18}$$

Note that this result is *independent of the radar wavelength.*

The frequency independence of the principal-plane sidelobes is illustrated in Fig. 11.25. For viewing angles away from normal incidence, the plate edges are the dominant sources of echo, and the sin $(x)/x$ pattern is the result of the individual edge contributions changing phase with respect to each other as the aspect angle changes. Noting from Table 11.1 that the radar echoes of straight edges perpendicular to the line of sight are independent of frequency, the result of Eq. (11.18) is to be expected.

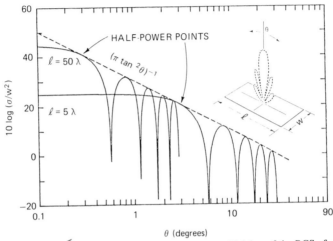

FIG. 11.25 The amplitudes of the principal-plane sidelobes of the RCS of a flat rectangular plate are independent of frequency. (*Courtesy of Walter W. Lund, Jr., The Boeing Company.*)

The physical optics formula for the RCS of a circular metallic disk is

$$\sigma = 16\pi \left| \frac{A \cos \theta}{\lambda} \cdot \frac{J_1(kd \sin \theta)}{kd \sin \theta} \right|^2 \qquad (11.19)$$

where A is the physical area of the disk, d is its diameter, and $J_1(x)$ is the Bessel function of the first kind of order 1. Equations (11.17) and (11.19) both reduce to the value listed in Table 11.1 for normal incidence.

The integral is somewhat more complicated to evaluate when the surface is singly or doubly curved. An exact evaluation can be performed for a circular cylinder and a spherical cap viewed along the axis of symmetry, but not for a truncated cone or a spherical cap seen along other than the axis of symmetry. Even so, the exact evaluation for the cylinder includes fictitious contributions from the shadow boundaries at the sides of the cylinder that do not appear in a *stationary phase approximation.*[32] The amplitude of the elemental surface patch contributions changes slowly over the surface of integration while the phase changes much more rapidly. As such, the net contribution in regions of rapid phase change is essentially zero and may be ignored. As the specular regions are approached, on the other hand, the phase variation slows down and then reverses as the specular point is crossed. This results in a nonzero specular contribution to the integral. The phase variation near the shadow boundaries is rapid; hence surface contributions there are ignored in a stationary phase evaluation, but an exact evaluation includes them because the shadow boundaries are the limits of integration. Because the actual surface field distributions do not suddenly drop to zero as the shadow boundary is crossed, as assumed by the theory, the shadow boundary contributions are spurious.[33,34] Therefore, a stationary phase approximation of the physical optics integral over closed curved surfaces tends to be more reliable than an exact evaluation.

With this in mind, the stationary phase result for a circular cylinder is

$$\sigma = ka\ell^2 \left| \frac{\sin (k\ell \sin \theta)}{k\ell \sin \theta} \right|^2 \tag{11.20}$$

where a is the radius of the cylinder, ℓ is its length, and θ is the angle off broadside incidence. Equation (11.20) includes only the contribution from the curved side of the cylinder and not its flat ends, which may be included by using the prescription of Eq. (11.19). Equation (11.20) may be used to estimate the RCS of a truncated right circular cone if the radius a is replaced by the mean radius of the cone and ℓ is replaced by the length of the slanted surface.

While the theory of physical optics offers a significant improvement over geometric optics for flat and singly curved surfaces, it suffers from other drawbacks. Although one obtains the proper result for most of the illuminated surface, the physical optics integral yields false contributions from the shadow boundaries, as noted above. Moreover, the theory shows no dependence on the polarization of the incident wave and yields different results when the receiver and the transmitter are interchanged. These effects contradict observed behavior. Finally, it errs by wider margins as the direction of observation moves farther away from the specular direction. As illustrated in Fig. 11.11, the theory is quite accurate at broadside incidence (the specular case), but the agreement between measurement and prediction becomes progressively worse as the scattering angle moves away from that direction. Keller's *geometrical theory of diffraction* (GTD) offers an improvement in both the polarization dependence and the predicted values in the wide-angle regions.[35,36]

GTD is a ray-tracing method that assigns an amplitude and phase to fields diffracted at smooth shadow boundaries and at surface discontinuities. Because the latter are much more significant in backscattering computations than the former, we focus here on edge diffraction. The theory assumes that a ray striking an edge excites a cone of diffracted rays, as in Fig. 11.26. The half angle of this *diffraction cone* is equal to the angle between the incident ray and the edge. Unless the point

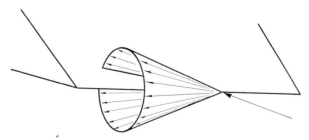

FIG. 11.26   The Keller cone of diffracted rays.

of observation lies on the diffraction cone, no value is assigned the diffracted field. The scattering direction in backscattering problems is the reverse of the direction of incidence, whence the diffraction cone becomes a disk, and the scattering edge element is perpendicular to the line of sight.

The amplitude of the diffracted field is given by the product of a *diffraction coefficient* and a *divergence factor*, and the phase depends on the phase of the edge excitation and on the distance between the observation point and the diffracting edge element. Two cases are recognized, depending whether the incident field is polarized parallel or perpendicular to the edge.

The diffracted field is given by the formula

$$E_d = \frac{\Gamma e^{iks} e^{i\pi/4}}{\sqrt{2\pi ks} \, \sin \beta} (X \mp Y) \tag{11.21}$$

where Γ is a divergence factor, X and Y are diffraction coefficients, β is the angle between the incident ray and the edge, and s is the distance to the observation point from the point of diffraction. The difference of the two diffraction coefficients is used when the incident electric field is parallel to the edge (TM polarization) and the sum when the incident magnetic field is parallel to the edge (TE polarization).

The divergence factor accounts for the decay in amplitude as the rays spread away from the edge element and includes the effects of the radius of the edge if it is curved, as at the end of a truncated cylinder, and the radius of curvature of the incident phase front.[37] The divergence factor for a two-dimensional edge (of infinite length) illuminated by a plane wave is $\Gamma = 1/s$. The diffraction coefficients are

$$X = \frac{\sin (\pi/n)/n}{\cos (\pi/n) - \cos [(\phi_i - \phi_s)/n]} \tag{11.22}$$

$$Y = \frac{\sin (\pi/n)/n}{\cos (\pi/n) - \cos [(\phi_i + \phi_s)/n]} \tag{11.23}$$

where ϕ_i and ϕ_s are the angles of the planes of incidence and scattering, as measured from one face of the wedge, and n is the exterior wedge angle normalized with respect to π; see Fig. 11.27. The three-dimensional result for an edge of finite length ℓ may be obtained by inserting Eqs. (11.22) and (11.23) in Eq. (11.21), using Eq. (11.21) for V_s/V_0 in Eq. (11.2), and then inserting Eq. (11.2) in Eq. (11.3).

Figures 11.28 and 11.29 compare measured and GTD-predicted RCS patterns of a right circular cone frustum. The theory replicates most of the pattern features for both polarizations but fails in three different aspect angle regions. These aspects are the specular directions of the flat surfaces at either end of the frustum (0 and 180° on the charts) and near the specular flash from the slanted side at 80°. The failure is due to a singularity in the diffraction coefficient Y along the reflection boundary, and a similar singularity occurs in the diffraction coefficient X along the shadow boundary, a situation encountered in forward scattering.

The singularities are overcome in the *physical theory of diffraction* (PTD) formulated by P. Ia. Ufimtsev.[39,40] (Although these publications may be difficult to find, we cite them here for completeness.) Like Keller, Ufimtsev relied on the (exact) canonical solution of the two-dimensional wedge problem, but he distin-

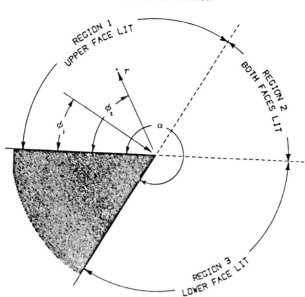

FIG. 11.27   Angles of incidence and scattering for wedge geometry.

guished between "uniform" and "nonuniform" induced surface currents. The uniform currents are the surface currents assumed in the theory of physical optics, and the nonuniform currents are associated with the edge itself (filamentary currents). The PTD result for two-dimensional problems may be represented as a linear combination of TM and TE polarizations,

$$E_s = E_0 f \frac{e^{ik\rho} e^{i\pi/4}}{\sqrt{2\pi k\rho}} \tag{11.24}$$

$$H_s = H_0 g \frac{e^{ik\rho} e^{i\pi/4}}{\sqrt{2\pi k\rho}} \tag{11.25}$$

where ρ is the distance to the far-field observation point and f and g are

$$f = \begin{cases} (X-Y)-(X_1 - Y_1) & 0 \le f_i < a\,2\,p \\ (X - Y) - (X_1 - Y_1) - (X_2 - Y_2) & \alpha - \pi \quad \phi_i \le p \\ (X - Y) - (X_2 - Y_2) & \pi \quad \phi_i \le a \end{cases} \tag{11.26}$$

$$g = \begin{cases} (X + Y) - (X_1 + Y_1) & 0 \le f_i \le a\,2\,p \\ (X + Y) - (X_1 + Y_1) - (X_2 + Y_2) & \alpha - \pi \quad \phi_i \le p \\ (X + Y) - (X_2 + Y_2) & \pi \quad \phi_i \le a \end{cases} \tag{11.27}$$

The subscripted coefficients are known as the *physical optics diffraction coefficients*,

$$X_1 = - \tan [(\phi_s - \phi_i)/2] \tag{11.28}$$

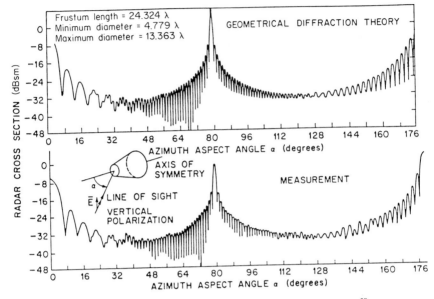

FIG. 11.28   RCS of a cone frustum, vertical polarization. (*Copyright 1966, IEEE.*[38])

$$Y_1 = -\tan[(\phi_s + \phi_i)/2] \tag{11.29}$$

$$X_2 = \tan[(\phi_s - \phi_i)/2] \tag{11.30}$$

$$Y_2 = -\tan[\alpha - (\phi_s + \phi_i)/2] \tag{11.31}$$

Because the PO diffraction coefficients depend on whether the upper face, the lower face, or both faces of the wedge are illuminated by the incident wave, the diffraction coefficients are combined differently in the three recognizable sectors defined in Eqs. (11.26) and (11.27). And because surface terms have been suppressed explicitly by the subtraction of the PO coefficients, the effects of surface currents (as distinguished from filamentary edge currents) must be accounted for independently. Those surface terms may be obtained, for example, by using geometrical optics or, paradoxically, the theory of physical optics after the edge terms themselves have been computed.

GTD and PTD are both based on the exact solution of the two-dimensional wedge problem, for which the directions of incidence and scattering are perpendicular to the edge. When extended to the case of oblique incidence, the direction of observation must lie along a generator of the Keller cone depicted in Fig. 11.26. If the edge is straight and of finite length, as in the three-dimensional world, Eq. (11.3) provides an approximation of the RCS. If the edge is curved, it may be regarded as a collection of infinitesimally short segments butted together, and the scattered fields may be computed via an integration of incremental fields diffracted by each element of the edge. This is the concept introduced by Mitzner,[41] and the summation of the fields diffracted by the edge elements implies an integral around the edge contour. (Although Mitzner's most significant results are embedded in a government document of limited distribution, we include this source in our references because of its significance.)

However, Mitzner sought the fields scattered in arbitrary directions, not just those along the local Keller cones, and for this purpose he developed his concept

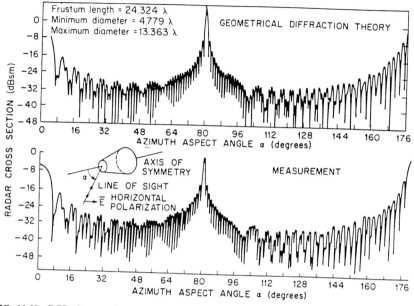

FIG. 11.29 RCS of a cone frustum, horizontal polarization. (*Copyright 1966, IEEE.*[38])

of the *incremental length diffraction coefficient*. Extending the example provided by Ufimtsev, he devised a set of diffraction coefficients for arbitrary directions of incidence and scattering. Not unexpectedly, those coefficients are more complicated than the X's and Y's appearing in Eqs. (11.22) and (11.23), and (11.28) through (11.31).[30,42]

Mitzner expressed his result as the diffracted electric-field components parallel and perpendicular to the plane of scattering in terms of the components of the incident electric field parallel and perpendicular to the plane of incidence. As such, the diffraction coefficients may be expressed as three separate pairs representing parallel-parallel, perpendicular-perpendicular, and parallel-perpendicular (or perpendicular-parallel) combinations. One member of each pair is due to the total surface current on the diffracting edge (including the assumed filamentary edge currents), and the other is due to the uniform physical optics currents. Mitzner subtracted one member of each pair from the other, thereby retaining the contributions from the filamentary currents alone.

The results have identically the form of Ufimtsev's expressions, in which the PO coefficients are subtracted from the non-PO coefficients. Thus, Mitzner's expression for the scattered field contains only the contributions from the filamentary edge currents. In applying his theory to scattering objects, therefore, the contributions of nonfilamentary induced surface currents must be accounted for separately, just as in Ufimtsev's physical theory of diffraction. When the directions of incidence and scattering become perpendicular to an edge, the perpendicular-parallel terms disappear and Mitzner's diffraction coefficients then reduce identically to Ufimtsev's.

Undertaking what he called a more rigorous evaluation of the fields induced on a wedge, Michaeli duplicated Mitzner's result for the total surface currents, confirming Mitzner's prior development, but he did not explicitly remove the PO surface-current contributions.[43] Thus, like Keller's X and Y, Michaeli's diffraction coefficients become singular in the transition regions of the reflection and shadow directions. Michaeli later investigated the removal of the singularities, the cleverest of which was the use of a skewed coordinate system along the wedge surfaces.[44,45]

While these methods of evaluating the fields scattered by edge elements may be applicable to smooth unbounded edges, they do not account for the discontinuities at corners where the edges turn abruptly in other directions. An attack on the problem has been suggested by Sikta et al.[46]

When applying these approximate high-frequency methods of estimating the fields scattered by complex objects, it is necessary to represent the object as a collection of surfaces having relatively simple mathematical descriptions. The actual surface profiles may be approximated by segments that have conveniently simple mathematical descriptions, such as flat plates, truncated spheroids, and truncated conic sections. The total RCS may be formed by summing the field contributions of the individual segments using the methods described above or whatever other tools are available. It is important to sum the field strengths of the individual contributions, complete with phase relationships, before squaring to obtain the total RCS as given by Eq. (11.1). This is tantamount to forming the coherent sum

$$\sigma = |\sum_p \sqrt{\sigma_p} e^{i\phi_p}|^2 \qquad (11.32)$$

where σ_p is the RCS of the pth contributor and ϕ_p is its relative phase angle, accounting for the two-way propagation of energy from the radar to the scattering feature and back again. If all phase angles are equally likely, one may form instead the noncoherent sum

$$\sigma = \sum_p \sigma_p \qquad (11.33)$$

The noncoherent RCS is meaningful only if a change in the aspect angle or a sweep in the instantaneous radar frequency does indeed result in a uniform distribution of phase angles. It is the average RCS formed over a time interval long enough to ensure the equal likelihood of all phase angles.

11.4 RCS MEASUREMENT TECHNIQUES

RCS measurements may be required for any of several reasons, ranging from scientific inquiry to verification of compliance with product specifications. There are no formal standards governing intrumentation and measurement methods, but informal standards of good measurement practice have been recognized for decades. Depending on the size of the test object, the frequencies to be used, and other test requirements, measurements may be made in indoor test facilities or on outdoor ranges. Because one is seldom interested in the RCS of an object for only one aspect angle, all static test ranges use turntables or rotators to vary the target aspect angle. Although the purpose of testing often governs how the measurements will be made, Refs. 47 and 48 are good overall guides for routine RCS testing.

General Requirements. The most important requirement for RCS measurements is that the test object be illuminated by a radar wave of acceptably uniform amplitude and phase. Good practice dictates that the amplitude of the incident wave deviate by no more than 0.5 dB over the transverse and longitudinal extent of the target and that the phase deviation be less than 22.5°. It is standard practice at some test ranges to physically probe the incident field at the onset of a test program to verify the amplitude uniformity of the incident wave.

The phase requirement is the basis of the far-field range criterion

$$R > 2D^2/\lambda \tag{11.34}$$

where R is the distance between the instrumentation radar and the test object and D is the maximum target dimension transverse to the line of sight. All other error sources being fixed, compliance with the far-field requirement is generally felt to yield data with an accuracy of 1 dB or better.[49] Figure 11.30 illustrates the far-field requirement for a variety of frequencies and target sizes.

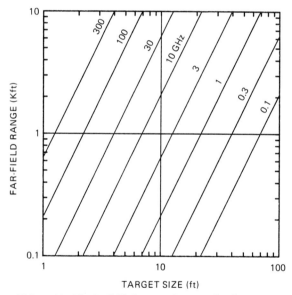

FIG. 11.30   The far-field distance. (*Reprinted with permission of Artech House, Inc.*[50])

Errors attributable to radar instrumentation should be held to 0.5 dB or less, which requires careful design and selection of components. The drift in system sensitivity should not exceed this value for the time it takes to record a single RCS pattern, which sometimes may approach an hour. The dynamic range of the system should be at least 40 dB, with 60 dB preferred. Linearity over this range should be 0.5 dB or better, and if not, steps should be taken to correct measured data via calibration of the receiver transfer function (gain characteristics).

RCS measurements should be calibrated by the *substitution method*, in which an object of known scattering characteristics is substituted for the target under test. Given the known (measured or calibrated) receiver gain characteristics, this establishes the constant by which a receiver output indication may be converted to an absolute radar cross-section value. Common calibration targets include metal spheres, right circular cylinders, flat plates, and corner reflectors. The radar cross sections of these objects may be calculated by using the expressions given in Sec. 11.3.

Because residual background reflections contaminate the desired target echo signal, they should be minimized by careful range design and operation. Interior walls in indoor test chambers must be covered with high-quality radar-absorbing material, and the surface of the ground on outdoor ranges should be smooth and free of vegetation. Target support structures should be designed specifically for low echo characteristics.

The effects of undesired background signals are illustrated in Fig. 11.31. Because the relative phase between the background signal and the target signal is unknown, two curves are shown; they correspond to perfect in-phase and out-of-phase conditions. If the background signal is equal to the target signal (ratio of 0 dB) and the two are in phase, the total received power is 4 times the power due to either one. This is the value shown at the upper left of the chart (6 dB). If the two are out of phase, they cancel each other and there is no signal at all (off the lower left of the chart). The chart shows that if the error due to background signals is to be 1 dB or less, the background must be at least 20 dB below the signal one intends to measure.

Three different kinds of support structures have been demonstrated to be useful in RCS measurements. They are the low-density plastic foam column, the string suspension harness, and the slender metal pylon. The echo from a plastic foam column arises from two mechanisms. One is a coherent surface reflection, and the other is a noncoherent volume contribution from the thousands of internal cells comprising the foam material.[51,52] The column should be designed so that its surfaces are never closer than 5 to 10° to the line of sight to the radar (depending on frequency), thereby minimizing the effect of the surface reflection. The noncoherent volume return is irreducible, however, and is not influenced by the orientation of the column. The volume return of suitable foam column support materials is of the order of -58 dBsm/ft^3 of material at 10 GHz.[53]

String suspension methods are best implemented indoors, where an overhead support point is normally available, although one documented design was seriously considered for outdoor use.[54] One of three configurations may be selected, all requiring a custom-made sling

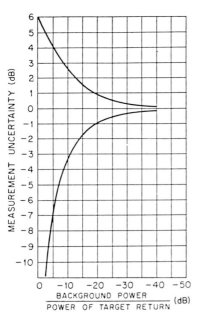

FIG. 11.31 Measurement error as a function of the relative background power level.

or harness to support the target. The first uses a single overhead support point and guy lines to a floor-mounted turntable to rotate the target. The second configuration suspends the target from an overhead turntable, reducing the guy lines and string loads at the expense of a more costly installation. The third configuration is the most costly, using a pair of turntables slaved together, one in the ceiling and one on the floor.

The echo signal from a string depends on the length and diameter of the string, its tilt angle with respect to the incident wave, and its dielectric constant. No matter what the tilt of the string, it will be presented normal to the line of sight twice in a complete rotation of the target and may cause a spike in the RCS pattern that could be erroneously attributed to the target unless otherwise accounted for. The RCS of a string rises with the fourth power of its diameter in the Rayleigh region (see Fig. 11.5), and for a given tensile strength the diameter rises only as the square root of the load to be supported. Thus, because the echo signal increases with the square of the load-carrying capacity, string suspension techniques are best suited for measurements of light objects or at low frequencies.

The metal target support pylon was first suggested in 1964,[54] but a practical implementation of the concept did not appear until 1976. The configuration of the pylon is sketched in Fig. 11.32, and it owes its electromagnetic performance to the sharpness of its leading edge and its tilt toward the radar (to the left in the diagram). Pylons as tall as 95 ft have been built, and it is customary to treat them with radar-absorbing material to suppress the echoes from the leading and trailing edges.

The obvious advantage of the metal pylon is its superior weight-carrying capability compared with that of strings and plastic foam columns. However, be-

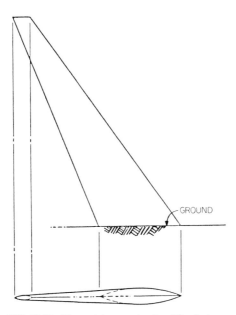

FIG. 11.32   The metal support pylon. The design is for an incident wave arriving from the left. (*Reprinted with permission of Artech House, Inc.*[55])

cause the top of the pylon is small, the rotation mechanism needed to vary the aspect angle of the target must be embedded in the target itself. This usually destroys the operational value of the target. Most of the rotators for these pylons are dual-axis, azimuth-over-elevation designs. When measurements are made with the azimuth rotation angle tilted back (away from the radar), parts of the target may sweep through the shadow cast by the top of the pylon, possibly degrading the measurements. One way to avoid this is to invert the target and tilt the rotation axis toward the radar instead of away from it. This requires the installation of the rotator in the top of the target as well as in the bottom. The unused internal cavities created for such installations must be concealed by covers or shields.

It is often necessary to measure scale models, which requires the application of scaling laws. Because nonconducting materials must be scaled differently than good conductors, it is not possible to satisfy all the scaling requirements for arbitrary targets composed of conducting and nonconducting materials. Most targets requiring scale-model testing, however, are dominantly metallic, for which the perfectly conducting scaling law is generally regarded as adequate.

When normalized with respect to the square of the wavelength, the RCS patterns of two perfectly conducting objects of identical shape but different size will be identical if the objects are the same number of wavelengths in size. If a model is one-tenth of full scale, for example, it should be measured at one-tenth of the full-scale wavelength (10 times the full-scale frequency). The RCS of the full-scale target may be obtained from the scale-model measurements by multiplying the scale-model RCS by the square of the ratio of the two frequencies. In this example, that factor is 10^2, or 20 dB.

Outdoor Test Ranges. Outdoor test ranges are required when test targets are too large to be measured indoors. The far-field criterion often requires that the range to the target be several thousand feet (see Fig. 11.30). Because the typical target height above the ground is a few dozen feet at best, the elevation angle to the target as seen from the radar is 1° at most and often less. At such low grazing angles the ground is strongly illuminated by the antennas, and unless the ground bounce can be suppressed, the target will be illuminated by a multipath field. In the design of an outdoor test range, therefore, a decision must be made whether to exploit the ground bounce or to attempt to defeat it. It is generally easier to exploit it than to eliminate it.

Test ranges designed to exploit the multipath effect may be asphalted to improve the ground reflection, although many ranges are operated over natural soil. Paving the range ensures uniformity in the characteristics of the ground plane from day to day and extends its operational usefulness to higher frequencies than might otherwise be possible. A conducting screen embedded in the asphalt may improve the reflection. Paving also reduces maintenance of the ground plane, such as might be required by periodical removal of vegetation and smoothing out windblown ridges in unstable soil.

The angle of incidence and the dielectric properties of asphalt and natural soil are such that the phase of the voltage reflection coefficient is within a few degrees of 180°. This being the case, one can usually choose a combination of target and antenna heights such that the wave reflected by the ground arrives at the target in phase with the wave propagated directly from the antennas. The indirect path should be a half wavelength longer than the direct path, resulting in the following rule for selecting the antenna and target heights:

$$h_a h_t = \lambda R/4 \qquad (11.35)$$

where h_a and h_t are the antenna and targets heights, respectively, and R is the range to the target.

Because most test ranges have turntables or target pylons installed at a few fixed locations relative to a permanent radar complex, the range R is usually restricted to a few preset values. The target is installed at a height h_t high enough to minimize spurious interactions with the ground, yet low enough to minimize the size and complexity of the target support structure. Therefore, it is the antenna height h_a that is most easily controlled and adjusted to optimize the location of the first lobe in the vertical multipath interference pattern. This is easily accomplished by mounting the radar antennas on carriages that can be raised or lowered along the side of a building or a tower provided for that purpose.

The ideal ground plane offers a theoretical sensitivity enhancement of 12 dB over identical measurements made in free space. The actual enhancement is usually significantly less than this, however, primarily because of the directivity of the antennas and imperfections in the ground plane. The directivity of the antennas precludes the target ever being squarely along the boresight of both the real antenna and its image in the ground plane at the same time, and the reflection coefficient of typical ground planes varies from 95 percent to as low as 50 percent or less. For all except very high and very low frequencies (millimeter wavelengths and VHF), typical sensitivities are of the order of 7 to 10 dB above free space instead of the ideal 12 dB.

When the range to the target is relatively short and tests must be performed over a wide range of frequencies, it is sometimes advantageous to attempt to defeat the ground-plane effect. One option is to install a berm shaped like an inverted V running between the radar and the target. The purpose of the slanted top of the berm is to deflect the ground-reflected wave out of the target zone. Another option is to install a series of low *radar fences* across the range. The design objective is to block ground-reflected rays from reaching the target from the radar, and vice versa, by shielding the specular zone on the ground from both. The near sides of the fences should be slanted to deflect energy upward, and may be covered with absorbing material. It is difficult, however, to prevent diffraction of radar energy from the top of the fences from reaching the target zone or to prevent target-diffracted signals from reaching the radar receiver via the same kind of mechanism.

Because of the large distances from the radar to the target on outdoor ranges, instrumentation radars typically develop peak signal powers ranging from 1 to 100 kW. Most of them emit simple pulsed waveforms whose pulses are from 0.1 to 0.5 μs wide, with pulse repetition rates of a few kilohertz. The radiated pulse should be wide enough to completely bracket the target but short enough to minimize background clutter contributions. To reduce measurement time and target exposure to weather or unauthorized observation, several instrumentation radars may be operated simultaneously. The radars may be triggered simultaneously or sequentially, depending on the particular requirements imposed at a given installation. Stepped- or swept-frequency waveforms may be employed to collect coherent test data for diagnostic purposes, as discussed in the next subsection.

Indoor Test Ranges. Indoor test ranges offer protection from weather and therefore more productive testing, but unless a very large facility is available, maximum target sizes are limited to a dozen feet or so. Because of the proximity of the walls, floor, and ceiling, they must be covered with high-quality absorbing material. The lower the intended frequency of operation, the more expensive the absorber becomes. Absorber reflectivity ratings of −50 dB are common among the materials used. This performance is usually achievable only with the pyramidal design.[56]

Early indoor chambers were rectangular in shape, and despite the installation of good absorbent materials on the walls, RCS measurements could be contaminated by wall reflections. The most sensitive part of the anechoic chamber is the rear wall, which receives 95 to 99 percent of the power radiated by the radar; hence one's best absorber should be reserved for the rear wall.[57] The floor, ceiling, and sidewalls also contribute errors, via a quadruplet of reflections not unlike those due to the ground plane of outdoor ranges. A remedy is the tapered anechoic chamber, which eliminates most of the sidewall reflections purely by means of geometrical control.[58,59,60]

Even targets of modest size cannot be measured at the far-field distance in indoor facilities because most chambers are not much more than 100 ft or so in length. It is possible, however, to provide the necessary uniformity of illumination by *collimating* the radiated beam. This can be done by inserting a lens between the radar and the target or by reflecting the radar beam off a collimating reflector. The concept is known as the *compact range* because a beam of parallel rays can be generated in a much shorter distance than would be possible without the collimating device.

Two successful lens designs have been documented.[61,62] The surface profiles of both were truncated hyperboloids of revolution, as suggested in Fig. 11.33, with the vertex facing the radar. While successful for the particular application they were designed for, these two lenses were too small (1.1 and 0.43 m in diameter) for most targets of interest.

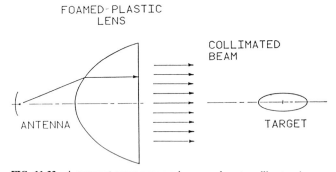

FIG. 11.33 A compact range uses a microwave lens to collimate a beam into parallel rays. The lens surface is a hyperboloid of revolution.

Because the flat rear face of the lens is parallel with the phase fronts there, it can be a significant source of undesired reflections. The rear-face reflections may be reduced by tilting the lens slightly, at the price of slight aberrations in the phase of the incident field. (Lenses also can be made with both surfaces curved, as is routinely done in the visible portion of the spectrum.) Lenses may be made of foamed plastic, and the dielectric constant of the material and the desired focal length determine the profile the lens must have. The uniformity required of material properties throughout the volume of lens, whether made of foamed or of solid plastic, has thus far discouraged the fabrication of larger versions.

The reflector offers a different way to collimate a beam. In contrast to the lens, which is placed between the radar and the test object, the radar and the test object remain on the same side of the reflector, as shown in Fig. 11.34. The reflector is typically an offset paraboloid, meaning that the paraboloidal surface does not include the vertex of the generating parabola. This permits the feed that excites the reflector to be placed out of the beam reflected toward the target. If the test object is held within one or two focal lengths of the reflector and if the reflector is excited by a suitably designed feed, the reflected wave is sensibly planar.[18,63]

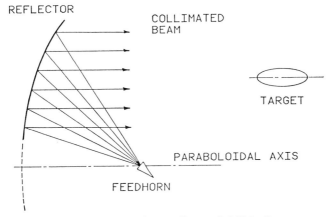

FIG. 11.34 A compact range using an offset paraboloidal reflector.

However, unless the edges of the reflector are carefully designed, the incident field in the target zone will be contaminated by fields diffracted from the edges of the reflector. The diffraction causes ripples in both the amplitude and the phase of the field distribution in the target zone. In some cases the effect is small enough to be ignored, but in high-quality installations the ripple may be objectionably large. Rolled-edge configurations that appear to have acceptable performance have been designed and tested.[64,65] The price paid for this improvement in performance is a much larger and more complicated reflector structure.

Targets measured indoors are placed much closer to the radar than those measured outdoors, and useful measurements may be made by using much less radiated power. Early indoor instrumentation radars relied on simple CW sources, and undesired chamber reflections were suppressed by a cancellation process. The procedure is to prepare the chamber for a measurement in every respect except for the installation of the target on its support fixture. A small sample of the transmitted signal is passed through a variable attenuator and a variable phase shifter and combined with the received signal. The amplitude and phase of the signal sample are then adjusted so as to cancel the signal received in the absence of the target.

The availability of low-cost, phase-locked, frequency-synthesized sources now makes it attractive to collect wideband RCS data, which contains far more target-scattering information than CW measurements made at single frequencies. When coherent RCS scattering data is suitably processed, it is possible to generate *radar imagery*, two-dimensional maps of the echo sources of test objects.[66]

Figure 11.35 is an example of such an image. The processing required to generate this image is a double Fourier transformation, one from the frequency domain to the time domain and the other from the angle domain to the cross-range domain. The frequency-time domain processing may be performed virtually in real time (a second or two for processing and display on a video screen), but the conversion from the angle domain to the cross-range domain must be performed offline. The fast Fourier transform (FFT) is invariably exploited to expedite the processing.

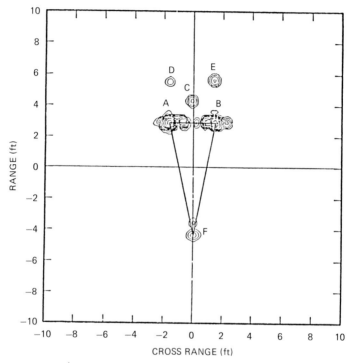

FIG. 11.35 Radar image of a wedge. (*Courtesy of D. L. Mensa, U.S. Navy Pacific Missile Test Center.*)

The resolution of the processed image in the time (range) domain is inversely proportional to the bandwidth of the emitted waveform. The resolution in the cross-range domain is inversely proportional to the aspect angle window over which the data is collected. Thus, the operating characteristics of the instrumentation system and the azimuthal data sampling rate must be decided before the data is collected. Because the cross-range coordinate of the resulting image is perpendicular to the axis of rotation of the target, it may be necessary to multiply that coordinate by a scale factor that effectively registers the generated image with, say, a plan view of the target.

The resulting data may be presented in the form of a contour map, as in Fig. 11.35, or in a gray-scale pixel format. The wedge target has been superposed in the figure for diagnostic purposes, and the particular attitude shown is for edge-on incidence. Images like this can be generated for any angle of incidence, provided the target has been rotated through a sector wide enough to yield the desired cross-range resolution and has been sampled at a sufficient number of angles over that sector, say, 64, 128, or 256 samples. In practice, the target is rotated continuously while the swept- or stepped-frequency data is collected. The angular speed must be slow enough that the phase of the return at the end of a frequency sweep due to target motion be within 22.5° from what it would have been had the target not moved.

Note that the leading edge of the wedge is an identifiable scatterer, although much less intense than the contributions from the base. In addition to the direct echoes A and B due to single diffraction from the base, there are additional contributions due to cross-base (multiple) diffraction. Contribution C is due to the excitation of one edge of the base by the other, which then diffracts the energy back to the source. This involves a single traverse across the base, and the phase delay due to the additional path length places the apparent source of scattering behind the actual base by half the width of the base. Contribution C appears along the centerline because the total propagation path from the source to the first edge, then to the second edge, and then back to the source is independent of the target rotation angle.

This is not the case for contributions D and E, which are due to double traverses of the base. That is, diffraction from one edge reaches the other edge across the base, which then diffracts some of the energy back toward the first edge in a second traverse. Because the diffraction responsible for contributions D and E crosses the base twice, D and E appear precisely one base width behind the actual base. Unlike contribution C, which lies along the centerline, D and E are displaced to the side, appearing directly behind the base edges. This is so because the phase of the excitation of the primary edge varies as the edge is moved toward or away from the source by the target rotation.

These ''ghost'' scatterers owe their existence to the way in which the data-processing system sorts the range and cross-range locations of scatterers. Down-range locations are sorted according to their processed time delays and cross-range locations according to their time-delay rates, whether due to real scatterers or to interactions between scatterers. Even though the contributions of some scattering centers may involve propagation in directions other than along the line of sight from the radar, the system has no way of discerning the fact. Therefore, despite the powerful diagnostic value of images like these, one must always be aware that multiple interactions between target elements can create scattering sources that are not where they appear to be.

11.5 ECHO REDUCTION

There are three motivations for controlling or attempting to control the radar cross section of objects that may fall in the beam of the radar. In civilian applications, the general objective is to enhance the radar echo of the object and thereby render it easier to detect and track. The enhanced signal improves safety and navigation, and metal corner reflectors have been marketed commercially for installation on small boats for this very purpose. Another civilian application is the reduction of the echoes of automobiles to minimize detection by police radars.[67] Because echo reduction is far more difficult and costly to achieve than echo enhancement, the civilian market for echo reduction devices has been weak.

Electronic warfare and electronic countermeasures have been by far the motivation for the echo reduction of military targets. Electronic countermeasures take many forms, including target masking or tracker system diversion by clouds of metal particles called *chaff*; the deliberate emission of signals that confuse or overwhelm the radar system, called *jamming*; and the reduction of the intrinsic echo of the target. The last is regarded a *passive* countermeasure because no signals are emitted.

These are not the only countermeasures available, of course. Detection may be reduced by the use of tactics, one of which is to fly close to the ground or the sea surface, where the incident radar beam is much weaker than at higher altitudes. The optical visibility of contrails from engine exhausts may be reduced by fuel additives. Infrared signals may be reduced by low-emissivity paints and coatings. Spread-spectrum radio transmission techniques reduce the overall RF power radiated by communications systems, thereby reducing detectability. It should be appreciated that radar echo reduction is only one of many countermeasures and that even this addresses only one of the many characteristics of the target that influence overall detectability.

There are only four ways to reduce the RCS of a body: shaping, absorbers, passive cancellation, and active cancellation. Because the cancellation schemes suffer limited bandwidth or are too complicated to implement, shaping and radar absorbers are by far the most effective methods of radar echo reduction.

Cancellation methods require loading the object with discrete antennalike elements (say, slots or dipoles) whose impedances must be chosen so as to cancel the returns from other parts of the body. The cancellation is imperfect and may even revert to a reinforcement when the frequency or the aspect angle shifts away from the nominal values for which the loads were designed. In some cases the real part of the required impedance is negative, implying that a source must be installed in the body. Difficulties in controlling self-oscillations and the phase and amplitude of the source output make shaping and absorbers much more attractive options.

By shaping we mean the intentional selection of target surfaces and features so as to minimize the amount of energy scattered back to the radar. Shaping includes specific design configurations, such as the placement of engine intakes, which can have large radar echoes, where they may be shielded from the incident wave by other parts of the target. The purpose of radar-absorbing materials is to soak up radar energy, which also minimizes the energy scattered back to the radar. Both methods have advantages and disadvantages.

Shaping. There is no point in considering shaping for echo reduction unless a specific threat direction can be identified in azimuth or elevation, or both. This is true because favorable surface orientations cannot be selected without knowing where the threat is likely to appear. If all directions are equally likely, then the advantage of choosing favorable surface orientations for one threat direction is canceled by an accompanying enhancement in another. In many situations, however, the general direction of the threat can be forecast.

Figure 11.36 illustrates the RCS reduction available by the use of shaping. The curves plotted there are based on theory and measurements and show how the nose-on (axial) RCS varies with the electrical size of each of the six rotationally symmetrical metallic bodies shown in Fig. 11.37. The diameters and projected areas of the objects are identical, and their volumes differ at most by a factor of 2. Except for the sphere, whose RCS is shown by the uppermost trace, all the objects have the same nose angle (40°), and of the six shapes the ogive exhibits the lowest RCS. Thus, at least along the axes of these particular bodies, the RCS can

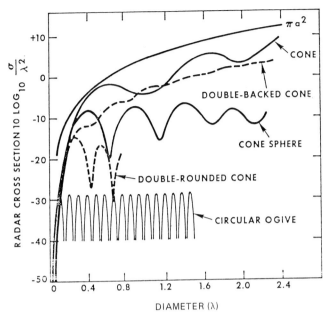

FIG. 11.36 RCS of a collection of bodies of revolution of similar size and projected area. (*Copyright 1964, IEEE.*[68])

be minimized by selecting the appropriate surface profile.

However, the attainment of low echoes over a range of aspect angles is usually accompanied by higher echo levels at other angles. Thus, the selection of an optimum shape should always include an evaluation of the variation of the RCS over a range of aspects wide enough to cover the anticipated threat directions. This implies the capability to measure the RCS patterns of a collection of objects with candidate surface profiles or the capability to predict those patterns, or both.

Two approaches may be taken in the application of shaping. One is to replace flat surfaces with curved surfaces and thereby eliminate narrow but intense specular lobes. While this does indeed reduce the magnitudes of specular echoes, it increases the general echo levels at nearby aspect angles. The other approach is to extend flat and singly curved surfaces so as to further narrow the specular lobe even if this increases its intensity. The logic of this approach is that the probability of detection is proportional to the average RCS over a range of solid angles of observation, and if the width of the lobe is narrow enough, its contribution to the average RCS may be less than if it were a wider but less intense lobe. The required RCS pattern levels of specific vehicle concepts should be established by means of mission analyses before one decides which shaping criterion is applicable.

Shaping is usually difficult to exploit or expensive to implement for vehicles or objects already in production. This is so because the vehicle configuration and profile have been selected and optimized for specific mission objectives; changes in the configuration are likely to impair the mission capabilities of the vehicle. If considered an option in the control of RCS, shaping must be included in the con-

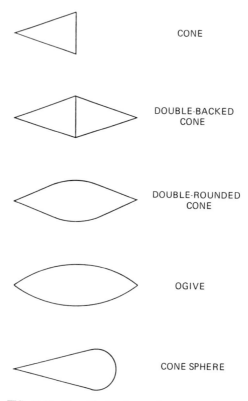

FIG. 11.37   The objects whose radar cross sections are plotted in Fig. 11.36.

ceptual design of the vehicle well before any production decisions are made. Furthermore, shaping is not very effective for bodies that are not electrically large.

Radar Absorbers.   The purpose of the radar absorber is to soak up incident energy and thereby reduce the energy scattered or reflected back to the radar. Most absorbers are designed to reduce specular reflections from metallic surfaces, but some have been designed for nonspecular scattering, primarily that due to surface traveling waves.

The simplest specular absorber is the Salisbury screen,[69] a thin resistive sheet mounted a quarter wavelength above a metal surface, shown schematically in Fig. 11.38. The transmission-line analogy is a lumped resistive element located a quarter wavelength toward the generator from a short circuit. Because the short circuit transforms to an open circuit at the lumped element, the effective impedance terminating the line is the resistive element itself. The reflection due to this termination becomes zero when the impedance of the element is identically the characteristic impedance of the line, Z_0. Because the impedance of free space is 377 ohms, the resistive sheet should have a resistivity of 377 ohms per square. Figure 11.39 charts the theoretical performance of the device for three values of resistivity. In the practical implementation of the design, the resistive layer is glued to a light plastic foam or honey-comb spacer backed by metal foil.

The Dällenbach layer is also a simple absorber. The material is uniform throughout its volume and is a mixture of compounds designed to have a specified index of refraction. That design may include materials with magnetic losses as well as carbon particles responsible for electric losses. The electric and magnetic susceptances (relative permittivity and relative permeability) therefore have imaginary components, resulting in an index of refraction with an imaginary component. The resulting imaginary part of the propagation constant attenuates waves traveling through the material.

Most of the commercial versions of Dällenbach absorbers are flexible and can be applied to modestly curved surfaces. The dielectric absorbers are typically made of a rubbery foam, sometimes urethane, impregnated with carbon particles. Impregnation may be performed by dipping a compressed slab of material in a graphite suspension bath, then wringing it out and drying it. Magnetic Dällenbach layers can be rolled from a mixture of natural or synthetic rubber

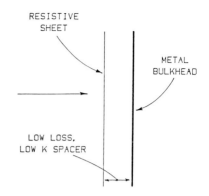

RESISTIVE SHEET

METAL BULKHEAD

LOW LOSS, LOW K SPACER

TRANSMISSION-LINE EQUIVALENT

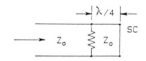

FIG. 11.38 The Salisbury screen and its transmission-line equivalent. Here K is the dielectric constant of the spacer between the resistive sheet and the metal backing plate.

loaded with carbonyl iron or ferrite powders. The lower the powder content, the more flexible the sheet but the less effective its electromagnetic performance. Dielectric and magnetic Dällenbach layers typically range from about 1 mm to 1 cm or more in thickness. The magnetic materials may have densities as high as 320 lb/ft^3.

The front face and the metal backing of the Dällenbach layer are the only sources of reflection. By using physically realizable materials, it is impossible to force either one of them to zero. The design objective, therefore, is to choose the electrical properties of the layer so that the two reflections tend to cancel each other. If the material properties are dominated by electric effects, the optimum layer thickness is near a quarter wavelength, and if they are dominantly magnetic, it is near a half wavelength. This is illustrated in Figs. 11.40 and 11.41.

The reflectivity of pure dielectric Dällenbach layers exhibits a pronounced resonant "notch" for a thickness of about a quarter wavelength (solid trace in Fig. 11.40). Pure magnetic Dällenbach layers probably don't exist, but if they did, they would have a similar notch for a thickness near a half wavelength (solid trace in Fig. 11.41). The performance of both materials deteriorates with increasing electrical thickness because the reflection due to the metal backing is attenuated too much to be effective in canceling the front-face reflection. Note that if the pure dielectric is augmented with some magnetic loss and the pure magnetic material is augmented with some dielectric loss, their bandwidths are improved considerably.

The quest for improved bandwidth motivated the development of absorbing

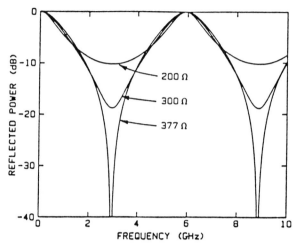

FIG. 11.39 Performance of the Salisbury screen. (*Reprinted with permission of Artech House, Inc.*[70])

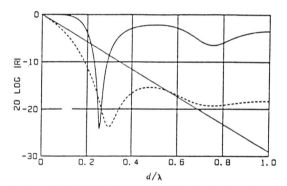

FIG. 11.40 Reflectivity curves for dominantly electric materials. The solid trace is for $|\epsilon_r| = 16$, $|\mu_r| = 1$, $\delta_\epsilon = 20°$, $\delta_\mu = 0°$. The dashed trace is for $|\epsilon_r| = 25$, $|\mu_r| = 16$, $\delta_\epsilon = 30°$, $\delta_\mu = 20°$. $\mu_r = |\mu_r| \exp(i\delta_\mu)$ and $\epsilon_r = |\epsilon_r| \exp(i\delta_\epsilon)$ are the complex permeability and permittivity of the material relative to that of free space. (*Copyright 1979, IEEE.*[71])

materials considerably more complex than the Salisbury screen and the Dällenbach layer. The natural extension of the Salisbury screen is the Jaumann absorber, which consists of a collection of resistive sheets stacked one above the other. Optimum performance is obtained when the spacing between sheets is fixed at a quarter wavelength (measured at the center of the desired band) and when the sheet resistivity varies from a high value at the outer sheet to a low value at the inner sheet. The theoretical performance of four designs is shown in Fig. 11.42; note that the bandwidth of the four-sheet design exceeds 100 percent at the −20 dB level.

Graded dielectric and magnetic absorbers exploit the same kind of layer-to-

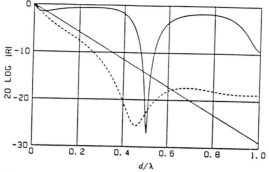

FIG. 11.41 Reflectivity curves for dominantly magnetic materials. The solid trace is for $|\mu_r| = 16$, $|\epsilon_r| = 1$, $\delta_\mu = 10°$, $\delta_\epsilon = 0°$. The dashed trace is for $|\mu_r| = 25$, $|\epsilon_r| = 16$, $\delta_\mu = 20°$, $\delta_\epsilon = 30°$. $\mu_r = |\mu_r| \exp(i\delta_\mu)$ and $\epsilon_r = |\epsilon_r| \exp(i\delta_\epsilon)$ are the complex permeability and permittivity of the material relative to that of free space. (*Copyright 1979, IEEE.*[71])

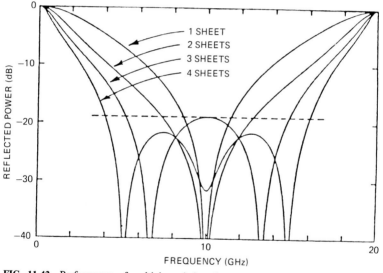

FIG. 11.42 Performance of multiple resistive sheets (Jaumann absorber). (*Reprinted with permission of Artech House, Inc.*[72])

layer variation used in the Jaumann absorber. Graded absorbers are made by cementing together stacks of Dällenbach layers, with the intrinsic impedance of each layer becoming smaller the closer the layer is to the metallic backing foil. Five or more layers have been used in the commercial production of graded dielectric absorbers, but commercial graded magnetic absorbers appear to have been limited to three layers. It is important in the design process to account for the actual thickness and electrical properties of the adhesive films used to bond the layers to each other.

The circuit analog sheet is an attempt to impart controllable reactive properties to the otherwise purely resistive sheet. The reactive part of the complex impedance of a sheet allows the designer to reduce the spacing between sheets or to improve the bandwidth of multisheet designs. This control is achieved by depositing a lossy coating on a thin, lossless, durable film in a prescribed pattern, an example of which is shown in Fig. 11.43.

FIG. 11.43 Example of a pattern deposited on a circuit analog sheet.

The particular pattern displayed is a rectangular collection of crosses. The shape, spacing, and thickness of the crosses govern the effective resistance of the sheet. Their size and shape govern the inductance of the sheet, and the shape of the crosses and their spacing govern the capacitance of the sheet. The variety of patterns is virtually unlimited, giving the designer enough parameters to develop absorbers that perform well under many different conditions. These might include, for example, arbitrary polarizations and angles of incidence.

Approximate expressions can be developed for the impedance of a sheet based on the characteristics of the deposited pattern, which simplifies the analysis of the performance of a collection of sheets. One of the attractive features of the circuit analog design approach is that it is possible to fabricate tough, thin, lightweight panels capable of withstanding, for example, the aerodynamic stresses of flight. On the other hand, circuit analog absorbers are costly.

The circuit analog sheet is a lossy version of a general class of patterns known as *frequency-selective surfaces* (FSS).[73] The FSS is usually a thin metallic pattern etched into or deposited onto a lossless substrate or film. As in the Jaumann and circuit analog absorbers, several layers may be cascaded to obtain the desired effect. The desired effect is usually to pass waves of a given range of frequencies (bandpass filtering) or all waves except those in a range of frequencies (bandstop filtering). Other designs use high-pass or low-pass filtering. Frequency-selective surfaces may be employed to conceal antennas, for example, by reflecting incident waves with undesired wavelengths while allowing others, say, communications or other radar signals, to pass through to the antenna beneath the FSS.

The pyramidal absorber used to suppress wall reflections in indoor chambers represents a particularly effective method of varying the effective impedance "seen" by an incident wave. The absorber is made of flexible, carbon-impregnated plastic foam cut in the form of pyramids. It exhibits optimum performance when the pyramids are pointed toward the direction of the incident wave, and the pyramids should be of the order of 3 to 6 wavelengths deep. Fire-retardant paint is usually applied to pyramidal absorbers to satisfy safety requirements, but at high frequencies the paint tends to degrade the performance of the material. Nevertheless, pyramidal absorbers of sufficient depth consistently turn in performances better than −50 dB. Because these absorbers do not rely on the cancellation of a front-face reflection by a rear-face reflection, they exhibit great bandwidth. In general, a pyramidal absorber with sharp tips and uniform bulk loss characteristics can have a bandwidth that exceeds 100:1.[60]

Nonspecular absorbers need not have the great thickness characterized by specular absorbers. Intended primarily for suppression of surface traveling-wave echoes, nonspecular materials have the opportunity to reduce the buildup of surface currents over several wavelengths *along the surface*. They are able to register quite respectable performances, therefore, simply because a thin layer attached to a metallic surface need not be very heavy. In this respect, the surface traveling-wave contribution due to long, smooth surfaces is one of the easiest to suppress. Even so, the thickness and geometric distribution of surface-wave absorbers should be varied for optimum performance.

11.6 SUMMARY

With the exception of insects, birds, and people, we have focused in this chapter on the radar echo characteristics of inanimate objects. Those characteristics are given by the radar cross section of the object, the abbreviation for which (RCS) is generally recognized. The RCS of an object depends on the size, shape, and composition of the object and on its orientation with respect to the direction of arrival and the polarization of the incident wave. The RCS itself may be measured at a suitably designed test facility or may be estimated by means of any of several analytical techniques. For estimation and prediction purposes, the RCS is proportional to the ratio of the scattered power density at the receiver to the incident power density at the target. It is even possible to define a bistatic scattering cross section when the radar receiver and transmitter are separated from one another.

Examples of the RCS of simple scattering features given in this chapter included edges and flat, singly curved and doubly curved structures. The echo of a flat plate is strong at normal incidence but decays as the plate is tilted; the larger the plate, the more rapid the angular decay. The echoes of reentrant structures are equally intense, but unlike those of flat plates, they remain at relatively high levels over a considerable range of incidence angles. This is true because reflected rays emerging from such structures, especially corner reflectors, tend to propagate back toward the source of the incident wave regardless of the direction of incidence. Particularly important reentrant structures are the intakes and exhausts of jet engines. Our examples included a chart of common scattering obstacles, ranging from insects to ships, and the range in their scattering amplitudes covered 120 dB (12 orders of magnitude).

The RCS of an object is an area, and that area may be expressed in terms of the square of a typical dimension of the object, or in absolute units, such as square meters, or in terms of the square of the wavelength of the incident wave. Examples cited from the literature included most of these units of area. Whatever the unit, however, the RCS is a gauge of the ratio of the scattered power density registered by a receiver to the power density of the incident wave at the target. Because a large metal sphere scatters incident energy nearly omnidirectionally, the RCS of an object may be expressed in terms of the area of an equivalent sphere that scatters the same power.

Methods of predicting the RCS of objects include exact solutions of the wave equation and (nearly) exact solutions of the integral equation. The wave equation is a second-order differential equation whose solutions are restricted to relatively simple geometrical shapes, including spheres, circular cylinders, and a few others. The solution of the integral equation is known as the method of moments and can be applied to bodies of arbitrary shape and composition. However, the method of moments is limited by computer resources (memory and execution time both) to practical objects not much more than a few wavelengths in size at most, and even then the computed results may contain spurious results. Nevertheless, both kinds of exact solution provide us the capability to understand the scattering mechanism and to explore the responses of simple and sometimes complex objects to the stimulus of an incident electromagnetic wave.

Few, if any, practical targets have shapes simple enough or small enough to fall within the province of the exact formulations, and one must turn to approximate methods of calculating the RCS. The bulk of the approximate methods lie in the optics regime in which the incident wavelength is significantly shorter than any characteristic target dimension. The optics methods include geometric optics, physical optics, the geometrical theory of diffraction, the physical theory of diffraction, the method of equivalent currents, and others. Happily enough, most of the optical approximations are acceptably accurate for objects as small as 2 wavelengths or so.

At the other end of the spectrum, where the scattering obstacle is small compared with the incident wavelength, analytical methods of RCS prediction are not as well developed. This is so because low-frequency expansions of the wave equation rapidly become very complicated beyond the first and second orders. Moreover, because the method of moments works quite well for electrically small bodies, the availability of this useful tool has led to the abandonment of low-frequency expansions as a way of estimating Rayleigh region RCS characteristics.

Because echo reduction and its effects are often the concern of modern radar designers, this chapter has included some of the methods by which that may be achieved. Although four methods of echo reduction actually are available, only two methods (shaping and radar absorbers) have been demonstrated to be effective, and only those two were discussed. The objective of shaping is to select the surface profiles of one's target so as to deflect reflected radar waves in any direction except back to the radar. When that is not effective or feasible, one may coat sensitive parts of one's target with radar-absorbing material.

These options are not to be considered lightly, for echo reduction comes at a very high price, including nonrecurring engineering costs, recurring maintenance costs, and reduced effectiveness of other mission functions. Finally, it was pointed out that in addition to passive echo suppression techniques, one has available many other options to avoid detection or to break track, including tactics and active electronic countermeasures. The system designer is well advised to consider all of them and not RCS reduction exclusively.

REFERENCES

1. Schneider, E. G.: Radar, *Proc. IRE*, vol. 34, pp. 528–578, August 1946.

2. Robertson, S. D.: Targets for Microwave Navigation, *Bell Syst. Tech. J.*, vol. 26, pp. 852–869, 1947.

3. Stratton, J. A.: "Electromagnetic Theory," McGraw-Hill Book Company, New York, 1941, pp. 414–420, 563–567.

4. Rheinstein, J.: Backscatter from Spheres: A Short-Pulse View, *IEEE Trans.*, vol. AP-16, pp. 89–97, January 1968.

5. Freeny, C. C.: Target Support Parameters Associated with Radar Reflectivity Measurements, *Proc. IEEE*, vol. 58, pp. 929–936, August 1965.

6. Chang, S. S., and V. V. Liepa: Measured Backscattering Cross Section of Thin Wires, *University of Michigan, Rad. Lab. Rept.* 8077-4-T, May 1967.

7. Peters, L., Jr.: End-Fire Echo Area of Long, Thin Bodies, *IRE Trans.*, vol. AP-6, pp. 133–139, January 1958.

8. Knott, E. F., and T. B. A. Senior: CW Measurements of Right Circular Cones, *University of Michigan, Rad. Lab. Rept.* 011758-1-T, April 1973.

9. Ross, R. A.: Radar Cross Section of Rectangular Plates, *IEEE Trans.*, vol. AP-14, pp. 329–335, May 1966.

10. Knott, E. F.: A Tool for Predicting the Radar Cross Section of an Arbitrary Corner Reflector, presented at IEEE Southeastcon '81 Conference, Huntsville, Ala., April 6–8, 1981; *IEEE Publ.* 81CH1650-1, pp. 17–20.

11. Knott, E. F.: RCS Reduction of Dihedral Corners, *IEEE Trans.*, vol. AP-25, pp. 406–409, May 1977.

12. Anderson, W. C.: Consequences of Non-Orthogonality on the Scattering Properties of Dihedral Reflectors, *IEEE Trans.*, vol. AP-35, pp. 1154–1159, October 1987.

13. Hajovsky, R. G., A. P. Deam, and A. H. LaGrone: Radar Reflections from Insects in the Lower Atmosphere, *IEEE Trans.*, vol. AP-14, pp. 224–227, March 1966.

14. Riley, J. R.: Radar Cross Section of Insects, *Proc. IEEE*, vol. 73, pp. 228–232, February 1985.

15. Vaughn, C. R.: Birds and Insects as Radar Targets: A Review, *Proc. IEEE*, vol. 73, pp. 205–227, February 1985.

16. Schultz, F. V., R. C. Burgener, and S. King: Measurements of the Radar Cross Section of a Man, *Proc. IRE*, vol. 46, pp. 476–481, February 1958.

17. Ridenour, L. N. (ed.): "Radar System Engineering," MIT Radiation Laboratory Series, vol. 1, McGraw-Hill Book Company, New York, 1947, p. 76.

18. Howell, N. A.: Design of Pulse Gated Compact Radar Cross Section Range, *1970 G-AP Int. Prog. & Dig., IEEE Publ.* 70c 36-AP, pp. 187–195, September 1970.

19. Olin, I. D., and F. D. Queen: Dynamic Measurement of Radar Cross Sections, *Proc. IEEE*, vol. 53, pp. 954–961, August 1965.

20. Skolnik, M. I.: "Introduction to Radar Systems," McGraw-Hill Book Company, New York, 1980, p. 45.

21. Skolnik, M. I.: An Empirical Formula for the Radar Cross Section of Ships at Grazing Incidence, *IEEE Trans.*, vol. AES-10, p. 292, March 1974.

22. Ramo, S., and J. R. Whinnery: "Fields and Waves in Modern Radio," 2d ed., John Wiley & Sons, New York, 1960, pp. 272–273.

23. Bowman, J. J., P. L. E. Uslenghi, and T. B. A. Senior (eds.): "Electromagnetic and Acoustic Scattering by Simple Shapes," North-Holland, Amsterdam, 1969, pp. 92–108.

24. Stratton, J. A.: Ref. 3, pp. 464–467.

25. Harrington, R. F.: "Field Computation by Moment Methods," Macmillan Company, New York, 1968.

26. Yaghjian, A. D.: Broadside Radar Cross Section of the Perfectly Conducting Cube, *IEEE Trans.*, vol. AP-33, pp. 321–329, March 1985.

27. Kleinman, R. E.: The Rayleigh Region, *Proc. IEEE*, vol. 53, pp. 848–856, August 1965.

28. Crispin, J. W., Jr., and K. M. Siegel (eds.): "Methods of Radar Cross Section Analysis," Academic Press, New York, 1968, pp. 144–152.

29. Ruck, G. T., D. E. Barrick, W. D. Stuart, and C. K. Krichbaum: "Radar Cross Section Handbook," vol. 1, Plenum Press, New York, 1970, pp. 50–59.

30. Knott, E. F.: A Progression of High-Frequency RCS Prediction Techniques, *Proc. IEEE*, vol. 73, pp. 252–264, February 1985.

31. Knott, E. F., J. F. Shaeffer, and M. T. Tuley: "Radar Cross Section," Artech House, Norwood, Mass., 1985, p. 123.

32. Knott, E. F., et al.: Ref. 31, pp. 124–129.

33. Senior, T. B. A.: A Survey of Analytical Techniques for Cross-Section Estimation, *Proc. IEEE*, vol. 53, pp. 822–833, August 1965.

34. Gupta, I. J., and W. D. Burnside: Physical Optics Correction for Backscattering from Curved Surfaces, *IEEE Trans.*, vol. AP-35, pp. 553–561, May 1987.

35. Keller, J. B.: Diffraction by an Aperture, *J. Appl. Phys.*, vol. 28, pp. 426–444, April 1957.

36. Keller, J. B.: Geometrical Theory of Diffraction, *J. Opt. Soc. Am.*, vol. 52, pp. 116–130, 1962.

37. Kouyoumjian, R. G., and P. H. Pathak: A Uniform Theory of Diffraction for an Edge in a Perfectly Conducting Surface, *Proc. IEEE*, vol. 62, pp. 1448–1461, November 1974.

38. Ross, R. A., and M. E. Bechtel: Radar Cross Section Prediction Using the Geometrical Theory of Diffraction, *IEEE Int. Antenna Propagat. Symp. Dig.*, pp. 18–23, 1966.

39. Ufimtsev, P. Ia.: Approximate Computation of the Diffraction of Plane Electromagnetic Waves at Certain Metal Boundaries, Part I: Diffraction Patterns at a Wedge and a Ribbon, *Zh. Tekhn. Fiz. (U.S.S.R)*, vol. 27, no. 8, pp. 1708–1718, 1957.

40. Ufimtsev, P. Ia.: Method of Edge Waves in the Physical Theory of Diffraction, *U.S. Air Force Systems Command, Foreign Technology Division Doc. FTD-HC-23-259-71*, 1971. (Translated from the Russian version published by Soviet Radio Publication House, Moscow, 1962.)

41. Mitzner, K. M: Incremental Length Diffraction Coefficients, *Northrop Corporation, Aircraft Div. Tech. Rept. AFAL-TR-73-296*, April 1974.

42. Knott, E. F.: The Relationship between Mitzner's ILDC and Michaeli's Equivalent Currents, *IEEE Trans.*, vol. AP-33, pp. 112–114, January 1985. [In the last term of Eq. (15), the dot preceding the minus sign should be deleted and β should be replaced by sin β; in Eq. (20), the sign of the first term on the right side must be reversed.]

43. Michaeli, A.: Equivalent Edge Currents for Arbitrary Aspects of Observation, *IEEE Trans.*, vol. AP-32, pp. 252–258, March 1984. (See also correction, vol. AP-33, p. 227, February 1985.)

44. Michaeli, A.: Elimination of Infinites in Equivalent Edge Currents, part I: Fringe Current Components, *IEEE Trans.*, vol. AP-34, pp. 912–918, July 1986.

45. Michaeli, A.: Elimination of Infinities in Equivalent Edge Currents, part II: Physical Optics Components, *IEEE Trans.*, vol. AP-34, pp. 1034–1037, August 1986.

46. Sikta, F. A., W. D. Burnside, T. T. Chu, and L. Peters, Jr.: First-Order Equivalent

Current and Corner Diffraction Scattering from Flat Plate Structures, *IEEE Trans.*, vol. AP-31, pp. 584–589, July 1983.

47. Mack, R. B.: Basic Design Principles of Electromagnetic Scattering Measurement Facilities, *Rome Air Development Center Rept. RADC-TR-*81-40, March 1981.

48. Dybdal, R. B.: Radar Cross Section Measurements, *Proc. IEEE*, vol. 75, pp. 498–516, April 1987.

49. Kouyoumjian, R. G., and L. Peters, Jr.: Range Requirements in Radar Cross Section Measurements, *Proc. IEEE*, vol. 53, pp. 920–928, August 1965.

50. Knott, E. F., et al.: Ref. 31, p. 327.

51. Plonus, M. A.: Theoretical Investigation of Scattering from Plastic Foams, *IEEE Trans.*, vol. AP-13, pp. 88–93, January 1965.

52. Senior, T. B. A., M. A. Plonus, and E. F. Knott: Designing Foamed-Plastic Materials, *Microwaves*, pp. 38–43, December 1964.

53. Knott, E. F., and T. B. A. Senior: Studies of Scattering by Cellular Plastic Materials, *University of Michigan, Rad. Lab. Rept.* 5849-1-F, April 1964.

54. Freeny, C. C.: Target Support Parameters Associated with Radar Reflectivity Measurements, *Proc. IEEE*, vol. 53, pp. 929–936, August 1965.

55. Knott, E. F., et al.: Ref. 31, p. 336.

56. Emerson, W. H.: Electromagnetic Wave Absorbers and Anechoic Chambers through the Years," *IEEE Trans.*, vol. AP-21, pp. 484–490, July 1973.

57. Solomon, L.: Radar Cross Section Measurements: How Accurate Are They? *Electronics*, vol. 35, pp. 48–52, July 20, 1962.

58. Emerson, W. H., and H. B. Sefton, Jr.: An Improved Design for Indoor Ranges, *Proc. IEEE*, vol. 53, pp. 1079–1081, August 1965.

59. King, H. E., F. I. Shimabukuro, and J. L. Wong: Characteristics of a Tapered Anechoic Chamber, *IEEE Trans.*, vol. AP-15, pp. 488–490, May 1967.

60. Dybdal, R. B., and C. O. Yowell: VHF to EHF Performance of a 90-Foot Quasi-Tapered Anechoic Chamber, *IEEE Trans.*, vol. AP-21, pp. 579–581, July 1973.

61. Mentzer, J. R.: The Use of Dielectric Lenses in Reflection Measurements, *Proc. IRE*, vol. 41, pp. 252–256, February 1953.

62. Olver, A. D., and A. A. Saleeb: Lens-Type Compact Antenna Range, *Electron. Lett.*, vol. 15, pp. 409–410, July 1979.

63. Johnson, R. C., H. A. Ecker, and R. A. Moore: Compact Range Techniques and Measurements, *IEEE Trans.*, vol. AP-17, pp. 568–576, September 1969.

64. Burnside, W. D., M. C. Gilreath, B. M. Kent, and G. L. Clerici: Curved Edge Modification of Compact Range Reflector, *IEEE Trans.*, vol. AP-35, pp. 176–182, February 1987.

65. Rudduck, R. C., M. C. Liang, W. D. Burnside, and J. S. Yu: Feasibility of Compact Ranges for Near-Zone Measurements, *IEEE Trans.*, vol. AP-35, pp. 280–286, March 1987.

66. Mensa, D. L.: "High Resolution Radar Imaging," Artech House, Norwood, Mass., 1981.

67. Bedard, Patrick: Stealth Cars, *Car & Driver*, vol. 26, pp. 140 et seq., December 1980.

68. Blore, W. E.: The Radar Cross Section of Ogives, Double-Backed Cones, Double-Rounded Cones, and Cone Spheres, *IEEE Trans.*, vol. AP-12, pp. 582–590, September 1964.

69. Salisbury, W. W.: Absorbent Body for Electromagnetic Waves, U.S. Patent 2,599,944, June 10, 1952.

70. Knott, E. F., et al.: Ref. 31, p. 242.

71. Knott, E. F.: The Thickness Criterion for Single-Layer Radar Absorbents, *IEEE Trans.*, vol. AP-27, pp. 698–701, September 1979.

72. Knott, E. F., et al.: Ref. 31, p. 248.

73. Pelton, E. L., and B. A. Munk: A Streamlined Metallic Radome, *IEEE Trans.*, vol. AP-22, pp. 799–803, November 1974.

CHAPTER 12
GROUND ECHO

Richard K. Moore
The University of Kansas

12.1 INTRODUCTION

Radar ground return is described by σ^0, the differential scattering cross section, or scattering coefficient (scattering cross section per unit area), rather than by the total scattering cross section σ used for discrete targets.[1] Since the total cross section σ of a patch of ground varies with the illuminated area and this is determined by the geometric radar parameters (pulse width, beamwidth, etc.), σ^0 was introduced to obtain a coefficient independent of these parameters.

Use of a differential scattering cross section implies that the return from the ground is contributed by a large number of scattering elements whose phases are independent. This is primarily because of differences in distance that, although small fractions of total distance, are many wavelengths. Superposition of power is possible for the computation of average returns. If this condition is not applicable to a particular ground target, the differential-scattering-cross-section concept has no meaning for that target. For example, a very-fine-resolution radar might be able to resolve a part of a car; the smooth surfaces on the car would not be properly represented by σ^0. On the other hand, a coarser radar might look at many cars in a large parking lot, and a valid σ^0 for the parking lot could be determined.

If a region illuminated at one time by a radar contains n scattering elements and the above criterion is satisfied so that power may be added, the radar equation becomes

$$P_r = \sum_i^n \frac{P_{ti} G_{ti} A_{ri} \sigma_i}{(4\pi R_i^2)^2} = \sum_i^n \frac{P_{ti} G_{ti} A_{ri} (\sigma_i / \Delta A_i) \Delta A_i}{(4\pi R_i^2)^2}$$

Here ΔA_i is an element of surface area, and P_{ti}, G_{ti}, and A_{ri} are values of P_t, G_t, and A_r appropriate for an element at the location of ΔA_i. The factor in parentheses in the numerator of the right-hand expression is the incremental scattering cross section for element $_i$, but this concept is meaningful only in an average. Thus the average power returned is given by

$$P_r = \sum_i^n \frac{P_{ti} G_{ti} A_{ri} \sigma^0 \Delta A_i}{(4\pi R_i^2)^2}$$

Here σ^0 has been used to denote the average value of $\sigma_i / \Delta A_i$. In this formulation, we may pass in the limit from the finite sum to the integral given by

$$\overline{P}_r = \frac{1}{(4\pi)^2} \int_{\text{Illuminated area}} \frac{P_t G_t A_r \sigma^0 dA}{R^4} \tag{12.1}$$

The bar over P_r implies the average value. This integral is not really correct, for there is a minimum size for real, independent scattering centers. Nevertheless, the concept is widely used and is applicable as long as the illuminated area is large enough to contain many such centers.

Figure 12.1 illustrates the geometry associated with Eq. (12.1). Note that, for a rectangular pulse, P_t is either zero or the peak transmitter power but, for other pulse shapes, the variation with t (or R) is significant. Actual pulses are often approximated by rectangular pulses with widths equal to their half-power widths. Real pulses cannot be rectangular after passing through real receiver bandwidths. The transmitting-antenna gain and receiving-antenna aperture are functions of the elevation and azimuth angles:

$$G_t = G_t(\theta, \phi) \qquad A_r = A_r(\theta, \phi) \tag{12.2a}$$

The differential scattering cross section itself is a function of both *look angle* and ground location:

$$\sigma^0 = \sigma^0(\theta, \phi, \text{location}) \tag{12.2b}$$

The integral of Eq. (12.1) must be inverted when σ^0 is measured. With narrow beams and short pulses the inversion is relatively easy, but with the wider beams and longer pulses used in many measurements the values obtained are sometimes poorly defined.

Some authors[2] use a scattering cross section per unit projected area rather than per unit ground area. Figure 12.2 illustrates by a *side view* the difference between ground area and projected area. The ground area is proportional to $\Delta\rho$, and the projected area is smaller. Thus,

$$\sigma^0 A = \gamma d(\text{projected area}) = \gamma \cos \theta \, dA \tag{12.3}$$

or $\qquad\qquad\qquad \sigma^0 = \gamma \cos \theta$

Since both γ and σ^0 are called scattering coefficients, readers of the literature must be especially careful to determine which is being used by a particular author.

Radar astronomers use a different σ:[3]

$$\sigma = \frac{\text{total return power from entire surface}}{\text{power returned from perfect isotropic sphere of same radius}} \tag{12.4}$$

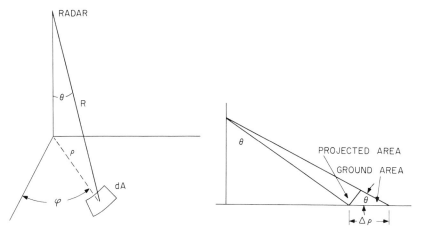

FIG. 12.1   Geometry of the radar equation.   **FIG. 12.2**   Ground area and projected area.

The resulting value for σ is usually much smaller than σ^0 for the planet at vertical incidence and is larger than the values of σ^0 near grazing incidence (return from the limb of the planet).

Relative Importance of Theory and Empiricism.   The theory of radar ground return has been the subject of many publications. The various theories, insofar as they can be confirmed by experiment, provide bases for judging the effects of variations in the dielectric properties of the ground, of the roughness of the ground and nature of vegetative or snow cover, of radar wavelength, and of angle of incidence. Viewed as aids to insight, radar ground-return theories can be extremely useful.

The validity of any ground-return theory must depend on the mathematical model used to describe the surface, as well as on the approximations required to obtain answers. Even the simplest ground surface, the sea, is extremely difficult to describe accurately; it is homogeneous to beyond the skin depth, contains relatively modest slopes, and (except for spray) has no part above another part of the surface. At grazing angles, shadowing of one wave by another might occur. Land surfaces are much more difficult to describe: Imagine an adequate mathematical description of the shape of a forest (when every leaf and pine needle must be described). Furthermore, land surfaces are seldom homogeneous either horizontally or with depth.

Since a true mathematical description of the ground surface appears out of the question, empirical measurements are necessary to describe the radar return from natural surfaces. The role of theory is to aid in interpreting these measurements and to suggest how they may be extrapolated.

Available Scattering Information.   Prior to 1972 the lack of coordinated research programs over the necessary long period resulted in only one really usable set of measurements, that at Ohio State University.[2,4] Since that time extensive measurements have been made from trucks and helicopters by the University of Kansas,[6,7] a group in the Netherlands,[8] and several groups in France.[9] These measurements concentrated especially on vegetation, with the Kansas measurements also including some work on snow and extensive work on sea ice. Most of these measurements were in the 10 to 80° range of incidence angles. Measurements near vertical are scarcer, while well-controlled experiments near grazing are very scarce indeed.

Airborne measurements are necessary to make larger scattering areas accessible. Although airborne programs for special purposes have been legion, curves of scattering coefficient versus angle for a known homogeneous area are scarce. The work at the MIT Radiation Laboratory[10] was early work by Philco Corporation.[11] Goodyear Aerospace Corporation,[12] General Precision Laboratory,[13] and the U.S. Naval Research Laboratory (NRL)[14–16] programs were important early. More recently, the Canada Centre for Remote Sensing (CCRS) has made numerous airborne scatterometer measurements,[17] especially over sea ice. The Environmental Research Institute of Michigan (ERIM),[18] CCRS, the European Space Agency (ESA),[19] and the Jet Propulsion Laboratory (JPL)[20] used imaging synthetic aperture radars (SARs) for some scattering measurements, but most were not well calibrated.

Results of most of these measurements are summarized in Ulaby, Moore, and Fung.[21] More complete summaries of the earlier work and near-grazing studies are in Long.[22] Many applications summaries are also in the "Manual of Remote Sensing."[23] Readers requiring more detailed information should consult these books.

12.2 PARAMETERS AFFECTING GROUND RETURN

Radar return depends upon a combination of system parameters and ground parameters:

Radar system parameters [Eqs. (12.1) and (12.2a and b)]

Wavelength

Power

Illuminated area

Direction of illumination (both azimuth and elevation)

Polarization

Ground parameters

Complex permittivity (conductivity and permittivity)

Roughness of surface

Inhomogeneity of subsurface or cover to depth where attenuation reduces wave to negligible amplitude

Different wavelengths are sensitive to different elements on the surface. One of the earliest known and most striking directional effects is the *cardinal-point* effect in return from cities: Radars looking in directions aligned with primary street grids observe stronger regular returns than radars at other angles. When radars are looking at a normal-incidence angle, horizontally polarized waves are reflected better by horizontal wires, rails, etc., than are vertically polarized waves.

If the geometry of two radar targets were the same, the returns would be stronger from the target with higher complex permittivity because larger currents (displacement or conduction) would be induced in it. Because identical geometries with differing permittivities do not occur in nature, this distinction is not easy to measure. Effective permittivity for ground targets is very strongly influenced by moisture content, since the relative permittivity of liquid water is from about 60 at X band to about 80 at S band and longer wavelengths whereas most dry solids have permittivities less than 8. Attenuation is also strongly influenced by moisture, since wet materials usually have higher conductivity than the same materials dry. Figures 12.3 and 12.4 show the effect of moisture content on properties of plants and of soil. The high permittivity of plants with much moisture means that radar return from crops varies as the plants mature, even when growth is neglected.

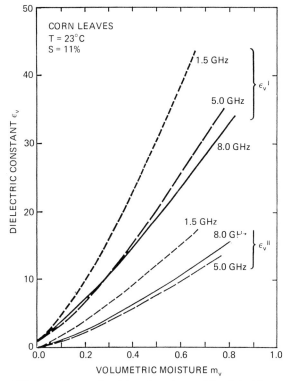

FIG. 12.3 Measured moisture dependence of the dielectric constant of corn leaves at 1.5, 5.0, and 8.0 GHz. S is the salinity of water content in parts per thousand, $\epsilon_v = \epsilon_v' - j\epsilon_v''$ is the complex dielectric constant in Fm^{-1}, and m_v is the volumetric moisture content in kg·m^{-3}. (*After Ulaby, Moore, and Fung.*[21])

The roughness of surfaces (especially natural ones) is difficult to describe mathematically but easy to understand qualitatively. Thus it is easy to see that a freshly plowed field is rougher than the same field after rain and wind have been at work on it. A forest is inherently rougher than either a field or a city. The dif-

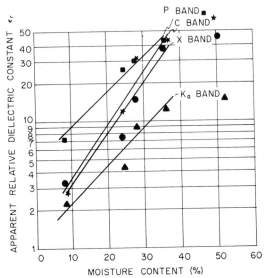

FIG. 12.4 Apparent relative dielectric constant versus moisture content (Richfield silt loam). (*After Lundien.*[24])

ference between the roughness of a city with flat walls interspersed with windowsills, with curbs, cars, and sidewalks, and the roughness of natural areas is harder to see.

Surfaces that are relatively smooth tend to reflect radio waves in accordance with the Fresnel-reflection direction,* and so they give strong backscatter only when the look angle is nearly normal to the surfaces. Rough surfaces, on the other hand, tend to reradiate nearly uniformly in all directions, and so they give relatively strong radar returns in any direction.

The problem of radar scatter is complicated because waves penetrate significant distances into many surfaces and vegetation canopies, and internal reflection and scatter contribute to the return. Measurements of attenuation for field crops[25,26] and grasses[27] show that most of the return is from the upper layers, with some contribution by the soil and lower layers if the vegetation is not very dense. Most of the signal returned from trees is usually from the upper and middle branches when the trees are in leaf,[28–32] although in winter the surface is a major contributor to the signal.

12.3 THEORETICAL MODELS AND THEIR LIMITATIONS

Descriptions of a Surface. Many theoretical models for radar return from the ground assume a rough boundary surface between air and an infinite homogeneous half space. Some include either vertical or horizontal homogeneities in the ground properties and in vegetative or snow covers.

---

*Angle of reflection equals angle of incidence.

Surface descriptions suitable for use in mathematical models are necessarily greatly idealized. Few natural grounds are truly homogeneous in composition over very wide areas. Descriptions of their detailed shape must be simplified if they are to be handled analytically, although computers permit the use of true descriptions. Very few surfaces have ever been measured to the precision appropriate for centimeter-wavelength radars; even for these there is no assurance that scattering boundaries do not exist within a skin depth beneath the surface. Surfaces containing vegetation and conglomerate rocks almost completely defy description.

Statistical descriptions of surfaces are used for most theories, since a theory should be representative of some kind of surface class, rather than of a particular surface, and since exact description is so difficult. The statistical descriptions themselves must be oversimplified, however. Many theories assume isotropic statistics, certainly not appropriate for plowed fields or gridded cities. Most theories assume some kind of model involving only two or three parameters (standard deviation, mean slope, correlation distance, etc.), whereas natural (or human-made) surfaces seldom are so simply described. The theories for vegetation and other volume scatterers have more parameters.

Simplified Models.   Early radar theories for ground return assumed, as in optics, that many targets could be described by a Lambert-law variation of intensity; that is, the differential scattering coefficient varies as $\cos^2 \theta$, with θ the angle of incidence. This "perfectly rough" assumption was soon found wanting, although it is a fair approximation for the return from many vegetated surfaces over the midrange of angles of incidence.

Clapp[10] described three models involving assemblies of spheres, with different spacings and either with or without a reflecting ground plane. These models yield variations from σ^0 independent of angle through $\sigma^0 \propto \cos \theta$ to $\sigma^0 \propto \cos^2 \theta$. Since the sphere models are highly artificial, only the resulting scatter laws need be considered. Most targets give returns that vary more rapidly over part of the incidence-angle regime than these models, although forests and similar rough targets of some depth sometimes give such slowly varying returns.

Since these rough-surface models usually fail to explain the rise in return near vertical incidence, other simplified models combine Lambert's law and other rough-surface scattering models with specular reflection at vertical incidence, and a smooth curve is drawn between the specular value and the rough-surface prediction.

Specular reflection is defined as reflection from a smooth plane and obeys the Fresnel reflection laws.[33] At normal incidence, the specular-reflection coefficient is therefore

$$\Gamma_R = \frac{\eta_g - \eta_0}{\eta_g + \eta_0}$$

where η_0, η_g are the intrinsic impedances of air and earth, respectively. The fraction of total incident power specularly reflected from a rough surface is[5]

$$e^{-2(2\pi\sigma_h/\lambda)^2}$$

where σ_h = standard deviation of surface height variations
         λ = wavelength

Since this proportion is down to 13.5 percent when $\sigma_h = \lambda/2\pi$ and to 1.8 percent when $\sigma_h = \lambda/(2\pi\sqrt{2})$, significant specular reflection is seldom found for the centimeter wavelengths usually used for radar. Nevertheless, a simplified model like this is convenient for some purposes.

Observation of reflected sunlight from rippled water, from roads, and from other smooth surfaces leads to the postulation of a facet theory.[34,35] The only sunlight reaching the observer from smooth surfaces such as water is that from facets for which angle of incidence equals angle of reflection. Thus the observed light may be described by methods of *geometric optics*.

When geometric optics is used to describe radar scatter, the surface of the ground is represented by small flat-plane segments. Radar return is assumed to occur only for facets oriented normal to the radar (normal orientation is required for backscatter so that the reflected wave returns to the source). Thus, if the slope distribution of such facets is known, the fraction normal to a given diverging beam can be established, and from this the return can be obtained. Geometric optics assumes zero wavelength, and so the results of such a theory are wavelength-independent, clearly not in accord with observation.

The facet model for radar return is extremely useful for qualitative discussions, and so modification to make it fit better with observation is appropriate. Two kinds of modification may be used, separately or jointly: considering the actual reradiation pattern of finite-size facets at finite wavelengths[36] and considering the effect of wavelength on establishing the effective number of facets.[37] Thus the scatter from a facet may actually occur in directions other than that requiring that angle of incidence equal angle of reflection. Figure 12.5 illustrates this. For large facets (compared with wavelength) most of the return occurs almost at normal incidence, whereas for small facets the orientation may be off normal by a considerable amount without great reduction in scatter. As the wavelength is increased, the category of a given facet changes from *large* to *small*; eventually the facet is smaller than a wavelength, and its reradiation pattern shape remains almost isotropic from that point. Many facets that would be separate at, say, a 1-cm wavelength are combined at a 1-m wavelength; the result may be a transition from rough- to smooth-surface behavior. Figure 12.6a shows a number of facets of different sizes contributing to a radar return.

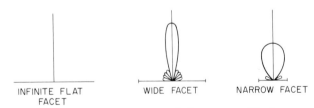

INFINITE FLAT   WIDE FACET   NARROW FACET
   FACET

FIG. 12.5 Normal-incidence reradiation patterns of facets.

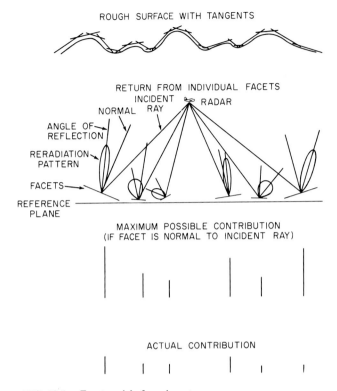

FIG. 12.6*a*   Facet model of a radar return.

Physical Optics Models. Theories based on applications of the Kirchhoff-Huygens principle have been thoroughly developed.[21,36,38–40] The Kirchhoff approximation is that the current flowing at each point in a locally curved (or rough) surface is the same as would flow in the same surface if it were flat and oriented tangent to the actual surface. This assumption permits construction of scattered fields by assuming that the current over a rough plane surface has the same magnitude as if the surface were smooth, but with phase perturbations set by the differing distances of individual points from the mean plane. For surfaces assumed to be azimuthally isotropic, the usual approach yields integrals of the form

$$\frac{1}{\cos^3 \theta} \int e^{-(2k\sigma_h \cos \theta)^2 [1 - \rho(\xi)]} J_0(2k\xi \cos \theta) \xi \, d\xi$$

where $\rho(\xi)$ = spatial autocorrelation function of surface heights
  θ = angle with vertical
  σ_h = standard deviation of surface heights
  k = $2\pi/\lambda$
  J_0 = first-order, first-kind Bessel function

The autocorrelation function of height with distance is seldom known for terrain, although it can be determined on a large scale by analysis of contour maps,[41] and it has been found for some areas by careful contouring at close intervals and subsequent analysis. Because of lack of knowledge of actual autocorrelations, most theory has been developed with artificial functions that are chosen more for their integrability than for their fit with nature; selection among them has been on the basis of which ones yield the best fit between theoretical and experimental scatter curves.

The correlation function first used[42] was gaussian:

$$\rho(\xi) = e^{-\xi^2/L^2} \tag{12.5}$$

where L is the *correlation length*. Not only is this a function that makes the integral analytically tractable, but it also gives exactly the same results as geometric optics.[43] Since it fails, like geometric optics, to explain frequency variation, it cannot be a truly representative correlation function, although it gives a scattering curve that fits several experimental curves near the vertical. The next most frequently used function is the exponential:

$$\rho(\xi) = e^{-|\xi|/L} \tag{12.6}$$

This has some basis in contour-map analysis;[41] the results fit both earth and lunar radar return over a wider range of angles than the gaussian[41,44] (but sometimes not as well near vertical). Furthermore, it has the merit that it exhibits frequency dependence. Resulting expressions for power (scattering coefficient) variations appear in Table 12.1.

TABLE 12.1 Scattering Coefficient Variation

Correlation coefficient	Power expression	Reference
$e^{-\xi^2/L^2}$	$\dfrac{K}{\sin \theta} e^{-(L^2/2\sigma_h^2)\tan^2 \theta}$	42
$e^{-|\xi|/L}$	$\dfrac{K\theta}{\cos^2 \theta \sin \theta} \left(1 + A \dfrac{\sin^2 \theta}{\cos^4 \theta}\right)^{-3/2}$	33

Small-Perturbation and Two-Scale Models. Recognition that existing models were inadequate for describing ocean scatter led to recognition that resonance of the signal with small structures on the surface has a powerful influence on the strength of the signal received.[45,46] Thus a small-perturbation method originally proposed by Rice[47] became the most popular way to describe ocean scatter. Its application to land scatter was not far behind.

The term *Bragg scatter* is often used to describe the mechanism for the small-perturbation model. The idea comes from the concept illustrated in Fig. 12.6b.

A single sinusoidal component of a complex surface is shown with an incoming radar wave at angle of incidence θ. The radar wavelength is λ, and the surface-component wavelength is Λ. When the signal travels an extra distance

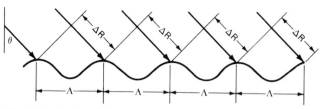

FIG. 12.6*b*   In-phase addition for Bragg scattering; $\Delta R = n\lambda/2$.

$\lambda = 2\Delta R$ between the source and two successive wave crests, the phase difference between the echoes from successive crests is 360°; so the echo signals all add in phase. If this condition is satisfied for a particular Λ and θ, it fails to be satisfied for others. Thus, this is a resonant selection for a given θ of a particular component of the surface Λ. The strength of the received signal is proportional to the height of this component and to the number of crests illuminated by the radar. If the surface has an underlying curvature, the number of illuminated crests satisfying the resonance criterion may be limited by the length of the essentially flat region; otherwise it is limited by the radar resolution.

The theoretical expression for the scattering coefficient is[48]

$$\sigma_{pq}^{\ 0} = 8k^4\sigma_1^2 \cos^4 \theta |\alpha_{pq}|^2 W(2k \sin \theta, 0) \qquad (12.7)$$

where p,q = polarization indices (H or V)
   k = $2\pi/\lambda$ (the radar wavenumber)
   α_{HH} = R_1 (Fresnel reflection coefficient for horizontal polarization)

$$\alpha_{VV} = (\epsilon_r - 1)\ \frac{\sin^2 \theta - \epsilon_r (1 + \sin^2 \theta)}{[\epsilon_r \cos \theta + (\epsilon_r - \sin^2 \theta)^{1/2}]^2}$$

where ϵ_r is the relative permittivity $\epsilon' - j\epsilon''$ and $\alpha_{VH} = \alpha_{HV} = 0$.

$W(2k \sin \theta, 0)$ is the *normalized roughness spectrum* (the Fourier transform of the surface autocorrelation function). It may be written as $W(K, 0)$, where K is the wavenumber for the surface. In terms of the wavelength on the surface Λ,

$$K = 2\pi/\Lambda$$

Thus the component of the surface that satisfies the Bragg resonance condition is

$$\Lambda = \lambda/2 \sin \theta \qquad (12.8)$$

The meaning of this is that the most important contributor to a surface return is the component of surface roughness with wavelength Λ. Even though other components may be *much* larger, the Bragg resonance makes this component more important. On the ocean this means that tiny ripples are more important than waves that are meters high; the same applies for land-surface scatter.

As originally developed, this theory was for perturbations to horizontal flat surfaces, but it was soon modified to handle surfaces with large-scale roughness. The large-scale roughness was assumed to cause a *tilting* of the flat surface to which the small-perturbation theory could be applied. The principal problem with this approach is deciding where in the spectrum lies the boundary between the larger components that do the tilting and smaller components that are Bragg-resonant. Many papers have been written to describe the evolution of this theory; for a complete summary, the reader is referred to Fung's development in Ref. 49.

Other Models.   The theory for volume scatter has led to many papers and continues to evolve. For a review of some of the approaches the reader should consult Fung's summary in Ref. 50 and papers by Kong, Lang, Fung, and Tsang. These models have been used reasonably successfully to describe scatter from vegetation,[51] snow,[52] and sea ice.[53] Models of straight vegetation such as wheat in terms of cylinders have had some success.[2] Corner-reflector effects have been used to describe strong returns from buildings at nonnormal incidence angles.[54] Other specialized models have been used for particular purposes.

Regardless of the model used and the approach applied to determining the field strength, theoretical work only guides understanding. Actual earth surfaces are too complex to be described adequately in any of the models, and the effects of signals that penetrate the ground and are scattered therein are too little known to permit its evaluation.

12.4 FADING OF GROUND ECHOES

The amplitude of ground echoes received by radars on moving vehicles fluctuates widely because of variations in phase shift for return from different parts of the illuminated area. In fact, even fixed radars frequently observe fluctuations in ground echoes because of motions of vegetation, automobiles, etc.

Regardless of the model used to describe a ground surface, signals are, in fact, returned from different positions not on a plane. As a radar moves past a patch of ground while illuminating it, the look angle changes, and this changes the relative distances to different parts of the surface; the result is that relative phase shift is changed. This is the same kind of relative-phase-shift change with direction that is present for an antenna array and results in the antenna pattern. For ground echo the distance is doubled; so the pattern of an echoing patch of length L has lobes of width $\lambda/2L$. This compares with λ/L for an antenna of the same cross-range length. Because the excitation of the *elements* of the scattering array is random, the scattering pattern in space also is random.

This fading phenomenon is usually described in terms of the doppler shift of the signal. Since different parts of the target are at slightly different angles, the signals from them experience slightly different doppler shifts. The doppler shift, of course, is simply the rate of change of phase due to motion. Thus the total rate of change of phase for a given target is

$$\omega = \omega_c + \omega_{di} = \frac{d\phi i}{dt} = \frac{d}{dt}(\omega_c t - 2kR_i) \tag{12.9}$$

where ω_c = carrier angular frequency
ω_{di} = doppler angular frequency for ith target
ϕ_i = phase for ith target
R_i = range from radar to ith target

The doppler shift can be expressed in terms of the velocity vector $\mathbf{v}$ as

$$\omega_{di} = -2k\frac{dR_i}{dt} = -2k\mathbf{v} \cdot \frac{\mathbf{R}_i}{R_i} = -2kv \cos (\mathbf{v}, \mathbf{R}_i) \qquad (12.10)$$

Hence the total field is given by

$$E = \sum_i A_i \exp\left\{ j\left[\omega_c t - \int_0^t 2k\mathbf{v} \cdot \frac{\mathbf{R}_i}{R_i}dt - 2kR_{i0} \right] \right\} \qquad (12.11)$$

where A_i is the field amplitude of the ith scatterer and R_{i0} is the range at time zero.

The only reason the scalar product is different for different scatterers is the different angle between the velocity vector and the direction to the scatterer. This results in a different doppler frequency for each scatterer. If we assume the locations to be random, as most theories do, the received signal is the same as one coming from a set of oscillators with random phases and unrelated frequencies. This same model of a group of randomly phased, different-frequency oscillators is used to describe noise; *thus the statistics of the fading signal and the statistics of random noise are the same.*

This means that the envelope of the received signal is a random variable with its amplitude described by a Rayleigh distribution. Such distributions have been measured for many ground-target echoes.[15] Although the actual distributions vary widely, no better description can be given for relatively homogeneous targets.

When a target is dominated by one large echo (such as a metal roof oriented to give a strong return), the distribution is better described by that for a sine wave in noise. If the large echo is considerably stronger than the mean of the remaining contributors to the return, this approaches a normal distribution about the value for the large echo. In practice, the distribution from large targets may be more complicated than either of the simple models described.

For reference, the two distributions are given:[55]

$$p(v)dv = \frac{v}{\psi_0} e^{-v^2/2\psi_0}dv \qquad \text{(Rayleigh)}$$

$$p(v)dv = \frac{v}{\psi_0^{1/2}} e^{-(v^2 + a^2)/2\psi_0} I_0\left(\frac{av}{\psi_0}\right)dv \qquad \text{(sine wave + noise)}$$

where v = envelope voltage
ψ_0 = mean square voltage
a = sine-wave peak voltage
$I_0(x)$ = Bessel function, first kind, zero order, imaginary argument

Fading-Rate Computations. Doppler frequency calculation is the easiest way to find fading rates. To compute the signal amplitude returned with a particular range of doppler shifts, all signals having such shifts must be summed. This requires knowledge of the contours of constant doppler shift (isodops) on the scattering surface. These contours must be established for each particular geometric arrangement. A simple example is presented here: horizontal motion over a plane earth. This is typical of an aircraft in ordinary cruising flight.

Consider travel in the y direction, with z vertical, and the altitude (fixed) $z = h$. Then

$$\mathbf{v} = \mathbf{1}_y v$$
$$\mathbf{R} = \mathbf{1}_x x + \mathbf{1}_y y - \mathbf{1}_z h$$

where $(\mathbf{1}_x, \mathbf{1}_y, \mathbf{1}_z)$ are unit vectors. Hence

$$v_r = \mathbf{v} \cdot \frac{\mathbf{R}}{R} = \frac{vy}{\sqrt{x^2 + y^2 + h^2}}$$

where v_r is the relative speed. Curves of constant relative speed are also curves of constant doppler shift. The equation of such a curve is

$$x^2 - y^2 \frac{v^2 - v_r^2}{v_r^2} + h^2 = 0$$

This is a hyperbola. The limiting curve for zero relative speed is a straight line perpendicular to the velocity vector. Figure 12.7 shows such a set of constant-doppler-shift contours.

The spectrum of fading can be calculated by a slight rearrangement of the radar equation (12.1). Thus, if $W_r(f_d)$ is the power received between frequencies f_d and $f_d + df_d$, the radar equation becomes

$$W_r(f_d)df_d = \frac{1}{(4\pi)^2} \underset{\substack{\text{Illuminated area} \\ \text{between } f_d \text{ and } f_d + df_d}}{\int \frac{P_t G_t A_r \sigma^0 \, dA}{R^4}} = \frac{df_d}{(4\pi)^2} \int \frac{P_t G_t A_r \sigma^0}{R^4} \left(-\frac{dA}{df_d} \right) \qquad (12.12)$$

This is an integral in which the area element between f_d and $f_d + df_d$ is expressed in terms of coordinates along and normal to the isodops. Such coordinates must be established for each particular case.

Figure 12.8 shows the geometry for horizontal travel. The coordinate σ is along the isodop, and η is normal to it. We can express Eq. (12.12) in terms of these coordinates as

$$W_r f_d = \frac{d\eta}{df_d} \left[\frac{\lambda^2}{(4\pi)^3} \right] \int_{\text{strip}} \left[\frac{P_t G^2 \sigma^0 d\xi}{R^4} \right] \qquad (12.13)$$

Note that P_t, the transmitted power, is nonzero in the integral only for the time it illuminates the ground. In pulse radars, only that part of the ground area providing signals back to the radar at a particular time can be considered to have finite P_t, and so the range of frequencies that can be present is limited by the pulse, as well as by the antennas and the maximum velocity.

Another example is shown in Fig. 12.9. This is the small illuminated area for a narrow-beam, short-pulse system. Here we can make linear approximations with-

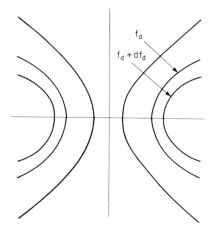

FIG. 12.7 Contours of constant doppler frequency shift on a plane earth due to horizontal motion.

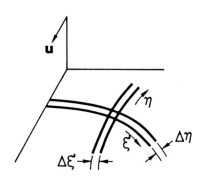

FIG. 12.8 Geometry of complex fading calculations. (*From Ulaby, Moore, and Fung.*[21])

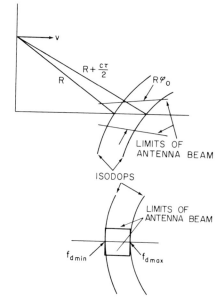

FIG. 12.9 Geometry of doppler-shift calculations for an airborne search radar.

out too much error. A pulse of length τ is transmitted from an antenna of beamwidth ϕ_0. For the simple illustration given here, we assume the pulse to be transmitted directly ahead of the horizontally moving vehicle. We may simplify the problem by assuming a rectangular illuminated area, $R\phi_0$ by $c\tau/(2 \sin \theta)$. Furthermore, the curvature of the isodops may be neglected, and so the doppler frequency is assumed to be the same for all maximum-range points and the same for all minimum-range points. With this assumption,

$$f_{d\max} = \frac{2v}{\lambda} \sin \theta_{\max}$$
$$f_{d\min} = \frac{2v}{\lambda} \sin \theta_{\min}$$

Thus the total width of the doppler spectrum is

$$\Delta f_d = \frac{2v}{\lambda}(\sin \theta_{\max} - \sin \theta_{\min})$$

For short pulses and angles away from vertical, this is

$$\Delta f_d \approx \frac{2v}{\lambda}\Delta\theta \cos \theta$$

In terms of pulse length, it becomes

$$\Delta f_d = \frac{vc\tau}{2h\lambda} \frac{\cos^3 \theta}{\sin \theta} \tag{12.14}$$

If the angular difference across the illuminated rectangle is small enough so that σ^0 is essentially constant, the doppler spectrum is a rectangle from $f_{\min}$ to $f_{\max}$.

In practice, antenna beams are not rectangular. The result is that the doppler spectrum for a side-looking radar like that of the example is not rectangular but rather has the shape of the antenna along-track pattern. Thus, if the antenna pattern in the along-track direction is $G = G(\beta)$, with β the angle off the beam center, we can express β in terms of the doppler frequency f_d as

$$\beta = f_d\lambda/2v$$

and the spectrum is

$$W(f_d) = \frac{\lambda^3 P_t \sigma^0 r_x}{2(4\pi)^3 R^3} G^2 \left[\frac{\lambda f_d}{2v}\right]$$

where r_x is the horizontal resolution in the range direction. Of course, the half-power beamwidth may be used as an approximation, resulting in the bandwidth given by Eq. (12.13).

Effect of Detection. The effect of detecting narrowband noise has been treated extensively in the literature. Here it is necessary only to show the postdetection spectrum of the preceding example and to consider the number of independently fading samples per second. Figure 12.10 shows the spectrum before and after detection. If square-law detection is assumed, the postdetection spectrum is the self-convolution of the predetection spectrum. Only the part that passes the low-pass filters in a detector is shown in the figure. The rectangular *RF spectrum* has become a triangular *video spectrum*.

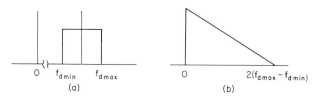

FIG. 12.10 Spectrum of fading from a homogeneous small patch (*a*) before and (*b*) after detection.

This spectrum describes the fading of the detector output for a CW radar. For a pulse radar, the spectrum is sampled by the PRF (pulse repetition frequency). If the PRF is high enough so that the entire spectrum can be reproduced (the PRF is higher than the Nyquist frequency, $2\Delta f_d$), the diagram indicated is that of the spectrum of the samples of a received pulse at a given range. Figure 12.11 shows a series of actual pulses, followed by a series of samples at range R_1. The spectrum of Fig. 12.10 is the spectrum of the envelope of samples at R_1 (after low-pass filtering). The spectrum of fading at a different range (or vertical angle) is different, in accord with Eq. (12.13).

For many purposes, the number of *independent* samples is important, since these may be treated by using the elementary statistics of uncorrelated samples. For continuous integration, the effective number of independent samples is[55]

$$N = \frac{\overline{P_e^2}T}{2\int_0^T \left[1 - \frac{x}{T}\right]R_{sf}(x)dx} \qquad (12.15)$$

where $\overline{P}_e$ is the mean envelope power, T is the integration (averaging) time, and $R_{sf}(t)$ is the autocovariance function for the detected voltage. For many practical purposes, if N is large, it may be approximated by

$$N \approx BT \qquad (12.16)$$

where B is the effective IF bandwidth. For the effect of short integration time, see Ref. 56.

Fading samples can, of course, also be independent because motion of the vehicle causes the beam to illuminate a different patch of ground. Thus, in a particular case, the independent-sample rate may be determined either by the motion of the illuminated patch over the ground or by the doppler effect, or by some combination of the two.

The number of independent samples determines the way in which the Rayleigh or other distributions may be applied. Thus, if 100 pulses give only 10 independent samples, the variance of the mean obtained by integrating these pulses is much greater than would be true if all 100 pulses were independent.

Doppler-based systems, such as doppler navigators and synthetic aperture radar systems, depend on the predetection spectrum for their operation, since they are coherent and do not use ordinary detection.

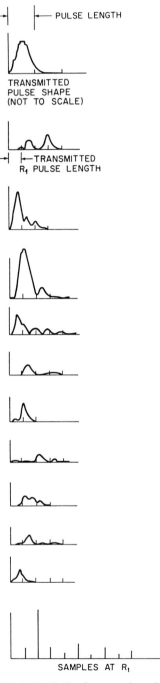

FIG. 12.11   Fading for successive pulses of a radar with ground target.

Moving-Target Surfaces. Sometimes clutter has internal motion. This can occur when fixed radars are used to observe movement of the sea and the land. On land, clutter motion is usually due to moving vegetation, although moving animals and machines create similar effects. The radar return from an assembly of scatterers like those of Fig. 12.8 can change because of motion of the individual scatterers just as it changes because of motion of the radar. Thus, if each scatterer is a tree, the waving of the trees as the wind blows causes relative phase shifts between the separate scatterers; the result is fading. For a fixed radar, this may be the only fading observed, except for very slow fading due to changes in refraction. For a moving radar, this motion of the target changes the relative velocities between target element and radar, so that the spectrum is different from that for a fixed surface. The width of the spectrum due to vehicle motion determines the ability of the radar to detect this target motion.

12.5 MEASUREMENT TECHNIQUES FOR GROUND RETURN

Special-purpose instrumentation radars and modified standard radars may be used to determine the ground return. Since the ground return is almost invariably due to scattering, these systems are termed *scatterometers*. Such systems may use CW signals with or without doppler processing, but they may also use both pulse and FM techniques. Scatterometers capable of measuring response over a wide range of frequencies are called *spectrometers*.[57] Various antenna patterns from pencil beams to fan beams may be used.

CW and FM-CW Systems. The simplest scatterometer uses a stationary CW radar. Such systems are not very flexible, but they are discussed here in some detail to illustrate calibration techniques that also apply to the more complex systems.

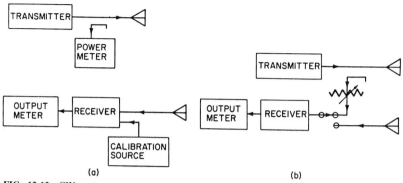

FIG. 12.12 CW-scatterometer-system block diagram. (a) Separate transmitter and receiver calibration. (b) Calibration of the ratio of received to transmitted power.

The CW scatterometer is shown in block form in Fig. 12.12. To evaluate σ^0 the ratio of transmitted to received power is required. The system in Fig. 12.12a measures transmitter power and receiver sensitivity separately. The transmitter feeds an antenna through a directional coupler so that a portion of the energy may be fed to a power meter. The receiver operates from a separate antenna (electrically isolated). The output of the receiver is detected, averaged, and displayed on a meter, oscilloscope, or other display or recorder. Its sensitivity must be checked by use of a calibration source. The calibrated signal may be fed through the receiver at a time when the transmitter is off. Figure 12.12b shows a similar arrangement in which the signal from the transmitter is attenuated a known amount and used to check the receiver. By comparing the output from the attenuated transmitter signal with that received from the ground, the scattering cross section may be determined without actually knowing the transmitted power and the receiver gain.

The calibrations shown in Fig. 12.12 are incomplete without knowledge of the antenna patterns and absolute gains. Since accurate gain measurements are difficult, absolute calibrations may be made by comparing received signals (with proper relative calibration) from the target being measured and from a *standard target*. Standard targets may be metal spheres, Luneburg-lens reflectors, metal plates, corner reflectors, or active radar calibrators (ARCs—actually repeaters).[58] Of the passive calibrators, the Luneburg-lens reflector is best, since it has a large cross section for its volume and has a very wide pattern so that alignment is not critical. Luneburg-lens reflectors are used for making strong radar targets of small vessels, and they may be obtained from companies that supply that market. For discussion of the relative merits of different passive calibration targets, see Ulaby, Moore, and Fung.[59]

The ideal receiver would respond linearly to its input, so that a single calibration at one input level would suffice for all levels. The usual receiver, however, has some nonlinearities due to detector properties and to saturation of its amplifiers by large signals. Figure 12.13 shows a typical input-output curve for a receiver. Two equal increments in input signal (Δ^i), as shown, produce different increments in output because of the nonlinearity of this curve. For this reason, receiver calibration must be performed over a range of input levels, and the nonlinearities must be compensated for in the data processing.

CW scatterometers depend on antenna beams to discriminate different angles

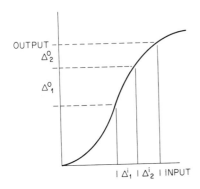

FIG. 12.13 Typical receiver input-output curve. Illustrated is the effect of non-linearity.

of incidence and different targets. Usually assumptions are made that the antenna pattern has constant gain within the actual 3 dB points and zero gain outside, but this clearly is not an accurate description. If large targets appear in the locations illuminated by the side of the main lobe or the minor lobes, their signals may contribute so much to the return that it is significantly changed. Since this changed signal is charged to the direction of the major lobe by the data reduction process, the resulting value for σ^0 is in error. Responses at vertical incidence frequently cause trouble, for vertical-incidence signals are usually fairly strong. Thus the antenna pattern must be accurately known and taken into account in the data analysis. A pattern with strong minor lobes may be simply inadmissible.

The scattering coefficient is determined by applying

$$P_r = \frac{P_t \lambda^2}{(4\pi)^3} \int_{\substack{\text{Illuminated} \\ \text{area}}} \frac{G_t^2 \sigma^0 dA}{R^4}$$

The integration is over whatever area is illuminated significantly, including the regions hit by the minor lobes. The usual assumption is that σ^0 is constant over the illuminated area, so that

$$P_r = \frac{P_t \lambda^2 \sigma^0}{(4\pi)^3} \int_{\substack{\text{Illuminated} \\ \text{area}}} \frac{G_t^2 dA}{R^4} \tag{12.17}$$

This assumption would be true only if the antenna confined the radiated energy to a very small spread of angles and to a fairly homogeneous region. The resulting expression is

$$\sigma^0 = \frac{(4\pi)^3 P_r}{P_t \lambda^2 \int_{\substack{\text{Illuminated} \\ \text{area}}} (G_t^2/R^4) dA} \tag{12.18}$$

Note that only the ratio of transmitted to received power is required, and so the technique of Fig. 12.12b is justified. Sometimes R, G_t, or both are assumed constant over the illuminated area, but such an approximation to Eq. (12.18) should be attempted only after checking its validity for a particular problem.

If the result of applying the technique of Eq. (12.18) to a set of measurements indicates that σ^0 probably did vary across the significantly illuminated area, this variation may be used as a first approximation to determine a function $f(\theta)$ describing the θ variation of σ^0, and a next-order approximation then becomes

$$\sigma^0 = \frac{(4\pi)^3 P_r}{P_t \lambda^2 \int\limits_{\substack{\text{Illuminated} \\ \text{area}}} [f(\theta)G_t^2/R^4] \, dA} \tag{12.19}$$

Proper scattering measurements demand an accurate and complete measurement of antenna gain G_t. This can be a very time-consuming and expensive process, particularly when the antenna is mounted on an aircraft or other metallic object. Nevertheless, complete patterns are a must for good scatter measurements.

Range-Measuring Systems. Radar's ability to separate returns from different ranges can be used advantageously along with directive antenna beams to simplify the scattering measurements. Most ranging scatterometers use either pulse modulation or FM, although more exotic modulations could also be used. The discussion here treats pulse systems, but since all other range-measuring systems can be reduced to equivalent pulse systems most results are general.

Figure 12.14 shows the way in which pulse measurement of range is used. Figure 12.14a shows a circular pencil beam. At angles near grazing, the illuminated patch set by the circular antenna pattern becomes rather long (the patch is an ellipse), and use of the pulse length to confine illumination to a part of the patch is helpful. Many systems that use beamwidth to set the measured area near vertical use range resolution for angles beyond, say, 60°.

Figure 12.14b shows an antenna pattern that takes better advantage of the possibilities of range measurement. A fan beam is used to illuminate a narrow strip along the ground, and the range resolution permits separating the returns from different angles by the time they return. This technique is especially effective at angles away from the vertical, for the resolution near the vertical is much poorer than near grazing. The simple approach assumes a constant gain across the beam and zero elsewhere:

$$G_t = 0 \qquad \phi_a < 2 \, f_0/2 \qquad \text{or} \qquad \phi_a > f_0/2$$
$$G_t = G_o \qquad - \phi_0/2 < f_a < f_0/2$$

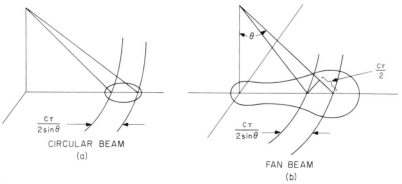

FIG. 12.14 Range resolution applied to scatterometry. (a) Improving one dimension of a circular-beam illumination pattern. (b) Use with a fan beam.

where ϕ_0 = beamwidth
ϕ_a = transverse angle with respect to antenna axis

With the further assumption that σ^0 is essentially constant and that the difference in range across a resolution element is negligible, the expression for σ^0 becomes

$$\sigma^0 = \frac{P_r(4\pi)^3 R^3 \sin\theta}{P_t \; \lambda^2 G_0 \phi_0 \, r_R}$$

where r_R is the short-range resolution.

Janza has reported details of calibration problems with a range-measuring pulsed radar scatterometer.[60,61]

CW-Doppler Scatterometers. A convenient way to measure the scattering coefficient at many angles simultaneously is with a CW system in which the relative velocities corresponding to different angles are separated by separating their doppler frequencies. The use of a fan beam with such a system permits the simultaneous measurement of scattering coefficients at points ahead of and behind the aircraft carrying the radar. Figure 12.15 shows this. The pattern of the antenna illumination on the ground is shown intersected by two isodops (lines of constant doppler frequency), with the width of the spectrum between them shown on the diagram. The distance between them can be seen to be

$$\Delta\rho = R(\sin\theta_2 - \sin\theta_1)$$

and
$$\Delta f_d = \frac{2v}{\lambda}(\sin\theta_2 - \sin\theta_1)$$

Thus the width of the element on the ground is related to the doppler frequency bandwidth by

$$\Delta\rho = \frac{R\lambda}{2v}(\Delta f_d)$$

Where this technique is applied to the radar equation and the following are assumed:

1. σ^0 constant in the illuminated area

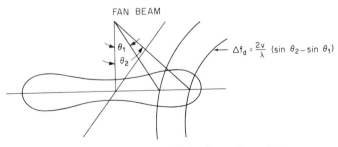

FAN BEAM

θ_1

θ_2

$\Delta f_d = \frac{2v}{\lambda}\ (\sin\theta_2 - \sin\theta_1)$

FIG. 12.15   Resolution in a fan-beam CW-doppler scatterometer.

2. Antenna gain constant over its beamwidth and zero elsewhere

3. Range variation across the small illuminated area negligible

$$P_r = \frac{P_t\lambda^2}{(4\pi)^3} \int \frac{G_t^2\sigma^0 dA}{R^4} = \frac{P_t\lambda^4\sigma^0 G_0^2\psi_0\Delta f_d}{2vR^2} \tag{12.21}$$

and so

$$\sigma^0 = \frac{P_r}{P_t} \frac{2vR^2}{\lambda^4 G_0^2\psi_0\Delta f_d} \tag{12.22}$$

Doppler scatterometers need not use fore-and-aft beams. The Seasat[62] and N-SCATT[63] spaceborne doppler scatterometers were designed with beams pointed (squinted) ahead and behind the normal to the ground track.

Independent Samples Required for Measurement Accuracy. The Rayleigh distribution describes the fading signal fairly well. If we assume a Rayleigh distribution of fading, the number of independent samples required for a given accuracy is shown in Fig. 12.16. The *range* defined in this figure is the range of mean values lying between 5 and 95 percent points on the distribution. This accuracy range is independent of any accuracy problems associated with calibration and knowledge of the antenna pattern.

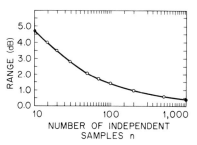

FIG. 12.16 Accuracy of averages for fading signals.

The precision of the measurement depends upon the number of independent samples, not on the total number of samples. The number of independent samples can be found from Eq. (12.15) or Eq. (12.16) after suitable analysis. This analysis assumes that only doppler fading contributes to independence but motion from one cell to another also adds independent samples. Thus, the total number of such samples is approximately the product of the number calculated from Eq. (12.13) and the number of ground cells averaged. Figure 12.17 shows some examples of the effect of the angle of incidence on the number of independent samples for a horizontally traveling scatterometer with a forward-pointed beam.

Study of the results obtained in this type of analysis indicates that, in regions where the scattering coefficient does not change rapidly with angle, the widest possible angular width (obtained by a longer pulse or a wider filter for a CW-doppler system) results in the maximum number of independent samples for a given distance traveled along the ground.

Near-Vertical Problem Most published radar return data purporting to include vertical incidence gives vertical-incidence scattering coefficients that are too small. This is a consequence of a fundamental problem in measuring near the vertical with a finite beamwidth or pulse length. Near-vertical radar returns from most targets drop off rapidly as the angle with the vertical is increased. Thus the

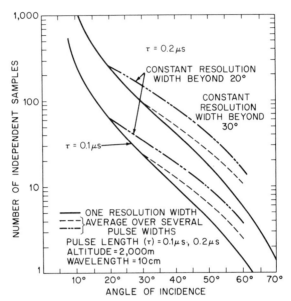

FIG. 12.17 Examples of the variation with angle of incidence of the number of independent samples for a scatterometer.

measuring beamwidth or pulse width usually encompasses signals from regions having values for σ^0 many decibels apart. Since the scattering coefficient varies much more rapidly near the vertical than at angles beyond 10 or 20° from the vertical, the problem is much more severe at the vertical. Furthermore, the problem is complicated at the vertical by the fact that the angular scale terminates there, so that a beam centered at the vertical illuminates weaker targets (σ^0) on both sides of its pattern, whereas a beam away from the vertical illuminates stronger signals on one side and weaker signals on the other.

Figure 12.18 shows what happens for a steeply descending curve of σ^0 versus θ. The radar return integral from Eq. (12.1) is a convolution integral; the figure shows the convolution of the beam pattern with the σ^0 curve. Clearly the average at the vertical is lower than it should be to indicate properly the variation of σ^0 near the vertical.

Figure 12.19 shows an example[64] based on the theoretical scattering coefficient for the sea derived from the spectra reported by the Stereo Wave Observation Project.[65] The effect of different beamwidths is clearly shown.

With a pulse or other range-measuring system, reported values are always in error because, as indicated above, it is almost impossible to resolve a narrow range of angles near the vertical.

Ground and Helicopter Scatterometers and Spectrometers. Many ground scattering measurements have been made with systems mounted on boom trucks and helicopters. Most of these are FM-CW systems[66,67] that use wide bandwidth to obtain extra independent samples rather than for fine resolution. Some use very wide bandwidth to obtain fine range resolution to locate sources of scattering.[68] Most have multiple-polarization capability, and some are capable of polarimetry because the phase of two received signals with orthogonal polarization can be measured.

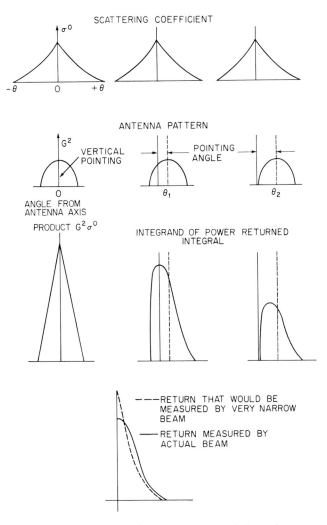

FIG. 12.18 How finite beamwidth causes a near-vertical error in measuring the scattering coefficient.

The basic elements of an FM-CW scatterometer are shown in Fig. 12.20. The swept oscillator must produce a linear sweep; this is easy with yttrium-iron-garnet (YIG)-tuned oscillators but requires linearizing circuits if tuning uses a varactor. If dual antennas are used (as shown), the overlap of the beams must be considered.[69] Single-antenna systems are sometimes used, with a circulator isolating transmitter and receiver; their performance is somewhat poorer than that of dual-antenna systems because of internal reflections and leakage through the circulator.

Two versions of the control and data-handling part of an FM-CW scatterometer are shown in Figs. 12.21 and 12.22. Figure 12.21 shows the common range-tracking scatterometer. This system can be used to measure

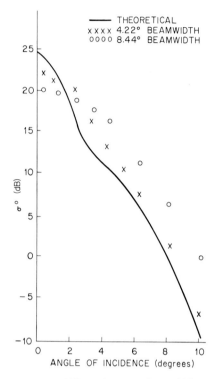

FIG. 12.19 Effect of antenna beamwidth on the measured scattering coefficient as a function of angle of incidence.

surface scattering coefficients when the distance between radar and target is changing, such as with a fixed radar observing the sea or a radar on a helicopter. If the scatterometer is mounted on a boom truck, the range tracker is not needed; but it is convenient because the range changes as the angle of incidence is changed. Figure 12.22 shows the kind of system that may be used to measure scattering from within a volume. By determining the spectrum of the return, the user can establish the scattering from different ranges. This system has been used in determining the sources of scatter in vegetation[25–27] and snow.

Ultrasonic waves in water can be used to simulate electromagnetic waves in air.[70–72] Because of the difference in velocity of propagation an acoustic frequency of 1 MHz corresponds with a wavelength of 1.5 mm. Such a wavelength is of a convenient size for many modeling measurements, and, of course, equipment in the 1-MHz region is in many ways easier to operate than equipment in the microwave region; certainly it is much easier to operate and less expensive than microwave equipment operating at a 1.5-mm wavelength.

Acoustic plane waves and electro-

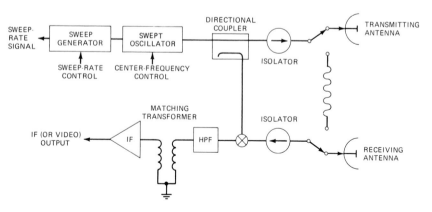

FIG. 12.20 Basic block diagram of an FM-CW scatterometer RF section.

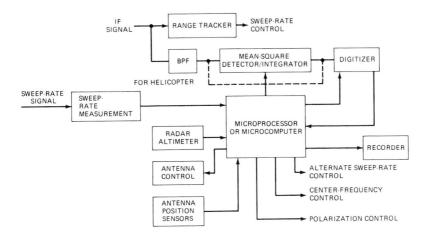

FIG. 12.21 Basic block diagram of an FM-CW range-tracking scatterometer: control and data-handling system.

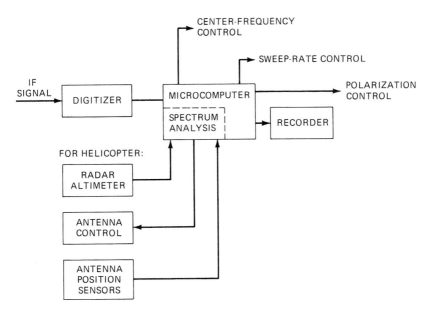

FIG. 12.22 Basic block diagram of an FM-CW range-discriminating scatterometer: control and data-handling system.

magnetic plane waves satisfy the same boundary conditions. When the scattering surfaces are not plane and when angles of incidence are rather oblique, the analogy between acoustic and electromagnetic waves is less valid.

Scattering Coefficients from Images. Radar images produced by real or synthetic aperture radars can be used for scattering coefficient measurement. Unfortunately, most such systems are uncalibrated; so the results are somewhat dubious, even on a relative basis when images are produced on different days. Relative calibration has been introduced into some systems.[12,18,20,73–75] Absolute calibration, which also serves as relative calibration in some cases, can be achieved by using strong reference targets, with the ARC repeaters especially suitable.[76] Another approach that has been used is to measure scattering from reference areas with a ground-based or helicopter system that is well calibrated and to compare the images to these measured values.[73,77]

Bistatic Measurements. Measurements of ground return when the receiver and transmitter are separated are comparatively rare. These measurements are very difficult to make from aircraft because it is necessary that both transmitter and receiver antennas look at the same ground point at the same time and that the signal be correlated with known antenna look angles. Furthermore, it is difficult to know the polarization, and the exact size and shape of the common area illuminated by the antenna beams are sometimes difficult to determine. For this reason, few bistatic measurements from aircraft have been reported in the literature.[78]

Laboratory bistatic measurements have been made by both the Waterways Experiment Station[24] and Ohio State University[2,4] groups using electromagnetic waves and by the University of Kansas[71] group using acoustic waves. Bistatic measurements of laser radiation have been made at Bell Telephone Laboratories,[79] and C-band measurements of buildings at the University of Kansas.[80]

Because of the antenna orientation problems, most electromagnetic bistatic measurements are only for forward scatter; that is, the receiver, transmitter, and target all lie in the same vertical plane. The acoustic measurements and optical measurements are easier to make over a wide range of angles and have been made with a fixed incidence angle and scatter directions covering the entire hemisphere.

Bistatic measurements call for additional calibration complications when made outside the laboratory because an absolute reference for both transmitter power and receiver sensitivity must be used. In the laboratory, however, it is possible to use techniques similar to those for monostatic measurements.

12.6 GENERAL MODELS FOR SCATTERING COEFFICIENT (CLUTTER MODELS)

Scatter measurements made during the 1970s allowed generation of models for average backscatter from large areas. In particular, these included measurements with the Skylab radiometer-scatterometer RADSCAT[83] and with truck-mounted microwave active spectrometers (MAS)[81] by the University of Kansas. Two different models were developed based on the same data, one a linear model and one a more complicated formulation. Here we present only the linear model.

These models are for *averages*, and the models do not include variations about the average. However, analysis of Shuttle Imaging Radar-B (SIR-B) data permits some estimates to be made of the variability to be expected for different sizes of illuminated footprint.

The general characteristics of radar backscatter over the range of angles of incidence have been known for decades. Figure 12.23 shows these. For like-polarized waves, one can break scatter into three angular regimes: near-vertical (the *quasi-specular region*), intermediate angles from 15 to about 80° (the *plateau region*), and near-grazing (the *shadow region*). Cross-polarized scatter does not have separate quasi-specular and plateau regions (the plateau extends to vertical), and too little is known to establish whether a shadow region exists.

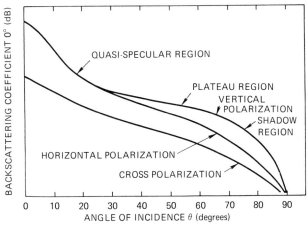

FIG. 12.23 General characteristics of scattering coefficient variation with angle of incidence. (*From Ulaby, Moore, and Fung.*[21])

For nearly every type of terrain, the measured data fits closely to the form

$$\sigma^0 = A_i e^{-\theta/\theta_i} \qquad (12.23a)$$

or
$$\sigma^0_{dB} = 10 \log A_i - 4.3434(\theta/\theta_i) \qquad (12.23b)$$

where A_i and θ_i differ for the near-vertical and midrange regions. Figure 12.24 shows an example of this variation. No theory gives exactly this result, but nearly all measurements fit such a model closely, and the model approximates most theoretical curves well over the relevant regions. This simple result means that simple clutter models may be developed and used although more complex models may be necessary for some remote-sensing applications.

The basis for the linear model[82] is a combination of the Skylab results over North America[83] and those from Kansas cropland measurements over three complete seasons with the microwave active spectrometer (MAS).[84] The 13.9-GHz Skylab RADSCAT had a ground footprint of from a 10-km circle at vertical to an ellipse of 20 by 30 km at 50°. The MAS had footprints at 50° ranging from 5.5 by 8.5 m at 1.1 GHz to 1.4 by 2.1 m at 17 GHz, but millions of measurements were averaged for the model. Because the Skylab data was at only one frequency and the responses for the two experiments were essentially the same at that

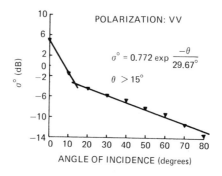

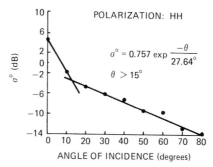

FIG. 12.24 Regression of average of all 1974 and 1975 13.8-GHz cropland data obtained with a microwave active spectrometer. (*From Moore, Soofi, and Purduski.*[82])

frequency, the frequency response shown in the model depends entirely on the MAS measurements.

The summer Skylab observations included deserts, grassland, cropland, and forests, whereas the Kansas measurements were only of cropland. However, early and late in the growing season the cropland was essentially bare, similar to the summer desert except for soil moisture content. During the height of the growing season the crops were dense enough so that scatter was similar to that from forests. Thus, the overall model seems representative of summer conditions averaged over all of North America.

The model takes the form

$$\sigma^0_{dB}(f,\theta) = A + B\theta + Cf + Df\theta \qquad 20° \le \theta \le 70° \qquad (12.24a)$$

where A, B, C, and D take on different values for different polarizations and above and below 6 GHz. The frequency response below 6 GHz is much more rapid than above 6 GHz. Moreover, at frequencies above 6 GHz the frequency response is independent of angle, so that $D = 0$. For lower frequencies, the frequency response is angle-dependent.

For angles less than 20°, only two points were available, 0° and 10°; so separate frequency regressions were run at each of these angles. The model for these angles is

$$\sigma^0_{dB}(f,\theta) = M(\theta) + N(\theta)f \qquad \theta = 0°, 10° \qquad (12.24b)$$

The frequency responses below 6 GHz differed for the two years; so the models have separate values of the constants for 1975 and 1976. The year 1976 was very dry in Kansas; so the 1975 values are probably more representative, but both are given here. Values of the constants are in Table 12.2. Figure 12.25 shows the clutter model for the midrange of angles as a function of frequency. The figure is only for vertical polarization because results are so similar for vertical and horizontal.

Ulaby developed a different, more complex model from the Kansas vegetation data.[85] This model fits curves rather than straight lines to the measured data. For most purposes the straight-line model is adequate, and it is much easier to use.

A straight-line model for snow-covered grassland similar to that for vegetation depends on a more limited data set.[86,87] The data was for only one season in Colorado when the snow was only about 50 cm deep. This means that the signal probably penetrated to the ground surface at frequencies below about 6 GHz. Nevertheless, the model indicates the kind of results to be expected for this important situation. Table 12.3 gives the resulting constants to use in Eq. (12.24a).

Snow scatter depends strongly on the free-water content of the upper layer of snow; so scatter is much lower from the wet daytime snow (where solar melting has commenced) than for the dry nighttime snow. Hence, different models must be used for day and night; compare the day and night measurements shown in Fig. 12.26. The difference between day and night scatter from snow is even more pronounced at 35 GHz, but the model does not include 35 GHz because no data exists between 17 and 35 GHz.

Although no specific clutter model has been developed for forest, results from the Skylab RADSCAT and Seasat scatterometer show that the Amazon rain forest scatters almost independently of the angle of incidence even near vertical.[88] The mean measured value at 33° was -5.9 ± 0.2 dB at 13.9 GHz. Observations with SIR-B indicated that this lack of angular variation of σ^0 also is present at 1.25 GHz, but lack of calibration prohibited learning the level of scatter at this frequency.

TABLE 12.2   Constants for Linear Scattering Model (Summer)*

Eq.	Polarization	Angular range,°	Frequency range, GHz	Constant A or M, dB	Angle slope B or N, dB	Frequency slope C, dB/GHz	Slope correction D, dB/ (° · GHz)
12.24a	V	20–60	1–6 (1975)	−14.3	−0.16	1.12	0.0051
	V	20–50	1–6 (1976)	−4.0	−0.35	−0.60	0.036
	V	20–70	6–17	−9.5	−0.13	0.32	0.015
	H	20–60	1–6 (1975)	−15.0	−0.21	1.24	0.040
	H	20–50	1–6 (1976)	−1.4	−0.36	−1.03	
	H	20–70	6–17	−9.1	−0.12	0.25	
12.24b	V and H	0	1–6 (1975)	7.6	. . .	−1.03	
	V and H	0	1–6 (1976)	6.4	. . .	−0.73	
	V and H	0	6–17	0.9	. . .	0.10	
	V and H	10	1–6 (1975)	−9.1	. . .	0.51	
	V and H	10	1–6 (1976)	−3.6	. . .	−0.41	
	V and H	10	6–17	−6.5	. . .	0.07	

*After Moore, Soofi, and Purduski.[82]

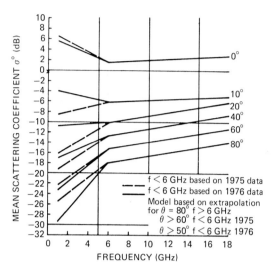

FIG. 12.25 General land-scattering-clutter model (vertical polarization). Horizontal polarization is very similar. (*From Moore, Soofi, and Purduski.*[82])

TABLE 12.3 Regression Results for Ground-Based Measurements of Snow-Covered Ground*

Time of day	Polarization	Frequency range, GHz	Constant A, dB	Angle slope B, dB/°	Frequency slope C, dB/ GHz	Slope correction D, dB/ (° · GHz)
Day	V	1–8	−10.0	−0.29	0.052	0.022
Day	V	13–17	0.02	−0.37	−0.50	0.021
Day	H	1–8	−11.9	−0.25	0.55	0.012
Day	H	13–17	−6.6	−0.31	0.0011	0.013
Night	V	1–8	−10.0	−0.33	−0.32	0.033
Night	V	13–17	−10.9	−0.13	0.70	0.00050
Night	H	1–8	−10.5	−0.30	0.20	0.027
Night	H	13–17	−16.9	−0.024	1.036	−0.0069

*After Moore, Soofi, and Purduski.[82]
NOTE: θ = 20 to 70°. Values of coefficients in this table also are considered those of the model.

The models described above are based on averages over very large areas. For this situation the variability from place to place is small, particularly in the midrange of angles. Figure 12.27 shows the mean and upper and lower decile values measured by the Skylab RADSCAT over North America. The larger variation near vertical apparently results from the effect of nearly specular reflection from water bodies. When the footprint is smaller, more variability occurs. This is shown in Fig. 12.28 from a study of the variation of scatter observed by SIR-B

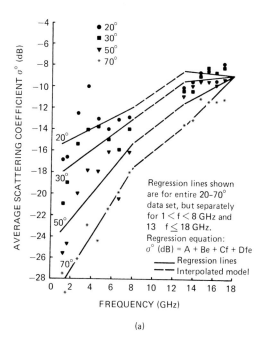

(a)

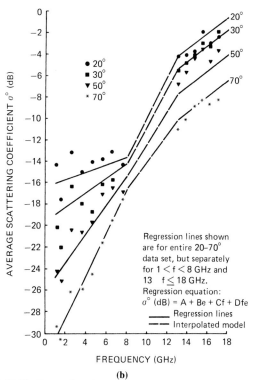

(b)

FIG. 12.26 Regressions for vertical-polarization clutter model for snow: (*a*) day and (*b*) night. Note the large differences. Horizontal polarization is similar. (*From Moore, Soofi, and Purduski.*[82])

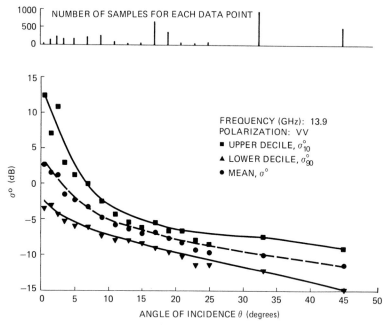

FIG. 12.27 Angular patterns of the mean, upper decile, and lower decile of Skylab scatterometer observations over North America during the summer season. (*From Moore et al., University of Kansas Remote Sensing Laboratory Technical Report 243-12, 1975.*)

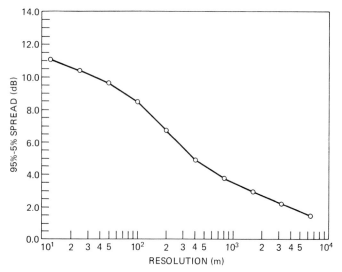

FIG. 12.28 90 percent range of pixel amplitude versus resolution.

with averages over different-sized footprints. For small footprints the scatter varies over a wide range, and system designers must account for this.

12.7 SCATTERING COEFFICIENT DATA

Numerous programs to gather scattering coefficient data existed prior to 1972, but sizable data collections with accompanying "ground truth" were rare. Since 1972, however, several major programs have changed the situation so that much information is now available. Indeed, this information is so widespread that an adequate summary of the literature is impossible. Hence, this section can only give highlights of the results and major programs. The reader should consult the three major compendia of such data for more information both on results and on bibliography[21,23,114] (note that information is spread through many chapters).

Some early scattering-coefficient-measurement programs worth mentioning include those of the Naval Research Laboratory,[15,16] Goodyear Aerospace Corporation,[12] Sandia Corporation (near-vertical data),[89,90] and particularly Ohio State University.[2,4] Since 1972 the largest program has been at the University of Kansas.[6,7,21,53,57,69,91] Extensive programs were also in France (Centre National d'Etudes Spatiales, Centre National d'Etudes des Télécommunications, Université Paul Sabatier),[9] the Netherlands,[8] Canada Centre for Remote Sensing (CCRS; especially sea ice),[17] and the University of Bern, Switzerland (snow).[92] Many of the results from these programs appear in digests of the International Geoscience and Remote Sensing Symposia (IGARSS; IEEE Geoscience and Remote Sensing Society) and journals such as *IEEE Transactions on Geoscience and Remote Sensing* and *on Ocean Engineering, International Journal of Remote Sensing, Remote Sensing of Environment*, and *Photogrammetric Engineering and Remote Sensing*.

Although calibrations for some of the older data were doubtful, summary presentations are not available for newer data. Accordingly, Fig. 12.29 shows an earlier summary based mostly on X-band data. One should be cautious in using this data, but the figure gives a feel for the overall variations. Figure 12.30 is a similar presentation for near-vertical data.[93] Calibration of the systems was good, but the antenna effect discussed in Sec. 12.5 makes the values from 0 to 5° low.

Effects of Roughness, Moisture Content, and Vegetation Cover. Scattering falls off more rapidly with angle for smooth surfaces than for rough surfaces. Since the roughness that affects radar must be measured in wavelength units, a surface smooth at long wavelengths may be rough at shorter ones. This is illustrated in Fig. 12.31,[94] which shows these effects with measurements from plowed fields. At 1.1 GHz the signal changed 44 dB between 0 and 30° for the smoothest field and only 4 dB for the roughest. At 7.25 GHz the smoothest field was rough enough to reduce the variation to 18 dB.

For most surfaces cross-polarized scatter is lower than like-polarized, often by about 10 dB. Cross-polarized scatter from smooth surfaces is much less near vertical than elsewhere. Figure 12.32[95] shows this effect. Cross-polarized returns from volume scatterers with elements that are large compared with a wavelength are stronger than for surfaces, sometimes being only 3 dB down.

Scatter depends on dielectric constant, which depends on moisture content. Thus scatter from wet soils at angles off vertical is usually much higher

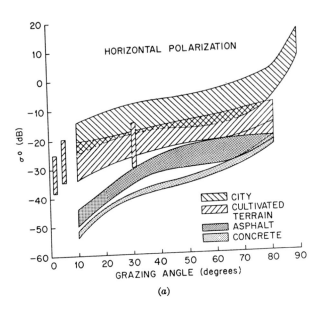

(a)

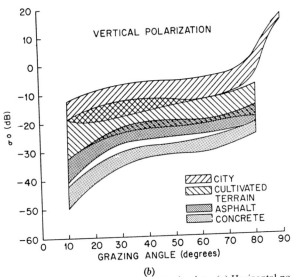

(b)

FIG. 12.29 Boundaries of measured radar data. (a) Horizontal polarization. (b) Vertical polarization. (*Courtesy of I. Katz.*)

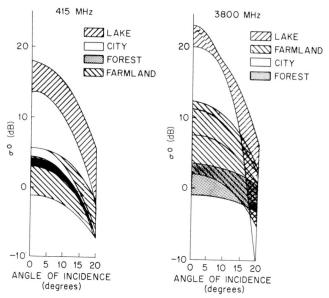

FIG. 12.30 Boundaries of measured radar return near vertical incidence, based on Sandia Corporation data. (*From Janza, Moore, and Warner.*[93])

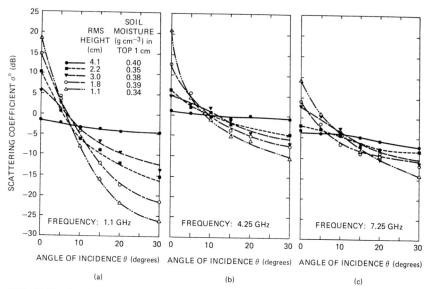

FIG. 12.31 Angular response of the scattering coefficient for five moist fields with different roughness at (*a*) 1.1 GHz, (*b*) 4.25 GHz, and (*c*) 7.25 GHz. (*From Ulaby, Moore, and Fung.*[21])

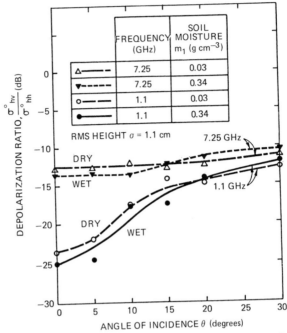

FIG. 12.32 Angular dependence of the depolarization ratio of a smooth surface. (*From Ulaby, Moore, and Fung.*[21])

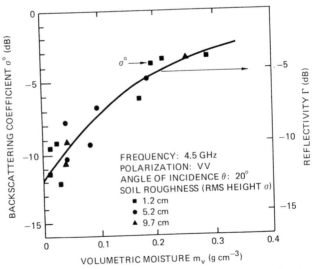

FIG. 12.33 Measured scattering coefficient $\sigma°$ (left scale) as a function of soil moisture content for three surface roughnesses. The solid curve is the reflectivity Γ (right scale) calculated on the basis of dielectric measurements. (*From LeToan.*[9])

than from dry soils. Figure 12.33 shows this.[9] The effect can be many decibels (9 dB in the figure).

Vegetation canopies over soil can contribute to scatter in the various ways shown in Fig. 12.34.[96] Figure 12.35[25] shows an example. Most of the scatter from the entire plant came from the top leaves, with enough attenuation there to reduce the scatter from stem, bottom leaves, and soil to measurable but negligible size. When those leaves were absent, the signals scattered from the soil and lower parts of the plant were about equal to each other and were much larger than when leaves were present.

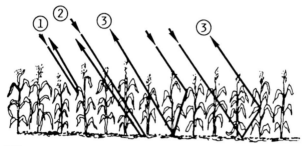

FIG. 12.34 Contributions to backscatter from a vegetation canopy over a soil surface. 1. Direct backscattering from plants. 2. Direct backscattering from soil (includes two-way attenuation by canopy). 3. Plant-soil multiple scattering. (*After Ulaby, Moore, and Fung.*[21])

Because volume scatter dominates for dense vegetation, especially trees, σ^0 is nearly independent of the angle of incidence. Figure 12.36[97] shows this with results from X-band imaging of a forest. The figure is a plot of γ rather than σ^0 ($\gamma = \sigma^0/\cos\theta$).

Soil Moisture. Figure 12.33 shows the size of the effect of soil moisture on σ^0. Soil moisture effects differ for different soils. Dobson and Ulaby[98] showed that this use of moisture expressed in percent of *field capacity* improved the fit between σ^0 and moisture content. Field capacity is a measure of how tightly the soil particles bind the water; the unbound water affects ϵ more. An empirical expression for field capacity is[99]

$$FC = 25.1 - 0.21S + 0.22C \text{ percent by weight}$$

where S and C are the percentages (by weight) of sand and clay in the soil. The soil moisture content in terms of field capacity is

$$m_f = 100m_g / FC \text{ percent}$$

with m_g the percent moisture in the soil by weight. When we use this measure, the relation between σ^0 in dB and m_f is linear even in the presence of moderate vegetation cover, as shown in Fig. 12.37.[100] The slope of this curve is somewhat different with vegetation cover than it is without, however. Although m_f is apparently at least as good as the volumetric moisture content for relating to σ^0, its use has been questioned.[101]

Soil moisture can affect a radar image, as has been demonstrated in imagery obtained from the Seasat L-band SAR.[102] A simulation experiment[103] showed

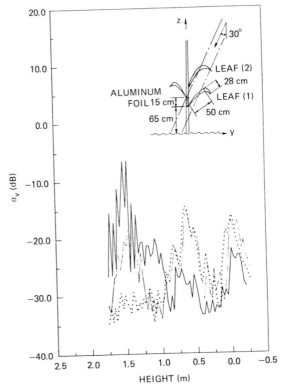

FIG. 12.35 FM-CW probing scatterometer measurements of a corn plant at 30°. The solid curve is the full plant; the dot-dash curve, leaf 1 removed; the dotted curve, leaf 2 removed. (*From Wu et al.*[25])

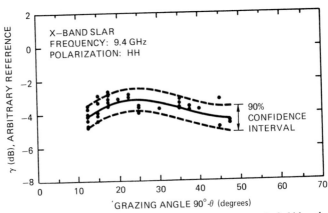

FIG. 12.36 Measured scattering variation of a forest parcel of old beech trees. Note use of δ (with an arbitrary reference) instead of σ^0 for the ordinate. (*From Hoekman.*[97])

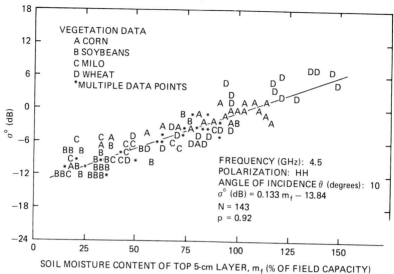

FIG. 12.37 4.5-GHz scattering coefficient versus soil moisture (percent of field capacity) for vegetation-covered soil. (*After Ulaby et al.*[100])

that one can estimate soil moisture within 20 percent for 90 percent of the pixels in an image. Moreover, it showed that resolutions between 100 and 1000 m were superior to finer resolutions for this purpose.

Vegetation. Backscatter from vegetation depends on many parameters and varies widely. Thus, although we can develop *average* models like those of Sec. 12.6, details are much more complex. The σ^0 varies with season, moisture content, state of growth, and time of day.

Figure 12.38[104] shows the seasonal variation for corn compared with a model presented in the reference. The much larger variation at 0^0 apparently results from the larger effect at vertical of the soil and consequently its moisture content. The rapid 12 dB swing between May 25 and June 1 results from drying of the soil. Even at $50°$, where attenuation through the canopy masks the soil effect, the seasonal variation exceeds 8 dB. Diurnal variations are relatively small but finite. They result both from plant moisture changes and from morphological changes (a corn plant actually lifts its leaves "to meet the sun"; morning glories close their flowers at night).

Most crops are planted in rows. This causes an azimuthal variation of σ^0, as shown in Fig. 12.39.[105] The modulation shown is the ratio of σ^0 looking parallel to the rows (more vegetation) to that looking normal. This phenomenon is much more pronounced at the lower frequencies, as shown.

Some general properties of vegetation scatter are visible in Fig. 12.40.[106] At low frequencies the decay with θ is rapid out to about $20°$ and then more gradual; most of the steep part results from surface echo. At higher frequencies the plant attenuation prevents a significant surface echo; so the angular variation is more uniform. Cross-polarized signals at vertical are negligible; so even at low frequencies the cross-polarized σ^0 varies uniformly. At both high and low frequencies it is about 10 dB below the like-polarized σ^0.

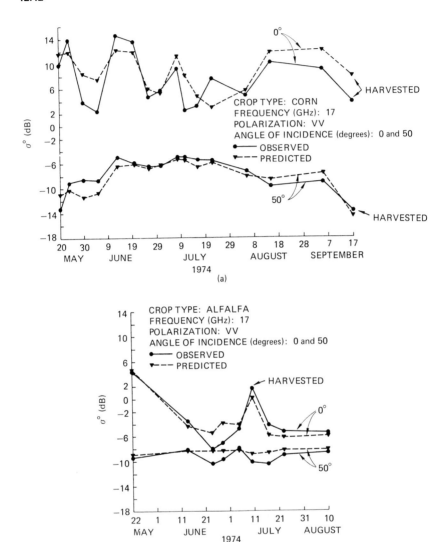

FIG. 12.38 Time variation of scatter from (*a*) corn and (*b*) alfalfa at incidence angles of vertical and 50°. (*From Attema and Ulaby.*[104])

Snow. When snow covers the ground, much of the scatter is from the snow rather than the underlying ground. Snow is both a volume-scattering and an attenuating medium. When the snow is dry, scatter comes from a large volume; when it is wet, the scattering volume is much less because of higher attenuation. As a result σ^0 decreases rapidly as the sun melts the top layer. Figure 12.41[86] illustrates how fast this can be and also shows that the effect is much greater at the higher frequencies where attenuation is greater. Figure 12.42[108]

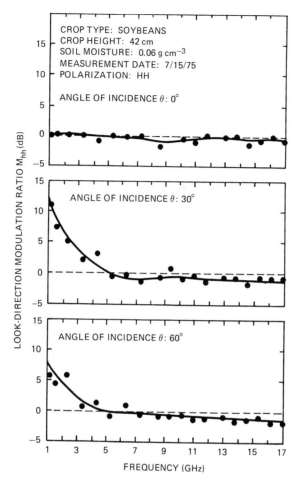

FIG. 12.39 Frequency response of the look-direction modulation ratio for a soybean field with horizontal polarization at incidence angles of 0, 30, and 60°. (*From Ulaby, Moore, and Fung.*[21])

shows the angular variation seen for snow-covered ground. Off-vertical scattering is much greater at higher frequencies. For the 58-cm depth shown, much of the scatter at 1.6 and 2.5 GHz is probably from the underlying surface.

Some reports state that there are *radar hot spots* in snow cover, particularly at 35 GHz. These reports result from improper interpretation of variations that are due to normal Rayleigh fading of the signal. Scatter from snow comes from many centers within the illuminated volume; so the conditions for Rayleigh fading are met. Measurements with suitable averaging in frequency or illumination angle demonstrate that snow-covered surfaces scatter essentially uniformly except for the effects of the multipath fading.

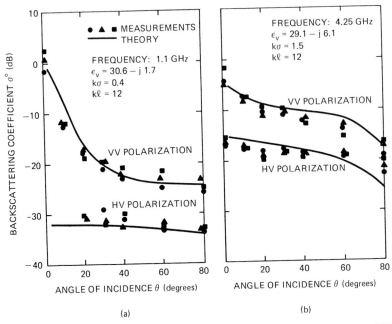

FIG. 12.40   Comparison of model calculations with measurements at (*a*) 1.1 GHz and (*b*) 4.25 GHz. (*From Eom and Fung.*[106])

Sea Ice.   Sea ice is a very complex medium. Ice observers characterize it in many different categories that depend on thickness, age, and history of formation.[109] Hence one cannot characterize its radar return in any simple way; in this sense it is like vegetation. The most important ice types from a radar point of view are first-year (FY 1 to 2 m thick), multiyear (MY >2 m thick), and a conglomeration of thinner types (< 1 m thick).

Like snow, sea ice influenced by solar melting and above freezing temperatures scatters microwaves very differently from the more normal cold-surface ice. In winter, the cold MY ice scatters much more than cold FY ice. In summer, σ^0 for MY ice decreases to about the same level as that of FY ice. Figure 12.43[110] shows this and typical angular responses. These curves are for 13.3 GHz, but the results would be similar at any frequency down to S band. Figure 12.44[91] shows the frequency variation of σ^0 for various kinds of ice. Shore-fast ice is grounded to the bottom at the shoreline; in this case it is probably MY. Gray ice is one of the types thinner than FY.

Kim[53] developed a theory that explains a wide range of sea-ice σ^0 measurements. From this and extensive data from the literature on ice properties, Fig. 12.45[91] shows the ranges of FY and MY scattering under winter conditions. Clearly higher frequencies are superior for identifying ice types to lower frequencies, and discrimination is not possible below about 5 GHz. At L band and below, the differences between MY and FY ice are small even in winter. This means that imaging radars can easily distinguish ice types by intensity alone at the higher frequencies in winter but not in summer. This fact is the basis for operational ice-monitoring systems by the Soviet Union [using the Toros K_u-band side-

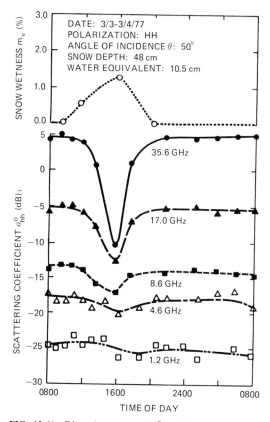

FIG. 12.41 Diurnal patterns of σ^0 and liquid-water content for snow at several frequencies. Note the extreme variation of the K_a band as the sun starts to melt the surface. (*From Stiles and Ulaby.*[107])

looking airborne radar (SLAR)][111] and Canada (using a modified X-band APS-94 SLAR and more recently the STAR-1 X-band SAR).

Snow cover on ice can mask ice scatter itself as with snow on land. Since the arctic is relatively dry, most areas have little snow, but snow does make distinguishing ice types difficult at times.

Programs to learn microwave properties of sea ice have been numerous because of the importance of arctic operations and meteorology. Microwave remote sensing is necessary to monitor ice properties in the arctic owing to the long winter night, frequent cloud cover, and inaccessibility.

12.8 IMAGING RADAR INTERPRETATION

Side-looking fine-resolution imaging radars with real or synthetic apertures produce images that closely resemble aerial photographs. Both shadows and differ-

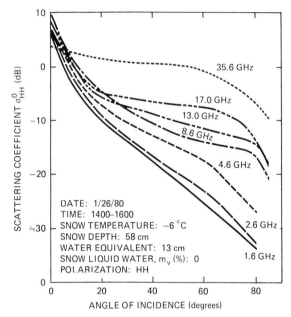

FIG. 12.42 Angular response of σ^0 of dry snow at different frequencies. Rapid falloff at lower frequencies apparently results from penetration to the smooth ground surface. (*From Stiles et al.*[108])

ences in σ^0 for different parts of the ground produce image-intensity variations like those in photographs. For this reason, photointerpreters can easily learn to interpret radar images. However, since the radar images are due to microwave reflectivity, not optical reflectivity, the interpreters must understand the differences and that the images at the different wavelengths are in fact complementary. Moreover, the geometrical distortions for radar images are those of a side-looking range measurement system, whereas those of aerial photos are those of a down-looking angle measurement system—a difference that the interpreter must understand. At low grazing angles, the distortions are small for radar, but at low incidence angles they are large. Moreover, the speckle in radar images is not present in photographs.

Modern imaging radars use digital recording, and the images are produced on film or manipulated digitally. Because the side-looking configuration produces a strip image, the output films are usually in the form of long strips. Most cameras produce separate images that are approximately square. Strip film cameras and optical-infrared scanners produce strip images like those of radars, but with different distortions because they are angle-, not range-, measuring devices.

Every pure and applied science that uses aerial photography can also use radar images. This is particularly useful in cloudy environments, but radar also is useful even in clear weather because performance is independent of time of day. Moreover, radar signatures of the ground differ from those in the visible and infrared. Radar has been applied to agriculture, forestry, geology, hydrology, urban geography, regional studies, oceanography, and ice mapping.

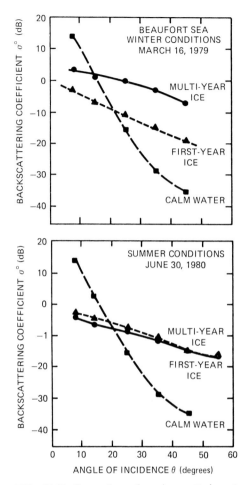

FIG. 12.43 Comparison of sea-ice scattering at 13.9 GHz in summer and winter. (*From Gray et al.*[110])

Fading complicates the interpretation of images by producing speckle. This means that averaging of the speckled image usually is necessary. Sometimes the processor does the averaging, and sometimes the interpreter does it mentally; it must be done to interpret an image. The image intensity for a single-look SAR follows the Rayleigh distribution. Most SAR processors sacrifice some spatial resolution by averaging, say, four pixels together. Transmitting more bandwidth than needed for range resolution accomplishes this purpose without loss of needed spatial resolution,[112] but it takes more transmitter power. Suitable frequency agility can accomplish the same result.

Tradeoffs exist between spatial resolution and measurement precision. The latter can be used to define a *gray-level resolution*.[113] One can then think of image resolution in terms of a volume:

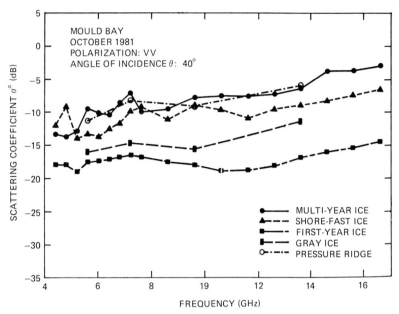

FIG. 12.44 Example of frequency response of $\sigma°$ for different kinds of sea ice. (*From Kim.*[53])

$$V = r_a r_y r_g$$

where r_a is the along-track resolution, r_y is the ground-range resolution, and r_g is the gray-level resolution. The referenced study showed that interpretability depends on V; so tradeoffs between the three elements of V are possible. Best results for a human interpreter occur when three independent samples of the fading are averaged. Ignoring this fading (speckle) can lead to erroneous conclusions on the spatial resolution needed for a given application.

Single-frequency, single-polarization radar images are useful. However, use of multiple polarizations (particularly including cross polarization) and multiple frequencies clearly increases their value. Different angles of incidence are most suitable for different applications. For example, soil moisture monitoring is best within 20° of vertical at frequencies near 5 GHz. Vegetation discrimination is better, however, at higher frequencies and angles of incidence.

Because the literature in this field is so massive, the radar engineer wishing to learn more about the subject should consult the "Manual of Remote Sensing,"[23] *Microwave Remote Sensing*,[21] especially Vol. III and Chap. 11 of Vol. II, and the journals outlined early in Sec. 12.7.

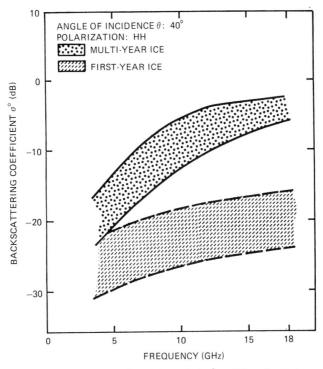

FIG. 12.45 Measurement-based theoretical σ^0 variations for first-year and multiyear sea ice. Ranges are determined by using known variations of ice characteristics. (*From Kim et al.*[91])

REFERENCES

1. Goldstein, H.: Sea Echo, in Kerr, D. E. (ed.): "Propagation of Short Radio Waves," MIT Radiation Laboratory Series, vol. 13, McGraw-Hill Book Company, New York, 1951, chap. 6.

2. Cosgriff, R. L., W. H. Peake, and R. C. Taylor: Terrain Scattering Properties for Sensor System Design, "Terrain Handbook II," Ohio State University, Eng. Exp. Sta. Antenna Lab., Columbus, 1959.

3. Moore, R. K.: Radar Scattering Cross-Section per Unit Area and Radar Astronomy, *IEEE Spectrum*, p. 156, April 1966.

4. Ruck, G., D. Barrick, W. Stuart, and C. Krichbaum: "Radar Cross Section Handbook," Plenum Press, New York, 1968.

5. Moore, R. K.: Resolution of Vertical Incidence Radar Return into Random and Specular Components, *University of New Mexico, Eng. Exp. Sta.*, Albuquerque, 1957.

6. Banhart, J. M. (ed.): Remote Sensing Laboratory Publication List 1964–1980, *University of Kansas, Remote Sensing Lab.*, Lawrence, 1981.

7. Banhart, J. M. (ed.): Remote Sensing Laboratory Publication List 1981–1983, *University of Kansas, Remote Sensing Lab.*, Vol. TR-103, Lawrence, 1984.

8. de Loor, G. P., P. Hoogeboom, and E. P. W. Attema: The Dutch ROVE Program, *IEEE Trans.*, vol. GE-20, pp. 3–11, 1982.

9. LeToan, T.: Active Microwave Signatures of Soil and Crops: Significant Results of Three Years of Experiments, *Dig. Int. Geosci. Remote Sensing Symp. (IGARSS '82), IEEE 82CH14723-6*, vol. I, 1982.

10. Clapp, R. E.: A Theoretical and Experimental Study of Radar Ground Return, *MIT Radiat. Lab. Rept.* 6024, Cambridge, Mass., 1946.

11. George, T. S.: Fluctuations of Ground Clutter Return in Airborne Radar Equipment, *Proc. IEE (London)*, vol. 99, pp. 92–99, 1952.

12. Reitz, E. A., et al.: Radar Terrain Return Study, Final Report: Measurements of Terrain Back-Scattering Coefficients with an Airborne X-Band Radar, *Goodyear Aerospace Corporation, GERA*-463, Phoenix, 1959.

13. Campbell, J. P.: Back-Scattering Characteristics of Land and Sea at X Band, *Proc. Natl. Conf. Aeronaut. Electron.*, 1958.

14. MacDonald, F. C.: The Correlation of Radar Sea Clutter on Vertical and Horizontal Polarization with Wave Height and Slope, *IRE. Conv. Rec.*, vol. 4, pp. 29–32, 1956.

15. Ament, W. S., F. C. MacDonald, and R. Shewbridge: Radar Terrain Reflections for Several Polarizations and Frequencies, *Proc. Symp. Radar Return, NOTS TP*2359, U.S. Naval Ordnance Test Station, China Lake, Calif., 1959.

16. Grant, C. R., and B. S. Yaplee: Backscattering from Water and Land at Centimeter and Millimeter Wavelengths, *Proc. IRE.*, vol. 45, pp. 972–982, 1957.

17. Livingstone, C. E. K., P. Singh, and A. L. Gray: Seasonal and Regional Variations of Active/Passive Microwave Signatures of Sea Ice, *IEEE Trans.*, vol. GE-25, pp. 159–173, 1987.

18. Larson, R. W., R. E. Hamilton, and F. L. Smith: Calibration of Synthetic Aperture Radar, *Dig. IGARSS '81*, pp. 938–943, 1981.

19. Haskell, A., and B. M. Sorensen: The European SAR-580 Project, *Dig. IGARSS '82, IEEE 82CH14723-6*, sess. WA-5, pp. 1.1–1.5, 1982.

20. Held, D. N.: The NASA/JPL Multipolarization SAR Aircraft Program, *Dig. IGARSS '85*, pp. 454–457, 1985.

21. Ulaby, F. T., R. K. Moore, and A. K. Fung: "Microwave Remote Sensing: Active and Passive," Addison-Wesley Publishing Company, Reading, Mass., vol. I, 1981; vol. II, 1982; Artech House, Norwood, Mass., vol. III, 1986.

22. Long, M. W.: "Radar Reflectivity of Land and Sea," 2d ed., Artech House, Norwood, Mass., 1983.

23. Colwell, R. N., D. S. Simonett, J. E. Estes, F. T. Ulaby, G. A. Thorley, et al.: "Manual of Remote Sensing," 2d ed., vols. I and II, American Society of Photogrammetry, Falls Church, Va., 1983.

24. Lundien, J. R.: Terrain Analysis by Electromagnetic Means: Radar Responses to Laboratory Prepared Soil Samples, *U.S. Army Waterways Exp. Sta., TR* 3-639, Vicksburg, Miss., 1966.

25. Wu, L. K., R. K. Moore, R. Zoughi, F. T. Ulaby, and A. Afifi: Preliminary Results on the Determination of the Sources of Scattering from Vegetation Canopies at 10 GHz, pts. I and II, *Int. J. Remote Sensing*, vol. 6, pp. 299–313, 1985.

26. Wu, L. K., R. K. Moore, and R. Zoughi: Sources of Scattering from Vegetation Canopies at 10 GHz, *IEEE Trans.*, vol. GE-23, pp. 737–745, 1985.

27. Zoughi, R., J. Bredow, and R. K. Moore: Evaluation and Comparison of Dominant Backscattering Sources at 10 GHz in Two Treatments of Tall-Grass Prairie, *Remote Sensing Environ.*, vol. 22, pp. 395–412, 1987.

28. Zoughi, R., L. K. Wu, and R. K. Moore: Identification of Major Backscattering Sources in Trees and Shrubs at 10 GHz, *Remote Sensing Environ.*, vol. 19, pp. 269–290, 1986.

29. Paris, J. F.: Probing Thick Vegetation Canopies with a Field Microwave Spectrometer, *IEEE Trans.*, vol. GE-24, pp. 886–893, 1986.

30. Wu, S. T.: Preliminary Report on Measurements of Forest Canopies with C-Band Radar Scatterometer at NASA/NSTL, *IEEE Trans., vol. GE-24, November 1986.*

31. Pitts, D. E., G. D. Badhwar, and E. Reyna: The Use of a Helicopter Mounted Ranging Scatterometer for Estimation of Extinction and Scattering Properties of Forest Canopies, *IEEE Trans.*, vol. GE-26, pp. 144–152, 1988.

32. Bernard, R., M. E. Frezal, D. Vidal-Madjar, D. Guyon, and J. Riom: Nadir Looking Airborne Radar and Possible Applications to Forestry, *Remote Sensing Environ.*, vol. 21, pp. 297–310, 1987.

33. Moore, R. K.: "Traveling Wave Engineering," McGraw-Hill Book Company, New York, 1960.

34. Schooley, A. H.: Upwind-Downwind Ratio of Radar Return Calculated from Facet Size Statistics of Wind Disturbed Water Surface, *Proc. IRE*, vol. 50, pp. 456–461, 1962.

35. Muhleman, D. O.: Radar Scattering from Venus and the Moon, *Astron. J.*, vol. 69, pp. 34–41, 1964.

36. Fung, A. K.: Theory of Cross Polarized Power Returned from a Random Surface, *Appl. Sci. Res.*, vol. 18, pp. 50–60, 1967.

37. Katz, I., and L. M. Spetner: Two Statistical Models for Radar Return, *IRE Trans.*, vol. AP-8, pp. 242–246, 1960.

38. Beckmann, P., and A. Spizzichino: "The Scattering of Electromagnetic Waves from Rough Surfaces," Macmillan Company, New York, 1963.

39. Beckmann, P.: Scattering by Composite Rough Surfaces, *Proc. IEEE*, vol. 53, pp. 1012–1015, 1965.

40. Fung, A. K., and H. J. Eom: An Approximate Model for Backscattering and Emission from Land and Sea, *Dig. IGARSS '81*, vol. I, pp. 620–628, 1981.

41. Hayre, H. S., and R. K. Moore: Theoretical Scattering Coefficients for Near-Vertical Incidence from Contour Maps, *J. Res. Nat. Bur. Stand.*, vol. 65D, pp. 427–432, 1961.

42. Davies, H.: The Reflection of Electromagnetic Waves from a Rough Surface, *Proc. IEE (London)*, pt. 4, vol. 101, pp. 209–214, 1954.

43. Fung, A. K., and R. K. Moore: The Correlation Function in Kirchhoff's Method of Solution of Scattering of Waves from Statistically Rough Surfaces, *J. Geophys. Res.*, vol. 71, pp. 2929–2943, 1966.

44. Evans, J. V., and G. H. Pettengill: The Scattering Behavior of the Moon at Wavelengths of 3.6, 68, and 784 Centimeters, *J. Geophys. Res.*, vol. 68, pp. 423–447, 1963.

45. Wright, J. W.: A New Model for Sea Clutter, *IEEE Trans.*, vol. AP-16, pp. 217–223, 1968.

46. Bass, F. G., I. M. Fuks, A. I. Kalmykov, I. E. Ostrovsky, and A. D. Rosenberg: Very High Frequency Radiowave Scattering by a Disturbed Sea Surface, *IEEE Trans.*, vol. AP-16, pp. 554–568, 1968.

47. Rice, S. O.: Reflection of Electromagnetic Waves by Slightly Rough Surfaces, *Commun. Pure Appl. Math.*, vol. 4, pp. 351–378, 1951.

48. Ref. 21, vol. II, p. 961.

49. Ref. 21, vol. II, chap. 12.

50. Ref. 21, vol. III, chap. 13.

51. Lang, R. H., and J. S. Sidhu: Electromagnetic Scattering from a Layer of Vegetation: A Discrete Approach, *IEEE Trans.*, vol. GE-21, pp. 62–71, 1983.

52. Fung, A. K.: A Review of Volume Scatter Theories for Modeling Applications, *Radio Sci.*, vol. 17, pp. 1007–1017, 1982.

53. Kim, Y. S.: Theoretical and Experimental Study of Radar Backscatter from Sea Ice, Ph.D. dissertation, University of Kansas, Lawrence, 1984.

54. Rydstrom, H. O.: Interpreting Local Geology from Radar Imagery, *Bull. Geol. Soc. Am.*, vol. 78, pp. 429–436, 1967.

55. Rice, S. O.: Mathematical Analysis of Random Noise, pt. I, *Bell Syst. Tech. J.*, vol. 23, pp. 282–332, 1944; pt. II, vol. 24, pp. 46–156, 1945.

56. Ref. 21, vol. II, pp. 487–492.

57. Ulaby, F. T., W. H. Stiles, D. Brunfeldt, and E. Wilson: 1-35 GHz Microwave Scatterometer, *Proc. IEEE/MTT-S, Int. Microwave Symp.*, IEEE 79CH1439-9 MIT-S, 1979.

58. Brunfeldt, D. R., and F. T. Ulaby: An Active Radar Calibration Target, *Dig. IGARSS '82*, IEEE 82CH14723-6, 1982.

59. Ref. 21, vol. II, pp. 766–779.

60. Janza, F. J.: The Analysis of a Pulse Radar Acquisition System and a Comparison of Analytical Models for Describing Land and Water Radar Return Phenomena, Ph.D. dissertation, University of New Mexico, Albuquerque, 1963.

61. Janza, F. J., R. K. Moore, and R. E. West: Accurate Radar Attenuation Measurements Achieved by Inflight Calibration, *IEEE Trans.*, vol. PGI-4, pp. 23–30, 1955.

62. Bracalente, E. M., W. L. Jones, and J. W. Johnson: The Seasat—A Satellite Scatterometer, *IEEE J.*, vol. OE-2, pp. 200–206, 1977.

63. Li, F. K., D. Callahan, D. Lame, and C. Winn: NASA Scatterometer on NROSS—A System for Global Observations on Ocean Winds, *Dig. IGARSS '84*, 1984.

64. Moore, R. K., and W. J. Pierson: Measuring Sea State and Estimating Surface Winds from a Polar Orbiting Satellite, *Proc. Int. Symp. Electromagn. Sensing of Earth from Satellites*, pp. R1–R26, 1965.

65. Cote, L. J., et al.: The Directional Spectrum of a Wind-Generated Sea as Determined from Data Obtained by the Stereo Wave Observation Project, *New York University Meteorol. Pap.*, vol. 2, no. 66, 1960.

66. Bush, T. F., and F. T. Ulaby: 8-18 GHz Radar Spectrometer, *University of Kansas, Remote Sensing Lab.*, vol. TR 177-43, Lawrence, September 1973.

67. Ref. 21, vol. II, pp. 779–791; vol. III, chap. 14.

68. Zoughi, R., L. K. Wu, and R. K. Moore: SOURCESCAT: A Very Fine Resolution Radar Scatterometer, *Microwave J.*, vol. 28, pp. 183–196, 1985.

69. Moore, R. K.: Effect of Pointing Errors and Range on Performance of Dual-Pencil-Beam Scatterometers, *IEEE Trans.*, vol. GE-23, pp. 901–905, 1985.

70. Edison, A. R.: An Acoustic Simulator for Modeling Backscatter of Electromagnetic Waves, Ph.D. dissertation, University of New Mexico, Albuquerque, 1961.

71. Parkins, B. E., and R. K. Moore: Omnidirectional Scattering of Acoustic Waves from Rough Surfaces of Known Statistics, *J. Acoust. Soc. Am.*, vol. 50, pp. 170–175, 1966.

72. Moore, R. K.: Acoustic Simulation of Radar Returns, *Microwaves*, vol. 1, no. 7, pp. 20–25, 1962.

73. Dobson, M. C., F. T. Ulaby, D. R. Brunfeldt, and D. N. Held: External Calibration of SIR-B Imagery with Area-Extended and Point Targets, *IEEE Trans.*, vol. GE-24, pp. 453–461, 1986.

74. Vaillant, D., and A. Wadsworth: Preliminary Results of Some Remote Sensing Campaigns of the French Airborne SAR VARAN-S, *Dig. IGARSS '86*, pp. 495–500, 1986.

75. Hirosawa, H., and Y. Matsuzaka: Calibration of Cross-Polarized SAR Imagery Using Dihedral Corner Reflectors, *Dig. IGARSS '86*, pp. 487–492, 1986.

76. Brunfeldt, D. R., and F. T. Ulaby: Active Reflector for Radar Calibration, *IEEE Trans.*, vol. GE-22, pp. 165–169, 1984.

77. Hartl, P., M. Reich, and S. Bhagavathula: An Attempt to Calibrate Air-Borne SAR Image Using Active Radar Calibrators and Ground-Based Scatterometers, *Dig. IGARSS '86*, pp. 501–508, 1986.

78. Larson, R. W., et al.: Bistatic Clutter Measurements, *IEEE Trans.*, vol. AP-26, pp. 801–804, 1978.

79. Renau, J., and J. A. Collinson: Measurements of Electromagnetic Backscattering from Known Rough Surfaces, *Bell Syst. Tech. J.*, vol. 44, pp. 2203–2226, 1965.

80. Kieu, D.: Effects of Tall Structures on Microwave Communication Systems, M.S. thesis, University of Kansas, Lawrence, 1988.

81. Stiles, W. H., D. Brunfeldt, and F. T. Ulaby: Performance Analysis of the MAS (Microwave Active Spectrometer) Systems: Calibration, Precision and Accuracy, *University of Kansas, Remote Sensing Lab.*, vol. TR 360-4, Lawrence, 1979.

82. Moore, R. K., K. A. Soofi, and S. M. Purduski: A Radar Clutter Model: Average Scattering Coefficients of Land, Snow, and Ice, *IEEE Trans.*, vol. AES-16, pp. 783–799, 1980.

83. Moore, R. K., et al.: Simultaneous Active and Passive Microwave Response of the Earth—The Skylab RADSCAT Experiment, *Proc. Ninth Int. Symp. Remote Sensing Environ.*, pp. 189–217, 1974.

84. Ref. 21. See summaries in vol. II, chap. 11, and vol. III, chap. 21.

85. Ulaby, F. T.: Vegetation Clutter Model, *IEEE Trans.*, vol. AP-28, pp. 538–545, 1980.

86. Stiles, W. H., and F. T. Ulaby: The Active and Passive Microwave Response to Snow Parameters, Part I: Wetness, *J. Geophys. Res.*, vol. 85, pp. 1037–1044, 1980.

87. Ulaby, F. T., and W. H. Stiles: The Active and Passive Microwave Response to Snow Parameters, Part II: Water Equivalent of Dry Snow, *J. Geophys. Res.*, vol. 85, pp. 1045–1049, 1980.

88. Birrer, I. J., E. M. Bracalante, G. J. Dome, J. Sweet, and G. Berthold: Signature of the Amazon Rain Forest Obtained with the Seasat Scatteromater, *IEEE Trans.*, vol. GE-20, pp. 11–17, 1982.

89. Edison, A. R., R. K. Moore, and B. D. Warner: Radar Return Measured at Near-Vertical Incidence, *IEEE Trans.*, vol. AP-8, pp. 246–254, 1960.

90. Bidwell, C. H., D. M. Gragg, and C. S. Williams: "Radar Return from the Vertical for Ground and Water Surface," Sandia Corporation, Albuquerque, N. Mex., 1960.

91. Kim, Y. S., R. K. Moore, R. G. Onstott, and S. P. Gogineni: Towards Identification of Optimum Radar Parameters for Sea-Ice Monitoring, *J. Glaciol.*, vol. 31, pp. 214–219, 1985.

92. Stotzer, E., V. Wegmuller, R. Huppi, and C. Matzler: Dielectric and Surface Parameters Related to Microwave Scatter and Emission Properties, *Dig. IGARSS '86*, pp. 599–609, 1986.

93. Janza, F. J., R. K. Moore, and B. D. Warner: Radar Cross-Sections of Terrain near Vertical Incidence at 415 Mc, 3800 Mc, and Extension of Analysis to X Band, *University of New Mexico, Eng. Exp. Sta.*, TR EE-21, Albuquerque, 1959.

94. Ref. 21, vol. III, Fig. 21.20, p. 1825.

95. Ref. 21, vol. III, Fig. 21.22, p. 1827.

96. Ref. 21, vol. III, Fig. 21.41, p. 1856.

97. Hoekman, D. H.: Radar Backscattering of Forest Stands, *Int. J. Remote Sensing*, vol. 6, pp. 325–343, 1985.

98. Dobson, M. C., and F. T. Ulaby: Microwave Backscatter Dependence on Surface Roughness, Soil Moisture and Soil Texture: Part III—Soil Tension, *IEEE Trans.*, vol. GE-19, pp. 51–61, 1981.

99. Schmugge, T. J.: Effect of Texture on Microwave Emission from Soils, *IEEE Trans.*, vol. GE-18, pp. 353–361, 1980.

100. Ulaby, F. T., A. Aslam, and M. C. Dobson: Effects of Vegetation Cover on the Radar Sensitivity to Soil Moisture, *University of Kansas, Remote Sensing Lab.*, TR 460-10, Lawrence, 1981.

101. Dobson, M. C., F. Kouyate, and F. T. Ulaby: A Reexamination of Soil Textural Effects on Microwave Emission and Backscattering, *IEEE Trans.*, vol. GE-22, pp. 530–535, 1984.

102. Ulaby, F. T., B. Brisco, and M. C. Dobson: Improved Spatial Mapping of Rainfall Events with Spaceborne SAR Imagery, *IEEE Trans.*, vol. GE-21, pp. 118–121, 1983.

103. Ulaby, F. T., M. C. Dobson, J. Stiles, R. K. Moore, and J. C. Holtzman: A Simulation Study of Soil Moisture Estimation by a Space SAR, *Photogramm. Eng. Remote Sensing*, vol. 48, pp. 645–660, 1982.

104. Attema, E., and F. T. Ulaby: Vegetation Modeled as a Water Cloud, *Radio Sci.*, vol. 13, pp. 357–364, 1978.

105. Ref. 21, vol. III, p. 1873.

106. Eom, H., and A. K. Fung: A Scatter Model for Vegetation Up to K_u-Band, *Remote Sensing Environ.*, vol. 15, pp. 185–200, 1984.

107. Stiles, W. H., and F. T. Ulaby: The Active and Passive Microwave Response to Snow Parameters, Part I: Wetness, *J. Geophys. Res.*, vol. 85, pp. 1037–1044, 1980.

108. Stiles, W. H., F. T. Ulaby, A. K. Fung, and A. Aslam: Radar Spectral Observations of Snow, *Dig. IGARSS '81*, pp. 654–668, 1981.

109. Bushuyev, A. V., N. A. Volkov, and V. S. Loshchilov: "Atlas of Ice Formations," Gidrometeoizdat, Leningrad, 1974. (In Russian with English annotations.)

110. Gray, A. L., R. K. Hawkins, C. E. Livingstone, L. D. Arsenault, and W. M. Johnstone: Simultaneous Scatterometer and Radiometer Measurements of Sea Ice Microwave Signatures, *IEEE J.*, vol. OE-7, pp. 20–32, 1982.

111. Loshchilov, V. S., and V. A. Voyevodin: Determining Elements of Drift of the Ice Cover and Movement of the Ice Edge by the Aid of the "Toros" Side Scanning Radar Station, *Probl. Arktiki Antarkt* (in Russian), vol. 40, pp. 23–30, 1972.

112. Moore, R. K., W. P. Waite, and J. W. Rouse: Panchromatic and Polypanchromatic Radar, *Proc. IEEE*, vol. 57, pp. 590–593, 1969.

113. Moore, R. K.: Tradeoff Between Picture Element Dimensions and Noncoherent Averaging in Side-Looking Airborne Radar, *IEEE Trans.*, vol. AES-15, pp. 696–708, 1979.

114. Ulaby, F. T., and M. C. Dobson: "Handbook of Radar Scattering Statistics for Terrain," Antech House, Norwood, Mass., 1989.

CHAPTER 13
SEA CLUTTER

Lewis B. Wetzel
Naval Research Laboratory

13.1 INTRODUCTION

For an operational radar, backscatter of the transmitted signal by elements of the sea surface often places severe limits on the detectability of returns from ships, aircraft and missiles, navigation buoys, and other targets sharing the radar resolution cell with the sea. These interfering signals are commonly referred to as *sea clutter* or *sea echo*. Since the sea presents a dynamic, endlessly variable face to the radar, an understanding of sea clutter will depend not only on finding suitable models to describe the surface scattering but on knowledge of the complex behavior of the sea as well. Fortunately, a close relationship between radar and oceanography has grown up in the remote-sensing community, leading to the accumulation of a large amount of useful information about scattering from the sea and how this scattering relates to oceanographic variables.

It would seem a simple matter to characterize sea clutter empirically by direct measurement of radar returns for a wide variety of both the radar and environmental parameters that appear to affect it. Parameters relating to the radar or its operating configuration, such as frequency, polarization, cell size, and grazing angle, may be specified by the experimenter, but the environmental parameters are quite another matter—for two reasons. First, it has not always been clear which environmental variables are important. For example, wind speed certainly seems to affect clutter levels, but correlation of clutter with, say, ships' anemometer readings has not been entirely satisfactory. The state of agitation of the surface (*sea state*) appears to have a strong effect, but it is a subjective measure, and its relation to the prevailing local winds is often uncertain. Moreover, it has been found that the temperatures of the air and the sea surface can affect the way in which the measured wind speed is related to the generation of clutter-producing waves, yet the importance of these effects were unappreciated over most of the history of sea clutter measurements; so air and sea temperatures were seldom recorded. Even if the importance of an environmental parameter has been recognized, it is often difficult to measure it with accuracy under real-sea conditions, and there are practical and budgetary limits to obtaining open-ocean measurements in sufficient variety to develop any really meaningful statistical models of clutter. Little wonder that many aspects of sea clutter remain frustratingly ill defined.

Before the late 1960s, most clutter data was collected in bits and pieces from isolated experiments, often with poor or incomplete ground truth. (For reviews of the older literature see, for example, Long,[1] Skolnik,[2] or Nathanson.[3]) Nevertheless, though much of the earlier clutter data was of limited scientific value, it did disclose some general trends, such as the tendency of clutter signal strength at low to intermediate grazing angles to increase with the grazing angle and with wind (or sea state) and generally to be greater for vertical polarization and in upwind-downwind directions.

It is commonly noted that, when viewed on an A scope, the appearance of sea clutter depends strongly on the size of the resolution cell, or *radar footprint*. For large cells it appears *distributed* in range and may be characterized by a surface-averaged cross section with relatively modest fluctuations about a mean value. As the size of the resolution cell is reduced, clutter takes on the appearance of isolated targetlike, or *discrete*, returns that vary in time. At these higher resolutions, the distributed clutter is often seen to consist of a dense sequence of discrete returns. When the discrete returns stand well out of the background, as they are seen to do for both polarizations but most clearly with horizontal polarization at small grazing angles, they are called *sea spikes* and are a common clutter contaminant in this radar operating regime.

Attempts to provide a theoretical explanation of the observed behavior of clutter signals trace essentially from the work pursued during World War II and described in the well-known MIT Radiation Laboratory book edited by Kerr.[4] Unfortunately, the scattering models developed during this period, along with most of those published over the following decade, failed to account for the behavior of sea backscatter in a very convincing way. In 1956, however, Crombie observed that at high-frequency (HF) wavelengths (tens of meters) scattering appeared to arise from a resonant interaction with sea waves of one-half of the incident wavelength, i.e., to be of the Bragg type.[5] Reinforced by the theoretical implications of various small waveheight approximations and wave tank measurements under idealized conditions, the *Bragg model* was introduced into the microwave regime by many workers in the mid-1960s.[6–8] This produced a revolution in thinking about the origins of sea clutter because it involved the sea wave *spectrum*, thus forging a link between clutter physics and oceanography in what became the field of *radio oceanography*. However, fundamental conceptual problems in applying the Bragg hypothesis in microwave scattering, along with recent questions about the validity of its predictions and the possibility of alternative scattering hypotheses, have reopened inquiry into the physical origins of sea scatter and how best to model it.[9–14] This being the case, speculation about physical models will be kept to a minimum in the sections on the empirical behavior of sea clutter. The problem of modeling sea scatter will be discussed separately in a later section.

13.2 DESCRIPTION OF THE SEA SURFACE

Close observation of the sea surface discloses a variety of features such as wedges, cusps, waves, foam, turbulence, and spray, as well as breaking events of all sizes and masses of falling water. Any or all of these might contribute to the scattering of electromagnetic waves responsible for sea clutter. The basic oceanographic descriptor of the sea surface, however, is the *wave spectrum*, which, while saying little about these features, contains a great deal of information about the sea surface in general and is central to the application of the Bragg scattering hypothesis. In view of the need to understand the sea surface in order to understand sea clutter and the prominence of the Bragg hypothesis in existing clutter models, some tutorial material describing the spectral characterization of the sea surface is included below.

There are basically two types of surface waves, *capillary* and *gravity*, depending on whether surface tension or gravity is the dominant restoring force. The transition between one and the other takes place at a wavelength of about 2 cm; so the smaller capillary waves supply the surface fine structure while gravity waves make up the larger and most visible surface structures. Waves have their origin ultimately in the wind, but this does not mean that the "local" wind is a particularly good indicator of what the wave structure beneath it will be. In order to arouse the surface to its *fully developed* or *equilibrium* state, the wind must blow for a sufficient time (*duration*) over a sufficient distance (*fetch*). That part of the wave structure directly produced by these winds is called *sea*. But waves propagate, so even in the absence of local wind, there can be significant local wave motion due to waves arriving from far away, perhaps from a distant storm. Waves of this type are called *swell*, and since the surface over which the waves travel acts as a low-pass filter, *swell* components often take the form of long-crested low-frequency sinusoids.

The Wave Spectrum. The wave spectrum which provides the primary oceanographic description of the sea surface appears in several forms. If the time history of the surface elevation is monitored at a fixed point, the resulting time series may be processed to provide a *frequency spectrum* $S(f)$ of the surface elevation, where $S(f)df$ is a measure of the *energy* (i.e., square of the waveheight) in the frequency interval between f and $f + df$. Wave spectra have been measured in the open ocean primarily for gravity waves down to wavelengths of about 1 m. Open-ocean measurements of capillary waves are especially difficult to perform.[15]

For a gravity wave, the frequency f and the wavenumber K are related by the dispersion relation

$$f = (\tfrac{1}{2}\pi)(gK)^{1/2} \tag{13.1}$$

where g is the acceleration of gravity and $K = 2\pi/\Lambda$, with Λ being the wavelength. Although each individual gravity wave obeys this relation, the waves at a point on the sea surface could come from any direction; so they are characterized by a two-dimensional propagation *vector* with orthogonal components K_x and K_y, where the K to be used in Eq. (13.1) is the magnitude $K = (K_x^2 + K_y^2)^{1/2}$.

The wavenumber spectrum associated with $S(f)$ is a function of the two components of K and is commonly written as $W(K_x, K_y)$. This is called the *directional wave spectrum* and expresses the asymmetries associated with winds, currents, refraction, isolated swell components, etc. For a given source of asymmetry like the wind, various parts of the spectrum will display different directional behaviors. For example, in a fully developed sea, the larger waves will tend to move in the direction of the wind while the smaller waves will be more isotropic. Directional spectra are more difficult to measure and are obtained by a variety of experimental methods, such as an array of wave staffs to measure surface heights over a matrix of points, a multiaxis accelerometer buoy, and stereo photography, and even by processing radar backscatter signals. However, a frequency spectrum measured at a point can contain no knowledge of wave direction; so a wavenumber spectrum $W(K)$ is often defined in terms of the frequency spectrum $S(f)$ by the relation

$$W(K) = S(f(K))(df/dK) \qquad (13.2)$$

with the relation between f and K given by Eq. (13.1). To account for the wind direction, $W(K)$ is sometimes multiplied by an empirical function of K and direction v relative to the (up)wind direction.

Oceanographers have not always been in complete agreement about the form of the frequency spectrum. Nonequilibrium wave conditions, inadequate sampling times, poor ground truth, etc., can contaminate the data set from which empirical spectra are derived. However, by careful selection of data from many sources, ensuring that only equilibrium (fully developed) sea conditions were represented and the wind was always measured at the same reference height, Pierson and Moskowitz[16] established an empirical spectrum that has proved popular and useful. It has the form

$$S(f) = Af^{-5}e^{-B(f_m/f)^4} \qquad (13.3)$$

where g is the acceleration of gravity, and $f_m = g/2\pi U$, corresponding to the frequency of a wave moving with a velocity equal to the wind speed U; A and B are empirical constants. This spectrum is illustrated in Fig. 13.1 for several wind

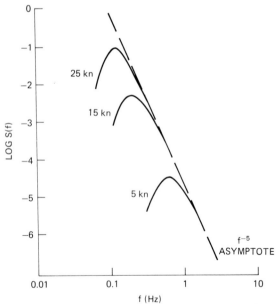

FIG. 13.1 Sea wave frequency spectra of the Pierson-Moskowitz type, representing fully developed seas.

speeds. The effect of increasing wind speed is simply to move the low-frequency cutoff to lower frequencies along the high-frequency f-minus-5 asymptote. (It should be noted that most of the oceanographers' spectra are based on measurements at relatively low frequencies and so cannot be taken seriously at frequencies above about 2 Hz. Nevertheless, these spectral forms are often used up to 20 Hz or greater in predicting radar clutter under the Bragg hypothesis.)

Converting this frequency spectrum into an isotropic wavenumber spectrum through Eq. (13.2) results in a spectrum of similar form, only with a K-minus-4 asymptote. Phillips[17] derived this asymptotic behavior on dimensional grounds, and a widely used simplification, obtained by replacing the smooth peak in Fig. 13.1 by a sharp cutoff, is generally referred to as the *Phillips spectrum* and in wavenumber space is written

$$
\begin{aligned}
W(K) &= 0.005/K^{-4} & K &> g/U^2 \\
&= 0 & K &< g/U^2
\end{aligned} \tag{13.4}
$$

where the cutoff wavenumber corresponds to the frequency f_m of the peak in Eq. (13.3). Opposed to this highly simplified form are increasingly complex spectra based on more careful empirical studies[18] as well as more sophisticated theoretical considerations.[19,20]

In discussing the characterization of the sea surface by its spectrum, it must be kept in mind that the spectrum is a highly averaged description of how the *energy* of the surface is distributed among the wavenumbers, or frequencies, of the waves present on it. Since the phases of these waves are lost, the spectrum gives no information about the morphology of the surface itself, i.e., about the complex surface features that are responsible for the scattered field. This point will be raised again in the section below on theories of sea clutter.

General Sea Descriptors. The shape of the curves in Fig. 13.1 suggests that the sea wave system has a relatively high Q; so it should be possible to get a rough idea of the behavior of the major waves on the surface by taking the values of *period* ($1/f$) and *wavelength* ($2\pi/K$) defined at the spectral peak. These values belong to a wave satisfying the dispersion relation Eq. (13.1) and having a phase velocity $C = 2\pi \times f/K$ equal to the wind speed U. By using Eq. (13.1), the period T and wavelength Λ thereby defined take the form

$$
T = 0.64U \qquad \Lambda = 0.64U^2 \tag{13.5}
$$

where U is in meters per second. Thus, for example, the largest waves in a fully developed sea for a 15-kn (7.5 m/s) wind will have a wavelength of about 120 ft (36 m) with a period of 5 s.

The statistical distribution of waveheights on the ocean surface is quite close to gaussian, with a mean square deviation that can be obtained by integrating the waveheight spectrum over all frequencies (or wavenumbers). For spectra resembling those in Fig. 13.1, the rms waveheight is given approximately by

$$
h_{\text{rms}} = 0.005U^2 \qquad \text{m} \tag{13.6}
$$

The rms waveheight contains contributions from all the waves on the surface, but very often it is the peak-to-trough height for the higher waves that is of major interest. This is certainly the case for a ship in a seaway or in the shadowing of the surface at low radar grazing angles. The *significant height*, or height of the one-third highest waves, provides such a measure. It is denoted by $H_{1/3}$ and is taken to be about 3 times the rms height given by Eq. (13.6). For a 15-kn wind, this is only about 3 ft, but for gale-force winds of 40 kn it rises to over 20 ft, which is a rather formidable sea.

Looking at the sea, an observer might describe what he or she sees in terms of a subjective *state of the sea*, e.g., "smooth," "rough," "terrifying!" If these descriptions are listed in order of severity and assigned numbers, these numbers define a *sea state*. A similar numerical scale exists for wind speeds, the *Beaufort wind scale*, with numbers about an integer higher than the corresponding sea state. But it is seldom used in reference to sea clutter.

There are, then, two numbers commonly used to indicate the activity of the sea surface: a subjective sea state and a measured wind speed. Only when the wind has sufficient *fetch* and *duration* to excite a *fully developed* sea, can a wave height be unambiguously associated with it. The surface descriptors generally used in connection with sea clutter—sea state, wind speed, and its associated equilibrium waveheight—are given in Table 13.1, with the wind speed in knots, the significant waveheight in feet, and the duration/fetch required for a fully developed sea in hours/nautical mile. It is of interest to note that the median wind speed over the world's oceans is about 15 kn, corresponding to sea state 3.

TABLE 13.1 Sea-Surface Descriptors

Sea state	Wind speed, kn	Waveheight $H_{1/3}$, ft	Duration/fetch, h/nmi
1 (smooth)	< 7	1	1/20
2 (slight)	7–12	1–3	5/50
3 (moderate)	12–16	3–5	15/100
4 (rough)	16–19	5–8	23/150
5 (very rough)	19–23	8–12	25/200
6 (high)	23–30	12–20	27/300
7 (very high)	30–45	20–40	30/500

13.3 EMPIRICAL BEHAVIOR OF SEA CLUTTER

Sea clutter is a function of many parameters, some of them showing a complicated interdependence; so it is not an easy task to establish its detailed behavior with a great deal of confidence or precision. For example, in a proper sea clutter measurement, the polarization, radar frequency, grazing angle, and resolution cell size will have been specified. Then the wind speed and direction must be measured at a reference altitude, and if the results are to be compared with those of other experimenters, the proper *duration* and *fetch* should be present to ensure standardization to equilibrium sea conditions. Since these measured winds are related to the wind structure at the surface through the atmospheric boundary layer, the shape of this layer must be determined by measuring the air and sea temperatures. To complicate the picture still further, it is becoming increasingly clear that sea backscatter has a strong dependence on the direction of the long waves, which include *swell*, in the measurement area; so ideally the *directional wave spectrum* should be measured as well. Obviously, it is unlikely that all these environmental parameters will be recorded with precision in every (or even *any*) sea clutter measurement; so considerable variability in the basic conditions under which sea clutter data is collected by different experimenters can be expected. It is of interest to note that in many of the reported measurements of sea clutter, particularly in the older literature, wide inconsistencies between wind speed and waveheight may be found. For example, a wind speed of 5 kn might be reported with waveheights of 6 ft, or 20-kn winds with 2-ft waves. These pairings are inconsistent with the values for an equilibrium sea described in Table 13.1 and indicate the unnoticed presence of heavy swell or highly nonequilibrium wind conditions, or both. Even with all the variables properly specified, recorded clutter data can be spread over a wide dynamic range, often as great as 40 dB at low grazing angles, so that clutter behavior is best described in terms of probability distribution functions.

Since sea clutter is generally viewed as a surface-distributed process, the basic clutter parameter is taken to be the normalized radar cross section (NRCS), σ^0, of the surface, commonly referred to as *sigma zero* and expressed in decibels relative to 1 m^2/m^2. It is obtained experimentally by dividing the measured radar cross section of an illuminated patch of the surface by a normalizing area; so differences in the definition of this area can lead to inconsistencies among various reports of NRCS measurements. Scattering from any distributed target involves the product of the transmitting and receiving system footprints integrated over the target. These footprints cover exactly the same area for a monostatic radar and will depend on the pulse and beamwidths, the range, and the grazing angle. If the footprints are assumed to be of the *cookie-cutter* type (constant amplitude falling sharply to zero at the half-power points), then the relation between the actual radar clutter cross section σ_c, as inferred from the received power via the radar equation, and the NRCS σ^0 is given by

$$\sigma^0 = \sigma_c/A_f \qquad (13.7)$$

where for a radar with an antenna beamwidth B and rectangular pulse of length T, viewing the surface at range R and grazing angle θ, the area A_f is either

$$A_f = \pi(BR)^2/4\sin\theta \qquad (13.8)$$

for beam-limited conditions [e.g., continuous-wave (CW) or long-pulse radar at high grazing angles] or

$$A_f = (c\tau/2)BR/\cos\theta \qquad (13.9)$$

for pulse-width-limited conditions (e.g., short-pulse radar at low grazing angles).

Real radars do not produce cookie-cutter footprints, however, since the antenna beam will have, say, a Bessel or gaussian profile and the pulse might be shaped. For this reason, an effective A must be obtained from a surface integration of the square of the actual amplitude profile of the footprint, which will always result in a smaller value of A than that defined by Eq. (13.8) or Eq. (13.9), and therefore in larger values of σ^0 as derived from measured values of σ_c by Eq. (13.7). Most experimenters use the half-power beamwidth in Eq. (13.8) or Eq. (13.9), with an error that is usually only 1 or 2 dB.

Dependence on Wind Speed, Grazing Angle, and Frequency. It was noted earlier that summaries of clutter measurements made before about 1970 may be found in several of the standard reference books on radar[2,3] and radar clutter.[1] Among the programs of this period, the most ambitious was that pursued in the late 1960s at the Naval Research Laboratory (NRL),[21] in which an airborne four-frequency radar (4FR), operating with both horizontal and vertical polarizations at UHF (428 MHz), L band (1228 MHz), C band (4455 MHz), and X band (8910 MHz), made clutter measurements upwind, downwind, and crosswind in winds from 5 to 50 kn for grazing angles between 5 and 90°. The system was calibrated against standard metal spheres, and wind speeds and waveheights were recorded in the measurement areas from instrumented ships.

Typically, samples of σ^0 for a given set of radar and environmental parameters are scattered over a wide range of values and in the NRL experiments were organized into probability distribution functions of the type shown in Fig. 13.2. The

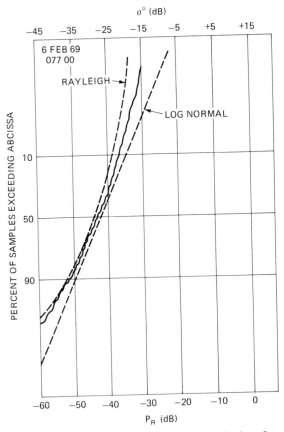

FIG. 13.2 An example of the probability distribution of sea clutter data. (*From Daley.*[21])

data, represented by the solid line, is plotted on normal probability paper with Rayleigh and log-normal distributions shown for comparison (dashed lines). The ordinate is the *percent of time by which the abscissa is exceeded*, and the abscissa is the value of σ^0 as defined by Eq. (13.7), with A taken from Eq. (13.8) or Eq. (13.9) as appropriate. This particular distribution is representative of clutter from a relatively large radar footprint (pulse length about 0.5 μs) measured at intermediate grazing angles (20 to 70°) for moderate wind speeds (about 15 kn). It is Rayleigh-like but shows a tendency toward log-normal behavior for the larger cross sections. From a detailed statistical analysis of the NRL 4FR data, Valenzuela and Laing[22] concluded that, for this data at least, the distributions of sea clutter cross sections were intermediate between the exponential (which is the power distribution corresponding to Rayleigh-distributed scattered-field amplitudes) and log-normal distributions.

Organizing the data samples into probability distributions makes the *median* (50 percent) value a convenient statistical measure of the clutter cross section. But many investigators process their data to provide the *mean* value, and since the conversion of a *median* to a *mean* requires knowledge of the probability distribution function, care must be taken to avoid ambiguity in comparing the measurements of different experimenters. The original analysis of the NRL 4FR data was based on *median* cross sections and the assumptions of the cookie-cutter antenna beam embodied in Eqs. (13.8) and (13.9).[21,23] In later presentations of this data,[24] the *median* values of σ^0 were replaced by *means*, raising them by about 1.6 dB, and the area A in Eq. (13.7) was redefined in terms of a more realistic tapered footprint, adding another 1 to 2 dB. This means that there is a difference of 3 to 4 dB between the earlier and later presentations of the same data, and since these results are widely used and quoted, it is important to ensure that the proper definition of σ^0 is being used when comparing them with clutter data that has been taken by other experimenters or in using these results in clutter predictions.

General Results. Being the first really comprehensive collection of clutter data over a wide range of radar frequencies, the 4FR program produced many plots showing the dependence of sea clutter on grazing angle, frequency, polarization, wind direction, and wind speed. However, comparison of these plots with others made both earlier and later shows the extent of the variations to be found in sea clutter measurements reported by different investigators for exactly the same set of parameters. This is seen clearly in Fig. 13.3*a* and *b*, which compares the grazing-angle dependence of X-band clutter data for wind speeds in the neighborhood of 15 kn obtained from four sources: NRL 4FR[24] (these are *mean* results for upwind directions and include the antenna corrections mentioned above), aircraft measurements by Masuko et al.[29] (also in the upwind direction), and summaries of the older data (pre-1970) taken from books on radar systems by Skolnik[2] and Nathanson.[3] The discrepancies between the different data sets can be accounted for, at least in part, as follows. The older data set was based on published measurements from various sources, and since there is no specification of wind direction, it may be assumed that it represents some kind of average of upwind, downwind, and crosswind directions. As will be seen below, this average is about 2 to 3 dB smaller than the upwind returns. Moreover, the *early* NRL 4FR data was used liberally in the older data summaries, and it was noted above that there is a difference of 3 to 4 dB between the early and later presentations of the same NRL 4FR data, the latter being used in Fig. 13.3*a* and *b*. With these

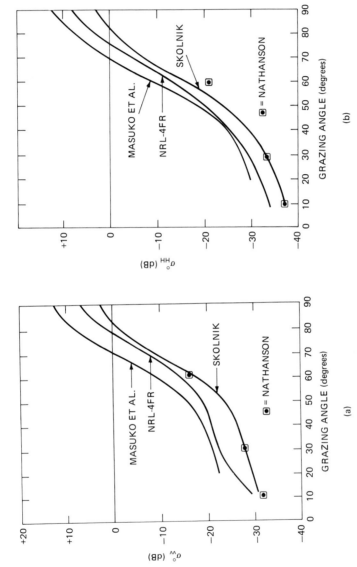

FIG. 13.3 Comparison of X-band clutter data from different sources for a nominal wind speed of 15 kn. (*a*) Vertical polarization. (*b*) Horizontal polarization. (*Based on data from Masuko et al.,[29] NRL 4FR,[21] Skolnik,[2] and Nathanson.[3]*)

corrections, the curves would show closer agreement. Nevertheless, it is clear that uncritical use of published clutter data could lead two radar systems designers to choose sea clutter estimates almost an order of magnitude apart for the same conditions.

The NRL 4FR data set is unique in that no other program has reported measurements made over so wide a range of frequencies, grazing angles, and wind speeds at the same time. Figure 13.4 shows the trends for both vertically and horizontally polarized sea clutter over a range of grazing angles down to 5°. The curves represent the centers of ±5 dB bands which contain the major returns for the three higher frequencies (L, C, and X bands—the UHF returns were a few decibels lower) and wind speeds above about 12 kn. The major differences in sea clutter for the two polarizations are seen to lie in the range of grazing angles between about 5 and 60°, where the horizontally polarized returns are smaller. This difference is found to be emphasized at both lower wind speeds and lower frequencies. The cross sections approach each other at high angles (>50°) and, for the higher microwave frequencies, at low angles (<5°) as well. In fact, for grazing angles less than a few degrees and moderate to strong wind speeds, several observers have reported that at X band and at the higher sea states the horizontally polarized returns often exceed the vertically polarized returns.[1,25,26]

The NRL 4FR system permitted transmission and reception on orthogonal polarizations so that data could be collected for cross-polarized sea clutter. These returns tended to have a weak dependence on grazing angle and were always smaller than either of the like-polarized returns, lying in the cross-hatched region shown on Fig. 13.4.

It is informative to compare measurements by different investigators in different parts of the world under similar wind conditions. Figure 13.5 displays measurements of vertically polarized sea clutter down to a grazing angle of 20° for wind speeds of about 15 kn from three independent experiments using airborne radars at C-, X-, and K-band frequencies.[27–29] While there is no assurance that all these measurements were made over fully developed seas, it is clear that there is a rather strong consistency among them, which reinforces the observation made in reference to Fig. 13.4 that the frequency dependence of sea clutter at intermediate grazing angles is weak at microwave frequencies from L to K band.

Dependence on Wind Speed. The relation between sea clutter and wind speed is complex and uncertain, since it has been found to depend on almost all the parameters that characterize sea clutter: frequency, grazing angle, polarization, the state of the sea surface, the direction and speed of the wind itself, and even on whether the measurements are made from an aircraft or a tower platform.[30]

A common way to organize clutter data is to seek the best straight-line fit (linear regression) between clutter cross sections in decibels and the log of the wind speed (or some other parameter). This, of course, *imposes* a power-law relation between the variables: $\sigma^0 \propto U^n$, where n is determined by the slope of the

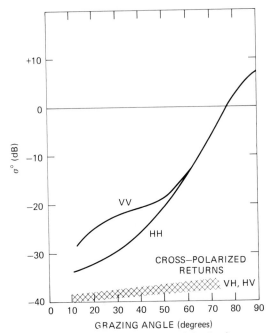

FIG. 13.4 General trends in clutter behavior for average wind speeds (about 15 kn) based on NRL 4FR data. Plots represent L-, C-, and X-band data within ±5 dB.

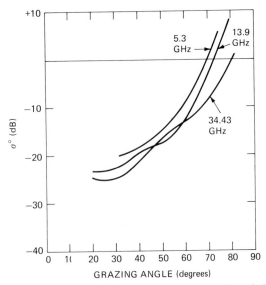

FIG. 13.5 Frequency dependence of sea clutter for wind speeds of about 15 kn: 5.3 GHz, Feindt;[27] 13.9 GHz, Schroeder;[28] 34.4 GHz, Masuko.[29]

line. An example is shown in Fig. 13.6.[31] On the other hand, the totality of the NRL 4FR results appeared to show saturation for wind speeds above about 20 kn, but the high and low- to moderate- wind-speed data was collected at different times in different places under different conditions of sea-surface development, and discrepancies between the two data sets for common wind speeds have weakened the evidence for saturation.[32] Other investigators deny that it is even possible to express wind dependence in the form of a power law, proposing the existence of a kind of threshold wind speed, below which clutter virtually vanishes and above which the clutter level rises toward a saturation value.[18] This is indicated by the curves in Fig. 13.7, where the straight lines correspond to various power laws. Once this possibility is raised, it is possible to find examples of data that appear to track such a curve while at the same time yielding a power law by linear regression, as illustrated in the tower data shown in Fig. 13.8.[31] This behavior is not uncommon.

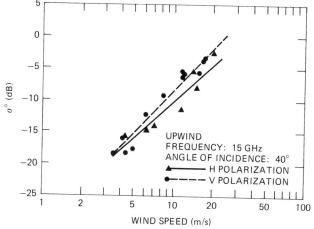

FIG. 13.6 Sea clutter from a tower platform with power-law wind-speed dependence defined by linear regression. (*From Chaudhry and Moore,*[31] *© 1984, IEEE.*)

Nevertheless, the imposition of a power-law relation provides a convenient way to visualize trends in the behavior of sea clutter with wind speed. The various aircraft measurements referred to above[27–29] as well as data from a tower in the North Sea[30,31] were all treated in this way, yielding plots of σ^0 as a function of wind speed and grazing angle of the form shown in Fig. 13.9a and b. Plots of this type give information about both the wind-speed and grazing-angle dependence of sea clutter for a given frequency, polarization, and wind direction. Figure 13.9a and b is based on a blend of radiometer-scatterometer (RADSCAT) data at 13.9 GHz[29] and measurements by Masuko et al. at 10 GHz,[29] both for upwind directions. Thus they can be viewed as representative of clutter behavior in the vicinity of X band, since the difference between the two frequencies is small. However, examination of the data points underlying these linear regressions show point scatter that sometimes resembles Fig. 13.6, sometimes Fig. 13.8, and sometimes neither; so the straight lines in these figures cannot be taken too seriously. In fact, it appears that there is no simple functional dependence of

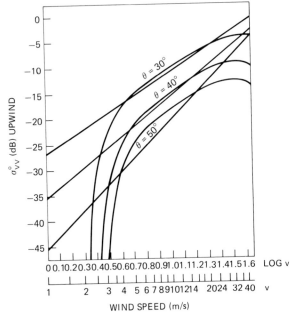

FIG. 13.7 A hypothetical wind-speed dependence of sea clutter (curved traces) compared with various power laws (straight lines). (*Derived from Pierson and Donelan.*[18])

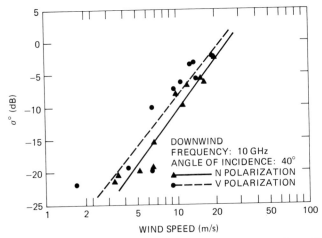

FIG. 13.8 Example of forcing a power-law fit (compare data with curves in Fig. 13.6). (*From Chaudhry and Moore,*[31] © *1984, IEEE.*)

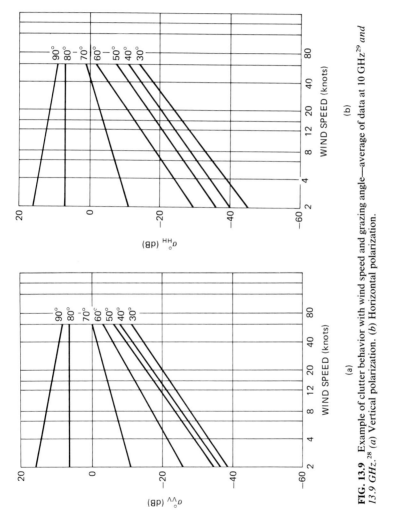

FIG. 13.9 Example of clutter behavior with wind speed and grazing angle—average of data at 10 GHz[29] and 13.9 GHz.[28] (a) Vertical polarization. (b) Horizontal polarization.

13.15

sea clutter on wind speed that can be established with any confidence from existing data, although most investigators would probably agree that the behavior of microwave sea clutter with wind speed at intermediate grazing angles can be roughly described as follows: for light winds (less than 6 to 8 kn) sea clutter is weak, variable, and ill defined; for intermediate winds (about 12 to 25 kn) it can be described roughly by a power law of the type found in Fig. 13.6; and for strong winds (above about 30 kn) there is a tendency for it to level off. In fact, the convergence of the lines in Fig. 13.9*a* and *b* with increasing wind speed suggests that the reflectivity of the sea surface is tending toward Lambert's law, for which there is no dependence on grazing angle, frequency, or polarization but only on surface *albedo*.

Dependence on Wind Direction. In several of the experiments referenced above, the dependence of sea backscatter on angle relative to the wind direction was found by recording the radar return from a spot on the surface while flying around it in a circle. Figure 13.10*a* and *b* gives an example of this behavior for grazing angles of about 45° and wind speeds close to 15 kn.[29] The figures contain results obtained independently by three different groups. The behavior shown here is representative of that found generally: sea clutter is strongest viewed upwind, weakest viewed crosswind, and of intermediate strength viewed downwind, the total variation being about 5 dB.

At High Grazing Angles. The top curve in Fig. 13.9*a* and *b* corresponds to clutter at a grazing angle of 90°, that is, for a radar looking straight down. On a strictly empirical basis, the clutter cross section at this angle is only weakly de

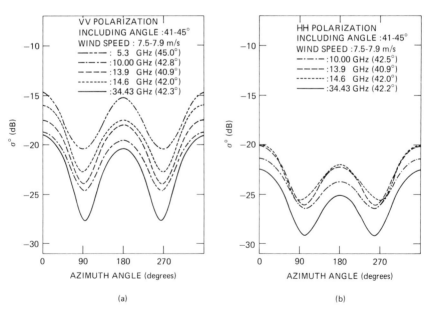

FIG. 13.10 Dependence of clutter on wind direction: nominal wind speed, 15 kn; grazing angle, 45°. (*From Masuko et al.,*[29] © *by the American Geophysical Union.*)

pendent on frequency, has a maximum of about +15 dB at zero wind speed (at least for the antenna beamwidths and experimental configurations reported), and falls off gradually as the wind picks up. Scattering at high grazing angles is commonly regarded as a form of specular scattering from tilted facets of the surface; so it is of interest to note that there appears to be a small range of angles in the neighborhood of 80° for which the cross section is almost completely independent of wind speed. Since these angles correspond to complements of the common rms sea slope angles of about 10°, it might be argued that as the wind increases, the clutter decrease due to increasing surface roughness is balanced at these angles by a clutter increase due to an increasing population of scattering facets. This line may therefore be regarded as the boundary separating the *specular* regime, where the cross section is decreased by surface roughness, from the *rough-surface* regime, where the cross section increases with surface roughness. It should further be noted that clutter measurements at these high grazing angles will be relatively sensitive to the averaging effects of wide antenna beamwidths, which could become a source of ambiguity in aircraft measurements at the lower radar frequencies.

At Low Grazing Angles. At low grazing angles, below mean sea slope angles of about 10°, sea clutter takes on a different character. The sharp clutter peaks known as sea spikes began to appear on A-scope presentations,[1,25,33] and the probability distributions assume a different form.[34] Figure 13.11 shows the presence of sea spikes in the time histories of returns from a fixed spot, measured from a tower in the Gulf of Mexico with a high-resolution X-band radar looking into an active sea at a 1.5° grazing angle.[33] The vertically polarized returns appear to be a bit broader, and while the horizontally polarized returns are more spiky, both polarizations display the sharp bursts characteristic of sea clutter at small grazing angles. The peak cross sections in these records are of the order of 10 m² and are roughly the same for the two polarizations, which is another characteristic of sea clutter at these angles. Interestingly, while the same measurements made in "calm" water looked virtually identical in every detail, peak cross sections were now only 10 cm², or 40 dB, less.

VERTICAL POLARIZATION

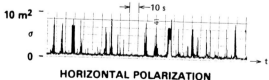

HORIZONTAL POLARIZATION

FIG. 13.11 Sea spikes at X band, 1.4° grazing angle, moderate to strong winds. Note equal amplitudes at the two polarizations. (*From Lewis and Olin.*[33])

Trizna has accumulated a considerable body of data from measurements of low-angle sea clutter using high-resolution (40-ns) shipboard radar in both the Atlantic and the Pacific oceans.[34] The probability distributions of the clutter cross sections were plotted in the manner of Fig. 13.12, which shows the distributions of horizontally polarized X-band data at a 3° grazing angle for low, medium, and high wind speeds (in order from left to right). The low-wind trace corresponds to a Rayleigh distribution, while the other straight-line segments are two-parameter Weibull distributions defined by different parameter pairs. It is clear that the behavior is different and considerably more complex than that shown in Fig. 13.2 for the higher grazing angles and wider pulses. Trizna interprets these distributions as follows: in each trace, the left-hand segment (lowest cross section) is actually receiver noise, recorded when the radar footprint lay in shadow; the middle section corresponds to distributed clutter, for reasons relating mainly to its weak dependence on resolution cell size; the right-hand section (highest cross section) describes the sea spikes, for reasons relating to the dependence on wind speed (similar to whitecap dependence) and the sheer size of the components (some individual absolute cross sections in excess of 1000 m²). For the higher wind speeds and fully developed seas encountered in the North Atlantic, the population of this sea-spike sector (the percentage of sea spikes) was found to grow as the 3.5th power of the wind speed, which, interestingly, is the same wind-speed dependence shown by the percentage of whitecaps seen on the surface.[36]

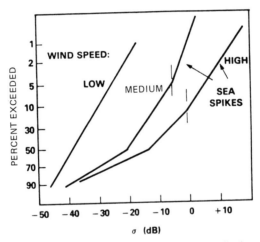

FIG. 13.12 Segmented clutter probability distributions at low grazing angles. (*Based on Trizna.*[34])

It should be kept in mind that, to the extent that the sea surface may be viewed as a stationary homogeneous process, as it generally is over the duration and spatial extent of any particular experimental event, the scattering cross section may be said to be *ergodic*, which means that the statistical results obtained by time averaging from a small cell are equivalent to a shorter time average from a larger cell, provided that the number of "samples" is the same in the two cases. For this reason, the statistical implications of experimental data can be properly compared only if the details of the sampling procedure are specified. However, the number of samples in the experimental results shown thus far have been sufficiently large that the differences between, for example, Figs. 13.2 and 13.12, may be considered real and related to differences in grazing angle rather than in resolution cell size. In fact, distributions closely resembling those in Fig. 13.12 were obtained much earlier from similar measurements with considerably broader pulse widths.[35]

At Very Low Grazing Angles. There is some evidence that sea clutter might drop off more sharply below a *critical angle* in the neighborhood of a degree or so (see Long[1]). This critical angle, or *critical range* for a radar at a fixed height, has been observed from time to time since first noted in early observations of sea clutter.[4] According to Katzin,[37] the critical angle occurs as a result of interference between direct and (perfectly) reflected rays at the scattering *targets* responsible for the clutter signal. While this simple picture can account for the *R*-minus-7 decay sometimes observed, a critical angle often fails to materialize, and when it does, it need not show an *R*-minus-7 decrease with range (or the equivalent fourth-power dependence on grazing angle).[1] An alternative explanation for this behavior, applicable at the higher microwave frequencies, has been suggested by Wetzel,[12,38] based on a *threshold-shadowing* model for upwind and downwind directions that implies a sharp decrease in the average cross section for grazing angles below a few degrees. In crosswind directions, with the radar looking along the troughs of the major waves, a much milder shadowing function will apply; so there should be a clear distinction between the upwind-downwind and crosswind behavior of sea clutter at very low grazing angles.

Examples of clutter behavior at these angles may be found in independent measurements at relatively high wind speeds by Hunter and Senior off the south coast of England[39] and by Sittrop off the west coast of Norway.[40] Their results for orthogonal directions relative to the wind are shown in Fig. 13.13, along with the predictions of a conventional shadowing function[41] and the threshold-shadowing function.[38] It would appear that a combination of conventional shadowing (which goes as the first power of the grazing angle) across the wind and threshold shadowing in upwind and downwind directions accounts for the observed behavior of this very low angle clutter quite well. The decay law for low-angle clutter should therefore depend on the viewing angle relative to the wind direction; so it might occur with powers between the first and the fourth. This is just what is observed.[42] It should be remarked, however, that shadowing at low grazing angles is a complex phenomenon (see below), and the physical origin or even the existence of a critical angle is still open to question. Moreover, there is relatively little good data on very low angle clutter for other than X-band frequencies; so the general behavior of sea clutter in this angular regime remains uncertain.

At HF and Millimeter-Wave Frequencies. All the measurements described above were made at *microwave* frequencies between UHF (428 MHz) and K_a band (35 GHz). High-frequency (HF) radars usually operate in the frequency range between about 5 and 30 MHz, corresponding to wavelengths between 60 and 10 m, respectively. Since the operation of such radars takes place either by the ground wave or over ionospheric (*sky-wave*) paths spanning great ranges, the grazing angles tend to be small (between 0 and 20°). For these wavelengths and grazing angles, initial measurements by Crombie indicated that the scattering from the sea surface was the result of Bragg scatter from sea waves of one-half the radar wavelength.[4] In the years since these early measurements, there has been considerable activity in the field of HF radar and HF clutter,[43,44] and the results can be summarized as follows: For vertical polarization, the major energy of the HF clutter signal appears in spectral lines displaced to either side of the

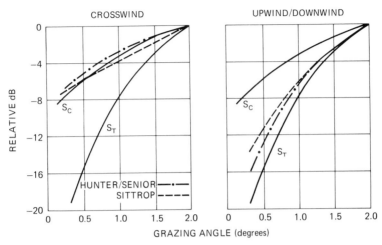

FIG. 13.13 Differential behavior of very low angle clutter for orthogonal wind directions: S_c is a conventional shadowing function;[41] S_T is a threshold-shadowing function.[38] *(Data from Hunter and Senior[39] and Sittrop.[40])*

carrier frequency by the frequency of sea waves having a wavelength equal to half the HF wavelength λ (in meters). The relative strengths of the plus and minus lines are determined by the proportion of advancing and receding Bragg-resonant wave components in the clutter cell. Provided the wind speed is greater than about $\sqrt{3\lambda}$ kn (with λ in meters) and the sea is fully developed, the clutter cross section σ^0 is about -27 dB and is relatively independent of wind speed and frequency. (The definition of σ^0 in HF radar is complicated by problems in properly defining antenna gains for ground-wave and sky-wave paths and by propagation effects due to the ionosphere.) The clutter spectrum tends to fill in around and between the lines as the wind picks up. For horizontal polarization (which is possible only over sky-wave paths), the cross section is much smaller and shows the characteristic fourth-power decay with decreasing grazing angle. For these HF wavelengths of tens of meters, the sea is relatively flat and the scattering laws are simple. A discussion of HF radar may be found in Chap. 24.

At the other end of the potentially useful radar spectrum, in the millimeter-wave band, the few published measurements of radar clutter lead to the conclusion that millimeter-wave backscatter behaves in much the same manner as backscatter at the lower microwave frequencies. This was suggested by the K-band curves shown in Fig. 13.5 for moderate wind speeds and further supported by some older shipboard data at frequencies between 9 and 49 GHz.[45] It should be noted that clutter signal paths lie close to the sea surface, where the atmospheric and water-vapor densities are highest. This means that at these higher frequencies the clutter signal will be strongly affected by the atmospheric absorption effects described in Chap. 2, and consequently the surface-related cross section inferred from the received signal strength in any given measurement will depend upon the path length. Moreover, the role of sea *spray* in both scattering and absorption will certainly be more important than at the lower microwave frequencies.

It is difficult to find clutter data at frequencies above K_a band, although *H*- and *V*-polarized returns at 95 GHz at a grazing angle of 1° were reported, both with values of close to -40 dB.[46,47] Interestingly, this is just the cross section measured at this angle by a number of investigators at X band (see Ref. 12), showing a similarity between the returns at these two widely spaced frequencies. However, at lower frequencies, at L band and below, there is a noticeable tendency for the cross section to fall off with decreasing grazing angles below about 15 to 20°.

The Spectrum of Sea Clutter. The scattering features producing sea clutter are associated with a surface subject to several types of motion. The features may themselves be moving with small group or phase velocities over this surface while the surface, in turn, is moved by the orbital velocities of the larger waves passing across it. Or the scatterers might be detached from the underlying surface, as in the plumes emitted at the crests of breaking waves, and move at speeds much greater than the orbital speeds.[48] At higher radar frequencies and in strong winds, the possibility of scattering from spray, advected by the wind field above the surface, must be considered. All this complex motion shows up in a doppler shift imparted to the scattered electromagnetic wave.

Surprisingly few measurements of microwave clutter spectra for real seas have been reported in the literature, and those few that exist can be separated into aircraft measurements of the spectral shape alone[49,50] and fixed-site shore measurements showing a shift in the spectral peak.[51,52] All these studies were performed at relatively low grazing angles (less than 10°), although Valenzuela and Laing include a few measurements up to 30°. Other measurements of sea clutter spectra include those made at much lower frequencies in the HF band, as described in the last section, those made under artificial conditions in the wave tanks,[53] whose application to real-sea conditions is uncertain, and other fixed-site measurements at high resolution and short averaging times, to be discussed later.

As it turns out, microwave sea clutter spectra have a rather simple form at the lower grazing angles. Figure 13.14 illustrates typical spectral behavior at the two polarizations, based on data collected by Pidgeon for C-band clutter looking upwind at a few degrees grazing.[51] The peak frequency of the upwind spectrum appears to be determined by the orbital velocity of the largest sea waves, plus a wind-dependent velocity increment containing, but not entirely explained by, wind-induced surface currents. The orbital velocity V_{orb} is taken to be that of the major waves and is obtained in terms of significant height $H_{1/3}$ and period "T" from the expression

$$V_{orb} = \pi H_{1/3} / "T" = 0.1U \qquad (13.10)$$

The approximate dependence on wind speed U was found by substituting $H_{1/3} = 3h_{rms}$ from Eq. (13.6), assuming a *fully developed sea*, and T from Eq. (13.5). To this there must be added a wind-drift velocity of about 3 percent of U and a fixed *scatterer* velocity (which appears to be about 0.25 m/s in the X- and C-band measurements[51,52,54]). Summing these components yields the virtual doppler velocity at the peak of the clutter spectrum for the particular case of a *vertically polarized, X- or C-band* radar looking *upwind* at *low grazing angles*:

$$V_{vir} \approx 0.25 + 0.13U \qquad m/s \qquad (13.11)$$

(As noted earlier, care must be taken whenever wind speed is used to parameterize a process that depends on waveheight. There is an unambiguous relation only for a fully developed sea in the absence of swell.) The remaining properties of the clutter spectrum can now be discussed in terms of V_{vir}. For example, the spectral peak for *horizontal* polarization follows a similar linear dependence on U, only with a coefficient lying somewhere between 0.17 and 0.20, as reflected by the sketch in Fig. 13.14. The (half-power) width of the clutter spectrum is roughly the same for both polarizations and is equal approximately to the upwind vertical velocity given in Eq. (13.11). For look directions away from upwind, the peak doppler follows a cosine dependence very closely, going to zero at crosswind aspects and turning negative downwind. Interestingly, the *bandwidth* of the spectrum remains relatively constant.

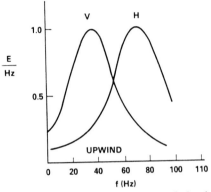

FIG. 13.14 Qualitative behavior of doppler spectra of sea clutter looking upwind at low grazing angles. (*Based on C-band measurements by Pidgeon.*[51])

The details of the clutter spectrum show little dependence on either the radar frequency or the grazing angle, at least for angles less than about 10°. In reviewing the results of measurements at four frequencies—UHF, L, C, and X bands—Valenzuela and Laing[50] noted a relatively weak tendency of clutter bandwidth to decrease with increases in frequency between the UHF and X bands and grazing angles between 5 and 30°. Since both of these variations entail a decrease in the size of the radar footprint on the surface, they might be due to a dependence on resolution cell dimensions, although the other workers found that the pulse length had little effect on clutter bandwidth for values between about 0.25 and 10 μs. The equivalence between time and space averaging in sea clutter measurements was discussed earlier, and in the case of clutter spectra the averaging times were all quite long (of the order of 10 to 20 min), which should be sufficient to stabilize the spectra for almost any resolution cell size.

Spectra obtained with *short* averaging times disclose something of the origins of the clutter spectrum. Figure 13.15 is a sequence of 0.2-s spectra obtained by Keller et al.[55] with a coherent vertically polarized X-band radar operating at a grazing angle of 35° and a resolution cell size of about 10 m². The zero-doppler reference in this figure was located arbitrarily at −16 Hz, and because of the high grazing angle the effects of both senses of the orbital velocities are seen, unlike the low-angle shadowed surface results shown in Fig. 13.14. The spread along each line is due to the small-scale wave motions on the surface, while the larger meanders are induced by the orbital velocities of the large waves moving through the measurement cell. The wind speed was 16.5 m/s, and a doppler shift of 100 Hz corresponds to a radial velocity of 1.6 m/s. The average clutter spectrum expected for this wind speed and grazing angle, with bandwidth obtained from Eq. (13.11), is sketched on the figure. The large spectral spike appearing in the center of the display is no doubt due to a wave breaking in or close to the measurement cell. The doppler velocity for this spike suggests a peak scatterer velocity of about half the wind speed, which would correspond to the group velocity of the longest waves on the surface. Although such events are relatively rare in a fixed area of 10 m², they should occur quite frequently within a large surveillance cell and might often have large scattering cross sections associated with them.

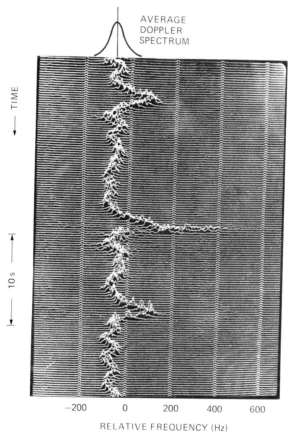

FIG. 13.15 Short-time averaged doppler spectra at X band for an intermediate grazing angle of 35°; spectra computed at 0.2-s intervals. (*From Keller et al.*[55])

Other Effects on Sea Clutter

Rain. Evidence of the effect of rain on sea clutter is mainly anecdotal; for example, radar operators report that sea clutter tends to decrease when it starts to rain. However, there has been little in the way of reliable, quantitative experimental information about the interaction between rain and wind-driven sea scatter. Laboratory measurements by Moore et al.[56] with artificial "rain" suggested that for light winds the backscatter level increased with the rain rate, while for heavy winds rain made little difference. In measurements in natural rain over Chesapeake Bay, Hansen[57] found that even a light rain (2 mm/h) changes the spectral character of sea clutter at moderate wind speeds (6 m/s) by introducing a significant high-frequency component. He also found some evidence in support of the radar operators, at least for the low grazing angles and horizontal polarizations with which most shipboard radars operate. Figure 13.16 compares the correlation function of sea clutter (X band, low grazing angle, H polarization) with and without rain for a 15-kn wind speed and a rain rate of 4 mm/h. The sharp decrease in correlation time in the presence of rain reflects the broadening of the clutter spectrum. Beyond this, there is virtually no quantitative information about the effect of rain on existing sea clutter.

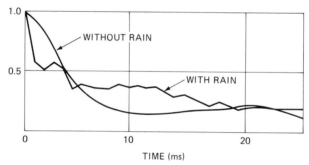

FIG. 13.16 Effect of rain on the correlation function of wind-driven sea clutter; X band, horizontal polarization, wind speed 15 kn, rain rate 4 mm/h. (*From Hansen.*[57])

The production of sea clutter by rain falling on a "calm" surface in the absence of wind was also investigated by Hansen, with the results shown in Fig. 13.17.[57] A high-resolution X-band radar (40-ns pulse, 1° beamwidth), operating at a grazing angle of about 3°, viewed the backscatter from a fixed spot on the windless surface of Chesapeake Bay as the rain steadily increased from 0 to 6 mm/h. The cross sections for vertical and horizontal polarizations were quite different for low rain rates but tended to merge at a rain rate of about 6 mm/h. The magnitude of this *splash* cross section rose to a σ^0 of about −40 dB, corresponding to wind-induced cross sections at this grazing angle for winds of about 10 kn. Further laboratory[58] and theoretical[59] studies have shown that the major scattering feature is the vertical *stalk* that emerges shortly after drop impact. Moreover, these studies suggest that the V-polarized returns from raindrop splashes should be relatively insensitive to the rain rate, while the H-polarized returns should show a strong dependence on both the rain rate and the drop-size distribution.

Propagation Effects. Another topic in sea clutter that has been largely unexplored is the role played by *propagation effects* within the atmospheric boundary layer lying over the sea surface. The effects of atmospheric absorption have been noted above in connection with millimeter-wave clutter. However, at very low grazing angles the ray paths joining the radar to the surface become very sensi-

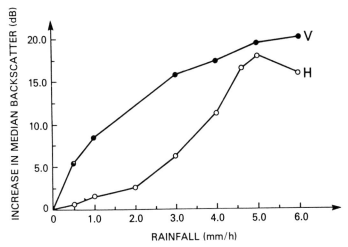

FIG. 13.17   Sea clutter produced by rain splashes alone on a calm surface. (*From Hansen.*[57])

tive to refractive inhomogeneities in the atmospheric boundary layer. Over distances approaching and beyond the conventional optical horizon, such perturbations could produce strong focus-defocus variations along the illumination profile[60] or a general rise in the local grazing angle.[38] Figure 13.18 gives an experimental example of the effect of *ducting* on very low angle sea clutter.[42] Since the grazing angle given as the abscissa is actually a plot of inverse range, the lifting of the cross section by ducting over an order-of-magnitude span of ranges is very likely due to a rise in the mean grazing angle produced by refraction in the evaporative layer.[38] Such effects should be suspected whenever the radar propagation path extends beyond the *optical* horizon.

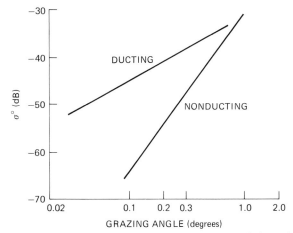

FIG. 13.18   Effect of ducting on low-angle clutter; wind speed about 10 kn.

Shadowing. The possibility of shadowing must be seriously considered whenever the sea is viewed at grazing angles smaller than the rms slope angle of the sea surface. Some examples were discussed earlier in connection with the behavior of sea clutter at low grazing angles in Fig. 13.13. In fact, the sharp falloff of the *nonconducting* data in Fig. 13.18 gives further evidence of the *threshold shadowing* mentioned there. However, the common idea of shadowing, along with all existing theories of a shadowed surface, rests on the geometrical optics concept of a sharp transition between light and darkness. By considering the implications of diffraction at the wave peaks, it is possible to determine the domain of radar frequencies and wind speeds over which the concepts of geometrical optics may be applied. This was done by Wetzel,[12] who showed in detail how diffraction, rather than shadowing, controls propagation into and out of the troughs of the waves under many of the usual frequencies and wind speeds encountered in practical radar operations at low grazing angles. For example, shadowing will take place at K_a band for any winds above 15 kn, yet will hardly ever occur at L-band frequencies.

Contaminants. The idea of pouring oil on troubled waters is a familiar one: the angry surface will smooth and subside. In another age, the survival-gear locker of every sailing ship would contain a bottle of oil to quiet the sea in a storm. Although the effectiveness of this procedure has always been somewhat controversial, there is no question that oil can produce a *slick* of smooth water at relatively low wind speeds. In fact, biological oils, produced by bacteria, algae, and plankton, can be found everywhere on the world's oceans and form natural slicks in those regions that combine the greatest oil concentration with the lowest wind speeds, e.g., close to continental shorelines.[61] Human-made contaminants can, of course, have the same effect. A layer of oil only 1 molecule thick will significantly affect the ability of the surface to support wave motions, but this layer must be continuous. The adjacent molecules then sense each other and form a film that is resistant to horizontal compression. The surface elasticity is changed, a type of longitudinal viscosity is introduced, and the surface becomes stabilized against the growth of short waves up to several inches in length.[62,63]

To the extent that radar sea clutter is produced by small-scale surface roughness (at grazing angles less than about 80°), the presence of oil on the surface should lead to a measurable decrease in clutter cross section. But, as noted above, the reduction of small wave motions requires the existence of a *continuous* monolayer; slick formation is a go–no-go process, and so slicks will tend to have relatively sharp boundaries. In operating the NRL 4FR system as a synthetic aperture radar to obtain images of the slicks produced by oil spills, Guinard found that the slicks were well defined, that it took very little oil to maintain a visible slick, that vertical polarization provided much greater contrast than did horizontal, and that the slicks were quenched by winds and currents.[64] Although signal strength was not recorded in this imaging experiment, later measurements at X and L bands by others[65] indicated that at the higher grazing angles (about 45°) the clutter reduction produced by the types of oil occurring in *natural* slicks was rather small, of the order of a few tenths of a percent. Since slicks are dispersed by the wind and associated wave action at wind speeds greater than about 10 kn, the effect of natural slicks on clutter may not be clear because they tend to occur in the regime of low wind speeds where the sea surface is already ill defined.

The celebrated *sun glitter* measurements by Cox and Munk[66] gave a quantitative measure of the effect of contaminants on the surface slopes in open water, showing that the wind-generated component of the rms slope of "oiled" waters is significantly smaller than that of "clean" water. The heavy human-made oils used in their experiment were effective in suppressing small-scale waves over a range of wind speeds well beyond those which would normally disperse the lighter natural oils; so the effect of oil spills on sea clutter should be expected to extend to the higher wind speeds. In fact, at these higher wind speeds the depression of radar backscatter by such oils at X and K_a bands can reach 10 to 20 dB at intermediate grazing angles between 30 and 60°.[67,68]

Currents. The most obvious effect of a current on sea clutter would be a shift in the peak of the doppler spectrum, similar to the contribution of the 3 percent wind-drift current mentioned in connection with Eq. (13.8). Another effect is related to the fact that the excitation of the surface-wave system depends on the *apparent* wind; so there can be significant differences in waveheight according as the wind is blowing with or against the current. According to Eq. (13.6), waveheight is proportional to the square of the wind speed; so in the Gulf Stream, for example, with a current of 4 kn flowing north, a 15-kn northerly blowing against the current will raise a sea 3 times as high as a 15-kn southerly blowing with the current. Even with no wind the presence of strong current shears can produce highly agitated surfaces. Shipboard observers have reported bands of roaring breakers passing by on an otherwise-smooth surface, presumably produced by powerful surface-current shears associated with large-amplitude internal waves.[69] In a more subtle way, currents are held responsible for synthetic aperture radar (SAR) images which contain the expression of bottom topography in shallow waters.[70] In each of the examples cited above, the current produces a change in the surface roughness, which can be expected to give rise to a change in sea clutter cross section.

Combined Effects. Some idea of the complexity within a clutter scene due to *other effects* may be obtained from Fig. 13.19, which shows a digitized PPI display of clutter in the Sargasso Sea, under light wind conditions and near a thermal oceanic front.[71] Although all possible contributing effects were not identified, observers noted the presence of human-made detritus organized by the currents at the edges of the thermal front, slicks probably of both natural and artificial origin, fronds of seaweed close to the surface, and the presence of light and variable winds. The dynamic range of the digitized (false-color) PPI was 30 dB, and some of the clutter contrasts, across what are obviously extremely sharp boundaries, were almost this great.

13.4 THEORIES OF SEA CLUTTER

The sea surface is so rich in potential scattering structures that in seeking to understand the phenomenology experimenters and theorists alike have proposed and have found support for almost any imaginable model. However, aside from providing an intellectual basis for "understanding" sea clutter phenomena, a theory of sea clutter should serve the practical purpose of providing accurate a priori predictions of all aspects of clutter behavior under all possible environmental conditions. At present, the theory of sea clutter does neither of these tasks very well and must be thought of as a book with the final chapters still to be written.

Before discussing the current theories of sea clutter, it is important to distinguish them from other so-called sea clutter models that are designed to provide a predictive capability. Some such models organize large quantities of empirical data by finding a multiple linear-regression formula relating the clutter cross section to a variety of parameters, such as grazing angle, wind speed, frequency, etc., all measured concurrently.[1,40] Even a multiparameter matrix tabulation

FIG. 13.19 Example of strong clutter contrasts under special environmental conditions; light winds in the Sargasso Sea; range 1 km. (*Provided by Trizna.*[71])

of the average values of all available data can be viewed as a model.[3] Data summarized in this way can often be useful to the system designer in establishing ballpark estimates of clutter levels under various conditions but is of little use in real-time predictions in a dynamic environment or in providing a basis for a physical understanding of sea backscatter.

In the development of models of sea backscatter based on physical theory, there are essentially two basic and distinct approaches. Historically, the first approach assumed the clutter return to have its origin in scattering *features*, or obstacles, actually present on or near the sea surface. For example, seeking to explain the difference between the cross sections for the two polarizations, Goldstein[72] applied a kind of rain model to the cloud of spray often seen above the surface of an active sea but failed to explain why the difference was the same whether or not spray was present. At least spray was an observable feature of the sea environment. Later features included smooth circular metallic disks,[37,73] arrays of semi-infinite planes,[74] and fields of hemispherical bosses,[75] to name a few. Obviously, the choice of these scattering obstacles related more to the pre-existence of convenient scattering solutions for these shapes than to insights gained from observing the sea. More recently, feature models have taken on a bit more reality by focusing on wedge shapes, as suggested by the Stokes waves and sharp crests observed on most natural water surfaces[12,25,76,77] and by the sloshes and plumes suggested by the properties of wave groups and the hydrodynamics of breaking waves.[12,48]

The other approach to theoretical modeling derives the scattered field from a global boundary-value problem (GBVP) in which the sea as a whole is considered a boundary surface whose corrugations are described by some kind of statistical process. An enormous literature is devoted to the theory of surface scatter from this point of view, stemming from the importance not only of radar sea scatter but radar ground scatter and sonar *reverberation* (the acoustic equivalent of radar clutter) from both the surface and the bottom of the sea. Since it is the GBVP approach that leads to the analytical expression of Bragg resonance scattering that has dominated the theory of sea backscatter since the late 1960s, a brief explanation of some of the central ideas is included below.

Theories Based on Global Boundary-Value Problems. Unfortunately, general formulations of the GBVP, though elegant, are of little practical value, and some kind of approximation is necessary to obtain useful quantitative results from them. The methods of approximation relate to the two methods of formulating the GBVP:

1. Small-amplitude approximations (sea wave heights much smaller than the radar wavelength) are used with Rayleigh's hypothesis, in which the boundary conditions are employed to match a spectrum of outgoing plane waves to the incident field.[8,78–80]

2. A general integral formulation based on Green's theorem is pursued either in a small-amplitude approximation[6,81] or under the assumptions of physical optics (surface curvatures much greater than the radar wavelength).[82–84]

In formulation 1, sometimes called the small-perturbation method (SPM) and associated most often with the work of Rice,[8] the surface displacements are assumed *everywhere* to be much smaller than the radar wavelength; so the method is directly applicable only to such cases as HF scattering, with wavelengths of tens of meters, at low to intermediate wind speeds, and with waveheights of a few meters at most. The solution is in the form of a power series in the ratio of sea waveheight to radar wavelength, and it predicts the first-order Bragg lines and second-order spectral filling around the lines that were mentioned in the earlier section on HF sea clutter.

On the other hand, the various integral formulations referenced above are usually formulated in very general terms, but in most of them σ^0 either appears in or can be put into a form represented schematically by the following simplified one-dimensional expression (see Ref. 10 or Ref. 83, for example):

$$\sigma^0(\psi) = Ak^2F_p(\psi)\int_{-\infty}^{\infty} dy e^{i2k_1 y}[e^{-4k_2^2h^2[1 - C(y)]} - e^{-4k_2^2h^2}] \quad (13.12)$$

where A is a constant; $k_1 = k \cos \psi$ and $k_2 = k \sin \psi$ where k is the radar wavenumber $(2\pi/\lambda)$; $F_p(\psi)$ is a function of polarization p, grazing angle ψ, and the electrical properties of seawater; h is the rms sea waveheight; and $C(y)$ is the surface correlation coefficient. Of course, the reduction of a complicated boundary-value problem to so simple a form requires assumptions about both the surface fields and the distribution of the sea heights, which is gaussian to a good approximation.[85] But except for the SPM approach mentioned above, where small ratios of h/λ are assumed right at the start, the results of virtually all existing GBVP theories derive from expressions resembling Eq. (13.12), for which there are no a priori restrictions on surface heights.

The statistical properties of the sea surface enter through the correlation coefficient $C(y)$ appearing under the integral sign in the exponential in the brackets, and by expanding this exponential Eq. (13.12) may be written

$$\sigma^0(\psi) = AF_p(\psi)e^{-4k_2^2h^2}\sum_{n=1}^{\infty}\frac{(4k_2^2)^n}{n!}W^{(n)}(2k_1) \qquad (13.13)$$

where

$$W^{(n)}(2k_1) = \int_{-\infty}^{\infty} d\tau\ e^{i2k_1\tau}[h^2C(\tau)]^n \qquad (13.14)$$

Interestingly, back in 1963 this expression was treated in some detail in the standard reference book by Beckmann and Spizzichino for the special case of a gaussian correlation function,[83] and while it becomes analytically intractable for more general correlation functions, it continues to show up in papers on rough-surface scattering.[10,86]

Bragg Scattering from Integral Formulations. In the limit of small ratios of rms waveheight to radar wavelength or, more specifically,

$$2kh \sin \psi - 1 \qquad (13.15)$$

only the first term in the series in Eq. (13.13) survives, and the cross section assumes the very simple form

$$\sigma^0(\psi) \not= 4\pi k^4 F_p'(\psi)\overset{(1)}{W} (2k \cos \psi) \qquad (13.16)$$

where the constant A has been made explicit and F' has absorbed a $\cos^2$ term from the series. $W^{(1)}$ is the Fourier transform of the surface correlation function, which makes it the sea wavenumber spectrum discussed in Sec. 13.2 evaluated at twice the (surface-projected) radar wavenumber, which is the *Bragg-resonant* wavenumber. Except possibly for the details of the angle factor F', Eq. (13.16) is equivalent to the result obtained by the SPM discussed above, and although it is sometimes felt that its derivation from a surface integral provides some potential for greater generalization, it carries with it all of the same restrictions.

Nevertheless, this direct, linear relation between radar cross section and the oceanographers' descriptor of the sea surface has had a powerful influence on thinking about the physical origins of sea clutter. It is intuitively appealing, and it offers a simple way both to predict radar clutter from measurements or forecasts of the sea spectrum and, inversely, to use radar backscatter measurements to provide remote sensing of the sea surface for oceanographic and meteorological applications. As noted earlier, HF radar clutter is described quite well by a Bragg model and in fact at most wind speeds is predicted with reasonable accuracy by Eq. (13.16). Moreover, this relation has been exploited with great success in providing long-wavelength sea spectra from HF clutter measurements.[87,88]

At microwave frequencies, however, the small-perturbation assumption on which the Bragg model rests is violated on any real sea surface. Consider, for example, an X-band radar (3-cm wavelength) looking at the sea with a grazing angle of 45°. The small-perturbation condition expressed by Eq. (13.15) means that the maximum departure of the sea surface from a flat plane must be much smaller than 7 mm. But for the median oceanic wind speed of 15 kn the rms wave height is 0.3 m, making $2kh \sin \psi = 84$. Thus the condition for using Eq. (13.16) is strongly violated, and the power series in Eq. (13.13) would contain thousands of terms, making it essentially useless whatever the form taken by Eq. (13.14). It appears, therefore, that the use of integral formulations in the practical solution of the sea clutter problem will require something more sophisticated.

Other Limiting Conditions. Instead of expanding the exponential in the integrand of Eq. (13.12), it should be possible, at least in principle, to replace $C(y)$ directly by the Fourier transform of $W(K)$ [the inverse of Eq. (13.14) for $n = 1$], thus providing a functional relationship between the radar cross section and the sea wave spectrum without the restrictions of a small-amplitude approximation. This approach involves extensive computations even to obtain limited results in individual cases, as shown in work by Holliday et al.[10] However, one interesting result of this approach was the discovery that the importance of the Bragg-resonant part of the spectrum appeared to decrease significantly with increasing wind speeds and frequencies, a conclusion that casts suspicion on the significance of Bragg scatter under conditions other than those for which Eq. (13.16) is valid.

In the opposite limit of large kh, another approximation is used. In this case the integrand in Eq. (13.12) is vanishingly small *except* for small values of y, for which the correlation coefficient $C(y)$ is close to unity. The first few terms in an expansion of $C(y)$ around $y = 0$ are

$$C(y) = 1 + C'(0)y + (\tfrac{1}{2})C''(0)y^2 \tag{13.17}$$

where $C(0)$ has been put equal to 1 in the first term. For gaussian-like correlation coefficients $C'(0) = 0$ and $C''(0) = -L^{-2}$, where L is the correlation length of the process. If Eq. (13.17) is substituted into Eq. (13.12) with these values, the resulting integral is in a standard form and the cross section becomes

$$\sigma^0(\psi) = A'F_p(\psi)(h/L)^{-1}\exp\left[-\cot^2\psi/4(h/L)^2\right] \tag{13.18}$$

which says that even under "very rough" conditions in which $kh \geq 1$, long, smoothly varying surfaces with small ratios (h/L) would produce very little clutter, while choppy, "noisy" surfaces with large ratios would scatter much more energetically. Unfortunately, this result has little value in making quantitative predictions of sea clutter from given environmental conditions.

If the basic integral formulation of the GBVP is solved in the physical optics approximation (large k) and the scattered field found by the method of stationary phase, the result is an expression commonly called the *specular return* because its origin may be traced to pieces of the surface that provide a reflection point for the incident wave.[83,90] This expression is written for a gaussian sea surface in the form

$$\sigma^0(\psi) = (|R|^2/s^2)\csc^4\psi\exp\left[-\cot^2\psi/s^2\right] \tag{13.19}$$

where s is the rms surface slope and R is the flat-surface reflection coefficient for normal incidence. This is the type of scattering alluded to in connection with the high-grazing-angle returns discussed in Sec. 13.3, and the tendency of σ^0 to round off for grazing angles close to $90°$ (see Figs. 13.3 and 13.4) may be ascribed to this mechanism.

From what has been said thus far, it can be seen that strict analytical solutions via the GBVP approach appear to run into dead ends: intractable formal expressions in the form of Eq. (13.12), small-amplitude approximations in the form of

Eq. (13.16) that make little sense for microwave scattering from real sea surfaces, or large-amplitude limits like Eqs. (13.18) and (13.19) that ultimately relate to the probability densities of specularly reflecting surface slopes. Judging from the success of Holliday et al. in exploiting expressions in the form of Eq. (13.12), the best hope of obtaining useful results for real seas from GBVP formalisms might well lie in further study of the surface correlation function.

The Composite-Surface Hypothesis. Since it is not clear how to extend straightforward GBVP solutions beyond the limiting approximations described above, a heuristic model was developed that viewed the sea as a carpet of Bragg scattering wavelets obeying the condition in Eq. (13.15), modulated by the motions of the larger waves on the surface.[89–91] It is imagined that the surface-wave spectrum can somehow be separated into two parts, one which contains the Bragg scattering wavelets and whose integrated rms waveheight satisfies the conditions of Eq. (13.16) and another which contains only the long waves that tilt and otherwise modulate the Bragg waves. Other assumptions include (1) that the correlation lengths of the short Bragg waves be long enough that a resonant interaction is possible but short enough that adjacent areas on the surface contribute to the total signal in random phase; and (2) that the long waves that tilt and modulate the short waves have radii of curvature sufficiently large that the curvature over the correlation length of the Bragg "patches" is small in some sense. Thus this model replaces the sea surface by an ensemble of "flat" Bragg scattering patches, tilted, heaved, and advected by the motion of the large-wave components. In its most elementary form, it interprets σ^0 (ψ) in Eq. (13.16) as the cross section of a patch with *local* grazing angle $\psi = \psi_0 + \alpha$, where α is the local wave slope and ψ_0 is the mean grazing angle. For the simple one-dimensional case, this quantity is averaged over the sea slope distribution $p(\alpha)$, yielding

$$\overline{\sigma^0}(\psi_0) = \int_{-\infty}^{\infty} \sigma^0(\psi_0 + \alpha)p(\alpha)d\alpha \qquad (13.20)$$

For a more general two-dimensional sea, the local grazing angle is a function of the slopes in and normal to the plane of incidence, while for each polarization p the angle function $F_p(\psi)$ in $\sigma^0(\psi)$ becomes a complex mixture of the angle functions of both polarizations.[89] Reference 91 contains a comprehensive discussion of this model.

Although the composite-surface model is often presented as if it emerged as a rigorous product of an integral formulation of the GBVP, it is not a scattering *theory* but instead a scattering *picture* assembled from a group of more or less plausible assumptions. Such models are never very satisfying, but with the failure of the more formal GBVP theories to provide a general framework for predicting and understanding sea clutter, this model has become the basis for most analytical approaches to backscatter from the sea.

Figure 13.20 is based on a comparison in Ref. 89 of the predictions of the Bragg model from the SPM [in the form of Eq. (13.16)] and the composite-surface model [in the form of Eq. (13.20)] with a sample of NRL 4FR data taken at high wind speeds (greater than 22 kn). The wave spectrum used was the Phillips spectrum given in Eq. (13.4). Historically, comparisons of this type have been used often to establish the Bragg scattering hypothesis as a useful clutter model,[89,91] and the agreement usually looks good, especially for the higher wind speeds. But it must be kept in mind that the small-amplitude approximation in Eq. (13.16) is *by itself* totally invalid for the extended sea surface; so it remains a curiosity how a totally invalid application of a scattering approximation can produce results as convincing as those shown in Fig. 13.20. The effect of using the composite-surface model is primarily to introduce an average over surface slopes that raises the horizontal return at the lower grazing angles. However, this effect would also occur if the scattering patch were populated not by Bragg wavelets but instead by some of the features described in the next section. Nevertheless, agreement between measurement and prediction of the type illustrated in Fig. 13.20 has kept alive belief in the Bragg scattering hypothesis in spite of the lack of a proper theory argued from first principles.

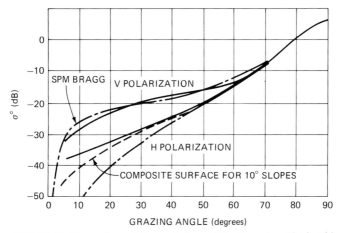

FIG. 13.20 Comparison of the predictions of the Bragg hypothesis with NRL 4FR data for higher wind speeds (>20 kn). (*Based on Valenzuela.*[89])

Scattering by Surface Features. The actual sea surface is a complex mixture of wedges, cusps, microbreakers, hydraulic shocks, patches of turbulence and gravity-capillary waves (both wind-driven and parasitic), punctuated on occasion by the sharp crest of a breaking wave, with plumes of water cascading down its face and a halo of spray above it. In other words, there is a large variety of scattering features associated with an active sea surface, any or all of which could contribute to the clutter signal.

The common Stokes wave[85] has a quasi-trochoidal structure that resembles a wedge on the surface, yet only recently were wedges studied as a possible source of sea clutter.[11,12,76,77] The scattering model is usually some variant of the familiar geometrical theory of diffraction (GTD),[92] which is strictly applicable to the clutter problem only when the edge of the wedge is normal to the plane of incidence. Nevertheless, the predictions gave both polarization and grazing-angle dependences over at least part of the ranges of these variables that were about as good as, and in some ways better than, the predictions of the Bragg or composite-surface models.[77]

One major problem with all models based on scattering features is the lack of any information about the shapes, sizes, orientations, speeds, and lifetimes of the features themselves. For this reason, the predictions of such models must contain arbitrary assumptions about these crucial parameters. However, there is often guidance from either observation or theory. For example, stability arguments prevent the interior angle of a wave crest from falling below 120°, which then becomes a convenient measure of the wedge angle in wedge-scattering models. In Fig. 13.21, the overall scale of wedge scattering as calculated by the GTD was adjusted to place the cluster of cross sections at the level of the experimental values; the polarization and grazing-angle dependence are shown by the circles. The dependence on these two parameters appears to be fairly well expressed by the wedge model, although it should be recalled that this model applies only to wedges whose edges are normal to the plane of incidence; so the application of GTD-based wedge models confronts difficulties just as important as those associated with the application of the Bragg model.

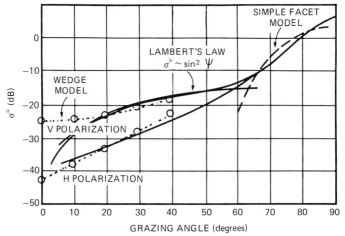

FIG. 13.21 Comparison of simple wedge and other models with NRL 4FR data (data the same as in Fig. 13.20).

Figure 13.21 also compares two very simple scattering models with the data in Fig. 13.20. Lambert's law, mentioned in connection with Fig. 13.9a and b, expresses the cross section in the form $\sigma^0 = A \sin^2 (\psi)$, where A is the surface albedo. Choosing $A = -17$ dB gives a fairly good match to the vertically polarized returns over a wide range of grazing angles. The *facet* model, expressed by Eq. (13.19) and often thought to describe clutter at the higher grazing angles, is shown for 20-kn seas.[89] The general behavior described by these two models seems to agree about as well as any other, although they too must employ arbitrary assumptions to obtain reasonable fits to the data; so the significance of this agreement is difficult to assess.

Another surface scattering feature was examined in connection with the complex behavior of sea spikes discovered by Lewis and Olin[33] (see Fig. 13.11). At low grazing angles, the most visible part of the surface consists of the higher elevations, particularly the peaks and faces of breaking waves. By using a plume model for the scattering elements associated with spilling breakers, a theory of scattering from breaking waves was developed that explained most of the observed behavior of sea spikes.[48] Of course, like all other models based on scattering features, it was necessary to make arbitrary assumptions about the sizes, shapes, and lifetimes of the scattering plumes. But these parameters were all inferred from observation of real sea surfaces, and the resulting predictions were surprisingly good.

Although scattering features have been introduced mainly in connection with low-grazing-angle sea clutter (Ref. 12 contains a detailed discussion), there is every reason to believe that feature scattering operates at all grazing angles. Considering the failure of scattering theories formulated as the GBVP to provide any predictions beyond those in certain limiting-case approximations and the precarious nature of the logical infrastructure of the Bragg hypothesis in microwave scattering, it is likely that a careful examination of the actual scattering features present on the sea surface might provide the basis for understanding sea clutter that has been sought for the last 40 years.

13.5 SUMMARY AND CONCLUSIONS

In the early days of radar, the importance of knowing the clutter environment led to many experiments under a variety of conditions. Variations in quality and completeness of ground truth, calibration of the equipment, and the competence of the experimenter led to results that often showed considerable inconsistency and suggested clutter behavior that was sometimes more a function of the vagaries of the experiment than of the physics of clutter. As data of increasingly better quality accumulated, it would seem that it should be possible to establish the behavior of sea clutter with increasing confidence. This has not always been so. For the higher 50 percent of wind speeds encountered over the world's oceans (above about 15 kn), microwave sea clutter at intermediate to high grazing angles has little dependence on frequency, and the effects of wind speed are uncertain, seeming to depend on polarization, wind direction, and grazing angle in confusing ways. The major areas of uncertainty, however, lie at any wind speed, whenever the grazing angle goes below a few degrees and the surface illumination begins to feel the effects of refraction and diffraction, and at any grazing angle, whenever the wind speed is less than about 10 kn, where peculiarities and uncertainties in the generation of surface roughness begin to emerge most strongly.

The jury is still out on the question of sea clutter theory. The most popular model, the composite-surface model, is actually an assemblage of assumptions supported by circumstantial evidence; there is still no clear reason why it should work as it does. Theories based on scattering by surface features are beginning to show promise, although the major problem of characterizing these features in a manner useful to quantitative predictions is still to be addressed. It will be interesting to see what progress will have been made in the theory of sea clutter by the publication of the next edition of this handbook.

REFERENCES

1. Long, M. W.: "Radar Reflectivity of Land and Sea," Artech House, Norwood, Mass., 1983.
2. Skolnik, M. I.: "Introduction to Radar Systems," McGraw-Hill Book Company, New York, 1980.

3. Nathanson, F. E.: "Radar Design Principles," McGraw-Hill Book Company, New York, 1969.

4. Kerr, D. E.: "Propagation of Short Radio Waves," McGraw-Hill Book Company, New York, 1951.

5. Crombie, D.: Doppler Spectrum of Sea Echo at 13.56 Mc/s, *Nature*, vol. 175, pp. 681–683, 1955.

6. Wright, J. W.: A New Model for Sea Clutter, *IEEE Trans.*, vol. AP-16, pp. 217–223, 1968.

7. Bass, F. G., I. M. Fuks, A. I. Kalmykov, I. E. Ostruvsky, and A. D. Rosenberg: Very High Frequency Radio Wave Scattering by a Disturbed Sea Surface, *IEEE Trans.*, vol. AP-16, pp. 554–568, 1968.

8. Barrick, D., and Q. Peake: A Review of Scattering from Surfaces with Different Roughness Scales, *Radio Sci.*, vol. 3, pp. 865–868, 1968.

9. Atlas, D., R. C. Beal, R. A. Brown, P. De Mey, R. K. Moore, C. G. Rapley, and C. T. Swift: Problems and Future Directions in Remote Sensing of the Oceans and Troposphere: A Workshop Report, *J. Geophys. Res.*, vol. 9(C2), pp. 2525–2548, 1986.

10. Holliday, D., G. St-Cyr, and N. E. Woods: A Radar Ocean Imaging Model for Small to Moderate Incidence Angles, *Int. J. Remote Sensing*, vol. 7, pp. 1809–1834, 1986.

11. Kwoh, D. S., and B. M. Lake: A Deterministic, Coherent, and Dual-Polarized Laboratory Study of Microwave Backscattering from Water Waves, part 1: Short Gravity Waves without Wind, *IEEE J. Oceanic Eng.*, vol. OE-9, pp. 291–308, 1984.

12. Wetzel, L. B.: Electromagnetic Scattering from the Sea at Low Grazing Angles, chap. 12 in Geernaert, G. L., and W. J. Plant (eds.): "Surface Waves and Fluxes: Current Theory and Remote Sensing," Reidel, Dordrecht, Netherlands, 1989.

13. Phillips, O. M.: Radar Returns from the Sea Surface—Bragg Scattering and Breaking Waves, *J. Phys. Oceanogr.*, vol. 18, pp. 1065–1074, 1988.

14. Middleton, D., and H. Mellin: Wind-Generated Solitons: A Potentially Significant Mechanism in Ocean Surface Wave Generation and Wave Scattering, *IEEE J. Oceanic Eng.*, vol. OE-10, pp. 471–476, 1985.

15. Tang, S., and O. H. Shemdin: Measurement of High-Frequency Waves Using a Wave Follower, *J. Geophys. Res.*, vol. 88, pp. 9832–9840, 1983.

16. Pierson, W. J., and L. Moskowitz: A Proposed Spectral Form for Fully Developed Seas Based on the Similarity Theory of S. A. Kitaigorodskii, *J. Geophys. Res.*, vol. 69, pp. 5181–5190, 1964.

17. Phillips, O. M.: "The Dynamics of the Upper Ocean," 2d ed., Cambridge University Press, Cambridge, England, 1977.

18. Pierson, W. J., Jr., and M. A. Donelan: Radar Scattering and Equilibrium Ranges in Wind-Generated Waves with Application to Scatterometry, *J. Geophys. Res.*, vol. 91(C5), pp. 4971–5029, 1987.

19. Kitaigorodskii, S. A.: On the Theory of the Equilibrium Range in the Spectrum of Wind-Generated Gravity Waves, *J. Phys. Oceanogr.*, vol. 13, pp. 816–827, 1983.

20. Phillips, O. M.: Spectral and Statistical Properties of the Equilibrium Range in Wind-Generated Gravity Waves, *J. Fluid Mech.*, vol. 156, pp. 505–531, 1985.

21. Daley, J. C., J. T. Ransone, J. A. Burkett, and J. R. Duncan: Sea Clutter Measurements on Four Frequencies, *Naval Res. Lab. Rept.* 6806, November 1968.

22. Valenzuela, G. R., and R. Laing: On the Statistics of Sea Clutter, *Naval Res. Lab. Rept.* 7349, December 1971.

23. Guinard, N. W., J. T. Ransone, Jr., and J. C. Daley: Variation of the NRCS of the Sea with Increasing Roughness, *J. Geophys. Res.*, vol. 76, pp. 1525–1538, 1971.

24. Daley, J. C.: Wind Dependence of Radar Sea Return, *J. Geophys. Res.*, vol. 78, pp. 7823–7833, 1973.

25. Kalmykov, A. I., and V. V. Pustovoytenko: On Polarization Features of Radio Signals Scattered from the Sea Surface at Small Grazing Angles, *J. Geophys. Res.*, vol. 81, pp. 1960–1964, 1976.

26. Katz, I., and L. M. Spetner: Polarization and Depression Angle Dependence of Radar Terrain Return, *J. Res. Nat. Bur. Stand., Sec. D*, vol. 64-D, pp. 483–486, 1960.

27. Feindt, F., V. Wismann, W. Alpers, and W. C. Keller: Airborne Measurements of the Ocean Radar Cross Section at 5.3 GHz as a Function of Wind Speed, *Radio Sci.*, vol. 21, pp. 845–856, 1986.

28. Schroeder, L. C., P. R. Schaffner, J. L. Mitchell, and W. L. Jones: AAFE RADSCAT 13.9-GHz Measurements and Analysis: Wind-Speed Signature of the Ocean, *IEEE J. Oceanic Eng.*, vol. OE-10, pp. 346–357, 1985.

29. Masuko, H., K. Okamoto, M. Shimada, and S. Niwa: Measurement of Microwave Backscattering Signatures of the Ocean Surface Using X Band and K_a Band Airborne Scatterometers, *J. Geophys. Res.*, vol. 91(C11), pp. 13065–13083, 1986.

30. de Loor, G. P., and P. Hoogeboom: Radar Backscatter Measurements from Platform Noordwijk in the North Sea, *IEEE J. Oceanic Eng.*, vol. OE-7, pp. 15–20, January 1982.

31. Chaudhry, A. H., and R. K. Moore: Tower-Based Backscatter Measurements of the Sea, *IEEE J. Oceanic Eng.*, vol. OE-9, pp. 309–316, December 1984.

32. Ulaby, F. T., R. K. Moore, and A. K. Fung: "Microwave Remote Sensing, Active and Passive, vol. III, Addison-Wesley Publishing Company, Reading, Mass., 1986, Sec. 20.2.

33. Lewis, B. L., and I. D. Olin: Experimental Study and Theoretical Model of High-Resolution Backscatter from the Sea, *Radio Sci.*, vol. 15, pp. 815–826, 1980.

34. Trizna, D.: Measurement and Interpretation of North Atlantic Ocean Marine Radar Sea Scatter, *Naval Res. Lab. Rept.* 9099, May 1988.

35. Trunk, G. V.: Radar Properties of Non-Rayleigh Sea Clutter, *IEEE Trans.*, vol. AES-8, pp. 196–204, 1972.

36. Wu, Jin: Variations of Whitecap Coverage with Wind Stress and Water Temperature, *J. Phys Oceanogr.*, vol. 18, pp 1448–1453, October 1988.

37. Katzin, M.: On the Mechanisms of Radar Sea Clutter, *Proc. IRE*, vol. 45, pp. 44–54, January 1957.

38. Wetzel, L. B.: A Model for Sea Backscatter Intermittency at Extreme Grazing Angles, *Radio Sci.*, vol. 12, pp. 749–756, 1977.

39. Hunter, I. M., and T. B. A. Senior: Experimental Studies of Sea Surface Effects on Low Angle Radars, *Proc. IEE*, vol. 113, pp. 1731–1740, 1966.

40. Sittrop, H.: X- and K_u-Band Radar Backscatter Characteristics of Sea Clutter, in Schanda, E. (ed.): *Proc. URSI Commission II Specialist Meeting on Microwave Scattering from the Earth*, Bern, 1974.

41. Smith, B. G.: Geometrical Shadowing of the Random Rough Surface, *IEEE Trans.*, vol. AP-15, pp. 668–671, 1967.

42. Dyer, F. B., and N. C. Currie: Some Comments on the Characterization of Radar Sea Echo, *Dig. Int. IEEE Symp. Antennas Propagat.*, July 10–12, 1974.

43. Barrick, D. E., J. M. Headrick, R. W. Bogle, and D. D. Crombie: Sea Backscatter at HF: Interpretation and Utilization of the Echo, *Proc. IEEE*, vol. 62, 1974.

44. Teague, C. C., G. L. Tyler, and R. H. Stewart: Studies of the Sea Using HF Radio Scatter, *IEEE J. Oceanic Eng.*, vol. OE-2, pp. 12–19, 1977.

45. Wiltse, J. C., S. P. Schlesinger, and C. M. Johnson: Back-Scattering Characteristics of the Sea in the Region from 10 to 50 KMC, *Proc. IRE*, vol. 45, pp. 220–228, 1957.

46. Ewell, G. W., M. M. Horst, and M. T. Tuley: Predicting the Performance of Low-Angle Microwave Search Radars—Targets, Sea Clutter, and the Detection Process, *Proc. OCEANS 79*, pp. 373–378, 1979.

47. Rivers, W. K.: Low-Angle Radar Sea Return at 3-MM Wavelength, Georgia Institute of Technology, Engineering Experiment Station, *Final Tech. Rept.*, Contract N62269-70-C-0489, November 1970.

48. Wetzel, L. B.: On Microwave Scattering by Breaking Waves, chap. 18 in Phillips, O. M., and K. Hasselmann (eds.): "Wave Dynamics and Radio Probing of the Ocean Surface," Plenum Press, New York, 1986, pp. 273–284.

49. Hicks, B. L., N. Knable, J. J. Kovaly, G. S. Newell, J. P. Ruina, and C. W. Sherwin: The Spectrum of X-Band Radiation Backscattered from the Sea Surface, *J. Geophys. Res.*, vol. 65, pp. 825–837, 1960.

50. Valenzuela, G. R., and R. Laing: Study of Doppler Spectra of Radar Sea Echo, *J. Geophys. Res.*, vol. 65, pp. 551–562, 1970.

51. Pidgeon, V. W.: Doppler Dependence of Sea Return, *J. Geophys. Res.*, vol. 73, pp. 1333–1341, 1968.

52. Mel'nichuk, Y. U., and A. A. Chernikov, Spectra of Radar Signals from Sea Surface for Different Polarizations, *Izv. Atmos. Oceanic. Phys.*, vol. 7, pp. 28–40, 1971.

53. Wright, J. W., and W. C. Keller: Doppler Spectra in Microwave Scattering from Wind Waves, *Phys. Fluids*, vol. 14, pp. 466–474, 1971.

54. Trizna, D.: A Model for Doppler Peak Spectral Shift for Low Grazing Angle Sea Scatter, *IEEE J. Oceanic Eng.*, vol. OE-10, pp. 368–375, 1985.

55. Keller, W. C., W. J. Plant, and G. R. Valenzuela: Observation of Breaking Ocean Waves with Coherent Microwave Radar, chap. 19 in Phillips, O. M., and K. Hasselmann (eds.): "Wave Dynamics and Radio Probing of the Ocean Surface," Plenum Press, New York, 1986, pp. 285–292.

56. Moore, R. K., Y. S. Yu, A. K. Fung, D. Kaneko, G. J. Dome, and R. E. Werp: Preliminary Study of Rain Effects on Radar Scattering from Water Surfaces, *IEEE J. Oceanic Eng.*, vol. OE-4, pp. 31–32, 1979.

57. Hansen, J. P.: High Resolution Radar Backscatter from a Rain Disturbed Sea Surface, *ISNR-84 Rec.*, Tokyo, Oct. 22–24, 1984.

58. Hansen, J. P.: A System for Performing Ultra High Resolution Backscatter Measurements of Splashes, *Proc. Int. Microwave Theory & Techniques Symp.*, Baltimore, 1986.

59. Wetzel, L. B.: On the Theory of Electromagnetic Scattering from a Raindrop Splash, *Naval Res. Lab. Mem. Rept.* 6103, Dec. 31, 1987.

60. Wetzel, L. B.: On the Origin of Long-Period Features in Low-Angle Sea Backscatter, *Radio Sci.*, vol. 13, pp. 313–320, 1978.

61. Garrett, W.: Physicochemical Effects of Organic Films at the Sea Surface and Their Role in the Interpretation of Remotely Sensed Imagery, in Herr, F. L., and J. Williams (eds.): *ONRL Workshop Proc.—Role of Surfactant Films on the Interfacial Properties of the Sea Surface*, pp. 1–18, Nov. 21, 1986.

62. Huhnerfuss, H., W. Alpers, W. D. Garrett, P. A. Lange, and S. Stolte: Attenuation of Capillary and Gravity Waves at Sea by Monomolecular Organic Surface Films, *J. Geophys. Res.*, vol. 88, pp. 9809–9816, 1983.

63. Scott, J. C.: Surface Films in Oceanography, in Herr, F. L., and J. Williams (eds.): *ONRL Workshop Proc.—Role of Surfactant Films on the Interfacial Properties of the Sea Surface*, pp. 19–40, Nov. 21, 1986.

64. Guinard, N. W.: Radar Detection of Oil Spills, *Joint Conf. Sensing of Environmental Pollutants*, Palo Alto, Calif., AIAA Pap. 71-1072, Nov. 8–10, 1971.

65. Hühnerfuss, H., W. Alpers, A. Cross, W. D. Garrett, W. C. Keller, P. A. Lange, W. J. Plant, F. Schlude, and D. L. Schuler: The Modification of X and L Band Radar Signals by Monomolecular Sea Slicks, *J. Geophys. Res.*, vol. 88, pp. 9817–9822, 1983.

66. Cox, C. S., and W. H. Munk: Statistics of the Sea Surface Derived from Sun Glitter, *J. Mar. Res.*, vol. 13, pp. 198–227, 1954.

67. Alpers, W., and H. Hühnerfuss: Radar Signatures of Oil Films Floating on the Sea Surface and the Marangoni Effect, *J. Geophys. Res.*, vol. 93, pp. 3642–3648, Apr. 15, 1988.

68. Masuko, H., and H. Inomata: Observations of Artificial Slicks by X and K_a Band Airborne Scatterometers, *Proc. Int. Geoscience and Remote Sensing Symp. (IGARSS'88)*, pp. 1089–1090, Edinburgh, Sept. 12–16, 1988. Published by European Space Agency, ESTEC, Noordwijk, Netherlands, 1988.

69. Perry, R. B., and G. R. Schimke: Large-Amplitude Internal Waves Observed off the Northwest Coast of Sumatra, *J. Geophys. Res.*, vol. 70, pp. 2319–2324, 1965.

70. Alpers, W., and I. Hennings: A Theory of the Imaging Mechanism of Underwater Bottom Topography by Real and Synthetic Aperture Radar, *J. Geophys. Res.*, vol. 89, pp. 10529–10546, 1984.

71. Trizna, D.: private communication.

72. Goldstein, H.: Frequency Dependence of the Properties of Sea Echo, *Phys. Rev.*, vol. 70, pp. 938–946, 1946.

73. Schooley, A. H.: Some Limiting Cases of Radar Sea Clutter Noise, *Proc. IRE*, vol. 44, pp. 1043–1047, 1956.

74. Ament, W. S.: Forward and Backscattering by Certain Rough Surfaces, *Trans. IRE*, vol. AP-4, pp. 369–373, 1956.

75. Twersky, V.: On the Scattering and Reflection of Electromagnetic Waves by Rough Surfaces, *Trans. IRE*, vol. AP-5, pp. 81–90, 1957.

76. Lyzenga, D. R., A. L. Maffett, and R. A. Schuchman: The Contribution of Wedge Scattering to the Radar Cross Section of the Ocean Surface, *IEEE Trans.* vol. GE-21, pp. 502–505, 1983.

77. Wetzel, L. B.: A Minimalist Approach to Sea Backscatter—The Wedge Model, *URSI Open Symp. Wave Propagat.: Remote Sensing and Communication*, University of New Hampshire, Durham, preprint volume, pp. 3.1.1–3.1.4, July 28–Aug. 1, 1986.

78. Rice, S. O.: Reflection of Electromagnetic Waves from Slightly Rough Surfaces, *Commun. Pure Appl. Math.*, vol. 4, pp. 361–378, 1951.

79. Peake, W. H.: Theory of Radar Return from Terrain, *IRE Nat. Conv. Rec.*, vol. 7, pp. 27–41, 1959.

80. Valenzuela, G. R.: Depolarization of EM Waves by Slightly Rough Surfaces, *IEEE Trans.*, vol. AP-15, pp. 552–559, 1967.

81. Bass, F. G., and I. M. Fuks: "Wave Scattering from Statistically Rough Surfaces," Pergamon Press, New York, 1979.

82. Eckart, C.: The Scattering of Sound from the Sea Surface, *J. Acoust. Soc. Am.*, vol. 25, pp. 566–570, 1953.

83. Beckmann, P., and A. Spizzichino: "The Scattering of Electromagnetic Waves from Rough Surfaces," Macmillan Company, New York, 1963.

84. Wetzel, L. B.: HF Sea Scatter and Ocean Wave Spectra, *URSI Spring Meet.*, National Academy of Sciences, Washington, April 1966.

85. Kinsman, B.: "Wind Waves," Prentice-Hall, Englewood Cliffs, N.J., 1965.

86. Fung, A. K., and G. W. Pan: A Scattering Model for Perfectly Conducting Random Surfaces: I. Model Development, *Int. J. Remote Sensing*, vol. 8, no. 11, pp. 1579–1593, 1987.

87. Barrick, D. E.: Extraction of Wave Parameters from Measured HF Sea Echo Doppler Spectra, *Radio Sci.*, vol. 12, pp. 415–424, 1977.

88. Maresca, J. W., and T. M. Georges: Measuring RMS Wave Height and the Scalar Ocean Wave Spectrum with HF Skywave Radar, *J. Geophys. Res.*, vol. 85, pp. 2759–2771, 1980.

89. Valenzuela, G. R.: Theories for the Interaction of Electromagnetic and Oceanic Waves—A Review, *Boundary-Layer Meteorol.*, vol. 13, pp. 61–85, 1978.

90. Kuryanov, B. F.: The Scattering of Sound at a Rough Surface with Two Types of Irregularity, *Sov. Phys. Acoust.*, vol. 8, pp. 252–257, 1963.

91. Plant, W. J.: Bragg Scattering of Electromagnetic Waves from the Air/Sea Interface, chap. 12 in Geernaert, G. L., and W. J. Plant (eds.): "Surface Waves and Fluxes: Current Theory and Remote Sensing," Reidel, Dordrecht, Netherlands, 1988.

92. Keller, J. B.: Diffraction by an Aperture, *J. Appl. Phys.*, vol. 28, pp. 426–444, 1957.

CHAPTER 14
CW AND FM RADAR

William K. Saunders
Formerly of Harry Diamond Laboratories

14.1 INTRODUCTION AND ADVANTAGES OF CW

The usual concept of radar is a pulse of energy being transmitted and its round-trip time being measured to determine target range. Fairly early it was recognized that a continuous wave (CW) would have advantages in the measurement of the doppler effect and that, by some sort of coding, it could measure range as well.

Among the advantages of CW radar are its apparent simplicity and the potential minimal spread in the transmitted spectrum. The latter reduces the radio interference problem and simplifies all microwave preselection, filtering, etc. A corollary is the ease in the handling of the received waveform, as minimum bandwidth is required in the IF circuitry. Also, with solid-state components peak power is usually little greater than average power; CW then becomes additionally attractive, particularly if the required average power is within the capability of a single solid-state component.

Another very apparent advantage of CW (unmodulated) radar is its ability to handle, without velocity ambiguity, targets at any range and with nearly any conceivable velocity. With pulse doppler or moving-target indication (MTI) radar this advantage is bought only with considerable complexity. An unmodulated CW radar is, of course, fundamentally incapable of measuring range itself. A modulated CW radar has all the unwanted compromises, such as between ambiguous range and ambiguous doppler, that are the bane of coherent pulsed radars. (See Chaps. 15 to 17.)

Since CW radar generates its required average power with minimal peak power and may have extremely great frequency diversity, it is less readily detectable by intercepting equipment. This is particularly true when the intercepting receiver depends on a pulse structure to produce either an audio or a visual indication. Police radars and certain low-level personnel detection radars have this element of surprise. Even a chopper receiver, in the simplest video version, may not give warning at sufficient range to prevent consequences.

It should not be concluded that CW radar has all these advantages without corresponding disadvantages. Spillover, the direct leakage of the transmitter and its accompanying noise into the receiver, is a severe problem. This was recognized fairly early by Hansen[1] and Varian[2] and others. In fact, the history of CW radar shows a continuous attempt to devise ingenious methods to achieve the desired sensitivity in spite of spillover.

14.2 DOPPLER EFFECT

Complete descriptions of the doppler phenomenon are given in most physics texts, and a discussion emphasizing radar is to be found in Skolnik[3] (chap. 3, pp. 68–69).

When the radar transmitter and receiver are colocated, the doppler frequency f_d obeys the relationship

$$f_d = \frac{2v_r f_T}{c}$$

where f_T = transmitted frequency
  c = velocity of propagation, 3×10^8 m/s
  v_r = relative (or radial) velocity of target with respect to radar

Thus when the relative velocity is 300 m/s, the doppler frequency at X band is about 20 kHz. Alternatively, 1 ft/s corresponds to 20 Hz at this frequency. Scaling is a convenient way to handle other microwave frequencies or velocities.

As in a pulse radar, a CW radar that uses a rapid rate of frequency modulation, in order to sample the doppler, must have this rate twice the highest expected doppler frequency if an unambiguous reading is to be obtained. If the rate falls below the doppler frequency itself, there are problems of blind speeds as well as ambiguities. (A blind speed is defined as a relative velocity that renders a target invisible.) This will be discussed in more detail in Sec. 14.10; see also Sec. 15.3.

14.3 UNMODULATED CW RADAR

Spectral Spreading. The following discussion concerns the larger CW radars used for target illumination in semiactive systems, for acquisition, or for warning. A highly simplified diagram is shown in Fig. 14.1. Insofar as the primary operation of the equipment is concerned, the transmitted signal may be considered an unmodulated CW, although small-amplitude amplitude modulation (AM) or small-deviation frequency modulation (FM) is sometimes employed to provide coding or to give a rough indication of range. The modulation frequency is chosen to lie above the doppler band of interest, and the circuitry is designed to degrade the basic noise performance as little as possible.

A spectrum-spreading problem is also posed by conical scan. In this case, the scan frequency will lie, it is hoped, below any doppler of interest, with the result that the conical-scan frequency will appear as small-amplitude sidebands on the doppler frequency when it is recovered in the equipment. In the material that follows, these secondary issues will be largely ignored, and an unmodulated CW transmission plus a receiver that introduces no intentional modulation will be assumed.

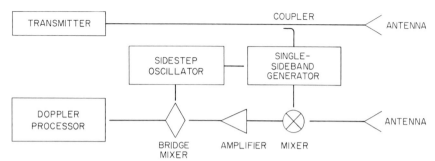

FIG. 14.1 Basic diagram of a CW radar.

Noise in Sources. The primary noise problem is in the microwave source itself. All klystrons, triodes, solid-state sources, etc., generate measurable noise sidebands in a band extending beyond any conceivable doppler frequency. Unless the source is unacceptably bad, these noise sidebands may be subdivided into pairs: those whose phase relationship to each other and the main carrier line is such as to represent amplitude modulation and those corresponding to a small index of frequency or phase modulation. The AM components, offset a given frequency from the carrier, are usually many decibels below the corresponding FM components. Moreover, balanced mixers, limiters, and other design techniques may be used to suppress AM noise. FM noise therefore is usually of greater concern in CW radar.

The FM noise of a good klystron amplifier driven by a klystron oscillator having either an active or a passive FM stabilizer is about 133 dB below the carrier in a 1-Hz bandwidth* 10 kHz removed from the carrier. The noise power decreases approximately as $1/f^2$ at larger offsets. The corresponding AM noise is 150 to 160 dB below the carrier. As the local-oscillator signal of a CW radar is usually derived from the transmitter, it too will show comparable amounts of noise. With noise this low relative to the carrier there is no concern that a significant portion of the target's energy will be lost in any practical filtering operation. What is of concern is the energy of the noise components carried into the receiver on the unwanted spillover and clutter signals. Moreover, should vibrations cause variations in the length or lengths of the spillover path the problem is more severe, since such effects introduce spectral lines that may fall directly into the doppler band of interest.

Noise from Clutter. Clutter, unwanted reflections from the ground, rain, etc., reflects the transmitted power and its noise sidebands back to the receiver. Suppose that in the foreground there is an industrial area having a clutter cross section of 0.1 m² per square meter of ground illuminated. Consider only an area located 2½ to 3½ km from the radar ($R \approx 35$ dB, where the range R is in decibels relative to 1 m) and illuminated with an antenna having 0.1 rad of beamwidth. There is a reflecting cross section σ of approximately

---

*In this chapter a 1-Hz bandwidth has been chosen as the reference. There has been no consistent practice in the literature, since it has been customary to use either the bandwidth of the system in question or that of the measuring equipment.

$$3 \times 10^3 \times 0.1 \times 10^3 \times 0.1 = 3.0 \times 10^4 \, m^2 \text{ or } 45 \text{ dB}$$

Suppose that there is present a target at a greater range and that a +22-dB signal-to-noise ratio in a 1-kHz bandwidth is needed to have a suitable probability of detection with an acceptable probability of false alarm (this includes allowance for a +10-dB receiver noise figure). If the target produces a 10-kHz doppler signal, the noise of the clutter signal must not exceed −144 dBm* + 22 dB = −122 dBm at the 10-kHz offset. If a transmitter is assumed in which the noise sidebands in a 1-kHz bandwidth are 103 dB below the carrier, the clutter signal must not exceed −19 dBm. Assume that the antenna gain is +30 dB and the transmitted power +60 dBm at X band. We have

$$(\text{power}) \quad (G^2) \quad (\lambda^2) \quad (64\pi^3) \quad (\sigma) \quad (R^4)$$

$$-19 \text{ dBm} > +60 \text{ dBm} + 60 \text{ dB} - 30 \text{ dB} - 33 \text{ dB} + 45 \text{ dB} - 140 \text{ dB}$$

$$> -38 \text{ dBm, or a 19 dB favorable margin}$$

It should be noted that we have assumed a very quiet transmitter and clutter centered about a 3-km range. (This ignores correlation; see below.)

It is convenient to express the FM-noise sidebands on the transmitter in another manner. Consider a single modulating frequency, or line, f_m in the noise having a peak frequency deviation ΔF_p. By the frequency-modulation formulas, the carrier has a peak amplitude of $J_0(\Delta F_p/f_m)$, and each of the nearest sidebands a peak amplitude of $J_1(\Delta F_p/f_m)$.[4] If the arguments are small, as they must be for our computations, the Bessel functions may be approximated by

$$J_0(X) \sim 1$$
$$J_1(X) \sim \frac{X}{2}$$

Hence the power ratio between the carrier and one of the first harmonic sidebands is $\Delta F_p^2/4f_m^2$. For the ratio of the power in both sidebands to that in the carrier (the quantity usually of interest),

$$\text{FM noise power} = \frac{\Delta F_p^2}{2f_m^2} = \frac{\Delta F_{rms}^2}{f_m^2} \qquad (14.1)$$

A 1-Hz peak deviation at a 10-kHz rate represents a double-sideband noise ratio of $\frac{1}{2}(\frac{1}{10^4})^2$ or −83 dB with respect to the carrier. The −103-dB (double sideband at 10 kHz) transmitter used as an example above has a peak deviation of 0.1 Hz. These numbers are all equivalent only when referred to a particular bandwidth, 1 kHz in this case. (The description in hertz is convenient only when the noise is expressed in the bandwidth of interest in a particular radar.)

The concept of a peak signal is properly associated only with a sine wave. With random noise the rms description is more meaningful. However, a noise power at a given frequency and bandwidth is equivalent to that which would be produced by a sine wave having a certain peak or rms value.[5]

To carry the computations further, the correlation effect must be discussed.

---

*Thermal noise in a 1-kHz band at 20°C.

When a transmitter is producing FM noise, it may be thought to be modulating in frequency at various rates and small deviations. Consider, for example, a particular one of these modulating frequencies. If it is a low frequency and the delay associated with the spillover or clutter is short, the returning signal finds the carrier at nearly the same frequency that it had at the time of transmission; that is, the decorrelation is small. Higher frequencies in the noise spectrum have greater decorrelation. Moreover, the effect is periodic with range: For any given sinusoidal modulating frequency the FM noise produced will increase as a function of range out to a given range and will then decrease. The zeros occur at the ranges $R = nc/2f_m$, where f_m is the frequency of the sinusoidal modulating component, c is the velocity of light, and n is any integer.

However, in general, one deals with noise rather than a sinusoidal component. For this reason, the discrete zeros indicated are seldom of interest, and for ease in the computations the signal is assumed to be decorrelated at a frequency f_l approaching $f_l = c/8R$. Some frequencies higher than f_l cause no problems at particular ranges, but nearby ones do so. Moreover, as the formulas below will show, the deviation of the recovered signal can be twice as great as that of the transmitter. (The returning signal may be swinging up while the transmitter is swinging down.)

The peak voltage of the first harmonic sideband in the IF spectrum of an FM signal mixed with itself after a time delay $T = 2R/c$ is[6]

$$v_{p1} = J_1\left(2\frac{\Delta F_p}{f_m} \sin \pi f_m T\right)$$

and the peak voltage of the carrier is

$$v_{p0} = J_0\left(2\frac{\Delta F_p}{f_m} \sin \pi f_m T\right)$$

In both formulas ΔF_p is the peak frequency deviation of the carrier. As before, $J_1(X) \sim x/2$, $J_0(X) \sim 1$, $X < 1$. Hence the ratio of the power in a single sideband to that in the carrier is

$$\frac{P_s}{P_c} = \left(\frac{\Delta F_p}{f_m} \sin \pi f_m T\right)^2$$

and in the pair of sidebands

$$\frac{P_{2s}}{P_c} = 2\left(\frac{\Delta F_p}{f_m} \sin \pi f_m T\right)^2 \tag{14.2}$$

The maximum value of this is $2(\Delta F_p/f_m)^2$, which is in agreement with the maximum deviation of the IF being twice as large as that of the transmitted frequency [Eq. (14.1)].

For smaller values of $f_m T$ the double-sideband power ratio is

$$\frac{P_{2s}}{P_c} = 2(\pi\Delta F_p T)^2 \qquad \pi f_m T < 1 \tag{14.3}$$

This is an interesting formula as it shows that when ΔF_p is constant, as it is with many klystrons, the correlated noise power is independent of frequency and directly dependent on range.

A convenient curve which gives the ratio of noise at the receiver to measured noise on the transmitter (Fig. 14.2) is based on the approximations of formulas (14.2) and (14.3). The dotted portion of the curve reflects only the approximations formula (14.2).

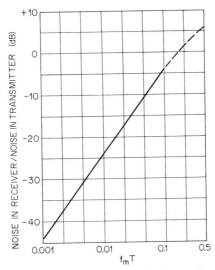

FIG. 14.2 Noise suppression by the correlation effect.

The problem started above can now be completed. The center of the clutter was taken at 3 km or a T of 2×10^{-5} s. The frequency of interest was 10^4 Hz. Hence $f_m T = 0.2$, which is beyond the region of noise correlation.

An unmodulated CW radar must contend with clutter almost down to zero range. Without the correlation effect, this would generally be impossible. For a given antenna beam the width of the illuminated clutter area decreases as R but there is a $1/R^4$ in the radar equation. The result is that the clutter return varies as $1/R^3$. The correlation effect shows that for a fixed noise frequency the correlated noise sidebands decrease as R^2 [Eq. (14.3), with $T = 2R/c$]. Hence there is an apparent rate of increase of $1/R$. The integral representing the clutter power appears to diverge, but two factors so far ignored have a decisive influence at very short ranges. The first is that the intersection of a beam emitted from an antenna of finite height and the earth is the interior of a hyperbola and not a sector, as implied above. The second is that at close ranges the clutter is in the Fresnel region of the antenna and the far-field gain formula no longer applies. In a more careful analysis, by using either of these factors, the integral may be shown to be convergent.

Shreve[7] has derived a formula for the double-sideband noise power. He took the boundary of the Fresnel region $R_F = D^2/\lambda$ as the lower limit of the integral.

His formula for the above parameters yields a value of -117 dBm for the correlated noise power from the clutter. This is for the extreme case of a single antenna at exact ground level looking into very severe clutter (0.1 m^2/m^2).

A more practical way to look at the problem is to note that clutter from very short ranges and spillover are almost equivalent phenomena. For a ground-based CW radar to operate at maximum sensitivity, two antennas must be employed; this reduces both the spillover and the near-in clutter since no close-in point can be in the main lobes of both transmitting and receiving beams. Moreover, as described below, spillover cancellation (and hence near-in clutter cancellation) is usually employed.

The discussion above assumes that the local oscillator employed in the radar is either derived from the transmitter or locked to it with a servo which has a frequency response sufficiently high to cover the doppler and noise band of interest.

Microphonism. Microphonism can cause the appearance of additional noise sidebands on the spillover and occasionally on the clutter signals. If the structures are sufficiently massive, the microphonism is greatest at the lower frequencies, where it can be counteracted by a feedthrough servo. To this end, however, it is most important that microwave components employed in the feedthrough nulling as well as in the remainder of the microwave circuitry be as rigid as possible.[8] It is customary to use a milled-block form of construction. In the rare cases where a single antenna plus duplexer or a pair of nested antennas has been used in an airborne high-power CW radar, the mechanical design problems have been all but insurmountable. Even in a ground-based radar, fans, drive motors, motor-generator sets, rotary joints, cavitation in the coolant, etc., are very troublesome.

Scanning and Target Properties. In addition to the spectral spreading caused by transmitter noise and by microphonism, there is a spreading of the CW energy by the target and by the scanning of the antenna. Generally, the spreading by even a rapidly scintillating aircraft target does not produce appreciable energy outside a normal doppler frequency bandwidth. The filter is usually set by the acquisition problem or the time on target rather than by the intrinsic character of the return signal. Rapid antenna scan, however, can cause an appreciable broadening of the spectrum produced by the clutter. Were it not for the particular shape of the typical antenna beam, the transients produced by clutter while scanning would be far more serious.

An approximate analysis assumes a gaussian two-way gain for the antenna, $G^2 = e^{-2.776\theta^2/\theta_B^2}$, where θ is measured from the axis of the beam and θ_B is the beamwidth between the half-power points of the antenna. We shall discuss a two-way pattern down 3 dB at $\pm\frac{1}{2}°$ ($\theta_B = 1°$). If the antenna scans $180°$ a second, we shall need the Fourier transform of e^{-at^2} with $a = 9 \times 10^4$. This has the form $Ae^{-\omega^2/3.6 \times 10^5}$, which is down to $\frac{1}{1000}$ (60 dB) of its peak when $\omega^2/(3.6 \times 10^5) = 6.9$, $\omega \approx 1575$, and $f \approx 250$ Hz.

Actual antenna patterns produce somewhat less favorable transients than the gaussian shape. Limiting in the receiver is equivalent to altering the shape of the beam.[9] For any antenna pattern there is a definite limitation on the scanning speed of a narrow-beam antenna. Actually, mechanical limitations usually prevent trouble except with the very slowest targets, but with nonmechanical scanning methods degradations may occur.

14.4 SOURCES

Master Oscillator Power Amplifier (MOPA) Chains. Requirements peculiar to CW radar are the use of extremely quiet tubes throughout the transmitter chain, very quiet power supplies, and, often, stabilization to reduce the total noise of the system. In theory, any of the methods for measuring FM and AM noise to be discussed in Sec. 14.5 might be modified to produce a noise-quieting servo. Practical considerations have resulted in a wide variety of additional schemes. The simplest is the introduction of a high-Q cavity between a klystron driver and the power amplifiers. Q's of 20,000 to 100,000 are normally employed. The action of the cavity is primarily that of an additional reactive element directly in parallel with the cavity in the klystron. With a high-quality reflex klystron having an FM noise 110 dB below the carrier in a 1-Hz band spaced 10 kHz from the carrier, use of the high-Q cavity as a passive stabilizer reduces the corresponding FM noise 130 to 135 dB below the carrier. The cost is a power loss of about 11 dB. The 20 to 25 dB noise improvement is obtained at most frequencies of interest. No noticeable improvement is made in the AM noise level at doppler frequencies by this technique.

It should be remembered that this results only in a stable driver, any noise generated in the power amplifier being unaffected. And, as noted above, unless the local oscillator is generated from the output of the power amplifier, which, of course, cannot be done in a pulse doppler system, this noise is uncorrelated. Fortunately, good power amplifiers driven by highly regulated supplies add extremely small amounts of excess, or additive, noise. (See Fig. 14.8.)

It is to be noted that the illuminator for the basic Hawk surface-to-air missile system used a magnetron as the transmitter rather than a MOPA chain. Cost, availability, lower weight, and lower high-voltage requirements were all factors in the choice. Later versions of Hawk use the inherently quieter klystrons. Banks[10] provides a definitive overview of the Hawk illuminator including both the noise degeneration loop (feedthrough nulling) and the transmitter stabilization microwave circuitry. An interesting feature of the latter is a spherical cavity that is far stiffer under vibration than the usual cylindrical cavities.

Active Stabilization. All the schemes for active stabilization on a MOPA chain depend on the use of a high-Q cavity as the reference element. The cavity must be isolated from the tube so that it functions as a measuring device without introducing the pulling that results from the frequency dependence of its susceptance.

For reflection and transmission cavities, useful equations adapted from Grauling and Healy[11] are given below. For a matched reflection cavity,

$$\Gamma = \frac{Z-1}{Z+1} = \frac{j\delta Q}{1+j\delta Q} = \frac{j2\delta Q_L}{1+j2\delta Q_L}$$

where Γ = reflection coefficient
  δ = $(f - f_0)/f_0$
  Q = unloaded Q of cavity
  Q_L = loaded Q of cavity
  Z = normalized impedance looking into cavity

The transmission cavity has similar characteristics except that both the carrier (V_c) and sidebands (V_{sb}) are passed. For a transmission cavity,

$$V_0 = \frac{V_1}{3 + j2\delta Q} = V_c + V_{sb}$$

(Both coupling coefficients have been assumed equal to unity.) V_1 is the input voltage and V_0 the output voltage. With a little algebra, it is seen that the frequency-dependent terms of Γ and V_{sb} are similar in form. In stabilization systems, $f - f_0$ is kept small and

Reflection: $\qquad \Gamma \approx j2\delta Q$

Transmission: $\qquad V_{sb} \approx 2(- 2j\delta Q)$

One might expect that the stabilization would be equally effective in reducing regardless of the frequency. This ignores two factors: the cavity has only a finite linear range, and larger values of f_m may produce sidebands that lie outside this range; for stability the servo that follows the cavity must have a response that rolls off at the higher frequencies.

The simplest stabilization bridges result directly from the character of the transmission and reaction cavities. The transmission-cavity bridge might have the arrangement of Fig. 14.3.

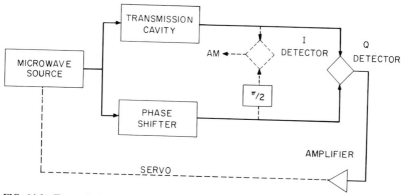

FIG. 14.3 Transmission bridge, video version.

The phase shifter is set so that the Q, or quadrature, detector receives signals in phase quadrature* and, to first order, is sensitive only to FM. Should there be a requirement for AM stabilization, it would normally be only at very low frequencies such as those introduced by the power supplies. Such a requirement could be met easily by adding a $\pi/2$ phase shift and an I (coherent amplitude-sensitive) detector, as indicated by the dotted blocks in Fig. 14.3. The discussion of the servo constants will be postponed until the end of this section.

---

*The most sensitive technique for adjusting the quadrature detector is to introduce intentional AM and null this by adjustment of the phase shifter. Maximizing the response to FM yields a less exact adjustment.

An obvious disadvantage of the transmission-cavity bridge is that the carrier is not suppressed in the microwave circuitry. Since the total input power is limited by fear of crystal damage and of exceeding the linear range in the mixing process, the intelligence signal power at a relatively low level is in competition with the thermal noise generated by the crystals.

The reflection cavity has an advantage in that a sizable portion of the carrier power is absorbed if the transmitter is kept tuned to the frequency of the cavity. This eliminates much of the saturation problem.

A particularly attractive arrangement was proposed by Marsh and Wiltshire[12] (Fig. 14.4). It is the basis of the earliest successful FM noise-measuring instruments and has been employed in stabilization as well.[13] It is the only bridge that removes most of the carrier power to avoid saturation of the mixer crystals. The key to its operation is a balancing element which matches exactly the reflection from the cavity at resonance.

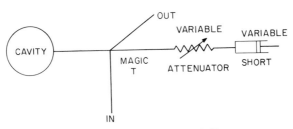

FIG. 14.4   Marsh and Wiltshire microwave bridge.

At resonance the cavity is nearly a perfect absorber, and the residual reflection is tuned out with the variable short and variable attenuator. As the frequency varies, the cavity produces a reactive component which alters the balance. The result, at least for small deviations, is a double-sideband suppressed-carrier signal. With care, the carrier may be suppressed as much as 40 dB with manually or statically balanced bridges and as much as 60 dB if either the cavity or the source is electronically or thermally tuned. The result is that 2 W of power may be handled in the manually tuned version, and up to 1000 W in the servo-tuned equipment. The balance of the circuitry is shown in Fig. 14.5.

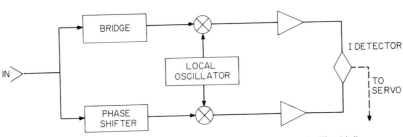

FIG. 14.5   IF circuitry for a Marsh and Wiltshire bridge (shown in Fig. 14.4).

As is seen, a properly phased second sample of the transmitter is processed in a parallel path and serves as the reinjected carrier to recover the sidebands at the I detector. Since the lower path is nondispersive and since the sidebands are small (a requisite for all the above quieting schemes), the large signal may be regarded as an essentially pure carrier in the reinjection process. The LO must be reasonably quiet, and the phase delay of the amplifiers must be matched.

Figure 14.6 gives the servo-loop gain, and Fig. 14.7 shows the three resulting curves: A, the FM noise on the free-running oscillator; B, the FM noise of the stabilized oscillator; and C, the expected theoretical improvement based on the servo gain and the noise analysis.

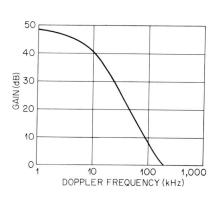

FIG. 14.6 Frequency control-loop gain.

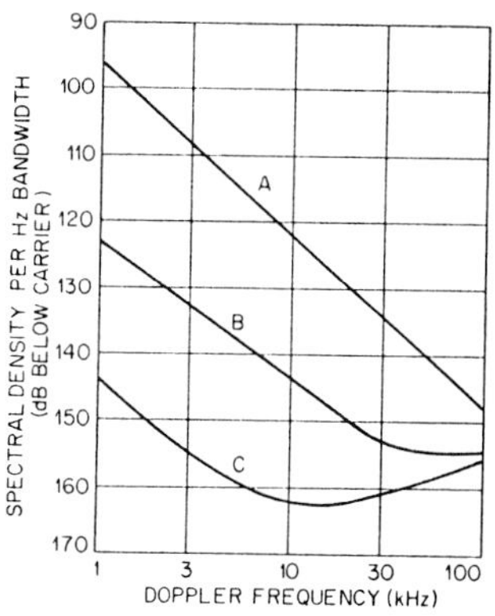

FIG. 14.7 Transmitter noise spectra: A, free-running FM; B, closed-loop FM; C, free-running FM divided by control-loop gain.

For details of the construction and a discussion of the results, see the first edition (1970) of the handbook or Ref. 12.

Stabilization of Power Oscillators. The active methods described above can be used for the stabilization of power oscillators as well as for the stabilization of drivers. The measurement bridges are unchanged, but servo circuits must be altered to operate at high voltage and, in some cases, supply considerable power, since multicavity klystrons and crossed-field devices such as magnetrons and amplitrons can be modulated only through their high-voltage high-current supplies. Moreover, the fact that these are essentially "stiff" devices imposes more stringent requirements on the design of the servo.

14.5 NOISE MEASUREMENT TECHNIQUE

Two basic types of noise measurement are of interest to the designer: primary-noise measurements to be made on drivers or power oscillators and additive-, or excess-, noise measurements to be made on amplifiers, multipliers, rotary joints, etc.

Although microwave cavities, such as used in the Marsh-Wiltshire bridge, were widely employed at one time, commercial instruments generally avoid them.[14] They accomplish this by comparing, in a phase detector, the source under test with either an external nearly duplicate source or an internal source supplied by a portion of the test equipment. If nearly duplicate sources are used, one is assured that at least one of them (not necessarily always the same one) is at least 3 dB quieter at each offset frequency than the phase noise indicated by the instrumentation. By using three essentially duplicate sources and measuring the phase noise generated by each pair at all the desired offset frequencies, one derives three sets of measurements. This leads to three equations with three unknowns, and the phase noise of each of the three sources can be derived as a function of frequency. If one of the internal sources supplied by the instrumentation is used, there is a distinct limitation owing to the phase-noise characteristic of that particular internal source. In general, this is the paramount limitation since the noise floor of the phase detection circuitry is usually well below that of the internal or, for that matter, the external reference sources. This assumes that the AM noise on both sources used in the test is well below the phase-modulation noise. The only safe course is first to measure the AM noise on any unknown source with a simple amplitude detector available with the instrumentation.

The instruments can provide a servo voltage to hold the two sources at the same frequency and in quadrature at the phase detector. If the sources are such that neither is readily voltage-tunable, then one source is chosen at a typical IF frequency away from the other, and an IF oscillator is locked, in the mean, to the difference frequency. This technique was originally employed in military test equipment designed to measure noise on radars in the field.[15,16] The instruments provide a wide range of internal frequencies through a combination of synthesizer techniques at the lower frequencies and combs going up to 18 GHz. The latter are created by using the harmonics of a step-recovery diode multiplier.

The signal coming from the phase detector is filtered to remove microwave frequencies and is then amplified in a low-noise baseband amplifier. The resulting phase noise can be measured by any of a variety of methods, including spectrum analyzers and analog wave analyzers. The most accurate and convenient method for the measurement of the lower frequency noise is the fast Fourier transform (FFT). The method is too time-consuming for analysis of the far-out phase noise.

With computer control of all of the components of the test equipment almost any desired measurement can be made, adjusted for filter shape, and printed out. There is even an option to remove spurs (spurious frequencies), occurring during the measurements, from the calculations and from the data. All this comes at a considerable cost, not all of which is monetary. One's ability to understand any laboratory technique and its inherent reliability are usually inversely related to complexity. For example, given enough equipment, each generating a plethora of internal signals, one is almost guaranteed spurs in any sensitive measurement. If the ultimate aim is a quiet transmitter, one attributes such spurs to gremlins in the test equipment at one's peril. A well-shielded screen room with a minimum of (well-understood) instrumentation in that screen room eliminates many variables.

In general, modern commercial instrumentation is vastly superior to the earlier cavity bridges for making routine measurements on low-power sources. It can measure phase noise very close to the carrier if the servo is tailored to force the two sources to track in the mean, without effectively locking them together, at the offset frequency of interest. By knowing the servo's characteristics, the instrumentation can adjust the data to reflect the phase noise actually present. The technique is limited, at small offset frequencies, by thermal noise that competes with the low values of phase deviation permitted by the servo. Even with this limitation, one is considerably better off than with a cavity bridge whose sensitivity falls off rapidly at small offsets. Without the aid of commercial phase-bridge instrumentation, it would have been difficult to develop the crystal sources having much reduced phase noise at close-in frequencies. (These have been key components in long-range airborne radars that are required to detect crossing targets immersed in clutter.) Nor are commercial instruments limited in their ability to measure phase noise at the larger offsets. They appear to have just two significant limitations. Cavity bridges are superior for development work on state-of-the-art sources, especially those that are difficult or expensive to produce in pairs, and in the measurement of high-power transmitters such as the Hawk illuminator. Compare curve *I* of Fig. 14.8, measured in the early 1960s, with curve *P* of the same figure, which is the measurement floor of typical commercial instrumentation.

An alternative to both cavity and source comparison techniques is the use of a delay line to provide a primary reference for the measurement of phase noise. A method that eliminates the noise contributed by the local oscillator is suggested in the appendix of Ref. 16. Unfortunately, the accuracy of any phase-noise measurement that depends on a delay line is proportional to the length of the delay. Long delays imply difficulty in maintaining sufficient signal amplitude to make satisfactory measurements. Incidentally, as noted above, many measurement techniques can be altered to provide a valid source-stabilization method. The delay line is an exception. When one attempts to servo-out phase noises at the

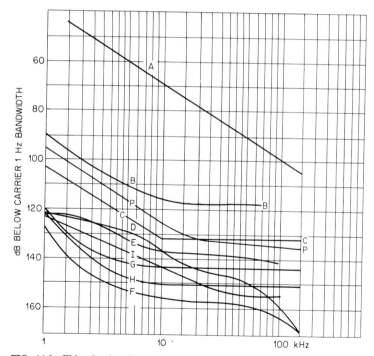

FIG. 14.8 FM noise in microwave sources. *A*, voltage-controlled *LC* oscillator multiplied to X band; *B*, crystal-controlled oscillator, step-recovery multiplier, to X band (*courtesy of D. Leeson*); *P*, noise floor at X band of 11729B/8640B combination (*courtesy of Hewlett-Packard*[14]); *C*, crystal oscillator (ST cut) multiplied to X band (*courtesy of Westinghouse Corporation*[18]); *D*, compact X-band klystron CW amplification (*Hughes*); *E*, compact X-band klystron pulsed amplification (*Hughes*); *F*, X-band klystron CW amplification (*Varian*); *G*, X-band klystron pulsed amplification (*Varian*); *H*, S-band electrostatically focused klystron amplifier (*Litton*); *I*, curve *B* in Fig. 14.7. Note that curves *D* to *H* are additive-noise measurements.

higher offset frequencies, one runs into the Nyquist restriction. For any fixed delay length, there is a corresponding offset frequency where the servo gain must go to zero or the whole system will become unstable.

Except for multipliers and dividers, the measurement of additive or excess phase noise* on components such as power amplifiers is considerably easier than the similar measurements on sources. All that is required is one moderately quiet source, a phase shifter, a phase detector, a suitable wave analyzer, and a method of calibration. The commercial instruments, described above, provide all this and much more. The reason that the source phase noise is not critical to the measurement is that it is common practice to add sufficient coaxial or microwave delay to equalize the two paths to the phase detector. Figure 14.2 indicates what such equalization (i.e., correlation) buys. As above, the amplitude modulation introduced by the source or the component under test must be checked first. For very demanding measurements, such as shown in curve F of Fig. 14.8, it might be well to consult the 1970 edition of this handbook and Ref. 17 of this chapter. For such work, a screen room is a must.

The measurement of multipliers is considerably more difficult since the two signals arriving at the phase detector must have the same frequency. This implies that two similar multipliers must enter the circuitry, and one has most of the problems associated with the measurement of sources. The only problem one is spared is the phase locking, which is usually required when working with sources. Fortunately, well-designed multipliers usually add little phase noise to a radar (above that to be expected from the increased FM deviation produced by the multiplication process). When 100 sources, consisting of a crystal oscillator plus a multiplier chain, were supplied by a subcontractor to a military radar program, the only ones that were unable to meet an extremely severe specification were ones that had substandard crystals in the oscillator.[16]

Because of the similarity of methods, measurement of noise in pulsed transmitters will merely be sketched. The measurement of pulsed sources is intrinsically much more difficult than the measurement of CW sources. The pulse structure produces very substantial AM that inevitably conflicts in direct and indirect ways with any attempt to measure the FM. In fact, it is possible to measure FM only up to half the repetition frequency, and then only by the use of rather sharp filters placed immediately following the Q detector. Measurements of -100 dB with respect to the carrier in a 1-Hz band 10 kHz from the carrier require excellent technique.

Similar problems occur in the measurement of the additive noise produced by pulse amplifiers. At Harry Diamond Laboratories some added sensitivity has been obtained by producing another pulse spectrum as similar as possible to that produced by the transmitter and subtracting this.

A suitable device for switching low-level signals is a PIN diode modulator, but with it there is difficulty in obtaining an exact reproduction of the pulse shape produced by a high-power amplifier. On the other hand, the rather large phase perturbation produced by the PIN diode modulator on the leading edge of the pulse is repeated pulse to pulse and produces spectral energy only at multiples of the repetition frequency, where measurements are impossible in any case.

Typical additive-noise measurements made on a variety of FM and CW sources at the Harry Diamond Laboratories[17] and elsewhere are shown in Fig. 14.8. The considerable improvement since 1970 in crystal-oscillator multiplier

---

*The adjective *residual* sometimes appears in the literature. It is not an apt choice in radar, where the source is usually the element that sets the phase-noise floor.

chains, especially below 5 kHz, can be seen by comparing curve B (1970) with curve C (1988). Even better performance is to be expected as research in this very important area continues. Although the curves are given to 150 kHz at most, one is often interested in FM noise out to $1/\tau$ (where τ is the pulse length). Solid-state sources, unlike klystrons, have white FM noise at the higher frequencies.[19] This noise folds down in the operation of a pulse doppler radar. The locked source method of measurement, mentioned on page 14.12, can be conveniently altered to measure the total folded noise. The servo is designed to remove noise from the phase detector in the correct proportion to account for the correlation effect. The output of the phase detector is then chopped at the radar's pulse repetition frequency and duty factor. The folding, thus produced, accurately reproduces the radar's demands on its source. The required $1/f^2$ frequency response of the servo is a convenient and stable choice. It is superior to a strictly narrowband loop followed by a shaped amplifier even for CW measurements.

14.6 RECEIVERS

RF Amplification. Although it is apparently attractive, low-noise RF amplification of the received signal has not been extensively employed in CW radar. Transistor amplifiers with excellent noise figures are available to K_u band. Traveling-wave tubes are expensive. In many cases, the determining factor in deciding against a low-noise RF amplifier has been the presence of spillover noise, clutter signals, and signals produced by electronic countermeasures that equal or exceed the noise contributed by conventional front ends.

Few modern receivers have been designed for CW target illuminators since many installations avoid the use of a separate receiver. With space in the nose of an aircraft at a premium, it is usual for airborne missile systems to employ a common antenna for both the tracking radar and the illuminator.[71] In some shipboard weapon systems, the illuminator does not require a receiver since the illuminator antenna is pointed to the direction of the target from information obtained by the tracking radar of the weapon control system rather than have the illuminator track the target itself.

Generation of the Local-Oscillator Signal. To realize adequate signal-to-noise performance, it is customary to perform the first amplification in a CW radar at an intermediate frequency such as 30 MHz. To obtain the necessary coherent local oscillator signal, various types of sidestep techniques are employed. These include modulators, balanced modulators, single-sideband generators (SSG), or phase-locked oscillators. The SSG is probably the most cumbersome, since the suppression of the carrier and unwanted sidebands is seldom better than 20 dB and filtering must be employed to suppress these signals further. The balanced modulator is much simpler and suppresses the carrier as well as the SSG. The filter needed for further carrier suppression usually suppresses the unwanted sideband to the desired level without added poles. The simple modulator is scarcely less complex than the balanced modulator and requires a sharper filter for the necessary added carrier suppression. Phase locking eliminates the need for high-frequency filtering altogether but requires a skillfully designed servo loop to impress the transmitter's FM noise faithfully on the local oscillator. It may also require a search mechanism to pull in initially. All these methods require the use of an oscillator at the intermediate frequency. The stability required

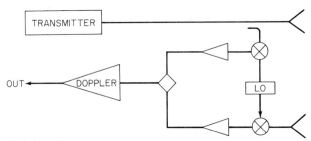

FIG. 14.9 Balanced receiver with a floating LO.

of this oscillator is not excessive since FM is contributed only in the ratio of the IF frequency to microwave frequency.

An alternative approach to the problem of an IF offset is the free-floating local oscillator employed by Harris et al.[13] and by O'Hara and Moore.[8] This has a similarity to the method used to introduce the local oscillator in the Marsh and Wiltshire bridge and requires twin IF amplifiers. The basic diagram is shown in Fig. 14.9. The simple appearance of the figure is deceptive. The local oscillator must normally be positioned by the AFC to hold the signal in the IF bands. As shown, the system folds the doppler frequencies. To avoid this, either quadrature techniques must be employed or a second sidestep be introduced into the reference channel. The latter is unattractive as it destroys the symmetry that may be required to assure uniformity of time delays to cancel the FM noise of the LO. Even in the simplest version the symmetry is far from complete, as the signal channel must handle signals over a wide range of amplitudes while the reference channel carries a signal of uniform amplitude.

IF Amplifier. Traditional low-noise IF amplifiers are usually employed. Because of the levels of clutter signals, ECM, and spillover signals that must be carried by the IF amplifier, it is usual to restrict the gain to no more than 40 dB. This establishes the noise figure and raises the signal to a value where microphonism is less serious without risking levels where saturation and the attendant intermodulation are problems.

Subcarriers. Although doppler filtering may be carried out at slightly higher levels, it is desirable to reject the signal produced by clutter and by spillover at the lowest level possible. Unfortunately, sufficiently high Q's are not available, even in quartz filters, to make it possible to reject clutter at, say, 30 MHz without diminishing the lower doppler frequencies as well.

The simplest method is to mix the signal from the IF amplifier with the signal used in the sidestep. This reduces the spillover signal to dc and the clutter signal to dc and very low frequencies. A multipole filter will suppress those unwanted signals with minor suppression of the very lowest dopplers. Unfortunately, this process folds the spectrum so that incoming targets are indistinguishable from outgoing targets and the random-noise sidebands accompanying each appear in the baseband amplifier. Even if one is prepared to accept the ambiguity, the 3 dB loss in the signal-to-noise ratio (*SNR*) is a matter of concern in a high-power radar.

There are two alternatives, both of which have been extensively employed. The first is a subcarrier band for the doppler intelligence which does not extend to dc but is centered at a frequency where either quartz or electromechanical filters have sufficient Q's to permit sharp filtering. (Values of 0.1 to 0.5 MHz or 1 to 5 MHz are suitable ranges for quartz filters; 0.1 to 0.5 MHz is proper for electromechanical filters.)

The second alternative is quadrature detection.[20] A suitable block diagram for this technique is shown in Fig. 14.10. A single 90° phase shift can be substituted for the +45° and −45° shifts at the constant frequency coming from the oscillator. The plus and minus 45° in the two signal paths are required to maintain a semblance of balance over a wide band of frequencies. A phasor diagram of the system (simplified by omitting the IF sidestep) is shown in Fig. 14.11. If the output from mixer 1 is

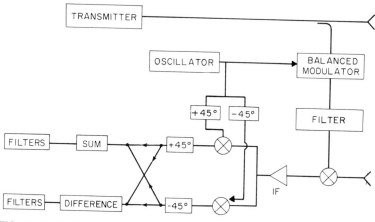

FIG. 14.10 Quadrature receiver.

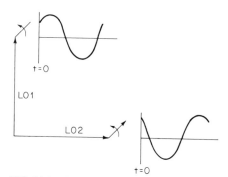

FIG. 14.11 Phasor diagram for a quadrature receiver.

advanced 90° and added to that of mixer 2, the signals will sum in the combiner and disappear in the differencer. A corresponding diagram would show that receding targets reinforce in the differencer and cancel in the combiner.

An advantage of the quadrature system is that the filter bands are completely symmetrical and all filter elements may be identical. Moreover, high-pass filters having steep slopes to reject clutter are somewhat easier to design near dc than they are at low IF. A disadvantage is the requirement to maintain balanced operation over the full range of doppler frequencies in the two second mixers and in the two 45° phase shifters in order to eliminate false targets.

Amplification. After the undesired clutter and spillover signals have been removed, substantial amplification can be achieved either at the second subcarrier frequency or, in the case of folded or quadrature systems, at the doppler frequency itself. It is customary to add additional filtering against the unwanted signals as interstage networks between the stages. The only requirement that must be met is that the combination of the amplification and the total filtering is such that the amplitude of the unwanted signals nowhere approaches the saturation level.

Doppler Filter Banks. Ideally no nonlinear operation occurs in the signal processing and amplification. There is still a coherent signal, and the band narrowing buys decibel for decibel in improved signal-to-noise ratio. Steinberg[21] has shown that, given a fixed doppler band, one pays no penalty in false-alarm rate for subdividing it. In a radar, then, it would be desirable to have the final doppler bandwidth limited only by the time on target. This might be possible in a rapidly scanning radar, but with tracking radars or illuminators the indicated bandwidths would be unrealistically narrow. Moreover, the target itself seldom produces a clearly defined doppler but, rather, a spread of frequencies by the scintillation and glint effects. Bandwidth may also have to be allowed for the coding frequencies which may accompany the doppler, such as those injected by conical scan. A typical circuit for an X-band radar might have a suitable bank of adjacent two-pole filters, each 1000 Hz wide, or an equivalent set of digital filters produced by an FFT.

Following each filter are a detector and a postdetection integrator whose time constant is matched to the time on target or, in the case of a tracker, the demands of the servo data rate. A threshold level is set in the circuitry following each detector; when this is exceeded, a voltage is generated and held until such time as it is read. In acquisition the threshold circuits are normally scanned by some type of readout mechanism. This is fundamentally a computer-type operation.

Doppler Trackers. Doppler filter banks are satisfactory for acquisition and for track-while-scan radars. They are not commonly used in tracking radars or illuminators to improve *SNR* since the use of a doppler tracker (speedgate) is far less complex. The usual speedgate circuit is identical with the AFC circuit in an FM radio. A voltage-controlled oscillator (VCO) is used to beat the signal to be analyzed to a convenient intermediate frequency. A narrowband amplifier at this frequency performs the filtering operation. The VCO is in turn controlled by the output of a discriminator connected to the amplifier. The input to the speedgate can either be the full doppler band, as in the folded or quadrature receiver, or be a subcarrier containing the full doppler intelligence. Although some clutter filtering may take place in the speedgate, earlier removal of unwanted signals is preferable. This is particularly important with the folded receiver in certain airborne situations in which the clutter has a substantial spread because mixer nonlinearities produce harmonics of the unwanted signals that may fall directly on the target signal in the speedgate.

Once in track, the speedgate follows the proper doppler component. The response is limited only by the bandwidth of the servo, which is designed to follow the expected target maneuvers. To acquire track, intelligence may be passed on an open-loop basis from the doppler filter bank, if one is available, or the VCO may have a sawtooth or triangular voltage applied to produce a programmed search. Search is stopped when the output registers the desired target. Coding signals may be employed to aid in the detection and the stopping.

It is usually necessary to restrict the VCO from moving to a frequency that will lock the speedgate on spillover or clutter. With ground-based systems the problem may be simply solved by fixed-limit stops placed on the search voltage. Airborne systems having clutter signals that vary in frequency require more sophisticated solutions.

Constant False-Alarm Rate (CFAR). A constant false-alarm rate in the presence of variable levels of noise is usually a requirement placed on any modern radar. It is very easily achieved in CW radars by the use of filter banks or FFTs. The energy reaching the filter banks is restricted either by automatic gain control (AGC) or, when feasible, by limiting, and the thresholds in the circuitry following the filter banks are properly set with respect to the level in the total band. In a typical setting technique, random noise is injected into the amplifier that drives the filter banks, and each threshold is set to achieve the desired false-alarm rate. The level of noise is then varied and the threshold rechecked. If the limiting is proper, the false-alarm rate should not change. However, target signals in the absence of noise are unaffected, as they do not change the total energy present in the broad doppler spectrum sufficiently to change the AGC level or reach the limiting level. Similar remarks apply to the speedgate as well.

14.7 MINIMIZATION OF FEEDTHROUGH

All major ground-based CW radars have two antennas to minimize spillover. Isolation may be improved further by the use of various absorbers or of an intentional feedthrough path that is adjusted in phase and amplitude to cancel spillover energy. In free space such solutions are all that would be required. When a radar scans across a rough ground plane, however, the energy reflected to the receiver antenna does not remain constant. A dynamic canceler is required. A diagram and description of one such device are given in Ref. 10. Other descriptions are to be found in Refs. 8 and 22.

All dynamic cancelers depend on synthesizing a proper amplitude and phase of a signal taken from the transmitter and using this to buck out the spillover signal. To achieve independence of the servos, the vector is synthesized in orthogonal rectangular coordinates. Figure 14.12 is a typical arrangement for use with a CW radar in which the local oscillator is derived by sidestepping the transmitter. Slight modification[8,13] is needed when the basic radar has the balanced method (Fig. 14.9) of generating the offset required for the first IF amplifier.

The servo amplifiers have response from dc to some frequency well below the doppler band of interest. They respond to the slow variations in the feedthrough signal without damage to the dopplers. For complete details of the mechanical design, see Ref. 8.

Harmer and O'Hara[22] show a variant of the equipment that may be used with a single antenna plus a duplexer. This would be very attractive, especially for an airborne radar that must fit into a small radome. Unfortunately, experience has shown that there is a limit to the transmitter power that may be employed in such

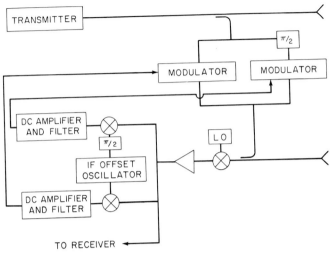

14.12 Feedthrough nulling bridge.

an arrangement. Beyond a modest level of power, the servo is unable to cancel out the −20 dB reflection from the antenna or duplexer sufficiently to prevent receiver degradation.

It should be noted that microwave-feedthrough cancellation is of principal value in preventing saturation and in minimizing the effects of AM noise. Because of the correlation effect, FM noise produced by spillover tends to cancel in the receiver. Near-in AM and FM noise produced by clutter is also beneficially reduced by the spillover servo, since, in nulling out the carrier, it automatically removes both sidebands, whatever their origin, as long as the decorrelation interval is short. Clutter signals from long ranges have both AM and FM noise that is essentially decorrelated, and feedthrough nulling of these signals may increase their deviation by a factor of 2 or their power by a factor of 4. See Eq. (14.3).

14.8 MISCELLANEOUS CW RADARS

There are several small CW radars for applications that require equipment of modest sensitivity. In all these the homodyne technique is employed, the transmitter itself serving as a local oscillator. The transmitter signal reaches the first mixer either by a direct connection or, more frequently, by controlled leakage.

CW Proximity Fuzes. The basic proximity fuze[23,24] is a CW homodyne device whose only range sensitivity is in the rise of the doppler voltage signals as ground is approached or in the behavior of the signal when the antenna pattern intercepts an aircraft. Commonly a single element is used as both oscillator and mixer-detector.

Characteristically, proximity fuzes use a common antenna for transmitting and receiving and hence suffer from a large leakage problem. The situation is tolerable only in the VHF band where the signals returned from the target (terrain or aircraft) are very large. Frequently a projectile body is used as an end-fed antenna although separate transverse dipole or loop antennas have been employed to avoid a null in the forward direction.

The principal problems with the device are those associated with requirements of small size, long shelf life, low cost, and reliability under high acceleration. Because of the very light weight of all solid-state circuitry with integrated components, complex circuits may be built that will allow proximity fuzes to withstand accelerations in excess of 100,000 g.

Police Radars. This is a straightforward application of the CW homodyne radar technique. Controlled leakage is used to supply the required LO signal to a single crystal mixer. The amplification takes place at the doppler frequency. At 10,525 MHz, one of the frequencies currently approved by the Federal Communications Commission (FCC), 50 m/h corresponds to 1570 Hz, which is in a convenient range.

A squelch circuit is used to prevent random or noisy signals from reaching the counter. Three amplifier levels relative to the squelch yield suitable gains for the detection of short-, medium-, or maximum-range automobiles. The output signal from the doppler amplifier is clipped, differentiated, and integrated. Each pulse from the differentiator makes a fixed contribution to the integrated signal, and the higher the frequency the greater the output. This dc value actuates a meter or a recording device marked directly in velocity. A tuning fork may be used to calibrate the equipment. Some equipments offer a burst mode which determines the speed of the vehicle before it can be altered.

14.9 FM RADAR

The material to follow is on the homodyne FM radar, i.e., a CW radar in which a microwave oscillator is frequency-modulated and serves as both transmitter and local oscillator. For additional material on FM radars, see Refs. 3 and 6. An excellent introduction to FM in general is contained in Ref. 4, Chap. 12.

There are three approaches to the analysis of this type of radar: the phasor diagram, the time-frequency plot, and Fourier analysis. One should have some facility with each. Perhaps the most useful attack for an FM radar having modest deviation is the phasor diagram. To construct the diagram, a large phasor is drawn to represent the carrier. This is taken as a reference and is considered stationary; higher frequencies are represented by phasors rotating counterclockwise and lower frequencies by phasors rotating clockwise. In applying the phasor method to FM homodyne radar, the instantaneous phase of the local oscillator (i.e., that of the transmitter) is taken as the reference phasor, and the returning signal or signals as the small phasor or phasors. The output from the mixer is proportional to the projection of the small phasor or phasors on the large one.

For example, consider an altimeter with triangular frequency modulation. In its phasor diagram (Fig. 14.13) the small phasor will, except at the turnarounds, swing either clockwise or counterclockwise at a uniform rate. If the swing is short (i.e., the range to the ground is short), then, depending on the phase, either of two situations results: Fig. 14.14a or b. In Fig. 14.14a twice as many cycles of difference frequency will be developed in unit time as in Fig. 14.14b. This leads to the so-called critical-distance problem in an FM altimeter. The situation will be covered more fully below; here the interest is in the phasor diagram and what it reveals.

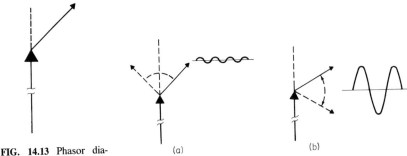

FIG. 14.13 Phasor diagram for an FM-CW radar.

(a)   (b)

FIG. 14.14   Phasor diagrams showing critical distance

The second method is the drawing of an instantaneous-frequency diagram. In these diagrams a curve is drawn in a time-frequency plane to represent each of the signals of interest. A typical plot, that for a sinusoidally modulated altimeter, is shown in Fig. 14.15. Curve *A* represents the frequency-time history of the transmitter (and local oscillator) and curves *B* and *C* that of returns from two different ranges. Note that the vertical distance between curves (e.g., curves *D* and *E*) yields a heuristic picture of the average frequency behavior of the difference signal from the mixer. This is somewhat naïve. Both the transmitted signal and the returned signal are periodic waves, as is their difference. Hence there cannot be a continuum of difference frequencies; there can be only harmonics of the fundamental modulation frequency. Diagrams such as Fig. 14.15 are most useful when the different frequencies indicated are several multiples of the repetition frequency. In this event, the many harmonic lines act almost like a continuum. Such a diagram would not be useful to discover the step error shown in the phasor diagram above.

Finally, there are mathematical approaches limited originally to those systems

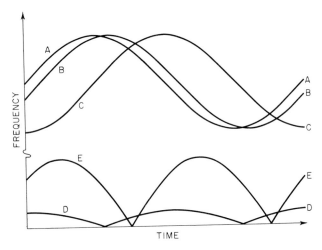

FIG. 14.15   Schematic diagram for a sinusoidally modulated FM.

employing one or more sinusoidal modulations. There exist exact analyses of triangular, sawtooth, dual triangular, dual sawtooth, and combinations of some of these with noise,[25] but heuristic techniques are usually a necessary starting point.

14.10 SINUSOIDAL MODULATION

Suppose one transmits an FM wave of the form

$$\mu_s = U_s \sin \left(\Omega_0 t + \frac{\Delta\Omega}{\omega_m} \sin \omega_m t \right)$$

where ω_m = modulation frequency
Ω_0 = carrier frequency
$\Delta\Omega/\omega_m$ = modulation index

An echo from a point target will have the form

$$\mu_e = U_e \left\{ \sin \left[\Omega_0(t - T) + \frac{\Delta\Omega}{\omega_m} \sin \omega_m(t - T) \right] + \phi \right\}$$

where ϕ = arbitrary phase angle produced on reflection
T = time delay of echo

To introduce the effect of doppler we let T be time-dependent: $T = T_0 + 2vt/c$, where v is the velocity of the echoing object and c the velocity of light. After the usual trigonometric manipulation, the difference μ_i takes the form

$$\mu_i = U_i \cos \left[\Omega_0 \left(T_0 + \frac{2vt}{c} \right) - \phi + D \cos \omega_m \left(t - \frac{T}{2} \right) \right]$$

$$D = \frac{2\Delta\Omega}{\omega_m} \sin \frac{\omega_m T}{2}$$

The reflection phase ϕ may generally be disregarded and μ_i expanded in a Fourier series.[6]

$$\mu_i = U_i \left(J_0(D) \cos \Omega_0 \left(T_0 + \frac{2vt}{c} \right) + \sum_{n \text{ odd}}^{\infty} (-1)^{(n+1)/2} J_n(D) \left\{ \sin \left[n\omega_m \left(t - \frac{T}{2} \right) \right. \right. \right.$$

$$\left. + \Omega_0 \left(T_0 + \frac{2vt}{c} \right) \right] - \sin \left[n\omega_m \left(t - \frac{T}{2} \right) - \Omega_0 \left(T_0 + \frac{2vt}{c} \right) \right] \right\}$$

$$+ \sum_{n \text{ even}}^{\infty} (-1)^{n/2} J_n(D) \left\{ \cos \left[n\omega_m \left(t - \frac{T}{2} \right) + \Omega_0 \left(T_0 + \frac{2vt}{c} \right) \right] \right.$$

$$\left. \left. \left. \right) \cos \left[n\omega_m \left(t - \frac{T}{2} \right) - \Omega_0 \left(T_0 + \frac{2vt}{c} \right) \right] \right\} \right)$$

[$J_n(D)$ is the nth-order Bessel function.]

In the usual case that a doppler frequency exists, μ_i is seen to be the signal represented by Fig. 14.16. No energy is present at the modulation frequency or its harmonics, but there are a pair of sidebands surrounding each harmonic of the modulation frequency and one just above dc. The amplitude of the pair bracketing the nth harmonic is proportional to $J_n(D)$, with D a multiplicative function of the modulation index, and a periodic function of the product of the modulating frequency and the time delay of the echo. The values of the appropriate $J_n(D)$ indicate the amplitude of the double-sideband signal that would be recaptured by an amplifier tuned to the region of a given harmonic of the modulation frequency. Figure 14.17 indicates some typical values that might occur in the systems to be discussed.

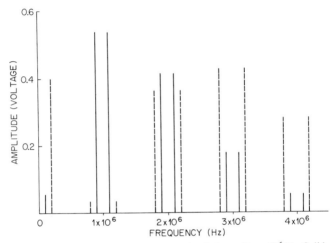

FIG. 14.16 Spectra of double-sideband signals for $\omega_m/2\pi = 10^6$ Hz. Solid lines: $D = 2.3$, doppler $= 10^5$ Hz; dotted lines: $D = 4$, doppler $= 2 \times 10^5$ Hz.

Other plots are easily constructed by using a table or graph of the lower-order Bessel functions. All such plots have the following characteristics: They are periodic in T, and the curve of the Bessel function is traced out to some argument and then back to the origin and so forth. There is a symmetry in the behavior of all the spectral components about the points $\omega_m T/2 = n\pi$; that is, the delay of the returning echo is equal to an integral number of periods of the modulation frequency. When only the amplitudes of spectral components are considered, additional points of symmetry occur at $\omega_m T/2 = \pi/2$, $3\pi/2$, etc. The behavior of the Bessel function, $J_n(X) \sim Ax^n$ for $x \ll 1$, indicates the character of the curve at those zeros produced by $D = 0$ and provides insight into the ability of the system to combat feedthrough.

In all the above it is assumed that the doppler does not exceed one-half of the modulation frequency, $\omega_m/2\pi$, for otherwise the doppler sidebands would

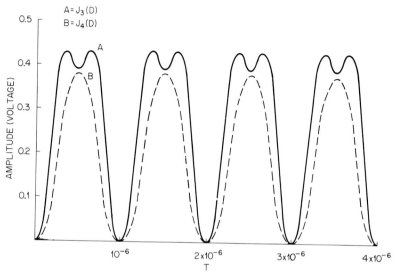

FIG. 14.17 Range response of third- and fourth-harmonic systems ($f_m = 10^6$; index = 2.4).

bear no easily discovered relationship to the modulation harmonics. There is an equivalent problem in pulse doppler radar when the PRF is not sufficiently high with respect to the doppler. The presence of the double sidebands around each harmonic causes certain problems in signal processing that will be discussed in Secs. 14.17 and 14.19. If it is desired to avoid double sidebands, an offset frequency, which is high with respect to the largest modulation harmonic of appreciable value, can be introduced. This eliminates the folding.

The sidebands around each of the harmonics of the modulation frequency require some examination. The formulas presented and Fig. 14.16 imply that the two sidebands about a given harmonic line are exactly equidistant from the frequency of the harmonic and that these distances are the same for all the lines associated with the different harmonics. This is not quite true; the analysis ignores $\omega_m T$ and ϕ and loses a small but sometimes critical second-order effect. If phase detection at a second mixer using a given harmonic of the modulation frequency as a reference is examined, it is seen that the two sidebands rotate in the phasor diagram at slightly different rates. Hence the line where they cross rotates, and the detector shifts between an I and a Q position. This means that a single phase detector will lose all sensitivity when the crossing line is exactly in the Q position. Since the crossing position is related to range, this means a null in the range response. It can be shown that a single phase detector adds a number of nulls in each repeat of the range response equal to the number of the harmonic; i.e., three extra nulls in a J_3 system, etc. By changing the phase of the reference, the nulls move; but they are always present. To avoid these nulls, single-sideband processing is required. Similar remarks apply to other types of modulation. One exception is sawtooth (see below). Section 14.12 has some discussion of phase rotation as well.

Double Sinusoidal Modulation. The mixing signal that results from the use of a modulation consisting of the sum of two sinusoidal components may be handled by similar mathematical techniques. In the presence of a low doppler frequency, pairs near the harmonics of the lower modulation frequency form about each harmonic of the higher one (Fig. 14.18). The lines actually represent the relative heights of the pairs of doppler sidebands which straddle the indicated lines. It is not feasible to show a 10-kHz doppler in the same scale as the 1-MHz harmonics of the first modulating frequency and the 40-kHz of the second. It can be shown that the doppler pair carried by the nth harmonic of the 40-kHz modulating frequency in the region of the mth harmonic of the 1-MHz modulating frequency has relative amplitudes equal to[26]

$$J_n\left(\frac{2\Delta\Omega_2}{\omega_2}\sin\frac{\omega_2 T}{2}\right) J_m\left(\frac{2\Delta\Omega_1}{\omega_1}\sin\frac{\omega_1 T}{2}\right)$$

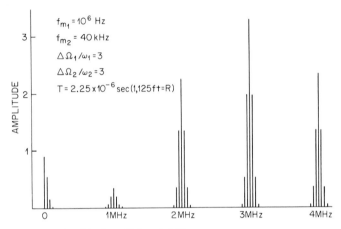

FIG. 14.18   Double sinusoidal modulation.

For values in Fig. 14.18 the sine in the first parentheses may be approximated by its argument, and J_n in turn approximated by the first term of its power series. For the plot, a value of T was chosen which yielded one of the many maxima of $J_3[(2\Delta\Omega_1/\omega_1)(\sin\omega_1 T/2)]$ and which, at the same time, displayed the contributions of the lower modulating frequency. It should be noted that such a simple diagram exists only if the doppler frequency is small compared with the lower modulating frequency and if the highest harmonic of the lower modulating frequency that is of interest (this is a function of T) does not exceed one-half of the spacing between the harmonics of the higher modulating frequency. If the dopplers are high, double modulations using widely different frequencies are usually not possible. It is possible to use a small amount of a nearby harmonic of the fundamental modulating frequency (see the discussion of the J_3 fuze in Sec. 14.17).

14.11   TRIANGULAR AND SAWTOOTH MODULATION

These modulations are usually handled by appeal to a time-frequency diagram. In most cases the interval surrounding the moment when the slope of the modulating function is discontinuous is not considered in the analysis. This is reasonable, as this period represents a relatively small portion of the recovered energy and one that is spread over a wide range of frequencies.

The usual approximation for either waveform is one based on similar triangles. In Fig. 14.19, A represents the wave transmitted by a typical altimeter, B the return from the ground, and C the center of the difference-frequency envelope at the mixer. As is customary, the distances α-β, β-γ are long compared with the spacing between A and B; that is, the wavelength of the modulating frequency is long compared with any altitude of interest. If ΔF represents the total frequency deviation, τ one-half of the period of the modulating frequency, and $T = 2R/c$ the delay, then $\Delta f = T\,\Delta F/\tau$. As noted above, this is only an approximation; no spectral line may exist at this exact frequency. Some typical numbers for an altimeter might be $\Delta F = 100$ MHz, $\tau = 1 \times 10^{-3}$ s (i.e., a 500-Hz deviation rate). At 10,000 ft, corresponding to a round-trip signal delay of $2 \times 10^4 \times 10^{-9}$ s, there would be a difference-frequency spectrum clustered near 2 MHz. The 500-Hz line structure usually can be ignored. Even at 50 ft the difference spectrum would center at 10 kHz, and the line structure would not be damaging.

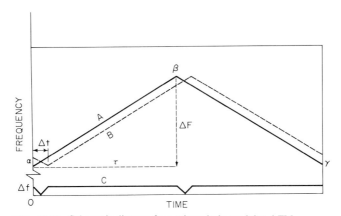

FIG. 14.19   Schematic diagram for a triangularly modulated FM.

So far doppler has been ignored. Altimeters working with clip-and-count circuits and almost no velocity component are not really affected. When one is working with higher-modulation frequencies and processing the energy in only a few of the lines with phase detectors, more careful analysis is required. If one examines curve B in Fig. 14.19 closely, it is seen that it does not run exactly parallel to or equidistant from curve A. Doppler produced by an approaching object raises curve B, bringing it closer to curve A on the upsweep and farther from A on the downsweep. More than this, the doppler impressed on B is somewhat greater when A is above its average frequency and less when A is below its average frequency. This produces a small lack of parallelism between the A and B curves. Curve C then shows a frequency reduced by the doppler during the upsweep and increased during the downsweep. The lower doppler sidebands around the harmonic lines are produced on the upsweep and the higher ones on the downsweep. As an approaching object comes closer, the zeros of the C curve shift to the left, which implies a phase shift with respect to any fixed reference, such as a harmonic of the modulation frequency. This leads to an alternate way of viewing the range behavior of a phase detector first noted in Sec. 14.10.

The calculations for a sawtooth modulation are similar. The only differences are in the much higher frequencies generated during the transient period and in the way in which any doppler comes into play. With a triangular modulation, a doppler less than the difference frequency ΔF averages out, since it increases the difference frequency on one sweep and decreases it on the other. With a sawtooth modulation, only an increase or a decrease is noted since the transients during the flyback are usually out of the passband of the difference amplifier. Moreover, there is a difference in the line structure since each repetition-rate line carries only one doppler sideband, not two.[27]

If either triangular or sawtooth modulation is regarded as a sample of an essentially constant mixer difference frequency, it is readily seen that the envelope of the spectral lines at the mixer has a $(\sin x)/x$ form. The scale of the $(\sin x)/x$ function is determined by the period of one constant-frequency portion with respect to the mean difference frequency. See curve C in Fig. 14.19. A comparable situation with a sinusoidal modulation produces a less compact spectrum and one that extends asymmetrically toward low harmonics.

Exact analyses of triangular and sawtooth modulations are now available in Tozzi.[25] By using the power of a mainframe computer, the Fresnel integrals which occur were evaluated, and the spectral components as a function of range were calculated. A sampling technique which represented each of the waveforms involved by upward of 50,000 to 100,000 samples was also employed to derive a sampled version of the waveform from the first mixer. It is now possible to analyze these waveforms without making the assumption that only the envelope of the spectral lines is of interest. This is of greatest value in designing very-short-range systems, such as microwave proximity fuzes, as will be discussed in Sec. 14.17.

14.12 NOISE MODULATION

A band-limited noise voltage may also be used to frequency-modulate a transmitter. In this case, it can be shown that a half-gaussian curve, peaked at zero, describes the dc of the mixer output as a function of the delay.[28] This is reasonable when the process is regarded as a correlation, for at zero delay it is a CW radar whereas with longer delays the signal decorrelates. The presence of doppler modifies the situation in that correlation occurs not at dc but at the doppler frequency. Used by itself, FM modulation by band-limited noise is not usually desirable, since maximum response occurs at zero range. This implies a problem with spillover.

14.13 CODED MODULATIONS

With the advent of cheap compact integrated circuits, particularly of nonlinear types such as bistables, it has become feasible to build shift registers to generate codes for small radars.[29,30] These codes, which may be regarded as a sequence of plus and minus 1s of length $2^n - 1$, have the desirable property that a code multiplied by itself and summed for zero delay has a value of $2^n - 1$, and if shifted modulo k, $0 < k < 2^n - 1$, the sum of the products does not exceed -1. The process is symbolized for a code of length 7. (In the correlation process, $+$ with $+$ equals $+$ and $+$ with $-$ equals $-$, and the terms are then summed.)

$$\frac{+\ +\ +\ -\ -\ +\ -\ +\ +\ +\ -\ -\ -\ +\ -}{+\ +\ +\ +\ +\ +\ +} = 7$$

$$\frac{+\ +\ +\ -\ -\ +\ -\ +\ +\ +\ -\ -\ -\ +\ -}{+\ +\ -\ +\ -\ -\ -} = -1$$

Figure 14.20 shows the correlation function of a 31-bit code as a function of delay.

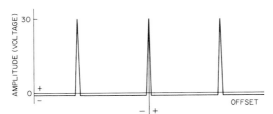

FIG. 14.20 Correlation function of a 31-bit shift-register code.

As used in CW radar, the code controls the phase, usually 0 or 180°, of the transmitted waveform. This is normally done by a diode or ferrite phase shifter which operates between the transmitter and the antenna. Subsequently, the returning waveform is decoded by comparing it with a suitably delayed version of the code. In some applications the delayed version of the code is available from the shift register (perhaps augmented); in other applications the delay of the code is accomplished by either an RF or a video delay line.

One interesting feature of the codes is their behavior when multiplied by either the unmodulated driver or the modulated transmitter. In both cases, there is generated a video signal with a bandwidth approximately that implied by the width of a single code bit. Use of the unmodulated RF signal in the mixer results in transference of the code to video without essential change in its structure. Use of a coded RF in the mixer scrambles the code. The result is that a target having sufficient extent to produce more than one delayed version of the original code yields echoes in the video that do not correspond to adjacent samples of the code. For example, if a 15-

bit code is modulated on RF and mixed with itself after a delay of 4 bits, the resulting video code will appear to be delayed 3 bits with respect to the original code. However, if the delay in the RF is 5 bits, the video will appear to be delayed 10 bits.[30] For ranging purposes, this requires a rewiring of the matrix or the generation and use of an equivalent product code at video.

Ambiguity becomes an issue when the total code duration becomes comparable with the duration of one cycle of the highest doppler that may be produced by the target.[31] In general, there are sidelobes in the doppler domain that rise to $1/\sqrt{N}$ instead of the $1/N$ realized with the stationary targets.

There are other types of binary codes as well as polyphase codes. An interesting example of the latter is found in Plantier,[32] who modified the work done by Lewis, Kretschmer, and Shelton[33] on generalized Frank polyphase codes to make it suitable for CW waveforms. However, binary shift-register codes have been used most often in CW radar. (See Ref. 34 and the bibliography there for more information.)

14.14 DUAL MODULATION

It is possible and frequently useful to employ combinations of modulating modes such as triangle plus sine, sine plus noise, triangle plus triangle, triangle plus noise, etc. In general, one of the modulating modes produces a large deviation, and the other is chosen to have a perturbation type of behavior on the mixer spectrum. This may be due to the low modulation index of the second signal or to a total deviation-time-delay product, ΔFT, which is small. For example, a device having sinusoidal modulation with an index of 3 and a mixer-amplifier combination that captures the third harmonic of the modulation frequency yields a range response that is periodic. The addition of a low-frequency triangle to the modulation function may be made to emphasize only the first peak. Similarly, a small sine-wave component may be added to what was basically a triangularly modulated altimeter to produce an ac servo signal.

Tozzi[25] has analyzed exactly triangle plus triangle, sawtooth plus sawtooth, and those combined modulations with modulation by noise, either random or pseudorandom. His aim was to use one modulation to reduce the range sidelobes produced by the other. This is an alternative to the method discussed in Sec. 14.17, which relies on a single modulating waveform, and shapes the range response by combining harmonics of the modulating frequency to produce the reference waveform used in a second mixer. Unfortunately Tozzi's extensive work in this area exists only in his thesis, but that document has evidently had considerable distribution; see, for example, the work of Randonat and Sautter, done in Germany, in Sec. 14.20.

14.15 LEAKAGE

All the material above assumes that only a wanted echo is processed in the first mixer. This is seldom the case. Direct leakage from transmitter to receiver usually produces a signal that is many orders of magnitude greater than the echo signal. Moreover, it is difficult to frequency-modulate a transmitter without producing incidental AM. The first mixer forms a signal representing all possible sums and differences between the echo, the leakage, and the unwanted AM components on both leakage and local oscillator signal. Some relief against AM on the local oscillator is provided by use of a balanced mixer—but only if the leakage signal at the mixer is smaller than the local oscillator signal. The other cross products are a source of considerable difficulty.

Much of the advantage of FM radar is the manner in which leakage is handled by the circuitry. For example, with a sinusoidal modulation having a deviation index approaching 3 and an amplifier following the mixer with a bandwidth sufficient only to carry the doppler lines surrounding the third harmonic of the modulation frequency, leakage is discriminated against rather substantially. All signals carried by the third harmonic are weighted by $J_3(D)$, where D is a function of the time delay, or distance. The time delay for leakage is normally small, and for those signals $J_3(D)$ may be approximated by $D^3/48$. This is evidently small for small D. A system capturing the first (or J_1) line would act less favorably since $J_1(D) \sim d/2$. This is a very complex question, particularly when the delay of the leakage may be substantial, the incidental AM components particularly high, or microphonism present to produce dopplerlike sidebands. It is at the very heart of any meaningful study of a given FM radar system.[6,35,36]

14.16 PERFORMANCE OF FM-CW SYSTEMS

From the previous paragraphs it should be realized that the small FM homodyne radar is seldom if ever limited by thermal noise in the first mixer. The badge of a novice in the FM radar field is a carefully worked out performance appraisal based only on the methods of Chap. 2. In general, it may be said that, with two antennas with good decoupling between them, a well-designed FM homodyne radar may come within 20 to 30 dB of the performance indicated by Chap. 2. A system designed to receive only distant echoes has considerable advantage over one such as a proximity fuze, that must cope with echoes having short delays.

14.17 SHORT-RANGE SYSTEMS AND MICROWAVE PROXIMITY FUZES

The earliest use of sinusoidal FM modulation in radar was for microwave proximity fuzes designed for some of the first antiaircraft guided missiles, such as Terrier, Sparrow 2, and Bomarc. These missiles had the space to carry far more complex electronics than the simple proximity fuzes described in Sec. 14.8. A small, rugged klystron was developed especially for this application and was driven by an ingenious one-tube circuit which served both to supply modulation voltage and to act as a servo to keep the klystron at mode center. The first amplifier was tuned to the third harmonic of the modulation frequency. Separate transmitting and receiving antennas were employed, and the 30 to 40 dB isolation achieved, coupled with the J_3 behavior, provided sufficient reduction of the leakage signal to permit a target of ¼ to 1 m^2 to be seen out to 100 to 150 ft. Because the deviation of the klystron was modest, with its amplitude modulation limited primarily to even harmonics, and because of the decoupling provided by the dual antennas, a simple amplitude detector sufficed for the second detector. The antennas formed a conical beam about the missile axis, with the angle of the cone chosen to be compatible with the fragment pattern of the warhead.[37]

If the third-order Bessel function modulation were designed to yield maximum signal return at, say, 100 ft, it would yield somewhat less sensitivity at close ranges than might have been desired. The problem was alleviated somewhat by multiple reflections between the target and the missile. It would have been possible to introduce a small amount of a second modulation at 3 times the frequency of the primary modulation. This would have produced an additional component in the third-harmonic amplifier, having J_1 behavior, i.e., a $1/R^2$ response, where R = range.

With missiles designed to operate against ground targets, a *setable-range* radar altimeter was desired. The first step toward the synthesis of a suitable transmitted waveform was made by Karr[38] and Horton.[28] Their starting point was random noise since the desire was to produce a signal that was not predictable for reasons of electronic counter-countermeasures. Beginning with the time-frequency plane, they noted that the slopes of the output waveform were directly related to the frequencies produced at the mixer, with time delay a multiplicative parameter. For fixed time delay, the probability density function of the absolute values of the derivatives of the output waveform (in the time-frequency plane) was linearly related to the probability density function of the mixer's difference frequencies. From this reasoning, a procedure was developed which started with gaussian noise and, by nonlinear processing, arrived at a correct modulating waveform to produce a desired difference-frequency spectrum.[6] This work was extended[39] to air target fuzes where the range response was required to increase as R^3 out to some predetermined range. (This function matches $1/R^3$ rather than the usual $1/R^4$ since the area of the target illuminated by a conical beam increases as R.) Alternatively, a triangular modulation might have been employed and the first amplifier shaped to match R^3. This would have been a difficult amplifier to design; for a better solution, see below.

When the exact analysis of triangular and sawtooth waveforms became available,[25] it was possible to treat these modulations without relying on the time-frequency plane. In particular, those cases where the lower harmonic lines were dominant could be analyzed. Carrying this work further, a group at the University of Florida analyzed systems, consisting of both a first mixer and second phase detector, where a combination of harmonics of the modulating frequency served as the reference. Phase detection against a single harmonic of the modulating frequency had been proposed as early as 1952[40] and, as noted in Ref. 36, was considered in doppler navigators. Some form of phase detection is also essential to filter out the results of unwanted AM carried by the leakage signal into the simpler systems that have solid-state sources, with perhaps unfavorable power-versus-frequency curves, and that cannot afford dual antennas. The Florida work extended the technique to combinations of harmonics of the modulation frequency. In a sense, this leads to an extension of the Woodward ambiguity function; i.e., two correlations, the first occurring in the microwave mixer and the second in another mixer where the signal passing through a moderately wideband IF amplifier is correlated with a waveform related to the modulation. More than this, they were able to show that, starting with a triangular, sawtooth, or any of several other modulations, a waveform for the second mixer could be synthesized to give a desired range response. This in effect partially sidesteps some of the limitations on the design of the transmitted waveform (Ref. 41, p. 125).

All this would be academic were it not for the vast strides made in integrated circuits. It is feasible to build a digital circuit on a very small chip that produces in sampled form any suitably band-limited waveform. The required samples are merely read out of a read-only memory (ROM). Since the desired waveform is normally symmetric, the necessary ROM capacity is halved. A simple example is shown in Fig. 14.21, where a shows the triangular modulating voltage; b, the reference waveform for the second mixer (a phase detector), given by $-\cos(6\omega_m t) + 0.8 \cos(8\omega_m t)$, the so-called cosine reference waveform; c, a second reference waveform, $-\sin(6\omega_m t) + 0.67 \sin(8\omega_m t)$ for $t < 0$, and $\sin(6\omega_m t) - 0.67 \sin(8\omega_m t)$ for $t > 0$, the sine, or quadrature, reference waveform;* d, the envelope of the range response using b (that using c is nearly the same); and e, the envelope of the response using only a sixth harmonic as a reference waveform (with slightly changed deviation to make the range peak match that of d). Except for the skew produced by the nearness to $R = 0$, the last is essentially $\sin(x)/x$, as predicted in Sec. 14.11. The skew would be far less pronounced if much higher harmonics representing a much longer range had been used. In comparing d with e, note the much improved range sidelobes in d, especially those in the very critical region beyond the range of interest.

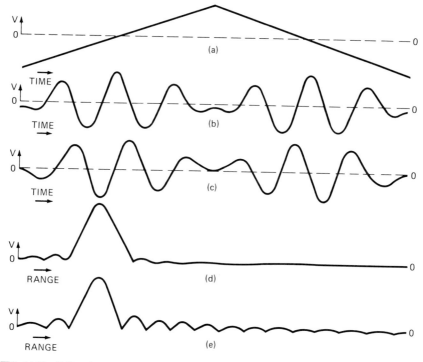

FIG. 14.21 Tailored range response. (See text for explanation.)

---

*A study of the waveforms arriving at the second mixer shows two distinct cases depending on the phase of the doppler. This is reflected in the appropriate reference curves b and c. The first is inherently symmetrical since it is a combination of cosines. The second has symmetry imposed on it by reversing the polarity of each of the sine waves at $t = 0$.

In addition to the analytical methods, including some which lead to a heuristic understanding of the system by short-range approximations that reduce the complexity of the mathematics, simulators have been built that permit the rapid test of a wide variety of modulating waveforms with their appropriate second-mixer waveforms.[42]

14.18 ALTIMETERS

Altimeters were the earliest FM-CW devices and among the most successful. The first models employed triangular modulation produced by various mechanical means, such as moving elements, preferably noncontacting, in the microwave cavity of a triode. Repeatability of modulation and mechanical reliability were difficult to achieve. As a bibliography of these devices alone would fill many pages, it appears best to discuss only principles and some variants.

All modern altimeters use wide-deviation FM at a low modulation frequency. The difference signal recovered by a homodyne receiver is spread over a very large number of closely spaced spectral lines. In general, the average frequency is deduced by clipping the signal and counting zero crossings. Under fairly general conditions it may be shown[43] that, as long as the delay and total deviation are kept fixed, the number of zeros in the difference wave is independent of the shape of the modulating waveform. Triangular modulation is usually employed, since this yields the most compact and easily amplified spectrum and the doppler effect balances out.

For very-low-altitude operation the step error, or critical distance, has been made an issue. This is a quantization in height $\Delta h = c/4\Delta F$, where c is the velocity of light and ΔF the total deviation. (This assumes a difference of one zero crossing, i.e., $\frac{1}{2}$ Hz.) For 100 MHz of deviation this represents $\frac{3}{4}$ m. The resulting irregularities in altimeter response are often not seen, however, since they may occur too rapidly to be followed by the meter.[44]

One of the other basic difficulties in altimetry has been a tendency to shift readings not because of the step error but because some distant terrain temporarily produces a greater signal at the limiter than the terrain directly beneath the aircraft. This cannot be counteracted by narrowing the antenna pattern since the pattern must be wide enough to accommodate attitude changes of the aircraft. Returns from rough water can be particularly deceptive. The water beneath the aircraft may not yield a large return whereas certain wave slopes illuminated at a considerable angle from the vertical may produce a large spectral reflection. The problem is intensified by the usual method of designing the passband of the difference-frequency amplifier. This has a nominal response increasing 6 dB per octave from the lowest to the highest frequencies of interest to compensate for the $1/(\text{altitude})^2$ behavior of the desired vertical incidence signal. One solution is a simplified frequency tracker which limits the upper-frequency response of the difference-frequency amplifier as a function of the frequency of the signal being tracked.

The Bendix ALA-52A altimeter[45] is the successor to the 51A described in the first edition of this handbook. Most of the basic philosophy is the same, but the instrumentation now depends heavily on digital techniques and there have been many significant changes in the hardware. The transmitter is a transistor modulated by a single varactor. The modulation is a 150-Hz triangular wave produced by digitally counting a stable oscillator and passing the counts through a digital-to-analog (D/A) converter. (For a technique that produces the microwave energy directly, see Ref. 46.) The reference voltage supplied to the D/A converter can be slightly varied to alter the scale and, hence, the nominal 130-MHz deviation. Key to the operation of the altimeter is a bulk-wave quartz delay that produces a 300-ft internal reference. This provides the processor with means to check the deviation and its linearity and to provide information to the circuitry that alerts the pilot to faults. Since the deviation of the transmitter signal produced by a varactor is inherently nonlinear and is also subject to unit-to-unit and temperature effects, two voltages are added to the modulation. The first is derived from a ladder of diodes that switch in and out associated resistors as the voltage equals specified values. The second is an adaptive voltage that is dependent on the frequency count provided by processing the signal from the internal-delay reference. (For an alternate solution to the nonlinear problem, see Tozzi.[47]) The action of this second circuit is speeded up by command of the processor when the internal delay indicates a serious nonlinearity in the modulation of the output microwave energy.

The preamplifier has a fundamental 6 dB per octave increase in gain and there is a 12 dB low-pass filter that moves, under command of the processor, to filter out the unwanted higher frequencies. The heart of the system is a digital microprocessor that, with the aid of other elements on the processor board, reads the output frequency from the tracking filter (via a smoothing circuit, originally a complex phase-lock loop but most recently a single integrated circuit comparitor), as well as the output from the internal reference, and produces a wide variety of control commands, such as the one to the adaptive linearization circuit, and fault indications. A second processor having a different architecture monitors the first processor and, if there is a disagreement, indicates a fault. This provides redundancy without the threat of common-mode failure.

Refinements of the circuitry include a gate that removes the lower frequencies produced at the turnaround of the modulation (Fig. 14.19) and a variation in the 12 dB per octave low-pass characteristic of the filter in the preamplifier when the equipment is searching below 100 ft. Part of the bias toward lower frequencies is removed to prevent the system from locking up on feedthrough. Even with all the precautions, including dual antennas, turnaround gating, and linearization, there is still unwanted energy in the 150-Hz lines at the lower frequencies.

The altimeter consumes but 30 W of prime power and weighs only 11 lb. For an inkling of the radar complexity that can be packaged in 11 lb, the service manual needed to describe it weighs 6½ lb.

The Collins ALT-55 altimeter[48] is designed primarily for use on the smaller aircraft employed by corporations and feeder airlines. It weighs, including antennas and indicator, only 8 lb. (The weight given for the ALA-52A did not include that for the antennas or the indicator.) The radiated power (350 mW), the modulation frequency (100 kHz), and the deviation (100 MHz) are all numbers roughly equivalent to those used by Bendix, but in the mechanization the Collins equipment is all analog and considerably less complex.

The transmitter frequency is generated by an oscillator running at 1433 MHz, which is then amplified and applied to a multiplying diode. Filters select the third harmonic to become the transmitted RF and the sixth harmonic to be used in a microwave delay circuit that provides the information to guarantee the proper deviation. The modulation voltage required to produce the FM is derived from a 100-Hz square wave that is integrated in an operational amplifier to derive a triangular waveform. The output signal from the delay line is mixed with the input signal, and the resulting difference frequency is counted. If it does not correspond to the 50-ft delay provided, then the modulation control circuitry operates to correct the deviation. It does this by injecting an additional square-wave voltage before the operational amplifier. Added to this square wave is the square-wave output derived from a second oscillator that varies or "wobbles" in frequency. This smooths out the step errors inherent in such a short delay.

The preamplifier circuit in the receiver has the usual 6 dB per octave increase in gain and a three-section low-pass filter that can be adjusted to roll off at any of four frequencies as commanded by an analog signal produced in the frequency-to-voltage converter (equivalent to a tracker). The smoothing of the output from the preamplifier is accomplished by a time-domain filter that takes the square wave from the preamplifier and produces a rectangular 40 percent up, 60 percent down wave. A so-called missing pulse detector notes sudden increases in the altitude frequency and works with the time-domain filter to remove the smoothing temporarily and reacquire the track.

All voltages supplied to the indicator are analog and are derived from circuitry that essentially transforms the rectangular 40–60 waves into uniform height pulses that are in turn integrated to produce a continuous voltage that is proportional to the frequency. Two such circuits are employed, and their sum forms the output voltage. The first produces 0.020 V/ft up to 500 ft, and the second adds to this 0.003 V/ft up to 2500 ft. The indicator scale clearly reflects this with 180° of rotation devoted to 0 to 500 ft and 60° of additional rotation allocated to 500 to 2500 ft. The area between 240° and approximately 360° has a vane that covers the needle and tells the pilot that there is insufficient return or a fault and that no information is currently available. There is also a check button that, when pressed, switches the output of the delay through the tracker and to the indicator, which should then read 50 ft. This vindicates most of the system but nothing beyond the multiplier diode in the transmitter. The choice of having the delay operate at twice the transmitted frequency, although having the advantage of doubling the deviation, makes it difficult to check to the front end as Bendix does.

Neither the Bendix nor the Collins mechanization is acceptable to the military because of their relatively wide-open front ends. A group in France has taken another approach that permits a much narrower first amplifier. It has built altimeters for both military and civilian aircraft by using a deviation that varies under control of the tracker to produce a difference frequency that is essentially constant with altitude. (This has much in parallel with the philosophy of the Improved Hawk missile with its so-called inverse receiver; see Chap. 19.) An earlier version of the altimeter using this technique is described in the previous edition of this handbook, and the bibliography there cites an excellent reference in English by Fouilloy. The only easily available reference to the group's more recent work is Tomas.[49] Inevitably its newer altimeters make much greater use of integrated circuits and digital techniques.

In addition to the FM-CW altimeters a wide variety of pulse radar altimeters has been developed. They dominate a large portion of both the commercial and the military markets. Literature on them may be obtained from Honeywell and others. There is not space to discuss them further, since this chapter is devoted primarily to FM-CW.

14.19 DOPPLER NAVIGATORS

The first doppler navigators were pulse systems conceived as logical extensions to clutter-referenced MTI radar. The transmitter was incoherent pulse to pulse, but since multiple beams were radiated, the effects of phase incoherence canceled out when signals having similar delay were mixed. One beam, for example, was radiated 45° downward and forward along the flight path, while another was radiated at a similar depression angle but rearward. The doppler difference as recovered in the subsequent mixing was indicative of the velocity of the aircraft along the flight path. (It was usual to employ an offset to prevent folding.) The use of the dual beams has a secondary advantage. It can be shown (Brown et al.,[50] pp. 781–782) that when an aircraft alternately climbs and dives, the average value of the doppler developed by differences between signals derived from symmetrical beams has less error than would be the case with a single beam.

If the aircraft axis is not parallel to the track on the ground, one must either rotate the fore-and-aft beams or compensate for the yaw. In either case at least one additional beam is required. More usually, two additional beams are used. The redundancy is used in the tracker to cross-check in each plane and requires that each equivalent pair yield similar answers. A common arrangement has four beams, all with the same depression, two pointed forward, one left, one right, in a symmetrical fashion, and two others pointed aft (Fig. 14.22).

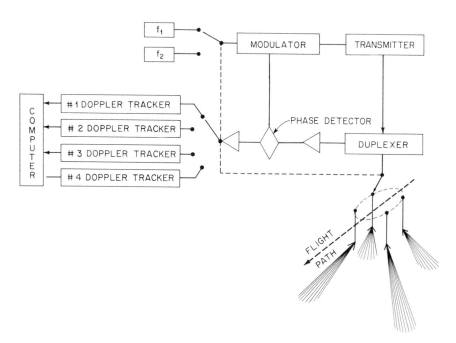

FIG. 14.22 Four-beam doppler navigator with individual trackers and alternating modulating frequencies.

Interest here is centered on the modern CW variants of the equipment. As a CW or FM-CW system is coherent by nature, there is no primary requirement to use pairs of beams to effect random-phase cancellation. Although one FM-CW device used differencing of beam pairs,* we shall consider just the one-beam type with its associated doppler tracker. It is to be understood that four trackers (or two pairs) are involved and that the transmitter-receiver is sequenced to each beam or beam pair in turn. A computer is required to derive the navigation information for presentation to the pilot.

The commercial airlines have shifted to nonradar forms of navigation and the high-altitude J_3 and J_4 systems described in the previous edition of the handbook are no longer produced. The emphasis has been on extremely small and light-weight all-solid-state equipments principally for helicopters. These all employ sinusoidal modulation and recover only the first harmonic, a J_1 system.[51] Since the velocities are low and more closely vertical antenna beams are employed, modulation frequencies in the 25- to 50-kHz range are practical. For the lower-altitude models the design can be such that the first zero in the range response falls well beyond the ranges of interest. The voltage entering a first-harmonic amplifier is approximately proportional to range (Sec. 14.10), and since the ground illuminated by the antennas increases as range squared, the net loss of signal is proportional to $1/R$ and the compensation of the system is quite good. The price paid is the increased feedthrough compared with the J_3 or J_4 system, and the design must take measures to suppress it. Since these systems are designed primarily for low altitudes, some relief is afforded by the relatively low sensitivity required. Another problem is that there is no assurance that a helicopter is moving forward with respect to the ground. One cannot assume that the doppler seen by the forward antennas is positive and that seen by the back antennas is negative. Doppler folding must be avoided. The preferred technique is some form of quadrature mixing (see Fig. 14.10 and accompanying text).

Various divisions of the Singer Corporation have produced doppler navigators for many years. A typical equipment is their prototype the SKD-2110.[52] This is a repackaged and reengineered version of the earlier navigators (AN/ASN-128) that are employed extensively on United States helicopters and those of many other countries. Over 2000 of the basic AN/ASN-128 version were manufactured.

The total weight of the SKD-2110 is but 11.5 lb, and an 8000-h mean time between failures (MTBF) should be achieved. (For comparison the earlier AN/ASN-128 weighed 29.2 lb and has shown a 2121-h MTBF.) The 2110 employs a Gunn diode as its oscillator. This is modulated with a varactor diode driven by a 30-kHz sine wave. The output power is about 50 mW at 13.3 GHz. A single microstrip antenna is used for both transmission and reception and is of a very lightweight construction. The four beams are at relatively small angles off vertical (about 15°), which brings the dopplers below the 30-kHz modulation frequency. The beamwidths and patterns are carefully designed to minimize the error that is present when the character of the terrain changes, as from land to water, or to rough water, etc. Beamwidths are about 3.7° in each axis.

---

*The use of differencing techniques, similar to those used with pulse systems, has the advantage that the degradation caused by low-frequency random FM on the klystron largely cancels out. Unfortunately all such systems contain a mixer where two signals, both having accompanying noise, are mixed. No completely satisfactory study of these "dirty" mixers with their noise-times-noise terms is known to the author. The alternative required for success with the single-beam tracker, which is to remove unwanted low-frequency FM from the transmitter, is a straightforward if difficult engineering task. It has become the usual approach.

The most important innovation is a leakage elimination filter that reduces the very large leakage components inherent in a single-antenna design. This takes the form of a servo which first measures the leakage components in both I and Q and then synthesizes a signal at sin $\omega_m t$ in proper phase and amplitude to cancel the unwanted signal. Fundamentally, it operates in the same manner as the feedthrough nulling bridge of Fig. 14.12. The difference is that the nulling is accomplished at an IF frequency rather than at a microwave frequency. The leakage generated from a 50-mW transmitter is no threat to the microwave crystals, and IF cancellation is perfectly effective as long as the gains are such that nonlinear saturation effects are avoided. A second requirement to be met is that the feedthrough on each of the four beams is different, and as the transmitter and receiver are shifted from beam to beam, the nulling bridge must be quickly readjusted.

In addition to the low-altitude equipments (below 15,000 ft) there is the SKD-2126[53] prototype usable to 40,000 ft. This employs dual antennas, two alternating modulation frequencies, 30 and 36.9 kHz (to eliminate the range hole), and a field-effect transistor (FET) amplifier to increase the output power to 300 mW. An FET front end is employed to reduce the contribution of $1/f$ noise in the necessarily wider IF amplifier centered at 33.5 kHz, and the antenna size and gain have been increased over those of the earlier systems. The weight is 23 lb.

The Canadians build a variety of J_1 systems for helicopter and drone applications.[54] Although they have built systems with a single antenna, most of their systems employ dual antennas, one for transmitting and one for receiving. They have also successfully mechanized a technique suggested by Fried[51] for deriving altitude from a phase measurement on the sidebands (see Sec. 14.10 where the phase rotation of sidebands is discussed). With this altitude information and information on the amplitude of the return signal they can draw inferences on the type of terrain beneath the helicopter and make a second-order correction in the doppler tracker to improve the accuracy of the track. For a full discussion of terrain effects, see Kayton and Fried.[55] The Canadians also employ a form of feedthrough cancellation in some of their systems.

One topic remains to be mentioned: that of the tracker required to follow the spread doppler spectrum. In theory this might be a speedgate similar to the one described in Sec. 14.6, but in practice sophisticated devices in wide variety have been employed: see, for example, Refs. 52 and 53.

14.20 *PERSONNEL DETECTION RADAR AND MISCELLANEOUS FM-CW SYSTEMS*

An example of a lightweight all-solid-state CW radar which uses a shift-register-type coded transmission and is designed for the detection of moving targets, including personnel, is described in Ref. 56. The equipment is extremely light, weighing, in various versions, from 2.2 to 3 lb. This includes a moderate-gain antenna and earphones for the operator, but not the batteries. As leakage is used in the first mixing operation, the code is initially scrambled (Sec. 14.13), but it is compared with a similarly scrambled code produced by a video mixing of the undelayed code and the code representing the desired 25-m range cell.

In addition to the primary coding, a secondary modulation is introduced. This secondary coding, in its undelayed version, is likewise a constituent of the video decoding signal. The signal returning from the target, however, carries both a doppler shift and a delayed version of the secondary modulation. The result is that the decoder produces correlation at the selected range, not at the doppler frequency but in the form of a double-sideband suppressed-carrier signal centered about the secondary modulation. Phase detection of this against the secondary modulation signal produces a doppler tone in the earphones whose amplitude is distance-dependent (as mentioned repeatedly in this chapter). By proper choice of the secondary modulation frequency and its phase at the phase detector, the system is designed to have a voltage sensitivity which increases linearly in range (to the approximation $\sin \theta \approx \theta$). This serves to deemphasize close-in returns just as a sensitivity time control (STC) deemphasizes close targets for a pulse radar. Such desensitization is very important with CW coded systems since the sidelobes on the code are down in voltage only $1/n$ (n = number of code bits), and this may be insufficient to cope with clutter which increases as $(\text{range})^{-3}$ at short ranges.

The ambiguity diagram for the double modulation undoubtedly has some additional spikes over those of the shift-register code alone, but as long as the dopplers of interest are low, these spikes are not of concern if they do not appreciably alter the response on the dc axis.

Although the above radar is no longer current, it does provide an excellent example of both pseudonoise techniques and dual modulation. Both of these areas have been heavily exploited, but there is space here for only some annotated references. The Naval Electronics Laboratory (now the Naval Ocean Systems Center) built a coded personnel detection radar with some assistance from the Technology Service Corporation.[57] Albanese and Klein[58] proposed a suitable staggered code for a CW radar that must detect high-speed targets and hence must have a relatively short code length to avoid doppler ambiguity. Fahey and O'Reilly[59] also describe a pseudo-noise-coded radar for high-speed targets. Sherif[60] has described a code plus a second modulation for a very-short-range system. Randonat and Sautter[61] have extended the work of Tozzi[25] to an asymmetrical triangular modulation for use in an automobile system. Their theoretical analysis employs transforms into the frequency plane similar to the work at the University of Florida (mentioned in Sec. 14.17). The evermore distant goal of solving the traffic problems brought about by the automobile has been a fertile field for FM-CW research since the late 1950s.[62–64] Meteorological studies[65–68] and industrial measurements[69,70] have extensively employed CW and FM-CW radar techniques. In short, the three applications discussed in Secs. 14.17 to 14.19 are only the tip of a very large iceberg—representing production equipment produced in quantity.

Although digital techniques such as the FFT[71] and D/A waveform generation have been mentioned, it is worth emphasizing that while they may be theoretically equivalent to analog filters, analog waveform generators, and so forth, there are significant practical differences. Many modern radars would exceed all weight and size constraints and show poor reliability were not digital techniques extensively employed. The system discussed in Sec. 14.17 is an obvious example. It would be difficult to envision a practical embodiment of such a system that had to depend on an analog equivalent of the ROM. Two other examples are systems[72] that could be built only with the advent of a very compact FFT, of an adaptive digital waveform generator, and of a small master computer. The second example describes a radar fuze that adapts the whole system to the velocity of the target while providing a continuous curve that is a best fit to the available discrete range measurements. This system and the AWG-9 employ interrupted CW (ICW) to reduce the effect of spillover. To complete a brief tour of this important area, see Ref. 71, pages 464–466 and 503–506, Nathanson,[73] pages 363 and 367–371, and the last paragraphs of Ref. 6. There is some difference between the results on page 369 of Ref. 73 and those on page 466 of Ref. 71, which undoubtedly arises from a difference in the two authors' viewpoints on eclipsing.

14.21 TAILORING THE RANGE RESPONSE OF FM-CW SYSTEMS

In the previous material there have been many examples of the range responses of various FM-CW systems: the range responses discussed in Sec. 14.17 and the J_1 response of the doppler navigators, for example. It might be well to summarize the various techniques that are available to produce a range response either for a wide band of ranges or for a specific narrow range cell. There are two obvious paths to a desired result. The first is to work on the modulation of the transmitted waveform. For example, a J_1 waveform without any modification yields an ideal response for a doppler navigator since the range response is approximately equal to R^2 in power, and that is just what is needed to compensate for the $1/R^2$ loss encountered as the helicopter rises in altitude. A little more complicated is the use of a bit of third harmonic with its peak in phase with the dominant first harmonic, as mentioned in Sec. 14.17, to modify the near response of a J_3 fuze. Next in complexity come true double modulations, such as one modulation, perhaps a sinusoidal one, and an additional lower frequency of random noise or pseudonoise. This has the property of reducing the periodic nature of the sine modulation and hence reducing the second and farther-out range peaks. There are double modulations where one chooses two relatively closely spaced frequencies of sinusoidal, sawtooth, or triangular modulation to shape the range response around a principal peak. In particular, the range sidelobes could be reduced, in much the same way that one reduces the sidelobes of an antenna.[25] There are also combinations of triangle plus pseudonoise, as found in some of the systems in Sec. 14.20, that use the pseudonoise for a fine range definition and the triangle for a coarse range discrimination. Finally there are the systems mentioned in Sec. 14.17 that do not even use any of the modulation waveforms discussed but shape a special one based on the theory between waveform slopes and difference frequencies.

The second attack is to take one of the standard modulating waveforms and work on the receiver to shape the desired range response. Some of the earliest work in this direction was to build two amplifiers, one capturing one harmonic of a modulating frequency and another a different harmonic, and then to take sums, differences, or ratios of the two outputs.[6] Since it was soon recognized that the taking of sums or differences (but not ratios or products) was essentially a linear process, a single amplifier followed by a phase detector switched with the sum of two harmonic waves could be substituted for the two amplifiers. This leads to the technique discussed at the end of Sec. 14.17. There is some parallelism between modulation waveform shaping and receiver response shaping. Many range responses can be attained by either technique and it is only a matter of simplicity which method is chosen. In theory one can combine both techniques, but usually this has not been done since one or the other technique usually has ample flexibility to produce the required range response.

In all this, one must ever be aware that certain waveforms and certain receivers are much more susceptible to leakage than, for example, a J_3 system. One pays a distinct price for increasing the short-range response in any FM-CW system. Only with considerable decoupling between separate transmitting and receiving antennas or with effective feedthrough canceling is a high level of near-in range response possible.

REFERENCES

1. Hansen, W. W.: in Ridenour, L. N. (ed.): "Radar System Engineering," MIT Radiation Laboratory Series, vol. 1, McGraw-Hill Book Company, New York, 1947, chap. 5.

2. Varian, R. H., W. W. Hansen, and J. R. Woodyard: Object Detecting and Locating System, U.S. Patent 2,435,615, Feb. 10, 1948.

3. Skolnik, M. I.: "Introduction to Radar Systems," 2d ed., McGraw-Hill Book Company, New York, 1980.

4. Arguimbau, L. B.: "Vacuum-Tube Circuits and Transistors," John Wiley & Sons, New York, 1956, pp. 504–513.

5. Raven, R. S.: Requirements on Master Oscillators for Coherent Radar, *Proc. IEEE,* vol. 54, pp. 237–243, February 1966. (Entire issue is devoted to frequency stability.)

6. Saunders, W. K.: Post War Developments in Continuous-Wave and Frequency-Modulated Radar, *IRE Trans.,* vol. ANE-8, pp. 7–19, March 1961; corrections on p. 105, September 1961.

7. Shreve, J. S.: Clutter Spreading Due to Oscillator Instability in Low Elevation Stationary Radars, *Harry Diamond Laboratories, Internal Rept.* R510-66-2, February 1966.

8. O'Hara, F. J., and G. M. Moore: A High Performance CW Receiver Using Feed-Thru Nulling, *Microwave J.,* vol. 6, pp. 63–71, September 1963.

9. Grasso, G.: Improvement Factor of a Nonlinear MTI in Point Clutter, *IEEE Trans.,* vol. AES-4, pp. 640–644, July 1968.

10. Banks, D. S.: Continuous Wave (CW) Radar, *Raytheon Electron. Prog.,* vol. XVII, no. 2, pp. 34–41, Summer 1975.

11. Grauling, C. H., Jr., and D. J. Healy III: Instrumentation for Measurement of the Short-Term Frequency Stability of Microwave Sources, *Proc. IEEE,* vol. 54, pp. 249–257, February 1966.

12. Marsh, S. B., and A. S. Wiltshire: The Use of a Microwave Discriminator in the Measurement of Noise Modulation on CW Transmitters, *Proc. IEE (London),* vol. 109B, pp. 665–667, May 1962.

13. Harris, I. R., et al.: Frequency and Amplitude Stabilization of Transmitters for CW Radar, *Proc. IEE (London),* vol. 111, pp. 1236–1240, July 1964.

14. Phase Noise Characterization of Microwave Oscillators, *Hewlett Packard Company, Product Note* 117298-1, March 1984.

15. Leeson, D. B., and W. K. Saunders: Test Equipment for Coherent Radar, *IEEE EASCON '74 Conf. Rec.,* pp. 221–228, June 1974.

16. Saunders, W. K.: Short Term Frequency Stability in Coherent Radar, *Twenty-sixth Annu. Frequency Control Symp.,* Atlantic City, N.J., pp. 21–28, June 1972.

17. Stable Transmitter Study, *Harry Diamond Laboratories, Rept.* 2146, February 1966; Phase II of Stable Transmitter Study, *Rept.* 6246, June 1966; Sann, K. H.: The Measurement of Near-Carrier Noise in Microwave Amplifiers, *IEEE Trans.,* vol. MTT-16, pp. 761–766, September 1968. (Issue is devoted to noise in microwave components).

18. Driscoll, M. M., and B. W. Kramer: Spectral Degradation in VHF Crystal Controlled Oscillators Due to Short Term Instability in the Quartz Resonator, *Proc. IEEE Ultrasonics Symp.,* pp. 340–345, October 1985; Driscoll, M. M., et al.: Low Noise Crystal Oscillators Using 50 Ohm Modular Sustaining Stages, *Proc. 40th Annu. Frequency Control Symp.,* pp. 329–335, May 1986; also, Driscoll, M. M.: Low Noise Microwave Signal Generation Using Bulk and Surface Acoustic Wave Resonators, *42d Annu. Frequency Control Symp.,* Baltimore, June 1988.

19. Leeson, D. B.: A Simple Model of Feedback Oscillator Noise Spectrum, *Proc. IEEE,* vol. 54, pp. 329–330, February 1966.

20. Kalmus, H. P.: Direction-Sensitive Doppler Device, *Proc. IRE*, vol. 45, pp. 689–700, June 1955; also U.S. Patent 2,934,756, April 1960.

21. Steinberg, B. D.: in Berkowitz, R. S. (ed.): "Modern Radar Analysis, Evaluation and System Design," John Wiley & Sons, New York, 1965, pt. VI, chap. 4.

22. Harmer, J. D., and W. S. O'Hara: Some Advances in CW Radar Techniques, *Proc. Nat. Conf. Mil. Electron.*, pp. 311–323, Washington, June 1961.

23. Hinman, W. S., Jr., and C. Brunetti: Radio Proximity-Fuze Development, *Proc. IRE*, vol. 34, pp. 976–986, December 1946.

24. Bonner, E. J.: The Radio Proximity Fuze, *Elec. Eng.*, vol. 66, pp. 888–893, September 1947.

25. Tozzi, L. M.: Resolution in Frequency Modulated Radars, Ph.D. thesis, University of Maryland, College Park, 1972.

26. Ismail, M. A. W.: A Study of the Double Modulated FM Radar, dissertation, Swiss Federal Institute of Technology, Zurich, 1955.

27. Peperone, S. J.: Sidelobe Suppression in a FM Altimeter, master's thesis, University of Maryland, College Park, May 1966; Withers, M. J.: Matched Filter for Frequency-Modulated Continuous Wave Radar Systems, *Proc. IEE (London)*, vol. 113, pp. 405–412, March 1966; Hymans, A. J., and J. Lait: Analysis of a Frequency-Modulated Continuous Wave Ranging System, *Proc. IEE (London)*, vol. 107B, pp. 365–372, July 1960.

28. Horton, B. M.: Noise-Modulated Distance Measuring Systems, *Proc. IRE*, vol. 47, pp. 823–828, May 1959.

29. Birdsall, T. G., and M. P. Ristenbatt: Introduction to Linear Shift Register Generated Sequences, *University of Michigan Research Institute, Tech. Rept.* 90, 1958.

30. Craig, S. E., W. Fishbein, and O. E. Rittenback: Continuous Wave Radar with High Range Resolution and Unambiguous Velocity Determination, *IRE Trans.*, vol. MIL-6, pp. 153–161 (especially pp. 153–157), April 1962.

31. Fowle, E. N.: The Design of Radar Signals, *Mitre Corporation, ESD-TR*-65-97, *AD* 617711, Bedford, Mass., June 1965.

32. Plantier, B.: Doppler Processing for Phase Coded Continuous Wave Radar, *IEE (London) RADAR '87 Conf. Publ.*, pp. 340–344, 1987.

33. Lewis, B. L., F. F. Kretschmer, and W. W. Shelton: "Aspects of Radar Signal Processing," Artech House, Norwood, Mass., 1986.

34. Chandler, J. P.: An Introduction to Pseudo-Noise Modulation, *Harry Diamond Laboratories, TM*-64-4 (and references therein), January 1964.

35. Pepper, W. H.: Detection of Modulated Signals, *Diamond Ordnance Fuze Laboratories, Internal Rept. M44.03-55-27*, November 1955.

36. Glegg, K. C. M.: A Low Noise CW Doppler Technique, *Proc. Nat. Conf. Aeronaut. Electron.*, Dayton, Ohio, pp. 133–144, 1958.

37. Kalmus, H. P., H. Goldberg, and M. Sanders: Low Noise Fuze, U.S. Patent 3,829,859, Aug. 13, 1974.

38. Private communication; see also P. W. Boesch and K. E. Hardinger: Low Altitude FM Altimeter, *Diamond Ordnance Fuze Laboratories, Tech. Rept.* 470, *AD* 225570, August 1959.

39. Levine, S.: The FLAT System, *Harry Diamond Laboratories, TR* 667, June 1959.

40. Goldberg, H., and M. Sanders: Fuze, U.S. Patent 3,872,792, Mar. 25, 1975.

41. Woodward, P. M.: "Probability and Information Theory, with Applications to Radar," McGraw-Hill Book Company, New York, 1953; reprinted by Artech House, Norwood, Mass.

42. Bartlett, M. C., and R. C. Johnson: A Digital IF Simulator for FM Ranging Systems, *University of Florida, Eng. Ind. Exp. Sta.*, Gainesville, March 1980.

43. Luck, D. G. C.: "Frequency Modulated Radar," McGraw-Hill Book Company, New York, 1949.

44. Sharpe, B. A.: Aircraft Radio Altimeters, *J. Inst. Navig.*, vol. 3, pp. 79–89 (especially p. 88), January 1950.

45. "Service Manual for ALA-52A Altimeter; Design Summary for the ALA-52A," Bendix Corporation, Fort Lauderdale, Fla., May 1982.

46. Griffiths, H. D., and W. D. Bradford: Digital Generation of Wideband FM Waveforms for Radar Altimeters, *IEE (London) Conf. RADAR '87 Publ.*, pp. 325–329, 1987.

47. Tozzi, L. M.: Linear Frequency Modulation of Oscillators, *Harry Diamond Laboratories, TR*-1586, Adelphi, Md., March 1972.

48. "ALT-55 Radio Altimeter System; Instruction Book," Collins (Rockwell International), Cedar Rapids, Iowa, October 1984.

49. Tomasi, J. P.: The Servoed Modulation FM-CW Radar Altimeters in Military Applications, *Proc. Mil. Microwaves Conf.*, pp. 421–425, London, 1978.

50. Brown, R. K., et al.: A Lightweight and Self-Contained Airborne Navigational System, *Proc. IRE*, vol. 47, pp. 778–807, May 1959.

51. Clegg, J. E., and J. W. Crompton: Low-Power CW Doppler Navigation Equipment, *IEE (London) Conv. Radio Aids Aeronaut. Mar. Navig.*, pp. 258–265, March 1958; Tollefson, R. D.: Application of FM Techniques to Doppler Radar Sensors, *Proc. Nat. Conf. Aeronaut. Electron.*, Dayton, Ohio, pp. 683–687, May 1959.; Fried, W. R.: An FM-CW Radar for Simultaneous Three Dimensional Velocity and Altitude Measurement, *IEEE Trans.*, vol. ANE-11, pp. 45–57, March 1964.

52. SKD-2110 Doppler Velocity Sensor/Navigator, *Kearfoot*, Little Falls, N.J., December 1983.

53. SKD-2126 Doppler Velocity Sensor (DVS) for Helicopters and Fixed-Wing Aircraft, *Kearfoot*, Little Falls, N.J., January, 1985.

54. Private communication and various brochures from Canadian Marconi Company, Avionics Division, Montreal.

55. Kayton, M., and W. R. Fried (eds.): "Avionics Navigation Systems," John Wiley & Sons, New York, 1969, chap. 6; Fried, W. R.: New Developments in Radar and Radio Sensors for Aircraft Navigation, *IEEE Trans.*, vol. AES-10, pp. 25–33, January 1974.

56. Frank, U. A., D. L. Kratzer, and J. L. Sullivan: The Two-Pound Radar, *RCA Eng.*, vol. 13, pp. 52–54, August–September 1967.

57. Bossert, L. H., and A. T. Roome: Lightweight Battlefield Surveillance Device, *Naval Ocean Systems Center, TR* 226, San Diego, May 1978; Strait, R. D.: Lightweight Battlefield Surveillance Device Technology Tradeoff Study, *Naval Ocean Systems Center, TR* 764, May 1982; Sampite, T. J.: Lightweight Battlefield Surveillance Radar (LBSR), FY 85 Year-End Report, *Naval Ocean Systems Center, TR* 1088, June 1986.

58. Albanese, D. F., and A. M. Klein: Pseudo-Random Code Waveform Design for CW Radar, *IEEE Trans.*, vol. AES-15, pp. 67–75, January 1979.

59. Fahey, M. D., and G. T. O'Reilly: A P-N Coded Radar, *Conf. Rec. IEEE-NTC*, vol. 1, pp. 18.4.1–18.4.3, 1978.

60. Sherif, A. S.: High Resolution CW Radar, *IEEE NAECON Conf. Rec.*, pp. 476–481, Dayton, Ohio, 1983; Novel Approach for Solution of CW Radar Problems Using Double Modulation, *IEEE EASCON '83 Conf. Rec.*, pp. 185–190, 1983; Experimental Results for Performance Evaluation of CW Radar with Double Modulation, *IEEE EASCON '83 Conf. Rec.*, pp. 191–196, 1983.

61. Randonat, U., and E. Sautter: Multitarget FM-CW Radar for Unambiguous Determi-

nation of Range and Doppler, *Nachrichtentech. Z.*, vol. 30, pp. 255–260, March 1977. (Summary in English, text in German.)

62. Nilssen, O. K.: A New Method of Range Measuring Doppler Radar, *IEEE Proc. Nat. Conf. Aeronaut. Electron.*, pp. 402–408, Dayton, Ohio, 1961.

63. Neininger, G.: An FM/CW Radar with High Resolution in Range and Doppler; Application for Anti-Collision Radar for Vehicles, *IEE (London) Conf. RADAR '77 Publ.* 155, pp. 526–530, 1977.

64. Seehausen, G.: 24 GHz FM-CW Radar for Detection of Information for Traffic Purposes, *IEEE MTT-S Int. Microwave Symp. Dig.*, pp. 251–253, 1984.

65. Chadwick, R. B., and R. G. Strauch: Processing of FM-CW Radar Signals from Distributed Targets, *IEEE Trans.*, vol. AES-15, pp. 185–189, January 1979.

66. Pasqualucci, F.: The Use of Pseudo Noise Phase Modulation in Meteorological Doppler Radars, *IEEE Aerosp. Appl. Conf. Dig.*, pp. 173–186, Vail, Colo., 1984.

67. Lighthart, L. P., and L. R. Nieuwkerk: Systems Aspects of a Solid State FM-CW Weather Surveillance Radar, *IEE (London) Conf. RADAR '87 Publ.*, pp. 112–115, 1987.

68. Shearman, E. D. R., et al.: An FMICW Ground-Wave Radar for Remote Sensing of Ocean Waves and Currents, *IEE (London) Conf. RADAR '87 Publ.*, pp. 598–605, 1987.

69. Whetton, C. P.: Industrial and Scientific Applications of Doppler Radar, *Microwave J.*, vol. 18, pp. 39–42, November 1975.

70. Cuthbert, L. G., et al.: Signal Processing in an FM-CW Radar for Detecting Voids and Hidden Objects in Building Materials, *Third Eur. Signal Process. Conf.*, vol. 2, pp. 1169–1172, The Hague, 1986.

71. Stimson, G. W.: "Introduction to Airborne Radar," Hughes Aircraft Company, El Segundo, Calif., 1983. (See especially Chaps. 18–20.)

72. Barrett, M., et al.: An X-Band FM-CW Navigation Radar, *IEE (London) Radar '87 Conf. Publ.*, pp. 448–452, 1987; also Saunders, W. K.: Velocity-Aided Range Acquisition and Tracking in Dual-Mode CW/FM-CW Radar, *IEE Proc.* (London), vol. 136, Part F, No. 4, August, 1989.

73. Nathanson, F. E.: "Radar Design Principles," McGraw-Hill Book Company, New York, 1969. (See also Fig. 17.19 and the accompanying text of the *Radar Handbook*.)

CHAPTER 15
MTI RADAR

William W. Shrader
V. Gregers-Hansen
Equipment Division
Raytheon Company

15.1 INTRODUCTION TO MTI RADAR

The purpose of moving-target indication (MTI) radar is to reject signals from fixed or slow-moving unwanted targets, such as buildings, hills, trees, sea, and rain, and retain for detection or display signals from moving targets such as aircraft. Figure 15.1 shows a pair of photographs of a PPI (plan position indicator) which illustrate the effectiveness of a properly working MTI system. The distance from the center to the edge of the PPI is 40 nmi. The range marks are at 10-nmi intervals. The picture on the left is the normal video display, showing the fixed-target returns. The picture on the right shows the MTI clutter rejection. The camera shutter was left open for three scans of the antenna; thus aircraft show up as a succession of three returns.

MTI radar utilizes the doppler shift imparted on the reflected signal by a moving target to distinguish moving targets from fixed targets. In a pulse radar system this doppler shift appears as a change of phase of received signals between consecutive radar pulses. Consider a radar which transmits a pulse of RF energy that is reflected by both a building (fixed target) and an airplane (moving target) approaching the radar. The reflected pulses return to the radar a certain time later. The radar then transmits a second pulse. The reflection from the building occurs in exactly the same amount of time, but the reflection from the moving aircraft occurs in less time because the aircraft has moved closer to the radar in the interval between transmitted pulses. The precise time that it takes the reflected signal to reach the radar is not of fundamental importance. What is significant is whether the time changes between pulses. The time change, which is of the order of a few nanoseconds for an aircraft target, is determined by comparing the phase of the received signal with the phase of a reference oscillator in the radar. If the target moves between pulses, the phase of the received pulses changes.

Figure 15.2 is a simplified block diagram of one form of a coherent MTI system. The RF oscillator feeds the pulsed amplifier, which transmits the pulses. The RF oscillator is also used as a phase reference for determining the phase of reflected signals. The phase information is stored in a PRI (pulse repetition interval) memory for the period between transmitted pulses, and it is also subtracted

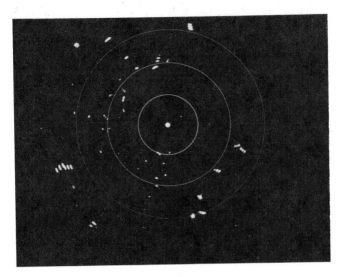

(b)

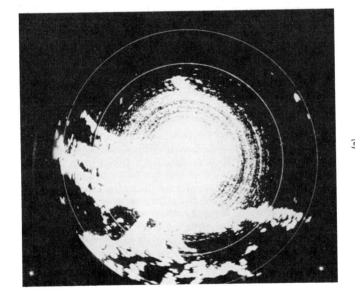

(a)

FIG. 15.1 (a) Normal video. (b) MTI video. These PPI photographs show how effective an MTI system can be. Aircraft appear as three consecutive blips in the right-hand picture because the camera shutter was open for three revolutions of the antenna. The PPI range is 40 nmi.

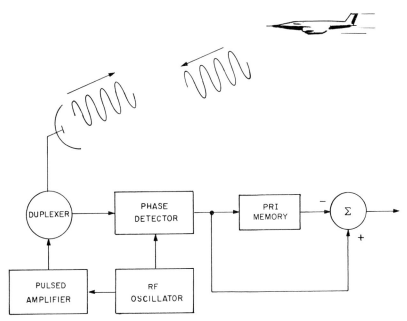

FIG. 15.2   Simplified block diagram of a coherent MTI system.

from the phase information from the previous transmitted pulse. There is an output from the subtractor only when a reflection has occurred from a moving target.

Moving-Target Indicator (MTI) Block Diagram. A block diagram of a complete MTI system is shown in Fig. 15.3. This block diagram represents an MTI system that uses a pulsed oscillator. It is not as sophisticated as MTI systems to be described later, but most of the practical considerations applying to any MTI system can be understood by examining this block diagram. The frequencies and 2500-μs interpulse period are typical for a 200-mi L-band radar.

The transmitter shown employs a magnetron. Because a magnetron is a pulsed oscillator that has no phase coherence between consecutive pulses, a phase reference must be established for each transmitted pulse. This is done by taking a sample of the transmitted pulse at a directional coupler, mixing this pulse with the stalo (stabilized local oscillator) and then using this pulse to phase-lock the coho (coherent oscillator). The coho then becomes the reference oscillator for the received signals. (The stability requirements for the coho and stalo will be described in Sec. 15.11.) The lock-pulse amplifier is gated off just before the end of the transmitted pulse because a magnetron emits a certain amount of noise during the fall of the high-voltage pulse applied to it, and this noise can prevent perfect locking of the coho.

The received signals are mixed with the stalo and amplified in a linear-limiting amplifier. (In some implementations the limiting is not deliberately provided. However, receiver saturation occurs at some signal level, and thus limiting inadvertently exists.)

The received signals are then compared in phase with the coho in a phase de-

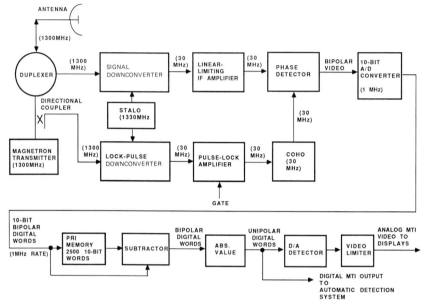

FIG. 15.3 MTI system block diagram.

tector. The output of the phase detector is a function of the relative phase of the signal and the coho, and it is also a function of the amplitude of the signal. At the output of the phase detector, the signal phase and amplitude information has been converted into bipolar video. The bipolar video received from a single transmitted pulse may appear as sketched in Fig. 15.4. If the point target is moving and if there is also a moving target in the region of strong clutter return, the superimposed bipolar video from several transmitted pulses may appear as in Fig. 15.5.

The remainder of the block diagram in Fig. 15.3 shows what is necessary for detecting the moving targets so that they may be displayed on a PPI or sent to an automatic target extractor. The bipolar video is converted to digital words in an analog-to-digital (A/D) converter. The A/D output is stored in a PRI memory and also subtracted from the memorized A/D output from the previous transmitted pulse.

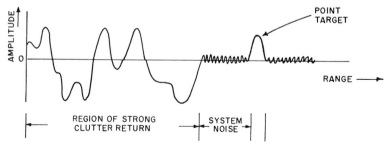

FIG. 15.4 Bipolar video: single sweep.

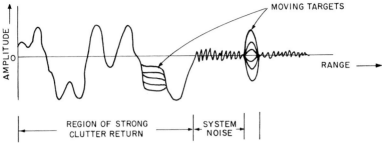

FIG. 15.5 Bipolar video: several sweeps.

The output of the subtractor is a digital bipolar signal that contains moving targets, system noise, and a small amount of clutter residue if clutter cancellation is not perfect. The absolute value of the signal is then converted to analog video in a digital-to-analog (D/A) converter for display on a PPI. The digital signal may also be sent to automatic target detection circuitry. The dynamic range (peak signal to rms noise) is limited to about 20 dB for a PPI display.

Moving-Target Detector (MTD) Block Diagram. In the moving-target detector (MTD) the basic MTI principle, as described above, is enhanced by increasing the linear dynamic range of the signal processor, using a number of parallel doppler filters followed by constant-false-alarm-rate (CFAR) processing, and adding one or more high-resolution clutter maps to suppress point clutter residues. With these additions a complete signal-processing system is obtained for suppressing clutter returns in a modern surveillance radar. A typical implementation of such an MTD processing system is shown in Fig. 15.6.

The MTD radar transmits a group of N pulses at a constant pulse repetition frequency (PRF) and at a fixed radar frequency. This set of pulses is usually referred to as the coherent processing interval (CPI) or pulse batch. Sometimes one or two additional fill pulses are added to the CPI in order to suppress range-ambiguous clutter returns, as might occur during periods of anomalous propagation. The returns received during one CPI are processed in the bank of N-pulse finite-impulse-response (FIR) filters. Then the radar may change PRF and/or RF

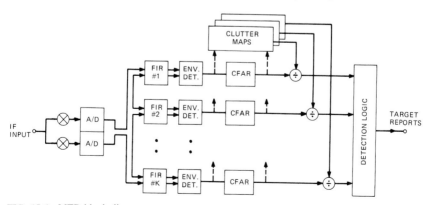

FIG. 15.6 MTD block diagram.

frequency and transmit another CPI of N pulses. Since most search radars are ambiguous in doppler, the use of different PRFs on successive coherent dwells will cause the target response to fall at different frequencies of the filter passband on the successive opportunities during the time on target, thus eliminating blind speeds.

Each doppler filter is designed to respond to targets in nonoverlapping portions of the doppler frequency band and to suppress sources of clutter at all other doppler frequencies. This approach maximizes the coherent signal integration in each doppler filter and provides clutter attenuation over a larger range of doppler frequencies than achievable with a single MTI filter. Thus one or more clutter filters may suppress multiple clutter sources located at different doppler frequencies. An example of the use of an MTD doppler filter bank against simultaneous land and weather clutter (Wx) is illustrated in Fig. 15.7. It can be seen that filters 3 and 4 will provide significant suppression of both clutter sources.

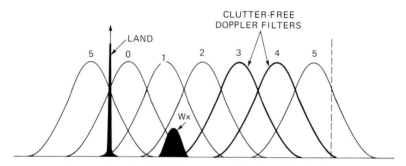

FIG. 15.7   Suppression of multiple clutter sources by using a doppler filter bank.

The output of each doppler filter is envelope-detected and processed through a cell-averaging CFAR processor to suppress residues due to range-extended clutter which may not have been fully suppressed by the filter.

As will be discussed later in this chapter, the conventional MTI detection system relies on a carefully controlled dynamic range in the IF section of the radar receiver in order to ensure that clutter residues at the MTI output are suppressed to the level of the receiver noise or below. This limited dynamic range, however, has the undesirable effect of causing additional clutter spectral broadening, and achievable clutter suppression is consequently reduced.

In the MTD one or more high-resolution clutter maps are used to suppress the clutter residues, after doppler filtering, to the receiver noise level (or, alternatively, to increase the detection threshold above the level of the residues). This in turn eliminates the need to restrict the IF dynamic range, which can then be set to the maximum value supported by the A/D converters. Thus, a system concept is obtained that provides a clutter suppression capability that is limited only by the radar system stability, the dynamic range of the receiver-processor, and the spectrum width of the returns from clutter. The concept of a high-resolution digital clutter map to suppress clutter residues is related to earlier efforts to construct analog area MTI systems using, for example, storage tubes.

In subsequent sections specific aspects of the design of an MTD system will be discussed. Thus Sec. 15.8 will discuss the design and performance of doppler filter banks, and a detailed discussion of clutter maps will follow in Sec. 15.14.

15.2 CLUTTER FILTER RESPONSE TO MOVING TARGETS

The response of an MTI system to a moving target varies as a function of the target's radial velocity. For the MTI system described above, the response, normalized for unity noise power gain, is shown in Fig. 15.8. Note that there is zero response to stationary targets and also to targets at ±89, ±178, ±267,...knots. These speeds, known as blind speeds, are where the targets move 0, ½, 1, 1½,...wavelengths between consecutive transmitted pulses. This results in the received signal being shifted precisely 360° or multiples thereof between pulses, which results in no change in the phase-detector output. The blind speeds can be calculated:

$$V_B = k\frac{\lambda f_r}{2} \quad k = \pm 0, 1, 2, \ldots \quad (15.1)$$

where V_B is the blind speed, in meters per second; λ is the transmitted wavelength, in meters; and f_r is the PRF, in hertz. A convenient set of units for this equation is

$$V_B \, (\text{kn}) = k\,\frac{0.29 f_r}{f_{\text{GHz}}} \quad k = \pm 0, 1, 2, \ldots \quad (15.2)$$

where f_r is the PRF, in hertz; and f_{GHz} is the transmitted frequency, in gigahertz. Note from the velocity response curve that the response to targets at velocities midway between the blind speeds is greater than the response for a normal receiver.

The abscissa of the velocity response curve can also be labeled in terms of

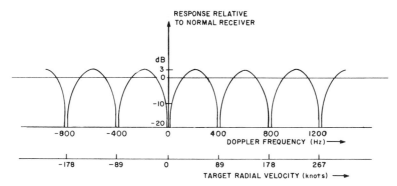

FIG. 15.8 MTI system response for 1300-MHz radar operating at 400 pps.

doppler frequency. The doppler frequency of the target can be calculated from

$$f_d = \frac{2V_R}{\lambda}$$ (15.3)

where f_d is the doppler frequency, in hertz; V_R is the target radial velocity, in meters per second; and λ is the transmitted wavelength, in meters. It can be seen from Fig. 15.8 that the doppler frequencies for which the system is blind occur at multiples of the pulse repetition frequency.

15.3 CLUTTER CHARACTERISTICS

Spectral Characteristics.   The spectrum of a pulsed transmitter transmitting a simple rectangular pulse of length τ is shown in Fig. 15.9. The spectral width of the (sin U)/U envelope is determined by the transmitted pulse width, the first nulls occurring at a frequency of $f_0 \pm 1/\tau$. The individual spectral lines are separated by a frequency equal to the PRF. These spectral lines fall at precisely the same frequencies as the blind speeds in Fig. 15.8. Thus a canceler will, in theory, fully reject signals with an ideal spectrum, as shown here. In practice, however, the spectral lines in clutter signals are broadened by motion of the clutter (such as windblown trees) and by motion of the antenna in a scanning radar. Barlow[1] stated that the returns from clutter have a gaussian spectrum, which may be characterized by its standard deviation σ_v. This spectral spread prevents perfect cancellation of clutter in the MTI system.

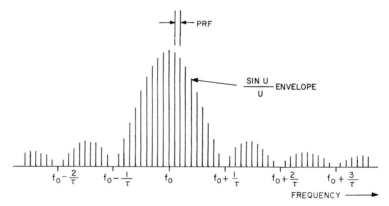

FIG. 15.9   Pulse transmitter spectrum.

Table 15.1 gives the standard deviation σ_v of the clutter spectrum in meters per second. More sophisticated and detailed clutter spectrum models exist,[6] but the gaussian model is usually an adequate model for understanding system limitations and obtaining good performance predictions.

Nathanson and Reilly[7] have shown that the clutter spectral width of rain is primarily due to a turbulence component and a wind-shear component (change in

TABLE 15.1   Summary of Standard Deviations of the Clutter Spectrum*

Source of clutter	Wind speed, kn	σ_v, m/s	Reference
Sparse woods	Calm	0.017	Barlow[1]
Wooded hills	10	0.04	Goldstein,[3] pp. 583–585
Wooded hills	20	0.22	Barlow[1]
Wooded hills	25	0.12	Goldstein,[3] pp. 583–585
Wooded hills	40	0.32	Goldstein,[3] p. 583
Sea echo		0.7	Wiltse et al.,[4] p. 226
Sea echo		0.75–1.0	Goldstein,[3] pp. 580–581
Sea echo	8–20	0.46–1.1	Hicks et al.,[5] p. 831
Sea echo	Windy	0.89	Barlow[1]
Chaff		0.37–0.91	Goldstein,[3] p. 472
Chaff	25	1.2	Goldstein,[3] p. 472
Chaff		1.1	Barlow[1]
Rain clouds		1.8–4.0	Goldstein,[3] p. 576
Rain clouds		2.0	Barlow[1]

*From Barton.[2]

wind velocity with altitude). Their measurements indicate that good average values are $\sigma_v = 1.0$ m/s for turbulence and $\sigma_v = 1.68$ m/(s/km) for wind shear. A convenient equation is $\sigma_v = 0.04R\theta_{el}$ m/s for the effects of wind shear, provided the rain fills the vertical beam, where R is the range to the weather, in nautical miles; and v_{el} is the one-way half-power vertical beamwidth, in degrees. Thus, for example, σ_v of rain viewed at 25 nmi with a vertical beamwidth of 4° would have a $\sigma_v = 4.1$ m/s, of which the shear component is dominant. Rain and chaff also have an average velocity, in addition to the spectral spread noted above, which must be taken into account when designing an MTI system.

The clutter spectral width in meters per second is independent of the radar frequency. The standard deviation of the clutter power spectrum σ_c, in hertz, is

$$\sigma_c = \frac{2\sigma_v}{\lambda} \quad \text{Hz} \tag{15.4}$$

where λ is the transmitted wavelength, in meters; and σ_v is the clutter standard deviation, in meters per second.

Antenna scanning also causes a spread of the clutter power spectrum because of amplitude modulation of the echo signals by the two-way antenna pattern.[2] The resulting clutter standard deviation is

$$\sigma_c = \frac{\sqrt{\ln 2}}{\pi} \times \frac{f_r}{n} = 0.265 \frac{f_r}{n} \quad \text{Hz} \tag{15.5}$$

where f_r is the PRF and n is the number of hits between the one-way 3 dB points of the antenna pattern. This equation was derived for a gaussian beam shape but is essentially independent of the actual beam shape or aperture illumination function used.

The clutter spectral spread due to scanning, normalized to the PRF, is

$$\sigma_c T = \frac{0.265}{n} \tag{15.6}$$

where $T = 1/\text{PRF}$ is the interpulse period.

Amplitude Characteristics. To predict the performance of an MTI system, the amplitude of the clutter signals with which a target must compete should be known. The amplitude of the clutter signals is dependent on the size of the resolution cell of the radar, the frequency of the radar, and the reflectivity of the clutter. The expected radar cross section of clutter can be expressed as the product of a reflectivity factor and the size of the volume or area of the resolution cell.

For surface clutter, as viewed by a surface-based radar,

$$\bar{\sigma} = A_c \sigma^0 = R\theta_{az}\frac{c\tau}{2}\sigma^0 \tag{15.7}$$

where $\bar{\sigma}$ is the average radar cross section, in square meters; A_c is the area of clutter illuminated, in square meters; R is the range to clutter patch, in meters; θ_{az} is the one-way half-power azimuthal beamwidth, in radians; c is the speed of propagation, 300 million m/s; τ is the half-power radar pulse length (after the matched filter), in seconds; and σ^0 is the average clutter reflectivity factor, in square meters per square meter.

For clutter that is airborne, such as chaff or rain,

$$\bar{\sigma} = V_c \eta = R\theta_{az}H\frac{c\tau}{2}\eta \tag{15.8}$$

where V_c is the volume of clutter illuminated, in cubic meters; H is the height of clutter, in meters (if clutter fills the vertical beam, then $H = R\theta_{el}$, where θ_{el} is the elevation beamwidth); and η is the clutter reflectivity factor, in square meters per cubic meter.

It should be noted that, for land clutter, σ^0 can vary considerably from one resolution cell to the next. A typical distribution of σ^0, taken from Barton,[8] is shown in Fig. 15.10. Typical values for σ^0 and η taken from the same reference are given in Table 15.2. Because of the imprecision in predicting σ^0 and η, these equations do not include an antenna beam-shape factor. For the measurement of the reflectivity of rain, references on radar meteorology present more precise equations.[9]

In addition to distributed clutter targets, there are many targets that appear as *points*, such as radio towers, water tanks, and buildings. These point targets typically have a radar cross section of 10^3 to 10^4 m².

Figure 15.11a shows a PPI display of all clutter observed with a surveillance radar with a 1.3° by 2-μs resolution cell in the mountainous region of Lakehead, Ontario, Canada. (The PPI range is set for 30 nmi.) Clutter that exceeds the minimum-discernible signal (MDS) level of the radar by 60 dB is shown in Fig. 15.11b. Note that the clutter in Fig. 15.11b is very spotty in character, including the strong fixed-point targets and returns from extended targets. It is significant that the extended targets are no longer very extended. The face of a mountain at 10 mi from 5 to 7 o'clock is only a line. If the MTI system were incapable of displaying an aircraft while it was over the mountain face, it would display the aircraft on the next scan of the antenna because the aircraft would have moved either farther or nearer. The PPI does not have a resolution that approaches the resolution of the signal-processing

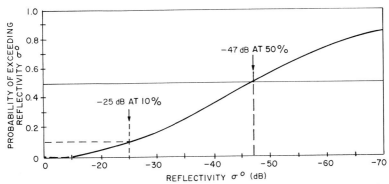

FIG. 15.10 Distribution of reflectivity for ground clutter typical of heavy clutter at S band.

circuits of this radar. Thus the apparent extended clutter has many weak areas not visible in these photographs, where targets could be detected by virtue of an MTI radar's interclutter visibility (defined in Sec. 15.4).

15.4 DEFINITIONS

Improvement Factor (I). The MTI improvement factor I is defined as "the signal-to-clutter ratio at the output of the clutter filter divided by the signal-to-clutter ratio at the input of the clutter filter, averaged uniformly over all target radial velocities of interest."[10] This definition accounts for both the clutter attenuation and the average noise gain of the MTI system. It is therefore a measure of the MTI system response to clutter relative to the average MTI system response to targets. An equivalent definition of improvement factor is $I = r_i/r_o$, where r_i is the input ratio of clutter to noise and r_o is the output ratio of clutter residue to noise. The use of I is encouraged instead of older terms, such as *cancellation ratio* and *clutter attenuation*, because these terms have not been consistently used in the literature and are not always normalized to the average canceler noise gain.

Signal-to-Clutter Ratio Improvement (I_{SCR}). For a system employing multiple doppler filters, such as the MTD, each filter will have a different improvement factor against the same clutter source. In this case it is preferable to define the performance against clutter in terms of the signal-to-clutter improvement (I_{SCR}) versus target doppler shift. This quantity is not included in the *IEEE Dictionary*,[10] but common usage defines the I_{SCR}, at each target doppler frequency, as the ratio of the signal-to-clutter ratio obtained at the output of the doppler filter bank (including all filters) to the signal-to-clutter ratio at the input of the filter bank. It should be noted that the signal-to-clutter improvement of any one filter is equal to the product of the MTI improvement factor of the filter as defined earlier and the coherent gain of the filter at the particular doppler frequency. The coherent gain of a doppler filter is equal to the increase in signal-to-thermal-noise ratio between the input and the output of the filter due to the coherent summation of individual target returns.

TABLE 15.2 Typical Values of Clutter Reflectivity*

Clutter	Reflectivity, λ, m η, $(m)^{-1}$	Conditions	Band λ, m	Clutter parameters for typical conditions			
				L 0.23	S 0.1	C 0.056	X 0.032
Land (excluding point clutter)	$\sigma^0 = \dfrac{0.00032}{\lambda}$ (worst 10 percent)	. . . .	σ^0 dB =	−29	−25	−22	−20
Point clutter	$\sigma = 10^4$ m²	. . . .	σ m² =	10^4	10^4	10^4	10^4
Sea (Beaufort scale K_B, angle E)	σ^0 dB $= -64 + 6K_B +$ $(\sin E)$dB $- \lambda$ dB	Sea state 4 (6-ft waves, rough); $E = 1°$	σ^0 dB =	−51.5	−47.5	−44.5	−42.5
Chaff (for fixed weight per unit volume)	$\eta = 3 \times 10^{-8}\lambda$	. . . .	η (m)$^{-1}$ =	7×10^{-9}	3×10^{-9}	1.7×10^{-9}	10^{-9}
Rain (for rate r, mm/h)	$\eta = 6 \times 10^{-14} r^{1.6} \lambda^{-4}$ (matched polarization)	$r = 4$ mm/h	η (m)$^{-1}$ =	2×10^{-10}	5×10^{-9}	5×10^{-8}	5×10^{-7}

*From Barton.[8]

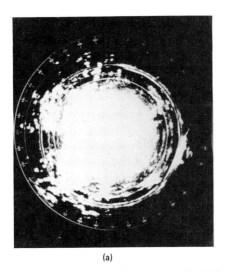

(a)   (b)

FIG. 15.11   PPI display, 30-nmi range, of (*a*) all clutter at a mountainous site and (*b*) clutter that exceeds the system noise level by 60 dB.

Subclutter Visibility (SCV).   The subclutter visibility (SCV) of a radar system is a measure of its ability to detect moving-target signals superimposed on clutter signals. A radar with 20 dB SCV can detect an aircraft flying over clutter whose signal return is 100 times stronger. The *IEEE Dictionary*[10] defines the subclutter visibility as "the ratio by which the target echo power may be weaker than the coincident clutter echo power and still be detected with specified detection and false alarm probabilities. Target and clutter powers are measured on a single pulse return and all target radial velocities are assumed equally likely." The SCV of two radars cannot necessarily be used to compare their performance while operating in the same environment, because the target-to-clutter ratio seen by each radar is proportional to the size of the radar resolution cell and may also be a function of frequency. Thus a radar with a 10-μs pulse length and a 10° beamwidth would need 20 dB more subclutter visibility than a radar with a 1-μs pulse and a 1° beamwidth for equal performance in a specified clutter environment.

The subclutter visibility of a radar, when expressed in decibels, is less than the improvement factor by the clutter visibility factor V_{oc} (see definition below).

Interclutter Visibility (ICV).   The interclutter visibility (ICV) of a radar is a measure of its capability to detect targets between points of strong clutter by virtue of the ability of the radar to resolve the areas of strong and weak clutter. A radar with high resolution makes available regions between points of strong clutter where the target-to-clutter ratio will be sufficient for target detection even though the SCV of the radar (based on average clutter) may be relatively low. A low-resolution radar averages the clutter over large resolution cells, most of which will contain one or more strong point targets, and thus the radar will have very little ICV. Because of the ICV capability of high-resolution radars, they tend to perform better in a clutter environment than would be predicted by using the average clutter amplitude characteristics of Sec. 15.3.[11] To achieve ICV, a mechanism must be furnished to provide CFAR operation against the residue from strong clutter. This CFAR is provided in the typical MTI system by IF limiting or, in the MTD implementation, through the use of high-resolution clutter maps.

Filter Mismatch Loss.   The maximum signal-to-noise ratio available from an N-pulse filter is N times the signal-to-noise ratio of a single pulse, assuming all pulses have equal amplitude. When weighting is applied to reject clutter and control the filter sidelobes, the peak output signal-to-noise ratio is reduced. The filter mismatch loss is the amount by which the peak-output signal-to-noise ratio is reduced by the use of the weighting. A three-pulse MTI filter using binomial weights has a filter mismatch loss of 0.51 dB. The mismatch loss for the binomial-weighted four-pulse canceler is 0.97 dB.

Clutter Visibility Factor (V_{oc}).   This factor is "the predetection signal-to-clutter ratio that provides stated probabilities of detection and false alarm on a display; in moving-target-indicator systems, it is the ratio after cancelation or doppler filtering."[10] The clutter visibility factor is the ratio by which the target signal must exceed the clutter residue so that target detection can occur without having the clutter residue result in false-target detections. The system must provide a threshold that the targets will cross and the clutter residue will not cross.

15.5 *IMPROVEMENT FACTOR CALCULATIONS*

Using Barton's approach (Ref. 2, pp. 210–219), the maximum improvement factor I against zero-mean clutter with a gaussian-shaped spectrum, for different implementations of the finite-impulse-response binomial-weight MTI canceler (see Sec. 15.7), is

$$I_1 \approx 2\left(\frac{f_r}{2\pi\sigma_c}\right)^2 \tag{15.9}$$

$$I_2 \approx 2\left(\frac{f_r}{2\pi\sigma_c}\right)^4 \tag{15.10}$$

$$I_3 \approx \frac{4}{3}\left(\frac{f_r}{2\pi\sigma_c}\right)^6 \tag{15.11}$$

where I_1 is the MTI improvement factor for the single-delay coherent canceler; I_2 is the MTI improvement factor for the dual-delay coherent canceler; I_3 is the MTI improvement factor for the triple-delay coherent canceler; σ_c is the rms frequency spread of the gaussian clutter power spectrum, in hertz; and f_r is the radar repetition frequency, in hertz. When the values of σ_c for scanning modulation [Eq. (15.5)] are substituted in the above equations for I, the limitation on I due to scanning is

$$I_1 \approx \frac{n^2}{1.39} \tag{15.12}$$

$$I_2 \approx \frac{n^4}{3.84} \tag{15.13}$$

$$I_3 \approx \frac{n^6}{16.0} \tag{15.14}$$

These relationships are shown graphically in Fig. 15.12. This derivation assumes a linear system. That is, it is assumed that the voltage envelope of the echo signals, as the antenna scans past a point target, is identical to the two-way antenna voltage pattern. This assumption of a linear system may be unrealistic for some practical MTI systems with relatively few hits per beamwidth, as discussed in Sec. 15.10.

The scanning limitation does not apply to a system that can step-scan, such as a phased array. Note, however, that sufficient pulses must be transmitted to initialize the filter before useful outputs may be obtained. For example, with a three-pulse binomial-weight canceler, the first two transmitted pulses initialize the canceler, and useful output is not available until after the third pulse has been transmitted. Feedback or infinite impulse response (IIR) filters would not be used with a step-scan system because of the long transient settling time of the filters.

The limitation on I due to internal-clutter fluctuations can be determined by substituting the appropriate value of σ_c into Eqs. (15.9) to (15.11). By letting $\sigma_c = 2\sigma_v/\lambda$, where σ_v is the rms velocity spread of the clutter, the limitation on I can be plotted for different types of clutter as a function of the wavelength λ and the pulse repetition frequency f_r. This is done for one-, two-, and three-delay binomial-weight cancelers in Figs. 15.13 to 15.15. The values of V_B given are the first blind speed of the radar (or where the first blind speed V_B would be for a staggered PRF system if staggering were not used). The improvement factor

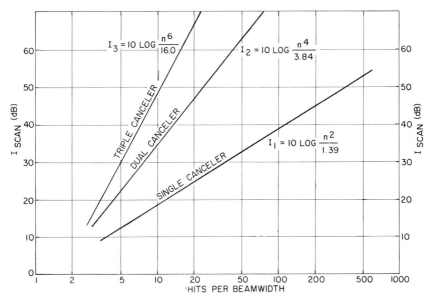

FIG. 15.12 Theoretical MTI improvement factor due to scan modulation; gaussian antenna pattern; n = number of pulses within the one-way half-power beamwidth.

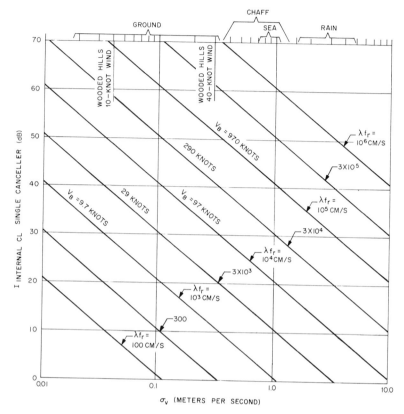

FIG. 15.13 MTI improvement factor as a function of the rms velocity spread of clutter for a two-pulse binomial-weight canceler.

shown in these figures for rain and chaff is based on the assumption that the average velocity of the rain and chaff has been compensated for so that the returns are centered in the canceler rejection notch. Unless such compensation is provided, the MTI offer little or no improvement for rain and chaff.

Two further limitations on I are the effect of pulse-to-pulse repetition-period staggering combined with clutter spectral spread from scanning and internal-clutter motion. These limitations, plotted in Figs. 15.16 and 15.17, apply to all cancelers, whether single or multiple. (The derivation of these limitations and a means of avoiding them by the use of time-varying weights are given in Sec. 15.9.)

15.6 OPTIMUM DESIGN OF CLUTTER FILTERS

The statistical theory of detection of signals in gaussian noise provides the required basis for the optimum design of radar clutter filters. Such theoretical results are important to the designer of a practical MTI or MTD system, in that

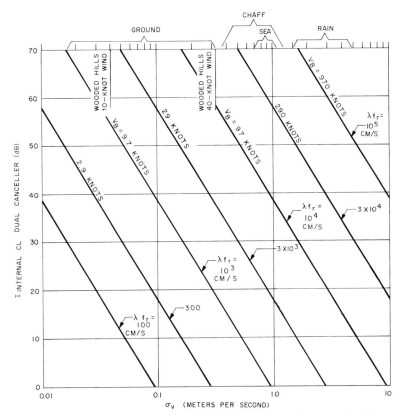

FIG. 15.14 MTI improvement factor as a function of the rms velocity spread of clutter for a three-pulse binomial-weight canceler.

they establish upper bounds on the achievable performance in a precisely specified clutter environment. It should be noted, however, that owing to the extreme variability of the characteristics of real clutter returns (power level, doppler shift, spectrum shape, spectral width, etc.) any attempt to actually approximate the performance of such optimum filters for the detection of targets in clutter requires the use of adaptive methods. The adaptive methods must estimate the unknown clutter statistics and subsequently implement the corresponding optimum filter. The design of such adaptive MTI systems is discussed in Sec. 15.13.

For a single radar pulse with a duration of a few microseconds, the doppler shift due to aircraft target motion is a small fraction of the signal bandwidth, and conventional MTI and pulse doppler processing are not applicable. It is well known that the classical single-pulse "matched" filter provides optimum radar detection performance when used in a white-noise background. Against clutter returns which have the same spectrum as the transmitted radar pulse, the matched filter is no longer optimum, but the potential improvement in the output signal-to-clutter ratio by designing a modified optimized filter is usually insignificant.

When the duration of the transmitted radar signal, whether CW or a repetitive

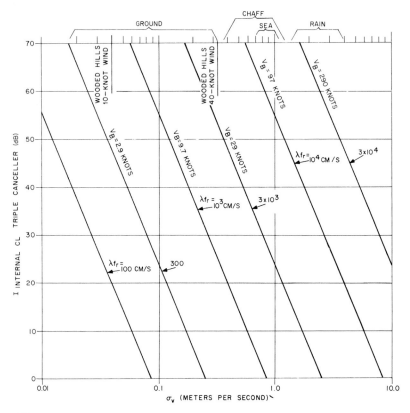

FIG. 15.15 MTI improvement factor as a function of the rms velocity spread of clutter for a four-pulse binomial-weight canceler.

train of N identical pulses, is comparable with or larger than the reciprocal of anticipated target doppler shifts, the difference between a conventional white-noise matched filter (or coherent integrator) and a filter optimized to reject the accompanying clutter becomes significant. The characteristics of the clutter are characterized by the covariance matrix Φ_C of the N clutter returns. If the power spectrum of the clutter is denoted $S_C(f)$ and the corresponding autocorrelation function is $R_C(t_i - t_j)$, then the elements of Φ_C are given by

$$\Phi_{ij} = R_C(t_i - t_j) \tag{15.15}$$

where t_i is the transmission time of the ith pulse. For example, for a gaussian-shaped clutter spectrum we have

$$S_C(f) = P_C \frac{1}{\sqrt{2\pi}\,\sigma_f} \exp\left[-\frac{(f - f_d)^2}{2\sigma_f^2} \right] \tag{15.16}$$

where P_C is the total clutter power, σ_f is the standard deviation of the clutter

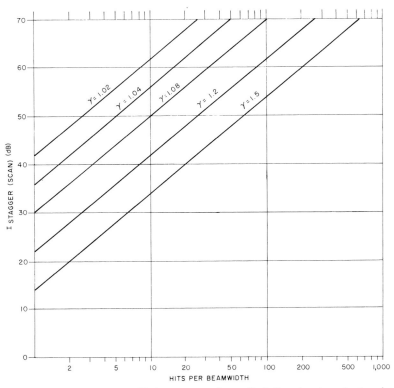

FIG. 15.16 Approximate MTI improvement factor limitation due to pulse-to-pulse repetition-period staggering and scanning (all canceler figurations). $I(\text{dB}) = 20 \log [2.5n/(\gamma - 1)]$; γ = maximum period/minimum period.

spectral width, and f_d is the average doppler shift of the clutter. The corresponding autocorrelation function is

$$R_C(\tau) = P_C \exp{(-4\pi\sigma_f^2\tau^2)} \exp{(-j2\pi f_d\tau)} \qquad (15.17)$$

For two pulses separated in time by the interpulse period T the complex correlation coefficient between two clutter returns is

$$\rho_T = \exp{(-4\pi\sigma_f^2 T^2)} \exp{(-j2\pi f_d T)} \qquad (15.18)$$

The second factor in this expression represents the phase shift caused by the doppler shift of the clutter returns.

For a known target doppler shift the received target return can be represented by an N-dimensional vector:

$$s = P_S \mathbf{f} \qquad (15.19)$$

where the elements of the vector $\mathbf{f}$ are $f_i = \exp{[j2\pi f_s t_i]}$. On the basis of this de-

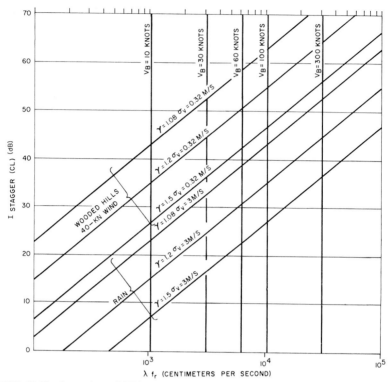

FIG. 15.17 Approximate MTI improvement factor limitation due to pulse-to-pulse staggering and internal-clutter motion (all canceler configurations). $I(dB) = 20 \log [0.33/(\gamma - 1)(\lambda f_r/\sigma_v)]$; γ = maximum period/minimum period.

scription of signal and clutter it has been shown[12] that the optimum doppler filter will have weights given by

$$w_{opt} = \Phi_C^{-1} s \tag{15.20}$$

and the corresponding signal-to-clutter improvement is

$$I_{SCR} = \frac{w_{opt}^T s \cdot s^{T*} w_{opt}^*}{w_{opt}^T \Phi_C w_{opt}^*} \tag{15.21}$$

where the asterisk denotes complex conjugation and superscript T is the transposition operator. An example where the optimum performance is determined for the case of clutter at zero doppler having a wide gaussian-shaped spectrum and a normalized width of $\sigma_f T = 0.1$ is shown in Fig. 15.18. In this case a coherent processing interval of CPI = nine pulses was assumed, and the limitation due to thermal noise was ignored by setting the clutter level at 100 dB above noise.

It should be kept in mind that Eq. (15.21) for the optimum weights will yield a different result for each different target doppler shift, so that a large number of

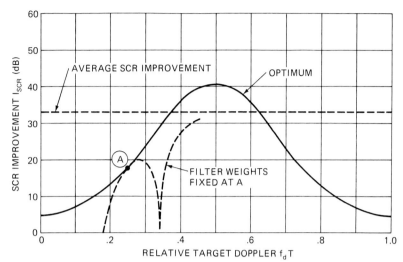

FIG. 15.18 Optimum signal-to-clutter ratio improvement (I_{SCR}) for gaussian-shaped clutter spectrum and a CPI of nine pulses; clutter-to-noise ratio, 100 dB.

parallel filters would be needed to approximate the optimum performance even when the clutter characteristics are known exactly. As an example, the response of the optimum filter designed for one particular target doppler frequency labeled as point A in Fig. 15.18 is shown in a broken line. At approximately ± 5 percent from the design doppler the performance starts to fall significantly below the optimum.

Also shown in Fig. 15.18 is a horizontal line labeled "average SCR improvement." This indicates the level corresponding to the average of the optimum SCR curve across one doppler interval and may be considered as a figure of merit for a multiple-filter doppler processor somewhat analogous to the MTI improvement factor defined for a single doppler filter. In Fig. 15.19 the optimum average I_{SCR} has been computed for several different values of the CPI as a function of the normalized spectrum width. These results may be used as a point of reference for practical doppler processor designs as discussed in Sec. 15.8. Note that for $\sigma_f T \approx 1$ the average SCR improvement is due only to the coherent integration of all the pulses in the CPI.

The implementation of a single MTI filter will result in a performance below that shown in Fig. 15.19. Further, it can be shown that the average SCR improvement calculated for a single filter is equal to the MTI improvement factor as defined in Sec. 15.4. The basis for obtaining the optimum MTI filter is again the covariance matrix of the clutter returns as given by Eq. (15.15). As shown by Capon,[13] the weights of the optimum MTI are found as the eigenvector corresponding to the smallest eigenvalue of the clutter covariance matrix and the MTI improvement factor is equal to the inverse of the smallest eigenvalue.

In Fig. 15.20 the improvement factor of an MTI using the optimum weights is compared with the binomial coefficient MTI for different values of the relative clutter spectral spread and shown as a function of the number of pulses in the CPI. These results again assume a gaussian-shaped clutter spectrum. For typical

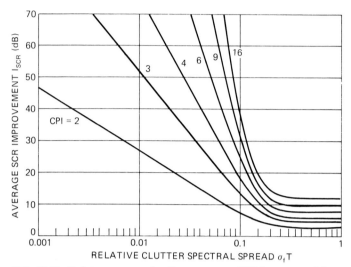

FIG. 15.19 Reference curve of optimum average SCR improvement for a gaussian-shaped clutter spectrum.

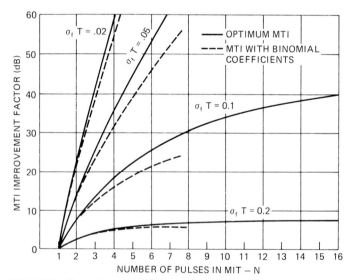

FIG. 15.20 Comparison of MTI improvement factor of binomial-weight MTI and optimum MTI against a gaussian-shaped clutter spectrum.

numbers of pulses in the MTI (three to five) the binomial coefficients are remarkably robust and provide a performance which is within a few decibels of the optimum. Again, it should be noted that any attempt to implement an MTI canceler, which performs close to the optimum, would require the use of adaptive techniques which estimate the clutter characteristics in real time. If the estimate is in error, the actual performance may fall below that of the binomial-weight MTI canceler.

15.7 MTI CLUTTER FILTER DESIGN

The MTI block diagram shown in Fig. 15.3 and discussed in detail in Sec. 15.2 uses a single-delay canceler. It is possible to utilize more than one delay and to introduce feedback and/or feedforward paths around the delays to change the MTI system response to targets of different velocities. Multiple-delay cancelers have wider clutter rejection notches than single-delay cancelers. The wider rejection notch encompasses more of the clutter spectrum and thus increases the MTI improvement factor attainable with a given clutter spectral distribution.

When a number of single-delay feedforward cancelers are cascaded in series, the overall filter voltage response is $k2^n \sin^n (\pi f_d T)$, where k is the target amplitude, n is the number of delays, f_d is the doppler frequency, and T is the interpulse period.[15] The cascaded single-delay cancelers can be rearranged as a transversal filter, and the weights for each pulse are the binomial coefficients with alternating sign: 1, -1 for two pulses; 1, -2, 1 for three pulses; 1, -3, 3, -1 for four pulses, etc. Changes of the binomial feedforward coefficients and/or the addition of feedback modify the filter characteristics. Within this chapter, reference to *binomial-weight cancelers* refers to cancelers with the $2^n \sin^n (\pi f_d T)$ transfer function.

Figures 15.21 to 15.23 represent typical velocity response curves obtainable from one-, two-, and three-delay cancelers. Shown also are the canceler configurations assumed, with appropriate Z-plane pole-zero diagrams. The Z plane is the comb-filter equivalent of the S plane,[16,17] with the left-hand side of the S plane transformed to the inside of the unit circle centered at $Z = 0$. Zero frequency is at $Z = 1 + j0$. The stability requirement is that the poles of the Z transfer function lie within the unit circle. Zeros may be anywhere.

These velocity response curves are calculated for a scanning radar system with 14.4 hits per beamwidth. An antenna beam shape of $(\sin U)/U$, terminated at the first nulls, was assumed. The shape of these curves, except very near the blind speeds, is essentially independent of the number of hits per beamwidth or the assumed beam shape.

The ordinate, labeled "response," represents the single-pulse signal-to-noise response of the MTI receiver relative to the signal-to-noise response of a normal linear receiver for the same target. Thus all the response curves are normalized with respect to the average gain for the given canceler configuration. The intersection at the ordinate represents the negative decibel value of I, the MTI improvement factor for a point clutter target processed in a linear system.

Because these curves show the signal-to-noise response for each output pulse from the MTI canceler, the inherent loss incurred in MTI processing due to the reduction of the effective number of independent pulses integrated[18] is not apparent. This loss may vary from 1½ to almost 3 dB, depending upon the number of pulses on target. In addition, if quadrature MTI channels (see Sec. 15.12) are not employed, there is an additional loss of 1½ to 3 dB, again depending upon the number of pulses on target.

The abscissa of these curves, V/V_B, represents the ratio of target velocity V to the blind speed $V_B = \lambda f_r/2$, where λ is the radar wavelength and f_r is the average

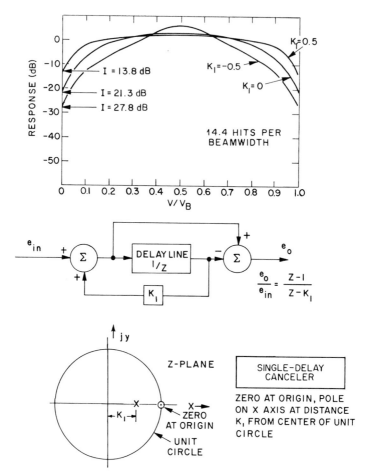

FIG. 15.21 One-delay canceler.

PRF of the radar. The abscissa can also be interpreted as the ratio of the doppler frequency to the average PRF of the radar.

The canceler configurations shown are not the most general feedforward, feedback networks possible but, rather, are practical configurations easy to implement. Many configurations are computationally equivalent. More flexibility in locating zeros and poles is achieved with delays in pairs as shown for the second and third delays of the triple canceler. (In this configuration, the zeros are constrained to the unit circle.)

The triple-canceler configuration is such that two of the zeros can be moved around the boundary of the unit circle in the Z plane. Moving the zeros gives a 4 or 5 dB increase in the MTI improvement factor for specific clutter spectral spreads, as compared with keeping all three zeros at the origin.[19]

It is interesting to note the width of the rejection notches for the different

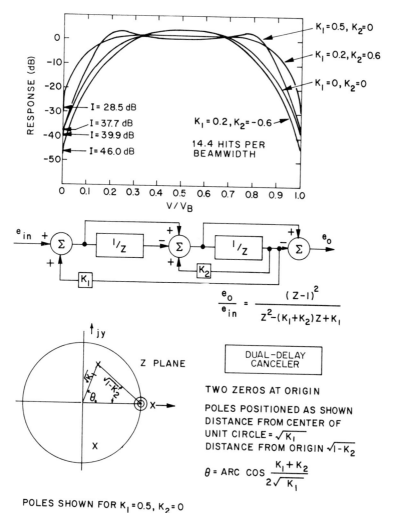

FIG. 15.22 Two-delay canceler.

binomial-weight canceler configurations. If the −6 dB response relative to normal response is used as the measuring point, the rejection is 24 percent of all target dopplers for the single canceler, 36 percent for the dual canceler, and 45 percent for the triple canceler. Consider the dual canceler, for example. Eliminating 36 percent of the dopplers means limiting the system to a long-term average of 64 percent single-scan probability of detection. Feedback can be used to narrow the rejection notch without much degradation of I. If feedback is used to increase the improvement factor, the single-scan probability of detection becomes worse.

Figure 15.24 shows the effect of feedback on I. These curves are calculated for a $(\sin U)/U$ antenna pattern terminated at the first nulls. The no-feedback curves shown here are almost indistinguishable from the theoretical curves derived for a

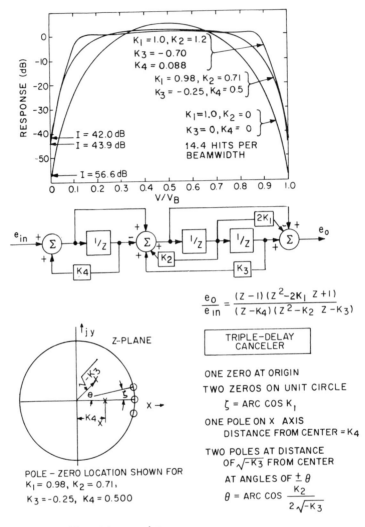

FIG. 15.23 Three-delay canceler.

gaussian pattern shown in Fig. 15.12. (One of the curves showing the effect of feedback on the triple canceler is not straight because two of the three zeros are not at the origin but have been moved along the unit circle the optimum amount for 14 hits per beamwidth. Thus, at 40 hits per beamwidth, these two zeros are too far removed from the origin to be very effective.)

In theory, it is possible to synthesize almost any velocity response curve with digital filters.[16] For each pair of poles and pair of zeros on the Z plane, two delay sections are required. The zeros are controlled by the feedforward paths, and the poles by the feedback paths.

Velocity response shaping can be accomplished by the use of feedforward only,

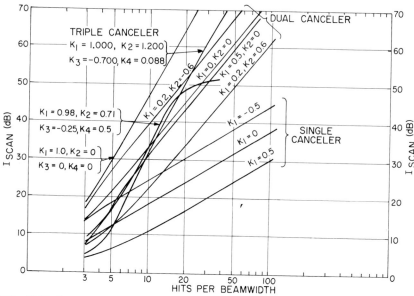

FIG. 15.24 Improvement factor limitation due to scanning for cancelers with feedback, computer-calculated for an assumed antenna pattern of $(\sin U)/U$ terminated at the first nulls.

without the use of feedback. The principal advantage of not using feedback is the excellent transient response of the canceler, an important consideration in a phased array or when pulse interference noise is present. If a phased array radar uses a feedback canceler, many pulses may have to be gated out after the beam has been repositioned before canceler transient ringing has settled to a tolerable level. An initialization technique has been proposed[20] to alleviate this problem, but it provides only partial reduction in the transient settling time. If feedforward only is used, only three or four pulses have to be gated out after moving the beam. The disadvantage of using feedforward for velocity response shaping is that an additional delay must be provided for each zero used to shape the response. Also, an inherent loss in improvement factor capability is caused by using zeros to shape the velocity response. This may or may not be significant, depending on the clutter spectral spread and the number of zeros available for cancellation. Figure 15.25 shows the velocity response and Z-plane diagram of a feedforward-only, shaped-response four-pulse canceler. Also shown are the velocity responses of a five-pulse feedforward canceler and a three-pulse feedback canceler. For the cancelers shown, the improvement factor capability of the three-pulse canceler is about 4 dB better than the shaped-response four-pulse feedforward canceler, independent of clutter spectral spread.

The five-pulse canceler response shown is a linear-phase[21] MTI filter described by Zverev.[22] The four zeros are located on the Z-plane real axis at $+1.$, $+1.$, -0.3575, and -2.7972. Much of the literature on filter synthesis describes linear-phase filters, but for MTI applications linear phase is of no importance. Almost identical filter responses can be obtained with nonlinear-phase filters that require fewer pulses, as shown in Fig. 15.25. Because only a fixed number of pulses is available during the time on target, none should be wasted. Thus one should choose the nonlinear-phase filter that uses fewer pulses.

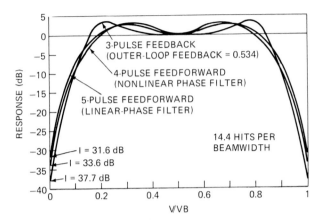

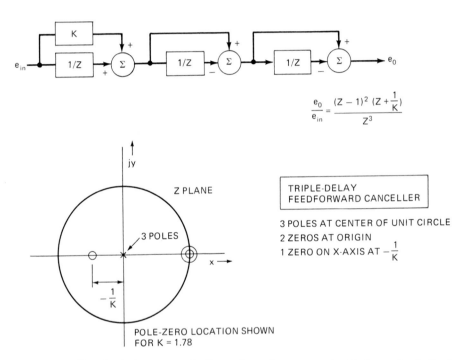

FIG. 15.25 Shaped-velocity-response feedforward cancelers compared with three-pulse feedback canceler. See text for five-pulse canceler parameters.

15.8 CLUTTER FILTER BANK DESIGN

As discussed in Sec. 15.1, the MTD uses a waveform consisting of coherent processing intervals (CPIs) of N pulses at the same PRF and RF frequency. The PRF and possibly the RF are changed from one CPI to the next. With this constraint only finite-impulse-response (FIR) filter designs are realistic candidates for the filter bank design. (Feedback filters require a number of pulses to settle after either the PRF or the RF is changed and thus would not be practical.)

The number of pulses available during the time when a surveillance radar beam illuminates a potential target position is determined by system parameters and requirements such as beamwidth, PRF, volume to be scanned, and the required data update rate. Given the constraint of the number of pulses on target, one must decide how many CPIs should occur during the time on target and how many pulses per CPI. The compromise is usually difficult. One wishes to use more pulses per CPI to enable the use of better filters, but one also wishes to have as many CPIs as possible. Multiple CPIs (at different PRFs and perhaps at different RF frequencies) improve detection and can provide information for true radial velocity determination.[23]

The design of the individual filters in the doppler filter bank is a compromise between the frequency sidelobe requirement and the degradation in the coherent integration gain of the filter. The number of doppler filters required for a given length of the CPI is a compromise between hardware complexity and the straddling loss at the crossover between filters. Finally the requirement of providing a high degree of clutter suppression at zero doppler (land clutter) sometimes introduces special design constraints.

When the number of pulses in a CPI is large (≥ 16), the systematic design procedure and efficient implementation of the fast Fourier transform (FFT) algorithm is particularly attractive. Through the use of appropriate weighting functions of the time-domain returns in a single CPI, the resulting frequency sidelobes can be readily controlled. Further, the number of filters (= the order of the transform) needed to cover the total doppler space (= the radar PRF) can be chosen independently of the CPI, as discussed below.

As the CPI becomes smaller (≤ 10), it will become important to consider special designs of the individual filters to match the specific clutter suppression requirements at different doppler frequencies in order to achieve better overall performance. While some systematic procedures are available for designing FIR filters subject to specific passband and stopband constraints, the straightforward approach for small CPIs is to use an empirical approach in which the zeros of each filter are adjusted until the desired response is obtained. An example of such filter designs is presented below.

Empirical Filter Design. An example of an empirical filter design for a six-pulse CPI follows. (The six pulses per CPI may be driven by system considerations, such as time on target.) Because the filter will use six pulses, only five zeros are available for the filter design: the number of zeros available is the number of pulses minus one. The filter design process consists of placing the zeros to obtain a filter bank response that conforms to the specified constraints. The example that follows was produced with an interactive computer program with which the zeros could be moved until the desired response was obtained. The assumed filter requirements are as follows:

1. Provide a response of −66 dB in the clutter rejection notch (relative to the peak target response) of the moving-target filters.

2. Provide a response of −46 dB for chaff rejection at velocities between ±20 percent of the ambiguous doppler frequency range.

3. Owing to hardware limitations, only five filters will be implemented.

4. Three of the five filters will reject fixed clutter and respond to moving targets. Two filters will respond to targets at zero doppler and its ambiguities. (With good fixed clutter rejection filters, it takes two or more coherent filters to cover the gap in response at zero velocity.)

With the above considerations, a filter bank can be constructed.

Figure 15.26a shows the filter designed to respond to targets in the middle of the doppler passband. The sidelobes near zero velocity are 66 dB down from the peak, thus providing good clutter rejection for clutter within 5 percent of zero doppler. The −46 dB sidelobe provides chaff rejection to ±16 percent. Because of the constraint of only 5 zeros available, this filter could not provide −46 dB rejection to ±20 percent.

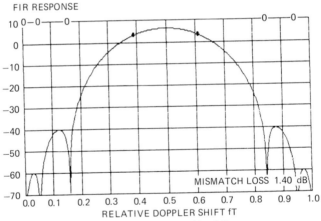

FIG. 15.26a Six-pulse filter for targets at $fT = 0.5$.

Figure 15.26b shows the filter that responds to targets as near as possible to zero doppler, while having zero-doppler response of −66 dB. Two zeros are placed near zero, providing −66 dB response to clutter at zero. The filter sidelobes between 0.8 and 1.0 doppler provide the specified chaff rejection of 48 dB. A mirror image of this filter is used for the third moving doppler filter. (The mirror-image filter has coefficients that are complex conjugates of the original filter coefficients.)

Figure 15.26c shows the first filter designed for response at zero doppler. Considerations here are that the straddling loss of the filter bank be minimized (this dictates the location of the peak), that the response to chaff at 0.8 doppler be down 46 dB, and that the mismatch loss be minimized. Minimizing the mismatch loss is accomplished by permitting the filter sidelobes between 0.3 and 0.8 to rise as high as needed (lower sidelobes in this range increase the mismatch loss). The second zero-doppler filter is the mirror image of this one.

FIR RESPONSE

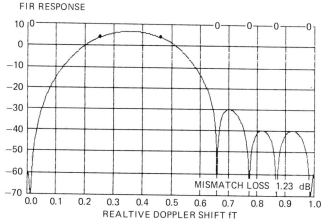

FIG. 15.26b Six-pulse filter for targets at $fT = 0.3$ that rejects fixed clutter.

FIR RESPONSE

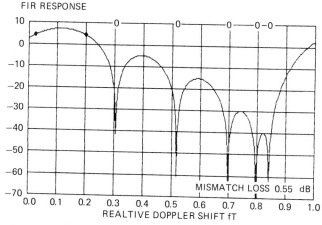

FIG. 15.26c Six-pulse filter that responds to targets at zero doppler but rejects chaff at $fT = 0.8$.

Figure 15.26d shows the composite response of the filter bank. Note that the filter peaks are fairly evenly distributed. The dip between the first zero-doppler filter and the first moving doppler filter is larger than the others, primarily because, under the constraints, it is impossible to move the first doppler filter nearer to zero velocity.

Chebyshev Filter Bank. For larger number of pulses in the CPI a more systematic approach to filter design is desirable. If a doppler filter design criterion is chosen that requires the filter sidelobes outside the main response to be below a specified level (i.e., providing a constant level of clutter suppression), while simultaneously minimizing the width of the filter response, a filter design based on the Dolph-Chebyshev distribution provides the optimum solution. Properties and design procedures based on the Dolph-Chebyshev distribution can be found in

FIR RESPONSE

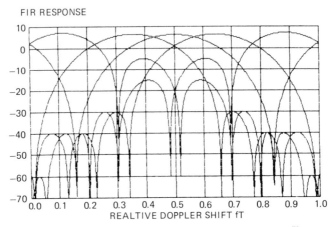

FIG. 15.26d Composite response of the bank for five six-pulse filters.

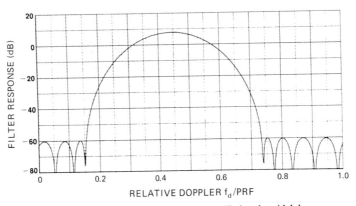

FIG. 15.27 Chebyshev FIR filter design with 68 dB doppler sidelobes.

the antenna literature. An example of a Chebyshev filter design for a CPI of nine pulses and a sidelobe requirement of 68 dB is shown in Fig. 15.27. The peak filter response can be located arbitrarily in frequency by adding a linear-phase term to the filter coefficients.

The total number of filters implemented to cover all doppler frequencies is a design option trading straddling loss at the filter crossover frequencies against implementation complexity. An example of a complete doppler filter bank implemented with nine uniformly spaced filters is shown in Fig. 15.28. The performance of this doppler filter bank against the clutter model considered in Fig. 15.18 is shown in Fig. 15.29. This graph shows the signal-to-clutter ratio improvement against clutter at zero doppler as a function of target doppler frequency. Only the response of the filter providing the largest improvement is plotted at each target doppler. For comparison the optimum curve from Fig. 15.18 is shown by a broken line and thus provides a direct assessment of how well the

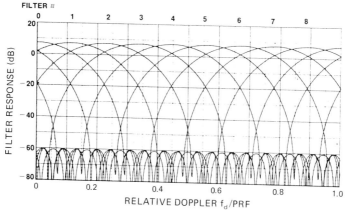

FIG. 15.28   Doppler filter bank of 68 dB Chebyshev filters. CPI = nine pulses.

Chebyshev filter design performs against a given clutter model. Also shown is the average SCR improvement for both the optimum and the Chebyshev filter bank.

Finally, Fig. 15.30 shows the average SCR improvement of the 68 dB Chebyshev doppler filter bank as well as the optimum curve (from Fig. 15.19) as a function of the relative spectrum spread of the clutter. Owing to the finite number of filters implemented in the filter bank, the average SCR improvement will change by a small amount if a doppler shift is introduced into the clutter returns. This effect is illustrated by the cross-hatched region, which shows upper and lower limits on the average SCR improvement for all possible clutter doppler shifts. For a smaller number of filters in the doppler filter bank this variation would be larger.

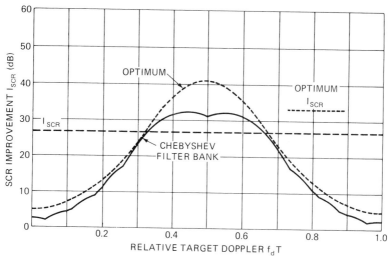

FIG. 15.29   SCR improvement of 68 dB Chebyshev doppler filter bank compared with the optimum.

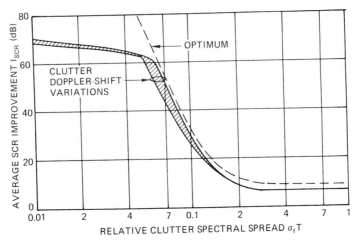

FIG. 15.30 Average SCR improvement for the 68 dB Chebyshev filter bank shown in Fig. 15.28. CPI = nine pulses. Optimum is from Fig. 15.19.

Fast Fourier Transform Filter Bank. For a large number of parallel doppler filters, hardware implementation can be significantly simplified through the use of the FFT algorithm. The use of this algorithm constrains all filters in the filter bank to have identical responses, and the filters will be uniformly spaced along the doppler axis. The number of filters implemented for a given size of the CPI can, however, be varied. For example, a larger number of filters can be realized by extending the received data with extra zero values (also known as zero padding) after the received returns have been appropriately weighted in accordance with the desired filter response (e.g., Chebyshev).

15.9 STAGGERED PRF

Stagger Design Procedures. The interval between radar pulses may be changed to shift the target velocities to which the MTI system is blind. The interval may be changed on a pulse-to-pulse, dwell-to-dwell (each dwell being a fraction of the beamwidth), or scan-to-scan basis. Each approach has advantages. The advantages of the scan-to-scan method are that the radar system is easier to build, and multiple-time-around clutter is canceled in a power amplifier MTI system. The transmitter stabilization necessary for good operation of an unstaggered MTI system costs money and weight. To stabilize the transmitter sufficiently for pulse-to-pulse or dwell-to-dwell stagger operation is considerably more difficult. Pulse-to-pulse staggering is used with MTI processing, while dwell-to-dwell staggering is used with filter bank processing.

For many MTI applications pulse-to-pulse staggering is essential. For example, if a binomial-weighted three-pulse canceler which has 36 percent–wide rejection notches is employed and if scan-to-scan pulse staggering is used, 36 percent of the desired targets would be missing on each scan owing to doppler considerations alone. This might be intolerable for some applications. With pulse-to-pulse staggering, good response can be obtained on all dopplers of interest on each scan. In addition, better velocity response can be obtained at some dopplers than either pulse interval will give on a scan-to-scan basis. This is so because pulse-to-pulse staggering produces doppler components in the passband of the MTI filter. Pulse-to-pulse staggering may degrade the improvement factor attainable, as shown in Figs. 15.16 and 15.17, but this degradation may not be significant, or it can be eliminated by the use of time-varying weights as described below. One further advantage of pulse-to-pulse staggering is that it may permit eliminating the use of feedback in the cancelers (used to narrow the blind-speed notches), which simplifies canceler design.

The optimum choice of the stagger ratio depends on the velocity range over which there must be no blind speeds and on the permissible depth of the first null in the velocity response curve. For many applications, a four-period stagger ratio is best, and a good set of stagger ratios can be obtained by adding the first blind speed (in V/V_B) to the numbers $-3, 2, -1, 3$ (or $3, -2, 1, -3$). Thus, in Fig. 15.31, where the first blind speed occurs at about $V/V_B = 28$, the stagger ratio is 25:30:27:31. (Alternating the long and short periods keeps the transmitter duty cycle as nearly constant as possible, as well as ensuring good response at the first null where $V = V_B$.) If using four interpulse periods permits the first null to be too deep, then five interpulse periods may be used, with the stagger ratio obtained by adding the first blind speed to the numbers $-6, +5, -4, +4, +1$. Figure 15.32 shows a velocity response curve for five pulse intervals. The depth of the first null can be predicted from Fig. 15.39, which is discussed later.

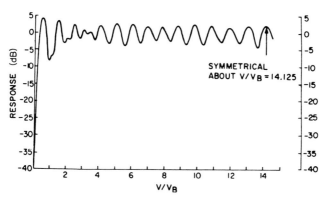

FIG. 15.31 Velocity response curve: dual canceler, no feedback, 25:30:27:31 pulse-interval ratio.

Figures 15.33 and 15.34 show two other velocity response curves calculated for four-pulse intervals. For a radar system with relatively few hits per beamwidth, it is not advantageous to use more than four or five different intervals because then the response to an individual target will depend on which part of the pulse sequence occurs as the peak of the beam passes the target. Random variation of the pulse intervals is not desirable (unless used as an electronic counter-countermeasure feature) because it permits the nulls to be deeper than the optimum choice of four- or five-pulse intervals.

When the ratio of pulse intervals is expressed as a set of relatively prime integers (i.e., a set of integers with no common divisor other than 1), the first true blind speed occurs at

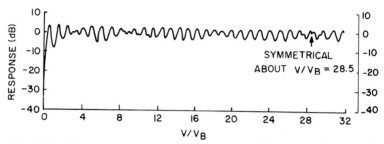

FIG. 15.32   Velocity response curve: three-pulse binomial canceler, 51:62:53:61:58 pulse-interval ratio.

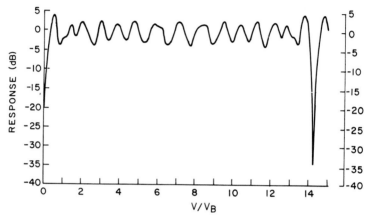

FIG. 15.33   Velocity response curve: three-pulse binomial canceler, 11:16:13:17 pulse-interval ratio.

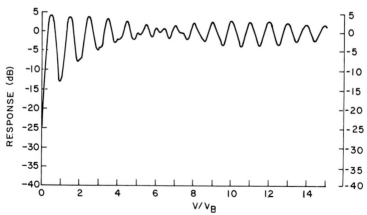

FIG. 15.34   Velocity response curve: three-pulse binomial canceler, 53:58:55:59 pulse-interval ratio. This response curve continues to $V/V_B = 53$ with no dips below 5 dB. The first blind speed is at $V/V_B = 56.25$.

$$\frac{V}{V_B} = \frac{R_1 + R_2 + R_3 + \cdots + R_N}{N} \tag{15.22}$$

where $(R_1, R_2, R_3, \ldots, R_N)$ are the set of integers and V_B is the blind speed corresponding to the average interpulse period. The velocity response curve is symmetrical about one-half of the value from Eq. (15.22). Thus, for the pulse-interval ratio 25:30:27:31, the first true blind speed occurs at $V/V_B = 28.25$, and the response curve is symmetrical about $V/V_B = 14.125$.

Feedback and Pulse-to-Pulse Staggering. When pulse-to-pulse staggering is employed, the effect of feedback is reduced. Staggering causes a modulation of the signal doppler at or near the maximum response frequency of the canceler. The amount of this modulation is proportional to the absolute target doppler so that, for an aircraft flying at V_B, the canceler response is essentially independent of the feedback employed. Figure 15.35 shows a plot of the effects of feedback on a dual-canceler system with 14.4 hits per beamwidth and a ratio of stagger intervals of 6:7:8. The feedback values employed are several of those used for the unstaggered velocity response plot in Fig. 15.22. If scan-to-scan pulse-interval staggering had been used instead of pulse-to-pulse, the no-feedback rms response for three scans at a target velocity of V_B would be -12.5 dB. The composite response for pulse-to-pulse staggering, however, is only -6 dB at V_B, thus illustrating the advantage of pulse-to-pulse staggering.

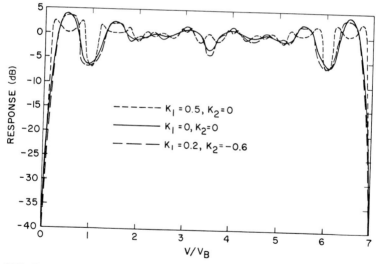

FIG. 15.35 Effect of feedback on the velocity response curve: dual canceler, 6:7:8 pulse-interval ratio.

Figure 15.36 shows the difference in response with a two-delay no-feedback canceler and with a three-delay canceler with a Chebyshev response (the same feedback used for the response in Fig. 15.23). These curves lie within about 3 dB of each other, except for velocities equal to $0.05V_B$ to $0.25V_B$. Since the radar signal from aircraft targets fluctuates considerably from scan to scan, for most

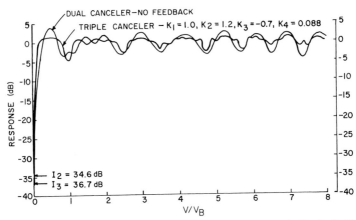

FIG. 15.36 Velocity response curves: triple canceler with no feedback; 12:16:13:18 pulse-interval ratio.

surveillance radar applications there is no practical advantage in employing the more complex canceler.

Figure 15.37 shows the difference in response between a five-pulse feedforward canceler and a three-pulse feedback canceler (the same such cancelers depicted in Fig. 15.25) for stagger ratios of 17:19:21:23. Although the feedback canceler response ripples more than the five-pulse feedforward canceler, no significant system advantage would accrue for the five-pulse canceler unless it is used in a batch mode, or step-scan mode, where the finite transient-response time

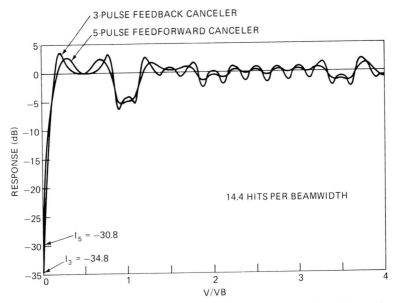

FIG. 15.37 Velocity response curves: five-pulse feedforward canceler and three-pulse feedback canceler of Fig. 15.25 for a 17:19:21:23 pulse-interval ratio.

becomes important (as discussed above, in Fig. 15.25 the four-pulse filter would be preferable to the five-pulse filter).

In some staggered systems, the use of feedback may serve a useful purpose: to increase the improvement factor of the system. Whether the feedback will help significantly depends on whether the stagger ratio or the scanning modulation limits the system performance more severely. If the limitation is primarily from scanning modulation, the feedback may be of some help.

Improvement Factor Limitations Caused by Staggering. When pulse-to-pulse staggering is used, it limits the attainable improvement factor owing to the unequal time spacing of the received clutter samples. The curves in Figs. 15.16 and 15.17, which have been referred to several times, give the approximate limitation on I caused by pulse-to-pulse staggering and either antenna scanning or internal clutter motion. They have been derived as explained below.

A two-delay canceler will perfectly cancel a linear waveform, $V(t) = c + at$, if it is sampled at equal time intervals independent of the constant c or the slope a. [Additional delay cancelers perfectly cancel additional waveform derivatives; e.g., a three-delay canceler will perfectly cancel $V(t) = c + at + bt^2$.] A stagger system with two pulse intervals samples the linear waveform at unequal intervals, and therefore there will be a voltage residue from the cancelers that is proportional to the slope a and inversely proportional to $\gamma - 1$, where γ is the ratio of the intervals. The apparent doppler frequency of the residue will be at one-half the average repetition rate of the system and thus will be at the frequency of maximum response of a no-feedback canceler.

The rate of change of phase or amplitude of clutter signals in a scanning radar is inversely proportional to the hits per beamwidth, n. Thus, with the use of a computer simulation to determine the proportionality constant, the limitation on I due to staggering is approximately

$$I \approx 20 \log \left(\frac{2.5n}{\gamma - 1} \right) \quad \text{dB} \tag{15.23}$$

which is plotted in Fig. 15.16.

These curves, which apply to all multiple-delay cancelers, give answers that are fairly close to the actual limitation that will be experienced for most practical stagger ratios. An example of the accuracy is as follows: A system with 14.4 hits per beamwidth, a four-pulse binomial weight canceler, and a 6:9:7:8 pulse-interval ratio has an improvement factor limitation of 36.5 dB due to staggering. The curve gives a limitation of 37.2 dB for this case. But note that if the sequence of pulse intervals is changed from 6:9:7:8 to 6:8:9:7, the actual limitation is 41.1 dB, which is 3.9 dB less than that indicated by the curve. This occurs because the primary modulation with a 6:9:7:8 pulse-interval ratio looks like a target at maximum-response speed, whereas the primary modulation with a 6:8:9:7 pulse-interval ratio looks like a target at one-half the speed of maximum response. Because it is desirable to average the transmitter duty cycle over as short a period as possible, the 6:9:7:8 pulse-interval ratio would probably be chosen for a practical system.

Once Eq. (15.23) for the limitation on I due to scanning and staggering is obtained, it is possible to determine the limitation on I due to internal-clutter motion and staggering. If

$$n = \frac{\sqrt{\ln(2)}}{2\pi} \times \frac{\lambda f_r}{\sigma_v} = 0.1325 \frac{\lambda f_r}{\sigma_v} \tag{15.24}$$

[from Eqs. (15.4) and (15.5)] is substituted into Eq. (15.23),

$$I = 20 \log \left(\frac{2.5}{\gamma - 1} \times \frac{0.1325 \lambda \, f_r}{\sigma_v} \right) = 20 \log \left(\frac{0.33 \lambda \, f_r}{(\gamma - 1)\sigma_v} \right) \qquad (15.25)$$

where λ is the wavelength, f_r is the average pulse repetition frequency, and σ_v is the rms velocity spread of scattering elements. This is plotted in Fig. 15.17 for rain and for wooded hills with a 40-kn wind. This limitation on the MTI improvement factor is independent of the type of canceler employed.

Time-Varying Weights. The improvement factor limitation caused by pulse-to-pulse staggering can be avoided by the use of time-varying weights in the canceler forward paths instead of binomial weights. The use of time-varying weights has no appreciable effect on the MTI velocity response curve. Whether the added complexity of utilizing time-varying weights is desirable depends on whether the stagger limitation is predominant. For two-delay cancelers, the stagger limitation is often comparable with the basic canceler capability without staggering. For three-delay cancelers, the stagger limitation usually predominates.

Consider the transmitter pulse train and the canceler configurations shown in Fig. 15.38. During the interval T_N when the returns from transmitted pulse P_N are being received, the two-delay canceler weights should be

$$A = 1 \qquad C = \frac{T_{N-2}}{T_{N-1}} \qquad B = -1 - C \qquad (15.26)$$

and the three-delay canceler weights should be

$$A = 1 \qquad C = 1 + \frac{T_{N-3} + T_{N-1}}{T_{N-2}}$$

$$B = -C \qquad D = -1 \qquad (15.27)$$

These weights have been derived by assuming that the cancelers should perfectly cancel a linear waveform $V(t) = c + at$, sampled at the stagger rate, independently of the values of the constant c or the slope a. [As mentioned at the beginning of this section, a multiple-delay canceler with binomial weights in an unstaggered system will perfectly cancel $V(t) = c + at$.]

The choice of $A = 1$ in both cases is arbitrary. In the three-delay canceler, setting $D = -1$ eliminates the opportunity for a second-order correction to cancel the quadratic term bt^2 which could be obtained if D were also time-varying. Computer calculations have shown that it is unnecessary to vary D in most practical systems.

Depth of First Null in Velocity Response. When selecting system parameters, it is useful to know the depth of the first few nulls to be expected in the velocity response curve. As discussed earlier, the null depths are essentially unaffected by feedback. They are also essentially independent of the type of canceler employed, whether single, dual, or triple, or of the number of hits per beamwidth. Figure 15.39 shows approximately what null depths can be expected versus the ratio of maximum to minimum interpulse period.

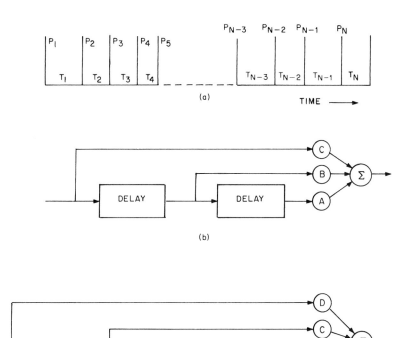

FIG. 15.38   Use of time-varying weights. (*a*) Pulse train. (*b*) Two-delay canceler. (*c*) Three-delay canceler.

15.10   IMPROVEMENT FACTOR RESTRICTION CAUSED BY LIMITING

Field measurements have indicated that the performance of many scanning multiple-delay MTI radar systems falls considerably short of the performance predicted above. This is true because the above theory is based on the assumption that the system is linear. As described earlier, many MTI systems use a linear-limiting amplifier preceding the canceler to adjust clutter residue to the level of thermal noise. Sometimes inadvertent clutter signal limiting occurs because of insufficient receiver dynamic range.

An example of how limiting the dynamic range adjusts the residue is shown in the MTI PPI photographs of Fig. 15.40. The range rings are at 5-mi intervals. A number of birds are shown on the display. The residue from clutter in the left photograph is solid out to 3 nmi and then decreases until it is almost entirely gone at 10 nmi. The MTI improvement factor in both pictures is 18 dB, but the input dynamic range (peak signal-to-rms noise) to the canceler was changed from 20 to 14 dB between the two pictures. An aircraft flying over the clutter in the first 5 mi in the left-hand picture could not be detected, no matter how large its radar cross

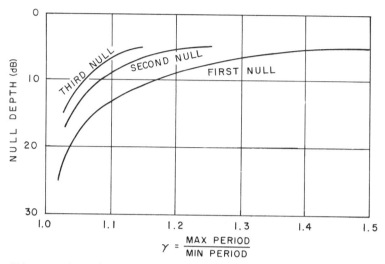

FIG. 15.39   Approximate depth of nulls in the velocity response curve for pulse-to-pulse staggered MTI.

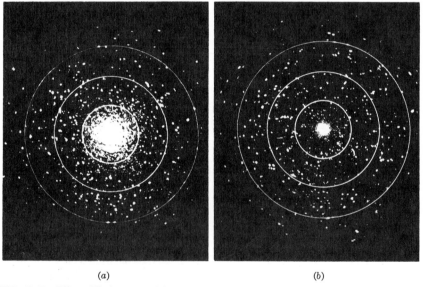

(a)   (b)

FIG. 15.40   Effect of limiters. (a) 18 dB improvement factor, 20 dB input dynamic range. (b) 18 dB improvement factor, 14 dB input dynamic range.

section. In the right-hand picture, the aircraft could be detected if the target-to-clutter cross-section ratio were sufficient.

Prior to the development of modern clutter maps for controlling false alarms caused by clutter residue, the use of IF limiting was essential for false-alarm control in an MTI radar. Such limiting, however, seriously affects the improvement obtainable with a scanning-limited, multiple-delay canceler because of the increased spectral spread of the clutter that exceeds the limit level. Part of the additional clutter spectral components comes from the sharp discontinuity in the envelope of returns as the clutter reaches the limit level.[24] A time-domain example of this phenomenon is shown in Fig. 15.41 for a radar with 16.4 hits per beamwidth. On the left is a point target that does not exceed the limit level; on the right is a point target that exceeds the limit level by 20 dB. Note that, for this example, I degrades by 12.8 dB for the dual canceler and by 26.5 dB for the triple canceler. The exact result of this calculation depends on the assumed shape of the antenna pattern. [In this case, a $(\sin U)/U$ pattern terminated at the first nulls was assumed.] There is a comparable spectral spread of limited distributed

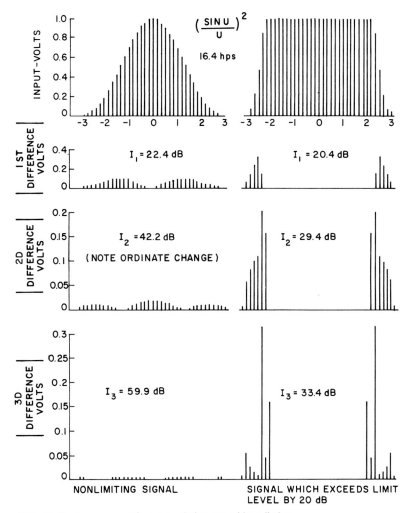

FIG. 15.41 Improvement factor restriction caused by a limiter.

clutter.[25,26] Figure 15.42a, b, and c, from Ref. 25, shows the expected improvement factor for two- three-, and four-pulse cancelers as a function of σ/L, the ratio of the rms clutter amplitude to the limit level. Hits per one-way half-power beamwidth are indicated by n.

When limiting is used for controlling false alarms, the dynamic range must be adjusted so that the residue from clutter is approximately equal to receiver noise. If this is not done, the clutter residue will be so strong that it will be impossible to detect desired targets in the clutter area. (It is possible to track a target through a clutter area with considerably more residue than permits initial detection of a target, but for surveillance and detection purposes the operators must have a clean display.)

Because MTI systems are built so that the dynamic range of signals into the canceler can be adjusted in the field to provide a good display, actual performance often falls far short of assumed performance without the user being aware of the difference.

With the proper employment of clutter maps, as described in Sec. 15.14, MTI radars can be operated with much larger dynamic ranges, and the degradation caused by IF limiters can be greatly reduced or eliminated.

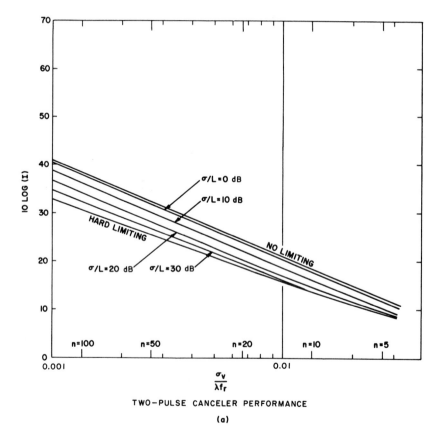

FIG. 15.42 Improvement factor restriction versus amount of limiting and clutter spectral spread. (a) Two-pulse canceler. (b) Three-pulse canceler. (c) Four-pulse canceler.

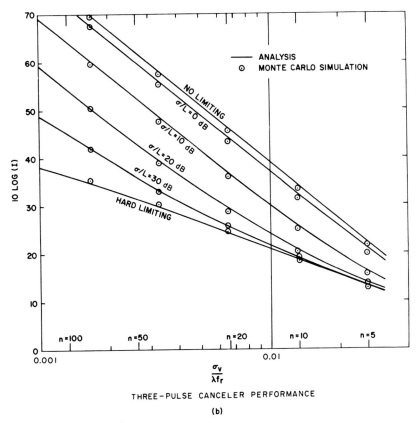

THREE-PULSE CANCELER PERFORMANCE

(b)

FIG. 15.42 *(Continued)*

15.11 *RADAR SYSTEM STABILITY AND A/D QUANTIZATION REQUIREMENTS*

System Instabilities. Not only do clutter motion and scanning affect the MTI improvement factor attainable, but system instabilities also place a limit on MTI performance. These instabilities come from the stalo and coho, from the transmitter pulse-to-pulse frequency change if a pulsed oscillator and from pulse-to-pulse phase change if a power amplifier, from the inability to lock the coho perfectly to the phase of the reference pulse, from time jitter and amplitude jitter on the pulses, and from quantization noise of the A/D converter. Weil has presented an excellent detailed discussion of these effects.[27,28]

Phase instabilities will be considered first. If the phases of consecutive received pulses relative to the phase of the coho differ by, say, 0.01 rad, a limitation of 40 dB is placed on the improvement factor possible. The 0.01-rad clutter vector change is equivalent to a target vector 40 dB weaker than the clutter su-

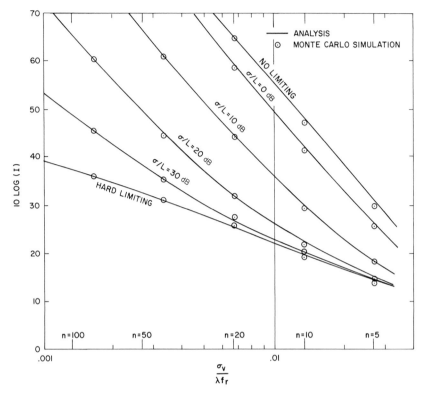

FOUR-PULSE CANCELER PERFORMANCE

(c)

FIG. 15.42   *(Continued)*

perimposed on the clutter, as shown in Fig. 15.43.

In the power amplifier MTI system shown in Fig. 15.44, pulse-to-pulse phase changes in the transmitted pulse can be introduced by the pulsed amplifier. The most common cause of a power amplifier introducing phase changes is ripple on

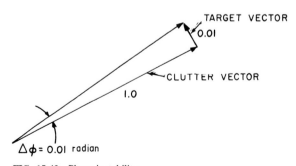

FIG. 15.43   Phase instability.

the high-voltage power supply. Other causes of phase instability include ac voltage on the transmitter tube filament and uneven power supply loading, such as that caused by pulse-to-pulse stagger.

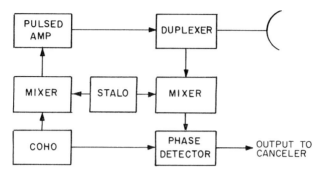

FIG. 15.44 Power amplifier simplified block diagram.

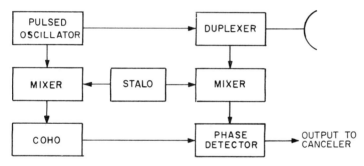

FIG. 15.45 Pulsed oscillator simplified block diagram.

In the pulsed oscillator system, shown in Fig. 15.45, pulse-to-pulse frequency changes result in phase run-out during the transmitted pulse. Phase run-out is the change of the transmitted pulse during the pulse duration with respect to the phase of the reference oscillator. If the coho locked perfectly to the end of the transmitted pulse, a total phase run-out of 0.02 rad during the transmitted pulse would then place an average limitation of 40 dB on the improvement factor attainable. Pulse-to-pulse frequency change in microwave oscillators is primarily caused by high-voltage power supply ripple. In the pulsed oscillator system, a pulse-to-pulse phase difference of 0.01 rad in locking the coho results in an improvement factor limitation of 40 dB.

The limitations on the improvement factor which are due to equipment instabilities in the form of frequency changes of the stalo and coho between consecutive transmitted pulses are a function of the range of the clutter. These changes are characterized in two ways. All oscillators have a noise spectrum. In addition, cavity oscillators, used because they are readily tunable, are microphonic, and thus their frequency may vary at an audio rate. The limitation on the improvement factor due to frequency changes is the difference in the number of radians that the oscillator runs through between the time of transmission and the time of reception of consecutive pulses. Thus the improvement factor will be limited to 40 dB if $2\pi\Delta fT = 0.01$ rad, where Δf is the oscillator frequency change between transmitted pulses and T is the transit time of the pulse to and from the target.

To evaluate the effects of oscillator phase noise on MTI performance, there are four steps. First, determine the single-sideband power spectral density of the phase noise as a function of frequency from the carrier.[29,30] Second, increase this spectral density by 6 dB. This accounts for a 3 dB increase because both sidebands of noise affect clutter residue and a 3 dB increase because the oscillator contributes noise during both transmitting and receiving. Third, adjust the spectral density for (a) correlation due to the range of the clutter of interest, (b) noise rejection due to the frequency response of the clutter filters, and (c) the frequency response of the receiver passband. Fourth, integrate the adjusted spectral density of the phase noise. The result is the limitation on the improvement factor due to the oscillator noise. Oscillator phase noise and adjustments to phase noise can all be approximated by straight lines on a decibel-versus-log frequency plot. The places where the straight lines intersect are called break frequencies.

The first adjustment, correlation due to range of the clutter of interest, reduces noise at the low frequencies by 20 dB per decade from the break frequency at $f = c/(2\pi R)$, where c is the speed of light and R is the clutter range. For the second adjustment, the response at low frequency of the clutter filters, cancelers with binomial weights have responses that fall off at 20 dB per decade for one delay, 40 dB per decade for two delays, 60 dB per decade for three delays, etc. The break frequencies for the start of the response falloff are $0.225f_r$ for one delay, $0.249f_r$ for two delays, $0.262f_r$ for three delays, and $0.271f_r$ for four delays. (If the clutter filters do not use binomial weights, the exact filter response must be used.) At frequencies higher than the start of the filter passband, the clutter filter is assumed to have unity gain because the average noise gain is unity.

For example, consider an oscillator (*oscillator* is assumed to include the complete microwave signal source, which typically includes a crystal oscillator and a multiplier chain) with single-sideband phase-noise spectral density as shown in Fig. 15.46. (One device for measuring phase noise is the Hewlett Packard 11729B Carrier Noise Test Set.[30]) The single-sideband noise is increased by 3 dB because both sidebands affect system stability and by an additional 3 dB because the oscillator introduces noise in both the transmitted signal (or coho locking signal if a magnetron transmitter is used) and the receiver downconversion process. Figure 15.47 shows the spectral modifications due to the system response. (a) The first modification accounts for correlation due to the range to the clutter of interest [assumed clutter range is $\approx$ 100 nmi (185.2 km); thus the break frequency is 3 $\times 10^8/(2\pi \times 185,200) = 365$ Hz]. (b) Second, a three-pulse binomial-weighted canceler is assumed with the radar operating at a PRF of 360 Hz. Thus the break frequency is $0.249 \times 360 = 90$ Hz. (c) Third, the receiver passband is assumed to extend from -500 kHz to $+500$ kHz with respect to the IF center frequency (1-MHz total passband) at the -3 dB points and determined by a two-pole filter. Thus the receiver passband response falls off at 40 dB per decade from the break frequency at 500 kHz as shown.

The modified phase-noise spectral density is shown in Fig. 15.48. The total noise power with respect to the carrier is determined by integration of the noise power under the curve. The equation for spectral power density of each segment as a function of frequency is

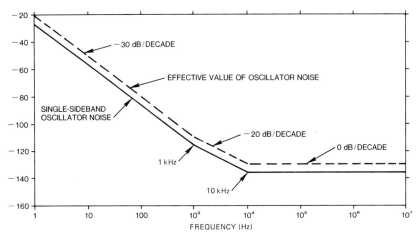

FIG. 15.46 Single-sideband phase-noise spectral density of a microwave oscillator and the effective noise density.

$$P(f) = P_{f1} \times 10^{\left[\frac{\text{SLOPE}}{10} \log\left(\frac{f}{f_1}\right)\right]} \tag{15.28}$$

where P_{f1} is spectral power density, in watts per hertz, at f_1 (for convenience, it is assumed that the carrier power is 1 W); SLOPE is the slope of the segment, in decibels per decade; and f_1 is the frequency where P_{f1} is specified.

For each segment of the spectrum with constant slope, this equation can be integrated by using Vigneri's method[31] or with calculators with an integrate function, such as the Hewlett Packard HP-15C. Table 15.3 gives the integration for the example. Note that the assumption is made that the carrier power is 1 W, so that, for example, -149.4 dBc/Hz becomes 1.148×10^{-15} W/Hz. When the integrated powers for all the segments are calculated, they are summed and then

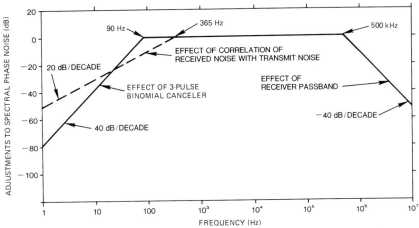

FIG. 15.47 Adjustments, based on system parameters (see text), to the phase noise of a microwave oscillator.

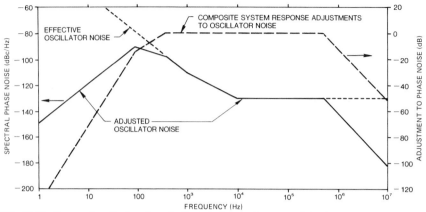

FIG. 15.48 Composite adjustments and adjusted phase-noise spectral density.

TABLE 15.3 Integration of the Phase-Noise Spectral Density of Fig. 15.46 with Adjustments of Fig. 15.47 as Shown in Fig. 15.48

Segment	f_1, Hz	f_2, Hz	Slope, decibels per decade	P_1, dBc/Hz	Integrated power, W	Integrated power, dBc
1	1	90	30.0	−149.4	0.188E-07	−77.25
2	90	365	−10.0	−90.8	0.105E-06	−69.80
3	365	1,000	−30.0	−96.9	0.323E-07	−74.91
4	1,000	10,000	−20.0	−110.0	0.900E-08	−80.46
5	10,000	500,000	0.0	−130.0	0.490E-07	−73.10
6	500,000	1,000,000	−40.0	−130.0	0.167E-07	−77.78
		Total integrated noise power			0.231E-06	− 66.37

converted back to dBc. The final answer, −66.37 dBc, is the limit on I that results from oscillator noise. The limit on I_{SCR} (dB) is I (dB) plus target integration gain (dB).

Time jitter of the transmitted pulses results in degradation of MTI systems. Time jitter results in failure of the leading and trailing edges of the pulses to cancel, the amplitude of each uncanceled part being $\Delta t/\tau$, where Δt is the time jitter and τ is the transmitted pulse length. The total residue power is $2(\Delta t/\tau)^2$, and therefore the limitation on the improvement factor due to time jitter is $I = 20 \log [\tau/(\sqrt{2}\,\Delta t)]$ (dB). This limit on the improvement factor is based on an uncoded transmitter pulse and on the assumption that the receiver bandwidth is matched to the duration of the transmitted pulse. In a pulse compression system, the receiver bandwidth is wider by the time-bandwidth ($B\tau$) product; thus the clutter residue power at each end of the pulse increases in proportion to the $B\tau$ product. The limit on I for a chirp pulse compression system is then $I = 20 \log [\tau/(\sqrt{2}\,\Delta t \sqrt{B\tau})]$. For pulse compression systems employing burst waveforms, the factor 2 in the preceding equation should be multiplied by the number of subpulses in the waveform. Thus, for example, the limit on I for a 13-pulse Barker code is

$$I = 20 \log [\tau/(\sqrt{2 \times 13} \, \Delta t \sqrt{13})] \quad \text{dB} \tag{15.29}$$

Pulse-width jitter results in one-half the residue of time jitter, and

$$I = 20 \log \frac{\tau}{\Delta PW \sqrt{B\tau}} \quad \text{dB} \tag{15.30}$$

where ΔPW is pulse-width jitter.

Amplitude jitter in the transmitted pulse also causes a limitation of

$$I = 20 \log \frac{A}{\Delta A} \quad \text{dB} \tag{15.31}$$

where A is the pulse amplitude and ΔA is the pulse-to-pulse change in amplitude. This limitation applies even though the system uses limiting before the canceler because there is always present much clutter that does not reach the limit level. With most transmitters, however, the amplitude jitter is insignificant after the frequency-stability or phase-stability requirements have been met.

Jitter in the sampling time in the A/D converter also limits MTI performance. If pulse compression is done prior to the A/D or if there is no pulse compression, this limit is

$$I = 20 \log \frac{\tau}{J \sqrt{B\tau}} \quad \text{dB} \tag{15.32}$$

where J is the timing jitter, τ is transmitted pulse length, and $B\tau$ is the time-bandwidth product. If pulse compression is done subsequent to the A/D converter, then the limitation is

$$I = 20 \log \frac{\tau}{JB\tau} \quad \text{dB} \tag{15.33}$$

The limitations on the attainable MTI improvement factor are summarized in Table 15.4. This discussion has assumed that the peak-to-peak values of these instabilities occur on a pulse-to-pulse basis, which is often the case in pulse-to-pulse staggered MTI operation. If it is known that the instabilities are random, the peak values shown in these equations can be replaced by the rms pulse-to-pulse values, which gives results essentially identical to Steinberg's results.[32]

If the instabilities occur at some known frequency, e.g., high-voltage power supply ripple, the relative effect of the instability can be determined by locating the response on the velocity response curve for the MTI system for a target at an equivalent doppler frequency. If, for instance, the response is 6 dB down from the maximum response, the limitation on I is about 6 dB less severe than indicated in the equations in Table 15.4.

If all sources of instability are independent, as would usually be the case, their individual power residues can be added to determine the total limitation on MTI performance.

Intrapulse frequency or phase variations do not interfere with good MTI operation provided they repeat precisely from pulse to pulse. The only concern is a

TABLE 15.4 Instability Limitations

Pulse-to-pulse instability	Limit on improvement factor
Transmitter frequency	$I = 20 \log [1/(\pi \, \Delta f \, \tau)]$
Stalo or coho frequency	$I = 20 \log [1/(2\pi \, \Delta f \, T)]$
Transmitter phase shift	$I = 20 \log (1/\Delta\phi)$
Coho locking	$I = 20 \log (1/\Delta\phi)$
Pulse timing	$I = 20 \log [\tau/(\sqrt{2}\Delta t \sqrt{B\tau})]$
Pulse width	$I = 20\log[\tau/(\Delta PW\sqrt{B\tau})]$
Pulse amplitude	$I = 20 \log (A/\Delta A)$
A/D jitter	$I = 20 \log [\tau(J \, \sqrt{B\tau})]$
A/D jitter with pulse compression following A/D	$I = 20 \log [\tau/(JB\tau)]$

where
$$
\begin{aligned}
\Delta f &= \text{interpulse frequency change} \\
\tau &= \text{transmitted pulse length} \\
T &= \text{transmission time to and from target} \\
\Delta\phi &= \text{interpulse phase change} \\
\Delta t &= \text{time jitter} \\
J &= \text{A/D sampling time jitter} \\
B\tau &= \text{time-bandwidth product of pulse} \\
&\quad \text{compression system } (B\tau = \text{unity for uncoded pulses}) \\
\Delta PW &= \text{pulse-width jitter} \\
A &= \text{pulse amplitude, V} \\
\Delta A &= \text{interpulse amplitude change}
\end{aligned}
$$

loss of sensitivity if phase run-out during the transmitted pulse or mistuning of the coho or stalo permits the received pulses to be significantly detuned from the intended IF frequency. If 1-rad phase run-out during the pulse is permitted, the system detuning may be as large as $1/(2\pi\tau)$ Hz with no degradation of MTI performance.

To give an example of interpulse-stability requirements, consider a 3000-MHz radar transmitting an uncoded 2-μs pulse and the requirement that no single system instability will limit the MTI improvement factor attainable at a range of 100 nmi to less than 50 dB, a voltage ratio of 316:1.

The rms pulse-to-pulse transmitter frequency change (if a pulsed oscillator) must be less than

$$
\Delta f = \frac{1}{316\pi\tau} = 504 \text{ Hz}
$$

which is a stability of about 2 parts in 10^7.

The rms pulse-to-pulse transmitter phase-shift change (if a power amplifier) must be less than

$$
\Delta\phi = \frac{1}{316} = 0.00316 \text{ rad} = 0.18°
$$

The stalo or coho frequency change must be less than

$$
\Delta f = \frac{1}{316 \, (2\pi) \, (100 \times 12.36 \times 10^{-6})} = 0.4 \text{ Hz}
$$

which is a short-term stability of 1 part in 10^{10} for the stalo (at about 3 GHz) and 1 part in 10^8 for the coho (assuming a 30-MHz IF frequency).

The coho locking (if a pulsed oscillator system) must be within

$$\Delta\phi = \frac{1}{316} = 0.00316 \text{ rad} = 0.18°$$

The pulse timing jitter must be less than

$$\Delta t = \frac{\tau}{316\sqrt{2}\sqrt{1}} = \frac{2 \times 10^{-6}}{316\sqrt{2}} = 4.5 \times 10^{-9} \text{ s}$$

The pulse-width jitter must be less than

$$\Delta PW = \frac{\tau}{316\sqrt{1}} = \frac{2 \times 10^{-6}}{316} = 6 \times 10^{-9} \text{ s}$$

The pulse amplitude change must be less than

$$\frac{\Delta A}{A} = \frac{1}{316} = 0.00316 = 0.3 \text{ percent}$$

The A/D sampling time jitter must be less than

$$J = \frac{\tau}{316\sqrt{1}} = \frac{2 \times 10^{-6}}{316} = 6 \times 10^{-9} \text{ s}$$

Of the above requirements, the only ones that may be difficult to meet are the stalo stability[33,34] and the coho-locking accuracy. However, in systems with large bandwidths (short compressed pulses) the timing jitter requirements become significant and may require special clock regeneration circuitry at key system locations.

Effect of Quantization Noise on Improvement Factor. Quantization noise, introduced in the A/D converter, limits the attainable MTI improvement factor. Consider a conventional video MTI system, as shown in Fig. 15.49. Because

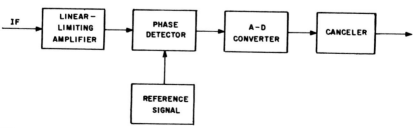

FIG. 15.49 Digital MTI consideration.

the peak signal level is controlled by the linear-limiting amplifier, the peak excursion of the phase-detector output is known, and the A/D converter is designed to cover this excursion. If the A/D converter uses N bits and the phase-detector output is from -1 to $+1$, the quantization interval is $2/(2^N - 1)$. The rms value of the signal-level deviation introduced by the A/D converter is $2/[(2^N - 1) \sqrt{12}]$. The limit on the MTI improvement factor that this imposes on a signal reaching the full excursion of the phase detector is found by substituting in the following equation from Table 15.4.

$$I = 20 \log \frac{A}{\Delta A} = 20 \log \left\{ \frac{1}{[(2^N - 1) \sqrt{3.0}]^{-1}} \right\} = 20 \log [(2^N - 1) \sqrt{3.0}]$$

$$(15.34)$$

Because two quadrature channels contribute independent A/D noise, the average limit on the improvement factor of a full range signal is

$$I = 20 \log \left[(2^N - 1) \sqrt{\frac{3.0}{2}} \right] = 20 \log [(2^N - 1) \sqrt{1.5}] \qquad (15.35)$$

If the signal does not reach the full excursion of the A/D converter, which is normally the case, then the quantization limit on I is proportionately more severe. For example, if the system is designed so that the mean level of the strongest clutter of interest is 3 dB below the A/D converter peak, the limit on I would be $20 \log [(2^N - 1) 0.75]$. (This is tabulated in Table 15.5.)

TABLE 15.5   Typical Limitation on I Due to A/D Quantization

Number of bits, N	Limit on MTI improvement factor I, dB
4	22.3
5	28.6
6	34.7
7	40.8
8	46.9
9	52.9
10	59.0
11	65.0
12	71.0

This discussion of A/D quantization noise has assumed perfect A/D converters. Many A/D converters, particularly under high-slew-rate conditions, are less than perfect. This in turn leads to system limitations more severe than predicted here. (See Sec. 15.12.)

Substituting the pulse-to-pulse rms deviation for ΔA in Eq. (15.34) was done on the assumption that the pulse-to-pulse quantization error is independent. Brennan and Reed[35] have calculated a *quieting effect* which occurs when the quantization interval is coarse compared with the clutter change between pulses (which results in a number of successive pulses out of the A/D converter having the same level), but this quieting effect will not occur with practical system parameters.

Pulse Compression Considerations. When an MTI system is used with pulse compression, the system target detection capability in clutter may be as good as a system transmitting the equivalent short pulse, or the performance may be no better than a system transmitting the same-length uncoded pulse. The kind of clutter environment, the system instabilities, and the signal processing utilized determine where the system performance will fall between the above two extremes. Unless provision is incorporated for coping with system instabilities, the MTI pulse compression system may fail to work at all in a clutter environment.

Ideally, a pulse compression receiver coupled with an MTI would appear as in Fig. 15.50a. If the pulse compression system were perfect, the compressed pulse would look as if the radar had transmitted and received a short pulse, and MTI processing could proceed as if the pulse compression had not existed. In practice,

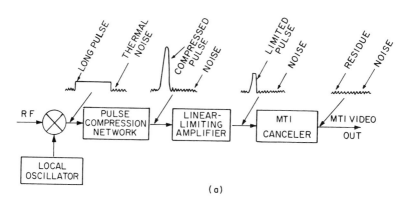

(a)

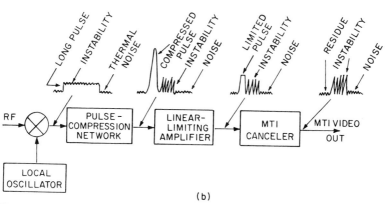

(b)

FIG. 15.50 Pulse compression with MTI. (a) Ideal but difficult-to-achieve combination. (b) Effect of oscillator on transmitter instabilities.

the compressed pulse will have time sidelobes from two basic causes. The first is system design, components that may be nonlinear with frequency, etc. These sidelobes will be stable; that is, they should repeat precisely on a pulse-to-pulse basis. The second cause of pulse compression sidelobes is system instabilities, such as noise on local oscillators, noise on transmitter power supplies, transmitter time jitter, and transmitter tube noise. These sidelobes are noiselike and are proportional to the clutter amplitude. For example, assume that the noiselike component of the sidelobes is down 40 dB from the peak transmitted signals. This noiselike component will not cancel in the MTI system, and therefore, for each clutter area that exceeds the system threshold by 40 dB or more, the residue will exceed the detection threshold. If the clutter exceeds the threshold by 60 dB, the residue from the MTI system will exceed the detection threshold by 20 dB, eliminating the effectiveness of the MTI. Figure 15.50b is a sketch of this effect.

One approach that has been successful in achieving the maximum MTI system performance attainable within the limits imposed by system and clutter instabilities is shown in Fig. 15.51. (Transmitter noise will be used in the following discussion to represent all possible system instabilities that create noiselike pulse compression time sidelobes.)

Limiter 1 is set so that the dynamic range at its output is equal to the range between peak transmitter power and transmitter noise in the system bandwidth. Limiter 2 is set so that the dynamic range at its output is equal to the expected MTI improvement factor. These limiter settings cause the residue due to transmitter noise and the residue due to other instabilities, such as quantization noise and internal-clutter motion, each to be equal to front-end thermal noise at the canceler output. This allows maximum sensitivity without an excessive false-alarm rate. The limiters are adjustable so that, when the system is placed in the field, they can be adjusted to take advantage of all the equipment and clutter stability that exists while precisely controlling the number of false alarms at the threshold output. Limiter 1 is a very efficient constant-false-alarm-rate device against transmitter noise because it suppresses the noise in direct proportion to the clutter signal strength but does not suppress at any time when the clutter signal is not strong. Thus a weak target that overlaps a strong piece of clutter by half of the uncompressed pulse length will be 50 percent suppressed, but the remaining 50 percent will be recompressed in the pulse compression network (with 6 dB amplitude loss) and can still be detected.

Although the limiter causes partial or complete suppression of some desired targets in the clutter areas, no targets are suppressed that could otherwise have been detected in the presence of transmitter noise at the system output if the limiter had not been used.

As a specific example, consider a system with a pulse compression ratio of about 30 dB and transmitter noise in the 15-MHz system noise bandwidth approximately 28 dB below the carrier power. (This is considered typical.[36]) Assume that the MTI canceler improvement factor is 30 dB, limited by clutter motion (internal-clutter spectral spread). With the above system parameters, a receiver

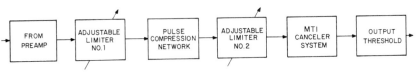

FIG. 15.51 Practical MTI pulse compression combination.

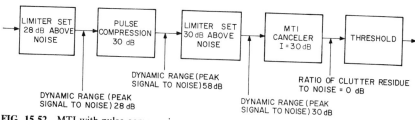

FIG. 15.52 MTI with pulse compression.

system that will provide the maximum obtainable performance should be adjusted as shown in Fig. 15.52. At the output of the pulse compression network, the transmitter noise will be equal to or less than thermal noise for either distributed clutter or point clutter. The peak clutter signals will vary from 28 dB above thermal noise for evenly distributed clutter to 58 dB above thermal noise for strong point clutter.

Because the MTI canceler is expected to attenuate clutter by 30 dB, the second limiter is provided to prevent the residue from strong clutter from exceeding the threshold. Without the second limiter, a strong-point reflector that was 58 dB above noise at the canceler input would have a residue 28 dB above noise at the canceler output. This would be indistinguishable from an aircraft target.

If the transmitter noise were 15 dB less than assumed above, the first limiter would be set 43 dB above thermal noise and much less target suppression would occur. Thus target detectability would improve in and near the strong clutter areas even though the MTI improvement factor was still limited to 30 dB by internal-clutter motion.

In summary, the noiselike pulse compression sidelobes and the duration of the uncompressed pulse dictate how effective a pulse compression MTI system can be. Systems have been built in which transmitter noise and long uncompressed pulses combined to make the systems incapable of detecting aircraft targets in or near land clutter. On the other hand, systems with low transmitter noise or with a short uncompressed pulse have proved satisfactory. Some existing pulse compression systems have not deliberately provided the two separate limiters described above, but the systems work because dynamic range is sufficiently restricted by circuit components. Other systems, such as those that deliberately hard-limit before pulse compression for CFAR reasons, do not have clutter residue problems but suffer from significant target suppression in the clutter areas.

An alternative to the use of limiters is the use of clutter maps. For the clutter maps to be successful in preventing detections from transmitter noise reflected from clutter and then dispersed in the pulse compression process, the clutter maps must be applied to the MTI filter outputs. Thus the clutter residue builds up in the maps, preventing detections on the residue in the filter outputs.

15.12 ANALOG-TO-DIGITAL CONVERSION CONSIDERATIONS

The accurate conversion of the radar IF signal into an equivalent digital representation is an important step in the implementation of a modern digital signal processor. This A/D conversion must preserve the amplitude and phase information contained in the radar returns with a minimum of error if the subsequent digital MTI or MTD processing is to provide the predicted level of performance. Figure 15.53 shows the block diagram of a typical I and Q phase detector circuit and the associated A/D converters and indicates the various error sources as discussed in detail below. Since most A/D converters will not provide predictable output codes when the input voltage exceeds their full-scale range, an amplitude limiter must be included at IF to ensure that it is impossible to drive the A/D converter beyond its maximum input value. More important, if limiting is allowed to occur at the A/D converter, severe harmonics will be introduced in the digital signal representation and the performance of a doppler filter bank will become unpredictable and most probably very poor. The two coherent phase detectors (balanced mixers) are driven by in-phase and quadrature CW signals from the coho. Harmonic outputs are removed by the low-pass filters. The resulting in-phase and quadrature representation of the amplitude and phase of the IF signal are then converted to an equivalent digital representation by the A/D converters, which are assumed to include the necessary sample-and-hold circuits. The error sources indicated in the diagram are as follows: (a) A phase detector nonlinearity which is caused by gain compression in the balanced mixer. This occurs if the power of the coho reference signal is not sufficiently higher than the largest input signal. (b) A quadrature phase error in the applied coho reference signals which distorts the reference coordinate system of the I and Q representation of the signal. (c) A gain error in one of the I or Q channels. This also distorts the reference coordinate system. (d) A dc offset at the input to the A/D converter. This is equivalent to a zero-doppler CW signal component. (e) The amplitude quantization due to the finite number of bits in the A/D converter. The effect of this quantization on the maximum improvement factor was discussed in Sec. 15.11.

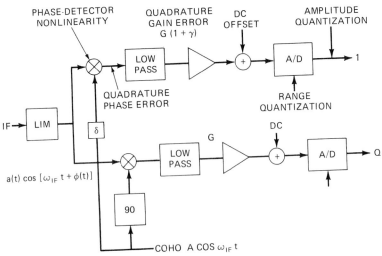

FIG. 15.53 Error sources in radar quadrature detection and A/D conversion.

Dynamic Range. The dynamic range of an A/D converter is limited by the number of bits in the output word. The quantization noise of an A/D converter has an rms value which is $N_Q = 1\sqrt{12}$ relative to the least significant bit (LSB). For an L-bit A/D converter the maximum amplitude which can be converted is proportional to 2^{L-1} (the most significant bit represents the sign), so that the maximum dynamic range is

$$DR_{max} = 2^{(L-1)} \sqrt{12} \qquad (15.36)$$

As an example, a 10-bit A/D converter will have a maximum dynamic range of 65.0 dB.

In a practical system the maximum dynamic range must be reduced because thermal noise at the input to the A/D must be well above the quantization noise in order to preserve the system noise figure and because the rms value of a sinusoidal signal is a factor of $\sqrt{2}$ below the peak value. The latter factor reduces the practical dynamic range by 3 dB. The rms noise level of system noise at the A/D input is usually set so that the rms value of the noise into the A/D converter is equal to the voltage increment corresponding to one to two levels. The number of levels corresponding to the rms noise level will be denoted k. Thus, the available dynamic range (voltage ratio) is

$$DR_{actual} = \frac{2^{(L-1)}}{\sqrt{2}} \frac{1}{k} \qquad (15.37)$$

since $2^{L-1}\sqrt{2}$ is the rms value of the maximum signal and k is the rms value of the noise. Thus for a 10-bit A/D converter and a noise level set at $k = 2.0$ the actual dynamic range would be 45.2 dB, almost 20 dB below the theoretical maximum.

From the value of the quantization noise ($N_Q = 1/12$) and the actual thermal noise at the output of the A/D converter, the noise figure degradation due to the quantization noise is found as

$$L_Q = 10 \log \left(\frac{k^2 + N_Q}{k^2} \right) \quad dB \qquad (15.38)$$

which for $k = 2.0$ is $L_Q = 0.09$ dB. For $k = 1.0$ the quantization loss is increased to $L_Q = 0.35$ dB, while the dynamic range is increased by 6 dB. The need to maintain an adequate level of noise in the signal processor to allow effective operation of CFAR processing circuits tends to favor a value of the input noise close to two levels ($k = 2.0$).

I and *Q* **Balance Requirements.** During the transformation of the amplitude and phase information in the IF signal into an equivalent representation in rectangular coordinates, dc offsets, amplitude imbalance, and phase errors are introduced. DC offsets which represent a spectral line at zero doppler at the A/D converter output can be compensated for quite readily. By implementing a long time-constant averaging circuit at the A/D converter output the dc component can be estimated and subtracted in each of the I and Q channels.

The effect of amplitude imbalance and phase errors is to generate spurious sidebands at the image doppler frequency corresponding to signals and clutter at the input. For clutter at zero doppler this sideband will also be located at zero doppler, and therefore it will not affect MTI performance. For doppler-shifted clutter returns which are suppressed by an MTI filter with an offset notch, the level of this sideband becomes a limiting factor on the improvement factor because it may be located at a doppler frequency at which the MTI filter has little or no attenuation. A calculation of the relative level of the spurious sideband caused by amplitude and phase imbalance has been made by Sinsky and Wang.[37] A graph showing the sideband level as a function of the quadrature error in amplitude or phase is shown in Fig. 15.54. In order to make this sideband less than −40 dB, the amplitude error must be less than 0.17 dB and the quadrature phase error less than 1.1°.

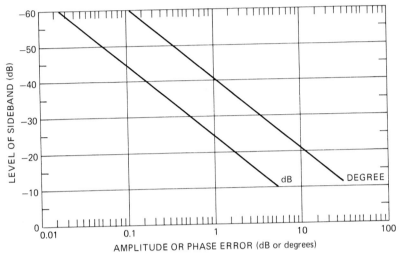

FIG. 15.54 Image sideband due to I and Q amplitude and phase errors.

Timing Jitter. The effect of timing jitter in the clock controlling the sample-and-hold circuit at the input to the A/D converter is equivalent to timing jitter on the transmitted pulse as discussed in Sec. 15.11. Its effect must therefore be included in the overall system stability budget for allowable timing jitter.

Linearity. For A/D converters with a large number of bits, the design of the coherent detectors needed to obtain the I and Q information becomes a compromise between noise figure and linearity. Any compression of the dynamic range at the limit of the input range will result in the generation of spurious outputs. If the nonlinearity is odd, symmetric odd harmonics will be generated. Calculations have shown that this sideband will be at about −48 dB when the dynamic-range compression is 0.1 dB at the maximum signal amplitude and about −31 dB when the dynamic-range compression reaches 1 dB. This effect is again important only when clutter at nonzero doppler has to be canceled in a filter with an offset rejection notch.

Accuracy. A/D converters are not necessarily perfect, as the discussion above has assumed. It is recommended that an A/D be evaluated with large signals at all frequencies within the receiver passband to establish that the quantization noise is as low as theoretically expected and that no spurious signals are produced. A/Ds with noise larger than theoretical are still usable, but it is necessary to consider their reduced performance in establishing system performance. For example, a noisy 14-bit A/D might be evaluated as being equivalent to a perfect 12-bit A/D converter.

15.13 ADAPTIVE MTI IMPLEMENTATION

When the doppler frequency of the returns from clutter is unknown at the radar input, special techniques are required to guarantee satisfactory clutter suppression. As discussed in Sec. 15.8, the doppler filter bank will usually be effective against moving clutter. This requires that the individual filters be designed with a low sidelobe level in the regions where clutter may appear and that each filter be followed by appropriate CFAR processing circuits to reject unwanted clutter residue. When clutter suppression is to be implemented with a single MTI filter, it is necessary to use adaptive techniques to ensure that the clutter falls in the MTI rejection notch. An example of such an adaptive MTI is TACCAR,[38] originally developed for airborne radars. In many applications the adaptive MTI will further have to take into account the situation where multiple clutter sources with different radial velocities are present at the same range and bearing.

Usually the doppler shift of clutter returns is caused by the wind field, and early attempts of compensating in the MTI have varied the coho frequency sinusoidally as a function of azimuth based on the average wind speed and direction. This approach is unsatisfactory because the wind field rarely is homogeneous over a large geographical area and because the wind velocity usually is a function of altitude due to wind shear (important for rain clutter and chaff). Against a single clutter source an implementation is required which permits the MTI clutter notch to be shifted as a function of range. An example of such an adaptive MTI implementation is shown in Fig. 15.55. The phase-error circuit compares the clutter return from one sweep to the next. Through a closed loop, which includes a smoothing time constant, the error signal controls a phase shifter at the coho output such that the doppler shift from pulse to pulse is removed. It should be noted that since the first sweep entering the MTI is taken as a reference, any phase shift run-out as a function of range will increase proportionally to the number of sweeps. Ultimately this run-out will exceed the speed of

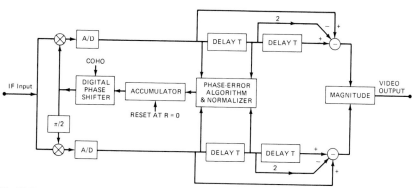

FIG. 15.55 Block diagram of closed-loop adaptive digital MTI.

response of the closed loop, and the MTI must be reset. This type of closed-loop adaptive MTI must therefore be operated for a finite set (batch) of pulses to ensure that this will not happen. Such batch-mode operation is also required if a combination of MTI operation and frequency agility is desired.

If a bimodal clutter situation is caused by the simultaneous presence of returns from land clutter and weather or chaff, an adaptive MTI can be implemented following a fixed-clutter-notch MTI section as illustrated in Fig. 15.56. The number of zeros used in the fixed- (zero-doppler) clutter-notch section of the MTI is determined by the required improvement factor and the spectral spread of the land clutter. Typically the fixed-notch MTI would use two or three zeros. For the adaptive portion of the MTI a fully digital implementation is shown in which the pulse-to-pulse phase shift of the clutter output from the first canceler is measured and averaged over a given number of range cells. This estimated phase shift is added to the phase shift which was applied to the data on the previous sweep, and this new phase shift is applied to the current data. The range averaging must be performed separately on the I and Q components of the measured phase in each range cell due to the 2π ambiguity of the phase representation itself. The accumulation of the applied phase shift from sweep to sweep, however, must be performed directly on the phase and is computed modulo 2π. The number of zeros of the adaptive MTI section is again determined by the required improvement factor and the expected spectral spread of the clutter. The phase shift is applied to the input data in the form of a complex multiply which again requires the transformation of the phase angle into rectangular coordinates. This transformation can easily be performed by a table lookup operation in a read-only memory.

When doppler shifts are introduced by digital means as described above, the accuracy of the I and Q representation of the original input data becomes an important consideration. Any dc offset, amplitude imbalance, quadrature phase error, or nonlinearity will result in the generation of undesired sidebands that will appear as residue at the canceler output. A discussion of A/D conversion considerations was presented in Sec. 15.12.

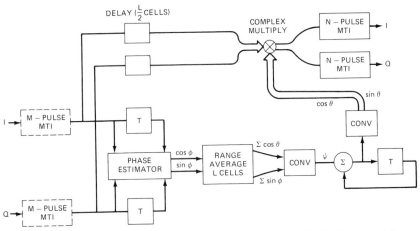

FIG. 15.56 Open-loop adaptive MTI for cancellation of simultaneous fixed and moving clutter.

In the adaptive MTI implementation described above the number of zeros allocated to each of the two cancelers was fixed, based on an a priori assessment of the clutter suppression requirement. The only variation possible would be to completely bypass one (or both) of the MTI cancelers if no land clutter or weather or chaff returns are received on a given radial. A more capable system can be implemented if the number of zeros can be allocated dynamically to either clutter source as a function of range. This leads to a fully adaptive MTI implementation using a more complex adaptation algorithm, as discussed below. Such an adaptive MTI may provide a performance close to the optimum discussed in Sec. 15.6.

In order to illustrate the difference in performance between such candidate MTI implementations, a specific example is considered next. For this example, land clutter returns are present at zero doppler with a normalized spectral spread of $\sigma_f T = 0.01$, and chaff returns are present at a normalized doppler offset of $f_d T = 0.25$ with a normalized spectral spread of $\sigma_f T = 0.05$. The power ratio of the land clutter to that of the chaff is denoted Q (dB). Thermal noise is not considered in this example. In both cases the total number of filter zeros was assumed to be equal to three. For the adaptive MTI with a fixed allocation of zeros, two zeros are located at zero doppler and the remaining zero is centered on the chaff returns. In the optimum MTI the zero locations are chosen so that the overall improvement factor is maximized. The results of this comparison are presented in Fig. 15.57, which shows the improvement factor for the optimum and the adaptive MTI as a function of the power ratio Q (dB). When Q is small so that chaff returns dominate, a significant performance improvement can be realized by using all MTI filter zeros to cancel the chaff returns. The performance difference for large values of Q is a result of an assumption made that the location of the third zero remains fixed at the chaff doppler frequency. In reality, the adaptive MTI would move its third zero to the land clutter as the land clutter residue starts to dominate the output of the first canceler. The zero locations of the optimum MTI are shown in Fig. 15.58 and can be seen to move between the land clutter at zero doppler toward the doppler of the chaff returns as the relative level of the land clutter becomes small.

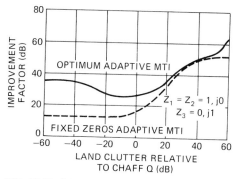

FIG. 15.57 Improvement factor comparison of optimum and adaptive MTI against fixed and moving clutter of ratio Q.

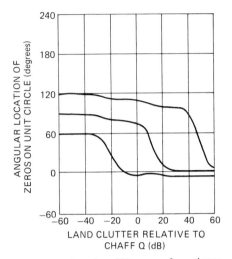

FIG. 15.58   Location of filter zeros for optimum MTI used against fixed and moving clutter.

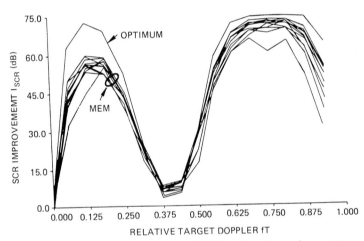

FIG. 15.59   Performance of an adaptive MTI based on the maximum-entropy spectral estimation theory.

The implementation of a fully adaptive MTI can be based on modern techniques of spectral estimation, as represented by the maximum-entropy method (MEM).[39] An example of the performance obtained through the use of an MTI based on the MEM principle is shown in Fig. 15.59. For this example bimodal clutter returns were assumed, consisting of land clutter at zero doppler 60 dB above noise having a relative spectral spread of $\sigma_f T = 0.01$ and 60 dB chaff returns at a normalized doppler frequency of $f_d T = 0.4$ with a relative spectral spread of $\sigma_f T = 0.06$. The total number of pulses processed in one coherent processing interval is CPI = 16. The MEM algorithm was implemented by using a lattice filter estimating seven poles of the clutter spectrum based on the 16 returns and by subsequently processing the 16 returns through a matching seven-zero FIR filter. The results of 12 independent simulation runs are shown in Fig. 15.59, and the optimum filter response is shown for reference.

15.14 CLUTTER MAP IMPLEMENTATION

In many MTI radar applications the clutter-to-noise ratio in the receiver will exceed the improvement factor limit of the system even when techniques such as sensitivity time control (STC), improved radar resolution, and reduced antenna gain close to the horizon are used to reduce the level of clutter returns. The resulting clutter residues after the MTI canceler must therefore be further suppressed to prevent saturation of the PPI display and/or an excessive false-alarm rate in an automatic target detection (ATD) system.

Against spatially homogeneous sources of clutter such as rain, sea clutter, or corridor chaff, a cell-averaging constant-false-alarm-rate (CFAR) processor following the MTI filter will usually provide good suppression of the clutter residues. Special features are sometimes added to the CFAR, such as greatest-of selection or two-parameter (scale and shape) normalization logic, in order to improve its effectiveness at clutter boundaries or if the probability distribution of the clutter amplitude is nongaussian. However, when the clutter returns are significantly nonhomogeneous, as is the case for typical land clutter returns, the performance of the cell-averaging CFAR will not be satisfactory and other means must be implemented to suppress the output clutter residues to the noise level.

The traditional solution to this problem has been to deliberately reduce the receiver dynamic range prior to the MTI filter to the same value as the maximum system improvement factor. Theoretically, then, the output residue should be at or below the normal receiver noise level, and no false alarms would be generated. In practice, the introduction of IF limiting against the ground clutter returns will result in an additional improvement factor restriction, as discussed in Sec. 15.10. Consequently, for the limited IF dynamic range to have the desired effect on the output residues, the limit level must be set 5 to 15 dB below the improvement factor limit of the linear system. The net result is that some of the clutter suppression capability of the MTI radar must be sacrificed in exchange for control of the output false-alarm rate.

Since returns from land clutter scatterers usually are spatially fixed and therefore appear at the same range and bearing from scan to scan, it has long been recognized that a suitable memory circuit could be used to store the clutter residues and remove them from the output residue on subsequent scans by either subtraction or gain normalization. This was the basic principle of the so-called area MTI, and many attempts have been made to implement an effective version of this circuit over an extended span of time. The main hindrance to its success has been the lack of appropriate memory technology, since the storage tube (long the only viable candidate) lacks in resolution, registration accuracy, simultaneous read-and-write capability, and stability. The development of high-capacity semiconductor memories is the technological breakthrough that has made the design of a working area MTI a reality. The *area MTI* is better known today as a *clutter map*, but both terms are used.

The clutter map may be considered as a type of CFAR where the reference samples, which are needed to estimate the level of the clutter (or clutter residue), are collected in the cell under test on a number of previous scans. Since aircraft targets usually move several resolution cells from one scan to the next, it is unlikely that the reference samples will be contaminated by a target return. Alternatively, by making the averaging time (in terms of past scans) long, the effect of an occasional target return can be minimized. While the primary purpose of the clutter map is to prevent false alarms due to discrete clutter or clutter residues which are at a fixed location, it may also be necessary to consider slowly moving point clutter in the clutter map design, either to suppress bird returns or because the radar is on a moving platform (e.g., a ship).

The memory of a clutter map is usually organized in a uniform grid of range and azimuth cells as illustrated in Fig. 15.60. Each map cell will typically have 8 to 16 bits of memory so that it will handle the full dynamic range of signals at its input and provide superclutter visibility when a target is flying over a point of clutter. The dimensions of each cell are a compromise between the required memory and several performance characteristics. These are the cutoff velocity of the map, its transient response, and the loss in sensitivity caused by the clutter map (similar to a CFAR loss). The minimum cell size will be constrained by the size of the radar resolution cell.

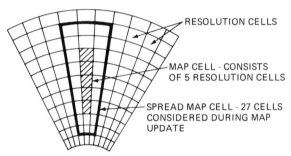

FIG. 15.60   Clutter map cell definition.

Each map cell is updated by the radar returns (or residues) falling within its borders (or in its vicinity) on several previous scans. To save memory the cells are usually updated by using a simple recursive (single-pole) filter of the form

$$y(i) = (1 - \alpha) \times y(i - 1) + \alpha x(i) \qquad (15.39)$$

where $y(i - 1)$ is the old clutter map amplitude, $y(i)$ is the updated clutter map amplitude, $x(i)$ is the radar output on the present scan, and the constant α determines the memory of the recursive filter. The test for detecting a target based on the output $x(i)$ is

$$x(i) \geq k_T y(i - 1) \qquad (15.40)$$

where the threshold constant k_T is selected to give the required false-alarm rate. Alternatively, the radar output can be normalized on the basis of the clutter map content to obtain an output $z(i) = x(i)/y(i - 1)$, which can be processed further if required. Analogously to the implementation of the cell-averaging CFAR processor, the amplitude $x(i)$ can be in linear, square-law, or logarithmic units.

The loss in detectability due to the clutter map is analogous to the CFAR loss analyzed in the literature for many different conditions. An analysis of the clutter map loss for single-hit detection using a square-law detector has been presented by Nitzberg.[40] These and other results can be summarized into a single universal curve of clutter map loss, L_{CM}, as a function of the ratio x/L_{eff}, as shown in Fig. 15.61, where x defines the required false-alarm probability according to $P_f = 10^{-x}$ and L_{eff} is the effective number of past observations averaged in the clutter map defined as

$$L_{eff} = \frac{2 - \alpha}{\alpha} \qquad (15.41)$$

For example, for $P_f = 10^{-5}$ and $\alpha = 0.125$ the clutter map loss is $L_{CM} = 1.8$ dB since $x = 5$ and $L_{eff} = 15$ for this case. Also shown in Fig. 15.61 is the curve for the conventional cell-averaging CFAR,[41] where all reference samples are equally weighted. If more than one noise amplitude is used to update the clutter map content on each scan, the value of L_{eff} should be increased proportionally.

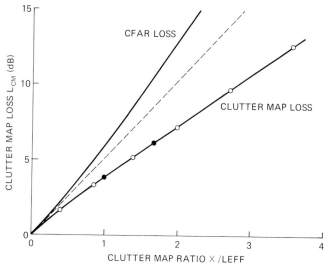

FIG. 15.61 Universal curve for determining detectability loss caused by the clutter map.

An analysis of the performance of typical implementations of clutter maps has been presented by Khoury and Hoyle.[42] From this reference a typical transient-response curve is shown in Fig. 15.62 for a Rayleigh fading point clutter source 20 dB above thermal noise, a filtering constant of $\alpha = 0.125$, and assuming four returns noncoherently integrated in each clutter map cell. The abscissa is in radar scans, and the ordinate is probability of detection of the point clutter source. Since the clutter point has the same amplitude statistics as thermal noise, the output false-alarm rate approaches $P_f = 10^{-6}$ asymptotically.

Against a slowly moving source of clutter (e.g., birds) the probability of detection may increase as the clutter source crosses the boundary between two clutter map cells. To prevent this a spreading technique can be used which will update each clutter map cell, not only with radar returns falling within its boundaries but also by using radar returns in adjacent cells in range and azimuth. Through the use of such spreading an additional degree of control over the clutter map velocity response can be achieved. An example of

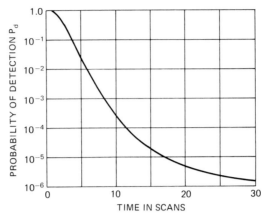

FIG. 15.62 Transient response of clutter map due to Swerling Case 2 point clutter model. (*From Khoury and Hoyle.*)[42]

the velocity response of a clutter map including such spreading is shown in Fig. 15.63. The range extent of the clutter map cell is 5 μs, the radar resolution cell is 1 μs, $n = 4$ pulses are noncoherently integrated, the filtering constant is $\alpha = 0.125$, the scan time is 5 s (12 r/min), and the $SNR = 20$ dB. On each scan, the clutter map cell is updated with the radar amplitudes in the five range cells falling within the clutter map cell and with the amplitude from one additional radar resolution cell before and after the clutter map cell. It is seen from Fig. 15.63 that the velocity response characteristic of the clutter map from stopband to passband is somewhat gradual in this particular implementation. This is partly due to the large size of the clutter map cell relative to the radar resolution. A finer-grain map with additional spreading would have a much better velocity response characteristic.

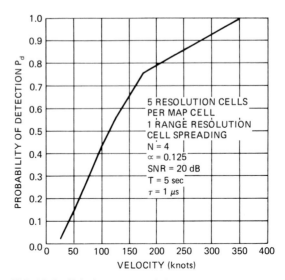

FIG. 15.63 Velocity response of clutter map.

A potential problem with the type of amplitude clutter map described in this section is the fact that a large target flying in front of a smaller target may cause enough buildup in the map to supress the small target. One way to overcome this problem in a system which includes automatic tracking would be to use the track prediction gate to inhibit updating of the clutter map with new (target) amplitudes.

15.15 CONSIDERATIONS APPLICABLE TO MTI RADAR SYSTEMS

MTI radar system design encompasses much more than receiver design. The entire radar system—transmitter, antenna, and operational parameters—must be keyed to function as part of an MTI radar. For example, excellent MTI circuitry will not perform satisfactorily unless the radar local oscillator is extremely stable, the transmitter has almost no pulse-to-pulse frequency or phase jitter, and the time on target is sufficient for coherent rejection of unwanted signals.

The design of MTI radar systems requires a number of compromises. One of the most serious problems is that desired targets often have radial velocities that are less than those of undesired targets. When a target flies past a radar, there is an interval when the doppler frequency passes through zero and the target is cancelled. This interval can be minimized by shaping the velocity response, but only at the expense of reducing the attainable improvement factor and increasing the response to undesired moving targets.

There are two classes of undesired moving targets that are particularly troublesome. These are typified by birds and automobiles. A large bird, such as a gull, crow, or buzzard, has a radar cross section of about 0.01 m^2 and flies 20 to 30 mi/h in still air. These flying speeds, combined with 10 to 20 mi/h winds, cause the birds often to fly at or near the velocity of the maximum MTI response. If there is one bird per square mile in the vicinity of the radar, there are over 1000 birds within 20 mi, which may saturate an automatic detection system and make a PPI display virtually useless. One approach to solving the bird problem that has been implemented on the air route surveillance radars[43-45] of the Federal Aviation Administration (FAA) is described below.

STC (sensitivity time control) can be employed to distinguish between small (0.01-m^2) targets and desired aircraft (1-m^2) targets. If the STC is programmed with a fourth-power law as a function of range, most of the bird returns can be eliminated from the display but desired targets are retained. The use of STC with a cosecant-squared antenna beam solves one problem but creates another: it also eliminates desired targets at high elevation angles where the antenna gain is low. The solution to this problem is to boost the antenna gain at high elevation angles to be considerably higher than the requirement for the cosecant-squared pattern. Not only does this compensate for the use of STC, but it also considerably enhances the target-to-clutter signal ratio for targets at high elevation angles, thus improving MTI performance. The penalty for this solution is a loss in the peak antenna gain which be achieved. An illustration of this approach is provided in Fig. 15.64, which shows the ARSR-2 antenna pattern and the corresponding free-space coverage. The loss in peak gain for this example, due to the boost of coverage at high angles, was about 2 dB.

The automobile and truck problem is more difficult to solve. The radar cross section of an automobile is as large as that from desired targets, and automobile speeds are well up in the response curves of most surveillance radars. If the radar is sited so that it can see many miles of highways, which is frequently the case in populated areas, automobiles are a major problem.

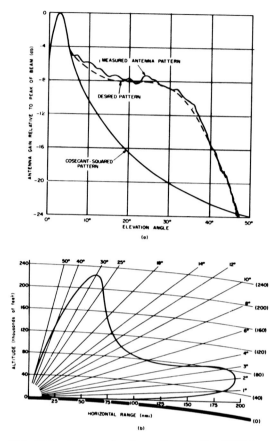

FIG. 15.64 Antenna elevation pattern for the ARSR-2 antenna. (*a*) Compared with the cosecant-squared pattern. (*b*) Free-space-coverage diagram.

One approach to solving the automobile problem is described in Refs. 46 and 47. On a horn-fed reflector antenna, a second horn is mounted beneath the primary horn to provide a high-angle receive-only beam. The energy is transmitted through the primary horn. The second horn is used for receiving for the first 20 mi, after which the receiver is electronically switched to the primary horn for the remainder of the interpulse interval. The use of the second beam at high angles greatly enhances the target-to-clutter signal ratio as well as the target-to-automobile signal ratio. This technique also greatly alleviates the bird problem because birds generally appear at low elevation angles.

REFERENCES

1. Barlow, Edward J.: Doppler Radar, *Proc. IRE*, vol. 37, pp. 340–355, April 1949.

2. Barton, David K.: "Radar System Analysis," Prentice-Hall, Englewood Cliffs, N.J., 1964.

3. Goldstein, Herbert: Sea Echo, the Origins of Echo Fluctuations, and the Fluctuations of Clutter Echoes, in Kerr, D. E. (ed.): "Propagation of Short Radio Waves," MIT Radiation Laboratory Series, vol. 13, McGraw-Hill Book Company, New York, 1951, secs. 6.6–6.21, pp. 560–587.

4. Wiltse, J. C., S. P. Schlesinger, and C. M. Johnson: Backscattering Characteristics of the Sea in the Region from 10 to 50 KMC, *Proc. IRE*, vol. 45, pp. 220–228, February 1957.

5. Hicks, B. L., N. Knable, J. J. Kovaly, G. S. Newell, J. P. Ruina, and C. W. Sherwin: The Spectrum of X-Band Radiation Back-Scattered from the Sea Surface, *J. Geophys. Res.*, vol. 65, pp. 828–837, March 1960.

6. Simkins, W. L., V. C. Vannicola, and J. P. Royan: Seek Igloo Radar Clutter Study, *Rome Air Development Center, Rept.* TR-77-338, October 1977. (DDC AD-A047 897.)

7. Nathanson, F. E., and J. P. Reilly: Radar Precipitation Echoes, *IEEE Trans.*, vol. AES-4, pp. 505–514, July 1968.

8. Barton, David K.: Radar Equations for Jamming and Clutter, *EASCON '67 Tech. Conv. Rec.*, Supplement to *IEEE Trans.*, vol. AES-3, pp. 340–355, November 1967.

9. Doviak, R. J., and D. S. Zrnić: "Doppler Radar and Weather Observations," Academic Press, Orlando, Fla., 1984.

10. "IEEE Standard Dictionary of Electrical and Electronics Terms," 3d ed., Institute of Electrical and Electronics Engineers, Inc., New York, 1984.

11. Barton, D. K., and W. W. Shrader: Interclutter Visibility in MTI Systems, *IEEE EASCON '69 Tech. Conv. Rec.*, pp. 294–297, October 1969.

12. Spafford, L.: Optimum Radar Signal Processing in Clutter, *IEEE Trans.*, vol. IT-14, pp. 734–743, September 1968.

13. Capon, J.: Optimum Weighting Functions for the Detection of Sampled Signals in Noise, *IEEE Trans.*, vol. IT-10, pp. 152–159, April 1964. (Reprinted in Ref. 14.)

14. Schleher, D. Curtis (ed.): "MTI Radar," Artech House, Norwood, Mass., 1978.

15. Skolnik, Merrill I.: "Introduction to Radar Systems," 2d ed., McGraw-Hill Book Company, New York, 1980.

16. White, W. D., and A. E. Ruvin: Recent Advances in the Synthesis of Comb Filters, *IRE Nat. Conv. Rec.*, vol. 5, pt. 2, pp. 186–200, 1957. (Reprinted in Ref. 14.)

17. Urkowitz, Harry: Analysis and Synthesis of Delay Line Periodic Filters, *IRE Trans.*, vol. CT-4, pp. 41–53, June 1957. (Reprinted in Ref. 14.)

18. Hall, W. M., and H. R. Ward: Signal-to-Noise Loss in Moving Target Indicator, *Proc. IEEE*, vol. 56, pp. 233–234, February 1968.

19. Shrader, W. W., and V. G. Hansen: Comments on "Coefficients for Feed-Forward MTI Radar Filters," *Proc. IEEE*, vol. 59, pp. 101–102, January 1971.

20. Fletcher, R. H., and D. W. Burlage: Improved MTI Performance for Phased Array in Severe Clutter Environments, *IEEE Conf. Publ. 105*, pp. 280–285, 1973.

21. Oppenheim, A. V., and R. W. Schafer: "Digital Signal Processing," Prentice-Hall, Englewood Cliffs, N.J., 1975.

22. Zverev, A. I.: Digital MTI Radar Filters, *IEEE Trans.*, vol. AU-16, pp. 422–432, September 1968.

23. Ludloff, A., and M. Minker: Reliability of Velocity Measurement by MTD Radar, *IEEE Trans.*, vol. AES-21, pp. 522–528, July 1985.

24. Grasso, G.: Improvement Factor of a Nonlinear MTI in Point Clutter, *IEEE Trans.*, vol. AES-4, pp. 640–644, November 1968.

25. Ward, H. R., and W. W. Shrader: MTI Performance Degradation Caused by Limiting, *EASCON '68 Tech. Conv. Rec.*, supplement to *IEEE Trans.*, vol. AE-4, pp. 168–174, November 1968.

26. Grasso, G., and P. F. Guarguaglini: Clutter Residues of a Coherent MTI Radar Receiver, *IEEE Trans.*, vol. AES-5, pp. 195–204, March 1969. (Reprinted in Ref. 14.)

27. Weil, T. A.: Applying the Amplitron and Stabilitron to MTI Radar System, *IRE Nat. Conv. Rec.*, vol. 6, pt. 5, pp. 120–130, 1958.

28. Weil, T. A.: An Introduction to MTI System Design, *Electron. Prog.*, vol. 4, pp. 10–16, May–June 1960.

29. Leeson, D. B., and G. F. Johnson: Short-Term Stability for a Doppler Radar: Requirements, Measurements and Techniques, *Proc. IEEE*, vol. 54, pp. 329–330, February 1966.

30. Hewlett-Packard Product Note 11729B-1, March 1984.

31. Vigneri, R., G. G. Gulbenkian, and N. Diepeveen: A Graphical Method for the Determination of Equivalent Noise Bandwidth, *Microwave J.*, vol. 11, pp. 49–52, June 1968.

32. Steinberg, Bernard D.: chaps. 1–4 in Berkowitz, R. S. (ed.): "Modern Radar: Analysis, Evaluation and System Design," pt. VI, John Wiley & Sons, New York, 1966, chaps. 1–4.

33. Stephenson, J. G.: Designing Stable Triode Microwave Oscillators, *Electronics*, vol. 28, pp. 184–187, March 1955.

34. Malling, L. R.: Phase Stable Oscillators for Space Communications, Including the Relationship between the Phase Noise, the Spectrum, the Short-Term Stability, and the *Q* of the Oscillator, *Proc. IRE*, vol. 50, pp. 1656–1664, July 1962.

35. Brennan, L. E., and I. S. Reed: Quantization Noise in Digital Moving Target Indications Systems, *IEEE Trans.*, vol. AES-2, pp. 655–658, November 1966. (Reprinted in Ref. 14.)

36. Bursweig, J., F. Hurt, and L. H. O'Brien: Intraspectral Noise and Transfer Functions of Pulsed Final Power Amplifiers, presented at *IEEE Electron. Devices Meet.*, Washington, Oct. 29, 1964.

37. Sinsky, A. I., and C. P. Wang: Error Analysis of a Quadrature Coherent Detection Processor, *IEEE Trans.*, vol. AES-10, pp. 880–883, November 1974.

38. Shrader, W. W.: MTI Radar, in Skolnik, M. I. (ed.): "Radar Handbook," McGraw-Hill Book Company, New York, 1970.

39. D'Addio, E., A. Farina, and F. A. Studer: The Maximum Entropy Method and Its Applications to Clutter Cancellation, *Riv. Tec. Selenia*, vol. 8, no. 3, pp. 15–24, 1983.

40. Nitzbert, R.: Clutter Map CFAR Anaysis, *IEEE Trans.*, vol. AES-22, pp. 419–421, July 1986.

41. Hansen, V. G.: Constant False Alarm Rate Processing in Search Radar, *IEE Conf. Publ.* 105, *Radar—Present and Future*, London, Oct. 23–25, 1973.

42. Khoury, E. N., and J. S. Hoyle: Clutter Maps: Design and Performance, *IEEE Nat. Radar Conf.*, Atlanta, 1984.

43. Shrader, W. W.: Antenna Considerations for Surveillance Radar Systems, *Proc. Seventh Ann. East Coast Conf. Aeronaut. Navig. Electron.*, October 1960.

44. Shrader, W. W.: Results of Antenna Pattern Considerations, *Proc. Eighth Ann. East Coast Conf. Aerosp. Navig. Electron.*, October 1961.

45. Shrader, W. W.: Reducing Clutter in Air Route Surveillance Radar, *Electronics*, pp. 37—41, Jan. 26, 1962.

46. Patrick, A. M.: Primary Radar in Air Traffic Control, *Interavia*, vol. 16, pp. 851—853, June 1961.

47. Mullholand, E. B., and F. R. Soden: Australia's ATC Radar Network, *Interavia*, vol. 22, pp. 511–513, Aprile 1967.

CHAPTER 16
AIRBORNE MTI

Fred M. Staudaher

Naval Research Laboratory

16.1 SYSTEMS USING AIRBORNE MTI TECHNIQUES

Airborne search radars were initially developed for the detection of ships by long-range patrol aircraft. During the latter part of World War II, airborne early-warning (AEW) radars were developed by the U.S. Navy to detect low-flying aircraft approaching a task force below the radar coverage of the ship's antenna. The advantage of the airborne platform in extending the maximum detection range for air and surface targets is apparent when one considers that the radar horizon is 12 nmi for a 100-ft antenna mast compared with 123 nmi for a 10,000-ft aircraft altitude.

Loss of picket ships due to kamikaze attacks led to the concept of the autonomous airborne detection and control station. This type of system was further developed as a barrier patrol aircraft for continental air defense.

The carrier-based E-2C aircraft (Fig. 16.1) uses AEW radar as the primary sensor in its airborne tactical data system. These radars with their extensive field of view are required to detect small aircraft targets against a background of sea and land clutter. Because of their primary mission of detecting low-flying aircraft, they cannot elevate their antenna beam to eliminate the clutter. These considerations have led to the development of airborne MTI (AMTI)[1-3] radar systems similar to those used in surface radars[1,4-6] discussed in the preceding chapter.

Airborne MTI radar systems have also been utilized to acquire and track targets in interceptor fire control systems. In this application the system has to discriminate against clutter only in the vicinity of a prescribed target. This allows the system to be optimized at the range and angular sector where the target is located. MTI is also used to detect moving ground vehicles by reconnaissance and tactical fighter-bomber aircraft. Because of the low target velocity, higher radar frequencies are employed to obtain a significant doppler shift. Since a strong clutter background is usually present, these systems can effectively utilize noncoherent MTI techniques.

The environment of high platform altitude, mobility, and speed coupled with restrictions on size, weight, and power consumption presents a unique set of

FIG. 16.1   E-2C airborne early-warning (AEW) aircraft showing rotodome housing the antenna.

problems to the designer of airborne MTI systems. This chapter will be devoted to considerations unique to the airborne environment.

16.2 COVERAGE CONSIDERATIONS

Search radars generally require 360° azimuthal coverage. This coverage is difficult to obtain on an aircraft since mounting an antenna in the clear presents major drag, stability, and structural problems. When extensive vertical coverage is required, the aircraft's planform and vertical stabilizer distort and shadow the antenna pattern. Analysis of tactical requirements may show that only a limited coverage sector is required. However, this sector usually has to be capable of being positioned over the full 360° relative to the aircraft's heading because of the requirements for coverage while reversing course, large crab angles when high winds are encountered, need to position ground track in relation to wind, nontypical operating situations, and operational requirements for coverage while proceeding to and from the station.

16.3 PLATFORM MOTION AND ALTITUDE EFFECTS ON MTI PERFORMANCE

MTI discriminates between airborne moving targets and stationary land or sea clutter. However, in the airborne case the clutter moves with respect to the airborne platform. It is possible to compensate for the mean clutter radial velocity by using techniques such as TACCAR (time-averaged-clutter coherent airborne radar).

As shown in Fig. 16.2, the apparent radial velocity of the clutter is $V_r = -V_g \cos \alpha$, where V_g is the ground speed of the platform and α is the angle subtended between the line of sight to a point on the earth's surface and the aircraft's velocity vector. Figure 16.3 shows the loci of constant radial velocity along the surface. In order to normalize the figure, a flat earth is assumed, and the normalized radial velocity $V_n = V_r/V_g$ is presented as a function of azimuth angle ψ and normalized ground range R/H, where H is the aircraft's altitude.

Instead of a single clutter doppler frequency corresponding to a constant radial velocity (V_B in Fig. 16.2) determined by the antenna pointing angle α_0, the radar sees a continuum of velocities. This results in a frequency spectrum at a particular range whose shape is determined by the antenna pattern that intersects the surface, the reflectivity of the clutter, and the velocity distribution within the beam. Furthermore, since V_r varies as a function of range at a particular azimuth ψ, the center frequency and spectrum shape vary as a function of range and azimuth angle ψ_0.

When the antenna is pointing ahead, the predominant effect is the variation of the center frequency corresponding to the change in α_0 with range. When the antenna is pointing abeam, the predominant effect is the velocity spread across the antenna beamwidth. These are classified as the slant-range effect and the platform-motion effect, respectively.

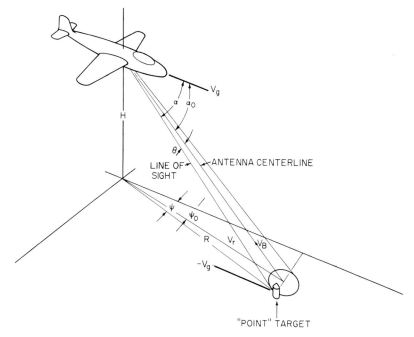

FIG. 16.2 Defining geometry: α_0 = antenna pointing angle; α = line-of-sight angle; θ = angle from antenna centerline; V_g = aircraft ground speed; V_r = radial velocity of point target; V_B = radial velocity along antenna centerline (boresight); ψ_0 = antenna azimuth angle; ψ = azimuth angle; R = ground range to point target; H = aircraft height.

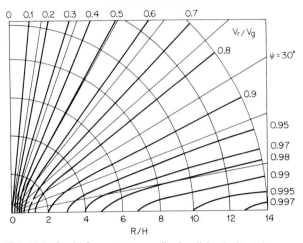

FIG. 16.3 Loci of constant normalized radial velocity V_r/V_g as a function of aircraft range-to-height ration R/H and azimuth angle ψ.

Effect of Slant Range on Doppler Offset. The antenna boresight velocity V_B is the ground-velocity component along the antenna centerline (boresight) and is given as $-V_g \cos \alpha_0$. If the clutter surface were coplanar with the aircraft, this component would be equal to $-V_g \cos \psi_0$ and would be independent of range. The ratio of the actual boresight velocity to the coplanar boresight velocity is defined as the normalized boresight-velocity ratio:

$$VBR = \frac{\cos \alpha_0}{\cos \psi_0} = \cos \phi_0 \qquad (16.1)$$

where ϕ_0 is the depression angle of the antenna centerline from the horizontal. Figure 16.4 shows the variation of the normalized boresight-velocity ratio as a

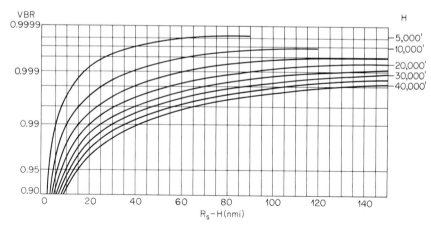

FIG. 16.4 Normalized boresight-velocity ratio VBR as a function of the difference between slant range R_s and aircraft altitude H for different aircraft altitudes.

function of slant range for a curved earth and different aircraft altitudes. The variation is fairly rapid for slant ranges less than 15 mi.

It is desirable to center the clutter spectrum in the notch (i.e., minimum-response region) of the AMTI filter in order to obtain maximum clutter rejection. This can be accomplished by offsetting the IF or RF frequency of the radar signal by an amount equal to the average doppler frequency of the clutter spectrum. Since the clutter center frequency varies with range and azimuth when the radar is moving, it is necessary for the filter notch to track the doppler-offset frequency, using an open- or closed-loop control system such as TACCAR, described below.

TACCAR. The MIT Lincoln Laboratory originally developed TACCAR to solve the AMTI radar problem. After many other approaches, it was recognized that if one used the clutter return rather than the transmit pulse to phase-lock the radar to the clutter filter, one could center the clutter in the filter stopband. The clutter phase varies from range cell to range cell owing to the distribution of the location of the scatterers in azimuth. Hence it is necessary to average the return for as long an interval as possible. Other processing features, such as phase comparison cancellation, were included in this radar (AN/APS-70). Today TACCAR is used to describe the centering of the returned clutter spectrum to the zero filter frequency. Since the technique compensates for drift in the various system elements and biases in the mean doppler frequency due to ocean currents, chaff, or weather clutter, it is used in shipboard and land-based radars as well as airborne radar.

A functional block diagram of an airborne radar employing TACCAR is shown in Fig. 16.5. The clutter error signal is obtained by measuring the pulse-to-pulse phase shift $\omega_d T_p$ of the clutter return. This provides a very sensitive error signal. The averaged error signal controls a voltage-controlled coherent master oscillator (COMO), which determines the transmitted frequency of the radar. The COMO is slaved to the system reference oscillator frequency via the automatic frequency control (AFC) loop shown in Fig. 16.5. This provides a stable reference in the absence of clutter. An input from the aircraft inertial navigation system and the

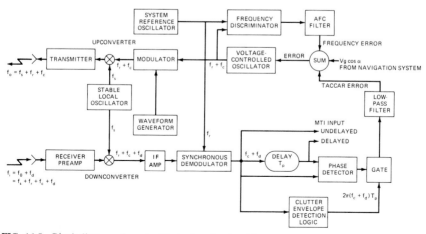

FIG. 16.5 Block diagram of a radar illustrating the signal flow path of the TACCAR control loop.

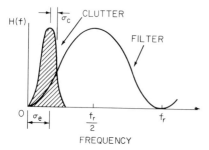

FIG. 16.6 Effect of doppler-offset error; f_r = PRF.

antenna servo provide a predicted doppler offset. These inputs allow the TACCAR system to provide a narrow-bandwidth correction signal.

Because of the noisy nature of the clutter signal, the need to have the control system bridge regions of weak clutter return, and the requirement not to respond to the doppler shift of a true target, the control system usually tracks the azimuth variation of a specific radar range interval. The maximum range of this interval is chosen so that clutter will be the dominant signal within the interval. The minimum range is chosen to exclude signals whose average frequency differs substantially from the frequency in the region of interest. For some applications it may be necessary to use multiple control loops, each one covering a specific range interval, or to vary the offset frequency in range. At any particular range the filter notch is effectively at one frequency and the center frequency of the clutter spectrum at another. The difference between these frequencies results in a doppler-offset error as shown in Fig. 16.6. The clutter spectrum will extend into more of the filter passband, and the improvement factor will be degraded.

Figure 16.7 shows the improvement factor for single- and double-delay cancelers as a function of the ratio of the notch-offset error to the pulse repetition frequency (PRF) for different clutter spectral widths. Fortunately, the platform-motion spectrum is narrow in the forward sector of coverage where offset error is

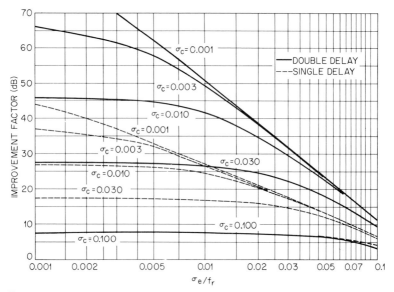

FIG. 16.7 Improvement factor I versus normalized doppler offset σ_e as a function of clutter spectrum width σ_c.

maximum. An offset error of one-hundredth of the PRF would yield 26 dB improvement for a double canceler with an input clutter spectrum whose width was 3 percent of the PRF. If the radar frequency were 10 GHz, PRF 1 kHz, and ground speed 580 kn, the notch would have to be held within 0.29 kn or $0.005V_g$.

Because of these requirements and the width of the platform-motion spectrum, stagger PRF systems must be chosen primarily on the basis of maintaining the stopband rather than flattening the passband. Similarly, higher-order delay-line filters (with or without feedback) are synthesized on the basis of stopband rejection. The limiting case is the narrowband filter bank where each individual filter consists of a small passband, the balance being stopband.

Platform-Motion Effect. To an airborne radar a clutter scatterer appears to have a radial velocity that differs from the antenna-boresight radial velocity at the same range by

$$
\begin{aligned}
V_e &= V_r - V_B \\
&= V_g \cos \alpha_0 - V_g \cos \alpha \\
&= V_g[\cos \alpha_0 - \cos (\alpha_0 + \theta)] \\
&= V_x \sin \theta + 2V_y \sin^2 \frac{\theta}{2}
\end{aligned}
\tag{16.2}
$$

for small values of θ and depression angle ϕ_0, where V_x is the horizontal component of velocity perpendicular to the antenna boresight and V_y is the component along the antenna boresight. θ is the azimuthal angle from the antenna boresight, or intersection of the vertical plane containing the boresight with the ground. The corresponding doppler frequency, when α_0 is a few beamwidths from ground track, is

$$
f_d = \frac{2V_x}{\lambda} \sin \theta \approx \frac{2V_x}{\lambda}\theta
\tag{16.3}
$$

This phenomenon results in a platform-motion clutter power spectrum which is weighted by the antenna's two-way power pattern in azimuth. The true spectrum may be approximated by a gaussian spectrum,

$$
H(f) = e^{-\frac{1}{2}(f_d/\sigma_{pm})^2} = e^{-(V_x\theta/\lambda\sigma_{pm})^2} \approx G^4(\theta)
\tag{16.4}
$$

$G^4(\theta)$, the two-way power pattern of the antenna, is 0.25 when $\theta = \theta_a/2$, where θ_a is the half-power beamwidth which can be approximated by λ/α, α being the effective horizontal aperture width. Thus

$$
e^{-\frac{1}{2}(V_x/a\sigma_{pm})^2} = 0.25
$$

or

$$
\sigma_{pm} = 0.6\frac{V_x}{a}
\tag{16.5}
$$

where V_x and a are in consistent units. This value is lower than ones derived by other authors.[4,5] However, it agrees with more exact analysis of antenna radiation patterns and experimental data analyzed by the author.

A more exact value of the parameter σ_{pm} may be obtained by matching a two-way power pattern of interest with the gaussian approximation at a specific point on the pattern, determining the standard deviation of θ by using statistical techniques, or fitting the pattern and using numerical methods. The calculation of the improvement factor I can be performed by averaging the resultant residue power, obtained by summing the signal phasors at specific values of θ, from null to null of the antenna pattern.

Figure 16.8 shows the effect of platform motion on the MTI improvement factor as a function of the fraction of the aperture displaced in the plane of the aperture per interpulse period T_p. A 5.4 percent displacement would reduce the double-delay improvement factor to 30 dB. This corresponds to a speed of 540 ft/s if the system has a PRF of 1000 Hz and a 10-ft antenna aperture. For a single-delay system the displacement would have to be held to 1.1 percent for a 30 dB performance limit.

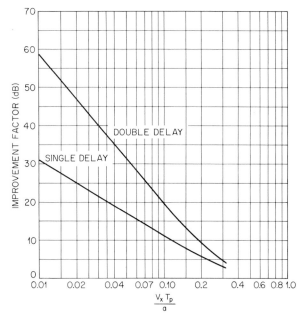

FIG. 16.8 Effect of platform motion on the MTI improvement factor as a function of the fraction of the horizontal antenna aperture displaced per interpulse period, $V_x T_p / a$.

16.4 PLATFORM-MOTION COMPENSATION ABEAM

The deleterious effects of platform motion can be reduced by physically or electronically displacing the antenna phase center along the plane of the aperture. This is referred to as the displaced phase center antenna (DPCA) technique.[7-11]

Physically Displaced Phase Center Antenna. In physical DPCA,[10,11] the

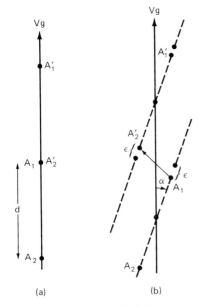

(a)　　　　　　　　(b)

FIG. 16.9 Physical DPCA defining geometry. (a) Perfect motion compensation, where A_1 and A_2 are the antenna phase centers for pulse 1 and the primed quantities are for pulse 2. (b) Imperfect motion compensation due to displacement error and alignment error α.

apertures of two side-looking antennas are aligned parallel with the aircraft longitudinal axis. Their phase centers are separated by the distance d. If the aircraft is moving at ground speed V_g, then the phase centers move V_gT_p during the interpulse period T_p.

In Fig. 16.9a the first pulse is transmitted and received on the forward antenna A_1. The second pulse is transmitted and received on the rear antenna A_2 during the next interpulse period. If $d = V_gT_p$, then the antenna used on the first pulse, A_1, will coincide with the antenna A_2 used on the second pulse. On a two-pulse-pair basis the signals received at A_1 and A_2 make it appear as if the antenna were stationary. There is actually a displacement with respect to the transmitter, but the signal path difference will be the same pulse to pulse. This will appear as a negligible range error.

Since it is difficult to change the spacing between antennas, the displacement is set by the design speed and PRF limits. Then the PRF is varied during operation to maintain the proper alignment.

If the antenna is not aligned with the flight path and if d is not equal to V_gT_p, then an error occurs between A_1 and A_2 as shown in Fig. 16.9b. The result is as though the aircraft were flying at a speed and heading such that the displacement $A_1 - A'_2$ occurs during an interpulse period. The TACCAR circuits could center the resultant spectrum at zero-doppler frequency. However, the cancellation will correspond to a value $V_xT_p = 2\epsilon$ in Fig. 16.8. If $2\epsilon/T_p$ is small enough, then the sidelobe clutter spectrum will be in the filter notch and will be canceled.

The two-antenna scheme is difficult to mechanize, and additional errors can occur if the antennas are mounted one above the other because of antenna field variations. These variations are caused by the difference in physical location due to vertical displacement, the effect of the different near-field environments, and fabrication errors. Furthermore, the PRF is effectively cut in half by having to receive during the unique alternate transmission-path configurations. A single-antenna scheme is possible by using an array with multiple feed structures that utilize a common set of elements and a switching network as shown in Fig. 16.10. The top row of switches connects the elements to the corporate feed or to the dummy loads. The bottom row of switches connects the corporate feed to the appropriate elements. The left six elements are active in the configuration illustrated. When the switches are placed in the alternate configuration, the right six elements are active. The subarrays are displaced from each other by two elements in the example.

Another variant is to have separate corporate feeds for the left and right subarrays of the antenna.[11] One of the subarrays or a central group of elements is

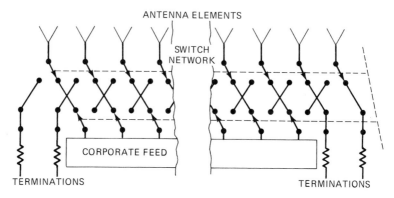

FIG. 16.10 Switching network to synthesize displaced subarrays, within an antenna array.

used as the transmit array to avoid high-power switching circuits. The separation of the subarrays d must equal $2V_gT_p$ to compensate for the transmit phase-center displacement. This allows $A_1 - A'_2$ to be paired and then A'_1 to be paired with A''_2, where A_1 is displaced from A_2 by d. This allows cancellation to be made every interpulse period, maintaining the effective PRF equal to the basic PRF.

The two-way patterns from the antenna at A_1 and A_2 must be nearly identical; otherwise, cancellation will be degraded. This degradation may be calculated by measuring the two antenna patterns, $G_1(\theta)$ and $G_2(\theta)$, and then calculating the correlation coefficient

$$\rho = \frac{[\int G_1^2(\theta)G_2^{*2}(\theta)d\theta]^2}{\int [G_1(\theta)]^4 d\theta \int [G_2(\theta)]^4 d\theta} \tag{16.6}$$

The resultant cancellation ratio is then

$$CR = 10 \log [1/(1 - \rho)] \tag{16.7}$$

If $G_2(\theta)$ is nearly identical to $G_1(\theta)$, then ρ is approximately equal to 1 and the cancellation ratio is large. When measuring G_1 and G_2, the array must be displaced for the second measurement to ensure that each subarray is in the same physical position on the antenna range.

Electronically Displaced Phase Center Antenna. Figure 16.11a shows the pulse-to-pulse phase advance of an elemental scatterer as seen by the radar receiver. The amplitude E_1 of the received signal is proportional to the two-way antenna field intensity. The phase advance is

$$2\eta = 2\pi f_d T_p = \frac{4\pi V_x T_p \sin \theta}{\lambda} \tag{16.8}$$

where f_d = doppler shift of scatterer [Eq. (16.3)]
T_p = interpulse period

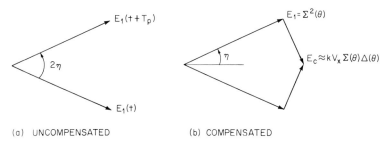

(a) UNCOMPENSATED   (b) COMPENSATED

FIG. 16.11   Phasor diagram showing the return from a point scatterer due to platform motion.

Figure 16.11b shows a method of correcting for the phase advance η. An idealized correction signal E_c is applied, leading the received signal by 90° and lagging the next received signal by 90°. For exact compensation the following relation would hold:

$$E_c = E_1 \tan \eta = \Sigma^2(\theta) \tan\frac{2\pi V_x T_p \sin \theta}{\lambda} \qquad (16.9)$$

This assumes a two-lobe antenna pattern similar to that in a monopulse tracking radar. Two receivers are used, one supplying a sum signal $\Sigma(\theta)$ and the other a difference signal $\Delta(\theta)$. The difference signal is used to compensate for the effects of platform motion.

If the system is designed to transmit the sum pattern $\Sigma(\theta)$ and receive both $\Sigma(\theta)$ and a difference pattern $\Delta(\theta)$, then at the design speed the received signal $\Sigma(\theta)\Delta(\theta)$ can be applied as the correction signal. The actual correction signal used to approximate E_c is $k\ \Sigma(\theta)\Delta(\theta)$, where k is the ratio of the amplification in the sum and difference channels of the receiver.

A uniformly illuminated monopulse array[12] has the difference signal Δ in quadrature with the sum and has the amplitude relationship

$$\Delta(\theta) = \Sigma(\theta) \tan \left(\frac{\pi W}{\lambda} \sin \theta\right) \qquad (16.10)$$

where W is the distance between the phase centers of the two halves of the antenna. Hence a choice of $W = 2V_x T_p$ and $k' = 1$ would ideally result in perfect cancellation.

In practice, a sum pattern is chosen based on the desired beamwidth, gain, and sidelobes for the detection system requirements. Then the difference pattern $\Delta(\theta)$ is synthesized independently, based on the relationship required at design radar platform speed and allowable sidelobes. The two patterns may be realized by combining the elements in separate corporate-feed structures.

Figure 16.12 shows the idealized improvement factor as a function of normalized aperture movement for a double-delay canceler. The improvement factor shown is the improvement factor for a point scatterer averaged over the null-to-null antenna beamwidth. In one case the gain ratio k' is optimized at each value of pulse-to-pulse displacement. In the other compensated case the optimum gain ratio k is approximated by the linear function of interpulse platform motion kV_x.

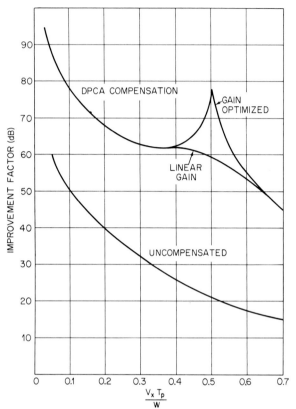

FIG. 16.12 MTI improvement factor I for DPCA compensation as a function of the fraction of the horizontal phase center separation W that the horizontal antenna aperture is displaced per interpulse period, $V_x T_p/W$. $W = 0.172a$.

A block diagram of the double-delay system is shown in Fig. 16.13. Since the transmitted pattern $\Sigma(\theta)$ appears in both channels, it is not shown. A single-delay system would not have the second delay line and subtractor. The normally required circuitry for maintaining coherence, gain and phase balance, and timing is not shown. The speed control V_x is bipolar and must be capable of reversing the sign of the $\Delta(\theta)$ signal in each channel when the antenna pointing angle changes from the port to the starboard side of the aircraft.

The hybrid amplifier shown has two input terminals which receive $\Sigma(\theta)$ and $j\Delta(\theta)$ and amplify the $\Delta(\theta)$ channel by kV_x relative to the $\Sigma(\theta)$ channel. The output terminals produce the sum and difference of the two amplified input signals. Since DPCA compensates for the complex signal, both amplitude and phase information must be retained. Therefore, these operations usually occur at RF or IF. Digital compensation can be used if synchronous detection and analog-to-digital (A/D) conversion are performed and the components are treated as complex phasors. Furthermore, the operations must be linear until the sum signal and difference signals have been processed by the hybrid amplifier. After this

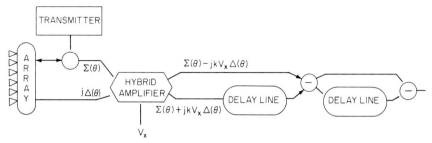

FIG. 16.13 Simplified double-delay DPCA mechanization.

single-pulse combination, the actual double cancellation can be performed by any of the processing techniques outlined in Chap. 15.

Power in the Antenna Sidelobes. Airborne systems are limited in their ability to reject clutter due to the power returned by the antenna sidelobes. The full 360° azimuthal pattern sees velocities from $-V_g$ to $+V_g$. The compensation circuits offset the velocity by an amount corresponding to the antenna boresight velocity V_B, but the total range of doppler frequencies corresponding to $2V_g$ is obtained because of echoes received via the sidelobes. For airborne systems with low and medium PRFs, these doppler frequencies can cover several multiples of the PRF so that the sidelobe power is folded into the filter. This limitation is a function of the antenna pointing angle, the MTI filter response, and the sidelobe pattern. If the sidelobes are relatively well distributed in azimuth, a measure of performance can be obtained by averaging the power returned by the sidelobes.

The limiting improvement factor due to sidelobes is

$$I_{\text{sl limit}} = \frac{K \int_{-\pi}^{\pi} G^4(\theta)\, d\theta}{\int_{\text{sl}} G^4(\theta) d\theta} \tag{16.11}$$

where the lower integral is taken outside the main-beam region. Main-beam effects would be included in the platform-motion improvement factor. The constant K is the noise normalization factor for the MTI filter. ($K = 2$ for single delay and 6 for double delay.) $G^4(\theta)$ is the two-way power of the antenna in the plane of the ground surface.

The DPCA performance described in the preceding subsection can be analyzed on the basis of radiation patterns or the equivalent aperture distribution function.[8] If the radiation pattern is used, the composite performance may be obtained either by applying the pattern functions over the entire 360° pattern or by combining the improvement factors for the DPCA main-beam and the sidelobe regions in the same manner as parallel impedances are combined:

$$\frac{1}{I_{\text{total}}} = \frac{1}{I_{\text{sl}}} + \frac{1}{I_{\text{DPCA}}} \tag{16.12}$$

If the aperture distribution is used, the sidelobe effects are inherent in the analysis. Care must be taken since if the array or reflector function is used without considering the weighting of the elemental pattern or the feed distribution, the inherent sidelobe pattern can obscure the main-beam compensation results.

16.5 SCANNING-MOTION COMPENSATION

Figure 16.14a shows a typical antenna main-beam radiation pattern and the response of a point scatterer for two successive pulses when the antenna is scanning. It is seen that the signals returned would differ by $\Delta G^2(\theta)$. This results in imperfect cancellation due to scanning. The average effect on the improvement factor can be obtained by integrating this differential effect over the main beams:

$$I_{\text{scan}} = \frac{2 \int_{-\theta_0}^{\theta_0} |G(\theta)|^2 d\theta}{\int_{-\theta_0}^{\theta_0} |G(\theta + T_p \dot{\theta}) - G(\theta)|^2 d\theta} \quad \text{for single-delay cancellation} \qquad (16.13a)$$

$$I_{\text{scan}} = \frac{6 \int_{-\theta_0}^{\theta_0} |G(\theta)|^2 d\theta}{\int_{-\theta_0}^{\theta_0} |G(\theta + T_p\dot{\theta}) - 2G(\theta) + G(\theta - T_p\dot{\theta})|^2 d\theta} \quad \text{for double-delay cancellation}$$

$$(16.13b)$$

where θ_0 = null of main beam
$G(\theta)$ = two-way voltage pattern

In order to treat scanning motion in the frequency domain, the apparent clutter velocity seen by the scanning antenna is examined to determine the doppler frequency. Each element of an array or incremental section of a continuous aperture can be considered as receiving a doppler-shifted signal due to the relative

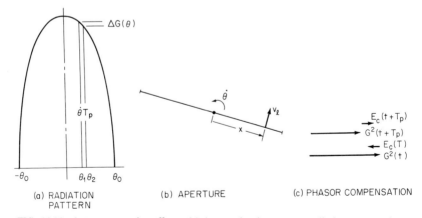

(a) RADIATION   (b) APERTURE   (c) PHASOR COMPENSATION
    PATTERN

FIG. 16.14 Antenna scanning effects. (a) As seen by the antenna radiation pattern, due to the apparent change in azimuth of the scatterer, $\theta_2 - \theta_1 = \dot{\theta} T_p$. (b) As seen by the aperture illumination function, due to the apparent motion, $v_l = x\dot{\theta}$, of the scatterer relative to the antenna element at position x. (c) Step-scan compensation of two received phasors.

motion of the clutter. The power received by the element is proportional to the two-way aperture power distribution function $F_2(x)$ at the element.

In addition to the velocity seen by all elements because of the motion of the platform, each element sees an apparent clutter velocity due to its rotational motion, as illustrated in Fig. 16.14b. The apparent velocity varies linearly along the aperture. Hence the two-way aperture distribution is mapped into the frequency domain. The resulting power spectrum due to the antenna scanning is

$$H(f) \not\approx F_2\left(\frac{\lambda f}{2\dot{\theta}}\right) \qquad 0 \leq f \leq \frac{a\,\dot{\theta}}{\lambda} \qquad (16.14)$$

where $\dot{\theta}$ = antenna rotation rate
a = horizontal antenna aperture

This spectrum can be approximated by a gaussian distribution with standard deviation

$$\sigma_c = 0.265\frac{f_r}{n} = 0.265\frac{\dot{\theta}}{\theta_a} \approx 0.265\frac{a\,\dot{\theta}}{\lambda} \qquad (16.15)$$

where λ and a are in the same units, θ_a is the one-way half-power beamwidth, and n is the number of hits per beamwidth. The approximation $\theta_a \approx \lambda/a$ is representative of antenna distribution yielding acceptable sidelobe levels.

It can be seen that the differential return is

$$\Delta G^2(\theta) \not\approx \frac{dG^2(\theta)}{d\theta}\Delta\theta = \frac{dG^2(\theta)}{d\theta}\dot{\theta}T_p \qquad (16.16)$$

This suggests[7,13] that a correction signal in the reverse sense to $\Delta G^2(\theta)$ be applied, as shown in Fig. 16.14c. Half the correction is added to one pulse and half subtracted from the other, so that

$$\text{Correction signal} = \frac{\Delta G^2(\theta)}{2} = \frac{\dot{\theta}T_p}{2}\frac{d\Sigma^2(\theta)}{d\theta} \qquad (16.17)$$

$$= \dot{\theta}T_p\Sigma(\theta)\frac{d\Sigma(\theta)}{d\theta}$$

where $\Sigma^2(\theta)$ was substituted for $G^2(\theta)$. The radar transmits a sum pattern $\Sigma(\theta)$ and receives on the difference pattern $\Delta(\theta)$, so that the received signal is proportional to the product of the two. If the signal received on the difference pattern is used as the correction, we have

$$E_c = \Delta(\theta)\Sigma(\theta) \qquad (16.18)$$

By comparing Eqs. (16.17) and (16.18), we see that, for E_c to approximate the correction signal, the difference patterns should be

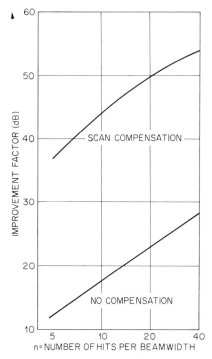

FIG. 16.15 MTI improvement factor for a step-scan compensation of a single-delay canceler as a function of the number of hits per beamwidth. The antenna pattern is $(\sin x)/x$.

$$\Delta(\theta) = \dot{\theta} T_p \frac{d\Sigma(\theta)}{d\theta} \qquad (16.19)$$

The derivative of the sum pattern is similar to a difference pattern in that it is positive at the main-beam null, $-\theta_0$, and decreases to zero on the antenna centerline and then goes negative until θ_0.

By referring to Fig. 16.13, one observes that the mechanization for scan compensation is fundamentally similar to the DPCA mechanization except that the difference signal is applied in phase with the sum signal and amplified by an amount determined by the antenna rotation per interpulse period.

The signals required, if the transmission signal $\Sigma(\theta)$ that appears in each channel is neglected, are $\Sigma(\theta) \pm \dot{\theta} T_p \Delta(\theta)$, where [is the ratio of the amplification in the two channels chosen to maximize the clutter rejection. The required difference-pattern slope is determined by the derivative of the scan pattern, which differs from the DPCA criterion. This technique is known as step-scan compensation because the system electronically points the antenna slightly ahead of and behind boresight each pulse so that a leading and lagging pair are taken from successive returns to obtain the effect of the antenna remaining stationary.

Figure 16.15 shows the improvement obtained by Dickey and Santa[7] for single-delay cancellation.

Compensation-Pattern Selection. Selection of the compensation pattern depends on the level of system performance required, the type of MTI filtering used, the platform velocity, scan rate, and the characteristics required by normal radar parameters such as resolution, distortion, gain, sidelobes, etc. For instance, an exponential pattern and its corresponding difference pattern are excellent for single-delay-cancellation DPCA but are unsatisfactory when double-delay cancellation is used. This is because the single-delay canceler requires the best match between the actual pattern and the required pattern near boresight, whereas double-delay cancellation requires the best match on the beam shoulder. Step-scan compensation usually requires the difference-pattern peaks to be near the nulls of the sum pattern to match.

Grissetti et al.[13] have shown that for step-scan compensation the improvement factor for single-delay cancellation increases as a function of the number of hits at

20 dB/decade; for the first-derivative*-type step-scan compensation, at the rate of 40 dB/decade; and with first- and second-derivative compensation, at the rate of 60 dB/decade. Hence, for a ground-based system that is limited by scan rate, one should improve the compensation pattern rather than use a higher-order MTI canceler. However, airborne systems are primarily limited by platform motion and require both better cancelers and compensation for operation in a land clutter environment. In the sea clutter environment the system is usually dominated by the spectral width of the velocity spectrum or platform motion rather than scanning. The applicability of DPCA or step-scan compensation in the latter case is dependent on the particular system parameters.

16.6 SIMULTANEOUS PLATFORM MOTION AND SCAN COMPENSATION

In AMTI systems having many hits per scan, scanning is a secondary limitation for an uncompensated double canceler. However, the performance of a DPCA system is significantly reduced when it is scanned. This is due to the scanning modulation on the difference pattern used for platform-motion compensation.

Since the DPCA applies the difference pattern in quadrature to the sum pattern to compensate for phase error and step scan applies the difference pattern in phase to compensate for amplitude error, it is possible to combine the two techniques by properly scaling and applying the difference pattern both in phase and in quadrature. The scaling factors are chosen to maximize the improvement factor under conditions of scanning and platform motion.

The relationships for a double-delay (three-pulse) AMTI are shown in the phasor diagram, Fig. 16.16. The phase advance between the first pair of pulses, received by the sum pattern Σ, is

$$2_{\eta 1} = \frac{4\pi T_p}{\lambda} \left[V_x \left(\sin\theta_2 - \sin\frac{\omega_r T_p}{2} \right) + V_y \left(\cos\frac{\omega_r T p}{2} - \cos\theta_2 \right) \right] \quad (16.20)$$

and the phase advance between the second pair of pulses is

$$2_{\eta 2} = \frac{4\pi T_p}{\lambda} \left[V_x \left(\sin\theta_2 + \sin\frac{\omega_r T_p}{2} \right) + V_y \left(\cos\frac{\omega_r T_p}{2} - \cos\theta_2 \right) \right] \quad (16.21)$$

where θ_2 is the direction of the clutter cell with respect to the antenna pointing angle when the second pulse is received and ω_r is the antenna scan rate. The subscripts on the received signals Σ_i and Δ_i indicate the pulse reception sequence.

The difference pattern Δ is used to generate an in-phase correction for scanning motion and a quadrature correction for platform motion. This process yields the set of resultant signals R_{ij}, where the subscript i denotes the pulse pair

---

*The compensation required by $\Delta G^2(\theta)/2$ can be determined from a Taylor's series expansion of $G^2(\theta)$. In the preceding discussion we used the first derivative. Using higher-order terms gives an improved correction signal.

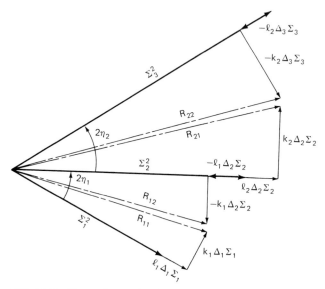

FIG. 16.16   Phasor diagram for simultaneous scanning and motion compensation.

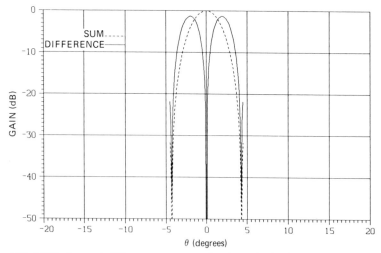

FIG. 16.17   Sum and difference patterns used to determine DPCA performance.

and the subscript j denotes the component of the pair. Since η_1 does not equal η_2, different weighting constants are required for each pulse pair. The values of k_1 for the quadrature correction of the first pulse pair, k_2 for the quadrature correction for the second pulse pair, ℓ_1 for the in-phase correction for the first pulse pair, and ℓ_2 for the in-phase correction for the second pulse pair are optimized by minimizing the integrated residue power over the significant portion of the antenna pattern, usually chosen between the first nulls of the main beam.

Figure 16.17 shows the sum and difference main-beam patterns for an aperture

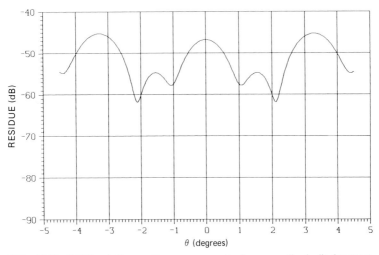

FIG. 16.18 DPCA clutter residue versus angle for normalized displacement $V_n = 0.04$ and normalized scanning motion $W_n = 0.04$.

20 wavelengths long. Figure 16.18 shows the residue for the case when the fraction of the horizontal aperture width a traveled per interpulse period T_p, $V_n = V_x T_p/a$, is equal to 0.04 and when the number of wavelengths that the aperture tip rotates per interpulse period, $W_n = a\omega_r T_p/2\lambda$, is equal to 0.04. The corresponding improvement factor is 52 dB.

The improvement factor is shown in Fig. 16.19 for a range of normalized platform motion V_n as a function of normalized scanning displacements W_n. The

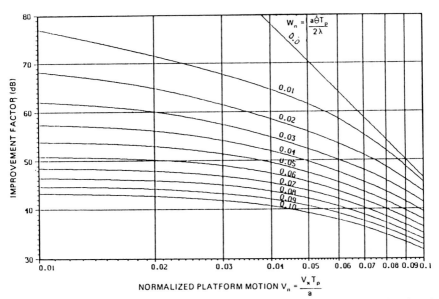

FIG. 16.19 DPCA improvement factor versus normalized platform motion V_n as a function of normalized scanning motion W_n.

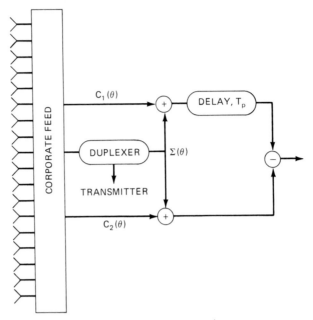

FIG. 16.20   Optimized DPCA phase compensation.

nonscanning case is shown as $W_n = 0$. The improvement factors were computed for the 20-wavelength aperture patterns shown in Fig. 16.17.

Andrews[14] has developed an optimization procedure for platform-motion compensation that rotates the phasors directly rather than by using a quadrature correction. The procedure determines the antenna feed coeficients for two compensation patterns, one of which, $C_1(\theta)$, is added to the sum pattern $\Sigma(\theta)$ and fed to the undelayed canceler path and the other, $C_2(\theta)$, is added to the sum pattern and fed to the delayed path as shown in Fig. 16.20. The procedure was developed for a single-delay canceler and a nonscanning antenna. Andrews used the procedure to minimize the residue power over the full antenna pattern, which includes the main-beam and sidelobe regions.

16.7 PLATFORM-MOTION COMPENSATION, FORWARD DIRECTION

The previous sections discussed the compensation for the component of platform motion parallel to the antenna aperture. TACCAR removes the average component of platform motion perpendicular to the aperture. Wheeler Laboratories (now Hazeltine Corporation) developed the Coincident Phase Center Technique (CPCT)[15] to remove the spectral spread due to the velocity component perpendicular to the aperture and due to the component parallel to the aperture. Removal of the component parallel to the aperture uses the DPCA pattern synthesis technique described in Ref. 8, which creates two similarly shaped illumination functions whose phase centers are physically displaced. Removal of the component perpendicular to the aperture is accomplished by a novel extension of this concept.

The first term of Eq. (16.2) for spectral width due to platform motion approaches zero as the antenna points ahead. However the second term of Eq. (16.2) dominates as the antenna approaches within a few beamwidths of the aircraft's ground track. In this region

$$f_d \approx \frac{4V_y}{\lambda} \sin 2\frac{\theta}{2} \approx \frac{V_y\theta^2}{\lambda} \qquad (16.22)$$

which yields a single-sided spectrum that is significantly narrower than the spectrum abeam. For moderate platform speeds and lower-frequency (UHF) radars this effect is negligible, and compensation is not required.

When it is necessary to compensate for this effect, the phase center of the antenna must be displaced ahead of the aperture and behind the aperture for alternate receive pulses so that the phase centers are coincident for a moving plat-

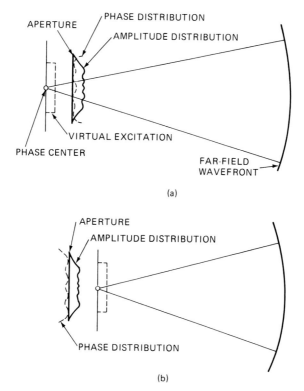

(a)

(b)

FIG. 16.21 CPCT concept showing displacement of the phase center (*a*) behind the physical aperture and (*b*) ahead of the physical aperture. (*Courtesy of Hazeltine Inc.*[15])

form. This technique can be extended to more than two pulses by using the necessary phase-center displacements for each pulse. In order to maintain the effective PRF, the displacement must compensate for the two-way transmission path. To accomplish this displacement, near-field antenna principles are utilized. A desired aperture distribution function is specified. The near-field amplitude and phase are calculated at a given distance from the origin. If this field is used as the actual illumination function, a virtual aperture is created with the desired distribution function at the same distance behind the physical antenna. Figure 16.21a[15] shows the phase and amplitude distribution required to form a uniform virtual distribution displaced behind the physical aperture. It can be shown that if the phase of the illumination function is reversed $\phi' = -\phi$, the desired virtual distribution function is displaced ahead of the aperture as shown in Fig. 16.21b.

In practice, performance is limited by the ability to produce the required illumination function. As the displacement increases, a larger physical aperture size is required to produce the desired virtual aperture size owing to beam spreading. This can be seen in Fig. 16.21. The effectiveness of the correction varies with elevation angle since the actual displacement along the line of sight varies with elevation angle. This effect is more pronounced at higher aircraft speeds and higher radar frequencies. A change in the magnitude of the correction factor or even the compensation pattern with range, height, and velocity could be utilized to retain performance.

Figure 16.22 illustrates the theoretical MTI performance of a CPCT system as a function of beam-pointing direction and interpulse motion normalized to the interpulse motion used to design the compensation patterns. (*Cancellation ratio*

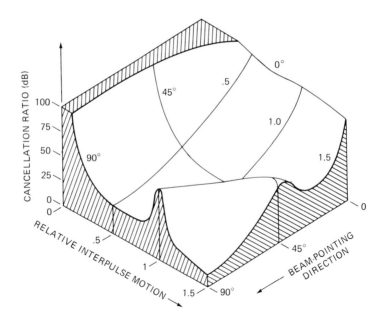

FIG. 16.22 CPCT cancellation ratio, in decibels, as a function of relative interpulse motion and beam-pointing direction. (*Courtesy of Hazeltine Inc.*[15])

is defined as the ratio of input clutter power to output clutter residue power.) The peak on the 90° axis is typical of the optimized DPCA performance illustrated in Figure 16.12.

16.8 SPACE-TIME ADAPTIVE MOTION COMPENSATION

Several methods have been described to compensate for antenna motion. All these techniques are applied in the radar design phase for a specific set of operational parameters. Controls (usually automatic) are provided to adjust weights for operational conditions around the design value.

The development of digital radar technology and economical high-speed processors allows the use of dynamic space-time adaptive array processing,[16] whereby a set of antenna patterns that displace the phase center of the array both along and orthogonal to the array are continually synthesized to maximize the signal-to-clutter ratio. Spatial adaptive array processing combines an array of signals received at the same instant of time that are sampled at the different spatial locations corresponding to the antenna elements. Temporal adaptive array processing combines an array of signals received at the same spatial location (e.g., the output of a reflector antenna) that are sampled at different instances of time, such as several interpulse periods for an adaptive MTI. Space-time adaptive array processing combines a two-dimensional array of signals sampled at different instances of time and at different spatial locations.

A basic block diagram of a radar incorporating space-time adaptive array processing is shown in Fig. 16.23. Circuits for auxiliary functions such as pulse compression, clutter gating, synchronization, and TACCAR are not shown. With the exception of the interchange of the corporate-feed and duplexing functions, the transmit channel is identical to that of any other radar. An individual duplexer is placed between each corporate-feed output and its corresponding antenna element. Provision could be included for electronic beam steering using high-power phase shifters or transmit modules with low-power beam steering.

On receive, each duplexer output is sent to its own digital receiver and adaptive processing module (APM), which provides a weighted undelayed signal that is combined with the outputs of the other adaptive processing modules to form an undelayed antenna beam. The weighted signal received on the previous pulse is combined with the corresponding outputs of the adaptive processing modules to form a delayed antenna beam. The two beams are then subtracted to produce the single-delay AMTI output. The output is then sent to the automatic detector for further processing and display. The output is also returned to the adaptive processing modules.

Figure 16.24 shows the block diagram of a typical digital receiver. The signal received from a single antenna element is amplified and converted to IF. The IF signal is further amplified and converted to baseband by using the synchronous demodulators. One of the bipolar video outputs, I, is the component that is in phase with the reference oscillator. The other bipolar video output, Q, is in quadrature with the reference oscillator. The two bipolar video signals are sampled for each range cell and converted to digital representation by the A/D converters. The output logic formats the I and Q values for transfer to the adaptive processing module. The I, Q pair of numbers is a phasor representing the instantaneous phase and amplitude of each range cell in rectangular coordinates.

Figure 16.25 shows a block diagram of the adaptive processing module used

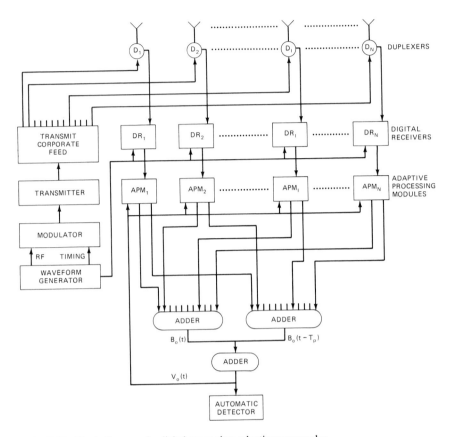

FIG. 16.23 Block diagram of a digital space-time adaptive array radar.

for space-time adaptive array processing. All components are digital processing blocks that can be implemented in various combinations of hardware and software. The complex value of the sampled signal $V_i(t)$ is multiplied by the complex adaptive weight W_{i1} to form the ith-channel input to the adder forming the undelayed antenna beam. The value is also routed to a buffer for storage. The previously stored value $V_i(t - T_p)$ is multiplied by the delayed channel weight W_{i2} to form the ith-channel input to the adder forming the delayed beam.

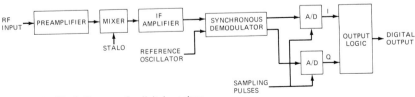

FIG. 16.24 Block diagram of a digital receiver.

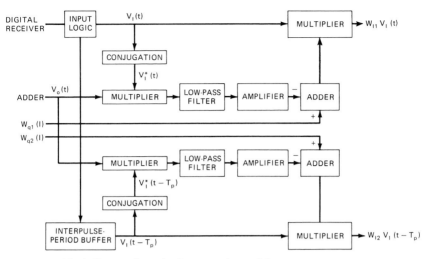

FIG. 16.25 Block diagram of an adaptive processing module.

The weights W_{i1} and W_{i2} are determined by the Howells-Applebaum algorithm.[17] The correlation between the input signal $V_i(t)$ and the output signal $V_o(t)$ is determined by multiplying $V_o(t)$ by the complex conjugate of the input signal $V^*_i(t)$ and averaging the resultant by passing it through a low-pass digital filter. This correlation is amplified and subtracted from the appropriate quiescent weight W_{q1i} or W_{q2i} to obtain the slowly varying weight W_{i1} or W_{i2}. The quiescent weight is the product of the antenna illumination factor for the ith element that will yield the desired antenna pattern and the MTI weight for the delayed or undelayed pulse. This would include the phase component required to steer the beam in a given direction. The closed-loop action will drive the weight so that the average correlation of the output V_o and each input V_i approximates the value of the quiescent weight.

$$(V^*_iV_o) \approx W_{q1} \quad \text{for} \quad i = 1 \text{ to } 2N$$

$$(V^*_i\Sigma W_jV_j) \approx W_{q1}$$

$$\Sigma(V^*_iV_j)W_j \approx W_{q1} \tag{16.23}$$

where ( ) indicates the time average. If we define the N values of each delayed variable by extending the subscript range over $N + 1$ to K, where $K = 2N$, and define $m_{ij} = (V^*_i V_j)$, we have the set of equations

$$m_{11}W_1 + W_{12}W_2 + \cdots + m_{1K}W_K = W_{q1}$$

$$M_{21}W_1 + m_{22}W_2 + \cdots + m_{2K}W_K = W_{q2}$$

$$\begin{array}{ccccc} \cdot & \cdot & \cdots & \cdot & \cdot \\ \cdot & \cdot & \cdots & \cdot & \cdot \\ \cdot & \cdot & \cdots & \cdot & \cdot \end{array}$$

$$m_{K1}W_1 + m_{K2}W_2 + \cdots + m_{KK}W_K = W_{qK}$$

which can be expressed in matrix notation as

$$MW = W_q \qquad (16.24)$$

This set of equations can be solved for the steady-state set of weights $W1$ to W_K, which can be expressed in matrix notation as the familiar equation

$$W = M^{-1}W_q \qquad (16.25)$$

These weights have been shown[17,18] to be the optimum set which maximizes the signal-to-interference ratio. Because of the smoothing required to keep the weights from jittering,[18] the weights adapt to their steady-state values in a time determined by the clutter power and the allowable steady-state variation in the weights. Other algorithms[19] can speed up the adaptation rate, but a more complex mechanization is required.

This process results in a delayed beam and an undelayed beam, whose phase centers are offset to compensate for platform motion. If jamming is present in the sidelobes or on the shoulder of the main beam, nulls will be formed in each jamming direction. If mixed clutter and jamming are present, the weights will adjust to maximize the signal-to-total-interference ratio. The signal is defined as a short-duration pulse return from the direction determined by the quiescent weights and at a doppler frequency corresponding to half of the PRF. Independent quiescent weights for each subchannel could be utilized to optimize the doppler response for another frequency.

Performance Capability of Space-Time Adaptive Arrays. The performance attainable from the space-time array is limited by the aircraft speed, the array alignment with respect to the aircraft ground track, and the system accuracies. A performance analysis for a 16-element, two-pulse space-time array is presented in Figs. 16.26 and 16.27. The antenna elements were spaced at a half wavelength and assumed to be omnidirectional. The clutter model was assumed to be homogeneous. The improvement factor is based on the fully adapted weights. The clutter-to-noise ratio was selected to limit performance to 92 dB.

The improvement factor for adaptive arrays is usually defined as the ratio of the signal-to-interference-power ratio at the output of the processor to the signal-to-interference ratio at the input of the processor. The signal is specified as coming from the direction and at the doppler frequency specified by the quiescent weights. In MTI systems, it is customary to define the improvement factor as the average response over the doppler interval. For a single-delay processor, the MTI improvement factor is 3 dB less than the adapted improvement factor shown in Figs. 16.28 and 16.27.

Figure 16.28 shows the improvement factor for an array pointing along the ground track of the aircraft as a function of motion expressed in terms of wavelengths per interpulse period. The dashed line shows the unadapted single-delay improvement factor for an antenna with a Dolph-Chebyshev aperture illumination that provides a 28 dB uniform peak sidelobe level. The solid line shows the adapted improvement factor to be 92 dB for a stationary antenna, reducing to 89 dB for 4 wavelengths per interpulse-period platform motion. Figure 16.27 shows

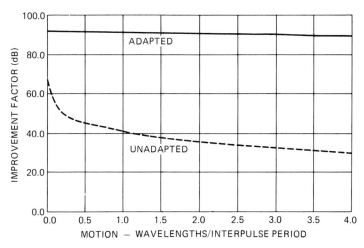

FIG. 16.26 Adapted and unadapted improvement factor as a function of normalized antenna motion per interpulse period; 16-element (half-wavelength spacing), two-pulse space-time adaptive processor; antenna array aligned to the ground track.

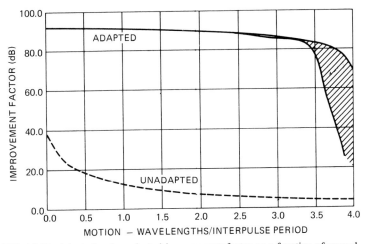

FIG. 16.27 Adapted and unadapted improvement factor as a function of normalized antenna motion per interpulse period; 16-element (half-wavelength spacing), two-pulse space-time adaptive processor; antenna array aligned to perpendicular to the ground track.

the performance when the antenna is pointing abeam. This is the standard DPCA case. The performance holds to 1.5 wavelengths per interpulse-period platform motion, then decreases slightly, and drops off sharply above 3.5 wavelengths per interpulse period. The shaded region is where the improvement factor varies within the limits. A peak occurs when the platform motion is a multiple of a quarter wavelength. The clutter-to-noise-ratio limitation results in cusping not being

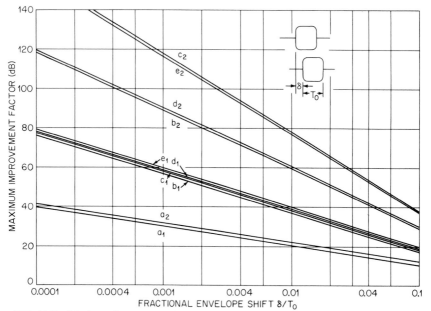

FIG. 16.28 Maximum improvement factor as a function of a fractional envelope shift for typical pulse envelope shapes. Subscript 1 indicates single delay; 2, double delay. Pulse shape a = rectangular; b = cosine; c = cosine-squared; d = triangular; e = gaussian.

visible at speeds less the 3.5 wavelengths per interpulse period. The performance at other angles is between these two cases, and peaking does not occur.

16.9 LIMITATION OF IMPROVEMENT FACTOR DUE TO PULSE ENVELOPE SHIFT

The doppler frequency that arises because of the radial component of aircraft motion results in an incremental phase shift between successive radar pulses. The envelope of the radar pulse is also delayed a corresponding amount. The TACCAR circuit usually compensates for the phase delay at IF by changing the phase of a CW reference oscillator. Hence the envelope of a single pulse is unaffected. The mismatch in the envelope delay time between successive pulses results in a residual signal, sometimes called *ranging noise*.

Figure 16.28 shows the effect of this residual on the performance of a single-delay canceler. The idealized rectangular pulse envelope gives a pessimistic picture of this effect. Most conventional representations of pulse shape give about the same performance, 20 dB per decade.

Figure 16.28 also shows the effect of the residual on a double cancellation system. In the case of double-delay cancellation the rectangular-pulse case shows only 1.8 dB improvement over the single-delay case. However, more realistic pulse representations show substantial improvement. The triangular and cosine representations have a rolloff of about 30 dB per decade; the smoother cosine-squared and gaussian representations roll off at 40 dB per decade.

16.10   EFFECT OF MULTIPLE SPECTRA

An airborne search-radar system may be operated at an altitude so that the radar horizon is approximately at the maximum range of interest. This results in sea or ground clutter being present at all ranges of interest. Other clutter sources such as rain and chaff may coexist with the surface clutter. In most instances these sources are moving at a speed determined by the mean wind aloft and have a mean doppler frequency significantly different from that of the surface clutter. If the MTI filter is tracking the surface clutter, the spectra of the sources with a different mean doppler frequency lie in the passband of the MTI filter. A 20-kn differential in an S-band system corresponds to 200 Hz, which would be at an optimum response in a 400-PRF system. A single-delay secondary canceler can be cascaded with either a single-delay or a double-delay primary canceler. The primary canceler tracks the mean surface velocity and rejects surface clutter. The single-delay canceler tracks the secondary source and rejects it. Since the pass and rejection bands of the two cancelers overlap, the MTI improvement factor for each clutter source is a function of their spectral separation.

Figure 16.29 shows the improvement factor for a double canceler which consists of two single cancelers, each tracking one of the spectra. It can be seen that, as the separation varies from 0 to one-half of the PRF, the performance degrades from that equivalent to a double canceler to the performance of a single canceler at half of the PRF.

The triple canceler has a double-delay canceler tracking the primary spectra and a single-delay canceler tracking the secondary spectra. The performance of the primary system varies from that of a triple canceler to a level less than that of a double canceler. The secondary-system performance varies from that of a triple canceler to a performance level lower than that of a single canceler.

16.11   DETECTION OF GROUND MOVING TARGETS

Vehicles and ships may have radial speeds that are significantly greater than the clutter velocity spectrum. This allows these targets to be detected. However, for an AEW system operating overland, ground traffic can saturate the tracking system. Furthermore, traffic density on major highways, target aspect changes causing strong scintillation, and shadowing by cultural features that occurs at low grazing angles can result in misassociation in the target-tracking system, causing false and runaway tracks. Runaway tracks are false tracks whose high speed causes them to rapidly move away from the true target position. They can associate with other reported positions or false alarms and thus perpetuate themselves. Since ground traffic is not of interest in the AEW case, these undesired targets are censored on the basis of highway grid maps, the small change in range during the antenna dwell, or the small velocity determined by a scan-to-scan processor or the tracking system. High-PRF pulse doppler radars use a low-frequency stopband filter to reject these velocity regions along with main-beam clutter. Low-PRF AEW radars bypass these censor circuits in the portion of the surveillance region that is over water. This allows slow-moving shipping targets to be detected.

Air-to-surface search radars, as well as airborne battlefield surveillance radars, are designed to maximize detection of slowly moving targets. Higher-

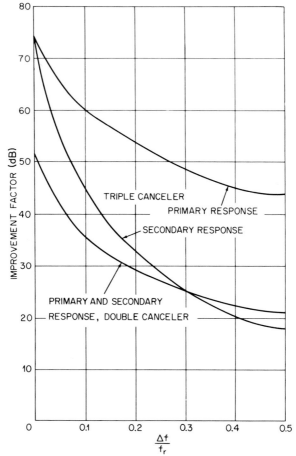

FIG. 16.29 MTI improvement factor for a double-notch canceler tracking two spectra as a function of the normalized spectra separation $\Delta f/f_r$. Normalized spectral width $\sigma_c/f_r = 0.01$.

frequency bands (X or K) are chosen to maximize the doppler shift. The PRF is chosen to optimize detection over the expected doppler frequency region of these targets. Since a strong clutter background is usually present, battlefield surveillance radars can effectively utilize noncoherent MTI techniques. However, the clutter spectrum is convolved with the target spectrum, which broadens the resultant target spectrum, thus widening the blind-speed zone and reducing doppler resolution. When the target phase coincides with the clutter phase, the targets are suppressed. In regions that are shadowed by hills or mountains, the targets are not detected.

Side-looking radars can produce a large number of pulses, thus increasing radar sensitivity. If a coherent radar is used, improved sensitivity and resolution can be obtained by using doppler filter banks or digital fast Fourier transform (FFT) processing. If the platform motion compared with the aperture length is sufficiently large, platform-motion compensation will be required.

Ship detection can be improved by rapidly scanning the antenna so that sea clutter is decorrelated and surface-target returns are integrated or leave a pattern of returns indicating their track. In some cases, frequency agility can also be utilized to decorrelate clutter and integrate ship target returns. Scan-to-scan video cancellation can be utilized for detecting moving targets overland if their scan-to-scan motion is of the order of the radar pulse width.

REFERENCES

1. Emerson, R. C.: Some Pulsed Doppler MTI and AMTI Techniques, *Rand Corporation Rept. R-274, DDC Doc. AD* 65881, Mar. 1, 1954. (Reprinted in Ref. 6.)

2. George, T. S.: Fluctuations of Ground Clutter Return in Airborne Radar Equipment, *Proc. IEE (London),* vol. 99, pt. IV, pp. 92–99, April 1952.

3. Dickey, F. R., Jr.: Theoretical Performance of Airborne Moving Target Indicators, *IRE Trans.,* vol. PGAE-8, pp. 12–23, June 1953.

4. Berkowitz, R. S. (ed.): "Modern Radar: Analysis, Evaluation and System Design," John Wiley & Sons, New York, 1966.

5. Barton, D. K.: "Radar Systems Analysis," Prentice-Hall, Englewood Cliffs, N.J., 1964.

6. Schlerer, D. C. (ed.): "MTI Radar," Artech House, Inc., Norwood, Mass., 1978.

7. Dickey, F. R., Jr., and M. M. Santa: Final Report on Anticlutter Techniques, *General Electric Company Rept. R65EMH37,* Mar. 1, 1953.

8. Anderson, D. B.: A Microwave Technique to Reduce Platform Motion and Scanning Noise in Airborne Moving Target Radar, *IRE WESCON Conv. Rec.,* vol. 2, pt. 1, pp. 202–211, 1958.

9. "Final Engineering Report on Displaced Phase Center Antenna," vol. 1, Mar. 26, 1956; vols. 2 and 3, Apr. 18, 1957, General Electric Company, Schenectady, N.Y.

10. Urkowitz, H.: The Effect of Antenna Patterns on Performance of Dual Antenna Radar Moving Target Indicators, *IEEE Trans.,* vol. ANE-11, pp. 218–223, December 1964.

11. Tsandoulis, G. N.: Tolerance Control in an Array Antenna, *Microwave J.,* pp. 24–35, October 1977.

12. Shroeder, K. G.: Beam Patterns for Phase Monopulse Arrays, *Microwaves,* pp. 18–27, March 1963.

13. Grissetti, R. S., M. M. Santa, and G. M. Kirkpatrick: Effect of Internal Fluctuations and Scanning on Clutter Attenuation in MTI Radar, *IRE Trans.,* vol. ANE-2, pp. 37–41, March 1955.

14. Andrews, G. A.: Airborne Radar Motion Compensation Techniques: Optimum Array Correction Patterns, *Naval Res. Lab. Rept.* 7977, Mar. 16, 1976.

15. Lopez, A. R., and W. W. Ganz: CPCT Antennas for AMTI Radar, vol. 2: Theoretical Study, *Air Force Avionics Lab. Rept. WL1630.22, AD* 51858, June 1970. (Not readily available.)

16. Brennan, L. E., J. D. Mallett, and I. S. Reed: Adaptive Arrays in Airborne MTI Radar, *IEEE Trans.,* vol. AP-24, pp. 607–615, September 1976.

17. Applebaum, S. P.: Adaptive Arrays, *IEEE Trans.,* vol. AP-24, pp. 585–598, September 1976.

18. Brennan, L. E., E. L. Pugh, and I. S. Reed: Control Loop Noise in Adaptive Array Antennas, *IEEE Trans.,* vol. AES-7, March 1971.

19. Monzingo, R. A., and T. W. Miller: "Introduction to Adaptive Arrays," John Wiley & Sons, New York, 1980.

CHAPTER 17
PULSE DOPPLER RADAR

William H. Long
David H. Mooney
William A. Skillman
Westinghouse Electric Corporation

17.1 CHARACTERISTICS AND APPLICATIONS

Nomenclature. For the purpose of this chapter, the term *pulse doppler* (PD) will be used for radars to which the following apply:

1. They utilize coherent transmission and reception; that is, each transmitted pulse and the receiver local oscillator are synchronized to a free-running, highly stable oscillator.

2. They use a sufficiently high pulse repetition frequency (PRF) to be ambiguous in range.

3. They employ coherent processing to reject main-beam clutter, enhance target detection, and aid in target discrimination or classification.

Applications. PD is applied principally to radar systems requiring the detection of moving targets in a severe clutter environment. Table 17.1 lists typical applications [1-10] and requirements. This chapter will deal principally with airborne applications, although the basic principles can also be applied to the ground-based case.

PRFs. Pulse doppler radars are generally divided into two broad PRF categories: medium and high PRF.[11] In a medium-PRF radar[12-14] the target and clutter ranges and velocities of interest are usually ambiguous, while in a high-PRF radar[15] the range is ambiguous but the velocity is unambiguous (or has at most a single velocity ambiguity as discussed later).

A low-PRF radar, commonly called a moving-target indicator (MTI),[16] is one in which the ranges of interest are unambiguous while the velocities are usually ambiguous. MTI radars are generally not categorized as pulse doppler radars, although the principles of operation are similar. A comparison of MTI and pulse doppler radars is shown in Table 17.2.

TABLE 17.1 Pulse Doppler Applications and Requirements

Radar application	Requirements
Airborne or spaceborne surveillance	Long detection range; accurate range data
Airborne interceptor or fire control	Medium detection range; accurate range, velocity data
Ground-based surveillance	Medium detection range; accurate range data
Battlefield surveillance (slow-moving target detection)	Medium detection range; accurate range, velocity data
Missile seeker	May not need true range information
Ground-based weapon control	Short range; accurate range, velocity data
Meteorological	High velocity and range data resolution
Missile warning	Short detection range; very low false-alarm rate

TABLE 17.2 Comparison of MTI and Pulse Doppler (PD) Radars

	Advantages	Disadvantages
MTI—low PRF	Can sort clutter from targets on basis of range. No range ghosts. Front-end STC suppresses sidelobe detections and reduces dynamic range requirements.	Low doppler visibility due to multiple blind speeds. Poor slow-moving target rejection. Cannot measure radial target velocity.
PD—medium PRF	Good performance at all target aspects. Good slow-moving target rejection. Measures radial velocity. Less range eclipsing than in high PRF.	Range ghosts. Sidelobe clutter limits performance. High stability requirements due to range folding.
PD—high PRF	Can be sidelobe clutter–free for some target aspects. Single doppler blind zone at zero velocity. Good slow-moving target rejection. Measures radial velocity. Velocity-only detection can improve detection range.	Sidelobe clutter limits performance. Range eclipsing. Range ghosts. High stability requirements due to range folding.

Pulse Doppler Spectrum. The transmitted spectrum of a pulse doppler radar consists of discrete lines at the carrier frequency f_0 and at sideband frequencies $f_0 \pm if_R$, where f_R is the PRF and i is an integer. The envelope of the spectrum is determined by the pulse shape. For the rectangular pulses usually employed, a $(\sin x)/x$ spectrum is obtained.

The received spectrum from a stationary target has lines that are doppler-shifted proportionally to the line of sight, or radial velocity, between the radar platform and the target. The two-way doppler shift is given by $f_d = (2V_R/\lambda) \cos \psi_0$, where λ is the radar wavelength, V_R is the radar platform velocity, and ψ_0 is

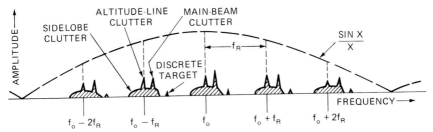

FIG. 17.1   Clutter and target frequency spectrum from a horizontally moving platform.

the angle between the velocity vector and the line of sight to the target. Illustrated in Fig. 17.1 is the received pulsed spectrum with returns from continuous clutter, such as the ground or clouds, and from discrete targets, such as aircraft, automobiles, tanks, etc.

Figure 17.2 shows the unfolded spectrum (i.e., no spectral foldover from adjacent PRF lines) in the case of horizontal motion of the radar platform, with a velocity V_R. The clutter-free region is defined as that portion of the spectrum in which no ground clutter can exist. (A clutter-free region usually does not exist with medium PRFs.) The sidelobe clutter region, $4V_R/\lambda$ in width, contains ground clutter power from the sidelobes of the antenna, although the clutter power may be below the noise level in part of the region. The main-beam region, located at $f_0 + (2V_R/\lambda) \cos \psi_0$, contains the strong return from the main beam of the antenna striking the ground at a scan angle of ψ_0, measured from the velocity vector. Rain and chaff clutter may also be large when the main beam illuminates a rain or chaff cloud. Motion due to winds may displace and/or spread the return in frequency.

Altitude-line clutter, which is due to ground clutter at near normal incidence directly below the radar platform, is at zero doppler if there is no vertical component of platform velocity. A discrete target return in the main beam is shown at $f_T = f_0 + (2V_R/\lambda) \cos \psi_0 + (2V_T/\lambda) \cos \psi_T$, where the target velocity is V_T, with an angle ψ_T between the target velocity vector and the radar target line of sight.

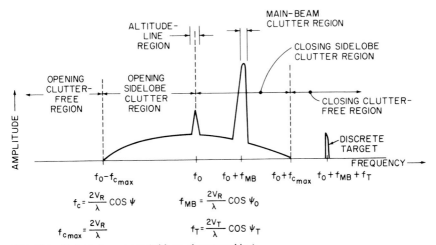

FIG. 17.2   Unfolded spectrum (with no clutter tracking).

The components of the spectrum shown in Fig. 17.2 will also vary with range as discussed later.

Figure 17.3 illustrates the various clutter doppler frequency regions as a function of the antenna azimuth and relative radar and target velocities, again for an unfolded spectrum. The ordinate is the radial, or line-of-sight, component of target velocity in units of radar platform velocity, so that the main-beam clutter region is at zero velocity and the sidelobe clutter region frequency boundaries vary sinusoidally with antenna azimuth. Thus, it shows the doppler regions in which the target can become clear of sidelobe clutter. For example, if the antenna azimuth angle is at zero, any head-on target ($V_T \cos \psi_T > 0$) is clear of sidelobe clutter, whereas if the radar is in trail behind the target ($\psi_T = 180°$ and $\psi_0 = 0°$), the target's radial velocity has to be greater than twice that of the radar to become clear of sidelobe clutter.

The sidelobe clear and clutter regions can also be expressed in terms of the aspect angle with respect to the target,[14] as shown in Fig. 17.4. Here, collision geometry is assumed in which the radar and target aircraft fly straight-line paths toward an intercept point; the look angle of the radar ψ_0 and the aspect angle of the target ψ_T are constant for a given set of radar and target speeds V_R and V_T, respectively. The center of the diagram is the target, and the angle to the radar on the circumference is the aspect angle. The aspect angle and look angles satisfy the equation $V_R \sin \psi_0 = V_T \sin \psi_T$, which is defined as a collision course. The target aspect angle is zero for a head-on condition and 180° for a tail chase. The aspect angle corresponding to the boundary between the sidelobe clutter region and the sidelobe clear region is a function of the relative radar-target velocity ratio and is shown in Fig. 17.4 for four cases. Case 1 is where the radar and target speeds are equal and the target can be seen clear of sidelobe clutter in a head-on

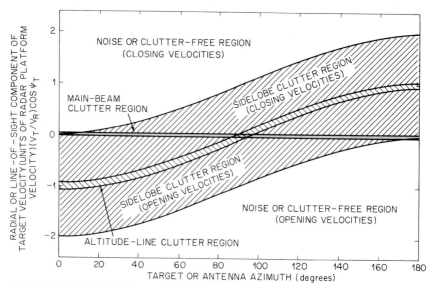

NOTE: WIDTH OF ALTITUDE–LINE AND MAIN–BEAM CLUTTER REGIONS VARIES WITH CONDITIONS; AZIMUTH IS MEASURED FROM RADAR PLATFORM VELOCITY VECTOR TO THE ANTENNA BORESIGHT OR TO THE LINE OF SIGHT TO THE TARGET; HORIZONTAL–MOTION CASE.

FIG. 17.3 Clutter and clutter-free regions as a function of target velocity and azimuth.

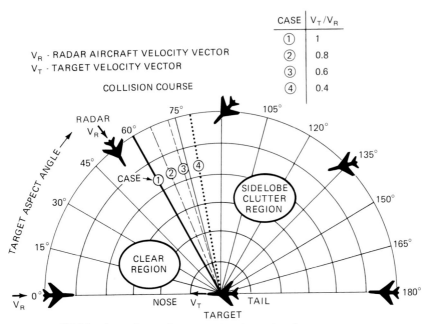

FIG. 17.4 Sidelobe clutter-clear regions versus target aspect angle.

aspect out to 60° on either side of the target's velocity vector. Similarly, Cases 2 to 4 show conditions where the target's speed is 0.8, 0.6, and 0.4 times the radar's speed, in which case the target can be seen clear of sidelobe clutter over a region of up to ±78.5° relative to the target's velocity vector. Again, these conditions are for an assumed collision course. As is evident, the aspect angle of the target clear of sidelobe clutter is always forward of the beam aspect.

Ambiguities and PRF Selection. Pulse doppler radars are generally ambiguous in either range or doppler, or both. The unambiguous range R_u is given by $c/2f_R$, where c is the speed of light and f_R is the PRF.

If the maximum target velocity to be observed is $\pm V_{T\max}$, then the minimum value of PRF, $f_{R\min}$, which is unambiguous in velocity (both magnitude and doppler sense, i.e., positive and negative), is

$$f_{R\min} = 4V_{T\max}/\lambda \qquad (17.1)$$

However, some pulse doppler radars employ a PRF which is unambiguous in velocity magnitude only, i.e., $f_{R\min} = 2V_{T\max}/\lambda$, and rely on detections in multiple PRFs during the time on target to resolve the ambiguity in doppler sense. These types of radars can be considered to be in the high-PRF category if the older definition of high PRF (no velocity ambiguity) is extended to allow one velocity ambiguity, that of doppler sense. The lower PRF eases the measurement of true range while retaining the high-PRF advantage of a single blind-speed region near zero doppler.

The choice between high and medium PRF involves a number of considerations, such as transmitter duty cycle limit, pulse compression availability, signal-processing capability, missile illumination requirements, etc., but often depends on the need for all-aspect target detectability. All-aspect coverage requires good performance in tail chase, where the target doppler is in the sidelobe clutter region near the altitude line. In a high-PRF radar, the range foldover may leave little clear region in the range dimension, thus degrading target detectability. By using a lower or medium PRF, the clear region in range is increased at the expense of velocity foldover for high-doppler targets that are in the clutter-free region in high PRF. For example, Fig. 17.5 shows the clutter-plus-noise-to-noise ratio in range doppler coordinates for a 12-kHz PRF at an altitude of 6000 ft showing the main-beam clutter, altitude line, and sidelobe clutter. The range dimension represents the unambiguous range interval R_u, and the frequency dimension represents the PRF interval. As is evident, there is a range doppler region in which the sidelobe clutter is below thermal noise and in which good target detectability can be achieved. The main-beam clutter is filtered out.

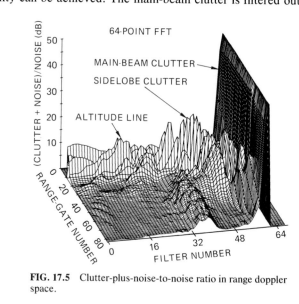

FIG. 17.5   Clutter-plus-noise-to-noise ratio in range doppler space.

Because the clutter is folded in both range and doppler with medium PRF, a number of PRFs may be required to obtain a satisfactory probability of sufficient detections to resolve the range and doppler ambiguities. The multiple PRFs move the relative location of the clear regions so that all-aspect target coverage is achieved. Since the sidelobe clutter generally covers the doppler region of interest, the ratio of the region with sidelobe clutter below noise relative to the total range-doppler space is a function of the radar altitude, speed, and antenna sidelobe level.

If a high-PRF waveform is used, the clear-range region disappears because the sidelobe clutter folds in range into the unambiguous range interval (assuming the target doppler is such that it still competes with the sidelobe clutter). However, in those doppler regions free of sidelobe clutter, as shown in Figs. 17.3 and 17.4, target detectability is limited only by thermal noise, independently of radar altitude, speed, and sidelobe level. This requires system stability sidebands to be well below noise for the worst-case main-beam clutter. Thus, although medium PRF provides all-aspect target coverage, the target is potentially competing with sidelobe clutter at all aspects, whereas with high PRF a target can become clear of sidelobe clutter at aspect angles forward of the beam aspect.

Basic Configuration. Figure 17.6 shows a representative configuration of a pulse doppler radar utilizing digital signal processing under the control of a central computer. Included are the transmitter suppression circuits, main-beam and sidelobe discrete rejection circuits, and ambiguity resolvers. The radar computer receives inputs from the on-board systems, such as the inertial unit and operator controls, and performs as a master controller for the radar. As such, it does the track loop and automatic gain control (AGC) loop filtering, antenna scan pattern generation, and clutter positioning as well as the target-processing functions (such as centroiding). In addition, the computer performs the multiple-target track functions when the radar is in a track-while-scan mode and may execute radar self-test and calibration routines. For simplicity only the search processing is shown.

Duplexer. The duplexer in a pulse doppler radar is usually a passive device such as a circulator which effectively switches the antenna between the transmitter and receiver. Considerable power may be coupled to the receiver since typically 20 to 25 dB isolation may be expected from ferrite circulators.

Receiver-Protector (R/P). The receiver-protector is a fast-response, high-power switch which prevents the transmitter output from the duplexer from damaging the sensitive receiver front end. Fast recovery is required to minimize desensitization in the range gates following the transmitted pulse.

RF Attenuator. The RF attenuator is used both for suppressing transmitter leakage from the R/P into the receiver (so that the receiver is not driven into saturation, which could lengthen recovery time after the transmitter is turned off) and for controlling the input signal levels into the receiver. The received levels are kept below saturation levels, typically with a clutter AGC in search and a target AGC in single-target track, to prevent spurious signals, which degrade performance, from being generated.

Clutter Positioning. A voltage-controlled oscillator (VCO), usually part of the stable local oscillator (stalo), is used to heterodyne main-beam clutter to zero frequency, or dc. With the clutter at dc the in-phase (I) and quadrature (Q) channel amplitude and phase-balance requirements are eased, as the images resulting from unbalance also fall near dc and can be filtered out along with the main-beam clutter.

Transmit Pulse Suppressor. Further attenuation of transmitter leakage is provided by the transmit pulse suppressor in the receiver IF, which is a gating device.

Signal Processing. The analog output of the receiver is downconverted to baseband (dc) via quadrature mixing. The in-phase and quadrature signals are passed through a matched filter and converted to digital words by an analog-to-digital (A/D) converter. Following the A/D is typically a delay-line clutter canceler and doppler filter bank for main-beam clutter rejection and coherent integration. The filter bank is usually realized by using the fast Fourier transform

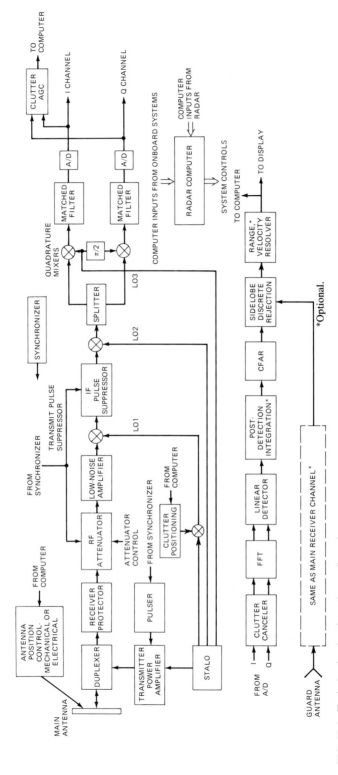

FIG. 17.6 Typical pulse doppler radar configuration.

(FFT) or by the discrete Fourier transform (DFT) for a small number of filters. Appropriate weighting is employed to reduce the filter sidelobes. The voltage envelope at the output of the FFT is formed by using an I/Q combining approximation. Postdetection integration (PDI) may be used where each range-gate–doppler-filter output is linearly summed over several coherent looks. The PDI output is compared with a detection threshold determined by a constant-false-alarm-rate (CFAR)[17–20] process.

Following the CFAR is the sidelobe discrete rejection logic, discussed in Sec. 17.2, and the range and velocity ambiguity resolvers (if used). The final detection outputs are passed to the radar display and computer.

17.2 PULSE DOPPLER CLUTTER

General. Clutter returns from various scatterers have a strong influence on the design of a pulse doppler radar as well as an effect on the probability of detection of point targets. Clutter scatterers include terrain, both ground and water, rain, snow, and chaff. Since the antennas generally used in pulse doppler radars have a single, relatively high-gain main beam, main-beam clutter may be the largest signal handled by the radar when in a down-look condition, which is a principal reason for the use of medium- and high-PRF pulse doppler radars. The narrow beam limits the frequency extent of this clutter to a relatively small portion of the doppler spectrum. The remainder of the antenna pattern consists of sidelobes which result in sidelobe clutter. This clutter is generally much smaller than the main-beam clutter but covers much more of the frequency domain. The sidelobe clutter from the ground directly below the radar, the altitude line, is frequently large owing to a high reflection coefficient at steep grazing angles, the large geometric area, and the short range. Range performance is degraded for targets in the sidelobe clutter region wherever the clutter is near or above the receiver noise level. Multiple PRFs may be used to move the target with respect to the clutter, thus avoiding completely blind ranges or blind frequencies due to high clutter levels. This relative motion occurs owing to the range and doppler foldover. If one PRF folds sidelobe clutter and a target to the same apparent range and doppler, a sufficient change of PRF will separate them.

Ground Clutter in a Stationary Radar. When the radar is fixed with respect to the ground, both main-beam and sidelobe clutter returns occur at zero-doppler offset, the transmit frequency. The sidelobe clutter is usually small compared with main-beam clutter as long as some part of the main beam strikes the ground. The clutter can be calculated as in a pulse radar, then folded in range as a function of the PRF.

Ground Clutter in a Moving Radar. When the radar is moving with a velocity V_R, the clutter is spread over the frequency domain as illustrated in Fig. 17.2 for the special case of horizontal motion. The foldover in range and doppler is illustrated in Fig. 17.7 for a medium-PRF radar where the clutter is

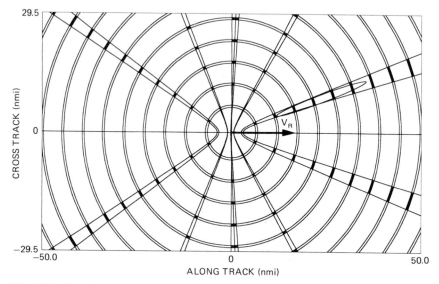

FIG. 17.7 Plan view of range-gate and doppler filter areas. Radar altitude, 10,000 ft; velocity, 1000 kn to right; dive angle, 10°; radar wavelength, 3 cm; PRF, 15 kHz; range-gate width, 6.67 μs; gate, 4; doppler filter, at 2 kHz; bandwidth, 1 kHz; beamwidth, 5° (circular); main-beam azimuth, 20°; depression angle, 5°.

ambiguous in both range and doppler. The radar platform is moving to the right at 1000 kn with a dive angle of 10°. The narrow annuli define the ground area that contributes to clutter in the selected range gate. The five narrow hyperbolic bands define the area that contributes to clutter in the selected doppler filter. The shaded intersections represent the area that contributes to the range-gate–doppler-filter cell. Each area contributes clutter power dependent on antenna gain in the direction of the area and the reflectivity of the area.

The main beam illuminates the elliptical area to the left of the ground track. Since this area lies entirely within the filter area, the main-beam clutter falls within this filter, and all other filters receive sidelobe clutter. Five range annuli are intersected by the main-beam ellipse; so the main-beam clutter in this range gate is the vector sum of the signals received from all five areas. Owing to this high degree of range foldover, all range gates will have approximately equal clutter.

If the main beam were scanned 360° in azimuth, the main-beam clutter would scan in frequency so that it would appear in the selected filter 10 times (twice for each hyperbolic band). In between, the filter would receive sidelobe clutter from all darkened intersections.

Clutter Return: General Equations. The clutter-to-noise ratio from a single clutter patch with incremental area dA at a range R is

$$C/N = \frac{P_{av}\, G_T\, G_R\, \lambda^2\, \sigma^0\, dA}{(4\pi)^3\, R^4\, L_c\, k\, T_s\, B_n} \tag{17.2}$$

where P_{av} = average transmit power
λ = operating wavelength
σ^0 = clutter backscatter coefficient
L_c = losses applicable to clutter
G_T = transmit gain in patch direction
G_R = receive gain in patch direction
k = Boltzmann's constant = 1.38054×10^{-23} W/(Hz/K)
T_s = system noise temperature, K
B_n = doppler filter bandwidth

The clutter-to-noise ratio from each radar resolution cell is the integral of Eq. (17.2) over the doppler and range extent of each of the ambiguous cell positions on the ground.[21-25] Under certain simplified conditions, the integration can be closed-form[25] while numeric integration may be used generally.

Sidelobe Clutter. The entire clutter spectrum can be calculated for each range gate by Eq. (17.2) if the antenna pattern is known in the lower hemisphere. In preliminary system design, the exact gain function may not be known, so that one useful approximation is that the sidelobe radiation is isotropic with a constant gain of G_{SL}.

Sidelobe Discretes. An inherent characteristic of airborne pulse doppler radars is that echoes from large objects on the ground (discretes), such as buildings, may be received through the antenna sidelobes and appear as though they were smaller moving targets in the main beam. This is a particularly severe problem in a medium-PRF radar, where all-aspect target performance is usually desired, as these returns compete with targets of interest. In a high-PRF radar, there is little if any range region clear of sidelobe clutter, such that the sidelobe clutter portion of the doppler spectrum is often not processed (since target detectability is severely degraded in this region). Further, in a high-PRF radar, especially at higher altitudes, the relative amplitudes of the distributed sidelobe clutter and the discrete returns are such that the discretes are not visible in the sidelobe clutter.

The apparent radar cross section (RCS), σ_{app}, of a sidelobe discrete with an RCS of σ is $\sigma_{app} = \sigma G_{SL}^2$, where G_{SL} is the sidelobe gain relative to the main beam. The larger-size discretes appear with a lower density than the smaller ones, and a model commonly assumed at the higher radar frequencies is as shown in Table 17.3. Thus, as a practical matter 10^6 m^2 discretes are rarely present, 10^5 m^2 sometimes, and 10^4 m^2 often.

Two mechanizations for detecting and eliminating false reports from sidelobe discretes are the guard channel and postdetection sensitivity time control (STC). These are discussed in the paragraphs which follow.

TABLE 17.3 Discrete Clutter Model

Radar cross section, m^2	Density, per mi^2
10^6	0.01
10^5	0.1
10^4	1

Guard Channel. The guard channel mechanization compares the outputs of two parallel receiving channels, one connected to the main antenna and the second to a guard antenna, to determine whether a received signal is in the main beam or the sidelobes.[26-28] The guard channel uses a broad-beam antenna that (ideally) has a pattern above the main-antenna sidelobes. A range-cell, doppler-filter by range-cell, doppler-filter comparison is made of the returns in both channels. Sidelobe returns are rejected (blanked) when they are larger in the guard receiver, and main-lobe returns are passed without blanking since they are larger in the main receiver.

A block diagram of a guard channel mechanization is shown in Fig. 17.8. After the CFAR circuits (which ideally would be identical in both channels), there are three thresholds: the main channel, guard channel, and main-to-guard-ratio threshold. The detection logic of these thresholds is also shown in Fig. 17.8.

The blanking which occurs because of the main-guard comparison affects the detectability in the main channel, the extent of which is a function of the threshold settings. The threshold settings are a tradeoff between false alarms due to sidelobe returns and detectability loss in the main channel. An example is shown in Fig. 17.9 for a nonfluctuating target, where the ordinate is the probability of detection in the final output and the abscissa is the signal-to-noise ratio (SNR) in the main channel. The quantity B^2 is the ratio of the guard channel SNR to the main channel SNR and is illustrated in Fig. 17.10. B^2 is small for a target in the main beam and large, 0 dB or so, for a target at the sidelobe peaks. In the example shown, there is a 0.5 dB detectability loss due to the guard blanking for targets in the main beam.

Ideally, the guard antenna gain pattern would exceed that of the main antenna at all angles in space (except for the main beam) to minimize detections through the sidelobes. If not, however, as illustrated in Fig. 17.10, returns through the sidelobe peaks above the guard pattern have a significant probability of detection in the main channel and would represent false detections.

Postdetection STC. A second approach to blanking sidelobe discretes is the postdetection STC,[29] the logic of which is shown in Fig. 17.11. Basically, the CFAR output data is correlated (resolved) in range 3 times. Each correlator calculates unambiguous range using M out of the N sets of detection data (e.g., three detections required out of eight PRFs). No doppler correlation is used since the doppler is ambiguous. The results of the first two correlations are used to blank all outputs which are likely to be sidelobe discretes from the final range correlator. Here, three range correlators are used in which the first, the A correlator, resolves the range ambiguities within some nominal range, say, 10 nmi, beyond which sidelobe discretes are not likely to be detected. A second correlator, the B correlator, resolves the range ambiguities out to the same range, but before a target can enter the B correlator, its amplitude is thresholded by a range-varying threshold (the STC threshold). A range-cell by range-cell comparison is made of the correlations in the A and B correlators, and if a range gate correlates in A and not in B, that gate is blanked out of the third correlator, the C correlator. The C correlator resolves the range ambiguities within the maximum range of interest.

The principle behind the postdetection STC approach is illustrated in Fig. 17.12, where the return of a target in the main beam and a large discrete target in the sidelobes is plotted versus unambiguous range (that is, after the range ambiguities have been resolved). Also shown are the normal CFAR threshold and the STC threshold versus range. As is evident, a discrete return in the sidelobes is

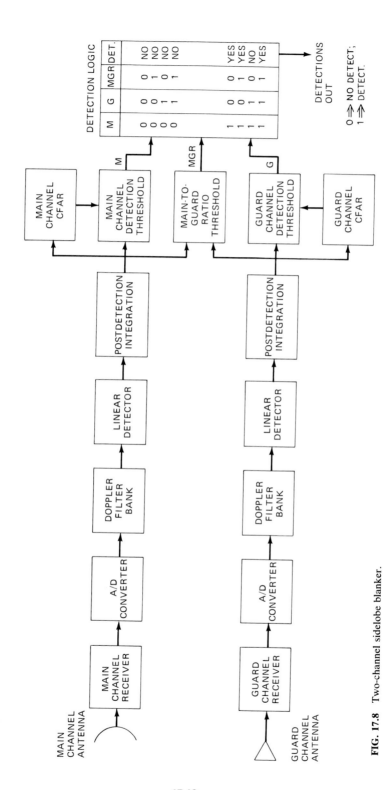

FIG. 17.8 Two-channel sidelobe blanker.

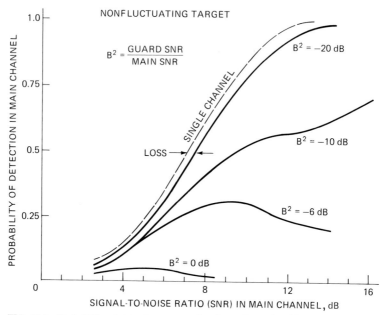

FIG. 17.9 Probability of detection versus signal-to-noise ratio with a guard channel.

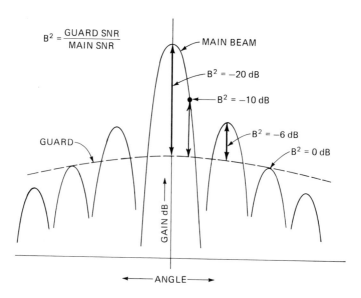

FIG. 17.10 Main and guard antenna patterns.

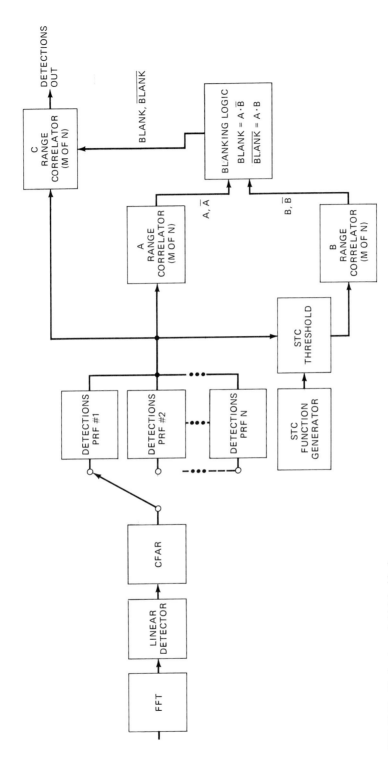

FIG. 17.11 Single-channel sidelobe blanker.

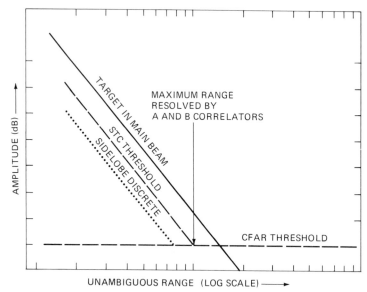

FIG. 17.12   Postdetection STC levels.

below the STC threshold, and a return in the main beam is above the threshold, such that the sidelobe discrete can be recognized and blanked from the final output but the target will not be blanked.

Main-Beam Clutter.   The   main-beam   clutter-to-noise   power   can   be approximated from Eq. (17.2) by substituting the intersected area for dA and summing over all intersections within the main beam.[30]

$$\frac{C}{N} = \frac{P_{av}\lambda^2 \, \theta_{az} \, (c\tau/2)}{(4\pi)^3 \, L_c \, K \, T_s \, B_n} \sum \frac{G_T \, G_R \, \sigma^0}{R^3 \cos \alpha} \tag{17.3}$$

The summation limits are the lower and upper edges of the smaller of the transmit and receive beams and

where θ_{az} = azimuth one-half power beamwidth, rad
τ = compressed pulsewidth
α = grazing angle at clutter patch

The remaining terms are as defined following Eq. (17.2).

Main-Beam Clutter Filtering.   In a pulse doppler radar utilizing digital signal processing, main-beam clutter is rejected by either a combination of a delay-line clutter canceler followed by a doppler filter bank or by a filter bank with low filter sidelobes. In either case, the filters around the main-beam clutter are blanked to minimize false alarms on main-beam clutter.

The choice between these options is a tradeoff of quantization noise and complexity versus the filter-weighting loss. If a canceler is used, filter weighting can be relaxed over that with a filter bank alone, since the canceler reduces the dynamic-range requirements into the FFT (if the main-beam clutter is the largest signal). Without a canceler, heavier weighting is needed to reduce sidelobes to a level so that the filter response to main-beam clutter is below the thermal-noise level. This weighting increases the filter noise bandwidth and hence increases the loss in signal-to-noise ratio.

The improvement factor for a DFT filter[31] is given by

$$I(K) = \frac{\left[\sum_{n=0}^{N-1} A_n^2\right]}{\sum_{n=0}^{N-1}\sum_{m=0}^{N-1} A_n A_m \exp\{-2[\pi(n-m)\sigma_c T]^2\}\cos[2\pi K(n-m)/N]} \qquad (17.4)$$

where A_i = DFT weight, $0 \le i \le N - 1$
N = number of points in DFT
σ_c = standard deviation of clutter spectrum
K = filter number ($K = 0$ is dc filter)
T = interpulse period

Here, the improvement factor for a filter (versus the more common definition applied to a delay-line canceler) is defined as the ratio of the total clutter power input to the filter to the clutter residue in that filter. Expressed another way, the improvement factor is the ratio of the clutter power out of a filter if it were centered over the clutter, and the clutter width reduced to zero, to the power out of the filter in actual operation.[32,33] Figure 17.13 shows the improvement factor of a 256-point, Dolph-Chebyshev weighted FFT as a function of the clutter width for various filter numbers in the filter bank.

If the main beam is pointed below the horizon and is greater than a beamwidth from 0° azimuth, the 6 dB clutter width due to platform motion Δf is

$$\Delta f = \frac{2V_R}{\lambda}\theta_B \sin \psi_0 \qquad (17.5)$$

where V_R = radar ground speed
ψ_0 = main-beam angle relative to velocity vector
θ_B = 3 dB one-way antenna beamwidth, rad
λ = RF wavelength

Clutter-Transient Suppression. When the PRF is changed for multiple-PRF ranging, or the slope is changed in linear FM ranging, or the RF carrier is changed, the transient change in the clutter return may cause degradation unless it is properly handled.[34] Since the clutter is usually ambiguous in range in a pulse doppler radar, the clutter power increases at each interpulse period (IPP) as clutter return is received from longer-range ambiguities, until the horizon is reached. This phenomenon is called *space charging.* Note that although an increasing number of clutter returns are received during the charging period, the vector sum may actually decrease owing to the random phase relations of the returns from different patches.

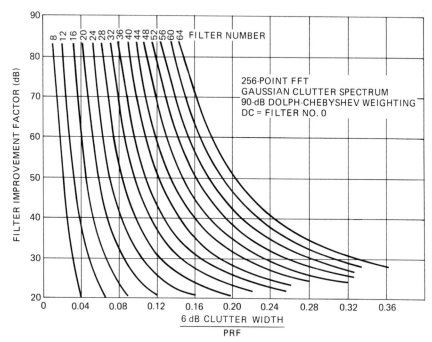

FIG. 17.13 Filter improvement factor versus clutter width.

If a clutter canceler is used, the output cannot begin to settle to steady-state value until space charging is complete. Some settling time must be allowed before signals are passed to the filter bank. Therefore, the coherent integration time available at each look is reduced from the total look time by the sum of the space charge time and the transient settling time. The canceler settling time can be eliminated by "precharging" the canceler with the steady-state input value.[35] This is done by changing the canceler gains so that all delay lines achieve their steady-state values on the first IPP of data.

If no canceler is used, signals can be passed to the filter bank after the space charge is complete, so that the coherent integration time is the total look time minus the space charge time.

Altitude-Line Clutter Filtering. The reflection from the earth directly beneath an airborne pulse radar is called altitude-line clutter. Because of specular reflection over smooth terrain, the large geometric area, and the relatively short range, this signal can be large. It lies within the sidelobe clutter region of the pulse doppler spectrum.

Because it can be much larger than diffuse sidelobe clutter and has a relatively narrow spectral width, altitude-line clutter is often removed by either a special CFAR which prevents detection of the altitude line or by a tracker-blanker which removes these reports from the final output. In the case of the tracker-blanker, a closed-loop tracker is used to position range and velocity gates around the altitude return and blank the affected range-doppler region.

17.3 *TIME GATING*

Time gating of the receiver permits blanking of transmitter leakage and its noise sidebands, elimination of excess receiver noise from competing with the signal, range gating for target tracking, and true range measurement, provided the ambiguity can be resolved.

Transmitted-Pulse Suppression. One major advantage of pulse doppler over CW systems is the time blanking of transmitter leakage so that receiver sensitivity is not degraded owing to saturation effects or to noise sidebands on the transmitter.

Harmonic Frequencies. Extreme care is required to prevent spurious signals from appearing in the system output. For example, if a 30-MHz IF receiver is being gated at a 110-kHz PRF, the 272d harmonic of the gating transient will fall at 29.92 MHz and the 273d at 30.03 MHz. Either of these harmonics may be within the doppler passband and therefore appear in the output. Although high-order harmonics of the gating transient are relatively small, they may be large compared with the signal since gating occurs early in the receiver.

Gating and Synchronization. One solution to the gating-harmonic problem is the use of balanced gating circuits and synchronization of the IF passband and the PRF so that the PRF harmonics all fall outside the useful portion of the passband. An alternative solution is to heterodyne the clutter to a frequency that is a multiple of the PRF so that the PRF harmonics are rejected with the clutter. However, such solutions preclude a variable-PRF system other than in discrete, accurately known steps.

Although synchronization of the PRF and the IF passband is usually necessary, synchronization at RF is not usually required. The harmful harmonics are of a much higher order and therefore are much smaller. In addition, the RF gating transients are usually further reduced in amplitude by the IF gating circuit.

Transmitter Leakage. The on-off ratio required for the overall transmitter blanking circuits is fairly large (more than can be obtained readily at RF without excessive insertion loss). Thus a combination RF and IF blanking system is usually employed. The transmitter leakage through the blanking circuits can be allowed to be as large as main-beam clutter if there is zero-doppler filtering to remove it. Alternatively, it must be a fraction of the noise power in a detection filter if there is no such filtering.

Range Gating. Range gating eliminates excess receiver noise from competing with the signal and permits target tracking and range measurement. Range gating is very similar to transmitted-pulse suppression. In a single-channel 0.5-duty-cycle system, one pulse-suppressor circuit serves both functions. In multiple-range-gated systems the range gates can serve both functions. If one circuit serves both functions, the on-off ratio must be adequate for pulse suppression, whereas if two are used, the range gate does not need as much rejection.

17.4 *RANGE-AMBIGUITY RESOLUTION*

Several methods of ranging are commonly employed in high PRF, while medium PRF is usually confined to multiple discrete PRF ranging.

High-PRF Ranging. Range-ambiguity resolution in high PRF is performed by modulating the transmitted signal and observing the phase shift of the modulation on the return echo. Modulation methods include varying the PRF, either continuously or in discrete steps; varying the RF carrier, with either linear or sinusoidal FM; or some form of pulse modulation such as pulse-width modulation (PWM), pulse-position modulation (PPM), or pulse-amplitude modulation (PAM). Of these modulation techniques, PWM and PPM may have large errors because of clipping of the received modulation by eclipsing or straddling (discussed in Sec. 17.7), and PAM is difficult to mechanize in both the transmitter and the receiver. Consequently, they will not be further considered here.

Multiple Discrete PRF Ranging. Ranging by use of several (usually two or three) fixed PRFs involves sequential measurement of the ambiguous range in each PRF, followed by comparison of the measurements to eliminate ambiguities.[36,37]

Figure 17.14 illustrates the principle of multiple-PRF ranging for a two-PRF, high-PRF radar. The PRFs are chosen to have a common submultiple frequency $1/T_u$. If the transmitted-pulse trains are compared in a coincidence detector, the common submultiple frequency is obtained. Similarly, if the received gates are compared in a coincidence detector, the same submultiple frequency shifted in time by the target range delay T_r is obtained. Measuring the time delay between the two sets of coincidence pulses yields the true target range. If desired, a three-PRF system can be mechanized similarly. The advantage obtained is the increased unambiguous range achievable.

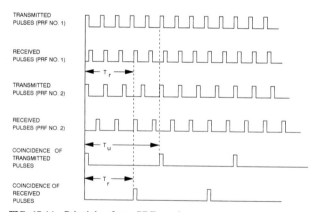

FIG. 17.14   Principle of two-PRF ranging.

In a surveillance radar a number of receiver gates are used to detect targets that may appear at any range within the interpulse period. Figure 17.15 illustrates a common method of spacing the gates for the general case where the gate spacing τ_s, the gate width τ_g, and the transmitted pulse τ_t are all unequal. Selecting $\tau_g > t_s$ reduces the range-gate straddle loss but increases the possibility of range ghosts. Selecting $\tau_t = \tau_g$ maximizes range performance.

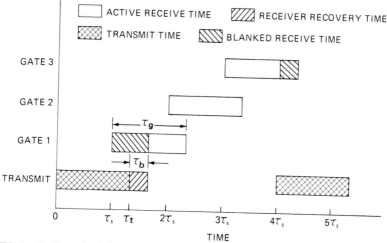

FIG. 17.15   Example of three range gates equally spaced in the interpulse period.

The PRFs are usually related by the ratios of closely spaced, relatively prime integers m_1, m_2, and m_3, as indicated in Table 17.4. Thus a three-PRF system using the seventh, eighth, and ninth submultiples of the range-gate clock frequency $f_c = 1/\tau_s$ as PRFs yields an unambiguous range of $7 \times 9 = 63$ times that of the middle PRF alone.

TABLE 17.4   Multiple-PRF Ranging Parameters

Item	Two-PRF	Three-PRF
Ranging parameters: $m_1 > m_2 > m_3$	m_1, m_2	m_1, m_2, m_3
Number of range-gated channels	$m_1 - 1$	$m_1 - 1$
PRFs		
f_{R1}	$1/m_1\tau_s$	$1/m_1\tau_s$
f_{R2} $(f_{R3} > f_{R2} > f_{R1} \geq f_{R\min})$	$1/m_2\tau_s$	$1/m_2\tau_s$
f_{R3}		$1/m_3\tau_s$
Unambiguous range ($R_{\max}$)	$m_2c/2f_{R1}$	$m_2m_3c/2f_{R1}$
Transmitter duty cycle, d	$\tau_t f_{R2}$	$\tau_t f_{R3}$
Ratio of highest and lowest PRF	m_1/m_2	m_1/m_3

NOTE: m_1, m_2, m_3 must be relatively prime integers.
τ_t = transmitted pulsewidth
τ_g = range-gate width
τ_b = blanking width due to receiver recovery
τ_s = range-gate spacing
f_c = range-gate clock = $1/\tau_s$

Figure 17.16 shows the maximum unambiguous range as a function of the minimum PRF, $f_{R\min}$, and the ranging parameter m_1, for the case where m_1, m_2, m_3 are consecutive integers. It is usually desirable to keep m_1 in the region from about 8 to 50. Thus the unambiguous range of a two-PRF system is seen to be rather limited, whereas three PRFs give much larger ranges. Some of the considerations influencing the choice of m_1 to this range are as follows:

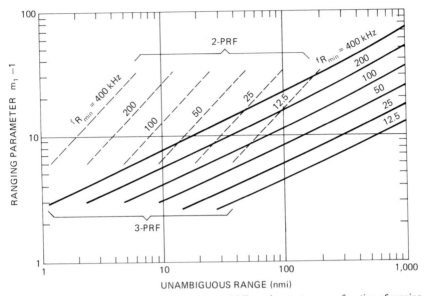

UNAMBIGUOUS RANGE (nmi)

FIG. 17.16 Unambiguous range for two- and three-PRF ranging systems as a function of ranging parameters, $m_1 - 1$, and minimum value of PRF for the case where m_1, m_2, m_3 are consecutive integers and m_1 is odd.

1. To minimize hardware, m_1 should be small since a maximum of $m_1 - 1$ range gates must be processed.

2. The probability of eclipse in at least one PRF is about $3/m_1$ for a three-PRF system, and so m_1 should be at least eight or higher since range cannot be measured if any PRF is eclipsed.

3. To get a long unambiguous range, m_1 should be large.

4. For good range resolution, τ_T must be small, which requires that m_1 be large. (Target range change during the dwell limits the minimum τ_t.)

5. To minimize the transmitter duty cycle and hence the average-power variation between PRFs, m_1 should be relatively large.

The Chinese remainder theorem is one means for calculating the true range from the several ambiguous measurements in a range-while-search system.[38] This approach permits a unique direct computation of the true-range cell number R_c from the three ambiguous-range cell numbers A_1, A_2, and A_3 (or two numbers for a two-PRF system). (The cell number is the range expressed in units of the pulse width and ranges from 0 to $m_i - 1$.) The theorem for a three-PRF system is expressed by the congruence

$$R_c \ (C_1 A_1 + C_2 A_2 + C_3 A_3) \ (\text{modulo } m_1 m_2 m_3) \qquad (17.6)$$

The smallest value of R_c that satisfies Eq. (17.6) is the remainder of the term within parentheses when divided by $m_1 m_2 m_3$ as many times as possible. Therefore, $0 \le R_c < m_1 m_2 m_3$. The constants C_1, C_2, and C_3 are related to m_1, m_2, and m_3 by the congruences

$$C_1 = b_1 m_2 m_3 \equiv 1 \ (\text{modulo } m_1) \tag{17.7}$$
$$C_2 = b_2 m_1 m_3 \equiv 1 \ (\text{modulo } m_2) \tag{17.8}$$
$$C_3 = b_3 m_1 m_2 \equiv 1 \ (\text{modulo } m_3) \tag{17.9}$$

where b_1 is the smallest positive integer which, when multiplied by $m_2 m_3$ and divided by m_1, gives unity as the remainder (and similarly for the other b's).

Once m_1, m_2, and m_3 have been chosen, the range can be computed from Eq.(17.6) by using the C values and the ambiguous-range cell numbers (A_1, A_2, A_3) in which the target is detected. For example, if $m_1 = 7$, $m_2 = 8$, $m_3 = 9$, then $b_1 = 4$, $b_2 = 7$, $b_3 = 5$, and the range is $R_c = (288 A_1 + 441 A_2 + 280 A_3)$ (modulo 504). If the target is in the first gate after the transmit pulse, $A_1 = A_2 = A_3 = 1$ and $R_c = (288 + 441 + 280)$ (modulo 504) = 1. An alternative to the Chinese remainder theorem is either a hard-wired correlator or a special-purpose computer that accepts detections from all PRFs and outputs all double or triple correlations.

Continuously Variable PRF Ranging. In a single-target tracking radar, the range ambiguity can be resolved by varying the PRF so that the target return is centered in the interpulse period. A high duty cycle, 0.333 to 0.5, may be used. Range R can then be calculated by

$$R = -\frac{\dot{R} f_R}{\dot{f}_R} \tag{17.10}$$

This method of range measurement has poor accuracy because of the errors involved in measuring the derivatives. An advantage of this technique is that the target return is never eclipsed by the transmitter pulse, thus improving tracking. It has a disadvantage, however, in that PRF harmonics can appear within the doppler band as spurious signals.

Linear-Carrier FM. Linear frequency modulation of the carrier can be used to measure range, especially in range-while-search applications. The modulation and demodulation to obtain range are the same as used in CW radar, but the transmission remains pulsed.

Assume that the dwell time is divided into two periods. In the first period, no FM is applied, and the doppler shift of the target is measured. In the second period, the transmitter frequency is varied linearly at a rate $\dot{f}$ in one direction. During the round-trip time to the target, the local oscillator has changed frequency so that the target return has a frequency shift, in addition to the doppler shift, that is proportional to range. The difference in the frequency Δf of the target return in the two periods is found, and the target range calculated from

$$R = \left| \frac{c \Delta f}{2 \dot{f}} \right| \tag{17.11}$$

The problem with only two FM segments during a dwell time is that, with more than a single target in the antenna beamwidth, range ghosts result. For example, with two targets present at different dopplers, the two frequencies observed during the FM period cannot be unambiguously paired with the two frequencies observed during the no-FM period. Thus, typical high-PRF range-while-search radars use a three-segment scheme in which there are no-

FM, FM-up, and FM-down segments. The range is found by selecting returns from each of the three segments that satisfy the relations

$$f_1 < f_0 < f_2 \tag{17.12}$$

$$f_1 + f_2 = 2f_0 \tag{17.13}$$

where f_0, f_1, and f_2 are the frequencies observed during the no-FM, FM-up, and FM-down segments, respectively. The range then is found from Eq. (17.11), where

$$\Delta f = f_2 - f_0 \quad \text{or} \quad (f_2 - f_1)/2 \quad \text{or} \quad f_0 - f_1 \tag{17.14}$$

An example is shown in Fig. 17.17.

Target	A	B
Range, nmi	10	20
Doppler frequency, kHz	21	29
FM shift, kHz	3	6
Observed frequencies		
f_0, no FM, kHz	21	29
f_1, FM up, kHz	18	23
f_2, FM down, kHz	24	35

Possible sets which satisfy the relations shown in Eqs. (17.12) and (17.13) are:

f_1	f_0	f_2	$2f_0$	$f_1 + f_2$	Target?	Range, nmi
18	21	24	42	42	Yes	10
18	21	35	42	53	No	
18	29	35	58	53	No	
23	29	35	58	58	Yes	20

FIG. 17.17 Three-slope FM ranging example. There are two targets, A and B; $\dot{f}$ = FM slope = 24.28 MHz/s.

If more than two targets are encountered during a dwell time, ghosts again result, as only $N - 1$ simultaneously detected targets can be resolved ghost-free where N is the number of FM slopes. This is not a severe problem in practice, however, for multiple targets in a single beamwidth are usually a transient phenomenon.

The accuracy of the range measurement improves as the FM slope increases since the observed frequency differences can be more accurately measured. However, the FM slope is limited by clutter-spreading considerations since during the FM periods the clutter is smeared in frequency and can appear in frequency regions normally clear of clutter. Range accuracies on the order of 1 or 2 mi can be reasonably achieved.

Sinusoidal-Carrier FM. This method is similar to that sometimes used in CW radar but retains the pulse transmission. It is particularly useful for tracking, either continuously or in a pause-to-range mode (discussed in Sec. 17.5). It is not suitable for range-while-search because of the relatively long time required to measure the phase shift of the sinusoidal modulation.

Medium-PRF Ranging. Multiple discrete PRF ranging, as discussed for high PRF, is also used for medium PRF except that the PRF selection criterion differs.[13] The technique of using closely spaced PRFs can be extended to medium PRF by employing three groups of three closely spaced PRFs, the groups being widely spaced to improve doppler visibility. The center PRF in each group is called the *major* PRF, and the adjacent ones the *minor* PRFs. Ranging is accomplished by requiring a detection in the major PRF and its adjacent minor PRFs and is effectively a detection criterion of exactly three detections out of three opportunities. This approach is attractive from a ghosting standpoint but suffers owing to the poor doppler visibility that results from having only three PRFs visible.

A better technique for medium PRF is to use seven or eight PRFs which cover nearly an octave in frequency and to require detections in at least three of these to declare a target report. The advantage is that doppler visibility is better than with the major-minor approach, and hence better range performance in sidelobe clutter is achieved (where some PRFs may be obscured by clutter). However, it is more susceptible to ghosting owing to the high doppler visibility. This problem is mitigated by also resolving the doppler ambiguities and using the true doppler for correlation to reject ghosts.

The basic accuracy of multiple-PRF ranging is on the order of the range-gate size (150 m/μs), but this can be improved to a fraction of the gate width by amplitude centroiding.

17.5 TARGET TRACKING

Target tracking can be performed on either a single target, using more or less conventional angle, range, and velocity tracking servo loops, or on multiple targets, using track-while-scan.

Single-Target Tracking. Angle tracking can be identical to a conventional pulse radar using monopulse, sequential lobing, or conical scan. Monopulse is more difficult to mechanize because of the problem of phase and amplitude matching of the multiple receiver channels, but the problem can be mitigated by using self-calibration routines controlled by the radar computer.

In a low-duty-cycle radar, range tracking is similar to pulse radar tracking in that split-gate tracking is used. In a high-duty-cycle radar, continuously variable PRF ranging or linear-FM ranging may be used. In a pulse doppler radar, the tracked range is usually ambiguous, so that provisions must be made to track through multiple interpulse periods and during eclipse (that is, when the target return overlaps the transmitted pulse).

Velocity (or doppler) tracking in a pulse doppler radar is carried out by forming a centroid on the target's doppler return in the filter bank. A closed-loop tracker then positions a doppler window around the tracked target such that returns which differ in doppler by more than a predetermined value are discarded by the tracker. In medium PRF, the PRF has to be adjusted to keep the doppler return away from the main-beam clutter notch as well as to avoid range eclipse.

Tracking through Eclipse. Because of the range ambiguities in medium and high PRF, the radar must cope with the loss of target each time that it passes through eclipse. Automatic tracking systems might recycle to the search mode if eclipse is not recognized and preventive measures taken.

The multiple-PRF true-ranging system is the most positive solution. Once true range has been determined so that the range ambiguity is resolved, PRF switching eliminates eclipsing. The onset of eclipse is detected by the range tracker by noting when the range gate begins to overlap the transmitted pulse. Then, before eclipse occurs, the PRF is switched to one of the other values. One of these values is certain to be uneclipsed, owing to the synchronization and relative PRF values. Since tracking is carried out in true range, no transient occurs and eclipse-free tracking continues indefinitely.

The continuously variable PRF system also permits eclipse-free tracking, but because of the spurious-signal problem it has not received much favor.

Other ranging systems do not permit eclipse-free tracking. FM ranging is not accurate enough to predict when eclipsing is about to occur. If the range is not accurately known, there is no way to anticipate an eclipse. In this case, an after-the-fact eclipse detection is made. A target-presence-detector circuit notes the absence of a signal and assumes that this is due to an eclipse. This circuit then commands the PRF to change value in an attempt to bring the target out of eclipse.

The problems with this approach are that target scintillation can cause PRF cycling and that there is no accurate way to predict which PRF will prevent eclipse. The latter problem can be reduced, if crude FM range data is available, by selecting the PRF from groups of PRFs, each group having values appropriate for a particular region of ranges. This reduces the time required in searching for an uneclipsed PRF.

Multiple-Target Tracking. Multiple-target tracking can be accomplished in several ways. One, track-while-scan, is to use the normal search mode with FM or multiple-PRF ranging and store the range, angle, and doppler of the reported detections in the computer. These detections are then used to form track files. The antenna scans in a normal search pattern, and a scan-to-scan correlation is made on the detections which update the track files. Although tracking accuracies are less than can be achieved in a single-target track, multiple targets can be tracked simultaneously over a large volume in space.

A second method of multiple-target tracking, pause-while-scan, particularly applicable to electronic scan antennas, is to scan in a normal search pattern, pause on each search detection, and enter a single-target track mode for a brief period. The advantage is that the resulting range, angle, and doppler measurements are more accurate than those made with a scanning antenna, but the time to search a volume in space is increased.

17.6 DYNAMIC-RANGE AND STABILITY REQUIREMENTS

Dynamic Range. Dynamic range as discussed here is the linear region above thermal noise over which the receiver and signal processor operate before any saturation (clipping) or gain limiting occurs. If saturations occur, spurious signals which degrade performance may be generated. For example, if main-beam clutter saturates, spurious frequencies can appear in the doppler passband normally clear of main-beam clutter and generate false-target reports. An AGC function is often employed to prevent saturations on either main-beam clutter in search or the target in single-target track mode. If saturations do occur in a range gate during an integration period, an option in a multiple-range-gated system is simply to blank detection reports from that gate.

The most stressing dynamic-range requirement is due to main-beam clutter when searching for a small low-flying target. Here, full sensitivity must be maintained in the presence of the clutter to maximize the probability of detecting the target.

The dynamic-range requirement of a pulse doppler radar, as determined by main-beam clutter, is a function not only of the basic radar parameters such as power, antenna gain, etc., but of radar altitude above the terrain and the radar cross section (RCS) of low-flying targets. As an example, Fig. 17.18 shows the maximum clutter-to-noise ratio (C/N_{max}) which appears in the ambiguous-range interval, i.e., after range folding, for a medium-PRF radar as a function of radar altitude and the range of the main-beam center. Note that the quantity plotted is the rms value of the clutter-to-noise ratio. A pencil-beam antenna pattern is as-

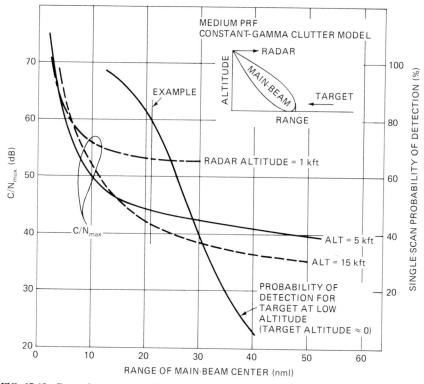

FIG. 17.18 Dynamic-range example.

sumed. At the longer ranges (small look-down angles), the clutter decreases with increasing altitude since range folding is less severe owing to less of the main beam intersecting the ground. At the shorter ranges, clutter increases with altitude since the clutter patch size on the ground increases. While Fig. 17.18 is for a medium-PRF radar, similar curves result for a high-PRF radar.

Also shown in Fig. 17.18 is the single-scan probability of detection P_d versus range for a given RCS target in a receiver with unlimited dynamic range. If it is desired to have the low-flying target reach at least, say, an 80 percent P_d before any gain limiting occurs, the dynamic-range requirement due to main-beam clutter is 53 dB at 1000 ft, 44 dB at 5000 ft, and 41 dB at 15,000 ft for this example. As is evident, the higher the desired probability of detection or the lower the radar altitude, the more dynamic range that is required. Further, if the specified target RCS is reduced, the dynamic-range requirement for the same desired P_d increases as the P_d-versus-range curve in Fig. 17.18 shifts to the left.

In a PD radar using digital signal processing, the dynamic range is most often limited by the A/D converters. The maximum signal level relative to thermal noise that can be processed linearly is related to the number of amplitude bits in the A/D by

$$\frac{S_{max}}{N} = 20 \log \left(\frac{2^{NAD} - 1}{noise} \right) \tag{17.15}$$

where S_{max}/N = maximum input level relative to noise, dB
NAD = number of amplitude bits in the A/D
noise = rms thermal-noise level at the A/D, quanta

From the relationships described above and assuming that the A/D limits the dynamic range, the A/D size can now be determined. An additional factor, that of a margin to allow for main-beam clutter fluctuations above the rms value, also needs to be considered. Since main-beam clutter time fluctuation statistics are highly dependent on the type of clutter being observed, such as sea clutter or clutter from an urban area, and are generally unknown, a value of 10 to 12 dB above the rms value is often assumed for the maximum level. Thus, the required number of amplitude bits in the A/D converter as determined by the main-beam clutter is

$$NAD \geq CEIL \left[\frac{(C/N)max \ (dB) + fluc \ margin \ (dB) + 20 \log \ (noise)(q)}{6} \right] \tag{17.16}$$

where CEIL is the next larger integer.

For the example cited in Fig. 17.18 where the maximum C/N is 53 dB at a 1000-ft altitude, and with a fluctuation margin of 10 dB and thermal noise at 1.414 quanta, the A/D requires at least 11 amplitude bits (plus a sign bit).

Stability Requirements. To achieve the theoretical clutter rejection and target detection and tracking performance of a pulse doppler system, the reference frequencies, timing signals, and signal-processing circuitry must be adequately stable.[39-42] In most cases, the major concern is with short-term stability rather than long-term drift. Long-term stability mainly affects velocity or range accuracy or spurious signals (due to PRF harmonics) but is relatively easy to make adequate. Short-term stability refers to variations within the round-trip radar echo time or during the signal integration time. The most severe stability requirements relate to the generation of spurious modulation sidebands on the main-beam clutter, which can appear as targets to the target detection circuitry. Thus, the ratio of main-beam clutter to system noise measured at the receiver output (C/N), including the fluctuation margin as discussed above, is the predominant parameter that determines stability requirements. However, at low spurious modulation frequencies, other constraints may become limiting.

Types of Spurious Modulation. The various spurious modulations that can appear on the received signal (clutter or target return) include both carrier and pulse modulation.

Carrier modulation can be amplitude modulation (AM), common FM, or independent FM. Common FM refers to identical modulation on both the transmitted signal and the receiver local-oscillator signal; independent FM appears on only one or the other or in the receiver following the first mixer.

Pulse modulation can be common pulse-position modulation (PPM), independent pulse-width modulation (PWM), or pulse droop. In these cases, common modulation refers to in-phase modulation on both the transmitter pulse and the receiver range gate or transmitter blanking gate. Independent modulation occurs if only one of these pulses is affected. Common PWM does not usually occur. However, if it does, it has requirements similar to those of common PPM. Also, pulse-amplitude modulation (PAM) can occur but is usually negligible when the other requirements are satisfied.

Sinusoidal Modulations. Any of these types of modulation may be caused by a sinusoidal disturbance, such as power supply ripple, line-frequency pickup, or sinusoidal vibration. Discrete sidebands at the modulation frequency and possibly higher harmonics will be introduced on the clutter and target signals. Since the pulse doppler receiver is basically a spectral analyzer, the radar requirements are most readily defined in terms of the allowable level of these modulation sidebands.

The predominant effect of these sidebands depends on the sideband frequency. For sideband frequencies greater than f_{min}, the sidebands on clutter signals fall outside the clutter rejection filter, where f_{min} is the minimum frequency separation of a detection filter from the edge of the main-beam clutter. This spread clutter must be kept below receiver noise; otherwise, it either will be detected as a target or will desensitize the receiver at these frequencies.

For sideband frequencies between $B_n/2$ and f_{min}, the concern is with either a target SNR loss or generation of false targets, where B_n is the receiver predetection-filter bandwidth. The SNR loss results from the sideband energy falling outside the detection filter. The false-target effect is due to modulation sidebands on a strong target appearing to the detection filters as weak targets.

For sideband frequencies below the reciprocal of the postdetection integration time T_1, sidebands per se cease to be of concern. However, the instantaneous-signal-frequency excursion should not be greater than the predetection bandwidth during the integration time, or a SNR loss results. Also, the signal amplitude should not be modulated significantly from one integration period to the next, or the sensitivity could change.

Sidebands in the immediate vicinity of the antenna lobing frequency or range track jitter frequency must be small enough to prevent excessive tracking error or noise.

TABLE 17.5 Allowable Deviation of Pulse or Carrier Sinusoidal Modulation as a Function of Modulating Frequency

| | | | | Maximum allowable deviation | | | | |
| | | | | Carrier modulation | | Pulse modulation | | |
Modulation frequency	Criterion	Factor Q	PAM-δ AM-M	Common FM δ_F	Independent FM δ_F	Independent PWM δ_W	Common PPM δ_P	Independent PPM δ_P
0 to $1/T_I$	Signal constant for T_I	0.1	$2Q$	$\cdots\cdots$	$\cdots\cdots$	$2\tau d_{min}Q$	$\dfrac{\tau d_{min}Q}{\pi f_m T_c}$	$2\tau d_{min}Q$
		$B_n/4\pi f_m T_I$						
$1/T_I$ to $1/\pi T_c$	Target SNR loss	$(0.1i!)^{1/i}$	No requirement	$Q/\pi f_m T_c$	$2Q$	No requirement (for natural sampling)		
	Clutter spreading	$[i!\sqrt{K_s/(C/N)}]^{1/i}$		$Q/\pi T_c$	$2f_m Q$			
$f_L - f_{bw}$ to $f_L + f_{bw}$	Tracking error	$K_M\sigma_E/\sqrt{2}(\theta\ or\ \tau)$	$2Q$	No requirement	No requirement	$2\tau d_{min}Q$	$\tau d_{min}Q$	$2\tau d_{min}Q$
$1/\pi T_c$ to $B_n/2$	Target SNR loss	$\cdots\cdots$	No requirement	$f_m(0.1i!)^{1/i}$	$2f_m(0.1i!)^{1/i}$	No requirement (for natural sampling)		
	Clutter spreading	$[i!\sqrt{K_s/(C/N)}]^{1/i}$		$f_m Q$	$2f_m Q$			
$B_n/2$ to f_{min}	Target SNR loss	$\cdots\cdots$	No requirement	$f_m(0.1i!)^{1/i}$	$2f_m(0.1i!)^{1/i}$	No requirement (for natural sampling)		
	Clutter spreading	$[i!\sqrt{K_{si}(C/N)}]^{1/i}$		$f_m Q$	$2f_m Q$			

5B_n to f_min	False targets	$\sqrt{K_s/(\max SNR)}$	2Q	$f_m\Omega$	$2f_m\Omega$	$\tau\Omega$	$2\tau Q$
f_{min} and higher	Clutter spreading	$\sqrt{K_s/(C/N)}$	2Q	$f_m\Omega$	$2f_m\Omega$	τQ	$2\tau Q$

M = peak fractional carrier amplitude modulation

δ_F = peak carrier frequency deviation, Hz

f_m = modulating frequency, Hz

T_c = two-way delay for main-beam clutter, s

δ_W = peak pulse-width deviation, s

d_{min} = minimum target signal duty cycle when partly eclipsed (say, $\tau f_R/5$)

δ_p = peak pulse-position deviation, s

K_x = safety factor of modulation power sidebands relative to system noise ($K_x \leq 1$)

f_L = tracking subcarrier frequency, Hz

f_{bw} = tracking-loop noise bandwidth, Hz

K_M = modulation sensitivity (fractional modulation/beamwidth or pulse-width error)

σ_E = rms tracking error due to modulation sidebands

θ = antenna beamwidth

max SNR = maximum target power SNR that does not cause automatic gain control or limiting

i = integer part of $[(B_n/2) + f_m]/f_m$

j = integer part of $(f_{min} + f_m)/f_m$

δ_Δ = peak fractional pulse-amplitude modulation

17.31

Applying all these constraints to the various types of modulations expected gives the sinusoidal modulation allowances of Table 17.5 where 0.1 dB tolerable loss was assumed in *SNR*. The various modulation-frequency regions (not sideband frequencies) corresponding to the predominant sideband effects (i.e., the criterion column in Table 17.5) are listed. A factor Q for each region is indicated; it is related to the maximum allowable deviation indicated in other columns. For example, for independent pulse-width modulation, for a modulation frequency greater than f_{min}, the maximum allowable deviation is

$$\delta_w = 2\tau Q = 2\tau\left[\frac{K_s}{(C/N)max}\right]^{1/2} \tag{17.17}$$

for acceptable clutter spreading. The safety factor K_s assures that the clutter sidebands will be buried in receiver noise.

Each tabulated value assumes that only one source of modulation is present. If multiple modulations are expected, an appropriate reduction factor must be provided so that the composite sidebands do not exceed the tolerable value.

In the receiver following the main-beam clutter filter, the clutter-spreading requirements are not germane. The *SNR* and false-target considerations then become the limiting factors at high-sideband frequencies as well as low-sideband frequencies.

Narrowband Noise. Although the requirements in Table 17.5 relate to a single-frequency sinusoidal modulation, they can be interpreted for narrowband-noise modulation. For this interpretation, the listed modulation values represent $\sqrt{2}$ times the allowable rms noise modulation referenced to a bandwidth of B_n. This interpretation is fairly good for modulating frequencies much greater than $B_n/2$, but for lower frequencies it is only a rough guide.

PPM and PWM. The pulse-position and pulse-width values are based on the assumption of natural sampling,[43] as is normally the case. In natural sampling, the deviation of the pulse edge is determined by the amplitude of the modulating signal at the time of occurrence of the pulse edge. Because of the range delay of target clutter signals and the range gating, PPM can be converted to PWM, which generates much larger sidebands. Most clutter signals are partially eclipsed or partially outside the range gate; therefore, the position of one edge of the pulse in the receiver is determined by the transmitted pulse, and the other edge is determined by the range gate or transmitter blanking pulse. Thus the original PPM can be converted into PWM as a function of modulating frequency and time delay to the clutter.

The deviation requirements in Table 17.5 were derived for the worst case, where the two common modulating signals are 180° out of phase or one edge of the received clutter pulse is completely eclipsed in the independent case.

Frequency Modulation. Common FM permits relaxed modulation requirements at low frequencies but 6 dB more severe requirements at high frequencies compared with independent FM.[44] This effect is caused by the range delay of the clutter. Since the transmitter and the receiver local oscillator are frequency-modulated in synchronism, the deviation of the IF difference signal is dependent on the range delay.

Droop. Although it does not cause new spectral sidebands and therefore was not included in Table 17.5, pulse droop on the transmitter modulator pulse is also of interest. This is because of the high phase-modulation sensitivity of traveling-wave tubes and klystrons typically used as RF amplifiers. A linear droop of the modulating voltage will serrodyne the RF signal, shifting the peak of the spectral envelope relative to the RF carrier frequency and reducing the useful signal power when passed through a filter matched to the pulse width. For small droop this loss is given by

$$\text{Fractional power loss} = \frac{1}{3}\left[\frac{\pi}{3.6}K_\phi\frac{\Delta V}{V}\right]^2 \qquad (17.18)$$

where K_ϕ = transmitter phase sensitivity (degrees phase change/percent voltage)
$\Delta V/V$ = fractional voltage droop on modulator pulse

Pulse-to-Pulse Random Modulation. In addition to the sinusoidal or narrowband-noise modulation, pulse-to-pulse random modulation may also be present. The predominant effect is clutter spreading noise into the detection filter. With the same notation as for Table 17.5, the factor Q is $[K_sf_R/(C/N)B_n]^{1/2}$, and the rms allowable fractional AM is equal to Q, as is the rms phase modulation (given in radians). The rms PPM or single-edge PWM allowable modulation (given in seconds) is τQ.

17.7 RANGE PERFORMANCE

Chapter 2 discusses the general radar range equation and the calculation of detection probability. This section extends those concepts to pulse doppler radars and includes a discussion of system losses and false-alarm probability. Generalized detection curves, which include multilook detection criteria, are presented.

Range Equation. In the doppler region where the signal does not fall in clutter, performance is limited only by system noise. The signal-to-noise ratio in the detection filter prior to postdetection integration for a target at range R is given by

$$SNR = \left(\frac{R_o}{R}\right)^4 \qquad (17.19)$$

$$R_o = \left(\frac{P_A\,G_T\,G_R\,\lambda^2\,\sigma_T}{(4\pi)^3k\,T_s\,B_n\,L}\right)^{1/4} \qquad (17.20)$$

where R_o = range at which $S/N = 1$
σ_T = target radar cross section
L = losses applicable to the target

The remaining terms are as defined following Eq. (17.2).

System Losses. Some of the losses inherent in, but not necessarily unique to, pulse doppler radars which employ digital signal processing are discussed below.

Quantization Noise Loss. This loss is due to the noise added by the A/D conversion process and to truncation due to finite word lengths in the signal-processing circuits which follow.[45]

CFAR Loss.   This is caused by an imperfect estimate of the detection threshold compared with the ideal threshold. The fluctuation in the estimate necessitates that the mean threshold be set higher than the ideal, hence a loss.

Doppler Filter Straddle Loss.   This loss is due to a target not always being in the center of a doppler filter. It is computed by assuming a uniformly distributed target doppler over one filter spacing and is a function of the FFT sidelobe weighting.

Amplitude-Weighting Loss.   This loss results from the increased noise bandwidth of the doppler filters that occurs because the filter sidelobe weighting. It can also be accounted for by an increase of the doppler filter noise bandwidth instead of as a separate loss.

Pulse Compression Mismatch Loss.   This is caused by the intentional mismatching of the pulse compression filter to reduce the time (range) sidelobes.

Guard Blanking Loss.   This is the detectability loss in the main channel caused by spurious blanking from the guard channel. (See Fig. 17.9.)

Eclipsing and Range-Gate Straddle Loss.   Because of eclipsing, the value of R_o, given by Eq. (17.20), may fall anywhere between zero and a maximum value, depending on the exact location of the target return in the interpulse period. When the PRF is high, so that many range ambiguities occur, the target range delay may be considered to be random from scan to scan, with a uniform distribution over the interpulse period. An approximate measure of performance in this case is found by first computing a detection curve averaged over target ambiguous ranges from zero to the range corresponding to the interpulse period. The loss is equal to the increase in signal-to-noise ratio required to obtain the same probability of detection with eclipsing or straddle as in the case when the transmit pulse is received by a matched gate with no straddle. Since the detection curve changes shape, the loss depends on the probability of detection selected. A less accurate approximation compares the average signal-to-noise ratio over the interpulse period with the signal-to-noise ratio of the matched case. In the case of M contiguous range gates of width τ that occupy the entire interpulse period except for the transmitted pulse also of width τ, the average eclipsing and straddle loss on a signal-to-noise-ratio basis is

$$\text{Eclipse and straddle loss} = \frac{Y}{3(M + 1)} \qquad \tau_t = \tau_g \qquad (17.21)$$

where $Y_1 = (1 - R)(2 + R)$ $M = 1$

$Y = (1 - R)(1 - R + 2X) + 2 + 1.75(M - 2)$ $M > 1, R \geqslant 0.618$

$Y = (1 - R)(1 + R + Z) + (Z - R)[Z(Z + X)]$
$\quad + (1 - Z)[Z(Z + 1) + 1] + 1 + 1.75(M - 2)$ $M > 1, R < 0.618$

$Z = 1/(1 + X)$

$X = \sqrt{1 - R}$

$R = \tau_b/\tau$

$\tau_b = $ width of first gate blanking

$\tau = $ width of transmitted pulse τ_t and receiver gate τ_g

$M = $ number of contiguous gates

The loss is plotted in Fig. 17.19. Signals cannot be received during the time τ_b following each transmit pulse owing to slow shutoff of the transmitter or recovery of the duplexer and/or receiver-protector.

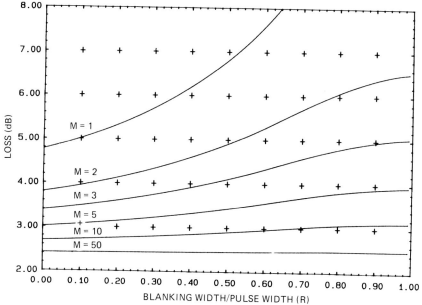

FIG. 17.19 Eclipsing and straddle loss for M contiguous range gates and equal transmitted pulse and range-gate widths as a function of the number of receive gates and blanking width.

Although Eq. (17.21) assumes contiguous range gates, the loss factor can be reduced by the use of overlapping gates at the expense of extra hardware and possibly more range ghosts.

Probability of False Alarm. PD radars often employ a multilook detection criterion to resolve range ambiguities such that during the time on target (dwell time) several PRFs are transmitted in successive looks and a threshold detection in more than one look is required for the radar to output a target report. For the case where a doppler filter bank in each range gate is used for coherent integration, possibly followed by a postdetection integrator, the probability of false alarm P_{FA} in each range gate–doppler filter required to obtain a given false report time T_{FR} is given approximately by

$$P_{FA} = \frac{1}{N_F}\left[\frac{0.693\,T_d}{\binom{n}{m} N_g\,T_{FR}}\right]^{1/m} \tag{17.22}$$

where N_F = number of independent doppler filters visible in the doppler passband (number of unblanked filters/FFT weighting factor)

n = number of looks in a dwell time

m = number of detections required for a target report (For example, 3 detections out of 8 PRFs is $m = 3$ and $n = 8$.)

T_d = total dwell time of the multiple PRFs including postdetection integration (if any) and any dead time

$\binom{n}{m}$ = binomial coefficient $n!/[m!(n - m)!]$

N_g = number of range gates in the output unambiguous-range interval (display range/range-gate size)

T_{FR} = false-report time [per Marcum's definition where the probability is 0.5 that at least one false report will occur in the false-report time (Ref. 46)]

Equation (17.22) is for the case where no doppler correlation is required for a target report. In the case where both range and doppler correlation are used, the required P_{FA} is

$$P_{FA} = \left[\frac{0.693\, T_d}{\binom{n}{m} N_{fu} N_g T_{FR} W^{m-1}} \right]^{1/m} \qquad (17.23)$$

where N_{fu} = number of independent doppler filters in the unambiguous doppler region and W = width (in filters) of the correlation window applied to detections following initial detection.

Probability of Detection. The probability-of-detection curves presented in Chap. 2 have been extended to include multilook detection criteria and are presented in a generalized fashion after Ref. 47; that is, they are presented in terms of gR/R_o, where the g factor is a function of the number of pulses integrated noncoherently and the probability of false alarm. For coherent integration $N = 1$ and g becomes

$$g = [- \ell n(P_{FA})]^{1/4} \qquad (17.24)$$

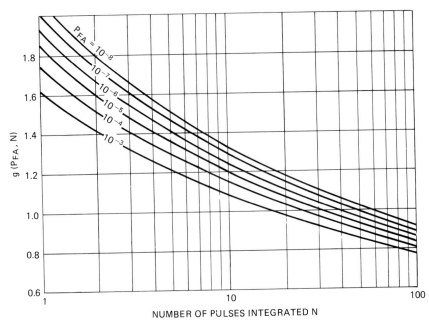

FIG. 17.20 Swerling's g factor as a function of P_{FA} and N (after Ref. 47).

The g factor is plotted in Fig. 17.20 for various false-alarm probabilities and number of pulses integrated.

The generalized results are based on the realization that Marcum's curves[46] are very similar over a wide range of parameters; one can reasonably use a single universal Marcum curve as shown in Fig. 17.21. This is patterned after the universal curve presented in Ref. 47, except that here it is for a nonfluctuating target. It is accurate to within 1 dB over the entire range (and closer over most of the range) of integration samples N from 1 to 100, probability of false-alarm values P_{FA} from 10^{-3} to 10^{-8}, and probability from 1 to 99 percent. To use Fig. 17.21, a value of $g(P_{FA},N)$ is first found from Fig. 17.20; this can then be used to convert gR/R_o values to R/R_o or signal-to-noise-ratio values.

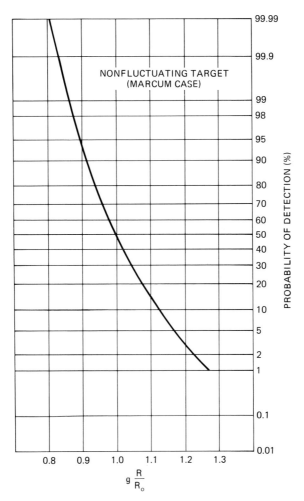

FIG. 17.21 Universal Marcum curve.

Generalized Curves. By using the universal Marcum curve, several detection cases have been generated and are shown in Fig. 17.22. These are all for a Swerling Case 1 target in which the target amplitude fluctuates independently from scan to scan but is constant within the dwell time. No losses other than the fluctuation loss have been included in these curves, so that any losses such as range-gate straddle and eclipsing can be accounted for in the computation of R_o.

The Swerling Case 1 single-scan detection curves can be closely approximated by

$$P_d = P_{FA}^{\left(\frac{1}{a + b \, SNR}\right)} \tag{17.25}$$

where P_d = single-scan detection probability
P_{FA} = probability of false alarm
SNR = signal-to-noise ratio = $(R_o/R)^4$

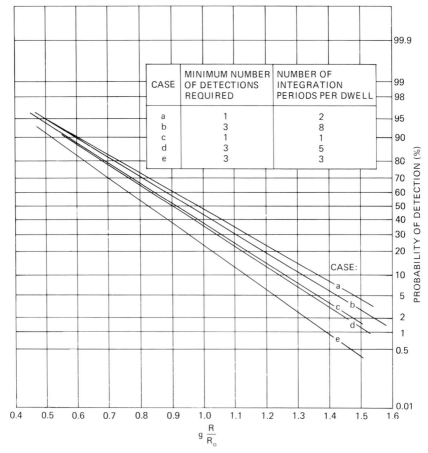

FIG. 17.22 Generalized single-scan probability of detection for a scan-to-scan, Swerling Case 1, Rayleigh fluctuating target.

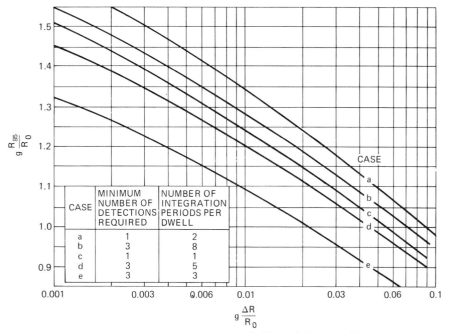

FIG. 17.23 Generalized 85 percent cumulative probability of detection for scan-to-scan, Swerling Case 1, Rayleigh fluctuating target.

The constants a and b can be found by substitution, using two pairs of P_d and gR/R_o values from Fig. 17.22, converting gR/R_o to SNR, and solving the resultant simultaneous equations.

The cumulative probability of detection (probability of detecting the target at least once in k scans), Pc_k, is defined as

$$Pc_k = 1 - \prod_{i=1}^{k}[1 - P_d(i)] \tag{17.26}$$

where $P_d(i)$ is the probability of detection on the ith scan. The cumulation may occur over a variable number of scans, such as when it begins at a range where $P_d(i)$ is approximately zero, or over a defined number of scans, where a 1-out-of-N acquisition criterion must be satisfied. The single-scan probability-of-detection curves shown in Fig. 17.22 have been used to compute the 85 percent cumulative probability of detection for the variable-scan case, shown in Fig. 17.23. ΔR is the change in range between successive scans for a fixed-velocity, radially moving target.

Clutter-Limited Case. The foregoing discussion assumed that the target fell in the noise-limited (i.e., clutter-free) part of the doppler band. If the target falls in the sidelobe clutter region, the range performance will be degraded, since the total power (system noise plus clutter) with which the target must compete is increased. The foregoing discussion can be applied to the sidelobe clutter region, however, by interpreting R_o as the range where the signal is equal to sidelobe clutter plus system noise.[48–50] The CFAR loss may also be higher owing to the increased variability of the threshold when the clutter varies over the target detection region.

REFERENCES

1. Skolnik, M. I.: Fifty Years of Radar, *Proc. IEEE*, vol. 73, pp. 182–197, February 1985.

2. Perkins, L. C., H. B. Smith, and D. H. Mooney: The Development of Airborne Pulse Doppler Radar, *IEEE Trans.*, vol. AES-20, pp. 290–303, May 1984.

3. Clarke, J., D. E. N. Davies, and M. F. Radford: Review of United Kingdom Radar, *IEEE Trans.*, vol. AES-20, pp. 506–520, September 1984.

4. Moaveni, M. K.: Corrections to "Radio Interference in Helicopter-Borne Pulse Doppler Radars," *IEEE Trans.*, vol. AES-14, p. 688, July 1978.

5. Moaveni, M. K.: Radio Interference in Helicopter-Borne Pulse Doppler Radars, *IEEE Trans.*, vol. AES-14, pp. 319–328, March 1978.

6. Ringel, M. B., D. H. Mooney, and W. H. Long: F-16 Pulse Doppler Radar (AN/APG-66) Performance, *IEEE Trans.*, vol. AES-19, pp. 147–158, January 1983.

7. Skillman, W. A.: Utilization of the E-3A Radar in Europe, *Proc. Mil. Electron. Def. Expo '78*, Wiesbaden, Germany, 1978.

8. Skillman, W. A.: Microwave Technology, Key to AWACS Success, *Proc. IEEE MTT/S*, Boston, 1983.

9. Clarke, J.: Airborne Early Warning Radar, *Proc. IEEE*, vol. 73, pp. 312–324, February 1985.

10. Doviak, R. J., D. S. Zrnic, and D. S. Sirmans: Doppler Weather Radar, *Proc. IEEE*, vol. 67, no. 11, pp. 1522–1553, 1979.

11. Stimson, G. W.: "Introduction to Airborne Radar," Hughes Aircraft Company, El Segundo, Calif., 1983, pt. 7.

12. Hovanessian, S. A.: Medium PRF Performance Analysis, *IEEE Trans.*, vol. AES-18, pp. 286–296, May 1982.

13. Aronoff, E., and N. M. Greenblatt: Medium PRF Radar Design and Performance, *20th Tri-Service Radar Symp.*, 1974. Reprinted in Barton, D. K.: "CW and Doppler Radars," vol. 7, Artech House, Norwood, Mass., 1978, sec. IV-7, pp. 261–276.

14. Long, W. H., and K. A. Harriger: Medium PRF for the AN/APG-66 Radar, *Proc. IEEE*, vol. 73, pp. 301–311, February 1985.

15. Goetz, L. P., and J. D. Albright: Airborne Pulse Doppler Radar, *IRE Trans.*, vol. MIL-5, pp. 116–126, April 1961. Reprinted in Barton, D. K.: "CW and Doppler Radars," vol. 7, Artech House, Norwood, Mass., 1978, sec. IV-3, pp. 215–225.

16. Skolnik, M. I.: "Introduction to Radar Systems," 2d ed., McGraw-Hill Book Company, New York, 1984, chap. 4.

17. Finn, H. M., and R. S. Johnson: Adaptive Detection Mode with Threshold Control as a Function of Spatially Sampled Clutter-Level Estimates, *RCA Rev.*, pp. 414–464, September 1968.

18. Steenson, B. O.: Detection Performance of a Mean-Level Threshold, *IEEE Trans.*, vol. AES-4, pp. 529–534, July 1968.

19. Rohling, H.: Radar CFAR Thresholding in Clutter and Multiple Target Situations, *IEEE Trans.*, vol. AES-19, pp. 608–621, July 1983.

20. Hansen, V. G.: Constant False Alarm Rate Processing in Search Radars, *Proc. IEEE Int. Radar Conf.*, pp. 325–332, London, 1973.

21. Farrell, J., and R. Taylor: Doppler Radar Clutter, *IEEE Trans.*, vol. ANE-11, pp.

162–172, September 1964. Reprinted in Barton, D. K.: "CW and Doppler Radars," vol. 7, Artech House, Norwood, Mass., 1978, sec. VI-2, pp. 351–361.

22. Helgostam, L., and B. Ronnerstam: Ground Clutter Calculation for Airborne Doppler Radar, *IEEE Trans.*, vol. MIL-9, pp. 294–297, July–October 1965.

23. Friedlander, A. L., and L. J. Greenstein: A Generalized Clutter Computation Procedure for Airborne Pulse Doppler Radars, *IEEE Trans.*, vol. AES-6, pp. 51–61, January 1970. Reprinted in Barton, D. K.: "CW and Doppler Radars," vol. 7, Artech House, Norwood, Mass., 1978, sec. VI-3, pp. 363–374.

24. Ringel, M. B.: An Advanced Computer Calculation of Ground Clutter in an Airborne Pulse Doppler Radar, *NAECON '77 Rec.*, pp. 921–928. Reprinted in Barton, D. K.: "CW and Doppler Radars," vol. 7, Artech House, Norwood, Mass., 1978, sec. VI-4, pp. 375–382.

25. Jao, J. K., and W. B. Goggins: Efficient, Closed-Form Computation of Airborne Pulse Doppler Clutter, *Proc. IEEE Int. Radar Conf.*, pp. 17–22, Washington, 1985.

26. Harvey, D. H., and T. L. Wood: Designs for Sidelobe Blanking Systems, *Proc. IEEE Int. Radar Conf.*, pp. 410–416, Washington, 1980.

27. Maisel, L.: Performance of Sidelobe Blanking Systems, *IEEE Trans.*, vol. AES-4, pp. 174–180, March 1968.

28. Finn, H. M., R. S. Johnson, and P. Z. Peebles: Fluctuating Target Detection in Clutter Using Sidelobe Blanking Logic, *IEEE Trans.*, vol. AES-7, pp. 147–159, May 1971.

29. Mooney, D. H.: Post Detection STC in a Medium PRF Pulse Doppler Radar, U.S. Patent 690,754, May 27, 1976.

30. Skillman, W. A.: "SIGCLUT: Surface and Volumetric Clutter-to-Noise, Jammer and Target Signal-to-Noise Radar Calculation Software and User's Manual," Artech House, Norwood, Mass., 1987.

31. Ziemer, R. E., and J. A. Ziegler: MTI Improvement Factors for Weighted DFTs, *IEEE Trans.*, vol. AES-16, pp. 393–397, May 1980.

32. Skillman, W. A.: "Radar Calculations Using the TI-59 Programmable Calculator," Artech House, Norwood, Mass., 1983, p. 308.

33. Skillman, W. A.: "Radar Calculations Using Personal Computers," Artech House, Norwood, Mass., 1984.

34. Ward, H. R.: Doppler Processor Rejection of Ambiguous Clutter, *IEEE Trans.*, vol. AES-11, July 1975. Reprinted in Barton, D. K.: "CW and Doppler Radars," vol. 7, Artech House, Norwood, Mass., 1978, sec. IV-11, pp. 299–301.

35. Fletcher, R. H., Jr., and D. W. Burlage: An Initialization Technique for Improved MTI Performance in Phased Array Radar, *Proc. IEEE*, vol. 60, pp. 1551–1552, December 1972.

36. Skillman, W. A., and D. H. Mooney: Multiple High-PRF Ranging, *Proc. IRE Conf. Mil. Electron.*, pp. 37–40, 1961. Reprinted in Barton, D. K.: "CW and Doppler Radars," vol. 7, Artech House, Norwood, Mass., 1978, sec. IV-1, pp. 205–213.

37. Hovanessian, S. A.: An Algorithm for Calculation of Range in Multiple PRF Radar, *IEEE Trans.*, vol. AES-12, pp. 287–289, March 1976.

38. Ore, O.: "Number Theory and Its History," McGraw-Hill Book Company, New York, 1948, pp. 246–249.

39. Goetz, L. P., and W. A. Skillman: Master Oscillator Requirements for Coherent Radar Sets, *IEEE-NASA Symp. Short Term Frequency Stability*, NASA-SP-80, November 1964.

40. Raven, R. S.: Requirements for Master Oscillators for Coherent Radar, *Proc. IEEE*, vol. 54, pp. 237–243, February 1966. Reprinted in Barton, D. K.: "CW and Doppler Radars," vol. 7, Artech House, Norwood, Mass., 1978, sec. V-1, pp. 317–323.

41. Gray, M., F. Hutchinson, D. Ridgely, F. Fruge, and D. Cooke: Stability Measurement

Problems and Techniques for Operational Airborne Pulse Doppler Radar, *IEEE Trans.*, vol. AES-5, pp. 632–637, July 1969.

42. Acker, A. E.: Eliminating Transmitted Clutter in Doppler Radar Systems, *Microwave J.*, vol. 18, pp. 47–50, November 1975. Reprinted in Barton, D. K.: "CW and Doppler Radars," vol. 7, Artech House, Norwood, Mass., 1978, sec. V-3, pp. 331–336.

43. Black, H. S.: "Modulation Theory," D. Van Nostrand Company, Princeton, N.J., 1953, p. 265.

44. Barton, D. K.: "Radar Systems Analysis," Prentice-Hall, Englewood Cliffs, N.J., 1964, p. 206.

45. Ziemer, R. E., T. Lewis, and L. Guthrie: Degradation Analysis of Pulse Doppler Radars Due to Signal Processing, *NAECON 1977 Rec.*, pp. 938–945. Reprinted in Barton, D. K.: "CW and Doppler Radars," vol. 7, Artech House, Norwood, Mass., 1978, sec. IV-12, pp. 303–312.

46. Marcum, J. I.: A Statistical Theory of Target Detection by Pulsed Radar, *IRE Trans.*, vol. IT-6, pp. 59–267, April 1960.

47. Swerling, P.: Probability of Detection for Fluctuating Targets, *IRE Trans.*, vol. IT-6, pp. 269–308, April 1960.

48. Mooney, D., and G. Ralston: Performance in Clutter of Airborne Pulse MTI, CW Doppler and Pulse Doppler Radar, *IRE Conv. Rec.*, vol. 9, pt. 5, pp. 55–62, 1961. Reprinted in Barton, D. K.: "CW and Doppler Radars," vol. 7, Artech House, Norwood, Mass., 1978, sec. V1-1, pp. 343–350.

49. Ringel, M. B.: Detection Range Analysis of an Airborne Medium PRF Radar, *IEEE NAECON Rec.*, Dayton, Ohio, pp. 358–362, 1981.

50. Holbourn, P. E., and A. M. Kinghorn: Performance Analysis of Airborne Pulse Doppler Radar, *Proc. IEEE Int. Radar Conf.*, pp. 12–16, Washington, 1985.

CHAPTER 18
TRACKING RADAR

Dean D. Howard
Locus, Inc., a subsidiary of Kaman Corp.

18.1 INTRODUCTION

A typical tracking radar has a pencil beam to receive echoes from a single target and track the target in angle, range, and/or doppler. Its resolution cell—defined by its antenna beamwidth, transmitter pulse length, and/or doppler bandwidth—is usually small compared with that of a search radar and is used to exclude undesired echoes or signals from other targets, clutter, and countermeasures. Electronic beam-scanning phased array tracking radars may track multiple targets by sequentially dwelling upon and measuring each target while excluding other echo or signal sources.

Because of its narrow beamwidth, typically from a fraction of 1° to 1 or 2°, a tracking radar usually depends upon information from a search radar or other source of target location to acquire the target, i.e., to place its beam on or in the vicinity of the target before initiating a track. Scanning of the beam within a limited angle sector may be needed to fully acquire the target within its beam and center the range-tracking gates on the echo pulse prior to locking on the target or closing the tracking loops.

The primary output of a tracking radar is the target location determined from the pointing angles of the beam and position of its range-tracking gates. The angle location is the data obtained from synchros or encoders on the antenna tracking axes shafts (or data from a beam-positioning computer of an electronic-scan phased array radar). In some cases, tracking lag is measured by converting tracking-lag-error voltages from the tracking loops to units of angle. This data is used to add to or subtract from the angle shaft position data for real-time correction of tracking lag.

There are a large variety of tracking-radar systems, including some that achieve simultaneously both surveillance and tracking functions. A widely used type of tracking radar and the one to be discussed in detail in this chapter is a ground-based system consisting of a pencil-beam antenna mounted on a rotatable platform which is caused by motor drive of its azimuth and elevation position to follow a target (Fig. 18.1). Errors in pointing direction are determined by sensing the angle of arrival of the echo wavefront and corrected by positioning the antenna to keep the target centered in the beam.

The principal applications of tracking radar are weapon control and missile-

FIG. 18.1 AN/FPQ-6 C-band monopulse precision tracking radar installation at the NASA Wallops Island Station, Va. It has a 29-ft-diameter antenna and a specified angle precision of 0.05 mrad rms.

18.2

range instrumentation. In both applications a high degree of precision and an accurate prediction of the future position of the target are generally required. The earliest use of tracking radar was in gunfire control. The azimuth angle, the elevation angle, and the range to the target were measured, and from the rate of change of these parameters the velocity vector of the target was computed and its future position predicted. This information was used to point the gun in the proper direction and to set the fuzing time. The tracking radar performs a similar role in providing guidance information and steering commands for missiles.

In missile-range instrumentation, the tracking-radar output is used to measure the trajectory of the missile and to predict future position. Tracking radar which computes the impact point of a missile continuously during flight is also important for range safety. Missile-range instrumentation radars are normally used with a beacon to provide a point-source target with high signal-to-noise ratio. Some of these systems achieve a precision of the order of 0.1 mil in angle and a range accuracy of 5 yd.

This chapter describes the conical-scan, sequential-lobing, and monopulse (both phase comparison and amplitude comparison) tracking-radar techniques, with the main emphasis on the amplitude-comparison monopulse radar.

18.2 SCANNING AND LOBING

The first technique used for angle tracking of targets by radar was to sense the target location with respect to the antenna axis by rapidly switching the antenna beam from one side of the antenna axis to the other, as in Fig. 18.2. The original tracking radars of this type, such as the SCR-268, used an array of radiating elements which could be switched in phase to provide two beam positions for the lobing operation. The radar operator observed an oscilloscope that displayed side by side the video returns from the two beam positions. When the target was on axis, the two pulses were of equal amplitude (Fig. 18.2a); when the target moved off axis, the two pulses became unequal (Fig. 18.2b). The radar operator, observing the existence of an error and its direction, could position the antenna to regain a balance between the two beam positions. This provided a manual tracking loop.

This lobing technique was extended to continuous rotation of the beam around the target (conical scan) as in Fig. 18.3.[1] Angle-error-detection circuitry is provided to generate error voltage outputs proportional to the tracking error and with a phase or polarity to indicate the direction of errors. The error signal actuates a servosystem to drive the antenna in the proper direction to reduce the error to zero.

Continuous beam scanning is accomplished by mechanically moving the feed of an antenna since the antenna beam will move off axis as the feed is moved off the focal point. The feed is typically moved in a circular path around the focal point, causing a corresponding movement of the antenna beam in a circular path around the target. A typical block diagram is shown in Fig. 18.4.[2] A range-tracking system is included which automatically follows the target in range, with range gates that turn on the radar receiver only during the time when the echo is expected from the target under track. Range gating excludes undesired targets and noise. The system also includes an automatic gain control (AGC) necessary to maintain constant angle sensitivity (volts of error-detector output per degree of

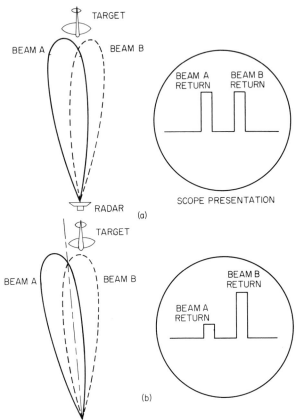

FIG. 18.2 Angle error sensing in one coordinate by switching the antenna beam position from one side of the target to the other. (*a*) Target located on the antenna axis. (*b*) Target at one side of the antenna axis.

error) independent of the amplitude of the echo signal. This provides the constant gain in the angle-tracking loops necessary for stable angle tracking.

The feed scan motion may be either a rotation or a nutation. A rotating feed turns as it moves with circular motion, causing the polarization to rotate. A nutating feed does not rotate the plane of polarization during the scan; it has a motion like moving one's hand in a circular path.

The radar video output contains the angle-tracking-error information in the envelope of the pulses, as shown in Fig. 18.5. The percentage modulation is proportional to the angle-tracking error, and the phase of the envelope function relative to the beam-scanning position contains direction information. Angle-tracking-error detection (error demodulation) is accomplished by a pair of phase detectors using a reference input from the scan motor. The phase detectors perform essentially as dot-product devices with sine-wave reference signals at the frequency of scan and of proper phases to obtain elevation error from one and

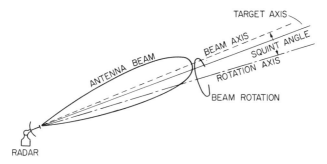

FIG. 18.3   Conical-scan tracking.

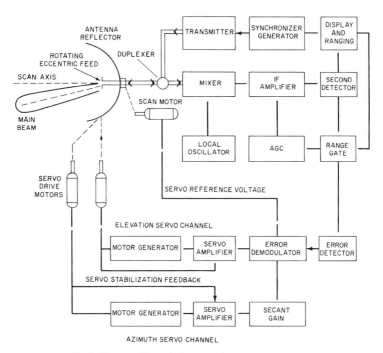

FIG. 18.4   Block diagram of a conical-scan radar.

azimuth error from the other. For example, the top scan position may be chosen as zero phase for a cosine function of the scan frequency. This provides a positive voltage output proportional to the angle error when the target is above the antenna axis. The reference signal to the second phase detector is generated with a 90° phase relation to the original reference. This provides a similar error voltage proportional to the azimuth-angle error and with polarity corresponding to the direction of error.

A secant correction (Fig. 18.4) is necessary in any conventional elevation-over-azimuth tracking radar where the elevation drive system rotates when the

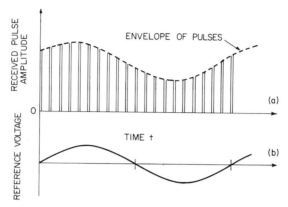

FIG. 18.5 (*a*) Angle error information contained in the envelope of the received pulses in a conical-scan radar. (*b*) Reference signal derived from the drive of the conical-scan feed.

antenna changes azimuth. A target flying a passing course by the radar will, at its closest point to the radar, cause the azimuth servo to drive faster at high elevation angles than at low elevation angles. In the extreme, when the target passes directly overhead, the azimuth drive would have to flip the antenna 180° at the instant when the target crosses overhead. This effect requires the azimuth-tracking loop gain to change approximately as the secant of elevation angle in order to maintain essentially constant overall azimuth loop gain. In practical tracking radars with the conventional elevation-over-azimuth mount the elevation angle is typically limited by this effect to a maximum of 85° since the servo bandwidth required for higher elevation angles exceeds practical limits.

A major parameter in a conical-scan radar is the size of the circle to be scanned relative to the beamwidth. Figure 18.6 shows a circle representing the 3 dB contour of the beam at one position of its scan. The half-power beamwidth is θ_B. The dashed circle represents the path described by the center of the beam as it is scanned. The radius of the dashed circle is β, the offset angle. The compromise that must be made in choosing β is between the loss of signal or antenna gain L_k (crossover loss) and the increase in angle sensitivity k_s of the angle-sensing circuits.

High angle-error sensitivity is desired to obtain higher voltage from the angle error detectors for a given true angle error relative to undesired voltages in the receiver output. The undesired receiver output includes angle errors caused by receiver thermal noise. For a given signal-to-noise ratio (*SNR*) the thermal-noise effects are inversely proportional to angle error sensitivity. Unfortunately, increasing β to increase k_s also increases the loss L_k, which reduces the *SNR*.

The relative values of k_s and L_k depend upon whether the target provides a beacon response that removes the transmitting modulation from the received signal or whether two-way skin tracking is performed. Two-way tracking gives a greater depth of modulation, or angle sensitivity, for a given β but doubles the loss in decibels. Figure 18.7 shows the loss of antenna gain and angle sensitivity as a function of β for the two cases.[2] The rms error caused by receiver thermal noise is inversely proportional to k_s and proportional to $\sqrt{L_k}$ (L_k expressed as power loss). The peaks of the dotted curves labeled $k_s/\sqrt{L_k}$ indicate the optimum offset angle β for minimizing receiver thermal-noise effects on angle tracking. However, the range-tracking system of the radar is affected by L_k only, and a β

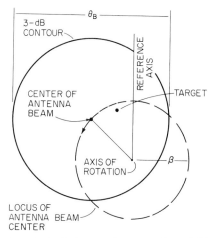

FIG. 18.6 Conical-scan-radar antenna beam 3 dB contour (solid circle) and path of rotation (dashed circle) of the beam center.

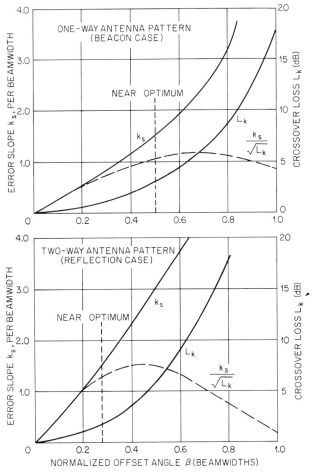

FIG. 18.7 Error slope k_s and crossover loss L_k.

of zero maximizes range-tracking performance. Therefore, values of β indicated by the vertical dashed lines are chosen smaller than optimum for angle tracking as a compromise between angle- and range-tracking performance.

The tracking-error information in beam-scanning tracking radars is a time fluctuation of the echo signal amplitude. Other sources of echo-signal-amplitude fluctuation such as target scintillation (Sec. 18.8) can cause false indications of tracking error. The undesired fluctuations that cause difficulty occur at about the same rate as the scan rate. Since target scintillation energy of aircraft is concentrated in the lower frequency range below approximately 100 Hz (particularly the troublesome propeller modulation), it is desirable to increase the scan rate as high as possible. The maximum practical rate is one-fourth of the pulse repetition frequency (PRF) so that four pulses provide a complete scan with one each up, down, right, and left. The maximum PRF and, consequently, maximum scan rate are limited by the maximum range of targets to be tracked by the radar. At a PRF of 1000 Hz the unambiguous range extends to about 80 nmi (at this range the echo is returning at the time when the radar is ready for its next transmission). The radar can track beyond this range by using *n*th-time-around tracking as described in Sec. 18.5.

High scan rates are difficult to achieve with mechanical scanning devices in a large antenna, and a variety of techniques to scan electronically have been used. In a small antenna such as that used in missile-homing heads, the dish rather than the feed may be tilted and rotated at high revolutions per minute (r/min) to achieve high scan rates. Scan rates of hundreds of r/min are frequently used—in some instances, as high as 2400 r/min, as in the AN/APN-58 target seeker. In the target seeker application the PRF can be high since the target is at short range. The coming of the jet aircraft caused additional problems for lobing systems because jet turbines cause significant modulations at high frequencies in regions near the maximum practical mechanical or electronic lobing rates. A further problem in scanning and lobing systems is a limitation on long-range tracking. At long ranges the time required for the radar signal to travel to the target and back becomes a significant portion of a scan cycle. For example, at a 100-Hz scan rate and the target at 460 mi, a signal transmitted on an up lobe will return as an echo when the antenna is looking on a down lobe, canceling the effect of the scan and the angle-error-sensing capability. In applications where this effect is significant, compensation can be provided if the range to the target is measured.

18.3 MONOPULSE (SIMULTANEOUS LOBING)

The susceptibility of scanning and lobing techniques to echo amplitude fluctuations was the major reason for developing a tracking radar that provides simultaneously all the necessary lobes for angle-error sensing. The output from the lobes may be compared simultaneously on a single pulse, eliminating any effect of time change of the echo amplitude. The technique was initially called *simultaneous lobing*, which was descriptive of the original designs. Later the term *monopulse* was used, referring to the ability to obtain complete angle error information on a single pulse. It has become the commonly used name for this tracking technique.

The original monopulse trackers suffered in antenna efficiency and complexity of microwave components since waveguide signal-combining circuitry was a relatively new art. These problems were overcome, and monopulse radar with off-the-shelf components can readily outperform scanning and lobing systems. The monopulse technique also has an inherent capability for high-precision angle measurement because its feed structure is rigidly mounted with no moving parts. This has made possible the development of pencil-beam tracking radars that meet missile-range instrumentation-radar requirements of 0.003° angle-tracking precision.

This chapter is devoted to tracking radar, but monopulse is used in other systems including homing devices, direction finders, and some search radars. However, most of the basic principles and limitations of monopulse apply for all applications. A more general coverage is found in Refs. 3 and 4.

Amplitude-Comparison Monopulse. A method for visualizing the operation of an amplitude-comparison monopulse receiver is to consider the echo signal at the focal plane of an antenna.[5] The echo is focused to a "spot" having a cross-section shape approximately of the form $J_1(X)/X$ for circular apertures, where $J_1(X)$ is the first-order Bessel function. The spot is centered in the focal plane when the target is on the antenna axis and moves off center when the target moves off axis. The antenna feed is located at the focal point to receive maximum energy from a target on axis.

An amplitude-comparison monopulse feed is designed to sense any lateral displacement of the spot from the center of the focal plane. A monopulse feed using the four-horn square, for example, would be centered at the focal point. It provides a symmetry so that when the spot is centered equal energy falls on each of the four horns. However, if the target moves off axis, causing the spot to shift, there is an unbalance of energy in the horns. The radar senses the target displacement by comparing the amplitude of the echo signal excited in each of the horns. This is accomplished by use of microwave hybrids to subtract outputs of pairs of horns, providing a sensitive device that gives signal output when there is an unbalance caused by the target being off axis. The RF circuitry for a conventional four-horn square (Fig. 18.8) subtracts the output of the left pair from the output of the right pair to sense any unbalance in the azimuth direction. It also subtracts the output of the top pair from the output of the bottom pair to sense any unbalance in the elevation direction.

The Fig. 18.8 comparator is the circuitry which performs the addition and subtraction of the feedhorn outputs to obtain the monopulse sum and difference signals. It is illustrated with hybrid-T or magic-T waveguide devices. These are four-port devices which, in basic form, have the inputs and outputs located at right angles to each other. However, the magic T's have been developed in convenient "folded" configurations for very compact comparator packages. The performance of these and other similar four-port devices is described in Ref. 3, Chap. 4.

The subtractor outputs are called difference signals, which are zero when the target is on axis, increasing in amplitude with increasing displacement of the target from the antenna axis. The difference signals also change 180° in phase from one side of center to the other. The sum of all four horn outputs provides a reference signal to allow angle-tracking sensitivity (volts per degree error) even though the target echo signal varies over a large dynamic range. AGC is necessary to keep the gain of the angle-tracking loops constant for stable automatic angle tracking.

Figure 18.9 is a block diagram of a typical monopulse radar. The sum signal, elevation difference signal, and azimuth difference signal are each converted to intermediate frequency (IF), using a common local oscillator to maintain relative phase at IF. The IF sum-signal output is detected and provides the video input to

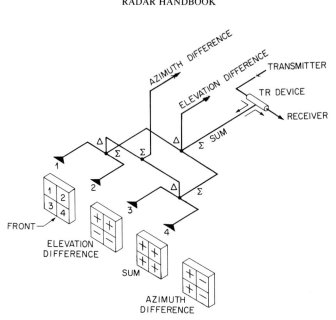

FIG. 18.8 Microwave-comparator circuitry used with a four-horn mono-pulse feed.

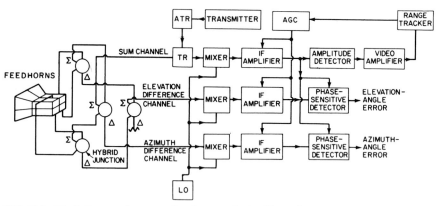

FIG. 18.9 Block diagram of a conventional monopulse tracking radar.

the range tracker. The range tracker determines the time of arrival of the desired target echo and provides gate pulses which turn on portions of the radar receiver only during the brief period when the desired target echo is expected. The gated video is used to generate the dc voltage proportional to the magnitude of the Σ signal or $|\Sigma|$ for the AGC of all three IF amplifier channels. The AGC maintains constant angle-tracking sensitivity (volts per degree error) even though the target echo signal varies over a large dynamic range by controlling gain or dividing by $|\Sigma|$. AGC is necessary to keep the gain of the angle-tracking loops constant for stable automatic angle tracking. Some monopulse systems, such as the two-channel monopulse, can provide instantaneous AGC or normalizing as described later in this section.

The sum signal at the IF output also provides a reference signal to phase detectors which derive angle-tracking-error voltages from the difference signal. The phase detectors are essentially a dot-product device producing the output voltage

$$e = \frac{|\Sigma| \, |\Delta|}{|\Sigma| \, |\Sigma|} \cos \theta \qquad \text{or} \qquad e = \frac{\Delta}{|\Sigma|} \cos \theta$$

where e = angle-error-detector output voltage
$|\Sigma|$ = magnitude of sum signal
$|\Delta|$ = magnitude of difference signal
θ = phase angle between sum and difference signals

The dot-product error detector is only one of a wide variety of monopulse angle error detectors described in Ref. 3, Chap. 7.

Normally, θ is either 0° or 180° when the radar is properly adjusted, and the only purpose of the phase-sensitive characteristic of the detector is to provide a plus or minus polarity corresponding to $\theta = 0°$ and $\theta = 180°$, respectively, giving direction sense to the angle-error-detector output.

In a pulsed tracking radar the angle-error-detector output is bipolar video; that is, it is a video pulse with an amplitude proportional to the angle error and whose polarity (positive or negative) corresponds to the direction of the error. This video is typically processed by a boxcar circuit which charges a capacitor to the peak video-pulse voltage and holds the charge until the next pulse, at which time the capacitor is discharged and recharged to the new pulse level. With moderate low-pass filtering, this gives a dc error voltage output employed by the servo amplifiers to correct the antenna position.

The three-channel amplitude-comparison monopulse tracking radar is the most commonly used monopulse system. The three signals may sometimes be combined in other ways to allow use of a two-channel or even a single-channel IF system as described later in this section.

Monopulse-Antenna Feed Techniques. Monopulse-radar feeds may have any of a large variety of configurations. For two-angle tracking such as azimuth and elevation, the feeds may include three or more apertures.[6] Single apertures are also employed by using higher-order waveguide modes to extract angle-error-sensing difference signals. There are many tradeoffs in feed design because optimum sum and difference signals, low sidelobe levels, omnipolarization capability, and simplicity cannot all be fully satisfied simultaneously. The term *simplicity* refers not only to cost saving but also to the use of noncomplex circuitry which is necessary to provide a broadband system with good boresight stability to meet precision-tracking requirements. (Boresight is the electrical axis of the antenna or the angular location of a signal source within the antenna beam at which the angle-error-detector outputs go through zero.)

Some of the typical monopulse feeds are described to show the basic relations involved in optimizing the various performance factors and how the more important factors can be optimized by a feed configuration but at the price of lower performance in other areas. Many new techniques have been added since the original four-horn square feed in order to provide good or excellent performance in all desired feed characteristics in a well-designed monopulse radar.

The original four-horn square monopulse feed is inefficient since the optimum feed size in the plane of angle measurement for the difference signals is approx-

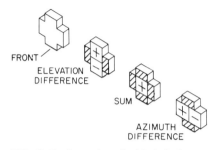

FIG. 18.10 Approximately ideal feed-aperture E-field distribution for sum and difference signals.

imately twice the optimum size for the sum signal.[7] Consequently, an intermediate size is typically used with a significant compromise for both sum and difference signals. The optimum four-horn square feed, which is subject to this compromise, is described in Ref. 3 as based on minimizing the angle error caused by receiver thermal noise. However, if sidelobes are a prime consideration, a somewhat different feed size may be desired.

The limitation of the four-horn square feed is that the sum- and difference-signal E fields cannot be controlled independently. If independent control could be provided, the *ideal* would be approximately as described in Fig. 18.10 with twice the dimension for the difference signals in the plane of error sensing than that for the sum signal.[7]

A technique used by the MIT Lincoln Laboratory to approach the ideal was the 12-horn feed (Fig. 18.11). The overall feed, as illustrated, is divided into small parts and the microwave circuitry selects the portions necessary for the sum and difference signals to approach the ideal. One disadvantage is that this feed requires a very complex microwave circuit. Also, the divided four-horn portions of the feed are each four element arrays which generate large feed sidelobes in the H plane because of the double-peak E field. Another consideration is that the 12-horn feed is not practical for focal-point-fed parabolas or reflectarrays because of its size. A focal-point feed is usually small to produce a broad pattern and must be compact to avoid blockage of the antenna aperture. In some cases the small size required is below waveguide cutoff, and dielectric loading becomes necessary to avoid cutoff.

A more practical approach to monopulse-antenna feed design uses higher-order waveguide modes rather than multiple horns for independent control of sum- and difference-signal E fields. This allows much greater simplicity and flexibility. A triple-mode two-horn feed used by RCA[7,8] retracts the E-plane septa to allow both the TE_{10} and TE_{30} modes to be excited and propagate in the double-width septumless region as illustrated in Fig. 18.12. At the septum the double-humped E field is represented by the combined TE_{10} and TE_{30} modes subtracting at the center and adding at the TE_{30}-mode outer peaks. However, since the two modes propagate at different velocities, a point is reached farther down the double-width guide where the two modes add in the center and subtract at the outer humps of the TE_{30} mode. The result is a sum-signal E field concentrated toward the center of the feed aperture.

This shaping of the sum-signal E field is accomplished independently of the difference-signal E field. The difference signal is two TE_{10}-mode signals arriving at the septum of Fig. 18.12 out of phase. At the septum it becomes the TE_{20} mode, which propagates to the horn aperture and uses the full width of the horn as desired. The TE_{20} mode has zero E field in the center of the waveguide where the septum is located and is unaffected by the septum.

The AN/FPS-16 radar feed used two retracted septum horns illustrated in Fig. 18.13. The TE_{20}-mode signals are added for the H-plane difference signal, the combined TE_{10} and TE_{30} modes are added for the sum signal, and they are subtracted for the E-plane difference signal. Since this is a focal-point feed, it is small in size (wavelengths) and RF currents tend to flow around the top and bottom

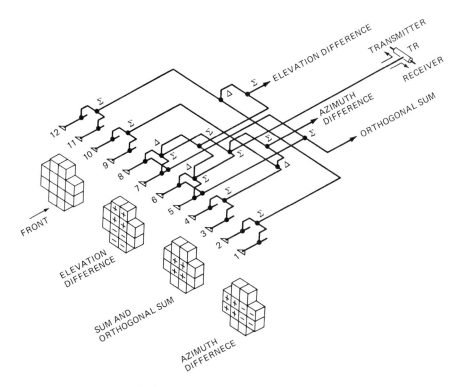

FIG. 18.11 Twelve-horn feed.

edges at the E-field peaks in the middle of the horns. This results in the need for the top and bottom matching stubs seen in Fig. 18.13.

A further step in feed development is the four-horn triple-mode feed illustrated in Fig. 18.14.[7] This feed uses the same approach as described above but with the addition of a top and bottom horn. This allows the E-plane difference signal to couple to all four horns and uses the full height of the feed. The sum signal uses only the center two horns to limit its E field in the E plane as desired for the ideal field shaping. The use of smaller top and bottom horns is a simpler method of concentrating the E field toward the center of the feed, where the full horn width is not needed.

The feeds described thus far are for linear-polarization operation. When circular polarization is needed in a paraboloid-type antenna, square or circular cross-section horn throats are used. The vertical and horizontal components from each horn are separated and comparators provided for each polarization. The sum and difference signals from the comparators are combined with 90° relative phase to obtain circular polarization. Use of the previously described feeds for circular polarization would require the waveguide circuitry to be prohibitively complex. Consequently, a five-horn feed is used as illustrated in Fig. 18.15.

The five-horn feed is selected because of the simplicity of the comparator which requires only two magic (or hybrid) T's for each polarization. The sum and difference signals are provided for the two linear-polarization components and, in an AN/FPQ-6 radar, are combined in a waveguide switch for selecting

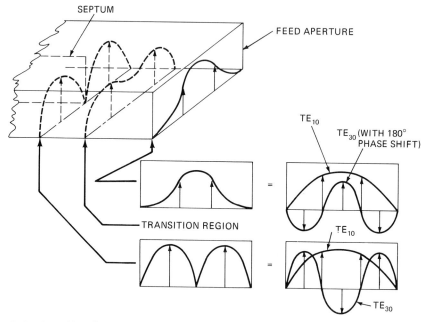

FIG. 18.12   Use of retracted septum to shape the sum-signal E field.

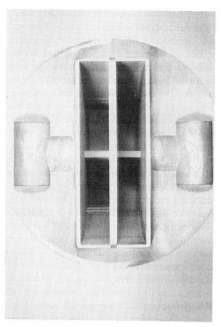

FIG. 18.13   AN/FPS-16 feed, front view. (*From S. M. Sherman, Ref. 3.*)

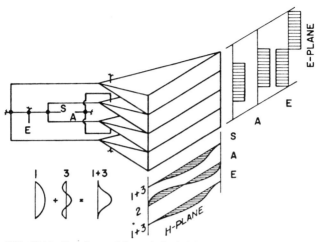

FIG. 18.14 Four-horn triple-mode feed. (*From P. W. Hannan, Ref. 7; copyright 1961, IEEE.*)

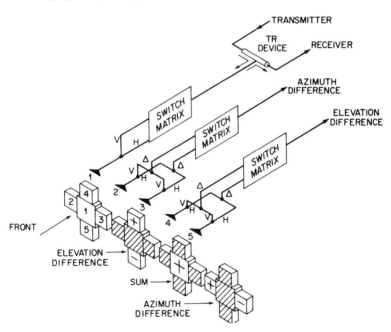

FIG. 18.15 Five-horn feed with coupling to both linear-polarization components, which are combined by the switch matrix to select horizontal, vertical, or circular polarization.

polarization. The switch selects either the vertical or the horizontal input component or combines them with a 90° relative phase for circular polarization. This feed does not provide optimum sum- and difference-signal E fields because the sum horn occupies space desired for the difference signals. Generally an undersized sum-signal horn is used as a compromise. However, the five-horn feed is a practical choice between complexity and efficiency. It has been used in several instrumentation radars including the AN/FPQ-6, AN/FPQ-10, AN/TPQ-18, and AN/MPS-36[9,10] and in the AN/TPQ-27 tactical precision-tracking radar.[11]

The multimode feed techniques can be expanded to other higher-order modes for error sensing and E-field shaping.[12,13,14] The difference signals are contained in unsymmetrical modes such as the TE_{20} mode for H-plane error sensing and combined TE_{11} and TM_{11} modes for E-plane error sensing. These modes provide the difference signals, and no comparators, as shown in Fig. 18.8, are used.[12] Generally, mode-coupling devices can give good performance in separating the symmetrical and unsymmetrical modes without significant cross-coupling problems.

Multiband monopulse feed configurations are practical and in use in several systems. A simple example is a combined X-band and K_a-band monopulse paraboloid antenna radar. Separate conventional feeds are used for each band, with the K_a-band feed as a Cassegrain feed and the X-band feed at the focal point.[15] The Cassegrain subdish is a hyperbolic-shaped grid of wires reflective to parallel polarization and transparent to orthogonal polarization. It is oriented to be transparent to the X-band focal-point feed behind it and reflective to the orthogonally polarized K_a-band feed extending from the vertex of the paraboloid.

Monopulse feed horns at different microwave frequencies can also be combined with horns interlaced. The multiband feed clusters will sacrifice efficiency but can satisfy multiband requirements in a single antenna.

AGC (Automatic Gain Control). To maintain a stable closed-loop servosystem for angle tracking, the radar must maintain essentially constant loop gain independent of target size and range. The problem is that monopulse difference signals from the antenna are proportional to both the angle displacement of the target from the antenna axis and the echo signal amplitude. For a given tracking error, the error voltage would change with echo amplitude and cause a corresponding change in loop gain.

AGC is used to remove the angle-error-detector-output dependence on echo amplitude and retain constant tracking loop gain. A typical AGC technique is illustrated in Fig. 18.16 for a one-angle coordinate tracking system. The AGC system detects the peak voltage of the sum signal and provides a negative dc voltage proportional to the peak signal voltage. The negative voltage is fed to the IF amplifier stages, where it is used to decrease gain as the signal increases. A high gain in the AGC loop is equivalent to dividing the IF output by a factor proportional to its amplitude.

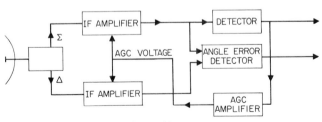

FIG. 18.16   AGC in monopulse tracking.

In a three-channel monopulse radar, all three channels are controlled by the AGC voltage, which effectively performs a division by the magnitude of the sum signal or echo amplitude. Conventional AGC with a control voltage is band-limited by filters, and the gain is essentially constant during the pulse repetition interval. Also, the AGC of the sum channel normalizes the sum echo pulse amplitude to similarly maintain a stable range-tracking servo loop.

The angle-error detector, assumed to be a product detector, has an output

$$|e| = k \, \frac{\Delta \, \Sigma}{|\Sigma||\Sigma|} \cos \theta$$

where $|e|$ is the magnitude of the angle error voltage. Phases are adjusted to provide 0 or 180° on a point-source target. The resultant is

$$|e| = \pm \, k \, \frac{\Delta}{|\Sigma|}$$

Complex targets can cause other phase relations as a part of the angle scintillation phenomenon.[3] The above error voltage proportional to the ratio of the difference signal divided by the sum signal is the desired angle-error-detector output, giving a constant angle error sensitivity.[3]

With limited AGC bandwidth, some rapid signal fluctuations modulate $|e|$, but the long-time-average angle sensitivity is constant. These fluctuations are largely from rapid changes in target reflectivity, $\sigma(t)$, that is, from target amplitude scintillation. The random modulation of $|e|$ causes an additional angle noise component that affects the choice of AGC bandwidth.

At very low signal-to-noise ratios ($SNR < 4$ dB) the AGC voltage is limited to a minimum value by the noise level. Therefore, as the signal decreases into the noise, the IF gain remains constant and the resultant angle sensitivity decreases. Consequently, the effect of thermal noise on tracking performance differs from the linear relations, which are accurate to within about 1 dB for an SNR of 4 dB or greater. Reference 2 discusses means for calculating thermal-noise effects for very low SNR conditions.

The AGC performance in conical-scan radars provides a similar constant angle error sensitivity. One major limitation in conical-scan radars is that the AGC bandwidth must be sufficiently lower than the scan frequency to prevent the AGC from removing the modulation containing the angle error information. The very low SNR effects on conical scan differ from the effects on monopulse, as discussed in detail in Ref. 2.

Phase-Comparison Monopulse. A second monopulse technique is the use of multiple antennas with overlapping (nonsquinted) beams pointed at the target. Interpolating target angles within the beam is accomplished, as shown in Fig. 18.17, by comparing the phase of the signals from the antennas (for simplicity a single-coordinate tracker is described). If the target were on the antenna boresight axis, the outputs of each individual aperture would be in phase. As the target moves off axis in either direction, there is a change in relative phase. The amplitudes of the signals in each aperture are the same so that the output of the angle error phase detector is determined by the relative phase only. The phase-detector circuit is adjusted with a 90° phase shift in one channel to give zero output when the target is on axis and an output increasing with increasing angular displacement of the target with a polarity corresponding to the direction of error.[3,4]

Typical flat-face corporate-fed phased arrays compare the output of halves of

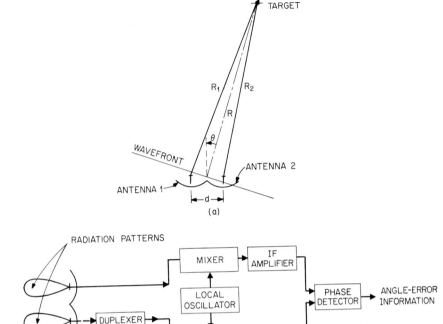

FIG. 18.17 (*a*) Wavefront phase relationships in a phase comparison monopulse radar. (*b*) Block diagram of a phase comparison monopulse radar (one angle coordinate).

the aperture and fall into the class of phase-comparison monopulse. However, the basic performance of amplitude- and phase-comparison monopulse is essentially the same.[3]

Figure 18.17 shows the antenna and receiver for one angular-coordinate tracking by phase-comparison monopulse. Any phase shifts occurring in the mixer and IF amplifier stages cause a shift in the boresight of the system. The disadvantages of phase-comparison monopulse compared with amplitude-comparison monopulse are the relative difficulty in maintaining a highly stable boresight and the difficulty in providing the desired antenna illumination taper for both sum and difference signals. The longer paths from the antenna outputs to the comparator circuitry make the phase-comparison system more susceptible to boresight change due to mechanical loading or sag, differential heating, etc.

A technique giving greater boresight stability combines the two antenna outputs at RF with passive circuitry to yield sum and difference signals, as shown in Fig. 18.18. These signals may then be processed as in a conventional amplitude-comparison monopulse receiver. The system shown in Fig. 18.18 would provide a relatively good difference-channel taper, having smoothly tapered *E* fields on each antenna. However, a sum-signal excitation with the two antennas provides a two-hump in-phase *E*-field distribution which causes high sidelobes since it looks like a two-element array. This problem may be reduced by allowing some aperture overlap but at the price of loss of angle sensitivity and antenna gain.

Monopulse Tracking with Phased Arrays. In general, phased-array tracking radars fall in either the amplitude- or the phase-comparison class, depending on the feed technique. Feedthrough lens arrays acting as an RF lens and reflectarrays acting like a parabolic reflector may use any of the described multihorn or multimode feeds, and the same general factors in optimizing a feed apply. Monopulse angle-error sensing can be accomplished in a corporate-feed array by using the two halves of the array (the top and bottom halves can also be used for elevation) as a phase-comparison tracker. Array antennas which use the two halves of the aperture for phase-comparison angle error sensing generally provide a good taper for the sum pattern, but the difference-signal E field across the array reaches a peak toward the center with a sudden 180° phase change. This sharp discontinuity at a maximum amplitude point causes undesired high sidelobes. Techniques such as use of separate feeds can provide the desired shaping of the difference-signal E-field distribution.

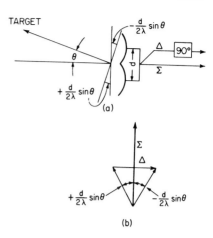

FIG. 18.18 (*a*) RF phase-comparison monopulse system with sum and difference outputs. (*b*) Vector diagram of the sum and difference signals.

The monopulse electronic-scan phased array is used in instrumentation radar to meet requirements for simultaneously tracking multiple targets.[16–18] An example is the Multiobject Tracking Radar (MOTR), AN/TPQ-39, built for the White Sands Missile Range for high-precision tracking. The MOTR provides a 60° cone of electronic pulse-to-pulse beam coverage plus mechanical pedestal movement to cover the hemisphere. High accuracy and efficiency can be maintained over the ± 30° electronic-scan coverage, and this coverage moves mechanically as needed to optimally move the electronic-scan coverage along with the target configuration to be tracked.[14,19]

One- and Two-Channel Monopulse. Monopulse radars may be constructed with fewer than the conventional three IF channels. This is accomplished by combining the sum and difference signals by some means so that they may be individually retrieved at the output. These techniques provide some advantages in AGC or other processing techniques but at the cost of *SNR* loss or of cross coupling between azimuth and elevation information.

A single-channel monopulse system called SCAMP (single-channel monopulse processor)[20] provides the desired constant angle error sensitivity by normalizing the difference signals with the sum signal in a single IF channel, as shown in Fig. 18.19. The signals are each converted from RF to different IF frequencies by separate local oscillators (LOs) of different frequencies for each signal. They are amplified in a single IF amplifier of sufficient bandwidth for all three signals at different frequencies. At the IF output the signals are hard-limited and separated by three narrowband filters. The signals are then converted to the same frequency by beating two of the signals with the frequency difference between their LOs and the LO of the third signal. The angle-error voltage is then determined by either a conventional phase detector or simply an amplitude detector.[14] The effect of AGC action and normalizing is performed by hard limiting which causes a weak-signal suppression of the difference signal similar to the effects of hard limiting on a weak signal in noise.[20]

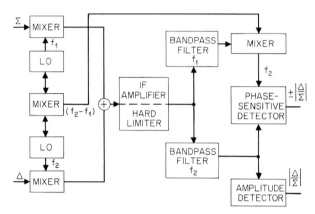

FIG. 18.19   Block diagram of SCAMP, a single-channel monopulse tracking system, demonstrating angle tracking in one angle coordinate. (The system is capable of tracking in both coordinates.) (*From W. L. Rubin and S. K. Kamen, Ref. 20.*)

The single-channel monopulse provides in effect an instantaneous AGC. Performance in the presence of thermal noise is about equal to that of the three-channel monopulse. However, the limiting process generates a significant cross-coupling problem,[21] causing a portion of the azimuth-error signal to appear in the elevation-angle-error-detector output and elevation error to appear in the azimuth-angle-error-detector output. Depending on the receiver configuration and the choice of IF frequencies, the cross modulation could cause serious errors and allow vulnerability to jamming. Reference 21 describes how the sum- and difference-channel bands in the wideband IF can be arranged to minimize cross coupling and suggests use of narrow banding to separate each signal before limiting in order to reduce jamming effects.

A two-channel monopulse receiver[22] may also be used by combining the sum and difference signals at RF, as shown in Fig. 18.20. The microwave resolver is a mechanically rotated RF coupling loop in circular waveguide. The azimuth and elevation difference signals are excited in this guide with E-field polarization oriented at 90°. The energy into the coupler contains both difference signals coupled as the cosine and sine of the angular position of the coupler, $\omega_s t$, where ω_s is the angular rate of rotation. The hybrid adds the combined difference signals Δ to the sum signal Σ. The $\Sigma + \Delta$ and $\Sigma - \Delta$ outputs each look like the output of a conical-scan tracker except that the modulation function differs by 180°. In case of failure of one channel, the radar can be operated as a scan-on-receive-only conical-scan radar with essentially the same performance as a conical-scan radar. The advantage of two channels with opposite-sense angle-error information on one with respect to the other is that signal fluctuations in the received signal are

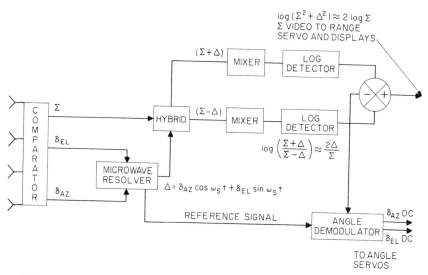

FIG. 18.20 Block diagram of a two-channel monopulse radar system. (*From R. S. Noblit, Ref. 22.*)

canceled in the postdetection subtractor at the IF output which retrieves the angle-error information. The log IF performs essentially as instantaneous AGC, giving the desired constant angle-error sensitivity of the difference signal normalized by the sum signal. The detected Δ information is a bipolar video where the error information is contained in the sinusoidal envelope. This signal is separated into its two components, azimuth- and elevation-error information, by an angle demodulation. The demodulator, using a reference from the drive on the rotating coupler, extracts the sine and cosine components from Δ to give the azimuth- and elevation-error signals. The two-channel monopulse technique is used in the AN/SPG-55 tracking radar and the AN/FPQ-10 missile-range instrumentation radar. The modulation caused by the microwave resolver is of concern in instrumentation radar applications because it adds spectral components in the signal which complicate the possible addition of pulse doppler tracking capability to the radar.

This system provides instantaneous AGC operation with only two IF channels and operation with reduced performance in case of failure of either channel. However, there is a loss of 3 dB *SNR* at the receiver inputs although this loss is partly regained by coherent addition of the Σ-signal information. The design of the microwave resolver must minimize loss through the device, and high precision is required to minimize cross coupling between azimuth and elevation channels. The resolver performance is improved by the use of ferrite switching devices to replace the mechanical rotating coupler.

Conopulse. Conopulse (also called scan with compensation) is a radar tracking technique that is a combination of monopulse and conical scan.[23,24] A pair of antenna beams is squinted in opposite directions from the antenna axis and rotated like a conical-scan-radar beam scan. Since they exist simultaneously, monopulse information can be obtained from the pair of beams. The plane in which monopulse information is measured rotates. Consequently, elevation and azimuth information is sequential and must be separated for use in each tracking coordinate. Conopulse provides the monopulse advantage of avoiding errors caused by amplitude scintillation, and it requires only two receivers. However, it has the disadvantage over conventional monopulse radar of lower angle data rate and the mechanical complexity of providing and coupling to a pair of rotating antenna feedhorns.

18.4 SERVOSYSTEMS FOR TRACKING RADAR

The servosystem of a tracking radar is the portion of the radar that receives as its input the tracking-error voltage and performs the task of moving the antenna beam in a direction that will reduce to zero the alignment error between the antenna axis and the target. For two-angle tracking with a mechanical-type antenna there are typically separate axes of rotation for azimuth and elevation and separate servosystems to move the antenna about each axis. A conventional servosystem is composed of amplifiers, filters, and a motor that moves the antenna in a direction to maintain the antenna axis on the target. Range tracking is accomplished by a similar function to maintain range gates centered on the received-echo pulses. This may be accomplished by analog techniques or by digital-counter registers that retain numbers corresponding to target range to provide a closed tracking loop digitally rather than mechanically.

Servosystems may contain hydraulic-drive motors, conventional electric motors geared down to drive the antenna, or direct-drive electric motors where the antenna mechanical axis shaft is part of the armature and the motor field is built into the supporting case. The direct drive is heavier for a given horsepower but eliminates gear backlash. The conventional motors may be provided in a duplicate drive with a small residual opposing torque to reduce backlash. Amplifier gain and filter characteristics as well as motor torque and inertia determine the velocity and acceleration capability or the ability to follow higher-order motions of the target.

It is desired that the antenna beam follow the center of the target as closely as possible, which implies that the servosystem should be capable of moving the antenna quickly. The combined velocity and acceleration characteristics of a servosystem can be described by the frequency response of the tracking loop, which is essentially a low-pass filter characteristic. Increasing the bandwidth increases the quickness of the servosystem and its ability to follow closely a strong, steady signal. However, a typical target causes scintillation of the echo signal, giving erroneous error-detector outputs, and at long range the echo is weak, allowing receiver noise to cause additional random fluctuations in the error-detector output. Consequently, a wide servo bandwidth which reduces lag errors allows the noise to cause erroneous motions of the tracking system. Therefore, for best overall performance it is necessary to limit the servo bandwidth to the minimum necessary to maintain a reasonably small tracking-lag error. There is an optimum bandwidth that minimizes the rms of the total erroneous outputs including both tracking lag and random noise, depending upon the target, its trajectory, and other radar parameters.

The optimum bandwidth for angle tracking is range-dependent. A target with typical velocity at long range has low angle rates and a low SNR, and a narrower servo passband will follow the target with reasonably small tracking lag while minimizing the response to receiver thermal noise. At close range the signal is strong, overriding receiver noise, but target angle scintillation errors proportional to the angular span of the target are large. A wider servo bandwidth is needed at close range to keep tracking lag within reasonable values, but it must not be wider than necessary or target scintillation errors become excessive.

The low-pass closed-loop characteristic of a servosystem is unity at zero frequency, typically remaining near this value up to a frequency near the low-pass cutoff, where it may peak up to higher gain, as shown in Fig. 18.21. The peaking is an indication of system instability but is allowed to be as high as tolerable, typically up to about 3 dB above unity gain to obtain maximum bandwidth for a given servomotor drive system. System A of Fig. 18.21 is a case of excessive peaking of about 8 dB. The effect of the peaking is observed by applying a step error input to the servosystem. The peaking of the low-pass characteristic results in an overshoot when the antenna axis moves to align with the target. High peaking causes a large overshoot and a return toward the target with an additional overshoot. In the extreme, as in system A of Fig. 18.21, the antenna zeros in on the target with a damped oscillation. An optimum system compromise between speed of response and overshoot, as in system B, allows the antenna to make a small overshoot with a reasonably rapid exponential movement back to the target. This corresponds to about 1.4-dB peaking of the closed-loop low-pass characteristic.

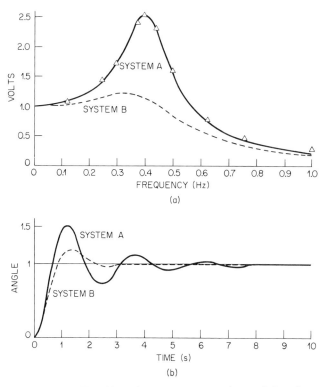

FIG. 18.21 (*a*) Closed-loop frequency-response characteristics of two servosystems. (*b*) Their corresponding time response to a step input.

To maximize the servo closed-loop bandwidth for a given motor drive system, tachometer feedback must be used. The tachometer provides a negative feedback voltage proportional to the servomotor speed. As the servomotor rotates the antenna toward a target, the error voltage decreases, dropping to zero when the antenna axis arrives at the target. However, the tachometer feedback provides a retarding motor torque opposing the system inertia that causes the overshoot, thus reducing the overshoot.

Resonances of the antenna and servosystem structure (the structure foundation is one of the most critical items) must be kept well above the bandwidth of the servosystem; otherwise the system can oscillate at the resonant frequency. A factor of at least 10 is desirable for the ratio of system resonant frequency to servo bandwidth. The high resonant frequency is difficult to obtain with a large antenna, such as the AN/FPQ-6 radar with a 29-ft dish, because of the large mass of the system. The ratio was pushed to a very minimum of about 3 to obtain a servo bandwidth of about 3.5 Hz. A smaller radar with a 12-ft dish, for example, can provide a servo bandwidth up to 7 or 8 Hz with conventional design.

A convenient method for calculating tracking error for a given target trajectory and servosystem is the use of the equation[25]

$$e(t) = \frac{\dot{\theta}(t)}{K_v} + \frac{\ddot{\theta}(t)}{K_a} - \frac{\dddot{\theta}(t)}{K_j} - \cdots$$

where $e(t)$ = tracking error as function of time
$\dot{\theta}(t)$ = angular velocity of target relative to radar versus time
$\ddot{\theta}(t)$ = angular acceleration
$\dddot{\theta}(t)$ = next higher order of angular motion (descriptively called *jerk*)
K_v = velocity constant of servosystem
K_a = acceleration constant
K_j = jerk constant of servosystem

A servosystem could be specified by selecting values for the three constants and, with a known target trajectory $\theta(t)$, the lag error calculated by using the

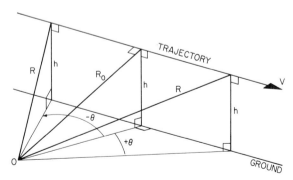

FIG. 18.22 Target trajectory on a passing course. (*From A. S. Locke, Ref. 25.*)

above equation. An example trajectory is shown in Fig. 18.22 for an aircraft flying a straight course past the radar with a minimum distance R_0 at altitude h. The figure shows that minimum range and maximum elevation angle occur when the radar is looking normal to the target path. The azimuth angle θ, with minimum range as the $0°$ azimuth reference point, starts at about $-90°$ and ends at about $+90°$, with its maximum rate of change $\dot{\theta}(t)$ at minimum range. For target parameters of a velocity of 500 kn, crossover range of 1000 yd, and altitude of 1500 ft, the derivatives of azimuth angle are $\dot{\theta}_{max} = 18.6°/s$, $\ddot{\theta}_{max} = 4°/s^2$, and $\dddot{\theta}_{max} = 4.2°/s^3$. Choosing example servosystem constants $K_v = 100$, $K_a = 111$, and $K_j = 1111$ results in a tracking-lag function

$$e(t) = \frac{\dot{\theta}(t)}{100} + \frac{\ddot{\theta}(t)}{111} - \frac{\dddot{\theta}(t)}{1111} \cdots$$

By substituting values for the derivatives, a time plot of tracking lag may be determined as shown in Fig. 18.23.

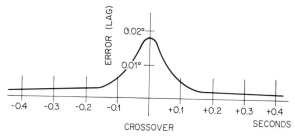

FIG. 18.23 Azimuth-tracking error for the passing-course target. (*From A. S. Locke, Ref. 25.*)

The range and elevation lag errors may be similarly determined by calculation of derivatives of $R(t)$ and $\phi(t)$, respectively, and use of the constants for these tracking systems. Elevation constants are similar to those for the azimuth system. The range tracker may be an inertialess electronic system with a double integration in the tracking loop. This is called a Type II system[25] with a $K_v = \infty$, causing velocity lag to be zero. The remaining significant lag components are the acceleration and jerk lags.

Electronically steerable arrays provide a means for inertialess angle tracking. However, because of this capability the system can track multiple targets by rapidly switching from one to another rather than continuously tracking a single target. The tracker simply places its beam at the location where the target is expected, corrects for the pointing error by converting error voltages (with a known angle error sensitivity) to units of angle, and moves to the next target. The system determines where the target was and, from calculations of target velocity and acceleration, predicts where it should be the next time the beam looks at the target. The lag error in this case is dependent on many factors, including the accuracy of the value of angle sensitivity used to convert error voltages to angular error, the size of the previous tracking error, and the time interval between looks.

18.5 TARGET ACQUISITION AND RANGE TRACKING

Range tracking is the process of continuously measuring the delay between the transmission of an RF pulse and the echo signal returned from a target. The range measurement is the most precise position-coordinate measurement of the radar; typically it can be within a few yards at hundreds-of-miles range. Range tracking usually provides the major means for discriminating the desired target from other targets (although doppler frequency and angle discrimination are also used) by performing range gating (time gating) to eliminate echoes of other targets from the error-detector outputs. The range-tracking circuitry is also used for acquiring a desired target. Range tracking requires not only that the time of travel of the pulse to and from the target be measured but that the return be identified as a target rather than noise and a range-time history of the target be maintained.

This discussion is for a typical pulse-type tracking radar. Range measurement may also be performed with CW radars using FM-CW, a frequency-modulated CW which is typically a linear-ramp FM. The target range is determined by the frequency difference between the echo-frequency ramp and the frequency of the ramp being transmitted. The performance of FM-CW systems with consideration of the doppler effect is described in Ref. 1.

Acquisition. The first function of the range tracker is acquisition of a desired target. Although this is not a tracking operation, it is a necessary first step before range tracking or angle tracking may take place in a typical radar. Some knowledge of target angular location is necessary for pencil-beam tracking radars to point their antenna beams in the direction of the target. This information, called *designation data*, may be provided by a search radar or some other source. It may be sufficiently accurate to place the pencil beam on the target or may require the tracker to scan a larger region of uncertainty. The range-tracking portion of the radar has the advantage of seeing all targets from close range out to the maximum range of the radar. It typically breaks this range into small increments, each of which may be simultaneously examined for the presence of a target. When beam scanning is necessary, the range tracker examines the increments for short periods such as 0.1 s, makes its decision about the presence of a target, and allows the beam to move to a new location if no target is present. This process is typically continuous for mechanical-type trackers which move the beam slowly enough that a target will remain well within the beamwidth for the short examination period of the range increments.

Target acquisition involves consideration of the S/N threshold and integration time needed to accomplish a given probability of detection with a given false-alarm rate similar to a search radar.[1] However, high false-alarm rates, as compared with values used for search radars, are used because the operator knows that the target is present, and operator fatigue from false alarms when waiting for a target is not involved. Optimum false-alarm rates are selected on the basis of performance of electronic circuits which observe range intervals to determine which interval has the target echo.

A typical technique is to set a voltage threshold sufficiently high to prevent most noise spikes from crossing the threshold but sufficiently low that a weak signal may cross. An observation is made after each transmitter pulse as to whether, in the range interval being examined, the threshold has been crossed. The integration time allows the radar to make this observation several times before deciding if there is a target present. The major difference between noise and a target is that noise spikes exceeding the threshold are random, but if a target is present the threshold crossings are more regular. One typical system simply counts the number of threshold crossings over the integration period, and if a crossing occurs for more than half the number of times that the radar has transmitted, a target is indicated as being present. If the radar pulse repetition frequency is 300 Hz and the integration time is 0.1 s, the radar will observe 30 threshold crossings if there is a strong, steady target. However, since the echo from a weak target combined with noise may not always cross the threshold, a limit may be set, such as 15 crossings, that must be exceeded during the integration period for a decision that a target is present. An example of expected performance on a nonscintillating target is a 90 percent probability of detection at a 2.5 dB-per-pulse SNR and a false-alarm rate of 10^{-5}. The AN/FPS-16 and AN/FPQ-6 instrumentation radars use these detection parameters with 10 contiguous gates of 1000 yd each for acquisition. The 10 gates give a coverage of a 5-nmi range interval.

Range Tracking. Once a target has been located, it is desired to follow the target in the range coordinate to provide continuous distance information or slant range to the target. Appropriate timing pulses provide range gating so that the angle-tracking circuits and AGC circuits look at only the short range interval, or time interval, when the desired echo pulse is expected. The range-tracking operation is carried out by a closed-loop tracker similar to the angle tracker. Error in centering the range gate on the target echo pulse is sensed, error voltages are generated, and circuitry is provided to respond to the error voltage by causing the gate to move in a direction to recenter on the target echo pulse.

The range-tracking error may be sensed in many ways. The most commonly used method is the early- and late-gate technique (Fig. 18.24). These gates are timed so that the early gate opens at the beginning of the main range gate and closes at the center of the main range gate. The late gate opens at the center and closes at the end of the main range gate. The early and late gates each allow the target video to charge capacitors during the time when the gates are open. The capacitors act as integrators; the early-gate capacitor charges to a voltage proportional to the area of the first half of the target video pulse, and the late-gate capacitor charges negatively proportionally to the later half of the target video. When the gates are properly centered about a symmetrical video pulse, the capacitors are equally charged. Summing their charge voltages yields a zero output.

When the gates are not centered about the target video so that the early gate extends past the center of the target video, the early-gate capacitor charged positively receives a greater charge. The late gate sees only a small portion of the pulse, resulting in a smaller negative charge. Summing the capacitor voltage results in a positive voltage output. Similarly, if the gates are early so that more of the target video falls in the late gate, the sum of the capacitor voltage results in a negative output. Over a range of errors of approximately $\pm \frac{1}{4}$ of the target video pulse width the voltage output is essentially a linear function of timing error and of a polarity corresponding to the direction of error.

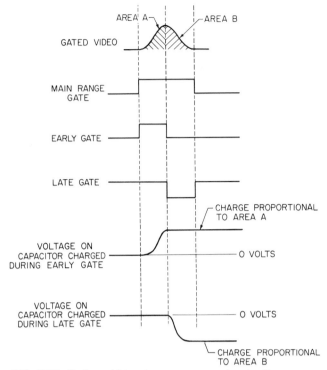

FIG. 18.24   Early- and late-gate range-error-sensing circuit.

Many radar range-tracking systems use sampling circuitry to take three to five samples in the vicinity of the echo video pulse. The amplitudes of the samples on the leading and lagging halves of the pulse are compared for range-error sensing similarly to the comparison of amplitudes in the early- and late-gate range trackers.

In some cases, leading- or lagging-edge range tracking is desired. This has been accomplished in some applications by simply adding a bias to move the error-sensing gates either to lead or to lag the center of the target. This provides some rejection by the gates of undesired returns that might occur near the target, such as echoes from other nearby targets. Threshold devices are also used as leading- or lagging-edge trackers by observing when the target video exceeds a given threshold level. The point of crossing of the threshold is used to trigger gating circuits to read out target range from timing devices or to generate a synthetic target pulse.

The range-tracking loop is closed by using the range-error-detector output to reposition range gates and correct range readout. One technique uses a high-speed digital counter driven by a stable oscillator. The counter is reset to zero at the time of the transmit pulse. Target range is represented by a number in a digital register, as shown in Fig. 18.25. A coincidence circuit senses when the digital counter reaches the number in the range register and generates the range gate, as indicated in the block diagram of Fig. 18.26. A range error sensed by the range-

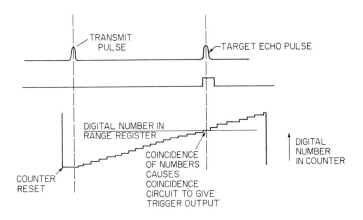

FIG. 18.25 Digital range tracker operation.

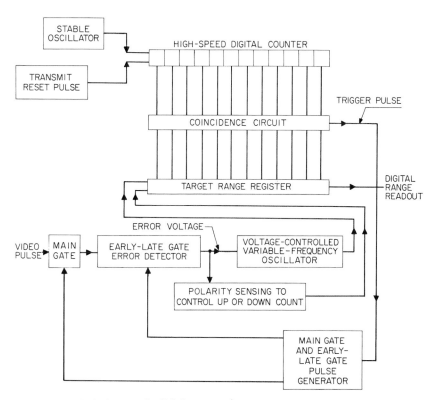

FIG. 18.26 Block diagram of a digital range tracker.

error detector results in an error voltage which drives a voltage-controlled variable-frequency oscillator to increase or decrease the count in the range register, depending upon the polarity of the error voltage. This changes the number in the range register toward the value corresponding to the range of the target. Range readout is accomplished by reading the number in the range register, where, for example, each bit may correspond to a 2-yd range. Another technique is to use a pair of oscillators.[26] The range gate is controlled by the beat frequency between the oscillators, where one is frequency-controlled by the range-error-detector-output voltage.

The electronic range tracker is inertialess, allowing any desired slew speed, and provides flexibility for conveniently generating acquisition gates for automatic-detection circuitry as well as transmitter trigger and pretrigger pulses. Tracking bandwidth is usually limited to that necessary for tracking to minimize loss of track to false targets and countermeasures. Many other electronic range-tracking techniques also offering most of these advantages are used.[2]

*n*th-Time-Around Range Tracking. To extend unambiguous range by reducing the PRF increases the acquisition time and reduces the data rate. A solution to this problem is called *n*th-*time-around tracking*, which avoids transmitting at the time that an echo is expected and which can resolve the range ambiguity. This allows the radar to operate at a high PRF and track unambiguously to long ranges where several pulses may be propagating in space to and from the target. The technique is useful only when a target is being tracked. During acquisition the radar must look at the region between transmitter pulses, and upon initial acquisition it closes the range- and angle-tracking loops without resolving the range ambiguity. The first step is to find in which range interval, or between which pair of transmitter pulses, the target is located. The zone *n* is determined by coding a transmitter pulse and counting how many pulses return before the coded pulse returns. The instrumentation radars provide *n*th-time-around tracking capability because beacons are used on rockets and space vehicles to provide sufficient signal level at long ranges.

To prevent the target echo from being blanked by a transmit pulse it is necessary to sense when the target is approaching an interference region and shift the region. This is accomplished by changing the PRF or alternately delaying groups of pulses equal to the number of pulses in propagation. This can be performed automatically to provide an optimum PRF shift or alternately delay pulse groups of the correct number of pulses.

18.6 SPECIAL MONOPULSE TECHNIQUES

High-Range-Resolution Monopulse. The use of high range resolution in monopulse radars offers the means for improving performance and extracting information about the target.[27] The basic approach is to provide sufficient range resolution to resolve the major scatterers of a target and, with monopulse processing, to locate each scatterer in range, azimuth, and elevation. This provides a three-dimensional (3D) radar image of a target's reflector locations and a fourth dimension, which is the reflector size measured by the amplitude of the echo from each scatterer.

The benefits include (1) greatly reduced target angle and range scintillation for applications requiring precise tracking of a point on a target, such as its center of gravity; (2) greatly reduced rain, sea clutter, and chaff backscatter, which are diminished proportionally to the range resolution; (3) a 3D target image plus a measure of reflector size (echo amplitude) for target recognition; and (4) a difficult waveform for some countermeasures to work against.[27]

Pulse compression will likely be necessary to maintain sufficient average power to meet radar range requirements. However, effective wideband surface acoustic-wave (SAW) pulse compression filters are available.[28] Also, if target detail is to be processed, high-speed sampling and digitizing techniques, which are available, will be needed.

Figure 18.27 demonstrates target-range-scintillation reduction by use of high

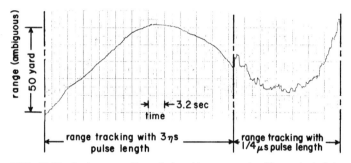

FIG. 18.27 Analog recording of closed-loop range-tracking output data demonstrating the major reduction in target range scintillation by 3-ns tracking as compared with 1/4-μs tracking. (*From D. D. Howard, Ref. 27; copyright 1975, IEEE.*)

range resolution. It shows a plot of radar measurement of the range of a target flying in approximately constant-range course. The true target wander is observed by the trend of the plotted range. For a 1/4-μs pulse length typical random fluctuations of about 3 to 4 yd rms are observed; however, with a 3-ns pulse length the range scintillation is essentially removed, although small errors from wander of center of gravity of the target will remain.

Measured high-range-resolution monopulse radar video outputs are shown in Fig. 18.28. This illustrates the monopulse sum-signal range-amplitude profile resolving major scatterers of a Super-Constellation aircraft. The monopulse difference-signal bipolar video polarity measures the direction, and amplitude measures the displacement of each reflector from the antenna axis (only the azimuth axis is shown). The average bipolar video results in a major reduction in rms error of the target center of gravity.

Dual-Band Monopulse. Dual-band monopulse can be efficiently accommodated on a single antenna to combine the complementary features of two RF bands.[15,29] A useful combination of bands is X band (9 GHz) and K_a band (35 GHz). The X-band operation provides the expected microwave performance with good radar range and precision tracking. Its weakness is the low-angle multipath region and the availability of electronic countermeasures in the band. The K_a band, although atmospheric- and rain-attenuation-limited, provides much greater accuracy in the low-angle multipath region and a second and more difficult band that the electronic-countermeasures techniques must cover.

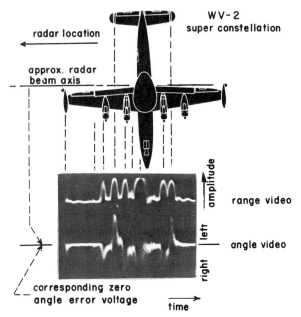

FIG. 18.28 High-range-resolution monopulse range and angle video from a Super-Constellation aircraft in flight. The radar operates at X band with a 1° beamwidth and a 3-ns pulse length. The angle video is a measure of the direction of the target from the antenna axis and the magnitude of the displacement.

A Naval Research Laboratory system called TRAKX (tracking radar at K_a and X bands) was designed for instrumentation radar applications to missile and training ranges.[15] Its purpose was to add precision tracking on targets essentially to "splash" and provide precision tracking at K_a band in an environment of X-band countermeasure operation.

A similar X-and-K_a-band system was developed by Hollandse Signaal- apparaten in the Netherlands for tactical application. The land-based version called FLY-CATCHER is part of a mobile anti-air-warfare system.[29] Another version, GOAL-KEEPER, is for shipboard anti-air-warfare application for the fire control of Gatling guns.[30] Both systems take full advantage of the two bands to provide precision tracking in multipath and electronic-countermeasures environments.

Mirror Antenna (Inverse Cassegrain). An antenna technique which uses a movable RF mirror for scanning the beam, called a mirror antenna or inverse Cassegrain, provides useful applications to monopulse radar. The technique uses a radome-supported wire-grid paraboloid which reflects parallel-polarized feed energy. The beam collimated by the paraboloid is reflected by a flat polarization rotation mirror. The energy polarization rotated by 90° can efficiently pass through the grid paraboloid. The advantages are as follows: (1) The mirror and its drive mechanism are the only moving parts for beam movement. The feed and radome-supported paraboloid remain fixed. (2) The beam movement is by a specular reflection, twice the angle of the mirror tilt.

This provides a compact structure for a given coverage requirement. (3) The normally lightweight mirror and the 2:1 beam displacement versus mirror tilt allow very rapid beam scan with low servo drive power.

The compactness and light weight are particularly attractive for airborne applications such as the Thompson-CSF Agave radar in the Super Étendards, which determines target range and designation data for the Exocet missile. It is a compact monopulse roll- and pitch-stabilized radar with 140° azimuth and 60° elevation scan.[31] The Israeli Elta subsidiary of Israeli Aircraft Industries developed an airborne tracking radar using this antenna technology for air-to-air combat and ground weapon delivery.[32]

A ground- or shipboard-based experimental mirror antenna system concept was developed with dual-band monopulse capability (3.0 GHz and 9.3 GHz). The objectives included high-speed beam movement for high-data-rate 3D surveillance and multitarget precision tracking.[33] The dual-band polarization-twist mirror design was accomplished with a two-layer mirror grid configuration.[34]

On-Axis Tracking. The best radar tracking performance is usually accomplished when the target is essentially on the radar antenna axis. Therefore, for maximum precision tracking it is desirable to minimize tracking lag and other error sources affecting the beam pointing. A technique called *on-axis* was developed to minimize radar axis deviation from the target by prediction and optimum filtering within the tracking loop.[14,35] The technique is particularly effective when the target trajectory is known approximately, such as when tracking satellites in orbit or a ballistic target. A computer in the tracking loop can cause the radar to follow an estimated set of orbital parameters, for example. It also performs optimum filtering of radar angle-error-detector output to generate an error trend from which it can update the assumed set of orbital parameters to correct the radar beam movement to a closer set of orbital parameters. By this means the radar antenna axis can be held on target with a minimum of error.

Improved tracking accuracy can also be provided on other targets where the approximate trajectory can be anticipated. However, performance of on-axis tracking is limited when tracking targets with unanticipated maneuvers.

18.7 SOURCES OF ERROR

There are many sources of error in radar tracking performance. Fortunately, most are insignificant except for very-high-precision tracking-radar applications such as range instrumentation, where the angle precision required may be of the order of 0.05 mrad (mrad, or milliradian, is one thousandth of a radian, or the angle subtended by 1-yd cross range at 1000-yd range). Many sources of error can be avoided or reduced by radar design or modification of the tracking geometry. Cost is a major factor in providing high-precision-tracking capability. Therefore, it is important to know how much error can be tolerated, which sources of error affect the application, and what is the most cost-effective means to satisfy the accuracy requirements.

Tracking radars track targets not only in angle but in range and sometimes in doppler. Consequently, the errors in each of these target parameters must be considered on most error budgets. The rest of this chapter will provide a guide for determining the significant error sources and their magnitudes.

It is important to recognize what the actual radar information output is. For a mechanically moved antenna, the angle-tracking output is usually obtained from the shaft position of the elevation and azimuth antenna axes. Absolute target location (relative to earth coordinates) will include the accuracy of the survey of the antenna pedestal site.

Phased array instrumentation radars, such as the MOTR (Multiobject Tracking Radar), provide electronic beam movement over a limited sector of about ±30° plus mechanical movement of the antenna to move the coverage sector.[16-19] The output will be mechanical shaft positions locating the normal to the array plus digital angle information from the electronic beam scan for each target.

18.8 TARGET-CAUSED ERRORS (TARGET NOISE)

Radar tracking of targets is most frequently performed by use of the echo signal reflected from the target. This is called *skin tracking* to differentiate it from *beacon tracking*, where a beacon or a transponder may be used to provide a stronger, point-source signal. Since most targets, such as aircraft, are complex in shape, the total echo signal is composed of the vector sum of a group of superimposed echo signals from the individual parts of the target, such as the engines, propellers, body, and wing edges. The motions of a target and parts of the target with respect to a radar cause the total echo signal to change with time, resulting in undesired fluctuations in the radar measurements of the parameters of the target. These fluctuations caused by the target only, excluding atmospheric effects and radar noise contributions, are called *target noise*.

This discussion of target noise is based largely on aircraft but is generally applicable to any target, including land targets of complex shape which are large with respect to a wavelength. The major difference is in the target motion, but the discussions are sufficiently general to be applied to any target situation.

The echo return from a complex target differs from that of a point source by the modulations that are produced by the changes in amplitude and relative phase of the returns from the individual elements. The word *modulations* is used in plural form because five types of modulation of the echo signal that are caused by a complex target affect radars. These are amplitude modulation, phase-front modulation (glint), polarization modulation, doppler modulation, and pulse time modulation (range glint). The basic mechanism by which the modulations are produced is the motion of the target, including random yaw, pitch, and roll, which causes the change in relative range of the various individual elements with respect to the radar.

Although the target motions may appear small, a change in relative range of the parts of a target of only one-half wavelength (because of the two-way radar path) causes a full 360° change in relative phase. At X band, this amounts to about ⅝ in, which is small even compared with the flexure between parts of an aircraft.

The five types of modulation caused by a complex target are discussed below.

Amplitude Noise. Amplitude noise is the change in echo signal amplitude caused by a complex-shaped target, excluding the effects of changing target range. It is the most obvious of the various types of echo signal modulation by a complex-shaped target and is readily visualized as the fluctuating sum of many small vectors changing randomly in relative phase. Although it is called noise, it may include periodic components. Amplitude noise typically falls into two categories: low frequency and high frequency. These categories overlap in some respects, but it is convenient to separate the noise into these two frequency ranges because they are generated by different phenomena and they are each significant to different functions of the radar.

Low-Frequency Amplitude Noise. The low-frequency amplitude noise is the time variation of the vector sum of the echoes from all the reflecting surfaces of the target. The time variation is visualized by considering the target as a rela- tively rigid body with normal random yaw, pitch, and roll motions. The small changes in relative range of the reflectors caused by this motion result in corre- sponding random changes in the relative phases. Consequently, the vector sum fluctuates randomly. Typically, target random motion is limited to small aspect changes such that the amplitudes of the echoes from the individual reflectors vary little over a period of seconds, and change in relative phase is the major contrib- utor. Exceptions are large flat surfaces with narrow reflection patterns.

An example of a target configuration is a distribution of reflecting surfaces that change in relative range with target motion. A typical pulse-to-pulse time func- tion is a slowly varying echo amplitude.[36] The low-frequency amplitude noise is seen to contribute the largest portion of the noise modulation density and is con- centrated mainly below about 10 Hz at X band. The amplitude-noise spectrum is similar for both large and small targets. This is so because the rate of relative range change is a function of both angular yaw and distance from the center of gravity of the aircraft. Thus a larger aircraft with slow yaw rates but greater wing- span generates a low-frequency noise spectrum similar to that of a small aircraft with higher yaw rates but smaller wingspan. However, the larger aircraft typi- cally has the broader noise spectrum because of the difference in distribution of dominant reflectors.

The radar RF frequency affects the low-frequency amplitude-noise spectrum shape where the spectrum width is closely proportional to the RF frequency (if the target span is assumed to be at least several wavelengths). The reason for this dependence is that the relative phase of the individual echo signals is a function of the number of wavelengths of change in relative range caused by the target random motion. Thus, with shorter wavelengths, a given relative range change will subtend more wavelengths, causing a higher phase rate, resulting in higher- frequency noise components. The rate of amplitude fluctuations of the envelope of the echo pulses is approximately proportional to the radar RF frequency.

A mathematical model of low-frequency amplitude noise of a typical aircraft is given by

$$A^2(f) = \frac{0.12B}{B^2 + f^2}$$

where $A^2(f)$ = (fractional modulation)^2/Hz
B = half-power bandwidth, Hz
f = frequency, Hz

The value of B falls typically between 1 and 2.5 Hz at X band, with the larger aircraft at the higher values because of the large reflectors, such as engines, spread out along the wings. These reflectors with the greater separation contrib- ute to the higher frequencies because their relative range change is larger for a given angular movement of the target. $A^2(f)$ is modulation power density such that the spectrum may be integrated over any frequency range to find the total noise power within a frequency band of interest. Taking the square root of the value of the integral gives the rms modulation.

High-Frequency Amplitude Noise. High-frequency amplitude noise consists of both random noise and periodic modulation. The random noise is largely a result of vibration and moving parts of the aircraft producing a relatively flat noise spectrum spread out to a few hundred hertz, depending on the type of aircraft. The rms noise density is typically a few percent modulation per $\sqrt{Hz}$.

The periodic modulation appearing as spikes in the spectrum of Fig. 18.29 is caused by rapidly rotating parts of an aircraft, such as the propellers. As the echo from a propeller blade changes with aspect when it rotates, a periodic modulation is produced. The background noise from the airframe is also observed. The spikes in the spectrum result from a fundamental modulation frequency related to the propeller r/min and number of blades. Since it is not usually sinusoidal, there are harmonic frequencies spread throughout the spectrum, as shown in Fig. 18.29 for the SNB, a small aircraft with two propeller engines. The location of these spikes is not dependent on RF frequency, as in the case of low-frequency amplitude noise, because the target controls the periodicity of the modulation, which is dependent only on the aircraft propeller rotation rate and number of blades. The high-frequency-noise modulation affects scan-type tracking radars, as described later, as well as giving some information as to the type of aircraft.

Effects of Amplitude Scintillation on Radar Performance. Amplitude noise affects all types of radars in probability of detection and tracking radar accuracy.[37-40] One effect on all types of tracking radars is from the interrelation between the low-frequency spectrum of amplitude noise, the AGC characteristics (which determine to what extent the slow fluctuations are smoothed), and the angle noise. The effects on angle noise are discussed later in this section, where it is stated that a fast-acting AGC is generally the preferred choice for maximizing overall tracking accuracy.

Another effect of amplitude noise on tracking radars concerns only conical-scan or sequentially lobed tracking radars because it is eliminated by monopulse techniques. Conical scan or sequential lobing to sense target direction depends upon measuring the amplitude of the signal for at least two different sequential antenna beam positions for each tracking axis. In azimuth tracking, for example, the antenna beam is displaced to the left of the target and then to the right. If the target were on the antenna axis, the signal would drop the same amount when the

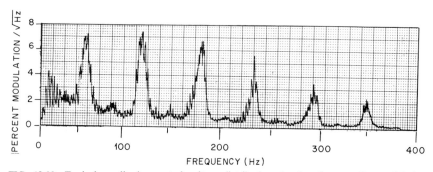

FIG. 18.29 Typical amplitude spectral voltage distribution showing the propeller modulation measured on a propeller-driven aircraft in flight. (*Fig. 4 from Dunn, Howard, and King, Ref. 37.*)

beam (assumed to be symmetrical) is moved an equal amount in either direction. The amplitudes for each beam position are subtracted in an angle error detector; hence the output is zero if the target is on the antenna axis and becomes finite, increasing positively or negatively as the target is moved off axis to the right or left.

Amplitude noise can cause the amplitude to change during the time taken to move the antenna beam from one position to the next. Even if the target is on axis, the amplitude at the two beam positions differs, thus causing an erroneous indication that the target is off axis. This effect is averaged out except for the noise spectral energy near the scan or lobing rate. For example, a periodic modulation spike near the scan rate will cause the tracking radar to drive its antenna in a circular motion around the target at a rate equal to the difference in frequency between the scan rate and the frequency of the spectral line. The direction clockwise or counterclockwise depends upon whether it is above or below the scan rate and whether the scan is clockwise or counterclockwise. The servosystem filters out all frequencies outside the frequency range between the scan rate plus the servo bandwidth and the scan rate minus the servo bandwidth. For a continuous noise background, the value of noise power density can be multiplied by this bandwidth, 2β, where β is the servo bandwidth, and an angle sensitivity constant which converts rms modulation to rms angle error.

An equation using this relation to calculate rms noise in scanning and lobing-type tracking radars caused by high-frequency amplitude noise[40] is

$$\sigma_s = \frac{\theta_B}{k_s}\sqrt{A^2(f_s)\beta}$$

where σ_s = rms angle error in same angular units as θ_B
$A(f_s)$ = rms fractional-modulation noise density in vicinity of scan rate
k_s = conical-scan error slope (k_s = 1.6 for system optimum[40])
θ_B = one-way antenna bandwidth
β = servo bandwidth, Hz

A sample calculation for an f_s of 120 Hz, where $A(f)$ from measured data taken on a large jet aircraft is approximately $0.018/\sqrt{Hz}$, θ_B is 25 mils, and β is 2 Hz, gives a σ_s of 0.42 mil rms.

In the case of a periodic modulation, where a spectral line falls within the band $f_s \pm \beta$, the rms noise is $\sigma_s = 0.67\theta_B A_f$, where A_f is the rms fractional modulation caused by the spectral line. The resultant rms tracking error σ_s will be periodic at the frequency $f_s - f_l$ where f_l is the frequency of the spectral line.

The effects of amplitude noise on target detection and acquisition are of concern in all types of radars,[2] particularly at long range where the signal is weak. The amplitude fluctuations can cause the signal to drop below the noise level for short periods of time, thus affecting the choice of thresholds, acquisition scan rate, and detection logic.[38-40]

Angle Noise (Glint).   Angle noise causes a change with time in the apparent location of the target with respect to a reference point on the target. This reference point is usually chosen as the *center of gravity* of the reflectivity distribution along the target coordinate of interest. The center of gravity is the long-time-averaged tracking point on a target. The term *glint* is sometimes used for angle noise, but it gives the false impression that the wander in the apparent position of a target always falls within the target span. The apparent

angular location of a target can fall at a point completely outside the
extremities of the target. This can be readily demonstrated both experimentally
and theoretically.[41,42] A pair of reflectors can be appropriately spaced to cause
a tracking radar with closed-loop tracking to align its antenna axis at a point
many times the reflector spacing away from the reflectors. If the reflectors are
stationary, the radar antenna will remain fixed on this large tracking error.
Figure 18.30 shows experimental data demonstrating this phenomenon with a
two-reflector target.

The angle-noise phenomenon affects all types of radars but is mainly of concern for tracking radars where precise target location is needed. To aid in visualizing why angle noise affects any radar-type angular-direction-sensing device,
the echo signal propagating in space was analyzed, showing that angle noise is
present in this propagating energy as a distortion of the phase front. Theoretical
plots of a distorted phase front from a dual source compare very closely with
photographs of the phase front from a dual vibrating source in a ripple-tank
experiment.[42] All radar angle-sensing devices sense, by one means or another,
the phase front of the signal, and indicate the target to be in a direction normal to
the phase front. Thus, the phase-front distortions affect all types of angle-sensing
radars.

Many measurements of angle noise have been made on a large variety of aircraft, and the results of theoretical studies have been verified. The theory and
measurements show that angle noise expressed in linear units of displacement,
such as yards, of the apparent position of the target from the center of gravity of
the target is independent of range (except for very short range). Therefore, the

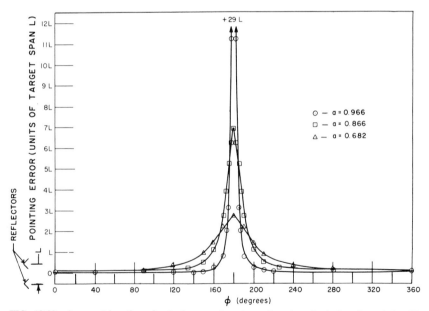

FIG. 18.30   Apparent location of a dual-source target as a function of relative phase ϕ for different values of relative amplitude a measured with a tracking radar. (*Fig. 5 from Howard, Ref.
42.*)

rms angle noise σ_{ang} is expressed in units of yards error measured at the target location. The results show that the rms value of angle noise σ_{ang} is equal to $R_0/\sqrt{2}$, where R_0 is the radius of gyration (taken along the angular coordinate of interest) of the distribution of reflecting areas of the target.[37] For example, if a target reflecting area has a $\cos^2(\pi\alpha/L)$ shaped distribution, where α is a variable and the target span is from $+L/2$ to $-L/2$, calculation of the radius of gyration divided by $\sqrt{2}$ gives a value of σ_{ang} of $0.19L$. Typical values of σ_{ang} on actual aircraft fall between about $0.15L$ and $0.20L$, depending on the distribution of the major reflecting areas such as engines, wing tanks, etc. A small aircraft, nose-on view, with a single engine and no significant reflectors attached to the wings will have a σ_{ang} near $0.1L$, whereas larger aircraft with an outboard engine and possibly wing tanks will have a σ_{ang} approaching the value of $0.2L$. Aircraft side view also tends toward the value of $0.2L$ because of a more continuous distribution of reflecting areas. Estimation of angle scintillation rms error in units of target span can be made by relating the approximate target distributions in Fig. 18.31 with actual aircraft configurations.

The value of σ_{ang} for a complex-shaped target is essentially a fixed value regardless of RF frequency, if a target span of at least several wavelengths is assumed, and is independent of the rate of random motion of the target. However, as described later, the spectral distribution of angle-noise power is directly affected by RF frequency, atmospheric turbulence, and other parameters.

Angle noise is typically gaussian-distributed. An example of measured distribution on an SNB aircraft is shown in Fig. 18.32. A relatively long time sample is needed, since short time samples of data can depart from the gaussian shape. Unusual targets may also depart from gaussian-distributed angle noise. Reference 37 gives data from two aircraft in formation which are gaussian-distributed when completely unresolved but change shape as the target comes close where the antenna begins to resolve the aircraft.

Although the rms value of angle noise is essentially a constant for a given target and aspect, the spectral shape of this energy is dependent on RF frequency and the random target motion. A typical spectrum shape is

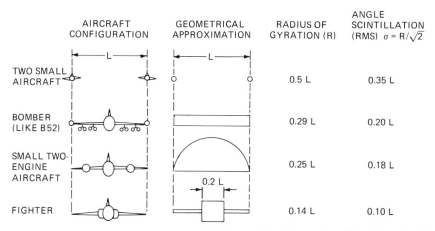

AIRCRAFT CONFIGURATION	GEOMETRICAL APPROXIMATION	RADIUS OF GYRATION (R)	ANGLE SCINTILLATION (RMS) $\sigma = R/\sqrt{2}$
TWO SMALL AIRCRAFT		0.5 L	0.35 L
BOMBER (LIKE B52)		0.29 L	0.20 L
SMALL TWO-ENGINE AIRCRAFT	0.2 L	0.25 L	0.18 L
FIGHTER		0.14 L	0.10 L

FIG. 18.31   RMS angle scintillation based on the theoretical relation to the radius of gyration of the distribution of reflecting areas of the target.

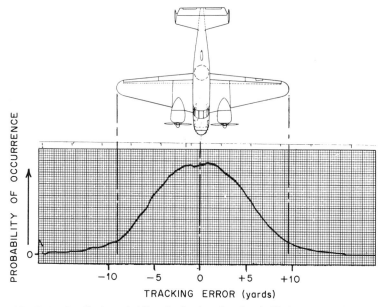

FIG. 18.32 Amplitude probability distribution of angle scintillation measured on an SNB-type aircraft.

$$N(f) = \sigma_{\text{ang}}^2 \frac{2B}{\pi(B^2 + f^2)}$$

where $N(f)$ = spectral noise power density, power/Hz
B = noise bandwidth, Hz
f = frequency, Hz

The values of B are proportional to RF frequency and dependent upon air turbulence effects on target motion and target aspect. An example of a measured amplitude scintillation spectrum is shown in Fig. 18.33. Typical values of B at X band in relatively turbulent air range from about 1.0 Hz for small aircraft to about 2.5 Hz for larger aircraft. B changes in proportion to RF frequency provided that the target span is at least a few wavelengths. Again long time samples are necessary to obtain a relatively smooth spectrum from measured data. For the above values of B, about 7 min of data was necessary to reach essentially the long-time-averaged characteristics. With only 1 min of data the noise power σ_{ang} would vary over 0.5 to 1.5 times the long-time-averaged σ_{ang}. At lower RF frequencies and in less turbulent atmosphere, B may be smaller, and proportionately longer time samples are necessary. Thus, for short time samples of radar performance, statistical variations must be expected.

To convert σ_{ang} expressed in linear units measured at the target to angular units for a radar at range r the following relation may be used:

$$\sigma_{\text{ang}} \text{ (angular mils)} = \frac{\sigma_{\text{ang}} \text{ (yd)}}{r \text{ (in thousands of yards)}}$$

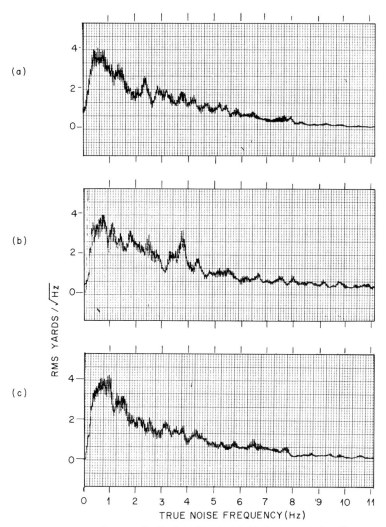

FIG. 18.33 Spectral-energy distribution of angle scintillation measured on the nose aspect of an SNB-type aircraft.

Since the angular errors caused by angle noise are inversely proportional to range, angle noise is mainly of concern at medium and close range. The resultant tracking noise can be reduced by lowering the servo bandwidth to reduce the radar's ability to follow the higher-frequency components of the noise. The amount of noise reduction may be estimated by comparing the area under a spectral-power-density plot of angle noise below the frequency corresponding to the radar servo bandwidth with the total area under the power-density plot. (The spectral-power-density plot may be obtained by squaring the ordinate values of a spectral-distribution plot such as Fig. 18.33.)

In general, it is recommended that the servo bandwidth be kept at a minimum to minimize angle noise. The minimum servo bandwidth is determined by the bandwidth required for the radar to follow the target trajectory. By Fourier transformation, a given target trajectory, in angle versus time, may be transformed into the power-density function of frequency. This frequency spectrum may be thought of as a signal where a matched filter would conform to the shape of the signal spectrum to maximize the signal-to-noise ratio of the antenna tracking function.

Multiple aircraft, such as a formation when unresolved by a radar, appear as a single large target to the radar. The value of σ_{ang} is large, reaching a theoretical maximum of $0.5L$, where L is the overall span of the formation. The worst case of aircraft formation is two aircraft where the reflecting areas are concentrated at the extremities of the overall target and σ_{ang} is close to the theoretical maximum. The value of σ_{ang} for any known formation may be readily calculated. For example, assume a formation of three small aircraft each with a wingspan of 20 yd; one aircraft is 26 yd to the left of the middle aircraft, and the third is 20 yd to the right. Each target has an average scattering cross section A, and the center of gravity with respect to the middle aircraft is

$$D_{cg} = \frac{A(-26) + A(0) + A(+20)}{3A} \quad \text{yd}$$

$D_{cg} = -2$ yd (with respect to the middle aircraft). The value R_0^2, the square of radius of gyration about D_{cg}, is obtained by summation of the products of the area of each target and the square of its distance from D_{cg}, plus the square of the value of R_0 for each target (about its own center of gravity) times its area. This summation is divided by the total area. For small targets with a 20-yd wingspan an individual σ_{ang} is typically 3 yd. Since $\sigma_{ang} = R_0/\sqrt{2}$, the R_0 for each target is about 4.2. Therefore, the R_0 for the formation is

$$R_0 = \sqrt{\frac{A(24)^2 + A(4.2)^2 + A(2)^2 + A(4.2)^2 + A(22)^2 + A(4.2)^2}{3A}}$$

$R_0 = 19.3$ yd, and $\sigma_{ang} = 13.6$ yd rms for the formation.

This method for calculating the rms value of angle scintillation can be used on any target, including single aircraft. Although reasonably good values of σ_{ang} can be determined for a single aircraft, accuracy is limited by the ability to estimate the scattering cross section of each reflecting area of the target.

The choice of AGC characteristics also affects the amount of angle noise followed by a tracking antenna. The AGC voltage is generated from the sum signal and follows the echo-signal-amplitude fluctuation. There is a degree of correlation between the angle-noise magnitude and echo-signal amplitude such that angle-noise peaks are generally accompanied by a dip or fade in amplitude. A slow AGC system which does not maintain constant signal level during rapid changes allows the signal level to drop during a rapid fade, reducing sensitivity (volts per degree angle error) during the larger angle-noise peaks. This results in a smaller rms tracking noise with a slow AGC system.[43–45]

However, this reasoning neglects an additional noise term, caused by the lack of full AGC action, which is proportional to tracking lag. A tracking lag causes a dc error voltage in the angle-detector output equal to angle error times the angle sensitivity. A slow AGC allows the amplitude noise to modulate the true tracking-error voltage, causing additional noise in angle tracking. Thus there will be an additional rms angle error proportional to tracking lag and dependent on the AGC time constant,[45] as illustrated in Fig. 18.34.

In general, a fast AGC is recommended because of the additional noise term caused by slow AGC and the possibility of large rms tracking errors, which can be considerably greater than the angle noise with a fast AGC. As previously discussed, angle noise is significant mainly at medium and close range where target angle rates are greatest. Thus, at close ranges where there is significant angle noise to be reduced by a slow AGC, the higher target angle rates also cause significant tracking lags. As seen in Fig. 18.34, a tracking lag of only one-half the target span will result in greater tracking noise in a slow AGC system, with the danger of much higher noise with greater lag. Therefore, for overall performance a fast AGC is recommended.

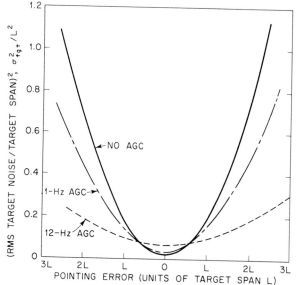

FIG. 18.34 Angle-scintillation noise power as a function of tracking error for three different AGC bandwidths. (*From Dunn, Howard, and King, Ref. 37.*)

Range Noise (Range Glint). Range noise, or random tracking errors in the range coordinate, caused by complex targets has received little attention in the past; however, with the increasing tracking-precision requirements, it has become a significant basic limitation in range tracking. Acquisition of a desired spectral line by a doppler tracking system is also limited by range noise. Coarse velocity information is obtained by differentiation of range to determine the desired spectral line for a pulse doppler tracker to acquire and track the spectral line. Range noise is a major limitation to the accuracy of velocity obtained from range rate and can prevent selection of the desired spectral line.

The range-tracking errors caused by a finite-size target and by multipath also cause significant angle-tracking errors in multilateration tracking systems which use high-precision range measurements from multiple locations to calculate target angle location. Multilateration systems depend upon very precise range measurements. Small range-tracking errors cause significant errors in calculated target angle. These errors must be fully understood to assess the performance of multilateration systems.

Target-caused range-tracking errors, similar to angle noise, are greater than the wander of the target center of gravity and can fall outside the target span.[46] Range-noise data was measured by using the split video range-error detector. Measurements were made on small and large aircraft and multiple aircraft.[47] Figure 18.35 shows typical samples of spectral-energy distributions and probability density functions for different target configurations. These characteristics follow very closely the relation of target angle noise to the target configuration radius of gyration along the range coordinate. For range tracking it is necessary to relate range noise to the target reflectivity distribution along the range coordinate. In general, the long-time-average value of the rms range error may be closely estimated by a value of 0.8 times the radius of gyration of the distribution of the reflecting areas in the range dimension based on many measurements of small, large, and multiple aircraft. Typically, in terms of target span along the range coordinate, the rms value will fall between 0.1 and 0.3 times the target span, being close to the 0.3 value for the tail view and nose view and 0.1 for the side view. The spectral shape may be closely estimated by using the same function of frequency as described for angle noise and the same value of bandwidth.

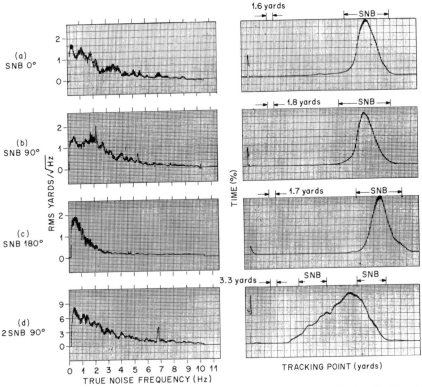

FIG. 18.35 Typical spectral-energy distributions for three views of the SNB aircraft: (*a*) nose view, (*b*) side view, (*c*) tail view, and (*d*) side view of two SNB aircraft flying in formation.

The error as a function of relative phase and amplitude of the target reflectors is similar to the angle glint phenomenon.

A beacon on a target can provide a point source (single-pulse response) to eliminate range error caused by the target. However, very stable circuitry is required to avoid pulse jitter and drift.

Doppler Scintillation. Doppler scintillation caused by a complex target may be divided into two phenomena:[48] (1) doppler spectral lines caused by moving parts of the aircraft such as propellers and jet turbine blades and (2) a continuous doppler spectrum spread by the motion of an aircraft in flight symmetrically above and below the average doppler of the target. A target typically has significant random yaw, pitch, and roll motion even on a "fixed" heading. Time plots of typical aircraft heading for an aircraft flying a "straight course" are observed to have typical random yaw motion, which causes the doppler scintillation and controls the spectra of other scintillation phenomena.

The doppler scintillation caused by the airframe of the target is a phenomenon related to the rate of change of angle noise previously described. The relation can be readily visualized by considering the phase-front-distortion concept of angle noise. Angle noise is the slope of the phase front of the echo of a complex target at any instant of time with respect to the flat (circular at near range) phase front that would have been radiated if there were a point source at the center of the target. The slope is a measure of angle error because the angle-tracking system indicates the direction to the target as the direction normal to the phase front. The doppler modulation is the time rate of phase change with respect to the phase that would exist if the target were a point source.

The distorted phase front from a complex target (two reflectors for this example) is caused to rotate[42] as the target yaws through a small change in aspect angle. As the distortion region of the phase front rotates past the radar, it causes a phase jump; the rate of this change of phase is a frequency. This frequency, at any instant, is referred to as instantaneous frequency. The phenomenon also occurs under multipath conditions in FM communications and has been demonstrated and analyzed in some detail.[49] If the target aspect angle rotated at a constant rate, the angle-noise time function and the instantaneous-doppler-frequency time function would be of identical shape.[42] However, where the target has typical random motion, the doppler modulation may be calculated for a given phase-front-distortion pattern and target random motion by differentiating the phase deviation (referenced to the phase that would be received if the target were a point source) with respect to time.

A typical example is analyzed where the angle scintillation is assumed to be a gaussian distribution approximation of data shown in Fig. 18.32. An example is calculated for a Boeing 707 aircraft nose view based on the relation of angle glint and doppler scintillation.[50] The calculated doppler spectrum is a spike shape peaked at the average target doppler. It compares closely with measurements of a Boeing 707 aircraft in flight by an X-band high-resolution doppler radar. Any constant turning rate or aspect change causes an additional spreading of the doppler spectrum.

Components of the target echo from rotating or moving parts of the target cause doppler lines at frequencies displaced from the airframe doppler spectrum. The periodic amplitude modulation causes pairs of doppler lines symmetrical about the doppler of the airframe velocity. Moving parts can also cause some pure frequency modulation which will result in a single set of doppler lines on one side of the airframe doppler spectrum.[48]

A major significance of the doppler modulation is its effect on doppler-measuring radars. A doppler tracking system which automatically tracks the frequency of a spectral line of the echo is subjected to two problems: (1) there is the possibility of locking onto a false line caused by moving parts of the target; and (2) when properly locked onto the airframe doppler spectrum, the doppler reading will be noisy as defined by the random fluctuations in instantaneous frequency as described by the spread of the doppler spectrum. Coherent beacons can provide a doppler-shifted response free of target-caused spectral spread and periodic modulations. A delay time is provided to separate the beacon response from the target skin echo.

Target doppler scintillation also offers useful information about the target configuration. Normal target motion will result in different doppler shifts for each major reflecting surface of a rigid-body target, and the shift will be a function of the displacement of the reflector from a reference point such as the center of rotation of the target random motions. Therefore, a high-resolution doppler system can resolve major reflectors and locate them in cross range as a function of the doppler difference from the reference reflector. This technique, called inverse synthetic aperture radar (ISAR), uses the target motion for the needed aspect change, instead of radar motion as used in conventional synthetic aperture radar, to obtain detailed cross-range target image information.[51–52]

18.9 OTHER EXTERNAL CAUSES OF ERROR

Multipath. Multipath angle errors are due to reflections of target echoes from other objects or surfaces causing echo energy to arrive by other than the direct return path. It is sometimes called low-angle error when applied to tracking of targets at small elevation angles over the earth or ocean surface.[53–57] Multipath errors are typically a special dual-source condition of angle noise as described in Fig. 18.36, where the target and its image reflected from a surface are the two sources. Over the ocean surface they are separated only in the elevation coordinate so that most of the error appears in the elevation-tracking channel. Where the error is severe, the residual crosstalk in the radar may cause some of the error to appear in the azimuth channel. Rough surfaces cause diffuse scattering, which can contribute errors to both azimuth and elevation tracking.[55] Other path geometry such as nonflat land or a building may cause a significant error to appear in the azimuth-tracking channel.

The major difficulty with low-angle tracking is that the target and its image are essentially coherent and their relative phase changes slowly and the angular error it causes is readily followed by an angle-tracking system. Furthermore, the paths are almost equal and in most cases they cannot be resolved by high-range-resolution techniques. Long time averages of the data do not in practice give target elevation; thus the multipath-angle problem has no simple solution and is generally minimized by using narrower-beam antennas.

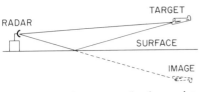

FIG. 18.36 Geometry of the radar multipath tracking condition where the reflection from a surface appears to the radar as an image below the surface.

When the target is at a low altitude, the multipath errors are severe, as observed in measured data shown in Fig. 18.37. This data is the multipath error of a 2.7° beamwidth S-band (3-GHz) tracking radar tracking an aircraft target with a beacon at 3300-ft altitude. The reference tracking data is from an AN/FPS-16 tracking radar with a 1.1° beamwidth at C band (5.7 GHz) simultaneously tracking the same target. There is a measurement bias error (observed in Fig. 18.37) of about 0.25°.

The tracking data of Fig. 18.37 is typical multipath error illustrating the phenomena from the region where the image enters the sidelobes to the region where it enters the main lobe. Three methods are used for predicting multipath errors, depending upon where the reflected image enters the antenna patterns.

At the far range, the image enters the antenna main lobe, and the error is essentially that of a two-reflector target glint error following approximately the equation[41]

$$e = 2h \frac{\rho^2 + \rho \cos \phi}{1 + \rho^2 + 2\rho \cos \phi}$$

where e = error, same units as h, measured at the target range relative to the
     target
   ρ = magnitude of surface reflection coefficient
   h = height of target
   ϕ = relative phase determined by geometry of direct and reflection signal
     paths as shown in Fig. 18.36

Although the fluctuations in ρ and ϕ alter the actual tracking from the theoretical, the equation gives a good indication of the errors to be expected.

At close range, when the radar main beam is above the image[40] and the image is seen by the difference-pattern sidelobes, the multipath errors are cyclic, almost sinusoidal, with an rms value predicted by the equation

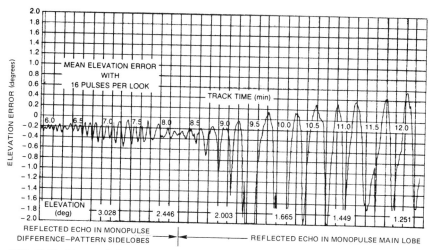

FIG. 18.37   Measured elevation-tracking error of an S-band radar using an AN/FPS-16 radar for a target elevation reference.

$$\sigma_E = \frac{\rho\theta_B}{\sqrt{8G_{se}(\text{peak})}}$$

where σ_E = rms elevation angle multipath error, same units as θ_B
$\quad\theta_B$ = one-way antenna beamwidth
$\quad\rho$ = reflection coefficient

and $G_{se}(\text{peak})$ is the power ratio of the tracking-antenna sum-pattern peak to the error-pattern peak sidelobe level at the angle of arrival of the image signal. The cyclic rate may be approximated by the equation

$$f_m = \frac{2hE}{\lambda}$$

where f_m = frequency of cyclic multipath error, rad/s
$\quad h$ = height of radar antenna
$\quad\lambda$ = wavelength, same units as h
$\quad E$ = rate of target elevation change as seen by radar, rad/s

The intermediate range is between the short-range region, where the image appears in the sidelobes, and the long-range region, where the image appears within the half-power beamwidth. The error is difficult to calculate in this region because it falls in the nonlinear error-sensing portion of the antenna patterns, and the radar response is strongly dependent upon the specific feed design and error-processing technique. However, Fig. 18.38[53,54] provides a practical means for determining approximate multipath-error values in this region. The curves are calculated multipath errors based on an assumed gaussian-shaped sum pattern and its derivative as the monopulse difference pattern. Figure 18.38 shows typical sidelobe multipath error for the higher-elevation targets and the linearly decreasing error versus target elevation, predicted by the above equation, for the very-low-elevation targets. The graph is normalized to radar beamwidth on both axes for convenient use with a wide variety of radars.

In the intermediate region, the error increases to a peak at target elevations of about 0.3 beamwidth. The peak value is dependent on several factors including surface roughness (which in part determines the value of ρ), servo bandwidth, and antenna characteristics in the region. The errors are severe and with *unsmoothed track* (wide servo bandwidth) the radar can break lock or lose track of the target.

When the surface is rough, corresponding to a reflection coefficient of about 0.3, the characteristic of the error versus elevation changes as observed in Fig. 18.38. The rough surface causes significant diffuse scattering rather than a mirror reflection. This changes the shape of the error curve and results in some residual elevation-angle error when the target elevation goes to zero. It also causes some significant azimuth error.[54]

Crosstalk Caused by Cross-Polarized Energy. Target echo energy cross-polarized to the radar antenna causes crosstalk (cross coupling) in tracking radars; i.e., azimuth error causes output from the elevation-error detector, and elevation error causes output from the azimuth-error detector (Ref. 14, Chap. 34). Generally this effect is negligible because the cross-polarized energy is usually less than the desired polarization, and it is typically reduced by about

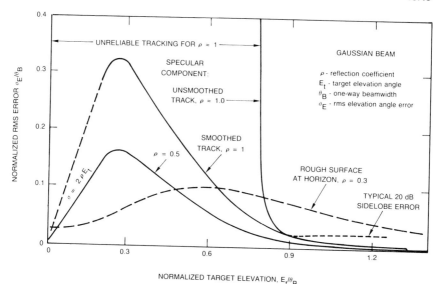

FIG. 18.38 Calculated rms multipath error σ_E versus target elevation E_t, both normalized to radar beamwidth θ_B.

20 dB by the antenna design. However, in special cases the resultant crosstalk can be very high and can cause a large tracking error and possible loss of track. For example, when tracking a linearly polarized beacon on a target, the target aspect might be such as to cause the return pulse energy to be almost fully cross-polarized.

Theoretically the coupling to cross-polarized energy is zero when the source is precisely on axis and increases with displacement from the axis.[58] Its effect on a tracking system is pure crosstalk so that a small tracking error in one tracking coordinate causes the antenna to move in the other coordinate. The error in the second coordinate then causes the antenna to move further from the source in the first coordinate. When there is no retarding effect, the cross-polarized energy causes the antenna to drive off target in one of the quadrants of the two-axis angle-tracking coordinate system, depending upon the direction of the initial error that moved the source off the precise on-axis position.[4,59]

A solution used with missile-range-instrumentation radar, where target aspect changes can cause a linearly polarized source to rotate to a cross-polarized position, is operation in a circularly polarized tracking mode. Coupling a linearly polarized signal to a circularly polarized antenna results in 3 dB signal loss, but it is independent of the direction of the linear polarization when rotated about a line in the direction toward the radar.

Troposphere Propagation. The troposphere is typically a nonhomogeneous medium for propagation and will cause random beam bending. Figure 18.39 illustrates the approximate relation of rms angle error to various atmospheric conditions.[2] The worst case is heavy cumulus clouds which form columns of shaded air that are cooler than the surrounding air and consequently of different dielectric constant. The result is typically a random beam bending as the radiated energy passes through these columns.

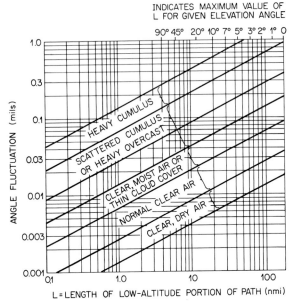

FIG. 18.39 Angle fluctuation versus path length for different tropospheres. (*From Final Report: Instrumentation Radar AN/FPS-16 (XN-2) by RCA under contract Bu Aer NOas 55-869c.*)

Figure 18.39 applies only for the portion of the beam that is within the troposphere. Once the beam goes above the troposphere (typically about 20,000 to 25,000 ft) there is no further beam bending.

The troposphere also affects target range measurement, but the errors are small, in the order of 1 to 2 ft maximum. However, even small errors of this magnitude will cause significant angle errors in multilateration systems which determine target angle by calculations with precision range measurements from separate locations.

18.10 INTERNAL SOURCES OF ERROR

Receiver Thermal Noise. The thermal noise of the receiver of a tracking radar causes erroneous outputs of the angle error detector which become significant at low *SNR*. The rms angle error σ_t, resulting from receiver thermal noise in a conical-scan radar is given by[2]

$$\sigma_t = \frac{1.4\theta_B}{k_s\sqrt{B\tau(S/N)(f_r/B_n)}}$$

where k_s = angle system error detection slope in Fig. 18.7
θ_B = antenna 3 dB beamwidth
S/N = signal-to-noise power ratio
f_r = pulse repetition frequency

β_n = servo bandwidth
B = receiver bandwidth
τ = pulse width

The equation shows that increasing $B\tau$ reduces σ_t. However, increasing $B\tau$ above the typically used value of about 1.2 for radars without pulse compression decreases the *SNR*. For optimum overall performance of the tracking radar, $B\tau$ should not exceed about 1.3. Similarly, increasing k_s by widening the offset angle (Fig. 18.7) reduces σ_t. However, crossover loss increases when the offset angle is increased, resulting in a loss in the *SNR*. Consequently, as described in Sec. 18.2, an offset angle for optimizing overall radar performance gives a k_s of about 1.5.

The angle error caused by receiver thermal noise in a monopulse tracking system is

$$\sigma_t = \frac{\theta_B}{k_m \sqrt{B\tau(S/N)(f_r/\beta_n)}}$$

where k_m is the angle-error-detector slope. The value of k_m is determined by the steepness of the antenna difference patterns, and a variety of values can be obtained, depending on the type of feed used. The values vary from 1.2 for the original four-horn square to a maximum of 1.9 for the MIT 12-horn feed. However, as described in the discussion on feeds, the 12-horn feed gives a lower antenna efficiency (0.58) than an optimum multimode monopulse feed, which can approach an efficiency of 0.75, although its slope is less, typically having a value of 1.7. Therefore, there is a tradeoff between slope and efficiency. A typical slope for monopulse radars is 1.57, representing a good modern four-horn feed design.

To compare the performance of conical scan and monopulse as affected by receiver thermal noise, the typical values are used. For conical scan, k_s is 1.5, giving a loss in *SNR* caused by the crossover (Fig. 18.7, see near optimum) of 3.1 dB for one-way (beacon) tracking and 2.0 dB for two-way echo tracking. For monopulse tracking, $k_m = 1.57$. The rms angle-tracking errors caused by receiver thermal noise are less in a monopulse radar by a factor of 1.8 for echo tracking and 2.1 for beacon tracking. This is a major improvement in angle-tracking performance by use of monopulse. Thus monopulse radar can precision-track a target to a greater distance (by a factor of 1.3 for echo tracking and 1.4 for beacon tracking) for a given allowable rms angle error than an equivalent conical-scan system having the same antenna size, transmitter power, and receiver noise figure.

For *SNR* <4 dB the relative performance changes somewhat because of noise-controlled AGC affecting angle sensitivity, servo bandwidth, and error-detector performance. The performance of conical scan relative to monopulse improves in the lower *SNR* regions, but monopulse retains its advantage for signals sufficient to maintain closed-loop tracking.

When there is a significant tracking lag or deliberate beam offset from the target, the error σ_{t0}, due to receiver noise for a given *SNR*, is given by the equation

$$\sigma_{t0} = \sigma_t \sqrt{L[1 + k(\theta_L/\theta_B)^2]}$$

where θ_L = lag angle, same units as θ_B
   L = antenna sum-pattern loss at angle θ_L

A similar range tracking error σ_{rt} results from receiver noise. The equation relating the error to SNR and system parameters is

$$\sigma_{rt} = \frac{\tau}{\sqrt{k_r\,(S/N)\,f_r/\beta_n}} \quad \text{ft (rms)}$$

where τ = pulse length, ft
   k_r = range-error-detector sensitivity (maximum value of 2.5 for a receiver where $B = 1.4$)
   S = signal power
   N = noise power

Other Internal Sources of Error. There are many other sources of internal errors which are small in a well-designed tracking radar. These include changes in relative phase and amplitude between monopulse receiver channels as a function of signal strength, RF frequency, detuning, and temperature. Also, pedestal bending from solar heating, nonorthogonality of pedestal axes, gearing backlash, bearing wobble, and many other factors contribute to errors. Table 18.1 lists the magnitude of these errors for the precision instrumentation radar AN/FPS-16.[2]
   Calibration is important to minimize internal errors.[14] When maximum performance is required, timely accurate calibration must be performed. For instrumen-

TABLE 18.1   Errors of Fixed rms Value

Error source	Angle error, mils rms	
	Bias	Noise
Boresight axis collimation	0.025	
Boresight axis drift	0.04	
Wind forces (50 mi/h)	0.02	0.012
Servo noise and unbalance	0.01	0.02
Subtotal: *radar-dependent* *tracking errors*	0.052	0.023
Leveling and north alignment	0.015	
Orthogonality of axes	0.02	
Mechanical deflections	0.01	
Thermal distortion	0.01	
Bearing wobble	. . .	0.005
Data gear error	. . .	0.03
Digital encoder error	. . .	0.025
Subtotal: *radar-dependent* *translation errors*	0.023	0.04
Tropospheric refraction	0.05	0.03
Total fixed error	0.078	0.054

tation radar, where the time of a tracking event is known, final calibration is performed just preceding the event to minimize drift errors.

18.11 SUMMARY OF SOURCES OF ERROR

Angle Measurement Errors. An inventory of sources of angle measurement errors is given in Table 18.2. This includes several sources that should be considered in addition to the radar-related sources.

Figure 18.40 is an example of measured tracking performance of an AN/FPS-16 radar tracking a 6-in metal sphere which provides a point-source target to eliminate target-caused errors. The data illustrates which error sources dominate at different regions of the radar range and their characteristics versus range.

TABLE 18.2 Inventory of Angle-Error Components*

Component	Bias	Noise

Radar-dependent tracking errors (deviation of antenna from target)	Boresight axis collimation
	Axis shift with:
	    RF and IF tuning
	    Receiver phase shift
	    Target amplitude
	    Temperature
	Wind force
	Antenna unbalance
	Servo unbalance
		Receiver thermal noise
		Multipath (elevation only)
		Wind gusts
		Servo electrical noise
		Servo mechanical noise

Radar-dependent translation errors (errors in converting antenna position to angular coordinates)	Leveling of pedestal North alignment Static flexure of pedestal and antenna Orthogonality of axes Solar heating	Dynamic deflection of pedestal and antenna Bearing wobble Data gear nonlinearity and backlash Data takeoff nonlinearity and granularity
Target-dependent tracking errors	Dynamic lag	Glint Dynamic lag variation Scintillation Beacon modulation
Propagation errors	Average refraction of troposphere Average refraction of ionosphere	Irregularities in tropospheric refraction Irregularities in ionospheric refraction
Apparent or instrumentation errors (for optical reference)	Telescope or reference instrument stability Film emulsion and base stability Optical parallax	Telescope, camera, or reference instrument vibration Film-transport jitter Reading error Granularity error Variation in optical parallax

*From D. K. Barton in R. S. Berkowitz (ed.), "Modern Radar," John Wiley & Sons, New York, 1965, chap. 7, p. 618.

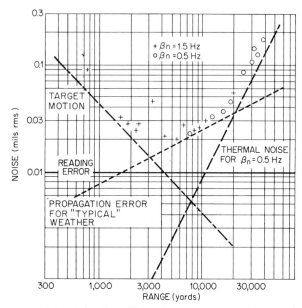

FIG. 18.40 Azimuth-tracking noise versus range using a 6-in metal sphere target supported inside a balloon to minimize target motion (estimated at 1.5 in rms). β_n = servo bandwidth.

Target-caused errors discussed in Sec. 18.8 include usual tracking events where the target extent is within the 3-dB beamwidth of the radar. However, a large target such as an aircraft formation may extend beyond the linear angle error region of the antenna patterns and eventually reach the point of resolution of the aircraft. The resultant angle-tracking error for large targets is illustrated by the example in Fig. 18.41. In Fig. 18.41a the typical gaussian-like glint error distribution is observed. With the wider separation of the aircraft the tracking-error distribution changes shape, becoming somewhat rectangular with a separation of aircraft as in Fig. 18.41b. At the widest separation, where the aircraft are almost resolved, as in Fig. 18.41c, the radar will track one aircraft until it fades and the other aircraft echo blossoms. Then the radar-tracking point will move to the other aircraft. The dwell on each target with random switching between the two aircraft causes the double-humped distribution of error.

Range Measurement Errors. The major sources of target range measurement errors are given in Table 18.3. Typical bias and noise errors in a precision-tracking radar are of a total rms value of 5 ft rms. Further details of range-error sources and their magnitudes are given in Ref. 2, Sec. 10.3.

18.12 ERROR REDUCTION TECHNIQUES

Multipath-Error Reduction. Very-low-altitude targets cause severe elevation-angle tracking errors, as described in Sec. 18.9, which may result in

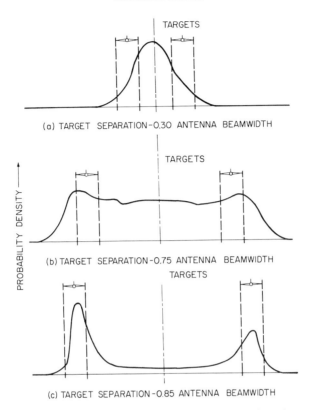

(a) TARGET SEPARATION-0.30 ANTENNA BEAMWIDTH

(b) TARGET SEPARATION-0.75 ANTENNA BEAMWIDTH

(c) TARGET SEPARATION-0.85 ANTENNA BEAMWIDTH

FIG. 18.41 Probability distribution of radar pointing when tracking two targets (where the left target is approximately 1.5 dB larger than the right target). Three different angular separations of the targets are (*a*) 0.30 antenna beamwidth, (*b*) 0.75 antenna beamwidth, and (*c*) 0.85 antenna beamwidth.

useless elevation tracking data and possible loss of track of the target. A variety of techniques have been developed to reduce these errors or their effects on radar tracking.[60-65] One simple approach to avoid loss of track in elevation is to open the elevation-tracking servo loop and place the antenna beam at about a half beamwidth above the horizon.[3,24] Azimuth closed-loop tracking may continue. Although the elevation-angle-error-detector output has large indicated angle errors, it is monitored to observe whether the target is maneuvering upward through the beam. A target rising through the beam will cause a positive angle-tracking error indication and the closed-loop elevation tracking resumed.

A very effective and direct approach to multipath-error reduction is to use a very narrow beam, usually accomplished by operating at short wavelengths such as the 8-mm (35-GHz, or K_a-band) region with usual microwave tracking aperture size.[29,30,61,62] This approach can reduce errors by two effects. First, as observed in Fig. 18.38, the angle where multipath errors become significant is proportional to beamwidth; therefore, the narrow beam reduces the altitude of the multipath region. Furthermore, as observed in Fig. 18.38, the magnitude of the elevation

TABLE 18.3   Inventory of Range-Error Components*

Component	Bias	Noise
Radar-dependent tracking errors	Zero range setting Range discriminator shift Servo unbalance Receiver delay	Receiver thermal noise Multipath Servo electrical noise Servo mechanical noise Variation in receiver delay
Radar-dependent translation errors	Range oscillator frequency Data takeoff zero setting	Range resolver error Internal jitter Data gear nonlinearity and backlash Data takeoff nonlinearity and granularity Range oscillator instability
Target-dependent tracking errors	Dynamic lag Beacon delay	Dynamic lag Glint Scintillation Beacon jitter
Propagation error	Average tropospheric refraction Average ionospheric refraction	Irregularities in tropospheric refraction Irregularities in ionospheric refraction

*From D. K. Barton, in R. S. Berkowitz (ed.), "Modern Radar," John Wiley & Sons, New York, 1965, chap. 7, p. 622.

multipath error reduces in direct proportion to beamwidth. The second advantage of shorter wavelengths is that even a relatively smooth sea, such as sea state 1, has waveheights of many wavelengths and appears rough, resulting in a small reflection coefficient.[66] This is observed in Fig. 18.38 to give small multipath errors. The 8-mm-wavelength monopulse capability may be effectively combined with a lower microwave band as described in Sec. 18.6 to take advantage of the complementary features of both bands.

There are several low-angle target-tracking techniques which modify microwave monopulse systems to reduce multipath errors. One is the complex-indicated-angle technique, which uses a conventional monopulse antenna but provides receivers for both the in-phase and quadrature monopulse difference signals for determination of target altitude.[3] The technique has ambiguities to be resolved and must maintain coherence between the monopulse sum and difference signals to provide high-accuracy target elevation angle. However, under practical multipath conditions, coherence is difficult to retain except with very-long-wavelength systems and a flat, relatively smooth ocean surface.

Several other techniques use additional antenna feed apertures or the equivalent to solve for target elevation. The conventional monopulse feed compares two apertures for each angle axis, which allows measurement of only the slope of the received phase front. Additional apertures provide phase-front curvature information from which target and image location can be derived.[64,65] Some systems perform adaptive processing of signals to effectively provide a control of tracking-pattern nulls on both the target and image.[65] In general, the techniques using additional apertures can significantly reduce multipath errors but are limited to the order of 0.1 to 0.2 beamwidths rms elevation-angle-error minimum in severe multipath.

Target Angle and Range Scintillation (Glint) Reduction. Target-caused errors in angle and range can be reduced by filtering, such as reducing tracking servo bandwidth. However, sufficient servo bandwidth must be retained to follow target trajectories. Unfortunately, target angle and range scintillation power density is normally concentrated below about 1 to 2 Hz when operating at microwave bands and falls within normally required servo bandwidths.

Target scintillation total noise power is relatively independent of frequency, but the spectral energy tends to spread upward in frequency as wavelength is reduced, resulting in lower noise power density in the servo passband. Therefore, operating at a shorter wavelength will result in lower target noise effects on closed-loop tracking.

Diversity techniques which can provide statistically independent samples of target scintillation offer a means for reducing target scintillation effects. The most practical technique is frequency diversity using pulse-to-pulse RF frequency change which will alter the phase relations between the echoes from dominant reflecting surfaces of the target.[67–69] The frequency change must be sufficient to cause enough change in relative phases of reflectors to result in a statistically independent sample of target scintillation at each new frequency. An approximate rule is a minimum frequency change of $1/\tau$, where τ is the radar range delay time between the leading and lagging extremities of the target. The rms angle and range target scintillation will be reduced by approximately $1/\sqrt{n}$, where n is the number of frequency steps provided.

Another means for reducing target scintillation is to use high range resolution sufficient to resolve the dominant reflecting surfaces of the target (see Sec. 18.6).[27] If the reflectors of the target are resolved and individually measured in angle and range at the error-detector outputs, the resulting video may be integrated to obtain a low-scintillation center of gravity or a precise leading-edge track on the closest reflector, for example. Figure 18.27 shows the range scintillation reduction measured with a 3-ns-pulse-length radar. Also, Fig. 18.28 shows measured monopulse sum- and difference-signal video from a Super-Constellation aircraft in track with a 3-ns-pulse-length radar. The separate video pulses from the resolved target reflectors are observed.

Reduction of Internally Caused Errors. Angle errors caused by receiver thermal noise as well as target scintillation are minimized by maintaining the target as close as possible to the tracking axis. The technique called *on-axis*, described in Sec. 18.6, is a means for placing a computer in the tracking loop to minimize lag and provide optimum angle error filtering.

Accurate system calibration also greatly reduces internal sources of error. Frequent calibration reduces drifts in component gain and phase and pedestal structure. Other internal sources of error with known characteristics can be automatically corrected to minimize their contamination of the output data.[14]

REFERENCES

1. Skolnik, M. I.: "Introduction to Radar Systems," 2d ed., McGraw-Hill Book Company, New York, 1980.

2. Barton, D. L.: "Radar Systems Analysis," Artech House, Norwood, Mass., 1977.

3. Sherman, S. M.: "Monopulse Principles and Techniques," Artech House, Norwood, Mass., 1986.

4. Leonov, A. I., and K. I. Formichev: "Monopulse Radar," Artech House, Norwood, Mass., 1986.

5. Dunn, J. H., and D. D. Howard: Precision Tracking with Monopulse Radar, *Electronics*, vol. 33, pp. 51–56, Apr. 22, 1960.

6. Peebles, P. Z., Jr.: Signal Processor and Accuracy of Three-Beam Monopulse Tracking Radar, *IEEE Trans.*, vol. AES-5, pp. 52–57, January 1969.

7. Hannan, P. W.: Optimum Feeds for All Three Modes of a Monopulse Antenna, I: Theory; II: Practice, *IEEE Trans.*, vol. AP-9, pp. 444–460, September 1961.

8. Final Report on Instrumentation Radar AN/FPS-16 (XN-2), *Radio Corporation of America, unpublished report NTIS 250500*, pp. 4-123–4-125.

9. Barton, D. K.: Recent Developments in Radar Instrumentation, *Astron. Aerosp. Eng.*, vol. 1, pp. 54–59, July 1963.

10. Nessmith, J. T.: Range Instrumentation Radars, *IEEE Trans.*, vol. AES-12, pp. 756–766, November 1976.

11. DiCurcio, J. A.: AN/TPQ-27 Precision Tracking Radar, *IEEE Int. Radar Conf. Rec.*, Arlington, Va., pp. 20–25, 1980.

12. Howard, D. D.: Single Aperture Monopulse Radar Multi-Mode Antenna Feed and Homing Device, *Proc. IEEE Int. Conv. Mil. Electron. Conf.*, pp. 259–263, Sept. 14–16, 1964.

13. Mikulich, P., R. Dolusic, C. Profera, and L. Yorkins: High Gain Cassegrain Monopulse Antenna, *IEEE G-AP Int. Antenna Propag. Symp. Rec.*, September 1968.

14. Johnson, R. C., and H. Jasik: "Antenna Engineering Handbook," 2d ed., McGraw-Hill Book Company, New York, 1984, chap. 34.

15. Cross, D., D. Howard, M. Lipka, A. Mays, and E. Ornstein: TRAKX: A Dual-Frequency Tracking Radar, *Microwave J.*, vol. 19, pp. 39–41, September 1976.

16. Hammond, V. W., and K. H. Wedge: The Application of Phased-Array Instrumentation Radar in Test and Evaluation Support, *Electron. Nat. Security Conf. Rec.*, Singapore, Jan. 17–19, 1985.

17. Bornholdt, J. W.: Instrumentation Radars: Technical Evaluation and Use, *Proc. Int. Telemetry Council*, November 1987.

18. Milway, W. B.: Multiple Target Instrumentation Radars for Military Test and Evaluation, *Proc. Int. Telemetry Conf.*, vol. XXI, 1985.

19. Stegall, R. L.,: Multiple Object Tracking Radar: System Engineering Considerations, *Proc. Int. Telemetry Council*, November, 1987.

20. Rubin, W. L., and S. K. Kamen: SCAMP—A Single-Channel Monopulse Radar Signal Processing Technique, *IEEE Trans.*, vol. MIL-6, pp. 146–152, April 1962.

21. Abel, J. E., S. F. George, and O. D. Sledge: The Possibility of Cross Modulation in the SCAMP Signal Processor, *Proc. IEEE*, vol. 53, pp. 317–318, March 1965.

22. Noblit, R. S.: Reliability without Redundancy from a Radar Monopulse Receiver, *Microwaves*, pp. 56–60, December 1967.

23. Sakamoto, H., and P. Z. Peebles, Jr.: Conopulse Radar, *IEEE Trans.*, vol. AES-14, pp. 199–208, January 1978.

24. Bakut, P. A., and I. S. Bol'shakov: "Questions of the Statistical Theory of Radar," vol. II, Sovetskoye Radio, Moscow, 1963, chaps. 10 and 11. (Translation available from NTIS, AD 645775, June 28, 1966.)

25. Locke, A. S.: "Guidance," D. Van Nostrand Company, Princeton, N.J., 1955, chap. 7.

26. Cross, D. C.: Low Jitter High Performance Electronic Range Tracker, *IEEE Int. Radar Conf. Rec.*, pp. 408–411, 1975.

27. Howard, D. D.: High Range-Resolution Monopulse Radar, *IEEE Trans.*, vol. AES-11, pp. 749–755, September 1975.

28. Weglein, R. D.: SAW Chirp Filter Performance above 1 GHz, *Proc. IEEE*, vol. 64, pp. 695–698, May 1976.

29. Malone, D. L.: FLYCATCHER, *Nat. Def.*, pp. 52–55, January 1984.

30. Hollandse Signaalapparaten B.V. advertisement, *Def. Electron.*, vol. 19, p. 67, April 1987.

31. Editor: Inside the Exocet: Flight of a Sea Skimmer, *Def. Electron.*, vol. 14, pp. 46–48, August 1982.

32. Editor: Special series: Israeli Avionics-2, *Aviat. Week Space Technol.*, pp. 38–49, Apr. 17, 1978.

33. Cross, D. C., D. D. Howard, and J. W. Titus: Mirror-Antenna Radar Concept, *Microwave J.*, vol. 29, pp. 323–335, May 1986.

34. Howard, D. D., and D. C. Cross: Mirror Antenna Dual-Band Light Weight Mirror Design, *IEEE Trans.*, vol. AP-33, pp. 286–294, March 1985.

35. Schelonka, E. P.: Adaptive Control Technique for On-Axis Radar, *Int. Radar Conf. Rec.*, pp. 396–401, 1975.

36. Olin, I. D., and F. D. Queen: Dynamic Measurement of Radar Cross Section, *Proc. IEEE*, vol. 53, pp. 954–961, August 1965.

37. Dunn, J. H., D. D. Howard, and A. M. King: Phenomena of Scintillation Noise in Radar-Tracking Systems, *Proc. IRE*, vol. 47, pp. 855–863, May 1959.

38. Swerling, P.: Probability of Detection of Fluctuating Targets, *IRE Trans.*, vol. IT-6, pp. 269–301, April 1960.

39. Skolnik, M. I.: "Introduction to Radar Systems," McGraw-Hill Book Company, New York, 1962, chap. 2.

40. Barton, D. K., "Modern Radar System Analysis," Artech House, Norwood, Mass., 1988, p. 388.

41. Merrill, G., D. J. Povejsil, R. S. Raven, and P. Waterman: "Airborne Radar," Boston Technical Publishers, 1965, pp. 203–207.

42. Howard, D. D.: Radar Target Angular Scintillation in Tracking and Guidance Systems Based on Echo Signal Phase Front Distortion, *Proc. Nat. Electron. Conf.*, vol. 15, October 1959.

43. Delano, R. H.: A Theory of Target Glint or Angle Scintillation in Radar Tracking, *Proc. IRE*, vol. 41, pp. 1778–1784, December 1953.

44. Delano, R. H., and I. Pfeffer: The Effects of AGC on Radar Tracking Noise, *Proc. IRE*, vol. 44, pp. 801–810, June 1956.

45. Dunn, J. H., and D. D. Howard: The Effects of Automatic Gain Control Performance on the Tracking Accuracy of Monopulse Radar Systems, *Proc. IRE*, vol. 47, pp. 430–435, March 1959.

46. Cross, D. C., and J. E. Evans: Target Generated Range Errors, *IEEE Int. Radar Conf. Rec.*, pp. 385–390, Arlington, Va., Apr. 21–23, 1975.

47. Povejsil, D. J., R. S. Raven, and P. Waterman: "Airborne Radar," D. Van Nostrand Company, Princeton, N.J., 1961, pp. 397–399.

48. Hynes, R., and R. E. Gardner: Doppler Spectra of S Band and X Band Signals, *IEEE Trans. Suppl.*, vol. AES-3, pp. 356–365, November 1967.

49. Corrington, M. S.: Frequency Modulation Distortion Caused by Multipath Transmission, *Proc. IRE*, vol. 33, pp. 879–891, December 1945.

50. Dunn, J. H., and D. D. Howard: Radar Target Amplitude, Angle, and Doppler Scintillation from Analysis of the Echo Signal Propagating in Space, *IEEE Trans.*, vol. MTT-16, pp. 715–728, September 1968.

51. Ausherman, A. A., A. Kozma, J. L. Walker, H. M. Jones, and E. C. Poggio: Development in Radar Imaging, *IEEE Trans.*, vol. AES-20, pp. 363–400, July 1984.

52. Dike, G., R. Wallenberg, and J. Potenza: Inverse SAR and Its Application to Aircraft Classification, *IEEE Int. Radar Conf. Rec.*, pp. 20–25, 1980.

53. Barton, D. K.: The Low-Angle Tracking Problem, *IEE Int. Radar Conf.*, London, Oct. 23–25, 1973.

54. Barton, D. K., and H. R. Ward: "Handbook of Radar Measurement," Prentice-Hall, Englewood Cliffs, N.J., 1969.

55. Barton, D. K.: Low-Angle Radar Tracking, *Proc. IEEE*, vol. 62, pp. 687–704, June 1974.

56. Barton, D. K.: "Radars," vol. 4: "Radar Resolution and Multipath Effects," Artech House, Norwood, Mass., 1978.

57. Howard, D. D., J. Nessmith, and S. M. Sherman: Monopulse Tracking Error Due to Multipath: Causes and Remedies, *EASCON Rec.*, pp. 175–182, 1971.

58. Jones, E. M. T.: Paraboloid Reflector and Hyperboloid Lens Antenna, *IRE Trans.*, vol. AP-2, pp. 119–127, July 1954.

59. Mitchell, R., et al.: Measurements of Performance of MIPIR (Missile Precision Instrumentation Radar Set AN/FPQ-6), *Final Rept., Navy Contract NOW*61-0428d, RCA, Missile and Surface Radar Division, Moorestown, N.J., December 1964.

60. Dax, P. R.: Accurate Tracking of Low Elevation Targets over the Sea with a Monopulse Radar, *IEE Radar Conf. Publ.* 105, "Radar—Present and Future," pp. 160–165, London, Oct. 23–25, 1973.

61. Howard, D. D.: Investigation and Application of Radar Techniques for Low-Altitude Target Tracking, *IEE Int. Radar Conf. Rec.*, London, Oct. 25–26, 1977.

62. Howard, D. D.: Environmental Effects on Precision Monopulse Instrumentation Tracking Radar at 35 GHz, *IEEE EASCON '79 Rec.*, October 1979.

63. McAulay, R. J., and T. P. McGarty: Maximum-Likelihood Detection of Unresolved Targets and Multipath, *IEEE Trans.*, vol. AES-10, pp. 821–829, November 1974.

64. White, W. D.: Techniques for Tracking Low-Altitude Radar Targets in the Presence of Multipath, *IEEE Trans.*, vol. AES-10, pp. 835–852, November 1974.

65. Peebles, P. Z., Jr.: Multipath Error Reduction Using Multiple Target Methods, *IEEE Trans.*, vol. AES-7, pp. 1123–1130, November 1971.

66. Nathanson, F. E.: "Radar Design Principles," McGraw-Hill Book Company, New York, 1969, p. 37.

67. Lind, G.: Reduction of Radar Tracking Errors with Frequency Agility, *IEEE Trans.*, vol. AES-4, pp. 410–416, May 1968.

68. Lind, G.: A Simple Approximation Formula for Glint Improvement with Frequency Agility, *IEEE Trans.*, AES-8, pp. 853–855, November 1972.

69. Barton, D. K.: "Frequency Agility and Diversity," vol. 6: "Radars," Artech House, Norwood, Mass., 1977.

CHAPTER 19
RADAR GUIDANCE OF MISSILES*

Alex Ivanov
Missile Systems Division
Raytheon Company

19.1 INTRODUCTION

Radar guided missiles represent one of the most widely used applications of the radar art, yet one about which much less has been published in the open literature than about other, more "conventional" radars. There is no nonmilitary use of this part of radar technology, and much of the detailed data is still classified. However, drawing solely on the unclassified published data permits at least a tutorial overview of radar guidance to be presented in this chapter.

Guided missiles can be characterized in several ways,[1–4] based on their mission, type of guidance, sensing wavelength, source of guidance energy, etc. The discussion here will narrow down to the particular radar homing types which form the vast majority of operational systems.

Based on their use, missile systems can be categorized as surface-to-surface, air-to-surface, surface-to-air, and air-to-air. The types of guidance are inertial, map-following, command, beam-riding, and homing. Types other than inertial can use the broadest range of the electromagnetic spectrum, from radio frequencies (RF) through infrared (IR) to the visible spectrum and beyond, to perform the guidance function.

Within these general categories, the surface-to-surface types [especially the intercontinental ballistic missile (ICBM) and the shorter-range ballistic] are usually inertially guided and fall outside the scope of this discussion. The primary exception is the antiship missile, which uses radar guidance and may be surface-(as well as air-) launched.[4,5] The main users of radar guidance are the air defense systems—surface-to-air or air-to-air. These also can employ IR or laser radars, but we shall restrict our discussion to the microwave radar category. Air-to-surface systems for use against ships, armored vehicles, or hard fixed targets

*The author is indebted to the many colleagues at Raytheon who reviewed various sections of this chapter and especially to David Barthuli, John Curley, and Al Williams for their many valuable comments and suggestions.

such as bridges, can use the visible spectrum (TV), IR, laser, or radar. Only the last-named category will be discussed.

Whether they are used against airborne or surface targets, guided missiles are intended to achieve a much higher accuracy than conventional artillery, which relies on open-loop prediction (even when optical or radar target tracking is employed for fire control). To achieve the required accuracy, a guidance system utilizes automatic closed-loop control by continuously sensing errors in the missile-to-target intercept geometry and translating them into corrective missile maneuvers designed to reduce miss distance to zero, although in practice a finite miss distance usually results.

Radar has been extensively used for command, beam-riding, and homing guidance. The simplest form of guidance is the beam rider. The target is tracked by a tracking radar (or, in early systems, by an operator using an optical sight with a radar slaved to it) which keeps the beam always pointed at the target. The missile itself does not perceive the target but detects its own position relative to the tracking beam. By keeping itself centered in the beam, it attempts, like the radar beam it rides, to pass through the target. In command guidance the target and the missile are tracked by separate radars (or by separate beams of a phased array radar). The missile itself does not perceive the target. Measured target and missile states are fed to a computer which calculates the missile trajectory required for intercept and develops the guidance commands which are continuously transmitted to the missile. In both these systems accuracy is inversely proportional to range from the radar, since a fixed angular error at the radar becomes an increasing linear error at increasing ranges.

Homing provides the highest accuracy at the cost of complexity of the missile-borne hardware. Whereas the beam rider and command systems require only a single receiver in the missile, to sense the beam or receive commands, the homing system perceives the target with its own radar (called the *seeker*), extracts tracking data from the received signal, and computes its own steering commands. As it closes on the target, a fixed angular error at the missile results in a decreasing linear error, providing the higher accuracy characteristic of homing guidance.

Homing systems can be further categorized on the basis of the source of the sensed radar energy into passive, semiactive, and active. Passive homing uses energy originating from the target (i.e., jamming or radar transmissions). An active homing system is a self-contained radar which transmits its own radar energy at the target and tracks the target-reflected energy. The semiactive system includes an external radar which illuminates the target while the missile receives and tracks the target-reflected energy to extract guidance information.

The waveforms used vary from noncoherent pulse (used in some early systems) to continuous wave (CW) and coherent pulse doppler (PD). The most widely used operational systems have, over the years, employed CW semiactive homing. Since the active systems differ only by virtue of the presence of the illuminator-transmitter on board the missile, a discussion of the semiactive system can be easily extended to cover the active type as well. Similarly, passive homing can be considered a subset of the semiactive.

Because of antenna size constraints, operating frequencies have generally been in the C, X, or K_u bands. The increased availability of components at higher frequencies has permitted operation at K_a and millimeter-wave frequencies in later-generation systems.

The nature of missile systems and the environment in which they are developed result in evolutionary changes rather than revolutionary innovations. Thus to understand today's systems we must understand how they became what they are. This chapter will, therefore, begin with a discussion of the CW semiactive system and trace the evolution of the CW seeker through several generations. Once these concepts have been explained, they will be extended to the active and passive systems. Functional operations (i.e., acquisition, tracking) and characteristics of subsystems will then be discussed.

19.2 OVERVIEW OF SEMIACTIVE CW SYSTEMS[6]

The basic semiactive system is conceptually illustrated in Fig. 19.1. The illuminator maintains the target within its radar beam throughout the engagement. The missile receives the target-reflected illumination in its front antenna and a sample of the directly received illumination (often through sidelobes of the illuminator antenna) in its rearward-looking reference (rear) antenna. The front and rear signals are coherently detected against each other, resulting in a spectrum which contains the doppler-shifted target signal at a frequency roughly proportional to closing velocity. A narrowband frequency tracker searches the spectrum, locks onto the target return, and extracts guidance information from it. The use of CW provides the capability of discriminating against clutter on the basis of doppler frequency and thus allows low-altitude operation.

Doppler Frequency Relationships. The geometry for the doppler frequency relationships in a general semiactive case is shown in Fig. 19.2. The doppler shift is a function of the transmitted frequency f_0 and radial velocity V_R [the component of velocity along the line of sight (LOS) from the source to the observer—either a receiver or a reflector].

$$f_{\text{dop}} = (f_0/c)V_R$$

where c = velocity of light.

In the geometry of Fig. 19.2, the *rear*, or reference, doppler is a function of the radial velocity of the missile with respect to the illuminator. The *front*, or target, doppler frequency, depends on the radial velocities of the illuminator, missile, and target. The resulting spectrum, when the front signal is coherently detected

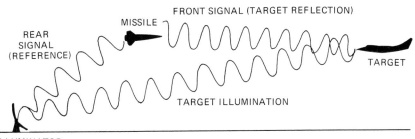

FIG. 19.1 Semiactive homing employs an illuminator to illuminate the target and provide a reference to the missile, which compares the reference with the reflected-target illumination to extract guidance information.

(mixed) against the rear, is the difference of the two.

$$f_{\text{rear}} = -\frac{f_0}{c}V_M \cos\theta + \frac{f_0}{c}V_I \cos A$$

$$f_{\text{front}} = \frac{f_0}{c}(V_T \cos\phi + V_T \cos\beta + V_M \cos\alpha + V_I \cos B)$$

$$f_d = f_{\text{front}} - f_{\text{rear}}$$

$$= \frac{f_0}{c}(V_M \cos\theta + V_M \cos\alpha + V_T \cos\phi + V_T \cos\beta + V_I \cos B - V_I \cos A)$$

For a stationary illuminator $V_I = 0$. Closing velocity is $V_C = V_M \cos\alpha + V_T \cos\beta$. For the head-on case, all angles become zero and $V_C = V_M + V_T$, with the result that $f_d = (f_0/c)2V_C$. The constant of proportionality f_0/c at X band (approximately 10 GHz) is 10 Hz/(ft/s), and hence the rule of thumb is that target doppler is 20 Hz for each foot per second of closing velocity at X band. Scaling is a convenient way to handle other frequency bands. Where more exact doppler frequencies must be known, the exact transmitted frequency f_0 should be used.

It is important to note that, in addition to the target, there exist large interfering signals within the spectrum of interest: clutter and feedthrough (spillover or leakage of the rear signal into the front receiver through backlobes of the front antenna). Because the frequency of the feedthrough is the same as the rear signal, for a system in which the front and rear signals are mixed directly (baseband conversion), feedthrough would occur at dc (zero frequency), with the approaching and receding spectra folded around it. As will be shown later, it is usually desirable to unfold the spectrum, to separate the approach and recede portions, so that feedthrough will occur at some arbitrary offset frequency. Figure 19.3 illustrates the latter case.

The clutter doppler can be calculated by using the same equations used for the target. Let the reflecting clutter patch be the target, with a velocity $V_T = 0$. Use the appropriate angles which relate the missile velocity vector to the missile-to-clutter patch LOS. Not only main-lobe but also sidelobe clutter must be considered.

The spectrum of Fig. 19.3a shows the case of a ground-to-air missile. For small look angles (α and θ in Fig. 19.2), the main-lobe clutter (MLC) occurs at a frequency corresponding to a velocity of approximately $2V_M$. Sidelobe clutter extends all the way from $2V_M$ to zero-doppler velocity (feedthrough) as the angle between the missile velocity vector and the reflecting clutter patch varies from 0° (head on) to 180° (backlobe clutter).

The air-to-air case of Fig. 19.3b differs in that the clutter spectrum extends below feedthrough due to the airborne illuminator's backlobe, which can produce a return from a clutter patch behind the aircraft, i.e., an angle of 180° with respect to the illuminator velocity vector.

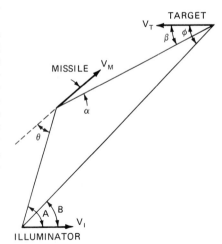

FIG. 19.2 Semiactive geometry. The radial velocity components of the three system elements contribute to the doppler shift of the received signal.

It should also be clarified what is meant by *approaching* and *receding* targets within the missile doppler spectrum. A target approaching the missile will yield a signal at a frequency above that of MLC (which corresponds to missile velocity). For the X-band case, let V_M = 2000 ft/s and V_T = 500 ft/s in level flight. MLC will then be at roughly 40 kHz and the target at 50 kHz. If the target were flying away from the missile at the same 500 ft/s velocity, its doppler frequency would be 30 kHz, or 10 kHz below MLC. However, the missile is still closing on this target at 1500 ft/s; so it is in the *approach* part of the spectrum, above feedthrough, even though it is an outbound, or receding, target. (Note that if the outbound target were faster than the missile, its doppler would be below feedthrough, but of course the missile would never catch up with it. Thus, it is a

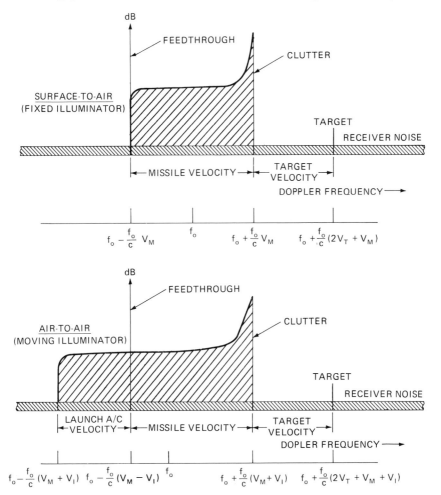

FIG. 19.3 Signal spectra for semiactive homing indicate the clutter and feedthrough (spillover) with which the target signal must compete. Both fixed and moving illuminator cases are shown. The frequencies shown are for maximum clutter extent; i.e., all angles shown in Fig. 19.2 are zero and all velocity vectors colinear.

meaningless case.) This is different for a ground-based radar, where the outbound target is in the *receding* part of the spectrum, below feedthrough. The significance of the above discussion is that while approaching targets are in a clutter-free region of the spectrum, the receding target lies in the sidelobe clutter and must compete directly with it: if the clutter in the detection cell exceeds the target, the target cannot be detected. This is of primary concern in look-down tail-chase air-to-air engagements.

One additional important factor must be noted. Although the target signal can be discriminated from the feedthrough and clutter on the basis of frequency (except for the receding target in sidelobe clutter), this is only true for spectrally pure signals. Noise on the transmitted signal and on any conversion oscillators within the missile will be spread throughout the doppler spectrum by the feedthrough and clutter and mask the target signal if the noise is not adequately controlled. Noise reduction is thus one of the key technologies required for good radar seeker performance.

Clutter and Feedthrough Considerations.[2,5,7]   The presence of clutter and feedthrough is one of the primary limitations on the performance achievable in a seeker and has been one of the main design drivers in the evolution of radar guided missiles. There are three main problems which must be addressed in connection with these large interfering signals. The first is the need to prevent lock on the clutter signal and its harmonics. Clutter in some geometries may be spectrally very narrow, resembling a target signal. Preventing lock is generally accomplished by limiting the portion of the doppler spectrum which is searched during the acquisition process, to exclude the clutter frequency. However, because clutter varies in frequency during missile flight, avoiding it can be a relatively complex problem, especially for slow-radial-velocity targets (small frequency separation from clutter). Feedthrough, on the other hand, is fixed in frequency and can thus be significantly attenuated with fixed filters and be easily avoided during search.

The second problem is often termed the *subclutter visibility* (SCV) or *subfeedthrough visibility* (SFV) problem. In essence, this refers to the maximum ratio of clutter (or feedthrough) to signal with which the system can operate. In its simplest form, this can be related to the dynamic range of the seeker receiver (i.e., its range of linear operation). One must consider not only possible suppression of the target signal by the clutter or feedthrough but also potential cross-modulation or intermodulation effects. As will be shown, gain normalization (automatic gain control, or AGC) is a key concern in achieving the required SCV.

The third problem is also related to SCV (and SFV) and is, as noted earlier, concerned with the spectral purity of the transmitter and the local oscillator. The spectrum of Fig. 19.3 will be broadened by noise, so that noise sidebands of feedthrough and clutter will appear at the target doppler frequency. In view of the magnitude of feedthrough and clutter, very low noise is required to prevent performance degradation (masking of the target). Maximum feedthrough levels can typically range from 80 to 100 dB above the target signal, while main-lobe clutter can be 40 to 80 dB greater than the target. However, the frequency separation between target and clutter is much smaller than between target and feedthrough; so, depending on the specific design and conditions, clutter may establish the more stringent noise requirement. Also, the effects of feedthrough noise can be reduced through cancellation in the missile receiver (see Sec. 19.4). Since amplitude-modulation (AM) noise in sources is generally well below frequency-modulation (FM) noise (20 dB is typical), the noise reduction techniques concentrate on FM noise.

Guidance Fundamentals.[7–12]   A detailed discussion of guidance is beyond the scope of this chapter. However, to understand the effect that the radar seeker's ability to measure the target's radar observables has on the performance of the missile (miss distance), a brief overview of guidance principles is presented here.

Virtually all homing systems employ some form of proportional navigation (PN), although modern control theory (such as Kalman filtering) has been used extensively to optimize performance of later-generation systems. The important fact to note is that PN can be accomplished with angle-only measurements and can thus become a fallback mode even if range or doppler (range rate) information—required for advanced guidance techniques—is unobtainable.

Proportional navigation is based on the fact that if two objects are closing on each other, they will collide if the LOS does not rotate in inertial space, as illustrated in Fig. 19.4. Any rotation of the LOS (i.e., an LOS rate) is indicative of a deviation from the collision course which must be corrected by a missile maneuver. In PN, the rate of rotation of the LOS is measured, and a lateral acceleration of the missile is commanded according to the equation

$$n_L = N' V_c \dot{\lambda}$$

where n_L = lateral acceleration
      N' = effective navigation ratio (constant, selected as discussed below)
      V_c = closing velocity
      $\dot{\lambda}$ = rate of change of the line of sight

The lateral acceleration ideally should be normal to the LOS; in practice the deflection of the missile control surfaces will result in acceleration normal to the missile velocity vector. The closing velocity can be estimated or, in the case of a doppler radar, measured (the target doppler is an approximate measure of V_c, as noted above). The LOS rate $\dot{\lambda}$ is measured by the seeker—this is the seeker's primary function. The value of N' is chosen to optimize performance in the face of initial errors or disturbances which would increase miss distance: heading error, target maneuver, system biases, and noise. For example, increasing values of N' cause early correction of collision course errors, reserving the missile's maneuver capability near intercept for countering target maneuvers and noise. Too high a value of N', however, results in too great a sensitivity to noise inputs, especially glint, which increases with decreasing range. In practice, N' values in the range of 3 to 5 are normally chosen.

The missile does not respond instantaneously to an LOS rate; rather, a finite response time, made up of several components, governs the process. This equivalent time constant, referred to as the guidance time constant τ_g, is a key parameter affecting miss distance.

Several time lags in series combine to produce τ_g. These are the track-loop time constant, the noise-filter time constant, and the autopilot-airframe response time. The antenna track-loop time constant can be eliminated as a contributor in certain configurations (LOS or LOS rate reconstruction[12]). The airframe aerodynamic response will vary with missile speed and altitude, and the autopilot must compensate for this variation. The final value of τ_g is a compromise between the rapid desired speed of response to counter target maneuvers and a long desired smoothing time to minimize glint. Moreover, the variations in τ_g brought about by *parasitic feedback* effects, such as radome aberration and imperfect antenna stabilization, must be controlled to avoid guidance loop instability.

A practical rule of thumb for a properly designed system is that a homing time

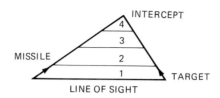

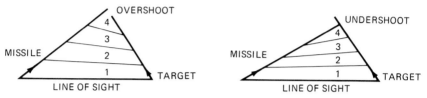

FIG. 19.4 Line-of-sight (LOS) motion of intercept. The line-of-sight rate is constant when missile and target are on an intercept (collision) trajectory.

of 10 τ_g will reduce miss distance to the asymptotically achievable value. This will therefore establish minimum range capability as well as set the requirement on the terminal guidance mode of a multimode missile.

The seeker's primary function is to generate an estimate of the inertial LOS rate. To accomplish this, it must track the target in angle and stabilize the antenna LOS against missile body motions, which could be erroneously interpreted as target motions. It is the accuracy of the resulting LOS rate estimate that will determine how well the missile performs.

The fundamental limit on achievable accuracy is the target's own angle noise (glint, scintillation, and depolarization). Other noise contributors must be minimized by proper design (i.e., maximize signal-to-noise ratio to minimize range-dependent noise, reduce the range-independent noise—servo and other instrumentation noise). Also, the correct angle measurement scale factor must be maintained over the full range of signal levels and over all look angles.

Finally, the effect of the radome must be considered. Because of aerodynamic considerations, the radome enclosing the gimballed antenna will be pointed rather than a hemisphere. Thus, at different gimbal (look) angles the radar signal will pass through a different portion of the radome, and the apparent LOS to the target will change with gimbal angle because of refraction (aberration). This results from different path lengths through the dielectric material (different curvature) as well as local differences in thickness or dielectric constant. A constant error would present no difficulty, since the tracking and guidance loops are driving the boresight error (LOS rate) to zero. It is the variations of the radome error with gimbal angle—radome error slope—which cause the problem by creating a feedback path.

Since the missile responds to a target LOS rate by maneuvering, the missile body orientation with respect to the observed LOS will change as a result of the maneuver. Thus the space-stabilized antenna, while maintaining track of the target, will move with respect to the radome, and the resulting change in the refraction angle will cause an apparent additional LOS rate, closing the feedback loop. The feedback can be either regenerative or degenerative, depending on the sign of the radome error slope (the direction of the radome error).

This phenomenon must be viewed in the context of the closed antenna tracking loop. Since for a constant LOS rate the residual boresight error is a constant, any radome error which tends to increase the apparent boresight error constitutes regenerative feedback. A radome error which makes the boresight error smaller is degenerative. To a first order, positive slopes (degenerative feedback) lower the guidance gain and lengthen the guidance time constant, making for a more sluggish response, while negative slopes (regenerative feedback) raise the guidance gain and shorten the guidance time constant to the point where missile instability could occur. The guidance design must avoid such an instability.

Target Illumination.[6,13] Target illumination for a CW semiactive missile system can be provided by a CW tracking radar, a CW transmitter slaved to another tracking radar, or a pulse or pulse doppler tracking radar at another frequency with the CW illumination injected into the antenna system from a separate CW transmitter.

The most capable of these configurations is the CW tracking illuminator. It is generally a two-dish radar because sufficient receiver-transmitter isolation cannot usually be achieved in a single-dish system. The CW tracking illuminator, since it uses the same radar signal to track the target as the missile utilizes for homing, sees essentially the same view of the target environment and can track targets at the same low altitudes as the missile seeker. The receiver portion of such an illuminator is very much like the seeker described in the following sections. The main differences stem from the much higher feedthrough levels in which the illuminator receiver operates and from the previously mentioned doppler spectrum differences (i.e., outbound targets are below feedthrough).

Alternatively, the illuminator can be the transmit-only portion of a radar slaved to a tracking radar—a mechanically scanned track-while-search (TWS) radar or a phased array which simultaneously maintains multiple target tracks with its electronically steered agile beam. In the third approach, where space constraints preclude use of separate antennas, such as in a fighter aircraft, a conventional pulse or PD radar tracks the target and the CW illumination is injected into the transmission port of the antenna from a separate transmitter.

Traditionally, the illuminator must continuously illuminate the target throughout the engagement. A system is therefore limited in its simultaneous-engagement capability by the number of available illuminators. A given illuminator must remain dedicated to its assigned target until the missile has achieved intercept; only then can it be reassigned to another. One of the primary reasons for active seekers is to remove this system firepower limitation, since each missile provides its own illumination. Another approach to avoiding the one-illuminator–one-target constraint is to use sampled data and time-share one illuminator (phased array or TWS) among several missiles.

19.3 SYSTEM EVOLUTION

Radar guided missiles have evolved through several generations since the first developments began in the closing years of World War II. The threat and the available technology have evolved over the years, and the systems have followed suit. As new requirements have been generated in response to more severe threats, new approaches using new technology have been developed. This section will attempt to trace some of these evolutionary developments.

Basic Semiactive Seeker.[2,6,7] The block diagram of Fig. 19.5 is representative of the earliest systems developed in the late 1940s and early 1950s. The simplest implementation of a CW missile seeker, it consists of a rear receiver, a front receiver, a signal processor (speedgate), and a tracking loop to control the gimballed front antenna. The missile also contains an autopilot to guide it and stabilize the airframe, a fuze to detonate the warhead at the optimum time, and a source of electrical and (in most missiles) hydraulic power.

The purpose of the rear receiver is to provide a coherent reference for detection of the front (target) signal. The rear signal, after conversion to IF, closes the automatic frequency control (AFC) loop around the microwave local oscillator (LO) and acts as the reference for the IF coherent detector. The target signal, received in the front antenna, is heterodyned to IF and amplified in a relatively wideband amplifier (typically 1 MHz or wider). It is then converted to baseband by mixing it with the rear signal in the balanced mixer (coherent detector). The doppler signal (now at baseband, with feedthrough at dc) is amplified in the video (doppler) amplifier, which has a bandwidth equal to the total range of possible doppler frequencies. It is then mixed with the speedgate LO, which is controlled by an AFC loop to keep the desired signal centered in the narrow speedgate (sometimes called the velocity gate or doppler tracker). Typical bandwidths range from 500 Hz to 2 kHz.[5]

Target acquisition is accomplished by sweeping the frequency of the speedgate LO over the designated portion of the doppler bandwidth. In essence, this sweeps the spectrum past the narrow frequency window of the speedgate. When a signal exceeds the detection threshold, the search is stopped and the signal is examined to verify that it is a coherent target rather than a false alarm due to noise. A valid target is then tracked in frequency, and guidance commands are extracted from it.

The front antenna conically scans the received beam. The resulting amplitude modulation of the received signal is recovered in the speedgate and resolved into the two orthogonal pitch and yaw gimbal axes. These pitch and yaw error signals are used to close the antenna tracking loop and to guide the missile.

The guidance error signal must be normalized (a constant scale factor of volts per degree off boresight is required) over the full dynamic range of target signal amplitudes (range, target size) in the presence of large feedthrough and clutter signals. Therefore, AGC in the receiver is necessary. Since the IF amplifier signal includes both the feedthrough and the clutter while the video amplifier includes the clutter, the specific AGC implementation must consider the degree to which these large interfering signals shall be allowed to control the gain for the target signal while preventing saturation on the interference. To maintain linear receiver operation over the large dynamic range is a major design challenge.

Unambiguous (Offset Video) Receiver. The basic receiver described above folds the spectrum around feedthrough, which occurs at dc. Although in the moving missile this does not produce an ambiguity, in the tracking illuminator the inbound targets must be distinguished from the outbound. This *unfolding* of the spectrum was initially achieved by use of a quadrature receiver (Chap. 14).

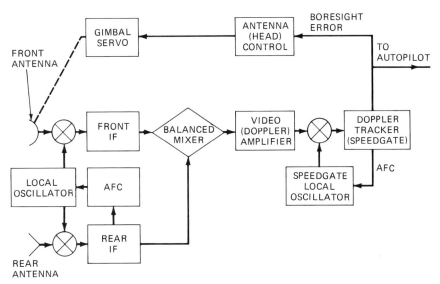

FIG. 19.5 This semiactive-seeker block diagram of a baseband conversion system is representative of the early-generation systems in which the rear (reference) and front signals are mixed
directly to extract the doppler-shifted target signal.

There are two additional drawbacks to the original configuration. Folding
the spectrum around feedthrough folds the receiver noise as well, resulting in
a 3 dB higher noise level (hence a 3 dB loss in sensitivity).[5] The other problem
stems from the fact that main-lobe clutter is the dominant signal in the doppler
spectrum. Clutter harmonics can be misclassified as targets and must therefore be avoided, thus limiting the usable range of target dopplers. For example, consider a clutter-to-signal ratio of 60 dB. A mere 0.1 percent second harmonic distortion would yield a clutter harmonic of the same magnitude as the
target. If a missile velocity of 2000 ft/s is assumed (40 kHz doppler at X band),
the harmonic would occur at 80 kHz, and the usable doppler spectrum, which
the speedgate would be able to search, could extend no further than 80 kHz (in
practice a safety margin of a few kilohertz would have to be maintained at
both ends of the search region, further limiting achievable performance). This
is illustrated in Fig. 19.6a.

However, by introducing a frequency offset before the coherent detector,[5] the
resulting spectrum will be as shown in Fig. 19.6b. This can be accomplished by
offsetting either the signal or the reference channel. Figure 19.7 shows the offset
reference configuration. Clutter harmonic distortion, noise foldover, and (for the
case of the illuminator) approach-recede ambiguity are eliminated. However,
feedthrough rejection now requires a complex notch filter at the relatively high
offset frequency rather than a simple high-pass filter. Also, clutter still controls
the gain normalization in the doppler amplifier.

For each frequency conversion, spurious higher-order mixer products must be
considered and kept out of the target spectrum. As additional conversions are
added, this task becomes increasingly difficult and seeker complexity grows. Ex-

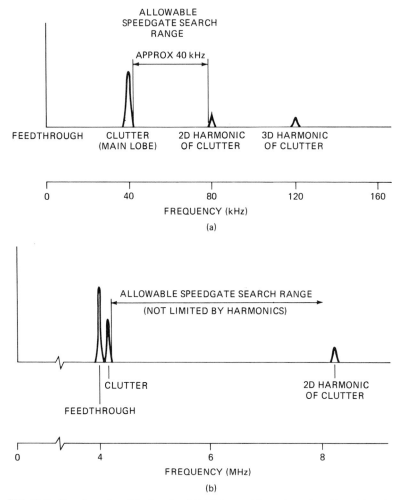

FIG. 19.6 Target spectra of the baseband (folded or ambiguous) receiver (*a*) and the off-set video (unambiguous) receiver (*b*) indicate the limitation which clutter harmonics impose on the achievable range of target velocities which can be handled.

tending the speedgate's frequency coverage to cope with faster targets and attempts to eliminate—or at least attenuate—clutter required additional conversions, which, even with the introduction of solid-state circuitry to replace vacuum tubes, resulted in prohibitive increases in size, complexity, and reduced reliability.[2,6]

Inverse Receiver.[2,6,14] The major breakthrough was the introduction of the *inverse receiver*, which gets its name from the fact that the bandwidth "funnel" of the conventional receiver (wide IF, narrower doppler amplifier, final narrowband speedgate) is inverted, with the final narrow banding (speedgating) placed right after the first conversion from microwave to IF. The crit-

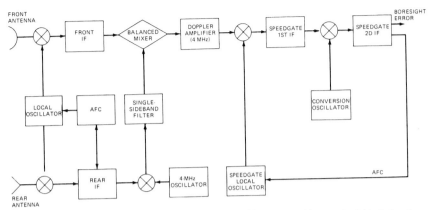

FIG. 19.7 Offset video receiver block diagram. It provides an unambiguous (unfolded) doppler spectrum by offsetting the rear reference before it is mixed with the front signal.

ical components necessary for the inverse receiver are highly selective filters at IF frequencies and low-noise tunable microwave sources.

The simplified block diagram of an inverse receiver is shown in Fig. 19.8. In the conventional receiver, the target signal must compete with feedthrough, clutter, and jamming until the final stages, with the dynamic-range requirements of the receiver and its AGC loops dictated by these large undesired signals. The inverse receiver, on the other hand, excludes them virtually at the input. The narrowband filter (usually a quartz crystal type), constituting the speedgate bandwidth, is placed in the IF after only a nominal amount of fixed preamplifier gain, sufficient to establish noise figure. One additional conversion is used in the receiver to avoid the problem of too much gain at one frequency. In the resulting two-conversion system, complexity is significantly reduced and unwanted signals

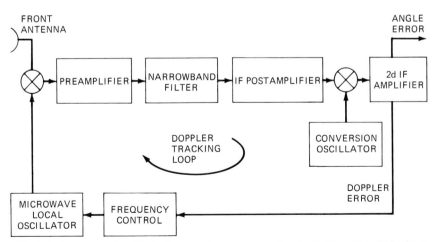

FIG. 19.8 Inverse-receiver block diagram. The narrow banding is placed very early in the receiver, inverting the bandwidth "funnel" of the conventional receiver and excluding interference from subsequent stages of the seeker.

are rejected very early in the signal path, thus reducing dynamic-range requirements and avoiding most possible sources of distortion.

The doppler tracking loop is closed through the microwave LO, which must, therefore, be tunable over the doppler frequency range of interest. This LO essentially fulfills the role of the speedgate LO in the conventional receiver of Fig. 19.5. The inverse receiver can be thought of as a double-conversion speedgate with the speedgate AFC loop closed around the microwave LO and the input to the speedgate being the microwave output of the seeker front antenna.

The IF spectrum at the mixer output will have the same form as Fig. 19.3. Sweeping the LO moves the spectrum past the narrowband filter to accomplish acquisition as in the conventional speedgate. Doppler tracking is similarly accomplished by controlling the LO frequency to keep the target in the narrow filter. The angle error signals required for guidance are extracted after the second IF amplifier. A single AGC loop (not shown), required to cope with only the target signal variations, is used to normalize the angle error signals.

Angle Tracking: Conical Scan to Monopulse.   This subsection assumes that the reader is familiar with the conical-scan and monopulse angle-tracking concepts described in Chap. 18.

Conical scan requires only a single channel and extracts the angle information which is contained in the amplitude and phase of the scan amplitude modulation by simple envelope detection. Conventional monopulse normally requires three complete channels, which must track in gain and phase to maintain the proper relationship between the sum and difference channel signals (the angle information is contained in the difference/sum ratio).[15] The complexity of monopulse, however, provides well-known performance advantages over conical scan.

Conical-scan processing requires that both the amplitude and the phase of the AM be preserved (at least one cycle of the scan is needed to make an angle measurement). The AGC which is required for gain normalization must therefore be slow enough not only to prevent it from following the scan AM envelope but to avoid any phase shift of the envelope[15] (since this would cause cross coupling between channels; i.e., a pitch error would couple into the yaw plane, and vice versa). Thus any externally generated amplitude fluctuations (propeller modulation, target fading noise, or jamming) at or near the scan frequency will be detected along with the target BSE and will result in noise or false data. In particular, this makes conical-scan systems susceptible to AM jamming at the scan frequency (the *spin frequency jammer*).[5]

The monopulse system extracts the angular information instantaneously by comparing the difference and sum channel signals. The gain normalization can therefore be made instantaneous (fast or instantaneous AGC), and the external amplitude variations, since they affect sum and difference channels by the same relative amount, are never detected as erroneous guidance signals.

The early systems all used conical scan for angle tracking because of its simplicity. The limited available volume and discrete-component tube technology of the period mandated a single-channel approach despite the performance limitations of conical scan. The inverse receiver permitted the performance of monopulse to be achieved with the single-channel simplicity of conical scan.[2,6,14]

Three identical mixers, preamplifiers and crystal filters, first process the three monopulse signals. Immediately after the narrowband filters, however, the difference channels are multiplexed with the sum channel at a moderate frequency (several kilohertz, much higher than the filter bandwidth). Interference at the multiplexing frequency is, therefore, prevented from passing through the filters. The modulated difference channels are combined with the sum signal, and the composite signal is processed in a single channel (just as a conical-scan signal). The AGC, required for gain normalization, is made faster than the bandwidth of the filters and thus acts as an instantaneous AGC. Its dynamic range has to cope only with target signal variations. The normalized monopulse error signals are then demultiplexed and used for closing the guidance loops. Frequency (AM or FM) or time multiplexing can be employed. If the frequency multiplexing is phased so as to produce AM sidebands, the processing is identical to that of conical scan.

Pulse Doppler (PD) Operation.[2–6] Semiactive systems using other than CW illumination have been employed. Some early systems employed noncoherent pulse waveforms, but they are not suitable for operation in clutter (except for very large target cross sections). Coherent PD systems, however, can approach the performance of CW.

The motivation for the use of PD in the seeker was to simplify the illuminator in air-to-air systems. For early-generation airborne radars, which employed a noncoherent pulse waveform, CW injection was the only practical solution. With the advent of coherent PD radars, an alternative way to achieve virtually CW operation without the penalty of the additional transmitter became available. This was to select a high-PRF (pulse repetition frequency), high-duty-cycle (30 to 50 percent) waveform and to use only the central line of the PD spectrum, both in the radar and in the seeker. This has sometimes been called *interrupted* CW (ICW).[3]

A high PRF is defined as one which is unambiguous in doppler. Thus when the receiver selects the central line, the spectrum is identical to the CW case. The radar receiver must be protected during transmission (duplexing and/or gating). In addition, the receiver may or may not use a range gate. If only the central-line power of the PD spectrum is used (no range gate), the resulting loss must be accepted. Use of a range gate matched to the pulse avoids this loss. In either case, the rest of the receiver and signal processing is the same as for a CW system.

The seeker implementation follows this same pattern. However, range gating in the seeker is generally not used with a high-duty-cycle system. The loss resulting from use of only the central line is essentially the duty cycle d_t. For a peak transmitted power P_t, the average power in the central line is $P_t(d_t)^2$, compared with the average power of the transmitted waveform of $P_t(d_t)$.

A low duty cycle (less than 10 percent) can also be used, but for this case the central-line power loss becomes prohibitive. Low-duty-cycle systems, therefore, must use range gating to optimize performance. In addition to retaining doppler resolution capability the range-gated system provides range resolution.

Active Seekers.[4,16] Active seekers can provide increased firepower as well as fire-and-forget (or launch-and-leave) operation. Thus, they have found application in both the air defense and the surface-target attack roles.

An active seeker is functionally the same as a semiactive seeker, with the exception that it carries along its own illuminator. Besides adding the transmitter, the other main difference in the active seeker configuration is elimination of the rear receiver, with the reference generated by offsetting the transmitter excitation (or drive) signal, as shown in Fig. 19.9.

Active seekers, since they use a single antenna both to transmit and to receive, cannot use CW because of the very limited isolation achievable. Noncoherent pulse or coherent PD waveforms have been employed, and either the central-line processing or the range-gated approach can be used for coherent operation.

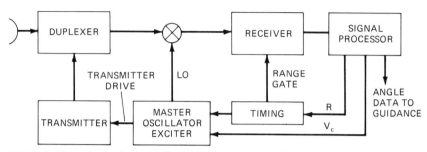

FIG. 19.9 The active seeker block diagram differs from the semiactive in that target illumination is provided by a self-contained transmitter.

Surface Targets.[4,5,17] Noncoherent pulse waveforms have been widely used in active seekers designed for attacking large-cross-section surface targets. For example, in antiship applications the slow target speed prevents effective doppler resolution from clutter, but the large target reflectivity provides an effective discriminant since the target return exceeds sea clutter by several orders of magnitude (large signal-to-clutter ratio). Even in antitank applications, such *contrast* discrimination of the target can be achieved if the size of the competing clutter patch can be reduced by the use of narrow-beamwidth antennas and narrow range gates. These noncoherent systems utilize low-duty-cycle short pulses or highly coded waveforms to achieve narrow range resolution. The resolution cell is determined by the range-gate duration in the range dimension and by antenna beamwidth in the cross-range (azimuth) dimension. The resulting surface clutter return, even for rough seas and fairly severe ground reflections, will contribute much less energy than the target echo even when the target fills only a small portion of the resolution cell. Thus the angle information derived will be primarily from the target, because of its large contrast with respect to the clutter background, and accurate homing can be achieved. It should be noted that radar cross sections of ships can be several thousand square meters, while those of tanks typically range from 25 to 125 m².[17]

For severe clutter and reduced target cross sections, coherent processing may be required. Although stationary or very slow-moving targets cannot be discriminated from clutter on the basis of doppler frequency, use of *doppler beam sharpening* or synthetic aperture techniques can reduce the effective size of the resolution cell and hence increase the signal-to-clutter ratio in the cell containing the target.[18] Angle tracking can thus take place to provide the required data for guidance.

Air Targets.[4,16] Although some early systems were designed to use noncoherent pulse waveforms, they were not suitable for low-altitude (high-clutter) operation against small-cross-section aircraft targets. Therefore, the typical air defense active missile employs some form of PD transmission and coherent processing to resolve and track the target in doppler (velocity) and sometimes also in range.

The active PD seeker can be thought of as a miniature airborne fire control radar. The waveform selection tradeoffs, especially the clutter-waveform interactions, are the same in the active seeker as in the airborne intercept (AI) radar. The unique problems of clutter ambiguities, eclipsing, range determination, etc., are the same as described in Chap. 17 and will not be repeated here. A key point is that the active seeker is a monostatic radar, whereas the semiactive system is bistatic. The doppler frequency relationships can be simply derived from the geometry of Fig. 19.2 by colocating the illuminator and missile and making the illuminator and missile velocity vectors coincident. The doppler spectrum will be like that of Fig. 19.3*b*.

Although it is highly desirable to select a high PRF (HPRF), which is unambiguous in doppler, it may be necessary in some system applications to use a medium PRF (MPRF) and operate with both range and velocity ambiguities (which must be resolved).[19] The tail-chase look-down air-to-air scenario is a key example. Since the target return must directly compete with the sidelobe clutter in the doppler resolution cell, it may be necessary to reduce the absolute amount of clutter contained in the cell. A positive signal-to-clutter ratio (*S/C*) is necessary to permit target visibility. One way to achieve this is to range-gate and thus reduce the size of the clutter patches, the return from which is accepted in the receiver. Reducing the PRF reduces the number of intervening range-ambiguous clutter patches which fold into the target doppler cell, further improving *S/C*.

To maximize range performance (not clutter-limited) average transmitter power must be the maximum practically achievable. Within the constraints of a tactical missile—small size, limited weight—this will tend to drive the design to higher-duty-cycle lower-peak-power systems. This is quite compatible with HPRF, where high-duty-cycle central-line processing has generally been used. If clutter is the limiting factor rather than receiver thermal noise, lower average power is acceptable—consistent with MPRF. The difficulty arises if the same system must achieve both long-range (noise-limited, approaching target) and tail-chase (clutter-limited) performance. Transmitter hardware constraints make it difficult to vary the waveform at will over a wide range of PRF and pulse width. Thus if an MPRF system is employed, the tendency will be toward long pulses (to keep average power high without increasing peak power). Therefore, to achieve good range resolution may require some form of pulse compression.

System Implementation.   Active seekers have used both conical scan and monopulse angle tracking, and the receivers have evolved from the conventional to the inverse configuration, just as with the semiactive.

Because of the limitation on achievable antenna size as well as transmitter power, the range performance of an active seeker will be considerably less than for a comparable size of semiactive seeker operating with a high-power large-antenna illuminator.[5] Thus active systems are used in short-range homing-all-the-way applications or as the terminal guidance mode of a multimode long-range system. For example, a midcourse mode employing inertial or command guidance can be used to bring the missile within the terminal guidance range (typically the last 10 guidance time constants or a few kilometers from intercept). The target coordinates in angle (antenna pointing) and range and/or velocity (doppler), provided by prelaunch data or by command updates during flight, initialize the seeker. The target uncertainty is searched by the seeker, and when the target is acquired, the missile transitions into the terminal phase of flight. Seeker operation then proceeds as for the semiactive case until target intercept.

Passive Seekers.   Three passive operating modes have been employed for missile guidance: antiradiation homing (ARH), home-on-jam (HOJ), and radiometric. ARH operation is used in missiles for attacking hostile radars, usually in an air-to-surface application (although air-to-air and surface-to-surface systems are also potential configurations). HOJ is an essential adjunct for semiactive and active systems to counter noise jamming.[4] The radiometric homing mode has been employed as a terminal guidance mode in millimeter-wave antitank missiles.[17]

Antiradiation Homing.[20]   ARH systems differ from the active or semiactive air defense or the ship or ground attack systems in that they are very wideband (octave bandwidth or wider). This need for wideband operation is the main driver in seeker design. The receiver configuration is very similar to the emitter location and identification systems often called ESM (electronic support measures).[21] However, the size and weight constraints of missile-borne hardware restrict usable approaches.

The parameters available to an ARH sensor include frequency, PRF, pulse width, angle (direction of arrival), and signal amplitude. Various combinations of these can be used to discriminate and select a specific emitter from among the multiplicity of signals present (estimates of 10^6 pulses per second have been cited as a "high-density" environment[21]). Signals must be initially sorted on the basis of frequency and then the pulse trains *deinterleaved* to select a particular emitter. Most radars will be detected primarily through their sidelobes, requiring reasonable sensitivity, but the dynamic range must be able to cope with main-lobe signals as well.

Broadband antennas can be gimballed or body-fixed. Since directional information must be determined on each received pulse, some form of monopulse antenna is required. Because the seeker may encounter any incident polarization, the antenna should have a uniform response to all senses of linear polarization.[20]

Four types of antenna systems are possible: an amplitude monopulse with four squinted beams, a three-channel phase-amplitude monopulse using four elements or apertures, an interferometer, and the two-channel polar monopulse using a four-arm dual-mode spiral.[22] The first three are conventional configurations except that each of the antenna elements is a broadband device such as a log-periodic type, conical log spiral, or cavity-backed planar spiral. The dual-mode spiral has been the preferred choice, since a single aperture generates all the direction-finding information (and thus makes full use of the limited available space), requires only two receiver channels, has excellent polarization characteristics, and is frequency-independent.[20]

The four arms of the spiral are fed by a mode-forming network to form a sum (Σ) and a delta (Δ) mode (hence the name *dual-mode*). The directional information is contained in the relative amplitude and phase of the Σ and Δ channels. The Δ/Σ ratio represents the magnitude of the BSE (the angle off axis in a cone of rotation about the boresight), while the relative phase indicates the direction on the cone of rotation. This polar information is then converted into the more conventional pitch and yaw coordinates.[23]

A variety of receiver types can be used to analyze the signal spectrum: wideband crystal video, instantaneous-frequency measurement (IFM), channelized, scanning superheterodyne, compressive (microscan), or Bragg cell (acoustooptic).[21,24] Size and weight limitations dictate that a single-channel approach be used in a seeker. The contradictory requirements of wide instantaneous bandwidth for rapid acquisition and narrow bandwidth for high sensitivity can be achieved by using switchable bandwidths, such as shown in the typical block diagram of Fig. 19.10. This also includes a compressive receiver which is ideal for CW signals. In this approach, the local oscillator is swept rapidly (*chirped*) to impress linear FM on the signal. A matched compressive delay-line filter then compresses the signal, producing a short pulse, the time of occurrence being indicative of RF frequency.[25,26]

Home-on-Jam.[3,4,27]   The HOJ mode is an essential part of a semiactive or active seeker. The use of wideband noise represents the earliest brute-force active jamming technique which can mask the desired target reflection. The jamming,

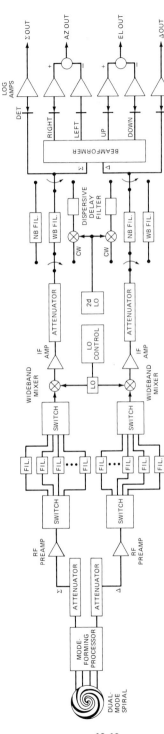

FIG. 19.10 Antiradiation homing (ARH) receiver. A typical configuration is shown, including a dual-mode spiral antenna, high first IF, switchable bandwidths, and compressive (microscan) CW processing. (*From Ref. 20.*)

19.19

however, is a powerful point source of radar energy which can provide more than adequate angle information for homing. All that is required is a means to allow the seeker angle track circuits to process the noise energy. When the jamming is such that tracking of the target *skin* return is not possible, the seeker switches to passive tracking of the received jamming energy. If the jammer-to-signal ratio (J/S) decreases to the point that skin track is again possible, this is given preference over HOJ. Also, provision must be made to allow switching between HOJ and skin if the jamming is intermittent. The criterion in all cases is that the mode which provides the better quality of guidance information should be given precedence.

Radiometric Homing. This mode utilizes the natural thermal radiation from targets for guidance. The very sensitive receiver detects the difference in radiation between the target and the ambient background. Use of this technique in millimeter-wave seekers against surface targets provides a terminal mode with significantly lower glint than the active or the semiactive radar mode.

Other System Configurations. Variations of the above seeker types as well as multimode combinations have been studied and in some cases implemented to take advantage of new and emerging technologies.

Sampled-Data Operation.[2,4,6] To overcome the limitation of tying up an illuminator for the duration of a semiactive engagement, a single radar can be time-shared among several missiles. This generally implies a phased array radar, although mechanically scanned track-while-scan (TWS) radars can provide this option in some cases.

The advent of phased array radars permitted a single transmitter to illuminate many targets by sequentially stepping its agile beam from one target to the next. The illumination was no longer continuous, and the missile would thus have to operate in a sampled-data mode, extracting information during the time that its target was illuminated (dwell time) and then holding the information until the next sample. The illumination waveform could be CW during the dwell time (interrupted or keyed CW), or PD could be employed with or without pulse compression.

For sampled-data operation, the primary difference is in the doppler acquisition scheme. Since target illumination occurs only in short bursts, the use of a sweeping gate for acquisition would result in excessively long acquisition times. The doppler uncertainty region must, therefore, be examined simultaneously by a bank of contiguous doppler filters. The illumination burst must be shaped or the received signal time-gated to prevent the spreading of clutter through the target doppler spectrum owing to the pulsed nature of the transmission. Finally, sample-and-hold circuits must be added in AFC, AGC, and angle track loops.

Either the conventional or the inverse receiver can be made to operate in the sampled-data mode. Sampled data can be used all the way to intercept or as a midcourse mode for an active terminal seeker. Lower data rates are allowable in midcourse than in terminal, providing an additional degree of freedom in the system design.

Retransmission Guidance.[13,28–31] Retransmission guidance, also known as TVM (target-via-missile or track-via-missile), was initially conceived as a simplification of missile-borne hardware, placing all the processing on the ground and making the seeker a simple repeater. In practice the repeater is limited in gain by the usual transmit/receive isolation (ring-around problem); so additional complexity must be added. At the same time, use of more complex pulse compression waveforms could be more easily accommodated by not requiring the sophisticated processing hardware to be missile-borne.

TVM is essentially a variation of semiactive homing. The target-reflected illumination is received in the missile, but instead of being processed on board it is retransmitted to the illuminating radar. Here the complex waveform is processed, guidance information extracted, and steering commands transmitted to the missile as in a command guidance system.

Multimode Systems.[2,4,16] The early CW semiactive systems were generally designed to home all the way from launch to intercept. In later-generation, more sophisticated systems, homing generally lasts for only the last few seconds of flight (typically 10 guidance time constants). In these systems, a midcourse phase (inertial, beam rider, or command) is employed to get the missile to an appropriate point on its trajectory, where it acquires the target (using prelaunch or in-flight commanded designation data) and enters the terminal (homing) phase of its flight. This is more efficient from the standpoint of both missile trajectory and radar power. The missile can fly out to longer ranges by a commanded or inertial up-and-over trajectory, spending less time in the denser air at low altitude. The radar power needed for illumination (semiactive or active) is sized by the terminal phase of flight, a fraction of the total intercept range. Midcourse commands impose much less severe demands on radar power since this is a one-way transmission path.

Combinations of semiactive or active radar with IR or ARH modes and the trend for operation at higher frequencies offer a large number of potential multimode seeker configurations.

19.4 SYSTEM FUNCTIONAL OPERATION

There are a number of necessary functions all of which must be successfully accomplished to permit a lethal intercept of the target by a guided missile. These begin with initial target detection and decision to engage and include missile launch, proper operation of the propulsion, guidance, and control systems through the flight, and fuzing and detonation of the warhead at intercept. We shall now consider the radar functions of target acquisition and tracking which provide the intelligence for guidance. Emphasis is on semiactive or active coherent operation unless noted otherwise.

Reference-Channel Operation.[2,5,6,32] Within the context of a coherent system, the seeker must have available as a reference a precise replica of the illuminating signal. In semiactive systems this has generally been provided by the rear (reference) receiver (although an alternative *on-board reference* approach is also possible). In active systems the reference is derived directly from the transmitter-exciter.

The reference must be spectrally pure (low-noise), and its frequency must accurately represent the illumination frequency to allow the target echo to fall within the bandwidth of the receiver. These requirements are relatively easy to meet in an active system because the same microwave source provides the reference and the transmitter exciter (drive) signal. In semiactive systems, particularly in the early-generation systems in which the transmitters were not crystal-controlled, providing a coherent reference posed a significant challenge.

The early-generation illuminator transmitters generally employed magnetrons or klystrons as power sources which, while possessing good short-term stability and low near-carrier FM noise, lacked the setability and long-term drift characteristics to allow precise frequency predictability. They could operate anywhere within relatively wide (compared with receiver bandwidth) frequency allocation bands. The missile LO, in conjunction with the rear receiver, thus has to search for, lock onto, and track the illuminator frequency throughout the flight to keep the LO properly positioned. Either an AFC or a phase-lock loop (PLL), closed around the LO, compensates for any frequency variations in the transmitter as well as for the rear doppler shift.

The rear acquisition process is similar to that of the speedgate. As shown in Fig. 19.11, the LO is swept in frequency over the illuminator uncertainty region. When the difference (mixer output) frequency falls within the rear IF bandwidth, the discriminator will generate an output, stopping the sweep and allowing the AFC loop to lock on. It will then continue to track the rear signal. Since the AFC loop is usually a first-order loop, a changing rear frequency will result in a frequency error (lag). Because the LO is common to both rear and front channels, however, the recovered doppler signal is not affected. A PLL would result in a phase error for a changing rear frequency but not in a frequency error. Some form of coding on the illuminator signal is often used to ensure that the correct signal is acquired by the rear loop.

With the development of crystal-controlled transmitters, the illumination frequency could be known precisely prior to missile launch. Thus, the microwave LO could be crystal-controlled and preset before launch. The requirement for tunability could thus be significantly reduced (no search would be required). The requirements for tunability and low FM noise are contradictory. It is much more difficult to design a low-noise oscillator with a wide tuning range than one with a fixed frequency (or a very small tuning range).

As noted, any modulation on the LO will spread clutter and feedthrough across the target doppler frequency band. If the modulation is noise, it will degrade sensitivity; if it consists of discrete frequencies (vibration-induced or resulting from power supply ripple), these can be interpreted as false targets. Thus the rear AFC (or PLL) loop is generally used as an FM noise degeneration loop around the microwave LO, with a bandwidth at least as wide as the target doppler band of interest.

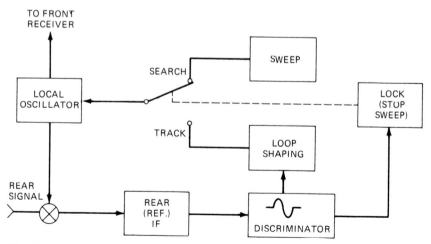

FIG. 19.11   The rear AFC loop uses open-loop search to acquire the rear signal and then switches to closed-loop tracking.

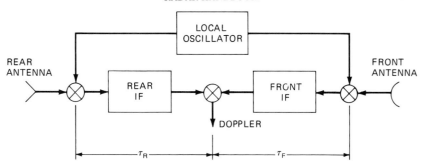

FIG. 19.12 Time-delay matched IFs provide cancellation of FM noise on the transmitter and missile local oscillator. If $\tau_R = \tau_F$, illuminator noise on feedthrough and local-oscillator noise are cancelled at the doppler output.

Transmitter noise will have the same effect as LO noise. In the case of the tracking CW illuminator, its own requirements will be more stringent than the missile-imposed ones. However, some other illuminators, such as those using CW injection, may have higher FM noise than allowable in high-feedthrough conditions. Here the use of time-delay matched front and rear receivers (see Fig. 19.12) can result in the cancellation of much of the transmitter FM noise present on the feedthrough when the front and rear signals are mixed in the doppler balanced mixer. At the same time, LO noise will similarly be reduced.

In systems in which the front and rear signals are not mixed directly, the total illuminator spectrum, as received at the rear receiver, can be transferred to the LO, which is then mixed with the front signal. If the proper phasing is maintained, feedthrough-noise cancellation can be achieved.

Use of a wideband loop, however, allows any corruption of the rear signal to be coupled onto the LO and thereby onto the front signal. These corrupting effects include rear signal level variations (fades), rear multipath, and rocket motor plume effects (attenuation and modulation).

If the LO is made sufficiently low-noise—for example, by use of a separate LO noise degeneration loop—and the illumination signal has long-term stability and is similarly noise-free, the rear loop need not be wide. A narrow-bandwidth loop, sufficient to control the LO frequency against relatively long-term variations (rather than high-frequency noise), greatly reduces the effects of rear signal degradation.

The ultimate step is the elimination of the rear receiver and its AFC (or PLL) loop by the use of an *on-board reference*. Use of a crystal-controlled illuminator transmitter and a crystal-controlled LO can provide the long-term oscillator stability which ensures that coherent detection of the doppler-shifted front signal still results. The use of fixed tuned oscillators makes it possible to achieve the low FM noise without elaborate degeneration loops.

Target Signal Detection.[2,5,6] Target detection, or *lock-on*, is a postlaunch function. Many missiles may not even be in a position to view the target before launch (e.g., inside a launch canister or a semisubmerged carry on an airplane). Those that might be able to lock before launch will break lock because of the launch shock, plume effects, or extremely high feedthrough or, in some cases, intentionally to avoid large interfering signals in the first seconds of flight. Thus the seeker will accomplish target detection in its front receiver at some prescribed time during flight (right after launch for a homing-all-the-way system or just prior to terminal for a multimode type).

The theory of detection of signals in noise has been covered extensively in the literature.[15,33-36] For the purposes of this discussion, some key differences between the search radar and the missile must be pointed out.

While the search radar must examine a large volume and decide whether a target is present, the missile begins its task with the knowledge that a target exists (else the missile would not have been committed to engage it) and must determine which resolution cell within the designated volume contains it. The missile must perform its detection within a finite, usually very limited time. Unlike the search radar which can find the target on the next or subsequent scan if it misses it the first time, the missile has no such luxury. The probability of finding the target must be very high—usually 95 percent or higher (requirements of 99 percent are not uncommon). The probability of false alarm can be moderately high; if a second decision criterion (verification) is employed, the initial detection threshold can be relatively low, ensuring a high probability of seeing the target. The penalty is the time wasted in verification, which is traded against total acquisition time. In practice, detection can be achieved with an independent sample S/N of 3 to 6 dB in the doppler resolution cell of a sweeping speedgate (and accumulation of several samples to yield the required cumulative detection probability).[5] The range calculation uses the standard radar range equation, but it is important to note that the conventional R^4 range product is $R_{IT}^2 R_{MT}^2$, where the two ranges— illuminator-to-target (R_{IT}) and missile-to-target (R_{MT})—are usually not equal in the semiactive case.

Acquisition with a sweeping speedgate is depicted in Fig. 19.13. The speedgate LO is some form of voltage-controlled oscillator (VCO). The linearity of its frequency-voltage characteristic must be very precisely controlled to allow precise sweep positioning and a constant sweep rate over the full doppler frequency range of interest. Also, since the VCO forms part of the AFC (or PLL) loop, control linearity is required to maintain loop gain constant over its total frequency coverage range. The extent of the sweep will differ from system to system, depending on the accuracy of the designation and on operating frequency; it may range from a few to several tens of kilohertz. The sweep rate is based on the doppler filter bandwidth and must be slow enough to ensure signal buildup in the filter. The rate and the extent of the sweep will determine the number of "looks" at the target during the available acquisition time and hence the cumulative detection probability. On the basis of the false-alarm rate and the verification time required for each alarm, the available search time will be decreased (i.e., fewer looks or independent samples will be available). These parameters can be traded off to achieve the optimum for a given system.

The actual detection process consists of programming the speedgate LO with a sawtooth or triangular sweep voltage. When the difference between the LO and the target doppler frequencies equals the speedgate filter center frequency, an output will be produced at the speedgate discriminator or at a separate amplitude detector. The output pulse is detected, and if it exceeds a preset threshold, the sweep is stopped for a few tens of milliseconds to determine whether the threshold crossing was a noise false alarm or a valid target return. This subsequent process of verification, or coherency check, examines the signal in the gate for persistence; a target will remain above the verification threshold (barring a fade), while noise will not.[5]

There are several possible methods of verification. The simplest is merely a measure of amplitude using a longer time constant than the pause circuit. A persistent signal will charge the capacitor of the long time constant and "hold lock." Another method is to apply a low-deviation, low-frequency FM (such that the

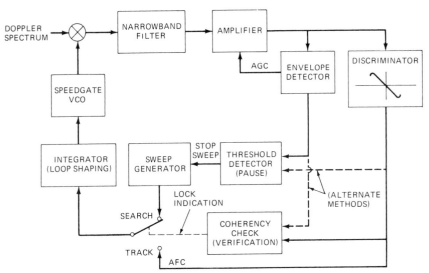

FIG. 19.13   A sweeping speedgate shows alternative methods of target detection and verification. After target lock-on the AFC loop keeps the target signal centered in the narrowband filter.

FM sidebands are contained within the speedgate bandwidth) either to the transmitted signal or to the speedgate LO. Detection of this FM at the discriminator output will hold lock. A signal which is present for several cycles of the modulating frequency will result in detection of the FM, while noise will not. Still another method is to attempt to phase-lock an oscillator to the speedgate output. If a coherent signal is present, the loop will lock on, and this fact is sensed to generate the verification signal. The method of FM'ing the transmitter was originally used as protection against locking on a vibration-induced false signal (microphonic) which could occur in the receiver. Such signals would not contain the correct FM and thus would be rejected.

Once the verification threshold has been crossed, the speedgate frequency-tracking loop is enabled (closed) and the speedgate is said to be *locked on*. Lock is held through target fades until the target signal reappears. If the target does not reappear after a preset delay, sweep resumes until the process of pause-verify is repeated. If the verification test fails after a pause, the initial detection is judged a false alarm and sweep resumes.

In sampled-data systems the illumination interval may not be long enough to allow time for a sweeping-gate acquisition system. In such applications detection must be accomplished in a bank of contiguous doppler filters which examine the band of uncertainty simultaneously. In the days of analog circuitry this was a costly, complex, bulky approach. With the advent of digital processing, such a parallel filter bank could be mechanized with a few digital chips executing a fast Fourier transform (FFT) algorithm. In range-gated PD systems, where each range gate must be followed by a doppler filter bank (i.e., the entire range-doppler uncertainty region must be searched), the digital-processing approach allows a system to be realized which would have been prohibitively complex and large in an analog configuration.

In a parallel filter bank acquisition system, the output of each filter is detected and thresholded to determine the presence of a target. Because there is a large number of independent samples during the acquisition interval, the false-alarm rate per filter must be correspondingly reduced to maintain the same overall false-alarm rate. Thus the threshold is higher than in the sweeping-speedgate case, and a higher S/N is required for detection, but a shorter total acquisition time results.

The above discussion is equally applicable to the conventional and the inverse receiver and to either a semiactive or an active seeker.

In noncoherent systems, when a range gate rather than a doppler gate is used, the range gate must be swept, or a bank of range gates must be used for target detection. This process is much like automatic detection in conventional pulse radar.

Target Signal Tracking. The detected signal must now be tracked to extract the information required for guidance. The primary data required is angle (or angle rate). The doppler and/or range gate is used to restrict the size of the resolution cell from which angle data is derived while excluding other targets, clutter, and interference. Although doppler (velocity) and range data (if available) are used in advanced guidance algorithms and for other missile logic decisions, predicted or assumed values can be substituted. There is no available substitute, however, for target angle data, without which homing cannot take place. Thus the primary function of the seeker is to extract the angle data sensed by the antenna and to maintain its fidelity.

The entire tracking process is very similar to that employed in a tracking radar. Thus only the aspects which are unique to the missile application will be detailed, and those which are common will be only briefly noted to provide continuity. A conventional tracking radar extracts target range, range rate, and angle data, which must be of high accuracy. This accuracy can be achieved with data smoothing since the data rate need not be extremely high, especially at long range. The missile, on the other hand, is closing on the target at high speed, and therefore its data rate must be much faster, consistent with the response time of the guidance and antenna tracking system and the range and doppler tracking loops. However, the S/N is consistently improving as the range decreases.

Doppler Tracking.[2,5,6,14,15] The simplest doppler tracking loop, shown in Fig. 19.13, is an AFC loop using a discriminator as the sensing element. Since this is a first-order loop, a constant acceleration (a linear frequency ramp input) will result in a frequency-tracking error. This error must be accommodated by the loop; hence the filter bandwidth is sized by the maximum acceleration which must be tracked. In conical-scan systems the scan sidebands will usually dictate the minimum speedgate bandwidth, but the acceleration lag is additive and must be taken into account. In monopulse systems the acceleration lag combined with the target spectrum itself will be the ultimate limit on the minimum usable bandwidth.

Use of a phase-lock loop eliminates the frequency error due to acceleration and allows a narrower velocity gate (narrower noise bandwidth, hence better sensitivity). However, a narrower bandwidth requires a slower sweep in acquisition and hence a longer acquisition time. Practical considerations will generally limit the minimum bandwidth achievable to about 100 Hz.[5]

Range Tracking.[15] Range-gated systems require range tracking as well. This is accomplished as in a conventional range-tracking radar by the use of a split range gate, which forms what is essentially a time discriminator. The error output closes the range-tracking loop to keep the target pulse centered in the range gate.

Range-gated pulse doppler systems track both in range and in doppler. The techniques are the same as used in PD tracking radars.

Angle Tracking.[2,5,6]   To provide the guidance data for homing requires measurement of the missile-to-target LOS rate. This implies that the seeker antenna boresight axis must be continuously aligned along the (approximate) LOS to the target so that the LOS rate can be measured. With a gimballed antenna this requires physically aiming the antenna; with a body-fixed antenna the LOS is steered electronically or merely calculated. In all cases the antenna LOS must be isolated from body motion. In a gimballed antenna this is usually achieved with gyros mounted on the antenna which are used in a stabilization loop. In body-fixed antennas, an inertial reference which senses body rates is used, and the body rates are subtracted from the antenna measurements to yield the required target angle data.

As noted earlier, radome aberration (refraction) and imperfect stabilization are error sources with which the guidance system must deal. Glint errors are a characteristic of the target and, to the extent that they are not reduced by techniques such as frequency agility, must also be accounted for by the guidance. The objective of the seeker is to make the measurement without any additional corruption of the angle data.

The basic antenna tracking loop is shown in Fig. 19.14 for a conical-scan seeker. The beam is scanned by spinning an offset feed or a subreflector in a Cassegrain antenna configuration. The spin motor is coupled to a reference generator, which provides two signals at the spin frequency in phase quadrature. The spinning beam amplitude modulates the target signal, with the modulation amplitude being proportional to the magnitude of the error and its phase representing the direction of the error off boresight. Phase comparison of the detected-signal AM envelope with the reference signals allows the direction of the error to be determined and resolved into the orthogonal pitch and yaw axes.

The antenna is mechanically steered in each axis by hydraulic or electric servos, driven by the pitch and yaw error signals. The loop in each plane drives the error signal to a null. A first-order loop will have a constant error for a constant-rate input. The detected boresight error will therefore be proportional to the LOS rate and can thus be used to implement the proportional-navigation guidance law.

In a monopulse system, the pitch and yaw errors are extracted by measurement of the Δ/Σ ratio. The servo loops function in the same manner, driving the errors to a null. The resultant errors in each plane are a measure of the LOS rate. A number of monopulse implementations are possible, including amplitude or phase monopulse in conventional three-channel or various two-channel configurations such as $(\Sigma \pm \Delta)$ or $(\Sigma \pm j\Delta)$.[37]

In the past the very limited space available in a missile put a premium on reducing the number of processing channels, both at RF and at IF, even if some performance compromises were involved. However, the increasing availability of high levels of integration in microwave and IF receiver circuits makes such considerations much less significant.

Receiver Processing.[2,5,6]   Extraction of the boresight error (angle data), as well as doppler and range data used to restrict the cell size from which the angle information is derived, requires extensive receiver processing. The very weak target signal must be amplified to a level sufficient to drive antenna servos, switch logic circuits, etc. Good noise figure must be achieved to maximize *S/N*, undesired signals and interference must be filtered, attenuated, or gated out, and the fidelity of the desired target data must be retained. Typical receiver gain may

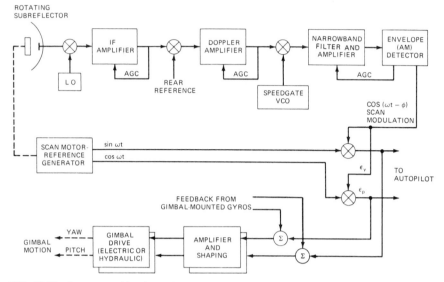

FIG. 19.14 Antenna tracking loop of an early-generation conical-scan seeker. Space stabilization is achieved by feedback from gimbal-mounted gyros.

range from 120 to 160 dB. The physical layout (isolation, shielding) cannot be overlooked as a potential source of distortion. The stability and sensitivity of the various closed loops must be maintained over a wide range of input conditions.

As noted, gain normalization is required in the receiver to maintain a constant guidance error gain. For most systems this implies the use of AGC. Depending on the receiver configuration, this may be carried out through various combinations in the IF, doppler amplifier, and speedgate. Each of these may have its own AGC loop, or the AGC may be derived later in the receiver and used to AGC an earlier section (for instance, AGC derived from the narrowband doppler amplifier output may be used to control the gain of the wideband IF). The considerations that dictate the specific implementation are the actual range of clutter and feedthrough levels and the degree to which they control the gain for the target signal. The dynamic range of these amplifiers and AGC loops forms one limitation on achievable SCV and SFV (subfeedthrough visibility).

The other concern is that of possible interaction between the large interfering signal and the small target which could cause false guidance commands. Mutual coupling or nonlinearities at large signal levels could couple amplitude fluctuations onto the target signal. Prevention of such nonlinearities is a major design objective, both in the circuit design and in the physical layout, since very high gain is concentrated in a very limited space.

Obviously, the inverse receiver, by rejecting interference early in the receiver, is much less susceptible to such problems and can achieve much better performance than the conventional (wide-front-end) configuration.

Monopulse systems can avoid the problems of conventional AGC by using instantaneous (or very fast) AGC, high instantaneous dynamic range, or, in some configurations, limiting or logarithmic receivers.

Performance Limitations. Even if all the preceding functions of acquisition and tracking are performed to perfection, this does not guarantee that the missile will score a target kill. There are factors not under the designer's control which will limit the achievable performance in terms of miss distance or lethality. Target noise has already been mentioned as a fundamental limit. Its effects can be minimized by choice of guidance time constant and navigation ratio and its magnitude reduced through frequency agility or high range or doppler resolution, but it cannot be totally eliminated. Similarly, clutter is an inescapable presence in low-altitude operation and a major design driver. Receiver design can ensure that clutter does not directly interfere with the target (for approaching geometries), and waveform selection (in active seekers) can reduce its magnitude in the receiver, but nonlinearities in the system (often the inevitable imperfections in the hardware) can result in corruption of the target data. The effects of multiple targets, multipath (which is a special case of the multiple target), and electronic countermeasures (ECM) are three additional factors which must be considered in the design to ensure that adequate performance is retained.

Multiple Targets.[38] A classic radar problem involves timely resolution of two closely spaced targets. In a tracking radar the presence of unresolved targets will degrade the quality of the data obtained; in a missile it can mean the difference between success and failure of its mission.

The case which is usually analyzed is that of two targets flying in close formation. It presents a common tactical situation as well as being a very stressing case.[39–42] Resolution can be accomplished in range, doppler, or angle, depending on what data is available. However, since range and doppler can be denied by simple noise jamming, angle is the only dimension in which eventual resolution can always be guaranteed, since the angular separation will at some point exceed the antenna beamwidth. The issue is whether resolution will occur early enough in the engagement to allow a lethal intercept against the resolved target (i.e., how many guidance time constants remain between resolution and intercept). Closing velocity, missile response time, target signature (jamming, nonjamming) and spacing, seeker antenna beamwidth, and type of angle processing are the factors which will determine the answer for each specific scenario.

For a monopulse seeker, nonjamming dual targets will result in the tracking of the stronger target, while dual noise-jamming targets will cause *centroid* tracking (i.e., the power-weighted centroid, or midway between two equal power jammers).[40] Centroid tracking would tend to make the missile fly down the middle, while for the nonjamming case target amplitude fluctuations can cause the guidance point to switch from one target to the other.

If the targets are separated by a distance of less than the effective range of the warhead, homing on the centroid would kill both, assuming the fuze and warhead function properly. If the spacing is large, so that resolution (separation greater than beamwidth) occurs early in the flight, the missile will have enough time to guide to the resolved target. Between these two extremes is the region where the effort to achieve early resolution has been concentrated. Clearly, a narrow beamwidth is a fundamental solution (as is a very large warhead). However, given the size limitation of missiles (diameters range from about 5 to 20 in), increasing the operating frequency is the only viable option.

Much effort has been devoted over the years to improving resolution in the doppler dimension by the use of narrow-bandwidth processing, such as phase-lock loops. The difficulty with such approaches has been the target amplitude fluctuation. Two identical targets will not present constant equal cross sections. If the resolved target fades, the loop may jump to the stronger unresolved target (i.e., true resolution does not occur until the final jump).[43,44]

Other approaches have been widely explored in the literature, and although improvements can be shown for particular cases, none have been found that will work against any number of targets and for all conditions.

Multipath (Image).[7,38] The case of the target image appearing below the surface, caused by multipath reflections from a smooth surface (especially the sea), is a special case of the multiple target. However, while in a dual-target situation it is of little consequence which target the missile resolves and homes on, this is not the case in multipath. The approach has been to minimize the reflected signal, and hence the *image* target, and to take steps to prevent the "noise" in the pitch (elevation) plane guidance channel from causing the missile to impact the surface in low-altitude intercepts.

In the case of the sea-skimming antiship missile, guidance in the elevation plane is often implemented with a radar altimeter, with target angle data used only for guidance in azimuth.[5]

For air defense missiles, this is not a viable solution. However, the choice of vertical polarization is commonly used to reduce the reflection coefficient at low grazing angles. Shaping of the trajectory so that the missile approaches the target with a dive angle which minimizes the reflection coefficient (Brewster angle) can provide additional improvement and prevent surface impact.[7,10]

Clearly, approaches which aid in the multiple-target case will be equally beneficial against multipath. Narrow antenna beamwidth and narrow doppler or range resolution are two such methods for reducing the image effects.

ECM.[27] One of the prime considerations in seeker design is the ECM environment in which the seeker must operate and the electronic counter-countermeasure (ECCM) solutions which must be implemented. These considerations are not substantially different from conventional radar ECCM, except that in the missile there is not an operator, who in the case of many conventional radars can play a significant ECCM role.

It is easy in the ECM-ECCM game to fall into the "what if" game with the *paper jammer* invented to exploit a specific implementation detail of the seeker and a responsive *paper ECCM* feature configured to solve it. Since the possibilities are limitless, a practical seeker design could never result. Since the enemy's ECM will never be under the designer's control, a specific defined threat should not be used to devise specific ECCM fixes. Rather, ECM should be considered in generic terms with robust ECCM techniques devised to counter generic classes of jamming. For example, use of monopulse is an ECCM technique which is inherently immune to amplitude-modulation ECM. For this reason, conical scan is no longer used except in the very simplest systems.[5,45]

ECM consists of both techniques and tactics. In the tactics category, jammers can be employed in standoff jammer (SOJ), escort, or self-screening jammer (SSJ) roles. The SOJ operates from ranges beyond the intercept capability of the missile and attempts to exploit geometry to mask the presence of penetrating attackers. SOJs are usually wideband to counter a number of defensive systems simultaneously. They cannot be engaged; therefore, the missile must be able to avoid or ignore SOJs (e.g., orient itself to place them in its antenna sidelobes) while engaging the attackers. SSJs are the targets that the missile must engage.

Escort jammers, flying with the attackers, are equivalent to SSJs from the missile standpoint and are targets to be engaged.

The generic ECM techniques can be broadly categorized as chaff, brute force (noise), and deception.[5] Chaff is just another form of clutter, and a coherent seeker with good SCV can inherently reject it. Noise jamming was the earliest form of active ECM and is still widely and effectively used. It can deny range and doppler measurement capability to any radar or seeker because it can always overpower the target echo given sufficient jammer power, and since the jammer transmission is a one-way path, it will have an R^2 advantage. However, the jammer when used in the SSJ role provides a beacon signal on which a missile can home by using its HOJ capability. Homing can be very effective providing certain precautions are taken (i.e., monopulse rather than conical scan).

The most difficult ECM to cope with is deception, which includes a wide variety of techniques. In general, this class of ECM attempts to generate false information to confuse the missile range, doppler, or angle measurement. It usually employs a repeater, which receives the illumination signal, amplifies it and applies some form of modulation, and then retransmits it at a level greater than the skin echo (but not overwhelmingly so) to persuade the seeker to accept it as the real target signal. Sophisticated signal processing is required to identify and counter such deceptive ECM. The details of this ongoing move and countermove struggle remain highly classified for obvious reasons.

The final measure of effectiveness of an ECM or ECCM technique cannot be based on a single jammer, missile, or launch platform but must consider the complete tactical environment. There will be a number of enemy aircraft, with and without ECM, and a number of different defensive systems, all interacting in a complex, shifting geometry. What works in one case may fail in another, and the result will be probabilistic.

19.5 SUBSYSTEMS AND INTEGRATION

This concluding section will consider the physical implementation of the missile subsystems and their integration into a complete missile. In block diagram or functional operation the missile seeker may seem very much like a conventional radar. The hardware, however, must meet a number of significant constraints which are unique to the missile application.

The missile environment is severe—shock, vibration, temperature extremes, high altitude, and humidity all must be endured while performance is maintained at the required level. Weight and volume are extremely limited. The real payload is the warhead; all else is excess baggage, essential to achieve the final desired result but at the expense of extra propulsion, structure, etc. A 1-lb reduction in seeker weight will save several times that in terms of missile weight, which is especially important in air-launched systems. The missile must be a very reliable, yet low-cost, producible piece of hardware. It is essentially a round of ammunition which must be produced in quantity, sit idle for prolonged periods, and then work unfailingly at a moment's notice. Yet it is one of the most sophisticated products of electronic technology.

The emphasis on reliability and producibility results in the requirement to employ proven, mature technology in the design of missile hardware. "Today's" state-of-the-art components rarely find their way into "today's" designs. This, coupled with the long development cycle, means that systems reach deployment containing old technology. Since their operational lifetime can be decades, the hardware in the field may contain components that long before were made obsolete by new developments.

Early-generation systems, designed in the early 1950s, used miniature and subminiature vacuum tubes and relays for switching. Systems of the 1960s utilized discrete transistors and first-generation (medium-scale integration) digital circuits for logic and switching. Many of these designs were still operational in the late 1980s and were expected to remain so for perhaps another decade. Microwave integrated circuits (MICs), analog ICs, and digital large-scale integration (LSI), often packaged in multichip hybrid packages, followed in the 1970s designs, while monolithic microwave integrated circuitry (MMIC) and very-high-speed integrated circuits (VHSIC) are the components for systems conceived in the 1980s.

Because of the fast pace of technology growth, the following subsections will concentrate only on the key subsystems and give a brief overview of the critical requirements and available design options. Space limitations do not allow more. Subsystems composed of IF and low-frequency analog circuitry and digital processors utilize standard components and techniques and will not be further described.

Radome.[46,47] The radome provides the necessary protection for the seeker antenna and must satisfy conflicting electromagnetic and aerodynamic requirements. From the standpoint of minimizing the radome slope (Sec. 19.2) the most desirable shape would be a hemisphere. This, however, is an unacceptable shape for a high-speed missile, and an aerodynamically shaped radome (usually some form of tangent ogive or similar configuration) with a fineness ratio (length/diameter) between 2:1 and 4:1 is chosen.[7,16]

The radome must withstand extreme aerodynamic heating as well as high mechanical stresses during flight. It must be able to fly through rain without erosion by impact with raindrops. At the same time it must maintain its electromagnetic properties, providing good transmission (low loss tangent) and, most important, a low error slope.

The radome design problem, once the shape has been dictated, becomes one of finding a material which can simultaneously meet both the mechanical and the electrical requirements. Based on electromagnetic considerations, an equivalent thickness of $\lambda/2$ is preferred for relatively narrowband applications. Mechanical strength limitations generally make thin-wall ($<\lambda/20$) radomes impractical.[16] Depending on operating frequency, thicker-wall radomes (in increments of $\lambda/2$) may be required to provide sufficient mechanical strength. The electrical properties of most concern to radome error slope are dielectric constant and its variation with temperature.

Materials which have been used include high-temperature glass or quartz fiber-reinforced polymers (i.e., fiberglass), refractory oxides such as alumina or fused silica, and ceramics such as Pyroceram. As missile speeds continue to increase, new materials such as silicon nitride are being introduced. Wideband radomes, for applications such as ARH, have generally utilized multilayer (sandwich) configurations, since no single material could achieve the performance required.

In order to achieve the required low error slopes (i.e., degrees of error per degree of gimbal angle change)—values of $\pm$ 0.05 to $\pm$ 0.01 deg/deg are typically specified[9,11]—precision grinding of radomes based on error slope measurements has been required. The initial manufacturing tolerances do not permit achievement of the required slope. With the advent of large-scale digital computing capability in missiles, however, the concept of measuring the radome error, storing the data in computer memory, and then electronically compensating the target angle data measured by the seeker becomes practical.

Antenna. The antenna is perhaps the most critical subsystem, since its design determines what the seeker will observe and thus governs the quality of the boresight error measurements. The angle processing (i.e., conical scanning or monopulse network) is generally considered an integral part of the antenna subsystem.

The key requirements are the highest possible gain, narrowest beamwidth, low sidelobes, the highest error sensitivity (i.e., monopulse slope), and, for ARH applications, wide bandwidth.[5,6] In active applications, the antenna must also be able to handle the transmitted power and be well matched to minimize any reflection of transmitter power back to the receiver.

Although most antennas are gimballed—the exceptions being interferometers and some body-fixed ARH types—much has been said about the possible application of electronically steered phased arrays, either conformal or mounted inside the radome. In particular, active arrays have been suggested as the answer for high-performance active seekers. For existing applications the gimballed antenna is quite capable of meeting requirements with much less hardware complexity and much lower cost. Until a reliable low-production-cost element [phase shifter or transmit-receive (TR) module] becomes available, the field will continue to be dominated by gimballed types.

The additional requirements imposed on gimballed antennas stem from the need to maximize available aperture and simplify the stabilization servo-loop requirements. To achieve this the antenna should be lightweight, have a low moment of inertia, yet be rugged.[16] To maximize aperture, the antenna should be thin so that its face is close to the center of rotation. A conventional front-fed parabola would have to be significantly smaller (because of its depth) to allow clearance at large gimbal angles than would a flat plate (see Fig. 19.15).

The first-generation conical-scan antennas were parabolas, often of the Cassegrain type. Monopulse seekers generally employ slotted waveguide or stripline flat plates which provide phase monopulse angle sensing by combining the outputs of the four quadrants. The small size, limited by missile diameters of 5 to 20 inches, results in a very small number of slots. The small number of slots does not allow much freedom for tapering amplitude distribution, thus making it difficult to achieve low sidelobes. Also, the importance of keeping the main beam as narrow as possible, to reduce the amount of clutter and jamming and to aid in multiple-target resolution, puts a limit on the amount of amplitude taper (hence beam broadening) that can be applied.[6]

Another important design decision is how much of the receiver should be on gimbal versus off gimbal. To minimize noise and maintain reasonable gain and phase tracking of monopulse channels, it is not practical to bring the low-level microwave signals off gimbal through flexible cables. Rotary joints (three or two channels, depending on monopulse-processing configuration) are required, or the conversion to IF can be accomplished on gimbal, and the IF signals brought off gimbal through cables. At the higher level and lower frequency of the IF the noise effects are significantly less than at microwave. However, the amount of added

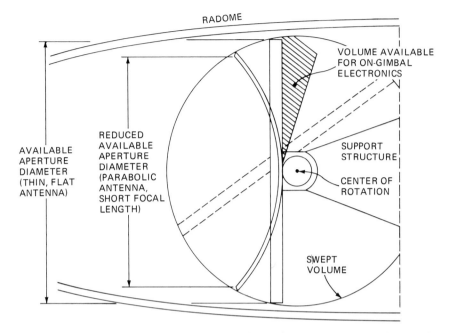

FIG. 19.15 Physical layout of gimballed antennas. The maximum antenna aperture is realized from a physically thin antenna mounted close to the center of rotation.

mass on gimbal must be held to a minimum, and cable torques or rotary-joint friction must be minimized to keep gimbal torque requirements reasonable and minimize nonlinear effects which can cause limit-cycle oscillation problems. Hydraulic servos have more than enough torque capability, but for electric drives available torque is limited and these effects must be considered. For active seekers, of course, loss and (for high powers) power-handling capability will necessitate the use of a rotary joint for routing the transmitter signal to the antenna.[16]

For many of the above reasons, a very attractive solution is the gimballed mirror (also called the inverse-Cassegrain) antenna.[48,49] As shown in Fig. 19.16, the antenna consists of a fixed feed and transreflector and a gimballed flat-plate twist reflector. Linearly polarized energy radiated from the feed is reflected by the parabolic transreflector, which is composed of a wire grid oriented to reflect the transmitted polarization. The reflected plane wave impinges on the twist reflector, which is essentially a quarter-wave plate. It reflects the incident energy while rotating its plane of polarization by 90°; so the transreflector grid is now transparent to the reflected polarization. This arrangement eliminates the need for rotary joints, allows all processing and receiver elements to be off gimbal, and has the added advantage that to achieve a given beam look angle θ the gimballed twist reflector has to move only θ/2. Since only the light twist reflector has to be moved, gimbal torques are minimized, and the 2:1 angle relationship makes for a simpler gimbal design. However, this also prevents LOS stabilization by means of gimbal-mounted gyros and requires remote stabilization by use of an inertial reference, as for any body-fixed antenna.

Interferometer antennas have been used in applications where the nose of the

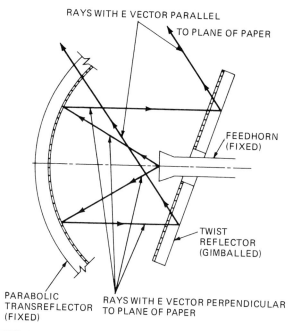

RAYS WITH E VECTOR PARALLEL
TO PLANE OF PAPER

FEEDHORN
(FIXED)

TWIST
REFLECTOR
(GIMBALLED)

PARABOLIC
TRANSREFLECTOR
(FIXED)

RAYS WITH E VECTOR PERPENDICULAR
TO PLANE OF PAPER

FIG. 19.16 The mirror scan antenna provides a mechanically steered beam with a fixed feed, requiring no rotary joints or flexible cables for the RF signals.

missile was unavailable, for example, in a ramjet-propelled missile with the air intake in the nose.[5,50] A typical configuration would employ two pairs of polyrod antennas on the missile circumference. Conceptually, the interferometer is a phase-monopulse antenna. However, the small element size yields a wide beamwidth and low gain. Also, the ambiguity among the several interferometer lobes must generally be resolved. Interferometers have been used in both semiactive and passive (ARH) systems. As discussed in Sec. 19.3, other wideband antennas have been used in body-fixed configurations. For example, conical spirals, because of their length, do not lend themselves to gimballing inside a radome over the required field of view.

Receiver. The front receiver converts the low-level microwave output of the antenna into a high-level signal at a frequency suitable for processing. This may be a baseband doppler signal which drives the speedgate in the early-generation systems or a serial bit stream driving a digital processor. In both cases, the signal must be amplified to a (relatively) constant amplitude, filtered, and frequency-translated (usually several times), while at the same time preserving the angle information on the target signal. Avoiding saturation and interference is thus a prime requirement. For monopulse systems, gain and phase tracking between channels is also required.

The rear receiver is a simpler version since it operates on a much stronger signal over a much smaller dynamic range. However, many of the filtering, receiver-protection, and frequency-conversion blocks are identical to those required in the front receiver.

The microwave portion of the receiver can range in complexity from merely a simple mixer to a fairly elaborate double-conversion system with RF pre-amplification and preselection as illustrated in Fig. 19.17 for a single channel.[6] The earliest systems used a balanced mixer to convert the microwave energy to IF, with no other filtering or amplification. The mixer was generally on the back of the antenna with a cable connection to the IF preamplifier located off gimbal. Use of a balanced mixer provides cancellation of AM noise on the LO signal. The driving requirement was minimum system noise figure.

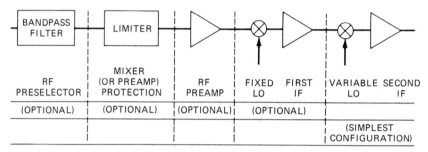

FIG. 19.17 This generalized receiver block diagram shows the range of possible implementations, from the simplest single-conversion mixer to the most elaborate double-conversion system with RF preamplification and receiver protection.

At the other end of the spectrum lies the double-conversion microwave receiver, with a high first IF in the UHF to 2-GHz range, followed by conversion to a conventional IF of 60 MHz or less. Most configurations lie somewhere between these extremes. Typically, some form of limiter precedes the first mixer to protect the mixer against burnout by high-power signals from friendly or hostile radars. RF preselection can consist of a fixed filter covering the operating bandwidth or a tunable preselector such as a yttrium iron garnet (YIG) filter.[51,52]

Some systems include an RF preamplifier. Low-noise field-effect transistor (FET) technology offers very low system noise figure (as low as 2 to 3 dB) with, however, reduced dynamic-range capability compared with a diode mixer. Although these low noise figures appear to provide significantly improved sensitivity (i.e., acquisition range) if the normally expected environment is dominated by ECM (as is likely in tactical situations), then it is jamming rather than thermal noise which sets the detection threshold, and the FET preamplifier, in reality, does not increase performance.[6] To maintain linearity in large-signal environments, the preamplifier may be bypassed with a switch, or additional attenuation may also be switched in (stepped AGC).

In a double-conversion receiver, a fixed LO is generally used at the first conversion, with the second conversion being used to close the doppler tracking loop in an inverse-receiver configuration. If only a single conversion is used, the doppler tracking loop in an inverse receiver will be closed through this mixer. With the conventional receiver, the LO is the tunable oscillator controlled by the rear AFC loop.

Other variations may use image-rejection mixers instead of bandpass filters to reject the image and avoid the noise foldover.[52]

For active seekers, a duplexer is required to provide receiver-transmitter isolation.

As the level of integration increases, the receiver can be thought of as a component rather than a subsystem, with most or all of it capable of being fabricated on a single chip.[53,54]

Low-Noise Frequency Reference. Perhaps the most difficult design challenge is the low-noise frequency reference. In a semiactive seeker, this subsystem acts as the LO, while in an active configuration it provides the transmitter RF drive. In either case, there is a primary oscillator which acts as the frequency reference for the seeker, with other required signals (such as LOs for other conversions in the system) derived from the primary oscillator by mixing or phase-lock techniques.

As described earlier, the conventional semiactive seeker's LO has evolved from a wide tuning range to essentially a fixed tuned (or slightly variable frequency) microwave source. It must have low near-carrier noise and also, based on the particular application, exhibit either stringent long-term frequency stability or have a continuous, linear, and repeatable tuning characteristic. The basic oscillator often incorporates a noise reduction technique to meet the low-noise requirement, which typically can be as low as 140 dBc/Hz (i.e., 140 dB below the carrier in a 1-Hz bandwidth) at a separation of 10 to 20 kHz from the carrier.[55]

The microwave LO in early-generation systems was usually a reflex klystron with mechanical coarse tuning (to cover the operating bandwidth) and electronic fine tuning. It required the rear loop to be used for noise degeneration.[6] Solid-state microwave sources utilized a lower-frequency VCO and frequency multipliers to produce the required frequency. In a typical configuration, a varactor-tuned transistor oscillator in the 1-GHz range provided the tunability, and a varactor or step-recovery diode was used as the multiplier.[32,56]

As the trend to more precise frequency control of transmitters allowed the missile LO to be crystal-controlled, basic crystal oscillators or voltage-controlled crystal oscillators (VCXOs) in the 5- to 100-MHz range were directly multiplied up to microwave frequencies or used to phase-lock a VCO. However, since the multiplication factors are high ($\times$ 200 for a 50-MHz to X-band multiplication), the noise on the fundamental oscillator must be correspondingly low to achieve the final requirement.[32,56]

Subsequent designs make use of surface acoustic-wave (SAW) devices (delay lines or resonators) as the feedback elements in oscillators.[55,57,58] SAW oscillators in the UHF to L-band frequency region require much lower multiplication ratios ($\times$ 10 to $\times$ 30 for an X-band LO) to get to the desired frequency; hence, they offer lower FM noise than crystal oscillators. Not only is the effect of multiplied noise correspondingly reduced, the hardware is significantly smaller and simpler. Use of multiple switched delay lines provides frequency agility. Use of SAW resonators as the feedback elements produces up to 20 dB lower FM noise than the delay-line implementation.[56] Another promising approach is the dielectric resonator oscillator (DRO).

For any of the above configurations, particular attention must be paid to vibration isolation. The normal missile vibration environment can transform a quiet oscillator into a noise generator if the component mounting, material selection, etc., are not stressed in the design.[55–57]

Signal Processing.[2,5,6] *Signal processing* refers to those hardware elements that perform target detection, range, doppler, and angle error extraction, and closure of the corresponding tracking loops. The division between the receiver and the signal processor is sometimes rather vague. In first-generation systems the speedgate was easily identifiable, but in the inverse receiver, the narrow banding (doppler signal processing) precedes most of the receiver gain; so the distinction is largely lost.

In modern digital implementations the key tradeoff is where to draw the line between analog and digital processing. Most required functions can be accomplished digitally; the decision to be made is whether the digital approach yields simpler hardware while maintaining the same performance.[59]

For example, a doppler filter bank implementation might consist of a number of narrow analog filters, a wide multipole bandpass analog *roughing filter* followed by a digital FFT, or a completely digital configuration.

The analog roughing filter may, however, significantly reduce the computational throughput requirements of the digital processor and thus be the selected approach.

Analog range gating, if used, must precede the narrow doppler filtering, since it is a wideband process (the narrow pulse shape must be preserved).[60] However, wide-bandwidth analog-to-digital conversion can provide for an all-digital range-doppler processor.

As the state of the art in analog-digital converters (ADCs) advances to higher speed and more bits of dynamic range, the digital boundary will move further forward in the seeker block diagram toward the receiver front end. The ideal limit would be an ADC at each antenna output port, but this goal is a long way from practical realization.

Transmitter. The transmitter is a critical element in active seeker design because in most applications it tends to dominate the weight/volume budget. Microminiaturization of other subsystems has reduced them to such a small fraction of their original size that the transmitter can easily represent 50 percent or more of the seeker weight/volume.

The key parameters are the average power, efficiency (since the transmitter size must consider the prime power source and thermal dissipation as well), any waveform limitations, weight/volume, power supply requirements, frequency, bandwidth, noise, and stability.

The major design decision is tube versus solid-state. Within each category numerous options are available. Depending on the operating frequency and the range requirement, coupled with missile size (i.e., available antenna aperture), the transmitter configuration is selected primarily on the basis of available power and power/weight, power/volume considerations. A significant factor in this tradeoff is the power supply voltage—in the 100-V range for solid-state and typically 10 kV or more for tubes.

A key difference for missile-borne transmitters is the operating time and the corresponding thermal design consideration. Unlike conventional radars which operate continuously and are designed for active cooling, active seekers operate for short periods (tens of seconds at most) and thus are amenable to passive cooling. What is required is sufficient thermal mass to dissipate the heat generated without exceeding thermal limits. This, of course, puts a premium on high efficiency, especially for high-power transmitters. For extended testing, active cooling provisions may be required.

For many surface-target applications, low-power (less than one to a few watts) solid-state transmitters are used. For air intercept of low-cross-section targets, where several hundred watts of average power are required, the choice has generally been a tube transmitter, especially at the higher frequencies. In applications requiring from tens to 100 to 200 W, tubes and solid-state are both viable candidates.[16]

Tube Transmitters.[61-63] Magnetrons, klystrons, and traveling-wave tubes (TWTs) have all been used as microwave power sources in active seeker applications. Magnetron oscillators were adequate for early-generation noncoherent pulse transmitters used against large-cross-section targets. The concept of injection-locking magnetrons has been advocated to provide coherent transmitter waveforms.[64-66] This approach sometimes includes combining the output of several magnetrons to achieve the required power.[64] A major shortcoming of the magnetron is its narrow bandwidth.

Klystrons can provide medium to high power in each of the commonly used operating bands, but again with low instantaneous bandwidth. A state-of-the-art published result from Varian describes a K_a-band 500 W average, 2.6 kW peak power klystron, which requires a 14 kV power supply.[63]

Conventional helix-type TWTs can provide several hundred watts at X band, with wide bandwidth. At higher frequencies, however, the reduced dimensions severely limit performance of helix tubes because of the difficulty of heat dissipation. Tens of watts is the maximum achieved average power at K_a band. The coupled-cavity TWT can provide high power at K_a and millimeter wavelengths,[67] but power supply voltages on the order of 30 kV are required, the tube is large (typical length, 16 in), and cost is high.[63]

The crossed-field amplifier (CFA) offers high power at high efficiency with much lower supply voltages and smaller size than a TWT. Bandwidth is moderate, but the lack of a grid makes modulation more difficult and limits how high a PRF can reasonably be achieved.[62]

Although tubes have been employed since the early-generation systems, numerous advances have occurred, spurred by the evolution of systems to higher (K_a and millimeter) frequencies. For missile applications, the high voltage requirements pose particular problems in packaging the power supply in the limited volume, especially for high-altitude operation. However, if the power requirements can only be met with a tube transmitter, these problems have to be faced and solutions devised regardless of the difficulty.

Solid-State Transmitters. A variety of solid-state devices have been applied to active seeker transmitters, both individually and in multidevice power combiners. The three main types are impact avalanche transit time (IMPATT) diodes, Gunn diodes, and FETs.[62,68]

For low-power systems, typified by air-to-surface short-range missiles, a single Gunn or IMPATT can suffice as the transmitter. In the medium- to high-power applications, the power of a number of devices must be combined to meet the requirement. IMPATTs provide higher power than Gunn diodes, with GaAs IMPATTs giving higher power and efficiency below about 40 GHz and silicon yielding better performance at higher frequencies. Gunn diodes have lower noise, making them well suited to injection-lock IMPATTs or as LOs or primary low-power CW sources.[69]

The highest solid-state power has been achieved by combining a number of IMPATT diodes in a resonant cavity (Kurokawa combiner). At X band 32 diodes have been combined to produce 313 W average power,[70] and approaches for combining as many as 64 diodes have been described.[71,72] The cavity results in a fairly narrow but adequate bandwidth (1 to 2 percent). The IMPATTs operate in an injection-locked mode at a modest gain (5 to 8 dB, depending on the number of diodes combined), thus requiring several stages to achieve the final power output. Commercially available GaAs diodes range from 10 W CW at X band to 3 W CW at K_a band; so combiner power would be correspondingly lower.[73]

The maximum achievable power depends on the number of diodes which can practically be combined. Physical size limits the number of diodes per cavity, but the power of several combiners can in theory be combined if the level of complexity is judged acceptable. IMPATT modulators must provide relatively high currents at low voltages (1.7 A at 100 V is typical for GaAs at X band).[73] IMPATT efficiencies tend to be lower than those of tube transmitters; so the thermal design is critical. GaAs FETs have generally exhibited lower power and higher efficiency than IMPATTs. Combining techniques for FETs have not matched the results of IMPATT combiners. As linear amplifiers, FETs have been used in the early stages of IMPATT transmitters. Discussion of active array approaches has centered on power FETs as the key components in the TR modules comprising such an antenna-transmitter.

Improvements in devices, combiners, and modulators continue to advance the capabilities of solid-state transmitters. Solid-state provides a lower-weight, more compact transmitter than a tube, but at considerable complexity. Reliability concerns tend to be somewhat offset by the concept of *graceful degradation*, since failure of one or a few devices will not result in total failure of the transmitter. For medium-power applications, solid-state will continue to be an attractive solution.

Power.   Prime power is generally provided by batteries, although electromechanical generators have been used as well. The design decision is whether to use multiple voltage taps to provide the various voltages or a single voltage battery and dc-to-dc converters. With the latter approach, a simpler battery results, but the converter chopping frequency is a potential source of noise in the system. The trend has been from silver zinc, used widely in earlier-generation systems, to thermal types in subsequent designs.

It has been found that isolating the higher-current-demand power supplies (such as those for the transmitter or digital computer) from the other lower-level subsystems significantly improves noise and interference problems. Additional voltage regulators and filtering are often necessary to reduce crosstalk between subsystems.

In addition to electrical power, some systems include pressurized hydraulic or pneumatic supplies for the antenna drives and/or control surface drives.

Integration.   A key to successful integration of the varied subsystems into a working missile seeker is functional partitioning. Efficient partitioning enables each subsystem to be specified, built, and tested as a functional entity and minimizes interconnections and system-level adjustments.

Perhaps the most important aspect is to preserve subsystem stand-alone testability, not requiring ''hand tailoring'' to match to an interfacing subsystem. As a rule, system-level adjustments should be avoided and subsystem interchangeability with zero adjustments should be the ultimate goal.

Varying test and maintenance philosophies have been applied to missiles. Extensive use of the built-in test is one approach, which results in periodic, sometimes frequent testing. Failure isolation and replacement of failed units are a consequence of this approach. However, subassembly repairs in the field have been found to introduce added problems, and such repairs are best made only at centralized depots or repair facilities.

A very successful approach has been the *wooden-round* concept,[28] in which a missile is normally not tested but treated like a round of ammunition. Excellent reliability results have been achieved with this no-field-test approach. Sample testing of missiles has been used to maintain confidence in the quality of deployed production lots. Reliability cannot be tested into hardware. If reliability is built in at the factory, there is no need for continual testing, and when the fire button is pushed, the missile should work properly with a high level of confidence.

REFERENCES

1. Phillips, T. L.: Anti-Aircraft Missile Guidance, *Electron. Prog.*, vol. II, pp. 1–5, March–April 1958.

2. Ivanov, A.: Improved Radar Designs Outwit Complex Threats, *Microwaves*, vol. 15, pp. 54–71, April 1976.

3. Gulick, J. F.: Overview of Missile Guidance, *IEEE EASCON Rec.*, pp. 194–198, April 1978.

4. Maurer, H. A.: Trends in Radar Missile Guidance, *Int. Def. Rev.*, vol. 15, special issue 14, "Air Defense Systems," pp. 95–102, 1982.

5. James, D. A.: "Radar Homing Guidance for Tactical Missiles," Halsted Press, a division of John Wiley & Sons, New York, 1986.

6. Ivanov, A.: Semi-Active Radar Guidance, *Microwave J.*, vol. 26, pp. 105–120, September 1983.

7. Fossier, M. W.: The Development of Radar Homing Missiles, *J. Guidance, Control, Dynamics*, vol. 7, pp. 641–651, November–December 1984.

8. Fossier, M. W., and B. Hall: Fundamentals of Homing Guidance, Raytheon Company unpublished seminar presentation, 1962.

9. Garnell, P.: "Guided Weapon Control Systems," 2d ed., Pergamon Press, Elmsford, N.Y., 1980.

10. Nesline, F. W.: Missile Guidance for Low-Altitude Air Defense, *Electron. Prog.*, vol. XXII, no. 3, pp. 18–26, Fall 1980.

11. Nesline, F. W., and P. Zarchan: Missile Guidance Design Tradeoffs for High-Altitude Air Defense, *J. Guidance, Control, Dynamics*, vol. 6, pp. 207–212, May–June 1983.

12. Nesline, F. W., and P. Zarchan: Line-of-Sight Reconstruction for Faster Homing Guidance, *J. Guidance, Control, Dynamics*, vol. 8, pp. 3–8, January–February 1985.

13. Long, J., and A. Ivanov: Radar Guidance of Missiles, *Electron. Prog.*, vol. XVI, no. 3, pp. 20–28, Fall 1974.

14. Jaffe, R. M.: Monopulse Radar Receiver, U.S. Patent 3,713,155, Jan. 23, 1973.

15. Skolnik, M. I.: "Introduction to Radar Systems," 2d ed., McGraw-Hill Book Company, New York, 1980.

16. Parker, D., and H. A. Maurer: The Era of Active RF Missiles, *Microwave J.*, vol. 27, pp. 24–36, February 1984.

17. Seashore, C. R.: MM-Wave Sensors for Missile Guidance, *Microwave J.*, vol. 26, pp. 133–144, September 1983.

18. Cherwek, R.: Coherent Active Seeker Guidance Concepts for Tactical Missiles, *IEEE EASCON Rec.*, pp. 199–202, Sept. 25–27, 1978.

19. Williams, F. D., and M. E. Radant: Airborne Radar and the Three PRFS, *Microwave J.*, vol. 26, pp. 129–135, July 1983.

20. McLendon, R., and C. Turner: Broadband Sensors for Lethal Defense Suppression, *Microwave J.*, vol. 26, pp. 85–102, September 1983.

21. Grant, P. M., and J. H. Collins: Introduction to Electronic Warfare, *Proc. IEE*, vol. 129, pt. F, pp. 113–132, June 1982.

22. Bullock, L. G., G. R. Oeh, and S. J. Sparagna: An Analysis of Wide-Band Microwave

Monopulse Direction-Finding Techniques. *IEEE Trans.*, vol. AES-7, pp. 188–201, January 1971.

23. Mosko, J.: An Introduction to Wideband, Two-Channel Direction-Finding Systems, *Microwave J.*, pt. I, vol. 27, pp. 91–106, February 1984; pt. II, vol. 27, pp. 105–122, March 1984.

24. Rappoli, F., and N. Stone: Receivers for Signal Acquisition, *Microwave J.*, vol. 20, pp. 29–33, January 1977.

25. Moule, G. L.: SAW Compressive Receivers for Radar Intercept, *Proc. IEE*, vol. 129, pt. F, pp. 180–196, June 1982.

26. Daniels, W. D., M. Churchman, R. Kyle, and W. Skudera: Compressive Receiver Technology, *Microwave J.*, vol. 29, pp. 175–185, April 1986.

27. Van Brunt, L. B.: "Applied ECM," vol. II, EW Engineering, Inc., Dunn Loring, Va., 1982.

28. Fossier, M. W.: Tactical Missile Guidance at Raytheon, *Electron. Prog.*, vol. XXII, no. 3, pp. 2–10, Fall 1980.

29. Carey, D., and W. Evans: The Patriot Radar in Tactical Air Defense, *Proc. IEEE EASCON*, pp. 64–70, Nov. 16–19, 1981.

30. The Patriot Surface-to-Air Missile System, *Mil. Technol.*, vol. 8, pp. 33–50, October 1984.

31. Graves, H. L.: The Patriot Air Defense System, *Electron. Prog.*, vol. XXVIII, pp. 4–15, Nov. 2, 1987.

32. Jerinic, G., N. Gregory, and W. Murphy: Low Noise Microwave Oscillator Design, *29th Ann. Frequency Control Symp.*, Fort Monmouth, N.J., May 1975.

33. Barton, D. K.: "Radar Systems Analysis," Prentice-Hall, Englewood Cliffs, N.J., 1964.

34. DiFranco, J. V., and W. L. Rubin: "Radar Detection," Prentice-Hall, Englewood Cliffs, N.J., 1968.

35. Nathanson, F. E.: "Radar Design Principles," McGraw-Hill Book Company, New York, 1969.

36. Meyer, D. P., and H. A. Mayer: "Radar Target Detection," Academic Press, New York, 1973.

37. Sherman, S. M.: "Monopulse Principles and Techniques," Artech House, Dedham, Mass., 1984.

38. Barton, D. K.: "Radars," vol. 4: "Radar Resolution and Multipath Effects," Artech House, Dedham, Mass., 1975.

39. Ducoff, M. R.: Closed Loop Angle Tracking of Unresolved Targets, *IEEE Int. Radar Conf.*, pp. 432–437, Apr. 28–30, 1980.

40. Kanter, I.: Varieties of Average Monopulse Responses to Multiple Targets, *IEEE Trans.*, vol. AES-17, pp. 25–28, January 1981.

41. Melino, C.: Average Monopulse Angle Tracking Response to Two Unresolved Sources, *IEEE Trans.*, vol. AES-23, pp. 634–643, September 1987.

42. McWhorter, L. V.: Response of Various Monopulse Seekers to a Multi-Source Environment, *IEEE Southeastcon Rec.*, pp. 698–709, Apr. 5–8, 1981.

43. Kliger, I., and C. Olenberger: Multiple Target Effects on Monopulse Signal Processing, *IEEE Trans.*, vol. AES-11, pp. 795–804, September 1975.

44. Kliger, I., and C. Olenberger: Phase-Lock Loop Jump Phenomenon in the Presence of Two Signals, *IEEE Trans.*, vol. AES-12, pp. 55–64, January 1976.

45. Johnston, S.: MM-Wave Radar: The New ECM/ECCM Frontier, *Microwave J.*, vol. 27, pp. 265–271, May 1984.

46. Rudge, A. W., G. A. E. Crone, and J. E. Summers: Radome Design and Performance: A Review, *Mil. Microwaves Conf.*, pp. 555–575, Oct. 22, 1980.

47. Fudge, D. L., and T. S. Moore: Ceramic Radome Development for Guided Weapons, *Mil. Microwaves Conf.*, pp. 584–591, Oct. 22, 1980.

48. Lewis, B. L., and J. P. Shelton: Mirror Scan Antenna Technology, *IEEE Int. Radar Conf.*, pp. 279–283, Apr. 28–30, 1980.

49. Cross, D. C., D. D. Howard, and J. W. Titus: Mirror-Antenna Radar Concept, *Microwave J.*, vol. 29, pp. 323–336, May 1986.

50. Feagler, E. R.: The Interferometer as a Sensor for Missile Guidance, *IEEE EASCON Rec.*, pp. 203–210, Sept. 25–27, 1978.

51. Daly, E., R. Sparks, and G. Spencer: Ganged Radio Frequency Filter, U.S. Patent 4,044,318, August 1977.

52. Briana, A. M.: Microwaves in Missiles, *Microwave J.*, vol. 22, pp. 41–42, March 1979.

53. Cardiasmenos, A. G.: Future Trends in Millimeter-Wave Receiver Design, *Mil. Electron./Countermeasures*, vol. 7, pp. 49–56, June 1981.

54. Seashore, C. R., and D. R. Singh: Millimeter-Wave ICs for Precision Guided Weapons, *Microwave J.*, vol. 26, pp. 51–67, June 1983.

55. Pedi, P., J. Loan, and E. McManus: The Role of SAW Oscillators in Military Radar Systems, *IEEE MTT-S 1983, Int. Microwave Symp.*, pp. 311–313, June 1983.

56. Cowan, D. A.: Voltage Controlled Oscillators for Missile Applications, *Mil. Microwaves Conf.*, London, Oct. 25–27, 1978.

57. Galani, Z., M. Bianchini, and R. DiBase: A Low Noise Frequency Agile X-Band Source, *IEEE MTT-S Dig.*, pp. 233–235, 1982.

58. Bianchini, M., J. B. Cole, R. DiBase, Z. Galani, R. W. Laton, and R. C. Waterman, Jr.: A Single-Resonator GaAs FET Oscillator with Noise Degeneration, *IEEE MTT-S Dig.*, pp. 270–273, 1984.

59. Hall, B. A., F. J. Langley, and K. O. Wefald: Computer Design Requirements for Digital Air-to-Air Missiles, *AIAA Guidance Control Conf.*, San Diego, August 1976.

60. Parode, L. C., and D. R. Holcomb: Range Gating System with Narrow Band Filtering, U.S. Patent 3,141,163, July 14, 1964.

61. Hicsimair, H., C. DeSantis, and N. Wilson: State-of-the-Art of Solid-State and Tube Transmitters, *Microwave J.*, vol. 26, pp. 46–48, October 1983.

62. Pallakoff, O. E.: Broadband, High-Power Devices, *Microwave J.*, vol. 28, pp. 69–90, February 1985.

63. Roach, B., G. Moen, and A. Acker: K_a-Band Radar—Which Tube to Choose, *MSN & CT*, vol. 17, pp. 92–96, July 1987.

64. Vyse, B., V. H. Smith, and M. O. White: The Use of Magnetrons in Coherent Transmitters, *Mil. Microwaves Conf.*, pp. 217–223, Oct. 20, 1982.

65. Vyse, B., V. H. Smith, and M. O. White: Magnetrons Offer Advantages in Coherent Transmitters, *MSN*, vol. 13, pp. 116–127, October 1983.

66. Best, W. S.: Coherent Transmitter Considerations Utilizing Injection Locked Magnetrons, *IEEE MTT-S Dig.*, pp. 356–358, 1984.

67. King, R. C. M.: The Design of Grid Pulsed, PPM Focused, Coupled-Cavity TWTs for Reliable Operation in Military Transmitters, *Microwave J.*, vol. 29, pp. 167–172, March 1986.

68. Ying, R. S.: Recent Advances in Millimeter Wave Solid-State Transmitters, *Microwave J.*, vol. 26, pp. 69–76, June 1983.

69. Kuno, H. J., and T. T. Fong: Solid-State MM-Wave Sources and Combiners, *Microwave J.*, vol. 22, pp. 47–85, June 1979.

70. Drubin, C. A., A. L. Hieber, G. Jerinic, and A. S. Marinilli: A 1 kW Peak, 300 W Avg Impatt Diode Injection Locked Oscillator, *IEEE MTT-S Dig.*, pp. 126–128, 1982.

71. Laton, R., S. Simoes, and L. Wagner: A Dual Diode TM_{020} Cavity for Impatt Diode Power Combining, *IEEE MTT-S Dig.*, pp. 129–131, 1982.

72. Wagner, L., R. Laton, and R. Wallace: The Effect of Dissimilar Impatt Diodes on Power Combining Efficiency, *IEEE MTT-S Dig.*, pp. 489–491, 1983.

73. Data Sheet—Gallium Arsenide Impatt Diodes, Raytheon Company, Research Division, 1987.

CHAPTER 20
HEIGHT FINDING AND 3D RADAR

David J. Murrow

General Electric Company

20.1 HEIGHT FINDING RADARS AND TECHNIQUES

Early Radar Techniques for Height Finding. Early radar techniques employed to find target height were classified according to whether or not the earth's surface was used in the measurement. The practice of using the earth's surface for height finding was quite common in early radar because antenna and transmitter technologies were limited to lower radio frequencies and broad elevation beams. The first United States operational shipborne radar, later designated CXAM and developed in 1939 by the U.S. Naval Research Laboratory (NRL), used the range of first detection of a target to estimate its height, based on a knowledge of the shape of the pattern near the horizon due to the primary multipath null. Later a refinement was made as the target traversed the higher-elevation multipath nulls or "fades." This technique, illustrated in Fig. 20.1a, was extensively employed on early shipborne radars, where advantage could be taken of the highly reflective nature of the sea surface. Of course, the technique was limited in performance by such uncontrollable factors as sea state, atmospheric refraction, target radar cross section, and target maneuvers.[1,2]

Reflections from the earth's surface were also used by other early contemporary ground-based radars, such as the British Chain Home (CH) series, which was employed in World War II for the defense of Britain. This radar was a pulsed high-frequency (HF) radar which made height measurements by comparing amplitudes of the (multipath-lobed) main beams of a pair of vertically mounted receiving antennas. Conceptualized in Fig. 20.1b, the technique was also utilized in early United States radars, notably, the Canadian-built United States radar SCR-588, and the United States–built SCR-527, both based on the British Type 7 radar design.[3]

One of the earliest and perhaps most direct form of radar height finding was to mechanically direct and hold a narrow-elevation-beam antenna pointed toward the target. The elevation angle of the target corresponds to the elevation readout on the antenna mount. In early radar systems employing this technique, an op-

20.1

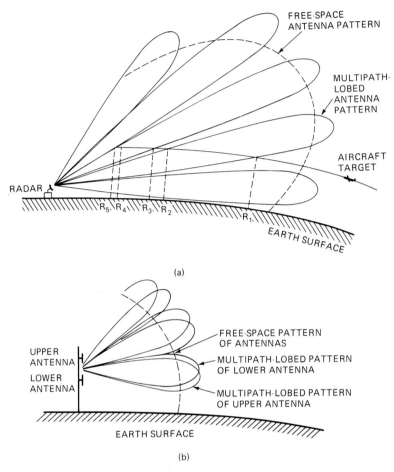

FIG. 20.1 Early radar height finding techniques. (*a*) Method of multipath nulls. (*b*) Amplitude comparison using multipath lobes.

erator would keep the antenna boresighted on the target with a handwheel while monitoring the target return strength. It was quickly learned that maximizing the signal strength of a target echo in a beam was not sensitive enough to provide the desired accuracies, and so alternative techniques were ultimately developed for this purpose. One of the first of these, called *lobe switching*, was first demonstrated in 1937 on a prototype of what later became the U.S. Army Signal Corps SCR-268 radar.[4] This radar was designed for directing antiaircraft gunfire and was the first production radar to use lobe-switching techniques to center the antenna on the target. Two separate identical beams, one above and one below the antenna boresight, are formed at the antenna on receive. By switching between the two beams and keeping the observed amplitudes equal, the SCR-268 elevation operator could keep the antenna boresighted on the target accurately.

If a dish antenna, which generates a narrow pencil-type beam in azimuth and elevation, is mechanically boresighted and trained at or in the vicinity of a target, allowing determination of its azimuth and elevation, the technique is called *searchlighting*. The searchlight technique was successfully employed on the British CMH radar and on the widely deployed United States SCR-584[5] as well as on the United States SCR-615 and the U.S. Navy SM radar. All these radars were S-band dish antenna radar systems. Some of these dish antennas employed conical scanning of a single beam to provide the elevation error signal required to accurately center the beam on the target. The accuracy of such a technique is very good but obviously is limited to one target at a time. Conical scanning and lobe switching are special cases of a general technique for developing off-boresight error signals called *sequential lobing*. The fundamental accuracy and limitations of the sequential-lobing technique are presented in Sec. 20.3. The searchlighting technique was the forerunner of modern-day tracking radars discussed in Chap. 18, many of which now employ monopulse techniques to develop off-boresight error signals. Obviously, techniques which require the antenna to be boresighted on the target are limited in simultaneous surveillance and height finding capability. Typically they make a measurement on a single target at a time and usually also require a designation at least in range and azimuth by an accompanying search radar. The concept of searchlighting and lobe switching is illustrated in Fig. 20.1c.

A widely used early radar dedicated to finding the height of a target in augmentation of a 2D surveillance set was the nodding antenna.* In this type of radar a horizontal fan beam, with a narrow elevation beamwidth, is mechanically scanned in elevation by rocking or "nodding" the entire antenna structure (Fig. 20.1d). As the radar beam traverses the target continuously transmitting pulses, the main-lobe target echoes that return are displayed to an operator by means of a range-height-indicator (RHI) type of display. This allowed the operator to precisely and directly estimate the target height of the target by a process termed *beam splitting*, referring to the process of estimating the center of the displayed target video. Although some nodding-antenna height finders had a slow azimuth rotation search mode, most relied on designations of azimuth from an operator. The operator would observe a detection by the 2D surveillance radar and then command a height determination by the height finder. The height finder would then slew to the commanded azimuth and obtain a height and range measurement. This method of operation was relatively slow and limited in multiple-azimuth target-tracking capability compared with 3D radars. These drawbacks seriously limited the continued use of the manual nodding-antenna height finder in military applications.

Several nodding-antenna height finders, notably the British Type 13 and the widely deployed United States AN/TPS-10, appeared in the mid- to late 1940s, when higher-frequency technology began to emerge.[6] The AN/TPS-10 X-band nodding-antenna height finder radar series was subsequently replaced by the AN/FPS-6, an S-band nodding-antenna radar also designed for the U.S. Army.[7] The AN/MPS-14 was a mobile version of the radar, and the AN/FPS-89 was an improved fixed-site version. The elevation beamwidth of the AN/FPS-6 was 0.9°, its azimuth beamwidth was 3.2°, and the entire antenna nodded at a rate of 20 to 30 nods per minute. The radar could scan in azimuth at a rate of 45° per second. It transmitted 2-μs pulses with pulse repetition frequencies (PRFs) from 300 to 400 Hz, and operated with a peak power of 4.5 MW.

---

*Nodding-antenna height finders have also been used for raid counting.

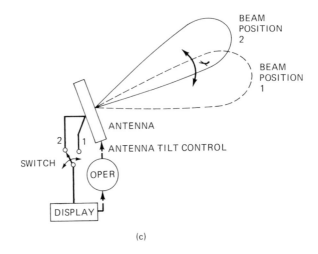

(c)

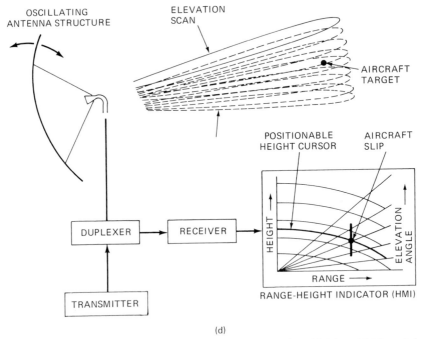

(d)

FIG. 20.1 *(Continued)* Early radar height finding techniques. (*c*) Searchlighting with lobe switching. (*d*) Nodding antenna.

The data rates of later versions of the nodding-antenna radar have been considerably improved over their predecessors. For example, the S600 series C-band nodding-antenna height finder is computer-controlled and -managed for maximum data rate, enabling it to obtain up to 22 height measurements per minute.[8]

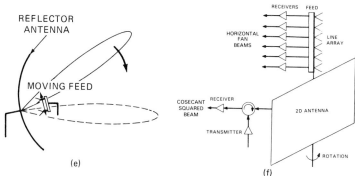

FIG. 20.1 (*Continued*) Early radar height finding techniques. (*e*) Electromechanical beam scanning. (*f*) 2D antenna with vertical line array.

Of course, it is possible to rapidly scan a horizontal fan beam in elevation by electromechanical means (Fig. 20.1*e*) instead of by mechanically rocking the entire antenna structure. Many nodding-beam-type height finders with ingenious means of beam scanning have been successfully deployed over the years. Two notable examples are the World War II SCI radar and the AN/SPS-8 shipborne radar, both of which used a Robinson-type electromechanical feed to rapidly scan the beam in elevation.

The development of higher-frequency microwave technology facilitated electrically larger apertures and correspondingly narrower beams, all in convenient physical sizes. Accompanying this evolution was a series of inventions of rapid electromechanical scanners based on geometric optic principles and developed for surveillance radar applications. These include the Robinson, delta *a* (Eagle), organ-pipe, Foster, Lewis, and Schwartschild scanners, along with a variety of polarization-switching mirror scanners. These scanners all utilized the motion of the feed structure of the antenna to control the incidence of illumination on the aperture, thereby scanning the beam. The reader may refer to a number of excellent sources for a detailed treatment of the method of operation of electromechanical scanners.[6,9,10] In principle, a relatively inexpensive volumetric 3D radar could be created by using an electromechanical feed and scanning a narrow pencil-type beam in elevation while rotating the antenna in azimuth. In practice this approach has not been employed because of the lack of waveform flexibility versus elevation angle imposed by the constant-rate scanning of the electromechanical feed.

During World War II, the British developed a very-high-frequency (VHF) phased array antenna height finder called the Variable Elevation Beam (VEB) radar. This 200-MHz radar utilized mechanically adjustable phase shifters to control the relative phase of nine groups of eight dipoles on a 240-ft mast. The resulting elevation beamwidth was approximately 1° in width and was phase-scanned over an elevation interval slightly more than 6° in extent.

A technique which has seen limited service for radar height finding is a vertical receive-only line array mounted on a conventional 2D surveillance radar as shown in Fig. 20.1*f*. The line array is processed to form a stack of receive horizontal fan beams, each of which is relatively narrow in elevation. Since the (narrow-azimuth) transmit beam is generated by the 2D radar antenna, the resulting stack of two-way beams is narrow in both azimuth and elevation. The stack of receive beams may be formed in a number of ways from the line array. One technique is the Butler matrix, an RF feed analog of the discrete Fourier transform. A second technique was designed for an experimental version of the FAA AHSR-1 S-band air traffic control.[11] The technique augmented the 2D surveillance aperture with a vertical receive-only line array of elements. Each beam of a vertical stack of horizontal fan beams was generated from the line array by combining energy coupled out of waveguide runs at the appropriate length from the element to produce a linear-phase gradient element to element. This produced a set of uniformly illuminated beams, each time-delay-steered to the desired elevation.

One of the early radars combining 2D surveillance with height finding was the AN/CPS-6B, which utilized the V-beam principle. The V-beam radar consisted of a primary and a secondary antenna aperture mounted on the same rotating shaft. The primary aperture operated as a conventional 2D radar, generating a vertical cosecant-squared fan beam which provided detections and the range and azimuth coordinates of targets in the surveillance volume. The secondary antenna aperture was similar to the primary aperture except that it rotated about an axis normal to the aperture. This produced a second fan beam which was tilted from the vertical plane. The two beams might be powered by the same or separate transmitters, but each beam had its own receiver. The tilted beam provided a second set of detections to the radar operator as the antenna rotated. The azimuth separation of the center of the two sets of detections corresponding to a single target, correlated by the operator using range, was directly proportional to the height of the target, to within flat-earth and normal propagation approximations. The concept of the V-beam radar is illustrated in Fig. 20.1g.

The V-beam radar has been referred to as a 3D radar by some authors.[3,7,10] Technically, however, it should not be classified as a true 3D radar because it lacks resolution in the elevation dimension. This shortcoming limits its use to low-density aircraft situations where it is unlikely to encounter two aircraft appearing at the same range and azimuth but at different heights.

The Japanese have developed a radar based on phase interferometry to find height in air traffic control applications,[12] but it also does not have resolution in the elevation dimension. The concept employs a set of four horizontal line arrays vertically displaced in a staggered fashion about a conventional 2D reflector-type antenna. The principle utilized by phase interferometry is that the phase difference between offset antennas is proportional to the sine of the angle of arrival of a received target echo as sketched in Fig. 20.1h.

Radars such as the VEB, the V-beam, the vertical line array plus 2D, the crossed-line array, and the phase interferometer plus 2D, which obtain simultaneous tricoordinate (range, azimuth, and elevation or height) measurements on a target but do not have significant resolution in the elevation dimension compared with their elevation coverage, might appropriately be termed 2½D radars.

Height Finding Techniques in 3D Radars. There are many types of radars that provide 3D information by simultaneously measuring the three basic position coordinates of a target (range, azimuth, and elevation). In this handbook, however, the convention is followed in which a *3D radar* is taken to be a surveillance radar whose antenna mechanically rotates in azimuth (to measure range and azimuth) and which obtains the elevation-angle measurement either by scanning one or more beams in elevation or by using contiguous, fixed-elevation beams.

Military interest in 3D radar stems from its ability to determine the height of a noncooperating target, along with its range and azimuth. Because of its better angu-

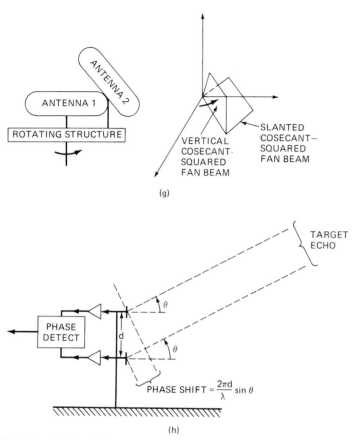

FIG. 20.1 *(Continued)* Early radar height finding techniques. (*g*) V-beam radar. (*h*) Phase interferometry.

lar resolution the 3D radar provides a higher-gain antenna and, arguably, a greater resistance to jamming and other forms of electronic countermeasures (ECM) than a combination of 2D and dedicated height finder. The counterargument points out that the 2D and height finder may be implemented in two separate frequency bands, forcing the jammer to spread out its energy, thereby diluting it.

Rotating 3D radars can be implemented as stacked-beam radars, frequency-scanned radars, phase-scanned radars, electromechanically scanned radars, and digital beamforming radars, according to how the elevation beams are formed and/or scanned in elevation.

Stacked-Beam Radars. Stacked-beam radars employ a vertical *stack* of simultaneously formed receive beams in elevation which are mechanically rotated in azimuth in order to perform search and tricoordinate target position estimation. The target is illuminated by a single transmit beam which is broad enough to cover the receive beam main lobes containing the target. Elevation-angle estimation may be accomplished in such a radar by an amplitude comparison technique, by which the amplitudes of the return at the target range in two or more adjacent simultaneous elevation receive beams are compared. The technique is thus a special case of the general technique of simultaneous lobing. Target height is determined via computer table lookup, using range and elevation entries. The sensitivity of this approach depends on the relative spacings of the beams in the stack, the aperture illumination used to form the beams, and the elevation angle of arrival of the target relative to the beam boresight placements, along with other equipment-related error sources.

It should be noted that the overall height finding performance of a stacked-beam radar is greatly influenced by the extent to which the designer includes anticlutter moving-target indication (MTI) and/or doppler processing in the beam stack, even in clear weather conditions. This is especially important for the lowest beam in the stack, as its main lobe intercepts the earth's surface, admitting surface clutter returns. However, it may also be important for all the beams in the stack, depending on the severity of the clutter. This is so because the elevation patterns of a stacked-beam radar are primarily one-way patterns, being dominated by the receiver elevation pattern. Thus, the elevation sidelobes protecting the radar upper beams from ground or sea clutter are the (one-way) elevation receive sidelobes. This is true because the transmit beam main lobe must be broad enough in elevation to cover all receive beams. This is different from the scanning pencil-beam radar, in which the product of the transmit and receive elevation sidelobes protects the radar upper beams from surface clutter. In benign surface clutter applications, it is economical to implement the stacked-beam radar without MTI or doppler processing in the beam stack, reserving this processing for a single cosecant-squared receive detection beam.

The AN/TPS-43 is an example of a widely deployed operational stacked-beam radar. Deployed in the 1970s, it is a transportable ground-based S-band radar which has been extensively used for air surveillance in the U.S. Air Force Tactical Air Command System (TACS). The radar employs a multiple-horn feed illuminating a reflector-type antenna rotating at 6 r/min to generate a stack of six receive beams in elevation. The original version of the radar utilized a linear-beam Twystron tube to generate approximately 4 MW of RF peak power in a 6.5-μs simple pulse. The radar is instrumented to a range of 240 nmi and operated with six PRFs averaging 250 Hz. The receive beams in the stack are 1.1 in azimuth and variable in elevation beamwidth in such a way that the six span the 20° of total elevation coverage. Subsequent versions of the radar provided pulse compression and improved MTI waveforms and processing.[13] The AN/TPS-75 is an upgraded version of this radar with a planar array low-sidelobe antenna.

Another example of a stacked-beam radar is the S713 Martello (Fig. 20.2*a*), an L-band transportable radar with an eight-beam stack. The Martello S713 radar employs IF processing to form the receive beam stack in elevation. In operation, a cosecant-squared transmit beam is formed and eight narrow beams are formed and processed on receive. A ninth receive beam, cosecant-squared in shape, is used for surveillance and detection. Azimuth and range are determined as in a conventional 2D radar. Height finding is accomplished by interpolating the received signal strengths in adjacent elevation beams of the stack to determine the target elevation angle. The array is 10.6 m high by 6.1 m wide and consists of 60 center-fed rows of 32 radiating elements, each equipped with 60 receivers to downconvert received RF to IF. The azimuth beam is 2.8° wide. The tube transmitter generates 3 MW of RF power at the peak of a 10-μs pulse and an average RF power of 8 kW. The radar is instrumented to 256 nmi and up to 30° in elevation and 100 kft in height. The antenna rotates at 6 r/min. A height accuracy of

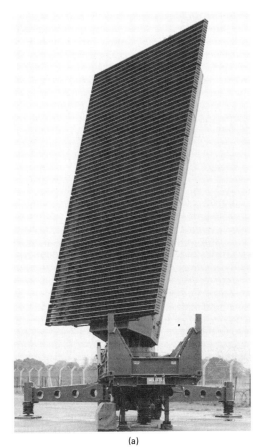

(a)

FIG. 20.2 Exemplar 3D radars. (*a*) S713 Martello stacked-beam 3D radar (*Courtesy Marconi Company*).

1000 ft on a small fighter aircraft at 100 nmi is claimed by the radar manufacturer. A solid-state transmitter version of the radar, the S273 with a shorter and wider array, is also available. This version offers a six- or eight-beam stack, with a 1.4° azimuth beamwidth but with wider beams in elevation covering to 20° total elevation. The solid-state transmitter consists of up to 40 modules generating 132 kW of total RF power at the peak of a 150-μs pulse and up to 5 kW of average power. The height accuracy claimed for the radar by its manufacturer is 1700 ft on a small fighter at 100 nmi.[14]

A radar which is a hybrid mix of stacked beams and phase steering is the RAT-31S, an S-band radar which phase-steers a stack of four beams in elevation to cover the surveillance volume. The radar employs monopulse to determine target height. It rotates a 13.2-ft-square array at 5 to 10 r/min while generating a stack of three receive beams covering 21° in elevation. The array is divided into three vertical sections. Each section of the array then generates its own beam, which is phase-steered over a designated section of the elevation coverage.[15]

Scanning Pencil-Beam Radars.   Another method of achieving 3D volumetric coverage suitable for high-air-traffic situations is the scanning pencil-beam radar. The most common radars in this class obtain high volumetric coverage by employing an antenna feed technique which electronically scans a narrow pencil-type beam through the elevation coverage as it rotates in azimuth, producing an azimuth-elevation scanning pattern similar to that of a TV raster scan.

For the air volume surveillance mission, electronic scanning provides flexibility and performance not readily available in electromechanical scanners. These advantages include (1) shorter volume surveillance frame times; (2) highly flexible computer-programmable waveform versus elevation time and energy management; (3) electronic compensation for moving platforms and mobile applications such as ground vehicles, ships, and aircraft; and (4) wide and flexible elevation coverage, including highly agile beam placement in elevation, programmable elevation coverage versus range and azimuth, good beam shape preservation over wide coverage, and flexible, precise control of beam placement versus azimuth, which is especially critical for low-elevation-beam performance.

Frequency Scanned Radars.   One of the earlier 3D radar techniques that has found application for the air surveillance mission is frequency scanning. Frequency-scanned arrays utilize the frequency-dependent phase characteristics of a length of transmission line, usually waveguide, to scan a pencil beam.[16] The waveguide is folded into a serpentine configuration on the side (or sides) or back of the array to provide output taps at the locations of the closely spaced antenna elements. A controlled change of transmit/receive RF frequency produces a different phase gradient across the aperture, electronically steering the beam to the desired elevation angle. Frequency scanning may be accomplished from pulse to pulse by changing the transmitter and receiver frequency sequentially from one pulse to the next, or "within" the pulse, by transmitting a chirp linear-frequency-modulated (LFM) pulse or sequence of contiguous subpulses each stepped in frequency, and by processing each of a stack of receive beams in elevation each at one of the subpulse frequencies.[17] The AN/SPS-39 S-band shipborne radar used a parabolic-cylinder reflector antenna fed by a line source to produce the change in phase with frequency necessary to scan its beam electronically in elevation. Upgraded with a planar array, this radar evolved into the AN/SPS-52[18] (Fig. 20.2b). The within-pulse approach was employed on the AR3D S-band surveillance radar. It transmits LFM pulses and extracts the target height via frequency discrimination in the receiver.[19]

The U.S. Marine Corps AN/TPS-32, the U.S. Navy shipborne AN/SPS-48, and the Series 320 radars are all examples of S-band 3D surveillance radars consisting of a small stack of frequency-scanned beams which are then step-scanned as a group to cover the elevation surveillance volume.[20,21]

The use of frequency-scanning beams in elevation as a height finding technique is a form of the general technique of sequential lobing, in which amplitudes from adjacent sequentially formed beams are compared to estimate the target elevation angle. The elevation-angle accuracy achievable in this class of radars is not as good as that of stacked-beam or phase-scanned monopulse radars, e.g., radars employing simultaneous lobing. There are several fundamental reasons for this. One is that because different frequencies are required in order to steer the beam, amplitude fluctuations in the target return are induced. These tend to dilute the quality of target angle information available in the multiple-beam target returns. The effect can be compensated by averaging out the target fluctuation effects by the use of noncoherent integration of multiple-frequency diversity subpulses in each beam. However, the diversity subpulse frequency separations

(b)

FIG. 20.2 *(Continued)* Exemplar 3D radars. (*b*) AN/SPS-52C shipboard frequency-scanned 3D radar (*Courtesy Hughes Aircraft Company*).

must be enough to induce target amplitude fluctuations while not causing too much beam steering—a difficult tradeoff in some applications. Sequential-lobing angle estimation techniques are also vulnerable to time-varying or amplitude-modulation jammers such as blinkers. The fact that RF frequencies correspond one to one with elevation angle in the frequency-scanned radar constrains it in the use of frequency agility for electronic counter-countermeasures (ECCM) purposes. It also tends to limit its flexibility in waveform time and energy management. The electronically steered phased array provides considerable relief to the designer from these limitations.

Phased Array Radars. Scanning or steering of a pencil-type narrow beam in elevation can be accomplished by means of electronically controlled phase shifters placed at the row feed outputs of an array antenna. This approach is the most flexible of the various 3D radar height finding techniques, allowing full use of the frequency band for purposes beyond beam scanning and allowing for complete independence of waveform and beam position. Height finding techniques which can be used with the phase-scanned array include a variety of coherent simultaneous-lobing (monopulse-multipulse* and phase-interferometry) techniques, as well as amplitude comparison sequential-lobing techniques. The phased array radar is becoming more commonplace in the present-day military marketplace, owing to an ever-escalating target and threat environment and dynamics.

The AN/TPS-59 L-band radar is an example of a long-range transportable 3D tactical radar with phase scanning to steer the beam in elevation. Developed for the U.S. Marine Corps, it is unique among air surveillance radars in that it was the first to employ an all-solid-state transmitter. The solid-state transmitter of this radar is distributed over the antenna aperture in the form of individual row transmitter units. The total transmit power is combined only in space, in the far-field collimated beam. The planar array antenna consists of 54 rows of horizontal stripline linear arrays. Each of the 54 rows contains its own solid-state modular transceiver consisting of a 1-kW nominal RF peak-power solid-state transmitter, integral power supply, low-noise receiver, phase shifter, duplexer, and logic control, all mounted on the antenna. The feed structure of the 15-ft by 30-ft planar array generates a full two-axis monopulse beam set on receive, consisting of a sum and two delta beams. An additional column feed provides a special low-angle height finding capability for the lowest angle beam positions. The feed generates a pair of squinted sum-type beams carefully placed in elevation and processed as a monopulse pair. The technique minimizes the effects of multipath. Its fundamental accuracy performance is considered in Sec. 20.3. The 1.6° by 3.2° monopulse beam set is electronically phased-scanned from -1 to 19°.[22–24] Fixed-site variants of this radar are the AN/FPS-117 SEEK IGLOO radar (Fig. 20.2c) and the GE-592 radar, both of which are distributed aperture solid-state and similar to the AN/TPS-59 but which employ a 24- by 24-ft array antenna with 44 rows and additional digital signal processing. The square-aperture array of the GE-592/FPS-117 radar generates a 2.2° azimuth by 2° elevation two-axis monopulse beam set.[25–28]

The HADR, deployed mainly in Europe for North Atlantic Treaty Organization (NATO) applications, is a ground-based 3D S-band phased array radar which also uses phase scanning in elevation and mechanical rotation in azimuth. The radar's 12.5-ft by 16-ft planar array rotates at 5 r/min while phase-scanning a single pencil-type 1.45° azimuth by 1.9° elevation beam through 12 long-range search beam positions and 4 MTI beam positions in elevation. The instrumented coverage of the radar is 250 nmi in range, 20° in elevation, 360 degrees° in azimuth, and 120 kft in height. Target height is estimated by using sequential lobing between contiguous beams in elevation.[29]

A significant example of a long-range airborne 3D surveillance radar is the AWACS (Airborne Warning and Control System) AN/APY-1 S-band radar used on the E-3A aircraft. Because of the limited vertical aperture extent of the AN/APY-1 radar, the elevation beam is relatively broad. Consequently the height accuracies achieved by the radar do not compare well with those of its surface-based counterparts.

---

*The term *multipulse* is used by the author to refer to a target angle estimation technique discussed in Sec. 20.3 for coherently combining received monopulse sum- and delta-channel (I,Q) target echo samples from multiple-pulse transmissions. This technique is to be distinguished from the combining of individual monopulse angle measurements from each of the multiple-pulse transmissions.

(c)

FIG. 20.2 *(Continued)* Exemplar 3D radars. *(c)* AN/FPS-117 fixed-site solid-state phase-scanned 3D radar *(Courtesy General Electric Company)*.

Digital Beamforming Radar.   A technology with considerable attractiveness for radar is digital beamforming. As a technique for finding target height, digital beamforming involves placing a receiver on each element of a vertical array of elements, or rows of elements. By digitally weighting and linearly combining the analog-to-digital (A/D) converted receiver outputs, a stack of receive beams or a single scanning receive beam in elevation can be generated. In this form, a digital beamforming radar is a special case of a stacked-beam radar, implemented in a technology that offers several advantages over conventional stacked-beam technology. The major advantage offered by digital beamforming technology is that of full adaptive control of the beam patterns for ECCM purposes. The major challenge faced by digital beamforming radar designers relative to height finding is to develop techniques to preserve monopulse ratios in the presence of adaptive array cancellation of jamming, including those in the main lobe. Monopulse beam pairs or stacked beams are easily generated digitally, but the accuracy of height finding depends on a precise, unambiguous knowledge of the relative patterns of the (adapted) height finding beams.

20.2 DERIVATION OF HEIGHT FROM RADAR MEASUREMENTS

Height in radar is always a derived rather than a measured quantity. This is true because a radar can only measure range and angle of arrival of target returns. Surface-based radars derive the height of a target from the range (time) of the echo return and elevation coordinate measurements. A radar on a ship, aircraft, or space satellite may be required to convert tricoordinate measurements relative to the antenna to an inertial reference system as part of the height calculation. The accurate calculation of height from radar measurements must provide for such effects as the location and orientation of the radar antenna in the desired reference coordinate system, the curvature of the earth, the refractive properties of the atmosphere, and the reflective nature of the earth's surface. Furthermore, if the target height is to be referenced to the local terrain, then the height of that possibly irregular terrain below the target must also be taken into account. The effects of some of the systematic internal equipment errors can also be partially offset by incorporating internal calibration measurements into the range and angle estimation algorithms.

Flat-Earth Approximation. For very-short-range targets, a sufficiently good estimate of target height is given by the flat-earth approximation:

$$h_T = h_a + R_T \sin \theta_T \qquad (20.1)$$

where h_a is the radar antenna height, R_T is the measured target range, θ_T is the measured or estimated target elevation angle, and h_T is the estimated target height.

Spherical Earth: Parabolic Approximation. A somewhat better approximation to target height which models the earth's curvature as parabolic in range can be derived by reference to Fig. 20.3b. For a radar located near the surface of the earth, it can be readily shown from the law of cosines that, to a first approximation,

$$h_T = h_a + R_T \sin \theta_T + R_T^2/2R_0 \qquad (20.2)$$

where r_0 is the radius of the earth and the other parameters are as defined above.

The height calculated with the above curved-earth algorithm exceeds that calculated by using the flat-earth algorithm, increasing quadratically with measured target range. The discrepancy reaches a value of about 88 ft for a measured target range of 10 nmi.

Spherical Earth: Exact Geometry. Again with reference to Fig. 20.3b, the exact target height can be calculated as follows:

$$h_T = [(R_0 + h_a)^2 + R_T^2 + 2(R_0 + h_a)R_T \sin \theta_T]^{1/2} - R_0 \qquad (20.3)$$

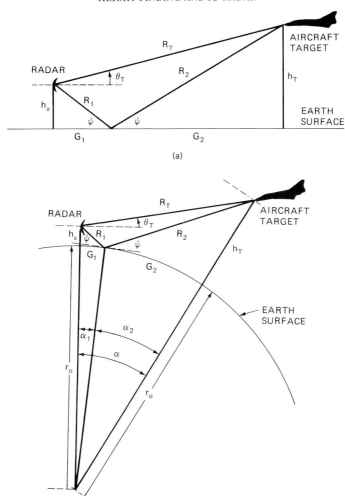

FIG. 20.3 Geometric considerations. (*a*) Flat-earth geometry. (*b*) Spherical-earth geometry.

Corrections for Atmospheric Refraction.* To further improve the accuracy of height computation, refraction of the radar beam along the ray path to the target must be taken into account. In free space, radio waves travel in straight lines. In the earth's atmosphere, however, electromagnetic waves are generally bent or refracted downward. The bending or refracting of radar waves in the atmosphere is caused by the variation with altitude of the index of refraction, which is defined as the ratio of the velocity of propagation in free space to the

---

*The data tables, figures, and portions of the following discussion on corrections for atmospheric refraction were extracted from and follow closely the original text of Burt Brown's Chap. 22, "Radar Handbook," 1st ed., edited by Merrill Skolnik.[10]

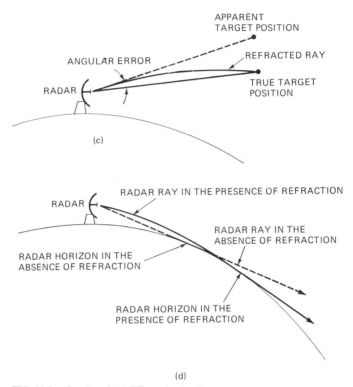

(c)

(d)

FIG. 20.3 *(Continued)* (c) Effect of refraction on radar horizon. (d) Angular error due to refraction.

velocity in the medium in question. One effect of refraction is to extend the radar distance to the horizon, as suggested in Fig. 20.3c. Another effect is the introduction of errors in the radar measurement of elevation angle. In the tropospheric portion of the atmosphere, the index of refraction n is a function of such meteorological variables as temperature, pressure, and water vapor and can be represented by[30]

$$(n - 1) \times 10^6 = N = \frac{77.6p}{T} + \frac{3.73 \times 10^5 e}{T^2} \tag{20.4}$$

where T = air temperature, K; p = barometric pressure, in millibars; and e = partial pressure of water vapor, in millibars. The parameter N is a scaled index of refraction termed *refractivity*.

Since the barometric pressure p and the water vapor content e decrease rapidly with height, the index of refraction normally decreases with increasing altitude. In a standard atmosphere, the index decreases at a rate of about 4.5×10^{-8} per meter of altitude. A typical value of the index of refraction at the surface of the earth is of the order of 1.0003. The Cosmic Ray Physics Laboratory (CRPL) standard atmosphere has been defined as one having an index of refraction of 1.000313 (or 313 N units for the refractivity N) and having an exponential decrease of refractive index with altitude.

$$N = N_s \exp (- ah) \qquad (20.5)$$

where $N_s = 313$ N units is the surface refractivity and $a = 0.04385$ per kft when h is in thousands of feet.

The classic method of accounting for atmospheric refraction in radar height computations is to replace the actual earth radius R_0 ($= 3440$ nmi) by an equivalent earth of radius $R_e = kR_0$ and to replace the actual atmosphere by a homogeneous atmosphere in which electromagnetic waves travel in straight lines rather than curved lines (Sec. 2.6). It may be shown by Snell's law in spherical geometry that the value of the factor k by which the earth's radius must be multiplied in order to plot the ray paths as straight lines is

$$k = \frac{1}{1 + R_0 \ (dn/dh)} \qquad (20.6)$$

where dn/dh is the rate of change of refractive index n with height. The vertical gradient of the refractive index dn/dh is normally negative. If, contrary to the CRPL standard atmosphere assumption, it is assumed that this gradient is constant with height, the value of k is 4/3. The use of the 4/3 effective earth's radius to account for the refraction of radio waves has been widely adopted in radio communications, propagation work, and radar.[31] The height calculated by using a 4/3 effective earth's radius is less than that calculated by using the actual earth radius, the difference increasing quadratically with measured target range, attaining a value of about 22 ft at 10 nmi.

The distance d to a horizon from a radar at height h_a may be shown from simple geometry to be approximately

$$d = \sqrt{2kR_0 h_a} \qquad (20.7)$$

where h_a is assumed to be small compared with R_0. For $k = 4/3$, the above expression reduces to a particularly convenient relationship if d and h_a are measured in nautical miles and feet respectively:

$$d(\text{nmi}) = 1.23\sqrt{h_a(\text{ft})} \qquad (20.8)$$

Refined computations of the angular deviations introduced when electromagnetic waves traverse a medium other than free space are discussed elsewhere.[32,33] Estimates of the height error for a target located at an altitude of 100,000 ft based on CRPL Reference Refractivity Atmosphere—1958, are contained in Table 20.1. It is noted that the magnitude of the height error is directly related to the surface refractivity and that above approximately 40° elevation angle the height error is independent of the surface refractivity. The height error is given in Table 20.2 as a function of slant range.

TABLE 20.1 Estimates of Height Error at an Altitude of 100,000 Ft Based on the CRPL Reference Refractivity Atmosphere—1958

Elevation angle,°	Height error, kft		
	Surface refractivity, $N_0 = 280$	Surface refractivity, $N_0 = 315$	Surface refractivity, $N_0 = 370$
1	9.14	11.12	14.73
2	5.65	6.75	8.63
4	2.63	3.08	3.82
6	1.44	1.68	2.06
8	0.89	1.03	1.26
10	0.60	0.69	0.84
15	0.28	0.32	0.39
20	0.16	0.18	0.22
40	0.04	0.04	0.05
70	0.01	0.02	0.02

TABLE 20.2 Estimate of Height Error at Slant Ranges of 100, 200, and 300 nmi Based on the CRPL Reference Refractivity Atmosphere—1958

Elevation angle,°	Height error, kft					
	Slant range, 100 nmi		Slant range, 200 nmi		Slant range, 300 nmi*	
	$N_0 = 280$	$N_0 = 370$	$N_0 = 280$	$N_0 = 370$	$N_0 = 280$	$N_0 = 370$
1	1.61	2.68	5.08	8.07	9.14	14.73
2	1.38	2.32	4.20	6.34		
4	1.12	1.73				
6	0.93	1.36				
8	0.78	1.13				

*Approximate slant range.

TABLE 20.3 Comparison of Heights Based on ⁴⁄₃-Earth's-Radius Principle with Heights Based on the Exponential Model*

Elevation angle,*	Slant range, 100 nmi			Slant range, 200 nmi			Slant range, 300 nmi		
	h_{exp}	$h^{4/3}$	Δh†	h_{exp}	$h^{4/3}$	Δh	h_{exp}	$h^{4/3}$	Δh
0.0	6.9	6.8	0.1	28.0	26.8	1.2	65.1	60.2	4.9
0.5	12.3	12.1	0.2	39.4	37.4	2.0	82.9	75.9	7.0
1.0	17.8	17.5	0.3	50.6	48.0	2.6			
2.0	28.6	28.1	0.5	72.8	69.1	3.7			
4.0	50.0	49.2	0.8						

*After Bauer, Mason, and Wilson.[34,35]
†$\Delta h = h_{exp} - h_{4/3}$; all heights in kilofeet.

A comparison of the heights based on a 4/3-earth's-radius principle with the heights based upon the exponential model is illustrated in Table 20.3. The data shows that, for a given elevation angle, the difference in height computation increases with slant range and that, for a given range, the height difference increases with elevation angle.

Compensation for Surface Refractivity Variation. For extremely accurate height calculations at long ranges, it is possible to correct for variations in surface refractivity in otherwise normal atmospheric refraction conditions. Such a technique is used, for example, in the General Electric series of solid-state 3D radars. The approach is to use offline ray tracing with an exponential model for atmospheric refraction. The resulting heights are then pretabulated as a function of elevation angle, range, and surface refractivity along with the partial derivatives of the height function with respect to the three above variables. These calculations are then stored in the radar computer database. In normal radar operation, the surface refractivity is measured periodically at the radar site, where it is used in conjunction with measured target elevation θ_T and range R_T to perform online table lookup of the tabulated height-refractivity parameters. The final height is computed by means of the interpolation

$$h_T = h_T (R_k, \theta_k, N_k) + \frac{\partial h_T}{\partial R_k}(R_T - R_k) + \frac{\partial h_T}{\partial \theta_k}(\theta_T - \theta_k) + \frac{\partial h_T}{\partial N_k}(N - N_k) \qquad (20.9)$$

where R_k, θ_k, and N_k are the closest stored values of range, elevation angle, and surface refractivity to the measured values.

Practical Corrections

Terrain Height Adjustments. If the height of the target above local terrain is to be obtained, the height relative to mean sea level must be corrected by the height of the terrain below the target. This involves calculation of ground range from target slant range and elevation angle and computer lookup of terrain height versus ground range and azimuth.

Platform Location, Orientation, and Stabilization. The calculation of target height with a radar on a moving platform, such as a ship, aircraft, or satellite (all of which are subject to uncertainties in location and orientation) is somewhat more complicated. Coordinate conversion of measured target range, azimuth, and elevation is necessary to determine the target height. Platform location and orientation must be sensed, and perhaps stabilized, and provided to the radar computer. Some of these quantities are also required for platform navigation and therefore may be available from the navigation gyros.

20.3 HEIGHT ACCURACY PERFORMANCE LIMITATIONS

The accuracy of the measurement of target height with a radar is conveniently expressed in terms of the root-mean-square error (rmse), i.e., the square root of the expected value of the square of the difference between the estimated target height and the actual target height. Because height is a derived quantity from the basic radar measurements of range and elevation angle, height accuracy can be expressed in terms of the rms errors associated with those measurements, as suggested in Table 20.4. The remainder of this section is devoted to analysis of the errors involved in the basic radar measurements, primarily elevation angle.

All radar measurements are in error because of the contamination of the received-signal echo with thermal noise. The common assumption about the nature of thermal noise, well justified by practical experience, is that it is a narrowband zero-mean gaussian random process. A particular pair of samples consisting of in-phase (I) and quadrature (Q) components can be properly

TABLE 20.4   Relationship of Target Height Error to Radar Range and Elevation Angle Measurement Errors*

Flat earth:
$$\sigma_h = (\sigma_R^2 \sin^2 \theta + R^2 \sigma_\theta^2 \cos^2 \theta)^{1/2}$$
Spherical earth: parabolic approximation:
$$\sigma_h = [\sigma_R^2 (R/R_e + \sin \theta)^2 + R^2 \sigma_\theta^2 \cos^2 \theta]^{1/2}$$
Spherical earth: exact geometry:
$$\sigma_h = \{[\sigma_R^2 (R^2 + (R_e + h_a)^2 \sin^2 \theta) + (R_e + h_a)^2 R^2 \sigma_\theta^2 \cos^2 \theta]/(R_e + h)^2\}^{1/2}$$

R_e = effective earth radius = kR_0
h_a = antenna height above earth surface
R = radar-measured target range
θ = radar-measured target elevation angle
h = radar-measured target height
σ_R = rmse of radar range measurement
σ_θ = rmse of radar elevation-angle measurement
σ_h = rmse of target height estimate

*Colocated and exactly known platform-antenna location and orientation and small measurement errors relative to target coordinate values are assumed.

viewed, therefore, as a single complex zero-mean gaussian random variable.* The rmse associated with the accuracy performance of a particular radar angle measurement technique, as limited solely by the presence of thermal noise on the technique, is termed herein the *fundamental accuracy* of the technique.

The fundamental accuracy of two general categories of elevation-angle estimation techniques is presented in this section: the sequential-lobing technique and the simultaneous-lobing technique. Other practical effects influencing the height accuracy of a radar system include: beam-pointing errors, pattern errors, channel mismatch errors, calibration, platform orientation and gyration, stabilization, compensation, ECM-ECCM and clutter errors, multipath, target fluctuations and thresholding effects, and multiple hit and channel combining.

Fundamental Accuracy of Sequential Lobing.   Sequential lobing is a technique used in radar for estimating the angle of arrival of electromagnetic radiation incident on an antenna by comparing the amplitudes of the received echoes in two or more sequentially formed or selected antenna beams. The technique is used for height finding by time—sequentially scanning a beam in elevation while transmitting and receiving pulses in each beam position. The pulse amplitudes in each beam position are envelope-detected and stored for use in a comparison with those from the other beams. The simplest form of radar sequential lobing compares the envelope-detected returns from a single pulse in each of two adjacent beams. The ratio of the detected pulse amplitude in one beam position to that in the other forms the basis for a table lookup or readout of target elevation angle. In early radar the readout was a calibrated dial or display. In 3D radars, computers provide elaborate lookup tables to relate the ratios to the target elevation angle.

The envelope-detected target returns in each of the beam positions are corrupted by thermal noise even in the most ideal of circumstances. This noise is due

---

*The term *complex* here refers to $I + jQ$. The complex sample has zero mean because of its random uniformly distributed phase on the interval $(0, 2\pi)$.

to thermally generated electronic noise in the radar receivers and to noiselike electromagnetic emissions from the sky and ground entering the antenna.

For a nonfluctuating target, a good approximation for the fundamental accuracy of the sequential-lobing technique for a large signal-to-noise ratio* is

$$\text{rmse} = \frac{1}{|\dot{f}|}\left(\frac{1+f^2}{2x}\right)^{1/2} \tag{20.10}$$

where, if θ is the target angle and $\hat{\theta}$ is its estimate, the rmse is defined by $[E(\hat{\theta}-\theta)^2]^{1/2}$. The various factors in the above rmse are defined as

$f = f(\theta) = G_2(\theta)/G_1(\theta) =$ ratio of two-way elevation beam power patterns
$\dot{f} = df/d\theta$; gives rmse in radians or milliradians (mrad)
$\dot{f} = df/d\,(\sin\theta)$; gives rmse in sines or millisines (msine)
$G_2(\theta) =$ two-way elevation beam power pattern in beam position 2
$G_1(\theta) =$ two-way elevation beam power pattern in beam position 1
$x =$ signal-to-noise ratio in beam position 1

Note that this result may be put into the familiar form

$$\text{rmse} = \frac{1}{K\sqrt{2x_0}} \tag{20.11}$$

where $K = \dfrac{|\dot{f}|\,|g_1|}{(1+f^2)^{1/2}}$ \hfill (20.12)

$g_1 =$ two-way normalized voltage pattern in beam position 1
$\quad = G_1^{1/2}/G_{10}^{1/2}$; $G_{10} =$ boresight antenna power gain in beam 1
and, $x_0 =$ boresight signal-to-noise ratio

The constant K in this form is a measure of the sensitivity of the angle estimation technique, in that the larger K is, the smaller are the rms errors and the better is the angle accuracy. A single value of the sensitivity (at beam peak or at crossover) is often used to compare techniques. However, care must be taken in this practice, because K is a strong function of the antenna beamwidth and of the target angle of arrival relative to the position of the beams in angle space. As expressed above, K includes the signal-to-noise-ratio dependence on the target angle through the two-way pattern of the beam. An alternative formulation could leave this pattern factor in the signal-to-noise ratio and omit it from the definition of K.

The fundamental accuracy of the sequential-lobing technique for a scanning beam generated by a uniformly illuminated aperture on transmit and receive is presented in Fig. 20.4a. Performance is presented in terms of a normalized ver-

---

*The signal-to-noise-ratio definition used here is E/N_o, where E is the received pulse energy and $N_o/2$ is the spectral power density of the interfering thermal noise.

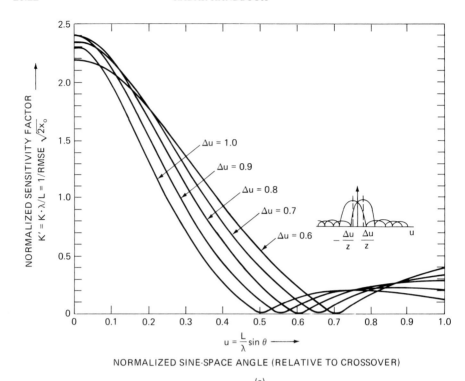

NORMALIZED SINE-SPACE ANGLE (RELATIVE TO CROSSOVER)

(a)

FIG. 20.4 Fundamental accuracy. (*a*) Two-beam sequential lobing: uniform (sin $\pi u/\pi u$) sum beams transmit and receive; separation = Δu; nonfluctuating target; N = one pulse per beam.

sion of the sensitivity factor $k = K\lambda/L$ versus a similarly normalized sine-space elevation angle of arrival, $u = L/\lambda \sin(\theta)$. The elevation angle θ is referenced to the crossover point halfway between the beam peaks. This normalization removes the aperture dimension L and transmit wavelength λ from the performance of the technique. The figure shows that the sensitivity factor peaks at crossover and is symmetrical about the crossover angle. The value of the normalized sensitivity factor at crossover depends on the two-way beam shape–aperture illumination and the step size between the beams. For a uniformly illuminated aperture whose beam is stepped by $\Delta u = 1$, between single-pulse transmission/reception, the normalized sensitivity factor associated with the two-beam sequential-lobing technique against a nonfluctuating target attains a value of approximately 2.15 at the crossover angle. Thus, for example, if the aperture height is 24 ft (7.3 m) and the radar is L band ($\lambda = 0.23$m), the appropriate normalization factor is $L/\lambda = 31.75$, so that the actual sequential-lobing sensitivity factor for a target at crossover is 68.25 V/(V · sine*), or 0.06825 V/(V · msine*). If the boresight

---

*A *sine* or a *millisine* is a unit of measure of the sine of an angle. For example, an angle of 0.7 rad (700 mrad) corresponds to sin (0.7) ˜ 0.64422 sine = 644.22 msines.

signal-to-noise ratio is 100 (20 dB), the rmse is easily calculated as rmse = 1.04 msines. Furthermore, if the beams have been electronically steered so that their crossover elevation angle is, say, 30° away from the antenna broadside, the angular accuracy can be calculated as rmse = 1.04/cos (30°) = 1.2 mrad.

If the beams are too closely spaced, sensitivity suffers because there isn't sufficient difference in the received echo strength to measure the angle of arrival accurately. On the other hand, if they are separated too far, there isn't sufficient signal-to-noise ratio in one of the beams for accuracy. It follows that there is a beam step size that optimizes accuracy by maximizing sensitivity and minimizing errors. This is clearly illustrated in Fig. 20.4a, which shows a maximization of the sensitivity factor for a beam spacing between $\Delta u = 0.8$ and $\Delta u = 0.9$. It can be shown analytically (for gaussian beam shapes) that the optimum spacing between beams is $\Delta u = 0.85$. It is also significant to note that the valid range of angular (u-space) coverage for the sequential-lobing technique using uniform illuminated sum beams is approximately $2 - \Delta u$, where Δu is the beam step size between transmission in u space. Within this region, the target is in the main lobe of each of the two beams, and its angle is uniquely and unambiguously determined by the ratio of the echo strengths in the two beams. Outside this region the target is in the sidelobes of at least one of the beams, and its angle of incidence cannot be unambiguously estimated with the technique. In practice, sidelobe responses to a target are eliminated by the use of sidelobe blanking. At the optimum beam spacing of $\Delta u = 0.85$, the valid angular coverage is $2 - 0.85 = 1.15$, or ± 0.575 about the crossover angle. Coverage can be increased from this value by decreased step size or by aperture weighting, but only at the expense of sensitivity.

It is possible to utilize more than two beam positions in the sequential-lobing estimation algorithm. In such an approach, performance is improved when the beam scans a small amount in angle between transmissions. As the beam steps past the target in elevation, it is possible to display intensity, or target echo strength, on a display or other indicator at the angular locations of the beam. A centroidal interpolation of the angular locations as weighted by the receive pulse amplitudes may be used to extract the target angle estimate. The mathematically equivalent estimation process is

$$\hat{\theta} = f^{-1} \frac{\displaystyle\sum_{k=1}^{N} |r_k|\theta_k}{\displaystyle\sum_{k=1}^{N} |r_k|} \tag{20.13}$$

where r_k = complex ($I_k + jQ_k$) sample at receiver–pulse matched-fiber output
θ_k = boresight elevation of beam position k
N = number of beam positions used in algorithm

and $\qquad f(\theta) = \displaystyle\sum_{k=1}^{N} \theta_k g_k \Big/ \sum_{k=1}^{N} g_k \qquad g_k = g_k(\theta) = G_k^{1/2} (\theta)/G_0^{1/2}$

where G_0 = boresight gain of beam. The fundamental accuracy of this multiple-beam position version of the sequential-lobing estimation algorithm is

$$\text{rmse} = \left[\frac{\sum_{k=1}^{N}(\theta_k - f)^2}{2\,\dot{f}^{\,2}x_0 \left(\sum_{k=1}^{N} g_k \right)^2} \right]^{1/2} \qquad (20.14)$$

This performance can be calculated at an arbitrary elevation angle between the beams θ, or it can be averaged over an elevation angle in a root-sum-square (rss) sense as

$$\overline{\text{rmse}} = \left(\frac{1}{\theta_s} \int_0^{\theta_s} \text{mse}(\theta)\, d\,\theta \right)^{1/2} \qquad (20.15)$$

where θ_s is the angular separation between the beam positions and $\text{mse}(\theta)$ is the mean square error.

Figure 20.4b displays this rss average accuracy for a nonfluctuating target versus signal-to-noise ratio for various beam spacings of a gaussian-shaped scanning beam. In general, the fundamental accuracy performance of the sequential-lobing technique is a function of the beam shapes and separations, the number of pulses

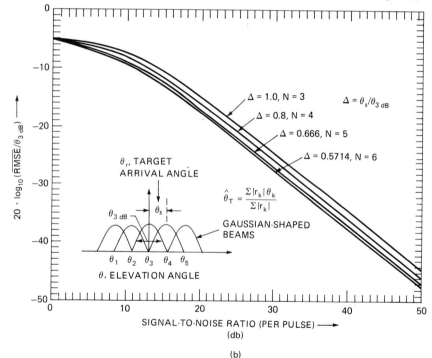

(b)

FIG. 20.4 *(Continued)* Fundamental accuracy. (*b*) Multiple-beam sequential lobing: nonfluctuating target.

noncoherently integrated in each beam, the type and amount of target echo pulse-to-pulse fluctuation within a beam, and of the prevailing beam-to-beam correlation in target fluctuation. For N pulse returns noncoherently integrated within each beam on a nonfluctuating target, the asymptotic rmse accuracy performance of two-beam sequential lobing is given by

$$\text{rmse} = \left(\frac{1 + f^2}{2f^2 Nx}\right)^{1/2} \tag{20.16}$$

where f = ratio of two-way elevation power patterns as before
$\quad\ N$ = number of pulses noncoherently integrated
$\quad\ x$ = per-pulse SNR

Since this form is the same as for a single pulse except for the factor of N in the denominator, numerical results may be obtained from Fig. 20.4a with appropriate scaling.

Fundamental Accuracy of Simultaneous Lobing. In the simultaneous-lobing method of angle estimation, two or more radar receive beams are simultaneously formed by the antenna and processed in parallel receive channels. A single transmit beam covers the angular region to be simultaneously processed on receive. Stacked beams, monopulse, and phase interferometry are all examples of the use of simultaneous lobing for target elevation angle estimation. While very different in implementation for a radar system, the fundamental accuracies of these techniques are all analyzed in a similar fashion, with approximate results that can be placed in the same form as Eq. (20.11). Because the receive beams in this technique are formed and processed simultaneously, the relative phase of the return between receive channels can, if desired, be used to aid the angle extraction accuracy. If it is used, the process is termed *phase-coherent* or simply *coherent*, and a close match in phase between receive channels must be maintained.

Monopulse. In general, the term *monopulse* refers to a radar technique to estimate the angle of arrival of a target echo resulting from a single-pulse transmission by using the amplitude and/or phase samples of the echo in a pair of simultaneously formed receive beams (Sec. 18.3). Historically the term has been associated with the simultaneous generation and processing of a *sum* receive beam and a *difference*, or *delta*, receive beam. These beams are so named because of the early and still common method used to form them, i.e., by adding and subtracting, respectively, the two halves of the antenna aperture. While this method is a relatively inexpensive way to produce a sum-difference beam pair, it is not necessarily the best way from a performance standpoint. Furthermore, it is unnecessarily constraining in many phased array applications, especially where the feeds account for a small fraction of the cost of the total radar. Typically a sum beam may be designed for good detectability and sidelobes. The delta beam is then optimized for accuracy performance, perhaps with other constraints. The defining characteristic of a sum beam is that it has approximately even symmetry about the beam boresight, while a delta beam has approximately odd symmetry about the same boresight. Without loss of generality, the delta beam may be assumed to be adjusted or calibrated to be in phase with the sum beam, in the sense that the ratio of the two patterns is real and odd about the beam boresight versus angle of arrival.

Monopulse techniques are classified according to the manner in which the incident radiation is sensed, i.e., according to antenna and beamforming techniques, and independently according to how the various beams and channels are subsequently processed and combined to produce a target angle estimate.[36,37] Amplitude comparison monopulse and phase comparison monopulse are categories of antenna-beamforming *sensing* techniques. In amplitude comparison monopulse, the antenna-beamformer generates a pair of sum and difference beams which, without loss of generality, may be assumed to be in phase, in the sense that their ratio is real. In phase comparison monopulse, two or more antennas or sets of radiating-receiving elements, physically separated in the elevation dimension, are used to generate two beams which have ideally identical patterns except for a phase difference which depends on the angle of incidence of the received target echo. Each of these techniques may be converted to the other, either in concept through mathematical sums and differences or physically through the use of passive RF hybrid combining devices. The fundamental accuracy performance of a phase comparison monopulse system is identical to that of an amplitude comparison monopulse system converted by this method, and vice versa. Therefore, the fundamental accuracy performance is addressed here from the conceptual viewpoint of amplitude comparison monopulse.

There are a variety of ways to implement monopulse processing on a sum-difference beam pair, depicted functionally in Fig. 20.5, some of which have a substantial impact on the fundamental monopulse accuracy performance. In each of these implementations, returns from a single transmission are received in simultaneously formed sum and difference beams and processed coherently. In the full-vector monopulse of Fig. 20.5a, two complex (I, Q) samples are fully utilized to calculate a complex monopulse ratio statistic. This calculated statistic, the *measured* monopulse ratio, provides the basis for a computer table lookup of the target angle of arrival relative to the null in the delta beam. The computer lookup function is simply a tabulated version of the assumed monopulse ratio consisting of the assumed delta beam antenna pattern to that of the assumed sum beam versus angle off-beam boresight. The tabulated monopulse ratio is inverted in the lookup process by entering the table with the measured monopulse ratio and finding the corresponding off-boresight angle. The full-vector monopulse processing in Fig. 20.5b differs somewhat from that in Fig. 20.5a, in that after low-noise amplification to establish the system noise figure, an RF quadrature hybrid device is used to combine the delta and sum beam signals 90° out of phase, i.e., as $\Sigma + j\Delta$. The purpose of this combining in the difference channel is to bring the signal strength in the difference channel to approximately the same amplitude at that in the sum channel. This causes unavoidable receiver nonlinearities to have nearly the same effect in the two channels, resulting in less degradation in accuracy performance attributable to receive-string nonlinearities. In the absence of nonlinearities, the two techniques in Fig. 20.5a and b are mathematically identical because

$$\text{Im}\left(\frac{\Sigma + j\Delta}{\Sigma}\right) = \text{Im}\left(1 + j\frac{\Delta}{\Sigma}\right) = \text{Im}\left|\frac{\Delta}{\Sigma}\right| \cos\phi \qquad (20.17)$$

Hence, they both provide the fundamental accuracy performance of full-vector monopulse processing, given by

$$\text{rmse} = \frac{\|w_\Delta - fw_\Sigma\|}{|\dot{f}|(2x)^{1/2}} \qquad (20.18)$$

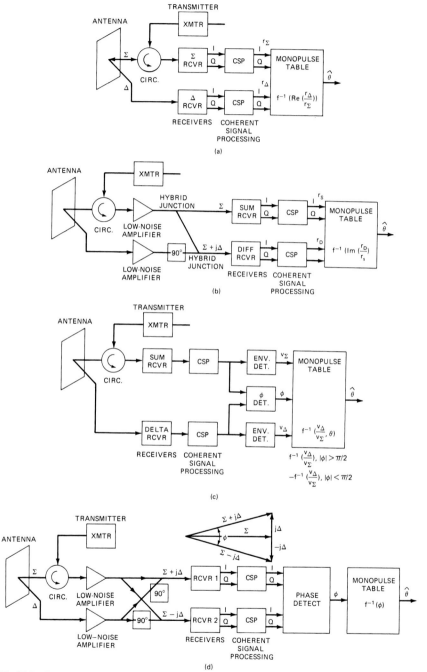

FIG. 20.5 Functional monopulse processing implementations. (*a*) Full-vector monopulse processing. (*b*) Full-vector monopulse with prehybrid combining. (*c*) Amplitude-only monopulse processing. (*d*) Phase-only monopulse processing.

where $f = f(\theta) = \Delta(\theta)/\Sigma(\theta)$, x = signal-to-noise ratio in the sum beam, and

Continuous aperture		*Discrete aperture array*	
$w_\Delta = w_\Delta(x)$	aperture illumination	$w_\Delta = (w_{\Delta n})$	$n = 1, N$ vectors of
$w_\Sigma = w_\Sigma(x)$	functions	$w_\Sigma = (w_{\Sigma n})$	array weights

$$\|w_\Delta\| = \left(\int_{-\infty}^{\infty} |w_\Delta(x)|^2 dx\right)^{1/2} \qquad \|w_\Delta\| = \left(\sum_1^N |w_{\Delta n}|^2\right)^{1/2}$$

$$\|w_\Sigma\| = \left(\int_{-\infty}^{\infty} |w_\Sigma(x)|^2 dx\right)^{1/2} \qquad \|w_\Sigma\| = \left(\sum_1^N |w_{\Sigma n}|^2\right)^{1/2}$$

$$\Delta(\theta) = \int_{-\infty}^{\infty} w_\Delta(x) \exp(j2\pi x (\sin\theta)) dx / \|w_\Delta\| \qquad \Delta(\theta) = \sum_1^N w_{\Delta n} \exp[j2\pi x_n(\sin\theta)]/\|w_\Delta\|$$

$$\Sigma(\theta) = \int_{-\infty}^{\infty} w_\Sigma(x) \exp(j2\pi x(\sin\theta)) dx / \|w_\Sigma\| \qquad \Sigma(\theta) = \sum_1^N w_{\Sigma n} \exp[j2\pi x_n(\sin\theta)]/\|w_\Sigma\|$$

Various authors have defined the *monopulse sensitivity factor* in different ways.[38] For the purposes of this chapter, the monopulse sensitivity factor is defined as the constant of proportionality required in the denominator of the rmse to convert the square root of twice the boresight signal-to-noise ratio in the beam to the rmse. Defined in this manner, the monopulse sensitivity factor has the desirable property of containing all target elevation angle-of-arrival information. The monopulse sensitivity factor for full-vector monopulse is

$$K = \frac{|\dot{f}| |g_\Sigma| |g_T|}{\|w_0 - fw_\Sigma\|} \tag{20.19}$$

where $g_\Sigma = g_\Sigma(\theta) = \Sigma(\theta)/\Sigma_0$ = sum-beam voltage pattern normalized to unity gain

$g_T = g_T(\theta) = G_T(\theta)/G_{T0}$ = transmit-beam voltage pattern normalized to unity gain

For orthogonal aperture illumination functions, where $\sum_{k=1}^N w_{\Sigma k} w^*_{\Delta k} = 0$ (usually the case in practice), this equation reduces to

$$K = \frac{|\dot{f}| |g_\Sigma| |g_T|}{(1 + f^2)^{1/2}} \tag{20.20}$$

This performance is presented graphically for several cases of interest in Fig. 20.6. A normalized sensitivity factor $k = K\lambda/L$ is plotted versus the u-space elevation angle of arrival $u = L/\lambda \sin\theta$, with θ referenced to the boresight of the sum and delta beams. The monopulse sensitivity factor peaks and is symmetrical about the boresight angle. The two curves in Fig. 20.6a correspond to uniformly illuminated sum beams. In one case, the delta beam is formed by subtracting the upper and lower halves of a uniformly illuminated aperture. For this delta beam illumination function, the normalized boresight monopulse sensitivity is $\pi/2 \backsim 1.57$. The sensitivity degrades off boresight, despite an increasing monopulse ratio slope, owing to a decreasing signal-to-noise ratio in the two channels and to

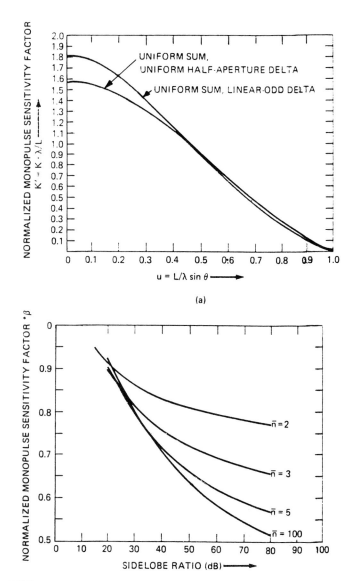

$^*\beta = \dfrac{RMSE_0}{RMSE}$; = accuracy with Taylor Σ, Bayliss Δ; $RMSE_0$ = accuracy with uniform Σ, linear-odd Δ.

(b)

FIG. 20.6   Fundamental accuracy of monopulse. (a) Vector monopulse accuracy: two common monopulse aperture illumination pairs. (b) Boresight vector monopulse sensitivities of Taylor sum and Bayliss delta aperture illuminations.

an increasing absolute value of the monopulse ratio $f(\theta)$. It can be shown that the boresight sensitivity of full-vector monopulse with a uniform sum beam can be maximized at a value of approximately 1.8 by employing a linear-odd aperture illumination function to generate the delta beam. The second curve in Fig. 20.6a illustrates performance for a linear-odd delta beam aperture illumination. The actual monopulse sensitivity factor can be calculated from the normalized sensitivity once the aperture height and RF wavelength have been specified. As an example, if $L/\lambda = 31.75$, the boresight monopulse sensitivity factor corresponding to the linear-odd delta beam illumination function is 0.05715 V/(V·msine). With a 20 dB signal-to-noise-ratio target return, this corresponds to a fundamental accuracy of 1.24 msines. For a uniform beam, the range of valid u-space angle coverage is approximately 2.0, corresponding to the sum-beam main-lobe null-to-null width. This is a principal advantage of monopulse because it allows reasonable spacings of the monopulse beams for coverage of large surveillance volumes. Coverage is increased with aperture weighting at the expense of monopulse sensitivity and fundamental accuracy. The effect on the boresight monopulse sensitivity of Taylor aperture weighting for the sum beam and Bayliss aperture weighting for the delta beam is illustrated in Fig. 20.6b. The sensitivity presented there is normalized by the sensitivity of the uniform-sum, linear-odd delta case, and is plotted for various values of the two parameters used to specify Taylor and Bayliss weighting, $\bar{n}$, and sidelobe ratio (SLR). It should be noted that not all combinations of $\bar{n}$ and SLR depicted in the figure constitute good aperture illumination design choices.

It is sometimes convenient and/or economical to perform coherent signal processing at RF or IF, by analog techniques, and then to carry out envelope and phase detection in the two channels. In amplitude-only monopulse the purpose of phase detection is solely to tell on which half of the beam the target return is incident. The angle off boresight is then determined via table lookup of the ratio of the envelope-detected signal strengths. The primary disadvantage of this approach is a degradation of accuracy at and near boresight relative to full-vector monopulse. It also provides less flexibility in coherent signal processing since it is analog instead of digital.

The fundamental accuracy performance of amplitude-only monopulse processing is degraded at boresight by the probability of incorrect phase detection, i.e., the probability of deciding that the target is below boresight when it is actually above, or vice versa. This probability is 0.5 at beam boresight, which results in boresight fundamental accuracy which is a factor of 2 worse than that of full vector monopulse. At off-boresight angles, the phase detection error probability depends on the signal-to-noise ratio. At angles far from the beam boresight, the signal-to-noise ratio diminishes, causing the error probability again to approach 0.5. A minimum-error probability–maximum-accuracy condition is reached for intermediate angles.

The last monopulse implementation illustrated (Fig. 20.5d) is termed *phase-only monopulse*. This processing is to be distinguished from the technique of phase interferometry, which has also been called by some authors[36,37] *phase-comparison monopulse*. In Fig. 20.5d, RF or IF hybrids are used to combine the sum and delta channels in quadrature, i.e., with a 90° phase shift. An accurate phase detector then detects the phase difference between the two channels. The underlying principle is that this phase difference will be in one-to-one correspondence with the delta-to-sum ratio, as illustrated in the vector diagram accompanying Fig. 20.5d. In phase-only monopulse, off-boresight accuracy is sacrificed to gain the benefit of identical amplitude signals in the two receiver-processor channels. If desired, the signals in the two channels may be hard-limited without affecting the fundamental accuracy of the phase-only monopulse processing. In principle, phase-only monopulse can be used to alleviate stringent receiver-processor dynamic-range requirements. However, other aspects of performance may suffer and should be examined carefully in the tradeoff process. Another advantage of phase-only monopulse, relative to vector and amplitude-only, is that the need for precise amplitude matching channel to channel is reduced.

Phase-only monopulse processing does not utilize the full target angle-of-arrival information available in the two beams. For this reason, its fundamental accuracy performance suffers. The fundamental accuracy of phase-only monopulse is identical to that of vector monopulse at boresight but degrades more rapidly off boresight. Full vector monopulse, using all the available information in the target returns, shows superior sensitivity at all target incidence angles. A uniformly illuminated aperture and beam and a uniform half-aperture difference beam are used for comparison of the three implementations.

In a radar which employs vector monopulse processing for height finding, it is possible to coherently precombine the returns from multiple pulses or subpulses in the simultaneous beams to form a single estimate of the target elevation angle of arrival, as suggested in Fig. 20.7. In this approach, the returns in the delta and sum channels are coherently cross-correlated pulse to pulse, and then the real part of the cross-correlation sum is normalized by a term determined by noncoherent integration in the sum channel to form the measured monopulse ratio. The same noncoherent sum used to normalize the measured monopulse ratio may be also used in the target decision logic for detection thresholding.

The rmse for multiple-pulse coherent monopulse differs from that of a single-pulse monopulse only by the square root of the number of pulses in the denominator. The results of Fig. 20.6, appropriately scaled, are applicable.

Stacked Beams. Stacked beams are another example of simultaneous lobing for target elevation-angle estimation. The processing of a pair of beams in the stack consists of an amplitude comparison table lookup. Its fundamental accuracy can also be placed in the form of Eq. (20.11).

In the stacked-beam radar, the transmit beam must be designed to cover all the beams within the stack and is therefore relatively wide in elevation beamwidth compared with that of a receive beam in the stack. A good approximation is that it is isotropic in elevation and thus is not a factor in the fundamental accuracy performance.

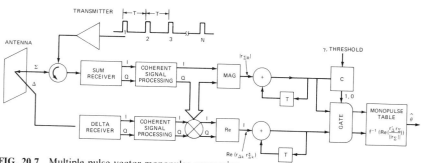

FIG. 20.7 Multiple-pulse vector monopulse processing.

The fundamental accuracy performance of a pair of uniformly illuminated beams in a stack is presented in Fig. 20.8 in terms of a normalized sensitivity factor $k = K\lambda/L$ versus normalized sine-space angle-of-arrival $u = L/\lambda \sin \theta$. The elevation u-space angle of arrival u of target energy is referenced to the crossover point halfway between the beams. Various beam separations in u space are illustrated. The sensitivity of the technique peaks at the crossover angle and is symmetrical about that angle, attaining a value at crossover which depends on the separation between the beams. A maximum crossover sensitivity of 1.95 is achieved for a u-space beam separation of 1.2. Coverage in u space provided by the uniform stacked-beam pair is approximately given by $2 - \Delta u$, where Δu is the u-space beam separation corresponding to a target in the main lobes of both beams. Outside this region, the target is in the sidelobes of one of the beams. In a stacked-beam radar, detections are made in a special cosecant-squared type of surveillance beam; so this condition is not sensed in the detection process. Thus, in order to eliminate the possibility of ambiguities, uniformly illuminated beams should be stacked at $\Delta u \leq 1$. The coverage of each beam pair may be increased by aperture weighting. In this case the beams may be stacked at greater separations but will possess reduced crossover sensitivity. The normalized crossover sensitivity associated with a pair of uniformly illuminated sum beams spaced at $\Delta u = 1$ is approximately 1.8. This corresponds to a fundamental accuracy of approximately 1.24 msines for a 24-ft aperture height at L band with a 20 dB target boresight signal-to-noise ratio.

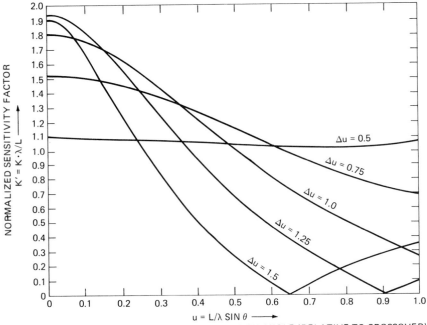

FIG. 20.8 Fundamental accuracy of stacked beams.

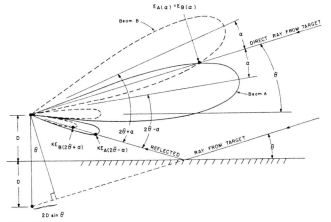

FIG. 20.9 Geometry for the analysis of elevation errors due to ground reflections in a simultaneous amplitude comparison radar.

Elevation Error Due to Surface Reflections.* One of the fundamental factors limiting the height accuracy in all height finding techniques that depend on elevation-angle measurements is the elevation-angle accuracy degradation due to the multipath from surface reflections. Such surface reflections vectorially combine with the direct-path signals entering the antenna to produce amplitude and phase variations which ordinarily cannot be separated from the direct-path signals. In general, the magnitude of such elevation-angle errors is such that, at low elevation angles where an appreciable portion of the antenna beam is directed into the ground, the elevation-angle errors are prohibitively large. Therefore, as a general rule, pencil-beam height finding radars and elevation-tracking radars cannot be expected to produce reliable elevation-angle data when their beams are pointed within about one beamwidth (-3 dB beamwidth) above the ground. At larger elevation angles, the magnitude of the elevation errors is a direct function of the ground-reflected relative field strength received in the respective negative-angle elevation sidelobe (i.e., the product of the relative sidelobe level and the ground-reflection coefficient).

Radar systems that employ a simultaneous amplitude comparison technique for target elevation-angle determination derive the elevation angle inside the radar beamwidth by measuring the ratio of simultaneous signal returns on two squinted received beams after having illuminated the target in some manner. The resulting elevation-angle data is independent of the manner in which the target is illuminated by the radar and is dependent only on the squinted *receiving*-antenna patterns.

In analyzing the elevation-angle errors due to ground reflections, we shall consider the case where the boresight crossover of a pair of squinted receiving beams (A and B) is oriented exactly on the target at elevation angle θ (a condition of zero error in the absence of ground reflections). See Fig. 20.9.

With pattern functions of beams A and B assumed to be identical and with the centerlines of beams A and B oriented at elevation angles of $\theta - \alpha$ and $\theta + \alpha$, respectively, the net received field strength at the feed points of beams A and B (relative to the peak of each beam) is then

---

*The material in this subsection was originally written by Burt Brown and appeared in Sec. 22.3 of the first edition of the handbook.

$$E_A = E_{A(\alpha)} + KE_{A(2\theta-\alpha)}e^{-j(\phi+2D\sin\theta)}$$

$$E_B = E_{B(\alpha)} + KE_{B(2\theta+\alpha)}e^{-j(\phi+2D\sin\theta)} \qquad (20.21)$$

where $\quad\quad K$ = amplitude of reflection coefficient

ϕ = phase of reflection coefficient

$E_{A(\alpha)} = E_{B(\alpha)}$ = relative received field strength of beams A and B from signal arriving along direct path at angle α from peaks of beams A and B

$E_{A(2\theta-\alpha)}$ = relative received field strength of beam A from reflected path at $2\theta - \alpha$ from peak of beam A

$E_{B(2\theta+\alpha)}$ = relative received field strength of beam B from reflected path at $2\theta + \alpha$ from peak of beam B

The magnitude of the off-boresight elevation-angle error due to ground reflections is a function of the ratio of the magnitudes of E_A and E_B, or

$$\text{Elevation error} = f\,\frac{|E_A|}{|E_B|} = f\,\frac{|E_{A(\alpha)} + KE_{A(2\theta-\alpha)}e^{-j(\phi+2D\sin\theta)}|}{|E_{B(\alpha)} + KE_{B(2\theta+\alpha)}e^{-j(\phi+2D\sin\theta)}|} \qquad (20.22)$$

When $KE_{A(2\theta-\alpha)}$ and $KE_{B(2\theta+\alpha)}$ are small compared with $E_{A(\alpha)}$, the maximum value of $f(|E_A|/|E_B|)$ is equal to

$$\frac{E_{A(\alpha)} + KE_{A(2\theta-\alpha)}}{E_{B(\alpha)} - KE_{B(2\theta+\alpha)}} \qquad (20.23)$$

To illustrate these effects, a specific example is cited where the amplitude comparison beams A and B are assumed to have the following characteristics: Antenna aperature (α) = 25.5λ

Receiving beam pattern function $\dfrac{\sin\,[\pi(\alpha/\lambda)\sin\theta]}{\pi(\alpha/\lambda)\sin\theta} = \dfrac{\sin\,(25.5\pi\sin\theta)}{25.5\pi\sin\theta}$

Beamwidth (at -3 dB points) = 2.0°

Squint angle (α) = 1.125°

Antenna height above ground = 50λ

The ground-reflection coefficient $Ke^{j\phi}$ is assumed to be $1.0e^{j\pi}$, which corresponds to horizontal polarization over an infinite conducting plane.

With these values, Fig. 20.10 shows a plot of the resultant relative field strengths in beams A and B as a function of the elevation boresight pointing angle θ. Note that in these curves, for each value of elevation angle θ, the antenna boresight is assumed to be directed exactly on the target. Thus, in the *absence* of ground reflections, the net field strengths in A and B would have been *equal* for all values of θ.

The corresponding off-boresight errors for the amplitude curves of Fig. 20.10 are shown in Fig. 20.11. In viewing the characteristics of the off-boresight error curves of Fig. 20.11, the important features to be noted are as follows:

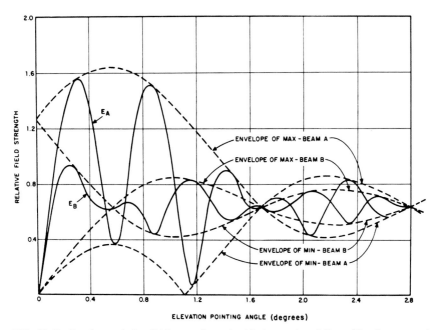

FIG. 20.10 Resultant relative field strength received in beams A and B resulting from ground reflections as a function of the elevation pointing angle of the beam crossover. Beamwidth is 2°; antenna height, 50λ; total squint angle, 2.25°; and reflection coefficient, 1.0.

1. The shape and amplitude of the *envelope of maximum errors* (dash curves) are dependent only on the antenna-beam pattern function in the elevation plane and the ground-reflection coefficient and are independent of the antenna electrical height above the ground.

2. The configuration of the error curve lying inside the envelope of maximum errors (i.e., the positions and spacing of the peak errors) is dictated chiefly by the antenna electrical height above the ground (in wavelengths) and, to a minor extent, by the phase angle of the antenna sidelobes that receive the reflected rays.

A similar analysis can be made of the errors due to surface reflections for a simultaneous phase comparison system, as found in Sec. 22.3 of the first edition of this handbook.

At high elevation pointing angles, where the errors are due only to surface reflections from the negative elevation-angle sidelobes, the errors contributed by various sidelobe levels in simultaneous amplitude comparison and phase comparison radars may be summarized approximately as follows: 0.2 to 0.3 beamwidth, when the surface-reflected sidelobe is 10 dB down from the peak; 0.07 to 0.10 beamwidth for −20 dB sidelobes; 0.025 to 0.035 for −30 dB sidelobes; 0.008 to 0.011 for −40 dB sidelobes; and about 0.003 beamwidth for −50 dB sidelobes illuminating the surface.

Low-Angle Squinted-Beam Height Finding. One height finding technique which has proved practical and effective against surface multipath is the so-called low-angle squinted-sum-beam height finding technique employed in the

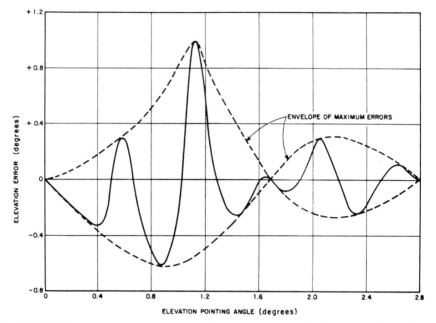

FIG. 20.11 Elevation errors due to ground reflection as a function of pointing angle for amplitude comparison system and uniform aperture distribution. Other conditions are as in Fig. 20.10.

TPS-59/GE-592/FPS-117 solid-state radar series.[27] The problem with conventional sum-delta monopulse in a surface multipath environment is that the delta-beam peak response is in the direction of the indirect-path reflection. The low-angle technique avoids this problem by using a pair of squinted beams on receive, as illustrated in Fig. 20.12. The lower beam, unweighted so as to generate as narrow a beam as permitted by the array aperture, is placed in elevation so that the indirect path is attenuated by the lower side of the beam. The upper beam, weighted to produce low sidelobes, is placed a degree or so above the lower beam so that the indirect-path echo is rejected by the sidelobes of the upper beam while the direct-path echo is received at high gain. This approach tends to minimize the amount of indirect-path energy in the two beams while maintaining coverage on the horizon.

The performance of the low-angle squinted-sum-beam technique also differs from that of a conventional stacked-beam pair, for two reasons. First, the two receive beams in the low-angle technique are not formed from identical aperture illuminations. Second, the transmit beam in the low-angle technique is narrow, reducing indirect multipath returns and to a lesser degree off-boresight signal-to-noise ratio.

The performance of a version of the low-angle squinted-sum-beam technique relative to that of a conventional monopulse is compared in Fig. 20.13 for a surface reflection coefficient of -1, approximately that of a smooth sea. The antenna in each case is situated at a height of twice the vertical aperture dimension above a flat earth. The monopulse case consists of a uniformly illuminated sum beam and a half-aperture uniform difference beam electronically phase-steered to $u_0 = 0.5$, processed, and implemented with full vector processing. The low-angle

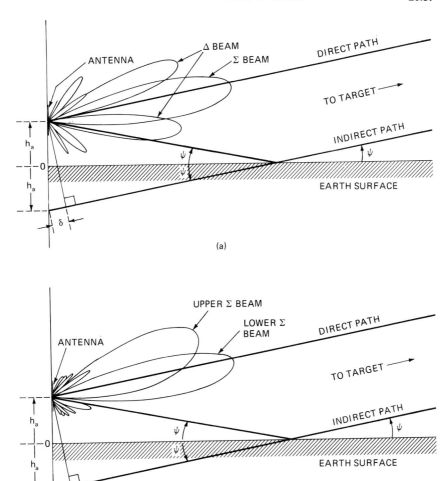

FIG. 20.12 Multipath impact on monopulse; δ = path difference. (*a*) Conventional sum-delta monopulse. (*b*) Squinted-sum-beam low-angle technique.

technique examined consists of a uniformly illuminated transmit/receive lower beam, accompanied by a weighted aperture upper beam. The lower beam is electronically phase-steered to $u_1 = 0.5$ while the upper beam is electronically phase-steered an additional $\delta u = 1.0$ (i.e., to $u_2 = 1.5$). This places the multipath largely in the sidelobe of the upper beams. In the monopulse case, the multipath introduces severe bias errors into the elevation-angle estimate with peaks on the order of 0.4 in u space. For a radar with an aperture of $L/\lambda = 32$, this corresponds to an rmse of 12.5 msines. The multipath bias errors dominate the total accuracy performance of the monopulse technique and, contrary to the behavior of the thermal errors, are not suppressed by high target signal-to-noise ratios. By

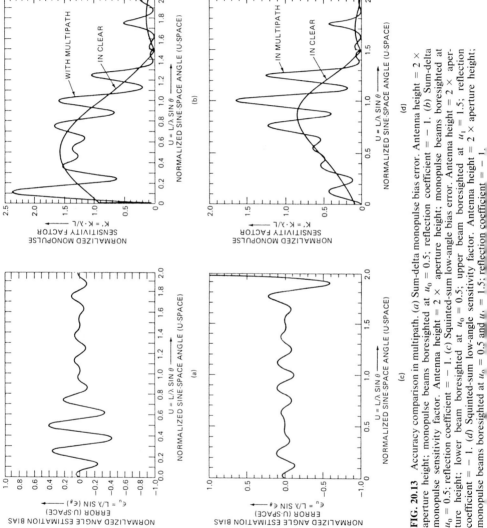

FIG. 20.13 Accuracy comparison in multipath. (*a*) Sum-delta monopulse bias error. Antenna height = 2 × aperture height; monopulse beams boresighted at $u_0 = 0.5$; reflection coefficient = − 1. (*b*) Sum-delta monopulse sensitivity factor. Antenna height = 2 × aperture height; monopulse beams boresighted at $u_0 = 0.5$; reflection coefficient = − 1. (*c*) Squinted-sum low-angle bias error. Antenna height = 2 × aperture height; lower beam boresighted at $u_0 = 0.5$; upper beam boresighted at $u_1 = 1.5$; reflection coefficient = − 1. (*d*) Squinted-sum low-angle sensitivity factor. Antenna height = 2 × aperture height; monopulse beams boresighted at $u_0 = 0.5$ and $u_1 = 1.5$; reflection coefficient = − 1.

contrast, the bias errors introduced by the multipath in the low-angle squinted-sum-beam technique are kept to peaks on the order of 0.15 in u space, corresponding to an rmse of approximately 4.7 msines for a radar of the same aperture. Further reduction might be possible with an optimization of the beam placement and aperture illumination functions.

REFERENCES

1. Skolnik, M. I.: Fifty Years of Radar, *Proc. IEEE*, vol. 73, pp. 182–197, February 1985.

2. Guerlac, H. E.: "Radar in World War II," Tomash Publishers, American Institute of Physics, Los Angeles, 1987.

3. Ridenour, L. N.: "Radar System Engineering," MIT Radiation Laboratory Series, vol. 1, McGraw-Hill Book Company, New York, 1947.

4. The SCR-268 Radar, *Electronics*, vol. 18, pp. 100–109, September 1945.

5. The SCR-584 Radar, *Electronics*, vol. 18, pp. 104–109, November 1945.

6. Schneider, E. G.: Radar, *Proc. IRE*, vol. 34, pp. 528–578, August 1946.

7. Brookner, E.: "Radar Technology," Artech House, Norwood, Mass., 1980, pp. 5–59.

8. Sutherland, J. W.: Marconi S600 Series of Radars, *Interavia*, vol. 23, pp. 73–75, January 1968.

9. Skolnik, M. I.: "Introduction to Radar Systems," 2d ed., McGraw-Hill Book Company, New York, 1980.

10. Brown, B. P.: Radar Height Finding, chap. 22 of Skolnik, M. I. (ed.): "Radar Handbook," 1st ed., McGraw-Hill Book Company, New York, 1970.

11. Simpson, T. J.: The Air Height Surveillance Radar and Use of Its Height Data in a Semi-Automatic Air Traffic Control System, *IRE Int. Conv. Rec.*, vol. 8., pt. 8, pp. 113–123, 1960.

12. Watanabe, M., T. Tamana, and N. Yamauchi: A Japanese 3-D Radar for Air Traffic Control, *Electronics*, p. 68, June 21, 1971.

13. AN/TPS-43E Tactical Radar System, brochure, Westinghouse Corporation.

14. The Martello High Power 3-D Radar System, brochure, Marconi Company.

15. RAT-31S 3D Surveillance Radar, brochure, Selenia Radar and Missile Systems Division, Rome.

16. Hammer, I. W.: Frequency-Scanned Arrays, chap. 13 of Skolnik, M. I. (ed.): "Radar Handbook," 1st ed., McGraw-Hill Book Company, New York, 1970.

17. Milne, K.: The Combination of Pulse Compression with Frequency Scanning for Three Dimensional Radars, *Radio Electron. Eng.*, vol. 28, pp. 89–106, August 1964.

18. Polmar, N.: "Ships and Aircraft of The U.S. Fleet," 14th ed., Naval Institute Press, Annapolis, Md., 1987, chap. 29, Electronic Systems.

19. AR-3D Mobile Air Defense Radar System, brochure, Plessey.

20. Pretty, R. T. (ed.): "Jane's Weapon Systems, 1981–1982," pp. 449–596.

21. Pfister, G.: The Series 320 Radar, Three Dimensional Air Surveillance Radar for the 1980's, *IEEE Trans.*, vol. AES-16, pp. 626–638, September 1980.

22. Lain, C. M., and E. J. Gersten: AN/TPS-59 System, *IEEE Int. Radar Conf. Rec.*, IEEE Publ. 75 CHO 938-1 AES, pp. 527–532, Apr. 21–23, 1975.

23. AN/TPS-59 Tactical Solid State Radar, brochure, General Electric Company.

24. AN/TPS-59: First Total Solid State Radar, *ADCOM Commun. Electron. Comput. Resources Dig.* (editorial article), October–November–December 1976.

25. AN/FPS-117 Minimally Attended Solid State Radar System, brochure, General Electric Company.

26. Gostin, J. J.: The GE592 Solid State Radar, *EASCON '80 Rec.*, pp. 197–203, IEEE Publ. 80 Ch 1578-4 AES, Sept. 29, 30, Oct. 1, 1980.

27. GE-592 Solid State Radar Systems, brochure, General Electric Company.

28. Klass, P. J.: Solid State 3D Radar for NATO Tested, *Aviat. Week Space Technol.*, May 21, 1979.

29. U.S. Air Force reports on Hughes Air Defense Radar, *Flight Int.*, Dec. 4, 1982.

30. Smith, E. K., and S. Weintraub: The Constraints in the Equation for Atmospheric Refractive Index at Radio Frequencies, *Proc. IRE*, vol. 41, pp. 1035–1037, August 1953.

31. Blake, L. V.: Ray Height Computation for a Continuous Nonlinear Atmospheric Refractive-Index Profile, *Radio Sci.*, vol. 3, pp. 85–92, January 1968.

32. Millman, G. H.: Atmospheric Effects on Radio Wave Propagation, in Berkowitz, R. S. (ed.): "Modern Radar Analysis, Evaluation and System Design," John Wiley & Sons, New York, 1965, pp. 315–377.

33. Bean, B. R., and E. J. Dutton: Radio Meteorology, *Nat. Bur. Stand. Monog.* 92, pp. 59–76, March 1966.

34. Bauer, J. R., W. C. Mason, and R. A. Wilson: Radio Refraction in a Cool Exponential Atmosphere, *MIT Lincoln Laboratory, Tech. Rept.* 186, August 1958.

35. Bauer, J. R., and R. A. Wilson: Precision Tropospheric Radio Refraction Corrections for Ranges from 10–500 Nautical Miles, *MIT Lincoln Laboratory, Rept.* 33G-0015, Feb. 20, 1961.

36. Sherman, S. M.: "Monopulse Principles and Techniques," Artech House, Norwood, Mass., 1985, chap. 5, chap. 12, pp. 345–348.

37. Rhodes, D. R.: "Introduction to Monopulse," McGraw-Hill Book Company, New York, 1959; reprinted by Artech House, Norwood, Mass., 1982.

38. Kinsey, R. R.: Monopulse Difference Slope and Gain Standards, *IRE Trans.*, vol. AP-10, pp. 343–344, May 1962.

CHAPTER 21
SYNTHETIC APERTURE RADAR

L. J. Cutrona
Sarcutron, Inc.

21.1 BASIC PRINCIPLES AND EARLY HISTORY

For airborne ground-mapping radar there has been continuous pressure and desire to achieve finer resolution. Initially, this finer resolution was achieved by the
application of "brute-force" techniques. Conventional radar systems of this type
were designed to achieve range resolution by the radiation of a short pulse and
azimuth resolution by the radiation of a narrow beam.

The range resolution problem and some of the pulse compression techniques
are discussed in Chap. 10. There it is shown that techniques are available for
achieving a resolution significantly finer than that corresponding to the pulse
width, provided a signal of sufficient bandwidth is transmitted. Since pulse compression is adequately treated in that chapter, the present chapter will discuss
pulse compression techniques only for cases in which the pulse compression
technique is intimately involved with synthetic aperture techniques. This is particularly true for configurations that perform both pulse compression and azimuth
compression simultaneously rather than with techniques that perform range compression and azimuth compression sequentially.

The basic technology discussed in this chapter is the exploitation of synthetic
aperture techniques for improving the azimuth resolution of a mapping radar to a
value significantly finer than that achievable by making use of the radiated
beamwidth.

Synthetic aperture radar (SAR) is based on the generation of an effective long
antenna by signal-processing means rather than by the actual use of a long physical antenna. In fact, only a single, relatively small, physical antenna is used in
most cases.

In considering a synthetic aperture, one makes reference to the characteristics
of a long linear array of physical antennas. In that case, a number of radiating
elements are constructed and placed at appropriate points along a straight line. In
the use of such a physical linear array, signals are fed simultaneously to each of
the elements of the array. Similarly, when the array is used as a receiver, the
elements receive signals simultaneously; in both the transmitting and the receiving modes, waveguide or other transmission-line interconnections are used, and
interference phenomena are exploited to get an effective radiation pattern.

The radiation pattern of a linear array is the product of two quantities if the radiating elements are identical. The radiation pattern of the array is the radiation pattern of a single element multiplied by an array factor. The array factor has significantly sharper lobes (narrower beamwidths) than the radiation patterns of the elements of the array. The half-power beamwidth β, in radians, of the array factor of such an antenna is given by

$$\beta = \frac{\lambda}{L} \tag{21.1}$$

In this expression, L is the length of the physical array, and λ is the wavelength.

In the synthetic antenna* case, only a single radiating element is used in most instances. This antenna is translated to take up sequential positions along a line. At each of these positions a signal is transmitted, and the radar signals received in response to that transmission are placed in storage. It is essential that the storage be such that both amplitude and phase of received signals are preserved.

After the radiating element has traversed a distance L_{eff}, the signals in storage resemble strongly the signals that would have been received by the elements of an actual linear array. Consequently, if the signals in storage are subjected to the same operations as those used in forming a physical linear array, one can get the effect of a long antenna aperture. This idea has resulted in the use of the term *synthetic aperture* to designate this technique.

In the case of an airborne ground-mapping radar system, the antenna usually is mounted to be side-looking, and the motion of the aircraft carries the radiating element to each of the positions of the array. These array positions are the location of the physical antenna at the times of transmission and reception of the radar signals.

The designer of a synthetic aperture radar has available a number of degrees of freedom that are not available to the designer of a physical linear array. These degrees of freedom derive from the fact that the signals in storage can be selected by range and that, if desired, a different operation can be performed on the signals at different ranges. One important operation of this type is that of *focusing*.

A physical linear array can be focused to a specific range. There will then be a depth of focus surrounding this range. However, most physical linear arrays are unfocused. This is sometimes stated by saying that the antenna is "focused at infinity." In a synthetic aperture radar, however, it is possible to focus each range separately by the proper adjustment of the phases of the received signals before the summation; this results in the effective synthetic aperture. Furthermore, if desired, a different weighting can be applied to each range, although usually the same type of weighting is used at all ranges.

There is another important difference between physical linear arrays and synthetic linear arrays. This difference results in the synthetic aperture having a resolution finer by a factor of 2 than that corresponding to a real linear array of the same length. Qualitatively, the following discussion indicates the physics resulting in this factor of 2. In a more general analysis, the factor 2 arises naturally.

In a physical linear array, the transmission of the signals results in an illumination of the target area. The angle selectivity of the linear array is provided only during the reception process. During this process, the differences in phase received by each element of the linear array give the antenna pattern. In the synthetic antenna radar, on the other hand, a single element radiates and receives signals. Consequently, the round-trip phase shift is effective in forming the effective radiation pattern. This relationship is written as

---

*The terms *synthetic antenna* and *synthetic aperture* are used interchangeably in this chapter.

SYNTHETIC APERTURE RADAR   **21.3**

$$\beta_{eff} = \frac{\lambda}{2L_{eff}} \qquad (21.2)$$

Here β_{eff} is the effective half-power beamwidth of the synthetic aperture, and L_{eff} is the length of the synthetic aperture.

A more detailed derivation of the resolution capability of a synthetic aperture radar will be given later in this chapter. The following derivation is that initially made by the author and his colleagues in the early days of synthetic aperture radar.

Let D represent the horizontal aperture of the physical antenna carried by an airborne ground-mapping radar. The width of the horizontal beam at range R gives the maximum value for the length of synthetic aperture that can be used at that range. Since the beamwidth of such an antenna is given by the ratio of the wavelength λ to its horizontal aperture D, the maximum length of this synthetic antenna aperture is given by

$$L_{eff} = \frac{R\lambda}{D} \qquad (21.3)$$

The linear resolution in azimuth δ_α is the product of the effective beamwidth given by Eq. (21.2) and the range R:

$$\delta_\alpha = \beta_{eff} R \qquad (21.4)$$

If Eqs. (21.2) and (21.3) are combined with Eq. (21.4), one obtains

$$\delta_\alpha = \frac{\lambda}{2L_{eff}} R = \frac{\lambda R}{2} \frac{D}{R\lambda} = \frac{D}{2} \qquad (21.5)$$

It will be noted that Eq. (21.5) indicates an azimuth linear resolution independent of both range and wavelength. Moreover, the result indicates that finer resolution is achievable with smaller rather than larger physical apertures. This spectacular result formed much of the motivation of the research in synthetic antenna radar.

The author was first exposed to the idea of a synthetic antenna radar in 1953, during a summer study which launched a program known as Project Michigan. During that summer, the ideas relating to synthetic antennas were presented by Dr. C. W. Sherwin,[1] then of the University of Illinois, Dr. Walt Hausz of the General Electric Company, and J. Koehler, at that time with Philco Corporation. Subsequently, it came to the author's attention that Carl Wiley and the Goodyear Aircraft Company had already undertaken some work and had made substantial progress in the synthetic-antenna area.

The Pioneer Award of the IEEE Aerospace and Electronic Systems Society was given to Carl Wiley in 1985 for his work in synthetic aperture radar. His remarks from that presentation are given in Ref. 2 and relate some of the early history of SAR.

Most of the early workers considered an unfocused synthetic antenna. However, Dr. Sherwin indicated that finer resolution should be achievable by using focusing because this technique removed what would otherwise be a restriction on the maximum length of synthetic antenna that could be used. The author and his colleagues at the University of Michigan undertook development of the focusing concept suggested by Dr. Sherwin.

21.2 FACTORS AFFECTING RESOLUTION OF A RADAR SYSTEM

In the following paragraphs a brief comparison of the conventional antenna, the unfocused synthetic antenna, and the focused antenna is given.[3,4] The language of synthetic apertures is used, and a comparison of the resolution capability for three cases is given. A more sophisticated derivation of simultaneous resolution in range and azimuth will be given later in this chapter.

Three cases are compared for their azimuth resolution capability: (1) the *conventional technique*, in which azimuth resolution depends upon the width of the radiated beam; (2) the *unfocused synthetic antenna technique*, in which the synthetic antenna length is made as long as the unfocused technique permits; and (3) the *focused synthetic antenna technique*, in which the synthetic antenna length is made equal to the linear width of the radiated beam at each range.

The linear azimuth resolution for the *conventional case* is given by

$$\text{Resolution}_{\text{conv}} = \frac{\lambda R}{D} \tag{21.6}$$

For the *unfocused case*, the resolution is

$$\text{Resolution}_{\text{unf}} = \frac{1}{2}\sqrt{\lambda R} \tag{21.7}$$

whereas for the *focused case*, the resolution is

$$\text{Resolution}_{\text{foc}} = \frac{D}{2} \tag{21.8}$$

where λ = wavelength of radar signal transmitted
D = horizontal aperture of antenna
R = radar range

Figure 21.1 is a plot of the resolution for each of these cases as a function of radar range. This plot is for an antenna aperture of 5 ft and a wavelength of 0.1 ft.

Conventional Technique. The conventional technique for achieving azimuth resolution has been that of radiating a narrow beam. In this case the *resolution* of a target depends upon whether the target is included within the half-power points of the radiated beam, although some techniques exist for resolving targets somewhat less than a beamwidth apart.

The computation of the linear azimuth resolution for the conventional case is well known. The appropriate expression is obtained by noting that the width of the radiated beam, in radians, is given by the ratio λ/D whereas the linear width of the beam at range R is the product of this beamwidth and range. These considerations lead to the result already written as Eq. (21.6).

A consideration from antenna theory is that Eq. (21.6) applies only to the far-

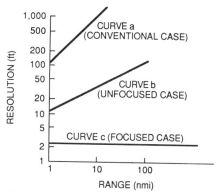

FIG. 21.1 Azimuth resolution for three cases: curve *a*, conventional; curve *b*, unfocused; curve *c*, focused.

field pattern of an antenna. The beginning of the far field occurs at a distance R_{min} for which

$$R_{min} \simeq \frac{D^2}{\lambda} \qquad (21.9)$$

It will be noted by substitution of Eq. (21.9) that the finest resolution achievable by the conventional technique is given by

$$\text{Minimum conventional resolution} = D \qquad (21.10)$$

The Unfocused Synthetic Aperture. The simpler of the synthetic antenna techniques is that which generates an unfocused synthetic aperture. In this case, the coherent signals received at the synthetic array points are integrated, with no attempt made to shift the phases of the signals before integration. This lack of phase adjustment imposes a maximum upon the synthetic antenna length that can be generated. This maximum synthetic antenna length occurs at a given range when the round-trip distance from a radar target to the center of the synthetic array differs by $\lambda/4$ from the round-trip distance between the radar target and the extremities of the synthetic aperture array.

The pertinent geometry is shown in Fig. 21.2. In this figure, R_0 represents the

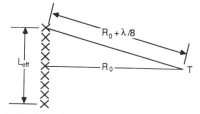

FIG. 21.2 Geometry for an unfocused synthetic antenna.

range from a radar target to the center of the array, and L_{eff} represents the maximum synthetic antenna length such that the distance from the target to the extremities of the synthetic aperture does not exceed $R_0 + \lambda/8$.

It is evident from this geometry that

$$\left(R_0 + \frac{\lambda}{8}\right)^2 = \frac{L_{eff}^2}{4} + R_0^2 \qquad (21.11)$$

If this expression is solved for L_{eff}, subject to the assumption that $\lambda/16$ is small compared with R_0, the result is

$$L_{eff} = \sqrt{R_0\lambda} \qquad (21.12)$$

Combination of Eqs. (21.2) and (21.12) gives

$$\beta_{eff} = \frac{1}{2}\sqrt{\frac{\lambda}{R_0}} \quad \text{rad} \qquad (21.13)$$

Multiplying this beamwidth by range results in the resolution given by Eq. (21.7).

It will be noted that for the unfocused case the transverse linear resolution is independent of the antenna aperture size, fineness of resolution is increased by the use of shorter wavelengths, resolution varies as the square root of λ, and the resolution deteriorates as the square root of range. A plot of Eq. (21.7) is given in Fig. 21.1.

The Focused Case. An expression for the resolution achievable in the focused case has been given as Eq. (21.8). It is significant that the azimuth resolution achievable for this case depends only upon the physical antenna aperture and that, in contradistinction to the conventional case, fine resolution requires the use of small rather than large antennas. Also significant is the fact that the achievable resolution for a given antenna size is independent both of the range and of the wavelength used. A graph of Eq. (21.8) is also shown in Fig. 21.1.

In order to achieve the resolution indicated by Eq. (21.8), the synthetic aperture length required is

$$L_{eff} = \frac{\lambda R}{D} \qquad (21.14)$$

The considerations used in arriving at Eq. (21.12) indicated that, unless additional processing were applied to the signals, antenna lengths such as those implied by Eq. (21.14) could not be achieved. The processing required is an adjustment of the phases of the signals received at each point of the synthetic antenna, which makes these signals of equal phase (cophase) for a given target. If this is done, the restrictions which limited the maximum antenna length to that given by Eq. (21.12) are no longer pertinent and the new limitation on the length of the synthetic antenna achievable becomes simply the linear width of the radiated beam at the range of the target.

In some cases, a resolution coarser than $D/2$ is sufficient. Then a fraction γ of the maximum focused synthetic antenna length can be used. For this case

$$L_{eff} = \frac{\gamma\lambda R}{D} \qquad (21.15)$$

and the achievable resolution is

$$\text{Resolution}_{foc} = \frac{D}{2\gamma} \qquad (21.16)$$

For situations in which the synthetic antenna length given by Eq. (21.15) is less than or equal to the synthetic antenna length for the unfocused case as given by Eq. (21.12), only a limited improvement in resolution is achievable for the focused case. However, if a resolution finer than that given by Eq. (21.7) is desired, focusing must be used. Focusing removes the restriction on synthetic aperture length that would otherwise apply.

21.3   RADAR SYSTEM PRELIMINARIES

Whether or not synthetic aperture generation is used, a number of components are required for a radar system. The use of a synthetic antenna and/or pulse compression places additional requirements on some of these components, especially with respect to coherence and stability.

It is the purpose of this section to present a block diagram of the portions of the radar system that precede the signal processor. A block diagram and several variants are described. The major variant is for the purpose of describing the transmitter-receiver portions of a radar system for the cases of synthetic antenna generation alone as compared with the case of synthetic antenna generation combined with pulse compression. The signal-processing operations will be discussed later.

The essential elements of a radar system useful in a synthetic aperture situation are shown in Fig. 21.3. The components that determine the radiated waveform are shown within the dotted lines in the upper left-hand corner of the diagram. This equipment consists of two stable oscillators. One of them is a local

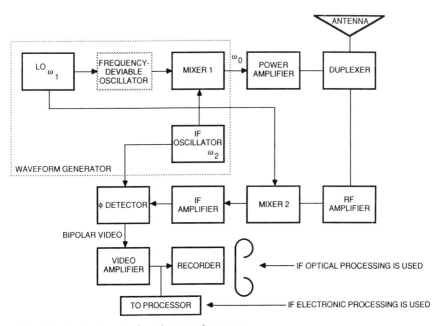

FIG. 21.3   Block diagram of a coherent radar system.

oscillator (LO) at radian frequency ω_2. The outputs of these oscillators are fed into mixer 1. In this mixer, a multiplicity of sum and difference frequencies is generated, and either the sum frequencies or the difference frequencies are selected and fed to the power amplifier.

If synthetic antenna generation without pulse compression is to be accomplished, the dotted component labeled "frequency-deviable oscillator" is not used, and the local oscillator is fed directly into mixer 1.

If pulse compression is to be combined with synthetic antenna generation, a frequency-deviable oscillator (FDO) is used for obtaining the desired waveform. In this case, the local oscillator is used to lock in the FDO. A ramp voltage is used to linearly frequency-modulate the FDO. This linearly frequency-modulated signal is then fed into mixer 1, instead of the LO signal, for the case of pulse compression. Waveforms other than linear frequency modulation may be used for pulse compression. In Fig. 21.3 the output of the video amplifier is fed to a recorder if optical processing is to be performed and/or to an electronic processor.

21.4 SIGNAL-PROCESSING THEORY

The theory of synthetic antenna generation combined with pulse compression is carried out below to show the information theoretic considerations involved and to indicate the operations necessary for achieving both azimuth synthetic antenna generation and pulse compression. A combined range-azimuth resolution function is derived. Following this treatment, an analysis of the signal-to-noise-ratio characteristics of a synthetic antenna compression radar are analyzed.

Detailed Resolution Analysis. The analysis will be carried out in terms of an ambiguity function whose properties indicate both the azimuth and the range resolution of the system. In the analysis, some conditions are stated for which the terms affecting range resolution can be factored from the elements affecting azimuth resolution, so that the resulting ambiguity function can be written as the product of two factors, one for range and one for azimuth.

Role of the Generalized Ambiguity Function. In this subsection, a definition of a generalized ambiguity function will be given, and its role in determining the resolution of a system will be interpreted.

To determine the generalized ambiguity function for a radar system, let a waveform $f(t)$ be radiated. We consider the operations performed upon the received signals with the objective of determining the radar reflectivity of the terrain being mapped. The function $f(t)$ may assume a variety of forms and may be a succession of short signals. If the quantity $\rho(x,y,z)$ represents the reflectivity of the terrain being mapped, the signal received by a radar system can be described by

$$s(t) = \iiint \rho(x,y,z) \, f\left[t - \frac{2R}{c}\right] dx \, dy \, dz \qquad (21.17)$$

The integration extends over the illuminated patch, and R is the range between a point (x,y,z) on the ground and the radar position $(vt,0,h)$. This equation shows that the received signal is the superposition of a large number of reflections within the illuminated pattern of the antenna and within the range gate which arrive simultaneously at the radar antenna.

The radar design problem is one of designing an operation on $s(t)$ to recover the reflectivity function $\rho(x,y,z)$. One such operation consists of passing the signals $s(t)$ through a matched filter. The operation of subjecting $s(t)$ to this matched filter is given by

$$e_0(R,R') = \int f^*\left[t - \frac{2R'}{c}\right]s(t)\,dt \qquad (21.18)$$

In this equation, the asterisk indicates complex conjugation, and R' indicates the range from the radar antenna to the specific point (x',y',z') at which the reflectivity is to be evaluated.

Substitution of Eq. (21.17) into Eq. (21.18) gives a fourfold integral for the output, namely,

$$e_0 = \int\int\int\int \rho(x,y,z)f\left[t - \frac{2R}{c}\right]f^*\left[t - \frac{2R'}{c}\right]dt\,dx\,dy\,dz \qquad (21.19)$$

If the order of integration can be inverted so that the integration with respect to t is performed first, one can define a quantity $\chi(x,y,z;x',y',z')$. This quantity is the generalized ambiguity function, given by

$$\chi(x,y,z;x',y',z') = \int f\left[t - \frac{2R}{c}\right]f^*\left[t - \frac{2R'}{c}\right]dt \qquad (21.20)$$

In terms of the generalized ambiguity function defined by Eq. (21.17), Eq. (21.19) can be rewritten as

$$e_0 = \int\int\int \chi(x,y,z;x',y',z')\rho(x,y,z)\,dx\,dy\,dz \qquad (21.21)$$

Equation (21.21) shows that the ambiguity function can be considered as a weighting function on $\rho(x,y,z)$. The output of the radar system, therefore, is the weighted average of ρ over a domain determined by the limits of integration. If the ambiguity function is localized at some point and is essentially zero at all other points, the output will be a good representation of the radar reflectivity at that point. Otherwise, the estimate of the reflectivity at a given point will be the weighted average given by Eq. (21.21).

Although no use will be made in this section of considerations determined by the limits of integration, it should be pointed out that Eq. (21.21) states that the output estimate of reflectivity in the radar system is a weighting of the reflectivity ρ by the product of the ambiguity function and the illumination function, whereby illumination function is meant the function that determines the distribution of signal energy over the plane. Ordinarily, the antenna illumination pattern, the pulse length, and the terms appearing in the radar equation determine this illumination function. In some cases, the ambiguity function χ has peaks at more than one point. If the illumination function excludes all but one of these peaks, an unambiguous system results.

Factorization of the Ambiguity Function. Let it be assumed that $f(t)$ can be written

$$f(t) = g(t)e^{i\omega_0 t} \tag{21.22}$$

In this equation, $g(t)$ is considered a complex function having both magnitude and phase, whereas ω_0 represents a carrier frequency. If $f(t)$ with the form given by Eq. (21.22) is used in Eq. (21.20), one obtains for the ambiguity function the expression

$$\chi = \int g\left[t - \frac{2R}{c}\right] g^*\left[t - \frac{2R'}{c}\right] e^{-i\omega_0(2R/c - 2R'/c)} \, dt \tag{21.23}$$

Let $f(t)$ consist of a sequence of transmissions. It is assumed that successive transmissions may be alike or that they may be different. Thus $f(t)$ will have the characteristics of being nonzero for a sequence of time intervals and of being zero otherwise. Further, let it be assumed that the exponential term in Eq. (21.23) varies slowly during each of these transmissions. This is equivalent to a statement that the electrical path length between a target and the radar changes by a small amount during each transmission. If this assumption is valid, the exponential term in Eq. (21.23) can be considered a constant during a given transmission, although it will vary between transmissions.

The integral that is the coefficient of the exponential term in Eq. (21.23) has the form of an autocorrelation function of g with itself. This autocorrelation function for g is given by Eq. (21.24). It will be noted that the autocorrelation function of g is a function of the difference in range $R - R'$. In Eq. (21.24) the integral is carried out over the times that $g(t - 2R/c)$ overlaps $g^*(t - 2R'/c)$ for a given transmission.

$$\phi_{gg} = \int g\left[t - \frac{2R}{c}\right] g^*\left[t - \frac{2R'}{c}\right] dt = \phi_{gg}\left[\frac{2R}{c} - \frac{2R'}{c}\right] \tag{21.24}$$

If the notation given by Eq. (21.24) is used, one obtains

$$\chi = \Sigma\phi_{gg}\left[\frac{2R}{c} - \frac{2R'}{c}\right] e^{-i\omega_0(2R/c - 2R'/c)} \tag{21.25}$$

Examination of Eq. (21.25) shows that if ϕ_{gg}, the autocorrelation function for g, is the same function for each member of the sequence of transmissions, then this element can be factored out and written outside the summation term of Eq. (21.25). The expression after this common term has been factored out is

$$\chi = \phi_{gg}\left[\frac{2R}{c} - \frac{2R'}{c}\right] \Sigma e^{-i\omega_0(2R/c - 2R'/c)} \tag{21.26}$$

The summation term in Eq. (21.26) gives the azimuth resolution of the system, whereas the term ϕ_{gg} gives the range resolution. It is evident from Eq. (21.26) that the autocorrelation function of g rather than g itself determines the range resolution of the system.

A variety of waveforms have been used to achieve range resolution. Among them the two most important are those in which $g(t)$ is a short pulse and those in which $g(t)$ is a linearly frequency-modulated short pulse (chirped signal). It is, of course, evident that any other waveform having a desirable autocorrelation function is equally possible for $g(t)$.

The form of $g(t)$ to be analyzed is that of the linearly frequency-modulated case.

Azimuth Resolution Factor of the Ambiguity Function.   The azimuth resolution capability of the system can be determined from an evaluation of the sum term written as the factor in Eq. (21.26). In this sum, R represents the range from the radar to an arbitrary point on the terrain being mapped, whereas R' represents the range to a specific point for which the reflectivity is to be estimated. The geometry appropriate to the synthetic antenna case is shown in Fig. 21.4. In this diagram, it is assumed that the aircraft carries a side-looking antenna and flies at height h and with velocity v along the x axis so that the aircraft location is given by

$$x = vt$$

Consider two points being mapped having coordinates $(0,y_0,0)$ and $(x',y_0,0)$. R_0 is defined by

$$R_0 = \sqrt{y_0^2 + h^2} \tag{21.27}$$

and R and R' can be written, respectively, as

$$R = \sqrt{R_0^2 + x^2} \approx R_0 + \frac{x^2}{2R_0} \tag{21.28}$$

$$R' = \sqrt{R_0^2 + (x - x')^2} \approx R_0 + \frac{(x - x')^2}{2R_0} \tag{21.29}$$

The second expression for Eqs. (21.28) and (21.29) is an approximation that is valid whenever the inequalities $x \ll R_0$, $(x - x') \ll R_0$ are satisfied. If the approximate forms for $R - R'$ as given by the approximate expressions in Eqs.

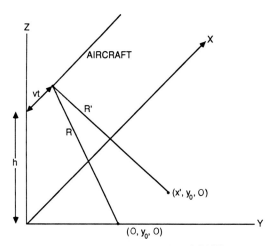

FIG. 21.4 Geometry for Eqs. (21.27) and (21.28).

(21.28) and (21.29) are used to evaluate the sum appearing in Eq. (21.26), one obtains

$$\Sigma e^{-i\omega_0(2R/c - 2R'/c)} = \Sigma e^{-i(2\omega_0/c)(2xx' - x'^2)/2R_0} \tag{21.30}$$

Thus far, the summation index has not been defined. To proceed further, it is necessary to indicate the summation index and its bounds. Let it be assumed that the transmissions radiated are a sequence of pulses with time intervals between pulses that are multiples of T. Then the variable x can be given as an integral multiple of the distance vT moved between successive transmissions. This relationship is

$$x = nvT \tag{21.31}$$

If Eq. (21.31) is substituted into Eq. (21.30), one obtains

$$\Sigma = e^{i(2\omega_0/c)x'^2/2R_0} \sum_{-N/2}^{N/2} e^{-i4\pi(x'/\lambda R_0)nvT} \tag{21.32}$$

In writing Eq. (21.32), the summation is carried over $N + 1$ terms. The synthetic antenna length implied by these limits is given by $L = NvT$.

Inasmuch as the summation terms in Eq. (21.32) are those corresponding to a geometric progression, the sum term can be immediately evaluated. The result is

$$\Sigma = e^{i(2\omega_0/c)x'^2/2R_0} \frac{\sin\,[(N + 1)4\pi x'vT/2\lambda R_0]}{\sin\,[4\pi x'vT/2\lambda R_0]} \tag{21.33}$$

The right-hand side of Eq. (21.33) gives the factor of the generalized ambiguity function that is responsible for the azimuth resolution. It will be noted that there are a phase term given by the exponential and a magnitude term given by the remaining terms in Eq. (21.33). The specific form of Eq. (21.33) is a consequence of the equal weighting of the signals. A weighting function can be used to shape the sidelobes in direct analogy with the use of such a technique in real antenna design.

Range Resolution Factor of the Ambiguity Function. This subsection considers the factor in the generalized ambiguity function, Eq. (21.25), that is responsible for range resolution. This factor, ϕ_{gg}, has been defined by Eq. (21.24). A specific form for $g(t)$ will be assumed, and an evaluation of ϕ_{gg} will be made for this specific waveform. The function $g(t)$ to be analyzed is that in which each radiation consists of a short, linearly frequency-modulated signal. An expression for $g(t)$ in this case is

$$g(t) = e^{i\alpha t^2} \tag{21.34}$$

The use of Eq. (21.34) in Eq. (21.24) results in

$$\phi_{gg} = e^{i\alpha[(2R/c)^2 - (2R'/c)^2]} \int_{-\tau/2}^{\tau/2} e^{-i\alpha[(4R/c)t-(4R'/c)t]} dt \tag{21.35}$$

$$\phi_{gg} = e^{i\alpha[(2R/c)^2 - (2R'/c)^2]} \tau \frac{\sin \{\alpha\tau[(2R/c) - (2R'/c)]\}}{\alpha\tau[(2R/c - 2R'/c)]} \tag{21.36}$$

Equation (21.36) gives the range resolution factor of the ambiguity function for a transmitted waveform of the type expressed by Eq. (21.34). It will be noted that this term consists of a phase term and an amplitude term.

Inasmuch as both the azimuth resolution factor and the range resolution factor have been evaluated, the generalized ambiguity function can be written:

$$\chi = e^{i\alpha[(2R/c)^2 - (2R'/c)^2]} \tau \frac{\sin \alpha\tau[(2R/c) - (2R'/c)]}{\alpha\tau(2R/c - 2R'/c)} e^{i(2\omega_0/c)(x'^2/2R_0)}$$

$$\times \frac{\sin [(N + 1)(2\pi x'vT)/(\lambda R_0]}{\sin [2\pi x'vT/\lambda R_0]} \tag{21.37}$$

In interpreting Eq. (21.37), it should be noted that there are phase terms and magnitude terms. One of the magnitude terms corresponds to the range resolution capability of the system; the other, to azimuth resolution. Quantitative expressions for the resolution in each of these coordinates will be obtained below.

The resolution terms in Eqs. (21.33), (21.36), (21.37) are of the form

$$\sin Nz/\sin z \tag{21.38}$$

with

$$z = 2\pi x'vT/\lambda R_0 \tag{21.39}$$

for the azimuth resolution case, and

$$\sin z/z \tag{21.40}$$

with

$$z = \alpha\tau[2R/c - 2R'/c] \tag{21.41}$$

for the range resolution case.

If the value

$$L = (N + 1)VT \tag{21.42}$$

is combined with Eq. (21.39) and one recognizes that

$$\alpha\tau = 2\pi B \tag{21.43}$$

where B is the chirp signal bandwidth, one can show that the azimuth resolution δ_a and the range resolution δ_r are given by

$$\delta_a = (1.4\lambda R_0)/(\pi L) \tag{21.44}$$

and

$$\delta_r = (1.4c)/(2\pi B) \tag{21.45}$$

Ambiguities. It is the objective of this subsection to make some observations regarding the possibility of multiple peaks (ambiguities) in the ambiguity function as given by Eq. (21.37) and its effect on system performance as given by Eq. (21.19).

The final term on the right-hand side of Eq. (21.37) is of the form

$$\frac{\sin [(N + 1)q]}{\sin q} \tag{21.46}$$

where the quantity q is defined by

$$q = \frac{2\pi x' vT}{\lambda R_0} \tag{21.47}$$

The azimuth resolution factor, therefore, has a peak whenever the quantity q takes on a value equal to an integral multiple of π rad. Thus the system is potentially ambiguous for values of x' that are solutions of

$$\frac{2\pi x' vT}{\lambda R_0} = m\pi \tag{21.48}$$

Actually, it is more meaningful to solve for the ratio x'/R_0. The angle v gives directions from the broadside at which angle ambiguities are potentially possible. This relationship is given as

$$\sin \theta = \frac{x'}{R_0} = \frac{\lambda}{D} \frac{D}{vT} \frac{m}{2} = \frac{m\beta D}{2vT} \tag{21.49}$$

In the last form of this equation, the numerator and the denominator have been multiplied by D, the horizontal aperture of the antenna, to express the result in terms of the radiated beamwidth $\beta = \lambda/D$.

Thus, the possibility of azimuth ambiguities arises as a natural consequence of the signals radiated and of the processing method. Ordinarily, these potential ambiguities in azimuth are suppressed by the illumination factor. The illumination pattern β is chosen so that the values of β corresponding to more than one value of m are not illuminated.

Possibilities also exist for ambiguities in range. The analysis carried out to the point of Eq. (21.37) was not sufficiently general to predict ambiguities in range. However, if reference is made to Eq. (21.24), it is evident that the autocorrelation function ϕ_{gg} will be periodic if $g(t)$ is periodic. Thus range ambiguities can also occur. In particular, range ambiguities will occur for ranges having a difference given by

$$\Delta R = \frac{cT}{2} \tag{21.50}$$

where T is the interpulse period.

To date, systems have been built which have avoided ambiguities by virtue of illuminating only the part of the ambiguity diagram that excludes all but one major peak. This technique has sometimes been referred to as *ambiguity avoidance*. For some sets of parameters, ambiguities cannot be avoided by using a radar with a single radiated beam. The use of multiple beams solves this problem. This topic is discussed in Sec. 21.5.

Signal-to-Noise-Ratio Considerations. It is the purpose of this subsection to derive expressions for signal-to-noise (S/N) ratio for radars in which pulse compression and synthetic antenna techniques are used. The signal-to-noise ratio for a radar system as a result of the reception of a single pulse is given by the well-known radar equation

$$\frac{S}{N} = \frac{P_t G_t A_r \sigma}{(4\pi)^2 R^4 k T_0 B F_n} \tag{21.51}$$

In a pulse compression radar, signal-to-noise improvement occurs in the ratio of the uncompressed pulse length τ_i to compressed pulse length τ_0.

In a radar that achieves its azimuth resolution by the generation of a synthetic antenna, there is an additional signal-to-noise improvement factor due to the integration of a number of pulses. The number of pulses integrated is equal to the product of the pulse repetition frequency (PRF) and the time necessary to generate the synthetic antenna. In turn, this time is equal to the ratio of synthetic length L to aircraft speed v.

An expression in which the product of both factors has been written is

$$\text{Improvement factor} = \frac{\tau_i}{\tau_0} \frac{\text{PRF } L}{v} \tag{21.52}$$

The length of synthetic antenna required to achieve azimuth resolution δ_{az} at range R and wavelength λ is given by

$$L = \frac{R\lambda}{2\delta_{az}} \tag{21.53}$$

The substitution of Eq. (21.53) into Eq. (21.52) gives for the improvement factor

$$\text{Improvement factor} = \left[\frac{\tau_i}{\tau_0}\right] \frac{\text{PRF } R\lambda}{2v\delta_{az}} \tag{21.54}$$

The signal-to-noise ratio including the improvement factor is obtained by multiplying together the expressions given by Eqs. (21.51) and (21.54). The result of this multiplication is

$$\frac{S}{N} = \frac{P_t G_t A_r \sigma}{(4\pi)^2 R^4 k T_0 B F_n} \frac{\tau_i}{\tau_0} \frac{\text{PRF } R\lambda}{2v\delta_{az}} \tag{21.55}$$

Although Eq. (21.55) contains the desired information, it is useful to modify the term somewhat by expressing the antenna gain in terms of the effective area of its aperture and of the wavelength. This expression is written

$$G_t = \frac{4\pi A_r}{\lambda^2} \qquad (21.56)$$

It is also desirable to collect together three terms in the numerator of Eq. (21.55), namely, P_t, the peak transmitted power; τ_i, the uncompressed pulse length; and the PRF. The product of these three factors gives the average power P_{av}. This relationship is written

$$P_{av} = P_t\tau_i\text{PRF} \qquad (21.57)$$

In the design of a radar system, the bandwidth B is chosen to be the reciprocal of τ_0. Hence the product of the bandwidth and the compressed pulse width is approximately equal to unity. This relationship is written

$$B\tau_0 \approx 1 \qquad (21.58)$$

Finally, it is useful to express the radar cross section σ in terms of the azimuth and range resolution, δ_{az} and δ_r, as well as in terms of the reflectivity of the terrain, ρ. The radar cross section is equal to the reflectivity of the terrain multiplied by the projected area. This projection accounts for the term $\sin\psi$. The expression for the radar cross section in terms of these parameters is given by

$$\sigma = \rho\delta_r\delta_{az} \sin\psi \qquad (21.59)$$

Substitution of Eqs. (21.56) to (21.59) for the corresponding quantities in Eq. (21.55) gives

$$\frac{S}{N} = \frac{P_t\sigma_i\text{PRF}4\pi A^2\rho\delta_r\delta_{az} (\sin\psi)\ R\lambda}{(4\pi)^2R^4\lambda^2kT_0F_n(B\tau_0)2v\delta_{az}} \qquad (21.60)$$

In writing Eq. (21.60), no cancellation of terms has been made. Canceling terms that appear in both the numerator and the denominator results in

$$\frac{S}{N} = \frac{P_{av}}{8\pi}\ \frac{A_r^2\rho\delta_r}{kT_0F_nR^3\lambda}\ \frac{\sin\psi}{v} \qquad (21.61)$$

This is the desired result.

Equation (21.61) does not take into account factors concerned with ambiguity avoidance. The inclusion of such effects is given in Ref. 5.

Equation (21.61) shows that the signal-to-noise ratio at the output of a radar that has used pulse compression and has generated a synthetic antenna has the following properties different from conventional radar:

1. The signal-to-noise ratio is proportional to the size of the range resolution element and is independent of the size of the azimuth resolution element.

2. The signal-to-noise ratio is inversely proportional to the third power of range.

3. The signal-to-noise ratio is inversely proportional to the wavelength.

4. The signal-to-noise ratio is inversely proportional to the speed of the aircraft.

Effect of Phase Errors. In actual equipment, phase errors arise from a number of sources. Some of the instabilities arise in oscillators and other electrical components of the radar, but other sources of phase error are inhomogeneities in the atmosphere or the result of uncompensated deviation of the aircraft from linear unaccelerated motion. A number of modifications of the synthetic antenna pattern result from such uncompensated phase errors. These modifications include beam canting, beam spreading, peak gain reduction, and redistribution of the ratio of energy in the main lobe to that in the sidelobes. An analytic formulation and Monte Carlo computer simulation of the effects of phase errors for normally distributed random phase errors and for three cross-correlation functions have been given by Greene and Moller.[6]

Signal Processing. The preceding subsections have discussed a number of aspects of radar signal processing. Also discussed has been the radar system up to the point of signal processing. As part of that analysis, the waveforms of signals at a number of points in the radar system have been described. It is the purpose of this subsection to discuss a number of aspects of signal processing that are common to all mechanizations.

Many fine-resolution radar systems employ both pulse compression and synthetic antenna generation.

Theoretic Aspects of Synthetic Aperture Generation. In generating a synthetic antenna, the returns from a number of spatial positions must be combined. In doing this, one usually wishes to apply weighting to the signals for synthetic antenna pattern sidelobe-level control; in the case of focused synthetic antennas, one also wishes to adjust the phases of the signals before combination.

In the preceding discussion, the signal was represented as a function of time. For present purposes it is preferable to consider the signals as a discrete sequence numbered from 1 to N. Let S_n represent the signal received when the physical antenna is at the nth position of the antenna array. Let W_n represent weighting applied to S_n, and let ϕ_n represent the phase adjustment required for focusing.

The operation of synthetic antenna generation then consists of taking the vector sum of the signals S_n, adjusting their phases, and multiplying by weighting factors. The sum of this operation is given by

$$\Sigma S_n e^{i\phi_n} W_n = \text{focused pattern} \qquad (21.62)$$

In the case of unfocused synthetic antenna generation the phase adjustments ϕ_n are not made. In this case the signal operations required have the form

$$\Sigma S_n W_n = \text{unfocused synthetic pattern} \qquad (21.63)$$

There are many mechanizations possible to carry out the operations indicated by Eqs. (21.62) and (21.63). Some of them are described below. Two common techniques are digital and optical in nature.

Discussions regarding optical and digital data processing are given in Refs. 7 through 11.

Optical Techniques.   The optical techniques involve the recording of the radar signals on a transparency, most frequently silver halide photographic film in any of a number of formats. Initially the successive range sweeps were placed parallel and side by side; later polar format was used. The growing understanding that we are really collecting a portion of the three-dimensional spectrum has led to use of a three-dimensional storage format.[8] This topic will be discussed further in Sec. 21.5.

The most frequently used optical processors are based on the tilted-plane optical processor described by Kozma, Leith, and Massey.[7] In this processor, both the input plane and the output plane are tilted (i.e., they are not perpendicular to the optical axis). The optical components are telescopic, and the powers of the elements in two perpendicular planes are unequal. The telescopic elements include both spherical and cylindrical elements.

The evolution of this processor is based in part upon the recognition that signal histories may be assigned focal lengths and behave to some degree as optical elements.

The azimuth along-track signals from a point target are similar to those of a zone plate. The focal length is proportional to target range. If pulse compression is used, all targets have associated with them zone plates in the range direction. These all have the same focal length. The recorded signals before processing are often referred to as *signal histories.*

Digital Processors.   Digital processing has emerged as the preferred means when the amount of data to be processed is not too great. Mechanization of SAR operations is often computation intensive.

In cases not based on polar format, correlation operations have been used. These operations are usually performed using frequency-plane equivalent of correlation.

$$\int f(q)\, g(q - x)\, dx = \int F(\omega)\, G(\omega)\, \exp\,(j\omega x)\, d\omega \qquad (21.64)$$

In other cases, such as polar format, Fourier transform operations are indicated and the use of the fast Fourier transform (FFT) plays an important role.

Associated with FFT processing is the fact that algorithms exist for processing two-dimensional data with sampling points in the rectangular arrays, whereas the sample points obtained are equally spaced on radial lines. This requires formatting operations on the data points to convert them to rectangular format.[12,13]

Imagery from synthetic aperture radars is shown in Figs. 21.5 and 21.6. These images were provided by the Environmental Research Institute of Michigan (ERIM).

21.5 ADDITIONAL SYSTEM CONSIDERATIONS

In this section a number of considerations peculiar to synthetic aperture radar are discussed. Some are additional performance requirements on the components of the system; some are concerned with system aspects.

Antenna.   The horizontal aperture of the antenna determines the finest azimuth resolution achievable in a single-beam synthetic aperture radar except for the searchlight mode. Moreover, in the signal processing it is assumed that the antenna gain is constant as a function of along-track position. Thus, it is necessary to have a degree of stabilization of antenna pointing so that the beam

FIG. 21.5 STAR-1 radar imagery, lower Lake St. Clair, upper Detroit River. Resolution, 20 ft (6 m). (*Courtesy Environmental Research Institute of Michigan.*)

rotation is some minor fraction of the beamwidth. In most cases, the antenna is side-looking, although in some cases the antenna is positioned to an angle off broadside and the system then operates in what is called the *squint* mode.

Receiver-Transmitter. The transmitter and receiver for synthetic antenna radars require maintenance of coherence of the radar signals. Consequently, there is emphasis on the stability of the oscillators and more rigorous requirements on the components. The output of coherent radar is the output of a synchronous demodulator rather than that of an envelope detector commonly used in radars. The output is bipolar video, in which a reference bias level corresponds to zero level of video output.

Storage and Recording. It is inherent that synthetic antenna radars and pulse compression radars require the storage of radar data, because the *data* for synthetic antenna generation does not occur simultaneously but is collected over some interval of time. Operations are then performed on these signals to achieve the selectivity of the radar. Moreover, each radar return participates in forming the output for a large number of points on the output map. The requirements for storage are therefore very large. Since a high volume of storage is required for a fine-resolution radar system, photographic storage is commonly used.

For digital processing, storage of the digital signals after analog-to-digital (A/D) conversion is required. The amount of this data can be great and often limits the area over which fine resolution can be obtained. A description of what needs to be done is given in Refs. 12 and 13.

In selecting a storage medium one must consider the rate at which information must be recorded, the amount of data to be recorded, and the rate at which the

FIG. 21.6 STAR-1 radar imagery, lower Detroit River. Resolution, 20 ft (6 m). (*Courtesy of Environmental Research Institute of Michigan.*)

storage must be read out for performing the azimuth compression and the pulse compression.

Motion Compensation. In generating the synthetic antenna the signal-processing equipment assumes that the radar flies along a straight line at constant speed. In practice, the vehicle carrying the antenna is subject to deviations from unaccelerated flight. Therefore, it is necessary to have auxiliary equipment to compensate for other than straight-line motion. Motion-compensation equipment must include sensors capable of detecting the deviation of the flight path from a linear path. The output of these sensors is used in a variety of ways. For motion compensation proper, the received-signal phase must be adjusted to compensate for the displacement of the real antenna from the location of the ideal synthetic antenna being generated.

A consideration of the geometry involved shows that the phase correction that must be applied is a function of depression angle. Consequently, the correction must be made as a function of range. The rate of change correction is very rapid at steep depression angles and becomes slower at shallow depression angles.

Squint Mode. In most examples of synthetic aperture radar, the beam is directed at right angles to the ground track of the aircraft. In some cases, however, it is desirable to "squint" the antenna beam so that an area either forward or aft of the aircraft is mapped. It is necessary to position the antenna beam so that the maximum of its radiation pattern points in the desired squint direction. Moreover, it is usually necessary to modify the signal processors to take into account the average doppler frequency shift that occurs when the antenna points in a direction other than normal to the flight path. It is, of course, also necessary to take the geometry of the squint mode into account in designing recorders and displays.

Spotlight Mode. In Sec. 21.2 Eq. (21.8) was derived for a radar in *strip-map* mode, i.e., for the case that the radar antenna is in a fixed orientation and the radar beamwidth $R\lambda/D$ is used as the length of synthetic aperture generated. One can increase the antenna length by use of *spotlight* mode. In this case the radar antenna is continuously pointed toward the region being imaged. For this case one can make a synthetic antenna length longer than $R\lambda/D$, or one can make several images and noncoherently integrate them. Spotlight mode also makes possible the use of higher antenna gain.

Effects of Motion Errors. In generating synthetic aperture antenna images, one needs to estimate the along-track and cross-track velocities of targets in order to derive the matched filter to use in imaging. If one has an error in the radial velocity of the target, one gets a rotation in target position. If one has an error in along-track velocity, a limit is set to the achievable resolution.

Multiple-Beam Radars. The analyses leading to Eqs. (21.5) and (21.8) are correct for synthetic aperture radars which radiate only one beam. However, system parameters sometimes dictate the use of multiple beams.

The use of multiple beams is motivated by several considerations, such as ambiguity avoidance and the achievement of higher antenna gain. The achievement of a larger area coverage rate is another but less likely motivation. A great deal of flexibility is possible. The multiple beams may be arranged in the azimuth direction or the range direction, or both. With the use of multiple beams, it is possible to achieve any desired combination of unambiguous range, resolution, and area rate. The antenna area and the number of beams are determined from the performance parameters.

A more complete analysis of multibeam systems is given in Refs. 14 through 17.

ISAR. *Inverse synthetic aperture radar* (ISAR) is the term used when the motion of the object being imaged is used instead of the motion of the radar. A more general case is that in which both the radar and the object are in motion. In ISAR, the target motion is often not known to the radar. Hence, a major part of the problem is determination of the target motion to generate the matched filter needed to generate an image. A number of techniques have been studied for providing data regarding both translation and rotation of moving objects. An example of such work is that of B. Steinberg.[18]

Three-Dimensional Spectrum. The analysis starting with Eqs. (21.15) and (21.16) contains the assumption that a matched filter is applied to the radar returns from each point. This can in fact be done. It would reduce the signal-processing load if a reference function could be applied over a region. This, too, can be done, but range walk and defocusing set limits to the size of a patch which can be handled in this manner. Of the methods that have been proposed, the use of polar format and its generalization and the collection of a portion of the three-dimensional spectrum of the scene being mapped are most significant.

If one starts with an expression such as Eq. (21.17), performs the integration with respect to τ, where τ is a dummy variable to replace t, and then makes a Fourier transform of the results, one gets

$$E_0(\omega) = \int \rho(x,y,z)|G(\omega)|^2 \exp\left[-j(2\omega/c)(2r/c - 2r'/c)\right] dx\, dy\, dz \quad (21.65)$$

In this equation $|G(\omega)|^2$ is the Fourier transform of the autocorrelation function of $g(t)$. Let the vector difference, or r, and r' be represented by q.

$$\mathbf{r} = \mathbf{r}' + \mathbf{q} \quad (21.66)$$

so that

$$r = r' + \frac{r' \cdot q}{r'} + \frac{q^2 + (r' \cdot q/r')^2}{2r'} \quad (21.67)$$

If one can neglect the last term, and if one writes

$$\hat{\mathbf{r}} = \mathbf{r}/r \quad (21.68)$$

for the unit vector along r, one can write

$$r - r' = \hat{\mathbf{r}} \cdot \mathbf{q} \quad (21.69)$$

Let a vector G be defined by

$$\mathbf{G} = (2\,\omega/c)\hat{\mathbf{r}} \quad (21.70)$$

then

$$E_0(\omega) = \int \rho(x,y,z)G(\omega)^2 \exp\left(-jG \cdot q\right) dq \quad (21.71)$$

One notes that except for the factor $|G(\omega)|^2$, Eq. (21.71) has the form of a three-dimensional spectrum of $\rho(\mathbf{q})$.

Equation (21.71) has been derived by Jack Walker.[8] Related developments involving use of the *projection-slice theorem* have been given by a number of other authors.[9,19]

In interpreting Eq. (21.71), it is useful to consider the point **r** as a general point on the object being imaged and **r'** as a reference point on the object. The vector **q** is then the vector from the reference point to all other points on the object, and the integration extends over the object.

Equation (21.71), being a three-dimensional spectrum of an object, requires that the data be taken with an origin of coordinates and a coordinate system fixed with respect to the object being imaged. Hence the effect of translation and rotation of a moving object must be compensated in order to image that object.

The three-dimensional spectrum is the proper format for storing radar data. The polar format is a special case in which the radar collection is performed in a plane. Use of the projection-slice theorem enables one to project the data along any direction. This promises to give images of *slices* of moving objects.

The projection of the radar data, followed by a two-dimensional Fourier transform, can be used to form images for the most general motion of both the radar and the moving object.

REFERENCES

1. Sherwin, C. W., P. Ruina, and R. D. Rawcliffe: Some Early Developments in Synthetic Aperture Radar Systems, *IRE Trans.*, vol. MIL-6, pp. 111–115, April 1962.

2. Wiley, C.: Pioneer Award acceptance remarks, *IEEE Trans.*, vol. AES-21, pp. 433–443, May 1986.

3. Cutrona, L. J., W. E. Vivian, E. N. Leith, and G. O. Hall: A High Resolution Radar Combat-Surveillance System, *IRE Trans.*, vol. MIL-5, pp. 127–131, April 1961.

4. Cutrona, L. J., and G. O. Hall: A Comparison of Techniques for Achieving Fine Azimuth Resolution, *IRE Trans.*, vol. MIL-6, pp. 119–121, April 1962.

5. Skolnik, M. I.: "Introduction to Radar Systems," 2d ed., McGraw-Hill Book Company, 1980, p. 522, Eqs. 14 and 16.

6. Greene, C. A., and R. T. Moller: The Effect of Normally Distributed, Random Phase Errors on Synthetic Array Gain Patterns, *IRE Trans.*, vol. MIL-6, pp. 130–139, April 1962.

7. Kozma, A., E. N. Leith, and N. G. Massey: Tilted Plane Optical Processor, *Appl. Opt.*, vol. 11, pp. 1766–1777, August 1972.

8. Walker, J. L.: Range Doppler Imaging of Rotating Objects, *IEEE Trans.*, vol. AES-16, pp. 23–52, January 1980.

9. Ausherman, D. A., A. Kozma, J. L. Walker, H. M. Jones, and E. C. Poggio: Developments in Radar Imaging, *IEEE Trans.*, vol. AES-20, pp. 363–400, July 1984.

10. Cutrona, L. J., E. N. Leith, C. J. Palermo, and L. J. Porcello: Optical Data Processing and Filtering System, *IRE Trans.*, vol. IT-6, pp. 386–400, June 1960.

11. McLeod, J.: The Axicon: A New Type of Optical Element, *J. Opt. Soc. Am.*, vol. 44, pp. 592–597, August 1954.

12. Ausherman, D. A.: Digital Image Processing, *Proc. SPIE*, vol. 528, pp. 118–133, Jan. 22–23, 1985.

13. Ausherman, D. A.: Digital versus Optical Techniques in Synthetic Aperture Radar (SAR) Data Processing, *Opt. Eng.*, vol. 19, pp. 157–167, March–April 1980.

14. Cutrona, L. J.: Means to Achieve Wide Swath Widths in Synthetic Aperture Radar, *Addendum to Proc. Synth. Aperture Radar Technol. Conf.*, pp. V-9-1–V-9-21, New Mexico State University, Las Cruces, Mar. 8–10, 1978.

15. Claassen, J. P., and J. Eckerman: A System Concept for Wide Swath Constant Incident Angle Coverage, *Proc. Synth. Aperture Radar Technol. Conf.*, Pap. VI-4, pp. VI-4-1–VI-4-19, New Mexico State University, Las Cruces, Mar. 8–10, 1978.

16. Cutrona, L. J.: Comparison of Sonar System Performance Achievable Using Synthetic-Aperture Techniques with the Performance Achievable by More Conventional Means, *J. Acoust. Soc. Am.*, vol. 58, pp. 336–348, August 1975.

17. Cutrona, L. J.: Additional Characteristics of Synthetic Aperture Sonar Systems and a Further Comparison with Non-Synthetic Aperture Sonar Systems, *J. Acoust. Soc. Am.*, vol. 61, pp. 1213–1217, May 1977.

18. Steinberg, B. D.: "Microwave Imaging with Large Antenna Arrays," John Wiley & Sons, New York, 1983.

19. Munson, D. C., Jr., J. D. O'Brien, and W. K. Jenkins: A Topographic Formulation of Spotlight-Mode Synthetic Aperture Radar, *Proc. IEEE*, vol. 71, pp. 917–925, August 1983.

CHAPTER 22
SPACE-BASED RADAR SYSTEMS AND TECHNOLOGY

Leopold J. Cantafio
Space and Technology Group TRW

22.1 INTRODUCTION

Significant developments have been made in space-based radar (SBR) systems and technology since the 1970 edition of the *Radar Handbook* was published. A new rendezvous radar was developed for the space shuttle and has become operational. The unmanned orbital maneuvering vehicle (OMV) will use a new low-cost rendezvous radar that is expected to be operational during the early 1990 time period. Synthetic aperture radar (SAR) types of SBR have been used for earth and planetary exploration. Altimeters have been used on many satellites. The technology of SBR subsystems has been developed in the areas of antennas, transmitters, receivers, solid-state transmit–receive (T/R) modules, signal processors, and prime power. This chapter will review SBR systems and technology with the intent to provide a description that is not too sketchy to be substantive. Therefore, selected systems and technology will be discussed. Several SBR systems for rendezvous, earth exploration, and planetary exploration missions will be described. Systems considerations such as the space environment, orbit selection, radar tradeoffs, advantages and disadvantages, and critical issues will be discussed. Many topics, such as electronic countermeasures, will have to be omitted. This chapter should be considered a status report on the new frontier for radar systems. A more comprehensive treatment of SBR can be found in "Spacebased Radar Handbook," written and edited by the author and published by Artech House.

22.2 SBR SYSTEMS CONSIDERATIONS

Types of SBR. There are three types of radar that have been and can be based in space. SBR that is typical of Type I is the small, short-range rendezvous radar such as those used on the Shuttle, Apollo, and Gemini programs.[1–4] Type II SBR includes the earth and planetary resources radar used for mapping, scatterometers, altimeters, and subsurface probing.[5–9] Side-looking SAR techniques are typical of mapping radars such as those used on the Seasat satellite in June 1978 and the Shuttle in November 1981 with the Shuttle Imaging Radar-A (SIR-A). Type III SBR includes the large phased array surveillance radar proposed for multimission defense, air traffic control, and disarmament functions.[10-14]

Type I SBR.   Gemini and Apollo programs demonstrated the first operational experience with the rendezvous maneuver. The successful performance of the rendezvous radars in these programs effectively opened the door to many possible missions that may be performed in space. The K_u-band integrated radar and communications subsystem (IRACS), designed for the space shuttle orbiter vehicle, demonstrated the rendezvous, satellite retrieval, and station-keeping missions. The maiden voyage for this radar was aboard *Challenger* (Shuttle) STS-7 on June 22, 1983.[15] During the STS-11 flight in February 1984, the K_u-band radar assisted in the checkout of the manned maneuvering unit (MMU) operations. The radar acquired and tracked mission specialist Robert Stewart in the MMU during his 300-ft sojourn into space. The radar measured the radar cross sections (RCS) of the MMU, which varied between 2.5 and 7.5 dBsm with acquisition at a range of 100 ft and track out to the maximum range of 308 ft. Average velocity during the mission was 0.7 ft/s.

The rendezvous radar provides the tracking function for a guidance system. The rendezvous phase of the mission begins after the radar acquires the target. Thereafter, the tracking function provides data on range, range rate, and the two components of the line-of-sight inertial rate. A digital guidance computer calculates relative velocity perpendicular to the line of sight, using range and angular rate data. The closing component of velocity is obtained from the doppler or by differentiation of radar range measurements. A simplified block diagram of a typical rendezvous guidance subsystem is shown in Fig. 22.1. The radar search and acquisition mode is initiated by the guidance computer. A relatively large solid angle is searched periodically until the target is acquired in range and angle. In order to maximize the probability of detection and acquisition, the kinematics are arranged such that a long search time is available before the target escapes from

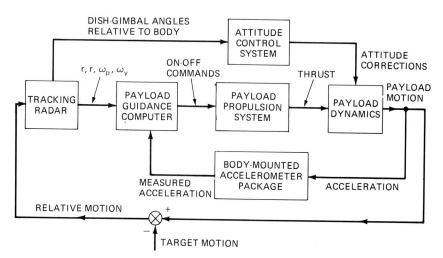

FIG. 22.1   Rendezvous guidance subsystem: simplified block diagram.[16]

TABLE 22.1   STS Rendezvous Radar Requirements*

Search	± 30° spiral scan
Acquisition	12 nmi on 0 dBsm SW-1; 300 nmi on +14 dBW transponder
Track	
Range	± 1 percent
Range rate	1 ft/s or 1 percent
Angle	8 mrad
Angle rate	0.14 mrad/s or 5 percent

*From Ref. 15.

the search sector. When detection has been accomplished, the search mode is stopped and the tracking mode is initiated by locking a tracking gate onto the target return and thereafter monopulse angle-tracking the antenna about an axis always directed toward the target. The tracking phase ends when rendezvous has been achieved within certain desired terminal accuracy on relative position and velocity. The typical requirements for the STS rendezvous radar are given in Table 22.1.[15]

At and immediately following acquisition, the relative velocity vector will generally lie in the direction of the instantaneous line of sight; however, there may be a substantial error equivalent to a relative velocity component perpendicular to the line of sight. The range at acquisition and the magnitude of the closing velocity are such that the rendezvous-phase duration can be several minutes. A reasonably long period is essential to an accurate rendezvous, since sufficient time must be allowed for smoothing the inherently noisy radar tracking data as well as for correcting measured errors. A period of as much as 10 to 20 min is still short compared with the overall mission duration. The effect of the differential earth gravity field has been shown by Hord[17] to be negligible for tracking-phase durations not exceeding 10 to 20 min. Furthermore, Wolverton[16] has shown that when the rendezvous time t_r is small compared with the product of the satellite orbital period T_0 and $(2\pi)^{-1}$, the orbital motion aspects of the rendezvous maneuver can be neglected.

Type II SBR.   Remote sensing of the earth from space began in 1960 with the launch of the first television and infrared observation satellite (Tiros) weather satellite. Remote sensing of the earth from space by radar began in 1975 with the launch of the GEOS-C by the National Aeronautics and Space Administration (NASA) and continued with the Seasat in 1978, the SIR-A on the Shuttle in 1981, and the SIR-B on the Shuttle STS-17 in 1984.

SEASAT-A SYSTEM.   The Seasat-A program was managed for the NASA Office of Applications by the California Institute of Technology Jet Propulsion Laboratory (JPL). The mission for Seasat-A was to demonstrate that measurements of ocean dynamics are feasible. The measurements included topography, surface winds, gravity waves, surface temperature, sea-ice extent and age, ocean features, and salinity. Precision of the geoid measurement was specified as ±10 cm.[18]

The Seasat-A satellite was launched at 6:12 P.M. PST on June 26, 1978. The orbital altitude was 783 km at apogee and 778 km at perigee. The retrograde polar orbit had an inclination angle of 108° and a period of 100.5 min. Three radar and two radiometer sensors were carried on the spacecraft. The coherent SAR, described in Sec. 22.3, operated at 1.275 GHz. The radar altimeter operated in the 12- to 14-GHz band and covered a 1.6-km swath directly below the spacecraft.

The wind scatterometer operated at 14.599 GHz and covered two swaths, each 400 km wide and offset on each side of the spacecraft. Four antennas were used to measure wind speed in the range from 4 to 28 m/s. The microwave radiometer had five frequency channels at 6.6 GHz, 10.6 GHz, 18 GHz, 21 GHz, and 37.6 GHz. A swath 1000 km wide, centered at the nadir, was covered. The visible and infrared (IR) radiometer covered a single swath 1800 km wide, symmetrical about the nadir.

Seasat-A collected data until Oct. 9, 1978, when a short circuit developed at the slip rings between the solar array and the power distribution bus.

The primary objectives of the SAR experiment on Seasat-A included (a) to obtain radar imagery of ocean wave patterns in deep oceans, (b) to obtain ocean wave patterns and water-land interaction data in coastal regions, and (c) to obtain radar imagery of sea and fresh-water ice and snow cover. The secondary objectives included (a) to obtain radar imagery of land surfaces; (b) to obtain data for mapping of the earth's surface; (c) to obtain data for estimates of land and sea surface roughness, ice type, differentiation of surface materials, vegetation, and landforms; (d) to obtain data for monitoring changes in the environment; (e) to obtain a demonstration of all-weather, day-night measurement capability; and (f) to obtain data useful for designing future high-resolution spaceborne radar systems.

GEOS-3. The Geodynamics Experimental Ocean Satellite (GEOS-3) was a remote-sensing satellite that contained five instruments in the experiment package.[19–21] These were (1) an SBR altimeter, (2) two C-band transponders, (3) an S-band transponder, (4) laser retroreflectors, and (5) a radio doppler system. The purpose of the GEOS-3 satellite was to perform experiments in support of the application of geodetic satellite techniques to geoscience investigations such as earth physics and oceanography. The SBR altimeter mission objective on the GEOS-3 satellite was to perform an in-orbit experiment that (1) determined the feasibility and utility of a space-borne radar altimeter to map the topography of the ocean surface with an absolute accuracy of ±5 m and with a relative accuracy of 1 to 2 m, (2) determined the feasibility of measuring waveheight, (3) determined the feasibility of measuring the deflection of the vertical at sea, and (4) contributed to the technology leading to a future operational altimeter satellite system with a 10-cm measurement capability.

The GEOS-C satellite (its designation was changed to GEOS-3 after successful orbit had been achieved) was launched on Apr. 9, 1975. The nominal orbit parameters were as follows: mean altitude, 843 km; inclination angle, 115°; eccentricity, 0.000; and period, 101.8 min. The GEOS-3 spacecraft was an eight-sided aluminum shell topped by a truncated pyramid. The satellite width was 132 cm (53 in), and the height was 81 cm (32 in); the weight of the GEOS-3 was 340 kg (750 lb).

Type III SBR. Before the design of a Type III SBR can begin, requirements for the surveillance radar systems must be specified. These requirements should include but not be limited to[22] (1) target radar cross section model, (2) target velocity and acceleration (maximum), (3) number of targets, (4) probability of detection, (5) probability of false alarm and false-alarm time, (6) track accuracy, (7) minimum target spacing, (8) designation error, (9) warning time, (10) length of detection fence, (11) revisit time, (12) clutter model, and (13) weather model. With these requirements as a minimum input to the design study, orbit selection can begin and parameter tradeoffs can be made. The influence of the space environment, interference, and clutter must be considered. Since the Shuttle (STS) can be a major launch vehicle for SBR, its capabilities should be examined. The advantages and disadvantages of large surveillance radar in space should also be considered.

Target characteristics and requirements for coverage, track data rate, and revisit rate are important parameters. The radar subclutter visibility capability, antenna size, scan rate, and grazing-angle limitations also determine the orbit selected for the SBR. The space environment itself can determine the selected orbit if the natural-radiation lifetime dosage that the SBR electronics receives is too large. Finally, there is the requirement to use the least number of satellites to keep total system cost to a minimum.

Considerations

Orbit Selection.   Many factors contribute to the selection of the orbit to be used for each type of SBR and particularly for a large surveillance-type SBR. The orbit parameters of period, altitude, and velocity are the first consideration. The velocity for a satellite in a circular orbit around the earth is given by[16]

$$V_c = \sqrt{\frac{\mu}{r}} \tag{22.1}$$

where r is the distance of the satellite from the center of the earth and μ is the product of the universal gravitational constant and the mass of the earth. The period of a satellite of the earth is given by[16]

$$T = \frac{2\pi\mu}{\sqrt{V_a^{\,3} V_p^{\,3}}} \tag{22.2}$$

where V_a is the velocity of the satellite at apogee and V_p is the velocity of the satellite at perigee. For a circular orbit, $V_a = V_p$ and the period of a circular orbiting satellite is

$$T_c = \frac{2\pi\mu}{V_c^{\,3}} \tag{22.3}$$

Table 22.2 shows selected calculations of circular-orbit velocity and period when the radius of the earth is 20.903 $(10)^6$ ft, μ is 1.4069 $(10)^{16}$ ft³/s², and 1 nmi is 6076.1 ft.

Many studies concerning the design of satellite constellations for optimal coverage have been made and reported.[23–28] Luders and Ginsberg[24] describe an analytical solution to the problem of achieving continuous coverage of latitudinally

TABLE 22.2   Selected Orbital Parameters

Altitude, nmi	Velocity, ft/s	Period, min
99	25,587	88
414	24,520	100
912	23,074	120
2,262	20,157	180
5,612	15,999	360
19,369	10,079	1,440

bounded zones of the globe. Emara and Leondes[27] solved the problem of simultaneous observations by at least four satellites by a constellation of the minimum number of satellites. Ballard[25] extended earlier work by Walker[23] and analyzed rosette constellations that provided the largest possible great-circle range between an observer anywhere on the earth's surface and the nearest subsatellite point. Single, double, triple, and quadruple visibility was provided by various constellations. Beste[26] designed satellite constellations that provided single and triple continuous coverage by the minimum number of satellites. All these studies determined coverage for satellites with sensors that observe only angles around the nadir. Electro-optical sensors and mapping radars are typical sensors that provide this coverage. However, these studies do not provide results for SBR surveillance sensors that must detect targets in clutter. These sensors typically have a *nadir hole* 20 to 30° off nadir in which the signal-to-clutter ratio (*SCR*) is too large for reliable detection. This is shown in Fig. 22.2 for a 50°

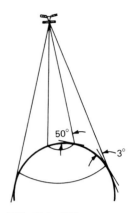

FIG. 22.2 SBR coverage and nadir hole.

maximum grazing angle and a 3° minimum grazing angle. The minimum grazing angle is a limit set by the atmospheric attenuation allocated in the SBR loss budget and the refraction angle error. To illustrate the different results that can be obtained, consider a requirement to provide continuous coverage of the earth from an orbital altitude of 10,371 km (5600 nmi). For a single sensor on each satellite with no grazing-angle limitations, a constellation of six satellites can provide the required continuous coverage from polar orbits. The satellites would be equally distributed in two orbital planes, using the study results given by Harney.[28] However, if the sensor in the SBR was limited to grazing angles between 3 and 60°, then the required coverage could be provided by a constellation of 10 satellites. This constellation consists of 1 satellite in each of 10 equally spaced orbit planes at an inclination of 49.4°, resembling the Walker 10/10/8 constellation.[25] If the grazing angles extend between 3 and 70°, then a 14-satellite constellation in a Walker 14/14/12 configuration provides a continuous global twofold coverage. The inclination angle of each orbital plane is 49.4°.

Space Environment. For a large phased array type of radar operating in space, the thermal and natural radiation environments have significant influence on the design of an SBR. Particular effects depend on the orbital altitude and the materials used in the structure.

THERMAL ENVIRONMENT EFFECTS. In general, distortion of a phased array antenna will cause a decrease in antenna gain. Figure 22.3 shows the effect of random phase errors caused by the distortion ϵ when the error correlation interval is large with respect to a wavelength. It is seen from Fig. 22.3 that a 2 dB loss in gain is obtained when the distortion is about one-tenth of a wavelength. Thus for a 50-m-diameter planar corporate phased array antenna operating at a wavelength of 10 cm, the rms distortion of the plane of the array must be held to less than 1 cm if a 2 dB loss in antenna gain is to be maintained.

Thermal distortion in a 70-ft (21.34-m) diameter parabolic reflector was studied[29] at synchronous orbit. Reflector performance comparisons were made for titanium and graphite composite materials. Generally the tolerances that must be held on reflector antennas are more severe than for phased arrays for the same performance. Figure 22.4 shows the results of the analysis. Performance of the

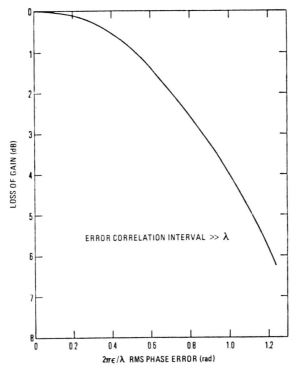

FIG. 22.3   Antenna loss of gain due to random phase errors.

graphite composite material is superior, giving an rms distortion of about 0.076 cm. If this is one-fiftieth of the wavelength, then the antenna could operate satisfactorily at a wavelength of 3.8 cm.

Consider a 70-m-diameter-lens phased array[30,31] at an altitude of 5600 nmi as shown in Fig. 22.5. The progress of the sun angle is shown. Simulations have predicted the following maximum and minimum temperatures for selected parts of the space-fed lens antenna:

Location	Temperature, K	
	Maximum	Minimum
Ground plane	264	224
Rim	182	160
Upper stays	231	186
Lower stays	217	201
Upper dipole plane	314	201
Lower dipole plane	274	220

By choosing the proper materials, the design of this class of antenna will experience low distortions compared with those allowable. Figure 22.6[32,33] shows the loss in relative gain for a 71-m-diameter space-fed antenna as a function of the

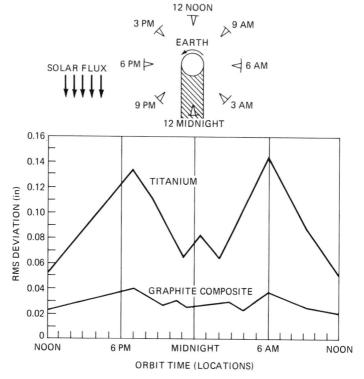

FIG. 22.4 Thermal distortion.[29]

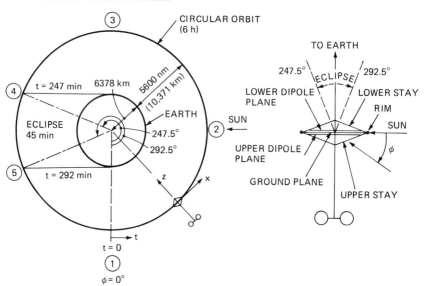

FIG. 22.5 SBR at 5600-nmi orbital altitude and the sun angle progression.[32]

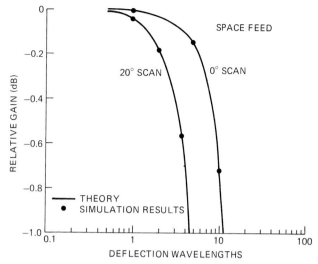

FIG. 22.6   Loss of gain due to distortion for a space-fed array.[32]

deflection or distortion in wavelengths. It is seen that the relative gain is down 1 dB when the distortion is about 5 wavelengths at a 20° scan angle.

RADIATION ENVIRONMENT EFFECTS.   SBRs can encounter in space particle radiation that may be due to both natural phenomena and nuclear detonations. The satellite must be designed to operate for a reasonable lifetime in the natural space environment. This environment is a function of orbital altitude. When the satellite is operating in midaltitude orbits, exposure to the earth's Van Allen belts will be predictable and its effect on radar electronics will be functions of the inherent hardness level of the components and the shielding used. (Reference 34 provides the trapped radiation data for proton and electron flux that has been measured as a function of altitude.) Figure 22.7[32] shows the total 5-year dose in rads (Si) that satellites in orbits between 350- and 6500-nmi altitudes will experience as a function of the aluminum shielding used. It appears that current technology in integrated-circuit hardening should produce a total dose hardness of about $5(10)^5$ rads (Si) for devices that are suitable for the SBR T/R modules. This hardness level is adequate for SBR deployment in many of the candidate orbits with a mission life in the natural environment of several years. A hardness of $5(10)^6$ rads (Si) which may be achievable is required for a 5-year mission life. Survival in a saturated nuclear environment typical of a high-altitude nuclear burst requires a hardness of 1 to $5(10)^7$ rads (Si), depending upon the specific orbit. The development and consistent fabrication of devices as hard as this are relatively uncertain.

Tradeoffs.   Obviously many tradeoffs can be made during the design of each type of SBR, depending upon the mission. In a dual-frequency surveillance and track radar performing an air traffic control (ATC) mission, as mentioned in Sec. 22.6, it is possible to trade off the length of the surveillance fence against the number of targets in track as functions of the track data rate and the radar-beam grazing angle. In a high-resolution-mapping radar mission it is possible to trade off the resolution against the orbital altitude as functions of radar wavelength and integration time. These trades are shown in detail in Refs. 12 and 13.

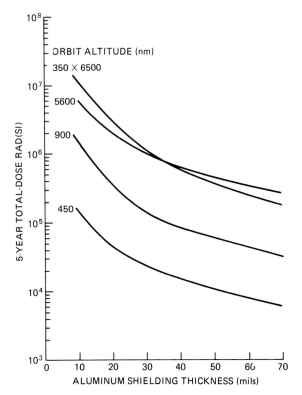

FIG. 22.7 Total dose versus shielding thickness for a 5-year mission.[32]

Clutter/Interference. SBR performance is significantly dependent upon clutter and interference, either intentional or unintentional. To illustrate the magnitude of the clutter problem, consider the ATC radar described in Sec. 22.6. When the grazing angle is 70° and the reflectivity of the ground is −15 dB, the main-beam clutter cross section is +57 dBsm. If the desired radar performance requires that a target with an RCS of +13 dBsm have an *SCR* of 25 dB, then the main-beam clutter cancellation ratio must be at least 69 dB. Therefore, SBR performance requires large clutter cancellation ratios. Reference 35 indicates that clutter cancellation ratios up to 90 dB can be obtained by using pulse doppler and displaced phase center antenna (DPCA) techniques.

Interference will enter the SBR antenna primarily through the sidelobes since the beamwidth is narrow. This interference can be either intentional noise jamming or unintentional from other radars. These effects can be reduced to acceptable levels if adaptive sidelobe cancellation techniques and sidelobe-blanking techniques are utilized.

Launcher Capabilities. The most probable launch vehicle for the SBR is the STS (shuttle). Therefore, STS capabilities to put various payloads that include one or more SBR satellites (and the propulsion systems to place them into the desired orbits) must be considered. Figure 22.8 shows the STS cargo weight as a function of orbit inclination angle for various circular orbital altitudes and orbital-

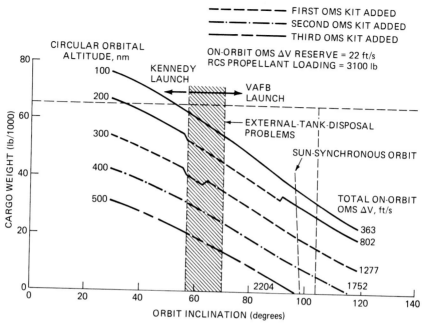

FIG. 22.8 STS (Shuttle) cargo weight versus inclination for various circular-orbit altitudes (delivery only—no rendezvous).

maneuvering-system (OMS) on-orbit velocity increments. It is seen that 64,000 lb can be delivered to a 100-nmi circular orbit inclined 50° from the Kennedy launch site in Florida. If each SBR weighs 9500 lb, then three SBR satellites can be placed into orbit along with 35,500 lb in propulsion for orbital transfer.

Advantages and Disadvantages of SBR Systems. When sensors are required for missions involving targets in space, ocean, and air and for missile defense missions, the use of SBRs should be considered. The advantages of such radars deployed in space compared with ground-based radars are described below.

1. Coverage in both space and time is limited only by the orbit selected and the number of satellites. Large-scale continuous observation can be obtained as shown in Figs. 22.9 and 22.10.[28] In Fig. 22.9 the required number of vehicles are shown as well as the number of orbit planes in which they are distributed to provide continuous coverage of the entire earth's surface from circular polar orbits. It is seen that six vehicles in two orbit planes can be used for vehicle altitudes greater than about 6000 nmi. There is no nadir hole in the satellite coverage. Figure 22.10 illustrates the special case of equatorial orbits and the number of vehicles required for continuous coverage. This situation is limited to the use of wide swaths that extend up to the specific latitudes indicated. It is seen that four vehicles can cover a 60° swath when the vehicles are at altitudes greater than about 6000 nmi. Temporal coverage is illustrated in Fig. 22.11, which shows the maximum time for viewing ground objects from a space vehicle if the objects are tracked.[28] It can be seen that a ground object can be observed for more than 7000 s when the orbital altitude is 6000 nmi.

2. When the SBR uses an electronic scanning antenna, it is possible to per-

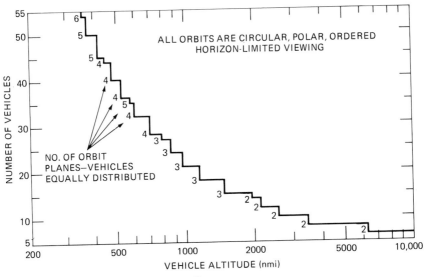

FIG. 22.9 Global coverage by polar orbits.[28]

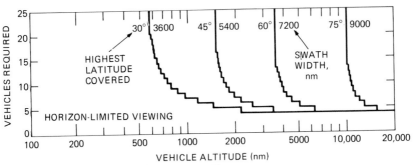

FIG. 22.10 Zonal coverage by equatorial orbits.[28]

form multiple missions. For example, a system of radar satellites can (*a*) search a fence formed completely around the continental United States (CONUS) to detect bombers at a distance from the coast, (*b*) search a fence over the poles to detect intercontinental ballistic missiles (ICBMs) before they can be detected by the Ballistic Missile Early Warning System (BMEWS), (*c*) monitor potential launch sites for space launches from any foreign country, (*d*) perform surveillance of ocean areas, (*e*) search a sea-launched ballistic missile (SLBM) detection fence, and (*f*) detect objects in space that appear to be threats to United States synchronous satellites. The number of missions is limited only by the weight and prime power available, but even these limitations can be overcome when the space shuttle is the planned launch vehicle. Therefore, the only real limitations are technology and cost.

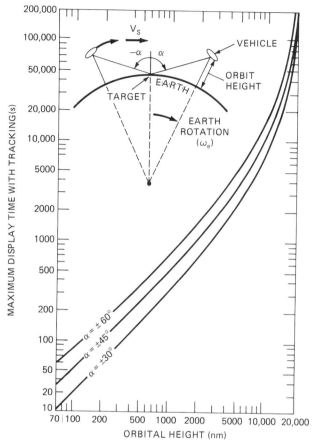

FIG. 22.11 Maximum time for viewing objects from a space vehicle if the objects are tracked.[28]

3. Atmospheric propagation problems can be minimized by proper selection of operating frequencies and favorable geometry selection.

4. No overseas stations are required if data is read out via relay satellites. Hence, the SBR system allows a country to be politically independent, and the loss of tracking stations in a foreign country has no impact on its system capabilities.

The factors that affect the pace of development of large radar systems in space are:

1. The technologies of large antenna structures in space, of large phased arrays in space, of large weights in space, and of large prime power systems in space are considered to be in their early stages.

2. The funds that can reasonably be spent on a space-based multimission radar system are to be determined. Even with the use of the Shuttle to reduce the cost per pound of payload into orbit, large investment costs are expected to be required for the SBR system.

22.3 SBR SYSTEM DESCRIPTIONS

The United States and the U.S.S.R. have deployed Type I and Type II space-based radars. This section describes some of these SBR systems.

STS Rendezvous Radar.[1,15,36] The Integrated Radar and Communications Subsystem (IRACS) was developed by Hughes Aircraft Company for use on the Space Transportation Systems (STS). The IRACS is a coherent range-gated pulsed doppler radar which searches for, acquires, and tracks other orbiting objects and provides the spatial measurement data needed to perform rapid and efficient rendezvous with those objects.

The IRACS performs both radar and communications functions for the STS. In the pulsed doppler radar mode it performs the rendezvous function just described. In the communications mode it searches for, acquires, and tracks the Tracking and Data Relay Satellite System (TDRSS) relay satellites to provide two-way communication between the space shuttle and ground tracking stations.

The IRACS hardware is subdivided into *deployed* and *inboard* assemblies. The deployed hardware is located within the Shuttle payload bay and is extended for operation through the open payload bay doors. Included in this hardware are the antenna reflector, feed, gimbals, drive motors, gyros, digital shaft encoders, rotary joints, transmitter, receiver, upconverter, first downconverter, and frequency synthesizer. The inboard hardware is located internally to the shuttle and includes the signal-processing, track-filtering, and control functions.

The K_u-band IRACS operates in the band of frequencies between 13.75 and 15.15 GHz, with radar operation between 13.75 and 14.0 GHz. There are two basic radar modes: a passive mode in which the target is noncooperative, in that no cross-section augmentation is present, and an active mode in which the target has an on-board transponder. The radar operates out to 12 nmi in the passive mode and out to 300 nmi, with a +14 dBm transponder, in the active mode. Submodes include an automatic search and angle and range track capability and external angle control operation. Under external angle control the antenna either is positioned by external slew commands or is referenced to inertial space or to the Shuttle axes. During automatic operation, angle, angle rate, range, and range rate measurements are made by the radar after track has been initiated. Under external angle control only range and range rate are measured.

The antenna is a 36-in-diameter center-fed parabola with 38.4 dB gain and 1.68° beamwidth. The five-element monopulse feed provides a sum and two orthogonal difference outputs. The difference outputs are time-multiplexed together into a *single* receiver difference channel for the angle-tracking operation. An auxiliary horn is monitored in the search mode, using the receiver difference channel, and compared with the main-antenna sum channel to prevent acquisition of large targets in the sidelobes of the main antenna. The auxiliary antenna has a peak gain which is about 20 dB less than that of the main antenna. Low-noise radio-frequency (RF) preamplifiers are used in the sum and difference channels. After amplification, at intermediate frequency (IF), the sum and difference channels are combined into a single receive channel for routing to the inboard electronics assemblies for further processing. The transmitter employs a traveling-wave tube (TWT) with 44 dB gain to amplify the coherent synthesizer output to 50 W of peak power. For short-range operation (down to 100 ft) the TWT is by-passed to reduce the power on the target. Five RF frequencies are used in the radar mode to decorrelate Swerling 1 (slowly fluctuating) target returns and improve detection. A 16-point digital Fourier transform (DFT) processor is employed to coherently integrate multiple pulse returns and to provide fine-resolution measurements of target relative velocity. The deployed assemblies weighed 135 lb, and the prime power was 460 W.

Seasat-A Synthetic Aperture Radar.[18,37] The Seasat-A was a focused SAR consisting of five subsystems: (1) spacecraft radar antenna, (2) spacecraft radar sensor, (3) spacecraft-to-ground data link, (4) ground data recorder and formatter, and (5) ground data processor. The antenna was a microstrip array of eight panels that were fed by a corporate-feed network and operated at 1275 MHz. Details of the Seasat-A antenna are discussed in Sec. 22.4. The solid-state radar transmitter generated a nominal peak power of 800 W with a linear frequency modulation (LFM) derived from a stable local oscillator (stalo). The antenna illuminated a 100-km-swath width at the surface of the earth with an antenna elevation beamwidth of 6° that was oriented at an angle of 20° with respect to the nadir. Upon reception of the reflected signal by the receiver in the radar sensor, the return signal was amplified by a sensitivity-time-controlled RF amplifier. This signal and a fraction of the radar stalo were then combined and transmitted to a ground station by an analog data link. At the ground station, the data line demodulator recovered the radar sensor stalo and the radar return signal. The recovered synchronously demodulated video radar signal was then converted into digital form by the radar data recorder and formatter subsystem. Upon conversion, the signal was buffered and recorded by a high-density magnetic tape recorder. Subsequently, the radar data processor converted the digital recorded data into a two-dimensional map of the radar cross section of the area observed by the antenna. The SAR system generated a 25-m-resolution radar map in elevation (across track) by time-gated compressed radar return signals and in azimuth (along track) by focusing the coherent radar returns during the data-processing interval in the earth-based signal processor. Total SAR on-orbit weight was 223 kg; required radar prime power was 624 W. Table 22.3 gives the characteristics of the Seasat SAR.

Shuttle Imaging Radar.[36] The technology developed for the Seasat-A SAR formed the basis for the shuttle imaging radar (SIR) series, SIR-A and SIR-B. Minor differences in the antenna will be discussed in Sec. 22.4. The L-band radar transmitter was utilized with slight bandwidth changes so that resolution was 40 m on SIR-A and 20 m on SIR-B. Swath width was 50 km for both radars. Orbital altitudes were 240 km and 220 km, respectively, so that radar range and incidence angles were different.

GEOS-C SBR System Characteristics.[8,19-21] The GEOS-C radar altimeter was a precision K_u-band (13.9-GHz) SBR altimeter developed primarily to measure ocean surface topography and sea state. It was a complex multimode radar system with two distinct radar gathering modes (global and intensive modes) and two corresponding self-test–calibration modes for use in on-orbit functional test and instrument calibration. The key performance features were its capability to (1) provide precise satellite-to-ocean surface-height measurements [precision

TABLE 22.3 Synthetic Aperture Radar

Antenna	
Type	Planar phased array (10.74 m × 2.16 m)
Beamwidth	1.1° azimuth, 6° elevation (1 dB points)
Look angle	20° depression, 90° with respect to the velocity vector
Gain	34.7 dB
Polarization	Horizontal
Weight	113 kg
Transmitter	
Type	Solid-state transistor
Efficiency	38 percent
RF carrier	1275 MHz
Peak power	800 W (nominal), 1125 W (maximum)
Pulse length	33.8 μs
PRF	1463, 1540, 1645 pps
Duty cycle	0.05 (maximum)
Average power	44.5 W (nominal), 62.6 W (maximum)
Waveform	Pulse, LFM, 19-MHz Bandwidth
Receiver	
Noise temperature	550 K
Bandwidth	22 MHz
System input noise	−127.42 dBW
AGC time constant	5 s
STC gain variation	9 dB
Stalo stability	3×10^{-10} in 5 ms
Recorder	25 kb/s digital
System weight	110 kg (excluding antenna)
Total prime power	624 W (maximum)
Resolution	25 m
Swath width	100 km
Swath length	2000 km per pass
Swath orientation	Right side of orbit path
Signal-to-noise ratio	9 dB (nominal)

of 50 cm in the global mode (GM) and 20 cm in the intensive mode (IM) at an output rate of one per second] for use in mapping the shape of the ocean surface and (2) provide data which can be processed to estimate peak-to-trough ocean waveheight (waveheights in the range of 2 to 10 m can be estimated to an accuracy of 25 percent). Several key areas of technology included in the design are (1) high-frequency logic circuitry with a 160-MHz clock and four-phase division for 1.56-ns resolution, (2) a wideband (100-MHz) linear FM pulse compression system with a compression ratio of 100:1 and a compressed pulse width of 12.5 ns, (3) high-speed sample-and-hold circuitry for accurate sampling of wideband (50-MHz) noisy video return signals, and (4) design and packaging of high-voltage (12-kV) power supplies for space application.

The instrument weighs 68 kg (150 lb) and occupies a volume of 0.119 m³ (4.2 ft³) including the antenna, which is a 0.6-m (24-in) diameter parabolic dish with a 2.6° beamwidth and a 36 dB gain. The instrument is packaged in two basic sections: an RF section and an attached electronics section, which are both mounted to a center-cylindrical disk baseplate with a diameter of 0.65 m (26 in). The major subsystems contained in the RF section are (1) the IM transmitter (chirp generator, upconverter, 1-W driver TWT and high-voltage power supply, 2-kW output TWT and high-voltage power supply), (2) the GM transmitter (a 2-kW peak-power magnetron and high-voltage power supply), (3) the RF switch assembly (RF switches, waveguide runs, calibrate attenuation path, and TR switch), and (4) the receiver front end (downconverter-preamplifier). The major subsystems contained in the attached electronics section are (1) the IF receiver (IF amplifiers, filters, pulse compressor, detectors), (2) the signal processor (AGC, acquisition, and tracking functions implemented with analog and digital circuitry on multilayer board assemblies), (3) the frequency synthesizer, (4) the mode control circuitry, (5) the calibrate-test circuitry, and (6) the low-voltage power supply. The nominal power required for operation was 71 W for the global mode and 126 W for the intensive mode (16 waveform samplers).

U.S.S.R. Cosmos 1500 Side-Looking Radar.[38,39]   The U.S.S.R. launched the Cosmos 1500 oceanographic satellite on Sept. 28, 1983, into a nominal 650-km polar orbit. The satellite was the first of a series intended to provide continuous world ocean observations for civil and military missions. The sensors provide side-looking radar (SLR), radiometric, and visual coverage of oceans and ice zones for land- and sea-based users through an operational distribution network.[38] Table 22.4 summarizes the parameters and performance of the real-beam SLR. The radar operates at a frequency of 9500 MHz with a magnetron transmitter that has a peak power output of 100 kW. The antenna is a slotted waveguide that is 11 m long and 4 cm high. Cosmos 1500 has demonstrated many significant capabilities, including (1) routine automatic picture transmission of SLR images of earth; (2) mapping of inhomogeneities of Antarctic and Greenland ice cover that were previously not detected; (3) radar images of polar regions of multiyear and first-year ice zones; (4) mapping of elongated zones of ice-cover continuity disturbances; (5) tracking of sea-ice drift by using a series of radar images of the same water area; (6) detection of oil slicks, wind fields, and currents; and (7) guidance of ships trapped in arctic ice during October–November 1983.

The orbit of Cosmos 1500 allowed complete earth coverage each 1.41 days for the optical sensors and each 5.9 days for the radar sensor. Subsequent launches of the Cosmos 1500 type of satellite have occurred.

22.4 TECHNOLOGY

The desire to develop large radars in space has stimulated progress in several new technologies such as (1) large deployable parabolic and phased array antennas, (2) lightweight, low-cost monolithic microwave integrated circuit (MMIC) transmit/receive modules, (3) high-level prime power systems, (4) efficient onboard signal processors, (5) large lightweight space structures, (6) lightweight, low-cost phase shifters, (7) radiation-hardened electronic devices, (8) materials with a low thermal coefficient of expansion, and (9) advanced calibration and self-test techniques. Some of these technologies are briefly reviewed here.

TABLE 22.4   Cosmos 1500 SBR Parameters and Performance

Type	Real-beam side-looking radar (460-km swath)
Frequency/wavelength	9500 MHz/3.15 cm
Antenna	
Type	Slotted waveguide
Size	11.085 m × 40 mm
No. of slots	480
Illumination	Cosine on a pedestal
Beamwidth	0.20° × 42°
Gain	35 dB
Sidelobes	−22 dB to −25 dB
Waveguide	Copper, 23 × 10-mm cross section
Polarization	Vertical
Swing angle	35° from nadir
Noise temperature	300 K
Transmitter	
Type	Magnetron
Power	100 kW peak, 30 W average
Pulse width	3 μs
PRF	100 pps
Loss	1.7 dB
Receiver	
Type	Superheterodyne
Noise power	−140 dBW
Loss	1.7 dB
Pulses integrated	8 noncoherent
LNA noise temperature	150 to 200 K
LNA gain	15 dB
Dynamic range	30 dB
IF	30 MHz ± 0.1 MHz
Input power	400 W
Range	700 km (minimum), 986 km (maximum)
SNR	0 dB on $\sigma^0 = -20$ dB

Antennas. The development of SBR is strongly dependent upon the technology of large space-deployable antennas. Large antennas must be used since the radar ranges are significantly greater than usual and the prime power in the radar is limited. The vacuum of space and the zero-g environment permit the deployment of antennas with low mass per unit antenna area. Antennas with large diameters, up to 1 km, have been discussed by United States developers.[40–48] In the U.S.S.R., antennas with diameters in the 1- to 10-km range have been discussed.[49] In addition to being large and deployable, the SBR antenna must maintain its desired shape whether it be parabolic or planar. As shown earlier (Fig. 22.3), small deviations can cause a significant loss in antenna gain. Stable configurations are obtained by using low-coefficient-of-thermal-expansion (CTE) materials. Characteristics of selected materials for stable RF systems are shown in Table 22.5. Data includes CTE, density, modulus, conductivity, and attenuation of WR 75 waveguide fabricated

TABLE 22.5   Potential Material Selection for Thermally Stable RF System

Material	Expansion coefficient, in/(in/°F) $\times 10^{-6}$	Density, lb/in^3	Young's modulus, $\times 10^6$ lb/in^2	Thermal conductivity, $\dfrac{Btu \cdot in}{h \cdot ft^2 \cdot °F}$	WR75 attenuation, dB/ft at 11.95 GHz
Aluminum	13.1	0.10	10	1513	0.049
Titanium	5.1	0.16	16	444	0.274
Invar	1.1	0.29	20	93	0.370
Beryllium	6.8	0.07	40–44	1138	0.082
Graphite/ epoxy	0.03	0.06	17–25	75 (axial), 7.3 (transverse)	1.560 (bare), 0.040 (coated)
Gold	6.8	0.70	. . .	2064	0.048
Copper	7.8	0.32	. . .	2944	0.040
Silver	11.0	0.38	. . .	3101	0.039
Rhodium	4.7	0.45	. . .	611	0.087
Kevlar®* 49	− 1.1 longitudinal, + 33 radial	0.052	19	0.334 (axial), 0.285 (transverse)	

*®Du Pont trademark.

out of each material. In the following discussion selected antenna designs are described to illustrate the state of the art in large antennas in space.

United States Space-Deployable Antennas.   A large space-deployable antenna that the United States deployed in space was the Lockheed-NASA ATS-6 parabolic reflector, launched in 1974. It was 9.1 m in diameter with a tolerance of 1.52 mm rms and a specific weight of 1.4 kg/m^2.[41,50] The ATS-6 antenna embodies the flex-rib technique. During the years subsequent to that launch, Lockheed has evolved flex-rib deployment technology to additional reflector designs, the polyconic and the maypole designs.[41]

Harris developed the radial-rib double-mesh design and in 1970 built a 12.5-ft-diameter antenna.[40] This was followed by the TDRSS 4.88-m-diameter antenna and three generic antenna designs including the radial-rib, TRAC, and hoop-column concepts. The weight-versus-diameter capabilities of these three designs are shown in Fig. 22.12.[47] As part of the NASA deployable antenna flight experiment (DAFE) design study, Harris estimated that a 50-m-diameter reflector assembly would have an overall weight of 819 kg (1805 lb).

The specific mass of this design is 0.417 kg/m^2, and the estimated surface error was 4 mm rms. In a parallel DAFE competition, the Grumman Aerospace Corporation designed a 50-m-diameter phased array lens antenna that would have a specific mass of 0.522 kg/m^2.

The DAFE studies were conducted for NASA Marshall Space Flight Center (MSFC) by Harris and Grumman in a competition during the period from August 1980 to September 1981. The primary objectives of the study were (1) to demonstrate, by a flight experiment, the capability to launch, deploy, retract, and return to earth a large (50-m-diameter) space frame; and (2) to verify, by flight experiment, the capability of the space frame to attain and maintain the dimensional

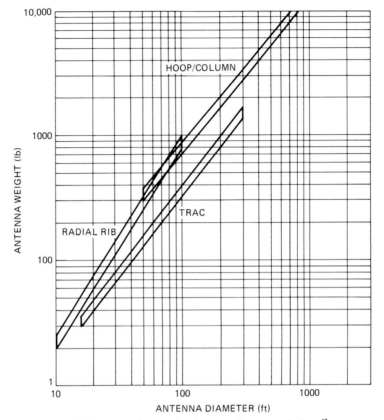

FIG. 22.12 Weight versus diameter for three generic antenna designs.[47]

precision required to operate as a spaceborne antenna. Both contractors devised orbiter-attached experiments that would maximize program outputs while minimizing orbiter and experiment risks. Although many flight configurations were designed, overall results were similar for both phased array and parabolic antennas. Both contractors also devised measurement techniques that would provide a 50-mil rms accuracy required for the measurement of antenna deformation.

General Dynamics has designed space-erectable antennas and parabolic graphite-epoxy reflectors for space applications.[43,46] A 2.44-m-diameter reflector was built and tested. It has a surface tolerance of 0.0635 mm rms and a specific mass of 4.4 kg/m². The space-erectable designs had a specific mass of 0.49 kg/m²; however, the tolerance was on the order of 10 mm rms. Therefore, the space-erectable antenna designs were configured primarily for relatively low-frequency operations.

TRW has developed an advanced antenna concept under work sponsored by JPL as part of the NASA large space systems technology (LSST) program.[51] The feasibility of stowing large, solid antenna reflectors in the Shuttle was examined. The antennas would be designed to operate in the 10- to 100-GHz range and main-

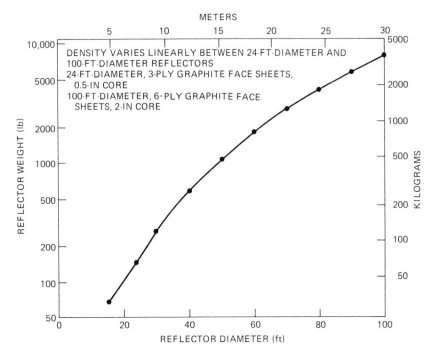

FIG. 22.13   TRW antenna reflector weight estimate.[51]

tain rms deviation on the order of 10^{-5}-diameter fabrication error. Thermal deviation for a 100-ft-diameter antenna was estimated to be 0.0034 in rms. The weight of antenna reflectors was estimated for diameters of 16 to 100 ft. Figure 22.13 shows the plot of reflector weight excluding the weight of feeds and subreflectors. The basic construction assumed a graphite-epoxy-aluminum honeycomb-sandwich configuration.

Antenna systems have been studied and fabricated under the LSST program. Lockheed Missiles and Space Company (LMSC) has demonstrated, in a simulated zero-gravity environment, the technology for a large space-deployable antenna.[52] LMSC fabricated a 22.5° sector of a 55-m-diameter *wrap-rib* parabolic antenna and deployed it in a ground-based zero-*g* facility. The surface of the antenna is a knit mesh of 1.2-mil gold-plated molybdenum wire that is contoured by graphite-epoxy ribs. Each rib weighs 9.1 kg, is 27.5 m long, and is lenticular in shape. This shape allows the ribs to collapse as they are wrapped around a central hub for stowage prior to deployment. The ribs resume their required structural shape as they unwind (under constraint), thereby stretching the mesh into the proper parabolic shape. The Lockheed development program was initiated to demonstrate the readiness of large-diameter offset reflector technology through development of ground-testable, flight-representative full-size hardware.

The Seasat-A antenna (designed by Ball) is a 10.74- by 2.16-m microstrip array that is deployed after orbit insertion. The operating wavelength is 23.5 cm. This antenna is very similar to the SIR-A. Both are significant developments in large deployable antennas.[6,37] The SIR-B antenna is similar except that it was mechan-

TABLE 22.6 Characteristics of Seasat, SIR-A, SIR-B, and SIR-C Antennas

	Seasat	SIR-A	SIR-B	SIR-C
Frequency	1275 MHz	1278 MHz	1282 MHz	1275 and 5300 MHz
Bandwidth	22 MHz	8 MHz	16 MHz	> 20 MHz
(1.5:1 VSWR)				
Gain	34.9 dB	33.6 dB	33.0 dB	37.0 dB (L band); 43.0 dB (C band)
Polarization	Horizontal linear	Horizontal linear	Horizontal linear	Horizontal linear and vertical linear
Beamwidths				
H plane	6.2°	6.2°	6.2°	Adjustable through amplitude and phase, 0.99° L band; 0.24° C band
E plane	1.1°	1.4°	1.1°	
Beam-pointing angle	20.5°	47°	15 to 60° (mechanical steering)	Tilted to 35°, then ±25° electronically steered
Size (deployed)	10.74 × 2.16 m	9.4 × 2.16 m	10.74 × 2.16 m	12.06 × 4.2 m
Size (folded)	1.34 × 2.16 m		4.1 × 2.16 m	4.1 × 4.2 m
Weight	103 kg	181 kg	306 kg	900 kg
Support structure	Graphite-epoxy 3D truss	Rigid aluminum 3D truss	Rigid aluminum 2D and 3D truss	Graphite-epoxy 2D truss
Fold mechanisms	Multifold (spring-loaded)	Fixed	Two folds (motor-driven)	Two folds (motor-driven)
Number of radiating elements	1024	896	1024	864 (L band); 5184 (C band)
Number of panels	8	7	8	9
Feed system	Microstrip, coaxial and suspended substrate	Microstrip, coaxial	Microstrip, coaxial	Microstrip, coaxial, waveguide
W/A, kg/m²	4.44	8.9145	13.1906	17.7683

ically steerable. The SIR-C antenna is electronically steerable and dual-frequency. Table 22.6 summarizes the RF and mechanical characteristics of the Ball Seasat-class antennas.

Ball Aerospace Systems Division designed an antenna for the low-altitude space-based radar (LASBR) mission.[53] The 13.8- by 63.6-m array is a direct extension of space-proven Seasat and SIR-A technology with stringent constraints of array two-way sidelobe and beam skirt performance. The design features a single-axis deployable truss fabricated from graphite-epoxy microstrip honeycomb panels and passive 3-bit hybrid phase shifters at each of the 49,152 elements. The loss and weight penalty (W/A = 4.02 kg/m^2) of a corporate-feed network is compensated for by using transmit and receive gain at each of 384 subpanels.

U.S.S.R. Cosmos 1500 Antenna. On Sept. 28, 1983, the U.S.S.R. launched the Cosmos 1500 satellite with an SLR for the all-weather probing of the surface and ice cover of the earth's seas and oceans.[38] (The SLR was mentioned in Sec. 22.3.) The antenna is a slotted-waveguide array of 480 slots with a length of 11.085 m and a height of 4 cm. The operational wavelength is 3.15 cm. The beamwidth of the antenna is 0.2 by 42°, providing a gain of 35 dB. The antenna is constructed out of a copper waveguide that measures 23 by 10 mm in cross section. The slots are in the wide wall, with variable spacing to provide a cosine on a pedestal amplitude distribution. Figure 22.14 shows the antenna during deployment. The five sections of the antenna are mated and held in position by spring-loaded locks on the ends that are operated by release mechanisms at the end of the deployment cycle. Helical springs are provided on the flange faces along the wide wall for electrically tight joints. Relative leakage power between sections is down by 50 dB. After deployment, the antenna can be rotated through 35° from the nadir.

Transmit/Receive Modules. During the advanced development stage of large phased array space-based radars, the use of small, low-cost, lightweight low-power T/R modules was proposed in active array configurations.[54] Goals for these T/R modules[31] included costs of less than $100 each in large mass production and a size of 1 in^2, using 0.5 to 1 W of power. Each module contains a phase shifter, drivers, logic switches, power amplifiers, low-noise receiver, and other components. They can also include a means for sensing and compensation of element displacement error. Further information on solid-state transmitters and transceiver module characteristics is found in Chap. 5.

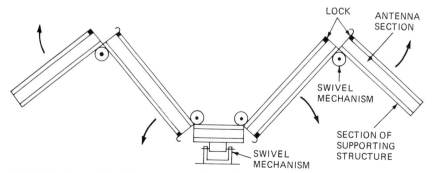

FIG. 22.14 Schematic of the antenna module for Cosmos 1500.

On-Board Processors.   The bandwidth requirements for the satellite communications data link can be reduced when an on-board processor is utilized. The major functions of an on-board processor in a large SBR can include pulse compression, doppler filtering, adaptive beamforming, calibration, range walk correction, video integration, constant false-alarm rate (CFAR), monopulse error signal, burst waveform weighting, sidelobe blanking, and editing out interference.

In addition to providing these functions, the processor must be low-powered, have low mass, operate for many years without manual repair, and have radiation-hardened memories. Technology using 16K random-access-memory (RAM) chips and very-large-scale-integration (VLSI) computers can reduce "typical" system power and mass from 3 kW and 2000 lb, respectively, to 400 W and 400 lb.[55] The Defense Advanced Research Projects Agency (DARPA) and others have been working on the development of an advanced on-board signal processor (AOSP) and have made considerable progress.[10] The concern here is to develop a very reliable and survivable on-board computer using gallium arsenide circuitry that can resist the electromagnetic pulse and other radiation effects produced by nuclear detonations. AOSP program technical goals include (1) prime power of 100 W, (2) weight of 100 lb, (3) volume of 2 ft^3, (4) 5-year life with a probability of survival of 95 percent, and (5) an input rate of 50 million words per second.[32,56]

Prime Power.   The performance of any SBR will ultimately be limited by the prime power system. The most frequently utilized source of prime power for satellites is the solar-battery configuration. High-efficiency GaAs solar cells have demonstrated efficiencies of 18 percent.[57] With the addition of other subsystems, including panels, rotary joints, slip rings, battery, power control, and distribution equipment, the specific power density of the prime power system is on the order of 13 to 24 W/kg. Solar-battery systems are limited and have several disadvantages that will be discussed later.

Space nuclear prime power systems offer certain advantages to SBR, and they have been launched into space by the United States since 1961, beginning with the SNAP-3A. Of the nuclear power systems that were placed into orbit between 1961 and 1977, only one was a nuclear reactor, SNAP-10A. Since then the technology has advanced[58–61] to the extent that it is estimated that an SP-100 type of nuclear reactor would have a mass of 2770 kg and a power output of 100 kW, thereby providing a specific power density of 36 W/kg.

Two baseline deployment configurations of solar-battery and nuclear prime power systems were designed with two power levels, 25 and 100 kW, for the same deployment altitude and are shown in Fig. 22.15.[61] It is seen that the solar systems are larger than the nuclear. As the power level increases, the increased size of the solar system becomes more pronounced. In comparing overall lengths, the 100-kW solar system is 2.4 times the length of the nuclear system. The weight of the solar system depends upon the orbital altitude and the operational requirement during eclipse. For a continuous-operation solar array at geosynchronous altitude, the 100-kW solar system weight is estimated to be 3970 kg.

In comparing advantages and disadvantages, the solar-battery system is based on known technology, and extrapolation to a larger power output is considered to be an engineering design task. The nuclear reactor design re-

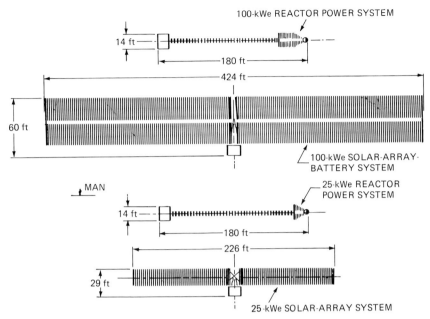

FIG. 22.15   Baseline prime power systems: deployed configurations.[61]

quires engineering development. The advantages of nuclear prime power systems include (1) reduced mass and size at the higher power levels; (2) no perturbation by natural background in low earth orbit (LEO) and geosynchronous orbit (GEO); (3) no need for alignment, gimballing, slip rings, and long-life batteries, suggesting that the nuclear system will have significantly enhanced reliability; (4) reduced effect on SBR antenna, i.e., multipath and sidelobes; (5) nuclear-hard compared with solar systems; (6) reduced optical and radar signature; (7) reduced cost by a factor of 3; (8) continuous availability of power; (9) no orientation requirements; (10) no maneuver limitations; (11) no power degradation, i.e., beginning-of-life–end-of-life (BOL-EOL) power level; and (12) no large, flexible structure.

The issue of safety was addressed in 1980 by the United Nations Working Group report,[59] which studied the safety of nuclear power sources (NPS) in space. That group reaffirmed the conclusion that NPS can be used safely in space. It placed responsibility on the launching nation to (1) conduct safety tests and evaluations consistent with international standards; (2) provide the United Nations with detailed design and test data of the NPS at launch time; and (3) when reentry of the NPS becomes reasonably certain, provide the United Nations with details of orbiting parameters, probable impact regions, power history, inventory of nuclear fuel, and radiation dosage at 1 m for survival sections. The working group noted that U-235-fueled reactors required 400 years' decay time to reduce fission product activity by a factor of 1000. It implied that a minimum orbit altitude of 300 nmi should be used.

It is obvious from a technical point of view that nuclear prime power systems should be utilized for large SBR systems whenever high power is required.

22.5 CRITICAL ISSUES

A succinct treatment of selected critical issues is given here. The critical issues in the development of SBR include (1) system cost, (2) system survivability or vulnerability, (3) system calibration, (4) antenna deployment and distortion, (5) onboard processing, and (6) nuclear prime power.

SBR System Costs. The author uses a cost-estimating ratio for SBR satellites of $64,000 per kilogram in 1988 dollars that is based upon informal study of many satellites that have been placed into orbit. Launch costs are not included; they depend upon the launch vehicle. Informal study of many satellite launches has resulted in the data shown in Fig. 22.16, which gives a launch cost for several types of vehicles when launched from two United States launch sites, the Eastern Test Range (ETR) and the Western Test Range (WTR). It can be seen that polar orbit costs are greater than launches from the ETR due east and that it is more economical (on a dollars-per-pound basis) to launch large payloads on STS and Titan class vehicles.

Survivability and Vulnerability. SBR system survivability and vulnerability must be demonstrated and tested. The natural space radiation environment will cause a significant total dose on a T/R module depending upon its shielding. Table 22.7 is a summary of the total dose for a 5-year period for circular orbits at altitudes of 450, 900, and 5600 nmi.[32] The T/R module in the analysis has an area of 1 in^2, and it has been assumed that the total dose values will be double those expected in order to account for the particle radiation that penetrates both sides of the module package. Some shielding may be provided by the chip substrate; however, this has been ignored.

22.6 SBR FUTURE POSSIBILITIES

Rendezvous Radar Missions. All satellite rendezvous missions have been performed by manned vehicles. In the foreseeable future, the majority of rendezvous missions might be conducted by unmanned vehicles such as the OMV. The planned list of missions for the OMV includes (1) large-observatory servicing at the shuttle, (2) payload placement, (3) payload retrieval, (4) payload reboost, (5) payload deboost to reentry, (6) payload viewing, (7) subsatellite mission, (8) multiple-payload mission, (9) in situ servicing mission, (10) STS transfer to space station, and (11) base support. Details of these missions may be found in the NASA OMV request for proposal.[62] The initial design of the OMV is modular so as to permit upgrading its capability to operate from the space station and to accommodate the following growth missions by the addition of appropriate kits or elements to the system: (1) logistic support, (2) debris collection mission, (3) extended on-orbit operation, (4) satellite buildup, (5) satellite refueling, (6) servicing mission, and (7) space station reboost.

A rendezvous radar that is low in cost and light in weight will be used to perform these future OMV missions. Such an OMV radar might have the major performance characteristics shown in Table 22.8. The rendezvous radar set (RRS) will be an X-band coherent, range-gated, pulse doppler radar with redundant

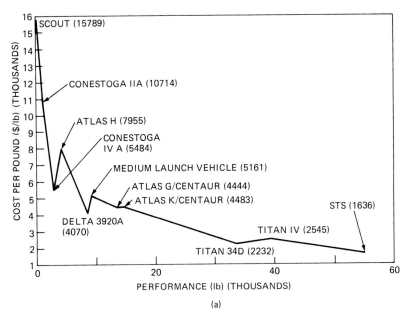

(a)

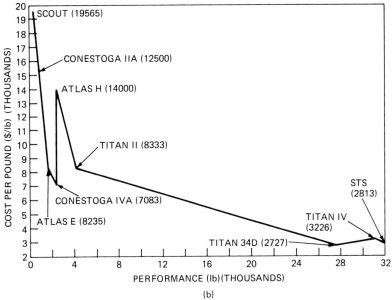

(b)

FIG. 22.16   Cost per pound to low earth orbit. (*a*) Eastern Test Range launch due east. (*b*) Western Test Range launch into polar orbit.

TABLE 22.7 Space Radiation Environment Summary*

SBR orbit, nmi	5-year total dose, rads (Si); aluminum shielding thickness		
	15 mils	25 mils	50 mils
450	$2(10)^5$	$6(10)^4$	$2(10)^4$
900	$2(10)^6$	$4(10)^5$	$2(10)^5$
5600	$6(10)^6$	$4(10)^6$	$(10)^6$

*From Ref. 32.

TABLE 22.8 OMV Radar Characteristics

Frequency	9.5 to 9.8 GHz
PRF	6.67 kHz
Pulse width	0.05, 0.2, 1.5, and 15 μs
Transmitter peak power	2 W (GaAs FET with $\approx$ 30-dB gain)
Receiver noise figure	< 4 dB, GaAs FET LNA
Antenna	Planar slotted array, linear polarization
Antenna size	14 by 15 by 1 in
Antenna gain and beamwidth	30.5 dB (at 9.65 GHz) and 5.0°
Search scan	± 20° cone, with 5-min scan time
Angle accuracy (3σ)	20 mrad
Range accuracy (3σ)	Greater of 20 ft or 2 percent of range
Range rate accuracy (3σ)	Greater of 0.1 ft/s or 2 percent of range rate
Deployed assembly weight	26 lb
Inboard assembly weight	50 lb (redundant total)
Electronics volume	$\approx$ 2 ft^3 (redundant total)
Prime power	< 60 W

electronics and redundant gimbal motor windings. The OMV system computer initiates the acquisition-search function to permit detection of a 1-m^2 Swerling 1 target at a 4.5-nmi range (with 99 percent probability of detection and a false-alarm rate of one alarm per hour.) Monopulse tracking is performed to within a minimum range of 35 ft. Peak power is programmed over a 50 dB range during the rendezvous maneuver to minimize the RF radiation intensity on sensitive targets. Pulse frequency agility is utilized; up to 30 carrier frequency changes in 10-MHz steps over the 300-MHz operating band are used to decorrelate Swerling 1 target fluctuations. At each dwell, 128 pulses are coherently integrated in the fast Fourier transform (FFT) processor prior to noncoherent integration of the FFT outputs. Up to 30 FFT outputs can be integrated.

Initial configuration of the Space Station will have limited tracking-system requirements that include tracking of cooperative vehicles within a 37-km control zone.[63] This is based upon the assumption that all vehicles will provide accurate position and velocity data to the space station tracking system by way of the space-to-space link. Automatic tracking of an extravehicular (EV) astronaut is not required. For growth configurations of the Space Station, the tracking system must expand to meet additional requirements. More co-orbiting vehicles, noncooperative or disabled vehicles, automatic tracking of EV astronauts, and sensors for berthing and docking operations will require additional tracking capabilities within the system. Some type of short-range radar may be required to track vehicles which do not have global positioning system (GPS) capability or vehicles which have been disabled. Results of preliminary tradeoffs on multiple-target tracking radars indicate[36] that either a K_a-band or an X-band phased array radar would be the preferred approach.

Remote-Sensing Missions.[64]   SBR will participate in many remote-sensing missions for observation of the earth and the planets. The SIR series is expected to have the capability to image the earth's surface by using all polarization states (*HH, VV,* and *HV*) and with multiple frequency bands.

A number of SBR SAR missions have been considered by various countries for various purposes. An example is the Canadian Radarsat, which employs a C-band SBR SAR primarily for monitoring polar ice dynamics for use in ship routing; the SAR will have a 200-km swath.

In planetary exploration areas, SBR imaging systems are key elements for the exploration of two bodies that are continuously cloud-covered, Venus and Titan. In the exploration of Venus during the late 1970s, a radar sensor on the Pioneer Venus Orbiter provided low-resolution (40 to 100 km) images of the planet. A U.S.S.R. Venera satellite[65] produced radar images of part of the northern hemisphere of Venus with a resolution of 1300 m. The United States Venus radar mission has the objective of providing global coverage with a resolution of 150 m. In the exploration of Titan, a satellite of Saturn, the larger distance to the earth will put a very tight limit on data rate transmission, which directly impacts mapping coverage and resolution. A radar can be placed into orbit around Saturn, and on selected orbits the spacecraft will fly by Titan. These flybys will be targeted so that during each flyby a different region of Titan will be mapped with an SBR SAR. The Titan radar mapper will have a very wide swath (600 to 800 km) to obtain a global map during the small number of flybys. Real aperture imaging will provide a resolution of 6 to 40 km. A synthetic aperture mode can be used to observe limited regions with a resolution of about 200 m.

Other missions using radar are planned for ocean scatterometry and altimeters. Scatterometers are used to obtain accurate measurement of global surface winds for oceanography and meteorology. Wind speed with errors of about 2 m/s and a wind direction error of less than 16° are sought. SBR altimeters expect to measure altitude with an error of 5 cm from a 1300-km polar orbit inclined 65° over the ocean. Over the solid surface of Mars, a 37-GHz altimeter on the Mars orbiter mission expects to gather global high-resolution topographic mapping data with a height resolution of 15 m.

The overall goal of the Earth Observing System (EOS) is to advance the scientific understanding of the entire earth system on the global scale through developing a deeper understanding of the components of that system, interactions among them, and how the system is changing.[66] International space station elements include the following satellites in polar and equatorial orbits (1) a NASA EOS platform at 824 km, sun-synchronous, 1:30 P.M. equator-crossing time, ascending-node orbit; (2) a European Space Agency (ESA) platform at 824 km, sun-synchronous, 10:00 A.M. equator-crossing time, descending-node orbit; and (3) the manned space station in a 335- to 460-km 28.5° inclined orbit. Instruments planned for the satellites include radar, radiometers, IR, optics, and ultraviolet (UV). These sensors will measure parameters such as winds, clouds, rain, liquid-moisture content, geologic parameters, ocean currents, etc. Radars will be used to make atmospheric and geological observations. Two of the radars proposed are the tropical-rain mapping radar (TRAMAR) and the land, ocean, and rain radar altimeter (LORRA).[67]

Global Air Traffic Surveillance.[12] ATC is increasingly a matter of global concern, and the explosive growth of aircraft density in, around, and between major metropolitan areas in Europe and North America is common knowledge. If a United Nations organization were responsible for ATC for 120 to 130 nations in the world, it is conceivable that as many as 84,000 commercial aircraft could require ATC in the twenty-first century. A rosette constellation of SBR satellites at an orbital altitude of 5600 nmi (10,371 km) in a 14/14/12 Walker orbit[23] inclined at 49.4° provides continuous worldwide visibility by at least two satellites simultaneously. (One satellite is deployed in each of 14 equally spaced orbit planes.) Each satellite provides radar coverage between grazing angles from 3 to 70°. The major subsystems in the satellite include (1) radar, (2) communications, (3) guidance and control, and (4) electrical power subsystems. Details of these subsystems are found in Ref. 12. The radar parameters are shown in Table 22.9. Each T/R module would have a peak power of 0.155 W and an average power of 15 mW and would weigh 5 g.

TABLE 22.9 Radar Parameters for Global Air Traffic Surveillance*

Antenna	
Type	Corporate-fed active phased array
Diameter	100 m
Frequency	2 GHz
Wavelength	0.15 m
Polarization	Circular
Number of elements	576,078
Number of modules	144,020
Element spacing	0.7244 wavelength
Beamwidth	1.83 mrad
Directive gain	66.42 dB
Maximum scan angle	22.4°
Receiver	
Type	Distributed solid-state monolithic T/R module
Bandwidth	500 kHz
System noise temperature	490 K
Compressed pulse width	2 μs
Transmitter	
Type	Distributed solid-state monolithic T/R module
Peak power	22.33 kW
Pulse width	2000 μs
Maximum duty	0.20
Frequency	2 GHz
Signal Processor	
Type	Digital
Input speed	50 million words per second

*From Ref. 12.

Military SBR Systems. Brookner and Mahoney[11] derived a satellite radar architecture for performing the basic surveillance missions for the fleet defense and air defense of the CONUS. The system was an L-band, corporate-fed phased array radar in orbit constellations of 3 to 12 satellites at altitudes from 600 to 2000 nmi. At the highest orbital altitude, a 10- by 30-m phased array that contained 15,000 radiating elements or modules was designed. The modules delivered an average power of 6 kW, and the radar required a prime power of 30 kW.

REFERENCES

1. Hughes Aircraft Company: "K$_u$-Band Integrated Radar and Communications Equipment for the Space Shuttle Orbiter Vehicle," preliminary design review, vol. 1, Mar. 14–24, 1978.

2. RCA Government and Commercial Systems, Aerospace Systems Division, Burlington, Mass.: "The Apollo LM Rendezvous Radar and Transponder," *Rept. LTM* 3300-14D, February 1971.

3. Quigley, W. W.: Gemini Rendezvous Radar, *Microwave J.*, pp. 39–45, June 1965.

4. Fenner, R. G., and R. F. Broderick: Spaceborne-Radar Applications, chap. 34 in Skolnik, M. I. (ed.): "Radar Handbook," McGraw-Hill Book Company, New York, 1970.

5. Elachi, C., et al.: Spaceborne Synthetic Aperture Imaging Radars: Applications, Techniques and Technology, *Proc. IEEE*, vol. 70, pp. 1174–1209, October 1982.

6. Elachi, C., and J. Granger: Spaceborne Imaging Radars Probe "in Depth," *IEEE Spectrum*, vol. 19, pp. 24–29, November 1982.

7. Williams, F. C., et al.: The Pioneer Venus Orbiter Radar, *1976 WESCON Sess. 4*, Los Angeles, Sept. 14–17, 1976.

8. Hofmeister, E. L., et al.: GOES-C Radar Altimeter, vol. 1, "Data Users Handbook," General Electric Company, Utica, N.Y., May 1976.

9. Soviet Radar Records Venus Surface Imager, *Aviat. Week Space Technol.*, vol. 119, p. 18, Oct. 24, 1983.

10. Ulsamer, E.: In Focus—Approach Set on Space Radars, *Air Force Mag.*, vol. 67, pp. 17–18, February 1984.

11. Brookner, E., and T. F. Mahoney: Derivation of a Satellite Radar Architecture for Air Surveillance," *IEEE EASCON '83 Conf. Rec.*, Washington, Sept. 19–21, 1983.

12. Cantafio, L. J., and J. S. Avrin: Satellite-Borne Radar for Global Air Traffic Surveillance, *IEEE ELECTRO '82 Prof. Sess. Rec.*, Boston, May 25–27, 1982.

13. Cantafio, L. J.: Space Based Radar Concept for the Proposed United Nations International Satellite Monitoring Agency, *Mil. Microwaves Conf.*, London, Oct. 24–26, 1984.

14. "The Implication of Establishing an International Satellite Monitoring Agency," UN Publ. Sales No. E.83.IX.3.

15. Griffin, J. W., et al.: K$_u$-Band—The First Year of Operation, *IEEE Int. Radar Conf. Rec.*, pp. 330–339, Arlington, Va., May 6–9, 1985. (IEEE Cat. No. 85CH2076-8.)

16. Wolverton, R. W., et al.: "Flight Performance Handbook for Orbital Operations," John Wiley & Sons, New York, 1961.

17. Hord, R. A.: Relative Motion in the Terminal Phase of Interception of a Satellite or Ballistic Missile, *NACA TN* 8399, September 1958.

18. Functional Requirements for the Seasat-A Synthetic Aperture Radar System, *Jet Propulsion Lab., FR No.* FM511774, rev. dated Aug. 2, 1976, Pasadena, Calif.

19. "Geodynamics Experimental Ocean Satellite Project of the Earth and Ocean Physics

Applications Program," NASA brochure, NASA Wallops Flight Center, Wallops Island, Va., 1975.

20. New Satellite to Measure Ocean Surface Topography and Sea State, *NASA News Release* 75-88, Washington, Mar. 31, 1975.

21. GEOS-C Mission Plan, NASA TK-6340-001, rev. 3, NASA Wallops Flight Center, Wallops Island, Va., Dec. 18, 1974.

22. Cantafio, L. J.: Satellite-Borne Radar, *Lecture IX, Adv. Radar Technol. Short Course*, scheduled by Technology Service Corporation, San Diego, Apr. 22, 1983.

23. Walker, J. G.: Continuous Whole Earth Coverage by Circular Orbit Satellite Patterns, *R. Aircr. Estab. Tech. Rept.* 77044, Mar. 24, 1977.

24. Luders, R. D., and L. J. Ginsberg: Continuous Zonal Coverage—A Generalized Analysis, *AIAA Pap.* 74-842, *AIAA Mech. Control of Flight Conf.*, Anaheim, Calif., Aug. 5–9, 1974.

25. Ballard, A. H.: Rosette Constellation of Earth Satellites, *IEEE Trans.*, vol. AES-16, pp. 656–673, September 1980.

26. Beste, D. C.: Design of Satellite Constellations for Optimal Continuous Coverage, *IEEE Trans.*, vol. AES-14, pp. 466–473, May 1978.

27. Emara, E. T., and C. T. Leondes: Minimum Number of Satellites for Three-Dimensional Continuous Worldwide Coverage, *IEEE Trans.*, vol. AES-13, pp. 108–111, March 1977.

28. Harney, E. D.: "Space Planners Guide," U.S. Air Force Systems Command, Publ. 0-774-405, 1965.

29. Fager, J. A.: Application of Graphite Composites to Future Spacecraft Antennas, *AIAA Pap.* 76-328, *Sixth Commun. Satellite Syst. Conf.*, Apr. 6–8, 1976.

30. Schultz, J. L., and P. Nosal: Space-Based Radar, *Horizons*, vol. 15, no. 1, p. 10, Grumman Aerospace Corporation, 1979.

31. Fawcette, J.: Large Radar Satellite Proposed, *Microwave Syst. News*, vol. 8, pp. 17–20, September 1978.

32. Mrstik, A. V., et al.: RF Systems in Space—Space-Based Radar Analysis, *General Research Corporation, RADC TR-83-91, Final Tech. Rept.*, vol. II, April 1983.

33. Ludwig, A. C., et al.: RF Systems in Space—Space Antennas Frequency (SARF) Simulation, *General Research Corporation, RADC TR-38-91, Final Tech. Rept.*, vol. I, April 1983.

34. Kendrick, J. B. (ed.): "TRW Space Data," 3d ed., 1967.

35. Barton, D. K.: A Half Century of Radar, *IEEE Trans.*, vol. MTT-32, pp. 1161–1169, September 1984.

36. Tu, K., et al.: Space Shuttle Communications and Tracking System, *Proc. IEEE.*, vol. 75, pp. 356–370, March 1987.

37. Brejcha, A. G., L. H. Keeler, and G. G. Sanford: The Seasat-A Synthetic Aperture Radar Antenna, *Synth. Aperture Radar Technol. Conf.*, Las Cruces, N. Mex., Mar. 8–10, 1978.

38. Kalmykov, A. I., et al.: Side-Looking Radar of Kosmos-1500 Satellite, *Issled. Zemli Kosmosa*, no. 3, May–June 1985.

39. Soviets Plan to Launch New Spacecraft, *Aviat. Week Space Technol.*, vol. 127, p. 27, Oct. 19, 1987.

40. Bearse, S. V.: Knitted Antenna Solving Knotty Problems, *Microwaves*, p. 14, March 1974.

41. Large Furlable Antenna Study, *Lockheed Missiles and Space Company, Rept. LMSC-D384797*, Jan. 20, 1975.

42. Cummings, Freeman, and Benz: Deployable Parabolic Antenna, U.S. Patent 3,789,375, Dec. 18, 1973, assigned to Rockwell Inc.

43. Fager, J. A., and R. Garriott: Large Aperture Expandable Truss Microwave Antenna, *IEEE Trans.*, vol. AP-17, pp. 452–458, July 1969.

44. Das, A., and J. A. Delaney: Spacecraft Phased Array Configurations, *IEEE Trans.*, vol. AP-17, pp. 522–524, July 1969.

45. Final Report—Spaceborne Radar Study, *Grumman Aerospace Corporation, AFSC-ESD contract F19628-74-R*-0140, *Rept.* 74-21AF-I, June 28, 1974.

46. Hagler, T.: Building Large Structures in Space, *Astronaut. Aeronaut.*, vol. 14, pp. 56–61, May 1976.

47. Deployable Antenna Flight Experiment—Preliminary Definition Study, *Harris Corporation, Third Q. Rev.*, June 24, 1981.

48. Deployable Antenna Flight Experiment Definition Study: Mid-Term Review, *Grumman Aerospace Corporation, NAS*-8-33932, Mar. 20, 1981.

49. Bujakes, V. I., et al.: Infinitely Built-Up Radio Telescope, *Pap. IAF-77-67, IAF XXVIII Cong.*, Prague, Sept. 25–Oct. 1, 1977.

50. Ulsamer, E.: ATS-6, NASA's Huge Transmitter in the Sky, *Air Force Mag.*, vol. 57, August 1974.

51. Archer, J. S.: Advanced Sunflower Antenna Concept Development, *LSST First Ann. Tech. Rev.*, NASA LRC, Nov. 7–8, 1979.

52. Lockheed Tests Large Space Antenna, *Aviat. Week Space Technol.*, vol. 120, p. 70, Apr. 30, 1984.

53. Larson, T. R., A Microstrip Honeycomb Array for the Low Altitude Space Based Radar Mission, *Ball Aerospace Systems Division, Rept. F*81-06, August 1981.

54. Final Report—Spaceborne Radar Study, *Grumman Aerospace Corporation, Rept.* 74-21-AF-1, *prepared for ADSC-ESD (XRS) contract F*19623-74-R-0140, June 28, 1974.

55. Thimlar, M. E., et al.: Future Space-Based Computer Processors, *Aerosp. Am.*, vol. 22, pp. 78–82, March 1984.

56. Works, G. A.: Advanced Onboard Signal Processor, *IEEE EASCON '80 Rec.*, p. 233, 1980.

57. GaAs Solar Cells, *Hughes Aircraft Company, presented at Space Power Conf.*, Los Angeles, Jan. 13–14, 1981.

58. Emigh, C. R.: Reactor Technology, January–March 1980, *Los Alamos Scientific Laboratory, Prog. Rept. LA*-8403-PR-UC-80, June 1980.

59. Buden, D., et al.: Space Nuclear Reactor Power Plants, *LASL Informal Rept. LA*-8223-MS-UC-33, January 1980.

60. Buden, D., et al.: Selection of Power Plant Elements for Future Reactor Space Electric Power Systems, *LASL Rept. LA*-7858, September 1979.

61. Kelley, J. H.: Minutes of Space Nuclear Power Service Working Group Meeting, Sept. 24, 1982, issued Oct. 8. 1982.

62. Orbital Maneuvering Vehicle, *NASA George C. Marshall Space Flight Center, Request for Proposal* 1-6-*pp*-01438, November 1985.

63. Dietz, R. H.: Space Station Communications and Tracking Systems, *Proc. IEEE*, vol. 75, pp. 371–382, March 1987.

64. Carver, K. R., C. Elachi, and F. T. Ulaby: Microwave Remote Sensing from Space, *Proc. IEEE*, vol. 73, pp. 970–996, June 1985.

65. Bogomolov et al.: Venera 15 and 16 Synthesized Aperture Radar in Orbit around Venus, *Izv. Vyssh. Uchebn. Zaved., Radiofiz.*, vol. 28, pp. 259–274, March 1985.

66. NASA Announcement of Opportunity: The Earth Observing System (EOS), *A.O.* No. *OSSA*-1-88, Jan. 19, 1988.

67. Lorra/Tramar Design—Feasibility Study, *Malibu Res. and TRW*, MRA p. 214-3, May 11, 1988.

CHAPTER 23
METEOROLOGICAL RADAR

Robert J. Serafin
National Center for Atmospheric Research*

23.1 INTRODUCTION

As this handbook is being written, dramatic changes are taking place in the field of radar meteorology. While the majority of radar engineers are familiar with current operational meteorological radars, few are aware of the advances that have been made in the past two decades. For example, doppler radar meteorology, using modern digital signal-processing techniques and display technology, has moved ahead so rapidly that the United States is now planning to replace its existing operational weather radar network with a next-generation doppler system (NEXRAD). This system will provide quantitative and automated real-time information on storms, precipitation, hurricanes, tornadoes, and a host of other important weather phenomena, with higher spatial and temporal resolution than ever before.[1] A second network of doppler radars, in airport terminal areas, will provide quantitative measurements of gust fronts, wind shear, microbursts, and other weather hazards for improving the safety of operations at major airports in the United States.[1,2] Next-generation doppler radars that use flat-plate antennas, color displays, and solid-state transmitters are now available for commercial aircraft. And many of these new technologies are being deployed in countries throughout the world.

In the research arenas, multiple-doppler radars are used for deriving three-dimensional wind fields.[3] Airborne doppler radar[4,5] has been used to duplicate these capabilities, thus providing for great mobility. Polarization diversity techniques[6] are used for discriminating ice particles from water, for improved quantitative precipitation measurement, and for detecting hail. And there is a new family of radars, ultrahigh-frequency (UHF) and very-high-frequency (VHF) fixed-beam systems that are being used to obtain continuous profiles of horizontal winds.[7] These examples are illustrative of the vitality of the field.

This chapter is intended to introduce the reader to meteorological radar and particularly those system characteristics that are unique to meteorological applications. In this regard, it should be noted that most meteorological radars appear

---
*The National Center for Atmospheric Research is sponsored by the National Science Foundation.
Special thanks are due to Victoria Holzhauer for her careful typing, assistance with figures, and editing of this manuscript. The author is also grateful to Richard Carbone and Jeffrey Keeler for their critical reviews.

similar to radars used for other purposes. Pulsed and pulsed doppler systems are common. Parabolic dish antennas, focal-point feeds, and low-noise solid-state receivers are used. Magnetrons, phase-locked magnetrons, klystrons, traveling-wave tubes, and other forms of transmitters are used.

The major distinction between meteorological radar and other kinds of radars lies in the nature of the targets. Meteorological targets are distributed in space and occupy a large fraction of the spatial resolution cells observed by the radar. Moreover, it is necessary to make quantitative measurements of the received signal's characteristics in order to estimate such parameters as precipitation rate, precipitation type, air motion, turbulence, and wind shear. In addition, because so many radar resolution cells contain useful information, meteorological radars require high-data-rate recording systems and effective means for real-time display.[8,9] Thus, while many radar applications call for discrimination of a relatively few targets from a clutter background, meteorological radars focus on making accurate estimates of the nature of the *weather clutter* itself. This poses some challenging problems for the radar system designer to address.

The discussion here will refer to a number of useful texts and references for the reader to use. However, Battan's text,[10] revised in 1973, deserves special mention for its clarity and completeness and remains a standard for courses in radar meteorology that are taught in universities around the world. Doviak and Zrnić[11] place special emphasis on doppler meteorological radar. Chapter 24 in the first "Radar Handbook," by Bean et al.,[12] addresses the problem of weather effects on radar. Finally, perhaps the broadest and most complete set of references on progress in the field can be found in the *Proceedings* and *Preprints* of the series of radar meteorology conferences sponsored by the American Meteorological Society (AMS). These documents can be found in most technical libraries and also can be obtained through the offices of the AMS in Boston.

23.2 THE RADAR RANGE EQUATION FOR METEOROLOGICAL TARGETS

The received power from distributed targets can be derived from any of a variety of expressions that are applicable to radar in general. A simple form, with which to begin, is given below:

$$P_r = \frac{\beta\sigma}{r^4} \qquad (23.1)$$

where β is a constant dependent upon radar system parameters, r is the range, and σ is the radar cross section.

It is in the calculation of σ for meteorological targets that the radar range equation differs from that for point targets. σ may be written

$$\sigma = \eta V \qquad (23.2)$$

where η is the radar reflectivity in units of cross-sectional area per unit volume and V is the volume sampled by the radar. η can be written as

$$\eta = \sum_{i=1}^{N} \sigma_i \qquad (23.3)$$

where N is the number of scatterers per unit volume and σ_i is the backscattering cross section of the ith scatterer. In general, the meteorological scatterers can take on a variety of forms, which include water droplets, ice crystals, hail, snow, and mixtures of the above.

Mie[13] developed a general theory for the energy backscattered by a plane wave impinging on spherical drops. This backscattered energy is a function of the wavelength, the complex index of refraction of the particle, and the ratio $2\pi\alpha/\lambda$, where α is the radius of the spherical particle and λ is the wavelength.

When the ratio $2\pi\alpha/\lambda - 1$, the Rayleigh approximation[10] may be applied, and σ_i becomes

$$\sigma_i = \frac{\pi^5}{\lambda^4}|K|^2 D_i^6 \tag{23.4}$$

where D_i is the diameter of the ith drop and

$$|K|^2 = \left|\frac{m^2 - 1}{m^2 + 2}\right|^2 \tag{23.5}$$

where m is the complex index of refraction. At temperatures between 0 and 20°C, for the water phase, and at centimeter wavelengths

$$|K|^2 \approx 0.93 \tag{23.6a}$$

and for the ice phase

$$|K|^2 \approx 0.20 \tag{23.6b}$$

Equation (23.3) can now be written as

$$\eta = \frac{\pi^5}{\lambda^4}|K|^2 \sum_{i=1}^{N} D_i^6 \tag{23.7}$$

and the radar reflectivity factor Z defined as

$$Z \sum_{i=1}^{N} D_i^6 \tag{23.8}$$

In radar meteorology, it is common to use the dimensions of millimeters for drop diameters D_i and to consider the summation to take place over a unit volume of size 1 m³. Therefore, the conventional unit of Z is in mm⁶/m³. For ice particles, D_i is given by the diameter of the water droplet that would result if the ice particle were to melt completely.

It is often convenient to treat the drop or particle size distribution as a continuous function with a number density $N(D)$, where $N(D)$ is the number of drops per unit volume, with diameters between D and $D + dD$. In this case, Z is given by the sixth moment of the particle size distribution,

$$Z = \int_0^\infty N(D)D^6 dD \tag{23.9}$$

If the radar beam is filled with scatterers, the sample volume V is given[10] approximately by

$$V \approx \frac{\pi \theta \phi r^2 c\tau}{8}$$ (23.10)

where θ and ϕ are the azimuth and elevation beamwidths, c is the velocity of light, and τ is the radar pulsewidth.

Substituting Eqs. (23.10), (23.2), and (23.4) into Eq. (23.1) gives

$$
\begin{aligned}
P_r &= \frac{\beta \pi}{r^4} \frac{\theta \phi r^2 c\tau}{8} \frac{\pi^5}{\lambda^4} |K|^2 \sum_{i=1}^{N} D_i^6 \\
&= \frac{\beta \pi^6 \theta \phi \, c\tau \, |K|^2}{8 \lambda^4 r^2} Z \\
&= \frac{\beta' Z}{r^2}
\end{aligned}
$$ (23.11)

This simple expression illustrates that the received power is a function only of β' (a constant dependent upon radar system parameters), is proportional to the radar reflectivity factor Z, and is inversely proportional to r^2.

In actual fact, the antenna gain is not uniform over the beamwidth, and the assumption of a uniform gain can lead to errors in the calculation of Z. Probert-Jones[14] took this into account, assumed a gaussian shape for the antenna beam, and derived the following equation for the received power:

$$P_r = \frac{P_t G^2 \lambda^2 \theta \phi \, c\tau}{512(2 \ln 2) \pi^2 r^2} \sum_{i=1}^{N} \sigma_i$$ (23.12)

where $2 \ln 2$ is the correction due to the gaussian-shaped beam.

By using the relationships in Eqs. (23.7) and (23.8), Eq. (23.12) can be written in terms of the reflectivity factor Z as

$$P_r = \frac{P_t G^2 \theta \phi c\tau \pi^3 |K|^2 Z}{512(2 \ln 2) r^2 \lambda^2}$$ (23.13)

One must be careful to use consistent units in Eq. (23.13). If meter-kilogram-seconds (mks) units are used, the calculation of Z from Eq. (23.13) will have dimensions of m^6/m^3. Conversion to the more commonly used units of mm^6/m^3 requires that the result be multiplied by the factor 10^{18}. Because Z values of interest can range over several orders of magnitude, a logarithmic scale is often used, where

$$dBZ = 10 \log Z$$ (23.14)

Equation (23.13) can be used to measure the reflectivity factor Z when the antenna beam is filled, when the Rayleigh approximation is valid, and when the scatterers are in either the ice or the water phase. Because all these conditions are not always satisfied, it is common to use the term Z_e, the effective reflectivity factor, in place of Z. When Z_e is used, it is generally understood that the above conditions are assumed. Practitioners in the field of radar meteorology often use Z_e and Z interchangeably, albeit incorrectly.

Finally, it is important to note the range of Z values that are of meteorological significance. In nonprecipitating clouds, Z values as small as -40 dBZ are of interest. In the optically clear boundary layer, Z values of the order -20 dBZ to 10 dBZ are of interest. In rain, Z may range from about 20 dBZ to as much as 60 dBZ, with a 55 to 60 dBZ rain being of the type that can cause severe flooding. Severe hailstorms may produce Z values higher than 70 dBZ. Operational radars are generally designed to detect Z values ranging from 10 to 60 dBZ, while research applications usually aim for the maximum dynamic range possible. In light of the above, operational radars often employ sensitivity time control (STC) to compensate for inverse r^2 dependence, but research radars usually do not use STC owing to the attendant loss of sensitivity at short ranges.

23.3 DESIGN CONSIDERATIONS

Three of the more significant factors that affect the design of meteorological radars are attenuation, range-velocity ambiguities, and ground clutter. The combination of these three, along with the need to obtain adequate spatial resolution, leads to a wavelength selection in the range of 3 to 10 cm for most meteorological applications.

Attenuation Effects. Attenuation has at least two negative effects on meteorological radar signals. First, because of attenuation it becomes difficult, if not impossible, to make quantitative measurements of the backscattered energy from precipitation which is at greater range (and at the same azimuth and elevation angles) than precipitation closer to the radar. This inability to precisely measure the backscattering cross section makes quantitative measurements of precipitation rates more difficult.

Second, if the attenuation due to precipitation or the intervening medium is sufficiently great, the signal from a precipitation cell behind a region of strong absorption may be totally obliterated, leading to potentially disastrous effects. One example of the potentially serious consequences of very strong absorption is the impact it might have on airborne storm avoidance radars, most of which are in the 3-cm band, although some use a 5-cm wavelength. Metcalf[15] has examined ground-based radar data from the storm that was responsible for the 1977 crash of Southern Airways Flight 242 in northwest Georgia. The crew had relied on its on-board radar for penetration of a severe storm. Metcalf shows strong evidence that the region penetrated by the aircraft, while appearing to be free of echo, had actually been obliterated because of severe attenuation. Severe storms can also produce very strong absorption at 5-cm wavelength, as noted by Allen et al.[16]

In some meteorological radar applications, it is desirable to attempt to measure attenuation along selected propagation paths. This is done because absorption is related to liquid-water content and can provide useful information for the detection of such phenomena as hail, in accordance with the dual-wavelength technique described by Eccles and Atlas.[17]

In the following subsections, quantitative expressions relating attenuation to precipitation are given. Much of this is taken from Bean, Dutton, and Warner.[12]

Battan's textbook[10] is also an excellent source for additional information on the absorbing properties of precipitation.

Attenuation in Clouds. Cloud droplets are regarded here as those water or ice particles having radii smaller than 100 μm, or 0.01 cm. For wavelengths of incident radiation well in excess of 0.5 cm, the attenuation becomes independent of the drop-size distribution. The generally accepted equations for attenuation by clouds usually show the moisture component of the equations in the form of the liquid-water content (grams per cubic meter). Observations indicate that the liquid-water concentration in clouds generally ranges from[18] 1 to 2.5 g/m³, although Weickmann and aufm Kampe[19] have reported isolated instances of cumulus congestus clouds with water contents of 4.0 g/m³ in the upper levels. In ice clouds, it rarely exceeds 0.5 and is often less than 0.1 g/m³. The attenuation due to cloud drops may be written[12]

$$K = K_1 M \qquad (23.15)$$

where K = attenuation, dB/km
K_1 = attenuation coefficient, dB/(km · g · m³)
M = liquid-water content, g/m³

$$M = \frac{4\pi\rho}{3} \sum_{i=1}^{N} a_i^3 \qquad (23.16)$$

$$K_1 = 0.4343 \frac{6\pi}{\lambda} \mathrm{Im}\left(-\frac{m^2 - 1}{m^2 + 2}\right) \qquad (23.17)$$

where the a_i are droplet radii, ρ is the density of water, and Im is the imaginary part. Values of K_1 for ice and water clouds are given for various wavelengths and temperatures by Gunn and East in Table 23.1.

Several important facts are demonstrated by Table 23.1. The decrease in attenuation with increasing wavelength is clearly shown. The values change by about an order of magnitude, for a change of λ from 1 to 3 cm. The data presented here also shows that attenuation in water clouds increases with decreasing temperature. Ice clouds give attenuations about two orders of magnitude smaller

TABLE 23.1 One-Way Attenuation Coefficient K_1 in Clouds in dB/(km · g · m³)*

Temperature, °C		Wavelength, cm			
		0.9	1.24	1.8	3.2
Water cloud	20	0.647	0.311	0.128	0.0483
	10	0.681	0.406	0.179	0.0630
	0	0.99	0.532	0.267	0.0858
	−8	1.25	0.684	0.34 (extrapolated)	0.112 (extrapolated)
Ice cloud	0	8.74×10^{-3}	6.35×10^{-3}	4.36×10^{-3}	2.46×10^{-3}
	−10	2.93×10^{-3}	2.11×10^{-3}	1.46×10^{-3}	8.19×10^{-4}
	−20	2.0×10^{-3}	1.45×10^{-3}	1.0×10^{-3}	5.63×10^{-4}

*After Gunn and East.[20]

than water clouds of the same water content. The attenuation of microwaves by ice clouds can be neglected for all practical purposes.[10]

Attenuation by Rain. Ryde and Ryde[21] calculated the effects of rain on microwave propagation and showed that absorption and scattering effects of raindrops become more pronounced at the higher microwave frequencies, where the wavelength and the raindrop diameters are more nearly comparable. In the 10-cm band and at shorter wavelengths the effects are appreciable, but at wavelengths in excess of 10 cm the effects are greatly decreased. It is also known that suspended water droplets and rain have an absorption rate in excess of that of the combined oxygen and water-vapor absorption.[22]

In practice, it has been convenient to express rain attenuation as a function of the precipitation rate R, which depends on both the liquid-water content and the fall velocity of the drops, the latter in turn depending on the size of the drops.

Ryde[23] studied the attenuation of microwaves by rain and deduced, by using Laws and Parsons[24] distributions, that this attenuation in decibels per kilometer can be approximated by

$$K_R = \int_0^{r_0} [R(r)]^\alpha dr \qquad (23.18)$$

where K_R = total attenuation, dB
 K = function of frequency[25]
 $R(r)$ = rainfall rate along path r
 r_0 = length of propagation path, km
 α = function of frequency[10]

Medhurst[26] shows that $\alpha = 1$ is a good assumption in many cases. The path loss per mile, according to Ryde, for the three carrier frequency bands of 4, 6, and 11 GHz, is shown in Fig. 23.1.

The greatest uncertainty in predictions of attenuation caused by rainfall, when theoretical formulas are used as a basis for calculation, is the extremely limited knowledge of drop-size distribution in rains of varying rates of fall under differing climatic and weather conditions. There is little evidence that a rain with a known rate of fall has a unique drop-size distribution, although studies on this problem seem to indicate that a certain most probable drop-size distribution can be attached to a rain of a given rate of fall.[27] Results of this study are shown in Table 23.2, which gives the percentage of total volume of rainfall occupied by raindrops of different diameters (centimeters) and varying rainfall rates (millimeters per hour). On the basis of these results, the absorption cross section of raindrops of different sizes is shown in Table 23.3. This table gives the decibel attenuation per kilometer in rains of different rates of fall for radio wavelengths between 0.3 and 10 cm.

Since the total-attenuation cross section[28] depends on the temperature (because of its effects on the dielectric properties of water), it is important to evaluate the attenuation of rains whose drops are at different temperatures from those in the preceding tables. Table 23.4 contains the necessary data relative to the change of attenuation with temperature and is to be used with Table 23.3.

To determine total attenuation caused by rainfall through a particular storm, something must be known about the nature of the storm itself and, consequently, about how its rainfall rates and drop sizes are distributed in three dimensions.

A systematic vertical variation of R, decaying with height above a measured surface value, seems to be appropriate in rainfall of a widespread (continuous) nature.[29] Widespread rainfall is usually triggered by a relatively large-scale mechanism, such as a frontal or monsoon situation. A vertical variation of R of the form

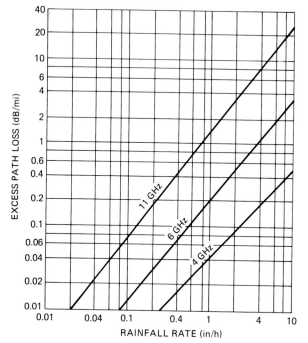

FIG. 23.1 Theoretical rain attenuation versus rainfall rate.

$$R = R_0 e^{-dh^2} \qquad (23.19)$$

can be assumed to be appropriate under continuous-rainfall conditions.[29] In Eq. (23.19), R_0 is the surface rainfall rate, h is the height above the earth's surface, and d is a constant, equal to about 0.2.

Convective-type precipitation, however, shows a quite different nature. The presence of the *virga* (precipitation aloft but evaporating before reaching the surface) associated with so many shower-type clouds indicates that Eq. (23.19) is not especially representative of shower rainfall. Dennis[30] has done considerable work in examining rainfall determinations in shower-type activity. His observations show that the reflectivity factor $Z(\text{mm}^6/\text{m}^3)$ of an element of a vertical slice taken through a spherical shower cell is well represented by a regression line of the form

$$Z = c_1(r_0 - r)^{c_2} \qquad (23.20)$$

In Eq. (23.20), r is the distance from the center of the cell of radius r_0, and c_1 and c_2 are positive constants.

Attenuation by Hail. Ryde[23] concluded that the attenuation caused by hail is one-hundredth of that caused by rain, that ice-crystal clouds cause no sensible attenuation, and that snow produces very small attenuation even at the excessive rate of fall of 5 in/h. However, the scattering by spheres surrounded by a concentric film of different dielectric constant does not give the same effect that

TABLE 23.2   Drop-Size Distribution*

Drop diameter D, cm	Precipitation rate p, mm/h							
	0.25	1.25	2.5	12.5	25	50	100	150
	Percentage of a given volume containing drops of diameter D							
0.05	28.0	10.9	7.3	2.6	1.7	1.2	1.0	1.0
0.10	50.1	37.1	27.8	11.5	7.6	5.4	4.6	4.1
0.15	18.2	31.3	32.8	24.5	18.4	12.5	8.8	7.6
0.20	3.0	13.5	19.0	25.4	23.9	19.9	13.9	11.7
0.25	0.7	4.9	7.9	17.3	19.9	20.9	17.1	13.9
0.30	.....	1.5	3.3	10.1	12.8	15.6	18.4	17.7
0.35	.....	0.6	1.1	4.3	8.2	10.9	15.0	16.1
0.40	.....	0.2	0.6	2.3	3.5	6.7	9.0	11.9
0.45	.....	.....	0.2	1.2	2.1	3.3	5.8	7.7
0.50	.....	.....	.....	0.6	1.1	1.8	3.0	3.6
0.55	.....	.....	.....	0.2	0.5	1.1	1.7	2.2
0.60	.....	.....	.....	.....	0.2	0.5	1.0	1.2
0.65	.....	.....	.....	.....	.....	0.2	0.7	1.0
0.70	.....	.....	.....	.....	.....	.....	.....	0.3

*From Burrows and Attwood.[27]

TABLE 23.3   Attenuation in Decibels per Kilometer for Different Rates of Rain Precipitation at Temperature 18°C*

Precipitation rate p, mm/h	Wavelength λ, cm								
	$\lambda = 0.3$	$\lambda = 0.4$	$\lambda = 0.5$	$\lambda = 0.6$	$\lambda = 1.0$	$\lambda = 1.25$	$\lambda = 3.0$	$\lambda = 3.2$	$\lambda = 10$
0.25	0.305	0.230	0.160	0.106	0.037	0.0215	0.00224	0.0019	0.0000997
1.25	1.15	0.929	0.720	0.549	0.228	0.136	0.0161	0.0117	0.000416
2.5	1.98	1.66	1.34	1.08	0.492	0.298	0.0388	0.0317	0.000785
12.5	6.72	6.04	5.36	4.72	2.73	1.77	0.285	0.238	0.00364
25.0	11.3	10.4	9.49	8.59	5.47	3.72	0.656	0.555	0.00728
50	19.2	17.9	16.6	15.3	10.7	7.67	1.46	1.26	0.0149
100	33.3	31.1	29.0	27.0	20.0	15.3	3.24	2.80	0.0311
150	46.0	43.7	40.5	37.9	28.8	22.8	4.97	4.39	0.0481

*From Burrows and Attwood.[27]

Ryde's results for dry particles would indicate.[23,31] For example, when one-tenth of the radius of an ice sphere of radius 0.2 cm melts, the scattering of 10-cm radiation is approximately 90 percent of the value that would be scattered by an all-water drop.

At wavelengths of 1 and 3 cm with $2a = 0.126$ (a = radius of drop), Kerker, Langleben, and Gunn[31] found that particles attained total-attenuation cross sections corresponding to all-melted particles when less than 10 percent of the ice particles was melted. When the melted mass reached about 10 to 20 percent, the attenuation was about twice that of a completely melted particle. These calculations show that the attenuation in the melting of ice immediately under the 0°C isotherm can be substantially larger than in the snow region just above and, under

TABLE 23.4   Correction Factor (Multiplicative) for Rainfall Attenuation*

Precipitation rate p, mm/h	λ, cm	0°C	10°C	18°C	30°C	40°C
0.25	0.5	0.85	0.95	1.0	1.02	0.99
	1.25	0.95	1.00	1.0	0.90	0.81
	3.2	1.21	1.10	1.0	0.79	0.55
	10.0	2.01	1.40	1.0	0.70	0.59
2.5	0.5	0.87	0.95	1.0	1.03	1.01
	1.25	0.85	0.99	1.0	0.92	0.80
	3.2	0.82	1.01	1.0	0.82	0.64
	10.0	2.02	1.40	1.0	0.70	0.59
12.5	0.5	0.90	0.96	1.0	1.02	1.00
	1.25	0.83	0.96	1.0	0.93	0.81
	3.2	0.64	0.88	1.0	0.90	0.70
	10.0	2.03	1.40	1.0	0.70	0.59
50.0	0.5	0.94	0.98	1.0	1.01	1.00
	1.25	0.84	0.95	1.0	0.95	0.83
	3.2	0.62	0.87	1.0	0.99	0.81
	10.0	2.01	1.40	1.0	0.70	0.58
150	0.5	0.96	0.98	1.0	1.01	1.00
	1.25	0.86	0.96	1.0	0.97	0.87
	3.2	0.66	0.88	1.0	1.03	0.89
	10.0	2.00	1.40	1.0	0.70	0.58

*From Burrows and Attwood.[27]

some circumstances, greater than in the rain below the melting level. Further melting cannot lead to much further enhancement, apparently, and may lead to a lessening of the reflectivity of the particle by bringing it to sphericity or by breaking up the particle. Melting of ice particles produces enhanced backscatter, and this effect gives rise to the radar-observed *bright band* near the 0°C isotherm.

Attenuation by Fog.   The characteristic feature of a fog is the reduction in visibility. Visibility is defined as the greatest distance in a given direction at which it is just possible to see and identify with the unaided eye (1) in the daytime a prominent dark object against the sky at the horizon and (2) at night a known, preferably unfocused, moderately intense light source.[32]

Although the visibility depends upon both drop size and number of drops and not entirely upon the liquid-water content, in practice the visibility is an approximation of the liquid-water content and therefore may be used to estimate radiowave attenuation.[33] On the basis of Ryde's work, Saxton and Hopkins[34] give the figures in Table 23.5 for the attenuation in a fog or clouds at 0°C temperature. The attenuation varies with the temperature because the dielectric constant of water varies with temperature; therefore, at 15 and 25°C the figures in Table 23.5 should be multiplied by 0.6 and 0.4, respectively. It is immediately noted that cloud or fog attenuation is an order of magnitude greater at 3.2 cm than at 10 cm. Nearly another order-of-magnitude increase occurs between 3.2 and 1.25 cm.

Range and Velocity Ambiguities.   The unambiguous doppler frequency or Nyquist frequency for a fixed pulse-repetition-frequency (PRF) radar is given by

$$\Delta f = \pm \text{ PRF}/2 \qquad (23.21)$$

TABLE 23.5 Attenuation Caused by Clouds or Fog
Temperature = 0°C*

Visibility, m	Attenuation, dB/km		
	λ = 1.25 cm	λ = 3.2 cm	λ = 10 cm
30	1.25	0.20	0.02
90	0.25	0.04	0.004
300	0.045	0.007	0.001

*From Saxton and Hopkins.[34]

where PRF is the pulse repetition frequency. The unambiguous range interval is given by

$$\Delta r = \frac{c}{2\text{PRF}} \qquad (23.22)$$

and the product $\Delta f \Delta r$ is simply

$$\Delta f \Delta r = \frac{c}{2} \qquad (23.23)$$

Since the doppler shift f and the target radial velocity v are linearly related by the expression

$$v = \frac{\lambda}{2} f \qquad (23.24)$$

it follows that the product of unambiguous velocity and unambiguous range is

$$\Delta v \Delta r = \frac{\lambda c}{4} \qquad (23.25)$$

and is maximized by maximizing λ, the transmitted wavelength.

Ground Clutter Effects. Many meteorological radar applications call for the detection of precipitation echoes in the presence of ground clutter. Airborne weather radars during takeoff or landing are particularly susceptible. Another application, in which ground clutter is serious, relates to the detection of low-level wind shear.

While ground clutter cannot be eliminated, its effects can be mitigated through careful design. The most straightforward approach is to use antennas with low sidelobes, particularly in elevation. A second approach is through the use of shorter wavelengths. Shorter wavelengths result in improved signal-to-clutter ratios owing to the fact that the backscattered weather signal power is inversely proportional to λ^4 while the ground clutter return is only weakly dependent on wavelength. If one assumes that the clutter signal is wavelength-independent and the antenna beamwidth is fixed, Eq. (23.13) may be used to show that the weather-signal-power to clutter-power ratio is inversely proportional to λ^2.

Typical Weather Radar Designs. There is no universal weather radar system design that can serve all purposes. Airborne weather radars are constrained by size and weight limitations. Ground-based radars may be constrained by cost considerations. Severe storm warning radars require long range and high unambiguous velocity, and they must penetrate very heavy rain, thus dictating long wavelengths. Radars designed for studies of nonprecipitating clouds may use short wavelengths[35,36] (8 mm or even 3 mm) in order to achieve sufficient sensitivity to detect small cloud particles of the order of 100 μm and smaller. And FM-CW radars[37] have been used to obtain very-high-range resolution for detection of very thin layers in the clear air.

However, most meteorological radars are conventional pulsed or pulsed doppler systems. Ground-based radars used for severe storm research or warning will normally use S-band (≈3 GHz) or C-band (≈5.5 GHz) transmitters. Airborne storm avoidance radars will use either C-band or X-band (≈10 GHz) transmitters.

A 1° beamwidth is commonly used for longer-range radars. Admittedly, this is somewhat arbitrary, but the choice of 1° is based upon several decades of experience. A 1° beam will provide resolution of 2 km at a range of 120 km. Because thunderstorms contain important spatial features, such as heavy precipitation shafts and updraft cores, with horizontal dimensions of the order 1 to 5 km, a 1° beam is reasonably well matched to the phenomena being observed. Shorter-range and airborne weather radars often employ beamwidths of between 2 and 3°.

Operational weather radars normally are capable of short- and long-pulse operation in the range of 0.5 μs to about 6 μs. Through pulse-width diversity, high resolution is obtained, usually at short range, while for long-range detection longer pulses provide increased sensitivity and tend to equalize the along-beam and cross-beam resolutions.

Equation (23.13) shows that the received power is directly proportional to the pulse width τ. The noise power N is conventionally given by

$$N = \kappa TB \tag{23.26}$$

where κ = Boltzmann's constant, 1.38×10^{-23} W/(H$_z$ · K)
       T = receiver noise temperature, K
       B = receiver noise bandwidth

For a matched receiver

$$B \approx \frac{1}{\tau} \tag{23.27}$$

The signal-to-noise ratio is therefore given by the proportionality

$$\frac{P_r}{N} \propto \frac{\tau}{\kappa TB} \approx \frac{\tau^2}{\kappa T} \tag{23.28}$$

Thus, for distributed targets and with the pulse volume filled with scatterers, the signal-to-noise ratio for a single pulse is proportional to the pulse width squared. This assumes that the peak power is unchanged and that the average power increases linearly with τ. If the transmitter's average power is fixed, the signal-to-noise ratio will be proportional to τ.

PRFs for meteorological radars range from as low as several hundred s^{-1} for long-range detection to several thousand s^{-1} for shorter-wavelength systems attempting to achieve high unambiguous velocities. Generally speaking, most meteorological doppler radars are operated in a single mode, compromising the radar's ability to unambiguously resolve either range or velocity. More recent designs, however, may use a dual pulse repetition period[38] (PRT) to resolve both range and velocity. Another approach[39] is to employ a transmitted-pulse sequence with random phases from pulse to pulse. Range ambiguities cannot be totally eliminated, but their effects can be significantly mitigated through these approaches.

To discuss design details of all types of meteorological radars is beyond the scope of this chapter. However, it will be useful to include some of the important characteristics of the NEXRAD radar, which illustrate the performance of a modern operational weather radar ca. 1989. Table 23.6 contains some of the more relevant NEXRAD design features.

TABLE 23.6   Some Relevant NEXRAD System Characteristics

Transmitted power (klystron)	700,000 W
Pulse width	1.6, 4.8 μs
Range (doppler mode)	230 km
Unambiguous velocity (doppler mode)	±50 m/s
Range (nondoppler mode)	460 km
Clutter rejection	50 dB
Beamwidth	1°
System sensitivity	(−8 dBZ at 50 km)

23.4 SIGNAL PROCESSING

It can be shown[8,11] that the received signal from meteorological targets is well represented by a narrowband gaussian process. This is a direct consequence of the fact that (1) the number of scatterers in the pulse volume is large ($>10^6$); (2) the pulse volume is large compared with the transmitted wavelength; (3) the pulse volume is filled with scatterers, causing all phases on the range from 0 to 2π to be returned; and (4) the particles are in motion with respect to one another due to turbulence, wind shear, and their varying fall speeds.

The superposition of the scattered electric fields from such a large number of particles (each with random phase) gives rise, through the central limit theorem, to a signal with gaussian statistics. Because the particles are in motion with respect to one another, there is also a doppler spread, often referred to as the variance of the doppler spectrum. Finally, since all the particles within the sample volume are moving with some mean or average radial velocity, there is a mean frequency of the doppler spectrum which is shifted from the transmitted frequency.

The power spectral density of a meteorological signal is depicted schematically in Fig. 23.2 and can be interpreted as follows. The received power is simply the integral under the curve and is given by

$$P_r = \int S(f)df = \int S(v)dv \tag{23.29}$$

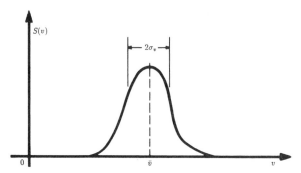

FIG. 23.2 The doppler spectrum. Received power, radial velocity, and spectrum width can be calculated and are directly related to meteorological variables.

where f and v are related by $f = (2/\lambda)v$.

The mean velocity is given by the first moment of the spectrum

$$\bar{v} \quad \frac{\displaystyle\int vS(v)dv}{\displaystyle\int S(v)dv} \tag{23.30}$$

The second central moment σ_v^2 is given by

$$\sigma_v^2 \quad \frac{\displaystyle\int (v - \bar{v})^2 S(v)dv}{\displaystyle\int S(v)dv} \tag{23.31}$$

where σ_v is the velocity width. Radar meteorologists refer to σ_v^2 as the spectrum variance because of its computational equivalence to the variance of a continuously distributed random variable. In short, $S(v)$ is analogous to a probability density function for v. The term *spectrum variance* will be used to refer to σ_v^2, and the term *spectrum width* to refer to σ_v. It is clear, therefore, that the doppler spectrum contains the information necessary to measure important signal parameters.

In the most general case, quadrature phase detection is used to obtain the real and imaginary parts of the complex signal envelope.[8] These are usually digitized in a large number of range gates (≈ 1000) at the radar's pulse repetition frequency. The resultant complex time series in each gate can then be processed by using a fast Fourier transform (FFT) to obtain an estimate of the doppler spectrum from which the mean velocity and spectrum variance can be obtained.

A more efficient estimation technique is described by Rummler.[40] This estimator makes use of the fact that the complex autocorrelation function of the signal has the general form

$$R(x) = P_r \rho(x) e^{j\frac{4\pi\bar{v}}{\lambda}x} \tag{23.32}$$

where $\rho(x)$ is the correlation coefficient and x is a dummy variable. It follows that $\bar{v}$, the mean velocity, is given by

$$\bar{v} = \frac{\lambda}{4\pi x}\text{arg}\ [R(x)] \tag{23.33}$$

It can also be shown that

$$\sigma_v^{\ 2} \approx \frac{\lambda^2}{8\pi^2 x^2}\left[1 - \frac{R(x)}{R(o) - N}\right] \tag{23.34}$$

where N is the noise power.

This estimator is widely used for mean-frequency estimation with doppler meteorological radars. The estimates are unbiased in the presence of noise when the doppler spectrum is symmetrical. Its greatest appeal, however, is due to its computational simplicity. For a pulsed radar, with a pulse repetition period (PRT) T, $R(T)$ is obtained from the simple expression[8]

$$R(T) = \frac{1}{N}\sum_{(k=0)}^{(N-1)} s_{k+1}s_k^* \tag{23.35}$$

where the s_k are the complex signal samples (sampled at the radar PRT) in a given range gate and s_k^* is the complex conjugate. It is clear that this algorithm requires only N complex multiplications for a time series of N samples while the FFT requires $N \log_2 N$. This *pulse-pair algorithm*, as it is often called, therefore not only is an excellent estimation technique but is less complex and costly than comparable FFT processors. In addition, FFT estimates of mean velocity and spectrum width are biased by receiver noise. If the FFT approach is used, the bias due to noise can be removed by estimating the noise threshold in the spectral domain and truncating the derived spectrum.[41]

For most applications, the pulse-pair processor has become the technique of choice. However, in some research applications it remains advantageous to have access to the full doppler spectrum. Very fast and programmable digital signal-processing chips make it possible for radar meteorologists to have their cake and eat it too. Flexibility due to programmability permits tailoring of the processor's characteristics to the application from day to day or even beam to beam and range gate to range gate. Until recently, most pulse-pair or FFT processors for meteorological radars have been hard-wired and therefore inflexible.

Measurement Accuracy. Because the received signals are sample functions from gaussian random processes, the doppler spectrum and its moments cannot be measured exactly in any finite period of time. Consequently, all measurements will be somewhat in error, with the error being a function of the properties of the atmosphere, the radar wavelength, and the time allocated to the measurement.

The theoretical development of signal estimator statistics is found in Denenberg, Serafin, and Peach[42] for the FFT technique. Doviak and Zrnić[11] cover the subject quite completely. Following are some useful expressions for the mean square error of mean power and mean velocity estimates.

Power Estimation. It is well known that for a gaussian process,[43] using square-law signal detection, samples of the mean power P_r of the process are exponentially distributed with variance P_r^2. Given a time T_0 allocated to the measurement and a signal bandwidth σ_f(Hz), there will be approximately $\sigma_f T_0$ independent samples of the square of the signal envelope. It follows, therefore, that an estimate $\hat{P}_r$ of the mean power for this process will have a variance or mean square error given by

$$\text{var}(\hat{P}_r) \approx \frac{P_r^2}{\sigma_f T_0} \tag{23.36}$$

Substituting for σ_f from the expression $\sigma_f = 2\sigma_v/\lambda$, where σ_v is the width of the doppler spectrum, Eq. (23.36) becomes

$$\text{var}(\hat{P}_r) \approx \frac{\lambda P_r^2}{2\sigma_v T_0} \tag{23.37}$$

This expression is valid for high signal-to-noise cases.

Velocity Estimation. Denenberg, Serafin, and Peach[42] give the following expression for the variance of mean-frequency estimates of the doppler spectrum

$$\text{var}(\hat{f}) = \frac{1}{P_r^2 T_0} \int f^2 S^2(f + \bar{f})\, df \tag{23.38}$$

This is an interesting result, showing that the variance of the estimate $\hat{f}$ is a function only of the shape of the doppler spectrum and the integration time T_0. If the spectrum has a gaussian shape, with variance σ_f^2, Eq. (23.38) becomes

$$\text{var}(\hat{f}) = \frac{\sigma_f}{4\sqrt{\pi}T_0} \tag{23.39}$$

Noting that $\text{var}(\hat{v}) = (\lambda/2)^2 \, \text{var}(\hat{f})$, we can write

$$\text{var}(\hat{v}) = \frac{\lambda\sigma_v}{8\sqrt{\pi}T_0} \tag{23.40}$$

If we multiply numerator and denominator by σ_v, Eq. (23.40) becomes

$$\text{var}(\hat{v}) = \frac{\lambda\sigma_v^2}{8\sqrt{\pi}\sigma_v T_0} = \frac{\sigma_v^2}{4\sqrt{\pi}\sigma_f T_0} \tag{23.41}$$

Thus, it is seen that the variance of the mean velocity estimate $\hat{v}$ is directly proportional to the variance of the doppler spectrum and inversely proportional to the number of independent samples. Note also that var $(\hat{v})$ is proportional to λ, indicating that, for the same processing time T_0 and for the same σ_v, the variance of the estimate can be reduced by reducing the wavelength, which increases the number of independent samples.

Equations (23.38), (23.39), (23.40), and (23.41) are applicable in high signal-to-noise-ratio cases. Zrnić[44] gives the following expression for the variance of the mean-frequency estimate $\hat{f}$ for the pulse-pair estimation technique and a gaussian-shaped spectrum

$$\text{var}(\hat{f}) = \frac{1}{8\pi^2 T_0 \rho^2(T)T}\left\{2\pi^{3/2}\sigma_f T + \frac{N^2}{S^2} + 2\frac{N}{S}[1 - \rho(2T)]\right\} \qquad (23.42)$$

where ρ is the correlation coefficient and N/S is the noise-to-signal ratio. Equation (23.42) applies to a single PRF with interpulse period T and assumes that all pulses in the interval T_0 are used in the estimation algorithm. It reduces exactly to Eq. (23.39) for large S/N and for narrow spectra, i.e., $\rho(T) \approx 1$. The reader is referred to Zrnić[44] for further details regarding the estimation of other moments of the doppler spectrum.

Processor Implementations. In nondoppler radars it is common to use log-video receivers along with sensitivity time control (STC) for inverse r^2 correction in order to achieve the widest dynamic range possible. For signal power estimation, the log-video signal is digitized and averaged or, in the most rudimentary of systems, may be used to modulate an analog PPI or other type of radarscope directly. Most modern meteorological radars, however, use digital averaging along with digital color displays for added quantitative precision. Note that when the logarithm is averaged, the estimate will be biased downward by as much as 2.5 dB.[45] This bias must be removed in order to accurately estimate the received signal power.

For doppler radars it has been common to use both linear and logarithmic receivers, with the log channel used for reflectivity estimation and the linear channel for doppler parameter estimation. This approach, however, often results in saturation of the linear channel and therefore some distortion of the doppler spectrum.[46]

Most modern designs now attempt to maintain linearity in the receiver through the use of a dynamic automatic gain control (AGC), whereby the receiver gain is adjusted from range gate to range gate through the use of rapidly switched attenuators. The estimate needed to select the proper attenuator may come from an independent log channel or may be based upon a short segment of the signal. Another approach[47] is to delay the signal for a period of the order of a microsecond while an estimate of signal strength can be made and the proper attenuator setting can be established. Clearly, such rapid switching in the receiver requires careful design in order to avoid the effects of switching transients. An approach that avoids transient effects is to use parallel IF strips, each with moderate dynamic range and fixed gains, and to sample the signal in the channel that is best matched to the signal strength.

In all these approaches, it is possible to achieve wide linear dynamic range of the order of 80 dB or greater and to use floating-point digital arithmetic. The reflectivity, mean doppler velocity, and spectrum width can all be estimated digitally from the floating-point linear channel samples.

23.5 OPERATIONAL APPLICATIONS

As has been demonstrated, meteorological radars measure backscattered power and radial velocity parameters. The challenge to the radar meteorologist is to translate these measurements, their spatial distributions, and their temporal evolution into quantitative assessments of the weather. The level of sophistication used in interpretation varies broadly, ranging from human interpretation of rudimentary gray-scale displays to computer-based algorithms and modern color-enhanced displays to assist human interpreters. Expert system approaches[48] that attempt to reproduce human interpretive logical processes can be employed effectively. Baynton et al.,[49] Wilson and Roesli,[50] and Serafin[1] all show how modern meteorological radars are used for forecasting the weather. The degree to which automation can be applied is evident in the NEXRAD radar system design, where the meteorological products shown in Table 23.7 will be automated.[51]

TABLE 23.7 NEXRAD Automated Products

Doppler radar data archive of storm phenomena
Precipitation analysis
Wind analysis
Tornado analysis
Fine-line analysis
Tropical cyclone analysis
Mesocyclone analysis
Thunderstorm analysis
Turbulence analysis
Icing analysis
Hail analysis
Freezing-melting analysis
Interpretive techniques
Multiple-radar mosaics

Precipitation Measurement. Among the more important parameters to be measured is rainfall, having significance to a number of water resource management problems related to agriculture, fresh-water supplies, storm drainage, and warnings of potential flooding.

The rainfall rate can be empirically related to the reflectivity factor[12] by an expression of the form

$$Z = aR^b \qquad (23.43)$$

where a and b are constants and R is the rainfall rate, usually in millimeters per hour. Battan[10] devotes three full pages of his book to the listing of dozens of Z-R relationships derived by investigators at various locations throughout the world, for various weather conditions and in all seasons of the year. The fact that no universal expression can be applied to all weather situations is not surprising when one notes that rainfall drop-size distributions are highly variable. For many conditions,[10] the drop-size distribution can be represented by an exponential function

$$N(D) = N_0 e^{-\Lambda D} \qquad (23.44)$$

where N_0 and Λ are constants. If $N(D)$ is known, the reflectivity factor can be calculated from Eq. (23.9). By using the terminal-fall speed data of Gunn and Kinzer,[52] the rainfall rate can also be obtained and Z directly related to R.

Clearly, a single-wavelength, single-polarization radar can measure only a single parameter Z and must assume Rayleigh scattering. Since the rainfall rate depends upon two parameters, N_0 and Λ, it is not surprising that Eq. (23.43) is nonuniversal. Despite this fact, Battan[10] lists four expressions as being "fairly typical" for the following four types of rain:

$$\text{Stratiform rain}^{53} \quad Z = 200\ R^{1.6} \qquad (23.45)$$

$$\text{Orographic rain}^{54} \quad Z = 31\ R^{1.71} \qquad (23.46)$$

$$\text{Thunderstorm rain}^{55} \quad Z = 486\ R^{1.37} \qquad (23.47)$$

$$\text{Snow}^{56} \quad Z = 2000\ R^2 \qquad (23.48)$$

Stratiform refers to widespread, relatively uniform rain. *Orographic* rain is precipitation that is induced or influenced by hills or mountains. In each of the above expressions, Z is in mm^6/m^3 and R is in mm/h. In Eq. (23.48), R is the precipitation rate of the melted snow.

For a more complete treatment of this topic, the reader is referred to Battan.[10] Wilson and Brandes[57] give a comprehensive treatment of how radar and rain-gauge data can be used to complement one another in measurements of precipitation over large areas. Bridges and Feldman[58] discuss how two independent measurements (reflectivity factor and attenuation) can be used to obtain both parameters of the drop-size distribution and therefore precisely determine the rainfall rate. Seliga and Bringi[59] show how the measurement of Z at horizontal and vertical polarization also can produce two independent measurements and therefore provide more accurate rainfall rate measurements. Zawadzki[60] argues, however, that other factors contribute far more to the variability of precipitation rate than does the drop-size distribution. He states, therefore, that dual-parameter estimation techniques are not likely to be successful in many cases. Wilson and Brandes[57] state that cumulative precipitation measurements with radar, in storm situations, can be expected to be accurate to a factor of 2 for 75 percent of the time. Accuracies over large areas can be improved to about 30 percent with the addition of a surface rain-gauge network. It is this author's opinion that no single topic in radar meteorology has received more attention than rainfall rate measurement. Although useful empirical expressions have evolved, a completely satisfactory approach remains to be discovered.

Severe Storm Warning. One of the primary purposes of weather radars is to provide timely warnings of severe weather phenomena such as tornadoes, damaging winds, and flash floods. Long-term forecasting of the precise location and level of severity of these phenomena, through numerical weather prediction techniques, is beyond the state of the art. Operational radars, however, can detect these phenomena and provide warnings (of up to 30 min) of approaching severe events; they can also detect the rotating mesocyclones in severe storms that are precursors to the development of tornadoes at the earth's surface.[61]

Tornado Detection. A single doppler radar can only measure the radial component of the vector wind field. Hence, exact measurements of vector winds at a point are generally not possible. However, rotating winds or vortices can be detected and their intensities measured by simply measuring the change in radial

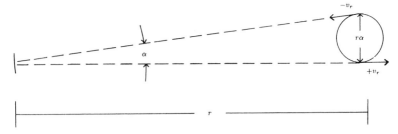

FIG. 23.3 Measurement of rotation or azimuthal shear in a mesocyclone. The azimuthal shear is given by $\Delta v/\Delta x = 2v_r/r\alpha$.

velocity with azimuth angle as shown in Fig. 23.3. The radar scans in azimuth and detects a couplet in radial velocity at constant range. The azimuthal shear is given simply by the expression

$$\frac{dv_r}{dx} \approx \frac{2v_r}{r\alpha} \tag{23.49}$$

where x is in the direction orthogonal to the radius r and d is the angle subtended by the circulation at range r.

Because mesocyclones, which spawn tornadoes, can be many kilometers in diameter, radars with 1° beams have the spatial resolution to detect mesocyclones at ranges in excess of 60 km. It should be clear that any mean translational motion would change the absolute values of the measured radial velocities but would not affect the shear measurement. Armstrong and Donaldson[62] were the first to use shear for severe storm detection. Azimuthal shear values of the order of 10^{-2} s^{-1} or greater and with vertical extent greater than the diameter of the mesocyclone are deemed necessary for a tornado to occur.[63]

Detection of the tornado vortex itself is not generally possible, since its horizontal extent may be only a few hundred meters. Detection of the radial shear, therefore, is not possible unless the tornado is close enough to the radar to be resolved by the beamwidth. In cases where the tornado falls entirely within the beam, the doppler spectral width[64] may be used to estimate tornadic intensity. In some cases, both a mesocyclone and its incipient tornado can be detected. Wilson and Roesli[50] show an excellent example of a tornado vortex signature (TVS) embedded within a larger mesocyclone.

Microbursts. Fujita and Caracena[65] first identified the microburst phenomenon as the cause of an airliner crash that took place in 1975. The microburst and its effects on an aircraft during takeoff or landing are depicted in Fig. 23.4. The microburst is simply a small-scale, short-duration downdraft emanating from a convective storm. This "burst" of air spreads out radially as it strikes the ground, forming a ring of diverging air about 0.3 to 1 km deep and of the order of 2 to 5 km in diameter. Aircraft, penetrating a microburst, experience first an increase in head wind and then a continuous, performance-robbing decrease in head wind, which can cause the plane to crash if encountered shortly before touchdown or just as the aircraft is taking off. More complete descriptions of microbursts and their effects on aviation safety are given by Fujita[66,67] and McCarthy and Serafin.[68]

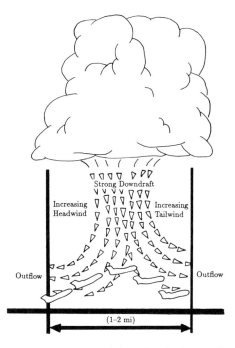

FIG. 23.4 Artist's depiction of a microburst and its effect on an aircraft during takeoff. The loss of airspeed near to the ground can be extremely hazardous.

Microburst detection, like tornado detection, is accomplished by estimating shear. However, in the case of the microburst, it is the radial shear of the radial velocity that is typically measured. Human interpretation of microburst signatures in color-enhanced radial velocity displays is easily accomplished with trained observers.[50] Radial velocity differences of 10 to 50 m/s are observed in microbursts. A radial velocity difference of 25 m/s over the length of a jet runway ($\approx$3 km) is of serious concern.

One principal problem concerning microbursts is their short lifetimes, which are of order 15 min. The duration of peak intensity is only 1 or 2 min. The Classify, Locate, and Avoid Wind Shear (CLAWS) project[69] in 1984 clearly demonstrated that a 2-min advance warning using doppler radar and human interpreters can be achieved. The use of doppler radars operationally, however, will require completely automated detection algorithms. A second major problem is ground clutter. Since the phenomenon occurs near the ground and oftentimes in very light or no precipitation, ground clutter mitigation is necessary.

C band seems to be the preferred operational frequency for several reasons. First, a C-band antenna will be physically smaller than an S-band antenna for the same beamwidth, an important consideration for use near airports. Second, since long-range detection is not of importance, attenuation effects are not of primary concern. Third, C band offers improved signal-to-clutter performance. X band is not the frequency of choice owing to more serious range-velocity ambiguities and the more severe attenuation that can occur in very heavy rain. It is expected that deployment of a national network of doppler radars near airports will begin in the early 1990s.

Hail. The NEXRAD radar will make use of a hail-detection algorithm similar to that discussed by Witt and Nelson.[70] This algorithm combines high reflectivity factor with echo height and upper-level radial velocity divergence to detect the occurrence of hail. Eventually, polarization diversity techniques may improve quantitative hail detection. Aydin, Seliga, and Balaji[71] propose a hail-detection technique using reflectivity measurements at orthogonal polarizations. This technique depends upon the fact that the ratio of horizontal to vertical reflectivity is unity (≈ 0 dB) when hail is present. This differs sharply from heavy rain, where this ratio can be as large as 6 dB. The combination of absolute reflectivity factor at horizontal polarization and ratio of reflectivities at horizontal and vertical polarizations (differential reflectivity) gives unique signatures for hail and heavy rain, each of which is characterized by high reflectivity factor. The difference in the differential reflectivity signatures is easily explained. Large raindrops assume pancakelike shapes as they fall and thus scatter back horizontally polarized electric fields more strongly than vertically polarized electric fields. Hailstones, while irregular in shape, appear to tumble while they fall and therefore exhibit no preferred orientation on average.

Wind Measurement. Lhermitte and Atlas[72] were the first to show how a single doppler radar can be used to measure vertical profiles of horizontal wind. This technique can be used if the precipitation and the wind are uniform in the region scanned by the radar. The method depends upon an analysis of the radial velocity measured during a complete scan in azimuth with elevation angle fixed. At any slant range r, the diameter of the region scanned is $r \cos \alpha$, and the height of the measurement is $r \sin \alpha$, where α is the elevation angle (see Fig. 23.5). If β is the azimuth angle, V_h is the horizontal wind speed, and V_f is the fall speed of the particles, the radial velocity at range r is given by

$$V_r(\beta) = V_h \cos \beta \cos \alpha + V_f \sin \alpha \qquad (23.50)$$

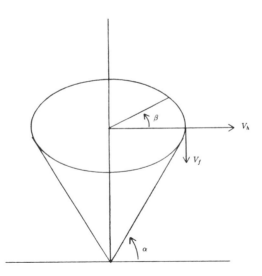

FIG. 23.5 Velocity-azimuth-display geometry for measuring horizontal wind with a single doppler radar. Measurement of the radial velocity for a complete azimuthal scan (β) permits measurement of horizontal winds.

A harmonic analysis can be used to obtain V_h, the horizontal wind speed, the wind direction, and V_f, the particle fall speed. The technique is referred to as the velocity-azimuth-display (VAD) technique. Browning and Wexler[73] later showed how the technique could be extended to measure other parameters of the wind field including divergence and deformation. Baynton et al.[49] show how the VAD can be applied in real time by using a color-enhanced radial velocity display.

Thunderstorm Prediction. Wilson and Schreiber[74] illustrate how modern meteorological doppler radar can be used to detect locations where new thunderstorm development is likely to occur. Modern radars have sufficient sensitivity to detect clear-air discontinuities in the lower 2 to 4 km of the atmosphere. Principally, this detection occurs in the summer months. The backscattering mechanism may be due to index-of-refraction inhomogeneities caused by turbulence in the lower layers and/or by insects. Wilson and Schreiber have found that about 90 percent of the thunderstorms that occur in the Front Range of the Rockies in the summertime develop over such boundaries. Since these boundaries can be detected before any clouds are present and because it is possible to infer the air mass convergence that is taking place along these boundaries through doppler measurements, more precise prediction of thunderstorm occurrence appears to be possible. From the radar designer's standpoint, such applications dictate the use of antennas with very low sidelobes and signal processors with significant ground clutter rejection capability. The NEXRAD radar system, with 50 or more dB of clutter rejection, is well suited to this eventual operational task.

23.6 RESEARCH APPLICATIONS

Operational meteorological radars are designed for reliability and simplicity of operation while providing the performance needed for operational applications. Research radars are considerably more complex, since cutting-edge research requires more detailed and more sensitive measurements of a multiplicity of variables simultaneously. In the research community, multiple-parameter radar studies, multiple-doppler radar network studies, and plans for airborne and spaceborne radars are all receiving considerable attention.

Multiple-Parameter Radar. It has been noted earlier that doppler radar provides a significant increase in the useful information that can be obtained from meteorological targets. The detection of hail, through the use of polarization diversity, adds additional information, and multiple wavelength provides yet another input related to the eventual interpretation of the size, water-phase state, and types of hydrometeors in all classes of clouds and precipitation. Very-short-wavelength radars are useful for probing newly developing clouds, while longer-wavelength radars are necessary for the study of severe storms. Researchers often need a wide range of these capabilities simultaneously. The capabilities desired of multiple-parameter meteorological radars are presented in the collection of papers edited by Hall.[6]

From the radar engineering standpoint, the challenge is considerable, requiring radar designers to develop fully coherent, polarization-diverse, and wavelength-diverse radars. Figure 23.6 is a photograph of the S (10-cm)- and X (3-cm)-band polarization-diverse doppler radar operated by the National Center for Atmospheric Research (NCAR). The system permits simultaneous measurements of the reflectivity factor on 2 wavelengths—the doppler parame-

FIG. 23.6 The CP-2, multiple-parameter radar at the National Center for Atmospheric Research, Boulder, Colorado. (*Courtesy of the National Center for Atmospheric Research.*)

ters on a single wavelength, S band, and polarization-diverse measurements at both wavelengths. The antenna beams are matched with approximately 1° beamwidths. The peak transmitted power at S band is 1 MW and 50 kW at X band. The pulse widths are approximately 1 μs, and the PRF is typically 1000 s^{-1}. The system is characteristic of the technologies currently in place in the research community in this field.

Multiple Radars. A single doppler radar measures only a single radial component of velocity. Lhermitte[3] was among the first to describe how two or more doppler radars could be used, scanning together, to obtain the full three-dimensional air motion fields in precipitation. This pioneering work led the way toward the use of networks of doppler radars for studies of individual clouds and larger-scale cloud systems. For the first time, it became possible to examine the three-dimensional structure of vector air motion in precipitation. Figure 23.7 illustrates an air motion field obtained by multiple-doppler radar observations in an individual convective storm cell. Shown are the horizontal

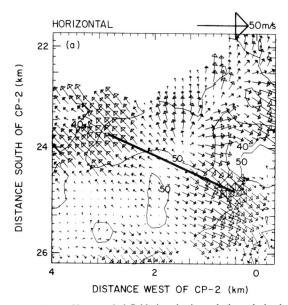

FIG. 23.7   Vector wind fields in a horizontal plane derived from dual-doppler radar observations of a summertime convective storm near Denver, Colorado. The divergent outflow is from a microburst. The dark, solid line is shown to indicate the length of a typical jet aircraft runway. (*Courtesy of the National Center for Atmospheric Research.*)

vector fields in a plane approximately 100 m above the surface. The phenomenon being measured is a low-level divergent outflow (or microburst). Figure 23.8 shows another example of air motion fields in a vertical plane orthogonal to an intense squall line in California.[75]

Rapid Scanning.   The use of multiple-doppler radars has provided dramatic new information on the internal winds in large precipitating systems— information that can be obtained in no other way. Despite the power of this technique, the spatial resolution in the derived three-dimensional motion fields is generally not better than of the order of 2 km. The reasons for this are several. The finite beamwidth limits the resolution available at longer ranges. At shorter ranges, the large solid angle that must be scanned in order to cover all regions of a storm requires total scanning times of the order of 3 to 5 min even for ideally situated storms. This is a consequence of the on-target time necessary for accurate radial velocity measurements. Finally, the storm itself is evolving and moving during this measurement time.

Some research applications require faster scanning. These applications include the study of finer-scale storm features, interactions between the kinematics and hydrometeor growth processes in the storms, and studies of electric-charge separation in clouds. Brook and Krehbiel[76] were the first to discuss a very-rapid-scanning radar (although nondoppler) for effectively obtaining snapshots of convective storms. Keeler and Frush[77] discuss design considerations for a rapid-scanning doppler radar. Any rapid-scanning approach generally must encompass

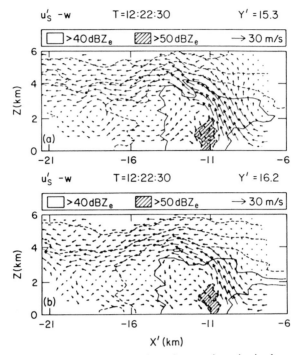

FIG. 23.8 Vertical cross sections of vector air motion in planes orthogonal to a California wintertime squall line.[75]

two features: (1) there must be relatively broadband transmissions to increase the independent samples available within the spatial resolution cell, thus reducing the dwell time; and (2) the antenna must be scanned—either very rapidly mechanically or electronically. An alternative approach might use several simultaneous beams and receivers.

Airborne and Space-Borne Radars. Hildebrand and Mueller[4] and Mueller and Hildebrand[5] have quantitatively demonstrated that it is possible for an airborne meteorological doppler radar to measure internal kinematic fields that are comparable to those obtained from ground-based systems. This powerful technique permits the use of a mobile platform, which therefore allows measurements over regions not accessible by ground-based systems. Moreover, the mobility of the aircraft permits longer-term observations of rapidly moving but long-lived storms and cloud systems. Figure 23.9 shows a photograph of the antenna of the airborne doppler radar mounted on the tail of the P-3 aircraft operated by the National Oceanic and Atmospheric Administration (NOAA). The antenna, covered by a cylindrical radome in flight, scans in range-height-indicator (RHI) mode, in vertical planes orthogonal to the aircraft fuselage. The aircraft is flown on orthogonal tracks in order to synthesize dual-doppler observations and therefore to obtain vector winds.

A point should be made here regarding the use of two doppler radars for measurements of three-dimensional winds. Since in principle two independent looks

FIG. 23.9 Tail-mounted doppler radar antenna on the P-3 research and reconnaissance aircraft operated by the National Oceanic and Atmospheric Administration (NOAA).

can measure only two components of vector air motion, the assumption of mass continuity is invoked. The equation of continuity ($\nabla \cdot \overline{V} = 0$) is used to obtain the third-dimensional component, where $\overline{V}$ is the vector air motion. The vertical air motion is calculated from vertical integration of the continuity equation.

Walther, Frush, and Hildebrand[78] describe a next-generation airborne doppler radar system that consists of two antennas mounted in the tail—one pointed forward from the orthogonal plane by an angle of the order of 30° and one pointed rearward, also by about 30°. With such a system, each antenna scans in a conical surface—one cone pointing forward, one rearward—thus permitting synthesis of a dual-doppler radar system along the aircraft track. Because the aircraft need not fly orthogonal tracks, the time required for measurements of cloud systems is dramatically reduced. Moreover, severe storms (which could otherwise not be penetrated along an orthogonal track) can be observed fully by an aircraft outside the regions of severe weather.

Among the more significant challenges facing researchers today is the need to make global measurements of precipitation. Understanding of the global climate requires that quantitative measurements of precipitation be made throughout the world, particularly in the tropics and over the oceans. Satellite observations appear to offer the only practical mechanism for obtaining these measurements. Meneghini and Atlas[79] describe a concept for a dual-wavelength radar for precipitation measurements from space.

Clear-Air Radars. Another form of doppler radar that has become popular in the research community is the so-called wind profiler. Wind profilers usually take the form of VHF and UHF fixed-beam systems, pointing vertically and at angles approximately 15° from the zenith. Such radars[7] can make doppler measurements throughout the range of altitudes from a few hundred meters to 15 km above the surface, depending upon the wavelength selected and the power-aperture product available. Very powerful radars of this type are referred to as Mesosphere, Stratosphere, Troposphere (MST) radars because of their ability to make measurements throughout most of these atmospheric regions. Powerful MST radars are operated at many laboratories around the world. Major facilities are located at Kiruna, Sweden; the Massachusetts Institute of Technology in Cambridge, Massachusetts; Arecibo, Puerto Rico; Jicamarca, Peru; and at the University of Kyoto in Japan.

These clear-air radars receive energy backscattered from index-of-refraction inhomogeneities due to atmospheric turbulence. The antenna systems usually take the form of phased arrays. Transmitters are generally in the form of high-powered, fully coherent transmitting tubes. One exception is at the University of Kyoto, where the antenna-transmitter system consists of more than 400 radiating elements, each with its own solid-state transmitter. This approach allows for full electronic scanning of the beam. A network of 400-MHz wind profilers in the central United States is also expected to use solid-state transmitters, but electronic scanning will not be possible.

The meteorological community is excited about these devices because of their ability to measure winds continuously. This capability permits the observation of smaller-scale temporal and spatial wind-field features than can be obtained from the global 12-hourly rawindsonde (balloon) networks. These smaller-scale measurements are important for understanding local and regional weather and for effective forecasting on these scales.

It is important to recognize that two-beam systems can measure horizontal winds if the wind field is uniform and if vertical velocities are negligible. A three-beam system can measure all three velocity components if the wind is uniform. Four- and five-beam systems allow one to determine the quality of the measurements by detecting the presence of nonuniformity. Carbone, Strauch, and Heymsfield[80] and Strauch et al.[81] address the issue of wind measurement error in detail.

The reader is referred to the review paper by Röttger and Larsen[82] for a thorough treatment of wind-profiler technology.

Synthetic Aperture Radar and Pulse Compression.   Metcalf and Holm[83] and Atlas and Moore[84] have considered the use of synthetic aperture radar (SAR) in order to obtain high-resolution measurements from mobile airborne or space-borne platforms. In general, both papers conclude that the cross-beam resolution possible is inherently limited by the decorrelation time of the targets due to their turbulent motion. Consequently, SAR offers little advantage over real aperture systems for meteorological applications from aircraft. However, space-borne systems can effectively use SAR because of the high speed of the orbiting spacecraft.

Pulse compression is not generally used for meteorological applications because peak power is not usually a limitation on system performance. Keeler and Frush,[77] however, point out that pulse compression can be of benefit in some rapid-scanning applications. In situations where signals are very weak (such as for MST applications), pulse compression is used to increase system sensitivity by increasing the average power of the system.

A note of caution is in order when considering pulse compression for meteorological radars. This relates to the matter of range sidelobes. Careful design is necessary to minimize these sidelobes, just as antenna sidelobes should be minimized, in order to mitigate the effects of interpretive errors caused by wide-dynamic-range distributed weather targets.

REFERENCES

1. Serafin, R. J.: New Nowcasting Opportunities Using Modern Meteorological Radar, *Proc. Mesoscale Analysis Forecast. Symp.*, pp. 35–41, European Space Agency, Paris, 1987.

2. McCarthy, J., J. Wilson, and T. T. Fujita: The Joint Airport Weather Studies (JAWS) Project, *Bull. Am. Meteorol. Soc.*, vol. 63, pp. 15–22, 1982.

3. Lhermitte, R. M.: Dual-Doppler Radar Observations of Convective Storm Circulations, *Preprints, 14th Conf. Radar Meteorol.*, pp. 139–144, American Meteorological Society, Boston, 1970.

4. Hildebrand, P., and C. Mueller: Evaluation of Meteorological Airborne Doppler Radar, Part I: Dual-Doppler Analyses of Air Motions, *J. Atmos. Ocean. Technol.*, vol. 2, pp. 362–380, 1985.

5. Mueller, C., and P. Hildebrand: Evaluation of Meteorological Airborne Doppler Radar, Part II: Triple-Doppler Analysis of Air Motions, *J. Atmos. Ocean. Technol.*, vol. 2, pp. 381–392, 1985.

6. Hall, M. (ed.): Special papers: Multiple Parameter Radar Measurements of Precipitation, *Radio Sci.*, vol. 19, 1984.

7. Strauch, R. G., D. A. Merritt, K. P. Moran, K. B. Earnshaw, and D. Van De Kamp: The Colorado Wind Profiling Network, *J. Atmos. Ocean. Technol.*, vol. 1, pp. 37–49, 1984.

8. Serafin, R. J., and R. Strauch: Meteorological Radar Signal Processing, in "Air Quality Meteorology and Atmospheric Ozone," American Society for Testing and Materials, Philadelphia, 1977, pp. 159–182.

9. Gray, G. R., R. J. Serafin, D. Atlas, R. E. Rinehart, and J. J. Boyajian: Real-Time Color Doppler Radar Display, *Bull. Am. Meteorol. Soc.*, vol. 56, pp. 580–588, 1975.

10. Battan, L. J.: "Radar Observation of the Atmosphere," University of Chicago Press, 1973.

11. Doviak, R. J., and D. S. Zrnić: "Doppler Radar and Weather Observations," Academic Press, Orlando, Fla., 1984.

12. Bean, B. R., E. J. Dutton, and B. D. Warner: Weather Effects on Radar, in Skolnik, M. (ed.): "Radar Handbook," McGraw-Hill Book Company, New York, 1970, pp. 24-1–24.40.

13. Mie, G.: Beiträge zur Optik trüber Medien, speziell kolloidaler Metallösungen [Contribution to the optics of suspended media, specifically colloidal metal suspensions], *Ann. Phys.*, vol. 25, pp. 377–445, 1908.

14. Probert-Jones, J. R.: The Radar Equation in Meteorology, *Q. J. R. Meteorol. Soc.*, vol. 88, pp. 485–495, 1962.

15. Metcalf, J. I.: Airborne Weather Radar and Severe Weather Penetration, *Preprints, 19th Conf. Radar Meteorol.*, pp. 125–129, American Meteorological Society, Boston, 1980.

16. Allen, R. H., D. W. Burgess, and R. J. Donaldson, Jr.: Severe 5-cm Radar Attenuation of the Wichita Falls Storm by Intervening Precipitation, *Preprints, 19th Conf. Radar Meteorol.*, pp. 87–89, American Meteorological Society, Boston, 1980.

17. Eccles, P. J., and D. Atlas: A Dual-Wavelength Radar Hail Detector, *J. Appl. Meteorol.*, vol. 12, pp. 847–854, 1973.

18. Donaldson, R. J., Jr.: The Measurement of Cloud Liquid-Water Content by Radar, *J. Meteorol.*, vol. 12, pp. 238–244, 1955.

19. Weickmann, H. K., and H. J. aufm Kampe: Physical Properties of Cumulus Clouds, *J. Meteorol.*, vol. 10, pp. 204–221, 1953.

20. Gunn, K. L. S., and T. W. R. East: The Microwave Properties of Precipitation Particles, *Q. J. R. Meteorol. Soc.*, vol. 80, pp. 522–545, 1954.

21. Ryde, J. W., and D. Ryde: "Attenuation of Centimeter Waves by Rain, Hail, Fog, and Clouds," General Electric Company, Wembley, England, 1945.

22. Bean, B. R., and R. Abbott: Oxygen and Water Vapor Absorption of Radio Waves in the Atmosphere, *Geofis. Pura Appl.*, vol. 37, pp. 127–144, 1957.

23. Ryde, J. W.: The Attenuation and Radar Echoes Produced at Centimetre Wavelengths by Various Meteorological Phenomena, in "Meteorological Factors in Radio Wave Propagation," Physical Society, London, 1946, pp. 169–188.

24. Laws, J. O., and D. A. Parsons: The Relationship of Raindrop Size to Intensity, *Trans. Am. Geophys. Union*, 24th Annual Meeting, pp. 452–460, 1943.

25. Schelleng, J. C., C. R. Burrows, and E. B. Ferrell: Ultra-Short-Wave Propagation, *Proc. IRE*, vol. 21, pp. 427–463, 1933.

26. Medhurst, R. G.: Rainfall Attenuation of Centimeter Waves: Comparison of Theory and Measurement, *IEEE Trans.*, vol. AP-13, pp. 550–564, 1965.

27. Burrows, C. R., and S. S. Attwood: "Radio Wave Propagation, Consolidated Summary Technical Report of the Committee on Propagation, NDRC," Academic Press, New York, 1949, p. 219.

28. Humphreys, W. J.: "Physics of the Air," McGraw-Hill Book Company, New York, 1940, p. 82.

29. Atlas, D., and E. Kessler III: A Model Atmosphere for Widespread Precipitation, *Aeronaut. Eng. Rev.*, vol. 16, pp. 69–75, 1957.

30. Dennis, A. S.: Rainfall Determinations by Meteorological Satellite Radar, *Stanford Research Institute, SRI Rept.* 4080, 1963.

31. Kerker, M., M. P. Langleben, and K. L. S. Gunn: Scattering of Microwaves by a Melting Spherical Ice Particle, *J. Meteorol.*, vol. 8, p. 424, 1951.

32. "Glossary of Meteorology," vol. 3, American Meteorological Society, Boston, 1959, p. 613.

33. Best, A. C.: "Physics in Meteorology," Sir Isaac Pitman & Sons, Ltd., London, 1957.

34. Saxton, J. A., and H. G. Hopkins: Some Adverse Influences of Meteorological Factors on Marine Navigational Radar, *Proc. IEE (London)*, vol. 98, pt. III, p. 26, 1951.

35. Pasqualucci, F., B. W. Bartram, R. A. Kropfli, and W. R. Moninger: A Millimeter-Wavelength Dual-Polarization Doppler Radar for Cloud and Precipitation Studies, *J. Clim. Appl. Meteorol.*, vol. 22, pp. 758–765, 1983.

36. Lhermitte, R.: A 94-GHz Doppler Radar for Cloud Observations, *J. Atmos. Ocean. Technol.*, vol. 4, pp. 36–48, 1987.

37. Richter, J. H.: High-Resolution Tropospheric Radar Sounding, *Proc. Colloq. Spectra Meteorol. Variables, Radio Sci.*, vol. 4, pp. 1261–1268, 1969.

38. Tang Dazhang, S. G. Geotis, R. E. Passarelli, Jr., A. L. Hansen, and C. L. Frush: Evaluation of an Alternating PRF Method for Extending the Range of Unambiguous Doppler Velocity, *Preprints, 22d Conf. Radar Meteorol.*, pp. 523–527, American Meteorological Society, Boston, 1984.

39. Laird, B. G.: On Ambiguity Resolution by Random Phase Processing, *Preprints, 20th Conf. Radar Meteorol.*, p. 327, American Meteorological Society, Boston, 1981.

40. Rummler, W. D.: Introduction of a New Estimator for Velocity Spectral Parameters, *Tech. Memo. MM*-68-4121-5, Bell Telephone Laboratories, Whippany, N.J., 1968.

41. Hildebrand, P. H., and R. H. Sekhon: Objective Determination of the Noise Level in Doppler Spectra. *J. Appl. Meteorol.*, vol. 13, pp. 808–811, 1974.

42. Denenberg, J. N., R. J. Serafin, and L. C. Peach: Uncertainties in Coherent Measurement of the Mean Frequency and Variance of the Doppler Spectrum from Meteorological Echoes, *Preprints, 15th Conf. Radar Meteorol.*, pp. 216–221, American Meteorological Society, Boston, 1972.

43. Davenport, W. B., Jr., and W. L. Root: "An Introduction to the Theory of Random Signals and Noise," McGraw-Hill Publishing Company, New York, 1958.

44. Zrnić, D. S.: Estimation of Spectral Moments for Weather Echoes, *IEEE Trans.*, vol. GE-17, pp. 113–128, 1979.

45. Marshall, J. S., and W. Hitschfeld: The Interpretation of the Fluctuating Echo for Randomly Distributed Scatterers, pt. I, *Can. J. Phys.*, vol. 31, pp. 962–994, 1953.

46. Frush, C.: Doppler Signal Processing Using IF Limiting, *Preprints, 20th Conf. Radar Meteorol.*, pp. 332–337, American Meteorological Society, Boston, 1981.

47. Mueller, E. A., and E. J. Silha: Unique Features of the CHILL Radar System, *Preprints, 18th Conf. Radar Meteorol.*, pp. 381–386, American Meteorological Society, Boston, 1978.

48. Campbell, S. D., and S. H. Olson: Recognizing Low-Altitude Wind Shear Hazards from Doppler Weather Radar: An Artificial Intelligence Approach, *J. Atmos. Ocean. Technol.*, vol. 4, p. 518, 1987.

49. Baynton, H. W., R. J. Serafin, C. L. Frush, G. R. Gray, P. V. Hobbs, R. A. Houze, Jr., and J. D. Locatelli: Real-Time Wind Measurement in Extratropical Cyclones by Means of Doppler Radar, *J. Appl. Meteorol.*, vol. 16, pp. 1022–1028, 1977.

50. Wilson, J., and H. P. Roesli: Use of Doppler Radar and Radar Networks in Mesoscale Analysis and Forecasting, *ESA J.*, vol. 9, pp. 125–146, 1985.

51. Bonewitz, J. D.: The NEXRAD Program—An Overview, *Preprints, 20th Conf. Radar Meteorol.*, pp. 757–761, American Meteorological Society, Boston, 1981.

52. Gunn, R., and Kinzer, G. D.: The Terminal Velocity of Fall for Water Droplets in Stagnant Air, *J. Meteorol.*, vol. 6, pp. 243–248, 1949.

53. Marshall, J. S., and W. M. K. Palmer: The Distribution of Raindrops with Size, *J. Meteorol.*, vol. 4, pp. 186–192, 1948.

54. Blanchard, D. C.: Raindrop Size Distribution in Hawaiian Rains, *J. Meteorol.*, vol. 10, pp. 457–473, 1953.

55. Jones, D. M. A.: 3 cm and 10 cm Wavelength Radiation Backscatter from Rain, *Proc. Fifth Weather Radar Conf.*, pp. 281–285, American Meteorological Society, Boston, 1955.

56. Gunn, K. L. S., and J. S. Marshall: The Distribution with Size of Aggregate Snowflakes, *J. Meteorol.*, vol. 15, pp. 452–466, 1958.

57. Wilson, J. W., and E. A. Brandes: Radar Measurement of Rainfall—A Summary, *Bull. Am. Meteorol. Soc.*, vol. 60, pp. 1048–1058, 1979.

58. Bridges, J., and J. Feldman: An Attenuation Reflectivity Technique to Determine the Drop Size Distribution of Water Clouds and Rain, *J. Appl. Meteorol.*, vol. 5, pp. 349–357, 1966.

59. Seliga, T. A., and V. N. Bringi: Potential Use of Radar Differential Reflectivity Measurements at Orthogonal Polarizations for Measuring Precipitation, *J. Appl. Meteorol.*, vol. 15, pp. 69–76, 1976.

60. Zawadzki, I.: Factors Affecting the Precision of Radar Measurements of Rain, *Preprints, 22d Conf. Radar Meteorol.*, pp. 251–256, American Meteorological Society, Boston, 1984.

61. Burgess, D., et al.: Final Report on the Joint Doppler Operational Project (JDOP), 1976–1978, *NOAA Tech. Memo. ERL NSSL*-86, 1979.

62. Armstrong, G. M., and R. J. Donaldson, Jr.: Plan Shear Indicator for Real-Time Doppler Identification of Hazardous Storm Winds, *J. Appl. Meteorol.*, vol. 8, pp. 376–383, 1969.

63. Donaldson, R. J., Jr.: Vortex Signature Recognition by a Doppler Radar, *J. Appl. Meteorol.*, vol. 9, pp. 661–670, 1970.

64. Zrnić, D. S., and R. J. Doviak: Velocity Spectra of Vortices Scanned with a Pulse Doppler Radar, *J. Appl. Meteorol.*, vol. 14, pp. 1531–1539, 1975.

65. Fujita, T., and F. Caracena: An Analysis of Three Weather-Related Aircraft Accidents, *Bull. Am. Meteorol. Soc.*, vol. 58, pp. 1164–1181, 1977.

66. Fujita, T.: "The Downburst," Satellite and Mesometeorology Research Project, Department of the Geophysical Sciences, University of Chicago, 1985.

67. Fujita, T.: "The DFW Microburst," Satellite and Meteorology Research Project, Department of the Geophysical Sciences, University of Chicago, 1986.

68. McCarthy, J., and R. Serafin: The Microburst: Hazard to Aviation, *Weatherwise*, vol. 37, no. 3, pp. 120–127, 1984.

69. McCarthy, J., J. Wilson, and M. Hjelmfelt: Operational Wind Shear Detection and Warning: The CLAWS Experience at Denver and Future Objectives, *Preprints, 23d Conf. Radar Meteorol.*, pp. 22–26, American Meteorological Society, Boston, 1986.

70. Witt, A., and S. P. Nelson: The Relationship between Upper-Level Divergent Outflow Magnitude as Measured by Doppler Radar and Hailstorm Intensity, *Preprints, 22d Conf. Radar Meteorol.*, pp. 108–111, American Meteorological Society, Boston, 1984.

71. Aydin, K., T. A. Seliga, and V. Balaji: Remote Sensing of Hail with a Dual Linear Polarization Radar, *J. Clim. Appl. Meteorol.*, vol. 25, pp. 1475–1484, 1986.

72. Lhermitte, R. M., and D. Atlas: Precipitation Motion by Pulse Doppler Radar, *Proc. Ninth Weather Radar Conf.*, pp. 218–223, American Meteorological Society, Boston, 1961.

73. Browning, K. A., and R. Wexler: A Determination of Kinematic Properties of a Wind Field Using Doppler Radar, *J. Appl. Meteorol.*, vol. 7, pp. 105–113, 1968.

74. Wilson, J. W., and W. E. Schreiber: Initiation of Convective Storms at Radar-Observed Boundary Layer Convergence Lines, *Mon. Weather Rev.*, vol. 114, pp. 2516–2536, 1986.

75. Carbone, R. E.: A Severe Frontal Rainband, Part I: Storm-Wide Hydrodynamic Structure, *J. Atmos. Sci.*, vol. 39, pp. 258–279, 1982.

76. Brook, M., and P. Krehbiel: A Fast-Scanning Meteorological Radar, *Preprints, 16th Conf. Radar Meteorol.*, pp. 26–31, American Meteorological Society, Boston, 1975.

77. Keeler, R. J., and C. L. Frush: Rapid-Scan Doppler Radar Development Considerations, Part II: Technology Assessment, *Preprints, 21st Conf. Radar Meteorol.*, pp. 284–290, American Meteorological Society, Boston, 1983.

78. Walther, C., C. Frush, and P. Hildebrand: The NCAR Airborne Doppler Radar, Part III: Overview of Radar Design Details, *Preprints, 23d Conf. Radar Meteorol.*, vol. I, pp. 155–158, American Meteorological Society, Boston, 1986.

79. Meneghini, R., and D. Atlas: Simultaneous Ocean Cross-Section and Rainfall Measurements from Space with a Nadir-Looking Radar, *J. Atmos. Ocean. Technol.*, vol. 3, pp. 400–413, 1986.

80. Carbone, R. E., R. Strauch, and G. M. Heymsfield: Simulation of Wind Profilers in Disturbed Conditions, *Preprints, 23d Conf. Radar Meteorol.*, vol. I, pp. 44–47, American Meteorological Society, Boston, 1986.

81. Strauch, R. G., B. L. Weber, A. S. Frisch, C. G. Little, D. A. Merritt, K. P. Moran, and D. C. Welsh: The Precision and Relative Accuracy of Profiler Wind Measurements, *J. Atmos. Ocean. Technol.*, vol. 4, pp. 563–571, 1987.

82. Röttger, J., and M. F. Larsen: Clear Air Radar Techniques, invited paper, *40th Conf.*

Radar Meteorol., American Meteorological Society, Boston, 1987. Published in Atlas, D. (ed.): "Radar Meteorology," American Meteorological Society, Boston, 1989.

83. Metcalf, J. I., and W. A. Holm: Meteorological Applications of Synthetic Aperture Radar, final report, Project A-2101, Engineering Experiment Station, Georgia Institute of Technology, 1979.

84. Atlas, D., and R. K. Moore: The Measurement of Precipitation with Synthetic Aperture Radar, *J. Atmos. Ocean. Technol.*, vol. 4, pp. 368–376, 1987.

CHAPTER 24
HF OVER-THE-HORIZON RADAR*

J. M. Headrick
Naval Research Laboratory

24.1 INTRODUCTION

Beyond-the-horizon ranges up to thousands of nautical miles can be achieved by radar operation in the high-frequency (HF) band (3 to 30 MHz). The longer-range performance is achieved by using sky-wave propagation. Ground-wave propagation over the sea is useful for short but still over-the-horizon distances. HF radar development over the past several decades has led to this capability,[1,2] and several operational systems are deployed.[3,4,5] Targets of interest are the same as for microwave radar and include aircraft, missiles, and ships. The wavelengths used are of the same order as ocean gravity waves, and this correspondence makes HF radar able to provide information on the waveheight directional spectrum and, by inference, surface winds and ocean currents.[6] In addition, this sensor is useful for observing various forms of high-altitude atmospheric ionization such as that due to aurora, meteors, and missile launches.[7,8] The wavelengths used and the nature of the transmission path make the spatial resolution coarse when compared with much-higher-frequency radars; however, the doppler resolution can be fine.

For effective radar operation, environmental parameters need to be determined in real time; transmission-path information is generally derived from adjunct vertical and oblique sounders as well as by using the radar itself as a sounder. An ionospheric electron-density model complex enough to enable adequate sounding interpretation is required. Ionospheric or transmission-path statistical forecasts are necessary for radar design and for development of the model, not for real-time operation. In addition, other users in the spectrum must be observed continuously and operating frequencies selected to avoid interference. In the context of transmission-path assessments for frequency selection, real time is measured as that interval in which there are no important changes in the ionosphere.

The magnitude and doppler distribution of the earth-surface backscatter is a major factor in setting system dynamic-range and signal-processing characteristics. Backscatter from the sea can be employed as a reference level and is a generally used diagnostic tool.

Figure 24.1 provides examples of radar coverage and spatial resolution appro-

---

* The comments, discussions and contributions by NRL colleagues J. Hudnall, J. McGeogh, R. Pilon, G. Skaggs, and J. Thomason are gratefully acknowledged.

priate for aircraft and ship targets. In the upper left the ray-path sketch shows that at higher-elevation radiation angles the rays escape, causing a skip zone of range coverage, that energy is returned to the earth until the reflection height horizon is reached, and that useful range coverage will lie between these limits. Although only one hop is shown, a multiplicity will exist, and energy may at times circle the earth. In the upper-right sketch, the ray paths show that different range extents are illuminated by using different operating frequencies, the longer ranges requiring the higher frequencies. In the example, 500 nmi is shown as the range coverage extent illuminated by one operating frequency. The trailing edge of an extent may vary as a function of radar parameters and target size, but the start is set by frequency selection and immediately follows the skip zone. Directly below is a plan view of the coverage area. It shows nine different areas that might be illuminated by a separate transmit beam of 8° width. It is likely that the ionosphere will vary across the 64° azimuth scan, so that if a single operating frequency is used, the range to the transmitter footprint will change with azimuth; however, in general a different operating frequency could be selected for each 8° to obtain the desired illumination. Each transmitter footprint is filled by 16 contiguous receive beams, each ½° wide. At the lower left, one illumination sector is shown divided into receive resolution cells with each cell being approximately 10 nmi on the side.

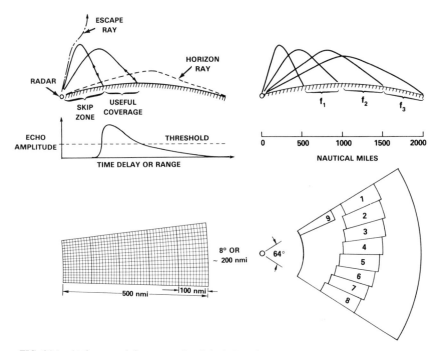

FIG. 24.1 At the upper left an example of single-hop sky-wave propagation with the useful coverage in range is sketched. On the upper right extended range coverage is obtained by using three frequencies. An azimuthal scan is shown in plan at the lower right, and at the lower left are given the receiver resolution cells in one transmitter footprint.

24.2 RADAR EQUATION

A form of the radar equation, Eq. (24.1), can be used to point to aspects of HF radar that are significantly different from radars that use higher frequencies. These differences include adaptation to environment, frequency and waveform selection, radar cross section, path losses, multipath effects, noise, interference, antenna gain, spatial resolution, and sky clutter.

$$\frac{S}{N} = \frac{P_{\mathrm{av}} \, G_t G_r T \lambda^2 \sigma F_p}{N_o \, L (4\pi)^3 R^4}$$
(24.1)

where S/N = output signal-to-noise ratio
P_{av} = average transmitted power, W
G_t = transmitter antenna gain
G_r = receiver antenna gain
T = effective processing time, s
λ = wavelength
σ = target radar cross section
F_p = propagation-path factor
N_o = noise power per hertz
L = transmission-path and system losses
R = distance between radar and target

These parameters are explained as follows:

1. *Antennas, G_t and G_r:* A common convention for HF-band radars is to include earth effects in the antenna performance characterization, and that convention will be used here. For example, a half-wave dipole in free space has a maximum gain over an isotrope of 2.15 dB. If that antenna is oriented vertically, just above but not touching a perfectly conducting earth, its maximum gain will be increased by a factor of 4, or 6 dB, to 8.15 dB at 0° elevation angle. Since the earth is never perfect, its conductivity and dielectric constant are factors in determining antenna performance. The electrical properties of the earth are a much stronger factor for vertical polarization than for horizontal; however, terrain features and surface roughness are important for both polarizations.

2. *Coherent processing time, T:* HF radar is generally a look-down type that has earth backscatter at the same ranges as targets. Doppler processing is used to separate targets from earth backscatter.

3. *Wavelength (λ):* The wavelength or operating frequency must be selected so that energy is refracted by the ionosphere to illuminate the desired area of the earth. The spectrum of the emissions must be constrained not to interfere with other users. Since both the ionosphere and the HF-band occupancy distributions are time-varying parameters, adaptive radar management is required.

4. *Radar cross section (RCS), σ:* The radar cross section of conventional targets will generally be a function of frequency, polarization, and aspect angle. Often some sort of average value is used in analysis. Clutter levels will be large relative to most targets and therefore are important in radar design. For the RCS of earth clutter, the surface scattering coefficient σ^0 is multiplied by the resolution cell size A and the cosine of the grazing angle. The important resolution cell size factors, receiver antenna beamwidth and spectral bandwidth, are not explicitly contained in Eq. (24.1)

5. *Propagation factors (F_p):* Several propagation phenomena, including Faraday polarization rotation, multipath, and ionospheric focusing, may need inclusion in the equation. With linearly polarized radiation, Faraday rotation will result in a polarization mismatch varying with time and distance. That is, the target-incident-energy polarization will rotate as distance changes. Since many targets have RCSs that vary with polarization, an important result is that the most favorable polarization will illuminate the target recurrently.

6. *Noise (N_0):* For radars operating in the HF band it is possible to design antennas and receivers with low enough noise figures that environmental noise is dominant.

7. *Losses (L):* The loss term contains the two-way losses along the path traversed including ionospheric absorption and ground-reflection losses as well as any radar system losses. Ionospheric losses, while predicted on a statistical basis, constitute a major unknown in real-time radar operation.

8. *Range (R):* The range in the equation is the distance along the virtual path between target and radar. The ionospheric reflection height needs to be used to convert this range to great-circle ground distance. The apparent range to a particular target may take on more than one value since multiple paths may exist.

24.3 TRANSMITTERS

Most of the radar designs and missions require transmitter average power levels between 10 kW and 1 MW. Antennas are generally arrays of elements, and the common trend is to drive each element with a separate amplifier. This approach permits beam steering at a low level in the amplifier chain. Power control and amplitude-shaping requirements indicate a linear amplifier design. Since the radar uses doppler filter signal processing to separate the targets from the clutter, the clutter returned on the phase and amplitude noise sidebands radiated by the transmitter should be below that of desired targets. This can put a stringent condition on the emitted signal-to-noise ratio of the transmitter. The signal-to-noise ratio of the initial signal synthesizer must meet the requirements. The lower-level signal amplification can generally be designed to add no noise. Mechanical vibration in the high-power amplifiers can add noise, and care must be exercised in the air or liquid coolant flow system design.

The active element in each final transmitter stage can be either a traditional vacuum tube[9] or a solid-state device.[10] If the radar is to perform wide-area surveillance, frequent frequency changes are required in order to cover the various range extents. In addition, relative phase or time-delay changes are required in each amplifier chain to accomplish azimuthal steering. Reference 11 contains a discussion on modern high-efficiency amplifier designs for broadcasting; while some of the techniques are useful in radar transmitters, the final amplifiers tend to be narrowband. A broad-bandwidth performance and a tolerance to a variable standing-wave ratio load are desired features in a wide-area surveillance radar. Since the antenna elements will be wideband, harmonic filters may be required. As an example, one transmitter and harmonic filter combination might have a 5- to 9-MHz passband and a stopband for 10 Mhz and higher frequencies; a second combination might pass up to 17 MHz and reject 18 Mhz and higher, and the design would continue in this manner to the highest frequency of operation.

24.4 ANTENNAS

A single antenna can be duplexed and used for both transmit and receive. The Naval Research Laboratory (NRL) magnetic-drum recording equipment (MADRE) antenna is an example.[1] This 100-m-wide by 40-m-high aperture provided sufficient gain and angular resolution for aircraft tracking in the upper part of the HF band: the frequencies used in daytime. A horizontal aperture of twice the width would provide similar resolution at night, where frequencies would be in the lower part of the HF band. However, it is desirable to have better azimuthal resolution for location accuracy and to reduce the clutter amplitude, and horizontal apertures of 3 km and even wider can be useful.[12,13] It is common practice to use separate transmit and receive antennas with the transmitter floodlighting many simultaneously formed receive beams. In elevation, desirable radiation angles run between 0 and 40°, with the specific angles depending on range and reflection height. Any sensitivity gained by directivity in elevation directly improves radar performance. This is in contrast with azimuth directivity for transmitting, where an increase in gain is accompanied by a decrease in area coverage. The AN/FPS 118 over-the-horizon (OTH) radar[3] does not employ steerable directivity in elevation but covers all necessary radiation angles with one broad elevation beam. This choice permits the antenna to have a relatively small vertical dimension. The AN/FPS 118 radar module uses a nominal 7.5° transmitter footprint filled with five receive beams; this combination is stepped as a surveillance barrier within a potential coverage area 500 to 1800 nmi in distance by 60° in width.

The antennas and power amplifiers used in HF broadcast station service have much in common with an HF radar transmitter-antenna combination; that is, the broadcast-service aim is to obtain a specified level of illumination over some selected area. The multiple-band and steerable broadcast antennas reviewed in the *IEEE Transactions on Broadcasting* Special Issue on Short-Wave Broadcasting,[11] those described in Johnson and Jasik,[14] pages 26–29 through 26–35, and those given in "Shortwave Antennas,"[15] Chaps. 13 and 14, are examples. It will be noticed that many of the broadcast-service antennas employ large vertical apertures. However, antennas used for radar have an added severe demand to minimize mechanical motion that would cause signal modulation, and this requirement is easier to meet with low-antenna-height designs.

Among the many factors that influence antenna design are:

1. The transmitter-receiver antenna gain product must be large enough to make its required contribution to sensitivity.

2. The receive beamwidth must be narrow enough to provide the required location accuracy.

3. The receive beamwidth must restrict clutter levels to values permitted by system dynamic range and slow-target detection requirements.

24.5 CLUTTER: THE ECHO FROM THE EARTH

Early in HF radar experiments it was noted that the clutter received by sky-wave paths represented a large signal and provided an indication of the earth-surface area illumination. Extensive observations made at the Naval Research Laboratory viewing alternately Atlantic Ocean areas and central United States areas indicated that, averaged over a wide area, sea clutter power levels were about an order of magnitude higher than those from an area of similar size in the central United States. Later observations indicated even less backscatter from ice-covered Greenland areas. The backscattered energy from land is topography-dependent; as an example, a city in the central plains provides a larger echo than its surroundings. In contrast with such abrupt changes, the seascattering coefficient σ^0 for HF radio waves is fairly uniform; that is, it changes gradually with range and azimuth. The sea echo power is proportional to the resolution cell area, to a good approximation. The nature of the sea echo needs some description. For the seascattering coefficient to be approximately constant, long time averaging (minutes) is required. Sky-wave transmission characteristics are notably variable; however, the sea echo can be used as an amplitude reference when care is exercised. An elementary description of sea echo behavior and some applications follows.

Ocean wave generation and propagation are a complex and not completely understood subject; a similar statement can be made about electromagnetic scattering from the waves. However, for the scale size of HF-band wavelengths the sea is a surface that is only slightly rough, and the method of Rice[16] can be used to explain reflection. Backscatter from the sea can be considered a resonant interaction. The disorganized-looking ocean waves are thought as being the sum of an infinite number of Fourier surfaces, each being a sinusoidally corrugated sheet with a different wavenumber and direction.[17] The principal backscatter for a grazing-incidence electromagnetic wave comes from that component sheet that has a wavelength equal to one-half of the radar wavelength and that is either directly approaching the radar or receding from it.[18,19] The surface scattering coefficient, or RCS per unit surface area (σ^0), is much larger for vertical than for horizontal polarization. The echo from the sea with horizontal polarization can be neglected at the smaller grazing angles. The water wavelengths between 10 and 100 m are gravity waves that in deep water follow the dispersive relation

$$v = \left[\frac{gL}{2\pi}\right]^{1/2} \tag{24.2}$$

where v = water-wave phase velocity
       g = acceleration of gravity
       L = water wavelength

Ocean waves are excited by the surface winds. If a wind blows at a constant velocity long enough and over sufficient fetch, a steady-state condition will be achieved where the wind provides just enough energy to the water to supply that lost in breaking and other dissipation. It will completely arouse the water waves with a phase velocity nearly equal to the wind speed. It will also completely arouse all the ocean waves of lower velocities. Figure 24.2a gives an example of the spectrum as derived from a measurement of waveheight versus time at a particular point. Pierson and Moskowitz have derived the following relation for a fully developed spectrum based upon empirical data.[20]

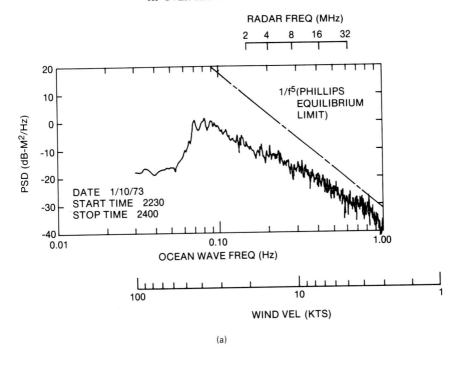

(a)

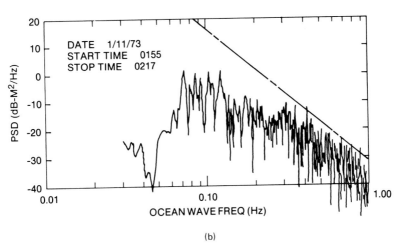

(b)

FIG. 24.2 (*a*) Ocean-waveheight power spectral density (PSD) is given as a function of water-wave frequency. The straight line is a saturation asymptote, and for this example the limit is approached at the higher frequencies. The scale marked "wind velocity" can be used to deduce that winds up to 40 kn have excited waves with frequencies as low as 0.08 Hz but that either the length of time or the fetch, or both, have not been sufficient for full development. In this example the individual spectral analyses, requiring about 100 s of data, have been averaged for 1½ h. (*b*) The spectral density is given as in *a* but is averaged over a shorter time, 23 min. Both higher peaks and deeper voids are seen. The scale across the top of *a* gives the radar operating frequency that experiences resonant backscatter, and it can be seen that at many frequencies this level is far below the long-term average.

$$S(\omega) = C_1 g^2 \omega^{-5} \exp\left[-C_2\left(\frac{g}{\omega v}\right)^4\right]$$ (24.3)

where S = waveheight squared per hertz
ω = water-wave angular frequency
v = wind speed
C_1, C_2 = constants

The exponential term approximates the decay in the spectrum above the wind maximum velocity. The ω^{-5} asymptotic term is a feature of major interest and use. This saturation effect suggests that the sea echo can provide an amplitude reference, and an important consequence is that with such a reference path losses may be estimated. The inference is that for vertically polarized waves from a radar, looking along the wind direction, σ^0 is constant for a specific radar operating frequency if the sea is fully developed. Further, following the analysis of Barrick,[19] σ^0 will have the same value for all water-wave frequencies along the wind direction where the sea is fully developed. In general, the scattering coefficient will be proportional to the resonant waveheight squared. The term *resonant* refers to the Fourier component of the surface where the water wavelength multiplied by the cosine of the grazing angle equals one-half of the radar wavelength. This type of scatter is frequently refered to as *Bragg scatter*. These models are helpful in developing an understanding of the sea surface and the radiowave interaction; however, a very long time average is implied when the water-wave spectrum is described as the product of two terms, one of which is frequency to the minus fifth power and the other an exponential with frequency in the argument. A single 100-s look at the waveheight-squared spectrum with a frequency resolution of $\frac{1}{100}$ Hz would yield a very jagged spectrum; Fig. 24.2*b* provides a short look to contrast with the long time average of Fig. 24.2*a*.

The variability and value of σ_{vv}^0 was examined with the San Clemente Island ground-wave radar.[21] The *vv* subscripts indicate that both transmission and reception were vertically polarized. This radar had several valuable and unique features: a transmission path out over the open sea, multiple-frequency operation in a repetition period, calibrated antennas, known transmitter power, and ground truth in the form of ocean-waveheight recordings. When looking into an approximately 20-kn wind, values of σ_{vv}^0 were found to be constant within a few decibels for operating frequencies where the water-wave spectrum was approximately fully developed; these observations provided a confirmation of Barrick's first-order theory.[19] It is emphasized that a constant level in scattering coefficient implies long averaging times. By using the antenna gain conventions stated earlier and a semi-isotropic sea directional spectrum, the value of σ_{vv}^0 was calculated as -29 dB, and the measured values were grouped between -7 and $+3$ dB of this value over a 5- to 20-MHz-frequency span. This experiment provided a direct measure of the sea-surface scattering coefficient and has exposed characteristics that should be considered when using σ^0 as a reference.

Some other features need consideration. Water-wave directions neither will be confined to the wind direction nor will they be semi-isotropic, but they will spread throughout 360°, with the spreading function depending upon frequency and other variables. Oceanographers generally treat the directional wave spectrum in the wind direction half plane only.[22] But HF radar has sufficient sensitivity to expose the waves running against the wind that have an RCS more than two orders of magnitude below those running with the wind. If the wave directional-spectrum distribution of Long and Trizna is used, the maximum value of σ^0 for a saturated sea is -27 dB in the upwind or downwind direction (longitudinal sea) and is -39 dB in the crosswind direction (transverse sea).[23] Figure 24.3 gives the nominal shape of the scattering coefficient for these two conditions. The statements so far have all related to vertical polarization, which will have the largest scattering coefficient. Sky-wave radar resolution cell sizes are generally large enough that Faraday rotation can be expected to cause illumination of a resolution cell to be over all polarization angles and consequently to reduce the average value of σ^0 by 3 or 4 dB, depending on the elevation radiation angle. If -30 dB is used for the effective scattering coefficient and a resolution cell is considered to be a square 10 nmi on a side, the surface area A is 85 dBsm and the RCS is $85 - 30 = 55$ dBsm. A 12 dB path enhancement will be effective owing to constructive multipath addition, which will increase the resolution cell effective RCS to 67 dBsm. The receiver and processor must be able to handle both the high-level signal due to this large RCS and those much smaller signals due to targets. An HF radar must be designed to accommodate such clutter levels even though they will not exist all the time or at any one time over all areas, especially at the lower operating frequencies.

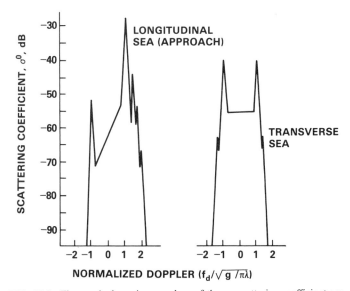

FIG. 24.3 The nominal maximum values of the seascattering coefficient are given as a function of doppler frequency for a coherent processing time of 51 s. A longitudinal, directly approaching or receding sea gives the largest value. A transverse, cross to the radar look direction, gives two peaks equal in height.

By scanning over a radar's total potential coverage area and using the ratio of the two resonant doppler frequency sea responses, a map of sea principal direction can be constructed.[23,24] By inference, surface winds can be mapped. Surface-wind maps are a regular contribution by the Jindalee radar.[25] Actual waveheight estimates can be made on the basis of the multiple scatter features of the sea echo spectra.[21,24,26–28] A major difficulty with these methods lies in eliminating corruption of the echo doppler spectrum by ionospheric path propagation.

Trizna has reported a method for estimating σ^0 from three simple and direct measurements made on the radar echo spectrum, and this method is relatively insensitive to propagation-path variation.[29,30] All the methods for estimating sea state or scattering coefficient require long coherent processing times plus incoherent averaging of a number of coherent times in order to achieve a distinct and repeatable amplitude-doppler signature. This type of radar operation will frequently be incompatible with other radar missions. However, the sea echo is a large signal, and it can be obtained with an adjunct oblique sounder operating in an appropriate radar mode.

In summary, the sea echo power in a resolution cell (1) is generally the largest in-band echo signal; (2) generally exists in the open ocean even in relative calm; (3) varies as the square of resonant waveheight, which is frequently limiting at the higher frequencies; and (4) varies with direction, being greatest for seas running toward or away from the radar.

24.6 RADAR CROSS SECTION

The RCS of the sea has been treated in Sec. 24.5. A number of natural scatterers have been described at the 1981 Symposium on the Effect of the Ionosphere on Radiowave Systems.[31] Findings on the nature of auroral echoes are given by Greenwald,[8] and an auroral echo-scattering model has been developed by Elkins.[32] Chapter 4 of Ref. 7 treats the RCS associated with rockets and their exhausts. Here attention will be confined to echoes from airborne and surface targets.

Aircraft and ships have dimensions that put them in the resonant scattering region. The smallest aircraft and cruise missiles will be in the Rayleigh scattering region for the lower half of the HF band. The RCS has aspect sensitivity but strongly depends upon the target's gross dimensions. For an aircraft the span of the wings, the fuselage length, the tail and elevator span, the vertical stabilizer and rudder height, and their relative locations are the features that determine the RCS. Target shaping of a scale size much less than a wavelength will have little effect. For bodies with high-conductivity surfaces the scattering cross section can be calculated by using numerical methods.[33] Facilities exist where good scale-model measurements can be made, and the Ohio State University compact range is an example.[34]

Rough but useful RCS estimates can be made by using the behavior of a few "canonical" shapes. Figure 24.4 is a family of plots giving RCS versus radar frequency for an oblong-shaped conducting body. The straight line marked 90° λ/2 dipole gives the RCS of a resonant, conducting half-wavelength rod, where the rod is parallel with the electric field. This geometry gives the maximum RCS for the rod. The upper scale of the abscissa gives the one-half-wavelength dimension of the frequency given on the lower scale. The curve marked 90° is the RCS of the oblong-shaped conducting body of 11-m length and 1-m thickness; again the target long dimension is aligned with the electric field. The maximum RCS coincides with the nominal half-wavelength dimension or with the first resonance. The curves marked 45, 15, and 0° give the RCS as the target is rotated to these angles in the plane that contains the electric vector. The little sketches give, at the left, the body shape, then the RCS patterns at nominal ½ wavelength, 1 wavelength, and ³⁄₂ wavelength in order to help visualize how the RCS will change as the as-

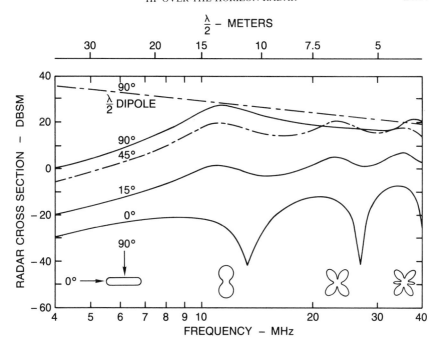

FIG. 24.4   The RCS is given versus frequency for an 11-m-long and 1-m-thick oblong conducting target. The E vector and the 11-m dimension are in the same plane; 0° (nose-on), 15°, 45°, and 90° (broadside) curves are given. The top dashed curve is for a resonant dipole at 90°. The little sketches at the first, second, and third resonances of the 11- by 1-m target show how the RCS behaves with an illumination aspect.

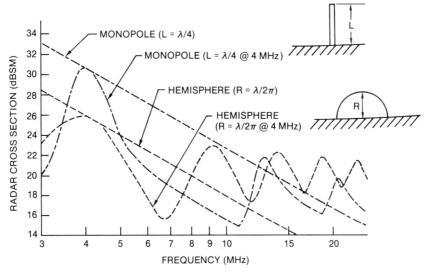

FIG. 24.5   The vertical-polarization RCS of a hemisphere and a monopole on a perfect conductor are plotted against the radar operating frequency.

pect angle is varied. For targets of other lengths with approximately the same shape factor, the response can be determined by sliding the curve along the $\lambda/2$ line and making the first resonance coincide with the line at the $\frac{1}{2}$-wavelength point. Aircraft with their wing-fuselage cruciform shape will have an RCS that varies with the aspect angle but not greatly. As has been mentioned, Faraday rotation will ensure periodic illumination with the most favorable polarization.

Figure 24.5 gives the vertical-polarization RCS of a rod and a hemisphere sticking out of a perfectly conducting surface. With these canonical shapes an estimate of RCS can be made for surface craft. For small vessels the mast height will be of most importance.[35] For surface targets where the maximum RCS is with vertical polarization, the 12 dB sky-wave RCS enhancement mentioned for the sea echo will occur.

24.7 NOISE AND INTERFERENCE

In the HF band, receiving systems can be designed such that external noise is dominant. The major source of noise at the lower frequencies is lightning discharges ionospherically propagated from all over the world (sferics). At the high end of the band, extraterrestrial or galactic noise may be larger than that due to sferics. Receive sites in an area of extensive electrical equipment use can find human-made noise dominant. The HF band is well occupied by other users, and channel selection can be such that other transmitter emissions constitute the background "noise" level.

The widely used source on noise is International Radio Consultative Committee (CCIR) Report 322.[36] This report is based upon measurements made at 16 locations throughout the world. The measurement and data analysis was performed to exclude individual collection site local thunderstorm contributions. Spaulding and Washburn[37] have added data from the U.S.S.R., and a revised CCIR report is available. The noise-level medians as a function of frequency are given in the form of worldwide maps by season and 4-h time block. Lucas and Harper[38] have provided a numerical representation of CCIR Report 322 useful for computer computations, and this has been revised by adding the work of Spaulding and Washburn. The numerical maps of median values are accompanied with decile values to indicate distributions over days of the season. These noise maps provide the level that an omnidirectional antenna would receive. The common method of use is to treat the noise as isotropic even though it must be azimuth- and elevation-angle-dependent. Examination of maps indicates that tropical rain forests and other regions of concentrated thunderstorm activity are major sources of noise. Also, examination suggests that a denser distribution of data collection sites would improve the data. It would be very useful to have noise source maps so that antenna patterns and explicit propagation effects could be taken into account. Some work toward source maps has been done by Ortenburger and Kramer,[39] but there is no known generally available database in this form.

Even though available noise description has limitations, it does provide a reference level for radar design. An HF radar is generally designed to take advantage of what the environment permits; that is, the receiver noise figure should be good enough to make environmental noise the limitation. An example of the CCIR Report 322 data will be discussed. Figure 24.6 was drawn from Lucas and Harper.[38] Noise power in a 1-Hz band relative to 1 W (dBW) is given as a func-

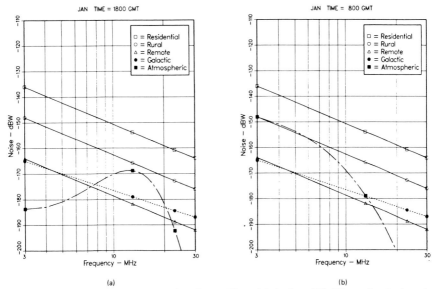

(a)   (b)

FIG. 24.6 Noise power per hertz is given for 38.65° north latitude and 76.53° west longitude and winter. (*a*) 1800 UTC is given as a daytime example. (*b*) 0800 UTC is given as a nighttime example. Universal time coordinated (UTC) and Greenwich mean time (GMT) are interchangeable in this treatment.

tion of frequency for three different sources of noise. The practice in use is to select the largest. This is a winter daytime example at a United States east coast location. The three straight lines are estimates of anthropogenic (formerly man-made) noise for three different types of sites. The shape of the anthropogenic curves is described by the equations

$$N_0 = -136 - 12.6 \ln (f/3) \qquad \text{residential}$$
$$N_0 = -148 - 12.6 \ln (f/3) \qquad \text{rural}$$
$$N_0 = -164 - 12.6 \ln (f/3) \qquad \text{remote}$$

where the frequency f is in megahertz and ln indicates the natural logarithm. The above trends with frequency approximate many measurements of human-made noise; however, ideally the curve would be based on measurements at the particular radar site. The galactic-noise curve should be selected when it is the largest and when there is a path through the ionosphere; the path will not exist for the lower operating frequencies in the daytime. The atmospheric noise rises from low frequencies to about 12 MHz and then rapidly falls. Figure 24.6*b* is for nighttime. All the curves are the same as in Fig. 24.6*a* except for atmospheric noise. At 10 MHz the night and day levels are the same; below 10 MHz the noise decreases with decreasing frequency in daytime and increases at night. Above 10 MHz daytime levels are greater than those at night. These effects can be partially explained by the very lossy long-range paths in day that attenuate the long-range noise at the lower frequencies and by there being few or no sky-wave paths to noise terrestrial sources at the higher frequencies at night. Later it will be seen that in general nighttime noise will be greater than daytime noise for sky-wave illumination of a selected range. The general trends of atmospheric noise in other seasons are similar to those of winter. However, there can be large differences in level at other locations on the earth.

Other effects that can control radar performance are sometimes mistaken for the *passive* noise discussed above. One of these is the spread-in-doppler clutter from localized high-density irregularities in sky ionization; this is sometimes refered to as *active* or multiplicative noise. The occurrence of this type of clutter is greater at night and is much more prevalent in the auroral zones and around the magnetic equator. As stated earlier, Elkins[32] has developed a model for HF auroral clutter that can be used to predict target obscuration when the transmission path is through the auroral region. Lucas has provided spread-F maps for inclusion in ionospheric models so that spread doppler clutter can be predicted.[40] Ionospheric irregularities that scatter back to the radar receiver occur much more often at night than by day at any latitude. Their effects can be reduced with spatial resolution.

Anyone engaged in extensive HF radar performance analyses should have the numerical description of noise maps in their computer data files. When only a few performance predictions are needed, CCIR Report 322 can be used manually.

24.8 SPECTRUM USE

The waveforms that can be effective for HF radar are in general similar to those used at the higher frequencies and are selected for similar reasons. However, the transmission path is dispersive, and waves experience polarization rotation with frequency; because of these effects, bandwidths are limited to the order of 100 kHz without correction. The more restrictive constraint on emissions is that of noninterference to and by other services.

Frequency spans in the HF spectrum are allocated for various types of service such as broadcasting, point-to-point communications, maritime mobile, aeronautical mobile, standard frequency and time, and amateur. The variability of the sky-wave transmission medium requires different operating frequencies at different times. A single point-to-point circuit can require as many as five different frequencies spread over a wide range if the circuit is to be reliable over all hours of the day and seasons of the year and through the solar activity cycle. If a radar is to perform surveillance over large areas by ionospheric refraction at all times of day, seasons, and degrees of solar activity, frequency channels distributed over a large part of the HF band are required, although only a single channel may be used at any one time. When the HF band is scanned with a spectrum analyzer at a particular hour, it can be seen that the gross features of occupancy are remarkably stationary over the days of a season. This is due to broadcast stations, fixed-service point-to-point transmitters, and many other spectrum users having regular schedules. Figure 24.7 examines one particular segment and time. These observations were made with 5-kHz-bandwidth filters. When narrower-bandwidth filters with steep skirts are used, several channels 5 to 10 kHz wide with no detectable users are generally found within any 1-MHz span. The maximum frequency that will still reflect energy back to the earth during the day may be twice that at night; therefore, the occupancy tends to be denser at night than during the day.

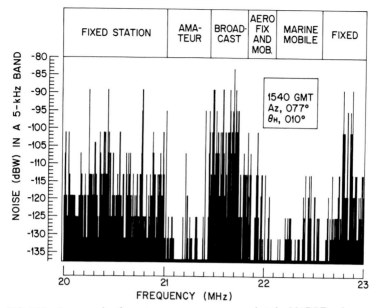

FIG. 24.7 An example of spectrum occupancy measured at the MADRE radar.

The part of the spectrum that is useful for sky-wave propagation is densely populated. Even out-of-band signal levels are a consideration in receiver front-end design, where it is convenient to have bandwidths much wider than that of the radar signal. There are a large number of broadcast stations that have 500-kW transmitters and antennas with more than 20-dB gain. Measurements made on the middle Atlantic coast of the United States show HF broadcast-band signals with strengths of 5 to 10 mV/m. These ambient levels must be accommodated in receiver design since a wideband front end is desirable for rapid and frequent frequency changes.

The practice in allocations for HF radar operation is to permit use of broad bands of the spectrum with a requirement to cause no interference to an existing service and to provide a lockout feature for channels that need protection. An integral part of an HF radar is a channel occupancy analyzer that provides a real-time description of spectrum availability.

24.9 SKY-WAVE TRANSMISSION MEDIUM

Solar radiation and particle emission bombardment are the cause of ionization in the earth's upper atmosphere. Even though there is no incident radiation at night, the ionization never completely decays; that is, there is always an ionosphere. The electron-density distribution is the major control over the propagation of HF radio waves. Ground illumination over the horizon is enabled by refraction in the ionosphere. When an oblique-incidence radio wave traverses a path where electron density is increasing with altitude, the ray is bent away from the vertical; if the gradient in density is sufficient, the wave will "reflect" back back to the earth, providing long-distance illumination. The lower the radio-wave frequency, the smaller the required gradient. Since some ionization in the upper atmosphere always exists, it is always possible to illuminate the earth over the horizon if there is freedom in frequency selection. Ionospheric outages do not exist in the sense that long-distance illumination is impossible. Path outages are due to deficiencies in frequency channel allocations and insufficient radiated power. Additional factors that can affect radar performance are ionization irregularities that degrade path quality and backscatter from spread-in-doppler ionization gradients that can obscure targets.

The solar activity that drives the ionization of the earth's atmosphere is variable on a diurnal, seasonal, and long-term basis with a superimposed random component. Current prediction and analysis methods depend upon a statistical description of the ionosphere. A large amount of vertical-incidence reflection-height versus frequency-sounding (ionosonde) data has been collected over several decades, and from this data most descriptions of the ionosphere are derived. Davies[41] and Chap. 10 of the "Handbook of Geophysics and the Space Environment"[42] can be read for information on ionospheric radio-wave transmission. The radar designer needs a statistical description that will permit matching the design to the required frequency span, power levels, and vertical radiation-angle gain. The radar operator needs a model with enough sophistication to permit full interpretation of the real-time soundings for both operating parameter selection and data analysis.

The regions of the ionosphere that are considered necessary to model for an understanding of transmission paths are as follows:

1. *D region:* This region occupies the lowest altitudes considered. It ranges from 50 to 90 km, where electron density rapidly increases with altitude in the daytime. The maximum ionization in the D region occurs near the subsolar point and will be greatest during periods of highest solar activity (sunspot maximum). The D region may not be explicit in some ionospheric models where its effects are accounted for with an empirically derived path-loss calculation. Most models have this nondeviative absorption as a median value plus a distribution.

2. *E region:* This ionization region extends between about 90 and 130 km in altitude with a maximum near 110 km when sunlit. In addition, there may be anomalous ionization referred to as sporadic E. This latter ionization layer is thin in altitude, may be either smooth or patchy, is seasonally and diurnally variable but not well correlated with solar activity, and has marked variation with latitude.

3. *F Region:* This is the highest-altitude region of interest for sky-wave propagation, and it is also the region of greatest electron density. In the daylight hours there may be two components that should be recognized, especially in summer. The F1 region lies between 130 and 200 km and, like the E region, is directly dependent upon solar radiation; it reaches maximum intensity about 1 h after local noon. The F2 region is variable in both time and geographical location. The altitudes of the F2 region peaks are considered to lie between 250 and 350 km in the middle latitudes. The F2-region ionization shows marked day-to-day variations and in general is not the regular sun follower that the E and F1 regions are. Most models have a statistical description of F2 maximum electron density (or critical frequency) in the form of a median and upper and lower decile values.

Goodman and Reilly discuss shortwave prediction methodologies on pages 230 to 237 of Ref. 11. The ionospheric models that have been extensively used in

HF radar performance analysis are in programs called ITSA-1, ITS-78, RADAR C, IONCAP, and AMBCOM.[43-47] Lucas[48] provides some detail on these models and their origins. In summary, they all draw on the large database of recorded ionospheric soundings made during the International Geophysical Year of 1957–1958 and the International Year of the Quiet Sun of 1964–1965. All the models rely heavily on data taken from the maximum and minimum solar activity years of one solar cycle, with supplement from an adjacent cycle of lower activity. Linear interpolation between the two extremes is used for conditions of intermediate solar activity. This limited database might be considered a serious deficiency since there can be considerable difference in the measure of solar activity from one cycle to another. However, while the degree of atmospheric ionization appears to be a strong function of position in the solar cycle, it has only a weak dependence on solar activity measure. The commonly used indices or measures of solar activity are the sunspot number (SSN) and the radiated-microwave flux density. On a yearly average basis these indices relate well to the median ionospheric description; however, short-term predictions are more difficult to relate to solar activity. In general, the models listed are part of a prediction method. Some of the prediction methods have not been well documented although widely distributed; also, users frequently "improve" upon a model and prediction method to suit their specific needs. As an example, the model RADAR C of Ref. 45 is the basic building block of Thomason, Skaggs, and Lloyd in NRL Report 8321;[49] however, they have added a D region, a collision-frequency distribution, an earth's magnetic field, a topside electron distribution, an auroral electron-density modification,[50] and other features that make the model more generally useful. The ionospheric model as described in NRL Report 8321 will be used for examples here.

A form for display of the data is that of vertical virtual height versus frequency sounding and a true-height plot. *Virtual height* is defined as the speed of light times the time delay for the ionospheric echo; *true height* is the actual distance to reflection height. The critical frequency is the highest frequency that is reflected. Reference 46 has a figure that illustrates how the stored median ionosphere compares with a set of actual soundings; it is shown here as Fig. 24.8. Each of these soundings was made at the same hour but on a different day of the month. In this figure each trace shows the virtual height of a vertical sounding versus probing frequency for the ordinary ray. When vertical soundings are made with an ionosonde that uses a linearly polarized antenna, the ionosphere will be birefractive and provide two traces that are called *ordinary* and *extraordinary*. With right-hand and left-hand circularly polarized antennas the two responses can be separated. Figure 24.9a and b gives an example in the mid-Atlantic off the east coast from the data file. Such plots can be obtained at a selected earth location for a specified level of solar activity (SSN), month, and time of day. Similarly to the illustration given in Fig. 24.8, if a collection of experimental soundings is taken at the geographical location, SSN, and times of day of Fig. 24.9 over a month, the medians will approximate the curves of Fig. 24.9 closely. In addition, the upper and lower decile values of the measured monthly families of F2 critical frequencies will deviate from the median by about ±25 percent. For radar performance calculations median values will be used. However, when designing a radar, distribution should be considered for lowest operating-frequency selection. Distributions are important in communications when a limited number of channels are assigned, but since the radar operating practice is to select a near-optimum frequency, distribution is not important. The model uses three parabolas to approximate the electron distribution with altitude. Electron density and

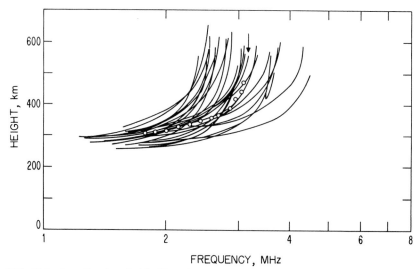

FIG. 24.8 A predicted vertical-incidence ionogram is compared with observed ionograms. The arrow points to the predicted 3.2-MHz critical frequency, and the little circles give points on the predicted median sounding. The measured median of this ensemble is about 3 MHz.

critical or plasma frequency are sometimes used almost interchangeably; when this occurs, the relation implied is $N_e = f^2/81$, where N_e is free-electron density, in numbers per cubic meter, and f is the frequency, in hertz. For the little tables shown in Figs. 24.9 and 24.10, FC is the critical frequency, in megahertz; HC is the height of maximum ionization or the nose of the parabola, in kilometers; and YM is the semithickness of the parabola, in kilometers. ES gives the sporadic-E distribution as M (median), L (lower), and U (upper) decile critical frequencies, in megahertz. This table shows the form of the stored data in the ionospheric model, and with it the various profiles can be generated. All these constants can be adjusted to fit diagnostic observations.

In Fig. 24.9 it is seen that the F2 critical is 6.5 MHz in the daytime and 4 MHz at night for summer. Figure 24.10 gives similar data but for winter, where the day and night comparison is from 8.4 MHz to 3.5 MHz. Figures 24.11 and 24.12 give summer and winter plasma-frequency contours for the same location versus time of day. These plots are provided to show how abrupt the night-to-day and day-to-night transitions are; during winter dawns the critical frequency changes from 2 to 5 MHz in 1 h. The radar-frequency management task is most difficult in these time periods; during most of the day and most of the night changes are relatively slow. This data indicates the median diurnal frequency variations required for a particular path: more than 2:1 in winter and somewhat less in summer. Day-to-day variability will impose greater extremes.

In this section the predictable and random variability of the transmission path has been indicated. When extensive and detailed radar performance calculations are to be made, a computer-stored base is required.

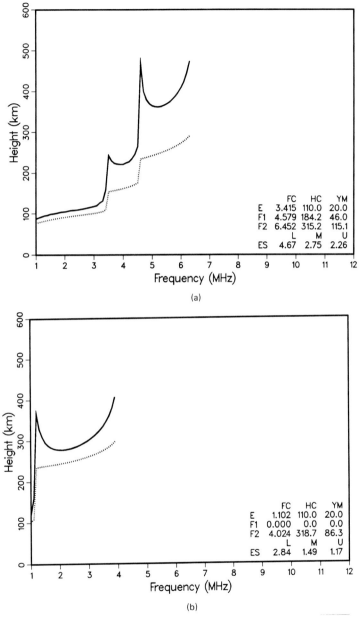

FIG. 24.9 The virtual (solid line) and true (dotted line) reflection heights are given for July, SSN = 50, and a mid-Atlantic-coast radar refraction area. (*a*) 1800 UTC is a daytime example, first hop. (*b*) 0800 UTC is a nighttime example.

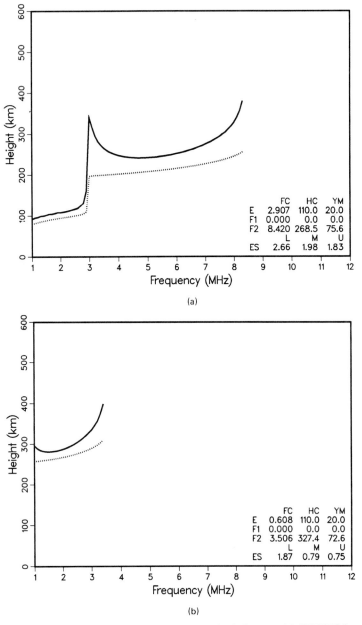

FIG. 24.10 Predicted ionograms as in Fig. 24.9, but in January. (*a*) 1800 UTC for day. (*b*) 0800 UTC for night.

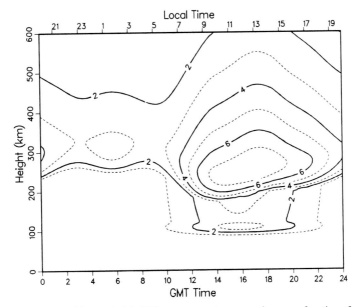

FIG. 24.11 Plasma- (critical-) frequency contours are given as a function of time of day for July; SSN = 50, latitude = 37.55°N, and longitude = 60.56°W.

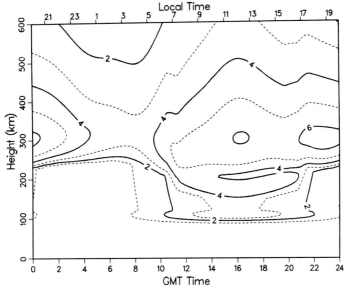

FIG. 24.12 Plasma-frequency contours are given as a function of time of day for January. Other controls are as in Fig. 24.11.

24.10 SKY-WAVE RADAR PERFORMANCE

The performance index used in this section will be the signal-to-noise ratio (*SNR*) indicated when using the transmission-medium model treated in Sec. 24.9 and the CCIR Report 322 noise discussed in Sec. 24.7. An additional need is a method to determine path effects. The detail with which path tracing is treated can vary greatly. A geometrical optics code that envolves integration along the ray as described by Jones and Stephenson[51] can provide paths in three dimensions, including delays and losses for both ordinary and extraordinary rays. When the details of electron distributions are uncertain, such comprehensive calculations are excessive. A number of other methods of path determination are available: for example, a path approximation technique goes with each of the ionospheric models mentioned in Sec. 24.9. For the performance exhibits treated here, the NRL-ITS Radar Performance Model will be used. This has been called RADAR C not quite correctly, but it is a lineal descendant of RADAR C.

NRL Memorandum Report 2500[52] describes the basic technique that will be used for path determinations. A simple closed-form virtual path trace, Snell's law for a spherically symmetric medium, is sequenced through elevation radiation angles in 1° increments. This process is incremented in 1-MHz steps over the radar's operating band. A vertical sounding of the ionosphere 700 km downrange has been used as the electron distribution for all one-hop paths, and a sounding 1400 km downrange is used for two-hop paths. Figures 24.9 and 24.10 gave a night and day example of the ionosphere 700 km downrange from a radar located at 38.65°N and 76.50°W looking east. Figure 24.13 gives constant plasma-frequency contours versus range from the radar for 0800 UTC, SSN 50, January

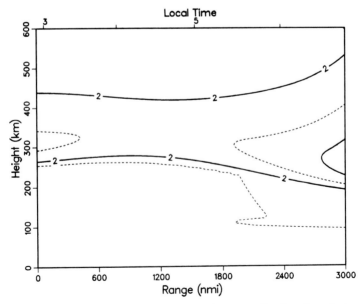

FIG. 24.13 Plasma-frequency contours are given, extending from the radar to a point east 3000 nmi downrange; January night example.

(night); and Fig. 24.14 for 1800 UTC (day). For the night case, the concentric spherical assumption from the 700-km downrange position will give paths that are slightly long for one-hop ranges. In the two-hop ranges the no-gradient assumption causes more distortion. In general, errors of this nature have little impact on performance prediction. However, near-real-time analysis for virtual range and azimuth correction to great-circle distance and bearing (grid registration) requires that tilt or gradient effects be taken into account. The daytime example has little horizontal gradient, and the simplifying assumption makes little difference. When better accuracy is desired, the correct vertical profile can be used for each radiation angle; also, gradients can be simulated by making the ionosphere nonconcentric with the earth. Both of these measures or something more complete should be used in radar performance assessment and management.

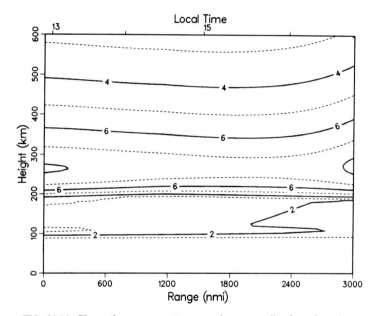

FIG. 24.14 Plasma-frequency contours are given, extending from the radar to a point east 3000 nmi downrange; January day example.

Figure 24.15 shows a performance prediction in the form of an oblique sounding. A typical sky-wave radar will be equipped with a vertical sounder and an oblique backscatter sounder for transmission-path analysis and to aid in radar-frequency management. Of course, the radar itself is an oblique sounder, but its sounding data is restricted to the frequency, waveform, and scan program of its primary surveillance task. An adjunct oblique sounder can present information in the form of Fig. 24.15 on earth backscatter echoes. In this prediction, SNR in decibels is plotted as a function of operating frequency and great-circle time delay or ground range. The numbers just above the abscissa (at 1-ms delay) are the noise powers in decibels below 1 W/Hz. For this plot the UTC time is 1800, SSN = 50, P_t = 200 kW, $G_t G_r$ = 50 dB, T = 1 s, and σ = 20 dBsm. Figure 24.16 gives the corresponding night plot. The shape of these displays is quite similar to what would be seen with a diagnostic oblique sounding; the levels would gener-

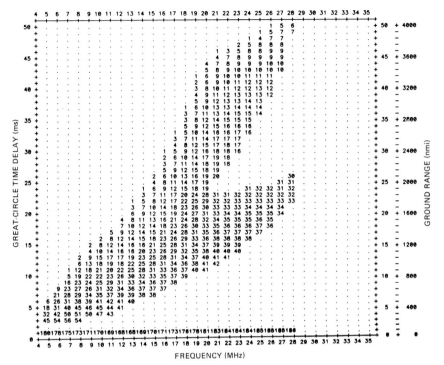

FIG. 24.15 *SNR* is given as a function of frequency and range in the form of a typical oblique backscatter sounding; January, 1800 GMT (day), SSN 50, location 38.65°N and 76.53°W, bearing 90°.

ally be greater since the resolution cell area times the surface scattering coefficient is generally much larger than 20 dBsm. Some of the night-day contrasts, such as available frequencies and difference in noise level for the same range, are evident. Also note that at night the 5-MHz lower frequency limit does not provide coverage closer than about 500 nmi. It should be remembered that this is a median SSN 50 calculation, and if consistent performance for ranges as close as 500 nmi is required during nights, a lower frequency limit should be selected to deal with periods of lower solar activity and the critical frequency distribution. The plots show that operation on a single frequency provides less than ±3 dB variation over a 500-nmi range interval. Also, if frequency selection had been made with a 2-MHz granularity instead of the 1 MHz used, the *SNR* would be reduced by only a decibel or so.

The performance-estimating aids that follow come from analyses as described above. After calculations as for oblique sounding, a range-ordered table of parameters is made. Parameter selections are made on the basis of the best *SNR* in each nominal 50-nmi interval, but the selection is adjusted to come from the adjacent lower frequency to avoid an optimistic bias. Then parameter plots are made as a function of range. The variables shown are losses, frequency, noise, and elevation radiation angle. The choice of range as the independent variable may seem artificial, but it is a useful approach for performance examination. With these curves the impact on radar *SNR* performance can be estimated for

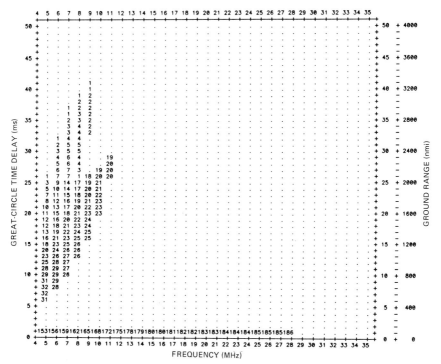

FIG. 24.16 *SNR* is given as a function of frequency and range in the form of an oblique sounding, as in Fig. 24.15, but for 0800 GMT (night).

selected antenna gain patterns, transmitted powers, target RCS, and coherent integration times (CIT). Figure 24.17 is an example for January with low solar activity in daytime. The R^4 loss is the fourth power of range to the target in meters given in decibels. $R^4 + L$ adds nondeviative absorption, deviative absorption, sporadic-E obscuration, and ground-reflection losses if there is more than one hop. The sharp increase in loss just before 2000 nmi is caused by transition from one to two hops; for two hops the lossy D region is transited twice as many times, ground-reflection loss is added, and required operation at a lower frequency increases loss. The jagged curve in the transition region is due to the parameter selection process; in radar operation the frequency would be selected to minimize transition effects. The frequency, radiation angle, and noise power per hertz that go with this site and look direction are also plotted. An example will be treated. It will be convenient to write the radar equation (24.1) in decibels:

$$SNR = P_{av} + G_t + G_r + \lambda^2 + T + \text{RCS} + F_p - (4\pi)^3 - (R^4 + L) - N$$

Select 1000 nmi as the range. Then the frequency is 17.5 MHz (wavelength = 17.1 m and $\lambda^2 = 25$ dB), noise power = -175 dB, and $R^4 + L = 261$ dB. Choose 53 dBW for P_{av}, 20 dB for G_t, 30 dB for G_r, 0 dBs for T, 20 dBsm for RCS, and 6 dB for F_p.

$$SNR = 53 + 20 + 30 + 25 + 0 + 20 + 6 - 33 - 261 - (-175) = 35 \text{ dB.}$$

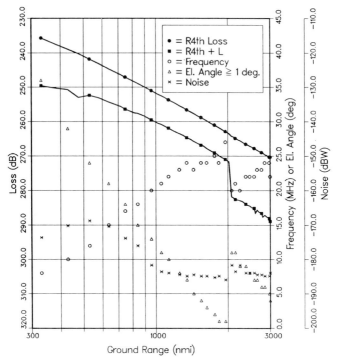

FIG. 24.17 Radar performance-controlling variables are given as a function of range; January, 1800 UTC, SSN 50.

Figure 24.18 shows the performance indicated with these assumptions for all ranges. A path factor enhancement of 6 dB has been chosen as an estimate of constructive multipath interference for an aircraft target that is effective in target detection. The beamwidth has been taken to be 5.7° and the surface scattering coefficient to be −35 dB, and with a 12 dB path enhancement the clutter level has been plotted. The clutter-to-noise ratio (*CNR*) at 1000 nmi is about 82 dB. For the constant beamwidth assumed, the clutter-to-signal ratio increases with range and is 47 dB at 1000 nmi. Large clutter-to-signal ratios are typical of HF radar; some form of doppler filtering is used to separate targets from clutter.

In Figs. 24.19 through 24.26 the performance-estimating curves are given for four seasons, night and day, and high and low solar activity. The permissible frequency selection is set between 5 and 28 MHz, and antenna radiation is not considered below an elevation angle of 1°. The analyses were made for a radar off the mid-Atlantic coast of the United States and should be a good approximation for any location where transmission paths are through the middle magnetic latitudes. This OTH performance presentation can be used to decide on the antenna patterns and powers required for specific targets and missions, or it can be used to exhibit periods of enhanced or degraded performance for an existing design. In looking at the performance index curves with radar range as the independent variable,

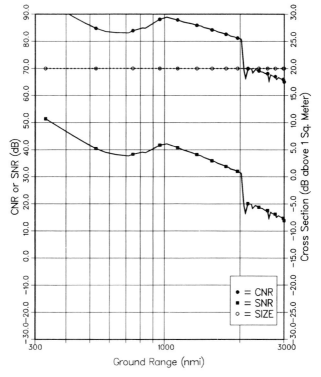

FIG. 24.18 A specific example of *SNR* and clutter-to-noise ratio (*CNR*) is determined by using Fig. 24.17. The target RCS is labeled "size" and is considered constant.

1. Summer shows much greater losses than other seasons.
2. Except for summer, night losses are only slightly less than day losses.
3. Night noise is much greater than day noise.
4. For a specific range, optimum frequencies vary by 3:1.

Several qualifiers should be kept in mind. At other geographic locations, the appropriate CCIR noise should be selected or, better yet, measured noise used. For radars that use auroral zone paths, specific analyses are required and target obscuration by spread-in-doppler clutter must be considered. The performance estimates from the figures assume that the radar design and waveforms are such that external noise is the control. The use of a single description for night and day gives a fair representation, but the transition from night to day is very abrupt and requires careful frequency management in radar operation. The ionospheric description that has been used is for what has been termed the *quiet ionosphere*; there will be a few hours per year when performance is very inferior to that predicted.

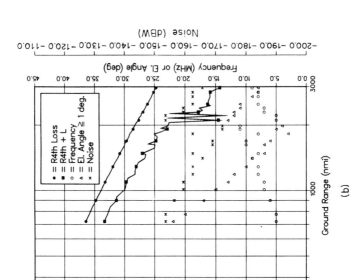

(b)

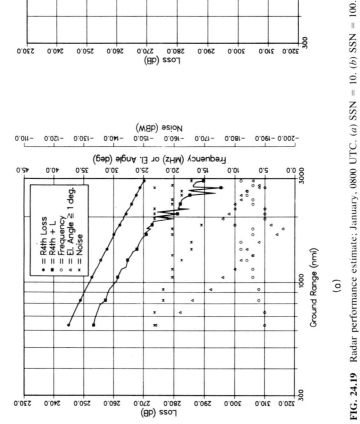

(a)

FIG. 24.19 Radar performance estimate; January, 0800 UTC. (*a*) SSN = 10. (*b*) SSN = 100.

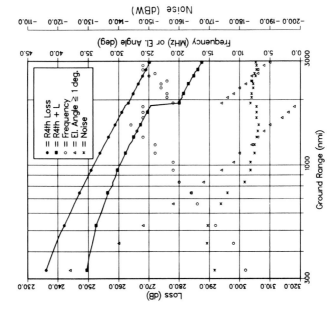

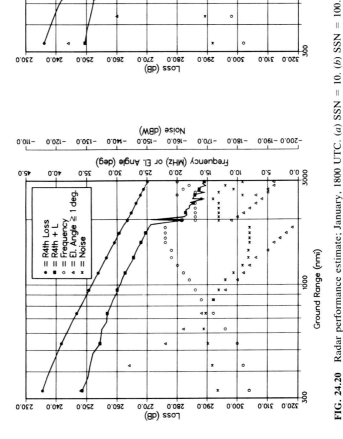

FIG. 24.20 Radar performance estimate; January, 1800 UTC. (*a*) SSN = 10. (*b*) SSN = 100.

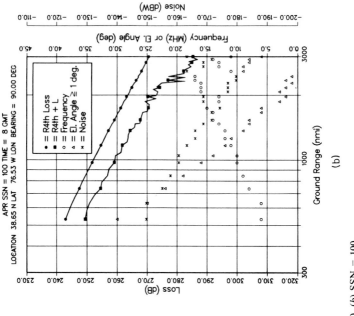

(b)

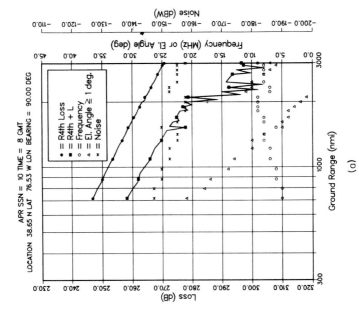

(a)

FIG. 24.21 Radar performance estimate; April, 0800 UTC. (*a*) SSN = 10. (*b*) SSN = 100.

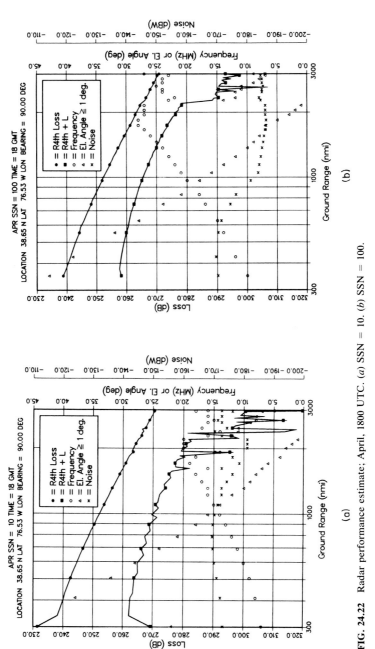

FIG. 24.22 Radar performance estimate; April, 1800 UTC. (a) SSN = 10. (b) SSN = 100.

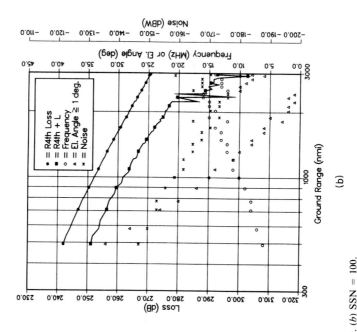

(b)

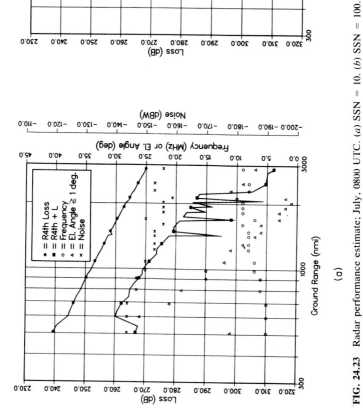

(a)

FIG. 24.23 Radar performance estimate; July, 0800 UTC. (*a*) SSN = 10. (*b*) SSN = 100.

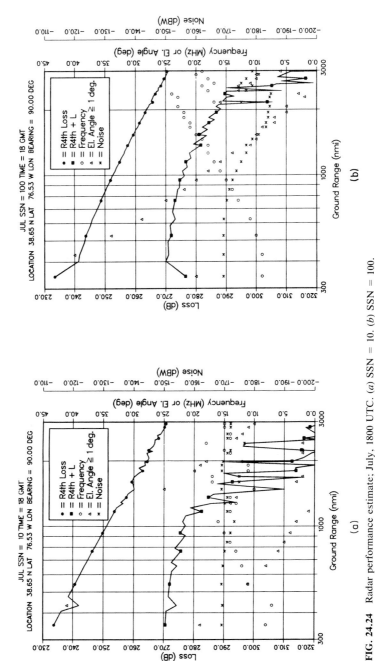

FIG. 24.24 Radar performance estimate; July, 1800 UTC. (a) SSN = 10. (b) SSN = 100.

24.33

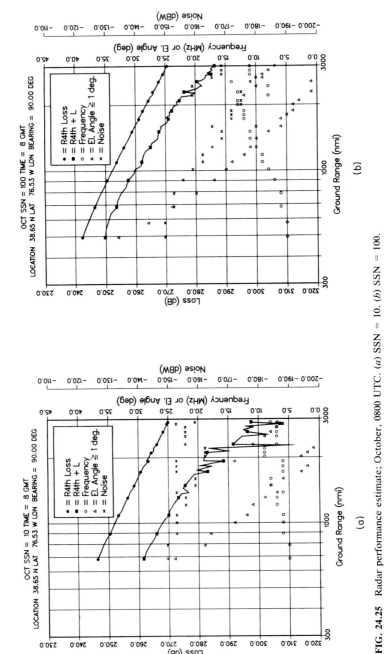

FIG. 24.25 Radar performance estimate: October, 0800 UTC. (*a*) SSN = 10. (*b*) SSN = 100.

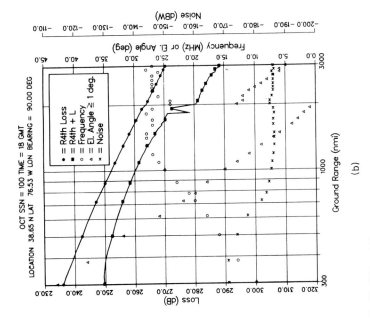

(b)

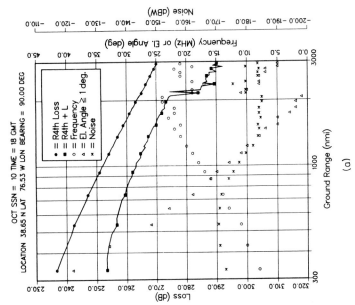

(a)

FIG. 24.26 Radar performance estimate; October, 1800 UTC. (*a*) SSN = 10. (*b*) SSN = 100.

24.11 RECEIVER-PROCESSOR

The prediction method indicated above is based upon long-term medians of vertical soundings, measured path losses, and observed noise. The statistical distributions are for a particular hour over days of the month or season. These kinds of statistics are insufficient to define the requirements for detection and tracking. An example will be given on the basis of a long dwell on a target using a constant frequency. The data will be used to indicate the required dynamic range and processor size and to show the input of the detection and tracking process. The amplitude levels in Fig. 24.27 are given in decibels relative to an arbitrary reference. Figure 24.27a gives a short time history of received power amplitude versus doppler frequency in one range gate. The waveform repetition frequency (wrf) was 20 Hz. Noise (N) samples were taken at wrf/2, target samples (T) on a target peak, and approach (A) and recede (R) on the resonant ocean wave peaks; N, T, A, and R are plotted in Fig. 24.27b. For this processing with a CIT of 12.8 s, the doppler filter bandwidth is a nominal 0.08 Hz, and at least 256 doppler filters should be used. The distance between the minimum noise points and the maximum clutter points is of the order of 100 dB, which indicates the dynamic range requirement if small targets are to be seen. For digital processing, an analog-to-digital (A/D) converter of at least 16-bit accuracy is in order. If data is processed

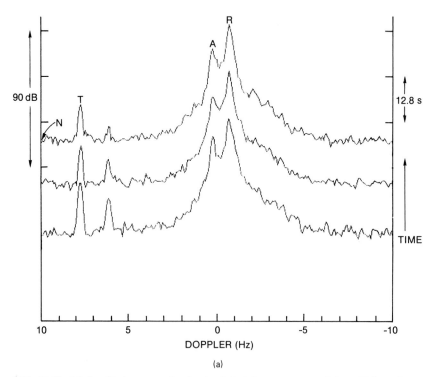

(a)

FIG. 24.27 (a) Amplitude versus doppler is plotted for a sequence of three 12.8-s coherent dwells. Indicated are a target marked T, the approach and recede resonant sea clutter peaks marked A and R, and the position for taking the noise sample marked N.

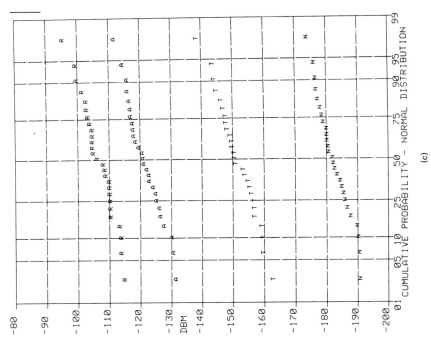

(c)

FIG. 24.27 (*Continued*) (*c*) Amplitude distributions of target, clutter peaks, and noise are plotted.

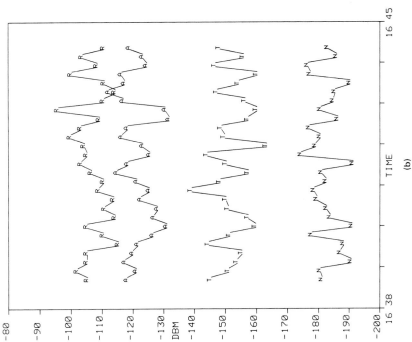

(b)

FIG. 24.27 (*Continued*) (*b*) Target, clutter peaks, and noise are plotted versus time for a longer sequence of data as in *a*.

24.37

for a radar operating as shown in Fig. 24.1, there will be a 500-nmi range extent at a 10-kHz resolution bandwidth requiring about 60 range gates; I and Q processing to expose unambiguously target dopplers between $\pm \mathrm{wrf}/2$ requires a sample rate of 40 kHz. For the Fig. 24.1 example, 16 simultaneous receiver-processor channels are required for the multiple receive beams. Figure 24.27c gives the corresponding power-level distributions. These approximately log-normal distributions are typical. The wide-area surveillance application makes automatic detection and tracking very desirable; the single transmitter footprint shown in Fig. 24.1 has 800 receive range-azimuth cells. Tracker requirements differ from those of other sensors in that it is generally necessary to have thresholds that permit many natural responses. It is common practice to defer target declaration until a track is recognized and thereby to reduce the false-alarm rate.

24.12 GROUND-WAVE RADAR PERFORMANCE

Ground-wave propagation as defined here will include all but sky-wave paths. That is, the paths or illumination considered are direct line-of-sight and by sea-

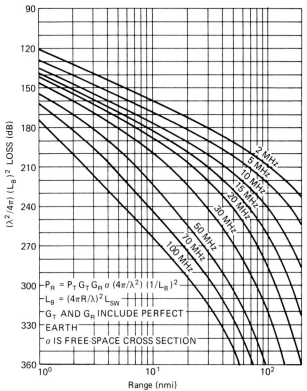

FIG. 24.28 Curves for estimating ground-wave radar performance are given as a function of range and are parametric in frequency. The surface is considered smooth, target and antenna heights are 2 m, conductivity is assumed to be 5 S/m, and the dielectric constant is 80.

surface reflection when radar and target are above the horizon and illumination in the penumbra and shadow region by a surface-attached wave. The feature that causes this propagation to be considered is that vertical polarization provides useful illumination down to the sea surface beyond the optical horizon. Figure 24.28 gives an example of ground-wave radar performance, parametric in frequency for the case in which both the radar antenna and the target are near the sea surface. These curves are for a smooth surface and use a ⁴⁄₃ earth radius to approximate atmospheric refraction effects. The propagation code is due to Berry

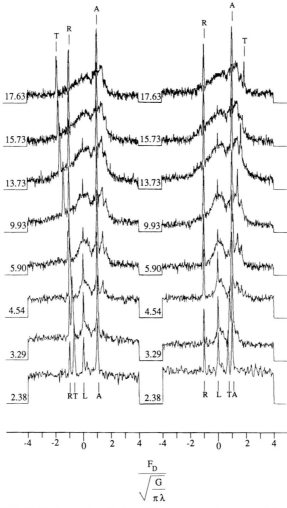

$$\frac{F_D}{\sqrt{\dfrac{G}{\pi\lambda}}}$$

FIG. 24.29 Examples of a target T are shown when approaching (right) and receding (left) in the presence of the sea echo. The format is of received power versus normalized doppler for seven radar operating frequencies. The resonant wave or Bragg peaks are marked A for approaching and R for receding. The peak at zero frequency is due to a stationary target in an antenna sidelobe.

and Chrisman,[53] and it is quite flexible, permitting antenna and target altitudes, surface conductivity and permittivity, polarization, and frequency to be specified. Sea roughness can be taken into account by using the work of Barrick.[54] Path-loss descriptors such as shown in Fig. 24.28 can be used to estimate performance. For example, consider a radar at 5 MHz with an average power of 10 kW (40 dBW), a transmit-receive antenna gain product of 15 dB, and a target at 100 nmi with an RCS of 20 dBsm; then the received power

$$P_r = 40 + 15 + 20 - 222 = -147 \text{ dBW}$$

By using the January nighttime noise as given in Fig. 24.6b,

$$SNR = P_r - N = -147 + 153 = 6 \text{ dB}$$

And if 10-s coherent processing time is used,

$$SNR = 16 \text{ dB}$$

Figure 24.29 provides a display of a 13-kn target and the sea echo as seen by a ground-wave radar. This is a family of received power versus doppler frequency plots over seven operating frequencies and for a target both approaching and receding. The abscissa units are in doppler normalized to the resonant wave or Bragg frequency; therefore, the resonant wave responses peak at ± 1. The amplitude range for each plot is 60 dB. The peaks at zero-doppler frequency are due to land in an antenna sidelobe. The target doppler coincides with a resonant line at 4.93 Mhz; it is between the lines below that frequency and outside them for frequencies above. The approach resonant wave peak is about 20 dB larger than the recede peak, indicating an almost direct sea into the radar. The processing used in developing these displays was 200-s CIT and 30-min averaging. This order of time for processing is appropriate for surface-target speeds and ground-wave radar; it also shows in doppler detail the form of the sea echo.

REFERENCES

1. Headrick, J. M., and M. I. Skolnik: Over-the-Horizon Radar in the HF Band, *Proc. IEEE*, vol. 62, pp. 664–673, June 1974.

2. Barnum, J. R.: Ship Detection with High Resolution HF Skywave Radar, *IEEE J. Ocean. Eng.*, vol. OE-11, pp. 196–210, April 1986.

3. Boutacoff, D. A.: Backscatter Radar Extends Early Warning Times, *Def. Electron.*, vol. 17, pp. 71–83, May 1985.

4. Sinnott, D. H.: The Jindalee Over-the-Horizon Radar System, *Conf. Air Power in the Defence of Australia*, Australian National University, Research School of Pacific Studies, Strategic and Defence Studies Centre, Canberra, July 14–18, 1986.

5. Wylder, J.: The Frontier for Sensor Technology, *Signal*, vol. 41, pp. 73–76, March 1987.

6. Guest editorial and invited papers in special issue on high-frequency radar for ocean and ice mapping and ship location: *IEEE J. Ocean. Eng.*, vol. OE-11, April 1986.

7. Kolosov, A. A. (ed.): "Fundamentals of Over-the-Horizon Radar," in Russian, *Radio i svyaz*, 1984. Also a translation by W. F. Barton, Artech House, Norwood, Mass., 1987.

8. Greenwald, R. A., K. B. Baker, R. A. Hutchins, and C. Hanuise: An HF Phased Array Radar for Studying Small-Scale Structure in the High Latitude Ionosphere, *Radio Sci.*, vol. 20, pp. 63–79, January–February 1985.

9. ITT Avionics Division, Electro-Physics Laboratories: EPL Model ATL-75 Transmitter for Radar and Communication, *IR&D Project Rept*. 274, *Results of Performance Measurements*, January 1975.

10. Hoft, D. J., and Fuat Agi: Solid State Transmitters for Modern Radar Applications, *CIE Int. Radar Conf. Rec.*, pp. 775–781, Nov. 4–7, 1986.

11. Guest editorial and invited papers in special issue on shortwave broadcasting, *IEEE Trans. Broadcast.*, vol. 34, June 1988.

12. Sweeney, L. E.: Spatial Properties of Ionospheric Radio Propagation as Determined with Half-Degree Azimuthal Resolution, *Stanford Electron. Lab. Tech. Rept.* 155 *SU-SEL-70-034*, Stanford University, June 1970.

13. Lynch, J. T.: Aperture Synthesis for HF Radio Signals Propagated via the F-Layer of the Ionosphere, *Stanford Electron. Lab. Tech. Rept.* 161 *SU-SEL-70-066*, Stanford University, September 1970.

14. Johnson, R. C., and H. Jasik (eds.): "Antenna Engineering Handbook," 2d ed., McGraw-Hill Book Company, New York, 1984.

15. Kurashov, A. G. (ed.): "Shortwave Antennas," 2d ed., in Russian, *Radio i svyaz*, January 1985.

16. Rice, S. O.: Reflection of Electromagnetic Waves from Slightly Rough Surfaces, in Kline, M. (ed.): "Theory of Electromagnetic Waves," Interscience Publishers, New York, 1951, pp. 351–378.

17. Pierson, W. J., G. Neumann, and R. W. James: Practical Methods for Observing and Forecasting Ocean Waves by Means of Wave Spectra and Statistics, *H.O. Pub.* 603, chap. 1, Hydrographic Office, U.S. Navy, 1960.

18. Crombie, D. D.: Doppler Spectrum of the Sea Echo at 13.56 Mcs, *Nature*, vol. 175, pp. 681–682, 1955.

19. Barrick, D. E.: First Order Theory and Analysis of MF/HF/VHF Scatter from the Sea, *IEEE Trans.*, vol. AP-20, pp. 2–10, January 1972.

20. Pierson, W. J., and L. Moskowitz: A Proposed Spectral Form for Fully Developed Wind Seas Based on the Similarity Theory of S. A. Kitaigordskii, *J. Geophys. Res.*, vol. 69, no. 24, pp. 5181–5190, 1964.

21. Barrick, D. E., J. M. Headrick, R. W. Bogle, and D. D. Crombie: Sea Backscatter at HF: Interpretation and Utilization of the Echo, *Proc. IEEE*, vol. 62, pp. 673–680, June 1974.

22. Clancy, R. M., J. E. Kaitala, and L. F. Zambresky: The Fleet Numerical Oceanography Center Global Spectral Ocean Wave Model, *Bull. Am. Meteorol. Soc.*, vol. 67, no. 5, May 1986.

23. Long, A. E., and D. B. Trizna: Mapping of North Atlantic Winds by HF Radar Sea Backscatter Interpretation, *IEEE Trans.*, vol. AP-21, pp. 680–685, September 1973.

24. Ahearn, J. L., S. R. Curley, J. M. Headrick, and D. B. Trizna: Tests of Remote Skywave Measurment of Ocean Surface Conditions, *Proc. IEEE*, vol. 62, pp. 681–686, June 1974.

25. Anderson, S. J.: Remote Sensing with the Jindalee Skywave Radar, *IEEE J. Ocean. Eng.*, vol. OE-11, pp. 158–163, April 1986.

26. Trizna, D. B., J. C. Moore, J. M. Headrick, and R. W. Bogle: Directional Sea Spectrum Determination Using HF Doppler Radar Techniques, *IEEE Trans.*, vol. AP-25, pp. 4–11, January 1977.

27. Barrick, D. E.: Extraction of Wave Parameters from Measured HF Radar Sea-Echo Spectra, *Radio Sci.*, vol. 12, no. 3, p. 415, 1977.

28. Lipa, B.: Derivation of Directional Ocean-Wave Spectra by Integral Inversion of Second-Order Radar Echoes, *Radio Sci.*, vol. 12, no. 3, p. 425, 1977.

29. Trizna, D. B.: Estimation of the Sea Surface Radar Cross Section at HF from Second-Order Doppler Spectrum Characteristics, *Naval Res. Lab. Rept.* 8579, May 1982.

30. Pilon, R. O., and J. M. Headrick: Estimating the Scattering Coefficient of the Ocean Surface for High-Frequency Over-the-Horizon Radar, *Naval Res. Lab. Memo. Rept.* 5741, May 1986.

31. Trizna, D. B., and J. M. Headrick: Ionospheric Effects on HF Over-The-Horizon Radar, in Goodman, J. M. (ed.): *Proc. Effect Ionosphere on Radiowave Syst.*, ONR/AFGL-sponsored, pp. 262–272, Apr. 14–16, 1961.

32. Elkins, T. J.: A Model for High Frequency Radar Auroral Clutter, *RADC Rept. TR-80-122*, March 1980.

33. Burke, G. J., and A. J. Poggio: Numerical Electromagnetic Code (NEC)—Method of Moments, *NOSC Tech. Doc.* 116, 1981.

34. Walton, E. K., and J. D. Young: The Ohio State University Compact Radar Cross Section Measurement Range, *IEEE Trans.*, vol. AP-32, pp. 1218–1223, November 1984.

35. Bogle, R. W., and D. B. Trizna: Small Boat Radar Cross Sections, *Naval Res. Lab. Memo. Rept.* 3322, July 1976.

36. CCIR (International Radio Consultative Committee): World Distribution and Characteristics of Atmospheric Radio Noise, *CCIR Rept.* 322, International Telecommunications Union, 1964.

37. Spaulding, A. D., and J. S. Washburn: Atmospheric Radio Noise: Worldwide Levels and Other Characteristics, *NTIA Rept.* 85-173, National Telecommunications and Information Administration, April 1985.

38. Lucas, D. L., and J. D. Harper: A Numerical Representation of CCIR Report 322 High Frequency (3–30 Mcs) Atmospheric Radio Noise Data, *Nat. Bur. Stand. Note* 318, Aug. 5, 1965.

39. Ortenburger, L. N., D. A. Schaefer, F. W. Smith, and A. J. Kramer: Prediction of HF Noise Directivity from Thunderstorm Probabilities, *GTE Sylvania Rept. EDL-M*1379, 1971.

40. Lucas, D. L.: Predictions of Backscatter Clutter Power in the Radar C Computer Program, final report on NRL Contract N0014-84-C-2451; CU5-36903, University of Colorado, Boulder, June 9, 1986.

41. Davies, K.: "Ionospheric Radio Propagation," *Nat. Bur. Stand. Monog.* 80, Apr. 1, 1965.

42. Jursa, A. S. (ed.): "Handbook of Geophysics and the Space Environment," Air Force Geophysics Laboratory, AFSC, U.S. Air Force, 1985.

43. Lucas, D. L., and G. W. Haydon: Predicting Statistical Performance Indexes for High Frequency Telecommunications Systems, *ESSA Tech. Rept. IER* 1 *ITSA* 1, U.S. Department of Commerce, 1966.

44. Barghausen, A. L., J. W. Finney, L. L. Proctor, and L. D. Schultz: Predicting Long-Term Operational Parameters of High-Frequency Sky-Wave Communications Systems, *ESSA Tech. Rept. ERL* 110-*ITS* 78, U.S. Department of Commerce, 1969.

45. Headrick, J. M., J. F. Thomason, D. L. Lucas, S. McCammon, R. Hanson, and J. Lloyd: Virtual Path Tracing for HF Radar Including an Ionospheric Model, *Naval Res. Lab. Memo. Rept.* 2226, March 1971.

46. Teters, L. R., J. L. Lloyd, G. W. Haydon, and D. L. Lucas: Estimating the Performance of Telecommunication Systems Using the Ionospheric Transmission Channel—

Ionospheric Communications Analysis and Prediction Program Users Manual, *Nat. Telecom. Inf. Adm. NTIA Rept.* 83-127, July 1983.

47. Hatfield, V. E.: HF Communications Predictions 1978 (An Economical Up-to-Date Computer Code, AMBCOM), *Solar Terrestrial Preduction Proc.*, vol. 4, in Donnelley, R. F. (ed.): "Prediction of Terrestrial Effects of Solar Activity," National Oceanic and Atmospheric Administration, 1980.

48. Lucas, D. L.: Ionospheric Parameters Used in Predicting the Performance of High Frequency Skywave Circuits, Interim Report on NRL Contract N00014-87-K-20009, Account 153-6943, University of Colorado, Boulder, Apr. 15, 1987.

49. Thomason, J., G. Skaggs, and J. Lloyd: A Global Ionospheric Model, *Naval Res. Lab. Rept.*, 8321, Aug. 20, 1979.

50. Miller, D. C., and J. Gibbs: Ionospheric Analysis and Ionospheric Modeling, *AFCRL Tech. Rept.* 75-549, July 1975.

51. Jones, R. M., and J. J. Stephenson: A Versatile Three-Dimensional Ray Tracing Computer Program for Radio Waves in the Ionosphere, *Office Telecom. Rept.* 75-76, October 1975.

52. Lucas, D. L., J. L. Lloyd, J. M. Headrick, and J. F. Thomason: Computer Techniques for Planning and Management of OTH Radars, *Naval Res. Lab. Memo. Rept.* 2500, September 1972.

53. Berry, L. A., and M. E. Chrisman: A FORTRAN Program for Calculation of Ground Wave Propagation Over Homogeneous Spherical Earth for Dipole Antennas, *Nat. Bur. Stand. Rept.* 9178, 1966.

54. Barrick, D. E.: Theory of HF and VHF Propagation across the Rough Sea, pts. 1 and 2, *Radio Sci.*, vol. 6, pp. 517–533, May 1971.

CHAPTER 25
BISTATIC RADAR

Nicholas J. Willis
Technology Service Corporation

25.1 CONCEPT AND DEFINITIONS

Bistatic radar employs two sites that are separated by a considerable distance. A transmitter is placed at one site, and the associated receiver is placed at the second site. Target detection is similar to that of monostatic radar: target illuminated by the transmitter and target echoes detected and processed by the receiver. Target location is similar to but more complicated than that of a monostatic radar: total signal propagation time, orthogonal angle measurements by the receiver, and some estimate of the transmitter location are required to solve the transmitter-target-receiver triangle, called the *bistatic triangle*. Continuous-wave (CW) waveforms can often be used by a bistatic radar because site separation, possibly augmented by sidelobe cancellation, provides sufficient spatial isolation of the *direct-path* transmit signal.

When separate transmit and receive antennas are at a single site, as is common in CW radars, the term *bistatic* is not used to describe such a system since the radar has characteristics of a monostatic radar. In special cases, the antennas can be at separate sites and the radar is still considered to operate monostatically. For example, an over-the-horizon (OTH) radar can have site separation of 100 km or more. But that separation is small compared with the target location of thousands of kilometers,[1,2] and the radar operates with monostatic characteristics.

When two or more receive sites with common spatial coverage are employed and target data from each site is combined at a central location, the system is called a *multistatic radar*. Thinned, random, distorted, and distributed arrays,[3–6] interferometric radars,[7–10] the radio camera,[11,12] and the multistatic measurement system[13,14] are sometimes considered a subset of multistatic radars. They usually combine data coherently from each receiver site to form a large receive aperture. Multiple transmitters can be used with any of these configurations. They can be located at separate sites or colocated with the receive sites. Three range-only monostatic radars combined in a radar net are sometimes called a *trilateration radar*. The trilateration concept applies to multistatic radars that measure target location by time-difference-of-arrival (TDOA) or differential doppler techniques.

The foregoing definitions are broad and traditional[1,15,16] but are by no means uniformly established in the literature. Terms such as *quasi-bistatic, quasi-*

monostatic, pseudo-monostatic, tristatic, polystatic, real multistatic, multi-bistatic, and *netted bistatic* have also been used.[17-20] They are usually special cases of the broad definitions given above.

Passive receiving systems, or electronic support measure (ESM) systems, often use two or more receiving sites. Their purpose is typically to detect, identify, and locate transmitters such as monostatic radars. They are also called *emitter locators*. Target location is by means of combined angle measurements from each site (e.g., triangulation), TDOA, and/or differential doppler measurements between sites. These systems usually are not designed to detect and process the echoes from targets illuminated by the transmitter. They can, however, be used with a bistatic or multistatic radar to identify and locate a suitable transmitter to initialize radar operations. Thus, while they have many requirements and characteristics common to multistatic radars, they are not radars and will not be considered here.

25.2 HISTORY

Early experimental radars in the United States, the United Kingdom, France, the Soviet Union, Germany, and Japan were of the bistatic type, where the transmitter and receiver were separated by a distance comparable to the target distance.[21-26] These bistatic radars used CW transmitters and detected a beat frequency produced between the direct-path signal from the transmitter and the doppler-frequency-shifted signal scattered by a moving target. This effect was called CW wave interference.[1] The geometry was similar to that of the forward-scatter (or near-forward-scatter) configuration, where the target position is near the baseline joining transmitter and receiver. Much of the early bistatic radar technology was derived from existing communications technology—separated sites, CW transmissions, and frequencies ranging from 25 to 80 MHz.[27] These early bistatic radars were typically configured as fixed, ground-based fences to detect the *presence* of aircraft: a major, emerging threat in the 1930s. The problem of extracting target *position* information from such radars could not readily be solved with techniques available at the time.[1]

Many of the early United States bistatic radar experiments were conducted by the Naval Research Laboratory (NRL).[1] In 1922 NRL researchers detected a wooden ship using a CW wave interference radar operating at 60 MHz. An NRL proposal for further work was rejected. In 1930 an aircraft was accidentally detected when it passed through a 33-MHz direction-finding beam received by an aircraft on the ground. Interest was revived, and in 1932 CW wave interference equipment detected an aircraft up to 80 km from the transmitter. In 1934 this work was disclosed in a patent, granted to Taylor, Young, and Hyland.[21]

In the Soviet Union an operational system, the RUS-1, evolved from an experimental bistatic CW radar.[24] By the time of the German invasion in 1941, 45 systems had been built and deployed to the Far East and the Caucasus. They were subsequently replaced by the RUS-2 and RUS-2C, both pulsed radars. The RUS-2 used two trucks, one for the transmitter and one for the receiver, separated by about 300 m to provide receiver isolation. Although the RUS-2 used two sites, separation was not sufficient to define the configuration as bistatic. The French also deployed a bistatic CW radar in a two-fence configuration prior to World War II, thus providing a coarse estimate of target course and speed.[24]

The Japanese deployed about 100 bistatic CW radar fences, called Type A, starting in 1941.[26] These remained in use until the end of World War II. Type A operated between 40 and 80 MHz with 3 to 400 W of transmitter power. Maximum detection ranges of up to 800 km on aircraft were achieved, with one system operating between Formosa (Taiwan) and Shanghai. Target location along the forward-scatter baseline was never achieved with this system.

A variation of these fence configurations was developed by the Germans during World War II.[17] They built a bistatic receiver, known as the Klein Heidelberg, that used a British Chain Home radar as the transmitter. The receiver gave warning of the onset of Allied bombing raids when the planes were over the English Channel, without endangering the German ground sites. This bistatic radar appears to be the first operational configuration to use a noncooperative transmitter.

The Chain Home radars themselves operated with separate transmitter and receiver sites, but again with separation small compared with target distance. However, they had a standard, reversionary mode in which, in the presence of electronic countermeasures (ECM) or a transmitter failure, a receiver site could operate with a transmitter at an adjacent site, hence becoming bistatic.[28]

The invention of the duplexer at NRL in 1936 provided a means of using pulsed waveforms with a common transmit and receive antenna. This single-site configuration is the familiar monostatic radar, and it greatly expanded the utility of radar, particularly for use by aircraft, ships, and mobile ground units. As a consequence bistatic radars became dormant.

It was not until the early 1950s that interest in bistatic radars was revived for aircraft detection.[1,29–31] The United States AN/FPS-23 was designed as a gap-filler fence for the Distant Early Warning (DEW) line in the arctic. It was installed in the mid-1950s but was later removed.[24] The Canadians also developed a bistatic radar for their McGill fence.[29,32] The United States Plato and Ordir ballistic missile detection systems were designed as the first multistatic radars; they combined range sum and doppler information from each receiver site to estimate target position. They were not deployed.[24,32]

The Azuza, Udop, and Mistram interferometeric radars, a variant of multistatic radars, were installed at the United States Eastern Test Range for precision measurement of target trajectories. They used a single CW transmitter, multiple receivers at separate, precisely located sites, and cooperative beacon transponders on the target.[9,10] The SPASUR, a satellite fence interferometric radar, was also implemented with a single CW transmitter and multiple receivers but with enough performance to detect satellite-skin echoes.[7,8]

A major development at this time was the semiactive homing missile seeker, in which the large, heavy, and costly transmitter could be off-loaded from the small, expendable missile onto the launch platform (Chap. 19). While these seekers are clearly a bistatic radar configuration, missile engineers have developed a different lexicon to describe their technology and operation, e.g., semiactive versus bistatic, illuminator versus transmitter, rear reference signal versus direct-path signal, etc. The missile and radar communities continue to go their separate ways.

In the 1950s and early 1960s bistatic radar system theory was codified.[15] Bistatic radar cross-section theory was developed, and measurements were taken.[33–41] Bistatic clutter measurements were also taken.[42,43] The name *bistatic radar* was coined by K. M. Siegel and R. E. Machol in 1952.[34]

Bistatic radars received renewed interest in the 1970s and 1980s as counters to retrodirective jammers and attacks by antiradiation missiles (ARMs). Retrodirective jamming levels can be reduced by selecting a geometry such that the receive site lies outside the jammer's main beam, which is directed at the transmit site. The effectiveness of an ARM attack can be reduced by removing the transmitter from the battle area into a "sanctuary," which is less vulnerable to attack. Several air defense field test programs explored these capabilities and the problems inherent in bistatic operation, such as time synchronization, coverage, and clutter suppression.[18,44–50]

Other bistatic radar concepts were identified and tested at this time, such as *clutter tuning* from an airborne transmitter and receiver.[51–53] One potential implementation of this concept allows the receiver to generate a synthetic aperture radar (SAR) map of modest resolution directly on its velocity vector—an impossible task for the monostatic SAR. Clutter tuning combined with the sanctuary concept protects the transmitter while allowing the receiver platform to fly toward the target with no radar emissions.

The concept of using a small bistatic receiver that "hitchhikes" off airborne radars was also developed and successfully tested.[54] It alerts and cues autonomous short-range air defense and ground surveillance systems to improve survivability and acquisition performance. This hitchhiking concept was extended to other transmitters of opportunity, including a commercial television station that served as a bistatic transmitter. Initial attempts to detect aircraft were only marginally successful.[55]

Bistatic radars using space-based transmitters and receivers that are either space-based, airborne, or ground-based have been studied.[3,56–59] Limited field tests were conducted by using a communication satellite as the transmitter and a ground-based receiver to detect aircraft.[58] Since the effective radiated power of the satellite was modest and the transmitter-to-target ranges were large, detection ranges were small, <4 km, unless a very large receive aperture was used.

A pulse doppler bistatic radar was developed and tested to protect military aircraft on the ground from intruders.[60] It was configured for near-forward-scatter operation. Five small portable transmitter-receiver units, typically separated by 65 m, were located around the aircraft, with one transmitter servicing an adjacent receiver. In field tests the radar detected moving targets, including high-speed vehicles and intruders creeping at 2 cm/s.

The Multistatic Measurement System (MMS) was installed at the United States Kwajalein Missile Range in 1980 to track ballistic missile skin echoes.[14] The TRADEX L-band and ALTAIR ultrahigh-frequency (UHF) monostatic radars are used to illuminate the targets, and the bistatic echoes, collected at two unmanned stations located about 40 km from the radars, are combined coherently at a central site. The system is projected to measure three-dimensional position and velocity with accuracies better than 4 m and 0.1 m/s, respectively, throughout reentry.[13]

Other multistatic radar concepts have been studied. They include the Doppler Acquisition System (DAS), which used multiple transmitters and receivers,[61] and Distributed Array Radar (DAR) concepts, with large[3] and small[5] spatial separation between receive sites. The DAS combines data from each site noncoherently; the DAR, coherently.

Bistatic radars have been analyzed, proposed, and in some cases developed for other than military applications. Such applications include high-resolution imaging at short ranges (in the near field of the antennas) for use by robotics in an industrial environment;[62] airport ground vehicle and aircraft collision warning and avoidance using a baseband bistatic radar;[63] planetary surface and environment measurements using a satellite-based transmitter and an earth-based receiver[64–67] or a planet-based transmitter and a satellite-based receiver;[68] geological probing of horizontally stratified, underground layers from a transmitter and receiver on the surface, usually operating at frequencies from 100 to 1000 MHz;[69] ocean wave spectral measurements (wavelength, frequency, and direction of travel) using a Loran-A system;[70] and detection and soundings of tropospheric layers, ionospheric layers, and high-altitude, clear-air atmospheric targets using ground-based sites.[16,71,72]

25.3 COORDINATE SYSTEM

A two-dimensional north-referenced coordinate system[73] is used throughout this chapter. Figure 25.1 shows the coordinate system and parameters defining bistatic radar operation in the x,y plane. This is sometimes called the *bistatic plane*.[74] The bistatic triangle lies in the bistatic plane. The distance L between the transmitter and the receiver is called the *baseline range* or simply the *baseline*. The angles θ_T and θ_R are, respectively, the transmitter and receiver look angles. They are also called angles of arrival (AOA) or lines of sight (LOS). Note that the bistatic angle $\beta = \theta_T - \theta_R$. It is also called the cut angle or the scattering angle. It is convenient to use β in calculations of target-related parameters and θ_T or θ_R in calculations of transmitter- or receiver-related parameters. Development of three-dimensional bistatic coordinate systems for some applications is available elsewhere.[16,46,48,75,76]

A useful relationship is that the bisector of the bistatic angle is orthogonal to

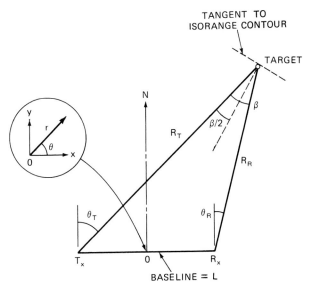

FIG. 25.1 Bistatic radar north coordinate system for two dimensions.[73]

the tangent of an ellipse with foci at the transmitter and receiver sites and passing through the target position. Such an ellipse is called an *isorange contour*. The tangent is often a good approximation to an isorange contour within the bistatic footprint, the area common to the transmit and receive beams.

Geometry often distinguishes bistatic from monostatic radar operation. In these distinguishing cases, equivalent monostatic operation is obtained by setting $L = O$, or $R_T = R_R$ and $\beta = 0$.

25.4 *RANGE RELATIONSHIPS*

Range Equation.[1,3,5,15,16,77,78] The range equation for a bistatic radar is derived in a manner completely analogous to that for a monostatic radar. With this analog, the bistatic radar maximum-range equation can be written as

$$(R_T R_R)_{\max} = \left(\frac{P_T G_T G_R \lambda^2 \sigma_B F_T^2 F_R^2}{(4\pi)^3 K T_s B_n (S/N)_{\min} L_T L_R} \right)^{1/2} \tag{25.1}$$

where R_T = transmitter-to-target range
$\quad\ R_R$ = receiver-to-target range
$\quad\ P_T$ = transmitter power
$\quad\ G_T$ = transmit antenna power gain
$\quad\ G_R$ = receive antenna power gain
$\quad\ \lambda$ = wavelength
$\quad\ \sigma_B$ = bistatic radar target cross section
$\quad\ F_T$ = pattern propagation factor for transmitter-to-target path
$\quad\ F_R$ = pattern propagation factor for target-to-receiver path
$\quad\ K$ = Boltzmann's constant
$\quad\ T_s$ = receive system noise temperature
$\quad\ B_n$ = noise bandwidth of receiver's predetection filter
$(S/N)_{\min}$ = signal-to-noise power ratio required for detection
$\quad\ L_T$ = transmit system losses (> 1) not included in other parameters
$\quad\ L_R$ = receive system loss (> 1) not included in other parameters

Equation (25.1) is related to the corresponding monostatic radar range equation by the following: $\sigma_M = \sigma_B$, $L_T L_R = L_M$, and $R_T^2 R_R^2 = R_M^4$. More specific formulations of the maximum-range equation, as given in Chap. 2, also apply to the bistatic radar case. Equation (25.1) is used in this chapter because it more clearly illustrates the utility of constant S/N contours (ovals of Cassini) and other geometric relationships. The right side of Eq. (25.1) is called the bistatic maximum-range product κ.

Ovals of Cassini. Equation (25.1), with $(R_T R_R)_{\max} = \kappa$, is the maximum-range oval of Cassini. It can be used to estimate the signal-to-noise S/N power ratio at any R_T and R_R simply by dropping the "max" and "min" designation for $(R_T R_R)$ and S/N respectively. Then when Eq. (25.1) is solved for S/N,

$$S/N = \frac{k}{R_T^2 R_R^2} \tag{25.2}$$

where S/N = signal-to-noise power ratio at ranges R_T, R_R, and

$$k = \frac{P_T G_T G_R \lambda^2 \sigma_B F_T^2 F_R^2}{(4\pi)^3 K T_S B_n L_T L_R} \tag{25.3}$$

The term k is the bistatic radar constant. The constants k and κ are related as

$$k = \kappa^2 (S/N)_{min} \tag{25.4}$$

Equation (25.2) represents one form of the ovals of Cassini. They can be plotted on the bistatic plane when R_T and R_R are converted to polar coordinates (r, θ), as shown on Fig. 25.1:

$$R_T^2 R_R^2 = (r^2 + L^2/4)^2 - r^2 L^2 \cos^2 \theta \tag{25.5}$$

where L is the baseline range. Figure 25.2 is such a plot for k arbitrarily set to $30L^4$.

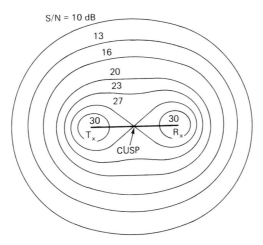

FIG. 25.2 Contours of constant signal-to-noise ratio, or ovals of Cassini, where the baseline $= L$ and $k = 30L^4$.

The ovals of Fig. 25.2 are contours of constant signal-to-noise ratio on any bistatic plane. They assume that an adequate line of sight (LOS) exists on the transmitter-to-target path and the receiver-to-target path and that σ_B, F_T, and F_R are invariant with r and θ, which is usually not the case. But this simplifying assumption is useful in understanding basic relationships and constraints. As S/N or L increases, the ovals shrink, finally collapsing around the transmitter and receiver sites. The point on the baseline where the oval breaks into two parts is called the cusp. The oval is called a lemniscate (of two parts) at this S/N. When $L = 0$, $R_T R_R = r^2$, which is the monostatic case where the ovals become circles.

Operating Regions. Ovals of Cassini define three distinct operating regions for a bistatic radar: receiver-centered region, transmitter-centered region, and receiver-transmitter-centered region, or simply the cosite region. Critical to the selection of these operating regions is the value of the bistatic radar constant k that is available. Many of the terms in Eq. (25.3) are transmitter-controlled. It is convenient to define three transmitter configurations that control k: dedicated, cooperative, and noncooperative. The *dedicated* transmitter is defined as being under both design and operational control of the bistatic radar system; the *cooperative* transmitter is designed for other functions but found suitable to support bistatic operations and can be controlled to do so; and the *noncooperative* transmitter, while suitable for bistatic operations, cannot be controlled. The bistatic receiver is sometimes said to hitchhike off a cooperative or noncooperative transmitter, usually a monostatic radar.

Table 25.1 summarizes useful bistatic radar applications permitted by operating regions and transmitter configurations. The two omitted entries on the "Transmitter-centered" row are operational constraints: a dedicated or cooperative transmitter can usually gather nearby data in a monostatic radar mode more easily than can a remote, bistatic receiver. The two omitted entries on the "Cosite" row are technical constraints: to generate a sufficiently large bistatic radar constant for cosite operation the transmitter design and operation must be optimized for bistatic radar use; hence the dedicated transmitter is often the only viable cosite configuration. Exceptions to this rule include exploiting HF groundwave propagation and occasional atmospheric ducting.

Isorange Contours. The transmitter-to-target-to-receiver range measured by a bistatic radar is the sum $(R_T + R_R)$. This sum locates the target somewhere on the surface of an ellipsoid whose two foci are the transmitter and receiver

TABLE 25.1 Bistatic Radar Applications

Bistatic radar operating regions	Range relationships	Transmitter configuration		
		Dedicated	Cooperative	Noncooperative
Receiver-centered	$R_T \gg R_R$ k small	• Air-to-ground attack (silent penetration) • Semiactive homing missile (lock on after launch)	• Short-range air defense • Ground surveillance • Passive situation awareness	• Passive situation awareness
Transmitter-centered	$R_R \gg R_T$ k small	. . . .	. . . .	• Intelligence data gathering • Missile launch alert
Cosite	$R_T \sim R_R$ k larger	• Medium-range air defense • Satellite tracking • Range instrumentation • Semiactive homing missile (lock on before launch) • Intrusion detection	. . . .	. . . .

sites. The intersection of the bistatic plane and this ellipsoid produces the familiar ellipses of constant range sum, or *isorange contours*.

Since the (constant range sum) isorange contours and the (constant S/N) ovals of Cassini are not colinear, the target's S/N will vary for each target position on the isorange contour. This variation can be important when target returns are processed over a bistatic range cell, defined by two concentric isorange contours with separation $\Delta R_B \approx c\tau/2 \cos(\beta/2)$, where τ = compressed pulse width. The S/N over an isorange contour, $(S/N)_i$ is

$$(S/N)_i = \frac{4k(1 + \cos\beta)^2}{[(R_T + R_R)^2 - L^2]^2} \tag{25.6}$$

where the denominator defines the isorange countour and the bistatic angle β defines the target's position on the isorange contour.

The maximum bistatic angle, β_{max}, on an isorange contour is $2\sin^{-1}[L/(R_T + R_R)]$, where $L/(R_T + R_R)$ is the eccentricity of the isorange contour. The minimum bistatic angle, β_{min}, is zero for all isorange contours. For example, when $L/(R_T + R_R)] = 0.95$, $\beta_{max} = 143.6°$ and $(S/N)_i$ at β_{max} is 20 dB less than at β_{min}.

25.5 AREA RELATIONSHIPS

Location.[1,15,16,18,46–48,73,79–83] Target position relative to the receive site (θ_R, R_R) is usually required in a bistatic radar. The receiver look angle θ_R is measured directly, or target azimuth and elevation measurements are converted directly to θ_R. Beam-splitting techniques can be used to increase the measurement accuracy.

The receiver-to-target range R_R cannot be measured directly, but it can be calculated by solving the bistatic triangle (Fig. 25.1). A typical solution in elliptical coordinates is[1]

$$R_R = \frac{(R_T + R_R)^2 - L^2}{2(R_T + R_R + L \sin\theta_R)} \tag{25.7}$$

The baseline L can be calculated from coordinates provided by a dedicated transmitter or measured by an emitter location system. The range sum $(R_T + R_R)$ can be estimated by two methods. In the direct method the receiver measures the time interval ΔT_{rt} between reception of the transmitted pulse and reception of the target echo. It then calculates the range sum as $(R_T + R_R) = c\Delta T_{rt} + L$. This method can be used with any transmitter configuration, given an adequate LOS between transmitter and receiver.

In the indirect method synchronized stable clocks are used by the receiver and (dedicated) transmitter. The receiver measures the time interval ΔT_{tt} between transmission of the pulse and reception of the target echo. It then calculates the range sum as $(R_T + R_R) = c\Delta T_{tt}$. A transmitter-to-receiver LOS is not required unless periodic clock synchronization is implemented over the direct path.

For the special case of a bistatic radar using the direct range sum estimation method, where $L \gg c\Delta T_{rt}$, Eq. (25.7) can be approximated as

$$R_R \simeq \frac{c\Delta T_{rt}}{1 + \sin \theta_R} \tag{25.8}$$

This approximation does not require an estimate of L. The error in Eq. (25.8) is less than 10 percent for $0° < \theta_R < 180°$ and $L > 4.6\, c\Delta T_{rt}$.

Other target location techniques are possible.[16,18] The transmitter beam-pointing angle θ_T can be used in place of θ_R. Unless the transmitter is also a monostatic radar tracking the target, target location accuracy is degraded, since beam splitting is sacrificed. A hyperbolic measurement system can be used, in which a receiver measures the difference in propagation times from two separate transmitters. The locus of target position now lies on a hyperbola, and the intersection of the receiver's AOA (angle of arrival) estimate with the hyperbola establishes the target position. Use of a third transmitter provides a full hyperbolic fix on the target. A theta-theta location technique uses the angles θ_T and θ_R and an estimate of L, where θ_T is typically provided by a monostatic radar, which acts as a cooperative bistatic transmitter.

For an elliptic location system, target location errors typically *increase* as the target approaches the baseline, ignoring S/N changes. The principal source of errors is the geometry inherent in Eq. (25.7). Additional errors occur when the direct range sum estimation method is used. They include interference from the direct-path signal (analogous to eclipsing), pulse instability, and multipath effects. Compounding the eclipsing problem is interference from range sidelobes when pulse compression is used by the transmitter. If linear FM pulse compression is used, Hamming or cosine-squared time-domain weighting by the receiver improves near-in sidelobe suppression by about 5 dB, when compared with the same type of frequency-domain weighting.[79]

For a hyperbolic location system, target location errors *decrease* as the target approaches the line joining the two transmitters. For a theta-theta location system, the error is a minimum when the target lies on the perpendicular bisector of the baseline with $\beta = 45°$ and increases elsewhere.[18] When successive data measurements (or redundant data) are available to a bistatic or multistatic radar, target state estimates can be made with Kalman or other types of filters.[80,81]

Coverage. Bistatic radar coverage, like monostatic radar coverage, is determined by both sensitivity and propagation. Bistatic radar sensitivity is set by the contour of constant $(S/N)_{min}$ and the oval of Cassini. Bistatic radar propagation requires a suitable path between the target and both sites and must include the effects of multipath, diffraction, refraction, shadowing, absorption, and geometry. The first five effects are usually included in the pattern propagation and loss factor terms of Eq. (25.1). The geometry effect is treated separately.

For given target, transmitter, and receiver altitudes the target must simultaneously be within LOS to both the transmitter and the receiver sites. For a smooth earth these LOS requirements are established by coverage circles centered at each site. Targets in the area common to both circles have an LOS to both sites as shown in Fig. 25.3. For a 4/3 earth model, the radius of these coverage circles, in kilometers, is approximated by[16]

$$r_R = 130(\sqrt{h_t} + \sqrt{h_R}) \tag{25.9}$$

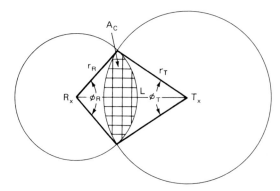

FIG. 25.3 Geometry for common-coverage area A_C.

$$\text{and } r_T = 130(\sqrt{h_t} + \sqrt{h_T}) \tag{25.10}$$

where h_t = target altitude, km
$\quad h_R$ = receive antenna altitude, km
$\quad h_T$ = transmit antenna altitude, km

If the receiver establishes synchronization via the direct-path link, then an adequate LOS is also required between transmitter and receiver. In this case $h_t = 0$ and $r_R + r_T \geq L$, where L is the baseline range. Thus,

$$L \leq 130(\sqrt{h_R} + \sqrt{h_T}) \tag{25.11}$$

If synchronization is accomplished by stable clocks, this LOS is not required and the system must satisfy only the requirements of Eqs. (25.9) and (25.10).

The common-coverage area A_C is shown in Fig. 25.3 as the intersection of the two coverage circles and is

$$A_C = \frac{1}{2}[r_R^2(\phi_R - \sin \phi_R) + r_T^2(\phi_T - \sin \phi_T)] \tag{25.12}$$

where ϕ_R and ϕ_T are shown on Fig. 25.3 and are

$$\phi_R = 2 \cos^{-1}\left(\frac{r_R^2 - r_T^2 + L^2}{2r_R L}\right) \tag{25.13}$$

$$\phi_T = 2 \cos^{-1}\left(\frac{r_T^2 - r_R^2 + L^2}{2r_T L}\right) \tag{25.14}$$

Terrain and other types of masking or shadowing degrade both monostatic and bistatic coverage. For ground-based bistatic transmitters and receivers the degradation can be severe.[84] For this reason some air defense bistatic radar concepts use an elevated or airborne transmitter.[44,45,48,54] As a general rule bistatic coverage is less than monostatic coverage in both single and netted configurations.

Clutter Cell Area.[42,51,59,73,85-89]   The main-lobe bistatic clutter cell area A_c is defined, in the broadest sense, as the intersection of the range resolution cell, the doppler resolution cell, and the bistatic main-beam footprint. The range and doppler resolution cells are defined by isorange and isodoppler contours, respectively. The bistatic footprint is the area on the ground, or clutter surface, common to the one-way transmit and receive beams, where the beamwidths are conventionally taken at the 3-dB points. Three clutter cell cases are usually of interest: beamwidth-limited, range-limited, and doppler-limited.

Beamwidth-Limited Clutter Cell Area.   The beamwidth-limited clutter cell area $(A_c)_b$ is the bistatic footprint. It has been evaluated for specific antenna pattern functions and specific geometries by numerical integration techniques.[42,85,86] At small grazing angles a two-dimensional approximation to $(A_c)_b$ is a parallelogram shown as the single-hatched area in Fig. 25.4 with area

$$(A_c)_b = \frac{R_R \Delta \theta_R R_T \Delta \theta_T}{\sin \beta} \tag{25.15}$$

where $R_R \Delta \theta_R$ is the cross-range dimension of the receive beam at the clutter cell, $R_T \Delta \theta_T$ is the corresponding dimension for the transmit beam, and $\Delta \theta_R$ and $\Delta \theta_T$ are, respectively, the 3 dB beamwidth of the receive and transmit beams. Respective transmit and receive beam rays are assumed to be parallel, which is a reasonable approximation when the range sum is much greater than the baseline range, $R_T + R_R \gg L$. The cell area is a minimum at $\beta = 90°$.

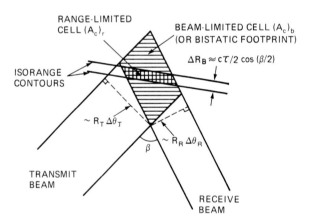

FIG. 25.4   Geometry for clutter cell areas.

Range-Limited Clutter Cell Area.   The range-limited clutter cell area $(A_c)_r$ has been evaluated at small grazing angles for all geometries of interest.[87] At small grazing angles and at large range sums ($R_T + R_R \gg$ L), a two-dimensional approximation to $(A_c)_r$ is a parallelogram shown as the double-hatched area in Fig. 25.4 with area

$$(A_c)_r = \frac{c\tau R_R \Delta \theta_R}{2 \cos^2 (\beta/2)} \tag{25.16}$$

where τ is the radar's compressed pulse width. The isorange contours are assumed to be straight lines within the bistatic footprint. For this example the cross-range dimension of the transmit beam $R_T\Delta\theta_T$ is greater than that of the receive beam $R_R\Delta\theta_R$, so that the clutter cell is determined by the intersection of the receive beam and the range cell. For a given geometry one or the other beam will usually determine the clutter cell area. In either case the cell area increases as β increases. For small range sums, the cell shape is trapezoidal or triangular at small β and is rhomboidal or hexagonal at large β.[87]

An exact expression for $(A_c)_r$ has been developed,[88] again for two dimensions, with one beam and the range cell determining the clutter cell area. Equation (25.16) gives results that are within a few percent of the exact results for $\beta < 90°$. The error increases significantly for $\beta \gg 90°$ and $\theta_R < -80°$.

Doppler-Limited Clutter Cell Area. The doppler-limited clutter cell area $(A_c)_d$ has been determined by numerical integration techniques when it is bounded by a range resolution cell.[51,89] No convenient algebraic expression has been developed for the cell area since the doppler cell size and orientation with respect to the baseline change as the transmitter and receiver velocity vectors and look angles change. In the special case where the transmitter and receiver velocity vectors are equal and the bistatic angle is large, the isorange and isodoppler contours are essentially parallel, creating very large clutter cell areas.[59]

25.6 DOPPLER RELATIONSHIPS

Figure 25.5 defines the geometry and kinematics for bistatic doppler when the target, transmitter, and receiver are moving. The target has a velocity vector of magnitude V and aspect angle δ referenced to the bistatic bisector. The transmitter and receiver have velocity vectors of magnitude V_T and V_R and aspect angles δ_T and δ_R referenced to the north coordinate system of Fig. 25.1, respectively.

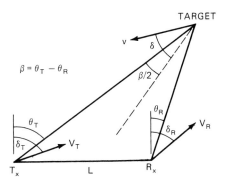

FIG. 25.5 Geometry for bistatic doppler.

Target Doppler. When the transmitter and receiver are stationary $(V_T = V_R = 0)$, the target's bistatic doppler at the receive site f_B is

$$f_B = (2V/\lambda)\cos\delta\cos(\beta/2) \tag{25.17}$$

When $\beta = 0°$, Eq. (25.17) reduces to the monostatic case, where δ is now the angle between the velocity vector and the radar-to-target LOS, which is colinear with the bistatic bisector. When $\beta = 180°$, the forward-scatter case, $f_B = 0$ for any δ. Equation (25.17) shows that:

- For a given δ, the magnitude of the bistatic target doppler is never greater than that of the monostatic target doppler when the monostatic radar is located on the bistatic bisector.

- For all β, when $-90° < \delta < +90°$, the bistatic doppler is positive; under this definition a closing target referenced to the bistatic bisector generates a positive, or *up*, doppler.

- For all β, when the target's velocity vector is normal to the bistatic bisector ($\delta = \pm 90°$) the bistatic doppler is zero; the vector is tangent to a range-sum ellipse drawn through the target position (a contour of zero target doppler).

- For all $\beta < 180°$, when the target's velocity vector is colinear with the bistatic bisector, the magnitude of the bistatic doppler is maximum; the vector is also tangent to an orthogonal hyperbola drawn through the target position, which is a contour of maximum target doppler.

Isodoppler Contours. When the target is stationary and the transmitter and receiver are moving (e.g., airborne), the bistatic doppler shift at the receiver site f_{TR} is

$$f_{TR} = (V_T/\lambda) \cos (\delta_T - \theta_T) + (V_R/\lambda) \cos (\delta_R - \theta_R) \qquad (25.18)$$

where the terms are defined on Fig. 25.5.

The locus of points for constant doppler shift on the earth's surface is called an *isodoppler contour*, or *isodop*. In the monostatic case and a flat earth, these isodops are conic sections in three dimensions and radial lines emanating from the radar in two dimensions. In the bistatic case the isodops are skewed, depending upon the geometry and kinematics. They are developed analytically for two dimensions and a flat earth by setting f_{TR} = constant in Eq. (25.18) and solving for θ_R (or θ_T if appropriate).

Figure 25.6 is a plot of bistatic isodops in a two-dimensional bistatic plane, i.e., where the transmitter and receiver are at zero or near-zero altitude, for the following conditions:[90]

$$V_T = V_R = 250 \text{ m/s}$$
$$\delta_T = 0°$$
$$\delta_R = 45°$$
$$\lambda = 0.03 \text{ m}$$

The dimension of the grid on the bistatic plane is arbitrary; that is, the isodops are invariant with scale. On the left and right sides of Fig. 25.6 the isodops approximate radial lines, which are pseudo-monostatic operating points.

25.7 TARGET CROSS SECTION[1,16,33–41,91–102,104,105]

The bistatic radar cross section (RCS) of a target σ_B is a measure, as is the monostatic radar cross section σ_M, of the energy scattered from the target in the

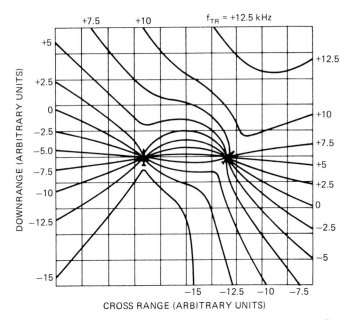

FIG. 25.6 Bistatic isodoppler contours for two dimensions and a flat earth.[90]

direction of the receiver. Bistatic cross sections are more complex than monostatic cross sections since σ_B is a function of aspect angle and bistatic angle.

Three regions of bistatic RCS are of interest: pseudo-monostatic, bistatic, and forward scatter (sometimes called near-forward scatter[99]). Each region is defined by the bistatic angle. The extent of each region is set primarily by physical characteristics of the target.

Pseudo-Monostatic RCS Region. The Crispin and Siegal monostatic-bistatic equivalence theorem applies in the pseudo-monostatic region:[36] for vanishingly small wavelengths the bistatic RCS of a sufficiently smooth, perfectly conducting target is equal to the monostatic RCS measured on the bisector of the bistatic angle. Sufficiently smooth targets typically include spheres, elliptic cylinders, cones, and ogives. Figure 25.7 shows the theoretical bistatic RCS of two perfectly conducting spheres as a function of bistatic angle.[1,92–95] For the larger sphere (near the optics region) the pseudo-monostatic region extends to $\beta = \sim 100°$, with an error of 3 dB. And even for the smaller sphere (in the resonance region) the pseudo-monostatic region extends to $\beta = \sim 40°$. Measurements[38] of a sphere with $a = 0.42 \lambda$, where a is the sphere radius, match within 3 dB the values for the smaller sphere in Fig. 25.7.

For targets of more complex structure, the extent of the pseudo-monostatic region is considerably reduced. A variation of the equivalence theorem developed by Kell[41] applies to this case: for small bistatic angles, typically less than 5°, the bistatic RCS of a complex target is equal to the monostatic RCS measured on the bisector of the bistatic angle at a frequency lower by a factor of cos ($\beta/2$).

Kell's complex targets are defined as an assembly of discrete scattering cen-

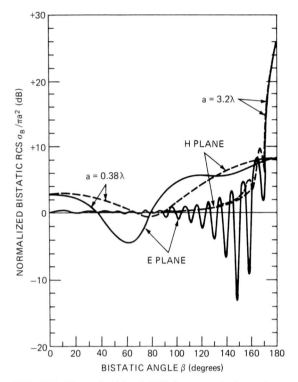

FIG. 25.7 Theoretical bistatic RCS for two perfectly conducting spheres, where a = sphere radius and λ = wavelength.[1,92-95]

ters (simple centers such as flat plates, reflex centers such as corner reflectors, skewed reflex centers such as a dihedral with corner $\neq 90°$ and stationary phase regions for creeping waves). When the wavelength is small compared with the target dimensions, these complex target models approximate many aircraft, ships, ground vehicles, and some missiles. The targets can be composed of conducting and dielectric materials.

The cos ($\beta/2$) frequency reduction term has little effect in Kell's pseudo-monostatic region, $0 < \beta < \sim 5°$, since a $5°$ bistatic angle corresponds to less than 0.1 percent shift in wavelength. At $\beta > 5°$ the change in radiation properties from discrete scattering centers is likely to dominate any cos ($\beta/2$) frequency reduction effect.[41] Thus the cos ($\beta/2$) term is often ignored.

Both versions of the equivalence theorem are valid when the positions of the transmitter and receiver are interchanged, given that the target-scattering media are reciprocal. Most media are reciprocal. Exceptions are gyrotropic media, such as ferrite materials and the ionosphere.[103]

Whenever the equivalence theorem is valid, Kell[41] provides a simple method for deriving bistatic RCS data from monostatic RCS data when plotted as a function of target aspect angle. Bistatic RCS data for the same polarization is obtained by translating along the target aspect angle axis by one-half of the desired bistatic angle. If monostatic RCS data is also available as a function of frequency, the monostatic curve for f sec ($\beta/2$), where f is the bistatic frequency, is used to estimate the bistatic RCS at f. As outlined earlier, this correction is usually small.

Bistatic RCS Region. The bistatic angle at which the equivalence theorem fails to predict the bistatic RCS identifies the start of the second, bistatic region. In this region the bistatic RCS diverges from the monostatic RCS. Kell[41] identified three sources of this divergence for complex targets and for a target aspect angle fixed with respect to the bistatic bisector. These sources are (1) changes in relative phase between discrete scattering centers, (2) changes in radiation from discrete scattering centers, and (3) changes in the existence of centers—appearance of new centers or disappearance of those previously present.

The first source is analogous to fluctuations in monostatic RCS as the target aspect angle changes, but now the effect is caused by a change in bistatic angle.[104] The second source occurs when, for example, the discrete scattering center reradiates, i.e., retroreflects, energy toward the transmitter and the receiver is positioned on the edge of or outside the retroflected beamwidth; thus the received energy is reduced. The third source is typically caused by shadowing, for example, by an aircraft fuselage blocking one of the bistatic paths—transmitter or receiver LOS to a scattering center.

In general, this divergence results in a bistatic RCS lower than the monostatic RCS for complex targets. Exceptions include (1) some target aspect angles that generate a low monostatic RCS and a high bistatic specular RCS at specific bistatic angles, (2) targets that are designed for low monostatic RCS over a range of aspect angles, and (3) shadowing that sometimes occur in a monostatic geometry and not in a bistatic geometry.[92]

Ewell and Zehner[97] measured the monostatic and bistatic RCS of coastal freighters at X band when both the transmitter and the receiver were near grazing incidence. The data was plotted as a ratio of bistatic to monostatic RCS, σ_B/σ_M. The measurements match Kell's model: of the 27 data points, 24 show bistatic RCS lower than monostatic RCS. The bistatic RCS reduction starts at about $\beta = 5°$ and trends downward to $\sigma_B/\sigma_M \approx -15$ dB at $\beta \approx 50°$. Most of the data points are in the region $5° < \beta < 30°$ where -2 dB $> \sigma_B/\sigma_M > -12$ dB.

Glint Reduction in the Bistatic RCS Region. A second effect can occur in the bistatic region. When the bistatic RCS reduction is caused by a loss or attenuation of large discrete scattering centers, for example through shadowing, target glint is often reduced. Target glint is the angular displacement in apparent phase center of a target return and is caused by the phase interference between two or more dominant scatters within a radar resolution cell. As the target aspect angle changes, the apparent phase center shifts, often with excursions beyond the physical extent of the target. These excursions can significantly increase the errors in angle tracking or measurement systems. When the returns from dominant scatterers are reduced in the bistatic region, the source and hence the magnitude of glint excursions are reduced. Limited measurements for tactical aircraft show that, for a 30° bistatic angle, peak glint excursions can be reduced by a factor of 2 or more, with most of the excursions contained within the physical extent of the target.[54]

Forward-Scatter RCS Region. The third bistatic RCS region, forward scatter, occurs when the bistatic angle approaches 180°. When $\beta = 180°$, Siegel[33] showed, based on physical optics, that the forward-scatter RCS, σ_F, of a target with silhouette (or shadow) area A is $\sigma_F = 4\pi A^2/\lambda^2$, where λ, the wavelength, is small compared with the target dimensions. The targets can be either smooth or complex structures and, from the application of Babinet's principle, can be totally absorbing.[37,91]

For $\beta < 180°$, the forward-scatter RCS rolls off from σ_F. The rolloff is approximated by treating the shadow area A as a uniformly illuminated antenna aperture. The radiation pattern of this *shadow aperture* is equal to the forward-scatter RCS rolloff when $(\pi - \beta)$ is substituted for the angle off the aperture normal. A sphere of radius a will roll off 3 dB at $(\pi - \beta) \simeq \lambda/\pi a$, when $a/\lambda \gg 1$.[15] Although the $a/\lambda \gg 1$ criterion is not satisfied in Fig. 25.7, the curve for $a = 3.2\lambda$ still exhibits this phenomenon: 3 dB reduction in σ_F at $\beta \simeq 174°$. (The value of σ_F at $\beta = 180°$ also matches $4\pi A^2/\lambda^2$ within 1 dB.) Figure 25.7 shows the rolloff approximating $J_0(x)/x$ down to $\beta \simeq 130°$, where J_0 is a Bessel function of zero order. A linear aperture of length D, with aspect angle orthogonal to the transmitter LOS, will roll off 3 dB at $(\pi - \beta) = \lambda/2D$, where $D/\lambda \gg 1$. The forward-scatter RCS rolloff continues, with sidelobes approximating $\sin x/x$ over the forward-scatter quadrant $(\beta > 90°)$.[105] For other aspect angles and targets with complex shadow apertures, calculation of the forward-scatter RCS rolloff usually requires computer simulation.

The forward-scatter RCS of more complex bodies has been simulated and measured; the bodies were both reflecting and absorbing.[34,37,38,92,98,100–102] Paddison et al.[100] report both measurements and calculations via computer simulation of forward-scatter RCS for a right circular aluminum cylinder at 35 GHz and bistatic angles up to 175.4°. Calculations were made via the method of moments,[106] and measurements were made by Delco.[98] A good match between measurements and calculations was obtained for targets with dimensions of the order of several wavelengths. A similar match to Delco measurements was obtained by Cha et al., using physical-theory-of-diffraction methods for targets that are larger than several wavelengths and the method of moments otherwise.[102]

Figure 25.8 shows calculations of a 16- by 1.85-cm cylinder with 992 facets at 35 GHz, for three fixed transmitter-to-target geometries: (*a*) near end on, (*b*) 45° aspect angle, and (*c*) broadside.[100] The broadside geometry shows the classic forward-scattering lobe from a rectangular aperture, with approximate $\sin x/x$ sidelobe rolloff out to $\beta \simeq 110°$. The three bistatic RCS regions are quite distinct: pseudo-monostatic at $\beta < 20°$, bistatic at $20° < \beta < 140°$, and forward scatter at $\beta > 140°$. The other two geometries show a similar but broader forward-scatter lobe, as is expected since the silhouette area and hence the shadowing aperture are smaller. The 45° aspect geometry is of interest because the RCS in the bistatic region is larger than the monostatic RCS for most bistatic angles. The large spike at $\beta = 90°$ is the bistatic specular lobe, analogous to the monostatic specular lobe in the broadside geometry. While Fig. 25.8 shows the clear dependency of bistatic RCS on both aspect and bistatic angle, it also serves to caution attempts to use oversimplified bistatic RCS models, especially in the bistatic region.

25.8 CLUTTER

The bistatic radar cross section of clutter σ_c is a measure, as is the monostatic radar clutter cross section, of the energy scattered from a clutter cell area A_c in the direction of the receiver. It is defined as $\sigma_c = \sigma_B^0 A_c$, where σ_B^0 is the scattering coefficient, or the clutter cross section per unit area of the illuminated sur-

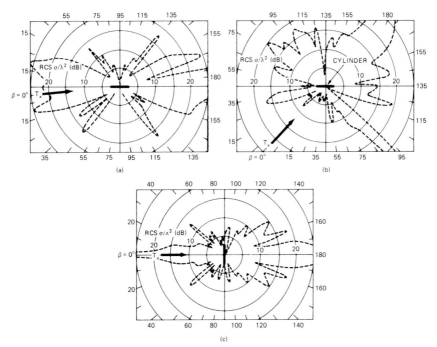

FIG. 25.8 Calculated bistatic RCS, replotted as a function of bistatic angle for a conducting cylinder, 16 by 1.85 cm at 35 GHz, *HH* polarization.[100] (*a*) Near end on. (*b*) 45° aspect angle. (*c*) Broadside.

face. The clutter cell area is given for beam- and range-limited cases in Sec. 25.5. This section considers measured and estimated values of $\sigma_B^{\,0}$, which vary as a function of the surface composition, frequency, and geometry. And, in contrast to the monostatic case, little measured data for $\sigma_B^{\,0}$ has been reported.[42,43,107–115]

The available database for terrain and sea clutter at microwave frequencies consists of six measurement programs, which are summarized in Table 25.2. The measurement angles shown in Table 25.2 are defined in Fig. 25.9, which is a clutter-centered coordinate system similar to those used in all the measurement programs. Because terrain and sea are reciprocal media, θ_i and θ_s are interchangeable in the subsequent data.[103] The Pidgeon data was analyzed by Domville[109] and Nathanson.[116] Vander Schurr and Tomlinson[117] analyzed the Larson and Cost data.

In addition to this database, bistatic reflectivity measurements have been made at optical[118] and sonic[119] wavelengths and of buildings,[120] airport structures,[121] and planetary surfaces.[66,122] In each of these measurements, the reflectivity data is expressed in terms of reflected power, not $\sigma_B^{\,0}$.[103]

The bistatic angle is calculated from the angles in Fig. 25.9 by the use of direction cosines:

$$\beta = \cos^{-1}(\sin \theta_i \sin \theta_s - \cos \theta_i \cos \theta_s \cos \phi) \qquad (25.19)$$

TABLE 25.2 Summary of Measurement Programs for Bistatic Scattering Coefficient, σ_B^0

Reference (date)	Organization	Author	Surface composition	Frequency	Polarization	Measurement angles (degrees)		
						θ_i	θ_s	ϕ
42 (1965)	Ohio State University (Antenna Laboratory)	Cost, Peake	Smooth sand, Loam, Soybeans, Rough sand	10 GHz	VV, HH, HV	5–30, 10–70	5–30, 5–90	0–145, 0, 180
			Loam with stubble, Grass	10 GHz	VV, HH, HV	5–70	5–90	0–180
43 (1966)	Johns Hopkins University (APL)	Pidgeon	Sea (sea states 1, 2, 3)	C band	VV, VH	0.2–3	10–90	180
107 (1967)			Sea (Beaufort, wind 5)	X band	HH	1–8	12–45	180
108 (1967)	GEC (Electronics) Ltd., England	Domville	Rural land, Urban land	X band	VV, HH	6–90*	6–180*	180, 165
109 (1968), 110 (1969)			Sea (20-kn wind), Sea (20-kn wind), Semidesert	X band, X band	VV, HH, VV, HH	~ 0–90*, ~ 0	~ 0–180*, ?	180, 165, 180, 165
111 (1977), 112 (1978)	University of Michigan (ERIM)	Larson, Heimiller	Grass with cement taxiway, Weeds and scrub trees	1.3 and 9.4 GHz	HH, HV, HH, HV	10, 40, 10, 15, 20	5, 10, 20, 5, 10, 20	0–180, 0–105
113 (1982), 114 (1984)	Georgia Institute of Technology (EES)	Ewell, Zehner	Sea (0.9-m, 1.2–1.8-m waveheights)	9.38 GHz	VV, HH	~ 0	~ 0	90–160
115 (1988)	University of Michigan (Department of Electrical Engineering and Computer Science)	Ulaby et al.	Visually smooth sand, Visually smooth sand, Rough sand, Gravel	35 GHz, 35 GHz	VV, HH, VH, HV, VV, HH, VH, HV	24, 30, 30	24, 30, 10–90	0–170, 0–170, 0–90

*Measured and interpolated data ranges.

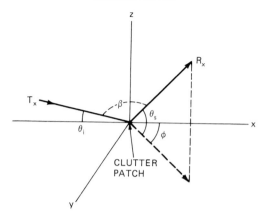

FIG. 25.9 Coordinate system for bistatic clutter measurements. θ_i = incident angle (in xz plane); θ_s = scattering angle (in plane containing z axis); ϕ = out-of-plane angle (in xy plane).

Two measurement sets are of interest: in plane, where $\phi = 180°$, and out of plane, where $\phi < 180°$. When $\phi = 180°$, $\beta = \theta_s - \theta_i$. In the monostatic case $\beta = 0$ and $\theta_s = \theta_i$ with $\phi = 180°$. Most of the data is taken at X band, with the most substantial in-plane database provided by Domville.[108–110] Because the database is sparse, mean values for σ_B^0 are usually given, with occasional standard deviations and probability distributions calculated.

In-Plane Land Clutter Scattering Coefficient. Figure 25.10 is a plot of Domville's X-band, vertically polarized data summary for rural land, consisting of open grassland, trees, and buildings.[108] Domville reports that since the data was a composite of different sources and averaged over different terrain conditions, differences of 10 dB in the values sometimes occurred. The spread in raw data within any data set ranged from 1 dB to 4.5 dB, however. The measured database consists of points near the lines $\theta_i = \theta_s$, $\theta_i = 90°$, and $\theta_s = 90°$ and of points along the specular ridge near the forward-scatter region. The remaining data points are interpolations.

Domville also summarized in-plane data for forest and urban areas.[108] The shape of constant σ_B^0 contours for all Domville's terrain types are similar. For urban areas σ_B^0 is generally 3 to 6 dB higher. The extent of the specular ridge is smaller, however. Because forest terrain is a more uniform scatterer, the cones extend into the forward quadrant ($\theta_s > 90°$). The ridge extent is smaller and its magnitude is about 16 dB below that of rural land. Other values of σ_B^0 for forest terrain are similar to those of rural land for $\theta_s < 90°$.

Domville reports[109] that at low θ_i no significant variation in σ_B^0 was observed for rural and forest terrain when measured at a small out-of-plane angle, $\phi = 165°$. Also at low θ_i no significant variation between horizontal, vertical, and crossed polarizations was observed for rural and forest terrain.

For semidesert, σ_B^0 was measured[110] at −40 dB for both horizontal and vertical polarization at $\theta_i < \sim 1°$ and for all $\theta_s > \sim 1°$. Crossed-polarization measurements were 5 to 10 dB lower. Also, σ_B^0 is reduced by about 0.3 dB/° as ϕ moves from 180 to 165°.

Although terrain conditions are different, the Cost in-plane data[42] matches the

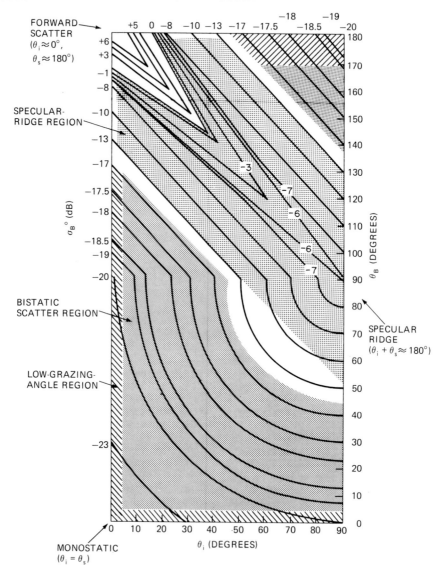

FIG. 25.10 X-band, vertically polarized, σ_B^0, in-plane ($\phi = 180°$) data summary for rural land.[108]

Domville data[108] within about 10 dB. The Cost data curves do not always approach the bistatic specular ridge monotonically even though the terrain conditions appear to be more uniform.

The in-plane Domville land clutter data can be divided into three regions: a low-grazing-angle region, where $\theta_i < \sim 3°$ or $\theta_s < \sim 3°$, the hatched area in Fig. 25.10; a specular-ridge region, where $140° \le (\theta_i + \theta_s) \le 220°$, the dotted area; and a bistatic scatter region, where (θ_i, θ_s) assume values shown as the shaded areas in Fig. 25.10. Each region can be modeled, by a *semiempirical process* (containing arbitrary constants that are adjusted to fit empirical data), as follows.

The low-grazing-angle and bistatic scatter regions are based on the constant-γ monostatic clutter model:

$$\sigma_M^{0} = \gamma \sin \theta_i \qquad (25.20)$$

where σ_M^{0} is the monostatic scattering coefficient, θ_i is the monostatic, or incident, angle on Fig. 25.9, and γ is a normalized reflectivity parameter. For farmland $\gamma \approx -15$ dB, and for wooded hills $\gamma \approx -10$ dB.[123]

The constant-γ bistatic-scatter-region model is developed by using a variation of the monostatic-bistatic equivalence theorem (Sec. 25.8), where $\sin \theta_i$ is replaced by the geometric mean of the sines of the incident and scattering angles, $(\sin \theta_i \sin \theta_s)^{1/2}$, in Eq. (25.20).[123] Hence

$$(\sigma_B^{0})_b = \gamma (\sin \theta_i \sin \theta_s)^{\frac{1}{2}} \qquad (25.21)$$

where $(\sigma_B^{0})_b$ is the scattering coefficient in the bistatic scatter region. Now γ can be estimated from Fig. 25.10 by using monostatic data, which is plotted along the line $\theta_i = \theta_s$. A value of $\gamma = -16$ dB in Eq. (25.20) fits the monostatic data within about 2 dB. Using $\gamma = -16$ dB in Eq. (25.21) yields a match within 3 dB to the bistatic data, including the small triangle in the forward quadrant.

The low-grazing-angle region is modeled by the sine of the arithmetic mean of the incident and scattering angles, $\sin [(\theta_i + \theta_s)/2]$. Hence

$$(\sigma_B^{0})_1 = \gamma \sin [(\theta_i + \theta_s)/2] \qquad (25.22)$$

where $(\sigma_B^{0})_1$ is the scattering coefficient in the low-grazing-angle region. The data match is again ~3 dB for $\gamma = -16$ dB, including the small quadrilateral in the upper right corner of Fig. 25.10. Since $(\theta_i + \theta_s)/2 = \theta_i + \beta/2$, Eq. (25.22) is an exact application of the monostatic-bistatic equivalence theorem. For very low grazing angles (θ_i or $\theta_s << \sim1°$), but excluding the specular-ridge region, the calculations for $(\sigma_B^{0})_1$ must be multiplied by the pattern propagation factors F_T^2 and F_R^2 and the loss terms L_T and L_R.[123]

The specular-ridge region is modeled for values of $(\sigma_B^{0})_s \leq 1$ by a variation of the Beckman and Spizzichino theory of forward scattering from rough surfaces:[124,125]

$$(\sigma_B^{0})_s = \exp [- (\beta_c/\sigma_s)^2] \qquad (25.23)$$

where $(\sigma_B^{0})_s$ = scattering coefficient in the specular-ridge region
$\quad\quad \sigma_s$ = rms surface slope
$\quad\quad \beta_c$ = angle between vertical and the bistatic bisector of θ_i and θ_s
$\quad\quad\;\; = |90 - (\theta_i + \theta_s)/2|$

For flat terrain $\sigma_s \approx 0.1$ rad. With a value of $\sigma_s = 0.17$ rad, Eq. (25.23) matches the specular ridge in Fig. 25.10 within 5 dB, for $(\sigma_B^{0})_s \leq 1$.

In-Plane Sea Clutter Scattering Coefficient. Limited in-plane sea clutter measurements have been taken.[43,107,109] The Domville data[109] contains a broad range of θ_i, θ_s measurement conditions but unfortunately estimates only wind conditions and not sea state. For vertical polarization, the Domville X-band data[109] and the Pidgeon C-band data[43] show spreads of about 10 dB, and their averages match within ±5 dB. For horizontal polarization, the Domville X-band data[109] and the Pidgeon X-band data[107] again show spreads of about 10 dB, but the match is only about ±10 dB.

In view of the limited database and the uncertainties in some of the measurement conditions, caution must be exercised in modeling this data. An approximate model is the direct application of the constant-γ monostatic clutter model, Eq. (25.20), when either θ_i or θ_s is held constant. Then for the region θ_i, $\theta_s > \sim 2°$ and $\theta_i + \theta_s < \sim 100°$, $\gamma = -20$ dB matches the available vertically polarized data within about 5 dB for a 20-kn wind ($\approx$ sea state 3 when fully developed).

Below about 2°, pattern propagation factors and losses affect the measurements. Values for σ_B^0 of -50 dB ±5 dB have been measured.[43] When the pattern propagation factors and losses are included in measurements, the data is sometimes called effective σ_B^0.[114] For $\theta_i + \theta_s > \sim 100°$, $\sigma_B^0 > 0$ dB, reaching $+10$ dB in the specular-ridge region. For horizontal polarization σ_B^0 is typically 1 to 5 dB lower,[109] but this difference is not significant compared with the data spread. Measured cross-polarized (VH) values for σ_B^0 are 10 to 15 dB lower than those for copolarized (VV) values at $\theta_i < 1°$ but only 5 to 8 dB lower at $\theta_i \approx 3°$.[43]

Out-of-Plane Scattering Coefficient. Limited out-of-plane land clutter measurements have been taken.[42,111,112,115] The Cost[42] and Ulaby[115] data shows reasonable correlation but only limited correlation with the Larson[111,112] data. There does not appear to be a satisfactory model of the available data.

However, general trends are apparent for all polarizations. First, σ_B^0 usually approaches a minimum as ϕ approaches 90°, with values 10 to 20 dB below the monostatic value ($\theta_i = \theta_s$, $\phi = 180°$). Second, out-of-plane σ_B^0 values are not significantly different (within ~ 5 dB) from in-plane σ_B^0 values for $\phi < \sim 10°$ and $\phi > \sim 140°$, i.e., angles close to in-plane conditions. The $\phi < \sim 10°$ limit is based on Cost, Ulaby, and Domville data; the $\phi > \sim 140°$ limit, on Ulaby and Larson data.

Ewell[113,114] measured horizontally and vertically polarized out-of-plane σ_B^0 for sea clutter at θ_i and θ_s near grazing incidence (θ_i, $\theta_s << 1°$). Visual estimates of sea conditions ranged from 0.9- to 1.8-m waveheight. Ratios of bistatic to monostatic scattering coefficients (median values) were calculated, with bistatic angles, $\beta \approx 180° - \phi$, ranging from 23° to 85°. The data implicitly included pattern propagation factors and losses. Since antenna heights were different, F_T, F_R, L_T, and L_R are expected to be different but were not measured. In all cases the measured bistatic to monostatic ratios were less than unity. In two cases they ranged from -2 dB to -12 dB, and in the third case they dropped from ~ -5 dB at $\beta = 23°$ to -20 to -25 dB at $\beta = 60°$. The trend was generally downward as β increased. Values for horizontal and vertical polarization showed no significant differences. For the most part both monostatic and bistatic data exhibited nearly log-normal amplitude distributions.

25.9 SPECIAL TECHNIQUES, PROBLEMS, AND REQUIREMENTS

Pulse Chasing.[49,73,126,127,129] The concept of pulse chasing has been proposed as a means to reduce the complexity and cost of multibeam bistatic receivers, which are one solution to the beam scan-on-scan problem. The simplest pulse-chasing concept replaces the multibeam receive system (*n* beams, receivers, and signal processors) with a single beam, receiver, and signal processor. As shown in Fig. 25.11, the single receive beam rapidly scans the volume covered by the transmit beam, essentially chasing the pulse as it propagates from the transmitter: hence the term *pulse chasing*. In addition to the usual requirements for solving the bistatic triangle, pulse chasing requires knowledge of θ_T and pulse transmission time,[126] which can be provided to the receive site by a data link. Alternatively, if the transmit beam scan rate and the pulse repetition frequency (PRF) are uniform, the receive site can estimate these parameters as the transmit beam passes by the receive site.[127]

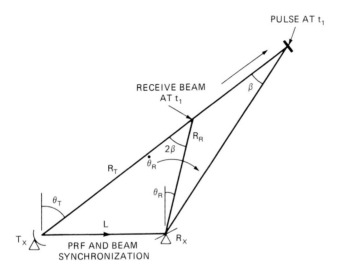

FIG. 25.11 Pulse chasing for the single-beam, continuous-scan case.

The receive beam-scanning rate must be at the transmitter's pulse propagation rate, modified by the usual geometric conditions. This rate, $\dot{\theta}_R$, is given by[73]

$$\dot{\theta}_R = c \tan (\beta/2)/R_R \qquad (25.24)$$

For typical geometries $\dot{\theta}_R$ can vary from 1°/μs to 0.01°/μs. These rates and rate changes require an inertialess antenna such as a phased array and fast diode phase shifters. Normally a phased array antenna used for surveillance is programmed to switch beams in increments of a beamwidth. Fractional shifts of a beamwidth can be achieved by changing the phase of a few (symmetric) pairs of phase shifters in the array. In this way a pseudo-continuous beam scan can be generated, with the required rates and rate changes.[128]

Because of pulse propagation delays from the target to the receiver, the pointing angle of the receive beam θ_R must lag the actual pulse position. For an instantaneous pulse position that generates a bistatic angle β, $\theta_R = \theta_T - 2\beta$. In terms of the bistatic triangle, the required receive beam-pointing angle is[73]

$$\theta_R = \theta_T - 2 \tan^{-1}\left(\frac{L \cos \theta_T}{R_T + R_R - L \sin \theta_T}\right) \tag{25.25}$$

The minimum receive beamwidth $(\Delta\theta_R)_m$ required to capture all returns from a range cell intersecting the common beam area is approximated by[73]

$$(\Delta\theta_R)_m \approx (c\tau_u \tan(\beta/2) + \Delta\theta_T R_T)/R_R \tag{25.26}$$

where τ_u is the uncompressed pulse width. The approximation assumes that respective rays from the transmit and receive beams are parallel. The approximation is reasonable when $(R_T + R_R) >> L$ or when $L >> c\tau_u$.

Other implementations of pulse chasing are possible. In one concept the n-beam receive antenna is retained and two receiver–signal processors (RSPs) are time-multiplexed across the n beams. One RSP steps across the even-numbered beams, and the other RSP steps across the odd-numbered beams, so that returns in beam pairs are processed simultaneously: (1,2), (2,3), (3,4), etc. This *leapfrog* sequence is required to capture all returns in the common-beam area.

A second concept uses two beams and two RSPs step-scanning over the volume covered by the n-beam antenna. It uses an identical leapfrog sequence. Both concepts relax the fractional beam scan requirements by either sampling or stepping the beams in units of a beamwidth. Since they both process returns across two beamwidths before switching, the beam dwell time T_b is approximately $2(\Delta\theta_R)_m R_R/c$ and the stepping rate is T_b^{-1}. The approximation assumes negligible phase-shift delays and settling times.

Beam Scan on Scan.   If high-gain scanning antennas are used by both the transmitter and the receiver in a bistatic surveillance radar, inefficient use is made of the radar energy, since only the volume common to both beams (the bistatic footprint) can be observed at any given time. Targets illuminated by the transmit beam outside the footprint are lost to the receiver. Four remedies to the beam scan-on-scan problem are possible: (1) fix the transmit beam for the time required to complete a surveillance frame by the receive beam, step the transmit beam one beamwidth and complete a second surveillance frame, and so forth until the transmit beam has stepped across the surveillance sector; (2) scan the transmit beam and use multiple simultaneous receive beams to cover the surveillance sector; (3) scan the transmit beam and chase the transmitted pulse with the receiver beam; and (4) broaden the transmit beamwidth to floodlight the surveillance sector, and scan the receive beam across the surveillance sector. The first and fourth remedies require a dedicated transmitter; the second and third do not.

The step-scan transmitter remedy increases the surveillance frame time by the number of required transmit beam steps. This increase is usually not acceptable for surveillance operations, and as a consequence the remedy is seldom considered. The multibeam-receiver remedy restores the frame time but increases receiver cost and complexity, since a multiple-beam antenna is required and an RSP must be used for each beam. The pulse-chasing remedy can remove requirements for multiple beams and RSPs, but at the penalty of using an inertialess (phased array) antenna with both complex and precise beam scheduling and/or multiplexing requirements, depending upon the configuration. The floodlight transmitter remedy removes all these complexities. It has the benefits of increasing data rates and simultaneously servicing multiple receivers. It incurs the penalty of a reduced S/N, directly as G_T is reduced; it also suffers increased sidelobe clutter levels. In short, there are no simple and inexpensive remedies to the beam scan-on-scan problem without suffering a penalty in surveillance performance.

Sidelobe Clutter. As with a monostatic radar, a bistatic radar must contend with sidelobe clutter. When both transmitter and receiver are ground-based and separated by a baseline range L, only ground clutter from regions having an adequate line of sight (LOS) to both the transmitter and the receiver will enter the receiver antenna sidelobes. That region is defined for a smooth earth as the common-coverage area A_C. From Eqs. (25.9), (25.10), and (25.12), when $h_t = 0$ and $L \geq r_R + r_T = 130 \ (\sqrt{h_R} + \sqrt{h_T})$, A_C for ground clutter is zero, and no sidelobe (or main-lobe) clutter enters the receive antenna. Targets with adequate LOS to both transmitter and receiver can be detected in a thermal-noise-limited background. This situation is analogous to a monostatic radar detecting targets at ranges greater than r_T.

This development applies to ocean scenarios but seldom is valid for land scenarios. In land scenarios variable terrain can decrease clutter levels by masking a clutter LOS when $L < r_R + r_T$ or increase clutter levels by generating a clutter LOS when $L > r_R + r_T$.

When the transmitter and/or the receiver is elevated or airborne, LOS restrictions are greatly reduced but not necessarily eliminated. Two clutter problems unique to bistatic radars are encountered in this situation. The first occurs when a floodlight transmit beam is used. To a first order, sidelobe clutter levels are reduced only by the one-way receive antenna sidelobes, in contrast to two-way sidelobe clutter reduction for a monostatic radar.

The second problem occurs when the transmitter and/or the receiver are moving, e.g., when airborne. Now the bistatic clutter doppler returns skew and spread, depending upon the geometry for each clutter patch and the kinematics of the transmit and receive platforms. Doppler skew is defined in terms of isodoppler contours, or isodops, given by Eq. (25.18) for two dimensions and a flat earth. The skew is range- and angle-dependent. The range-dependent skewing effect is not present in an airborne monostatic radar. Clutter spread in a particular sidelobe range cell is centered on the doppler skew present in the range cell.

These skewing and spreading effects, along with increased clutter levels, can greatly complicate the ability of a bistatic radar to detect targets in clutter. Remedies include conventional doppler filtering and high time-bandwidth waveforms; the judicious use of masking when available; control of the geometry, especially when a dedicated or cooperative transmitter is available; design of very low receive (and transmit when possible) antenna sidelobe levels; sidelobe blanking of discrete clutter returns; range or range-doppler averaging in the constant false-alarm rate (CFAR) unit for homogeneous clutter; and spatial excision of clutter returns. One implementation of this last technique relies on knowledge of the geometry and kinematics to predict the clutter doppler and spread in a given area. Then a filter or gate is set to excise main-beam clutter returns in that area. The amount of range-doppler space excised by this procedure can be as high as 8 percent.[45]

Time Synchronization. Time synchronization is required between the bistatic transmitter and receiver for range measurement. Timing accuracies on the order of a fraction of the transmitter's (compressed) pulse width are typically desired over the duration of an operation. Time synchronization can be accomplished directly by receiving a signal from the transmitter, demodulating the signal if necessary, and using the demodulated signal to synchronize a clock in the receiver. The transmitter signal can be sent via landline, via a communication link, or directly at the transmitter's RF if an adequate line of sight (LOS) exists between transmitter and receiver. If an adequate LOS is not available, it can be sent via a scatter path, where the scatterer has adequate LOS to both the transmitter and the receiver.[49] In this case, the scatterer must lie in the common-coverage area, as defined by Eq. (25.12). Transmission via tropospheric scatter can also be used in special cases.[50] In all these direct time synchronization schemes, implementation is straightforward, much like the initial synchronization process in communication systems. They can also be used for any type of transmitter pulse repetition interval (PRI) modulation: stable, staggered, jittered, and random. With time synchronization established, target range is calculated via Eq. (25.7) or similar methods.

For stable PRIs, time synchronization can be accomplished indirectly by using identical stable clocks at the transmitter and receiver sites. The clocks can be synchronized periodically, for example, whenever the transmitter and receiver are within LOS or located together if one or both are mobile. Direct time synchronization methods can be used for this task. Alternatively, the stable clocks can be slaved to a second source, such as Navstar GPS or loran C.[50,82,130] Indirect time synchronization can also be employed with a dedicated or cooperative transmitter using random PRIs if a random code sequence is established a priori and is known by the receive site.

In direct time synchronization, the required clock stability between updates is, to a first order, $\Delta\tau/T_u$, where $\Delta\tau$ is the required timing accuracy and T_u is the clock update interval. The update interval typically ranges from a minimum of the transmitter's interpulse period to a maximum of the transmitter's antenna scan period. The former usually requires a dedicated link between transmitter and receiver; the latter can be implemented whenever the transmit beam scans past the receive site, given an adequate LOS, and is sometimes called *direct breakthrough*.[130] Temperature-controlled crystal oscillators can often satisfy these requirements.

However, when direct-breakthrough time synchronization is used, multipath and other propagation anomalies, as well as radio-frequency interference (RFI), will degrade the accuracy of updating. Errors of ± 1 μs have been measured when a direct LOS is available.[131] They increase to ± 5 μs over a tropospheric propagation path.[130]

Since two clocks are used in indirect time synchronization, clock stability is, to a first order, $\Delta\tau/2T_u$. For T_u on the order of hours, atomic clocks are usually required to satisfy this requirement. Temperature-controlled crystal oscillators, an integral part of atomic clocks, will usually satisfy short-term (< 1 s) stability requirements. If the stable clocks are slaved to a second source, estimated timing accuracies of 0.5 μs for loran-C and <0.1 μs for Navstar GPS are reported.[82]

Phase Synchronization and Stability. As with monostatic radars, doppler or MTI processing can be used by the bistatic receiver to reject clutter or chaff. If noncoherent MTI is acceptable for clutter rejection, the bistatic receiver can use a clutter reference, exactly as a monostatic radar would, given that clutter patches are illuminated by the transmitter.

In one bistatic noncoherent MTI implementation, called *phase priming*, an oscillator at the receiver was phase-synchronized at the PRF rate with a small sample of close-in clutter returns.[132] Phase coherence was obtained within about 10 μs and extended over 1 ms. The process was found insensitive to the clutter signal level but quite sensitive to pulse-to-pulse phase fluctuations in the clutter signal.

If coherent processing is required, phase synchronization can be established with methods similar to those used for time synchronization: directly by phase-locking the receiver to the transmit signal or indirectly by using identical stable clocks in the transmitter and the receiver. Phase accuracy, or stability, requirements are the same as those for coherent processing by a monostatic radar: from 0.01λ to 0.1λ, or $3.6°$ to $36°$ of RF phase over a coherent processing interval,[133] with 0.01λ representing more typical design requirements.

Direct phase locking can be implemented as in direct time synchronization: via landline, communication link, or at the transmitter's RF. If a direct RF link is used, adequate transmitter-to-receiver LOS is again required. It is also subject to multipath and to phase reversals if coherent operation is required across transmitter antenna sidelobes. However, this latter problem can be overcome by a Costas loop for phase reversals near $180°$.[134] An extension of direct-path phase locking is the use of the direct-path signal as a reference signal in a correlation processor.[135]

For direct-path phase locking, clock stability is $\Delta\phi/2\pi f\Delta T$, where $\Delta\phi$ is the allowable rms sinusoidal phase error, f is the transmitter frequency, and ΔT is the difference in propagation time between the transmitter-target-receiver path and the transmitter-receiver (direct) path.[133] As with time synchronization, this requirement can usually be satisfied by a temperature-controlled crystal oscillator.

For matched stable clocks in the transmitter and receiver, phase stability is usually required over a coherent processing time T. Thus clock stability is $\Delta\phi/2\pi fT$. Again, atomic clocks are usually required, with crystal oscillators used for short-term stability. However, when $T < \sim 1$ s, integral crystal oscillators are usually acceptable. Quadratic phase errors caused by long-term drift in the stable clocks are usually smaller than allowable short-term sinusodial phase errors and can often be ignored.

In most types of SAR images, the integrated sidelobe ratio (ISLR) is an important criterion for image quality. It is a measure of the energy from a particular target that appears at image locations other than that corresponding to the target. Typically a -30- to -40-dB ISLR allocation for clock, or stable local oscillator (stalo), phase noise is desired.[53] When a single stalo is used, as in the monostatic case, these levels can be achieved for long coherent integration terms ($T > 10$ s) since low-frequency components of the phase noise are partially canceled in the demodulation process. However, since both bistatic phase synchronization techniques use two stalos, these low-frequency components do not cancel, resulting in higher ISLRs. Thus the bistatic SAR image quality, in terms of doppler or azimuth sidelobes, is degraded for coherent integration times greater than about 1 s at X band[53] unless very-high-quality clocks are used.

When direct-path phase locking is used by a bistatic SAR, the required motion-compensation phase shift (to track the target phase) must correct for relative motion between transmitter and receiver. When matched stable clocks are used, this correction is not required.[133]

Either time or phase errors can dominate synchronization requirements, depending upon the range and doppler accuracies needed. While all these requirements usually can be met, implementation is more complicated, time-consuming, and costly when compared with a monostatic system, which uses one clock for both time and phase synchronization.

REFERENCES

1. Skolnik, M. I.: "Introduction to Radar Systems," McGraw-Hill Book Company, New York, 1980.

2. *Microwave Syst. News Commun. Technol.*, vol. 18, p. 60, February 1988.

3. Heimiller, R. C., J. E. Belyea, and P. G. Tomlinson: Distributed Array Radar, *IEEE Trans.*, vol. AES-19, pp. 831–839, 1983.

4. Steinberg, B. D.: "Principles of Aperture and Array System Design—Including Random and Adaptive Arrays," John Wiley & Sons, New York, 1976.

5. Steinberg, B. D., and E. Yadin: Distributed Airborne Array Concepts, *IEEE Trans.*, vol. AES-18, pp. 219–226, 1982.

6. Steinberg, B. D.: High Angular Microwave Resolution from Distorted Arrays, *Proc. Int. Comput. Conf.*, vol. 23, 1980.

7. Easton, R. L., and J. J. Fleming: The Navy Space Surveillance System, *Proc. IRE*, vol. 48, pp. 663–669, 1960.

8. Mengel, J. T.: Tracking the Earth Satellite, and Data Transmission by Radio, *Proc. IRE*, vol. 44, pp. 755–760, June 1956.

9. Merters, L. E., and R. H. Tabeling: Tracking Instrumentation and Accuracy on the Eastern Test Range, *IEEE Trans.*, vol. SET-11, pp. 14–23, March 1965.

10. Scavullo, J. J., and F. J. Paul, "Aerospace Ranges: Instrumentation," D. Van Nostrand Company, Princeton, N.J., 1965.

11. Steinberg, B. D., et al.: First Experimental Results for the Valley Forge Radio Camera Program, *Proc. IEEE*, vol. 67, pp. 1370–1371, September 1979.

12. Steinberg, B. D.: Radar Imaging from a Distributed Array: The Radio Camera Algorithm and Experiments, *IEEE Trans.*, vol. AP-29, pp. 740–748, September 1981.

13. Salah, J. E., and J. E. Morriello: Development of a Multistatic Measurement System, *IEEE Int. Radar Conf.*, pp. 88–93, 1980.

14. Multistatic Mode Raises Radar Accuracy, *Aviat. Week Space Technol.*, pp. 62–69, July 14, 1980.

15. Skolnik, M. I.: An Analysis of Bistatic Radar, *IRE Trans.*, vol. ANE-8, pp. 19–27, March 1961.

16. Caspers, J. M.: Bistatic and Multistatic Radar, chap. 36 in Skolnik, M. I. (ed.): "Radar Handbook," McGraw-Hill Book Company, New York, 1970.

17. Ewing, E. F.: The Applicability of Bistatic Radar to Short Range Surveillance, *IEE Conf. Radar 77, Publ.* 155, pp. 53–58, London, 1977.

18. Ewing, E. F., and L. W. Dicken: Some Applications of Bistatic and Multi-Bistatic Radars, *Int. Radar Conf.*, pp. 222–231, Paris, 1978.

19. Farina, A., and E. Hanle: Position Accuracy in Netted Monostatic and Bistatic Radar, *IEEE Trans.*, vol. AES-19, pp. 513–520, July 1983.

20. Hanle, E.: Survey of Bistatic and Multistatic Radar, *Proc. IEE*, vol. 133, pt. F, pp. 587–595, December 1986.

21. Taylor, A. H., L. C. Young, and L. A. Hyland: U.S. Patent 1,981,884, System for Detecting Objects by Radio, Nov. 27, 1934.

22. Williams, A. F.: The Study of Radar, "Research Science and Its Application in Industry," vol. 6, Butterworth Scientific Publications, London, 1953, pp. 434–440.

23. Watson-Watt, Sir R.: "The Pulse of Radar," Dial Press, New York, 1959.

24. Skolnik, M. I.: Fifty Years of Radar, *Proc. IEEE*, vol. 73, pp. 182–197, February 1985.

25. Guerlac, H. E.: "Radar in World War II," vols. I and II, Tomask/American Institute of Physics, New York, 1987.

26. Price, A.: "The History of US Electronic Warfare," vol. 1, The Association of Old Crows, 1984.

27. Barton, D. K.: Historical Perspective on Radar, *Microwave J.*, vol. 23, p. 21, August 1980.

28. Summers, J. E., and D. J. Browning: An Introduction to Airborne Bistatic Radar, *IEE Colloq. Ground Airborne Multistatic Radar*, pp. 2/1–2/5, London, 1981.

29. Eon, L. G.: An Investigation of the Techniques Designed to Provide Early Warning Radar Fence for the Air Defense of Canada, *Defense Research Board (Canada), Rept. TELS* 100, Dec. 1, 1952.

30. Sloane, E. A., J. Salerno, E. S. Candidas, and M. I. Skolnik: A Bistatic CW Radar, *MIT Lincoln Laboratory Tech. Rept.* 82, AD 76454, Lexington, Mass., June 6, 1955.

31. Skolnik, M. I., J. Salerno, and E. S. Candidas: Prediction of Bistatic CW Radar Performance, *Symp. Radar Detection Theory, ONR Symp. Rept. ACR*-10, pp. 267–278, Washington, Mar. 1–2, 1956.

32. Skolnik, M. I.: private communication, September 1986.

33. Siegel, K. M., et al.: Bistatic Radar Cross Sections of Surfaces of Revolution, *J. Appl. Phys.*, vol. 26, pp. 297–305, March 1955.

34. Siegel, K. M.: Bistatic Radars and Forward Scattering, *Proc. Nat. Conf. Aeronaut. Electron.*, pp. 286–290, May 12–14, 1958.

35. Schultz, F. V., et al.: Measurement of the Radar Cross-Section of a Man, *Proc. IRE*, vol. 46, pp. 476–481, February 1958.

36. Crispin, J. W., Jr., et al.: "A Theoretical Method for the Calculation of Radar Cross Section of Aircraft and Missiles," *University of Michigan, Radiation Lab. Rept.* 2591-1-*H*, July 1959.

37. Hiatt, R. E., et al.: Forward Scattering by Coated Objects Illuminated by Short Wavelength Radar, *Proc. IRE*, vol. 48, pp. 1630–1635, September 1960.

38. Garbacz, R. J., and D. L. Moffett: An Experimental Study of Bistatic Scattering from Some Small, Absorber-Coated, Metal Shapes, *Proc. IRE*, vol. 49, pp. 1184–1192, July 1961.

39. Andreasen, M. G.: Scattering from Bodies of Revolution, *IEEE Trans.*, vol. AP-13, pp. 303–310, March 1965.

40. Mullin, C. R., et al.: A Numerical Technique for the Determination of the Scattering Cross Sections of Infinite Cylinders of Arbitrary Geometric Cross Section, *IEEE Trans.*, vol. AP-13, pp. 141–149, January 1965.

41. Kell, R. E.: On the Derivation of Bistatic RCS from Monostatic Measurements, *Proc. IEEE*, vol. 53, pp. 983–988, August 1965.

42. Cost, S. T.: "Measurements of the Bistatic Echo Area of Terrain of X-Band," *Ohio State University, Antenna Lab. Rept.* 1822-2, May 1965.

43. Pidgeon, V. W.: Bistatic Cross Section of the Sea, *IEEE Trans.*, vol. AP-14, pp. 405–406, May 1966.

44. Lefevre, R. J.: Bistatic Radar: New Application for an Old Technique, *WESCON Conf. Rec.*, pp. 1–20, San Francisco, 1979.

45. Fleming, F. L., and N. J. Willis: Sanctuary Radar, *Proc. Mil. Microwaves Conf.*, pp. 103–108, London, Oct. 22–24, 1980.

46. Forrest, J. R., and J. G. Schoenenberger: Totally Independent Bistatic Radar Receiver with Real-Time Microprocessor Scan Correction, *IEEE Int. Radar Conf.*, pp. 380–386, 1980.

47. Pell, C., et al.: An Experimental Bistatic Radar Trials System, *IEE Colloq. Ground Airborne Multistatic Radar*, pp. 6/1–6/12, London, 1981.

48. Schoenenberger, J. G., and J. R. Forrest: Principles of Independent Receivers for Use with Co-operative Radar Transmitters, *Radio Electron. Eng.*, vol. 52, pp. 93–101, February 1982.

49. Soame, T. A., and D. M. Gould: Description of an Experimental Bistatic Radar System, *IEE Int. Radar Conf. Publ.* 281, pp. 12–16, 1987.

50. Dunsmore, M. R. B.: Bistatic Radars for Air Defense, *IEE Int. Radar Conf. Publ.* 281, pp. 7–11, 1987.

51. Lorti, D. C., and M. Balser, Simulated Performance of a Tactical Bistatic Radar System, *IEEE EASCON 77 Rec. Publ.* 77 CH1255-9, pp. 4-4A–4-40, Arlington, Va., 1977.

52. Tactical Bistatic Radar Demonstrated, *Def. Electron.*, no. 12, pp. 78–82, 1980.

53. Auterman, J. L.: Phase Stability Requirements for a Bistatic SAR, *Proc. IEEE Nat. Radar Conf.*, pp. 48–52, Atlanta, March 1984.

54. Bistatic Radars Hold Promise for Future Systems, *Microwave Syst. News*, pp. 119–136, October 1984.

55. Griffiths, H. D., et al.: Television-Based Bistatic Radar, *Proc. IEE*, vol. 133, pt. F, pp. 649–657, December 1986.

56. Tomiyasu, K., Bistatic Synthetic Aperture Radar Using Two Satellites, *IEEE EASCON Rec.*, pp. 106–110, Arlington, Va., 1978.

57. Lee, P. K., and T. F. Coffey: Space-Based Bistatic Radar: Opportunity for Future Tactical Air Surveillance, *IEEE Int. Radar Conf.*, pp. 322–329, Washington, 1985.

58. Hsu, Y. S., and D. C. Lorti: Spaceborne Bistatic Radar—An Overview, *Proc. IEE*, vol. 133, pt. F, pp. 642–648, December 1986.

59. Anthony, S., et al.: Calibration Considerations in a Large Bistatic Angle Airborne Radar System for Ground Clutter Measurements, *Proc. IEEE Nat. Radar Conf.*, pp. 230–234, Ann Arbor, Mich., Apr. 20, 1988.

60. Walker, B. C., and M. W. Callahan: A Bistatic Pulse-Doppler Intruder-Detection Radar, *IEEE Int. Radar Conf.*, pp. 130–134, 1985.

61. Dawson, C. H.: Inactive Doppler Acquisition Systems, *Trans. AIEE*, vol. 81, pp. 568–571, January 1963.

62. Detlefsen, J.: Application of Multistatic Radar Principles to Short Range Imaging, *Proc. IEE*, vol. 133, pt. F, December 1986.

63. Nicholson, A. M., and G. F. Ross: A New Radar Concept for Short Range Application, *IEEE Int. Radar Conf.*, 1975.

64. Tyler, G. L.: The Bistatic Continuous-Wave Radar Method for the Study of Planetary Surfaces, *J. Geophys. Res.*, vol. 71, pp. 1559–1567, Mar. 15, 1966.

65. Tyler, G. L., et al.: Bistatic Radar Detection of Lunar Scattering Centers with Lunar Orbiter 1, *Science*, vol. 157, pp. 193–195, July 1967.

66. Pavel'yev, A. G., et al.: The Study of Venus by Means of the Bistatic Radar Method, *Radio Eng. Electron. Phys. (U.S.S.R.)*, vol. 23, October 1978.

67. Zebker, H. Z., and G. L. Tyler: Thickness of Saturn's Rings Inferred from Voyager 1 Observations of Microwave Scatter, *Science*, vol. 113, pp. 396–398, January 1984.

68. Tang, C. H., et al.: Measurements of Electrical Properties of the Martian Surface, *J. Geophys. Res.*, vol. 82, pp. 4305–4315, September 1977.

69. Zhou Zheng-Ou, et al.: A Bistatic Radar for Geological Probing, *Microwave J.*, pp. 257–263, May 1984.

70. Peterson, A. M., et al.: Bistatic Radar Observation of Long Period, Directional Ocean-Wave Spectra with Loran-A, *Science*, vol. 170, pp. 158–161, October 1970.

71. Doviak, R. J., et al.: Bistatic Radar Detection of High Altitude Clear Air Atmospheric Targets, *Radio Sci.*, vol. 7, pp. 993–1003, November 1972.

72. Wright, J. W., and R. I. Kressman: First Bistatic Oblique Incidence Ionograms Between Digital Ionosondes, *Radio Sci.*, vol. 18, pp. 608–614, July–August 1983.

73. Jackson, M. C.: The Geometry of Bistatic Radar Systems, *IEE Proc.*, vol. 133, pt. F, pp. 604–612, December 1986.

74. Davies, D. E. N.: Use of Bistatic Radar Techniques to Improve Resolution in the Vertical Plane, *IEE Electron. Lett.*, vol. 4, pp. 170–171, May 3, 1968.

75. McCall, E. G.: Bistatic Clutter in a Moving Receiver System, *RCA Rev.*, pp. 518–540, September 1969.

76. Crowder, H. A.: Ground Clutter Isodops for Coherent Bistatic Radar, *IRE Nat. Conv. Rec.*, pt. 5, pp. 88–94, New York, 1959.

77. Dana, R. A., and D. L. Knapp: The Impact of Strong Scintillation on Space Based Radar Design, I: Coherent Detection, *IEEE Trans.*, vol. AES-19, July 1983.

78. Pyati, V. P.: The Role of Circular Polarization in Bistatic Radars for Mitigation of Interference Due to Rain, *IEEE Trans.*, vol. AP-32, pp. 295–296, March 1984.

79. McCue, J. J. G.: Suppression of Range Sidelobes in Bistatic Radars, *Proc. IEEE*, vol. 68, pp. 422–423, March 1980.

80. Buchner, M. R.: A Multistatic Track Filter with Optimal Measurement Selection, *IEE Radar Conf.*, pp. 72–75, London, 1977.

81. Farina, A.: Tracking Function in Bistatic and Multistatic Radar Systems, *Proc. IEE*, vol. 133, pt. F, pp. 630–637, December 1986.

82. Retzer, G.: Some Basic Comments on Multistatic Radar Concepts and Techniques, *IEE Colloq. Ground Airborne Multistatic Radar*, pp. 3/1–3/3, London, 1981.

83. Hoisington, D. B., and C. E. Carroll: "Improved Sweep Waveform Generator for Bistatic Radar," U.S. Naval Postgraduate School, Monterey, Calif., August 1975.

84. Kuschel, H.: Bistatic Radar Coverage—A Quantification of System and Environmental Interferences, *IEE Int. Radar Conf. Publ.* 281, pp. 17–21, 1987.

85. Barrick, D. E. Normalization of Bistatic Radar Return, *Ohio State University, Res. Found. Rept.* 1388-13, Jan. 15, 1964.

86. Peake, W. H., and S. T. Cost: The Bistatic Echo Area of Terrain of 10 GHz, *WESCON 1968*, sess. 22/2, pp. 1–10.

87. Weiner, M. M., and P. D. Kaplan: Bistatic Surface Clutter Resolution Area at Small Grazing Angles, *MITRE Corporation, RADC-TR*-82-289, *AD A*123660, Bedford, Mass., November 1982.

88. Moyer, L. R., C. J. Morgan, and D. A. Rugger: An Exact Expression for the Resolution Cell Area in a Special Case of Bistatic Radar Systems, *Trans. IEEE*, vol. AES-25, July 1989.

89. Lorti, D. C., and J. J. Bowman: Will Tactical Aircraft Use Bistatic Radar?, *Microwave Syst. News*, vol. 8, pp. 49–54, September 1978.

90. Moyer, L. R. (TSC): private communication, February 1988.

91. Kock, W. I.: Related Experiments with Sound Waves and Electromagnetic Waves, *Proc. IRE*, vol. 47, pp. 1200–1201, July 1959.

92. Siegel, K. M., et al.: RCS Calculation of Simple Shapes—Bistatic, chap. 5, "Methods of Radar Cross-Section Analysis," Academic Press, New York, 1968.

93. Weil, H., et al.: Scattering of Electromagnetic Waves by Spheres, *University of Michigan, Radiat. Lab. Stud. Radar Cross Sections X, Rept.* 2255-20-T, contract AF 30(602)-1070, July 1956.

94. King, R. W. P., and T. T. Wu: "The Scattering and Diffraction of Waves," Harvard University Press, Cambridge, Mass., 1959.

95. Goodrich, R. F., et al.: Diffraction and Scattering by Regular Bodies—I: The Sphere, *University of Michigan, Dept. Electr. Eng. Rept.* 3648-1-T, 1961.

96. Matsuo, M., et al.: Bistatic Radar Cross Section Measurements by Pendulum Method, *IEEE Trans.*, vol. AP-18, pp. 83–88, January 1970.

97. Ewell, G. W., and S. P. Zehner: Bistatic Radar Cross Section of Ship Targets, *IEEE J. Ocean. Eng.*, vol. OE-5, pp. 211–215, October 1980.

98. Radar Cross-Section Measurements, *General Motors Corporation, Delco Electron. Div. Rept. R*81-152, Santa Barbara, Calif., 1981.

99. Bachman, C. G.: "Radar Targets," Lexington Books, Lexington, Mass., 1982, p. 29.

100. Paddison, F. C., et al.: Large Bistatic Angle Radar Cross Section of a Right Circular Cylinder, *Electromagnetics*, vol. 5, pp. 63–77, 1985.

101. Glaser, J. I.: Bistatic RCS of Complex Objects Near Forward Scatter, *IEEE Trans.*, vol. AES-21, pp. 70–78, January 1985.

102. Cha, Chung-Chi, et al.: An RCS Analysis of Generic Airborne Vehicles' Dependence on Frequency and Bistatic Angle, *IEEE Nat. Radar Conf.*, pp. 214–219, Ann Arbor, Mich., Apr. 20, 1988.

103. Weiner, M. M. (MITRE Corporation): private communication, April 1988.

104. Pierson, W. A., et al.: The Effect of Coupling on Monostatic-Bistatic Equivalence, *Proc. IEEE*, pp. 84–86, January 1971.

105. Barton, D. K.: "Modern Radar System Analysis," Artech House, Norwood, Mass., 1988, pp. 121–123.

106. Burk, G. J., and A. J. Foggio: "Numerical Electromagnetic Code (NEC)—Method of Moments," Naval Ocean Systems Center, San Diego, 1981.

107. Pidgeon, V. W.: Bistatic Cross Section of the Sea for Beaufort 5 Sea, *Science Technol.*, vol. 17, American Astronautical Society, San Diego, 1968, pp. 447–448.

108. Domville, A. R.: The Bistatic Reflection from Land and Sea of X-Band Radio Waves, pt. I, *GEC (Electronics) Ltd., Memo. SLM* 1802, Stanmore, England, July 1967.

109. Domville, A. R.: The Bistatic Reflection from Land and Sea of X-Band Radio Waves, pt. II, *GEC (Electronics) Ltd., Memo SLM* 2116, Stanmore, England, July 1968.

110. Domville, A. R.: The Bistatic Reflection from Land and Sea of X-Band Radio Waves, pt. II—Suppl., *GEC-AEI (Electronics) Ltd., Memo. SLM* 2116 (*Suppl.*), Stanmore, England, July 1969.

111. Larson, R. W., et al.: Bistatic Clutter Data Measurements Program, *Environmental Research Institute of Michigan, RADC-TR*-77-389, AD-A049037, November 1977.

112. Larson, R. W., et al.: Bistatic Clutter Measurements, *IEEE Trans.*, vol. AP-26, pp. 801–804, November 1978.

113. Ewell, G. W., and S. P. Zehner: Bistatic Sea Clutter Return Near Grazing Incidence, *IEE Radar Conf. Publ.* 216, pp. 188–192, October 1982.

114. Ewell, G. W., Bistatic Radar Cross Section Measurements, chap. 7 in Currie, N. C. (ed.): "Technology of Radar Reflectivity Measurement," Artech House, Norwood, Mass., 1984.

115. Ulaby, F. T., et al.: Millimeter-Wave Bistatic Scattering from Ground and Vegetation Targets, *IEEE Trans.*, vol. GE-26, no. 3, May 1988.

116. Nathanson, F. E., "Radar Design Principles," McGraw-Hill Book Company, New York, 1969.

117. Vander Schurr, R. E., and P. G. Tomlinson: Bistatic Clutter Analysis, *Decision-Science Applications, Inc., RADC-TR*-79-70, April 1979.

118. Sauermann, G. O., and P. C. Waterman: Scattering Modeling: Investigation of Scattering by Rough Surfaces, *MITRE Corporation, Rept. MTR*-2762, *AFAL-TR*-73-334, January 1974.

119. Zornig, J. G., et al.: Bistatic Surface Scattering Strength at Short Wavelengths, *Yale University, Dept. Eng. Appl. Sci. Rept. CS*-9, AD-A041316, June 1977.

120. Bramley, E. N., and S. M. Cherry: Investigation of Microwave Scattering by Tall Buildings, *Proc. IEE*, vol. 120, pp. 833–842, August 1973.

121. Brindly, A. E., et al.: A Joint Army/Air Force Investigation of Reflection Coefficient at C and K_u Bands for Vertical, Horizontal and Circular System Polarizations, *IIT Research Institute, Final Rept., TR*-76-67, *AD-A*031403, Chicago, July 1976.

122. Tang, C. H., et al.: Bistatic Radar Measurements of Electrical Properties of the Martian Surface, *J. Geophys. Res.*, vol. 82, pp. 4305–4315, September 1977.

123. Barton, D. K.: Land Clutter Models for Radar Design and Analysis, *Proc. IEEE*, vol. 73, pp. 198–204, February 1985.

124. Beckman, P., and A. Spizzichino: "The Scattering of EM Waves from Rough Surfaces," Pergamon Press, New York, 1963.

125. Nathanson, F., Technology Service Corporation: private communication, May 1988.

126. Hanle, E.: Pulse Chasing with Bistatic Radar-Combined Space-Time Filtering, in Schussler, H. W. (ed.): "Signal Processing II: Theories and Applications," Elsevier Science Publishers B.V., North Holland, pp. 665–668.

127. Schoenenberger, J. G., and J. R. Forrest: Principles of Independent Receivers for Use with Co-operative Radar Transmitters, *Radio Electron. Eng.*, vol. 52, pp. 93–101, February 1982.

128. Frank, J., and J. Ruze, Beam Steering Increments for a Phased Array, *IEEE Trans.*, vol. AP-15, pp. 820–821, November 1967.

129. Freedman, N.: Bistatic Radar System Configuration and Evaluation, *Raytheon Company, Independ. Dev. Proj.* 76D-220, *Final Rept.* ER76-4414, Dec. 30, 1976.

130. Bovey, C. K., and C. P. Horne, Synchronization Aspects for Bistatic Radars, *IEE Int. Radar Conf. Publ.* 281, pp. 22–25, 1987.

131. Schoenenberger, J. G., et al.: Design and Implementation of a UHF Band Bistatic Radar Receiver, *IEE Colloq. Ground Airborne Multistatic Radar*, pp. 7/1–7/3, London, 1981.

132. Griffiths, H. D., and S. M. Carter: Provision of Moving Target Indication in an Independent Bistatic Radar Receiver, *Radio Electron. Eng.*, vol. 54, pp. 336–342, July–August 1984.

133. Kirk, Jr., J. C.: Bistatic SAR Motion Compensation, *IEEE Int. Radar Conf.*, pp. 360–365, 1985.

134. Costas, J. P.: Synchronous Communications, *Proc. IRE*, vol. 44, pp. 1713–1718, December 1956.

135. Retzer, G.: A Concept for Signal Processing in Bistatic Radar, *IEEE Int. Radar Conf.*, pp. 288–293, 1980.

INDEX

ABOUT THE EDITOR IN CHIEF

Merrill I. Skolnik, known worldwide for his leadership in radar research and development, has been affiliated with the Johns Hopkins Radiation Laboratory, Sylvania, MIT Lincoln Laboratory, the Research Division of Electronic Communications Inc., the Institute for Defense Analyses, and the U.S. Naval Research Laboratory. He received his doctorate in electrical engineering from Johns Hopkins University, where he also earned B.E. and M.S.E. degrees. He is the author of the leading college textbook on radar, *Introduction to Radar Systems* (McGraw-Hill), now in its second edition, and the editor of *Radar Applications*.

He is a member of the National Academy of Engineering, a Fellow of the IEEE, and has served as editor of the *Proceedings of the IEEE*.